Lateral Electromagnetic Waves

Ronold W.P. King
Margaret Owens
Tai Tsun Wu

Lateral Electromagnetic Waves

Theory and Applications to Communications, Geophysical Exploration, and Remote Sensing

Springer-Verlag
New York Berlin Heidelberg London Paris
Tokyo Hong Kong Barcelona Budapest

Ronold W.P. King
Margaret Owens
Tai Tsun Wu
Gordon McKay Laboratory
Harvard University
Cambridge, MA 02138 USA

With 197 illustrations.

Library of Congress Cataloging-in-Publication Data
King, Ronold Wyeth Percival, 1905-
Lateral electromagnetic waves : theory and applications to communications, geophysical exploration, and remote sensing / Ronold W.P. King, Margaret Owens, Tai Tsun Wu.
p. cm.—
Includes bibliographical references and index.
ISBN 0-387-97679-5.
1. Electromagnetic waves. I. Wu, Tai Tsun. II. Owens, Margaret. III. Title. IV. Series.
QC661.K525 1992
539.2—dc20 91-27759

Printed on acid-free paper.

Production managed by Christin R. Ciresi; manufacturing supervised by Robert Paella.
Photo composed copy prepared from the authors TEX files.
Printed and bound by Edwards Brothers, Ann Arbor, MI.
Printed in the United States of America.

9 8 7 6 5 4 3 2 1

ISBN 0-387-97679-5 Springer-Verlag New York Berlin Heidelberg
ISBN 3-540-97679-5 Springer-Verlag Berlin Heidelberg New York

To
King Ching and Wei Van Wu
Graham and Tena Owens
Mary M. King

It is strange but true; for truth is always strange,
stranger than fiction.

Byron

Preface

The propagation of waves along and across the boundary between two media with different characteristic velocities is much more complicated when the source is on or near the boundary than when it is far away and the incident waves are plane. Examples of waves generated by localized sources near a boundary are the electromagnetic waves from the currents in a dipole on the surface of the earth and the seismic waves from a slip event in a fault in the earth's crust like the San Andreas fault in California. Both involve a type of surface wave that is called a *lateral wave* in electromagnetics and a *head wave* in seismology. Since the two are analogous and the latter is more easily visualized, it is conveniently used here to introduce and describe this important type of surface wave using the data of Y. Ben-Zion and P. Malin ("San Andreas Fault Zone Head Waves Near Parkfield, CA," Science **251**, 1592–1594, 29 March 1991).

The San Andreas fault is a vertical discontinuity in the surface of the earth in California. Near Parkfield, the average pressure-wave velocity in the granite (Region 1) west of the fault is $v_1 = 5.5$ km/sec; in the Franciscan rock (Region 2) east of the fault the velocity is $v_2 = 5.3$ km/sec. When a slip event occurs at some depth in the fault, this becomes the source of outward traveling spherical waves in each medium. These pressure (P) waves reach a receiver located in each region along radial lines or rays, as shown in Fig. 1. They are labeled P. Also reaching the receivers in each medium are pressure waves that have traveled upward to the air surface where they are reflected downward at the appropriate angle to the receiver. These are labeled PR. The times of arrival of these waves are consistent with

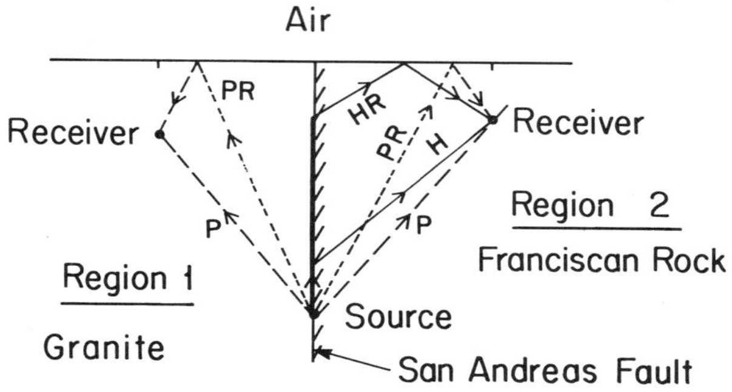

Figure 1. Spherical P-waves and head waves generated by a slip event at the source.

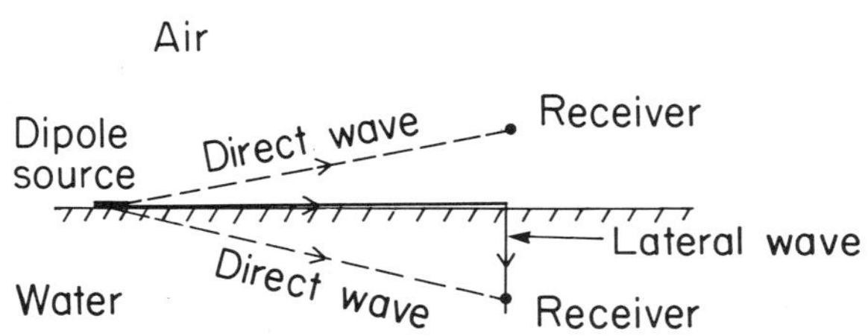

Figure 2. Spherical direct waves and lateral waves generated by a horizontal dipole on the air–water surface.

the distances traveled for each medium and their characteristic velocities. In the faster medium 1, these direct and once-reflected P waves are the only ones that reach the receiver. In the slower medium 2, there are not only the P and PR waves but also two additional waves, the direct and reflected head (H) waves. These have the following origin: As the spherical P waves travel in Region 1 along the boundary of Region 2, they continuously excite refracted waves that travel into Region 2 at the critical angle determined by the ratio of velocities v_1/v_2. The head wave that travels directly from the boundary to the receiver has its apparent origin at the appropriate point on that boundary, as shown in Fig. 1 where it is denoted by H. A second head wave that is reflected from the air–earth boundary also reaches the receiver in Region 2. Its path is shown in Fig. 1. It is denoted by HR. The first wave to arrive at the receiver in Region 1 is the direct P wave; it is followed by the PR wave. In Region 2 the first wave to arrive at the receiver is the direct head wave (H). The second to arrive is the P wave, the third is the HR wave, and the last is the PR wave. In seismology it is clearly important to understand these waves since the location of the source can be determined correctly only by tracing back the rays of the P waves in each region to their point of intersection.

Lateral waves in electromagnetic theory behave much like head waves in seismology. For example, a horizontal dipole on the surface of a lake sends out spherical radio waves into the air (Region 1) and the lake (Region 2). These travel, respectively, with the velocity of light $v_1 = c$ in the air and the velocity $v_2 = c/9$ in the lake water. They can be observed by receivers located in each region, as shown in Fig. 2. As the spherical wave travels along the surface of the water in air, it continuously generates waves called lateral (L) waves that propagate into the water at the critical angle determined by the ratio of velocities $v_1/v_2 = 9$. This is sufficiently great so that the angle of refraction in Region 2 is practically 90°. The receiver in the water observes the lateral wave first since it travels a long distance in air and then a fairly short distance in water. The direct wave travels entirely in water at a much slower speed.

This book is concerned with the rigorous analytical derivation of the complete set of electromagnetic waves—spherical waves and lateral waves—generated by horizontal or vertical dipoles near or on the plane boundary between two half-spaces with quite different characteristic wave velocities. Formulas for these fields are first derived in the form of general integrals which represent the *exact solution*. These integrals are then evaluated subject to a number of conditions of which the principal one is the requirement that the ratio of wave velocities be quite large.

Since there are a number of different types of electromagnetic surface waves, Chapter 1 is concerned with their individual characterization in a historical sequence. Chapter 2 reviews the basic electromagnetic theory including the fields of unit electric and magnetic dipoles in air and the transmission and reflection of plane waves across plane boundaries between electrically different media. The main problem of the book is formulated in Chapter 3 for the vertical dipole and in Chapter 5 for the horizontal dipole located near the plane boundary between two electrically different media. These chapters in conjunction with the associated appendices contain the major part of the complicated mathematical analysis. Chapter 4 for the vertical dipole and Chapters 6–8 for the horizontal dipole are devoted to the verification and applications of the newly derived formulas. These can be read without understanding the long and detailed derivation of the formulas used in them. Chapter 9 deals with the added complication introduced by anisotropy in one of the regions. Chapter 10 is concerned with the propagation of the lateral waves along a plane boundary on which vertical discontinuities are erected.

The theory of multilayered regions is formulated in general in Chapter 11 and applied in Chapter 12 specifically to the propagation of lateral waves in the three-layered region consisting of the ocean crust below a layer of sediment when excited by a horizontal dipole in the sea. The application of the theory to a three-layered region consisting of air, ice, and sea water or of air, dielectric substrate, and highly conducting base when the dipole is in the air is carried out in Chapter 15.

Chapters 13 and 14 are devoted to the propagation of electromagnetic pulses in the presence of a plane boundary. In Chapters 16–18, antennas useful in the generation of lateral waves are analyzed. The book concludes with a set of appendices that deal systematically with the evaluation of the complicated Bessel-function integrals that are encountered in the several chapters.

The original motivation for writing this book can be found in Chapter 11, titled "Antennas near Planar Surfaces," in the book *Antennas in Matter* by King and Smith (MIT Press, 1981). The exact general integrals for the complete electromagnetic field in the earth or sea generated by a horizontal electric dipole also in the earth or sea below the air surface are contained in this chapter together with their evaluation by numerical methods for a wide range of parameters. The graphical representation of the calculated fields and their correlation in specific ranges with approximate formulas in the book *Dipole Radiation in the Presence of a Conducting Half-Space* by A. Baños, Jr. (Pergamon Press, Oxford, England, 1966) exhibited regularities that suggested that there should be relatively simple continuous formulas to describe them and the associated exact integrals. Subsequently, in a lecture course at Harvard University, the theoretical foundation for the evaluation of the integrals for the electromagnetic field in a new and general form was developed. The first surprisingly simple integrated formula derived by this method immediately suggested that there must exist an interference pattern for the radial electric field along the boundary of a dielectric like fresh water. A comparison of the field calculated from the new formula and the numerical evaluation of the exact integrals at sufficiently small intervals showed the interference pattern to exist and the new formula to have a high degree of accuracy. This encouraging discovery provided the challenge to reduce to relatively simple, physically meaningful form the complicated phenomena associated with electromagnetic-wave propagation in the presence of boundaries, including especially the lateral wave.

Except for the important contribution of a major part of Chapter 9 by Wei-Yan Pan during his residence at Harvard as a Visiting Scholar, the important experimental work of Michael F. Brown described in Chapters 6 and 10, several figures in Chapter 12 from the Ph.D. thesis of John M. Dunn, and calculations and figures in Chapter 15 from the Masters thesis of David R. Gulla, the contents of the book are the result of research by the authors. Computations for many of the figures were carried out by Barbara S. Sandler. Most of the figures were drawn by William Minty and Marie Youk-See; photographic work was carried out by Armand Dionne. The authors are pleased to acknowledge the understanding and guidance of the editorial staff of Springer-Verlag New York. The researches reported in this book were supported in part by the Department of Defense under Joint Services Electronics Program contract N00014-84-K-0465 and grant N00014-89-J-1023 and an Office of Naval Research contract N00014-79-K-0419 monitored by Dr. Jack Heacock.

Although complete in itself, this book is restricted to electromagnetic waves in the presence of *plane* boundaries. What happens to a lateral wave generated by a vertical dipole on the surface of the *spherical* earth? The currents generated by a vertical monopole on an infinite perfectly conducting plane travel outward to infinity without radiating. If the monopole is erected on a perfectly conducting sphere, the outward-traveling currents radiate. When the plane is imperfectly conducting or dielectric, a lateral wave is generated that continuously transfers energy into the conducting or dielectric region. When the sphere is the imperfectly conducting earth, the outward-traveling currents are associated with both a lateral wave and a radiation field. A formulation of this problem and its analytical solution are a challenge for the future. Upon its solution depends an accurate understanding of long-range radio communication and over-the-horizon radar.

Harvard University
Cambridge, Massachusetts

Ronold W. P. King
Margaret Owens
Tai T. Wu

Contents

List of Symbols

$\mathbf{A}$	vector potential, 28
$A_e(\Theta)$	field factor of electric type, 9
$A_m(\Theta)$	field factor of magnetic type, 10
a	radius of cylindrical antenna, 126
a	radius of scattering cylinder, 270
a_e	effective radius, 603
$\mathbf{B}$	magnetic (flux density) vector, 26
B	driving-point susceptance, 594
$\mathbf{B}^c$	correction term for reflected magnetic field, 57
$\mathbf{B}^d$	direct-field component of magnetic field, 57
$\mathbf{B}^i$	reflected magnetic field due to ideal image, 57
$\mathbf{B}^L$	lateral-wave component of magnetic field, 57
B^r	magnetic far field, 98
B^s	surface-wave component of magnetic field, 91
$\mathbf{B}^{\text{inc}}$	incident magnetic field, 35
B_θ	transverse magnetic field, 5
b	outer radius of insulating sheath, 249
$C_2(z)$	real part of Fresnel integral, 15
c	velocity of light, 485
c_L	capacitance per unit length, 609
D	depth of ideal reflecting layer, 316
d	depth of scattering rod in Region 1, 254

d	depth of source dipole in Region 1, 17
d	height of source dipole in air, 8
d'	height of horizontal dipole over dielectric-coated conducting plane, 522
d_s	skin depth, 136
$\mathbf{E}$	electric vector, 26
$\mathbf{E}^c$	correction term for reflected electric field, 57
$\mathbf{E}^d$	direct-field component of electric field, 57
$\mathbf{E}^i$	reflected electric field due to ideal image, 57
E^i	incident electric field, 363
$\mathbf{E}^L$	lateral-wave component of electric field, 57
E^r	electric far field, 98
E^S	once-reflected spherical-wave electric field, 316
E^s	scattered electric field, 368
E^s	surface-wave component of electric field, 91
E^t	total electric field, 259
$\mathbf{E}^{\mathrm{tr}}$	transmitted electric field, 35
$\mathbf{E}^{\mathrm{inc}}$	incident electric field, 35
$\mathbf{E}^{\mathrm{ref}}$	reflected electric field, 35
E^{scat}	scattered electric field, 256
E_z	axial electric field, 5
$e(\rho, 0, z)$	normalized scattered electric field, 368
$\mathrm{erf}(z)$	error function, 15
$\mathcal{F}(p)$	Fresnel-integral function, 62
f	operating frequency, 69
f_c	cutoff frequency, 567
f_r	plane-wave reflection coefficient, 21; *see* f_{er}, f_{mr}
f_{er}, f_{et}	plane-wave reflection and transmission coefficients of electric type, 9, 40
f_{mr}, f_{mt}	plane-wave reflection and transmission coefficients of magnetic type, 10, 40
$f(\rho; k_1, k_2)$	radial function, 62
G	driving-point conductance, 594
g_L	conductance per unit length, 609
$g(\rho; k_1, k_2)$	radial function, 62
$\mathbf{H}$	magnetic vector, 169
H_θ	transverse magnetic field, 5
h	half-length of dipole, 254
h	half-length of scattering cylinder, 270
h	length of horizontal section of wave antenna, 245
h	length of monopole, 126
h_e	effective half-length of dipole, 109

h_m	maximizing length of insulated antenna, 250
h_{em}	maximum effective length, 251
$h_e(\phi_0)$	complex effective length of wave antenna, 247
$h(\rho; k_1, k_2)$	radial function, 162
I_s	current induced along buried conductor, 255
I_x	current on horizontal antenna, 247
I_z	current on vertical antenna, 126
I_z	total axial current, 5
$I\Delta l$	electric moment of dipole of length Δl, 126
$\mathbf{J}$	volume density of conduction current, 27
$\mathbf{J}^e$	impressed volume density of current, 27
J_z	volume density of axial current, 5
$\mathbf{K}_m$	magnetic surface current, 199
K_N	normalized surface density of current, 381
$\mathbf{k}$	wave vector, 35
k	complex wave number, 26
k_c	wave number of copper conductor, 592
k_d	wave number of dielectric sheath, 608
k_e, k_h	spheroidal coordinates, 2
k_{I}	complex wave number of current in bare antenna with finite conductivity, 597
k_i	wave number of insulating sheath, 255
k_L	complex wave number of current in wave antenna, 246
k_L, k_T	wave numbers for one-dimensionally anisotropic half-space, 326
k_L'	complex wave number for induced current on buried conductor, 255
$k(\omega)$	effective wave number of microstrip transmission line, 572
L	height of cylindrical metal reflecting wall, 380
l	half-length of buried conductor, 254
l	length of bare monopole terminating insulated antenna, 274
l	series inductance per unit length, 609
l	thickness of dielectric layer, 522
l	thickness of sediment layer, 407
l	thickness of substrate of microstrip, 560
l_e	effective length of bare monopole, 275
l_n	thickness of nth layer of layered half-space, 398
$\mathbf{M}$	volume density of magnetization, 27
$\mathbf{M}^e$	impressed volume density of magnetization, 27
m^e	magnetization or magnetic moment of dipole, 28

N	index of refraction, 9
$\mathbf{P}$	volume density of polarization, 27
P	power radiated by dipole, 104
$\mathbf{P}^e$	impressed volume density of polarization, 27
P_a	portion of power remaining in air, 104
P_g	portion of power transferred into earth, 104
P_{in}	input power to antenna, 614
P_0	total radiated power of isolated dipole, 109
P_+	upward-directed power of isolated dipole, 108
P_-	downward-directed power of isolated dipole, 109
p	numerical distance, 62
p^e	polarization or electric moment of dipole, 27
p_a	normalized power in air, 110
p_g	normalized power in earth, 110
p_S	Sommerfeld numerical distance, 15
q	charge per unit length, 2
$\mathcal{R}$	ratio of once-reflected spherical wave to lateral wave, 318
R	driving-point resistance, 599
R	plane-wave reflection coefficient for normal incidence, 198
R^e	external or radiation resistance, 599
R^i	internal resistance, 599
R^r	radiation resistance, 109
R_s	surface resistance, 134
r, Θ, Φ	spherical coordinates, 1
r^e	series external resistance per unit length, 609
r^i	internal resistance per unit length, 596
r_0	d.c. resistance per unit length, 274
r_{12}	series mutual resistance per unit length, 609
$\mathbf{S}$	Poynting vector, 1
$\mathbf{S}_E$, $\mathbf{S}_M$	Poynting vectors of electric and magnetic type, 206
$S_2(z)$	imaginary part of Fresnel integral, 15
T_e, T_m	plane-wave transmission coefficients for electric field of electric and magnetic type, 316
t	time, 1
$U(t)$	Heaviside or step function, 484
$U(a, z)$	parabolic cylinder function, 505, 506
V_0^e	emf of generator, 246
$V(a, z)$	parabolic cylinder function, 506
$V(t, x)$	Gaussian voltage pulse, 572
v	velocity of pulse in dielectric, 552

v_p	phase velocity, 36
v_s	phase velocity along surface, 6
X	driving-point reactance, 599
x, y, z	rectangular coordinates, 51
x^e	series external reactance per unit length, 609
x^i	internal reactance per unit length, 596
x_{12}	series mutual reactance per unit length, 609
Y_e	effective driving-point admittance, 603
Y_{sm}	surface admittance of magnetic type, 40
y_L	shunt admittance per unit length, 609
y_{sm}	normalized surface admittance of magnetic type, 10, 40
Z	driving-point impedance, 598
Z_c	characteristic impedance, 246
Z_c'	characteristic impedance of buried conductor, 255
Z_g	impedance of generator, 246
Z_h	terminating impedance at $x = h$, 247
Z_L	load impedance, 624
Z_l	terminal impedance of bare monopole in sea water, 274
Z_s	surface impedance, 6
Z_{se}	surface impedance of electric type, 40
Z_{in}	driving-point impedance, 247
Z_0	driving-point impedance, 127
Z_0	terminating impedance at $x = 0$, 247
z	cylindrical coordinate, 29
z	depth of point of observation in Region 1, 17
z^e	series external impedance per unit length, 609
z^i	internal impedance per unit length, 596
z_m'	maximum depth of penetration of lateral wave, 209
z_{se}	normalized surface impedance of electric type, 9, 40
z_{12}	series mutual impedance per unit length, 609
α	attenuation constant, 26
α_{I}	attenuation constant of current in bare antenna with finite conductivity, 597
β	real wave number or phase constant, 26
β_{I}	phase constant of current in bare antenna with finite conductivity, 597
Γ	gamma function, 113
Γ_m	reflection coefficient of magnetic type, 136
γ	propagation constant, 53
$\gamma_e, \gamma_L; \gamma_T$	propagation constants for one-dimensionally anisotropic half-space, 328

Δ	normalizing factor, 592
Δl	length of infinitesimal dipole, 126
δ_s	skin depth, 26
$\delta(t)$	δ-function, 459
ϵ	permittivity, 26, 27
$\tilde{\epsilon}$	complex permittivity, 26
ϵ', ϵ''	real and imaginary parts of ϵ, 27
ϵ_e	real effective permittivity, 27
ϵ_L, ϵ_T	permittivities of one-dimensionally anisotropic half-space, 325
ϵ_r	relative permittivity, 26
ϵ_0	permittivity of free space, 26
$\epsilon_{1r\ \mathrm{eff}}$	relative effective permittivity of microstrip, 572
ζ	wave impedance, 28
ζ_0	wave impedance of free space, 28
η	radiation efficiency, 105
η	wave admittance, 10
η_0	wave admittance of free space, 10
Θ	spherical coordinate, 1
Θ_c	critical angle for total internal reflection, 21
Θ_1	angle of refraction, 36
Θ_{2B}	Brewster angle, 41
Θ_{2r}	angle of reflection, 36
θ_h	complex terminal function, 610
λ	wavelength, 26
μ	permeability, 27
μ_0	permeability of free space, 27
Π_{2z}	Hertz potential, 482
ρ	radial distance from source dipole, 17
ρ, ϕ, z	cylindrical coordinates, 29
ρ^e	impressed volume density of charge, 27
ρ_b	radius of discontinuous region, 363
ρ_h	real part of terminal function, 610
ρ_m	radial distance to maximum depth of penetration of lateral wave, 208
σ	conductivity, 26, 27
σ', σ''	real and imaginary parts of σ, 27
σ_c	conductivity of copper conductor, 592
σ_d	conductivity of dielectric sheath, 608

σ_e	real effective conductivity, 27
σ_L, σ_T	conductivities of one-dimensionally anisotropic half-space, 324
Φ	spherical coordinate, 1
Φ_h	imaginary part of terminal function, 610
ϕ	cylindrical coordinate, 29
ϕ	scalar potential, 28
ψ_{er}, ψ_{mr}	phase angles of reflection coefficients, 9, 10
ω	angular frequency, 5

1
Historical and Technical Overview of Electromagnetic Surface Waves; Introduction to Lateral Waves

1.1 Electromagnetic Waves

When an oscillating infinitesimal electric dipole with unit electric moment $Ih = 1$ ampere meter is located in air far from all material media, it generates an electromagnetic field in the form of outward-traveling spherical waves. At sufficiently large electrical distances from the source [$kr \gg 1$, where $k = \omega(\mu_0 \epsilon_0)^{1/2}$ is the wave number], its component electric and magnetic fields in the spherical coordinates r, Θ, Φ are

$$E_\Theta(r,t) = -cB_\Phi(r,t) = -\frac{\omega\mu_0}{4\pi r} \sin\Theta \sin(\omega t - kr), \tag{1.1.1}$$

where Θ is measured from the positive z-axis and $c = \omega/k$ is the velocity of light. In (1.1.1), the argument of the sine function is, in general, $\omega t - kr +$ constant; the origin of the time variable t has been chosen such that this constant is absorbed in ωt. The E_Θ and B_Φ components are mutually perpendicular and both are perpendicular to the radial direction of propagation which is also the direction of the Poynting vector $\mathbf{S}(r,t) = \mu_0^{-1}\, \mathbf{E}(r,t) \times \mathbf{B}(r,t)$. The expanding spherical surfaces of constant phase are defined by

$$\omega t - kr = \omega\left(t - \frac{r}{c}\right) = \text{constant}. \tag{1.1.2}$$

Each surface travels radially outward with the velocity

$$\frac{dr}{dt} = \frac{\omega}{k} = c. \tag{1.1.3}$$

The distance between adjacent spherical surfaces of the same phase at any given instant is

$$r_2 - r_1 = \frac{2\pi}{k} = \lambda, \tag{1.1.4}$$

where λ is the wavelength.

At a sufficiently large radial distance from the source and in a volume V that is dimensionally small compared to this distance, the spherical surfaces of constant phase are approximately plane and $1/r$ is approximately constant at $1/r_0$. In V at $\Theta = \pi/2$ and $\Phi = 0$, the unit vectors are $\hat{\mathbf{r}} \to \hat{\mathbf{x}}$, $\hat{\boldsymbol{\Theta}} \to -\hat{\mathbf{z}}$, $\hat{\boldsymbol{\Phi}} \to \hat{\mathbf{y}}$. The field (1.1.1) becomes

$$E_z(x,t) = cB_y(x,t) = \frac{\omega\mu_0}{4\pi r_0}\sin(\omega t - kx). \tag{1.1.5}$$

This is a plane wave traveling in the positive x-direction with the velocity of light.

Close to the infinitesimal dipole source where $kr \lesssim 1$, the electromagnetic field is much more complicated. In addition to the two transverse components $E_\Theta(r,t)$ and $B_\Phi(r,t)$, there is a radial component $E_r(r,t)$ in the direction of propagation. The phase and amplitude relations among the three components are not simple, so that the picture of outward-traveling spherical surfaces of constant phase is not readily extended to small electrical distances even though quite simple analytical formulas for $\mathbf{E}(r,t)$ and $\mathbf{B}(r,t)$ are readily available—see, for example, eqs. (2.2.25)–(2.2.27) of Chapter 2. The near field is more interesting when the source is a very thin half-wave dipole with a sinusoidally distributed current (King 1956, Chap. 5; King and Prasad 1986, Sec. 8.5). In this case, the spheroidal coordinates k_e, k_h, and Φ are convenient since the surfaces of constant phase are spheroids with the ends of the dipole as foci. They expand with the velocity of light along the axes of the dipole and, of course, with greater phase velocities in other directions including the equatorial plane. As they expand, they become more spherical. Tangent to a spheroidal surface are the components $E_e \to E_\Theta$ and $B_\Phi \to B_\Phi$ as the surface becomes spherical. E_e and B_Φ are in the same phase at all points on an expanding spheroidal surface. Perpendicular to such a surface is the component E_h which approaches E_r as it expands. Everywhere on a given spheroidal surface, E_h differs in phase by 90° from E_e and B_Φ and, therefore, the electric field is elliptically polarized. The simple wave picture of expanding surfaces of constant phase is continuous from the surface of the half-wave dipole where the spheroid degenerates to be the straight line between the foci to large distances where the spheroid is a great sphere. On the dipole, E_e and B_Φ are in phase with the sinusoidally distributed electric current I and E_h is in phase with the cosinusoidally distributed electric charge per unit length q.

If an infinitesimal dipole source is embedded in an infinite lossless dielectric medium other than air, the electromagnetic field is like that in

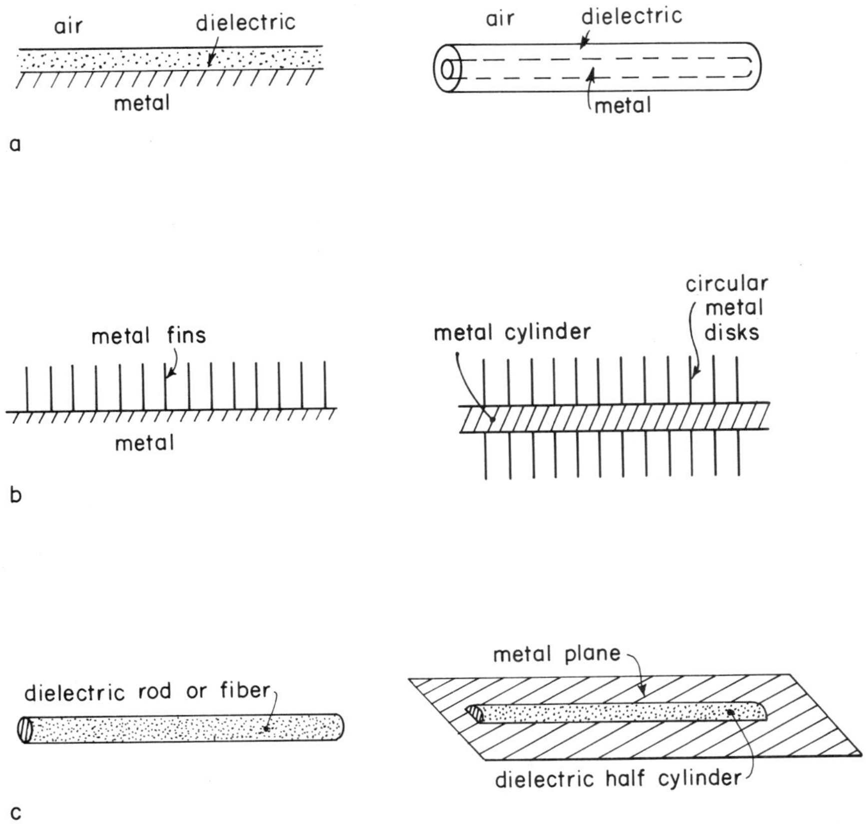

Figure 1.2.1. (a) Dielectric-coated metal plane and cylinder; (b) corrugated metal plane and cylinder; (c) dielectric rod and dielectric image line.

or surfaces. In general, it is quite complicated even when the metal surface or surfaces are treated as perfect conductors. The dielectric rod or optical fiber involves no metal and the electromagnetic field is distributed in both the dielectric rod and the surrounding air. For single-mode operation, most of the field—which includes three components of the electric field and three components of the magnetic field—is confined in the air quite close to the dielectric rod.

A quite different type of electromagnetic wave travels along a smooth cylindrical conductor in air when this carries an alternating current. The associated field was first analyzed by Sommerfeld (1899). Outside the conductor the electric field is elliptically polarized with a larger radial and smaller axial component in the direction of propagation, as indicated in Fig. 1.2.2. The latter component generates a cylindrical wave that travels radially inward, but is highly attenuated. The attenuation constant is equal to the phase constant; together they determine the radial amplitude and air but the wave number, which remains real, is different and given by $k = \omega(\mu\epsilon)^{1/2}$. If the medium is not a perfect dielectric, the wave number is complex in the form $k = \beta + i\alpha = (\omega^2\mu\epsilon + i\omega\mu\sigma)^{1/2}$, where σ is the electric conductivity of the dielectric. The time dependence used throughout this book is $\exp(-i\omega t)$. [For readers who are more at home with the time dependence $\exp(j\omega t)$, it is only necessary to substitute everywhere $i \rightarrow -j$.] For this case of the lossy dielectric, the outward-traveling spherical wave is exponentially attenuated at large distances from the source. It has the form $\exp(ikr)/r = [\exp(-\alpha r)/r]\exp(i\beta r)$. When $\alpha \ll \beta$, it is still possible to approximate a plane wave at large radial distances and in sufficiently small volumes. On the other hand, when the medium is highly lossy so that $\sigma \gg \omega\epsilon$, then $\alpha \sim \beta$ and it is generally not possible to approximate a plane wave.

When the dipole source is not in an infinite homogeneous medium but in one of two electrically different half-spaces, the nature of the electromagnetic field is greatly modified, especially near the boundary surface. If the incident field is treated like a plane wave arriving at the boundary at an arbitrary angle, the boundary conditions require a reflected field in the half-space of incidence and a refracted field in the second half-space. For the actually incident spherical wave, the boundary conditions cannot be satisfied with incident, reflected, and refracted waves. Except when the field incident on the boundary surface is a plane wave, the boundary conditions are satisfied only if an electromagnetic *surface wave* of a type usually called a *lateral wave* is added to the incident, reflected, and refracted waves. Indeed, the principal component of the electromagnetic field along the boundary is often the lateral wave. It is not a plane wave and its unusual properties and many useful applications are the subject of this book. In order to relate them to and distinguish them from other types of electromagnetic waves that travel along surfaces, it is appropriate to describe some of these.

1.2 Electromagnetic Waves Along Surfaces

The propagation of electromagnetic waves in the presence of a boundary between two quite different materials occurs in a variety of forms that depend on the nature of the boundary surface and the properties of the two media. Perhaps the best known electromagnetic surface waves are those associated with surface waveguides. These include dielectric-coated metal planes and cylinders (Fig. 1.2.1a), corrugated metal planes and cylinders (Fig. 1.2.1b), and dielectric rods or optical fibers and half-cylinders of dielectric on metal planes—known as dielectric image lines (Fig. 1.2.1c). In all of these the electromagnetic field is guided along the boundary surface

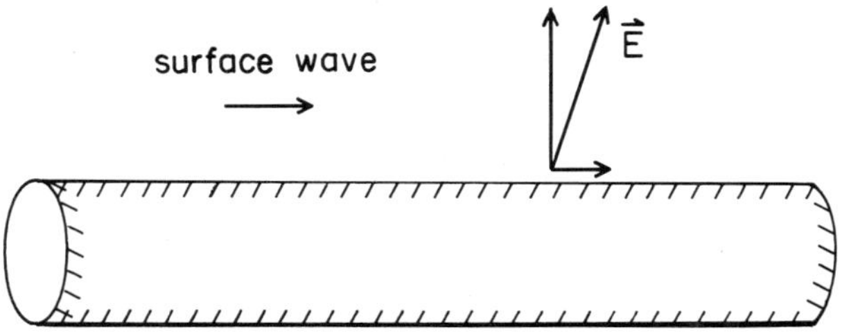

Figure 1.2.2. Electric field on surface of metal conductor.

phase distribution commonly referred to as skin effect and shown in Fig. 1.2.3. The volume density $J_z(\rho, z)$ of axially directed current and the axial electric field $E_{1z}(\rho, z)$ at the cross section z in the conductor are given by

$$\sigma_1 E_{1z}(\rho, z) = J_z(\rho, z) = \frac{I_z(a, z)}{\pi a^2} \frac{k_1 a}{2} \frac{J_0(k_1 \rho)}{J_1(k_1 a)}, \tag{1.2.1}$$

when $I_z(a, z)$ is a slowly varying function of z in the sense that $|I_z(a, z)| \gg a|dI_z(a, z)/dz|$. In (1.2.1), $I_z(a, z)$ is the total axial current given by

$$I_z(a, z) = \int_0^a J_z(\rho, z)\, \rho\, d\rho, \tag{1.2.2}$$

and J_0 and J_1 are Bessel functions of the first kind. The complex wave number for the field in the conductor is

$$k_1 = \beta_1 + i\alpha_1 \sim (1 + i)\left(\frac{\omega \mu_1 \sigma_1}{2}\right)^{1/2}, \tag{1.2.3}$$

when $\sigma_1 \gg \omega\epsilon$. Here σ_1 is the conductivity, ω the angular frequency, and μ_1 the permeability. (Unless otherwise indicated, all media will be assumed to be nonmagnetic so that $\mu_1 = \mu_0 = 4\pi \times 10^{-7}$ H/m.) When the electrical radius $\beta_1 a$ of the conductor is sufficiently great, (1.2.1) can be approximated as follows:

$$\sigma_1 E_{1z}(\rho, z) = J_z(\rho, z) \sim \frac{k_1 I_z(a, z)}{2\pi a}\left(\frac{a}{\rho}\right)^{1/2} e^{-\beta_1(a-\rho)} e^{i\beta_1(a-\rho)}; \qquad \beta_1 \rho \gtrsim 10. \tag{1.2.4}$$

Note that $\beta_1 a \geq \beta_1 \rho$. At radial distances not too close to the axis, this represents a wave traveling radially inward with amplitude given by $\exp[-\beta_1(a - \rho)]/\rho^{1/2}$ and radial phase velocity $v_1 = \omega/\beta_1 = (2\omega/\mu_1\sigma_1)^{1/2}$.

The magnetic field on the surface of the conductor is

$$H_{1\theta}(a, z) = \frac{B_{1\theta}(a, z)}{\mu_1} = \frac{I_z(a, z)}{2\pi a}. \tag{1.2.5}$$

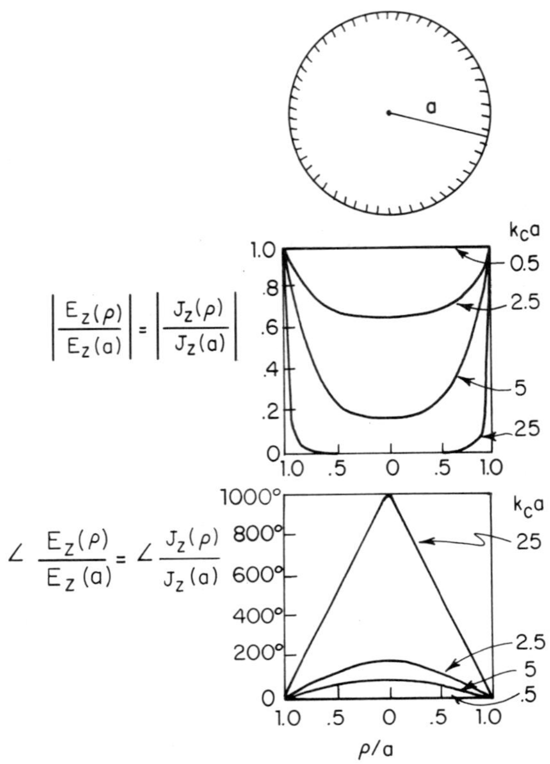

Figure 1.2.3. Radial distribution of axial current density and axial electric field in cylindrical conductor.

The ratio of the axial electric to the transverse magnetic field is known as the surface impedance. It is

$$Z_s = \frac{E_{1z}(a,z)}{H_{1\theta}(a,z)} = \frac{2\pi a E_{1z}(a,z)}{I_z(a,z)} = \frac{k_1}{\sigma_1}\frac{J_0(k_1 a)}{J_1(k_1 a)}. \tag{1.2.6}$$

For large and positive $\alpha_1 a$, $J_0(k_1 a)/J_1(k_1 a) \to -i$ and

$$Z_s = -\frac{ik_1}{\sigma_1} = \frac{\beta_1(1-i)}{\sigma_1} = (1-i)\left(\frac{\omega\mu_1}{2\sigma_1}\right)^{1/2}. \tag{1.2.7}$$

The relation

$$E_{1z}(a,z) = Z_s H_{1\theta}(a,z) \tag{1.2.8}$$

is known as the impedance boundary condition when $a \to \infty$ and the boundary surface between the conductor and air is planar. The corresponding boundary relation for $B_{1\theta}(a,z)$ is

$$E_{1z}(a,z) = v_s B_{1\theta}(a,z); \qquad v_s = \frac{Z_s}{\mu_1} = (1-i)\left(\frac{\omega}{2\mu_1\sigma_1}\right)^{1/2}. \tag{1.2.9}$$

The analogous phenomenon for a plane boundary between air and a conductor was discussed by Zenneck (1907) and related to the propagation of radio waves over the surface of the earth. With the existence and significance of the ionosphere still unknown, Zenneck sought to explain long-distance propagation of radio waves in terms of a surface wave that decays less rapidly than the $1/r$ decrease for space waves. The properties of radial cylindrical surface waves (Zenneck waves) as possible solutions of Maxwell's equations are well described by Barlow and Cullen (1953). They derive formulas for the field along the boundary between a conducting or dielectric half-space (Region 1, $z \geq 0$; ϵ_1, σ_1) with the complex wave number

$$k_1 = \beta_1 + i\alpha_1 = \omega\left[\mu_0\left(\epsilon_1 + \frac{i\sigma_1}{\omega}\right)\right]^{1/2}, \tag{1.2.10}$$

and air (Region 2, $z \leq 0$; $\epsilon_2 = \epsilon_0$, $\sigma_2 = 0$) with the wave number

$$k_2 = \omega(\mu_0\epsilon_0)^{1/2}. \tag{1.2.11}$$

Here ϵ_j, $j = 1, 2$, is the permittivity of Region j, and $\mu_2 = \mu_1 = \mu_0$. If the condition

$$|k_1|^2 \gg k_2^2 \quad \text{or} \quad |k_1|^2 \geq 9k_2^2 \tag{1.2.12}$$

is imposed, the following considerably simplified formulas (which serve present purposes) are obtained in the cylindrical coordinates ρ, ϕ, z:

In Region 1,

$$B_{1\phi} = \mu_0 A e^{ik_1 z} H_1^{(1)}(k_2\rho), \tag{1.2.13}$$

$$E_{1\rho} = -\frac{\omega}{k_1} B_{1\phi}, \tag{1.2.14}$$

$$E_{1z} = \frac{i\omega\mu_0 k_2}{k_1^2} A e^{ik_1 z} H_0^{(1)}(k_2\rho); \tag{1.2.15}$$

in Region 2,

$$B_{2\phi} = \mu_0 A e^{-i(k_2^2/k_1)z} H_1^{(1)}(k_2\rho), \tag{1.2.16}$$

$$E_{2\rho} = -\frac{\omega}{k_1} B_{2\phi}, \tag{1.2.17}$$

$$E_{2z} = \frac{i\omega\mu_0}{k_2} A e^{-i(k_2^2/k_1)z} H_0^{(1)}(k_2\rho), \tag{1.2.18}$$

where H_0 and H_1 are Hankel functions and A is an amplitude that depends on the nature of the unspecified source at the origin. Note that the field decays as $\exp(-\alpha_1 z)$ in the conducting or dielectric half-space, and more slowly as $\exp[-\alpha_1(k_2^2/|k_1|^2)z]$ in the air. It is confined to a relatively thin layer on each side of the surface $z = 0$. When $k_2\rho \gtrsim 10$, the following asymptotic forms of the Hankel functions are good approximations:

$$H_0^{(1)}(k_2\rho) \sim \left(\frac{\pi k_2\rho}{2}\right)^{-1/2} \exp\left[i\left(k_2\rho - \frac{\pi}{4}\right)\right], \tag{1.2.19}$$

$$H_1^{(1)}(k_2\rho) \sim \left(\frac{\pi k_2\rho}{2}\right)^{-1/2} \exp\left[i\left(k_2\rho - \frac{3\pi}{4}\right)\right]. \tag{1.2.20}$$

It follows that when $k_2\rho \gtrsim 10$, the entire surface-wave field decreases as $\rho^{-1/2}$ with increasing radial distance ρ. This is the essential characteristic of the Zenneck wave.

Sommerfeld (1909, 1926) derived general integrals for the complete electromagnetic field of a vertical dipole on the surface of the earth (represented as a homogeneous, isotropic half-space) and verified that this quite complicated solution actually contained the surface-wave term with its $\rho^{-1/2}$ rate of decrease. This seemed to confirm the possibility of attributing long-distance radio transmission to the Zenneck surface wave. However, in a reexamination of his theory Sommerfeld (1935) concluded that the surface-wave term could not be separated from the other terms in the complete solution for the field of a vertical electric dipole. He also stated that there were no possible circumstances under which the Zenneck surface wave could exist alone or even form the principal part of the complete representation of the outward-traveling waves from the dipole. Nevertheless, the existence or nonexistence of the Zenneck surface wave continued as the controversial subject of a long series of papers. Since the details are well summarized by Baños (1966), they are not repeated here. Useful additional discussions are in Hill and Wait (1978) and Wait and Hill (1979).

1.3 Surface Waves in the Field of a Vertical Electric Dipole over the Earth; The Formula of Norton

Even though the surface-wave term in the formula for the field of a vertical electric dipole over the earth has no application to phenomena now attributed to the ionosphere, it is nevertheless a significant term in the radiation field. This arises from the fact that the field in the absence of the surface-wave term vanishes along the surface of the earth—in the spherical coordinates r, Θ, Φ when $\Theta = \pi/2$ (King 1956, Sec. 7-18). This is readily understood from the formula for the field of an electrically short vertical dipole with unit electric moment ($I\Delta l = 1$ A m) at a height d in air (Region 2) over the earth (Region 1) (see Fig. 1.3.1a). It is

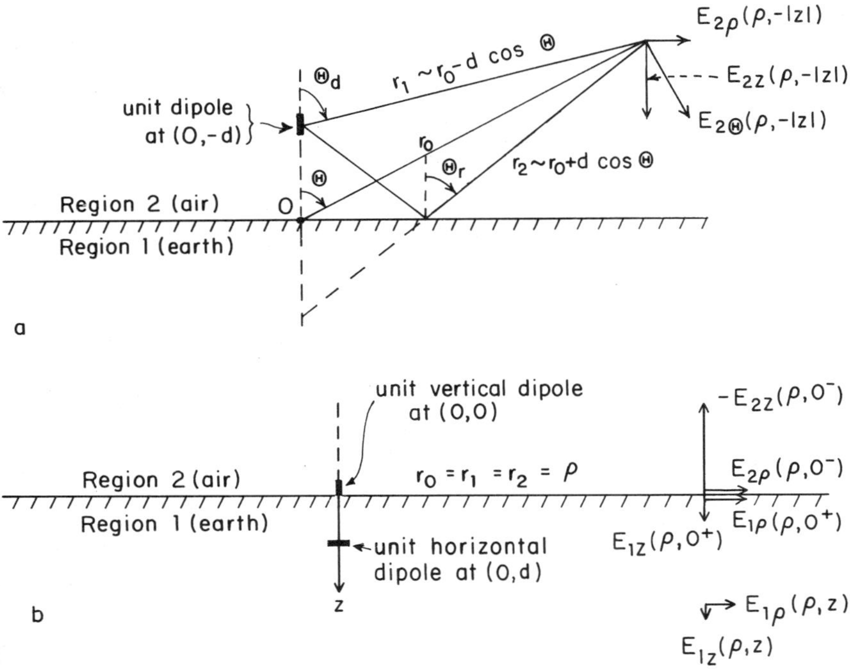

Figure 1.3.1. (a) Vertical dipole with unit electric moment ($I\Delta l = 1$ A m) at height d over earth. (b) Unit vertical dipole at surface, unit horizontal dipole at depth d in earth.

$$E_{2\Theta} = -\frac{i\omega\mu_0}{4\pi}\frac{e^{ik_2r_0}}{r_0}A_e(\Theta)\sin\Theta, \tag{1.3.1a}$$

$$A_e(\Theta) = e^{-ik_2d\cos\Theta} + f_{er}e^{ik_2d\cos\Theta}, \tag{1.3.1b}$$

where Θ is measured from the vertical axis and $A_e(\Theta)$ is the field factor. The plane-wave reflection coefficient of electric type (electric field in the plane of incidence) is $f_{er} = |f_{er}|\exp(i\psi_{er})$. It is defined as follows in terms of the square of the complex index of refraction

$$N^2 = \left(\frac{k_1}{k_2}\right)^2 = \frac{\tilde{\epsilon}_1}{\epsilon_2} = \epsilon_{1r} + \frac{i\sigma_1}{\omega\epsilon_0}, \tag{1.3.2}$$

$$f_{er} = \frac{N^2\cos\Theta - (N^2-\sin^2\Theta)^{1/2}}{N^2\cos\Theta + (N^2-\sin^2\Theta)^{1/2}} = \frac{\cos\Theta - z_{se}(\Theta)}{\cos\Theta + z_{se}(\Theta)}, \tag{1.3.3}$$

where $z_{se}(\Theta)$ is the normalized surface impedance of electric type, viz.,

$$z_{se}(\Theta) = \frac{Z_{se}(\Theta)}{\zeta_2} = \frac{(N^2-\sin^2\Theta)^{1/2}}{N^2}, \tag{1.3.4a}$$

$$\zeta_2 = \zeta_0 = \left(\frac{\mu_0}{\epsilon_0}\right)^{1/2} \sim 120\pi\ \Omega. \tag{1.3.4b}$$

(See Section 2.1 for a more accurate value of ζ_0.) The corresponding formulas for an electrically small horizontal loop (vertical magnetic dipole) with unit magnetic moment ($I\Delta S = 1$ A m^2) at a height d over the earth is

$$E_{2\Phi} = \frac{\omega\mu_0}{4\pi}\frac{e^{ik_2 r_0}}{r_0}A_m(\Theta)\sin\Theta, \tag{1.3.5a}$$

$$A_m(\Theta) = e^{-ik_2 d\cos\Theta} + f_{mr}e^{ik_2 d\cos\Theta}. \tag{1.3.5b}$$

The plane-wave reflection coefficient of magnetic type (electric field perpendicular to the plane of incidence) is $f_{mr} = |f_{mr}|\exp(i\psi_{mr})$. It is defined as follows:

$$f_{mr} = \frac{\cos\Theta - (N^2 - \sin^2\Theta)^{1/2}}{\cos\Theta + (N^2 - \sin^2\Theta)^{1/2}} = \frac{\cos\Theta - y_{sm}(\Theta)}{\cos\Theta + y_{sm}(\Theta)}, \tag{1.3.6}$$

where $y_{sm}(\Theta)$ is the normalized surface admittance of magnetic type, viz.,

$$y_{sm}(\Theta) = \frac{Y_{sm}(\Theta)}{\eta_2} = (N^2 - \sin^2\Theta)^{1/2}, \tag{1.3.7a}$$

$$\eta_2 = \eta_0 = \left(\frac{\epsilon_0}{\mu_0}\right)^{1/2} \sim \frac{1}{120\pi}\ \text{S}. \tag{1.3.7b}$$

Note that for normal incidence with $\Theta = 0$,

$$Z_s = Z_{se}(0) = \frac{\zeta_2}{N} = \frac{\zeta_2 k_2}{k_1} = \frac{\omega\mu_0}{k_1}, \tag{1.3.8a}$$

$$Y_s = Y_{sm}(0) = \eta_2 N = \frac{\eta_2 k_1}{k_2} = \frac{k_1}{\omega\mu_0}. \tag{1.3.8b}$$

When Region 1 is a good conductor, $k_1 = (i\omega\mu_0\sigma_1)^{1/2}$ and $Y_s^{-1} = Z_s = (-i\omega\mu_0/\sigma_1)^{1/2} = (1-i)(\omega\mu_0/2\sigma_1)^{1/2}$ in agreement with (1.2.7). When $\Theta = \pi/2$, $f_{er} = f_{mr} = -1$; when $\Theta = 0$, $f_{er} = -f_{mr} = (N-1)/(N+1)$. Thus, the magnitude of f_{er} has values at or near 1 when $\Theta = 0$ and $\pi/2$ and a deep minimum between $\Theta = 0$ and $\pi/2$. Graphs of both $|f_{er}|$, ψ_{er}, and $|f_{mr}|$, ψ_{mr} are shown in Figs. 1.3.2a and 1.3.2b for a range of values of N^2 appropriate to soil and water, respectively, where ψ_{er} and ψ_{mr} are the phases of f_{er} and f_{mr}. The complete field factors $A_e(\Theta)\sin\Theta$ and $A_m(\Theta)\sin\Theta$ are shown in Fig. 1.3.3 for electric and magnetic dipoles on the surface of the earth ($d = 0$). It is seen that the fields of electric type have maxima between $\Theta = 60°$ and $90°$, depending on the value of N, but vanish at $\Theta = \pi/2$ for all finite N. Only for $N = i\infty$, i.e., for a perfect conductor, does the maximum for the electric dipole occur at $\Theta = \pi/2$. It is virtually $\pi/2$ for a metal surface. For a magnetic dipole on a perfect conductor, the field at $\Theta = \pi/2$ is identically zero. The corresponding field

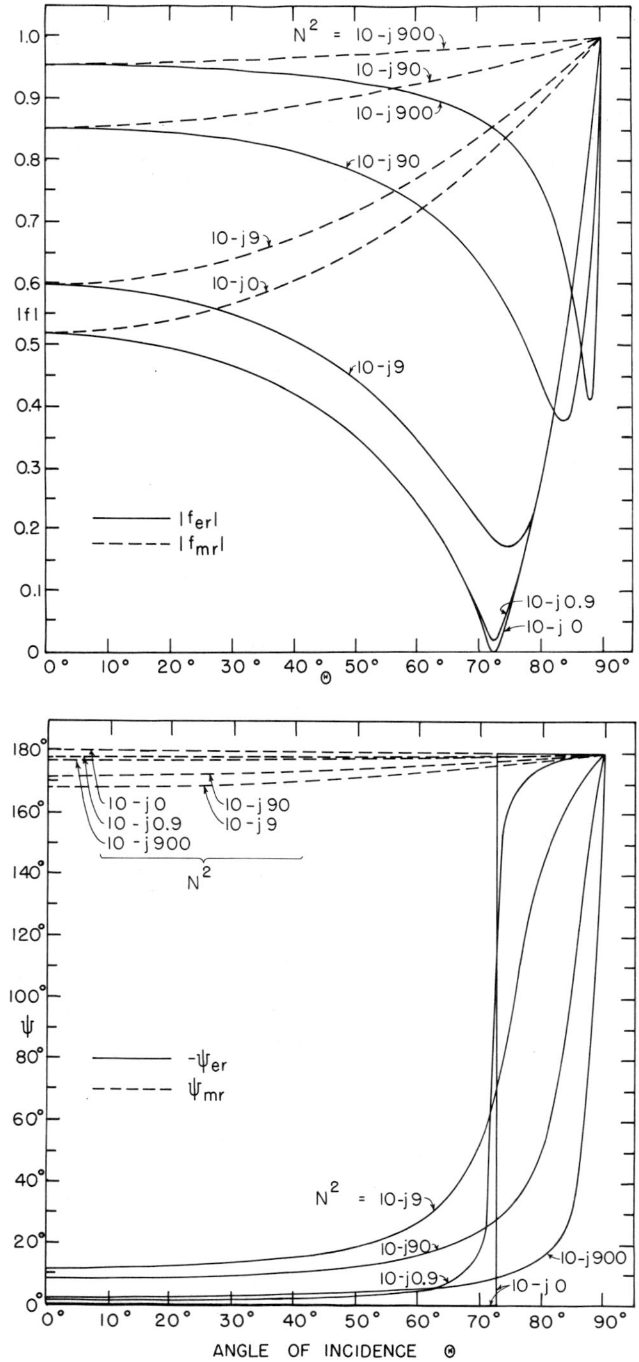

Figure 1.3.2a. Reflection coefficients f_{er} and f_{mr}; $\epsilon_{1r} = 10$. Note that $j = -i$.

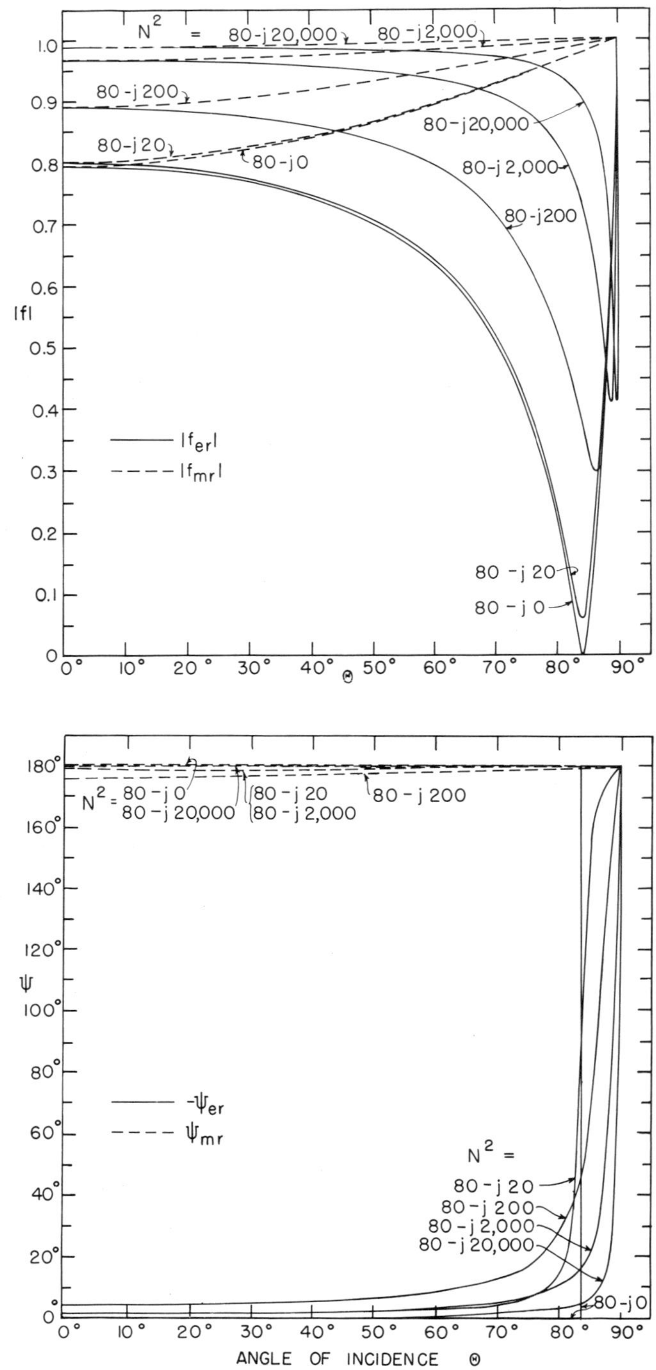

Figure 1.3.2b. Reflection coefficients f_{er} and f_{mr}; $\epsilon_{1r} = 80$. Note that $j = -i$.

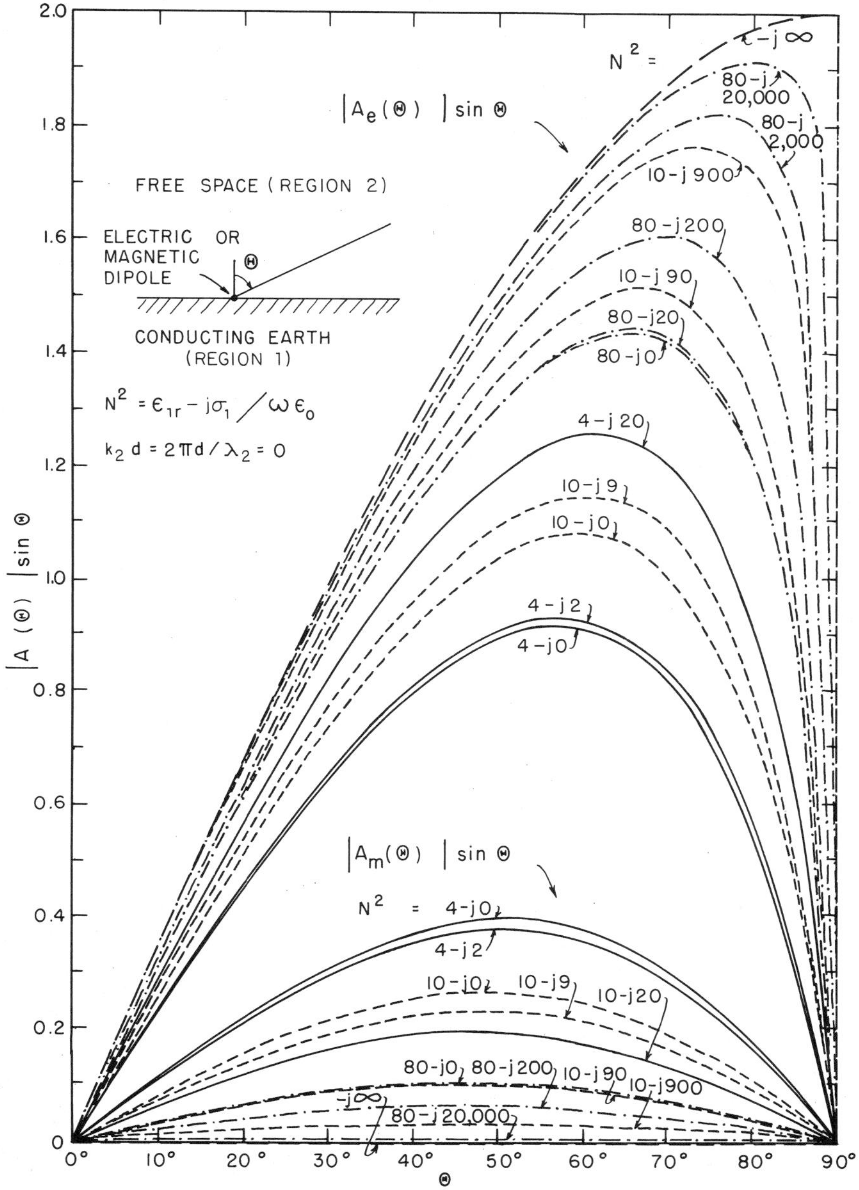

Figure 1.3.3. Field factors of vertical electric and magnetic dipoles on conducting earth; $d = 0$. Note that $j = -i$.

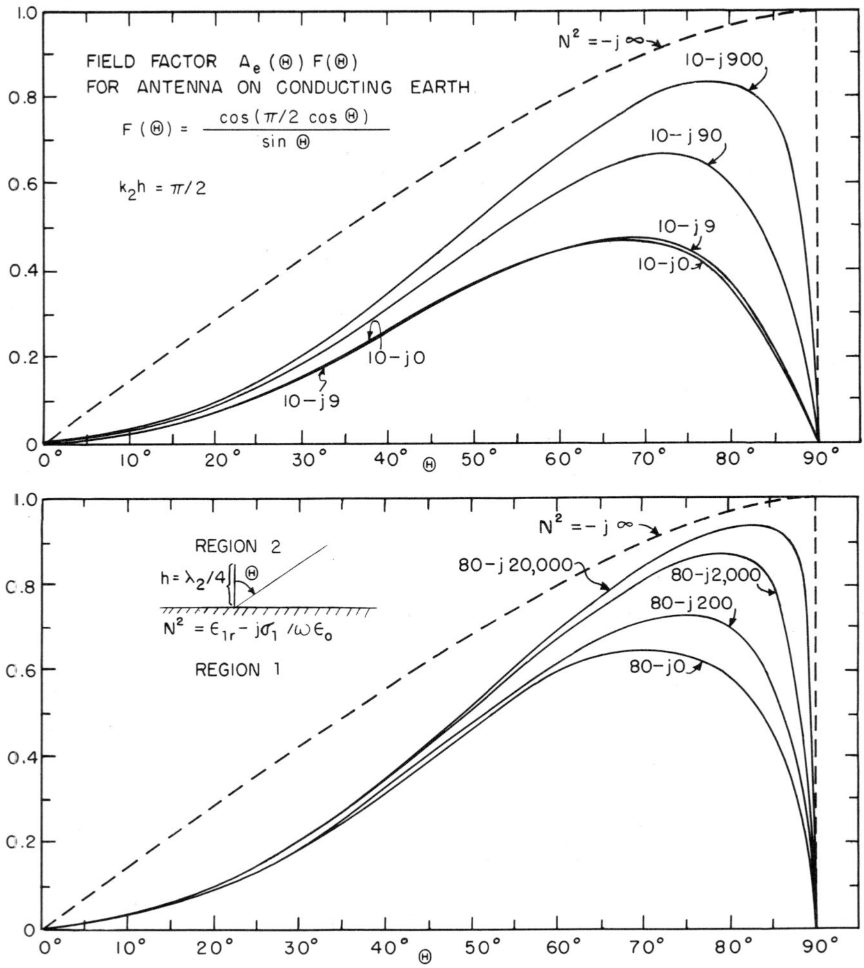

Figure 1.3.4. Vertical field pattern of quarter-wave monopole on a conducting earth or water; $\epsilon_{1r} = 10$ and 80. Note that $j = -i$.

patterns for an electric quarter-wave monopole on the earth are shown in Fig. 1.3.4. They are seen to vanish at $\Theta = \pi/2$ except when the surface of the earth is replaced by a perfect conductor. Note that the larger N^2, the closer to $\Theta = \pi/2$ is the maximum of the field pattern and the steeper is the drop to zero at $\Theta = \pi/2$. This means that a vertical electric receiving antenna located at a sufficient height can be exposed to a near maximum field. Actually, the field at $\Theta = \pi/2$ is not zero as indicated by the space-wave terms consisting of direct and reflected parts, but is determined by a surface-wave term that must be added to (1.3.1a). It is significant only when Θ is near or at $\pi/2$.

In order to make Sommerfeld's complicated solution with its complex integrals convenient for engineering use, Norton (1936), with the help of expressions for the ground-wave potential due to Van der Pol and Niessen (1930) and the electric field due to Wise (1931), reduced the surface-wave term for the electric dipole to formulas and graphs. These involved approximations that restricted their ranges of validity. Norton then compared the fields obtained from them with the available experimental results in order to determine these ranges. He was careful to point out that "when the theory is used to predict the field intensity in other frequency and distance ranges than those for which it has been checked ..., there is the possibility of error due to leaving out some factor which becomes important in these new situations." A helpful discussion and correction of Norton's work has been provided by R. J. King (1969).

Norton's surface-wave term in the field of a unit ($I\Delta l = 1$ A m) vertical electric dipole in Region 2 (air) on the boundary ($d = 0$) with Region 1 (earth) is (see Fig. 1.3.1b)

$$E_{2z}^s(\rho, 0) = \frac{i\omega\mu_0}{2\pi} \frac{e^{ik_2\rho}}{\rho} F_e, \tag{1.3.9}$$

where

$$F_e = 1 + i(\pi p_S)^{1/2} e^{-p_S}[1 - \operatorname{erf}(-ip_S^{1/2})], \tag{1.3.10}$$

and Sommerfeld's *numerical distance* is defined by

$$p_S = |p_S|e^{ib} = ik_2\rho\left(\frac{k_2^2}{2k_1^2}\right) = ip. \tag{1.3.11}$$

The error function is

$$\operatorname{erf}(z) = 2\pi^{-1/2}\int_0^z e^{-t^2}\,dt. \tag{1.3.12}$$

Norton provided graphs and tables of F_e as a function of the magnitude and argument of the numerical distance p_S. Their quantitative significance is considered later. A useful alternative form of (1.3.9) is in terms of the Fresnel integral,

$$C_2(z) + iS_2(z) = \int_0^z (2\pi t)^{-1/2} e^{it}\,dt. \tag{1.3.13}$$

With it, (1.3.9) becomes

$$E_{2z}^s(\rho, 0) = \frac{\omega\mu_0}{2\pi k_2} e^{ik_2\rho}\left[\frac{ik_2}{\rho} - \frac{k_2^3}{k_1}\left(\frac{\pi}{k_2\rho}\right)^{1/2} e^{-ip}\mathcal{F}(p)\right], \tag{1.3.14}$$

with

$$\mathcal{F}(p) = \tfrac{1}{2}(1 + i) - C_2(p) - iS_2(p). \tag{1.3.15}$$

The leading term for very large values of $k_2\rho$, i.e., when $k_2\rho \gg \pi|k_1^2/k_2^2|$, is

$$E_{2z}^s(\rho,0) \sim -\frac{\omega\mu_0}{2\pi}\frac{k_1^2}{k_2^3}\frac{e^{ik_2\rho}}{\rho^2}. \tag{1.3.16}$$

For small and intermediate values of $k_2\rho$, i.e., when $k_2\rho \ll \pi|k_1^2/k_2^2|$,

$$\begin{aligned} E_{2z}^s(\rho,0) &\sim \frac{\omega\mu_0}{2\pi k_2}e^{ik_2\rho}\left[\frac{ik_2}{\rho} - \frac{1}{2}(1+i)\frac{k_2^3}{k_1}\left(\frac{\pi}{k_2\rho}\right)^{1/2}\right] \\ &\sim \frac{i\omega\mu_0}{2\pi}\frac{e^{ik_2\rho}}{\rho}. \end{aligned} \tag{1.3.17}$$

The Zenneck surface-wave term appears explicitly in (1.3.17) where it is the term that varies with radial distance as $\rho^{-1/2}$. However, as predicted by Sommerfeld (1935), it is not the dominant term since it is multiplied by the small factor k_2/k_1. As ρ increases, the valid range of (1.3.17) is passed and (1.3.14) must be used. Finally, for sufficiently large values of ρ—where it might be expected that the $\rho^{-1/2}$ term would dominate over the ρ^{-1} term—the Fresnel integral contributes major factors such that in the second term in (1.3.14) the factor $\exp(-ip)$ is canceled and the term then includes a term $(-ik_2/\rho)$, which exactly cancels the first term in (1.3.14), and a ρ^{-2} term, which gives (1.3.16). The $\rho^{-1/2}$ term disappears when $k_2\rho$ is sufficiently large. Thus, the Zenneck surface-wave term, while formally present, is only a relatively small and unimportant part of (1.3.14).

The problem treated by Sommerfeld and evaluated by Norton concerned propagation over a plane earth. An extension of this work to the spherical earth is contained in a treatise by Bremmer (1949) and a more recent paper by Hill and Wait (1980).

1.4 General Integrals and Approximate Formulas

Following the work of Norton, which addressed primarily the problem of surface-wave communication between vertical transmitting and receiving antennas in the air on the surface of the earth, interest turned to the challenging problem of communicating with submerged submarines. Extensive studies were carried out relating especially to the electromagnetic fields generated by horizontal and vertical electric and magnetic dipoles in the sea. This work included an extensive series of papers by Wait (1971) and reports by Baños and Wesley which culminated in the exhaustive book by Baños (1966). In this, the entire problem is reformulated; general integrals are obtained for the Hertz potentials and formal integrals for the components of the electromagnetic field in both regions. Baños' exposition, like the earlier work by Wait (1961, 1971), follows the original work of Sommerfeld (1909,

1926, 1935) in representing the electromagnetic field in terms of derivatives of the relevant Hertz potentials. These are generally so complicated that they provide neither insight into the physical nature of the surface waves nor practical convenience in the determination of quantitative results.

In order to overcome the handicap of complexity, Baños made very extensive investigations of the possibility of deriving useful approximate formulas. He suggested general restrictions for their evaluation which are equivalent to the following:

$$|k_1| \geq 3|k_2|; \qquad |k_1\rho| \geq 3; \qquad \rho \geq 5z; \qquad \rho \geq 5d, \tag{1.4.1}$$

where d is the depth of the source dipole in Region 1 and z is the depth of the point of observation in Region 1; ρ is the radial distance from the source at $(0,0,d)$. These conditions are not very restrictive. However, Baños found that with only these restrictions, the formulas for the components of the electromagnetic field were much "too cumbersome" for practical use. He, like Wait (1961, 1971), imposed much more severe conditions in order to obtain simpler expressions. Unfortunately, the illuminating, quite simple formulas he derived are valid only in three generally nonoverlapping ranges. His simplest expressions (Baños 1966, Chap. 7) are identified with the restricted ranges defined essentially as follows:

$$\text{Asymptotic or far field:} \qquad 1 \ll |k_2\rho(k_2^2/2k_1^2)|, \tag{1.4.2}$$

$$\text{Intermediate field:} \qquad |k_2\rho(k_2^2/2k_1^2)| < 1 < |k_2\rho|, \tag{1.4.3}$$

$$\text{Near field:} \qquad |k_2\rho| < 1 < |k_1\rho|. \tag{1.4.4}$$

These conditions were imposed by Baños in addition to (1.4.1). They correspond roughly to the three ranges defined by Wait (1961).

Baños' approximate, simple formulas for the cylindrical components of the lateral-wave electric field at (ρ, ϕ, z) in Region 1 due to a horizontal electric dipole with unit electric moment ($I\Delta l = 1$ A m) located at $(0,0,d)$ also in Region 1 are as follows:

Asymptotic field: $\quad |k_2\rho| \gg |2k_1^2/k_2^2|$

$$E_{1\rho}^L \sim \frac{\omega\mu_0}{2\pi k_2}\,\frac{\cos\phi}{\rho^2}\,e^{i[k_2\rho + k_1(z+d)]}, \tag{1.4.5}$$

$$E_{1\phi}^L \sim \frac{\omega\mu_0 k_2}{2\pi k_1^2}\,\frac{\sin\phi}{\rho^2}\,e^{i[k_2\rho + k_1(z+d)]}, \tag{1.4.6}$$

$$E_{1z}^L \sim -\frac{\omega\mu_0}{2\pi k_1}\,\frac{\cos\phi}{\rho^2}\,e^{i[k_2\rho + k_1(z+d)]}. \tag{1.4.7}$$

Intermediate field: $\quad 1 < |k_2\rho| < |2k_1^2/k_2^2|$

$$E_{1\rho}^L \sim -\frac{i\omega\mu_0 k_2^2}{2\pi k_1^2}\,\frac{\cos\phi}{\rho}\left[1 + \frac{ik_2}{k_1}\left(\frac{i\pi k_2\rho}{2}\right)^{1/2}\right]e^{i[k_2\rho + k_1(z+d)]}, \tag{1.4.8}$$

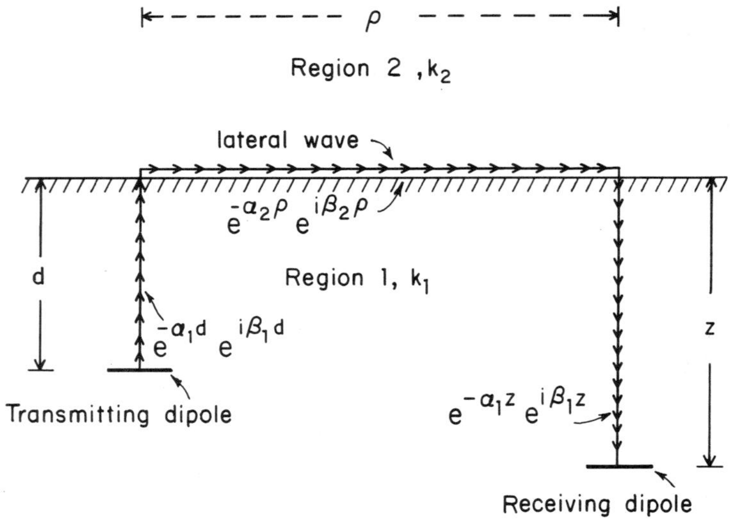

Figure 1.4.1. Apparent path of lateral wave and its relation to the mathematical representation $\exp\{-[\alpha_2\rho + \alpha_1(z+d)]\}\exp\{i[\beta_2\rho + \beta_1(z+d)]\}$.

$$E_{1\phi}^L \sim \frac{\omega\mu_0 k_2}{\pi k_1^2}\,\frac{\sin\phi}{\rho^2}\left[1 + \frac{ik_2}{2k_1}\left(\frac{i\pi k_2\rho}{2}\right)^{1/2}\right] e^{i[k_2\rho + k_1(z+d)]}, \tag{1.4.9}$$

$$E_{1z}^L \sim \frac{i\omega\mu_0 k_2^3}{2\pi k_1^3}\,\frac{\cos\phi}{\rho}\left[1 + \frac{ik_2}{k_1}\left(\frac{i\pi k_2\rho}{2}\right)^{1/2}\right] e^{i[k_2\rho + k_1(z+d)]}. \tag{1.4.10}$$

Near field: $\quad k_2/k_1 < |k_2\rho| < 1$

$$E_{1\rho}^L \sim \frac{i\omega\mu_0}{2\pi k_1^2}\,\frac{\cos\phi}{\rho^3}\, e^{i[k_2\rho + k_1(z+d)]}, \tag{1.4.11}$$

$$E_{1\phi}^L \sim \frac{i\omega\mu_0}{\pi k_1^2}\,\frac{\sin\phi}{\rho^3}\, e^{i[k_2\rho + k_1(z+d)]}, \tag{1.4.12}$$

$$E_{1z}^L \sim -\frac{\omega\mu_0 k_2^2}{2\pi k_1^3}\,\frac{\cos\phi}{\rho^2}\, e^{i[k_2\rho + k_1(z+d)]}. \tag{1.4.13}$$

The principal lateral-wave characteristic of these formulas is the exponential term which indicates a plane wave traveling vertically from the source a distance d in Region 1 to the boundary, then radially along the boundary in Region 2 a distance ρ, and finally vertically a distance z in Region 1 to the point of observation, as shown in Fig. 1.4.1. Extensive graphs of $|E_{1\rho}|$, $|E_{1\phi}|$, and $|E_{1z}|$ calculated from Baños' approximate formulas (1.4.5)–(1.4.13) for sea water, lake water, and dry earth are in Figs. 11.7.1a through 11.7.3c of King and Smith (1981) for wide ranges of the radial distance and frequency. Wait's formulas (1961) with his conditions are as follows:

Asymptotic field: $|k_2\rho| \gg 1; \quad |k_1| \gg |k_2|; \quad d^2 \ll \rho^2; \quad z^2 \ll \rho^2$

$$E_{1\rho}^L \sim -\frac{i\omega\mu_0 k_2^2}{2\pi k_1^2}\frac{\cos\phi}{\rho} \times [1 + i(\pi p_S)^{1/2} e^{-p_S}\,\mathrm{erfc}(-ip_S^{1/2})]e^{i[k_2\rho + k_1(z+d)]}, \tag{1.4.14}$$

$$E_{1\phi}^L \sim \frac{\omega\mu_0 k_2}{\pi k_1^2}\frac{\sin\phi}{\rho^2}\, e^{i[k_2\rho + k_1(z+d)]}, \tag{1.4.15}$$

$$E_{1z}^L \sim \frac{i\omega\mu_0 k_2^3}{2\pi k_1^3}\frac{\cos\phi}{\rho} \times [1 + i(\pi p_S)^{1/2} e^{-p_S}\,\mathrm{erfc}(-ip_S^{1/2})]e^{i[k_2\rho + k_1(z+d)]}. \tag{1.4.16}$$

Near field: $|k_2\rho| \ll |2k_1^2/k_2^2|; \quad |k_1| \gg |k_2|; \quad d^2 \ll \rho^2; \quad z^2 \ll \rho^2$

$$E_{1\rho}^L \sim \frac{i\omega\mu_0}{2\pi k_1^2}\frac{\cos\phi}{\rho^3}(1 - ik_2\rho - k_2^2\rho^2)e^{i[k_2\rho + k_1(z+d)]}, \tag{1.4.17}$$

$$E_{1\phi}^L \sim \frac{i\omega\mu_0}{\pi k_1^2}\frac{\sin\phi}{\rho^3}(1 - ik_2\rho)e^{i[k_2\rho + k_1(z+d)]}, \tag{1.4.18}$$

$$E_{1z}^L \sim -\frac{\omega\mu_0 k_2^2}{2\pi k_1^3}\frac{\cos\phi}{\rho^2}(1 - ik_2\rho)e^{i[k_2\rho + k_1(z+d)]}. \tag{1.4.19}$$

Quasi-static field: $|k_2| \to 0; \quad z = d = 0$

$$E_{1\rho}^L \sim \frac{i\omega\mu_0}{\pi k_1^2}\frac{\cos\phi}{\rho^3}, \tag{1.4.20}$$

$$E_{1\phi}^L \sim \frac{i\omega\mu_0}{2\pi k_1^2}\frac{\sin\phi}{\rho^3}, \tag{1.4.21}$$

$$E_{1z}^L \sim 0. \tag{1.4.22}$$

These formulas are included here, without discussion, for later reference. In (1.4.14) and (1.4.16), p_S is Sommerfeld's numerical distance; it is given in (1.3.11).

Electromagnetic waves are highly attenuated in sea water where, except at frequencies above 1 GHz, $\alpha_1 \sim \beta_1 \sim (\omega\mu_0\sigma_1/2)^{1/2}$. Only at very low frequencies is α_1 small enough and $\exp(-\alpha_1 r)$ large enough at practical distances r to permit communication with submerged submarines. Accordingly, this problem was studied primarily in terms of near-field or quasi-static formulas which are valid when $\beta_1 r \ll 1$. An extensive treatment of the quasi-static approximation has been given by Bannister and Dube (1978).

1.5 Numerical Methods with High-Speed Computers

When high-speed computers became available, attention was directed to the numerical evaluation of the Sommerfeld integrals in order to determine the electromagnetic field of horizontal electric dipoles embedded in the earth or sea near their surface with air. The early calculations of Siegel and King (1970, 1971) applied specifically to insulated traveling-wave antennas in the ocean. The problem was studied more generally by Lytle, Miller, and Lager (1976), Bubenik (1977), and Rahmat-Samii, Mittra, and Parhami (1981). Extensive systematic calculations of the fields of horizontal electric dipoles in air-bounded media ranging from sea water to dry earth over a wide range of frequencies were carried out by King and Sandler (1977), King, Sandler, and Shen (1980), and King and Smith (1981). Similar calculations for rock-bounded sea water were reported by King, deBettencourt, and Sandler (1979). All of these have been compared with calculations from Baños' approximate formulas in order to establish the validity of the specified ranges. Extensive measurements of the electromagnetic fields of horizontal electric antennas in sea water (Siegel and King 1973) and lake water (Shen, King and Sorbello 1976) have been made and compared with both the appropriate numerical results and with Baños' approximate formulas. More recently numerically evaluated fields of horizontal dipoles on the sea bottom have been reported by Young and Cox (1981) and Chave and Cox (1982).

Although numerical methods and high-speed computers have proved to be very valuable tools in the evaluation of fields that depend on the complex integrals of Sommerfeld, they do have serious limitations. These are a consequence of the unavailability of analytical expressions that give insight into and understanding of the complicated fields. Even very extensive and systematic evaluations and tabulations like those in King and Smith (1981) failed to reveal the physically fascinating and important interference patterns generated by the surface wave and direct wave near a horizontal dipole in various half-spaces bounded by air. These are described and discussed later. Similarly, the numerical calculations of Young and Cox (1981), while correctly calculating the desired specific electric field on the sea bottom, failed to disclose the reasons for the very different behavior of the components or the orientation of the antenna required for simple physical interpretation of the results. This is also discussed later. Computer-based numerical evaluations of the fields of dipoles near an interface are powerful but essentially blind. They are a most valuable supplement to, but no substitute for, analytical formulas when these are both accurate and conceptually simple. It is the purpose of this book to derive and apply such formulas subject only to the general restrictions (1.4.1)—or in some cases only $|k_1| \geq 3|k_2|$—which are almost always satisfied in practice.

1.6 Lateral Waves

The particular type of electromagnetic surface wave that is generated by horizontal and vertical dipoles on or near the boundary between air and earth (soil, lake water, sea water) or the sea floor and sea water is generally known as a lateral wave. It has been studied extensively and very generally in its relation to total reflection phenomena not only in electromagnetics and optics, but also in acoustics and seismic phenomena. Introductions to this approach to lateral waves are given in a treatise by Brekhovskikh (1980) and an excellent summary by Tamir in Chapter 13 of a book edited by Boardman (1982). These works contain extensive lists of references.

As pointed out by Tamir, a lateral wave is generated "by a bundle of rays or a beam of *bounded extent* at an angle Θ close to Θ_c," the critical angle for total internal reflection. "A lateral wave is absent if the incident field is an infinite plane wave, so that the realistic finite width of the incident field is instrumental in producing the lateral wave. In this context, the lateral wave manifests itself by (a) producing a lateral shift of the bounded reflected field, and (b) exciting a weak field that accompanies the reflected field. Neither of these two effects is accounted for by geometrical optics." Tamir provides a mathematical classification of fields supported by a boundary in terms of an assumed incident field in a two-dimensional situation. This is done in terms of the contours of integration in the complex plane and distinguishes between surface-wave modes, leaky waves, radiation modes, and the lateral wave.

Both Brekhovskikh (1980) and Tamir in Boardman (1982) discuss the incident spherical wave and the incident cylindrical wave due to a line source. The latter investigates the simpler case of the line source at a depth d in Region 1 ($k_1 = Nk_2$, $|N| > 1$) illustrated in Fig. 1.6.1. The asymptotic or far field observed at $P(\rho, z)$ is given by

$$E_1(\rho, z) = E_1^{\text{inc}}(\rho, z) + E_1^{\text{ref}}(\rho, z) + E_1^{L}(\rho, z), \tag{1.6.1}$$

where the direct field is

$$E_1^{\text{inc}}(\rho, z) = \frac{Ae^{ik_1r_1}}{(k_1r_1)^{1/2}}. \tag{1.6.2}$$

Here, A is proportional to the intensity of the line source and $r_1 = [(z - d)^2 + \rho^2]^{1/2}$. The reflected field is

$$E_1^{\text{ref}}(\rho, z) = Af_r\,\frac{e^{ik_1r_2}}{(k_1r_2)^{1/2}}, \tag{1.6.3}$$

where $r_2 = [(z+d)^2+\rho^2]^{1/2}$ and f_r is the plane-wave reflection factor which has the forms

$$f_r \equiv f_{er} = \frac{N^2\cos\Theta - (N^2 - \sin^2\Theta)^{1/2}}{N^2\cos\Theta + (N^2 - \sin^2\Theta)^{1/2}} \tag{1.6.4}$$

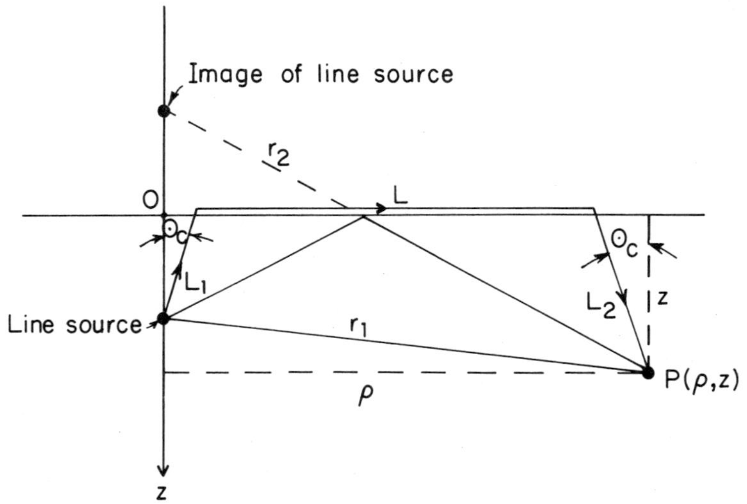

Figure 1.6.1. Path of lateral wave from line source (after Tamir).

for E-polarization, or

$$f_r \equiv f_{mr} = \frac{\cos\Theta - (N^2 - \sin^2\Theta)^{1/2}}{\cos\Theta + (N^2 - \sin^2\Theta)^{1/2}} \tag{1.6.5}$$

for H-polarization. The lateral wave in (1.6.1) is

$$E_1^L = A\,\frac{2iN^2}{N^2-1}\,\frac{\exp\{i[k_1(L_1+L_2)+k_2L]\}}{(k_2L)^{3/2}} \tag{1.6.6}$$

for E-polarization, or

$$E_1^L = A\,\frac{2i}{N^2-1}\,\frac{\exp\{i[k_1(L_1+L_2)+k_2L]\}}{(k_2L)^{3/2}} \tag{1.6.7}$$

for H-polarization. The distances L_1 and L_2 in Region 1 and L in Region 2 are shown in Fig. 1.6.1. When the critical angle of incidence is small, $\Theta_c \sim 0$, $L_1 \sim d$, $L_2 \sim z$, and $L \sim \rho$.

A comparison of the asymptotic lateral-wave field (1.4.7) for the infinitesimal dipole with (1.6.6) or (1.6.7) for the line source shows that they are entirely comparable. The $1/\rho^2$ dependence for an outgoing spherical wave corresponds to the $1/L^{3/2}$ dependence for an outgoing cylindrical wave. Baños' formulas assume $|k_1| \gg |k_2|$ so that the critical angle of incidence Θ_c is very small and the wave proceeds almost vertically in Region 1 with $L_1 \sim d$, $L_2 \sim z$, and $L \sim \rho$.

1.7 Lateral Waves and the Geoelectromagnetic Investigation of the Earth's Crust and Upper Mantle

Since electrical conductivity is an important parameter that characterizes the composition and physical state of the earth's interior, its experimental determination as a function of depth can provide information about the present state of the earth's crust and upper mantle. In order to measure the conductivity in the earth, geophysicists have made use of natural electromagnetic fields generated by time-varying currents in the ionosphere and magnetosphere. Their nature is summarized in Chapter 1, "Geoelectromagnetic Fields," in the book by Rokityansky (1982). The low frequencies of such natural fields allow them to penetrate deeply into the earth where they induce eddy currents with amplitudes and distributions that are directly related to the magnitude and distribution of the conductivity.

Geophysicists assume that over limited areas on the earth the currents in the ionosphere can be approximated by linear distributions of dipoles that are infinite in extent, i.e., by line sources (Vanyan 1967, p. 127) located above the earth's surface. This means that measurements and calculations of incident fields on and in the earth are at points effectively under the source currents. With these currents of the order of 80 to 100 km high, the field incident on the earth is approximated by a spherical or locally plane wave. Even if the angle of incidence in the air is not exactly zero, the inhomogeneous field traveling into the conducting earth or sea will have amplitude fronts that are parallel to the boundary surface and, because of the large index of refraction, phase fronts that are also essentially parallel to the earth's surface. This means that the assumption of a vertically downward-traveling plane wave is a generally good one. It is the basis of the analytical formulation of Rokityansky (1982). Note that the lateral wave excited by the line source along the surface of the earth or sea (as discussed in Section 1.6) is not involved so long as attention is directed exclusively to the region under the line source.

The primary electromagnetic field can also be generated by so-called "controlled sources" which consist either of an insulated wire grounded at both ends (horizontal electric dipole) or of an insulated horizontal wire loop (vertical magnetic dipole). Antennas of these types are located in the air just above the surface of the earth or sea, or in the sea just above the sea floor. Along the vertical axis through the center of the horizontal dipole source, the plane-wave approximation used with the natural sources may be adequate; but for the fields radially outward along the boundary between the air and the earth or sea or between the sea and the oceanic crust, the lateral waves generated by the dipole cannot be ignored. There is no mention of lateral waves in the work of Vanyan (1967) and only casual mention in the treatise by Kaufman and Keller (1983) where, in considering the "relationships between the field measured on the sea floor and the electrical properties of a two-layer sequence beneath the sea floor," the following statement is made: "It is obvious that where the depth of the sea is relatively small, there will be an effect from electromagnetic energy travelling from the source (on the sea floor) upwards to the sea surface, along the surface, and finally, back down to the observation site." This is a lateral wave and its effect is ignorable when the sea is sufficiently deep. However, no reference is made to the equally obvious and highly significant lateral wave that proceeds from the source to the sea floor, along this *in the crust*, and finally back up to the observation point in the sea! The existence of such a surface wave is recognized by Young and Cox (1981, p. 1043) when they state: "The transmitter and receiver were placed on the sea floor. The signals traveled from the transmitter to the receiver in a down-under-up path thereby usefully sampling the electrical conductivity properties of the crustal and upper mantle rocks in the region between them." But their "best fit" model does not mention the lateral wave in the interpretation which includes only waves reflected from eight layers at different depths up to 33 km. Analyses carried out in later chapters derive the *complete* field generated by a horizontal dipole in both the sea and the adjacent oceanic crust, including the lateral wave along the boundary in the crust and the spherical waves in both regions.

The theoretical physical foundations for geoelectromagnetic investigations are properly begun with Maxwell's equations. Most treatments by geophysicists including, for example, the recent books by Rokityansky (1982) and Kaufman and Keller (1983), and the basic paper by Chave and Cox (1982), immediately specialize these equations by setting

$$\nabla \times \mathbf{B}_j = \mu_j \sigma_j \mathbf{E}_j, \tag{1.7.1}$$

instead of

$$\nabla \times \mathbf{B}_j = \mu_j (\sigma_j - i\omega\epsilon_j) \mathbf{E}_j, \tag{1.7.2}$$

for all regions, $j = 1, 2, \ldots, N$, that may be involved. This is, of course, a valid approximation when

$$\frac{\omega\epsilon_j}{\sigma_j} \ll 1; \qquad k_j \sim (i\omega\mu_j\sigma_j)^{1/2}. \tag{1.7.3}$$

At the low frequencies of interest in most geophysical studies on the sea floor or in the interior of the oceanic crust, this condition is very well satisfied. But to apply (1.7.1) and formulas derived from it to a dielectric like air by setting $\sigma = 0$ (Kaufman and Keller 1983, p. 232) is questionable in any general sense since $\omega\epsilon/\sigma \to \infty$ except when ω is identically zero, i.e., for direct currents. When both σ_j and $\omega\epsilon_j$ are zero, $k_j = 0$ and the important surface wave is eliminated.

Actually, the severe condition (1.7.3) is quite unnecessary and *simple integrated* expressions for the complete electromagnetic field of vertical and horizontal electric dipoles near the boundary between the half-spaces 1 and 2 can be derived whenever the two wave numbers are significantly different, that is, when

$$|k_1|^2 \gg |k_2|^2 \quad \text{or} \quad |k_1| \geq 3|k_2|, \tag{1.7.4}$$

where $k_j^2 = \omega^2\mu_j\epsilon_j + i\omega\mu_j\sigma_j$, $j = 1, 2$. This condition can be satisfied when one or both regions are dielectrics or conductors at high and low frequencies.

Geophysicists use the terms "quasi-stationary" and "quasi-static" to identify fields calculated subject to (1.7.3) with no restriction on distance from the source. In this book, in agreement with the usual terminology in electromagnetics (e.g., Kraichman 1970, p. 3-15; King and Prasad 1986, p. 275), these terms are used synonymously with "near field" for the field when $|kr| < 1$, where $k = (\omega^2\mu_j\epsilon_j + i\omega\mu_j\sigma_j)^{1/2}$ is the wave number of the region in which the field is defined and r is the radial distance from the source to the point of observation. When r is fixed, the frequencies for which $|kr| < 1$ define the "quasi-static" or "quasi-stationary" state; when the frequency is fixed, the radial distances which satisfy $|kr| < 1$ contain the "near field."

2
Electromagnetic Preliminaries

2.1 Maxwell's Equations; Parameters

With an assumed time dependence of $\exp(-i\omega t)$, the basic equations that govern the electromagnetic field in all homogeneous, isotropic media are

$$\nabla \times \mathbf{B} = \mu \mathbf{J}^e - \frac{ik^2}{\omega}\mathbf{E}; \qquad \nabla \cdot \mathbf{B} = 0, \tag{2.1.1}$$

$$\nabla \times \mathbf{E} = i\omega \mathbf{B}; \qquad \nabla \cdot \mathbf{E} = \frac{\rho^e}{\tilde{\epsilon}}, \tag{2.1.2}$$

where

$$k = \beta + i\alpha = \omega(\mu\tilde{\epsilon})^{1/2} \tag{2.1.3}$$

is the complex wave number and

$$\tilde{\epsilon} = \epsilon + \frac{i\sigma}{\omega} \tag{2.1.4}$$

is the complex permittivity. In (2.1.4), σ is the conductivity and $\epsilon = \epsilon_0 \epsilon_r$ is the permittivity with $\epsilon_0 = 8.854 \times 10^{-12}$ F/m. The real part of the complex wave number is the real wave number or phase constant β. The associated wavelength λ and phase velocity v are

$$\lambda = \frac{2\pi}{\beta}; \qquad v = \frac{\omega}{\beta}. \tag{2.1.5}$$

The imaginary part of the complex wave number is the attenuation constant α. The associated skin depth is

$$\delta_s = \frac{1}{\alpha}. \tag{2.1.6}$$

The forms (2.1.1) and (2.1.2) of Maxwell's equations assume that all material media are linearly polarizing, magnetizing, and conducting, so that the volume densities of polarization **P**, magnetization **M**, and the conduction current **J** obey the following linear constitutive relations:

$$\mathbf{P} = (\epsilon - \epsilon_0)\mathbf{E}; \qquad -\mathbf{M} = (\mu^{-1} - \mu_0^{-1})\mathbf{B} = 0; \qquad \mathbf{J} = \sigma \mathbf{E}. \tag{2.1.7}$$

The permittivity ϵ and conductivity σ are, in general, complex, due to polarization and conduction lag. Hence,

$$\epsilon = \epsilon' + i\epsilon''; \qquad \sigma = \sigma' + i\sigma'', \tag{2.1.8}$$

and the real effective permittivity and conductivity are

$$\epsilon_e = \epsilon' - \frac{\sigma''}{\omega}; \qquad \sigma_e = \sigma' + \omega\epsilon''. \tag{2.1.9}$$

Note that ϵ and σ are also, in general, functions of the frequency. Since magnetizable media are rarely involved in the applications of lateral waves, it is assumed throughout this book that $\mathbf{M} = 0$ and

$$\mu = \mu_0 = 4\pi \times 10^{-7} \text{ H/m}. \tag{2.1.10}$$

In (2.1.1), $\mathbf{J}^e$ is the volume density of externally maintained or impressed electric current in the localized regions occupied by sources such as electric dipoles or antennas embedded in the ambient medium. $\mathbf{J}^e$ may be an impressed conduction current with an associated impressed volume density of charge ρ^e that satisfies the equation of continuity

$$\nabla \cdot \mathbf{J}^e - i\omega\rho^e = 0. \tag{2.1.11}$$

Alternatively, $\mathbf{J}^e$ may be an equivalent polarization current in which case $\mathbf{J}^e = -i\omega\mathbf{P}^e$, $\rho^e = -\nabla \cdot \mathbf{P}^e$, where $\mathbf{P}^e$ is the impressed volume density of polarization. Similarly, $\mathbf{M}^e$ is the impressed volume density of magnetization in a localized region, $\mathbf{J}^e = \nabla \times \mathbf{M}^e$ is the associated magnetization current, and $\rho^e = 0$. It is realized in practice by a small loop antenna. For idealized infinitesimal x-directed dipoles, the electric moment is the integral

$$I_x l = \int_\tau J_x^e \, d\tau, \tag{2.1.12}$$

where I_x is the uniform current in an equivalent dipole with the length l. Alternatively,

$$I_x l = -i\omega \int_\tau P_x^e \, d\tau = -i\omega p_x^e, \tag{2.1.13}$$

where p_x^e is the polarization or electric moment of the dipole. A convenient but fictitious magnetic dipole with the length l and fictitious uniform "magnetic current" V_x (consisting of fictitious moving magnetic poles) can be introduced by analogy with (2.1.13). Thus,

$$V_x l = i\omega\mu_0 \int_\tau M_x^e \, d\tau = i\omega\mu_0 m_x^e, \tag{2.1.14}$$

where m_x^e is the magnetization or magnetic moment of the dipole. For the practically realizable, electrically small, equivalent loop antenna

$$m_x^e = I_\theta S, \tag{2.1.15}$$

where I_θ is the current around a circular loop with area S. It follows that in the practical realization of the magnetic dipole by a loop antenna,

$$V_x l = i\omega\mu_0 I_\theta S. \tag{2.1.16}$$

In the calculation of fields it is usually convenient to use idealized *unit* dipoles. That is,

$$I_x l = 1 \text{ A m}; \qquad V_x l = 1 \text{ V m}. \tag{2.1.17}$$

The field of an actual, electrically small antenna is obtained from the field of the unit dipole by multiplying the field of the unit dipole by the equivalent electric or magnetic moments, $I_x l$ or $V_x l$.

It is at times convenient to introduce the wave impedance

$$\zeta = \frac{\omega\mu_0}{k} = \left(\frac{\mu_0}{\tilde{\epsilon}}\right)^{1/2}. \tag{2.1.18}$$

For air,

$$\begin{aligned} \zeta_0 &= \frac{\omega\mu_0}{k_0} = \left(\frac{\mu_0}{\epsilon_0}\right)^{1/2} = \mu_0 c \\ &= 4\pi \times 10^{-7} \times 299\,792\,458 \sim 120\pi\ \Omega. \end{aligned} \tag{2.1.19}$$

2.2 The Vector and Scalar Potentials; The Electromagnetic Field of a Unit Electric Dipole

The vector potential $\mathbf{A}$ and scalar potential ϕ in the Lorentz gauge are usually defined as follows:

$$\nabla \times \mathbf{A} = \mathbf{B}, \tag{2.2.1}$$

$$-\nabla\phi = \mathbf{E} - i\omega\mathbf{A}, \tag{2.2.2}$$

$$\nabla \cdot \mathbf{A} = \frac{ik^2\phi}{\omega} \quad \text{(Lorentz condition)}. \tag{2.2.3}$$

It follows that the vector potential satisfies the second-order equation:

$$\nabla^2\mathbf{A} + k^2\mathbf{A} = -\mu_0\mathbf{J}^e. \tag{2.2.4}$$

An x-directed unit infinitesimal electric dipole at the origin of coordinates in an infinite medium is defined by

$$\mathbf{J}^e = \hat{\mathbf{x}}\delta^3(\mathbf{r}), \tag{2.2.5}$$

where $\delta^3(\mathbf{r})$ is the three-dimensional delta function. With (2.2.5), the solution of (2.2.4) is

$$\mathbf{A} = \mu_0 \hat{\mathbf{x}} \frac{e^{ikr}}{4\pi r}, \tag{2.2.6}$$

where $r = (x^2+y^2+z^2)^{1/2}$. The exponential $\exp(ikr) = \exp(-\alpha r)\exp(i\beta r)$ is chosen in preference to $\exp(-ikr) = \exp(\alpha r)\exp(-i\beta r)$ because it satisfies the radiation condition—which is equivalent to requiring an outward-traveling wave at large distances from the source.

The complete electromagnetic field is now readily evaluated with (2.2.1) to calculate $\mathbf{B}$ and with (2.1.1) to calculate $\mathbf{E}$. Specifically,

$$\mathbf{B} = \nabla \times \mathbf{A}, \tag{2.2.7}$$

$$\mathbf{E} = \frac{i\omega}{k^2}\nabla \times \mathbf{B} = \frac{i\omega}{k^2}\nabla \times \nabla \times \mathbf{A}. \tag{2.2.8a}$$

Alternatively, from (2.2.2) and (2.2.3),

$$\mathbf{E} = -\nabla\phi + i\omega\mathbf{A} = \frac{i\omega}{k^2}(\nabla\nabla\cdot\mathbf{A} + k^2\mathbf{A}). \tag{2.2.8b}$$

In Cartesian coordinates the components are

$$B_x = 0, \tag{2.2.9}$$

$$B_y = \frac{\partial A_x}{\partial z} = \frac{z}{r}\frac{\partial A_x}{\partial r} = \frac{\mu_0}{4\pi}\frac{z}{r}\left(\frac{ik}{r} - \frac{1}{r^2}\right)e^{ikr}, \tag{2.2.10}$$

$$B_z = -\frac{\partial A_x}{\partial y} = -\frac{y}{r}\frac{\partial A_x}{\partial r} = -\frac{\mu_0}{4\pi}\frac{y}{r}\left(\frac{ik}{r} - \frac{1}{r^2}\right)e^{ikr}, \tag{2.2.11}$$

$$\begin{aligned} E_x &= \frac{i\omega}{k^2}\left(\frac{\partial B_z}{\partial y} - \frac{\partial B_y}{\partial z}\right) \\ &= \frac{i\omega\mu_0}{4\pi k^2}\left[\frac{k^2}{r} + \frac{ik}{r^2} - \frac{1}{r^3} - \frac{x^2}{r^2}\left(\frac{k^2}{r} + \frac{3ik}{r^2} - \frac{3}{r^3}\right)\right]e^{ikr}, \end{aligned} \tag{2.2.12}$$

$$E_y = -\frac{i\omega}{k^2}\frac{\partial B_z}{\partial x} = -\frac{i\omega\mu_0}{4\pi k^2}\frac{xy}{r^2}\left(\frac{k^2}{r} + \frac{3ik}{r^2} - \frac{3}{r^3}\right)e^{ikr}, \tag{2.2.13}$$

$$E_z = \frac{i\omega}{k^2}\frac{\partial B_y}{\partial x} = -\frac{i\omega\mu_0}{4\pi k^2}\frac{xz}{r^2}\left(\frac{k^2}{r} + \frac{3ik}{r^2} - \frac{3}{r^3}\right)e^{ikr}. \tag{2.2.14}$$

In cylindrical coordinates (ρ, ϕ, z),

$$\rho = (x^2 + y^2)^{1/2}; \qquad \phi = \tan^{-1}\left(\frac{y}{x}\right), \tag{2.2.15a}$$

$$\cos\phi = \frac{x}{\rho}; \qquad \sin\phi = \frac{y}{\rho}, \tag{2.2.15b}$$

$$E_\rho = E_x \cos\phi + E_y \sin\phi; \qquad E_\phi = -E_x \sin\phi + E_y \cos\phi. \tag{2.2.16}$$

Hence, for the x-directed electric dipole,

$$E_\rho = -\frac{i\omega\mu_0}{4\pi k^2}\left[\frac{2ik}{r^2} - \frac{2}{r^3} - \frac{z^2}{r^2}\left(\frac{k^2}{r} + \frac{3ik}{r^2} - \frac{3}{r^3}\right)\right] e^{ikr}\cos\phi, \tag{2.2.17}$$

$$E_\phi = -\frac{i\omega\mu_0}{4\pi k^2}\left(\frac{k^2}{r} + \frac{ik}{r^2} - \frac{1}{r^3}\right) e^{ikr}\sin\phi, \tag{2.2.18}$$

$$E_z = -\frac{i\omega\mu_0}{4\pi k^2}\frac{\rho z}{r^2}\left(\frac{k^2}{r} + \frac{3ik}{r^2} - \frac{3}{r^3}\right) e^{ikr}\cos\phi, \tag{2.2.19}$$

$$B_\rho = \frac{\mu_0}{4\pi}\frac{z}{r}\left(\frac{ik}{r} - \frac{1}{r^2}\right) e^{ikr}\sin\phi, \tag{2.2.20}$$

$$B_\phi = \frac{\mu_0}{4\pi}\frac{z}{r}\left(\frac{ik}{r} - \frac{1}{r^2}\right) e^{ikr}\cos\phi, \tag{2.2.21}$$

$$B_z = -\frac{\mu_0}{4\pi}\frac{\rho}{r}\left(\frac{ik}{r} - \frac{1}{r^2}\right) e^{ikr}\sin\phi. \tag{2.2.22}$$

The field of a z-directed electric dipole in cylindrical coordinates is calculated with

$$\mathbf{J}^e = \hat{\mathbf{z}}\delta^3(\mathbf{r}) \tag{2.2.23}$$

from

$$\mathbf{A} = \mu_0 \hat{\mathbf{z}}\,\frac{e^{ikr}}{4\pi r} \tag{2.2.24}$$

and (2.2.7) and (2.2.8). It is

$$E_\rho = -\frac{\omega\mu_0}{4\pi k^2}\frac{\rho z}{r^2}\left(\frac{ik^2}{r} - \frac{3k}{r^2} - \frac{3i}{r^3}\right) e^{ikr}, \tag{2.2.25}$$

$$E_z = \frac{\omega\mu_0}{4\pi k^2}\left[\frac{ik^2}{r} - \frac{k}{r^2} - \frac{i}{r^3} - \frac{z^2}{r^2}\left(\frac{ik^2}{r} - \frac{3k}{r^2} - \frac{3i}{r^3}\right)\right] e^{ikr}, \tag{2.2.26a}$$

$$= \frac{\omega\mu_0}{4\pi k^2}\left[\frac{2k}{r^2} + \frac{2i}{r^3} + \frac{\rho^2}{r^2}\left(\frac{ik^2}{r} - \frac{3k}{r^2} - \frac{3i}{r^3}\right)\right] e^{ikr}, \tag{2.2.26b}$$

$$B_\phi = -\frac{\mu_0}{4\pi}\frac{\rho}{r}\left(\frac{ik}{r} - \frac{1}{r^2}\right) e^{ikr}. \tag{2.2.27}$$

2.3 The Electromagnetic Field of a Unit Magnetic Dipole

With $\mathbf{J}^e = \nabla \times \mathbf{M}^e$ in (2.1.1), $\rho^e = 0$ in (2.1.2), and a prime added to $\mathbf{E}$ and $\mathbf{B}$ to distinguish the field of a magnetic dipole from that of an electric dipole, (2.1.1) and (2.1.2) become

$$\nabla \times (\mathbf{B}' - \mu_0 \mathbf{M}^e) = -\frac{ik^2}{\omega}\mathbf{E}'; \qquad \nabla \cdot \mathbf{B}' = 0, \tag{2.3.1}$$

$$\nabla \times \mathbf{E}' = i\omega \mathbf{B}'; \qquad \nabla \cdot \mathbf{E}' = 0. \tag{2.3.2}$$

Now let

$$\mathbf{K} = \mathbf{B}' - \mu_0 \mathbf{M}^e; \qquad \mathbf{L} = \mathbf{E}'. \tag{2.3.3}$$

Then,

$$\nabla \times \mathbf{L} = i\omega\mu_0 \mathbf{M}^e + i\omega \mathbf{K}; \qquad \nabla \cdot \mathbf{L} = 0, \tag{2.3.4}$$

$$\nabla \times \mathbf{K} = -\frac{ik^2}{\omega}\mathbf{L}; \qquad \nabla \cdot \mathbf{K} = 0. \tag{2.3.5}$$

A comparison of (2.3.4) with (2.1.1) and (2.3.5) with (2.1.2) shows that these equations are the same with

$$\mathbf{J}^e = i\omega\mu_0 \mathbf{M}^e; \qquad \frac{\mathbf{B}}{\mu_0} = \mathbf{L}; \qquad \tilde{\epsilon}\,\mathbf{E} = -\mathbf{K}. \tag{2.3.6}$$

With (2.3.6) and (2.2.5), an x-directed infinitesimal magnetic dipole at the origin of coordinates in an infinite medium is defined by

$$i\omega\mu_0 \mathbf{M}^e = \hat{\mathbf{x}}\delta^3(\mathbf{r}). \tag{2.3.7}$$

At all points outside the infinitesimal dipole source, $\mathbf{M}^e = 0$ so that (2.3.3) with (2.3.6) gives

$$\mathbf{E}' = \frac{\mathbf{B}}{\mu_0}; \qquad \mathbf{B}' = -\tilde{\epsilon}\,\mathbf{E}, \tag{2.3.8}$$

where the prime indicates a field generated by a unit magnetic dipole, and the absence of a prime the field generated by a unit electric dipole.

It follows that the field of a magnetic dipole with unit moment is given by the field of a comparable electric dipole with the substitutions (2.3.8). Specifically, the field of an x-directed magnetic dipole as obtained from (2.2.9)–(2.2.14) at all points outside the dipole is

$$E'_x = 0, \tag{2.3.9}$$

$$E'_y = \frac{1}{4\pi}\frac{z}{r}\left(\frac{ik}{r} - \frac{1}{r^2}\right)e^{ikr}, \tag{2.3.10}$$

$$E'_z = -\frac{1}{4\pi}\frac{y}{r}\left(\frac{ik}{r} - \frac{1}{r^2}\right)e^{ikr}, \tag{2.3.11}$$

$$B'_x = -\frac{i}{4\pi\omega}\left[\frac{k^2}{r} + \frac{ik}{r^2} - \frac{1}{r^3} - \frac{x^2}{r^2}\left(\frac{k^2}{r} + \frac{3ik}{r^2} - \frac{3}{r^3}\right)\right]e^{ikr}, \tag{2.3.12}$$

$$B'_y = \frac{i}{4\pi\omega}\frac{xy}{r^2}\left(\frac{k^2}{r} + \frac{3ik}{r^2} - \frac{3}{r^3}\right)e^{ikr}, \tag{2.3.13}$$

$$B'_z = \frac{i}{4\pi\omega}\frac{xz}{r^2}\left(\frac{k^2}{r} + \frac{3ik}{r^2} - \frac{3}{r^3}\right)e^{ikr}. \tag{2.3.14}$$

The field of a y-directed magnetic dipole with $i\omega\mu_0\mathbf{M}^e = \hat{\mathbf{y}}\delta^3(\mathbf{r})$ is obtained from (2.3.9)–(2.3.14) with the sequence: $x \to y,\ y \to z,\ z \to x$. At points outside the dipole it is

$$E'_x = -\frac{1}{4\pi}\frac{z}{r}\left(\frac{ik}{r} - \frac{1}{r^2}\right)e^{ikr}, \tag{2.3.15}$$

$$E'_y = 0, \tag{2.3.16}$$

$$E'_z = \frac{1}{4\pi}\frac{x}{r}\left(\frac{ik}{r} - \frac{1}{r^2}\right)e^{ikr}, \tag{2.3.17}$$

$$B'_x = \frac{i}{4\pi\omega}\frac{xy}{r^2}\left(\frac{k^2}{r} + \frac{3ik}{r^2} - \frac{3}{r^3}\right)e^{ikr}, \tag{2.3.18}$$

$$B'_y = -\frac{i}{4\pi\omega}\left[\frac{k^2}{r} + \frac{ik}{r^2} - \frac{1}{r^3} - \frac{y^2}{r^2}\left(\frac{k^2}{r} + \frac{3ik}{r^2} - \frac{3}{r^3}\right)\right]e^{ikr}, \tag{2.3.19}$$

$$B'_z = \frac{i}{4\pi\omega}\frac{yz}{r^2}\left(\frac{k^2}{r} + \frac{3ik}{r^2} - \frac{3}{r^3}\right)e^{ikr}. \tag{2.3.20}$$

The field for the y-directed magnetic dipole in cylindrical coordinates (ρ, ϕ, z) is obtained from (2.3.15)–(2.3.20) with the relations:

$$E'_\rho = E'_x\cos\phi + E'_y\sin\phi = E'_x\frac{x}{\rho} + E'_y\frac{y}{\rho}, \tag{2.3.21}$$

$$E'_\phi = -E'_x\sin\phi + E'_y\cos\phi = -E'_x\frac{y}{\rho} + E'_y\frac{x}{\rho}. \tag{2.3.22}$$

Similar relations apply to the magnetic field. Note that E'_z and B'_z are unchanged but $(x/r) = (x/\rho)(\rho/r) = (\rho/r)\cos\phi$, $(y/r) = (y/\rho)(\rho/r) = (\rho/r)\sin\phi$, and $x^2 + y^2 = \rho^2$. The result is

$$E'_\rho = -\frac{1}{4\pi}\frac{z}{r}\left(\frac{ik}{r} - \frac{1}{r^2}\right)e^{ikr}\cos\phi, \tag{2.3.23}$$

$$E'_\phi = \frac{1}{4\pi}\frac{z}{r}\left(\frac{ik}{r} - \frac{1}{r^2}\right)e^{ikr}\sin\phi, \tag{2.3.24}$$

$$E'_z = \frac{1}{4\pi}\frac{\rho}{r}\left(\frac{ik}{r} - \frac{1}{r^2}\right)e^{ikr}\cos\phi, \tag{2.3.25}$$

$$B'_\rho = \frac{i}{4\pi\omega}\left[\frac{2ik}{r^2} - \frac{2}{r^3} - \frac{z^2}{r^2}\left(\frac{k^2}{r} + \frac{3ik}{r^2} - \frac{3}{r^3}\right)\right]e^{ikr}\sin\phi, \tag{2.3.26}$$

$$B'_\phi = -\frac{i}{4\pi\omega}\left(\frac{k^2}{r} + \frac{ik}{r^2} - \frac{1}{r^3}\right)e^{ikr}\cos\phi, \tag{2.3.27}$$

$$B'_z = \frac{i}{4\pi\omega}\frac{\rho z}{r^2}\left(\frac{k^2}{r} + \frac{3ik}{r^2} - \frac{3}{r^3}\right)e^{ikr}\sin\phi. \tag{2.3.28}$$

The field of a z-directed vertical magnetic dipole with $i\omega\mu_0\mathbf{M}^e = \hat{\mathbf{z}}\delta^3(\mathbf{r})$ is obtained from (2.2.25)–(2.2.27) with (2.3.8). Thus, in cylindrical coordinates,

$$B'_\rho = \frac{1}{4\pi\omega}\frac{\rho z}{r^2}\left(\frac{ik^2}{r} - \frac{3k}{r^2} - \frac{3i}{r^3}\right)e^{ikr}, \tag{2.3.29}$$

$$B'_z = -\frac{1}{4\pi\omega}\left[\frac{ik^2}{r} - \frac{k}{r^2} - \frac{i}{r^3} - \frac{z^2}{r^2}\left(\frac{ik^2}{r} - \frac{3k}{r^2} - \frac{3i}{r^3}\right)\right]e^{ikr}, \tag{2.3.30}$$

$$E'_\phi = -\frac{1}{4\pi}\frac{\rho}{r}\left(\frac{ik}{r} - \frac{1}{r^2}\right)e^{ikr}. \tag{2.3.31}$$

2.4 The Reflection and Transmission of Plane Waves at the Plane Boundary Between Electrically Different Media

Before beginning the analysis of the electromagnetic field generated by an electric dipole near the boundary between two electrically different half-spaces, it is useful to review the reflection and refraction of *plane waves* at such a boundary. The electromagnetic fields generated by dipoles near plane boundaries do not consist of plane waves. However, in special circumstances plane waves are useful in the interpretation or evaluation of dipole fields. But these are usually not the familiar homogeneous plane waves with coincident amplitude and phase fronts. Also relevant are inhomogeneous waves with surfaces of constant amplitude that do not coincide with the surfaces of constant phase. Such waves characterize the refracted field in an imperfect dielectric like the earth or sea when a homogeneous plane wave is incident on its plane boundary from a perfect dielectric like air. In multiply-layered regions, inhomogeneous waves may be incident from an imperfect dielectric on the plane boundary with another medium the properties of which may range from those of a very good dielectric to those of a conductor. The nature of both the reflected and refracted fields is of interest. The reflection and refraction of homogeneous and inhomogeneous plane waves at plane boundaries are considered in a logical sequence in the following sections.

Specifically, and without loss of generality, let a plane wave be incident on the plane boundary $z = 0$ from Region 2, as shown in Fig. 2.4.1. In general,

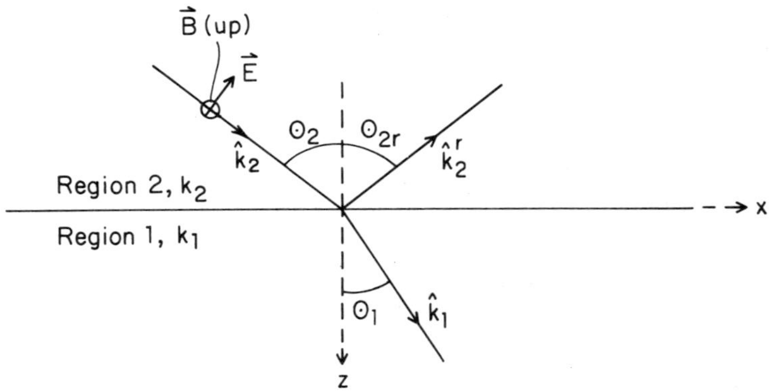

Figure 2.4.1. Plane wave incident on Region 1; E-polarization.

the wave numbers of the two half-spaces are complex so that $k_1 = \beta_1 + i\alpha_1$ for Region 1 ($z \geq 0$) and $k_2 = \beta_2 + i\alpha_2$ for Region 2 ($z \leq 0$). It is convenient to distinguish four cases that depend on whether the regions are lossy. These are as follows:

Case A: Both regions are lossless, so that $\alpha_1 = \alpha_2 = 0$; $k_1 = \beta_1$ and $k_2 = \beta_2$ are real.

Case B: Region 1 is lossy, Region 2 lossless, so that $\alpha_1 > 0$, $\alpha_2 = 0$; $k_1 = \beta_1 + i\alpha_1$, $k_2 = \beta_2$ is real.

Case C: Region 1 is lossless, Region 2 lossy, so that $\alpha_1 = 0$, $\alpha_2 > 0$; $k_1 = \beta_1$ is real, $k_2 = \beta_2 + i\alpha_2$.

Case D: Both regions are lossy, so that $\alpha_1 > 0$, $\alpha_2 > 0$; $k_1 = \beta_1 + i\alpha_1$, $k_2 = \beta_2 + i\alpha_2$.

The cases B and C differ in that in B the wave originates in a lossless Region 2 ($\alpha_2 = 0$); in C it originates in a lossy Region 2 ($\alpha_2 > 0$). The four cases are treated in order in the following sections.

2.5 The Reflection and Transmission of Plane Waves at the Boundary Between Two Lossless Half-Spaces (Case A)

A plane wave with arbitrary polarization of the electromagnetic field is incident from a perfect dielectric half-space [Region 2, $z \leq 0$, with the real wave number $k_2 = \omega(\mu_0\epsilon_2)^{1/2}$] on the plane boundary $z = 0$ of a similarly perfect dielectric half-space [Region 1, $z \geq 0$, with the real wave number $k_1 = \omega(\mu_0\epsilon_1)^{1/2}$]. The angle of incidence is Θ_2, as shown in Fig. 2.4.1. Let the coordinate axes be chosen so that the incident wave vector $\mathbf{k}_2$ is in the

xz-plane, i.e.,

$$\hat{\mathbf{y}} \cdot \mathbf{k}_2 = 0. \tag{2.5.1}$$

Then the spatial dependence of the incident electromagnetic field is

$$\mathbf{E}_2^{\text{inc}}(\mathbf{r}) = \mathbf{E}_2^{\text{inc}}(0)e^{i\mathbf{k}_2\cdot\mathbf{r}} = \mathbf{E}_2^{\text{inc}}(0)e^{i(k_{2x}x+k_{2z}z)}; \qquad z \le 0. \tag{2.5.2}$$

The wave vector is

$$\mathbf{k}_2 = \hat{\mathbf{x}}k_{2x} + \hat{\mathbf{z}}k_{2z} = \hat{\mathbf{x}}k_2 \sin\Theta_2 + \hat{\mathbf{z}}k_2\cos\Theta_2, \tag{2.5.3}$$

where k_{2x}, k_{2z}, and Θ_2 are all real. That is,

$$\operatorname{Im} k_{2x} = \operatorname{Im} k_{2z} = \operatorname{Im}\Theta_2 = 0. \tag{2.5.4}$$

The associated magnetic field is

$$\mathbf{B}_2^{\text{inc}}(\mathbf{r}) = \mathbf{B}_2^{\text{inc}}(0)e^{i\mathbf{k}_2\cdot\mathbf{r}} = \mathbf{B}_2^{\text{inc}}(0)e^{i(k_{2x}x+k_{2z}z)}; \qquad z \le 0, \tag{2.5.5}$$

with

$$\mathbf{B}_2^{\text{inc}}(0) = -\frac{i}{\omega}\,\mathbf{k}_2 \times \mathbf{E}_2^{\text{inc}}(0). \tag{2.5.6}$$

Because of translational invariance in the x- and y-directions, the reflected and transmitted (or refracted) fields are also plane waves. They are

$$\mathbf{E}_2^{\text{ref}}(\mathbf{r}) = \mathbf{E}_2^{\text{ref}}(0)e^{i\mathbf{k}_2^r\cdot\mathbf{r}} = \mathbf{E}_2^{\text{ref}}(0)e^{i(k_{2x}^r x+k_{2z}^r z)}; \qquad z \le 0, \tag{2.5.7}$$

and

$$\mathbf{E}_1^{\text{tr}}(\mathbf{r}) \equiv \mathbf{E}_1(\mathbf{r}) = \mathbf{E}_1(0)e^{i\mathbf{k}_1\cdot\mathbf{r}} = \mathbf{E}_1(0)e^{i(k_{1x}x+k_{1z}z)}; \qquad z \ge 0. \tag{2.5.8}$$

The corresponding components of the magnetic field are similar. It is an immediate consequence of Maxwell's equations (2.1.1) and (2.1.2) that, not only

$$k_{2x}^2 + k_{2z}^2 = k_2^2 \tag{2.5.9}$$

as implied in (2.5.3), but also

$$(k_{2x}^r)^2 + (k_{2z}^r)^2 = k_2^2; \qquad k_{1x}^2 + k_{1z}^2 = k_1^2. \tag{2.5.10}$$

The boundary conditions require the continuity of the tangential components of the electric and magnetic fields across the boundary at $z = 0$. This implies in particular that

$$k_{2x} = k_{2x}^r = k_{1x}. \tag{2.5.11}$$

This relation is of fundamental importance and actually has a generality that far transcends the special conditions in this section, as explained in later sections of this chapter.

Let the angles of reflection and refraction, Θ_{2r} and Θ_1, be defined in the usual manner, as shown in Fig. 2.4.1. That is,

$$k_{2x}^r = k_2 \sin \Theta_{2r}; \qquad k_{2z}^r = -k_2 \cos \Theta_{2r}, \tag{2.5.12}$$

$$k_{1x} = k_1 \sin \Theta_1; \qquad k_{1z} = k_1 \cos \Theta_1. \tag{2.5.13}$$

It then follows from (2.5.11) that

$$k_{2z}^r = -k_{2z}; \qquad \Theta_{2r} = \Theta_2; \qquad k_1 \sin \Theta_1 = k_2 \sin \Theta_2. \tag{2.5.14}$$

This is Snell's law for isotropic linear media.

It follows from (2.5.4) and (2.5.11) that the quantities k_{2x}, k_{2z}, k_{2x}^r, k_{2z}^r, k_{1x}, and Θ_2 are real. However, k_{1z} and Θ_1 may or may not be real. They are both real when

$$k_1 \geq k_2 \sin \Theta_2. \tag{2.5.15}$$

In this case,

$$\begin{aligned} k_{1z} &= (k_1^2 - k_{1x}^2)^{1/2} \\ &= (k_1^2 - k_1^2 \sin^2 \Theta_1)^{1/2} = (k_1^2 - k_2^2 \sin^2 \Theta_2)^{1/2}. \end{aligned} \tag{2.5.16}$$

It follows that, with $k_{1z} = \beta_{1z} + i\alpha_{1z}$, (2.5.16) in the range (2.5.15) gives

$$k_{1z} = \beta_{1z}; \qquad \alpha_{1z} = 0. \tag{2.5.17}$$

In this case the electric field is given by

$$\mathbf{E}_2(\mathbf{r}) = \mathbf{E}_2^{\text{inc}}(\mathbf{r}) + \mathbf{E}_2^{\text{ref}}(\mathbf{r}), \tag{2.5.18}$$

where

$$\mathbf{E}_2^{\text{inc}}(\mathbf{r}) = \mathbf{E}_2^{\text{inc}}(0) e^{ik_2(x \sin \Theta_2 + z \cos \Theta_2)}; \qquad z \leq 0, \tag{2.5.19}$$

$$\mathbf{E}_2^{\text{ref}}(\mathbf{r}) = \mathbf{E}_2^{\text{ref}}(0) e^{ik_2(x \sin \Theta_2 - z \cos \Theta_2)}; \qquad z \leq 0, \tag{2.5.20}$$

and by

$$\begin{aligned} \mathbf{E}_1^{\text{tr}}(\mathbf{r}) &= \mathbf{E}_1(0) e^{ik_1(x \sin \Theta_1 + z \cos \Theta_1)} \\ &= \mathbf{E}_1(0) e^{ik_2 x \sin \Theta_2} e^{ik_1 z \cos \Theta_1}; \qquad z \geq 0. \end{aligned} \tag{2.5.21}$$

In (2.5.21), $\Theta_1 = \sin^{-1}[(k_2/k_1) \sin \Theta_2]$. The formulas (2.5.18)–(2.5.21) (and similar ones for the magnetic field) represent the familiar incident and reflected plane waves in Region 2 with equal angles of incidence and reflection ($\Theta_{2r} = \Theta_2$) and a refracted wave proceeding in the direction of Θ_1 determined by the index of refraction and Snell's law, i.e.,

$$N_{12} \equiv \frac{k_1}{k_2} = \left(\frac{\epsilon_1}{\epsilon_2}\right)^{1/2} = \frac{v_{p2}}{v_{p1}} = \frac{\sin \Theta_2}{\sin \Theta_1}, \tag{2.5.22}$$

where $v_{pj} = \omega/k_j = (\mu_0 \epsilon_j)^{-1/2}$, $j = 1, 2$, is the phase velocity. The field in both regions can be resolved into a progressive or running component proceeding in the positive x-direction and given by $\exp(ik_2 x \sin\Theta_2)$. In Region 2 ($z \leq 0$), the component perpendicular to the boundary is a standing wave defined by $[\mathbf{E}_2^{\text{inc}}(0)\exp(ik_2 z\cos\Theta_2) + \mathbf{E}_2^{\text{ref}}(0)\exp(-ik_2 z\cos\Theta_2)]$; in Region 1 ($z \geq 0$), the component in the z-direction is a progressive wave given by $\exp(ik_1 z\cos\Theta_1)$.

The component wave number k_{1z} and the angle of refraction Θ_1 are not real when

$$k_1 < k_2 \sin\Theta_2. \tag{2.5.23}$$

Clearly, the angle

$$\Theta_{2\text{cr}} \equiv \sin^{-1}\left(\frac{k_1}{k_2}\right) \tag{2.5.24}$$

is a *critical angle*. When the angle of incidence exceeds the critical angle, i.e., when $\Theta_2 > \Theta_{2\text{cr}}$, (2.5.23) is satisfied and (2.5.10) gives

$$\begin{aligned} k_{1z} &= (k_1^2 - k_{1x}^2)^{1/2} \\ &= (k_1^2 - k_1^2\sin^2\Theta_1)^{1/2} = (k_1^2 - k_2^2\sin^2\Theta_2)^{1/2} \\ &= i(k_2^2\sin^2\Theta_2 - k_1^2)^{1/2}, \end{aligned} \tag{2.5.25}$$

which is a *pure imaginary*. The sign of the square root is chosen so that $\operatorname{Im} k_{1z} > 0$. With $k_{1z} = \beta_{1z} + i\alpha_{1z}$, it follows from (2.5.25) that

$$k_{1z} = i\alpha_{1z}; \qquad \beta_{1z} = 0. \tag{2.5.26}$$

The angle Θ_1 is readily obtained with (2.5.13). Specifically,

$$\Theta_1 = \cot^{-1}\left(\frac{k_{1z}}{k_{1x}}\right) = \cot^{-1}\left(\frac{i\alpha_{1z}}{k_{2x}}\right) = -i\coth^{-1}\left(\frac{\alpha_{1z}}{k_{2x}}\right). \tag{2.5.27}$$

It is a pure imaginary.

With (2.5.26), (2.5.8) becomes

$$\mathbf{E}_1^{\text{tr}}(\mathbf{r}) = \mathbf{E}_1(0)e^{-\alpha_{1z}z}e^{ik_{1x}x}; \qquad z \geq 0. \tag{2.5.28}$$

Similarly,

$$\mathbf{B}_1^{\text{tr}}(\mathbf{r}) = \mathbf{B}_1(0)e^{-\alpha_{1z}z}e^{ik_{1x}x}; \qquad z \geq 0. \tag{2.5.29}$$

Thus, the electromagnetic field in Region 1 ($z \geq 0$) decays exponentially perpendicularly from the boundary $z = 0$ and there is no refracted wave since $\beta_{1z} = 0$. This phenomenon is known as *total internal reflection.* It can occur only when the real wave number of the region containing the incident field exceeds that of the region containing the transmitted field—in the case at hand, when $k_2 > k_1$. The transmitted field in Region 1, $z \geq 0$, now consists of only a wave traveling parallel to the boundary in the positive x-direction. It is given by (2.5.28) and (2.5.29). It has a plane wave front that decreases as $\exp(-\alpha_{1z}z)$ when $z \geq 0$. Since it is significant only close to the boundary, it is properly called a *surface wave.* Actually, it is the bounding edge of the x-directed plane wave traveling in Region 2. This extends to the boundary at $z = 0$ with constant amplitude and then decreases exponentially across the boundary in Region 1.

So far, the discussion in this section is entirely independent of the polarization of the incident plane electromagnetic wave. In order to relate the reflected and transmitted fields to the incident field, it is necessary to express the complete field as the linear superposition of two component parts—the one of electric type or E-polarized with the electric field in the plane of incidence, the second of magnetic type or H-polarized with the magnetic field in the plane of incidence. When the xz-plane is the plane of incidence, the entire field is y-invariant, and Maxwell's equations separate into two independent sets which are equivalent to

$$\frac{\partial^2 B_y}{\partial x^2} + \frac{\partial^2 B_y}{\partial z^2} + k^2 B_y = 0; \quad E_x = -\frac{i\omega}{k^2}\frac{\partial B_y}{\partial z}; \quad E_z = \frac{i\omega}{k^2}\frac{\partial B_y}{\partial x}, \tag{2.5.30}$$

$$\frac{\partial^2 E_y}{\partial x^2} + \frac{\partial^2 E_y}{\partial z^2} + k^2 E_y = 0; \quad B_x = \frac{i}{\omega}\frac{\partial E_y}{\partial z}; \quad B_z = -\frac{i}{\omega}\frac{\partial E_y}{\partial x}. \tag{2.5.31}$$

The field of electric type defined by (2.5.30) includes $\mathbf{E}_2 = \hat{\mathbf{x}}E_{2x} - \hat{\mathbf{z}}E_{2z}$ and $\mathbf{B}_2 = \hat{\mathbf{y}}B_{2y}$; the field of magnetic type consists of $\mathbf{B}_2 = -\hat{\mathbf{x}}B_{2x} + \hat{\mathbf{z}}B_{2z}$ and $\mathbf{E}_2 = \hat{\mathbf{y}}E_{2y}$. The components in both regions are as follows:

Electric type:

$$B_{2y}(x,z) = B_{2y}^{\text{inc}}(0,0)e^{ik_2 x \sin\Theta_2} \times \left[e^{ik_2 z\cos\Theta_2} + f_{er}(\Theta_2)e^{-ik_2 z\cos\Theta_2}\right]; \qquad z \le 0, \tag{2.5.32}$$

$$E_{2x}(x,z) = \frac{\omega}{k_2}B_{2y}^{\text{inc}}(0,0)e^{ik_2 x \sin\Theta_2} \times \left[e^{ik_2 z\cos\Theta_2} - f_{er}(\Theta_2)e^{-ik_2 z\cos\Theta_2}\right]\cos\Theta_2; \qquad z \le 0, \tag{2.5.33}$$

$$E_{2z}(x,z) = -\frac{\omega}{k_2}B_{2y}^{\text{inc}}(0,0)e^{ik_2 x \sin\Theta_2} \times \left[e^{ik_2 z\cos\Theta_2} + f_{er}(\Theta_2)e^{-ik_2 z\cos\Theta_2}\right]\sin\Theta_2; \qquad z \le 0, \tag{2.5.34}$$

$$B_{1y}(x,z) = f_{et}(\Theta_2)B_{2y}^{\text{inc}}(0,0)e^{ik_1(x\sin\Theta_1 + z\cos\Theta_1)}; \qquad z \ge 0, \tag{2.5.35}$$

$$E_{1x}(x,z) = \frac{\omega}{k_1}f_{et}(\Theta_2)B_{2y}^{\text{inc}}(0,0)e^{ik_1(x\sin\Theta_1 + z\cos\Theta_1)}\cos\Theta_1; \qquad z \ge 0, \tag{2.5.36}$$

$$E_{1z}(x,z) = -\frac{\omega}{k_1}f_{et}(\Theta_2)B_{2y}^{\text{inc}}(0,0)e^{ik_1(x\sin\Theta_1 + z\cos\Theta_1)}\sin\Theta_1; \qquad z \ge 0. \tag{2.5.37}$$

Magnetic type:

$$E_{2y}(x,z) = E_{2y}^{\text{inc}}(0,0)e^{ik_2 x \sin\Theta_2} \times \left[e^{ik_2 z\cos\Theta_2} + f_{mr}(\Theta_2)e^{-ik_2 z\cos\Theta_2}\right]; \qquad z \le 0, \tag{2.5.38}$$

$$B_{2x}(x,z) = -\frac{k_2}{\omega} E_{2y}^{\rm inc}(0,0) e^{ik_2 x \sin\Theta_2} \times [e^{ik_2 z\cos\Theta_2} - f_{mr}(\Theta_2) e^{-ik_2 z \cos\Theta_2}] \cos\Theta_2; \quad z \le 0, \tag{2.5.39}$$

$$B_{2z}(x,z) = \frac{k_2}{\omega} E_{2y}^{\rm inc}(0,0) e^{ik_2 x \sin\Theta_2} \times [e^{ik_2 z\cos\Theta_2} + f_{mr}(\Theta_2) e^{-ik_2 z \cos\Theta_2}] \sin\Theta_2; \quad z \le 0, \tag{2.5.40}$$

$$E_{1y}(x,z) = f_{mt}(\Theta_2) E_{2y}^{\rm inc}(0,0) e^{ik_1(x\sin\Theta_1 + z\cos\Theta_1)}; \quad z \ge 0, \tag{2.5.41}$$

$$B_{1x}(x,z) = -\frac{k_1}{\omega} f_{mt}(\Theta_2) E_{2y}^{\rm inc}(0,0) e^{ik_1(x\sin\Theta_1 + z\cos\Theta_1)} \cos\Theta_1; \quad z \ge 0, \tag{2.5.42}$$

$$B_{1z}(x,z) = \frac{k_1}{\omega} f_{mt}(\Theta_2) E_{2y}^{\rm inc}(0,0) e^{ik_1(x\sin\Theta_1 + z\cos\Theta_1)} \sin\Theta_1; \quad z \ge 0. \tag{2.5.43}$$

When the boundary conditions requiring continuity of the tangential components of the electric and magnetic fields are applied at $z = 0$, the reflection and transmission coefficients can be evaluated. They can be expressed in a number of ways. One useful form is in terms of the wave numbers and their ratio, the index of refraction. Specifically, with

$$N_{12} = \frac{k_1}{k_2}, \tag{2.5.44}$$

$$f_{er}(\Theta_2) = \frac{B_{2y}^{\rm ref}(0,0)}{B_{2y}^{\rm inc}(0,0)} = \frac{k_2^{-1}\cos\Theta_2 - k_1^{-1}\cos\Theta_1}{k_2^{-1}\cos\Theta_2 + k_1^{-1}\cos\Theta_1} = \frac{N_{12}^2\cos\Theta_2 - (N_{12}^2 - \sin^2\Theta_2)^{1/2}}{N_{12}^2\cos\Theta_2 + (N_{12}^2 - \sin^2\Theta_2)^{1/2}}, \tag{2.5.45}$$

$$f_{et}(\Theta_2) = \frac{B_{1y}^{\rm tr}(0,0)}{B_{2y}^{\rm inc}(0,0)} = \frac{2k_2^{-1}\cos\Theta_2}{k_2^{-1}\cos\Theta_2 + k_1^{-1}\cos\Theta_1} = \frac{2N_{12}^2\cos\Theta_2}{N_{12}^2\cos\Theta_2 + (N_{12}^2 - \sin^2\Theta_2)^{1/2}}, \tag{2.5.46}$$

$$f_{mr}(\Theta_2) = \frac{E_{2y}^{\rm ref}(0,0)}{E_{2y}^{\rm inc}(0,0)} = \frac{k_2\cos\Theta_2 - k_1\cos\Theta_1}{k_2\cos\Theta_2 + k_1\cos\Theta_1} = \frac{\cos\Theta_2 - (N_{12}^2 - \sin^2\Theta_2)^{1/2}}{\cos\Theta_2 + (N_{12}^2 - \sin^2\Theta_2)^{1/2}}, \tag{2.5.47}$$

$$f_{mt}(\Theta_2) = \frac{E_{1y}^{\rm tr}(0,0)}{E_{2y}^{\rm inc}(0,0)} = \frac{2k_2\cos\Theta_2}{k_2\cos\Theta_2 + k_1\cos\Theta_1} = \frac{2\cos\Theta_2}{\cos\Theta_2 + (N_{12}^2 - \sin^2\Theta_2)^{1/2}}. \tag{2.5.48}$$

An alternative representation of the reflection and transmission coefficients is in terms of generalized surface impedances and admittances. With the wave impedances of the two media given by

$$\zeta_1 = \left(\frac{\mu_0}{\epsilon_1}\right)^{1/2} = \frac{\omega\mu_0}{k_1}; \qquad \zeta_2 = \left(\frac{\mu_0}{\epsilon_2}\right)^{1/2} = \frac{\omega\mu_0}{k_2}, \tag{2.5.49}$$

the surface impedance of electric type and surface admittance of magnetic type are defined as follows:

$$\begin{aligned} Z_{se}(\Theta_2) = \frac{\mu_0 E_{2x}(0,0)}{B_{2y}(0,0)} &= -\frac{i\omega\mu_0[\partial B_{2y}(x,z)/\partial z]_{z=0}}{k_2^2 B_{2y}(0,0)} \\ &= \frac{\zeta_2(N_{12}^2 - \sin^2\Theta_2)^{1/2}}{N_{12}^2}, \end{aligned} \tag{2.5.50}$$

$$\begin{aligned} Y_{sm}(\Theta_2) = -\frac{B_{2x}(0,0)}{\mu_0 E_{2y}(0,0)} &= \frac{[\partial E_{2y}(x,z)/\partial z]_{z=0}}{i\omega\mu_0 E_{2y}(0,0)} \\ &= \frac{(N_{12}^2 - \sin^2\Theta_2)^{1/2}}{\zeta_2}. \end{aligned} \tag{2.5.51}$$

The normalized surface impedance and admittance as seen from Region 2 are

$$z_{se}(\Theta_2) = \frac{Z_{se}(\Theta_2)}{\zeta_2} = \frac{(N_{12}^2 - \sin^2\Theta_2)^{1/2}}{N_{12}^2}, \tag{2.5.52}$$

$$y_{sm}(\Theta_2) = Y_{sm}(\Theta_2)\zeta_2 = (N_{12}^2 - \sin^2\Theta_2)^{1/2}. \tag{2.5.53}$$

In terms of these quantities the reflection and transmission coefficients are

$$f_{er}(\Theta_2) = \frac{\cos\Theta_2 - z_{se}(\Theta_2)}{\cos\Theta_2 + z_{se}(\Theta_2)}, \tag{2.5.54}$$

$$f_{et}(\Theta_2) = \frac{2\cos\Theta_2}{\cos\Theta_2 + z_{se}(\Theta_2)}, \tag{2.5.55}$$

$$f_{mr}(\Theta_2) = \frac{\cos\Theta_2 - y_{sm}(\Theta_2)}{\cos\Theta_2 + y_{sm}(\Theta_2)}, \tag{2.5.56}$$

$$f_{mt}(\Theta_2) = \frac{2\cos\Theta_2}{\cos\Theta_2 + y_{sm}(\Theta_2)}. \tag{2.5.57}$$

A third useful form of the reflection and transmission coefficients is in terms of the components of the propagation vectors $\mathbf{k}_1 = \hat{\mathbf{x}}k_{1x} + \hat{\mathbf{z}}k_{1z} = \hat{\mathbf{x}}k_1\sin\Theta_1 + \hat{\mathbf{z}}k_1\cos\Theta_1$ and $\mathbf{k}_2 = \hat{\mathbf{x}}k_{2x} + \hat{\mathbf{z}}k_{2z} = \hat{\mathbf{x}}k_2\sin\Theta_2 + \hat{\mathbf{z}}k_2\cos\Theta_2$. The formulas are

$$f_{er} = \frac{k_1^2 k_{2z} - k_2^2 k_{1z}}{k_1^2 k_{2z} + k_2^2 k_{1z}}; \qquad f_{et} = \frac{2k_1^2 k_{2z}}{k_1^2 k_{2z} + k_2^2 k_{1z}}, \tag{2.5.58}$$

$$f_{mr} = \frac{k_{2z} - k_{1z}}{k_{2z} + k_{1z}}; \qquad f_{mt} = \frac{2k_{2z}}{k_{1z} + k_{2z}}. \tag{2.5.59}$$

At normal incidence, $\Theta_2 = 0$ and the formulas for electric and magnetic types reduce to

$$\begin{aligned} f_{er}(0) &= -f_{mr}(0) \\ &= \frac{k_1 - k_2}{k_1 + k_2} = \frac{N_{12} - 1}{N_{12} + 1} = \frac{1 - z_{se}(0)}{1 + z_{se}(0)} = -\frac{1 - y_{sm}(0)}{1 + y_{sm}(0)}, \end{aligned} \tag{2.5.60}$$

$$\begin{aligned} f_{et}(0) &= \frac{k_1}{k_2} f_{mt}(0) \\ &= \frac{2k_1}{k_1 + k_2} = \frac{2N_{12}}{N_{12} + 1} = \frac{2}{1 + z_{se}(0)} = \frac{2k_1}{k_2[1 + y_{sm}(0)]}. \end{aligned} \tag{2.5.61}$$

Also,

$$z_{se}(0) = y_{sm}^{-1}(0) = N_{12}^{-1} = \frac{k_2}{k_1}, \tag{2.5.62a}$$

$$Z_{se}(0) = Y_{sm}^{-1}(0) = \frac{\omega\mu_0}{k_1} = \zeta_1. \tag{2.5.62b}$$

At the boundary $z = 0$ and at normal incidence, the relations (2.5.50) and (2.5.51) in the form

$$\mathbf{E}_2 = -\mu_0^{-1}\hat{\mathbf{z}} \times \mathbf{B}_2 Z_{se}(0) = -\frac{\omega}{k_1}\hat{\mathbf{z}} \times \mathbf{B}_2 \tag{2.5.63}$$

are known as the impedance boundary condition.

The electromagnetic field of electric type has quite different properties from the field of magnetic type. Most significant is the vanishing of the reflected field in Region 2 when $k_2 > k_1$ and the angle of incidence Θ_2 has a particular value Θ_{2B} known as the *Brewster angle.* This is readily determined from (2.5.45) and (2.5.46) with $\Theta_2 = \Theta_{2B}$ when

$$f_{er}(\Theta_{2B}) = 0 \quad \text{or} \quad \tan\Theta_{2B} = N_{12} = \frac{k_1}{k_2} = \left(\frac{\epsilon_1}{\epsilon_2}\right)^{1/2}, \tag{2.5.64a}$$

$$f_{et}(\Theta_{2B}) = 1. \tag{2.5.64b}$$

When Region 2 is air and Region 1 has the values $\epsilon_{1r} = 2$, 4, 10, and 81, $\Theta_{2B} = 54^\circ.7$, $63^\circ.4$, $72^\circ.5$, and $83^\circ.7$, respectively. At the Brewster angle, the two half-spaces are perfectly matched, and

$$\sin\Theta_1 = \frac{k_2}{k_1}\sin\Theta_{2B} = \frac{k_2}{(k_1^2 + k_2^2)^{1/2}} = \cos\Theta_{2B}. \tag{2.5.65a}$$

Hence,

$$\Theta_1 = \frac{\pi}{2} - \Theta_{2B}, \tag{2.5.65b}$$

and the propagation vectors $\mathbf{k}_2^r$ for the reflected field and $\mathbf{k}_1$ for the transmitted field are mutually perpendicular. Since $\mathbf{E}_1$ and $\mathbf{P}_1 = (\epsilon_1 - \epsilon_0)\mathbf{E}_1$ are perpendicular to $\mathbf{k}_1$, they are parallel to $\mathbf{k}_2^r$. Thus, the oscillating dipoles that determine the volume density of polarization $\mathbf{P}_1$ in Region 1 have their axes—along which they do not radiate—parallel to $\mathbf{k}_2^r$. Thus, they generate no field in the direction of reflection.

2.6 The Reflection and Transmission of Plane Waves at the Boundary of a Dissipative Half-Space (Case B)

The next step in the study of the reflection and transmission of plane waves at plane boundaries is to generalize the considerations of the preceding section to those of Case B in which Region 1 is dissipative, i.e., $k_1 = \beta_1 + i\alpha_1$ with $\alpha_1 > 0$. Mathematically, this generalization can be carried out very simply by analytic continuation. More explicitly, every equation in Section 2.5 is valid not only for Case A (with Regions 1 and 2 both perfect dielectrics), but also for Case B [with Region 2 a perfect dielectric (k_2 real) and Region 1 dissipative (k_1 complex)] except equation (2.5.24) which cannot hold when k_1 is complex and $\Theta_{2\mathrm{cr}}$ is real.

Since analytic continuation does not always clarify the physical interpretation, a more detailed discussion is desirable. Cases A and B are similar in the sense that Θ_2, k_{2x}, k_{2z}, k_{2x}^r, k_{2z}^r, and k_{1x} are all real. They differ in that k_{1z} is either purely real or purely imaginary in Case A, but complex in Case B. In general, a plane wave with the form

$$e^{i\mathbf{k}\cdot\mathbf{r}} = e^{i(k_x x + k_y y + k_z z)} \tag{2.6.1}$$

is called an inhomogeneous plane wave when k_x, k_y, and k_z are complex. Such a wave has two characteristic directions, the one with direction $\hat{\mathbf{n}}$ of increasing phase, the other with direction $\hat{\mathbf{n}}'$ of decreasing amplitude. With

$$\mathbf{k} = \hat{\mathbf{x}}k_x + \hat{\mathbf{y}}k_y + \hat{\mathbf{z}}k_z, \tag{2.6.2}$$

$$k_j = \beta_j + i\alpha_j; \qquad j = x,\, y,\, z, \tag{2.6.3}$$

and

$$\alpha_j > 0; \qquad \operatorname{Im} k > 0, \tag{2.6.4}$$

the plane wave (2.6.1) becomes

$$e^{i\mathbf{k}\cdot\mathbf{r}} = e^{-(\alpha_x x + \alpha_y y + \alpha_z z)}\, e^{i(\beta_x x + \beta_y y + \beta_z z)}. \tag{2.6.5}$$

The directions of increasing phase and decreasing amplitude have the direction cosines

$$\hat{\mathbf{n}} = \frac{(\beta_x, \beta_y, \beta_z)}{(\beta_x^2 + \beta_y^2 + \beta_z^2)^{1/2}}, \tag{2.6.6}$$

$$\hat{\mathbf{n}}' = \frac{(\alpha_x, \alpha_y, \alpha_z)}{(\alpha_x^2 + \alpha_y^2 + \alpha_z^2)^{1/2}}. \tag{2.6.7}$$

Note that in the limiting case when the components of $\mathbf{k}$ are all real and the α's are all zero, the amplitude is a constant and $\hat{\mathbf{n}}'$ is not defined.

In general, the two real unit vectors $\hat{\mathbf{n}}$ and $\hat{\mathbf{n}}'$ are distinct and the direction of propagation of the surfaces of constant phase differs from the direction of decreasing amplitude. The angle between these directions is limited by (2.6.2) and (2.6.4) which require that

$$\hat{\mathbf{n}} \cdot \hat{\mathbf{n}}' > 0, \tag{2.6.8}$$

i.e., the angle between $\hat{\mathbf{n}}$ and $\hat{\mathbf{n}}'$ must be less than $\pi/2$ for all lossy media. For the limiting case of a loss-free medium with the components of $\mathbf{k}$ not all real, $\hat{\mathbf{n}}\cdot\hat{\mathbf{n}}' = 0$, which means that the direction of increasing phase is always perpendicular to the direction of decreasing amplitude. For example, this is true of the wave in Region 1 of Section 2.5 when $\Theta_2 > \Theta_{2\mathrm{cr}}$ and total internal reflection occurs. In this case, $\hat{\mathbf{n}} = \hat{\mathbf{x}}$, $\hat{\mathbf{n}}' = \hat{\mathbf{z}}$.

When Region 1 is lossy, the field has the form

$$e^{i\mathbf{k}\cdot\mathbf{r}} = e^{-\alpha_{1z}z} e^{i(\beta_{1x}x + \beta_{1z}z)} \tag{2.6.9}$$

since $k_{1x} = \beta_{1x}$ is real but $k_{1z} = \beta_{1z} + i\alpha_{1z}$ is complex. Note that the sign of k_{1z} has been chosen so that $\beta_{1z} > 0$ *and* $\alpha_{1z} > 0$. This choice is made so that

$$n_z > 0 \quad \text{and} \quad n_z' > 0 \tag{2.6.10}$$

(in fact, $n_z' = 1$) and this choice is possible because k_{1x} is real. This assures that

$$\operatorname{Im} k_{1z}^2 = \operatorname{Im}(k_1^2 - k_{1x}^2) = \operatorname{Im} k_1^2 > 0. \tag{2.6.11}$$

Thus, no matter how the direction of propagation in Region 1 is interpreted, it must have a component in the positive z-direction because of (2.6.10).

For the lossy Region 1 of interest here, $\mathbf{k}_1 = \hat{\mathbf{x}}k_{1x} + \hat{\mathbf{z}}k_{1z}$ with $k_{1x} = k_1 \sin\Theta_1 = k_2 \sin\Theta_2 = k_{2x}$, a real quantity. Thus, $k_{1x} = \beta_{1x}$ and $\alpha_{1x} = 0$. On the other hand,

$$k_{1z} = (k_1^2 - k_{1x}^2)^{1/2} = (k_1^2 - k_2^2 \sin^2\Theta_2)^{1/2}. \tag{2.6.12}$$

Since $k_1 = \beta_1 + i\alpha_1$, this becomes

$$k_{1z} = (\beta_1^2 - \alpha_1^2 - k_2^2 \sin^2\Theta_2 + 2i\alpha_1\beta_1)^{1/2}. \tag{2.6.13}$$

This can be separated into real and imaginary parts as follows:

$$k_{1z} = (\beta_1^2 - \alpha_1^2 - k_2^2 \sin^2\Theta_2)^{1/2}(1 + ip_1)^{1/2}, \tag{2.6.14}$$

$$p_1 = \frac{2\alpha_1\beta_1}{\beta_1^2 - \alpha_1^2 - k_2^2 \sin^2\Theta_2}. \tag{2.6.15}$$

The real and imaginary parts are conveniently expressed in terms of the real functions

$$f(p_1) = \cosh(\tfrac{1}{2}\sinh^{-1} p_1) = \{\tfrac{1}{2}[(1+p_1^2)^{1/2}+1]\}^{1/2}, \tag{2.6.16}$$

$$g(p_1) = \sinh(\tfrac{1}{2}\sinh^{-1} p_1) = \frac{p_1}{|p_1|}\{\tfrac{1}{2}[(1+p_1^2)^{1/2}-1]\}^{1/2}. \tag{2.6.17}$$

Thus,

$$k_{1z} = (\beta_1^2 - \alpha_1^2 - k_2^2\sin^2\Theta_2)^{1/2}[f(p_1) + ig(p_1)]. \tag{2.6.18}$$

It is now possible to distinguish two ranges and the boundary between them. Thus,

(a) $\beta_1^2 - \alpha_1^2 > k_2^2 \sin^2\Theta_2$:

$$\begin{aligned} k_{1z} &= \beta_{1z} + i\alpha_{1z} \\ &= (\beta_1^2 - \alpha_1^2 - k_2^2\sin^2\Theta_2)^{1/2}[f(p_1) + ig(p_1)], \end{aligned} \tag{2.6.19a}$$

(b) $\beta_1^2 - \alpha_1^2 = k_2^2 \sin^2\Theta_2$:

$$k_{1z} = \beta_{1z} + i\alpha_{1z} = (2i\alpha_1\beta_1)^{1/2} = (1+i)(\alpha_1\beta_1)^{1/2}, \tag{2.6.19b}$$

(c) $\beta_1^2 - \alpha_1^2 < k_2^2 \sin^2\Theta_2$:

$$\begin{aligned} k_{1z} &= \beta_{1z} + i\alpha_{1z} \\ &= (k_2^2\sin^2\Theta_2 - \beta_1^2 + \alpha_1^2)^{1/2}[if(|p_1|) + g(|p_1|)]. \end{aligned} \tag{2.6.19c}$$

Since $f(p_1) > g(p_1)$, it follows that (a) is the range of transmission into Region 1 since the phase constant β_{1z} is greater than the attenuation constant α_{1z}. Similarly, (c) is the range of attenuation in Region 1 since $\alpha_{1z} > \beta_{1z}$. The boundary between these two ranges is (b) with $\alpha_{1z} = \beta_{1z}$. In the limit $\alpha_1 = 0,\ f(p_1) = 1,\ g(p_1) = 0$ and

(a) $\beta_1^2 > k_2^2 \sin^2\Theta_2$:

$$k_{1z} = \beta_{1z} + i\alpha_{1z} = (\beta_1^2 - k_2^2\sin^2\Theta_2)^{1/2} = \beta_{1z}; \qquad \alpha_{1z} = 0,$$

(b) $\beta_1^2 = k_2^2 \sin^2\Theta_2$:

$$k_{1z} = \beta_{1z} + i\alpha_{1z} = 0,$$

(c) $\beta_1^2 < k_2^2 \sin^2\Theta_2$:

$$k_{1z} = \beta_{1z} + i\alpha_{1z} = i(k_2^2\sin^2\Theta_2 - \beta_1^2)^{1/2} = i\alpha_{1z}; \qquad \beta_{1z} = 0.$$

Clearly, range (a) is that specified by (2.5.17); the associated transmitted field in Region 1 is (2.5.21) and a similar expression for the magnetic field. The range (c) is that defined by (2.5.26); the associated attenuated field in Region 1 is given by (2.5.28) and (2.5.29). It corresponds to total internal reflection in Region 2. The boundary range (b) is characterized by $k_{1z} = 0$ or $\alpha_{1z} = \beta_{1z} = 0$.

The real angle of transmission into Region 1 is the angle ψ_1 between the normal $\hat{\mathbf{n}}_1$ to the surface of constant phase and the z-axis in Region 1. It is given by

$$\psi_1 = \sin^{-1} \frac{\beta_{1x}}{(\beta_{1x}^2 + \beta_{1z}^2)^{1/2}}. \tag{2.6.20}$$

For range (a), this becomes

$$\psi_1 = \sin^{-1} \frac{k_2 \sin \Theta_2}{[k_2^2 \sin^2 \Theta_2 + (\beta_1^2 - \alpha_1^2 - k_2^2 \sin^2 \Theta_2) f^2(p_1)]^{1/2}}. \tag{2.6.21a}$$

For range (c),

$$\psi_1 = \sin^{-1} \frac{k_2 \sin \Theta_2}{[k_2^2 \sin^2 \Theta_2 + (k_2^2 \sin^2 \Theta_2 - \beta_1^2 + \alpha_1^2) g^2(|p_1|)]^{1/2}}. \tag{2.6.21b}$$

When $\alpha_1 = 0$, $f(p_1) = 1$ and (2.6.21a) reduces to

$$\text{range (a):} \quad \psi_1 = \Theta_1 = \sin^{-1}\left(\frac{k_2}{k_1} \sin \Theta_2\right); \qquad \alpha_1 = 0, \tag{2.6.22}$$

which is Snell's law for perfect dielectrics. The critical angle for total internal reflection is defined by (2.6.22) with $\Theta_1 = \pi/2$ or $\Theta_2 = \Theta_{2\text{cr}} = \sin^{-1}(k_1/k_2)$. The corresponding expression when $\alpha_1 > 0$ is given by (2.6.21a) with $\psi_1 = \pi/2$. This gives

$$k_2 \sin \Theta_2 = [k_2^2 \sin^2 \Theta_2 + (\beta_1^2 - \alpha_1^2 - k_2^2 \sin^2 \Theta_2) f^2(p_1)]^{1/2}. \tag{2.6.23}$$

With (2.6.15) and (2.6.16), this becomes

$$\beta_1^2 - \alpha_1^2 - k_2^2 \sin^2 \Theta_2 + [(\beta_1^2 - \alpha_1^2 - k_2^2 \sin^2 \Theta_2)^2 + 4\alpha_1^2 \beta_1^2]^{1/2} = 0. \tag{2.6.24}$$

This has no solution for $\sin \Theta_2$ except when $\alpha_1 = 0$. This means that there is no angle of incidence Θ_2 for which the real angle of refraction is $\psi_1 = \pi/2$.

When $\alpha_1 = 0$ and $g(|p_1|) = 0$ in (2.6.21b), the result is

$$\text{range (c):} \quad \psi_1 = \sin^{-1}(1) = \frac{\pi}{2}. \tag{2.6.25}$$

This is the angle of transmission in Region 1 when the angle of incidence in Region 2 is equal to or greater than the critical angle. The wave in Region 1 is a surface wave that travels along the boundary.

The largest possible angle of transmission ψ_1 occurs when the angle of incidence is $\Theta_2 = \pi/2$ at grazing incidence. In this case, (2.6.21a,b) give

$$\text{range (a):} \quad \sin^2 \psi_{1\max} = k_2^2 [k_2^2 + (\beta_1^2 - \alpha_1^2 - k_2^2) f^2(p_1)]^{-1}, \tag{2.6.26a}$$

$$\text{range (c):} \quad \sin^2 \psi_{1\max} = k_2^2 [k_2^2 + (k_2^2 - \beta_1^2 + \alpha_1^2) g^2(|p_1|)]^{-1}. \tag{2.6.26b}$$

With (2.6.15)–(2.6.17), the following results are obtained:

$$\text{range (a):} \quad \sin \psi_{1\max} = k_2 \big(\tfrac{1}{2}\{\beta_1^2 - \alpha_1^2 + k_2^2 + [(\beta_1^2 - \alpha_1^2 - k_2^2)^2 + 4\alpha_1^2 \beta_1^2]^{1/2}\}\big)^{-1/2}, \tag{2.6.27a}$$

$$\text{range (c):}\quad \sin\psi_{1\max} = k_2\left(\tfrac{1}{2}\{k_2^2+\beta_1^2-\alpha_1^2 + [(k_2^2-\beta_1^2+\alpha_1^2)^2+4\alpha_1^2\beta_1^2]^{1/2}\}\right)^{-1/2}. \tag{2.6.27b}$$

When $\alpha_1 = 0$ and $\Theta_2 = \pi/2$, these give, respectively, (2.6.22) and (2.6.25). For all values of $\alpha_1 > 0$,

$$\text{range (a):}\quad \psi_{1\max} < \sin^{-1}\left(\frac{k_2}{\beta_1}\right), \tag{2.6.28a}$$

$$\text{range (c):}\quad \psi_{1\max} < \frac{\pi}{2}. \tag{2.6.28b}$$

The last condition confirms the statement following (2.6.24). It indicates that there is always a transmitted wave in Region 1 that does not travel along the boundary when $\alpha_1 > 0$.

Since

$$\beta_{1x} = (\beta_{1x}^2+\beta_{1z}^2)^{1/2}\sin\psi_1; \qquad \beta_{1z} = (\beta_{1x}^2+\beta_{1z}^2)^{1/2}\cos\psi_1, \tag{2.6.29}$$

where ψ_1 is defined in (2.6.20), the field transmitted into Region 1 is given by (2.5.8),

$$\begin{aligned}\mathbf{E}_1^{\text{tr}}(\mathbf{r}) &= \mathbf{E}_1(0)e^{i(k_{1x}x+k_{1z}z)}\\ &= \mathbf{E}_1(0)e^{-\alpha_{1z}z}\,e^{i(\beta_{1x}^2+\beta_{1z}^2)^{1/2}(x\sin\psi_1+z\cos\psi_1)},\end{aligned} \tag{2.6.30}$$

where $\beta_{1x} = \beta_{2x} = k_2\sin\Theta_2$, and where α_{1z} and β_{1z} are defined in (2.6.19a-c) for the different ranges. This differs from the corresponding field transmitted into a perfect dielectric Region 1 only in the appearance of the exponential attenuation factor $\exp(-\alpha_{1z}z)$. It represents surfaces of constant phase traveling in the direction of ψ_1. The real wavelength is $\lambda_1 = 2\pi(\beta_{1x}^2+\beta_{1z}^2)^{-1/2}$; the real phase velocity is $v_{p1} = \omega(\beta_{1x}^2+\beta_{1z}^2)^{-1/2}$ in the direction of ψ_1. The amplitude factor $\exp(-\alpha_{1z}z)$ changes none of this. However, the amplitude of the wave is no longer unity independent of x and z. It decreases exponentially with distance as the surfaces of constant phase advance, but the distance is not the radial distance $(x^2+z^2)^{1/2}$ of travel but the vertical distance z traversed as the phase front advances radially. An alternative, illuminating manner of interpreting the inhomogeneous field is as an exponentially attenuated wave,

$$e^{-\alpha_{1z}z}\,e^{i\beta_{1z}z} = e^{-\alpha_{1z}z}\,e^{i(\beta_{1x}^2+\beta_{1z}^2)^{1/2}z\cos\psi_1}, \tag{2.6.31}$$

traveling vertically down into Region 1 from the boundary at $z = 0$. This downward traveling wave simultaneously moves in the positive x-direction as the unattenuated wave

$$e^{i\beta_{1x}x} = e^{i(\beta_{1x}^2+\beta_{1z}^2)^{1/2}x\sin\psi_1} = e^{ik_2x\sin\Theta_2}. \tag{2.6.32}$$

2.7 The Reflection and Transmission of Plane Waves Incident from a Lossy Half-Space (Cases C and D)

It remains to discuss Cases C and D of Section 2.4 which deal with a wave incident from a lossy half-space, i.e., $\alpha_2 > 0$, where Region 2 with the complex wave number $k_2 = \beta_2 + i\alpha_2$ is the region of incidence. Strictly speaking, a plane wave incident in a lossy medium is not physically meaningful because the electromagnetic field increases exponentially without limit as the "source" is approached. However, in the limited context of one or more infinite parallel planar boundaries, it is useful to make use of such an incident plane wave.

In Case C, $\alpha_2 > 0$ in the region of incidence, $\alpha_1 = 0$ in the loss-free region of transmission and refraction. These are the reciprocal conditions of those in Case B. Since k_{2x}, k_{2z}, and k_{1x} are all real in Case B, the corresponding conditions for Case C include k_{1x} and k_{2x} real. That is,

$$\operatorname{Im} k_{1x} = \operatorname{Im} k_{2x} = 0. \tag{2.7.1}$$

The important conclusion is reached that

$$k_{2x} \text{ is real} \tag{2.7.2}$$

for Cases A, B, C, and also D. In other words, k_{2x} is always to be taken to be real. More generally, for an arbitrary scattering plane, not necessarily the xy-plane, the transverse components

$$k_{2T} = k_{2T}^r = k_{1T} \text{ are always real.} \tag{2.7.3}$$

Once this crucial point is realized, the computation of the reflection and refraction of homogeneous or inhomogeneous plane waves is straightforward. In particular, the relations (2.5.32)–(2.5.59) are all true in general. The angles Θ_1 and Θ_2 are such that in the general formula $k_1 \sin\Theta_1 = k_2 \sin\Theta_2$ each side is real; the angles and the wave numbers are, in general, not real.

Several further comments on the crucial point (2.7.3) may be useful:

(a) In the study of lateral waves and more generally of geometry with translational invariance in two directions, it is natural and often necessary to use Fourier transforms in these two directions. When this is carried out without deforming the contours of integration, (2.7.3) is automatically satisfied. Therefore, no generalization beyond (2.7.3) is needed for this rather extensive class of electromagnetic problems.

(b) It must be remembered that an incident plane wave in a lossy medium must be treated with great care. In the case at hand, the class of acceptable plane waves in a lossy medium depends on the orientation of the plane boundaries between the two or more regions. Specifically, if an incident plane wave in Region 2 with the propagation vector $\mathbf{k}_2 = \hat{\mathbf{z}}k_{2_z} + \mathbf{k}_{2T}$ satisfies the condition

$$\hat{\mathbf{z}} \times \mathbf{k}_2 \text{ is real,} \tag{2.7.4}$$

that is, k_{2T} is real, this condition is, in general, no longer satisfied when the normal $\hat{\mathbf{z}}$ to the boundary is changed. The condition (2.7.4) can be satisfied for all $\hat{\mathbf{z}}$ only when Region 2 is lossless and k_2 is real. Clearly, (b) is closely related to (a).

(c) If the condition that k_{1_x} and k_{2_x} are both real is not satisfied, the transmitted wave has the form

$$e^{i\mathbf{k}_1\cdot\mathbf{r}} = e^{i[(\beta_{1_x}+i\alpha_{1_x})x+(\beta_{1_z}+i\alpha_{1_z})z]}. \tag{2.7.5}$$

If this is to be valid in Region 1 when this is loss-free so that k_1 is real, it follows that

$$\mathbf{k}_1 = \hat{\mathbf{x}}k_{1_x} + \hat{\mathbf{z}}k_{1_z}, \tag{2.7.6a}$$

with

$$\begin{aligned} k_1 &= (k_{1_x}^2 + k_{1_z}^2)^{1/2} \\ &= [(\beta_{1_x}^2 - \alpha_{1_x}^2 + \beta_{1_z}^2 - \alpha_{1_z}^2) + 2i(\alpha_{1_x}\beta_{1_x} + \alpha_{1_z}\beta_{1_z})]. \end{aligned} \tag{2.7.6b}$$

Since k_1 is real, it is necessary that

$$\alpha_{1_x}\beta_{1_x} + \alpha_{1_z}\beta_{1_z} = 0. \tag{2.7.7}$$

Since (2.6.10) implies that $\alpha_{1_z} > 0$ and $\beta_{1_z} > 0$, it follows that

$$\alpha_{1_x}\beta_{1_x} < 0, \tag{2.7.8a}$$

or

$$\operatorname{Im} \mathbf{k}_{1T} \cdot \mathbf{k}_{1T} < 0. \tag{2.7.8b}$$

This is a possible generalization of (2.7.1), but appears to have no useful physical application.

A specific example of Case D when both the region of incidence (Region 2, $z \le 0$) and the region of refraction and transmission (Region 1, $z \ge 0$) are dissipative is readily studied. The field in Region 2 must be an inhomogeneous wave with the following components if it is of electric type:

$$B_{2y}(x,z) = B_{2y}^{\text{inc}}(0,0)e^{ik_{2x}x}(e^{ik_{2z}z} + f_{er}e^{-ik_{2z}z}); \qquad z \le 0, \tag{2.7.9}$$

$$E_{2x}(x,z) = \frac{\omega k_{2z}}{k_2^2}\, B_{2y}^{\text{inc}}(0,0)e^{ik_{2x}x}(e^{ik_{2z}z} - f_{er}e^{-ik_{2z}z}); \qquad z \le 0, \tag{2.7.10}$$

$$E_{2z}(x,z) = -\frac{\omega k_{2x}}{k_2^2}\, B_{2y}^{\text{inc}}(0,0)e^{ik_{2x}x}(e^{ik_{2z}z} + f_{er}e^{-ik_{2z}z}); \qquad z \le 0, \tag{2.7.11}$$

$$B_{1y}(x,z) = f_{et} B_{2y}^{\text{inc}}(0,0) e^{ik_{1x}x} e^{ik_{1z}z}; \qquad z \geq 0, \tag{2.7.12}$$

$$E_{1x}(x,z) = \frac{\omega k_{1z}}{k_1^2} f_{et} B_{2y}^{\text{inc}}(0,0) e^{ik_{1x}x} e^{ik_{1z}z}; \qquad z \geq 0, \tag{2.7.13}$$

$$E_{1z}(x,z) = -\frac{\omega k_{1x}}{k_1^2} f_{et} B_{2y}^{\text{inc}}(0,0) e^{ik_{1x}x} e^{ik_{1z}z}; \qquad z \geq 0. \tag{2.7.14}$$

Here,

$$f_{er} = \frac{k_1^2 k_{2z} - k_2^2 k_{1z}}{k_1^2 k_{2z} + k_2^2 k_{1z}}; \qquad f_{et} = \frac{2k_1^2 k_{2z}}{k_1^2 k_{2z} + k_2^2 k_{1z}}, \tag{2.7.15}$$

which are identical with (2.5.58) for Case A. Also, Snell's law applies, viz.,

$$k_{1x} = k_{2x} \quad \text{or} \quad k_1 \sin\Theta_1 = k_2 \sin\Theta_2, \tag{2.7.16}$$

where k_{1x} and k_{2x} are real, but $k_1 = \beta_1 + i\alpha_1$, $k_2 = \beta_2 + i\alpha_2$, Θ_1, and Θ_2 are, in general, all complex. Also,

$$\begin{aligned} k_{1z} = \beta_{1z} + i\alpha_{1z} &= (k_1^2 - k_{1x}^2)^{1/2} \\ &= (\beta_1^2 - \alpha_1^2 - k_{1x}^2 + 2i\alpha_1\beta_1)^{1/2}, \end{aligned} \tag{2.7.17}$$

$$\begin{aligned} k_{2z} = \beta_{2z} + i\alpha_{2z} &= (k_2^2 - k_{2x}^2)^{1/2} \\ &= (\beta_2^2 - \alpha_2^2 - k_{2x}^2 + 2i\alpha_2\beta_2)^{1/2}. \end{aligned} \tag{2.7.18}$$

The real angles of propagation of the surfaces of constant phase in the two regions are given by

$$\psi_2 = \sin^{-1} \frac{\beta_{2x}}{(\beta_{2x}^2 + \beta_{2z}^2)^{1/2}}, \tag{2.7.19}$$

$$\psi_1 = \sin^{-1} \frac{\beta_{1x}}{(\beta_{1x}^2 + \beta_{1z}^2)^{1/2}}, \tag{2.7.20}$$

where $\beta_{1x} = k_{1x}$ and $\beta_{2x} = k_{2x}$ are real. Snell's law for the real angles ψ_1 and ψ_2 is

$$\beta_{1x} = \beta_{2x} \quad \text{or} \quad (\beta_{1x}^2 + \beta_{1z}^2)^{1/2} \sin\psi_1 = (\beta_{2x}^2 + \beta_{2z}^2)^{1/2} \sin\psi_2. \tag{2.7.21}$$

The formulas (2.7.9)–(2.7.14) show that in Region 2 there is a component wave traveling vertically downward. It is exponentially attenuated and given by

$$e^{-\alpha_{2z}z} e^{i\beta_{2z}z}; \qquad z \leq 0. \tag{2.7.22}$$

At the boundary, $z = 0$, it is partly reflected as a vertically upward traveling, exponentially attenuated wave given by

$$f_{er} e^{\alpha_{2z}z} e^{-i\beta_{2z}z}; \qquad z \leq 0, \tag{2.7.23}$$

and partly transmitted into Region 1 as the transmitted wave given by

$$f_{et} e^{-\alpha_{1z}z} e^{i\beta_{1z}z}; \qquad z \geq 0. \tag{2.7.24}$$

In Region 2 the downward and upward traveling waves combine to form a standing wave. Since Region 1 is unbounded downward, the downward traveling wave in Region 1 is an exponentially attenuated progressive wave. The entire distribution of vertical component waves in both regions travels horizontally in the positive x-direction as an unattenuated progressive wave given by

$$e^{ik_{2x}x} = e^{ik_{1x}x}. \tag{2.7.25}$$

The vertical attenuated plane waves combine with the horizontal unattenuated plane waves to produce a single inhomogeneous wave with surfaces of constant phase that travel in the directions of the real angles ψ_2 and ψ_1, respectively in the two regions, and horizontal surfaces of constant amplitude that move vertically.

If Region 1 is lossless so that $\alpha_1 = 0$ and k_1 is real (Case C), the downward traveling component wave in Region 1 is unattenuated provided that $k_1 > k_{2x}$. It combines with the unattenuated horizontal wave to produce a single homogeneous plane wave traveling in the direction of $\psi_1 = \Theta_1$ as required by Snell's law

$$\beta_1 \sin\psi_1 \equiv k_1 \sin\Theta_1 = k_2 \sin\Theta_2, \tag{2.7.26}$$

where now k_1 and Θ_1 are real, k_2 and Θ_2 both complex. Thus, an inhomogeneous wave in the lossy Region 2 generates a homogeneous refracted wave in the loss-free Region 1.

3
The Electromagnetic Field of a Unit Vertical Electric Dipole in the Presence of a Plane Boundary

3.1 Maxwell's Equations and Their Transforms

The electromagnetic field generated by a vertical electric dipole near the boundary between two quite different material half-spaces has been known in terms of general integrals for many years. Following Sommerfeld, it has been expressed in terms of derivatives of the Hertz potential. In this form it is difficult to visualize the nature of the electromagnetic field. A somewhat better insight is obtained from the explicit integrals for the components of the electromagnetic field. These can be obtained by carrying out the differentiation of the Hertz potential or directly from Maxwell's equations. The latter course is more direct and is followed here.

The geometry and notation underlying the analysis are shown in Fig. 3.1.1. The vertical dipole with unit electric moment (i.e., $I\Delta l = 1$ A m) is located on the downward-directed z-axis at a distance d from the origin of coordinates on the interface, i.e., the xy-plane. The electromagnetic field is to be determined at an arbitrary point (x, y, z) in rectangular or (ρ, ϕ, z) in cylindrical coordinates. The lower half-space ($z \geq 0$) is Region 1, the upper half-space ($z \leq 0$) is Region 2. The two regions are characterized by the complex wave numbers $k_j = \beta_j + i\alpha_j = \omega(\mu_j\tilde{\epsilon}_j)^{1/2}$, where $\tilde{\epsilon}_j = \epsilon_j + i\sigma_j/\omega$ and $\epsilon_j = \epsilon_0\epsilon_{jr}$ with $j = 1, 2$. It is assumed that both regions are nonmagnetic so that $\mu_1 = \mu_2 = \mu_0 = 4\pi \times 10^{-7}$ H/m. The time dependence $\exp(-i\omega t)$ is used.

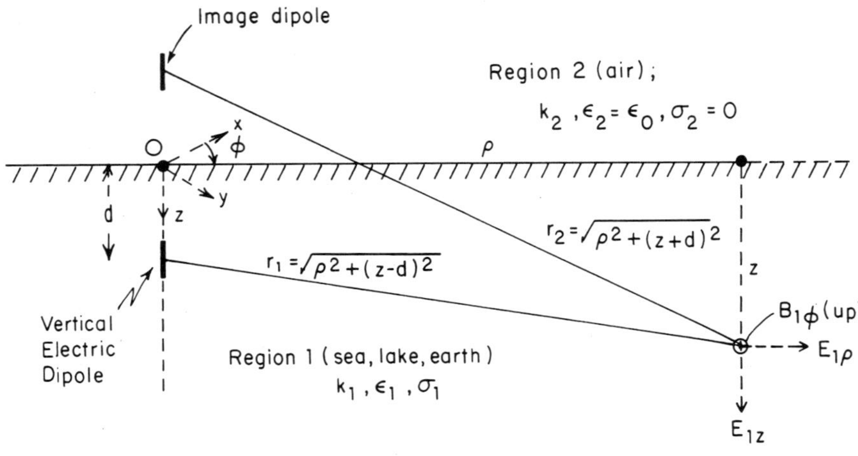

Figure 3.1.1. Vertical electric dipole at depth d in Region 1. Electromagnetic field calculated at (ρ, z).

Maxwell's equations for the two regions ($j = 1, 2$) are

$$\nabla \times \mathbf{E}_j = i\omega \mathbf{B}_j, \tag{3.1.1}$$

$$\nabla \times \mathbf{B}_j = -\frac{ik_j^2}{\omega}\,\mathbf{E}_j + \mu_0 \mathbf{J}, \tag{3.1.2}$$

where

$$\mathbf{J} = \hat{\mathbf{z}} J_z; \qquad J_z = \delta(x)\delta(y)\delta(z-d). \tag{3.1.3}$$

Note that with rotational symmetry, $B_{jz} = 0$. To obtain the explicit formulas for the components of the field, it is convenient to make use of the following Fourier transforms:

$$\mathbf{E}(x,y,z) = \frac{1}{(2\pi)^2}\int_{-\infty}^{\infty} d\xi \int_{-\infty}^{\infty} d\eta\, e^{i(\xi x + \eta y)}\bar{\mathbf{E}}(\xi,\eta,z), \tag{3.1.4}$$

$$\mathbf{B}(x,y,z) = \frac{1}{(2\pi)^2}\int_{-\infty}^{\infty} d\xi \int_{-\infty}^{\infty} d\eta\, e^{i(\xi x + \eta y)}\bar{\mathbf{B}}(\xi,\eta,z), \tag{3.1.5}$$

$$\mathbf{J}(x,y,z) = \frac{1}{(2\pi)^2}\int_{-\infty}^{\infty} d\xi \int_{-\infty}^{\infty} d\eta\, e^{i(\xi x + \eta y)}\bar{\mathbf{J}}(\xi,\eta,z), \tag{3.1.6}$$

where

$$\bar{J}_z(\xi,\eta,z) = \delta(z-d). \tag{3.1.7}$$

With (3.1.4)–(3.1.7), Maxwell's equations are transformed into

$$i\eta \bar{E}_{jz} - \frac{\partial \bar{E}_{jy}}{\partial z} = i\omega \bar{B}_{jx}, \qquad \frac{\partial \bar{E}_{jx}}{\partial z} - i\xi \bar{E}_{jz} = i\omega \bar{B}_{jy},$$

$$i\xi \bar{E}_{jy} - i\eta \bar{E}_{jx} = 0; \tag{3.1.8}$$

$$\frac{\partial \bar{B}_{jy}}{\partial z} = \frac{ik_j^2}{\omega}\, \bar{E}_{jx}, \qquad \frac{\partial \bar{B}_{jx}}{\partial z} = -\frac{ik_j^2}{\omega}\, \bar{E}_{jy},$$

$$i\xi \bar{B}_{jy} - i\eta \bar{B}_{jx} = -\frac{ik_j^2}{\omega}\, \bar{E}_{jz} + \mu_0 \delta(z-d). \tag{3.1.9}$$

It follows from (3.1.8) that

$$\bar{E}_{jy} = \frac{\eta}{\xi}\, \bar{E}_{jx}; \qquad \bar{B}_{jy} = -\frac{\xi}{\eta}\, \bar{B}_{jx}. \tag{3.1.10}$$

Suitable combinations of the above equations yield ordinary differential equations for $\bar{B}_{jx}$ and $\bar{B}_{jy}$. For example, the equation for $\bar{B}_{jx}$ is

$$\left(\frac{d^2}{dz^2} + \gamma_j^2\right) \bar{B}_{jx} = -i\eta\mu_0 \delta(z-d), \tag{3.1.11}$$

with

$$\gamma_j = (k_j^2 - \xi^2 - \eta^2)^{1/2}; \qquad j = 1,\, 2. \tag{3.1.12}$$

The square root is to be taken so that γ_j is in the first quadrant.

The equation for $\bar{B}_{jy}$ is obtained from (3.1.11) with (3.1.10). Also,

$$\bar{E}_{jx} = -\frac{i\omega}{k_j^2} \frac{\partial \bar{B}_{jy}}{\partial z} = \frac{i\omega}{k_j^2} \frac{\xi}{\eta} \frac{\partial \bar{B}_{jx}}{\partial z}, \tag{3.1.13}$$

$$\bar{E}_{jy} = \frac{i\omega}{k_j^2} \frac{\partial \bar{B}_{jx}}{\partial z}, \tag{3.1.14}$$

$$\bar{E}_{jz} = \frac{\omega}{\eta k_j^2} \left(\frac{d^2}{dz^2} + k_j^2\right) \bar{B}_{jx}. \tag{3.1.15}$$

3.2 Solution of the Transformed Equations

The solutions of (3.1.11) are

$$\bar{B}_{1x}(z) = C_1 e^{i\gamma_1 z} - \frac{\eta\mu_0}{2\gamma_1}\, e^{i\gamma_1 |z-d|}; \qquad z \geq 0, \tag{3.2.1}$$

$$\bar{B}_{2x}(z) = C_2 e^{-i\gamma_2 z}; \qquad z \leq 0. \tag{3.2.2}$$

In (3.2.1) and (3.2.2) terms proportional, respectively, to $\exp(-i\gamma_1 z)$ and $\exp(i\gamma_2 z)$ have been omitted. The explanation is the following. If the media are lossy, i.e., if $\operatorname{Im}\gamma_1 > 0$ and $\operatorname{Im}\gamma_2 > 0$, these omitted terms cannot be present because they lead to exponentially large fields as $z \to \infty$ for (3.2.1) and $z \to -\infty$ for (3.2.2). If one or both media are lossless, then these terms should still be absent. This follows from either the limit of the loss approaching zero or from the Sommerfeld radiation condition.

The boundary condition, $\bar{B}_{1x}(0) = \bar{B}_{2x}(0)$, leads to

$$C_1 = C_2 + \frac{\eta\mu_0}{2\gamma_1}\, e^{i\gamma_1 d}, \tag{3.2.3}$$

so that

$$\bar{B}_{1x}(z) = C_2 e^{i\gamma_1 z} + \frac{\eta\mu_0}{2\gamma_1}\,(e^{i\gamma_1(z+d)} - e^{i\gamma_1|z-d|}). \tag{3.2.4}$$

With (3.1.13), (3.2.2), and (3.2.4), it follows that

$$\bar{E}_{1x}(z) = -\frac{\omega}{k_1^2}\left(\frac{\gamma_1\xi}{\eta}\, C_2 e^{i\gamma_1 z} + \frac{\xi\mu_0}{2}\,(e^{i\gamma_1(z+d)} \mp e^{i\gamma_1|z-d|})\right), \tag{3.2.5}$$

where the upper sign is for $z > d$, the lower sign for $0 \le z \le d$, and

$$\bar{E}_{2x}(z) = \frac{\omega}{k_2^2}\,\frac{\gamma_2\xi}{\eta}\, C_2 e^{-i\gamma_2 z}; \qquad z \le 0. \tag{3.2.6}$$

The boundary condition, $\bar{E}_{1x}(0) = \bar{E}_{2x}(0)$, leads to

$$C_2 = -\frac{\mu_0\eta k_2^2 e^{i\gamma_1 d}}{k_1^2\gamma_2 + k_2^2\gamma_1}. \tag{3.2.7}$$

When this value of C_2 is substituted in (3.2.2), (3.2.4), (3.2.5), and (3.2.6), the final solutions for $\bar{B}_{1x}(z)$, $\bar{B}_{2x}(z)$, $\bar{E}_{1x}(z)$, and $\bar{E}_{2x}(z)$ are obtained. The corresponding solutions for $\bar{B}_{1y}(z)$, $\bar{B}_{2y}(z)$, $\bar{E}_{1y}(z)$, and $\bar{E}_{2y}(z)$ follow from (3.1.10). Finally, $\bar{E}_{1z}(z)$ and $\bar{E}_{2z}(z)$ are obtained with (3.1.15). The complete results for the two regions are summarized below. The following shorthand notation is used:

$$N \equiv k_1^2\gamma_2 + k_2^2\gamma_1. \tag{3.2.8}$$

Region 1, $z \ge 0$:

$$\bar{B}_{1x}(\xi,\eta,z) = -\mu_0\eta\left(\frac{e^{i\gamma_1|z-d|}}{2\gamma_1} - \frac{e^{i\gamma_1(z+d)}}{2\gamma_1} + \frac{k_2^2 e^{i\gamma_1(z+d)}}{N}\right), \tag{3.2.9}$$

$$\bar{B}_{1y}(\xi,\eta,z) = -\frac{\xi}{\eta}\,\bar{B}_{1x}(\xi,\eta,z), \tag{3.2.10}$$

$$\bar{B}_{1z}(\xi,\eta,z) = 0, \tag{3.2.11}$$

$$\begin{aligned}\bar{E}_{1x}(\xi,\eta,z) = -\frac{\omega\mu_0\xi}{k_1^2}\Bigg(&\mp\frac{e^{i\gamma_1|z-d|}}{2} + \frac{e^{i\gamma_1(z+d)}}{2} \\ &- \frac{k_2^2\gamma_1 e^{i\gamma_1(z+d)}}{N}\Bigg); \qquad \begin{cases} z > d \\ 0 \le z \le d \end{cases},\end{aligned} \tag{3.2.12}$$

$$\bar{E}_{1y}(\xi,\eta,z) = \frac{\eta}{\xi}\,\bar{E}_{1x}(\xi,\eta,z), \tag{3.2.13}$$

$$\bar{E}_{1z}(\xi,\eta,z) = -\frac{\omega\mu_0(\xi^2+\eta^2)}{k_1^2}\left(\frac{e^{i\gamma_1|z-d|}}{2\gamma_1} - \frac{e^{i\gamma_1(z+d)}}{2\gamma_1} + \frac{k_2^2 e^{i\gamma_1(z+d)}}{N}\right). \tag{3.2.14}$$

Region 2, $z \le 0$:

$$\bar{B}_{2x}(\xi,\eta,z) = -\mu_0\eta\,\frac{k_2^2 e^{i\gamma_1 d}e^{-i\gamma_2 z}}{N}, \tag{3.2.15}$$

$$\bar{B}_{2y}(\xi,\eta,z) = -\frac{\xi}{\eta}\,\bar{B}_{2x}(\xi,\eta,z), \tag{3.2.16}$$

$$\bar{B}_{2z}(\xi,\eta,z) = 0, \tag{3.2.17}$$

$$\bar{E}_{2x}(\xi,\eta,z) = -\omega\mu_0\xi\,\frac{\gamma_2 e^{i\gamma_1 d}e^{-i\gamma_2 z}}{N}, \tag{3.2.18}$$

$$\bar{E}_{2y}(\xi,\eta,z) = \frac{\eta}{\xi}\,\bar{E}_{2x}(\xi,\eta,z), \tag{3.2.19}$$

$$\bar{E}_{2z}(\xi,\eta,z) = -\omega\mu_0(\xi^2+\eta^2)\,\frac{e^{i\gamma_1 d}e^{-i\gamma_2 z}}{N}. \tag{3.2.20}$$

Note that, in agreement with the boundary condition, $\tilde{\epsilon}_1 E_{1n} = \tilde{\epsilon}_2 E_{2n}$, on the normal component of the electric field,

$$k_1^2\bar{E}_{1z}(\xi,\eta,0) = k_2^2\bar{E}_{2z}(\xi,\eta,0). \tag{3.2.21}$$

3.3 General Integrals for the Field When the Dipole Is in Region 1

The substitution of (3.2.9)–(3.2.20) into (3.1.4) and (3.1.5) yields the general integrals for the components of the electric and magnetic fields in Cartesian coordinates. When these are converted to cylindrical coordinates with the relations

$$x = \rho\cos\phi; \quad y = \rho\sin\phi; \qquad \xi = \lambda\cos\phi'; \quad \eta = \lambda\sin\phi', \tag{3.3.1}$$

$$\rho^2 = x^2 + y^2; \qquad \lambda^2 = \xi^2 + \eta^2, \tag{3.3.2}$$

$$d\xi\,d\eta = \lambda\,d\phi'\,d\lambda; \qquad \xi x + \eta y = \lambda\rho\cos(\phi - \phi'), \tag{3.3.3}$$

$$E_\rho = E_x\cos\phi + E_y\sin\phi, \tag{3.3.4}$$

$$B_\phi = -B_x\sin\phi + B_y\cos\phi, \tag{3.3.5}$$

it is found that

$$B_{jz}(\rho, z) = 0; \quad B_{j\rho}(\rho, z) = 0; \quad E_{j\phi}(\rho, z) = 0; \quad j = 1,\, 2. \tag{3.3.6}$$

With the integral representations of the Bessel functions, viz.,

$$J_0(\lambda\rho) = \frac{1}{2\pi}\int_0^{2\pi} e^{i\lambda\rho\cos(\phi-\phi')}\, d\phi', \tag{3.3.7}$$

$$J_1(\lambda\rho) = -\frac{i}{2\pi}\int_0^{2\pi} e^{i\lambda\rho\cos(\phi-\phi')}\cos(\phi-\phi')\, d\phi', \tag{3.3.8}$$

it follows that, for Region 1, $z \geq 0$,

$$B_{1\phi}(\rho, z) = \frac{i\mu_0}{2\pi}\int_0^\infty \left(\frac{e^{i\gamma_1|z-d|}}{2\gamma_1} - \frac{e^{i\gamma_1(z+d)}}{2\gamma_1} + \frac{k_2^2 e^{i\gamma_1(z+d)}}{N}\right) J_1(\lambda\rho)\lambda^2\, d\lambda, \tag{3.3.9}$$

$$E_{1\rho}(\rho, z) = \frac{i\omega\mu_0}{2\pi k_1^2}\int_0^\infty \left(\pm\frac{e^{i\gamma_1|z-d|}}{2} - \frac{e^{i\gamma_1(z+d)}}{2} + \frac{k_2^2\gamma_1 e^{i\gamma_1(z+d)}}{N}\right) J_1(\lambda\rho)\lambda^2\, d\lambda; \quad \begin{cases} z > d \\ 0 \leq z \leq d \end{cases}, \tag{3.3.10}$$

$$E_{1z}(\rho, z) = -\frac{\omega\mu_0}{2\pi k_1^2}\int_0^\infty \left(\frac{e^{i\gamma_1|z-d|}}{2\gamma_1} - \frac{e^{i\gamma_1(z+d)}}{2\gamma_1} + \frac{k_2^2 e^{i\gamma_1(z+d)}}{N}\right) J_0(\lambda\rho)\lambda^3\, d\lambda; \tag{3.3.11}$$

for Region 2, $z \leq 0$,

$$B_{2\phi}(\rho, z) = \frac{i\mu_0 k_2^2}{2\pi}\int_0^\infty \frac{e^{i\gamma_1 d}e^{-i\gamma_2 z}}{N} J_1(\lambda\rho)\lambda^2\, d\lambda, \tag{3.3.12}$$

$$E_{2\rho}(\rho, z) = -\frac{i\omega\mu_0}{2\pi}\int_0^\infty \frac{\gamma_2 e^{i\gamma_1 d}e^{-i\gamma_2 z}}{N} J_1(\lambda\rho)\lambda^2\, d\lambda, \tag{3.3.13}$$

$$E_{2z}(\rho, z) = -\frac{\omega\mu_0}{2\pi}\int_0^\infty \frac{e^{i\gamma_1 d}e^{-i\gamma_2 z}}{N} J_0(\lambda\rho)\lambda^3\, d\lambda, \tag{3.3.14}$$

where $N = k_1^2\gamma_2 + k_2^2\gamma_1$ with $\gamma_j = (k_j^2 - \lambda^2)^{1/2}$, $j = 1,\, 2$. The expressions (3.3.9)–(3.3.14) are the desired general integrals for the electromagnetic field in both half-spaces due to a vertical dipole at $z = d$ in Region 1.

3.4 Integrated Formulas for the Electromagnetic Field of a Vertical Electric Dipole at the Depth d in Region 1

In order to evaluate the integrals in (3.3.9)–(3.3.11) for the components of the electromagnetic field at (ρ, z) in Region 1 when the vertical electric dipole is also in Region 1 at $(0, d)$, it is convenient to examine separately four different parts. These are the direct field, the reflected field due to an ideal image, the lateral-wave field, and a correction for the reflected field when this is not accurately that of an ideal image. Specifically, let

$$\mathbf{E}_1(\rho, z) = \mathbf{E}_1^d(\rho, z) + \mathbf{E}_1^i(\rho, z) + \mathbf{E}_1^L(\rho, z) + \mathbf{E}_1^c(\rho, z), \tag{3.4.1}$$

$$\mathbf{B}_1(\rho, z) = \mathbf{B}_1^d(\rho, z) + \mathbf{B}_1^i(\rho, z) + \mathbf{B}_1^L(\rho, z) + \mathbf{B}_1^c(\rho, z). \tag{3.4.2}$$

The direct field is the entire field that would be observed at (ρ, z) if there were no boundary and medium 1 filled all of space. It involves only the wave number k_1 and the distance $r_1 = [\rho^2 + (z-d)^2]^{1/2}$. With the integrals tabulated in Appendix A, it is given by

$$\begin{aligned} B_{1\phi}^d(\rho, z) &= \frac{i\mu_0}{4\pi} \int_0^\infty \frac{e^{i\gamma_1|z-d|}}{\gamma_1} J_1(\lambda\rho)\lambda^2\, d\lambda \\ &= \frac{i\mu_0}{4\pi} \frac{\rho}{2} I_7(\rho, z-d) \\ &= -\frac{\mu_0}{4\pi} e^{ik_1 r_1} \left(\frac{ik_1}{r_1} - \frac{1}{r_1^2}\right)\left(\frac{\rho}{r_1}\right), \end{aligned} \tag{3.4.3}$$

$$\begin{aligned} E_{1\rho}^d(\rho, z) &= \frac{i\omega\mu_0}{4\pi k_1^2} \int_0^\infty \pm e^{i\gamma_1|z-d|} J_1(\lambda\rho)\lambda^2\, d\lambda; \qquad \begin{cases} z > d \\ 0 < z < d \end{cases} \\ &= \frac{i\omega\mu_0}{4\pi k_1^2} \frac{\rho}{2} I_{22}(\rho, z-d) \\ &= -\frac{\omega\mu_0}{4\pi k_1^2} e^{ik_1 r_1} \left(\frac{ik_1^2}{r_1} - \frac{3k_1}{r_1^2} - \frac{3i}{r_1^3}\right)\left(\frac{\rho}{r_1}\right)\left(\frac{z-d}{r_1}\right), \end{aligned} \tag{3.4.4}$$

$$\begin{aligned} E_{1z}^d(\rho, z) &= -\frac{\omega\mu_0}{4\pi k_1^2} \int_0^\infty \frac{e^{i\gamma_1|z-d|}}{\gamma_1} J_0(\lambda\rho)\lambda^3\, d\lambda \\ &= -\frac{\omega\mu_0}{4\pi k_1^2} \frac{1}{2} [I_7(\rho, z-d) + I_8(\rho, z-d)] \\ &= \frac{\omega\mu_0}{4\pi k_1^2} e^{ik_1 r_1} \left[\frac{ik_1^2}{r_1} - \frac{k_1}{r_1^2} - \frac{i}{r_1^3} \right. \\ &\qquad \left. - \left(\frac{z-d}{r_1}\right)^2 \left(\frac{ik_1^2}{r_1} - \frac{3k_1}{r_1^2} - \frac{3i}{r_1^3}\right)\right]. \end{aligned} \tag{3.4.5}$$

Actually the field of a vertical electric dipole in a homogeneous infinite medium is well known and given in Chapter 2. It is, of course, the same as derived in (3.4.3)–(3.4.5).

The ideal reflected field is the field at (ρ, z) due to an image antenna at $(0, -d)$ when there is no boundary and the properties of medium 1 characterize all space. The electric moment of the image is the negative of that of the source. The ideal reflected field involves only the wave number k_1 and the distance $r_2 = [\rho^2 + (z+d)^2]^{1/2}$. It is

$$\begin{aligned} B_{1\phi}^i(\rho, z) &= -\frac{i\mu_0}{4\pi} \int_0^\infty \frac{e^{i\gamma_1(z+d)}}{\gamma_1} J_1(\lambda\rho)\lambda^2 \, d\lambda \\ &= -\frac{i\mu_0}{4\pi} \frac{\rho}{2} I_7(\rho, z+d) \\ &= \frac{\mu_0}{4\pi} e^{ik_1 r_2} \left(\frac{ik_1}{r_2} - \frac{1}{r_2^2} \right) \left(\frac{\rho}{r_2} \right), \end{aligned} \tag{3.4.6}$$

$$\begin{aligned} E_{1\rho}^i(\rho, z) &= -\frac{i\omega\mu_0}{4\pi k_1^2} \int_0^\infty e^{i\gamma_1(z+d)} J_1(\lambda\rho)\lambda^2 \, d\lambda \\ &= -\frac{i\omega\mu_0}{4\pi k_1^2} \frac{\rho}{2} I_{22}(\rho, z+d) \\ &= \frac{\omega\mu_0}{4\pi k_1^2} e^{ik_1 r_2} \left(\frac{ik_1^2}{r_2} - \frac{3k_1}{r_2^2} - \frac{3i}{r_2^3} \right) \left(\frac{\rho}{r_2} \right) \left(\frac{z+d}{r_2} \right), \end{aligned} \tag{3.4.7}$$

$$\begin{aligned} E_{1z}^i(\rho, z) &= \frac{\omega\mu_0}{4\pi k_1^2} \int_0^\infty \frac{e^{i\gamma_1(z+d)}}{\gamma_1} J_0(\lambda\rho)\lambda^3 \, d\lambda \\ &= \frac{\omega\mu_0}{4\pi k_1^2} \frac{1}{2} [I_7(\rho, z+d) + I_8(\rho, z+d)] \\ &= -\frac{\omega\mu_0}{4\pi k_1^2} e^{ik_1 r_2} \left[\frac{ik_1^2}{r_2} - \frac{k_1}{r_2^2} - \frac{i}{r_2^3} \right. \\ &\qquad \left. - \left(\frac{z+d}{r_2} \right)^2 \left(\frac{ik_1^2}{r_2} - \frac{3k_1}{r_2^2} - \frac{3i}{r_2^3} \right) \right]. \end{aligned} \tag{3.4.8}$$

The integrated exact formulas again agree with the field of the electric dipole given in Chapter 2.

The rest of the field is defined by the following expressions in which the integrals with appropriate constant factors are represented by the three functions $F_{1m}(\rho, z+d)$ with $m = \phi$, ρ, and z:

$$B_{1\phi}^L(\rho, z) + B_{1\phi}^c(\rho, z) = \frac{i\mu_0}{2\pi} F_{1\phi}(\rho, z+d);$$

$$F_{1\phi}(\rho, z+d) = k_2^2 \int_0^\infty \frac{e^{i\gamma_1(z+d)}}{N} J_1(\lambda\rho)\lambda^2 \, d\lambda, \tag{3.4.9}$$

$$E_{1\rho}^L(\rho, z) + E_{1\rho}^c(\rho, z) = \frac{i\omega\mu_0}{2\pi k_1^2} F_{1\rho}(\rho, z+d);$$

$$F_{1\rho}(\rho, z+d) = k_2^2 \int_0^\infty \frac{\gamma_1 e^{i\gamma_1(z+d)}}{N} J_1(\lambda\rho)\lambda^2\, d\lambda, \tag{3.4.10}$$

$$E_{1z}^L(\rho, z) + E_{1z}^c(\rho, z) = -\frac{\omega\mu_0}{2\pi k_1^2} F_{1z}(\rho, z+d);$$

$$F_{1z}(\rho, z+d) = k_2^2 \int_0^\infty \frac{e^{i\gamma_1(z+d)}}{N} J_0(\lambda\rho)\lambda^3\, d\lambda. \tag{3.4.11}$$

These expressions involve both wave numbers k_1 and k_2.

The integrals (3.4.9)–(3.4.11) can be evaluated subject to the inequality $|k_1|^2 \gg |k_2|^2$ which is usually adequately approximated by

$$|k_1| \geq 3|k_2|. \tag{3.4.12}$$

This condition can be used to separate each of the integrals into several integrable parts. Some of these integrate into formulas that involve $\exp(ik_1r_2)$, that is, propagation entirely in Region 1. Others involve both $\exp[ik_1(z+d)]$ and $\exp(ik_2\rho)$, that is, propagation partly in Region 1 and partly in Region 2. Considerable simplification in the derivation of the integrated formulas for (3.4.9)–(3.4.11) is achieved with the conditions

$$\rho \geq 5|z|; \qquad \rho \geq 5d, \tag{3.4.13}$$

which are not really restrictive in most applications when Region 1 is the earth or sea. With them, the approximations

$$r_1 = [\rho^2 + (z-d)^2]^{1/2} \sim \rho; \qquad r_2 = [\rho^2 + (z+d)^2]^{1/2} \sim \rho \tag{3.4.14}$$

are made where appropriate. One integral is reduced to simplest form subject to the condition

$$|k_1\rho| \geq 3. \tag{3.4.15}$$

The integrals (3.4.9)–(3.4.11) all have the same denominator, $N = k_1^2\gamma_2 + k_2^2\gamma_1$. With $|k_1|^2 \gg |k_2|^2$, the leading part of each integral is obtained with the denominator approximated by $k_1^2\gamma_2$. It follows that $1/N$ is conveniently and exactly written in the form

$$\frac{1}{N} = \frac{1}{k_1^2\gamma_2} + \left(\frac{1}{k_1^2\gamma_2 + k_2^2\gamma_1} - \frac{1}{k_1^2\gamma_2}\right). \tag{3.4.16}$$

With this representation, the integrals (3.4.9)–(3.4.11) can be expressed as follows:

$$F_{1\phi}(\rho, z+d) = \frac{k_2^2}{k_1^2} \int_0^\infty \gamma_2^{-1} e^{i\gamma_1(z+d)} J_1(\lambda\rho)\lambda^2\, d\lambda + G_{1\phi}(\rho, z), \tag{3.4.17}$$

$$F_{1\rho}(\rho, z+d) = \frac{k_2^2}{k_1^2} \int_0^\infty \frac{\gamma_1}{\gamma_2} e^{i\gamma_1(z+d)} J_1(\lambda\rho)\lambda^2\, d\lambda + G_{1\rho}(\rho, z), \tag{3.4.18}$$

$$F_{1z}(\rho, z+d) = \frac{k_2^2}{k_1^2} \int_0^\infty \gamma_2^{-1} e^{i\gamma_1(z+d)} J_0(\lambda\rho)\lambda^3\, d\lambda + G_{1z}(\rho, z). \tag{3.4.19}$$

Since contributions to the integrals in the functions $G_{1m}(\rho, z)$ come primarily when λ is of the order of magnitude of k_2, it is a good approximation to set $\gamma_1 = (k_1^2 - \lambda^2)^{1/2} \sim k_1$, so that

$$G_{1m}(\rho, z) \sim G_{1m}(\rho) e^{ik_1(z+d)}; \qquad m = \phi,\ \rho,\ z. \tag{3.4.20}$$

With this approximation,

$$G_{1\phi}(\rho, z) \sim e^{ik_1(z+d)} G_{1\phi}(\rho);$$

$$G_{1\phi}(\rho) = k_2^2 \int_0^\infty \left(\frac{1}{k_1^2\gamma_2 + k_2^2\gamma_1} - \frac{1}{k_1^2\gamma_2} \right) J_1(\lambda\rho)\lambda^2\, d\lambda, \tag{3.4.21}$$

$$G_{1\rho}(\rho, z) \sim e^{ik_1(z+d)} G_{1\rho}(\rho);$$

$$G_{1\rho}(\rho) = k_2^2 \int_0^\infty \left(\frac{1}{k_1^2\gamma_2 + k_2^2\gamma_1} - \frac{1}{k_1^2\gamma_2} \right) \gamma_1 J_1(\lambda\rho)\lambda^2\, d\lambda, \tag{3.4.22}$$

$$G_{1z}(\rho, z) \sim e^{ik_1(z+d)} G_{1z}(\rho);$$

$$G_{1z}(\rho) = k_2^2 \int_0^\infty \left(\frac{1}{k_1^2\gamma_2 + k_2^2\gamma_1} - \frac{1}{k_1^2\gamma_2} \right) J_0(\lambda\rho)\lambda^3\, d\lambda. \tag{3.4.23}$$

The physical significance of the approximation (3.4.20) can be understood more completely after the detailed properties of the lateral wave have been determined. It is helpful at this point to note that when a dipole is at the depth d in the medium with the larger wave number (Region 1), total internal reflection at the boundary limits the field transmitted into Region 2 to that within a relatively small solid angle above the source. This means that it travels approximately vertically upward in Region 1 as implied in the factor $\exp(ik_1 d)$. Similarly, in terms of locally plane inhomogeneous waves any field in Region 2 incident on the boundary is transmitted into Region 1 with the amplitude determined by $\exp(-\alpha_1 z)$ where α_1 is the imaginary part of $k_1 = \beta_1 + i\alpha_1$. The phase fronts are not horizontal planes and depend on the angle of incidence. However, since they are bent toward the normal on entering Region 1 with $|k_1| \geq 3|k_2|$, their direction is not far from perpendicular. Thus, the factor $\exp(ik_1 z) = \exp(-\alpha_1 z)\exp(i\beta_1 z)$ is certainly a good approximation so long as $\beta_1 z$ is not too large.

The real test of (3.4.20) is in quantitative comparisons with the numerical evaluation of the relevant integrals for critical ranges of the parameters and variables. Such comparisons have been made in detail and are described and discussed in Chapter 6. The agreement is very good, and this is the experimental confirmation of the mathematically well-founded approximation in (3.4.20).

The integral in (3.4.17) is evaluated in (D35) with (D14) of Appendix D; the integral in (3.4.18) is evaluated in (D22) with (D1) of Appendix D. The integral in (3.4.19) is evaluated with (D34) and (D35) in Appendix D. The integrals in (3.4.21)–(3.4.23) are evaluated in (E60)–(E62) of Appendix E. When these integrated values are combined in (3.4.17)–(3.4.19), the following formulas are obtained:

$$\begin{aligned}F_{1\phi}(\rho, z+d) &= \frac{k_2^2}{k_1^2}\frac{\rho}{2} I_R(\rho, z+d) + e^{ik_1(z+d)} G_{1\phi}(\rho)\\ &= \frac{ik_2^2}{k_1^2}\Bigg\{ e^{ik_1(z+d)} e^{ik_2\rho} \left[\frac{ik_2}{\rho} - \frac{1}{\rho^2} \right.\\ &\qquad \left. - \frac{k_2^3}{k_1}\left(\frac{\pi}{k_2\rho}\right)^{1/2} e^{-ip}\mathcal{F}(p)\right]\\ &\qquad - ie^{ik_1r_2}\left(\frac{z+d}{\rho}\right)\left(\frac{ik_1}{\rho} - \frac{3}{2\rho^2}\right)\Bigg\},\end{aligned} \tag{3.4.24}$$

$$\begin{aligned}F_{1\rho}(\rho, z+d) &= \frac{k_2^2}{k_1^2}\frac{\rho}{2} I_A(\rho, z+d) + e^{ik_1(z+d)} G_{1\rho}(\rho)\\ &= \frac{ik_2^2}{k_1}\Bigg\{ e^{ik_1(z+d)} e^{ik_2\rho} \left[\frac{ik_2}{\rho} - \frac{1}{\rho^2} \right.\\ &\qquad \left. - \frac{k_2^3}{k_1}\left(\frac{\pi}{k_2\rho}\right)^{1/2} e^{-ip}\mathcal{F}(p)\right]\\ &\qquad - ie^{ik_1r_2}\left(\frac{1}{\rho^2} + \frac{3i}{2k_1\rho^3}\right)\Bigg\},\end{aligned} \tag{3.4.25}$$

$$\begin{aligned}F_{1z}(\rho, z+d) &= \frac{k_2^2}{k_1^2}\frac{1}{2}\left[I_Q(\rho, z+d) + I_R(\rho, z+d)\right] + e^{ik_1(z+d)} G_{1z}(\rho)\\ &= -\frac{k_2^2}{k_1^2}\Bigg\{ e^{ik_1(z+d)} e^{ik_2\rho} \left[\frac{ik_2^2}{\rho} - \frac{k_2}{\rho^2} - \frac{i}{\rho^3} \right.\\ &\qquad \left. - \frac{k_2^4}{k_1}\left(\frac{\pi}{k_2\rho}\right)^{1/2} e^{-ip}\mathcal{F}(p)\right]\\ &\qquad - ie^{ik_1r_2}\left(\frac{z+d}{\rho}\right)\left(\frac{ik_1^2}{\rho} - \frac{k_1}{2\rho^2} + \frac{7i}{8\rho^3}\right)\Bigg\}.\end{aligned} \tag{3.4.26}$$

With these formulas and subject to (3.4.12) and (3.4.13), the field defined in (3.4.9)–(3.4.11) can be written in the following compact form. The lateral-wave and correction terms are shown separately since they are readily identified; the correction term involves $\exp(ik_1r_2)$.

$$\begin{aligned}B_{1\phi}^L(\rho, z) &= -\frac{\mu_0 k_2^2}{2\pi k_1^2} e^{ik_1(z+d)} e^{ik_2\rho} f(\rho; k_1, k_2);\\ B_{1\phi}^c(\rho, z) &= \frac{i\mu_0 k_2^2}{2\pi k_1^2}\left(\frac{z+d}{\rho}\right)\left(\frac{ik_1}{\rho} - \frac{3}{2\rho^2}\right) e^{ik_1r_2},\end{aligned} \tag{3.4.27}$$

$$\begin{aligned}E_{1\rho}^L(\rho, z) &= -\frac{\omega\mu_0 k_2^2}{2\pi k_1^3} e^{ik_1(z+d)} e^{ik_2\rho} f(\rho; k_1, k_2);\\ E_{1\rho}^c(\rho, z) &= \frac{i\omega\mu_0 k_2^2}{2\pi k_1^3} e^{ik_1r_2}\left(\frac{1}{\rho^2} + \frac{3i}{2k_1\rho^3}\right),\end{aligned} \tag{3.4.28}$$

$$E_{1z}^L(\rho, z) = \frac{\omega\mu_0 k_2^3}{2\pi k_1^4} e^{ik_1(z+d)} e^{ik_2\rho} g(\rho; k_1, k_2);$$

$$E_{1z}^c(\rho, z) = -\frac{i\omega\mu_0 k_2^2}{2\pi k_1^4}\left(\frac{z+d}{\rho}\right)\left(\frac{ik_1^2}{\rho} - \frac{k_1}{2\rho^2} + \frac{7i}{8\rho^3}\right) e^{ik_1 r_2}, \tag{3.4.29}$$

where the radial functions $f(\rho; k_1, k_2)$ and $g(\rho; k_1, k_2)$ are defined as follows:

$$f(\rho; k_1, k_2) = \frac{ik_2}{\rho} - \frac{1}{\rho^2} - \frac{k_2^3}{k_1}\left(\frac{\pi}{k_2\rho}\right)^{1/2} e^{-ip}\mathcal{F}(p), \tag{3.4.30}$$

$$g(\rho; k_1, k_2) = \frac{ik_2}{\rho} - \frac{1}{\rho^2} - \frac{i}{k_2\rho^3} - \frac{k_2^3}{k_1}\left(\frac{\pi}{k_2\rho}\right)^{1/2} e^{-ip}\mathcal{F}(p). \tag{3.4.31}$$

Here

$$p \equiv \frac{k_2^3\rho}{2k_1^2} \tag{3.4.32}$$

is a convenient quantity introduced by Sommerfeld (1935) in the form $ik_2^3\rho/2k_1^2$ and called by him the *numerical distance.* The function $\mathcal{F}(p)$ is

$$\begin{aligned}\mathcal{F}(p) &= \int_p^\infty \frac{e^{it}}{(2\pi t)^{1/2}}\,dt \\ &= \int_0^\infty \frac{e^{it}}{(2\pi t)^{1/2}}\,dt - \int_0^p \frac{\cos t + i\sin t}{(2\pi t)^{1/2}}\,dt \\ &= \tfrac{1}{2}(1+i) - C_2(p) - iS_2(p),\end{aligned} \tag{3.4.33}$$

where $C_2 + iS_2$ is the Fresnel integral defined and tabulated in Abramowitz and Stegun (1972, pp. 300, 321, and 322).

The functions $f(\rho; k_1, k_2)$ and $g(\rho; k_1, k_2)$ defined in (3.4.30) and (3.4.31) constitute the complete radial dependence of the lateral-wave field when multiplied by $\exp(ik_2\rho)$. This is a surprisingly simple representation of the equivalent general integrals. The only part that is not completely transparent is the Fresnel-integral term which is complicated by the fact that the argument is, in general, complex. However, when the argument is sufficiently great, specifically when

$$|p| \ge 4 \quad \text{or} \quad |k_2\rho| \ge 8\left|\frac{k_1^2}{k_2^2}\right|, \tag{3.4.34}$$

the Fresnel integral $\mathcal{F}(p)$ cancels both the exponential factor and the square root preceding it in (3.4.30) and (3.4.31) and the remaining expression includes the term $-ik_2/\rho$. The simple result subject to (3.4.34) is

$$f(\rho; k_1, k_2) \sim g(\rho; k_1, k_2) = -\frac{1}{\rho^2}\left(\frac{k_1^2}{k_2^2} + 1\right) \sim -\frac{k_1^2}{k_2^2\rho^2}. \tag{3.4.35}$$

The relation (3.4.35) is readily shown with the help of formulas (7.1.3), (7.1.23), and (7.3.22) in Abramowitz and Stegun (1972, pp. 297, 298, and 301). These relate the Fresnel integral to an asymptotic expansion of the error function. Specifically, with

$$C_2(p) + iS_2(p) = \tfrac{1}{2}(1+i) - \sqrt{\tfrac{1}{2}i}\, e^{ip} w(\sqrt{ip}), \tag{3.4.36}$$

$$w(x) = e^{-x^2}\operatorname{erfc}(-ix) \sim \frac{i}{x\sqrt{\pi}}\left(1 + \frac{1}{2x^2} + \frac{3}{(2x^2)^2} + \cdots\right), \tag{3.4.37}$$

it follows that with (3.4.33) and (3.4.32),

$$\mathcal{F}(p) = ie^{ip}(2p\pi)^{-1/2}\left(1 - \frac{i}{2p}\right) = ie^{ip}\,\frac{k_1}{k_2}\,(\pi k_2\rho)^{-1/2}\left(1 - \frac{ik_1^2}{k_2^3\rho}\right). \tag{3.4.38}$$

When this is substituted in (3.4.30) and (3.4.31), the result is (3.4.35).

3.5 The Polarization of the Electromagnetic Field of a Vertical Electric Dipole on the Boundary in Region 1; Wave Tilt

When the vertical dipole is in Region 1 but on the boundary (i.e., when $d \to 0+$ so that $r_1 = r_2$), the direct and reflected fields given respectively by (3.4.3)–(3.4.5) and (3.4.6)–(3.4.8) cancel and the entire field is given by (3.4.27)–(3.4.29) with $d = 0$. It is

$$B_{1\phi}(\rho, z) = -\frac{\mu_0 k_2^2}{2\pi k_1^2}\left[f(\rho; k_1, k_2)e^{ik_2\rho}e^{ik_1 z} - \frac{iz}{\rho}\left(\frac{ik_1}{\rho} - \frac{3}{2\rho^2}\right)e^{ik_1 r_0}\right], \tag{3.5.1}$$

$$E_{1\rho}(\rho, z) = -\frac{\omega\mu_0 k_2^2}{2\pi k_1^3}\left[f(\rho; k_1, k_2)e^{ik_2\rho}e^{ik_1 z} - i\left(\frac{1}{\rho^2} + \frac{3i}{2k_1\rho^3}\right)e^{ik_1 r_0}\right], \tag{3.5.2}$$

$$E_{1z}(\rho, z) = \frac{\omega\mu_0 k_2^2}{2\pi k_1^4}\left[k_2 g(\rho; k_1, k_2)e^{ik_2\rho}e^{ik_1 z} - \frac{iz}{\rho}\left(\frac{ik_1^2}{\rho} - \frac{k_1}{2\rho^2} + \frac{7i}{8\rho^3}\right)e^{ik_1 r_0}\right], \tag{3.5.3}$$

where

$$r_0 = (\rho^2 + z^2)^{1/2} \text{ in phases;} \qquad r_0 \sim \rho \text{ in amplitudes.} \tag{3.5.4}$$

The field on the boundary in Region 1 is given by (3.5.1)–(3.5.3) with $z = 0$. The field on the boundary in Region 2 when the dipole is on the boundary in Region 1 is then most easily obtained with the boundary conditions. Thus,

$$B_{2\phi}(\rho, 0) = B_{1\phi}(\rho, 0) = -\frac{\mu_0 k_2^2}{2\pi k_1^2} f(\rho; k_1, k_2) e^{ik_2\rho}, \tag{3.5.5}$$

$$E_{2\rho}(\rho, 0) = E_{1\rho}(\rho, 0) = -\frac{\omega\mu_0 k_2^2}{2\pi k_1^3}\left[f(\rho; k_1, k_2)e^{ik_2\rho} - i\left(\frac{1}{\rho^2} + \frac{3i}{2k_1\rho^3}\right)e^{ik_1\rho}\right], \tag{3.5.6}$$

$$E_{2z}(\rho, 0) = \frac{k_1^2}{k_2^2} E_{1z}(\rho, 0) = \frac{\omega\mu_0 k_2}{2\pi k_1^2} g(\rho; k_1, k_2) e^{ik_2\rho}. \tag{3.5.7}$$

When Region 2 is air or rock and Region 1 is earth or sea water, respectively, the attenuation in the latter is much greater than in the former, so that the reflected-field correction term in (3.5.6) is usually negligible at quite short distances from the source and

$$\mathbf{E}_2(\rho, 0) = -\frac{\omega\mu_0 k_2^2}{2\pi k_1^3} f(\rho; k_1, k_2) e^{ik_2\rho}\left(\hat{\boldsymbol{\rho}} - \hat{\mathbf{z}}\,\frac{k_1 g(\rho; k_1, k_2)}{k_2 f(\rho; k_1, k_2)}\right). \tag{3.5.8}$$

Similarly in Region 1, from (3.5.2) and (3.5.3),

$$\mathbf{E}_1(\rho, z) = -\frac{\omega\mu_0 k_2^2}{2\pi k_1^3} f(\rho; k_1, k_2) e^{ik_2\rho} e^{ik_1 z}\left(\hat{\boldsymbol{\rho}} - \hat{\mathbf{z}}\,\frac{k_2 g(\rho; k_1, k_2)}{k_1 f(\rho; k_1, k_2)}\right). \tag{3.5.9}$$

Again, because of the high attenuation in Region 1, the terms multiplied by $\exp(ik_1 r_0) = \exp(-\alpha_1 r_0)\exp(i\beta_1 r_0)$ are negligible at the distances of interest. At distances $|k_2\rho| \geq 5$, $g(\rho; k_1, k_2) \sim f(\rho; k_1, k_2)$ so that the ratio $g(\rho; k_1, k_2)/f(\rho; k_1, k_2) \sim 1$. At these distances the time-dependent values of the electric field are the following:

In Region 2 at $(\rho, 0)$,

$$\mathbf{E}_2(\rho, 0; t) = \mathrm{Re}\left[K(\rho, 0)e^{-i\theta_K(\rho,0)} \times \left(\hat{\boldsymbol{\rho}} e^{-i\omega t} - \hat{\mathbf{z}}\left|\frac{k_1}{k_2}\right| e^{-i(\omega t - \theta_1 + \theta_2)}\right)\right]; \tag{3.5.10}$$

in Region 1 at (ρ, z),

$$\mathbf{E}_1(\rho, z; t) = \mathrm{Re}\left[K(\rho, z)e^{-i\theta_K(\rho,z)} \times \left(\hat{\boldsymbol{\rho}} e^{-i\omega t} - \hat{\mathbf{z}}\left|\frac{k_2}{k_1}\right| e^{-i(\omega t + \theta_1 - \theta_2)}\right)\right], \tag{3.5.11}$$

where

$$K(\rho,z)e^{-i\theta_K(\rho,z)} = -\frac{\omega\mu_0 k_2^2}{2\pi k_1^3} f(\rho;k_1,k_2)e^{ik_2\rho}e^{ik_1 z},$$
$$\theta_1 = \tan^{-1}\left(\frac{\alpha_1}{\beta_1}\right); \qquad \theta_2 = \tan^{-1}\left(\frac{\alpha_2}{\beta_2}\right). \tag{3.5.12}$$

These equations reduce, respectively, to

$$\left|\frac{k_2}{k_1}\right|^2 E_{2z}^2(\rho,0;t) - 2\left|\frac{k_2}{k_1}\right| E_{2z}(\rho,0;t)E_{2\rho}(\rho,0;t)\cos(\theta_1-\theta_2)$$
$$+ E_{2\rho}^2(\rho,0;t) - K^2(\rho,0)\sin^2(\theta_1-\theta_2) = 0 \tag{3.5.13}$$

for Region 2, and

$$\left|\frac{k_1}{k_2}\right|^2 E_{1z}^2(\rho,z;t) - 2\left|\frac{k_1}{k_2}\right| E_{1z}(\rho,z;t)E_{1\rho}(\rho,z;t)\cos(\theta_1-\theta_2)$$
$$+ E_{1\rho}^2(\rho,z;t) - K^2(\rho,z)\sin^2(\theta_1-\theta_2) = 0 \tag{3.5.14}$$

for Region 1.

These are the equations of ellipses with their major axes rotated with respect to the z-axis by the following wave-tilt angles:

In Region 2,

$$\psi_2 = \frac{1}{2}\tan^{-1}\left(\frac{2|k_2/k_1|\cos(\theta_1-\theta_2)}{1-|k_2^2/k_1^2|}\right); \tag{3.5.15}$$

in Region 1,

$$\psi_1 = \frac{1}{2}\tan^{-1}\left(\frac{2|k_1/k_2|\cos(\theta_1-\theta_2)}{1-|k_1^2/k_2^2|}\right) = \frac{\pi}{2} - \psi_2, \tag{3.5.16}$$

where

$$\cos(\theta_1-\theta_2) = \cos\theta_1\cos\theta_2 + \sin\theta_1\sin\theta_2 = \frac{\beta_1\beta_2+\alpha_1\alpha_2}{|k_1|\,|k_2|}. \tag{3.5.17}$$

Since $|k_2|^2 \ll |k_1|^2$, these are well approximated by

$$\psi_2 \sim \frac{1}{2}\tan^{-1}\left(2\left|\frac{k_2}{k_1}\right|\cos(\theta_1-\theta_2)\right)$$
$$= \frac{1}{2}\tan^{-1}\left(\frac{2(\beta_1\beta_2+\alpha_1\alpha_2)}{|k_1|^2}\right), \tag{3.5.18}$$
$$\psi_1 \sim \frac{\pi}{2} - \psi_2 = \frac{\pi}{2} - \frac{1}{2}\tan^{-1}\left(\frac{2(\beta_1\beta_2+\alpha_1\alpha_2)}{|k_1|^2}\right). \tag{3.5.19}$$

The condition $|k_1| \geq 3|k_2|$ ensures that the argument of the arctangent is small so that

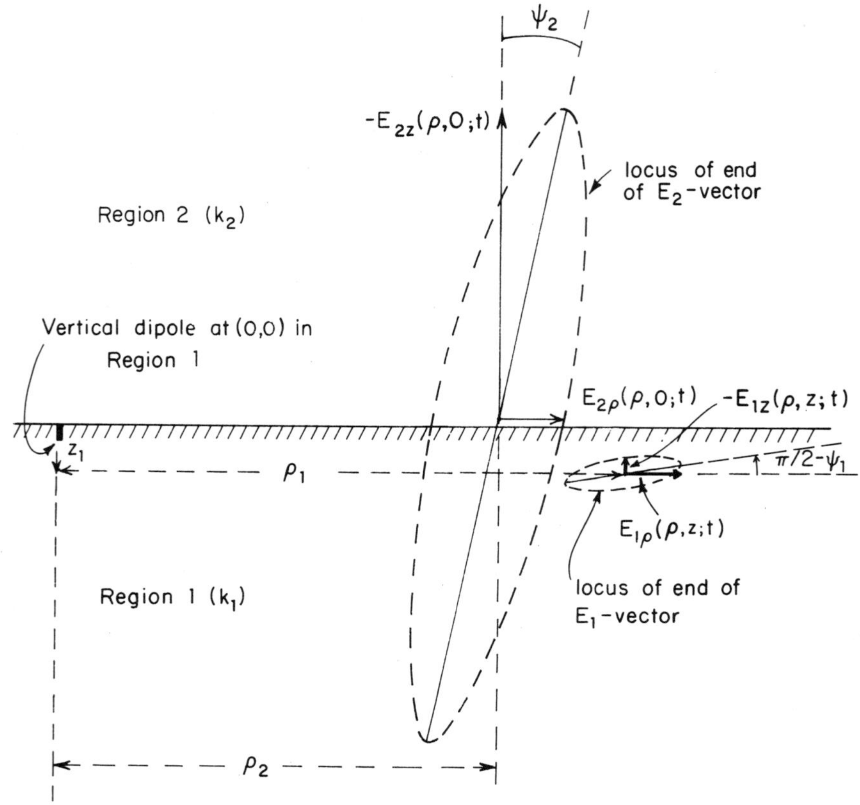

Figure 3.5.1. Elliptical loci of endpoints of E-vectors at $(\rho_2, 0)$ in Region 2 and at (ρ_1, z_1) in Region 1.

$$\psi_2 \sim \frac{\beta_1\beta_2 + \alpha_1\alpha_2}{|k_1|^2}; \qquad \psi_1 \sim \frac{\pi}{2} - \frac{\beta_1\beta_2 + \alpha_1\alpha_2}{|k_1|^2}. \tag{3.5.20}$$

These equations show that the electric field $\mathbf{E}_2(\rho, 0; t)$ is elliptically polarized with the component $E_{2z}(\rho, 0; t)$ much greater than the component $E_{2\rho}(\rho, 0; t)$, as shown in Fig. 3.5.1. Similarly, the electric field $\mathbf{E}_1(\rho, z; t)$ in Region 1 is also elliptically polarized but with $E_{1\rho}(\rho, z; t)$ much greater than $E_{1z}(\rho, z; t)$. This is also shown in Fig. 3.5.1. Note that

$$E_{2z}(\rho, 0-) : E_\rho(\rho, 0) : E_{2z}(\rho, 0+) \sim \left|\frac{k_1}{k_2}\right| : 1 : \left|\frac{k_2}{k_1}\right|. \tag{3.5.21}$$

3.6 The Complete Electromagnetic Field in Region 1 of the Vertical Electric Dipole in Region 1

When the vertical electric dipole is located at a depth d along the z-axis in Region 1, the electromagnetic field in Region 1 at all locations (ρ, z) not excluded by the conditions

$$|k_1| \geq 3|k_2|; \qquad \rho \geq 5z, \qquad \rho \geq 5d; \qquad |k_1\rho| \geq 3, \tag{3.6.1}$$

is given by

$$\begin{aligned} E_{1\rho}(\rho, z) = -\frac{\omega\mu_0}{2\pi k_1^2}\Bigg\{\frac{k_2^2}{k_1}\bigg[& f(\rho; k_1, k_2)e^{ik_2\rho}e^{ik_1(z+d)} \\ & - ie^{ik_1r_2}\left(\frac{1}{\rho^2} + \frac{3i}{2k_1\rho^3}\right)\bigg] \\ & + \frac{e^{ik_1r_1}}{2}\left(\frac{ik_1^2}{r_1} - \frac{3k_1}{r_1^2} - \frac{3i}{r_1^3}\right)\left(\frac{\rho}{r_1}\right)\left(\frac{z-d}{r_1}\right) \\ & - \frac{e^{ik_1r_2}}{2}\left(\frac{ik_1^2}{r_2} - \frac{3k_1}{r_2^2} - \frac{3i}{r_2^3}\right)\left(\frac{\rho}{r_2}\right)\left(\frac{z+d}{r_2}\right)\Bigg\}, \end{aligned} \tag{3.6.2}$$

$$\begin{aligned} E_{1z}(\rho, z) = \frac{\omega\mu_0}{2\pi k_1^2}\Bigg\{\frac{k_2^2}{k_1^2}\bigg[& k_2 g(\rho; k_1, k_2)e^{ik_2\rho}e^{ik_1(z+d)} \\ & - ie^{ik_1r_2}\left(\frac{z+d}{\rho}\right)\left(\frac{ik_1^2}{\rho} - \frac{k_1}{2\rho^2} + \frac{7i}{8\rho^3}\right)\bigg] \\ & + \frac{e^{ik_1r_1}}{2}\bigg[\frac{ik_1^2}{r_1} - \frac{k_1}{r_1^2} - \frac{i}{r_1^3} \\ & \qquad - \left(\frac{z-d}{r_1}\right)^2\left(\frac{ik_1^2}{r_1} - \frac{3k_1}{r_1^2} - \frac{3i}{r_1^3}\right)\bigg] \\ & - \frac{e^{ik_1r_2}}{2}\bigg[\frac{ik_1^2}{r_2} - \frac{k_1}{r_2^2} - \frac{i}{r_2^3} \\ & \qquad - \left(\frac{z+d}{r_2}\right)^2\left(\frac{ik_1^2}{r_2} - \frac{3k_1}{r_2^2} - \frac{3i}{r_2^3}\right)\bigg]\Bigg\}, \end{aligned} \tag{3.6.3}$$

$$\begin{aligned} B_{1\phi}(\rho, z) = -\frac{\mu_0}{2\pi}\Bigg\{\frac{k_2^2}{k_1^2}\bigg[& f(\rho; k_1, k_2)e^{ik_2\rho}e^{ik_1(z+d)} \\ & - ie^{ik_1r_2}\left(\frac{z+d}{\rho}\right)\left(\frac{ik_1}{\rho} - \frac{3}{2\rho^2}\right)\bigg] \\ & + \frac{e^{ik_1r_1}}{2}\left(\frac{ik_1}{r_1} - \frac{1}{r_1^2}\right)\left(\frac{\rho}{r_1}\right) \end{aligned}$$

$$- \frac{e^{ik_1 r_2}}{2}\left(\frac{ik_1}{r_2} - \frac{1}{r_2^2}\right)\left(\frac{\rho}{r_2}\right)\Bigg\}. \tag{3.6.4}$$

In these formulas $r_1 = [\rho^2 + (z-d)^2]^{1/2}$, $r_2 = [\rho^2 + (z+d)^2]^{1/2}$, and $f(\rho; k_1, k_2)$ and $g(\rho; k_1, k_2)$ are defined in (3.4.30), (3.4.31), and (3.4.35). The direct- and image-field terms are obtained from (3.4.3)–(3.4.8). If the conditions $\rho^2 \gg z^2$ and $\rho^2 \gg d^2$ are also applied to the direct and image fields, so that $r_1 \sim r_2 \sim \rho$ in amplitudes, $\rho/r_1 \sim \rho/r_2 \sim 1$, and the terms in $(z-d)^2/r_1^2$ and $(z+d)^2/r_2^2$ are negligible, the simpler results are

$$\begin{aligned} E_{1\rho}(\rho, z) = -\frac{\omega\mu_0}{2\pi k_1^2}\Bigg\{ \frac{k_2^2}{k_1}\Bigg[& f(\rho; k_1, k_2) e^{ik_2\rho} e^{ik_1(z+d)} \\ & - i e^{ik_1 r_2}\left(\frac{1}{\rho^2} + \frac{3i}{2k_1\rho^3}\right)\Bigg] \\ & + \frac{1}{2}\left(\frac{z-d}{\rho} e^{ik_1 r_1} - \frac{z+d}{\rho} e^{ik_1 r_2}\right) \\ & \times \left(\frac{ik_1^2}{\rho} - \frac{3k_1}{\rho^2} - \frac{3i}{\rho^3}\right)\Bigg\}, \end{aligned} \tag{3.6.5}$$

$$\begin{aligned} E_{1z}(\rho, z) = \frac{\omega\mu_0}{2\pi k_1^2}\Bigg\{ \frac{k_2^2}{k_1^2}\Bigg[& k_2 g(\rho; k_1, k_2) e^{ik_2\rho} e^{ik_1(z+d)} \\ & - i e^{ik_1 r_2}\left(\frac{z+d}{\rho}\right)\left(\frac{ik_1^2}{\rho} - \frac{k_1}{2\rho^2} + \frac{7i}{8\rho^3}\right)\Bigg] \\ & + \frac{1}{2}\left(e^{ik_1 r_1} - e^{ik_1 r_2}\right)\left(\frac{ik_1^2}{\rho} - \frac{k_1}{\rho^2} - \frac{i}{\rho^3}\right)\Bigg\}, \end{aligned} \tag{3.6.6}$$

$$\begin{aligned} B_{1\phi}(\rho, z) = -\frac{\mu_0}{2\pi}\Bigg\{ \frac{k_2^2}{k_1^2}\Bigg[& f(\rho; k_1, k_2) e^{ik_2\rho} e^{ik_1(z+d)} \\ & - i e^{ik_1 r_2}\left(\frac{z+d}{\rho}\right)\left(\frac{ik_1}{\rho} - \frac{3}{2\rho^2}\right)\Bigg] \\ & + \frac{1}{2}\left(e^{ik_1 r_1} - e^{ik_1 r_2}\right)\left(\frac{ik_1}{\rho} - \frac{1}{\rho^2}\right)\Bigg\}. \end{aligned} \tag{3.6.7}$$

When the vertical dipole is located just across the boundary in Region 2 with $d \sim 0$ and $r_1 = r_2 = r_0 = (\rho^2 + z^2)^{1/2}$, the components of the field in Region 1 are

$$\begin{aligned} E_{1\rho}(\rho, z) = -\frac{\omega\mu_0}{2\pi k_1}\Bigg[& f(\rho; k_1, k_2) e^{ik_2\rho} e^{ik_1 z} \\ & - i e^{ik_1 r_0}\left(\frac{1}{\rho^2} + \frac{3i}{2k_1\rho^3}\right)\Bigg], \end{aligned} \tag{3.6.8}$$

$$\begin{aligned} E_{1z}(\rho, z) = \frac{\omega\mu_0}{2\pi k_1^2}\Bigg[& k_2 g(\rho; k_1, k_2) e^{ik_2\rho} e^{ik_1 z} \\ & - \frac{iz}{\rho} e^{ik_1 r_0}\left(\frac{ik_1^2}{\rho} - \frac{k_1}{2\rho^2} + \frac{7i}{8\rho^3}\right)\Bigg], \end{aligned} \tag{3.6.9}$$

$$B_{1\phi}(\rho,z) = -\frac{\mu_0}{2\pi}\left[f(\rho;k_1,k_2)e^{ik_2\rho}e^{ik_1z} - \frac{iz}{\rho}\,e^{ik_1r_0}\left(\frac{ik_1}{\rho} - \frac{3}{2\rho^2}\right)\right]. \tag{3.6.10}$$

Note that moving the dipole from $d = 0$ in Region 1 to $d = 0$ in Region 2 leaves the surface wave unchanged in form but multiplied by the large factor k_1^2/k_2^2. In most applications for which k_1 is complex, the terms multiplied by $\exp(ik_1r_0)$ in (3.6.8)–(3.6.10) are negligible at moderate distances from the source.

The field in Region 2 on the boundary when the dipole is at $(0,0)$ in Region 2 is

$$E_{2\rho}(\rho,0) = -\frac{\omega\mu_0}{2\pi k_1}\left[f(\rho;k_1,k_2)e^{ik_2\rho} - ie^{ik_1\rho}\left(\frac{1}{\rho^2} + \frac{3i}{2k_1\rho^3}\right)\right], \tag{3.6.11}$$

$$E_{2z}(\rho,0) = \frac{\omega\mu_0}{2\pi k_2}\,g(\rho;k_1,k_2)e^{ik_2\rho}, \tag{3.6.12}$$

$$B_{2\phi}(\rho,0) = -\frac{\mu_0}{2\pi}\,f(\rho;k_1,k_2)e^{ik_2\rho}. \tag{3.6.13}$$

Formulas for the complete electromagnetic field anywhere in Region 2 when the vertical dipole is located at $z = -d$ in Region 2 are derived in Chapter 4.

In order to present a complete and unified picture of the properties of the several components of the electromagnetic field of both vertical and horizontal dipoles and compare them with measurements made in a model tank, the frequency $f = 600$ MHz was chosen with Region 1 salt water ($\sigma_1 = 3.5$ S/m, $\epsilon_{1r} = 80$, $k_1 = \beta_1 + i\alpha_1 = 129.4 + i64.1 = 144.4e^{i0.46}$ m^{-1}) and Region 2 air ($k_2 = 4\pi$ m^{-1}). Graphs of the magnitudes of the components of the electric field on the boundary due to a vertical electric dipole also on the boundary are shown in Fig. 3.6.1. Curve ① is the vertical electric field in air on the surface at $(\rho,0)$ due to a vertical dipole in air on the surface at $(0,0)$. It is denoted by $|E_{2z}(\rho,0)|_{2v}$ where the first two subscripts indicate the z-component in Region 2, the second two subscripts a vertical dipole in Region 2 also on the surface. The circled points along the curve were obtained from Norton's *ground wave term* (1.3.9) using his tabulation of the function F_e defined in (1.3.10). The associated radial component is represented by Curve ②. It is computed from (3.6.11). It is smaller than the z-component (represented by Curve ①) by the factor k_2/k_1. Curve ③ represents $|E_{1z}(\rho,0)|_{2v}$, the z-component of the electric field in Region 1 (sea water) due to a vertical dipole on the boundary in Region 2 (air) and $|E_{2z}(\rho,0)|_{1v}$, the z-component of the field in Region 2 (air) due to a vertical dipole on the boundary in Region 1 (sea water). It is smaller than $|E_{2z}(\rho,0)|_{2v}$ shown by Curve ① by the very small factor k_2^2/k_1^2. The magnetic field $|B_{2\phi}(\rho,0)|_{2v} = |B_{1\phi}(\rho,0)|_{2v}$ is not shown since it is given by $B_{1\phi}(\rho,0) = B_{2\phi}(\rho,0) = (k_1/\omega)E_{2\rho}(\rho,0)$. Curve ② for $20\log_{10}|E_{2\rho}(\rho,0)|$

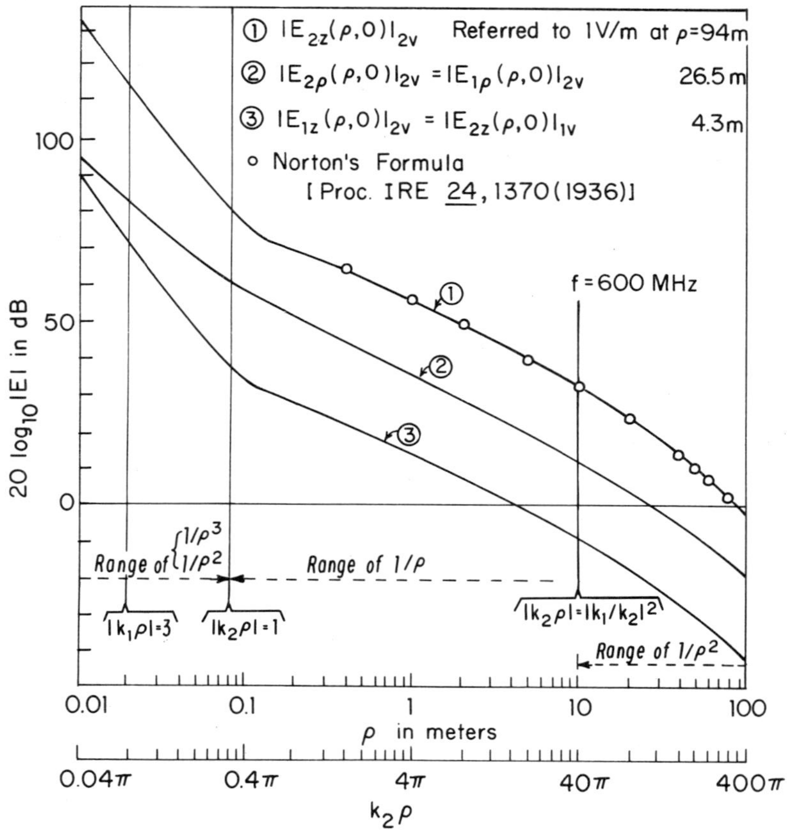

Figure 3.6.1. Components of electric field $E(\rho, 0)$ in V/m of unit vertical electric dipole ($I\Delta l = 1$ A m) at boundary between Region 1 (sea water, $\epsilon_{1r} = 80$, $\sigma_1 = 3.5$ S/m) and Region 2 (air).

need only be reduced by $20 \log_{10} |k_1/\omega| = 11$ dB to obtain the graph for $20 \log_{10} |B_{2\phi}(\rho, 0)|$.

Curve ② in Fig. 3.6.1 for $|E_{2\rho}(\rho, 0)|_{2v}$ is reproduced in Fig. 3.6.2 in order to intercompare the contributions by the several terms in

$$E_{2\rho}(\rho, 0) = -\frac{\omega\mu_0}{2\pi k_1}\left[\frac{ik_2}{\rho} - \frac{1}{\rho^2} - \frac{k_2^3}{k_1}\left(\frac{\pi}{k_2\rho}\right)^{1/2} e^{-ip}\mathcal{F}(p)\right] e^{ik_2\rho}, \quad (3.6.14)$$

$$\sim \frac{\omega\mu_0 k_1}{2\pi k_2^2}\frac{e^{ik_2\rho}}{\rho^2}; \qquad |p| = \left|\frac{k_2^3\rho}{2k_1^2}\right| \geq 4. \quad (3.6.15)$$

Specifically, the simple terms $|(\omega\mu_0/2\pi k_1)[(ik_2/\rho) - (1/\rho^2)]|$ and the Fresnel-integral term $|(\omega\mu_0 k_2^3/2\pi k_1^2)(\pi/k_2\rho)^{1/2} e^{-ip}\mathcal{F}(p)|$—which conforms precisely to the Zenneck wave $|(\omega\mu_0 k_2^3/2\pi k_1^2)(\pi/2k_2\rho)^{1/2}|$ in the range

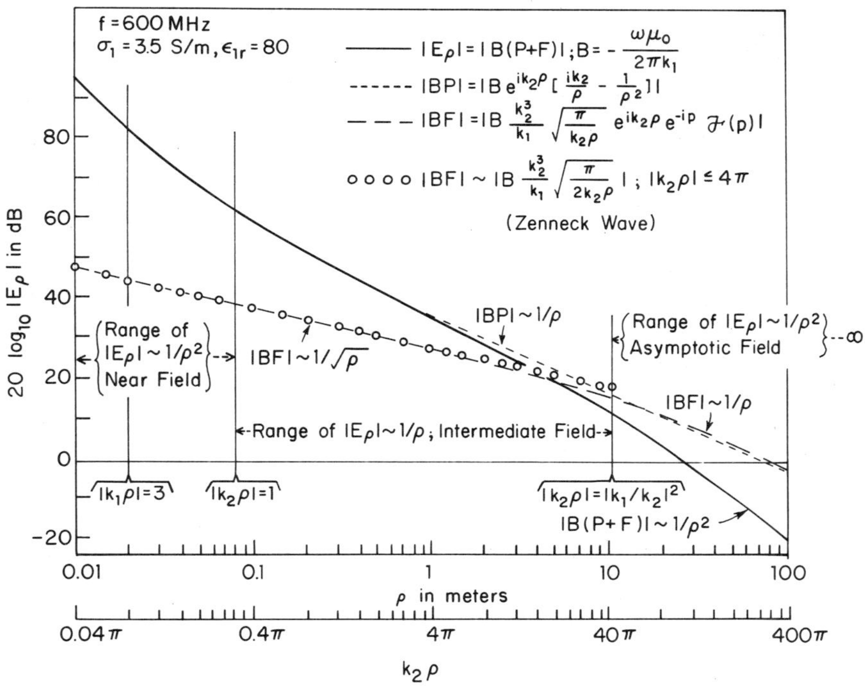

Figure 3.6.2. Radial electric field $E_{2\rho}(\rho, 0) = E_{1\rho}(\rho, 0)$ in V/m on boundary between sea water and air due to unit vertical electric dipole ($I\Delta l = 1$ A m) at surface in air. (Referred to 1 V/m at $\rho = 25$ m.)

$|k_2\rho| \leq 4\pi$—are compared with the complete formula (3.6.14) and with its asymptotic form (3.6.15).

The two vertical lines $|k_2\rho| = 1$ and $|k_2\rho| = |k_1/k_2|^2$ are shown in both Fig. 3.6.1 and Fig. 3.6.2. These are the approximate demarcation lines between the near field (which varies approximately as $1/\rho^2$ for $|E_\rho|$, $1/\rho^3$ for $|E_z|$), the intermediate field (which varies approximately as $1/\rho$), and the asymptotic field (which varies approximately as $1/\rho^2$).

3.7 The Field in Region 2 for the Vertical Electric Dipole in Region 1

The complete field at all points in Region 1 due to a vertical electric dipole in Region 1 at $(0, d)$ is given in the preceding section. From this it is possible to obtain the field in Region 2 along the boundary $z = 0$ with the help of the boundary conditions. Thus,

$$\begin{aligned} E_{2\rho}(\rho,0) &= E_{1\rho}(\rho,0) \\ &= -\frac{\omega\mu_0}{2\pi k_1^2}\left\{\frac{k_2^2}{k_1}\left[f(\rho;k_1,k_2)e^{ik_2\rho}e^{ik_1 d} - ie^{ik_1 r_d}\left(\frac{1}{\rho^2}+\frac{3i}{2k_1\rho^3}\right)\right]\right. \\ &\qquad \left. -\frac{\rho d}{r_d^2}\left(\frac{ik_1^2}{r_d}-\frac{3k_1}{r_d^2}-\frac{3i}{r_d^3}\right)e^{ik_1 r_d}\right\}, \end{aligned} \tag{3.7.1}$$

$$\begin{aligned} E_{2z}(\rho,0) &= \frac{k_1^2}{k_2^2}E_{1z}(\rho,0) \\ &= \frac{\omega\mu_0}{2\pi k_1^2}\left[k_2 g(\rho;k_1,k_2)e^{ik_2\rho}e^{ik_1 d}\right. \\ &\qquad \left. -\frac{id}{\rho}e^{ik_1 r_d}\left(\frac{ik_1^2}{\rho}-\frac{k_1}{2\rho^2}+\frac{7i}{8\rho^3}\right)\right], \end{aligned} \tag{3.7.2}$$

$$\begin{aligned} B_{2\phi}(\rho,0) &= B_{1\phi}(\rho,0) \\ &= -\frac{\mu_0 k_2^2}{2\pi k_1^2}\left[f(\rho;k_1,k_2)e^{ik_2\rho}e^{ik_1 d} - \frac{id}{\rho}e^{ik_1 r_d}\left(\frac{ik_1}{\rho}-\frac{3}{2\rho^2}\right)\right], \end{aligned} \tag{3.7.3}$$

where $r_d = (\rho^2 + d^2)^{1/2}$. The functions $f(\rho; k_1, k_2)$ and $g(\rho; k_1, k_2)$ are defined in (3.4.30), (3.4.31), and (3.4.35).

The complete field at all points in Region 2 not on the boundary is also of interest. In particular, it is important to determine the depth of penetration of the lateral-wave field that travels along the boundary and the nature of the field that travels into Region 2 away from the boundary. The field at all points in Region 2 is given by the general integrals (3.3.12)–(3.3.14). They are rewritten below for convenient discussion. Note that the integral for $E_{2\rho}(\rho, z)$ is expanded with the identity $k_1^2\gamma_2/N = 1 - k_2^2\gamma_1/N$ in order to obtain an integrable form.

$$B_{2\phi}(\rho,z) = \frac{i\mu_0 k_2^2}{2\pi}\int_0^\infty \frac{e^{i\gamma_1 d}e^{-i\gamma_2 z}}{N}J_1(\lambda\rho)\lambda^2\,d\lambda; \qquad z \le 0, \tag{3.7.4}$$

$$E_{2\rho}(\rho,z) = -\frac{i\omega\mu_0}{2\pi}\int_0^\infty \frac{\gamma_2 e^{i\gamma_1 d}e^{-i\gamma_2 z}}{N}J_1(\lambda\rho)\lambda^2\,d\lambda; \qquad z \le 0, \tag{3.7.5a}$$

$$\begin{aligned} &= -\frac{i\omega\mu_0}{2\pi k_1^2}\left(\int_0^\infty e^{i\gamma_1 d}e^{-i\gamma_2 z}J_1(\lambda\rho)\lambda^2\,d\lambda\right. \\ &\qquad \left. -k_2^2\int_0^\infty \frac{\gamma_1}{N}e^{i\gamma_1 d}e^{-i\gamma_2 z}J_1(\lambda\rho)\lambda^2\,d\lambda\right); \\ &\qquad z \le 0, \end{aligned} \tag{3.7.5b}$$

$$E_{2z}(\rho,z) = -\frac{\omega\mu_0}{2\pi}\int_0^\infty \frac{e^{i\gamma_1 d}e^{-i\gamma_2 z}}{N}J_0(\lambda\rho)\lambda^3\,d\lambda; \qquad z \le 0, \tag{3.7.6}$$

where $N = k_1^2\gamma_2 + k_2^2\gamma_1$ and $\gamma_j = (k_j^2 - \lambda^2)^{1/2}$, $j = 1, 2$. The evaluation of these integrals can be carried out directly subject to the conditions specified in (3.4.12) and (3.4.13). With the condition $|k_1^2| \gg |k_2^2|$, it is evident that the first term in $N = k_1^2\gamma_2 + k_2^2\gamma_1$ is the larger one. This permits the separation of three of the four integrals in (3.7.4)–(3.7.6) into two parts by writing

$$\frac{1}{N} = \frac{1}{N_0} + \left(\frac{1}{N} - \frac{1}{N_0}\right), \tag{3.7.7}$$

where

$$N_0 \equiv k_1^2\gamma_2 \equiv k_1^2(k_2^2 - \lambda^2)^{1/2}. \tag{3.7.8}$$

The integrals are

$$\begin{aligned} B_{2\phi}(\rho, z) = \frac{i\mu_0}{2\pi}\bigg[\frac{k_2^2}{k_1^2}\int_0^\infty \frac{e^{i\gamma_1 d}e^{-i\gamma_2 z}}{\gamma_2} J_1(\lambda\rho)\lambda^2\,d\lambda \\ + k_2^2\int_0^\infty \left(\frac{1}{N} - \frac{1}{N_0}\right)e^{i\gamma_1 d}e^{-i\gamma_2 z}J_1(\lambda\rho)\lambda^2\,d\lambda\bigg], \end{aligned} \tag{3.7.9}$$

$$\begin{aligned} E_{2\rho}(\rho, z) = -\frac{i\omega\mu_0}{2\pi k_1^2}\bigg[\int_0^\infty e^{i\gamma_1 d}e^{-i\gamma_2 z}J_1(\lambda\rho)\lambda^2\,d\lambda \\ - \frac{k_2^2}{k_1^2}\int_0^\infty \frac{\gamma_1}{\gamma_2}\,e^{i\gamma_1 d}e^{-i\gamma_2 z}J_1(\lambda\rho)\lambda^2\,d\lambda \\ - k_2^2\int_0^\infty \gamma_1\left(\frac{1}{N} - \frac{1}{N_0}\right)e^{i\gamma_1 d}e^{-i\gamma_2 z}J_1(\lambda\rho)\lambda^2\,d\lambda\bigg], \end{aligned} \tag{3.7.10}$$

$$\begin{aligned} E_{2z}(\rho, z) = -\frac{\omega\mu_0}{2\pi k_1^2}\bigg[\int_0^\infty \frac{e^{i\gamma_1 d}e^{-i\gamma_2 z}}{\gamma_2} J_0(\lambda\rho)\lambda^3\,d\lambda \\ + k_1^2\int_0^\infty \left(\frac{1}{N} - \frac{1}{N_0}\right)e^{i\gamma_1 d}e^{-i\gamma_2 z}J_0(\lambda\rho)\lambda^3\,d\lambda\bigg]. \end{aligned} \tag{3.7.11}$$

The first integrals in (3.7.9) and (3.7.11) and the first two integrals in (3.7.10) can be evaluated with the procedure described in Appendix D in conjunction with (D18) and (D19). Specifically, for the present application, let

$$f(k_1, k_2; \lambda) = \int_0^\infty e^{i\gamma_1 d}e^{-i\gamma_2 z}J_1(\lambda\rho)\lambda^2\,d\lambda; \qquad z \le 0. \tag{3.7.12}$$

Appendix D gives

$$f(k_1, k_2; \lambda) \sim f_1(k_1, k_2; \lambda) + f_2(k_1, k_2; \lambda) - f_3(k_1, k_2; \lambda), \tag{3.7.13}$$

where

$$f_1(k_1, k_2; \lambda) = f(k_1, k_2; \lambda \sim k_1),$$
$$f_2(k_1, k_2; \lambda) = f(k_1, k_2; \lambda \sim k_2), \tag{3.7.14}$$

and

$$f_3(k_1, k_2; \lambda) = f_1(k_1, k_2; \lambda \sim k_2) = f_2(k_1, k_2; \lambda \sim k_1). \tag{3.7.15}$$

The expanded formulas for the first integrals denoted by a superscript (0) are

$$\begin{aligned}
B_{2\phi}^{(0)}(\rho, z) &= \frac{i\mu_0 k_2^2}{2\pi k_1^2} \int_0^\infty \gamma_2^{-1} e^{-i\gamma_2 z} e^{i\gamma_1 d} J_1(\lambda\rho)\lambda^2\, d\lambda \\
&\sim \frac{i\mu_0 k_2^2}{2\pi k_1^2} \left(-ie^{k_1 z} \int_0^\infty J_1(\lambda\rho) e^{i\gamma_1 d} \lambda\, d\lambda \right. \\
&\qquad + e^{ik_1 d} \int_0^\infty \gamma_2^{-1} e^{-i\gamma_2 z} J_1(\lambda\rho)\lambda^2\, d\lambda \\
&\qquad \left. + ie^{k_1 z} e^{ik_1 d} \int_0^\infty J_1(\lambda\rho)\lambda\, d\lambda \right) \\
&= \frac{i\mu_0 k_2^2}{2\pi k_1^2} \left(-ie^{k_1 z} \frac{\rho}{2} I_{20}(k_1; \rho, d) + e^{ik_1 d} \frac{\rho}{2} I_7(k_2; \rho, -z) \right. \\
&\qquad \left. + ie^{k_1 z} e^{ik_1 d} \frac{\rho}{2} I_{20}(\rho, 0) \right) \\
&= -\frac{\mu_0 k_2^2}{2\pi k_1^2} \left[e^{ik_1 d} e^{ik_2 r_0} \left(\frac{ik_2}{r_0} - \frac{1}{r_0^2} \right) \left(\frac{\rho}{r_0} \right) \right. \\
&\qquad \left. - ie^{k_1 z} e^{ik_1 r_d} \left(\frac{d}{\rho} \right) \left(\frac{ik_1}{\rho} - \frac{3}{2\rho^2} \right) \right]; \quad z \le 0,
\end{aligned} \tag{3.7.16}$$

$$\begin{aligned}
E_{2\rho}^{(0)}(\rho, z) &= -\frac{i\omega\mu_0}{2\pi k_1^2} \int_0^\infty e^{-i\gamma_2 z} e^{i\gamma_1 d} J_1(\lambda\rho)\lambda^2\, d\lambda \\
&\sim -\frac{i\omega\mu_0}{2\pi k_1^2} \left(e^{k_1 z} \int_0^\infty e^{i\gamma_1 d} J_1(\lambda\rho)\lambda^2\, d\lambda \right. \\
&\qquad + e^{ik_1 d} \int_0^\infty e^{-i\gamma_2 z} J_1(\lambda\rho)\lambda^2\, d\lambda \\
&\qquad \left. - e^{k_1 z} e^{ik_1 d} \int_0^\infty J_1(\lambda\rho)\lambda^2\, d\lambda \right) \\
&= -\frac{i\omega\mu_0}{2\pi k_1^2} \left(e^{k_1 z} \frac{\rho}{2} I_{22}(k_1; \rho, d) + e^{ik_1 d} \frac{\rho}{2} I_{22}(k_2; \rho, -z) \right. \\
&\qquad \left. - e^{k_1 z} e^{ik_1 d} \frac{\rho}{2} I_{22}(\rho, 0) \right) \\
&= -\frac{\omega\mu_0}{2\pi k_1^2} \left[e^{ik_1 d} e^{ik_2 r_0} \left(\frac{ik_2^2}{r_0} - \frac{3k_2}{r_0^2} - \frac{3i}{r_0^3} \right) \left(\frac{\rho z}{r_0^2} \right) \right.
\end{aligned}$$

$$- e^{k_1 z} e^{ik_1 r_d}\left(\frac{ik_1^2}{r_d} - \frac{3k_1}{r_d^2} - \frac{3i}{r_d^3}\right)\left(\frac{\rho d}{r_d^2}\right)\Big]; \qquad z \le 0, \qquad (3.7.17)$$

$$\begin{aligned}
E_{2z}^{(0)}(\rho, z) &= -\frac{\omega\mu_0}{2\pi k_1^2}\int_0^\infty \gamma_2^{-1} e^{-i\gamma_2 z} e^{i\gamma_1 d} J_0(\lambda\rho)\lambda^3\, d\lambda \\
&\sim -\frac{\omega\mu_0}{2\pi k_1^2}\Bigg(-ie^{k_1 z}\int_0^\infty e^{i\gamma_1 d} J_0(\lambda\rho)\lambda^2\, d\lambda \\
&\qquad + e^{ik_1 d}\int_0^\infty \gamma_2^{-1} e^{-i\gamma_2 z} J_0(\lambda\rho)\lambda^3\, d\lambda \\
&\qquad + ie^{k_1 z} e^{ik_1 d}\int_0^\infty J_0(\lambda\rho)\lambda^2\, d\lambda\Bigg) \\
&= -\frac{\omega\mu_0}{2\pi k_1^2}\Bigg(-ie^{k_1 z}\,\frac{1}{2}\left[I_{20}(k_1;\rho,d) + I_{21}(k_1;\rho,d)\right] \\
&\qquad + e^{ik_1 d}\,\frac{1}{2}\left[I_7(k_2;\rho,-z) + I_8(k_2;\rho,-z)\right] \\
&\qquad + ie^{k_1 z} e^{ik_1 d} I_{28}(\rho, 0)\Bigg) \\
&= \frac{\omega\mu_0}{2\pi k_1^2}\Bigg\{e^{ik_1 d} e^{ik_2 r_0}\Bigg[\frac{ik_2^2}{r_0} - \frac{k_2}{r_0^2} - \frac{i}{r_0^3} \\
&\qquad - \left(\frac{ik_2^2}{r_0} - \frac{3k_2}{r_0^2} - \frac{3i}{r_0^3}\right)\left(\frac{z^2}{r_0^2}\right)\Bigg] \\
&\qquad - ie^{k_1 z} e^{ik_1 r_d}\left(\frac{d}{\rho}\right)\left(\frac{ik_1^2}{\rho} - \frac{k_1}{2\rho^2} + \frac{7i}{8\rho^3}\right)\Bigg\}; \qquad z \le 0, \qquad (3.7.18)
\end{aligned}$$

where $r_0 = (\rho^2 + z^2)^{1/2}$. The integrated values are readily obtained from Appendix A.

The second integral in (3.7.10) can be evaluated with the same procedure. In this case,

$$\begin{aligned}
&\int_0^\infty \frac{\gamma_1}{\gamma_2}\, e^{i\gamma_1 d} e^{-i\gamma_2 z} J_1(\lambda\rho)\lambda^2\, d\lambda \\
&\quad \sim -ie^{k_1 z}\int_0^\infty \gamma_1 e^{i\gamma_1 d} J_1(\lambda\rho)\lambda\, d\lambda \\
&\qquad + k_1 e^{ik_1 d}\int_0^\infty \gamma_2^{-1} e^{-i\gamma_2 z} J_1(\lambda\rho)\lambda^2\, d\lambda \\
&\qquad + ik_1 e^{k_1 z} e^{ik_1 d}\int_0^\infty J_1(\lambda\rho)\lambda\, d\lambda. \qquad (3.7.19)
\end{aligned}$$

With reference to Appendix A, the first integral on the right is (A13a) with $z = d$, the second integral is (A7a), and the third integral is (A20b). The combined results are

$$\begin{aligned}
&\int_0^\infty \frac{\gamma_1}{\gamma_2}\, e^{i\gamma_1 d} e^{-i\gamma_2 z} J_1(\lambda\rho)\lambda^2\, d\lambda \\
&\quad = -ie^{k_1 z}\,\frac{\rho}{2}\, I_{13}(k_1;\rho,d) + k_1 e^{ik_1 d}\,\frac{\rho}{2}\, I_7(k_2;\rho,-z) \\
&\qquad + ik_1 e^{k_1 z} e^{ik_1 d}\,\frac{\rho}{2}\, I_{20}(\rho,0) \\
&\quad \sim -ik_1 e^{k_1 z}\left[\frac{e^{ik_1 d}}{\rho^2} + ie^{ik_1 r_d}\left(\frac{1}{\rho^2} + \frac{3i}{2k_1\rho^3}\right)\right] \\
&\qquad - k_1 k_2 e^{ik_1 d}\left(\frac{\rho}{r_0}\right)\left(\frac{1}{r_0} + \frac{i}{k_2 r_0^2}\right) e^{ik_2 r_0} + ik_1\,\frac{e^{k_1 z} e^{ik_1 d}}{\rho^2}; \\
&\qquad\qquad z \le 0, \quad (3.7.20)
\end{aligned}$$

so that

$$\begin{aligned}
E_{2\rho}^{(2)}(\rho,z) = \frac{i\omega\mu_0 k_2^2}{2\pi k_1^3}&\left[e^{k_1 z} e^{ik_1 r_d}\left(\frac{1}{\rho^2} + \frac{3i}{2k_1\rho^3}\right)\right. \\
&\left. - e^{ik_1 d} e^{ik_2 r_0}\left(\frac{k_2}{r_0} + \frac{i}{r_0^2}\right)\left(\frac{\rho}{r_0}\right)\right]. \qquad (3.7.21)
\end{aligned}$$

The last integrals in (3.7.9)–(3.7.11) contain the factor

$$\frac{1}{N} - \frac{1}{N_0} = \frac{1}{k_1^2\gamma_2 + k_2^2\gamma_1} - \frac{1}{k_1^2\gamma_2}. \qquad (3.7.22)$$

In these it is clear that the major contributions must come from the range $\lambda \sim O(k_2)$. Therefore, $\gamma_1 = (k_1^2 - \lambda^2)^{1/2} \sim k_1$ and the integrals reduce to

$$\begin{aligned}
B_{2\phi}^{(1)}(\rho,z) &= \frac{i\mu_0 k_2^2}{2\pi}\, e^{ik_1 d} G_{2\phi}(\rho,z), \\
G_{2\phi}(\rho,z) &= \int_0^\infty \left(\frac{1}{N} - \frac{1}{N_0}\right) e^{-i\gamma_2 z} J_1(\lambda\rho)\lambda^2\, d\lambda; \qquad z \le 0, \quad (3.7.23) \\
E_{2\rho}^{(1)}(\rho,z) &= \frac{i\omega\mu_0 k_2^2}{2\pi k_1}\, e^{ik_1 d} G_{2\phi}(\rho,z); \qquad z \le 0, \quad (3.7.24)
\end{aligned}$$

$$\begin{aligned}
E_{2z}^{(1)}(\rho,z) &= -\frac{\omega\mu_0}{2\pi}\, e^{ik_1 d} G_{2z}(\rho,z), \\
G_{2z}(\rho,z) &= \int_0^\infty \left(\frac{1}{N} - \frac{1}{N_0}\right) e^{-i\gamma_2 z} J_0(\lambda\rho)\lambda^3\, d\lambda; \qquad z \le 0. \quad (3.7.25)
\end{aligned}$$

The two functions, $G_{2\phi}$ and G_{2z}, are evaluated in Appendix E. The results are

$$B_{2\phi}^{(1)}(\rho,z) = \frac{\mu_0 k_2^5}{2\pi k_1^3}\, e^{ik_1 d} e^{ik_2 r_0}\left(\frac{\pi}{k_2\rho}\right)^{1/2} e^{-iP}\mathcal{F}(P); \qquad z \le 0, \quad (3.7.26)$$

$$E_{2\rho}^{(1)}(\rho,z) = \frac{\omega\mu_0 k_2^5}{2\pi k_1^4}\, e^{ik_1 d} e^{ik_2 r_0}\left(\frac{\pi}{k_2\rho}\right)^{1/2} e^{-iP}\mathcal{F}(P); \qquad z \le 0, \quad (3.7.27)$$

$$E_{2z}^{(1)}(\rho, z) = -\frac{\omega\mu_0 k_2^4}{2\pi k_1^3} e^{ik_1 d} e^{ik_2 r_0} \left(\frac{\pi}{k_2\rho}\right)^{1/2} e^{-iP}\mathcal{F}(P); \quad z \leq 0, \quad (3.7.28)$$

where

$$P = \frac{(R-Z)^2}{R}; \qquad R = \frac{k_2^3\rho}{2k_1^2}; \qquad Z = \frac{k_2^2 z}{2k_1}, \qquad z \leq 0, \qquad (3.7.29)$$

and $\mathcal{F}(P)$ is given in (3.4.33). Note that the function P defined in (3.7.29) reduces to the function p defined in (3.4.32) when $z \sim 0$.

The complete field is the sum of (3.7.16)–(3.7.18), (3.7.21), and (3.7.26)–(3.7.28). The results are listed first in formulas (a). When the conditions $\rho \geq 5|z|$ and $\rho \geq 5d$ are also used in the exact direct- and image-field terms, so that $r_0 \sim \rho$ and $r_d \sim \rho$ in amplitudes, $P \sim R$, and terms of the order z^2/ρ^2 are neglected, these expressions reduce to the following simpler forms in formulas (b):

$$\begin{aligned} B_{2\phi}(\rho, z) &= B_{2\phi}^{(0)}(\rho, z) + B_{2\phi}^{(1)}(\rho, z) \\ &= -\frac{\mu_0 k_2^2}{2\pi k_1^2}\left\{ e^{ik_1 d} e^{ik_2 r_0}\left[\left(\frac{ik_2}{r_0} - \frac{1}{r_0^2}\right)\left(\frac{\rho}{r_0}\right) \right.\right. \\ &\qquad \left. - \frac{k_2^3}{k_1}\left(\frac{\pi}{k_2\rho}\right)^{1/2} e^{-iP}\mathcal{F}(P)\right] \\ &\qquad \left. - ie^{k_1 z} e^{ik_1 r_d}\left(\frac{d}{\rho}\right)\left(\frac{ik_1}{\rho} - \frac{3}{2\rho^2}\right)\right\}; \quad z \leq 0, \quad (3.7.30\text{a}) \\ &\sim -\frac{\mu_0 k_2^2}{2\pi k_1^2}\left[e^{ik_1 d} e^{ik_2 r_0} f(\rho; k_1, k_2) \right. \\ &\qquad \left. - ie^{k_1 z} e^{ik_1 r_d}\left(\frac{d}{\rho}\right)\left(\frac{ik_1}{\rho} - \frac{3}{2\rho^2}\right)\right]; \quad z \leq 0, \quad (3.7.30\text{b}) \end{aligned}$$

$$\begin{aligned} E_{2\rho}(\rho, z) &= E_{2\rho}^{(0)}(\rho, z) + E_{2\rho}^{(1)}(\rho, z) + E_{2\rho}^{(2)}(\rho, z) \\ &= -\frac{\omega\mu_0}{2\pi k_1^2}\left(\frac{k_2^2}{k_1}\left\{ e^{ik_1 d} e^{ik_2 r_0}\left[\left(\frac{ik_2}{r_0} - \frac{1}{r_0^2}\right)\left(\frac{\rho}{r_0}\right) \right.\right.\right. \\ &\qquad \left. - \frac{k_2^3}{k_1}\left(\frac{\pi}{k_2\rho}\right)^{1/2} e^{-iP}\mathcal{F}(P)\right] \\ &\qquad \left. - ie^{k_1 z} e^{ik_1 r_d}\left(\frac{1}{\rho^2} + \frac{3i}{2k_1\rho^3}\right)\right\} \\ &\qquad - e^{k_1 z} e^{ik_1 r_d}\left(\frac{ik_1^2}{r_d} - \frac{3k_1}{r_d^2} - \frac{3i}{r_d^3}\right)\left(\frac{\rho d}{r_d^2}\right) \\ &\qquad \left. + e^{ik_1 d} e^{ik_2 r_0}\left(\frac{ik_2^2}{r_0} - \frac{3k_2}{r_0^2} - \frac{3i}{r_0^3}\right)\left(\frac{\rho z}{r_0^2}\right)\right); \\ &\qquad z \leq 0, \quad (3.7.31\text{a}) \end{aligned}$$

$$
\begin{aligned}
&\sim -\frac{\omega\mu_0}{2\pi k_1^2}\left\{\frac{k_2^2}{k_1}\left[e^{ik_1 d}e^{ik_2 r_0} f(\rho;k_1,k_2)\right.\right.\\
&\qquad\left. - ie^{k_1 z}e^{ik_1 r_d}\left(\frac{1}{\rho^2}+\frac{3i}{2k_1\rho^3}\right)\right]\\
&\qquad - e^{k_1 z}e^{ik_1 r_d}\left(\frac{ik_1^2}{\rho}-\frac{3k_1}{\rho^2}-\frac{3i}{\rho^3}\right)\left(\frac{d}{\rho}\right)\\
&\qquad\left. + e^{ik_1 d}e^{ik_2 r_0}\left(\frac{ik_2^2}{\rho}-\frac{3k_2}{\rho^2}-\frac{3i}{\rho^3}\right)\left(\frac{z}{\rho}\right)\right\};\\
&\qquad\qquad z \le 0, \qquad (3.7.31\text{b})
\end{aligned}
$$

$$
\begin{aligned}
E_{2z}(\rho,z) &= E_{2z}^{(0)}(\rho,z) + E_{2z}^{(1)}(\rho,z)\\
&= \frac{\omega\mu_0}{2\pi k_1^2}\left\{e^{ik_1 d}e^{ik_2 r_0}\left[\frac{ik_2^2}{r_0}-\frac{k_2}{r_0^2}-\frac{i}{r_0^3}\right.\right.\\
&\qquad - \left(\frac{ik_2^2}{r_0}-\frac{3k_2}{r_0^2}-\frac{3i}{r_0^3}\right)\left(\frac{z^2}{r_0^2}\right)\\
&\qquad\left. - \frac{k_2^4}{k_1}\left(\frac{\pi}{k_2\rho}\right)^{1/2} e^{-iP}\mathcal{F}(P)\right]\\
&\qquad\left. - ie^{k_1 z}e^{ik_1 r_d}\left(\frac{d}{\rho}\right)\left(\frac{ik_1^2}{\rho}-\frac{k_1}{2\rho^2}+\frac{7i}{8\rho^3}\right)\right\};\\
&\qquad\qquad z \le 0, \qquad (3.7.32\text{a})
\end{aligned}
$$

$$
\begin{aligned}
&\sim \frac{\omega\mu_0}{2\pi k_1^2}\left[e^{ik_1 d}e^{ik_2 r_0} k_2 g(\rho;k_1,k_2)\right.\\
&\qquad\left. - ie^{k_1 z}e^{ik_1 r_d}\left(\frac{d}{\rho}\right)\left(\frac{ik_1^2}{\rho}-\frac{k_1}{2\rho^2}+\frac{7i}{8\rho^3}\right)\right];\\
&\qquad\qquad z \le 0, \qquad (3.7.32\text{b})
\end{aligned}
$$

where $r_0 = (\rho^2 + z^2)^{1/2}$ and $r_d = (\rho^2 + d^2)^{1/2}$; P, R, and Z are given in (3.7.29); $\mathcal{F}(P)$ is in (3.4.33); and $f(\rho; k_1, k_2)$ and $g(\rho; k_1, k_2)$ are defined in (3.4.30), (3.4.31), and (3.4.35). Note that, in the range $\rho \ge 5|z|$ in which the Fresnel-integral terms are valid, $k_2\rho + Z^2/R = k_2(\rho + z^2/2\rho) \sim k_2 r_0$. Note also that the terms multiplied by $\exp(k_1 z) = \exp(-k_1|z|)$ are not part of the propagating field in Region 2; they are exponentially attenuated upward from the boundary.

Subject to the condition $|k_1|^2 \gg |k_2|^2$, these formulas are good approximations throughout Region 2 except near the source. The condition $\rho \ge 5|z|$ limits the range to near the boundary. For most applications involving vertical dipoles in the earth or sea, this is the only range of interest. Where this is not the case, the restriction can be removed as shown in Sections 3.9 and 4.2. Note that these formulas satisfy the boundary conditions at $z = 0$: $B_{2\phi}(\rho, 0) = B_{1\phi}(\rho, 0)$, $E_{2z}(\rho, 0) = (k_1^2/k_2^2)E_{1z}(\rho, 0)$, and $E_{2\rho}(\rho, 0) = E_{1\rho}(\rho, 0)$.

3.8 Alternative Derivation of the Field in Region 2 for the Vertical Electric Dipole in Region 1

An alternative calculation of the field in Region 2 is of interest since it gives additional physical insight. It can also be applied to situations in which the direct calculation of the relevant integrals cannot be carried out. As a first step, let the integrals be written down for the field generated in Region 2 by a unit vertical dipole at the origin on the boundary in Region 2. The three components are given by (3.4.9)–(3.4.11) with the subscripts 1 and 2 interchanged, z replaced by $-z$, ϕ by $-\phi$, $I\Delta l = 1$ A m by $I\Delta l = -1$ A m, and $d = 0$. That is,

$$B_{2\phi}(\rho, z) = \frac{i\mu_0 k_1^2}{2\pi} \int_0^\infty \frac{e^{-i\gamma_2 z}}{N} J_1(\lambda\rho)\lambda^2\, d\lambda; \qquad z \leq 0, \tag{3.8.1}$$

$$E_{2\rho}(\rho, z) = -\frac{i\omega\mu_0 k_1^2}{2\pi k_2^2} \int_0^\infty \frac{\gamma_2 e^{-i\gamma_2 z}}{N} J_1(\lambda\rho)\lambda^2\, d\lambda; \qquad z \leq 0, \tag{3.8.2}$$

$$E_{2z}(\rho, z) = -\frac{\omega\mu_0 k_1^2}{2\pi k_2^2} \int_0^\infty \frac{e^{-i\gamma_2 z}}{N} J_0(\lambda\rho)\lambda^3\, d\lambda; \qquad z \leq 0. \tag{3.8.3}$$

When these formulas are compared with (3.3.12)–(3.3.14) for the field in Region 2 due to a vertical dipole at $z = d$ in Region 1, they are found to differ only by the factor $(k_2^2/k_1^2)\exp(i\gamma_1 d)$. In the evaluation of the G_1 integrals for Region 1, the approximation $\gamma_1 \sim k_1$ is made so that $\exp[i\gamma_1(z+d)] \sim \exp[ik_1(z+d)]$. If the same approximation is made here, the field in Region 2 is given by (3.8.1)–(3.8.3) multiplied by $(k_2^2/k_1^2)\exp(ik_1 d)$. Thus, the unit dipole *at* $z = d$ *in Region 1* generates a field in Region 2 that is approximately the same as the field of a vertical dipole with moment $(k_2^2/k_1^2)\exp(ik_1 d)$ located at the origin *on the boundary* $z = 0$ *in Region 2*. The approximation is good for B_ϕ and E_z, but is inadequate for E_ρ at and near the boundary $z = 0$.

The field of an isolated vertical dipole is well known. As a first approximation, let the boundary with Region 1 be treated as that of a perfect conductor along which $E_{2\,\text{tang}} = 0$. The field of a vertical dipole and image with moment $(k_2^2/k_1^2)\exp(ik_1 d)$ located at $(0,0)$ in Region 2 is

$$B_{2\phi}(\rho, z) = -\frac{\mu_0 k_2^2}{2\pi k_1^2}\, e^{ik_1 d}\left(\frac{ik_2}{r_0} - \frac{1}{r_0^2}\right)\left(\frac{\rho}{r_0}\right) e^{ik_2 r_0}; \qquad z \leq 0, \tag{3.8.4}$$

$$E_{2\rho}(\rho, z) = -\frac{\omega\mu_0 k_2}{2\pi k_1^2}\, e^{ik_1 d}\left(\frac{ik_2}{r_0} - \frac{3}{r_0^2} - \frac{3i}{k_2 r_0^3}\right)\left(\frac{\rho z}{r_0^2}\right) e^{ik_2 r_0};$$
$$z \leq 0, \tag{3.8.5}$$

$$E_{2z}(\rho, z) = \frac{\omega\mu_0 k_2}{2\pi k_1^2}\, e^{ik_1 d}\left[\frac{ik_2}{r_0} - \frac{1}{r_0^2} - \frac{i}{k_2 r_0^3}\right.$$

$$- \left(\frac{ik_2}{r_0} - \frac{3}{r_0^2} - \frac{3i}{k_2 r_0^3} \right) \left(\frac{z^2}{r_0^2} \right) \Bigg] e^{ik_2 r_0}; \qquad z \leq 0, \quad (3.8.6)$$

where $r_0 = (\rho^2 + z^2)^{1/2}$. This is the field in Region 2 except near the boundary where the boundary condition $E_{2\,\mathrm{tang}} \sim 0$ is not accurate. The correct boundary condition is

$$\mathbf{E}_{1\,\mathrm{tang}}(\rho, 0) = \mathbf{E}_{2\,\mathrm{tang}}(\rho, 0) = -\frac{\omega}{k_1}\, \hat{\mathbf{z}} \times \mathbf{B}_{2\,\mathrm{tang}}(\rho, 0) \qquad (3.8.7a)$$

or, with (3.8.4)–(3.8.6),

$$E_{1\rho}(\rho, 0) = E_{2\rho}(\rho, 0) = \frac{\omega}{k_1}\, B_{2\phi}(\rho, 0), \qquad (3.8.7b)$$

where $z \leq 0$ is Region 2. Use can be made of this accurate relation to extend the range of (3.8.4)–(3.8.6) to the boundary. The condition

$$|n^2| = \left| \frac{k_2^2}{k_1^2} \right| \ll 1 \qquad (3.8.8)$$

is assumed.

The correcting formulation begins with Maxwell's equations in cylindrical coordinates. With $E_\phi = B_\rho = B_z = 0$ and $\partial/\partial\phi = 0$, these reduce to

$$\frac{\partial B_\phi}{\partial z} = \frac{ik^2}{\omega}\, E_\rho, \qquad (3.8.9)$$

$$\frac{1}{\rho} \frac{\partial}{\partial \rho} (\rho B_\phi) = -\frac{ik^2}{\omega}\, E_z, \qquad (3.8.10)$$

$$\frac{\partial E_\rho}{\partial z} - \frac{\partial E_z}{\partial \rho} = i\omega B_\phi. \qquad (3.8.11)$$

These equations are to be applied to the components of the field (3.8.4)–(3.8.6) after the variables ρ and z have been changed according to

$$R = \frac{k_2^3 \rho}{2k_1^2} = \frac{n^2 k_2 \rho}{2}; \qquad Z = \frac{k_2^2 z}{2k_1} = \frac{n k_2 z}{2}, \qquad (3.8.12)$$

or

$$k_2 \rho = \frac{2R}{n^2}; \qquad k_2 z = \frac{2Z}{n}. \qquad (3.8.13)$$

Since the correction to the formulas (3.8.4)–(3.8.6) is to apply only near the boundary where $\rho^2 \gg z^2$, it is adequate to set $r_0 = (\rho^2 + z^2)^{1/2} \sim \rho$ in amplitudes and $r_0 \sim \rho + z^2/2\rho$ in phases. In order to simplify the evaluation, the correction is derived only for the leading terms.

When expressed in terms of the new variables R and Z, and with

$$A = \frac{i\omega\mu_0 k_2^3 n^2}{4\pi k_1^2}\, e^{ik_1 d}, \qquad (3.8.14)$$

the leading terms in (3.8.4)–(3.8.6) become

$$B_{2\phi} = A\frac{k_2}{\omega}e^{i2R/n^2}\tilde{B}_{2\phi}; \qquad \tilde{B}_{2\phi} = -\frac{1}{R}e^{iZ^2/R}, \tag{3.8.15}$$

$$E_{2\rho} = Ae^{i2R/n^2}n\tilde{E}_{2\rho}; \qquad \tilde{E}_{2\rho} = -\frac{Z}{R^2}e^{iZ^2/R}, \tag{3.8.16}$$

$$E_{2z} = Ae^{i2R/n^2}\tilde{E}_{2z}; \qquad \tilde{E}_{2z} = \frac{1}{R}e^{iZ^2/R}. \tag{3.8.17}$$

Evidently, from (3.8.15) and (3.8.17),

$$\tilde{E}_{2z} = -\tilde{B}_{2\phi}. \tag{3.8.18}$$

Also, from (3.8.7b) with (3.8.15) and (3.8.16),

$$\tilde{E}_{2\rho} = \tilde{B}_{2\phi} \quad \text{at } Z = 0. \tag{3.8.19}$$

With (3.8.13) and (3.8.18), Maxwell's equations (3.8.9)–(3.8.11) become

$$\frac{\partial \tilde{B}_{2\phi}}{\partial Z} - 2i\tilde{E}_{2\rho} = 0, \tag{3.8.20}$$

$$\frac{\partial \tilde{B}_{2\phi}}{\partial R} + \frac{\tilde{B}_{2\phi}}{R} = 0, \tag{3.8.21}$$

$$\frac{\partial \tilde{E}_{2\rho}}{\partial Z} + \frac{\partial \tilde{B}_{2\phi}}{\partial R} = 0. \tag{3.8.22}$$

These can be combined to give the following equation for $\tilde{B}_{2\phi}$:

$$\left(\frac{1}{R} + 2\frac{\partial}{\partial R}\right)\tilde{B}_{2\phi} - \frac{i}{2}\frac{\partial^2 \tilde{B}_{2\phi}}{\partial Z^2} = 0. \tag{3.8.23}$$

The relevant boundary condition is (3.8.19) and the required behavior for sufficiently large values of Z is

$$\left.\begin{aligned} \tilde{E}_{2\rho} &\to -\frac{Z}{R^2}e^{iZ^2/R} \\ \tilde{B}_{2\phi} &\to -\frac{1}{R}e^{iZ^2/R} \end{aligned}\right\} \quad \text{far from the boundary.} \tag{3.8.24}$$

The solution of (3.8.23) subject to (3.8.19) and (3.8.24) is carried out with the substitution

$$F \equiv F(R, Z) = R^{1/2}\tilde{B}_{2\phi}. \tag{3.8.25}$$

With this substitution, (3.8.23) becomes

$$\frac{\partial^2 F}{\partial Z^2} + 4i\frac{\partial F}{\partial R} = 0. \tag{3.8.26}$$

Also, with (3.8.20),

$$\tilde{E}_{2\rho} = -\frac{i}{2}\frac{\partial \tilde{B}_{2\phi}}{\partial Z} = -\frac{i}{2}R^{-1/2}\frac{\partial F}{\partial Z}, \tag{3.8.27}$$

so that, with (3.8.19), (3.8.27) becomes

$$\left[\frac{\partial F}{\partial Z} - 2iF\right]_{Z=0} = 0. \tag{3.8.28}$$

With (3.8.24),

$$F \to -R^{-1/2}e^{iZ^2/R} \quad \text{far from the boundary.} \tag{3.8.29}$$

The solution of (3.8.27) subject to (3.8.28) and (3.8.29) remains to be carried out.

Since far from the boundary the solution is (3.8.29), the solution near the boundary can be obtained in the form

$$F_0 = F + R^{-1/2}e^{iZ^2/R}, \tag{3.8.30}$$

where $F_0 \sim 0$ far from the boundary. Since

$$\left[\frac{\partial^2}{\partial Z^2} + 4i\frac{\partial}{\partial R}\right]\frac{e^{iZ^2/R}}{R^{1/2}} = 0,$$

it follows that

$$\frac{\partial^2 F_0}{\partial Z^2} + 4i\frac{\partial F_0}{\partial R} = 0. \tag{3.8.31}$$

Also, with (3.8.28),

$$\left[\frac{\partial F_0}{\partial Z} - 2iF_0\right]_{Z=0} = -2iR^{-1/2}. \tag{3.8.32}$$

A solution of (3.8.31) with (3.8.32) can be obtained with the Fourier transform

$$\bar{F}_0(\xi, Z) = \int_0^\infty F_0(R, Z)e^{-i\xi R}\,dR. \tag{3.8.33}$$

When (3.8.31) and (3.8.32) are Fourier transformed, they become

$$\frac{\partial^2 \bar{F}_0}{\partial Z^2} - 4\xi\bar{F}_0 = 0; \qquad \bar{F}_0 = A(\xi)e^{2\xi^{1/2}Z}, \tag{3.8.34}$$

$$\left[\frac{\partial \bar{F}_0}{\partial Z} - 2i\bar{F}_0\right]_{Z=0} = -2(\pi i)^{1/2}\xi^{-1/2}. \tag{3.8.35}$$

With (3.8.34) and (3.8.35),

$$A(\xi) = -\frac{(\pi i)^{1/2}}{\xi^{1/2}(\xi^{1/2} - i)}, \tag{3.8.36}$$

so that

$$\bar{F}_0(\xi, Z) = -\frac{(\pi i)^{1/2}e^{2\xi^{1/2}Z}}{\xi^{1/2}(\xi^{1/2} - i)} \tag{3.8.37}$$

and

$$F_0(R,Z) = -\left(\frac{i}{4\pi}\right)^{1/2} \int_{-\infty}^{\infty} \frac{e^{2\xi^{1/2}Z}}{\xi^{1/2}(\xi^{1/2}-i)}\, e^{i\xi R}\, d\xi. \tag{3.8.38}$$

When this value is substituted in (3.8.30), the solution for F is obtained. It is

$$F(R,Z) = -R^{-1/2}e^{iZ^2/R} - \left(\frac{i}{4\pi}\right)^{1/2} \int_{-\infty}^{\infty} \frac{e^{2\xi^{1/2}Z}}{\xi^{1/2}(\xi^{1/2}-i)}\, e^{i\xi R}\, d\xi. \tag{3.8.39}$$

With (3.8.39), (3.8.25) gives

$$\tilde{B}_{2\phi} = -\frac{e^{iZ^2/R}}{R} - \left(\frac{i}{4\pi R}\right)^{1/2} I(R,Z), \tag{3.8.40}$$

where

$$I \equiv I(R,Z) = \int_{-\infty}^{\infty} \frac{1}{\xi^{1/2}(\xi^{1/2}-i)}\, e^{2\xi^{1/2}Z} e^{i\xi R}\, d\xi. \tag{3.8.41}$$

This integral can be expressed in terms of the Fresnel integral. As a first step, it is to be noted that

$$\begin{aligned}
\frac{1}{2}\frac{\partial}{\partial Z}\left(e^{-2iZ} I\right) &= \left(\frac{1}{2}\frac{\partial I}{\partial Z} - iI\right) e^{-2iZ} \\
&= e^{-2iZ} \int_{-\infty}^{\infty} \xi^{-1/2} e^{2\xi^{1/2}Z} e^{i\xi R}\, d\xi \\
&= 2\left(\frac{\pi i}{R}\right)^{1/2} e^{-2iZ} e^{iZ^2/R}.
\end{aligned} \tag{3.8.42}$$

Since $I(\infty) = 0$, it follows that, for $Z \le 0$, (3.8.42) gives

$$I = 4\left(\frac{\pi i}{R}\right)^{1/2} e^{2iZ} \int_{-\infty}^{Z} e^{-2iZ'} e^{iZ'^2/R}\, dZ'. \tag{3.8.43}$$

Now let the variable be changed to $Z'' = R - Z'$. Then

$$I = 4\left(\frac{\pi i}{R}\right)^{1/2} e^{2iZ} e^{-iR} \int_{R-Z}^{\infty} e^{iZ''^2/R}\, dZ''. \tag{3.8.44}$$

Finally, let $t = Z''^2/R$ so that

$$I = 4\left(\frac{\pi i}{R}\right)^{1/2} e^{2iZ} e^{-iR} \left(\frac{\pi R}{2}\right)^{1/2} \int_{P}^{\infty} \frac{e^{it}}{(2\pi t)^{1/2}}\, dt, \tag{3.8.45}$$

where

$$P = \frac{(R-Z)^2}{R}; \qquad Z \le 0, \tag{3.8.46}$$

and R and Z are given in (3.8.12). Also,

$$\int_P^\infty \frac{e^{it}}{(2\pi t)^{1/2}}\,dt = \tfrac{1}{2}(1+i) - C_2(P) - iS_2(P) \equiv \mathcal{F}(P), \tag{3.8.47}$$

where

$$C_2(P) + iS_2(P) = \int_0^P \frac{e^{it}}{(2\pi t)^{1/2}}\,dt \tag{3.8.48}$$

is the Fresnel integral. It follows that

$$I = 4\pi\left(\frac{i}{2}\right)^{1/2} e^{2iZ}e^{-iR}\left[\tfrac{1}{2}(1+i) - C_2(P) - iS_2(P)\right] \tag{3.8.49}$$

and

$$\tilde{B}_{2\phi} = -\frac{e^{iZ^2/R}}{R}\,(1+\mathcal{G}), \tag{3.8.50}$$

with

$$\mathcal{G} \equiv i(2\pi R)^{1/2}e^{-i(R-Z)^2/R}\mathcal{F}[(R-Z)^2/R]; \qquad Z \le 0. \tag{3.8.51}$$

When (3.8.50) is substituted in (3.8.15), this becomes

$$B_{2\phi} \sim -A\,\frac{k_2}{\omega}\,e^{i2R/n^2}\,\frac{e^{iZ^2/R}}{R}\,(1+\mathcal{G}). \tag{3.8.52}$$

With (3.8.18) and (3.8.50), (3.8.17) becomes

$$E_{2z} \sim Ae^{i2R/n^2}\,\frac{e^{iZ^2/R}}{R}\,(1+\mathcal{G}). \tag{3.8.53}$$

With (3.8.27) and (3.8.50),

$$\tilde{E}_{2\rho} \sim -\frac{e^{iZ^2/R}}{R}\left(\frac{Z}{R}+1+\mathcal{G}\right); \qquad Z \le 0. \tag{3.8.54}$$

When this is substituted in (3.8.16), the result is

$$E_{2\rho} \sim -Ae^{i2R/n^2}e^{iZ^2/R}\,\frac{n}{R}\left(\frac{Z}{R}+1+\mathcal{G}\right); \qquad Z \le 0. \tag{3.8.55}$$

When some of the variables R and Z are changed back to ρ and z, the corrected leading $1/\rho$ terms are

$$B_{2\phi}(\rho,z) \sim -\frac{\mu_0 k_2^2}{2\pi k_1^2}\,e^{ik_1 d}e^{ik_2 r_0}\,\frac{ik_2}{\rho}\,(1+\mathcal{G}), \tag{3.8.56}$$

$$E_{2\rho}(\rho,z) \sim -\frac{\omega\mu_0 k_2}{2\pi k_1^2}\,e^{ik_1 d}e^{ik_2 r_0}\,\frac{ik_2}{\rho}\left(\frac{z}{\rho}+\frac{k_2}{k_1}\,(1+\mathcal{G})\right), \tag{3.8.57}$$

$$E_{2z}(\rho,z) \sim \frac{\omega\mu_0 k_2}{2\pi k_1^2}\,e^{ik_1 d}e^{ik_2 r_0}\,\frac{ik_2}{\rho}\,(1+\mathcal{G}), \tag{3.8.58}$$

where the approximation $k_2\rho + Z^2/R = k_2(\rho + z^2/2\rho) \sim k_2 r_0$, $r_0 = (\rho^2 + z^2)^{1/2}$, has been made. These can be combined with the complete expressions (3.8.4)–(3.8.6) with care to satisfy (3.8.7b). This requires the addition of the term $-1/\rho^2$ in $E_{2\rho}(\rho, z)$. The result is

$$B_{2\phi}(\rho, z) = -\frac{\mu_0 k_2^2}{2\pi k_1^2}\, e^{ik_1 d} e^{ik_2 r_0} \times \left[\left(\frac{ik_2}{r_0} - \frac{1}{r_0^2}\right)\left(\frac{\rho}{r_0}\right) + e^{ik_2 r_0}\,\frac{ik_2}{\rho}\,\mathcal{G}\right]; \qquad z \le 0, \tag{3.8.59}$$

$$E_{2\rho}(\rho, z) = -\frac{\omega\mu_0 k_2}{2\pi k_1^2}\, e^{ik_1 d} e^{ik_2 r_0} \times \left[\left(\frac{ik_2}{r_0} - \frac{3}{r_0^2} - \frac{3i}{k_2 r_0^3}\right)\left(\frac{\rho z}{r_0^2}\right) + \left(\frac{k_2}{k_1}\right)\left(\frac{ik_2}{\rho}\,(1 + \mathcal{G}) - \frac{1}{\rho^2}\right)\right]; \qquad z \le 0, \tag{3.8.60}$$

$$E_{2z}(\rho, z) = \frac{\omega\mu_0 k_2}{2\pi k_1^2}\, e^{ik_1 d} e^{ik_2 r_0} \times \left[\frac{ik_2}{r_0} - \frac{1}{r_0^2} - \frac{i}{k_2 r_0^3} - \left(\frac{ik_2}{r_0} - \frac{3}{r_0^2} - \frac{3i}{k_2 r_0^3}\right)\left(\frac{z^2}{r_0^2}\right) + \frac{ik_2}{\rho}\,\mathcal{G}\right]; \qquad z \le 0. \tag{3.8.61}$$

Here

$$\mathcal{G} = i(2\pi R)^{1/2} e^{-iP} \mathcal{F}(P), \tag{3.8.62}$$

P is defined in (3.8.46), R and Z in (3.8.12), and $\mathcal{F}(P)$ in (3.4.33). Note that when $Z = 0$,

$$\mathcal{G} \to i(2\pi R)^{1/2} e^{-iR} \mathcal{F}(R) = \frac{ik_2^2 \rho}{k_1}\left(\frac{\pi}{k_2\rho}\right)^{1/2} e^{-iR} \mathcal{F}(R). \tag{3.8.63}$$

The formulas (3.8.59)–(3.8.61) agree with (3.7.30)–(3.7.32) obtained by direct integration. They give the complete electromagnetic field in Region 2 [not too near the equivalent source and, for $E_{2\rho}(\rho, z)$, not too near the boundary $z = 0$] due to a unit vertical dipole at $z = d$ in Region 1. The validity of the method of evaluation of (3.8.59)–(3.8.61) is thus established.

3.9 Generalization of the Field in Region 2

For possible applications that involve the field throughout Region 2, the condition $\rho \geq 5|z|$ (which restricts the formulas to a small angle above the boundary) can be removed. Actually this condition has been used only in the derivation of the Fresnel-integral terms. A more general treatment—necessary for Section 2 in Chapter 4—can easily be applied. With it, the restriction $\rho \geq 5|z|$ is eliminated to obtain formulas valid everywhere in Region 2. The necessary changes are limited to the terms with superscript (1) in (3.7.23)–(3.7.25). The integrals $G_{2\phi}(\rho, z)$ and $G_{2z}(\rho, z)$ are generalized as follows:

$$G_{2\phi}(\rho, z) \to -\frac{i}{k_1^2} T(\rho, z), \tag{3.9.1}$$

$$G_{2z}(\rho, z) \to \frac{k_2}{k_1^2} \frac{\rho}{r_0} T(\rho, z), \tag{3.9.2}$$

where

$$T(\rho, z) = \frac{k_2^3}{k_1} \left(\frac{\pi}{k_2 r_0} \right)^{1/2} e^{-iP_0} \mathcal{F}(P_0), \tag{3.9.3}$$

with

$$P_0 = S_0 \left(\frac{S_0 - Z}{R} \right)^2 = \frac{k_2^3 r_0}{2k_1^2} \left(\frac{k_2 r_0 - k_1 z}{k_2 \rho} \right)^2, \tag{3.9.4a}$$

$$S_0 = \frac{k_2^3 r_0}{2k_1^2}; \qquad R = \frac{k_2^3 \rho}{2k_1^2}; \qquad Z = \frac{k_2^2 z}{2k_1}, \qquad z \leq 0, \tag{3.9.4b}$$

and $r_0 = (\rho^2 + z^2)^{1/2}$. The Fresnel-integral term $T(\rho, z)$ is derived in Appendix F. Its relation to $G_{2\phi}(\rho, z)$ and $G_{2z}(\rho, z)$ is shown in the concluding section of Appendix F.

The new formulas—restricted only by $k_1 \geq 3|k_2|$—are

$$\begin{aligned} B_{2\phi}(\rho, z) = -\frac{\mu_0 k_2^2}{2\pi k_1^2} \Bigg\{ & e^{ik_2 r_0} e^{ik_1 d} \Bigg[\left(\frac{\rho}{r_0} \right) \left(\frac{ik_2}{r_0} - \frac{1}{r_0^2} \right) \\ & - \frac{k_2^3}{k_1} \left(\frac{\pi}{k_2 r_0} \right)^{1/2} e^{-iP_0} \mathcal{F}(P_0) \Bigg] \\ & - i e^{k_1 z} e^{ik_1 r_d} \left(\frac{d}{\rho} \right) \left(\frac{ik_1}{\rho} - \frac{3}{2\rho^2} \right) \Bigg\}; \qquad z \leq 0, \end{aligned} \tag{3.9.5}$$

$$\begin{aligned} E_{2\rho}(\rho, z) = -\frac{\omega \mu_0}{2\pi k_1^2} \Bigg(\frac{k_2^2}{k_1} \Bigg\{ & e^{ik_1 d} e^{ik_2 r_0} \Bigg[\left(\frac{ik_2}{r_0} - \frac{1}{r_0^2} \right) \left(\frac{\rho}{r_0} \right) \\ & - \frac{k_2^3}{k_1} \left(\frac{\pi}{k_2 r_0} \right)^{1/2} e^{-iP_0} \mathcal{F}(P_0) \Bigg] \end{aligned}$$

$$- ie^{k_1 z} e^{ik_1 r_d} \left(\frac{1}{\rho^2} + \frac{3i}{2k_1 \rho^3} \right) \Bigg\}$$

$$- e^{k_1 z} e^{ik_1 r_d} \left(\frac{ik_1^2}{r_d} - \frac{3k_1}{r_d^2} - \frac{3i}{r_d^3} \right) \left(\frac{\rho d}{r_d^2} \right)$$

$$+ e^{ik_1 d} e^{ik_2 r_0} \left(\frac{ik_2^2}{r_0} - \frac{3k_2}{r_0^2} - \frac{3i}{r_0^3} \right) \left(\frac{\rho z}{r_0^2} \right) \Bigg);$$

$$z \le 0, \quad (3.9.6)$$

$$E_{2z}(\rho, z) = \frac{\omega \mu_0}{2\pi k_1^2} \Bigg\{ e^{ik_2 r_0} e^{ik_1 d} \Bigg[\frac{ik_2^2}{r_0} - \frac{k_2}{r_0^2} - \frac{i}{r_0^3}$$

$$- \left(\frac{ik_2^2}{r_0} - \frac{3k_2}{r_0^2} - \frac{3i}{r_0^3} \right) \left(\frac{z^2}{r_0^2} \right)$$

$$- \frac{k_2^4}{k_1} \left(\frac{\pi}{k_2 r_0} \right)^{1/2} \left(\frac{\rho}{r_0} \right) e^{-iP_0} \mathcal{F}(P_0) \Bigg]$$

$$- ie^{k_1 z} e^{ik_1 r_d} \left(\frac{d}{\rho} \right) \left(\frac{ik_1^2}{\rho} - \frac{k_1}{2\rho^2} + \frac{7i}{8\rho^3} \right) \Bigg\};$$

$$z \le 0. \quad (3.9.7)$$

It is readily verified that when $\rho \ge 5|z|$ so that $r_0 \sim \rho$, these formulas reduce exactly to (3.7.30)–(3.7.32).

4
Applications of the Theory of the Vertical Dipole Near the Boundary Between Two Half-Spaces

4.1 Limitations of the Theory

The relatively simple formulas developed in the preceding chapter for the electromagnetic field generated by a vertical electric dipole near the plane boundary between electrically different regions are complete and accurate within the specified conditions. When applied to practical situations such as propagation over the earth or sea, the geometry of two half-spaces with a plane boundary may not be adequate. When large distances are involved, the curvature of the earth must be considered. Unfortunately, there are no formulas available for the spherical geometry comparable to those in Chapter 3 for the plane boundary. The best that is available are methods for correcting the fields over a plane earth so that they may be applied approximately to the actual spherical earth (Norton 1941; CCIR 1978).

A further complicating factor in problems relating to wave propagation over the surface of the earth is the presence of the ionosphere (Watt 1967). In effect, this changes the two-media problem to one that involves three media, viz., the earth and the ionosphere separated by air. Wave propagation in this waveguide-like configuration involves multiple reflections at the two boundaries and the superposition of several different modes. Since a lateral wave traveling along the surface of the earth extends upward into the air only a very small distance compared to the distance of travel, its general form may remain largely unaffected by a superimposed field that involves multiple reflections and waveguide modes. The three-layer problem is discussed in Chapter 11.

In the following applications of the theory of lateral waves generated by a vertical dipole, no account is taken of either the curvature of the earth or of the presence of the ionosphere.

4.2 Radio Communication over the Surface of the Earth; Electromagnetic Field in Air Due to a Vertical Electric Dipole at Height d in Air

In Chapter 1 the electromagnetic far field of a vertical electric dipole on the surface of the earth was obtained as the superposition of the direct field and the reflected field calculated with the plane-wave reflection coefficient. It was shown that these two components exactly cancel at all points along the surface (except when the earth is replaced by a perfect conductor). Formulas for the actual field are given in Chapter 3, eqs. (3.6.11)–(3.6.13), specifically for the field due to a dipole on the boundary in air. These show that the entire far field on the boundary $z = 0$ is due to the lateral wave and has a $1/\rho^2$ dependence [see (3.4.35)]. Radio communication between vertical antennas on the earth makes use of the vertical component $E_{2z}(\rho, 0)$ represented in Fig. 3.6.1. It is the purpose of this section to generalize these results to obtain the complete electromagnetic field (not just the far field) at any point in the air due to a vertical dipole at any height d over the earth.

The electromagnetic field of a unit infinitesimal vertical dipole at $z = -d$ in Region 2 is obtained from (3.3.9)–(3.3.11) with an interchange of the subscripts 1 and 2, the replacement of z by $-z$, and ϕ by $-\phi$. To simplify the notation, it is convenient to introduce $z' = -z$ and $\phi' = -\phi$. The formulas for Region 2 with $z' \geq 0$ are

$$B_{2\phi'}(\rho, z') = \frac{i\mu_0}{2\pi} \int_0^\infty \left(\frac{e^{i\gamma_2|z'-d|}}{2\gamma_2} - \frac{e^{i\gamma_2(z'+d)}}{2\gamma_2} + \frac{k_1^2 e^{i\gamma_2(z'+d)}}{N} \right) J_1(\lambda\rho)\lambda^2\, d\lambda, \tag{4.2.1}$$

$$E_{2\rho}(\rho, z') = \frac{i\omega\mu_0}{2\pi k_2^2} \int_0^\infty \left(\pm\frac{e^{i\gamma_2|z'-d|}}{2} - \frac{e^{i\gamma_2(z'+d)}}{2} + \frac{k_1^2\gamma_2 e^{i\gamma_2(z'+d)}}{N} \right) J_1(\lambda\rho)\lambda^2\, d\lambda;$$

$$\begin{cases} z' > d \\ 0 \leq z' \leq d \end{cases}, \tag{4.2.2}$$

$$E_{2z'}(\rho, z') = -\frac{\omega\mu_0}{2\pi k_2^2} \int_0^\infty \left(\frac{e^{i\gamma_2|z'-d|}}{2\gamma_2} - \frac{e^{i\gamma_2(z'+d)}}{2\gamma_2} + \frac{k_1^2 e^{i\gamma_2(z'+d)}}{N} \right) J_0(\lambda\rho)\lambda^3 \, d\lambda. \tag{4.2.3}$$

For Region 1, $z' \leq 0$, it follows from (3.3.12)–(3.3.14) that

$$B_{1\phi'}(\rho, z') = \frac{i\mu_0 k_1^2}{2\pi} \int_0^\infty \frac{e^{i\gamma_2 d} e^{-i\gamma_1 z'}}{N} J_1(\lambda\rho)\lambda^2 \, d\lambda, \tag{4.2.4}$$

$$E_{1\rho}(\rho, z') = -\frac{i\omega\mu_0}{2\pi} \int_0^\infty \frac{\gamma_1 e^{i\gamma_2 d} e^{-i\gamma_1 z'}}{N} J_1(\lambda\rho)\lambda^2 \, d\lambda, \tag{4.2.5}$$

$$E_{1z'}(\rho, z') = -\frac{\omega\mu_0}{2\pi} \int_0^\infty \frac{e^{i\gamma_2 d} e^{-i\gamma_1 z'}}{N} J_0(\lambda\rho)\lambda^3 \, d\lambda, \tag{4.2.6}$$

where

$$N \equiv k_1^2\gamma_2 + k_2^2\gamma_1, \tag{4.2.7}$$

with $\gamma_j = (k_j^2 - \lambda^2)^{1/2}$, $j = 1, 2$. As indicated by the negative sign of the second term in formulas (4.2.1)–(4.2.3), the image is negative. This is appropriate when the dipole and the point of observation are in the earth or sea (Region 1) (Fig. 4.2.1a). The mathematically equivalent formulas with a positive image—which is physically more appropriate when the dipole and the point of observation are in the air (Region 2) (Fig. 4.2.1b)—are given by

$$B_{2\phi'}(\rho, z') = \frac{i\mu_0}{2\pi} \int_0^\infty \left[\frac{e^{i\gamma_2|z'-d|}}{2\gamma_2} + \frac{e^{i\gamma_2(z'+d)}}{2\gamma_2} + k_1^2 e^{i\gamma_2(z'+d)} \left(\frac{1}{N} - \frac{1}{N_0} \right) \right] J_1(\lambda\rho)\lambda^2 \, d\lambda, \tag{4.2.8}$$

$$E_{2\rho}(\rho, z') = \frac{i\omega\mu_0}{2\pi k_2^2} \int_0^\infty \left[\pm\frac{e^{i\gamma_2|z'-d|}}{2} + \frac{e^{i\gamma_2(z'+d)}}{2} + k_1^2 e^{i\gamma_2(z'+d)}\gamma_2 \left(\frac{1}{N} - \frac{1}{N_0} \right) \right] J_1(\lambda\rho)\lambda^2 \, d\lambda; \quad \begin{cases} z' > d \\ 0 \leq z' \leq d \end{cases}, \tag{4.2.9}$$

$$E_{2z'}(\rho, z') = -\frac{\omega\mu_0}{2\pi k_2^2} \int_0^\infty \left[\frac{e^{i\gamma_2|z'-d|}}{2\gamma_2} + \frac{e^{i\gamma_2(z'+d)}}{2\gamma_2} + k_1^2 e^{i\gamma_2(z'+d)} \left(\frac{1}{N} - \frac{1}{N_0} \right) \right] J_0(\lambda\rho)\lambda^3 \, d\lambda, \tag{4.2.10}$$

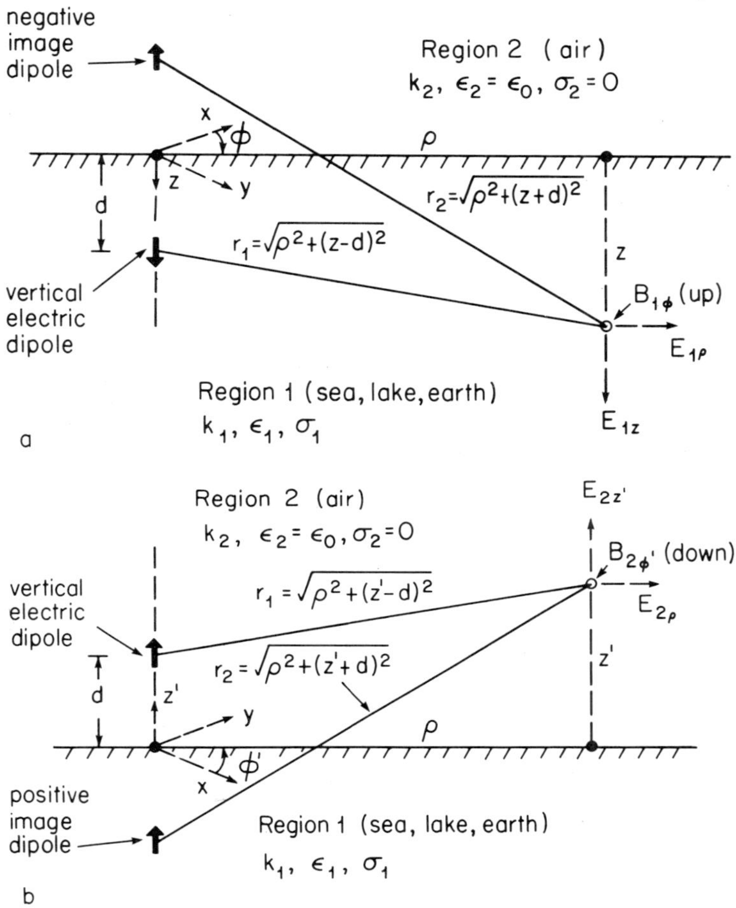

Figure 4.2.1. (a) Vertical electric dipole at depth d in Region 1. Electromagnetic field calculated at (ρ, z). (b) Vertical electric dipole at height d in Region 2. Field calculated at (ρ, z').

where

$$N_0 = k_1^2 \gamma_2. \tag{4.2.11}$$

As with (3.3.9)–(3.3.11), the first two terms in (4.2.8)–(4.2.10) are, respectively, the direct field and the image field. The third term represents the surface wave. Thus, with

$$B_{2\phi'}(\rho, z') = B_{2\phi'}^d(\rho, z') + B_{2\phi'}^i(\rho, z') + B_{2\phi'}^s(\rho, z'), \tag{4.2.12}$$

$$E_{2\rho}(\rho, z') = E_{2\rho}^d(\rho, z') + E_{2\rho}^i(\rho, z') + E_{2\rho}^s(\rho, z'), \tag{4.2.13}$$

$$E_{2z'}(\rho, z') = E_{2z'}^d(\rho, z') + E_{2z'}^i(\rho, z') + E_{2z'}^s(\rho, z'), \tag{4.2.14}$$

the exact direct-field components are

$$
\begin{aligned}
B^d_{2\phi'}(\rho, z') &= \frac{i\mu_0}{4\pi} \int_0^\infty \frac{e^{i\gamma_2|z'-d|}}{\gamma_2} J_1(\lambda\rho)\lambda^2\, d\lambda \\
&= \frac{i\mu_0}{4\pi} \frac{\rho}{2} I_7(k_2; \rho, z'-d) \\
&= -\frac{\mu_0}{4\pi} e^{ik_2 r_1} \left(\frac{ik_2}{r_1} - \frac{1}{r_1^2} \right) \left(\frac{\rho}{r_1} \right),
\end{aligned} \tag{4.2.15}
$$

$$
\begin{aligned}
E^d_{2\rho}(\rho, z') &= \frac{i\omega\mu_0}{4\pi k_2^2} \int_0^\infty \pm e^{i\gamma_2|z'-d|} J_1(\lambda\rho)\lambda^2\, d\lambda; \qquad \begin{cases} z' > d \\ 0 \le z' \le d \end{cases} \\
&= \frac{i\omega\mu_0}{4\pi k_2^2} \frac{\rho}{2} I_{22}(k_2; \rho, z'-d) \\
&= -\frac{\omega\mu_0}{4\pi k_2^2} e^{ik_2 r_1} \left(\frac{ik_2^2}{r_1} - \frac{3k_2}{r_1^2} - \frac{3i}{r_1^3} \right) \left(\frac{\rho}{r_1} \right) \left(\frac{z'-d}{r_1} \right),
\end{aligned} \tag{4.2.16}
$$

$$
\begin{aligned}
E^d_{2z'}(\rho, z') &= -\frac{\omega\mu_0}{4\pi k_2^2} \int_0^\infty \frac{e^{i\gamma_2|z'-d|}}{\gamma_2} J_0(\lambda\rho)\lambda^3\, d\lambda \\
&= -\frac{\omega\mu_0}{4\pi k_2^2} \frac{1}{2} [I_7(k_2; \rho, z'-d) + I_8(k_2; \rho, z'-d)] \\
&= \frac{\omega\mu_0}{4\pi k_2^2} e^{ik_2 r_1} \left[\frac{ik_2^2}{r_1} - \frac{k_2}{r_1^2} - \frac{i}{r_1^3} \right. \\
&\qquad \left. - \left(\frac{z'-d}{r_1} \right)^2 \left(\frac{ik_2^2}{r_1} - \frac{3k_2}{r_1^2} - \frac{3i}{r_1^3} \right) \right],
\end{aligned} \tag{4.2.17}
$$

where $r_1 = [\rho^2 + (z'-d)^2]^{1/2}$. The exact image fields are

$$
\begin{aligned}
B^i_{2\phi'}(\rho, z') &= \frac{i\mu_0}{4\pi} \int_0^\infty \frac{e^{i\gamma_2(z'+d)}}{\gamma_2} J_1(\lambda\rho)\lambda^2\, d\lambda \\
&= \frac{i\mu_0}{4\pi} \frac{\rho}{2} I_7(k_2; \rho, z'+d) \\
&= -\frac{\mu_0}{4\pi} e^{ik_2 r_2} \left(\frac{ik_2}{r_2} - \frac{1}{r_2^2} \right) \left(\frac{\rho}{r_2} \right),
\end{aligned} \tag{4.2.18}
$$

$$
\begin{aligned}
E^i_{2\rho}(\rho, z') &= \frac{i\omega\mu_0}{4\pi k_2^2} \int_0^\infty e^{i\gamma_2(z'+d)} J_1(\lambda\rho)\lambda^2\, d\lambda \\
&= \frac{i\omega\mu_0}{4\pi k_2^2} \frac{\rho}{2} I_{22}(k_2; \rho, z'+d) \\
&= -\frac{\omega\mu_0}{4\pi k_2^2} e^{ik_2 r_2} \left(\frac{ik_2^2}{r_2} - \frac{3k_2}{r_2^2} - \frac{3i}{r_2^3} \right) \left(\frac{\rho}{r_2} \right) \left(\frac{z'+d}{r_2} \right),
\end{aligned} \tag{4.2.19}
$$

$$
\begin{aligned}
E^i_{2z'}(\rho, z') &= -\frac{\omega\mu_0}{4\pi k_2^2} \int_0^\infty \frac{e^{i\gamma_2(z'+d)}}{\gamma_2} J_0(\lambda\rho)\lambda^3\, d\lambda \\
&= -\frac{\omega\mu_0}{4\pi k_2^2} \frac{1}{2} [I_7(k_2; \rho, z'+d) + I_8(k_2; \rho, z'+d)]
\end{aligned}
$$

$$
\begin{aligned}
&= \frac{\omega\mu_0}{4\pi k_2^2}\, e^{ik_2 r_2}\left[\frac{ik_2^2}{r_2} - \frac{k_2}{r_2^2} - \frac{i}{r_2^3}\right.\\
&\qquad\left. - \left(\frac{z'+d}{r_2}\right)^2\left(\frac{ik_2^2}{r_2} - \frac{3k_2}{r_2^2} - \frac{3i}{r_2^3}\right)\right],
\end{aligned}
\tag{4.2.20}
$$

where $r_2 = [\rho^2 + (z'+d)^2]^{1/2}$.

The surface-wave terms are

$$
B_{2\phi'}^s(\rho, z') = \frac{i\mu_0 k_1^2}{2\pi}\, G_{2\phi'}(\rho, z'+d), \tag{4.2.21a}
$$

$$
G_{2\phi'}(\rho, z'+d) = \int_0^\infty \left(\frac{1}{N} - \frac{1}{N_0}\right) e^{i\gamma_2(z'+d)} J_1(\lambda\rho)\lambda^2\, d\lambda, \tag{4.2.21b}
$$

$$
E_{2\rho}^s(\rho, z') = \frac{i\omega\mu_0 k_1^2}{2\pi k_2^2}\int_0^\infty \gamma_2\left(\frac{1}{N} - \frac{1}{N_0}\right) e^{i\gamma_2(z'+d)} J_1(\lambda\rho)\lambda^2\, d\lambda \tag{4.2.22a}
$$

$$
\begin{aligned}
&= -\frac{i\omega\mu_0}{2\pi k_1^2}\int_0^\infty \frac{\gamma_1}{\gamma_2}\, e^{i\gamma_2(z'+d)} J_1(\lambda\rho)\lambda^2\, d\lambda\\
&\quad - \frac{i\omega\mu_0}{2\pi}\int_0^\infty \gamma_1\left(\frac{1}{N} - \frac{1}{N_0}\right) e^{i\gamma_2(z'+d)} J_1(\lambda\rho)\lambda^2\, d\lambda
\end{aligned}
\tag{4.2.22b}
$$

$$
\begin{aligned}
&= -\frac{i\omega\mu_0}{2\pi k_1^2}\int_0^\infty \frac{\gamma_1}{\gamma_2}\, e^{i\gamma_2(z'+d)} J_1(\lambda\rho)\lambda^2\, d\lambda\\
&\quad - \frac{i\omega\mu_0 k_1}{2\pi}\, G_{2\phi'}(\rho, z'+d)\\
&= \frac{i\omega\mu_0}{2\pi k_1}\left[\left(\frac{\rho}{r_2}\right)\left(\frac{k_2}{r_2} + \frac{i}{r_2^2}\right) e^{ik_2 r_2}\right.\\
&\qquad\left. - e^{-k_1(z'+d)} e^{ik_1\rho}\left(\frac{1}{\rho^2} + \frac{3i}{2k_1\rho^3}\right)\right]\\
&\quad - \frac{i\omega\mu_0 k_1}{2\pi}\, G_{2\phi'}(\rho, z'+d),
\end{aligned}
\tag{4.2.22c}
$$

$$
E_{2z'}^s(\rho, z') = -\frac{\omega\mu_0 k_1^2}{2\pi k_2^2}\, G_{2z'}(\rho, z'+d), \tag{4.2.23a}
$$

$$
G_{2z'}(\rho, z'+d) = \int_0^\infty \left(\frac{1}{N} - \frac{1}{N_0}\right) e^{i\gamma_2(z'+d)} J_0(\lambda\rho)\lambda^3\, d\lambda. \tag{4.2.23b}
$$

The first integral in (4.2.22b) is evaluated in (3.7.20) with some changes in notation. Specifically, in (3.7.20) set $d = 0$ and replace $-z$ by $z'+d$ so that r_0 becomes r_2. In the second integral in (4.2.22b), the significant contributions to the integrand occur when $\lambda \sim k_2$ so that $\gamma_1 = (k_1^2 - \lambda^2)^{1/2} \sim k_1$. The integrals $G_{2\phi'}(\rho, z'+d)$ and $G_{2z'}(\rho, z'+d)$ are evaluated systematically in Appendix F. The results are summarized in the last section. Note that the integrated formulas are subject only to the condition

$$
|k_1^2| \gg |k_2^2| \quad \text{or} \quad |k_1| \geq 3|k_2|; \tag{4.2.24a}
$$

the restrictions

$$\rho^2 \gg |z|^2; \quad \rho^2 \gg d^2, \tag{4.2.24b}$$

have been eliminated so that the integrated formulas are valid for all values of ρ and z. The results are

$$B^s_{2\phi'}(\rho, z') = \frac{\mu_0 k_2^3}{2\pi k_1} e^{ik_2 r_2} \left(\frac{\pi}{k_2 r_2}\right)^{1/2} e^{-iP_2} \mathcal{F}(P_2), \tag{4.2.25}$$

$$E^s_{2\rho}(\rho, z') = \frac{\omega\mu_0}{2\pi k_1} e^{ik_2 r_2} \left[\left(\frac{\rho}{r_2}\right)\left(\frac{ik_2}{r_2} - \frac{1}{r_2^2}\right) - \frac{k_2^3}{k_1}\left(\frac{\pi}{k_2 r_2}\right)^{1/2} e^{-iP_2}\mathcal{F}(P_2)\right], \tag{4.2.26}$$

$$E^s_{2z'}(\rho, z') = -\frac{\omega\mu_0 k_2^2}{2\pi k_1} e^{ik_2 r_2} \left(\frac{\pi}{k_2 r_2}\right)^{1/2} \left(\frac{\rho}{r_2}\right) e^{-iP_2}\mathcal{F}(P_2), \tag{4.2.27}$$

where

$$P_2 = S_2\left(\frac{S_2 + Z' + D}{R}\right)^2 = \frac{k_2^3 r_2}{2k_1^2}\left(\frac{k_2 r_2 + k_1(z'+d)}{k_2\rho}\right)^2, \tag{4.2.28a}$$

$$S_2 = \frac{k_2^3 r_2}{2k_1^2}; \qquad R = \frac{k_2^3 \rho}{2k_1^2}; \qquad Z' = \frac{k_2^2 z'}{2k_1}; \qquad D = \frac{k_2^2 d}{2k_1}, \tag{4.2.28b}$$

and $r_2 = [\rho^2 + (z'+d)^2]^{1/2}$. Also,

$$\mathcal{F}(P_2) = \int_{P_2}^{\infty} \frac{e^{it}}{(2\pi t)^{1/2}}\, dt = \tfrac{1}{2}(1+i) - C_2(P_2) - iS_2(P_2), \tag{4.2.29}$$

where $C_2(P_2) + iS_2(P_2)$ is the Fresnel integral. Note that, in (4.2.22c), the term multiplied by $\exp[-k_1(z'+d)]$ is needed to satisfy the boundary condition but does not contribute to the propagating field; it has been omitted in (4.2.26).

The complete field in Region 2 is the sum of the direct, reflected, and surface-wave parts as defined in (4.2.12)–(4.2.14). Note that when the dipole is below the surface in the earth—as shown in Fig. 4.2.1a—the surface wave is known as a lateral wave. When the dipole is moved across the boundary into the air—as shown in Fig. 4.2.1b—the surface wave is generally called the Norton surface wave. This is a matter of terminology, not structure, since the only difference in the field when $d \to 0$ (dipole in air) and $d \to 0$ (dipole in earth) is a factor k_1^2/k_2^2. Subject to condition (4.2.24a) on the surface-wave terms, the complete field is

$$B_{2\phi'}(\rho, z') = -\frac{\mu_0}{2\pi}\left[\frac{e^{ik_2 r_1}}{2}\left(\frac{\rho}{r_1}\right)\left(\frac{ik_2}{r_1} - \frac{1}{r_1^2}\right) + \frac{e^{ik_2 r_2}}{2}\left(\frac{\rho}{r_2}\right)\left(\frac{ik_2}{r_2} - \frac{1}{r_2^2}\right)\right.$$

$$- e^{ik_2r_2}\frac{k_2^3}{k_1}\left(\frac{\pi}{k_2r_2}\right)^{1/2} e^{-iP_2}\mathcal{F}(P_2)\Bigg], \tag{4.2.30}$$

$$\begin{aligned}E_{2\rho}(\rho, z') = -\frac{\omega\mu_0}{2\pi k_2}\Bigg\{&\frac{e^{ik_2r_1}}{2}\left(\frac{\rho}{r_1}\right)\left(\frac{z'-d}{r_1}\right)\left(\frac{ik_2}{r_1} - \frac{3}{r_1^2} - \frac{3i}{k_2r_1^3}\right)\\ &+ \frac{e^{ik_2r_2}}{2}\left(\frac{\rho}{r_2}\right)\left(\frac{z'+d}{r_2}\right)\left(\frac{ik_2}{r_2} - \frac{3}{r_2^2} - \frac{3i}{k_2r_2^3}\right)\\ &- \frac{k_2}{k_1}e^{ik_2r_2}\Bigg[\left(\frac{\rho}{r_2}\right)\left(\frac{ik_2}{r_2} - \frac{1}{r_2^2}\right)\\ &\qquad - \frac{k_2^3}{k_1}\left(\frac{\pi}{k_2r_2}\right)^{1/2} e^{-iP_2}\mathcal{F}(P_2)\Bigg]\Bigg\},\end{aligned} \tag{4.2.31}$$

$$\begin{aligned}E_{2z'}(\rho, z') = \frac{\omega\mu_0}{2\pi k_2}\Bigg\{&\frac{e^{ik_2r_1}}{2}\Bigg[\frac{ik_2}{r_1} - \frac{1}{r_1^2} - \frac{i}{k_2r_1^3}\\ &\qquad - \left(\frac{z'-d}{r_1}\right)^2\left(\frac{ik_2}{r_1} - \frac{3}{r_1^2} - \frac{3i}{k_2r_1^3}\right)\Bigg]\\ &+ \frac{e^{ik_2r_2}}{2}\Bigg[\frac{ik_2}{r_2} - \frac{1}{r_2^2} - \frac{i}{k_2r_2^3}\\ &\qquad - \left(\frac{z'+d}{r_2}\right)^2\left(\frac{ik_2}{r_2} - \frac{3}{r_2^2} - \frac{3i}{k_2r_2^3}\right)\Bigg]\\ &- e^{ik_2r_2}\frac{k_2^3}{k_1}\left(\frac{\pi}{k_2r_2}\right)^{1/2}\left(\frac{\rho}{r_2}\right)e^{-iP_2}\mathcal{F}(P_2)\Bigg\},\end{aligned} \tag{4.2.32}$$

where $r_1 = [\rho^2 + (z'-d)^2]^{1/2}$ and $r_2 = [\rho^2 + (z'+d)^2]^{1/2}$. Note that when Region 1 is a perfect conductor, $\sigma_1 \to \infty$ and $k_1 \to \infty$, so that the surface-wave terms with k_1 in the denominator vanish and the entire field is given by the direct and image terms. This is the complete field, not just the far field which is given by the $1/r$ terms.

The field on the boundary where $z' = 0$ is

$$\begin{aligned}B_{2\phi'}(\rho, 0) = -\frac{\mu_0}{2\pi}e^{ik_2r_d}\Bigg[&\left(\frac{\rho}{r_d}\right)\left(\frac{ik_2}{r_d} - \frac{1}{r_d^2}\right)\\ &- \frac{k_2^3}{k_1}\left(\frac{\pi}{k_2r_d}\right)^{1/2} e^{-iP_d}\mathcal{F}(P_d)\Bigg],\end{aligned} \tag{4.2.33}$$

$$\begin{aligned}E_{2\rho}(\rho, 0) = \frac{\omega\mu_0}{2\pi k_1}e^{ik_2r_d}\Bigg[&\left(\frac{\rho}{r_d}\right)\left(\frac{ik_2}{r_d} - \frac{1}{r_d^2}\right)\\ &- \frac{k_2^3}{k_1}\left(\frac{\pi}{k_2r_d}\right)^{1/2} e^{-iP_d}\mathcal{F}(P_d)\Bigg],\end{aligned} \tag{4.2.34}$$

$$E_{2z'}(\rho, 0) = \frac{\omega\mu_0}{2\pi k_2}e^{ik_2r_d}\Bigg[\frac{ik_2}{r_d} - \frac{1}{r_d^2} - \frac{i}{k_2r_d^3}$$

$$- \left(\frac{d}{r_d}\right)^2 \left(\frac{ik_2}{r_d} - \frac{3}{r_d^2} - \frac{3i}{k_2 r_d^3}\right)$$
$$- \frac{k_2^3}{k_1}\left(\frac{\pi}{k_2 r_d}\right)^{1/2}\left(\frac{\rho}{r_d}\right) e^{-iP_d}\mathcal{F}(P_d)\Bigg], \tag{4.2.35}$$

where

$$P_d = S_d\left(\frac{S_d + D}{R}\right)^2 = \frac{k_2^3 r_d}{2k_1^2}\left(\frac{k_2 r_d + k_1 d}{k_2 \rho}\right)^2; \qquad S_2 = \frac{k_2^3 r_d}{2k_1^2}, \tag{4.2.36}$$

R and D are defined in (4.2.28b), and $r_d = (\rho^2 + d^2)^{1/2}$.

Of particular interest is the field at any point in Region 2 (air) when the dipole is also in Region 2 but on the surface of Region 1 (earth or sea), i.e., $d \to 0$ from above. The determination of this field was the original purpose of the many investigations in the period from Sommerfeld to Norton. This field is given by (4.2.30)–(4.2.32) with $d = 0$. It is

$$B_{2\phi'}(\rho, z') = -\frac{\mu_0}{2\pi} e^{ik_2 r_0}\left[\left(\frac{\rho}{r_0}\right)\left(\frac{ik_2}{r_0} - \frac{1}{r_0^2}\right)\right.$$
$$\left. - \frac{k_2^3}{k_1}\left(\frac{\pi}{k_2 r_0}\right)^{1/2} e^{-iP_0}\mathcal{F}(P_0)\right], \tag{4.2.37}$$

$$E_{2\rho}(\rho, z') = -\frac{\omega\mu_0}{2\pi k_2} e^{ik_2 r_0}$$
$$\times \left\{\left(\frac{\rho z'}{r_0^2}\right)\left(\frac{ik_2}{r_0} - \frac{3}{r_0^2} - \frac{3i}{k_2 r_0^3}\right)\right.$$
$$- \frac{k_2}{k_1}\left[\left(\frac{\rho}{r_0}\right)\left(\frac{ik_2}{r_0} - \frac{1}{r_0^2}\right)\right.$$
$$\left.\left. - \frac{k_2^3}{k_1}\left(\frac{\pi}{k_2 r_0}\right)^{1/2} e^{-iP_0}\mathcal{F}(P_0)\right]\right\}, \tag{4.2.38}$$

$$E_{2z'}(\rho, z') = \frac{\omega\mu_0}{2\pi k_2} e^{ik_2 r_0}$$
$$\times \left[\frac{ik_2}{r_0} - \frac{1}{r_0^2} - \frac{i}{k_2 r_0^3}\right.$$
$$- \left(\frac{z'}{r_0}\right)^2\left(\frac{ik_2}{r_0} - \frac{3}{r_0^2} - \frac{3i}{k_2 r_0^3}\right)$$
$$\left. - \frac{k_2^3}{k_1}\left(\frac{\pi}{k_2 r_0}\right)^{1/2}\left(\frac{\rho}{r_0}\right) e^{-iP_0}\mathcal{F}(P_0)\right], \tag{4.2.39}$$

where

$$P_0 = S_0\left(\frac{S_0 + Z'}{R}\right)^2 = \frac{k_2^3 r_0}{2k_1^2}\left(\frac{k_2 r_0 + k_1 z'}{k_2 \rho}\right)^2; \qquad S_0 = \frac{k_2^3 r_0}{2k_1^2}, \tag{4.2.40}$$

R and Z' are defined in (4.2.28b), and $r_0 = (\rho^2 + z'^2)^{1/2}$.

The field in (4.2.37)–(4.2.39) is often convenient in the spherical coordinates r_0, Θ, Φ. With the formulas

$$E_{2r}(r_0,\Theta) = E_{2\rho}(\rho,z')\sin\Theta + E_{2z'}(\rho,z')\cos\Theta, \tag{4.2.41a}$$

$$E_{2\Theta}(r_0,\Theta) = E_{2\rho}(\rho,z')\cos\Theta - E_{2z'}(\rho,z')\sin\Theta, \tag{4.2.41b}$$

$$B_{2\Phi}(r_0,\Theta) = B_{2\phi'}(\rho,z'), \tag{4.2.41c}$$

together with $\sin\Theta = \rho/r_0$ and $\cos\Theta = z'/r_0$, the components of the field are

$$B_{2\Phi}(r_0,\Theta) = -\frac{\mu_0}{2\pi}\,e^{ik_2r_0}\left[\sin\Theta\left(\frac{ik_2}{r_0}-\frac{1}{r_0^2}\right) - \frac{k_2^3}{k_1}\left(\frac{\pi}{k_2r_0}\right)^{1/2} e^{-iP_0}\mathcal{F}(P_0)\right], \tag{4.2.42}$$

$$\begin{aligned}E_{2r}(r_0,\Theta) &= \frac{\omega\mu_0}{2\pi k_2}\,e^{ik_2r_0}\\ &\times\left\{\cos\Theta\left[\frac{2}{r_0^2}+\frac{2i}{k_2r_0^3} - \sin\Theta\,\frac{k_2^3}{k_1}\left(\frac{\pi}{k_2r_0}\right)^{1/2} e^{-iP_0}\mathcal{F}(P_0)\right]\right.\\ &\quad\left. + \sin\Theta\,\frac{k_2}{k_1}\left[\sin\Theta\left(\frac{ik_2}{r_0}-\frac{1}{r_0^2}\right) - \frac{k_2^3}{k_1}\left(\frac{\pi}{k_2r_0}\right)^{1/2} e^{-iP_0}\mathcal{F}(P_0)\right]\right\},\end{aligned} \tag{4.2.43}$$

$$\begin{aligned}E_{2\Theta}(r_0,\Theta) &= -\frac{\omega\mu_0}{2\pi k_2}\,e^{ik_2r_0}\\ &\times\left\{\sin\Theta\left[\frac{ik_2}{r_0}-\frac{1}{r_0^2}-\frac{i}{k_2r_0^3} - \sin\Theta\,\frac{k_2^3}{k_1}\left(\frac{\pi}{k_2r_0}\right)^{1/2} e^{-iP_0}\mathcal{F}(P_0)\right]\right.\\ &\quad\left. - \cos\Theta\,\frac{k_2}{k_1}\left[\sin\Theta\left(\frac{ik_2}{r_0}-\frac{1}{r_0^2}\right) - \frac{k_2^3}{k_1}\left(\frac{\pi}{k_2r_0}\right)^{1/2} e^{-iP_0}\mathcal{F}(P_0)\right]\right\},\end{aligned} \tag{4.2.44}$$

where P_0 is given in (4.2.40). At the boundary where $z' = 0$ and $\Theta = \pi/2$, these formulas reduce to

$$-\frac{\omega}{k_1}\,B_{2\Phi}(r_0,\pi/2) = E_{2r}(r_0,\pi/2) = \frac{\omega\mu_0}{2\pi k_1}\,e^{ik_2\rho}f(\rho;k_1,k_2), \tag{4.2.45}$$

$$E_{2\Theta}(r_0,\pi/2) = -\frac{\omega\mu_0}{2\pi k_2}\,e^{ik_2\rho}g(\rho;k_1,k_2), \tag{4.2.46}$$

where

$$g(\rho; k_1, k_2) = f(\rho; k_1, k_2) - \frac{i}{k_2\rho^3}$$
$$= \frac{ik_2}{\rho} - \frac{1}{\rho^2} - \frac{i}{k_2\rho^3} - \frac{k_2^3}{k_1}\left(\frac{\pi}{k_2\rho}\right)^{1/2} e^{-iR}\mathcal{F}(R), \tag{4.2.47}$$

and $R = k_2^3\rho/2k_1^2$. Note that at sufficient distance where

$$k_2\rho \geq \frac{8|k_1^2|}{k_2^2} \quad \text{or} \quad |R| \geq 4, \tag{4.2.48}$$

the Fresnel-integral term becomes

$$-\frac{k_2^3}{k_1}\left(\frac{\pi}{k_2\rho}\right)^{1/2} e^{-iR}\mathcal{F}(R) \to -\frac{ik_2}{\rho} - \frac{k_1^2}{k_2^2\rho^2}, \tag{4.2.49}$$

so that

$$g(\rho; k_1, k_2) \sim f(\rho; k_1, k_2) = -\frac{1}{\rho^2}\left(\frac{k_1^2}{k_2^2} + 1\right) \sim -\frac{k_1^2}{k_2^2\rho^2}. \tag{4.2.50}$$

Evidently, the Fresnel-integral term cancels the ik_2/ρ term and leaves the term $-k_1^2/k_2^2\rho^2$. The entire far field along the boundary—as defined by (4.2.48)—is given by (4.2.45) and (4.2.46) with $g(\rho; k_1, k_2)$ and $f(\rho; k_1, k_2)$ replaced by $-k_1^2/k_2^2\rho^2$. That is,

$$\frac{\omega}{k_2} B_{2\Phi}^r(r_0, \pi/2) = -\frac{k_1}{k_2} E_{2r}^r(r_0, \pi/2)$$
$$= E_{2\Theta}^r(r_0, \pi/2) = \frac{\omega\mu_0 k_1^2}{2\pi k_2^3} \frac{e^{ik_2\rho}}{\rho^2}. \tag{4.2.51}$$

The large-argument approximation of the Fresnel-integral terms in (4.2.42)–(4.2.44) can also be used. This is

$$T \equiv \frac{k_2^3}{k_1}\left(\frac{\pi}{k_2 r_0}\right)^{1/2} e^{-iP_0}\mathcal{F}(P_0)$$
$$\to T^r = \frac{k_2^3}{k_1}\left(\frac{\pi}{k_2 r_0}\right)^{1/2} \frac{1}{(2\pi P_0)^{1/2}}\left(\frac{1}{2P_0} + i\right), \tag{4.2.52}$$

when $|P_0| \geq 4$. With (4.2.40), this becomes

$$T^r = \frac{k_1^2\rho^3}{k_2^2 r_0^5[1 + (k_1 z'/k_2 r_0)]^3} + \frac{ik_2\rho}{r_0^2[1 + (k_1 z'/k_2 r_0)]}. \tag{4.2.53}$$

In the transformations (4.2.49) and (4.2.52) use is made of the relation

$$C_2(p) + iS_2(p) = \tfrac{1}{2}(1+i) - \sqrt{\tfrac{1}{2}i}\, e^{ip} w(\sqrt{ip}), \tag{4.2.54}$$

and the asymptotic expansion of the error function,

$$w(x) = e^{-x^2}\,\mathrm{erfc}(-ix) \sim \frac{1}{x\sqrt{\pi}}\left(1 + \frac{1}{2x^2} + \frac{3}{(2x^2)^2} + \cdots\right). \tag{4.2.55}$$

In spherical coordinates with $\rho = r_0 \sin\Theta$ and $z' = r_0 \cos\Theta$,

$$T^r(r_0, \Theta) = \frac{ik_2 \sin\Theta}{r_0[1 + (k_1/k_2)\cos\Theta]} + \frac{k_1^2 \sin^3\Theta}{k_2^2 r_0^2 [1 + (k_1/k_2)\cos\Theta]^3}. \tag{4.2.56}$$

Note that when $\Theta = \pi/2$,

$$T^r(r_0, \pi/2) = \frac{k_1^2}{k_2^2 r_0^2} + \frac{ik_2}{r_0}; \qquad r_0 = \rho. \tag{4.2.57}$$

With these formulas the far-field expressions for the electromagnetic field of a unit vertical dipole in air on the surface $z' = 0$ of the earth can be defined subject to the far-field condition $|k_2^3 r_0/k_1^2| > 8$. They necessarily include the $1/r_0^2$ term from the Fresnel integral. Note that the additional $1/r_0^2$ term is retained in $E_{2r}(r_0, \Theta)$ since the $1/r_0$ terms cancel.

$$B_{2\Phi}^r(r_0, \Theta) = -\frac{\mu_0}{2\pi}\, e^{ik_2 r_0}\left(\frac{ik_2}{r_0}\sin\Theta - T^r(r_0, \Theta)\right), \tag{4.2.58}$$

$$\begin{aligned} E_{2r}^r(r_0, \Theta) = -\frac{\omega\mu_0}{2\pi k_2}\, e^{ik_2 r_0}\Bigg[&\cos\Theta\left(T^r(r_0, \Theta)\sin\Theta - \frac{2}{r_0^2}\right) \\ &- \sin\Theta\,\frac{k_2}{k_1}\left(\frac{ik_2 \sin\Theta}{r_0} - T^r(r_0, \Theta)\right)\Bigg], \end{aligned} \tag{4.2.59}$$

$$\begin{aligned} E_{2\Theta}^r(r_0, \Theta) = -\frac{\omega\mu_0}{2\pi k_2}\, e^{ik_2 r_0}\Bigg[&\sin\Theta\left(\frac{ik_2}{r_0} - T^r(r_0, \Theta)\sin\Theta\right) \\ &- \cos\Theta\,\frac{k_2}{k_1}\left(\frac{ik_2 \sin\Theta}{r_0} - T^r(r_0, \Theta)\right)\Bigg]. \end{aligned} \tag{4.2.60}$$

These expressions give the complete far field of the unit vertical dipole. They can be rearranged with (4.2.56) to separate the $1/r_0$ and $1/r_0^2$ terms as follows:

$$\begin{aligned} B_{2\Phi}^r(r_0, \Theta) = -\frac{\mu_0}{2\pi}\, e^{ik_2 r_0}\Bigg[&\frac{ik_2}{r_0}\left(\frac{k_1 \sin\Theta\cos\Theta}{k_2 + k_1\cos\Theta}\right) \\ &- \frac{k_1^2 \sin^3\Theta}{k_2^2 r_0^2 [1 + (k_1/k_2)\cos\Theta]^3}\Bigg], \end{aligned} \tag{4.2.61}$$

$$\begin{aligned} E_{2r}^r(r_0, \Theta) = \frac{\omega\mu_0}{2\pi k_2}\, &e^{ik_2 r_0} \\ \times &\left\{\frac{k_2}{k_1}\left[\frac{ik_2 \sin^2\Theta}{r_0} - \left(1 + \frac{k_1}{k_2}\cos\Theta\right)T^r(r_0, \Theta)\sin\Theta\right]\right. \\ &\left. + \frac{2}{r_0^2}\cos\Theta\right\} \end{aligned}$$

$$= \frac{\omega\mu_0}{2\pi k_2} e^{ik_2r_0}\left(\frac{2}{r_0^2}\cos\Theta - \frac{k_1 \sin^4\Theta}{k_2 r_0^2[1 + (k_1/k_2)\cos\Theta]^2}\right), \tag{4.2.62}$$

$$E_{2\Theta}^r(r_0, \Theta) = -\frac{\omega\mu_0}{2\pi k_2} e^{ik_2r_0}\left[\frac{ik_2}{r_0}\left(\frac{k_1 \sin\Theta\cos\Theta}{k_2 + k_1\cos\Theta}\right)\right.$$
$$\left. - \frac{k_1^2[\sin^2\Theta - (k_2/k_1)\cos\Theta]\sin^3\Theta}{k_2^2 r_0^2[1 + (k_1/k_2)\cos\Theta]^3}\right]. \tag{4.2.63}$$

Note that, just as with the isolated dipole, $E_{2r}^r(r_0, \Theta)$ can have no $1/r_0$ terms. Unlike the isolated dipole for which $E_{2r}^r(r_0, \pi/2) = 0$, $E_{2r}^r(r_0, \pi/2)$ in (4.2.62) is a relatively very large maximum value. It is contributed entirely by the $1/r_0^2$ terms of the surface wave.

Graphs of the magnitude of $E_{2\Theta}^r(r_0, \Theta)$ for a unit vertical dipole as calculated from (4.2.63) are shown in Fig. 4.2.2 for four different Regions 1. These are (a) salt water with $\sigma_1 = 4$ S/m, $\epsilon_{1r} = 80$; (b) wet earth with $\sigma_1 = 0.4$ S/m, $\epsilon_{1r} = 12$; (c) dry earth with $\sigma_1 = 0.04$ S/m, $\epsilon_{1r} = 8$; and (d) lake water with $\sigma_1 = 0.004$ S/m, $\epsilon_{1r} = 80$. The graph for a perfectly conducting earth ($\sigma_1 \to \infty$) is also shown in dashed line. The frequency is $f = 10$ MHz and the radial distance is $r_0 = 500$ km. The magnitude of $E_{2\Theta}^r(r_0, \pi/2)$ is due entirely to the surface wave. The numerical values of the maxima of the field, of the angle $\Theta_{\max}$ where this occurs, and of the field with $\Theta = \pi/2$ are also presented in the figure. The range in which the surface wave gradually cancels the $1/r_0$ terms of the direct and image fields and contributes the ultimately dominant $(k_1/k_2r_0)^2$ term as Θ increases is from $\Theta_{\max}$ to 90°. This important range is only 11°.5 over salt water and close to 25° over dry earth or lake water. Its principal part is shown in Fig. 4.2.3 with $80° \leq \Theta \leq 90°$ on a linear scale and $|E_{2\Theta}^r(r_0, \Theta)|$ on a logarithmic scale.

It is significant to note that the $1/r_0$ term in (4.2.63) is identically the far field given in (1.3.1a,b) in terms of the plane-wave reflection coefficient f_{er} in (1.3.3). That is, with $d = 0$,

$$-\frac{i\omega\mu_0}{4\pi}\frac{e^{ik_2r_0}}{r_0}(1 + f_{er})\sin\Theta = -\frac{i\omega\mu_0}{4\pi}\frac{e^{ik_2r_0}}{r_0}\frac{2k_1\cos\Theta}{k_1\cos\Theta + k_2}\sin\Theta$$
$$= -\frac{\omega\mu_0}{2\pi k_2}e^{ik_2r_0}\frac{ik_2}{r_0}\left(\frac{k_1\sin\Theta\cos\Theta}{k_2 + k_1\cos\Theta}\right). \tag{4.2.64}$$

In the formula (1.3.3) for the reflection coefficient, $\sin^2\Theta$ is neglected compared to (k_1^2/k_2^2) since $k_2^2 \ll |k_1^2|$.

Graphs of the magnitude of $E_{2r}^r(r_0, \Theta)$ as calculated from (4.2.62) are shown in Fig. 4.2.4 for the same four different Regions 1. Note that $E_{2r}^r(r_0, \Theta)$ is composed entirely of $1/r_0^2$ and negligible $1/r_0^3$ terms. The scales for the four regions are adjusted so that the surface-wave values at $\Theta = \pi/2$ are equal. The actual numerical values are given in tabular form. Note that when the four scales are the same, the fields at $\Theta = 0°$ are all equal.

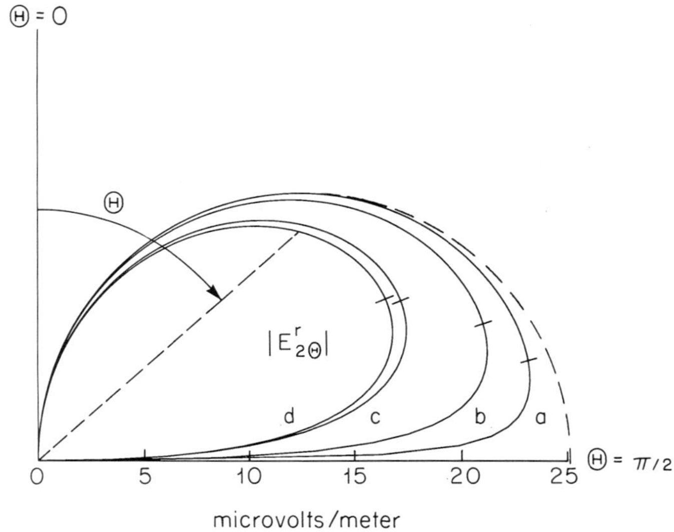

	σ_1	ϵ_{1r}	$\Theta_{\max}$	$E^r_{2\Theta}(r_0,\Theta)_{\max}$	$E^r_{2\Theta}(r_0,\pi/2)$
(a)	4.0 S/m	80	78°.5	2.36×10^{-5} V/m	1.73×10^{-6} V/m
(b)	0.4	12	73°.0	2.20×10^{-5}	1.73×10^{-7}
(c)	0.04	8	66°.0	1.87×10^{-5}	1.74×10^{-8}
(d)	0.004	80	65°.5	1.80×10^{-5}	1.93×10^{-8}

Figure 4.2.2. Complete field of $|E^r_{2\Theta}(r_0,\Theta)|$ for vertical dipole in air on boundary between air and (a) sea water, (b) wet earth, (c) dry earth, and (d) lake water. Frequency $f = 10$ MHz; radial distance $r_0 = 500$ km. The dashed curve is for $\sigma_1 = \infty$.

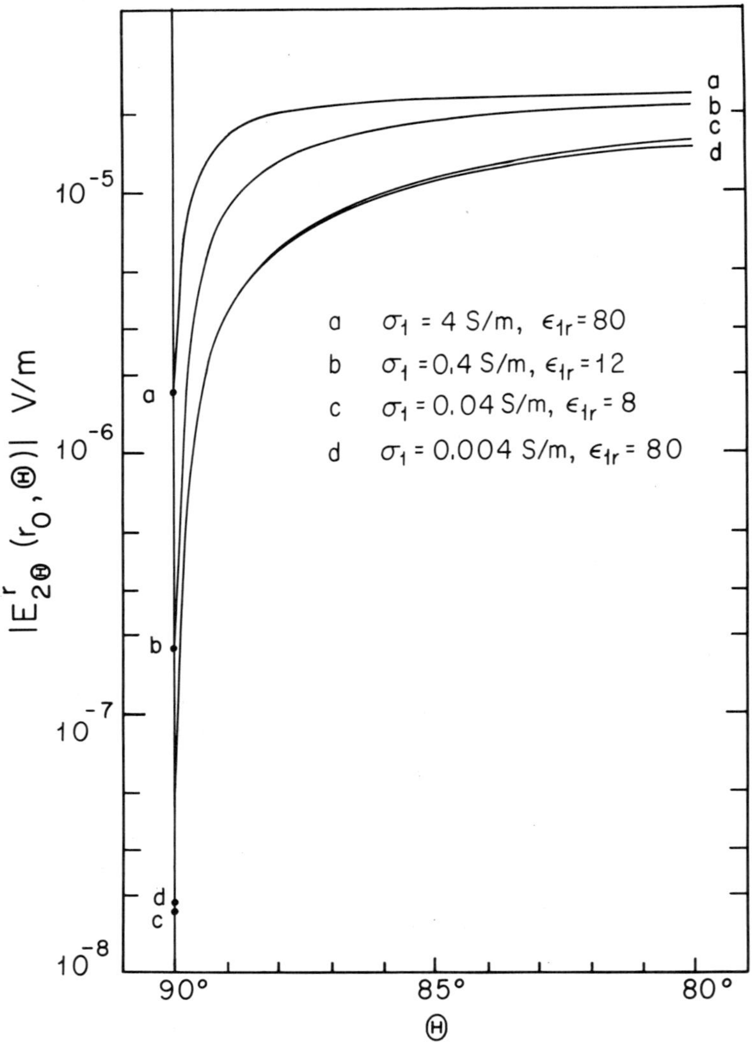

Figure 4.2.3. Enlarged section of Fig. 4.2.2 near $\Theta = 90^\circ$.

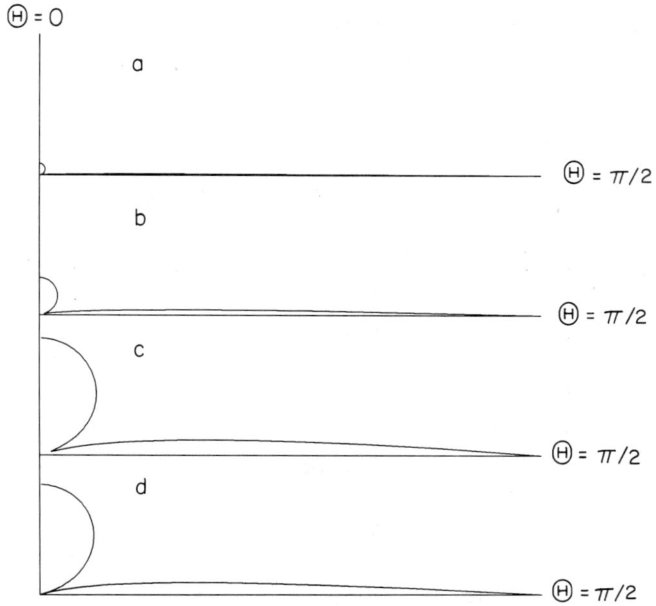

	σ_1	ϵ_{1r}	$E^r_{2r}(r_0, \pi/2)$
(a)	4.0 S/m	80	2.04×10^{-8} V/m
(b)	0.4	12	6.44×10^{-9}
(c)	0.04	8	2.04×10^{-9}
(d)	0.004	80	2.15×10^{-9}

Figure 4.2.4. Complete field of $|E^r_{2r}(r_0, \Theta)|$ for vertical dipole in air on boundary between air and (a) sea water, (b) wet earth, (c) dry earth, and (d) lake water. Frequency $f = 10$ MHz; radial distance $r_0 = 500$ km.

It is of interest to mention that when the integrals (4.2.21b) and (4.2.23b) are evaluated subject to the conditions (4.2.24b) as well as (4.2.24a), the Fresnel-integral terms are multiplied by the exponential factor $\exp[ik_2(\rho + z'^2/2\rho)]$ instead of $\exp(ik_2 r_0)$. If this is accepted as accurate instead of as an approximation of $\exp(ik_2 r_0)$, it leads to an apparent interference pattern due to the superposition of the direct and surface waves. The more general evaluation of (4.2.21b) and (4.2.23b) with no restrictions on ρ and z' shows that the quantity $\rho + z'^2/2\rho$ is actually an approximation of $r_0 = (\rho^2 + z'^2)^{1/2}$ that is valid only when $z'^2/\rho^2 \ll 1$ or Θ is close to 90°. However, the relatively simple picture of the surface waves contained in (4.2.61)–(4.2.63) and Figs. 4.2.2 and 4.2.4 is valid only at large distances from the source—beyond the Sommerfeld numerical distance, i.e., $|k_2^3 r_0/2k_1^2| \geq 4$. Nearer the source than this, the surface-wave term T in (4.2.52) does not have the simple form (4.2.53) so that the surface wave and its interaction with the direct wave are complicated. However, at distances much smaller than the numerical distance, the surface-wave term T is negligible.

4.3 Division of Radiated Power Between Air and Earth Half-Spaces

The power P radiated by a vertical dipole at a height d in air (see Fig. 4.2.1b) over a plane earth can be separated into two parts: a part P_a that remains in the upper half-space (Region 2, air; $z' \geq 0$), and a part P_g that is transferred into the lower half-space (Region 1, earth; $z' \leq 0$). These are defined in terms of the vertical component of the Poynting vector, viz.,

$$S_{2z'}(\rho, z') = \frac{1}{2\mu_0} E_{2\rho}(\rho, z') B^*_{2\phi'}(\rho, z'); \qquad z' \geq 0, \tag{4.3.1}$$

where the asterisk denotes the complex conjugate. Thus,

$$\left.\begin{matrix} P_a \\ P_g \end{matrix}\right\} = \pm \operatorname{Re} 2\pi \int_0^\infty S_{2z'}(\rho, z'_\pm)\rho\, d\rho. \tag{4.3.2}$$

In (4.3.2), z'_+ means $z' > d$ and z'_- means $0 \leq z' \leq d$. The components of the electromagnetic field are given by (4.2.8)–(4.2.10). They are

$$B_{2\phi'}(\rho, z') = \frac{i\mu_0}{4\pi} \int_0^\infty \gamma_2^{-1}[e^{i\gamma_2|z'-d|} + e^{i\gamma_2(z'+d)}(1 - G_{21})] \times J_1(\lambda\rho)\lambda^2\, d\lambda, \tag{4.3.3}$$

$$E_{2\rho}(\rho, z') = \frac{i\omega\mu_0}{4\pi k_2^2} \int_0^\infty [\pm e^{i\gamma_2|z'-d|} + e^{i\gamma_2(z'+d)}(1 - G_{21})] \times J_1(\lambda\rho)\lambda^2\, d\lambda; \qquad \begin{cases} z' > d \\ 0 \leq z' \leq d \end{cases}, \tag{4.3.4}$$

$$E_{2z'}(\rho, z') = -\frac{\omega\mu_0}{4\pi k_2^2} \int_0^\infty \gamma_2^{-1}[e^{i\gamma_2|z'-d|} + e^{i\gamma_2(z'+d)}(1 - G_{21})] \times J_0(\lambda\rho)\lambda^3\, d\lambda, \tag{4.3.5}$$

where, with the time dependence $e^{-i\omega t}$,

$$k_1 = \beta_1 + i\alpha_1 = \omega\mu_0^{1/2}\left(\epsilon_1 + \frac{i\sigma_1}{\omega}\right)^{1/2}; \qquad k_2 = \omega\mu_0^{1/2}\epsilon_0^{1/2}, \tag{4.3.6}$$

$$G_{21} = 2k_2^2\gamma_1 N^{-1};$$
$$N = k_1^2\gamma_2 + k_2^2\gamma_1; \qquad \gamma_j = (k_j^2 - \lambda^2)^{1/2}, \quad j = 1, 2. \tag{4.3.7}$$

Radio engineers are interested in a quantity called the *radiation efficiency* and defined as follows:

$$\eta = \frac{P_a}{P_a + P_g}. \tag{4.3.8}$$

This is the fraction of the power radiated by the dipole that remains in the upper half-space. Since it appears superficially obvious that radio transmission between antennas in the air depends on the power that remains in the upper half-space, it seems reasonable to assume that the most desirable antenna is one that has the highest radiation efficiency. Actually, this ignores the important fact that power ultimately transferred to the earth at large radial distances is available in the air at all shorter distances.

When (4.3.3) and (4.3.4) are substituted in (4.3.1), the following result is obtained:

$$\begin{aligned} S_{2z'}(\rho, z'_\pm) = \frac{\omega\mu_0}{32\pi^2 k_2^2} & \int_0^\infty f_\rho(\gamma_2, z'_\pm) J_1(\lambda\rho)\lambda^2 \, d\lambda \\ & \times \int_0^\infty f^*_{\phi'}(\gamma_2^*, z'_\pm) J_1(l\rho) l^2 \, dl. \end{aligned} \tag{4.3.9}$$

For $z' = z'_+ > d$,

$$f_\rho(\gamma_2, z'_+) = e^{i\gamma_2(z'-d)} + (1 - G_{21}) e^{i\gamma_2(z'+d)}, \tag{4.3.10}$$

$$f^*_{\phi'}(\gamma_2^*, z'_+) = \frac{1}{\gamma_2^*} [e^{-i\gamma_2^*(z'-d)} + (1 - G^*_{21}) e^{-i\gamma_2^*(z'+d)}]. \tag{4.3.11}$$

For $z' = z'_-$, $0 \le z' \le d$,

$$f_\rho(\gamma_2, z'_-) = -e^{i\gamma_2(d-z')} + (1 - G_{21}) e^{i\gamma_2(d+z')}, \tag{4.3.12}$$

$$f^*_{\phi'}(\gamma_2^*, z'_-) = \frac{1}{\gamma_2^*} [e^{-i\gamma_2^*(d-z')} + (1 - G^*_{21}) e^{-i\gamma_2^*(d+z')}]. \tag{4.3.13}$$

Note that $\gamma_2 = (k_2^2 - \lambda^2)^{1/2}$ when $0 \le \lambda \le k_2$ and $\gamma_2 = i(\lambda^2 - k_2^2)^{1/2} = i\nu_2$ when $k_2 < \lambda$. In this latter range, $\gamma_2^* = -i\nu_2$, $G^*_{21} = 2k_2^2\gamma_1^*/N^*$, $\gamma_1^* = (k_1^{*2} - \lambda^2)^{1/2}$.

When (4.3.9) is substituted into (4.3.2), this becomes

$$\begin{aligned} \left.\begin{matrix} P_a \\ P_g \end{matrix}\right\} = \pm \operatorname{Re} \frac{\omega\mu_0}{16\pi k_2^2} & \int_0^\infty d\lambda\, \lambda^2 \int_0^\infty dl\, l^2 \int_0^\infty d\rho\, \rho \\ & \times [J_1(\lambda\rho) J_1(l\rho)] f_\rho(\gamma_2, z'_\pm) f^*_{\phi'}(\gamma_2^*, z'_\pm). \end{aligned} \tag{4.3.14}$$

Here the integration with respect to ρ can be carried out with the following orthogonality relation:

$$\int_0^\infty J_1(at) J_1(bt) t \, dt = \frac{1}{b} \delta(b - a), \tag{4.3.15}$$

where $\delta(b-a)$ is the delta function. (A more rigorous derivation of this formula makes use of the Cartesian components of the field. The cylindrical components are convenient here.) With (4.3.15), (4.3.14) becomes

$$\left.\begin{matrix} P_a \\ P_g \end{matrix}\right\} = \pm \operatorname{Re} \frac{\omega\mu_0}{16\pi k_2^2} \int_0^\infty f_\rho(\gamma_2, z'_\pm) f^*_{\phi'}(\gamma_2^*, z'_\pm)\lambda^3\, d\lambda, \tag{4.3.16}$$

where $f_\rho(\gamma_2, z'_\pm)$ and $f^*_{\phi'}(\gamma_2^*, z'_\pm)$ are defined in (4.3.10)–(4.3.13).

The power radiated upward with $z' = z'_+$ involves the following product:

$$\begin{aligned} f_\rho(\gamma_2, z'_+) f^*_{\phi'}(\gamma_2^*, z'_+) &= \frac{1}{\gamma_2^*}\,[e^{i\gamma_2(z'-d)} + (1-G_{21})e^{i\gamma_2(z'+d)}] \\ &\quad \times [e^{-i\gamma_2^*(z'-d)} + (1-G_{21}^*)e^{-i\gamma_2^*(z'+d)}]. \end{aligned} \tag{4.3.17}$$

In the range $0 \le \lambda \le k_2$, $\gamma_2^* = \gamma_2$ is real so that

$$\begin{aligned} f_\rho(\gamma_2, z'_+) f^*_{\phi'}(\gamma_2^*, z'_+) &= \gamma_2^{-1}[2 - (G_{21} + G_{21}^*) + G_{21}G_{21}^* \\ &\quad + 2\cos 2\gamma_2 d - (G_{21}e^{2i\gamma_2 d} + G_{21}^* e^{-2i\gamma_2 d})]. \end{aligned} \tag{4.3.18}$$

In the range $\lambda > k_2$, $\gamma_2 = i\nu_2$ and $\gamma_2^* = -i\nu_2$, so that

$$\begin{aligned} f_\rho(\gamma_2, z'_+) f^*_{\phi'}(\gamma_2^*, z'_+) &= \frac{i}{\nu_2}\,[e^{-\nu_2(z'-d)} + (1-G_{21})e^{-\nu_2(z'+d)}] \\ &\quad \times [e^{-\nu_2(z'-d)} + (1-G_{21}^*)e^{-\nu_2(z'+d)}] \\ &= \frac{ie^{-2\nu_2 z'}}{\nu_2}\,[2(1+\cosh 2\nu_2 d) \\ &\quad + (G_{21}G_{21}^* - G_{21} - G_{21}^*)e^{-2\nu_2 d} \\ &\quad - (G_{21} + G_{21}^*)]. \end{aligned} \tag{4.3.19}$$

When (4.3.18) and (4.3.19) are substituted into (4.3.16) with the upper signs, the result is

$$\begin{aligned} P_a = \operatorname{Re} \frac{\omega\mu_0}{16\pi k_2^2} \Bigg(\int_0^{k_2} &\gamma_2^{-1}[2(1+\cos 2\gamma_2 d) + G_{21}G_{21}^* - G_{21} - G_{21}^* \\ &- G_{21}e^{2i\gamma_2 d} - G_{21}^* e^{-2i\gamma_2 d}]\lambda^3\, d\lambda \\ + i\int_{k_2}^\infty &\frac{e^{-2\nu_2 z'}}{\nu_2}\,[2(1+\cosh 2\nu_2 d) \\ &+ (G_{21}G_{21}^* - G_{21} - G_{21}^*)e^{-2\nu_2 d} \\ &- (G_{21} + G_{21}^*)]\lambda^3\, d\lambda \Bigg). \end{aligned} \tag{4.3.20}$$

In this expression, the two terms in square brackets are real so that the entire second integral is a pure imaginary and can be omitted. This leaves

$$P_a = \operatorname{Re} \frac{\omega\mu_0}{16\pi k_2^2} \int_0^{k_2} \gamma_2^{-1}(4\cos^2 \gamma_2 d + |G_{21}|^2$$

$$- 2G_{21} - 2G_{21}e^{2i\gamma_2 d})\lambda^3\,d\lambda. \tag{4.3.21a}$$

In the indicated range of integration, G_{21} is real so that the several terms can be combined to give

$$P_a = \frac{\omega\mu_0}{16\pi k_2^2}\int_0^{k_2} \gamma_2^{-1}[4(1-G_{21})\cos^2\gamma_2 d + G_{21}^2]\lambda^3\,d\lambda. \tag{4.3.21b}$$

The power transferred into the lower half-space involves the following product:

$$f_\rho(\gamma_2, z'_-)f^*_{\phi'}(\gamma_2^*, z'_-) = \frac{1}{\gamma_2^*}\,[-e^{i\gamma_2(d-z')} + (1-G_{21})e^{i\gamma_2(d+z')}]$$
$$\times\,[e^{-i\gamma_2^*(d-z')} + (1-G_{21}^*)e^{-i\gamma_2^*(d+z')}]. \tag{4.3.22}$$

In the range $0 \le \lambda \le k_2$, $\gamma_2^* = \gamma_2$ is real so that

$$f_\rho(\gamma_2, z'_-)f^*_{\phi'}(\gamma_2^*, z'_-) = \gamma_2^{-1}(G_{21}G_{21}^* - G_{21} - G_{21}^* - G_{21}e^{2i\gamma_2 z'}$$
$$+ G_{21}^*e^{-2i\gamma_2 z'} + 2i\sin 2\gamma_2 z'). \tag{4.3.23}$$

In the range $\lambda > k_2$, $\gamma_2 = i\nu_2$, $\gamma_2^* = -i\nu_2$, and

$$f_\rho(\gamma_2, z'_-)f^*_{\phi'}(\gamma_2^*, z'_-) = \frac{i}{\nu_2}\,e^{-2\nu_2 d}[-2\sinh 2\nu_2 z'$$
$$+ e^{-2\nu_2 z'}(G_{21}G_{21}^* - G_{21} - G_{21}^*)$$
$$- G_{21} + G_{21}^*]. \tag{4.3.24}$$

When (4.3.23) and (4.3.24) are substituted into (4.3.16) with the lower signs, the result is

$$P_g = -\operatorname{Re}\frac{\omega\mu_0}{16\pi k_2^2}\Bigg(\int_0^{k_2}\gamma_2^{-1}(G_{21}G_{21}^* - G_{21} - G_{21}^* - G_{21}e^{2i\gamma_2 z'}$$
$$+ G_{21}^*e^{-2i\gamma_2 z'} + 2i\sin 2\gamma_2 z')\lambda^3\,d\lambda$$
$$+ i\int_{k_2}^{\infty}\frac{e^{-2\nu_2 d}}{\nu_2}\,[-2\sinh 2\nu_2 z'$$
$$+ e^{-2\nu_2 z'}(G_{21}G_{21}^* - G_{21} - G_{21}^*)$$
$$- G_{21} + G_{21}^*]\lambda^3\,d\lambda\Bigg). \tag{4.3.25}$$

Here $G_{21}e^{2i\gamma_2 z'} - G_{21}^*e^{-2i\gamma_2 z'}$ and $G_{21} - G_{21}^*$ are pure imaginaries. Also, $G_{21} - G_{21}^* = 2i\operatorname{Im}G_{21}$. It follows that

$$P_g = -\frac{\omega\mu_0}{16\pi k_2^2}\Bigg(\operatorname{Re}\int_0^{k_2}\gamma_2^{-1}(|G_{21}|^2 - 2G_{21})\lambda^3\,d\lambda$$
$$+ 2\operatorname{Im}\int_{k_2}^{\infty}\nu_2^{-1}G_{21}e^{-2\nu_2 d}\lambda^3\,d\lambda\Bigg). \tag{4.3.26a}$$

As shown in Appendix B [see formulas (B1)–(B4)], the second integral in (4.3.26a) can be expanded and the parts that have no imaginary part omitted. Also, G_{21} is real in the range of integration of the first integral in (4.3.26a). Hence,

$$P_g = -\frac{\omega\mu_0}{16\pi k_2^2}\left(\int_0^{k_2} \gamma_2^{-1}(G_{21}^2 - 2G_{21})\lambda^3\, d\lambda \right.$$
$$\left. - 4k_1^2k_2^2 \int_{k_2}^{k_1} \frac{\gamma_1}{k_1^4\nu_2^2 + k_2^4\gamma_1^2}\, e^{-2\nu_2 d}\lambda^3\, d\lambda\right). \tag{4.3.26b}$$

Note that both P_a and P_g depend on the height d of the dipole but are independent of the location z' of the planes of integration within the ranges $0 \le z'_- \le d$ and $z'_+ > d$.

The total power radiated is

$$P_a + P_g = \frac{\omega\mu_0}{8\pi k_2^2}\left(\int_0^{k_2} \gamma_2^{-1}[2(1 - G_{21})\cos^2\gamma_2 d + G_{21}]\lambda^3\, d\lambda \right.$$
$$\left. + 2k_1^2k_2^2 \int_{k_2}^{k_1} \frac{\gamma_1}{k_1^4\nu_2^2 + k_2^4\gamma_1^2}\, e^{-2\nu_2 d}\lambda^3\, d\lambda\right). \tag{4.3.27}$$

G_{21} is defined in (4.3.7).

Two special cases are of interest. The first is the homogeneous region with $k_1 = k_2$. The upward-directed power with $z' > d$ now involves the functions

$$f_\rho(\gamma_2, z'_+) = e^{i\gamma_2(z'-d)}, \qquad f^*_{\phi'}(\gamma_2^*, z'_+) = \frac{1}{\gamma_2^*}\, e^{-i\gamma_2^*(z'-d)}, \tag{4.3.28}$$

so that

$$P_+ = \operatorname{Re} \frac{\omega\mu_0}{16\pi k_2^2}\int_0^\infty \frac{1}{\gamma_2^*}\, e^{i\gamma_2(z'-d)}e^{-i\gamma_2^*(z'-d)}\lambda^3\, d\lambda. \tag{4.3.29}$$

In the range $0 \le \lambda \le k_2$, $\gamma_2^* = \gamma_2$ is real; in the range $\lambda > k_2$, $\gamma_2 = i(\lambda^2 - k_2^2)^{1/2} = i\nu_2$ and $\gamma_2^* = -i\nu_2$. It follows that

$$P_+ = \operatorname{Re} \frac{\omega\mu_0}{16\pi k_2^2}\left(\int_0^{k_2} \frac{\lambda^3\, d\lambda}{(k_2^2 - \lambda^2)^{1/2}} \right.$$
$$\left. + ie^{-2\nu_2(z'-d)}\int_{k_2}^\infty \frac{\lambda^3\, d\lambda}{(\lambda^2 - k_2^2)^{1/2}}\right)$$
$$= \operatorname{Re} \frac{\omega\mu_0}{16\pi k_2^2}\left(\frac{(k_2^2 - \lambda^2)^{3/2}}{3} - k_2^2(k_2^2 - \lambda^2)^{1/2}\right)_0^{k_2}$$
$$= \frac{\omega\mu_0 k_2}{24\pi}. \tag{4.3.30}$$

Note that the second integral above is real so that, when multiplied by i, it is a pure imaginary that contributes nothing to the real part.

The downward-directed power involves the functions

$$f_\rho(\gamma_2, z'_-) = -e^{i\gamma_2(d-z')}, \qquad f^*_{\phi'}(\gamma_2^*, z'_-) = \frac{1}{\gamma_2^*}\, e^{-i\gamma_2^*(d-z')}, \tag{4.3.31}$$

so that

$$\begin{aligned} P_- &= \mathrm{Re}\,\frac{\omega\mu_0}{16\pi k_2^2}\int_0^\infty \frac{1}{\gamma_2^*}\, e^{i\gamma_2(d-z')}e^{-i\gamma_2^*(d-z')}\lambda^3\, d\lambda \\ &= \frac{\omega\mu_0 k_2}{24\pi}. \end{aligned} \tag{4.3.32}$$

The total radiated power is

$$\begin{aligned} P_0 &= P_+ + P_- \\ &= \frac{\omega\mu_0 k_2}{12\pi} = \frac{k_2^2\zeta_0}{12\pi} = 10k_2^2 \text{ watts/ampere meter.} \end{aligned} \tag{4.3.33a}$$

For a dipole with the electric moment $2Ih_e$, where h_e is the effective half-length,

$$P_0 = 10k_2^2 \times 4I^2h_e^2 = 40k_2^2h_e^2I^2. \tag{4.3.33b}$$

The well-known formula for the radiation resistance is

$$R^r = \frac{2P_0}{I^2} = 80k_2^2h_e^2. \tag{4.3.34}$$

The second special case is the dipole over a perfectly conducting half-space or a dielectric half-space with infinite permittivity, so that $k_1 \to \infty$. This means that

$$G_{21} = \frac{2k_2^2\gamma_1}{k_1^2\gamma_2 + k_2^2\gamma_1} \to 0, \tag{4.3.35}$$

so that, from (4.3.26), $P_g = 0$ and, from (4.3.21),

$$P_t = P_a = \frac{\omega\mu_0}{4\pi k_2^2}\,\mathrm{Re}\int_0^{k_2}\frac{\cos^2\gamma_2 d}{\gamma_2}\,\lambda^3\, d\lambda. \tag{4.3.36}$$

This is readily integrated with the change of variable,

$$\begin{aligned} &u = \gamma_2 d = d(k_2^2 - \lambda^2)^{1/2}; && u^2 = d^2(k_2^2 - \lambda^2), \\ &u\,du = -d^2\lambda\, d\lambda; && \lambda^2 = k_2^2 - \frac{u^2}{d^2}; \\ &\text{when } \lambda = 0,\ \ u = k_2 d; && \text{when } \lambda = k_2,\ \ u = 0. \end{aligned}$$

Hence,

$$P_a = \frac{\omega\mu_0}{4\pi k_2^2}\,\mathrm{Re}\int_0^{k_2 d}\frac{\cos^2 u}{(u/d)}\left(k_2^2 - \frac{u^2}{d^2}\right)\frac{u}{d^2}\,du$$

$$= \frac{\omega\mu_0 k_2}{12\pi}\left(1 + 3\,\frac{\sin 2k_2 d - 2k_2 d\cos 2k_2 d}{(2k_2 d)^3}\right). \tag{4.3.37}$$

When the dipole is on the boundary so that $d = 0$, it constitutes a monopole with the electric moment Ih_e on a perfectly conducting half-space. The power radiated by the monopole is

$$P_m = \frac{\omega\mu_0 k_2}{12\pi}\, I^2 h_e^2 \times 2 = \frac{\omega\mu_0 k_2}{6\pi}\, I^2 h_e^2 = 20 k_2^2 h_e^2 I^2. \tag{4.3.38}$$

[Note that as $k_2 d \to 0$, $\sin 2k_2 d \to 2k_2 d - (2k_2 d)^3/6$; $\cos 2k_2 d \to 1 - (2k_2 d)^2/2$.]

The well-known formula for the radiation resistance is

$$R^r = \frac{2P_m}{I^2} = 40 k_2^2 h_e^2. \tag{4.3.39}$$

The next step is to evaluate the integrals (4.3.21) and (4.3.26) for P_a and P_g and to study the analytical formulas obtained in the limit as $d \to 0$. Since for this purpose it is unimportant whether the lower half-space—earth—is a perfect dielectric or a conducting region, the analytically simpler case of a perfect dielectric is investigated. The powers in the air and earth are conveniently normalized with respect to the power radiated by the isolated dipole with the same electric moment. The normalized powers are

$$p_a = \frac{P_a}{P_0} = \frac{3}{4k_2^3}\,\mathrm{Re}\int_0^{k_2} \gamma_2^{-1}(4\cos^2\gamma_2 d + |G_{21}|^2 - 2G_{21} - 2G_{21}e^{2i\gamma_2 d})\lambda^3\,d\lambda, \tag{4.3.40}$$

$$p_g = \frac{P_g}{P_0} = -\frac{3}{4k_2^3}\left(\mathrm{Re}\int_0^{k_2} \gamma_2^{-1}(|G_{21}|^2 - 2G_{21})\lambda^3\,d\lambda + 2\,\mathrm{Im}\int_{k_2}^{\infty} \nu_2^{-1} G_{21} e^{-2\nu_2 d}\lambda^3\,d\lambda\right). \tag{4.3.41}$$

In these formulas, $\gamma_2 = (k_2^2 - \lambda^2)^{1/2} = i\nu_2 = i(\lambda^2 - k_2^2)^{1/2}$; $G_{21} = 2k_2^2\gamma_1 N^{-1}$, $N = k_1^2\gamma_2 + k_2^2\gamma_1$. In the range, $k_2 \le \lambda \le \infty$, $N = ik_1^2\nu_2 + k_2^2\gamma_1$. These formulas can be expressed as follows:

$$p_a = 1 + 3\left(\frac{\sin 2k_2 d - 2k_2 d\cos 2k_2 d}{(2k_2 d)^3}\right) + J_{1a} + J_{2a} + J_{3a} \tag{4.3.42}$$

with

$$J_{1a} = -\,\mathrm{Re}\,\frac{3}{k_2}\int_0^{k_2} \frac{\gamma_1}{\gamma_2}\,\frac{1}{N}\,\lambda^3\,d\lambda, \tag{4.3.43}$$

$$J_{2a} = \mathrm{Re}\; 3k_2\int_0^{k_2} \frac{\gamma_1^2}{\gamma_2}\,\frac{1}{|N^2|}\,\lambda^3\,d\lambda, \tag{4.3.44}$$

$$J_{3a} = -\operatorname{Re} \frac{3}{k_2} \int_0^{k_2} \frac{\gamma_1}{\gamma_2} \frac{1}{N} e^{2i\gamma_2 d} \lambda^3 \, d\lambda; \tag{4.3.45}$$

$$p_g = J_{1g} + J_{2g} + J_{3g} = -(J_{1a} + J_{2a}) + J_{3g} \tag{4.3.46}$$

with

$$J_{3g} = -\operatorname{Im} \frac{3}{k_2} \int_{k_2}^{\infty} \frac{\gamma_1}{\nu_2} \frac{1}{N} e^{-2\nu_2 d} \lambda^3 \, d\lambda; \tag{4.3.47}$$

and

$$p_t = p_a + p_g = 1 + 3\left(\frac{\sin 2k_2 d - 2k_2 d \cos 2k_2 d}{(2k_2 d)^3} \right) + J_{3a} + J_{3g}. \tag{4.3.48}$$

The integrals involved in (4.3.42)–(4.3.48) are evaluated in Appendix G, subject to the condition $k_1^2 \gg k_2^2$. With

$$\operatorname{Si} x = \int_0^x \frac{\sin u}{u} \, du \quad \text{and} \quad \operatorname{Cin} x = \int_0^x \frac{1 - \cos u}{u} \, du,$$

the results are

$$\begin{aligned} p_a = 1 &+ 3\left(\frac{\sin 2k_2 d - 2k_2 d \cos 2k_2 d}{(2k_2 d)^3} \right) \\ &- \frac{3k_2}{k_1} \Bigg\{ \left(1 - \frac{3k_2^2}{k_1^2}\right) \ln\left(1 + \frac{k_1}{k_2}\right) - \frac{3}{2} + \frac{3k_2}{k_1} \\ &\quad + \left(1 - \frac{k_2^2}{k_1^2}\right) \cos \frac{2k_2^2 d}{k_1} \Bigg[\ln\left(1 + \frac{k_1}{k_2}\right) \\ &\qquad\qquad - \operatorname{Cin} 2k_2 d \left(1 + \frac{k_2}{k_1}\right) + \operatorname{Cin} \frac{2k_2^2 d}{k_1} \Bigg] \\ &\quad + \left(1 - \frac{k_2^2}{k_1^2}\right) \sin \frac{2k_2^2 d}{k_1} \left[\operatorname{Si} 2k_2 d \left(1 + \frac{k_2}{k_1}\right) - \operatorname{Si} \frac{2k_2^2 d}{k_1} \right] \\ &\quad + \frac{1 - \cos 2k_2 d}{4k_2^2 d^2} - \left(1 - \frac{k_2}{k_1}\right) \frac{\sin 2k_2 d}{2k_2 d} \Bigg\}, \end{aligned} \tag{4.3.49}$$

$$\begin{aligned} p_g = \frac{3k_2}{k_1} &\Bigg[\left(1 - \frac{3k_2^2}{k_1^2}\right) \ln\left(1 + \frac{k_1}{k_2}\right) - \frac{3}{2} + \frac{3k_2}{k_1} \\ &- \cos \frac{2k_2^2 d}{k_1} \left(\gamma + \ln \frac{2k_2^2 d}{k_1} - \operatorname{Cin} \frac{2k_2^2 d}{k_1} \right) \\ &- \sin \frac{2k_2^2 d}{k_1} \left(\operatorname{Si} \frac{2k_2^2 d}{k_1} - \frac{\pi}{2} \right) + \left(1 - \cos \frac{2k_2^2 d}{k_1}\right) E_1(2k_1 d) \\ &+ \gamma + \ln 2k_1 d - 1 + \ln 2 - \frac{\pi}{2} \int_0^{2k_1 d} [I_1(z) - L_1(z)] \frac{dz}{z} \Bigg] \\ + \frac{k_1}{k_2} &\left(1 - \frac{3\pi}{4k_1 d} [I_2(2k_1 d) - L_2(2k_1 d)] \right). \end{aligned} \tag{4.3.50}$$

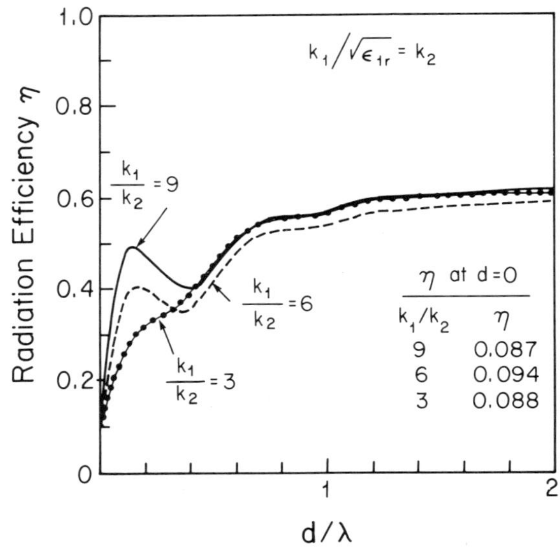

Figure 4.3.1. Radiation efficiency η of vertical dipole at height d in air (k_2) over dielectric half-space (k_1).

Here I is a modified Bessel function and L is a Struve function.

When the dipole is on the surface in the air, $d = 0$ and the above formulas reduce to

$$p_a = 2 - \frac{6k_2}{k_1}\left[\left(1 - \frac{2k_2^2}{k_1^2}\right)\ln\left(1 + \frac{k_1}{k_2}\right) - 1 + \frac{2k_2}{k_1}\right], \tag{4.3.51}$$

$$p_g = \frac{3k_2}{k_1}\left[\left(1 - \frac{3k_2^2}{k_1^2}\right)\ln\left(1 + \frac{k_1}{k_2}\right)\right.$$
$$\left. + 2\ln\frac{k_1}{k_2} - \frac{5}{2} + \ln 2 + \frac{3k_2}{k_1}\right] + \frac{k_1}{k_2}. \tag{4.3.52}$$

Complete graphs of the radiation efficiency $\eta = p_a/(p_a + p_g)$ are shown in Fig. 4.3.1 with $k_1/k_2 = 9$, 6, and 3. In all cases, the frequency is assumed to be sufficiently high ($\omega\epsilon_1 \gg \sigma_1$) so that Region 1 is a good dielectric with negligible conductivity. These results are in good agreement with the corresponding numerically evaluated ones of Hansen (1972) except near and at $d = 0$. The ratio η remains finite for all values of k_1/k_2 and does not vanish as indicated by Hansen.

Graphs of p_a, p_g, and η as functions of k_1/k_2 in the range $1 \leq k_1/k_2 \leq 15$ are shown in Fig. 4.3.2 specifically when $d = 0$. The range $2 \leq k_1/k_2 \leq 15$ was calculated from (4.3.51) and (4.3.52). The point at $k_1 = k_2$ was obtained directly from the integrals (4.3.40) and (4.3.41). With $d = 0$ and $G_{21} = 1$, these reduce to an elementary integral that gives $p_a = p_g = 0.5$. In Fig. 4.3.3, the graph for $d = 0$ is extended to larger values of k_1/k_2 and the limit $k_1/k_2 = \infty$ is indicated where η rises discontinuously to 1. The

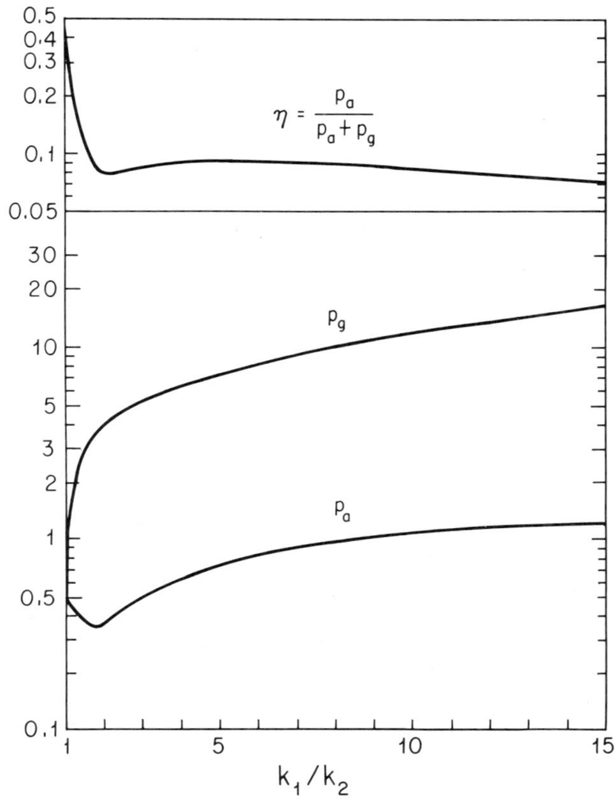

Figure 4.3.2. Normalized powers into air (p_a) and earth (p_g) and radiation efficiency (η) of vertical dipole on surface of dielectric half-space (k_1) at $d = 0$ in air (k_2).

constant value $\eta = 1$ for $d = \infty$ is also shown.

It appears from (4.3.50) and (4.3.52) that $p_g \to \infty$ when $k_1 \to \infty$ due to the presence of the term multiplied by k_1/k_2. Actually this is not the case because formulas (4.3.50) and (4.3.52) are not valid when $k_1 \to \infty$. This is evident from the following asymptotic expansions. In (4.3.50) one of the crucial terms is $L_2(2k_1d) - I_2(2k_1d)$. With formulas (9.6.6) and (12.2.6) in Abramowitz and Stegun (1972, pp. 375 and 498), it follows that

$$L_2(2k_1d) - I_2(2k_1d) \sim \frac{1}{\pi}\sum_{m=0}^{\infty} \frac{(-1)^{m+1}\Gamma(m+\frac{1}{2})}{\Gamma(\frac{5}{2}-m)(k_1d)^{2m-1}}$$
$$= \frac{k_1d}{\pi}\sum_{m=0}^{\infty} \frac{(-1)^{m+1}\Gamma(m+\frac{1}{2})}{\Gamma(\frac{5}{2}-m)(k_1^2d^2)^m}$$

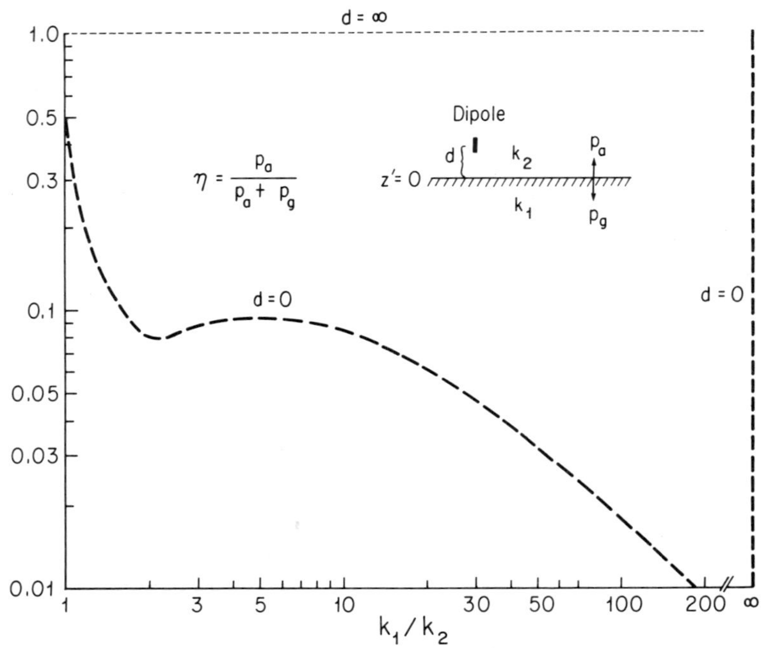

Figure 4.3.3. Fraction of power η in upper half-space with dipole at $d = 0$ and $d = \infty$.

$$= \frac{k_1 d}{\pi}\left(-\frac{\Gamma(\frac{1}{2})}{\Gamma(\frac{5}{2})} + \sum_{m=0}^{\infty} \frac{(-1)^m \Gamma(m+\frac{3}{2})}{\Gamma(\frac{3}{2}-m)(k_1^2 d^2)^{m+1}}\right). \tag{4.3.53a}$$

With $\Gamma(\frac{1}{2})/\Gamma(\frac{5}{2}) = \sqrt{\pi}/(3\sqrt{\pi}/4) = 4/3$, it follows that

$$\begin{aligned}
&\frac{k_1}{k_2}\left(1 - \frac{3\pi}{4k_1 d}\left[I_2(2k_1 d) - L_2(2k_1 d)\right]\right) \\
&\quad \sim \frac{k_1}{k_2}\left(1 - 1 + \frac{3}{4}\sum_{m=0}^{\infty} \frac{(-1)^m \Gamma(m+\frac{3}{2})}{\Gamma(\frac{3}{2}-m)(k_1^2 d^2)^{m+1}}\right) \\
&\quad = \frac{3k_2}{4k_1}\frac{1}{k_2^2 d^2}\sum_{m=0}^{\infty} \frac{(-1)^m \Gamma(m+\frac{3}{2})}{\Gamma(\frac{3}{2}-m)(k_1^2 d^2)^{m}}.
\end{aligned} \tag{4.3.53b}$$

Clearly, this *decreases* with increasing k_1/k_2 for any nonzero value of $k_1 d$.

A large-argument approximation of the second crucial term in (4.3.50), viz.,

$$\frac{\pi}{2}\int_0^{2k_1 d} \left[I_1(z) - L_1(z)\right] \frac{dz}{z}$$

$$= \frac{\pi}{2}\left(\int_0^{2k_1 d} [I_0(z) - L_0(z)]\, dz - [I_1(z) - L_1(z)]\right),$$

can also be obtained with formulas (12.2.6) and (12.2.8) in Abramowitz and Stegun (1972, p. 498). These give

$$\frac{\pi}{2}\int_0^{2k_1 d} [I_1(z) - L_1(z)]\, \frac{dz}{z}$$

$$\sim \ln 4k_1 d + \gamma - \sum_{m=1}^{\infty} \frac{(2m)!(2m-1)!}{(m!)^2(4k_1 d)^{2m}} + \frac{1}{2}\sum_{m=0}^{\infty} \frac{(-1)^{m+1}\Gamma(m+\frac{1}{2})}{\Gamma(\frac{3}{2}-m)(k_1 d)^{2m}}$$

$$= \gamma + \ln 4k_1 d - \sum_{m=1}^{\infty} \frac{(2m)!(2m-1)!}{(m!)^2(4k_1 d)^{2m}} - \frac{1}{2}\frac{\Gamma(\frac{1}{2})}{\Gamma(\frac{3}{2})}$$

$$+ \frac{1}{2}\sum_{m=0}^{\infty} \frac{(-1)^m\Gamma(m+\frac{3}{2})}{\Gamma(\frac{1}{2}-m)(k_1^2 d^2)^{m+1}}$$

$$= \gamma + \ln 2k_1 d + \ln 2 - 1 + \frac{1}{2k_1^2 d^2}\sum_{m=0}^{\infty} \frac{(-1)^m\Gamma(m+\frac{3}{2})}{\Gamma(\frac{1}{2}-m)(k_1^2 d^2)^m}$$

$$- \sum_{m=1}^{\infty} \frac{(2m)!(2m-1)!}{(m!)^2(4k_1 d)^{2m}}. \qquad (4.3.53c)$$

Here the only term that increases without limit as $k_1 \to \infty$ is $\ln 2k_1 d$. However, the entire expression is multiplied in (4.3.50) by k_2/k_1 and

$$\lim_{k_1\to\infty} 2k_2 d\, \frac{\ln 2k_1 d}{2k_1 d} = 2k_2 d \lim_{x\to\infty} \frac{\ln x}{x} = 2k_2 d \lim_{x\to\infty} \frac{1}{x} = 0. \qquad (4.3.53d)$$

Evidently, the critical quantity in the evaluation of the power into the earth is $k_1 d$. The simultaneous limits $d \to 0$ and $k_1 \to \infty$ are indeterminate. In (4.3.50), the limit $d \to 0$ leads to (4.3.52), but this is not valid with $k_1 \to \infty$ because the quantity $k_1 d$ has been set equal to zero.

The limit $k_1 \to \infty$ has been carried out in conjunction with (4.3.36), and the subsequent application of the limit $d \to 0$ led to (4.3.38).

The graphs in Fig. 4.3.1 show that the radiation efficiency of a vertical electric dipole—as defined in (4.3.8)—drops to a small value as the height d of the dipole is reduced to zero. This means that most of the power radiated by the dipole is ultimately transferred into the earth. Does this mean that the dipole on the earth's surface is an undesirable antenna for transmission to arbitrarily located receivers in the air or on the surface of the earth? In particular, is it undesirable for the transmission and reception of low-angle radiation in over-the-horizon radar? Just what does the radiation efficiency defined in (4.3.8) actually mean in terms of the electromagnetic field generated or received by a vertical electric dipole on the surface of the earth?

The electromagnetic field of a dipole when it is on or close to the boundary ($d \sim 0$) but still in the air is given in (4.2.37)–(4.2.39). The field in the air on the surface $z = 0$ of the earth is

$$B_{2\phi'}(\rho,0) = -\frac{\mu_0}{2\pi}\, e^{ik_2\rho}\left[\frac{ik_2}{\rho} - \frac{1}{\rho^2} - \frac{k_2^3}{k_1}\left(\frac{\pi}{k_2\rho}\right)^{1/2} e^{-iR}\mathcal{F}(R)\right], \tag{4.3.54}$$

$$E_{2\rho}(\rho,0) = \frac{\omega\mu_0}{2\pi k_1}\, e^{ik_2\rho}\left[\frac{ik_2}{\rho} - \frac{1}{\rho^2} - \frac{k_2^3}{k_1}\left(\frac{\pi}{k_2\rho}\right)^{1/2} e^{-iR}\mathcal{F}(R)\right], \tag{4.3.55}$$

$$\begin{aligned} E_{2z'}(\rho,0) = \frac{\omega\mu_0}{2\pi k_2}\, e^{ik_2\rho}\Bigg[\frac{ik_2}{\rho} - \frac{1}{\rho^2} - \frac{i}{k_2\rho^3} \\ - \frac{k_2^3}{k_1}\left(\frac{\pi}{k_2\rho}\right)^{1/2} e^{-iR}\mathcal{F}(R)\Bigg], \end{aligned} \tag{4.3.56}$$

where $R = k_2^3\rho/2k_1^2$. At very large distances from the source where $|R| \geq 4$, the Fresnel-integral term assumes a simple asymptotic form. This is given in (4.2.52) and (4.2.53). When $z' = d = 0$, this becomes

$$T^r = \frac{ik_2}{\rho} + \frac{k_1^2}{k_2^2\rho^2}. \tag{4.3.57}$$

With (4.3.57), (4.3.54)–(4.3.56) reduce to

$$\frac{k_1}{k_2}\, E_{2\rho}(\rho,0) = -\frac{\omega}{k_2}\, B_{2\phi'}(\rho,0) = E_{2z'}(\rho,0) = -\frac{\omega\mu_0}{2\pi k_2}\,\frac{k_1^2}{k_2^2}\,\frac{e^{ik_2\rho}}{\rho^2} \tag{4.3.58}$$

when

$$k_2\rho \geq \frac{8|k_1^2|}{k_2^2}. \tag{4.3.59}$$

Of primary interest is the far-field component $E_{2\Theta}^r(r_0,\Theta)$ as given by (4.2.63). This has a large $1/r_0$ term except near $\Theta = \pi/2$ where it vanishes. At $\Theta = \pi/2$, the asymptotic form of the Fresnel-integral term given in (4.2.56) is

$$T^r = \frac{ik_2}{r_0} + \frac{k_1^2}{k_2^2 r_0^2}; \qquad r_0 = (\rho^2 + z'^2)^{1/2} = \rho \tag{4.3.60}$$

so that

$$E_{2\Theta}^r(r_0,\pi/2) = \frac{\omega\mu_0}{2\pi k_2}\,\frac{k_1^2}{k_2^2}\,\frac{e^{ik_2 r_0}}{r_0^2}; \qquad r_0 = \rho. \tag{4.3.61}$$

When $|k_1 z'| \ll k_2 r_0$, $\sin\Theta = \rho/r_0 \sim 1$, $\cos\Theta = z'/r_0 \ll 1$, and

$$E_{2\Theta}^r(r_0,\Theta) \sim -\frac{\omega\mu_0}{2\pi k_2}\,\frac{k_1}{k_2}\,\frac{e^{ik_2 r_0}}{r_0^2}\left(ik_2 z' - \frac{k_1}{k_2}\right). \tag{4.3.62}$$

The $1/r_0$ term in (4.2.63) is the far-field contribution by the plane-wave reflection coefficient; the $1/r_0^2$ term is the surface wave. Graphs of

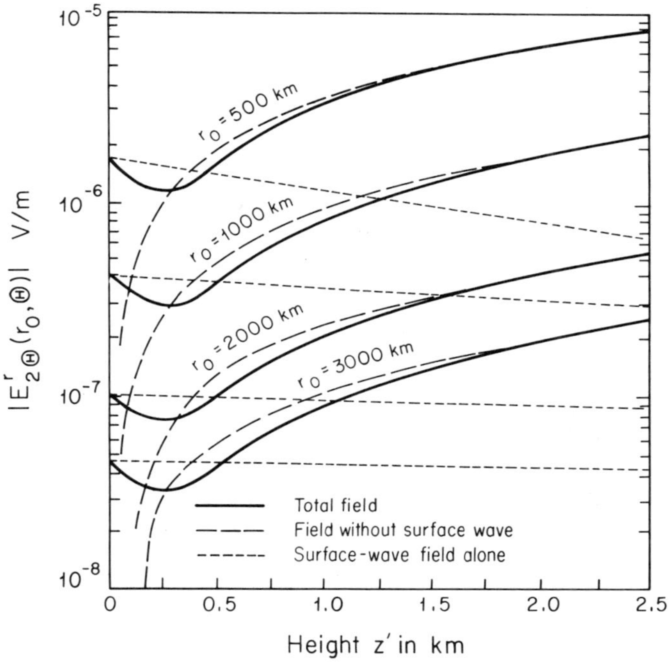

Figure 4.3.4. Magnitude of $E^r_{2\Theta}(r_0, \Theta)$ at height z' in air over sea water in the range $z' \ll r_0$. [Note that $|E^r_{2\Theta}(r_0, \Theta)| \sim 1/r_0^2$ to agree with (4.3.62).]

$|E^r_{2\Theta}(r_0, \Theta)|$ are in Fig. 4.2.2 in a conventional polar plot, and in Fig. 4.2.3 in a linear/logarithmic plot for the range $80° \leq \Theta \leq 90°$ for the low-angle radiation. Figure 4.3.4 shows $|E^r_{2\Theta}(r_0, \Theta)|$ when $z' \ll r_0$ as a function of z' specifically when Region 1 is sea water. The magnitudes of the field without the surface-wave term and of the field due to the surface-wave term alone are shown separately. Note that both terms decrease with distance as $1/r_0^2$ for any fixed z'.

For engineering applications it may be useful to generalize the far-field formulas (4.2.61) and (4.2.63) so that they can be used for all values of k_1 without the restriction $|k_1| \geq 3k_2$. This is readily done for the $1/r_0$ terms simply by substituting the exact formula for the plane-wave transmission coefficient f_{er} for the approximate one that appears in (4.2.61) and (4.2.63). That is,

$$\tfrac{1}{2}(1 + f_{er}) = \frac{n^2 \cos\Theta}{n^2 \cos\Theta + (n^2 - \sin^2\Theta)^{1/2}}; \qquad n^2 = \frac{k_1^2}{k_2^2} \tag{4.3.63}$$

is substituted in the $1/r_0$ terms for

$$\tfrac{1}{2}(1 + f_{er}) \sim \frac{k_1 \cos\Theta}{k_2 + k_1 \cos\Theta} \tag{4.3.64}$$

(which is obtained from the exact formula when $\sin^2\Theta \le 1$ is neglected compared to n^2). With this substitution, (4.2.61) and (4.2.63) have the following forms:

$$B^r_{2\Phi}(r_0, \Theta) = -\frac{\mu_0}{2\pi}\, e^{ik_2 r_0} \times \left[\frac{ik_2}{r_0}\left(\frac{(k_1^2/k_2^2)\sin\Theta\cos\Theta}{(k_1^2/k_2^2)\cos\Theta + [(k_1^2/k_2^2) - \sin^2\Theta]^{1/2}}\right) - \frac{k_1^2 \sin^3\Theta}{k_2^2 r_0^2 [1 + (k_1/k_2)\cos\Theta]^3}\right], \tag{4.3.65}$$

$$E^r_{2\Theta}(r_0, \Theta) = -\frac{\omega\mu_0}{2\pi k_2}\, e^{ik_2 r_0} \times \left[\frac{ik_2}{r_0}\left(\frac{(k_1^2/k_2^2)\sin\Theta\cos\Theta}{(k_1^2/k_2^2)\cos\Theta + [(k_1^2/k_2^2) - \sin^2\Theta]^{1/2}}\right) - \frac{k_1^2}{k_2^2 r_0^2}\left(\frac{\sin^3\Theta\,[\sin^2\Theta - (k_2/k_1)\cos\Theta]}{[1 + (k_1/k_2)\cos\Theta]^3}\right)\right]. \tag{4.3.66}$$

These formulas are valid subject to the far-field condition

$$k_2 r_0 \ge \left|\frac{8k_1^2}{k_2^2}\right|. \tag{4.3.67}$$

When $k_1 = k_2$, this condition approximates the usual far-field requirement $k_2 r_0 \gg 1$ so that the terms $1/(k_2^2 r_0^2)$ and $1/(k_2^3 r_0^3)$ can be neglected compared to $1/k_2 r_0$. When $k_1 = k_2$, the far field in (4.3.65) and (4.3.66) becomes

$$E^r_{2\Theta}(r_0, \Theta) = \frac{\omega}{k_2}\, B^r_{2\Phi}(r_0, \Theta) = -\frac{\omega\mu_0}{4\pi k_2}\, e^{ik_2 r_0}\, \frac{ik_2}{r_0} \sin\Theta, \tag{4.3.68}$$

which is the far field of the isolated dipole in air. Note that when $k_1 = k_2$, the $1/r_0$ terms do not vanish when $\Theta = \pi/2$ but have maxima so that *all* $1/r_0^2$ terms are negligible.

The accuracy of the surface-wave term is reduced when $|k_1/k_2|$ is smaller than 3, but it should still give the right order of magnitude at $\Theta = \pi/2$ until, as k_1 approaches k_2, it becomes negligibly small. The fact that it does not vanish when $k_1 = k_2$ is irrelevant since r_0 must be chosen large enough so that it—along with all the other $1/r_0^2$ terms—is negligible.

The Poynting vector in Region 2 (air) is defined by

$$\mathbf{S}_2(\rho, z') = \hat{\mathbf{z}}' S_{2z'}(\rho, z') + \hat{\boldsymbol{\rho}} S_{2\rho}(\rho, z') = \frac{1}{2\mu_0}\,[\mathbf{E}_2(\rho, z') \times \mathbf{B}_2^*(\rho, z')], \tag{4.3.69}$$

$$S_{2z'}(\rho, z') = \frac{1}{2\mu_0}\, E_{2\rho}(\rho, z') B^*_{2\phi'}(\rho, z'), \tag{4.3.70a}$$

$$S_{2\rho}(\rho, z') = -\frac{1}{2\mu_0} E_{2z'}(\rho, z') B^*_{2\phi'}(\rho, z'). \tag{4.3.70b}$$

The slope of the locus of the Poynting vector is

$$\frac{dz'}{d\rho} = \frac{\operatorname{Re} S_{2z'}(\rho, z')}{\operatorname{Re} S_{2\rho}(\rho, z')}. \tag{4.3.71}$$

The general equation of the locus obtained with (4.3.70a,b) and the components of the field as given in (4.2.37)–(4.2.39) is complicated and no explicit solution is available. In the near and intermediate range, an approximate formula is given in Section 4.5 that indicates that the Poynting vector at a point $(\rho, 0)$ on the boundary has followed a curved path from the dipole at $(0, 0)$ upward into the air and back down to the earth at $(\rho, 0)$. For present purposes, it is adequate to determine its slope as it reaches the surface. This is readily determined from the field in both the intermediate and far ranges. In the former, defined by $k_2\rho < |k_1^2|/k_2^2$, the Fresnel term contributes negligibly and (4.3.54)–(4.3.56) give

$$\begin{aligned} S_{2z'}(\rho, 0) &= \frac{1}{2\mu_0} E_{2\rho}(\rho, 0) B^*_{2\phi'}(\rho, 0) \\ &\sim -\frac{\omega\mu_0}{8\pi^2 k_1}\left(\frac{ik_2}{\rho} - \frac{1}{\rho^2}\right)\left(-\frac{ik_2}{\rho} - \frac{1}{\rho^2}\right) \\ &\sim -\frac{\omega\mu_0}{8\pi^2 k_1}\frac{k_2^2}{\rho^2}, \end{aligned} \tag{4.3.72}$$

$$\begin{aligned} S_{2\rho}(\rho, 0) &= -\frac{1}{2\mu_0} E_{2z'}(\rho, 0) B^*_{2\phi'}(\rho, 0) \\ &\sim \frac{\omega\mu_0}{8\pi^2 k_2}\left(\frac{ik_2}{\rho} - \frac{1}{\rho^2} - \frac{i}{k_2\rho^3}\right)\left(-\frac{ik_2}{\rho} - \frac{1}{\rho^2}\right) \\ &\sim \frac{\omega\mu_0}{8\pi^2 k_2}\frac{k_2^2}{\rho^2}. \end{aligned} \tag{4.3.73}$$

In the far field, where $k_2\rho \geq 8|k_1^2|/k_2^2$, (4.3.58) gives

$$S^r_{2z'}(\rho, 0) = -\frac{\omega\mu_0}{8\pi^2 k_1}\frac{(k_1 k_1^*)^2}{k_2^4}\frac{1}{\rho^4}, \tag{4.3.74}$$

$$S^r_{2\rho}(\rho, 0) = \frac{\omega\mu_0}{8\pi^2 k_2}\frac{(k_1 k_1^*)^2}{k_2^4}\frac{1}{\rho^4}. \tag{4.3.75}$$

In each range,

$$\frac{dz'}{d\rho} = \frac{\operatorname{Re} S_{2z'}(\rho, 0)}{\operatorname{Re} S_{2\rho}(\rho, 0)} \sim -\frac{k_2}{\beta_1}, \tag{4.3.76}$$

where β_1 is the real part of $k_1 = \beta_1 + i\alpha_1$. Since the boundary conditions at $z' = 0$ are

$$E_{1\rho}(\rho, 0) = E_{2\rho}(\rho, 0); \quad B_{1\phi'}(\rho, 0) = B_{2\phi'}(\rho, 0);$$

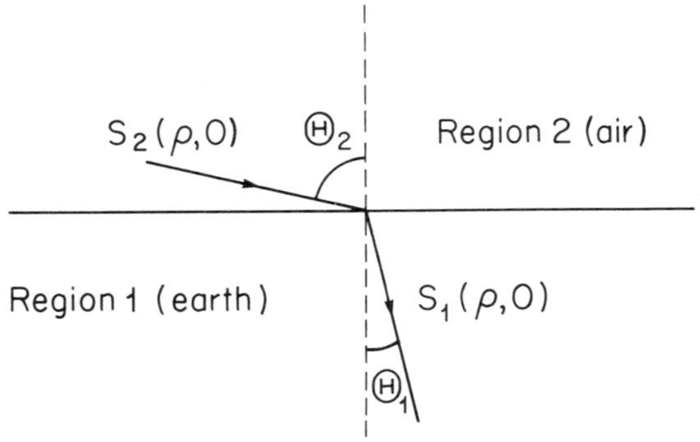

Figure 4.3.5. Poynting vector at point on air–earth boundary in far zone.

$$E_{1z'}(\rho,0) = \frac{k_2^2}{k_1^2}\, E_{2z'}(\rho,0), \tag{4.3.77}$$

it follows that

$$S_{1z'}(\rho,0) = S_{2z'}(\rho,0); \qquad S_{1\rho}(\rho,0) = \frac{k_2^2}{k_1^2}\, S_{2\rho}(\rho,0). \tag{4.3.78}$$

Hence, the slope of the Poynting vector in the earth just below the surface is

$$\left(\frac{dz'}{d\rho}\right)_{z'=0-} = -\frac{k_2}{\beta_1}\cdot\frac{\beta_1^2}{k_2^2} = -\frac{\beta_1}{k_2}. \tag{4.3.79}$$

The Poynting vector in both regions at a point on their boundary in the intermediate or the far field is shown in Fig. 4.3.5 with $\beta_1/k_2 = 2$. This rather small value is chosen to permit a clearer diagram. Actual values of β_1/k_2 are much greater. It is seen that the Poynting vector arrives at the point of observation $(\rho, 0)$ traveling almost horizontally, i.e., with a large angle of incidence in the air, and enters the earth to travel almost vertically downward, i.e., with a small angle of refraction. Since $S_{2\rho}(\rho,0)$ is greater than $S_{2z'}(\rho,0)$ by a factor of k_1/k_2, most of the power continues to travel radially along the surface in the air. However, a small fraction is continuously transferred into the earth where it travels almost vertically downward. This is a complicated phenomenon that includes incident, reflected (reradiated), and refracted fields and associated powers. It is important to note that in the far field only the surface wave is involved in this transfer of power into the earth.

A study of the fraction of power that remains in the air (upper half-space) and the fraction that is transferred into the earth (lower half-space) when the radiating source is a vertical dipole at a height d in the air shows that as $d \to 0$ the power in the air is reduced to a deep minimum but not zero. A study of the properties of the electromagnetic field generated by the same dipole shows that the power transferred into the earth is associated with the surface wave that travels outward along the boundary in the air.

Furthermore, when $d \sim 0$, the entire far field along that boundary is due to the surface wave.

An antenna with high radiation efficiency in the sense defined by (4.3.8) necessarily must generate a very weak surface wave. On the other hand, an antenna that generates a strong low-angle field with a significant value along the boundary surface, $\Theta = 90°$, necessarily generates a strong surface wave that transfers all the power associated with it into the earth. Therefore, it has a low radiation efficiency. This suggests that the concept of radiation efficiency as defined by (4.3.8) is not a useful figure-of-merit. The surface-wave part of all low-angle radiation is associated with power that is transferred into the earth. It is of major importance in over-the-horizon radar which requires an antenna that generates a strong spherical wave for reflection from the ionosphere and a strong surface wave for use when reflections from the ionosphere are interrupted by geomagnetic storms.

4.4 Communication with Submarines—Unit Vertical Dipole

Radio communication with submerged submarines is made difficult by the high attenuation of electromagnetic waves traveling through salt water with its high conductivity in the range $\sigma_1 \sim 2.8-4$ S/m. For a dipole in the sea water, this is evident from the direct- and reflected-wave terms in (3.6.2)–(3.6.4) which have the exponential factors $\exp(ik_1 r_1) = \exp(-\alpha_1 r_1) \times \exp(i\beta_1 r_1)$ and $\exp(ik_1 r_2) = \exp(-\alpha_1 r_2)\exp(i\beta_1 r_2)$, where $r_1 = [\rho^2 + (z-d)^2]^{1/2}$ and $r_2 = [\rho^2 + (z+d)^2]^{1/2}$. For a dipole on the surface, $d = 0$ and $r_1 = r_2 = r_0 = (\rho^2 + z^2)^{1/2}$. With $\alpha_1 \sim (\omega\mu_0\sigma_1/2)^{1/2}$ for sea water, it is essential that both the direct path through the salt water and the attenuation constant α_1 be kept as small as possible. The latter requires a low frequency. The shortest distance of transmission through sea water to a submerged submarine is vertically down. Because of the large index of refraction of sea water, an electromagnetic wave from a source in the air, such as an aircraft, the space shuttle, or a satellite, will be refracted almost vertically down from any angle of incidence.

A less obvious and more useful location for the antenna is just above or just below the surface of the sea at an arbitrary distance from the location of the submarine. The surface wave generated by such an antenna travels horizontally along the surface *in the air*, then vertically downward to the submarine. All three components of the electromagnetic wave have the factor $\exp(ik_2\rho)\exp(ik_1 z) = \exp(-\alpha_1 z)\exp[i(k_2\rho + \beta_1 z)]$ for a dipole on the surface in the air or $\exp(ik_2\rho)\exp[ik_1(z+d)] = \exp[-\alpha_1(z+d)]\exp\{i[k_2\rho + \beta_1(z+d)]\}$ for a dipole at the depth d below the surface in the salt water. Clearly, the entire exponential attenuation insofar as the location of the submarine is concerned is $\exp(-\alpha_1 z)$ for transmission vertically down—the same as for transmission from an aircraft. For the much less effective vertical dipole at the depth d in the sea, the attenuation factor $\exp(-\alpha_1 d)$ is also involved so that the distance $s = z + d$ is relevant.

The attenuation constant $\alpha_1 = (\omega\mu_0\sigma_1/2)^{1/2}$ for sea water is shown in Fig. 4.4.1 as a function of the frequency. Also shown is the attenuation factor $\exp(-\alpha_1 s)$ in the form $20 \log_{10}[\exp(-\alpha_1 s)] = -8.686\alpha_1 s$, where $s = z$ for the dipole on the surface ($d = 0$) and $s = z + d$ for the dipole at the depth d. The advantage of a low frequency at increasing depths is clear.

In order to determine the optimum frequency for long-distance communication with submarines, it is not sufficient to minimize the decrease in signal strength with depth. It is also necessary to consider the decrease in amplitude with radial distance. This is not exponential in air, but is determined by the radial functions $f(\rho; k_1, k_2)$ and $g(\rho; k_1, k_2)$ defined in (3.4.30) and (3.4.31). Actually, since these differ only in the near field and since $|E^L_{1z}(\rho, z)/E^L_{1\rho}(\rho, z)| \sim |k_2/k_1|$, it is evident that $E^L_{1\rho}(\rho, z)$ is the larger component of the electric field at all points in the sea that are not very close to the dipole. With

$$\frac{\omega\mu_0}{2\pi} = \frac{k_2\zeta_0}{2\pi} = 60k_2 \ \Omega/\text{m}, \tag{4.4.1}$$

(3.6.8) gives

$$|E_{1\rho}(\rho, 0)| = 60\left|\frac{k_2}{k_1} f(\rho; k_1, k_2)\right| \ \text{V/m} \tag{4.4.2}$$

for the unit electric dipole in air on the surface of the ocean. Graphs of $20 \log_{10} |E_{1\rho}(\rho, 0)|$ as a function of the radial distance ρ are shown in Fig. 4.4.2. On each curve the approximate boundary $k_2\rho = 1$ between the near field where $|E_{1\rho}(\rho, 0)| \sim 1/\rho^2$ and the intermediate field where $|E_{1\rho}(\rho, 0)| \sim 1/\rho$ is indicated, as is the approximate boundary $k_2\rho = |k_1/k_2|^2$ between the intermediate and the asymptotic fields. In the latter, $|E_{1\rho}(\rho, 0)| \sim 1/\rho^2$. These boundaries are important because the desirable range is the intermediate one where the field decreases less rapidly with the radial distance. The limits of the intermediate range together with the associated amplitudes of the electric field at the air–sea water interface are listed for a selection of frequencies in Table 4.4.1. For possible radial distances from 50 to 5000 km, the intermediate range requires frequencies between about 1 and 100 kHz.

The optimum frequency for a vertical dipole at $z = 0$ in air involves both the radial distance ρ and the depth z of the submarine. If the former is limited to the intermediate range, $f(\rho; k_1, k_2) = ik_2/\rho$ so that, with (3.6.8),

$$|E_{1\rho}(\rho, z)| = \frac{60k_2^2}{|k_1|\rho} e^{-\alpha_1 z}. \tag{4.4.3}$$

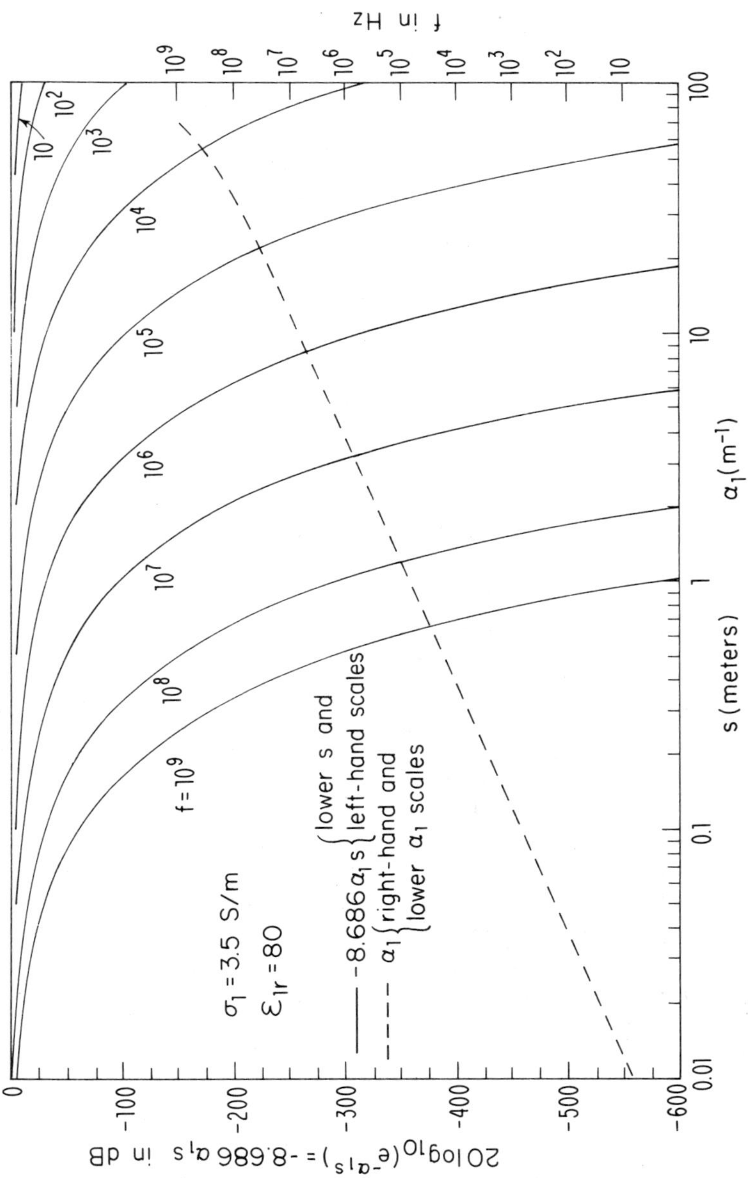

Figure 4.4.1. Decrease in amplitude with depth $s = z + d$ in sea water; attenuation constant α_1 as a function of frequency.

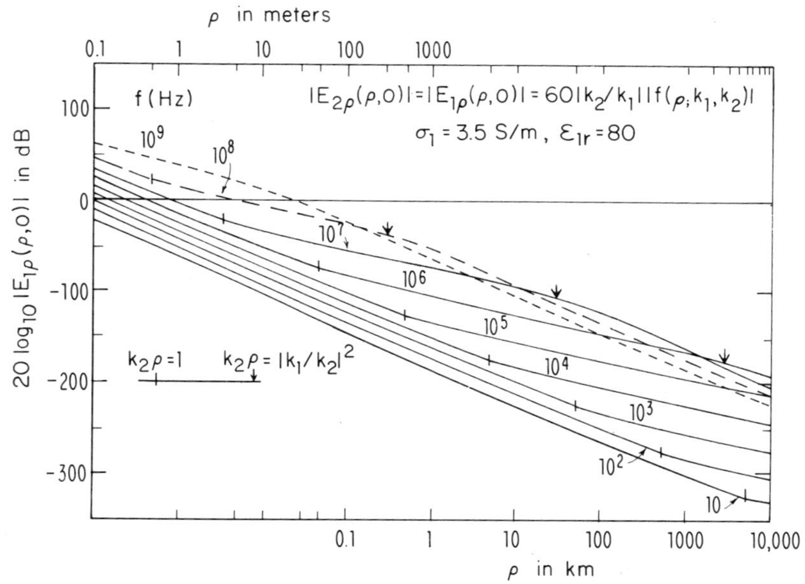

Figure 4.4.2. Magnitude of $E_{1\rho}(\rho, 0)$ in V/m due to vertical unit electric dipole in air above the surface of sea water.

Table 4.4.1. Intermediate Ranges in Distance and Amplitude for Vertical Dipole in Air

f	Intermediate Range	$20 \log_{10} \lvert E_{1\rho}(\rho, 0) \rvert$
10	5000 to beyond 10,000 km	-327 to -335 dB
10^2	500 to beyond 10,000 km	-277 to -306
10^3	50 to beyond 10,000 km	-227 to -276
10^4	5 to beyond 10,000 km	-177 to -246
10^5	500 m to beyond 10,000 km	-127 to -216
10^6	50 m to 3000 km	-77 to -176
10^7	5 m to 30 km	-27 to -107
10^8	0.5 to 300 m	23 to -38

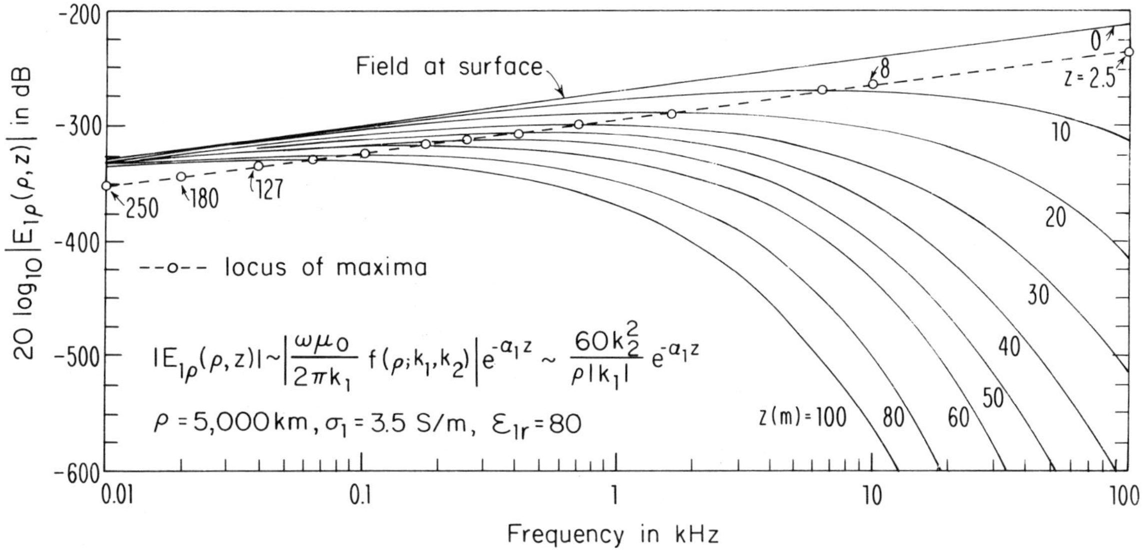

Figure 4.4.3. Magnitude of $E_{1\rho}(\rho, z)$ in V/m at depth z and $\rho = 5000$ km due to vertical electric dipole in air on the surface of sea water as a function of the frequency with z as parameter.

For the frequencies available in the intermediate range, the wave number of sea water is well approximated by $k_1 = \beta_1 + i\alpha_1 = (1+i)(\omega\mu_0\sigma_1/2)^{1/2}$. It follows that α_1 in (4.4.3) is given by $\alpha_1 = (\omega\mu_0\sigma_1/2)^{1/2}$.

The maximum value of $|E_{1\rho}(\rho, z)|$ with respect to the frequency is obtained from the equation $\partial|E_{1\rho}(\rho, z)|/\partial f = 0$. With (4.4.3), this gives the following maximizing condition:

$$\alpha_1 z = 3$$

or

$$fz^2 = \frac{9}{\pi\mu_0\sigma_1} = 6.5 \times 10^5 \text{ m}^2/\text{sec} \quad (\sigma_1 = 3.5 \text{ S/m}). \tag{4.4.4}$$

This formula is independent of the radial distance ρ but only within the intermediate range specified in Table 4.4.1. Graphs of $20 \log_{10}|E_{1\rho}(\rho, z)|$, as given by (4.4.3), are shown in Fig. 4.4.3 for $\rho = 5000$ km as a function of frequency in the range from 0.01 to 100 kHz with the depth z as the parameter. The field at the surface ($z = 0$) is the straight line at the top; the dashed line is the locus of the very broad maximum field at the maximizing frequency for each depth z. It is seen that the greater the depth z, the lower must be the frequency if the maximum possible field for the particular value of z is to be available. On the other hand, the magnitude of the available maximum decreases continuously with frequency. Thus, for a field no smaller than -325 dB for the unit dipole (-215 dB for the practical Cutler antenna described in Section 4.5), a frequency of 20 kHz is useful only to depths $z \sim 20$ m and a frequency of 1 kHz to depths down to $z = 50$ m.

Note that the graphs in Fig. 4.4.3 for $\rho = 5000$ km are readily applied to any other radial distance ρ within the intermediate range by adding $20 \log_{10}(5000/\rho \text{ in km})$ to the ordinate.

4.5 Communication with Submarines—A Practical Antenna

The electromagnetic fields determined in Section 4.4 are generated by idealized vertical dipoles with unit electric moments, $I\Delta l = 1$ A m, where I is the uniform current in amperes and Δl is the length (infinitesimal) in meters. Practical vertical monopole antennas on the earth are usually bare metal cylinders of length h and radius $a \ll h$. For antennas with the electrical lengths $k_2 h \leq \pi/2$, the distribution of current for calculating the electric moment is adequately approximated by

$$I_z(z) = \frac{V_0^e}{Z_0} \frac{\sin k_2(h-z)}{\sin k_2 h}, \tag{4.5.1}$$

where $z \geq 0$ and Z_0 is the driving-point impedance. For calculating the field in the equatorial plane, the antenna can be represented by an equivalent monopole with the uniform current $I_z(0)$ and an effective length h_e given by

$$I_z(0)h_e = \int_0^h I_z(z)\,dz. \tag{4.5.2}$$

The electrical effective length is

$$k_2 h_e = \int_0^h \frac{\sin k_2(h-z)}{\sin k_2 h}\,k_2\,dz = \tan\left(\frac{k_2 h}{2}\right). \tag{4.5.3}$$

[Note that when $I_z(z)$ is complex, h_e is also complex.] For an electrically short antenna with $k_2^2 h^2 \leq 0.3$, $h_e \sim h/2$. If the antenna is sufficiently top-loaded so that the current distribution is approximately uniform in amplitude,

$$h_e \sim h. \tag{4.5.4}$$

The field in the equatorial plane of a vertical monopole is well approximated by the field of a unit dipole multiplied by $I_z(0)h_e$.

A practical vertical antenna designed to excite lateral waves along the surface of the sea is the VLF antenna at the U.S. Naval Radio Station in Cutler, Maine (Watt 1967). It operates in the frequency range from 14 to 28.5 kHz and consists of two base-driven vertical monopoles erected on a peninsula with an extensive radial ground system that ends in the ocean. The distance between the two antennas is only 0.089 to 0.192 wavelength, so that the field pattern does not differ significantly from a circle when the two antennas are driven in phase. The antennas are 298 m high, but the six very large diamond-shaped panels of conductors that serve as the top load for each element are arranged in an umbrella fashion with an outer height of 243 m. The effective height of the array with dual operation is only $h_e = 147$ m at $f = 14$ kHz and $h_e = 151$ m at $f = 28.5$ kHz. These correspond, respectively, to electrical lengths of $k_2 h_e = 0.043$ and $k_2 h_e = 0.090$. The associated radiation resistances are $R_r = 40k_2^2 h_e^2 \sim 0.074$ and 0.32 Ω. With a total current of 2000 A, the effective electric moment is $I_z(0)h_e \sim 3.0 \times 10^5$ A m in the frequency range from 14 to 28.5 kHz. The associated radiated power is $P_r = I_z^2(0)R_r = 0.30$ MW at $f = 14$ kHz and 1.28 MW at $f = 28.5$ kHz. The total input power is of the order of 2 MW.

An increase in electric moment from 1 A m to 3×10^5 A m corresponds to an increase in the electric field by $20\log_{10}(3\times10^5) \sim 110$ dB. This means that the minimum level of -325 dB assumed for the unit dipole is raised to -215 dB by the Cutler antenna. This corresponds to a field intensity of 1.75×10^{-5} μV/m. If the submarine has a horizontal trailing-wire antenna with an effective length for reception of the order of 600 m, the required voltage sensitivity of the receiver is of the order of 0.01 μV.

It may be concluded that for communicating with submarines at radial distances up to 5000 km and depths of 10 to 20 m, top-loaded, electrically very short vertical monopoles like those at Cutler, Maine, operate effectively at frequencies in the 15- to 30-kHz range with currents up to 2000 A. If useful signals are to be received at greater depths, e.g., up to 40 or 50 m, much greater currents and much higher and more directive antennas are required. In view of the already very large physical size of the Cutler array, no significant increase seems practical. Due to corona, a substantially greater driving voltage and resulting current is also ruled out. A reduction in frequency from 15–30 kHz to the order of 1 kHz to make communication to depths up to 50 m possible necessarily requires an increase in the dimensions of a top-loaded antenna by at least a factor of 10 in order to maintain a comparable electrical effective length and radiation resistance. This is not realistic. Note that possible reflections from the ionosphere (which might substantially increase the signal) and the effect of the earth's curvature (which reduces the signal) have not been considered.

4.6 The Measurement of the Conductivity of the Oceanic Crust with a Vertical Electric Dipole as the Source—Theory

Another interesting application of the lateral-wave field of the vertical electric dipole near a boundary is to the measurement of the conductivity of the oceanic crust (Region 2). For this purpose the dipole is best located in the sea (Region 1) on or near the sea floor ($d \sim 0$), as shown in Fig. 4.6.1. Interest is usually in measurements at low frequencies in a range for which not only

$$\omega\epsilon_1 \ll \sigma_1; \qquad k_1 = \beta_1 + i\alpha_1 \sim (i\omega\mu_0\sigma_1)^{1/2} = (1+i)\left(\frac{\omega\mu_0\sigma_1}{2}\right)^{1/2}, \tag{4.6.1}$$

but also

$$\omega\epsilon_2 \ll \sigma_2; \qquad k_2 = \beta_2 + i\alpha_2 \sim (i\omega\mu_0\sigma_2)^{1/2} = (1+i)\left(\frac{\omega\mu_0\sigma_2}{2}\right)^{1/2}. \tag{4.6.2}$$

Measurements are made (usually of the magnetic field) on the sea floor ($z = 0$) within radial distances ρ from the source that are well within the intermediate or even near-field ranges. That is,

$$|k_2\rho| < \left|\frac{k_1^2}{k_2^2}\right|; \qquad |k_1|^2 \gg |k_2|^2. \tag{4.6.3}$$

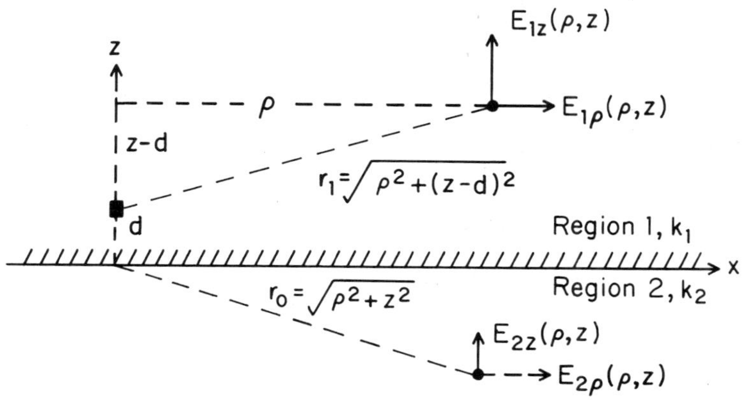

Figure 4.6.1. Vertical dipole at $(0, d)$ in Region 1.

The magnetic field in this range is given by (3.6.7) with (3.4.30), and with the Fresnel-integral term omitted because of (4.6.3). With (4.6.1) and (4.6.2), it follows that $k_2^2/k_1^2 = \sigma_2/\sigma_1$ and

$$B_{1\phi}(\rho, 0) = -\frac{\mu_0\sigma_2}{2\pi\sigma_1}\left[\left(\frac{ik_2}{\rho} - \frac{1}{\rho^2}\right)e^{ik_2\rho}e^{ik_1d} - \frac{id}{\rho}\left(\frac{ik_1}{\rho} - \frac{3}{2\rho^2}\right)e^{ik_1r_d}\right]. \tag{4.6.4}$$

When the vertical antenna extends from the sea floor to a height h in the salt water (and the depth of the sea is large compared to the skin depth), the effective height for a dipole at $z = 0$ on the sea floor, i.e., with $d = 0$, and with a current of uniform amplitude $I_z(z) = I$ is

$$h_e = \int_0^h e^{ik_1z}\,dz = ik_1^{-1}(1 - e^{ik_1h}), \tag{4.6.5}$$

and the magnetic field on the sea floor is

$$B_{1\phi}(\rho, 0) = -\frac{\mu_0\sigma_2}{2\pi\sigma_1}\,Ih_e e^{ik_2\rho}\left(\frac{ik_2}{\rho} - \frac{1}{\rho^2}\right). \tag{4.6.6}$$

Note that in this case the dipole is electrically long in sea water so that (4.6.5) differs from (4.5.2). Formula (4.6.6) is general and complete, subject only to the condition $|k_1/k_2|^2 = \sigma_1/\sigma_2 \geq 9$. At frequencies and radial distances that satisfy the condition

$$|k_2\rho| \leq 0.5, \tag{4.6.7}$$

(4.6.6) reduces approximately to the following quasi-static form:

$$B_{1\phi}(\rho, 0) \sim \frac{\mu_0\sigma_2}{2\pi\sigma_1}\frac{Ih_e}{\rho^2}. \tag{4.6.8}$$

The field in the oceanic crust, Region 2, is obtained from (3.7.30)–(3.7.32) with $\exp(ik_1 d)$ (for the unit dipole at $z = d$) replaced by Ih_e with h_e defined in (4.6.5) (for the insulated antenna with uniform current I extending from the sea floor to $z = h$ in the sea). In the range of interest, the Fresnel-integral terms are negligible as are the terms with $\exp(k_1 z)$ as a factor except at $z = 0$. The relevant components of the field are

$$B_{2\phi}(\rho, z) = -\frac{\mu_0 k_2^2}{2\pi k_1^2}\, Ih_e e^{ik_2 r_0}\left(\frac{ik_2}{r_0} - \frac{1}{r_0^2}\right)\left(\frac{\rho}{r_0}\right); \qquad z \leq 0, \quad (4.6.9)$$

$$E_{2\rho}(\rho, z) = -\frac{\omega\mu_0 k_2}{2\pi k_1^2}\, Ih_e e^{ik_2 r_0}\left[\left(\frac{k_2}{k_1}\right)\left(\frac{ik_2}{r_0} - \frac{1}{r_0^2}\right)\left(\frac{\rho}{r_0}\right) + \left(\frac{ik_2}{r_0} - \frac{3}{r_0^2} - \frac{3i}{k_2 r_0^3}\right)\left(\frac{\rho z}{r_0^2}\right)\right]; \qquad z \leq 0, \quad (4.6.10)$$

$$E_{2z}(\rho, z) = \frac{\omega\mu_0 k_2}{2\pi k_1^2}\, Ih_e e^{ik_2 r_0}\left[\frac{ik_2}{r_0} - \frac{1}{r_0^2} - \frac{i}{k_2 r_0^3} - \left(\frac{ik_2}{r_0} - \frac{3}{r_0^2} - \frac{3i}{k_2 r_0^3}\right)\left(\frac{z^2}{r_0^2}\right)\right]; \qquad z \leq 0. \quad (4.6.11)$$

In order to understand the significance with respect to the properties of the oceanic crust (Region 2) of measurements of the field in the sea (Region 1) on the sea floor, it is necessary to determine the depth of penetration into the crust. The formula (4.6.9), which gives the complete magnetic field at the points of measurement, suggests that the wave emanating from the vertical antenna simply propagates along the sea floor but entirely in the crust. There is no indication of any depth.

The path taken by the electromagnetic wave as it proceeds from the antenna to the point of observation can be determined from the locus of the Poynting vector in Region 2. It is defined as

$$\begin{aligned}\mathbf{S}_2(\rho, z) &= \frac{1}{2\mu_0}\,[\mathbf{E}_2(\rho, z) \times \mathbf{B}_2^*(\rho, z)] \\ &= \frac{1}{2\mu_0}\,[\hat{\mathbf{z}} E_{2\rho}(\rho, z) B_{2\phi}^*(\rho, z) - \hat{\boldsymbol{\rho}} E_{2z}(\rho, z) B_{2\phi}^*(\rho, z)]; \qquad z \leq 0. \quad (4.6.12)\end{aligned}$$

The asterisk indicates the complex conjugate. The slope of the locus of the Poynting vector is

$$\frac{dz}{d\rho} = \frac{\operatorname{Re} S_{2z}(\rho, z)}{\operatorname{Re} S_{2\rho}(\rho, z)}. \qquad (4.6.13)$$

This leads to a very complicated equation in general. Fortunately, it reduces to simpler form in the near field and in part of the intermediate field at frequencies that are sufficiently low so that (4.6.1) and (4.6.2) apply and $k_2/k_1 = (\sigma_2/\sigma_1)^{1/2}$. With $\rho \geq 5|z|$ so that $r_0 = (\rho^2 + z^2)^{1/2} \sim \rho$ in amplitudes and with $|k_2^2\rho^2| \ll 1$, the $1/r_0$ and z^2/r_0^2 terms in (4.6.9)–(4.6.11) are negligible and

$$B_{2\phi}(\rho, z) \sim \frac{\mu_0 k_2^2}{2\pi k_1^2} \frac{Ih_e}{\rho^2} e^{ik_2 r_0}, \tag{4.6.14}$$

$$E_{2\rho}(\rho, z) \sim \frac{\omega\mu_0 k_2^2}{2\pi k_1^2} Ih_e e^{ik_2 r_0} \times \left[\frac{3z}{\rho}\left(\frac{1}{k_2\rho^2} + \frac{i}{k_2^2\rho^3}\right) + \left(\frac{k_2}{k_1}\right)\left(\frac{1}{k_2\rho^2}\right)\right], \tag{4.6.15}$$

$$E_{2z}(\rho, z) \sim -\frac{\omega\mu_0 k_2^2}{2\pi k_1^2} Ih_e e^{ik_2 r_0}\left(\frac{1}{k_2\rho^2} + \frac{i}{k_2^2\rho^3}\right). \tag{4.6.16}$$

It follows that

$$S_{2z}(\rho, z) = \frac{1}{2\mu_0} E_{2\rho}(\rho, z) B_{2\phi}^*(\rho, z) = \frac{\omega\mu_0 k_2^4 I^2 h_e^2}{8\pi^2 k_1^4}\left[\frac{3z}{\rho}\left(\frac{\beta_2(1-i)}{|k_2^2|\rho^2} + \frac{1}{|k_2^2|\rho^3}\right) + \left(\frac{k_2}{k_1}\right)\left(\frac{\beta_2(1-i)}{|k_2^2|\rho^2}\right)\right]\left(\frac{1}{\rho^2}\right), \tag{4.6.17a}$$

$$\operatorname{Re} S_{2z}(\rho, z) = \frac{\omega\mu_0 k_2^4 I^2 h_e^2}{8\pi^2 k_1^4 |k_2^2|\rho^5}\left(\frac{3z}{\rho}(\beta_2\rho + 1) + \frac{k_2}{k_1}\beta_2\rho\right). \tag{4.6.17b}$$

Similarly,

$$S_{2\rho}(\rho, z) = -\frac{1}{2\mu_0} E_{2z}(\rho, z) B_{2\phi}^*(\rho, z) = \frac{\omega\mu_0 k_2^4 I^2 h_e^2}{8\pi^2 k_1^4}\left(\frac{\beta_2(1-i)}{|k_2^2|\rho^2} + \frac{1}{|k_2^2|\rho^3}\right)\left(\frac{1}{\rho^2}\right), \tag{4.6.18a}$$

$$\operatorname{Re} S_{2\rho}(\rho, z) = \frac{\omega\mu_0 k_2^4 I^2 h_e^2}{8\pi^2 k_1^4 |k_2^2|\rho^5}(\beta_2\rho + 1). \tag{4.6.18b}$$

When (4.6.17b) and (4.6.18b) are inserted in (4.6.13), this becomes

$$-\frac{dz'}{d\rho} = \frac{(-3z'/\rho)(\beta_2\rho + 1) + (k_2/k_1)\beta_2\rho}{\beta_2\rho + 1} = -\frac{3z'}{\rho} + \frac{k_2}{k_1}\frac{\beta_2\rho}{\beta_2\rho + 1}, \tag{4.6.19}$$

where, since z is negative in Region 2, it is convenient to introduce $z' = -z$. Equation (4.6.19) can be solved with the substitutions $z' = s\rho^2$ and $(dz'/d\rho) = 2\rho s + \rho^2(ds/d\rho)$. Thus, with $\delta = \beta_2/\beta_1 = (\sigma_2/\sigma_1)^{1/2}$,

$$\rho\left(2s + \rho\frac{ds}{d\rho}\right) = 3\rho s - \frac{\delta\beta_2\rho}{1 + \beta_2\rho} \quad \text{or} \quad \rho\frac{ds}{d\rho} = s - \frac{\delta\beta_2}{1 + \beta_2\rho}, \tag{4.6.20a}$$

or

$$\frac{ds}{s-\delta\beta_2} = \frac{d\rho}{\rho(1+\beta_2\rho)}. \tag{4.6.20b}$$

The solution is

$$\ln(s-\delta\beta_2) = \ln\left(\frac{C\rho}{1+\beta_2\rho}\right), \tag{4.6.21}$$

where C is a constant of integration. The solution with $s = z'/\rho^2$ is

$$z' = \rho^2\left(\delta\beta_2 + \frac{C\rho}{1+\beta_2\rho}\right). \tag{4.6.22}$$

Since $z' = 0$ at both the origin $\rho = 0$ and the radial distance $\rho = \rho_0$ where the lateral wave returns to the boundary surface, it follows that $C = -\delta\beta_2(1+\beta_2\rho_0)/\rho_0$ and

$$z' = \delta\beta_2\rho^2\left(1 - \frac{\rho(1+\beta_2\rho_0)}{\rho_0(1+\beta_2\rho)}\right). \tag{4.6.23}$$

This equation defines the locus of the Poynting vector from the equivalent dipole source at $\rho = 0$, $z' = 0$ to the point of observation at $\rho = \rho_0$, $z' = 0$. With any given value of $\beta_2\rho_0$ in the range

$$|k_2\rho_0| \le 0.6 \quad \text{or} \quad \beta_2\rho_0 \le 0.42, \tag{4.6.24}$$

the locus is readily evaluated for each ρ_0.

Since the ratio $(1+\beta_2\rho_0)(1+\beta_2\rho)^{-1}$ ranges only from 1.42 at $\rho = 0$ to 1 at $\rho = \rho_0$, the characteristics of the locus are more conveniently evaluated from the simple form

$$z' = \delta\beta_2\rho^2\left(1 - \frac{\rho}{\rho_0}\right), \tag{4.6.25}$$

and subsequently corrected with the factor $(1+\beta_2\rho_0)(1+\beta_2\rho)^{-1}$ if greater accuracy is required. Thus, the maximum penetration into Region 2 at $z' = z'_m$, $\rho = \rho_m$ is determined from

$$\frac{dz'}{d\rho} = 0 = 2\rho_m - \frac{3\rho_m^2}{\rho_0} \quad \text{or} \quad \rho_m = \frac{2\rho_0}{3} = 0.67\rho_0 \tag{4.6.26}$$

and

$$z'_m = \delta\beta_2\rho_m^2\left(1 - \frac{\rho_m}{\rho_0}\right) = \delta\beta_2\left(\frac{2\rho_0}{3}\right)^2\left(1 - \frac{2}{3}\right) = 0.15\delta\beta_2\rho_0^2. \tag{4.6.27}$$

The corrected values are obtained with the substitution of (4.6.26) on the right in (4.6.23). Thus,

$$\frac{dz'}{d\rho} = 0 = 2\rho_{mc} - \frac{3\rho_{mc}^2}{\rho_0}\left(\frac{1+\beta_2\rho_0}{1+2\beta_2\rho_0/3}\right)$$

or

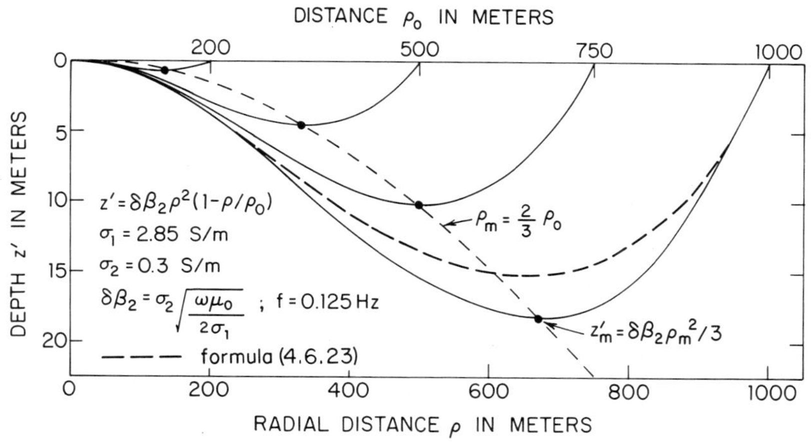

Figure 4.6.2. Locus of Poynting vector in half-space model of the oceanic crust; $k_1 = (i\omega\mu_0\sigma_1)^{1/2}$, $k_2 = (i\omega\mu_0\sigma_2)^{1/2}$; $S_{1\rho}(\rho_0, 0) = (\sigma_2/\sigma_1)S_{2\rho}(\rho_0, 0)$, $S_{1z}(\rho_0, 0) = S_{2z}(\rho_0, 0)$.

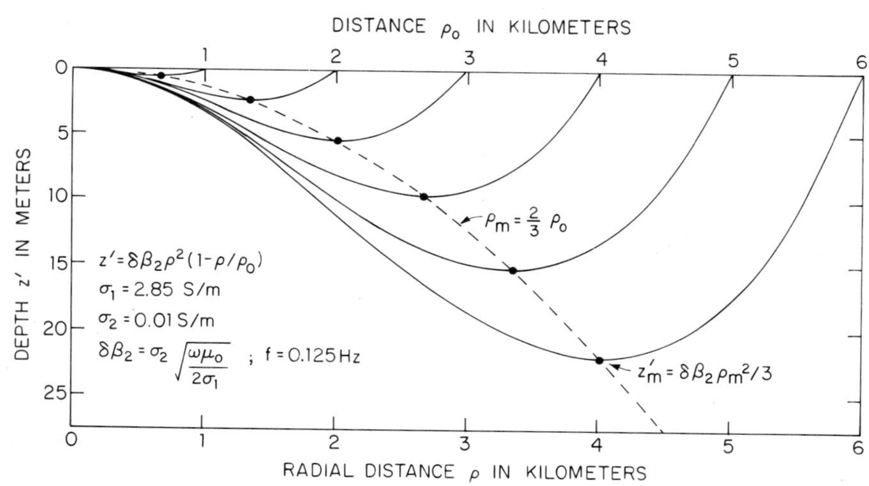

Figure 4.6.3. Locus of Poynting vector in half-space model of the oceanic crust; $k_1 = (i\omega\mu_0\sigma_1)^{1/2}$, $k_2 = (i\omega\mu_0\sigma_2)^{1/2}$; $S_{1\rho}(\rho_0, 0) = (\sigma_2/\sigma_1)S_{2\rho}(\rho_0, 0)$, $S_{1z}(\rho_0, 0) = S_{2z}(\rho_0, 0)$.

$$\rho_{mc} = \frac{2\rho_0}{3}\left(\frac{1 + 2\beta_2\rho_0/3}{1 + \beta_2\rho_0}\right). \tag{4.6.28}$$

Similarly,

$$z'_{mc} = \delta\beta_2\left(\frac{2\rho_0}{3}\right)^2\left[1 - \frac{2}{3}\left(\frac{1 + \beta_2\rho_0}{1 + 2\beta_2\rho_0/3}\right)\right]. \tag{4.6.29}$$

With $\beta_2\rho_0 = 0.42$,

$$\rho_{mc} = \frac{2\rho_0}{3}\left(\frac{1.28}{1.42}\right) = 0.60\rho_0, \tag{4.6.30}$$

$$z'_{mc} = \delta\beta_2\rho_0^2\left(\frac{4}{9}\right)\left(1 - \frac{2}{3}\cdot\frac{1.42}{1.28}\right) = 0.12\delta\beta_2\rho_0^2. \tag{4.6.31}$$

These represent the maximum departure from the approximate values in (4.6.26) and (4.6.27) obtained from the simple formula (4.6.25).

Note that with $\delta = \beta_2/\beta_1$, it follows that

$$\delta\beta_2 = \frac{\beta_2^2}{\beta_1} = \sigma_2 R_s, \tag{4.6.32}$$

where

$$R_s = \left(\frac{\omega\mu_0}{2\sigma_1}\right)^{1/2} \tag{4.6.33}$$

is the surface resistance.

Graphs of the loci defined by (4.6.25) are shown in Figs. 4.6.2 and 4.6.3 for parameters that apply to actual measurements made on the sea floor. Figure 4.6.2 applies to the sea–sediment boundary with $\sigma_1 = 2.85$ S/m, $\sigma_2 = 0.3$ S/m, and $f = 0.125$ Hz. For a radial range of 1 km, the maximum depth of penetration is about 18 m. The dashed curve is calculated from the more accurate formula (4.6.23). It is seen that it differs from the approximate curve calculated from the simple formula (4.6.25) primarily in yielding a somewhat smaller depth of penetration. For many purposes the simpler formula should be adequate. Figure 4.6.3 applies to a sea–rock boundary with $\sigma_1 = 2.85$ S/m, $\sigma_2 = 0.01$ S/m, and $f = 0.125$ Hz. For a radial range of 5 km, the maximum depth of penetration of the lateral wave into the rock is only 15.3 m.

4.7 The Measurement of the Conductivity of the Oceanic Crust Using a Vertical Dipole—Comparison of Theory with Measurement

An interesting set of measurements has been reported by Edwards et al. (1985) for determining the conductivity of the oceanic crust below the Bute

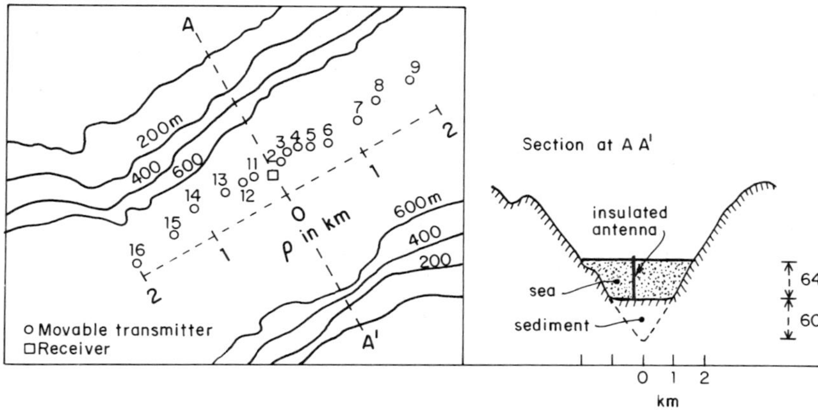

Figure 4.7.1. Location of measurements by Edwards et al. (1985).

Inlet, B.C. (Canada). It provides a useful vehicle for the application and discussion of the theory developed in Section 4.6. The site of the measurements has the cross section sketched on the right in Fig. 4.7.1. The sea water (Region 1, $\sigma_1 = 2.85$ S/m) has a depth $D_1 = 640$ m in a channel that is almost 4 km wide at the surface and 2 km wide on the floor. Below the water is a layer of sediment (Region 2, σ_2) that fills a trough that is estimated from the slopes of the rocky sides to have a central depth near $D_2 = 600$ m. The conductivity of the sediment was estimated from measurements of the moisture content of sample cores to range from 1/11th to 1/5th that of the salt water so that $\sigma_2 \sim 0.26$ to 0.57 S/m. The country rock (Region 3, σ_3) below the sediment has a conductivity of the order of $\sigma_3 \sim 0.01$ S/m. Above the sea water is, of course, air (Region 0).

The antenna used by Edwards et al. (1985) was an insulated conductor that extended from the sea floor to the air surface so that $h = D_1 = 640$ m. At each end the antenna was conductively connected to the salt water by bare metal rods. Its consecutive locations in the Bute Inlet are indicated by the numbered circles in Fig. 4.7.1. At the very low frequency of $f = 0.125$ Hz, the wave number of the salt water is $k_1 = (i\omega\mu_0\sigma_1)^{1/2} = 1.68 \times 10^{-3} e^{i\pi/4}$ m^{-1} and the electrical length of the antenna is $\beta_1 h = 0.76$. This is short enough so that a current with the uniform amplitude I is maintained over its entire length. The electric moment of an approximately equivalent infinitesimal dipole at $d = 0$ is Ih_e, where h_e is the effective length of the antenna. For an antenna of length h in an ocean with a depth D_1 that is much greater than both the skin depth and the length of the antenna, the effective length is

$$h_e = \int_0^h e^{ik_1 z}\, dz = ik_1^{-1}(1 - e^{ik_1 h}). \tag{4.7.1}$$

With $k_1 = 1.68 \times 10^{-3} e^{i\pi/4}$ m^{-1} and $h = 640$ m, $h_e = 438 e^{i0.335}$ m. Since the antenna extends from the sea floor to the surface with $h = D_1 = 640$ m and this is actually less than the skin depth $d_s = \alpha_1^{-1} = 847$ m, the presence of the sea–air boundary must be taken into account. This is readily done since, at the very low frequency used, the reflection coefficient at the air–sea boundary for the magnetic field of a wave incident from the sea is extremely close to -1. It follows that the boundary surface can be approximated by a magnetic wall on which the tangential magnetic field vanishes. That is, $B_{1\phi}(\rho, h) = 0$. This means that the entire air half-space can be replaced by an inverted image of the lower half-space containing the sea and oceanic crust. The antenna thus has an image with reversed current so that the effective length becomes

$$h_e = \int_0^h e^{ik_1 z}\, dz - \int_h^{2h} e^{ik_1 z}\, dz = ik_1^{-1}(1 - e^{ik_1 h})^2. \tag{4.7.2}$$

With $k_1 = 1.68 \times 10^{-3} e^{i\pi/4}$ m^{-1} and $h = 640$ m, this gives $h_e = 321 e^{-i0.12}$ m. In order to maintain the zero value of the magnetic field at $z = h$—now the midpoint in the ocean with image which has a combined depth of $2h$—the lateral wave traveling in the oceanic crust at $z < 0$ must also have an image that travels along the image of the crust at $z > 2h$. The contribution of the image wave to the field at $(\rho, 0)$ on the sea floor is identical with that of the actual wave but with opposite sign and with an additional distance $2h$ of travel down in the sea water from the plane $z = 2h$ to the plane $z = 0$. The total magnetic field observed in the sea on the sea floor, that is, at $(\rho, 0)$ in Region 1, includes the lateral wave traveling vertically upward into the sea from the sediment, the image wave traveling vertically down, and multiple reflections of these. The magnetic field at any point z is given by

$$B_{1\phi}(\rho, z) = B_{1\phi}^{\text{inc}}(\rho, 0)\left(\frac{e^{ik_1 z} - e^{ik_1(2h - z)}}{1 + \Gamma_m e^{i2k_1 h}}\right), \tag{4.7.3}$$

where

$$\Gamma_m = -\frac{k_1 - k_2}{k_1 + k_2} \equiv -\frac{\sqrt{\sigma_1} - \sqrt{\sigma_2}}{\sqrt{\sigma_1} + \sqrt{\sigma_2}} \tag{4.7.4}$$

is the reflection coefficient of magnetic type at the sea–sediment boundary. Note that $B_{1\phi}(\rho, h) = 0$ as required. With $\sigma_2 = 0.4$ S/m and $\sigma_1 = 2.85$ S/m, $\Gamma_m = -0.45$. The field on the sea floor at $z = 0$ is given by

$$B_{1\phi}(\rho, 0) = B_{1\phi}^{\text{inc}}(\rho, 0)\left(\frac{1 - e^{i2k_1 h}}{1 + \Gamma_m e^{i2k_1 h}}\right). \tag{4.7.5}$$

The complete effective length as modified by the effect of the air–sea boundary is

$$h_e = \frac{i}{k_1}\,\frac{(1 - e^{ik_1 h})^2(1 - e^{i2k_1 h})}{1 + \Gamma_m e^{i2k_1 h}}. \tag{4.7.6}$$

With $\Gamma_m = -0.45$, $k_1 = 1.68 \times 10^{-3} e^{i\pi/4}$ m^{-1}, and $h = 640$ m,

$$h_e = 326 e^{-i0.241} \text{ m}. \tag{4.7.7}$$

With $\Gamma_m = 0$, $h_e = 329e^{-i0.341}$ m. Clearly, only the first reflection is significant.

It is to be noted that even though all of the multiple reflections of the lateral waves reaching the point of observation at $(\rho, 0)$ in the sea are included, the antenna generating these waves includes only the section of length $2h$ in the sea with image and images of equal length in the oceanic crust and its image. This gives a total length of $6h$ with a uniform current that alternates in sign in successive lengths h. The effect of multiple reflections on the waves leaving the antenna are not included and cannot be provided by image antennas unless the reflection coefficient of the crust is close to ± 1—which it is not. When σ_2 lies between 0.26 and 0.57 S/m and $\sigma_1 = 2.85$ S/m, the reflection coefficient, $-(\sqrt{\sigma_1} - \sqrt{\sigma_2})/(\sqrt{\sigma_1} + \sqrt{\sigma_2})$, ranges from -0.53 to -0.38. Formula (4.7.6) is evidently accurate so long as contributions from image antennas beyond $4k_1h$ are negligible. That is, it is necessary that $|\exp(i4k_1h)| = \exp(-4\alpha_1 h) \ll 1$. With $\alpha_1 = 1.188 \times 10^{-3}$ m^{-1} and $h = 640$ m, $\exp(-4\alpha_1 h) = 0.02$. It follows that the magnitude of the lateral wave and its image should be correct within 2% when (4.7.6) is used. Note, however, that (4.7.6) is not useful in the limit of zero frequency. An alternative formula that has application in this limit is given and discussed in Section 4.8.

Measurements of the horizontal magnetic field on the sea floor were made by Edwards et al. (1985) with the transmitting antenna at the numbered locations relative to the stationary receiving antenna (magnetometer), as shown on the left in Fig. 4.7.1. The measured field is shown by the numbered crosses in Fig. 4.7.2 as a function of the radial distance from the base of the transmitting antenna.

The values of the wave numbers k_j, $j = 0, 1, 2, 3$, for the air, sea water, sediment, and rock with the suggested values of σ_j are listed in Table 4.7.1 at $f = 0.125$ Hz. Also given are the ratios $\delta = k_2/k_1 = (\sigma_2/\sigma_1)^{1/2}$; the ranges of electrical distance $|k_2\rho|$ corresponding to the actual ranges of measurement, $0.1 \le \rho \le 2$ km; the maximum allowable ranges of ρ to satisfy the condition for the neglect of the far-field terms; and the associated maximum depths of penetration of the Poynting vector.

In general, the electromagnetic field in a four-layered region cannot be determined from formulas for the field in a two-layered region. In the case at hand, the properties of the layers at the very low frequency and the locations of the transmitting antenna and the magnetometer provide a special situation which makes the two-layer theory useful for determining the conductivity of the layer of sediment from measurements of the magnetic field on the sea floor. However, the idealization that all boundaries are parallel plane surfaces is required.

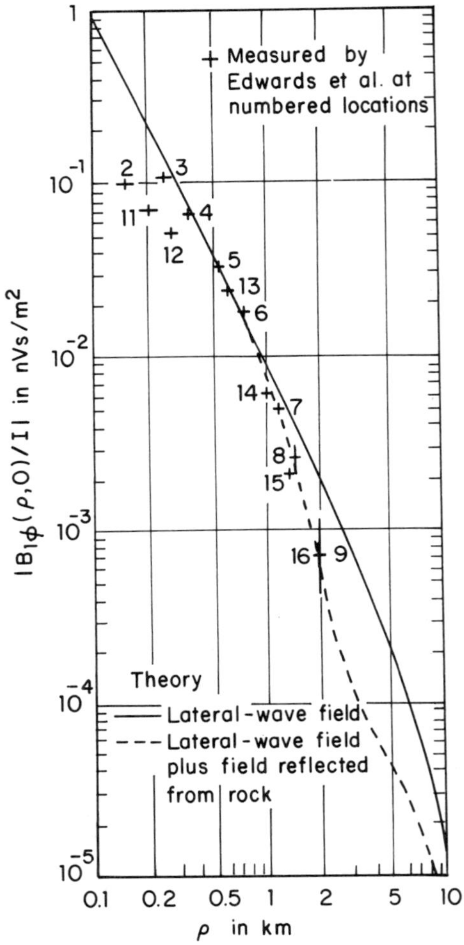

Figure 4.7.2. Magnetic field measured by Edwards et al. and field calculated from (4.7.9) and (4.7.18); $f = 0.125$ Hz; $\sigma_1 = 2.85$ S/m, $\sigma_2 = 0.4$ S/m, $\sigma_3 = 0.01$ S/m; $D_1 = 640$ m, $D_2 = 600$ m.

The measured field is in the sea water (Region 1) at its boundary with the sediment (Region 2). Theory shows that if the sea were unbounded upward and the sediment unbounded downward, the direct and reflected fields of a vertical antenna on the boundary would cancel so that the entire magnetic field along the boundary is the lateral-wave field. However, theory also shows that for the radial distances involved in the measurements the maximum depth of penetration z'_m of the lateral wave into the sediment is a small fraction of the depth D_2 of that sediment. This means that the lateral-wave field $B^L_{1\phi}(\rho, 0)$ is the same when $D_2 \gg z'_m$ as when $D_2 = \infty$. But it may not be the entire field. There may be a significant reflected field

Table 4.7.1. Numerical Values of Parameters

σ_2 in S/m	0.26	0.3
k_2 in m^{-1}	$5.06 \times 10^{-4} e^{i\pi/4}$	$5.44 \times 10^{-4} e^{i\pi/4}$
$\delta = (\sigma_2/\sigma_1)^{1/2}$	0.30	0.324
Range: $0.1 \le \rho \le 2$ km	$0.05 \le \lvert k_2\rho\rvert \le 1$	$0.054 \le \lvert k_2\rho\rvert \le 1.09$
Near field: $\rho \le 0.6\lvert k_2^{-1}\rvert$	1.19 km	1.10 km
$z'_m = 0.15\delta\beta_2\rho^2$	22.7 m	22.6 m

σ_2 in S/m	0.4	0.57
k_2 in m^{-1}	$6.44 \times 10^{-4} e^{i\pi/4}$	$7.50 \times 10^{-4} e^{i\pi/4}$
$\delta = (\sigma_2/\sigma_1)^{1/2}$	0.375	0.45
Range: $0.1 \le \rho \le 2$ km	$0.064 \le \lvert k_2\rho\rvert \le 1.28$	$0.075 \le \lvert k_2\rho\rvert \le 1.5$
Near field: $\rho \le 0.6\lvert k_2^{-1}\rvert$	0.93 km	0.80 km
$z'_m = 0.15\delta\beta_2\rho^2$	22.5 m	22.9 m

$$\begin{aligned} &f = 0.125 \text{ Hz}, && k_0 = 0.26 \times 10^{-8} \text{ m}^{-1};\\ &\sigma_1 = 2.85 \text{ S/m}, && k_1 = 1.68 \times 10^{-3} e^{i\pi/4} \text{ m}^{-1};\\ &\sigma_3 \sim 0.01 \text{ S/m}, && k_3 = 0.99 \times 10^{-4} e^{i\pi/4} \text{ m}^{-1};\\ &h = 640 \text{ m}, && h_e = 326 e^{-i0.241} \text{ m} \end{aligned}$$

from the sediment–rock boundary that travels upward in the sediment to the point of observation in the sea. Furthermore, since the vertical antenna extends to the air–sea surface, its currents also generate a lateral wave that propagates along that boundary and down in the sea to the point of measurement on the sea floor. Thus,

$$B_{1\phi}(\rho,0) = B_{1\phi}^L(\rho,0) + B_{1\phi}^A(\rho,0) + B_{1\phi}^T(\rho,0), \tag{4.7.8}$$

where

$$B_{1\phi}^L(\rho,0) = -\frac{\mu_0 k_2^2}{2\pi k_1^2}\, I h_e e^{ik_2\rho}\left(\frac{ik_2}{\rho} - \frac{1}{\rho^2}\right) \tag{4.7.9}$$

is the lateral wave along the sediment–sea boundary and

$$B_{1\phi}^A(\rho,0) = -\frac{\mu_0 k_0^2}{2\pi k_1^2}\, I h_e e^{ik_1 D_1} e^{ik_0\rho}\left(\frac{ik_0}{\rho} - \frac{1}{\rho^2}\right) \tag{4.7.10}$$

is the lateral wave that travels along the air–sea boundary a radial distance ρ and then proceeds vertically down the depth D_1 of the sea to the point of measurement on the sea floor. $B_{1\phi}^T(\rho,0)$ is the field that propagates down

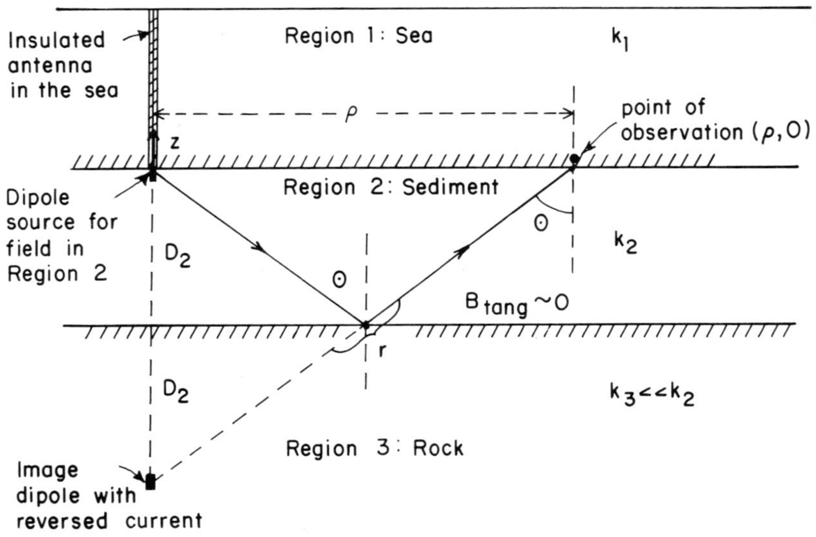

Figure 4.7.3. Schematic diagram locating the equivalent source dipole for the field in Region 2 at $(0, 0)$, its image with reversed current at $(0, -2D_2)$, and the point of observation in Region 1 at $(\rho, 0)$.

in the sediment as a spherical wave given by (4.6.9)–(4.6.11), is partially reflected at the rock boundary, and then travels upward to the sediment–sea boundary where it is partly reflected, partly transmitted to the point of observation in the sea. This is shown schematically in Fig. 4.7.3.

It is readily shown that $B_{1\phi}^A(\rho, 0)$ is negligible compared with $B_{1\phi}^L(\rho, 0)$. The ratio $B_{1\phi}^A(\rho, 0)/B_{1\phi}^L(\rho, 0)$ is dominated by the ratio

$$\frac{k_0^2}{k_2^2} = \frac{\omega^2\mu_0\epsilon_0}{i\omega\mu_0\sigma_2} = -\frac{i\omega\epsilon_0}{\sigma_2} = -i1.74 \times 10^{-11},$$

when $f = 0.125$ Hz and $\sigma_2 = 0.4$ S/m. This is so small that $B_{1\phi}^A(\rho, 0)$ can be neglected. It follows that (4.7.8) reduces to

$$B_{1\phi}(\rho, 0) = B_{1\phi}^L(\rho, 0)\left(1 + \frac{B_{1\phi}^T(\rho, 0)}{B_{1\phi}^L(\rho, 0)}\right). \tag{4.7.11}$$

The field $B_{1\phi}^T(\rho, 0)$ can be determined in a step-by-step manner. Consider first the spherical wave traveling downward in the sediment from the equivalent vertical dipole at $(0, 0)$ in the sediment. This is incident on the sediment–rock boundary. If on this surface it is treated as a locally plane wave, the reflection coefficient is

$$f_{er} = \frac{(\sigma_3/\sigma_2)\cos\Theta - [(\sigma_3/\sigma_2) - \sin^2\Theta]^{1/2}}{(\sigma_3/\sigma_2)\cos\Theta + [(\sigma_3/\sigma_2) - \sin^2\Theta]^{1/2}}, \tag{4.7.12}$$

where the angle of incidence is Θ and $\cos\Theta = D_2/[D_2^2 + (\rho/2)^2]^{1/2}$. With $\sigma_3 \sim 0.01$ S/m and $\sigma_2 \sim 0.4$ S/m, the quantity $\sigma_3/\sigma_2 \sim 0.024$ is very small. It follows that, except for very small angles, $\sin^2\Theta$ is significantly greater than σ_3/σ_2 and

$$f_{er} \sim -1. \tag{4.7.13}$$

This value is independent of the precise value of σ_3/σ_2 so long as this is small. With (4.7.13), the reflected field is the same as the field of an image of the source dipole and located at $(0, -2D_2)$ but with reversed electric moment, i.e., with the moment $-Ih_e$. The magnetic field of this image dipole incident on the sea floor from below, i.e., from the sediment, is

$$B_{2\phi}^R(\rho, 0) = \frac{\mu_0 k_2^2}{2\pi k_1^2}\, Ih_e e^{ik_2 r}\left(\frac{ik_2}{r} - \frac{1}{r^2}\right)\left(\frac{\rho}{r}\right), \tag{4.7.14}$$

where $r = (4D_2^2 + \rho^2)^{1/2}$ is the distance from the image dipole at $(0, -2D_2)$ to the point $(\rho, 0)$ on the boundary but *in the sediment.* The point of observation is at $(\rho, 0)$ *in the sea.* The field at this point is given by (4.7.14) multiplied by the transmission coefficient. That is,

$$B_{1\phi}^T(\rho, 0) = B_{2\phi}^R(\rho, 0) f_{et}, \tag{4.7.15}$$

where f_{et} is the transmission coefficient for $B_{2\phi}^R(\rho, 0)$ which is transverse to the plane of incidence. It is given by

$$\begin{aligned} f_{et} &= \frac{2k_1^2 \cos\Theta}{k_1^2\cos\Theta + k_2(k_1^2 - k_2^2\sin^2\Theta)^{1/2}} \\ &= \frac{2(\sigma_1/\sigma_2)\cos\Theta}{(\sigma_1/\sigma_2)\cos\Theta + [(\sigma_1/\sigma_2) - \sin^2\Theta]^{1/2}}. \end{aligned} \tag{4.7.16}$$

Since $k_2^2/k_1^2 = \sigma_2/\sigma_1 = 0.4/2.85 = 0.14$ is quite small and significant reflections occur only when $\rho > 2D_2$ and ρ/r is not a small fraction, a satisfactory approximation is

$$f_{et} \sim \frac{2\cos\Theta}{\cos\Theta + (\sigma_2/\sigma_1)^{1/2}} = \frac{4D_2}{2D_2 + r(\sigma_2/\sigma_1)^{1/2}} \tag{4.7.17}$$

since $\cos\Theta = 2D_2/r$. The Brewster angle at which the entire field is transmitted into the sea occurs with $\rho = 3.5$ km when $D_2 = 600$ m. It follows with (4.7.9), (4.7.14), (4.7.15), and (4.7.17) in (4.7.11) that

$$\begin{aligned} B_{1\phi}(\rho, 0) = B_{1\phi}^L(\rho, 0)\Bigg[1 - e^{ik_2(r-\rho)}\left(\frac{ik_2 r - 1}{ik_2\rho - 1}\right)\left(\frac{\rho}{r}\right)^3 \\ \times \left(\frac{4D_2}{2D_2 + r(\sigma_2/\sigma_1)^{1/2}}\right)\Bigg], \end{aligned} \tag{4.7.18}$$

where $r = (4D_2^2 + \rho^2)^{1/2}$. This is the total magnetic field at the point of observation $(\rho, 0)$ on the sea floor due to the lateral wave traveling in the sediment along a locus like those shown in Fig. 4.7.2 and the reflection from the rock layer at the depth D_2.

The magnitude of the lateral-wave magnetic field given by (4.7.9) has been calculated with $\sigma_1 = 2.85$ S/m, $D_1 = 640$ m, $\sigma_2 = 0.4$ S/m, $D_2 = 600$ m, and $\sigma_3 = 0.01$ S/m over a radial range from $\rho = 0.1$ to $\rho = 10$ km. It is shown in Fig. 4.7.2. The complete magnetic field including the reflection from the rock layer as calculated from (4.7.18) is shown in dashed line in Fig. 4.7.2. The reflected field is seen to produce a significant decrease in the field due to the lateral wave in a radial range centered around $\rho = 3.5$ km. At this distance, the angle of incidence on the sediment–sea boundary of the diagonally upward-traveling reflected field from the sediment–rock boundary is the Brewster angle of $\Theta = 71°$ (see Fig. 4.7.3).

When compared with the magnetic field measured by Edwards et al. (1985) in the Bute Inlet, B.C., at the numbered locations shown in Fig. 4.7.1, the theoretical curve is seen to be in excellent agreement over the entire range. This indicates that the conductivity of the sediment must have an average value very near $\sigma_2 = 0.4$ S/m at least over distances of the order of 23 m (Table 4.7.1) below the sea floor actually traversed by the lateral wave that reaches the point of observation. This value differs from $\sigma_2 = 0.52$ S/m calculated by Edwards et al. from the same data by a method that is not given in detail. However, with $\omega\epsilon_2 = 0$ and $k_2 = (i\omega\mu_0\sigma_2)^{1/2}$, formula (20) in Edwards, Nobes, and DeLaurier (1984) is essentially the same as (4.7.9).

4.8 A Note on the Effective Length of the Antenna

Formula (4.7.9) for the magnetic field on the sea floor corresponds to formula (20) given by Edwards, Nobes, and DeLaurier (1984) but their effective length is

$$h_e = \frac{i}{k_1}\left(\frac{1 - 2e^{ik_1h} + e^{i2k_1h}}{1 - e^{i2k_1h}}\right) = \frac{i}{k_1}\left(\frac{1 - e^{ik_1h}}{1 + e^{ik_1h}}\right). \tag{4.8.1}$$

Their formula is obtained from the exact integral in the low-frequency limit $\omega \to 0$ with the condition that the reflection coefficient of the crust be close to -1.

It is interesting to note that (4.8.1) is exactly the effective length of a *semi-infinite* array of images extending from the sea–air boundary to $+\infty$, with the reflection coefficient of all *images* of the sea–crust boundary set equal to -1. No reflections from the actual sea–crust boundary at $z = 0$ are included, only reflections from its images in the sea–air boundary. That is,

$$h_e = \left(\int_0^h - \int_h^{2h} + \int_{2h}^{3h} - \int_{3h}^{4h} + \cdots \right) e^{ik_1 z}\, dz$$
$$= \frac{i}{k_1}(1 - e^{ik_1 h})(1 - e^{ik_1 h} + e^{i2k_1 h} - e^{i3k_1 h} + \cdots) \tag{4.8.2}$$

or

$$h_e = \frac{i}{k_1}\left(\frac{1 - e^{ik_1 h}}{1 + e^{ik_1 h}}\right). \tag{4.8.3}$$

This is exactly (4.8.1). The field generated by this array violates the boundary condition that $B_{1\phi}(\rho, z) = 0$ when $z = h,\ 3h,\ 5h, \ldots$, i.e., on the sea–air boundary and its images, yet assumes the reflection coefficient for each of these boundaries to be -1. The lateral wave along the actual sea crust boundary is preserved because the reflection coefficient for it has its true value and is not set equal to -1 as for all images of the same boundary. However, the lateral waves along the image boundaries are all canceled. If the contribution from the image lateral wave in (4.7.6) is set equal to zero [the term $\exp(i2k_1 h)$ in the numerator] and Γ_m is set equal to -1, (4.7.6) is identically (4.8.1).

So long as $\exp(-4\alpha_1 h) \ll 1$, formula (4.7.6) should be more accurate than (4.8.1). However, in the limit $\omega \to 0$, (4.7.6) has no application and (4.8.1) reduces to $h/2$. With $k_1 = 1.68 \times 10^{-3} e^{i\pi/4}$ m^{-1} and $h = 640$ m, (4.8.1) gives

$$h_e = 318 e^{i1.0}. \tag{4.8.4}$$

This differs from $h_e = 326$ given by (4.7.6) by 2.5%. Since the first reflection from the sea–air boundary makes the only really significant contribution, the treatment of higher-order reflections is largely irrelevant.

5
The Electromagnetic Field of a Horizontal Electric Dipole in the Presence of a Plane Boundary

5.1 Introduction

The vertical electric dipole is an effective and very useful transmitter when erected in air on the surface of the earth or ocean and operated in the range of frequencies used in broadcast radio communication. Antenna lengths of a quarter- or half-wavelength in air are practical and directional arrays are readily constructed. The entire far field along the surface of the earth or sea is a lateral wave often called the Norton surface wave. At lower frequencies required for communication with submerged submarines, the vertical dipole becomes impractical because it cannot be made high enough to have a reasonable electrical length and radiation resistance. When embedded in the earth or sea (Region 1, wave number k_1) near a boundary with air, rock, or sediment (Region 2, wave number k_2), the effectiveness of the vertical dipole is greatly diminished because its field is multiplied by the very small factor $(k_2/k_1)^2$. This restricts its application to the near range and part of the intermediate range unless enormous amounts of power are supplied.

Instead of arranging an antenna perpendicular to a boundary, it can be located horizontally near the interface in either medium. The structural properties of horizontal antennas are very different from those of vertical ones. As a consequence, the length of the element is not a limitation even at very low frequencies. This means that electrically long traveling-wave antennas can be constructed for use on either side of a boundary. The electromagnetic fields generated by electrically short and long horizontal antennas have many useful applications in subsurface and close-to-the-surface communication, radar, and geophysical prospecting and diagnostics both from the air–earth boundary and from the boundary between the sea and the oceanic crust. The first step in the determination of such fields is a detailed study of the electromagnetic field on both sides of a boundary surface due to a horizontal unit electric dipole located near or on the boundary surface. The procedure to be followed is similar to that used for the vertical dipole, but the analysis is more complicated because all six components of the electromagnetic field are involved and not only the three, E_ρ, E_z, and B_ϕ.

The first analysis of the field generated by a unit horizontal electric dipole near the plane boundary between two electrically different half-spaces was in the dissertation of von Hörschelmann (1911), a student of Sommerfeld. As with the vertical electric dipole, the electromagnetic field is expressed in terms of derivatives of general integrals of the Hertz potential; in this case two components are required. This formulation was also used by other investigators including, especially, Baños and Wesley (1953, 1954), Wait (1961), and Baños (1966). More recently, integrals for the components of the electromagnetic field were derived directly from Maxwell's equations by King and Smith (1981). For completeness this last derivation is summarized here.

5.2 Maxwell's Equations and Their Transforms

Consider the x-directed horizontal electric dipole shown in Fig. 5.2.1 at the point $z = d$ on the downward-directed z-axis in the half-space (Region 1) defined by $z \geq 0$. Region 2 is defined by $z \leq 0$. The wave numbers of the two regions are $k_j = \beta_j + i\alpha_j = \omega(\mu_0\tilde{\epsilon}_j)^{1/2}$, where $\tilde{\epsilon}_j = \epsilon_j + i\sigma_j/\omega$ and $j = 1, 2$. It is assumed that both regions are nonmagnetic so that $\mu_1 = \mu_2 = \mu_0$.

Maxwell's equations in the two regions are

$$\nabla \times \mathbf{E}_j = i\omega\mathbf{B}_j, \tag{5.2.1}$$

$$\nabla \times \mathbf{B}_j = \mu_0(-i\omega\tilde{\epsilon}_j\mathbf{E}_j + \hat{\mathbf{x}}J_x), \tag{5.2.2}$$

where

$$J_x = \delta(x)\delta(y)\delta(z - d) \tag{5.2.3}$$

is the volume density of current in the dipole normalized to have a unit electric moment ($I\Delta l = 1$ A m). They are to be solved for $\mathbf{E}_j$ and $\mathbf{B}_j$, $j = 1, 2$, subject to the boundary conditions

$$E_{1x}(x, y, 0) = E_{2x}(x, y, 0),$$

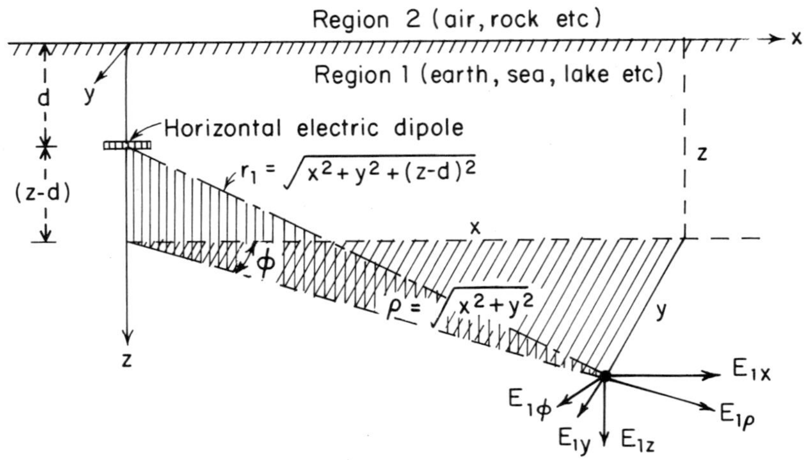

Figure 5.2.1. Schematic diagram of x-directed horizontal electric dipole at $(0, 0, d)$ and electric field at (x, y, z) or (ρ, ϕ, z).

$$E_{1y}(x, y, 0) = E_{2y}(x, y, 0),$$

$$k_1 E_{1z}(x, y, 0) = k_2 E_{2z}(x, y, 0); \tag{5.2.4}$$

$$\mathbf{B}_1(x, y, 0) = \mathbf{B}_2(x, y, 0). \tag{5.2.5}$$

The translational invariance of the boundary in the x- and y-directions suggests the use of the transform

$$\mathbf{E}(x, y, z) = \frac{1}{(2\pi)^2} \int_{-\infty}^{\infty} d\xi \int_{-\infty}^{\infty} d\eta \; e^{i(\xi x + \eta y)} \bar{\mathbf{E}}(\xi, \eta, z), \tag{5.2.6}$$

and a similar one for $\mathbf{B}(x, y, z)$. With (5.2.3), the transform of $J_x(x, y, z)$ is

$$\bar{J}_x(\xi, \eta, z) = \delta(z - d). \tag{5.2.7}$$

The transformed equations (5.2.1) and (5.2.2) in Cartesian coordinates are

$$i\eta \bar{E}_{jz} - \frac{\partial}{\partial z} \bar{E}_{jy} = i\omega \bar{B}_{jx}, \tag{5.2.8}$$

$$\frac{\partial}{\partial z} \bar{E}_{jx} - i\xi \bar{E}_{jz} = i\omega \bar{B}_{jy}, \tag{5.2.9}$$

$$i\xi \bar{E}_{jy} - i\eta \bar{E}_{jx} = i\omega \bar{B}_{jz}, \tag{5.2.10}$$

$$\begin{aligned} i\eta \bar{B}_{jz} - \frac{\partial}{\partial z} \bar{B}_{jy} &= -i\omega\mu_0 \tilde{\epsilon}_j \bar{E}_{jx} + \mu_0 \delta(z - d) \\ &= -\frac{ik_j^2}{\omega} \bar{E}_{jx} + \mu_0 \delta(z - d), \end{aligned} \tag{5.2.11}$$

$$\frac{\partial}{\partial z} \bar{B}_{jx} - i\xi \bar{B}_{jz} = -i\omega\mu_0 \tilde{\epsilon}_j \bar{E}_{jy} = -\frac{ik_j^2}{\omega} \bar{E}_{jy}, \tag{5.2.12}$$

$$i\xi \bar{B}_{jy} - i\eta \bar{B}_{jx} = -i\omega\mu_0\tilde{\epsilon}_j \bar{E}_{jz} = -\frac{ik_j^2}{\omega}\,\bar{E}_{jz}. \tag{5.2.13}$$

With these equations the y- and z-components of both $\bar{\mathbf{E}}_j$ and $\bar{\mathbf{B}}_j$ can be expressed in terms of $\bar{E}_{jx}$ and $\bar{B}_{jx}$. The results are

$$\bar{E}_{jy} = \frac{1}{k_j^2 - \xi^2}\left(-\xi\eta\bar{E}_{jx} + i\omega\,\frac{\partial}{\partial z}\,\bar{B}_{jx}\right), \tag{5.2.14}$$

$$\bar{E}_{jz} = \frac{1}{k_j^2 - \xi^2}\left(i\xi\,\frac{\partial}{\partial z}\,\bar{E}_{jx} + \eta\omega\bar{B}_{jx}\right), \tag{5.2.15}$$

$$\bar{B}_{jy} = \frac{1}{k_j^2 - \xi^2}\left(-\frac{ik_j^2}{\omega}\,\frac{\partial}{\partial z}\,\bar{E}_{jx} - \xi\eta\bar{B}_{jx}\right), \tag{5.2.16}$$

$$\bar{B}_{jz} = \frac{1}{k_j^2 - \xi^2}\left(-\frac{k_j^2\eta}{\omega}\,\bar{E}_{jx} + i\xi\,\frac{\partial}{\partial z}\,\bar{B}_{jx}\right). \tag{5.2.17}$$

The substitution of (5.2.14)–(5.2.17) into (5.2.8) and (5.2.11) leads to the following ordinary differential equations for $\bar{E}_{jx}$ and $\bar{B}_{jx}$:

$$\left(\frac{d^2}{dz^2} + \gamma_j^2\right)\bar{E}_{jx} = \frac{\omega\mu_0(k_j^2 - \xi^2)}{ik_j^2}\,\delta(z-d), \tag{5.2.18}$$

$$\left(\frac{d^2}{dz^2} + \gamma_j^2\right)\bar{B}_{jx} = 0, \tag{5.2.19}$$

where

$$\gamma_j^2 = k_j^2 - \xi^2 - \eta^2 \tag{5.2.20}$$

and

$$\gamma_j = (k_j^2 - \xi^2 - \eta^2)^{1/2}. \tag{5.2.21}$$

5.3 Solution of the Differential Equations for $\bar{E}_x$ and $\bar{B}_x$

Consider first the equation (5.2.19) for $\bar{B}_{jx}$ with $j = 1$ and $z > 0$. The general solution in Region 1, $z > 0$, is

$$\bar{B}_{1x} = C_1' e^{-i\gamma_1 z} + C_1 e^{i\gamma_1 z}, \tag{5.3.1}$$

where, from (5.2.21) with $k_1 = \beta_1 + i\alpha_1$,

$$\gamma_1 = (\beta_1^2 - \alpha_1^2 - \xi^2 - \eta^2 + 2i\alpha_1\beta_1)^{1/2}. \tag{5.3.2}$$

When $\xi^2 + \eta^2 > \beta_1^2 - \alpha_1^2$,

$$\begin{aligned}\gamma_1 &= i(\xi^2 + \eta^2 - \beta_1^2 + \alpha_1^2)^{1/2}(1 - ip_1)^{1/2} \\ &= i(\xi^2 + \eta^2 - \beta_1^2 + \alpha_1^2)^{1/2}[f(p_1) - ig(p_1)],\end{aligned} \tag{5.3.3}$$

where $p_1 = 2\alpha_1\beta_1/(\xi^2 + \eta^2 - \beta_1^2 + \alpha_1^2)$ is positive and real, and $f(p_1) = \cosh(\frac{1}{2}\sinh^{-1} p_1)$ and $g(p_1) = \sinh(\frac{1}{2}\sinh^{-1} p_1)$. It follows that

$$\begin{aligned}\pm i\gamma_1 z = &\mp z(\xi^2 + \eta^2 - \beta_1^2 + \alpha_1^2)^{1/2} f(p_1) \\ &\pm iz(\xi^2 + \eta^2 - \beta_1^2 + \alpha_1^2)^{1/2} g(p_1).\end{aligned} \tag{5.3.4}$$

Clearly, with $z > 0$, $\exp(-i\gamma_1 z) \to \infty$ as $z \to \infty$, which is an unacceptable solution. Therefore, $C_1' = 0$ and

$$\bar{B}_{1x} = C_1 e^{i\gamma_1 z}; \qquad z \geq 0. \tag{5.3.5}$$

This solution was obtained for $\xi^2 + \eta^2 > \beta_1^2 - \alpha_1^2$; by continuity it also holds when $\xi^2 + \eta^2 < \beta_1^2 - \alpha_1^2$.

When the same sequence of steps is followed for Region 2, $z < 0$, and the boundary condition $\bar{B}_{1x} = \bar{B}_{2x}$ at $z = 0$ is applied, it follows that

$$\bar{B}_{2x} = C_1 e^{-i\gamma_2 z}; \qquad z < 0. \tag{5.3.6}$$

The differential equation (5.2.18) for $\bar{E}_{2x}$ is homogeneous and a solution is obtained just as for $\bar{B}_{2x}$. It is

$$\bar{E}_{2x} = C_2 e^{-i\gamma_2 z}; \qquad z < 0. \tag{5.3.7}$$

The δ-function on the right side of (5.2.18) makes the solution for $\bar{E}_{1x}$, $z > 0$, more complicated. The particular integral for the equation is

$$\bar{E}_{1x_p} = -\frac{\omega\mu_0(k_1^2 - \xi^2)}{2\gamma_1 k_1^2}\, e^{i\gamma_1|z-d|}, \tag{5.3.8}$$

so that

$$\bar{E}_{1x} = C' e^{-i\gamma_1 z} + C e^{i\gamma_1 z} - \frac{\omega\mu_0(k_1^2 - \xi^2)}{2\gamma_1 k_1^2}\, e^{i\gamma_1|z-d|}; \qquad z \geq 0. \tag{5.3.9}$$

As in the solution (5.3.1) for $\bar{B}_{1x}$, the constant C' must vanish since $\exp(-i\gamma_1 z) \to \infty$ as $z \to \infty$. The boundary condition $\bar{E}_{1x} = \bar{E}_{2x}$ at $z = 0$ gives

$$C = C_2 + \frac{\omega\mu_0(k_1^2 - \xi^2)}{2\gamma_1 k_1^2}\, e^{i\gamma_1 d}, \tag{5.3.10}$$

so that

$$\bar{E}_{1x} = C_2 e^{i\gamma_1 z} - \frac{\omega\mu_0(k_1^2 - \xi^2)}{i\gamma_1 k_1^2}\, e^{i\gamma_1 d} \sin\gamma_1 z; \qquad 0 \leq z \leq d, \tag{5.3.11}$$

$$\bar{E}_{1x} = \left(C_2 - \frac{\omega\mu_0(k_1^2 - \xi^2)}{i\gamma_1 k_1^2} \sin\gamma_1 d \right) e^{i\gamma_1 z}; \qquad d \leq z. \tag{5.3.12}$$

The solutions (5.3.5) and (5.3.6) for $\bar{B}_x$ and (5.3.7), (5.3.11), and (5.3.12) for $\bar{E}_x$ satisfy the boundary conditions $E_{1x} = E_{2x}$, $B_{1x} = B_{2x}$ at $z = 0$. The continuity of E_y, kE_z, B_y, and B_z remains to be assured. Since these four additional boundary conditions are not independent, only two need to be enforced in order to determine the arbitrary constants C_1 and C_2 in the formulas for $\bar{E}_x$ and $\bar{B}_x$. Specifically, the conditions $E_{2y} = E_{1y}$, $B_{2y} = B_{1y}$ at $z = 0$ lead to the following equations with (5.2.14) and (5.2.16):

$$\frac{1}{k_2^2 - \xi^2}\left(-\xi\eta C_2 + \omega\gamma_2 C_1 \right) = \frac{1}{k_1^2 - \xi^2}\left(-\xi\eta C_2 - \omega\gamma_1 C_1 \right), \tag{5.3.13}$$

$$\begin{aligned} &\frac{1}{k_2^2 - \xi^2}\left(-\frac{k_2^2}{\omega}\gamma_2 C_2 - \xi\eta C_1 \right) \\ &= \frac{1}{k_1^2 - \xi^2}\left(\frac{k_1^2}{\omega}\gamma_1 C_2 + \mu_0(k_1^2 - \xi^2)e^{i\gamma_1 d} - \xi\eta C_1 \right). \end{aligned} \tag{5.3.14}$$

These reduce to

$$\left(\frac{\gamma_1}{k_1^2 - \xi^2} + \frac{\gamma_2}{k_2^2 - \xi^2} \right) C_1 + \left(\frac{1}{k_1^2 - \xi^2} - \frac{1}{k_2^2 - \xi^2} \right) \frac{\xi\eta C_2}{\omega} = 0, \tag{5.3.15}$$

$$\left(\frac{1}{k_1^2 - \xi^2} - \frac{1}{k_2^2 - \xi^2} \right) \xi\eta C_1 - \left(\frac{\gamma_1 k_1^2}{k_1^2 - \xi^2} + \frac{\gamma_2 k_2^2}{k_2^2 - \xi^2} \right) \frac{C_2}{\omega} = \mu_0 e^{i\gamma_1 d}, \tag{5.3.16}$$

or, in matrix form,

$$\begin{bmatrix} \left(\frac{\gamma_1}{k_1^2 - \xi^2} + \frac{\gamma_2}{k_2^2 - \xi^2} \right) & \left(\frac{1}{k_1^2 - \xi^2} - \frac{1}{k_2^2 - \xi^2} \right)\xi\eta \\ \left(\frac{1}{k_1^2 - \xi^2} - \frac{1}{k_2^2 - \xi^2} \right)\xi\eta & -\left(\frac{\gamma_1 k_1^2}{k_1^2 - \xi^2} + \frac{\gamma_2 k_2^2}{k_2^2 - \xi^2} \right) \end{bmatrix} \begin{bmatrix} C_1 \\ \frac{C_2}{\omega} \end{bmatrix} = \begin{bmatrix} 0 \\ \mu_0 e^{i\gamma_1 d} \end{bmatrix}. \tag{5.3.17}$$

The determinant of the 2×2 matrix in (5.3.17) appears to be quite complicated. Actually it is surprisingly simple. With $\gamma_j^2 = k_j^2 - \xi^2 - \eta^2$ and extensive manipulation, it reduces to

$$\det[\ \ ] = -\frac{MN}{(k_1^2 - \xi^2)(k_2^2 - \xi^2)}, \tag{5.3.18}$$

where

$$M \equiv \gamma_1 + \gamma_2; \qquad N \equiv k_1^2\gamma_2 + k_2^2\gamma_1. \tag{5.3.19}$$

The constants C_1 and C_2 are

$$C_1 = -\frac{\xi\eta(k_1^2 - k_2^2)}{MN}\mu_0 e^{i\gamma_1 d}, \tag{5.3.20}$$

$$C_2 = -\frac{\gamma_1(k_2^2 - \xi^2) + \gamma_2(k_1^2 - \xi^2)}{MN} \omega\mu_0 e^{i\gamma_1 d}. \tag{5.3.21}$$

When these expressions are substituted in (5.3.5)–(5.3.7), (5.3.11), and (5.3.12), the solutions $\bar{B}_{jx}$ and $\bar{E}_{jx}$, $j = 1, 2$, are obtained. The further substitution of $\bar{B}_{jx}$ and $\bar{E}_{jx}$ in (5.2.14)–(5.2.17) gives $\bar{E}_{jy}$, $\bar{E}_{jz}$, $\bar{B}_{jy}$, and $\bar{B}_{jz}$.

5.4 The Components of the Electromagnetic Field of the Horizontal Dipole as General Integrals

The six components of the electromagnetic field in Cartesian coordinates are obtained in the form of double infinite integrals when the Fourier transforms derived in Section 5.3 are substituted in formulas like (5.2.6). In Region 1 with $0 \leq z \leq d$ they are

$$E_{1x} = -\frac{\omega\mu_0}{4\pi^2}\int_{-\infty}^{\infty} d\xi \int_{-\infty}^{\infty} d\eta \; e^{i(\xi x + \eta y)} \times \left(\frac{\gamma_1(k_2^2 - \xi^2) + \gamma_2(k_1^2 - \xi^2)}{MN} e^{i\gamma_1 z} + \frac{k_1^2 - \xi^2}{i\gamma_1 k_1^2} \sin\gamma_1 z\right) e^{i\gamma_1 d}, \tag{5.4.1}$$

$$E_{1y} = \frac{\omega\mu_0}{4\pi^2}\int_{-\infty}^{\infty} d\xi \int_{-\infty}^{\infty} d\eta \; \xi\eta e^{i(\xi x + \eta y)} \left(\frac{\sin\gamma_1 z}{i\gamma_1 k_1^2} + \frac{e^{i\gamma_1 z}}{N}\right) e^{i\gamma_1 d}, \tag{5.4.2}$$

$$E_{1z} = \frac{\omega\mu_0}{4\pi^2}\int_{-\infty}^{\infty} d\xi \int_{-\infty}^{\infty} d\eta \; \xi e^{i(\xi x + \eta y)} \left(\frac{\gamma_2}{N} e^{i\gamma_1 z} - \frac{\cos\gamma_1 z}{k_1^2}\right) e^{i\gamma_1 d}, \tag{5.4.3}$$

$$B_{1x} = -\frac{\mu_0}{4\pi^2}\int_{-\infty}^{\infty} d\xi \int_{-\infty}^{\infty} d\eta \; \xi\eta e^{i(\xi x + \eta y)} \frac{k_1^2 - k_2^2}{MN} e^{i\gamma_1(z+d)}, \tag{5.4.4}$$

$$B_{1y} = \frac{\mu_0}{4\pi^2}\int_{-\infty}^{\infty} d\xi \int_{-\infty}^{\infty} d\eta \; e^{i(\xi x + \eta y)} \times \left(\cos\gamma_1 z + \frac{k_2^2\eta^2 - k_1^2(k_2^2 - \xi^2 + \gamma_1\gamma_2)}{MN} e^{i\gamma_1 z}\right) e^{i\gamma_1 d}, \tag{5.4.5}$$

$$B_{1z} = \frac{\mu_0}{4\pi^2}\int_{-\infty}^{\infty} d\xi \int_{-\infty}^{\infty} d\eta \; \eta e^{i(\xi x + \eta y)} \left(\frac{\sin\gamma_1 z}{i\gamma_1} + \frac{e^{i\gamma_1 z}}{M}\right) e^{i\gamma_1 d}. \tag{5.4.6}$$

In Region 1 with $d \leq z$,

$$E_{1x} = -\frac{\omega\mu_0}{4\pi^2}\int_{-\infty}^{\infty} d\xi \int_{-\infty}^{\infty} d\eta \; e^{i(\xi x + \eta y)} \times \left(\frac{\gamma_1(k_2^2 - \xi^2) + \gamma_2(k_1^2 - \xi^2)}{MN} e^{i\gamma_1 d} + \frac{k_1^2 - \xi^2}{i\gamma_1 k_1^2} \sin\gamma_1 d\right) e^{i\gamma_1 z}, \tag{5.4.7}$$

$$E_{1y} = \frac{\omega\mu_0}{4\pi^2}\int_{-\infty}^{\infty} d\xi \int_{-\infty}^{\infty} d\eta\ \xi\eta e^{i(\xi x+\eta y)}\left(\frac{\sin\gamma_1 d}{i\gamma_1 k_1^2} + \frac{e^{i\gamma_1 d}}{N}\right)e^{i\gamma_1 z}, \tag{5.4.8}$$

$$E_{1z} = \frac{\omega\mu_0}{4\pi^2}\int_{-\infty}^{\infty} d\xi \int_{-\infty}^{\infty} d\eta\ \xi e^{i(\xi x+\eta y)}\left(\frac{\gamma_2}{N}\, e^{i\gamma_1 d} + \frac{\sin\gamma_1 d}{i\gamma_1 k_1^2}\right)e^{i\gamma_1 z}, \tag{5.4.9}$$

$$B_{1x} = -\frac{\mu_0}{4\pi^2}\int_{-\infty}^{\infty} d\xi \int_{-\infty}^{\infty} d\eta\ \xi\eta e^{i(\xi x+\eta y)}\, \frac{k_1^2 - k_2^2}{MN}\, e^{i\gamma_1(z+d)}, \tag{5.4.10}$$

$$B_{1y} = \frac{\mu_0}{4\pi^2}\int_{-\infty}^{\infty} d\xi \int_{-\infty}^{\infty} d\eta\ e^{i(\xi x+\eta y)}$$
$$\times\left(i\sin\gamma_1 d + \frac{k_2^2\eta^2 - k_1^2(k_2^2 - \xi^2 + \gamma_1\gamma_2)}{MN}\, e^{i\gamma_1 d}\right)e^{i\gamma_1 z}, \tag{5.4.11}$$

$$B_{1z} = \frac{\mu_0}{4\pi^2}\int_{-\infty}^{\infty} d\xi \int_{-\infty}^{\infty} d\eta\ \eta e^{i(\xi x+\eta y)}\left(\frac{\sin\gamma_1 d}{i\gamma_1} + \frac{e^{i\gamma_1 d}}{M}\right)e^{i\gamma_1 z}. \tag{5.4.12}$$

In Region 2 where $z \leq 0$,

$$E_{2x} = -\frac{\omega\mu_0}{4\pi^2}\int_{-\infty}^{\infty} d\xi \int_{-\infty}^{\infty} d\eta\ e^{i(\xi x+\eta y)}$$
$$\times\left(\frac{\gamma_1(k_2^2 - \xi^2) + \gamma_2(k_1^2 - \xi^2)}{MN}\right)e^{i(\gamma_1 d - \gamma_2 z)}, \tag{5.4.13}$$

$$E_{2y} = \frac{\omega\mu_0}{4\pi^2}\int_{-\infty}^{\infty} d\xi \int_{-\infty}^{\infty} d\eta\ \xi\eta e^{i(\xi x+\eta y)}\, \frac{e^{i(\gamma_1 d - \gamma_2 z)}}{N}, \tag{5.4.14}$$

$$E_{2z} = -\frac{\omega\mu_0}{4\pi^2}\int_{-\infty}^{\infty} d\xi \int_{-\infty}^{\infty} d\eta\ \xi e^{i(\xi x+\eta y)}\, \frac{\gamma_1}{N}\, e^{i(\gamma_1 d - \gamma_2 z)}, \tag{5.4.15}$$

$$B_{2x} = -\frac{\mu_0}{4\pi^2}\int_{-\infty}^{\infty} d\xi \int_{-\infty}^{\infty} d\eta\ \xi\eta e^{i(\xi x+\eta y)}\, \frac{k_1^2 - k_2^2}{MN}\, e^{i(\gamma_1 d - \gamma_2 z)}, \tag{5.4.16}$$

$$B_{2y} = \frac{\mu_0}{4\pi^2}\int_{-\infty}^{\infty} d\xi \int_{-\infty}^{\infty} d\eta\ e^{i(\xi x+\eta y)}$$
$$\times\left(\frac{\gamma_2}{M} + \frac{\xi^2(k_1^2 - k_2^2)}{MN}\right)e^{i(\gamma_1 d - \gamma_2 z)}, \tag{5.4.17}$$

$$B_{2z} = \frac{\mu_0}{4\pi^2}\int_{-\infty}^{\infty} d\xi \int_{-\infty}^{\infty} d\eta\ \eta e^{i(\xi x+\eta y)}\, \frac{e^{i(\gamma_1 d - \gamma_2 z)}}{M}. \tag{5.4.18}$$

This completes the formulation of the electromagnetic field at all points in two half-spaces with different electrical properties when the source is a horizontal infinitesimal dipole with unit electric moment located at a distance d from and parallel to the boundary surface. A more convenient form for the evaluation of the integrals is in the cylindrical coordinates ρ, ϕ, z. These are defined by

$$x = \rho\cos\phi; \quad y = \rho\sin\phi; \qquad \xi = \lambda\cos\phi'; \quad \eta = \lambda\sin\phi', \tag{5.4.19}$$

where the Cartesian transform variables ξ and η are expressed in terms of the cylindrical transform variables λ and ϕ'. (Note that λ is not a wavelength.) It follows that

$$\rho = (x^2 + y^2)^{1/2}; \qquad \phi = \tan^{-1}\left(\frac{y}{x}\right),$$
$$\lambda = (\xi^2 + \eta^2)^{1/2}; \qquad \phi' = \tan^{-1}\left(\frac{\eta}{\xi}\right). \tag{5.4.20}$$

Also,

$$\gamma_j = (k_j^2 - \xi^2 - \eta^2)^{1/2} = (k_j^2 - \lambda^2)^{1/2}; \qquad j = 1,\, 2, \tag{5.4.21}$$
$$\xi x + \eta y = \lambda\rho\cos(\phi - \phi'); \qquad d\xi\, d\eta = \lambda\, d\phi'\, d\lambda. \tag{5.4.22}$$

The cylindrical components of the electric field are given by

$$E_\rho = E_x \cos\phi + E_y \sin\phi; \qquad E_\phi = -E_x \sin\phi + E_y \cos\phi. \tag{5.4.23}$$

The component E_{1x} in the range $d \leq z$ is

$$\begin{aligned}
E_{1x} &= -\frac{\omega\mu_0}{4\pi^2}\int_0^{2\pi} d\phi' \int_0^\infty d\lambda\, \lambda\, e^{i\lambda\rho\cos(\phi-\phi')} \\
&\quad \times \left(\frac{\gamma_1(k_2^2 - \lambda^2\cos^2\phi') + \gamma_2(k_1^2 - \lambda^2\cos^2\phi')}{MN}\, e^{i\gamma_1 d} \right. \\
&\qquad \left. + \frac{k_1^2 - \lambda^2\cos^2\phi'}{i\gamma_1 k_1^2} \sin\gamma_1 d \right) e^{i\gamma_1 z} \\
&= -\frac{\omega\mu_0}{4\pi}\int_0^\infty d\lambda\, \lambda \left(\frac{e^{i\gamma_1 d}}{M} + \frac{k_1^2}{i\gamma_1 k_1^2}\sin\gamma_1 d \right) e^{i\gamma_1 z} \\
&\qquad \times \int_0^{2\pi} d\phi'\, e^{i\lambda\rho\cos(\phi-\phi')} \\
&\quad + \frac{\omega\mu_0}{4\pi}\int_0^\infty d\lambda\, \lambda^3 \left(\frac{e^{i\gamma_1 d}}{N} + \frac{\sin\gamma_1 d}{i\gamma_1 k_1^2} \right) e^{i\gamma_1 z} \\
&\qquad \times \int_0^{2\pi} d\phi'\, e^{i\lambda\rho\cos(\phi-\phi')} \cos^2\phi'.
\end{aligned} \tag{5.4.24}$$

A similar expression for E_{1y} is readily obtained. The integral form of the Bessel function can now be introduced. It is

$$J_n(\lambda\rho) = \frac{i^{-n}}{2\pi}\int_0^{2\pi} e^{i\lambda\rho\cos\theta}\, e^{in\theta}\, d\theta. \tag{5.4.25}$$

With (5.4.25), it follows that

$$\frac{1}{2\pi}\int_0^{2\pi} e^{i\lambda\rho\cos(\phi-\phi')}\, d\phi' = J_0(\lambda\rho), \tag{5.4.26}$$
$$\frac{1}{2\pi}\int_0^{2\pi} e^{i\lambda\rho\cos(\phi-\phi')} \cos^2\phi'\, d\phi' = \tfrac{1}{2}[J_0(\lambda\rho) - J_2(\lambda\rho)\cos 2\phi]. \tag{5.4.27}$$

These expressions can be used in (5.4.24) and the analogous expression for E_{1y} to replace the ϕ' integrals. The resulting formulas can then be substituted in (5.4.23) to obtain an expression for $E_{1\rho}$. The application of standard differential and functional equations for Bessel functions and the convenient notation

$$P \equiv \frac{\gamma_2 - \gamma_1}{M} = \frac{\gamma_2 - \gamma_1}{\gamma_2 + \gamma_1}; \qquad Q \equiv \frac{k_1^2\gamma_2 - k_2^2\gamma_1}{N} = \frac{k_1^2\gamma_2 - k_2^2\gamma_1}{k_1^2\gamma_2 + k_2^2\gamma_1} \tag{5.4.28}$$

lead to the formula for $E_{1\rho}$. Similar procedures yield the associated formulas for the other components. They are listed as follows for Region 1, $z \geq 0$:

$$\begin{aligned} E_{1\rho} = -\frac{\omega\mu_0}{4\pi k_1^2}\cos\phi \Bigg[\int_0^\infty \Bigg(k_1^2 J_0(\lambda\rho) \\ - \frac{\lambda^2}{2}\left[J_0(\lambda\rho) - J_2(\lambda\rho)\right]\Bigg)\gamma_1^{-1}e^{i\gamma_1|z-d|}\lambda\, d\lambda \\ + \int_0^\infty \Bigg(\frac{\gamma_1 Q}{2}\left[J_0(\lambda\rho) - J_2(\lambda\rho)\right] \\ - \frac{k_1^2 P}{2\gamma_1}\left[J_0(\lambda\rho) + J_2(\lambda\rho)\right]\Bigg)e^{i\gamma_1(z+d)}\lambda\, d\lambda\Bigg], \end{aligned} \tag{5.4.29}$$

$$\begin{aligned} E_{1\phi} = \frac{\omega\mu_0}{4\pi k_1^2}\sin\phi \Bigg[\int_0^\infty \Bigg(k_1^2 J_0(\lambda\rho) \\ - \frac{\lambda^2}{2}\left[J_0(\lambda\rho) + J_2(\lambda\rho)\right]\Bigg)\gamma_1^{-1}e^{i\gamma_1|z-d|}\lambda\, d\lambda \\ + \int_0^\infty \Bigg(\frac{\gamma_1 Q}{2}\left[J_0(\lambda\rho) + J_2(\lambda\rho)\right] \\ - \frac{k_1^2 P}{2\gamma_1}\left[J_0(\lambda\rho) - J_2(\lambda\rho)\right]\Bigg)e^{i\gamma_1(z+d)}\lambda\, d\lambda\Bigg], \end{aligned} \tag{5.4.30}$$

$$E_{1z} = \frac{i\omega\mu_0}{4\pi k_1^2}\cos\phi \int_0^\infty \left(\pm e^{i\gamma_1|z-d|} + Qe^{i\gamma_1(z+d)}\right)J_1(\lambda\rho)\lambda^2\, d\lambda, \tag{5.4.31}$$

$$\begin{aligned} B_{1\rho} = -\frac{\mu_0}{4\pi}\sin\phi \Bigg[\pm\int_0^\infty J_0(\lambda\rho)e^{i\gamma_1|z-d|}\lambda\, d\lambda \\ + \int_0^\infty \Bigg(\frac{Q}{2}\left[J_0(\lambda\rho) + J_2(\lambda\rho)\right] \\ - \frac{P}{2}\left[J_0(\lambda\rho) - J_2(\lambda\rho)\right]\Bigg)e^{i\gamma_1(z+d)}\lambda\, d\lambda\Bigg], \end{aligned} \tag{5.4.32}$$

$$B_{1\phi} = -\frac{\mu_0}{4\pi}\cos\phi\left[\pm\int_0^\infty J_0(\lambda\rho)e^{i\gamma_1|z-d|}\lambda\,d\lambda + \int_0^\infty\left(\frac{Q}{2}\left[J_0(\lambda\rho) - J_2(\lambda\rho)\right] - \frac{P}{2}\left[J_0(\lambda\rho) + J_2(\lambda\rho)\right]\right)e^{i\gamma_1(z+d)}\lambda\,d\lambda\right], \tag{5.4.33}$$

$$B_{1z} = \frac{i\mu_0}{4\pi}\sin\phi\int_0^\infty (e^{i\gamma_1|z-d|} - Pe^{i\gamma_1(z+d)})\gamma_1^{-1}J_1(\lambda\rho)\lambda^2\,d\lambda. \tag{5.4.34}$$

Where two signs appear, the upper one is for $z > d$ and the lower is for $0 \leq z \leq d$.

The corresponding formulas for Region 2, $z \leq 0$, are

$$E_{2\rho} = -\frac{\omega\mu_0}{4\pi}\cos\phi\int_0^\infty\left(M^{-1}[J_0(\lambda\rho) + J_2(\lambda\rho)] + \frac{\gamma_1\gamma_2}{N}\left[J_0(\lambda\rho) - J_2(\lambda\rho)\right]\right)e^{i(\gamma_1 d - \gamma_2 z)}\lambda\,d\lambda, \tag{5.4.35}$$

$$E_{2\phi} = \frac{\omega\mu_0}{4\pi}\sin\phi\int_0^\infty\left(M^{-1}[J_0(\lambda\rho) - J_2(\lambda\rho)] + \frac{\gamma_1\gamma_2}{N}\left[J_0(\lambda\rho) + J_2(\lambda\rho)\right]\right)e^{i(\gamma_1 d - \gamma_2 z)}\lambda\,d\lambda, \tag{5.4.36}$$

$$E_{2z} = -\frac{i\omega\mu_0}{2\pi}\cos\phi\int_0^\infty \frac{\gamma_1}{N}J_1(\lambda\rho)e^{i(\gamma_1 d - \gamma_2 z)}\lambda^2\,d\lambda, \tag{5.4.37}$$

$$B_{2\rho} = \frac{\mu_0}{4\pi}\sin\phi\int_0^\infty\left(\frac{\gamma_2}{M}\left[J_0(\lambda\rho) - J_2(\lambda\rho)\right] + \frac{k_2^2\gamma_1}{N}\left[J_0(\lambda\rho) + J_2(\lambda\rho)\right]\right)e^{i(\gamma_1 d - \gamma_2 z)}\lambda\,d\lambda, \tag{5.4.38}$$

$$B_{2\phi} = \frac{\mu_0}{4\pi}\cos\phi\int_0^\infty\left(\frac{\gamma_2}{M}\left[J_0(\lambda\rho) + J_2(\lambda\rho)\right] + \frac{k_2^2\gamma_1}{N}\left[J_0(\lambda\rho) - J_2(\lambda\rho)\right]\right)e^{i(\gamma_1 d - \gamma_2 z)}\lambda\,d\lambda, \tag{5.4.39}$$

$$B_{2z} = \frac{i\mu_0}{2\pi}\sin\phi\int_0^\infty M^{-1}J_1(\lambda\rho)e^{i(\gamma_1 d - \gamma_2 z)}\lambda^2\,d\lambda. \tag{5.4.40}$$

These formulas give the complete field in both regions in the form of general integrals.

5.5 Integrated Formulas for the Electromagnetic Field in Regions 1 and 2 of the Horizontal Electric Dipole at the Depth d in Region 1

In order to evaluate the integrals in (5.4.29)–(5.4.34) for the electromagnetic field in Region 1 ($z \geq 0$) due to a horizontal electric dipole at $z = d$ also in Region 1, it is convenient to rearrange them so that they represent the sum of the direct field, the ideal reflected field or field of an ideal image, and the rest of the field. This last includes the lateral-wave field and correction terms for the reflected field to take account of the fact that it is not actually that of an ideal image. The rearranged integrals are

$$E_{1\rho} = -\frac{\omega\mu_0}{4\pi k_1^2}\cos\phi\,[F_{\rho 0}(\rho, z-d) + F_{\rho 0}(\rho, z+d) + F_{\rho 1}(\rho, z+d)], \tag{5.5.1}$$

where

$$F_{\rho 0}(\rho, z-d) = \int_0^\infty \left(\frac{\gamma_1}{2}\,[J_0(\lambda\rho) - J_2(\lambda\rho)] + \frac{k_1^2}{2\gamma_1}\,[J_0(\lambda\rho) + J_2(\lambda\rho)]\right) e^{i\gamma_1|z-d|}\lambda\,d\lambda, \tag{5.5.2}$$

$$F_{\rho 0}(\rho, z+d) = \int_0^\infty \left(\frac{\gamma_1}{2}\,[J_0(\lambda\rho) - J_2(\lambda\rho)] + \frac{k_1^2}{2\gamma_1}\,[J_0(\lambda\rho) + J_2(\lambda\rho)]\right) e^{i\gamma_1(z+d)}\lambda\,d\lambda, \tag{5.5.3}$$

$$F_{\rho 1}(\rho, z+d) = \int_0^\infty \left(\frac{\gamma_1}{2}\,(Q-1)[J_0(\lambda\rho) - J_2(\lambda\rho)] - \frac{k_1^2}{2\gamma_1}\,(P+1)[J_0(\lambda\rho) + J_2(\lambda\rho)]\right) \times e^{i\gamma_1(z+d)}\lambda\,d\lambda; \tag{5.5.4}$$

$$E_{1\phi} = \frac{\omega\mu_0}{4\pi k_1^2}\sin\phi\,[F_{\phi 0}(\rho, z-d) + F_{\phi 0}(\rho, z+d) + F_{\phi 1}(\rho, z+d)], \tag{5.5.5}$$

where

$$F_{\phi 0}(\rho, z-d) = \int_0^\infty \left(\frac{\gamma_1}{2}\,[J_0(\lambda\rho) + J_2(\lambda\rho)] + \frac{k_1^2}{2\gamma_1}\,[J_0(\lambda\rho) - J_2(\lambda\rho)]\right) e^{i\gamma_1|z-d|}\lambda\,d\lambda, \tag{5.5.6}$$

$$F_{\phi 0}(\rho, z+d) = \int_0^\infty \left(\frac{\gamma_1}{2}\,[J_0(\lambda\rho) + J_2(\lambda\rho)]\right.$$

$$+ \frac{k_1^2}{2\gamma_1}[J_0(\lambda\rho) - J_2(\lambda\rho)]\Big) e^{i\gamma_1(z+d)}\lambda\,d\lambda, \tag{5.5.7}$$

$$\begin{aligned} F_{\phi 1}(\rho, z+d) = \int_0^\infty \Big(&\frac{\gamma_1}{2}(Q-1)[J_0(\lambda\rho) + J_2(\lambda\rho)] \\ &- \frac{k_1^2}{2\gamma_1}(P+1)[J_0(\lambda\rho) - J_2(\lambda\rho)]\Big) \\ &\times e^{i\gamma_1(z+d)}\lambda\,d\lambda; \end{aligned} \tag{5.5.8}$$

$$E_{1z} = \frac{i\omega\mu_0}{4\pi k_1^2}\cos\phi\,[F_{z0}(\rho, z-d) + F_{z0}(\rho, z+d) + F_{z1}(\rho, z+d)], \tag{5.5.9}$$

where

$$F_{z0}(\rho, z-d) = \pm\int_0^\infty J_1(\lambda\rho)e^{i\gamma_1|z-d|}\lambda^2\,d\lambda; \qquad \begin{cases} z > d \\ 0 \le z \le d \end{cases}, \tag{5.5.10}$$

$$F_{z0}(\rho, z+d) = \int_0^\infty J_1(\lambda\rho)e^{i\gamma_1(z+d)}\lambda^2\,d\lambda, \tag{5.5.11}$$

$$F_{z1}(\rho, z+d) = \int_0^\infty (Q-1)J_1(\lambda\rho)e^{i\gamma_1(z+d)}\lambda^2\,d\lambda; \tag{5.5.12}$$

$$B_{1\rho} = -\frac{\mu_0}{4\pi}\sin\phi\,[G_{\rho 0}(\rho, z-d) + G_{\rho 0}(\rho, z+d) + G_{\rho 1}(\rho, z+d)], \tag{5.5.13}$$

$$B_{1\phi} = -\frac{\mu_0}{4\pi}\cos\phi\,[G_{\phi 0}(\rho, z-d) + G_{\phi 0}(\rho, z+d) + G_{\phi 1}(\rho, z+d)], \tag{5.5.14}$$

where

$$\begin{aligned} G_{\rho 0}(\rho, z-d) &= G_{\phi 0}(\rho, z-d) \\ &= \pm\int_0^\infty J_0(\lambda\rho)e^{i\gamma_1|z-d|}\lambda\,d\lambda; \qquad \begin{cases} z > d \\ 0 \le z \le d \end{cases}, \end{aligned} \tag{5.5.15}$$

$$G_{\rho 0}(\rho, z+d) = G_{\phi 0}(\rho, z+d) = \int_0^\infty J_0(\lambda\rho)e^{i\gamma_1(z+d)}\lambda\,d\lambda, \tag{5.5.16}$$

$$\begin{aligned} \left.\begin{matrix} G_{\rho 1}(\rho, z+d) \\ G_{\phi 1}(\rho, z+d) \end{matrix}\right\} = \int_0^\infty \Big(&\frac{1}{2}(Q-1)[J_0(\lambda\rho) \pm J_2(\lambda\rho)] \\ &- \frac{1}{2}(P+1)[J_0(\lambda\rho) \mp J_2(\lambda\rho)]\Big) \\ &\times e^{i\gamma_1(z+d)}\lambda\,d\lambda; \end{aligned} \tag{5.5.17}$$

$$B_{1z} = \frac{i\mu_0}{4\pi}\sin\phi\,[G_{z0}(\rho, z-d) - G_{z0}(\rho, z+d) + G_{z1}(\rho, z+d)], \tag{5.5.18}$$

where

$$G_{z0}(\rho, z-d) = \int_0^\infty \gamma_1^{-1}J_1(\lambda\rho)e^{i\gamma_1|z-d|}\lambda^2\,d\lambda, \tag{5.5.19}$$

$$G_{z0}(\rho, z+d) = \int_0^\infty \gamma_1^{-1} J_1(\lambda\rho) e^{i\gamma_1(z+d)} \lambda^2 \, d\lambda, \tag{5.5.20}$$

$$G_{z1}(\rho, z+d) = -\int_0^\infty (P-1)\gamma_1^{-1} J_1(\lambda\rho) e^{i\gamma_1(z+d)} \lambda^2 \, d\lambda. \tag{5.5.21}$$

In the expressions for the components of the electromagnetic field, the integrals $F_{m0}(\rho, z-d)$ and $G_{m0}(\rho, z-d)$, $m = \rho, \phi, z$, represent the direct field in Region 1 of the horizontal dipole at $(0, 0, d)$; $F_{m0}(\rho, z+d)$ and $G_{m0}(\rho, z+d)$ represent the ideal reflected field or the field of an ideal image dipole at $(0, 0, -d)$ when there is no boundary and all space contains medium 1. The rigorous expressions for these fields can be obtained from the appropriate integrals in Appendix A and, more conveniently, from the formulas in Chapter 2. The formulas are written first in exact form. When combined with other terms that presuppose the conditions

$$\rho \geq 5d; \qquad \rho \geq 5|z|, \tag{5.5.22a}$$

they are also written with terms in $(z \pm d)^2/r^2$ omitted and with the approximations

$$\begin{aligned} r_1 &= [\rho^2 + (z-d)^2]^{1/2} \sim \rho, \\ r_2 &= [\rho^2 + (z+d)^2]^{1/2} \sim \rho \end{aligned} \tag{5.5.22b}$$

in amplitudes but with no simplification in phases.

$$\begin{aligned} F_{\rho 0}(\rho, z \pm d) = -e^{ik_1 r} \Bigg[&\frac{2k_1}{r^2} + \frac{2i}{r^3} \\ &+ \frac{(z \pm d)^2}{r^2} \left(\frac{ik_1^2}{r} - \frac{3k_1}{r^2} - \frac{3i}{r^3} \right) \Bigg], \end{aligned} \tag{5.5.23}$$

$$F_{\phi 0}(\rho, z \pm d) = -e^{ik_1 r} \left(\frac{ik_1^2}{r} - \frac{k_1}{r^2} - \frac{i}{r^3} \right), \tag{5.5.24}$$

$$F_{z0}(\rho, z \pm d) = -e^{ik_1 r} \left(\frac{\rho}{r} \right) \left(\frac{z \pm d}{r} \right) \left(\frac{k_1^2}{r} + \frac{3ik_1}{r^2} - \frac{3}{r^3} \right), \tag{5.5.25}$$

$$\begin{aligned} G_{\rho 0}(\rho, z \pm d) &= G_{\phi 0}(\rho, z \pm d) \\ &= -e^{ik_1 r} \left(\frac{z \pm d}{r} \right) \left(\frac{ik_1}{r} - \frac{1}{r^2} \right), \end{aligned} \tag{5.5.26}$$

$$G_{z0}(\rho, z \pm d) = -e^{ik_1 r} \left(\frac{\rho}{r} \right) \left(\frac{k_1}{r} + \frac{i}{r^2} \right). \tag{5.5.27}$$

In these formulas, $r = r_2 = [\rho^2 + (z+d)^2]^{1/2}$ for the upper sign and $r = r_1 = [\rho^2 + (z-d)^2]^{1/2}$ for the lower sign.

The evaluation of the integrals $F_{m1}(\rho, z+d)$ and $G_{m1}(\rho, z+d)$ is carried out in Section 5.6, subject to the condition

$$|k_1| \geq 3|k_2|. \tag{5.5.28}$$

The results are

$$F_{\rho 1}(\rho, z+d) = 2e^{ik_1(z+d)}e^{ik_2\rho}\left[\frac{ik_2^2}{\rho} - \frac{k_2}{\rho^2} - \frac{i}{\rho^3} - \frac{k_2^4}{k_1}\left(\frac{\pi}{k_2\rho}\right)^{1/2} e^{-ip}\mathcal{F}(p)\right] + 2e^{ik_1r_2}\left[\frac{k_1}{r_2^2} + \frac{i}{r_2^3} + \left(\frac{z+d}{\rho}\right)\left(\frac{ik_1}{\rho^2} - \frac{3}{2\rho^3}\right)\right], \tag{5.5.29}$$

$$F_{\phi 1}(\rho, z+d) = 4e^{ik_1(z+d)}e^{ik_2\rho}\left[\frac{k_2}{\rho^2} + \frac{i}{\rho^3} + \frac{ik_2^3}{2k_1\rho}\left(\frac{\pi}{k_2\rho}\right)^{1/2} e^{-ip}\mathcal{F}(p)\right] + 4e^{ik_1r_2}\left(\frac{ik_1^2}{2r_2} - \frac{k_1}{r_2^2} - \frac{i}{r_2^3}\right) + 2ie^{ik_1r_2}\left(\frac{z+d}{r_2}\right)\left(\frac{ik_1^2}{\rho} - \frac{3k_1}{2\rho^2} - \frac{5i}{8\rho^3}\right) - 2e^{ik_1r_2}\left(\frac{z+d}{r_2}\right)^2\left(\frac{ik_1^2}{r_2} - \frac{3k_1}{r_2^2} - \frac{3i}{r_2^3}\right), \tag{5.5.30}$$

$$F_{z1}(\rho, z+d) = \frac{2k_2^2}{k_1}\, e^{ik_1(z+d)}e^{ik_2\rho}\left[\frac{k_2}{\rho} + \frac{i}{\rho^2} + \frac{ik_2^3}{k_1}\left(\frac{\pi}{k_2\rho}\right)^{1/2} e^{-ip}\mathcal{F}(p)\right] - \frac{2k_2^2}{k_1}\, e^{ik_1r_2}\left(\frac{1}{\rho^2} + \frac{3i}{2k_1\rho^3}\right), \tag{5.5.31}$$

$$G_{\rho 1}(\rho, z+d) = \frac{2}{k_1}\, e^{ik_1(z+d)}e^{ik_2\rho}\left[\frac{2k_2}{\rho^2} + \frac{2i}{\rho^3} + \frac{ik_2^3}{k_1\rho}\left(\frac{\pi}{k_2\rho}\right)^{1/2} e^{-ip}\mathcal{F}(p)\right] + 2e^{ik_1r_2}\left[\frac{i}{\rho^2} - \frac{7}{2k_1\rho^3} + \left(\frac{z+d}{r_2}\right)\left(\frac{ik_1}{r_2} - \frac{5}{r_2^2} - \frac{12i}{k_1r_2^3}\right)\right], \tag{5.5.32}$$

$$G_{\phi 1}(\rho, z+d) = \frac{2i}{k_1}\, e^{ik_1(z+d)}e^{ik_2\rho}\left[\frac{k_2^2}{\rho} + \frac{ik_2}{\rho^2} - \frac{1}{\rho^3} + \frac{ik_2^4}{k_1}\left(\frac{\pi}{k_2\rho}\right)^{1/2} e^{-ip}\mathcal{F}(p)\right] + \frac{e^{ik_1r_2}}{k_1}\left[\frac{2}{\rho^3} + \frac{3i}{k_1\rho^4} + 2\left(\frac{z+d}{r_2}\right)\left(\frac{k_1}{r_2^2} + \frac{3i}{r_2^3}\right)\right], \tag{5.5.33}$$

$$G_{z1}(\rho, z+d) = -\frac{2ik_2^2}{k_1^2}\, e^{ik_1(z+d)}e^{ik_2\rho}\left(\frac{1}{\rho^2} + \frac{3i}{k_2\rho^3} - \frac{3}{k_2^2\rho^4}\right) + 2ie^{ik_1r_2}\left(\frac{\rho}{r_2}\right)\left[\frac{1}{r_2^2} + \frac{3i}{k_1r_2^3} - \frac{3}{k_1^2r_2^4}\right.$$

$$+\left(\frac{z+d}{\rho}\right)\left(\frac{k_1}{\rho}+\frac{3i}{2\rho^2}-\frac{9}{8k_1\rho^3}\right)$$

$$+\left(\frac{z+d}{r_2}\right)^2\left(\frac{ik_1}{r_2}-\frac{6}{r_2^2}-\frac{15i}{k_1r_2^3}\right)\Bigg],\tag{5.5.34}$$

with

$$\mathcal{F}(p)=\tfrac{1}{2}(1+i)-C_2(p)-iS_2(p);\qquad p\equiv\frac{k_2^3\rho}{2k_1^2},\tag{5.5.35}$$

and where

$$C_2(p)+iS_2(p)=\int_0^p\frac{e^{it}}{(2\pi t)^{1/2}}\,dt\tag{5.5.36}$$

is the Fresnel integral.

When the functions $F_{m0}(\rho,z-d)$, $F_{m0}(\rho,z+d)$, $F_{m1}(\rho,z+d)$ and $G_{m0}(\rho,z-d)$, $G_{m0}(\rho,z+d)$, $G_{m1}(\rho,z+d)$ are combined according to (5.5.1), (5.5.5), (5.5.9), (5.5.13), (5.5.14), and (5.5.18), the following formulas are obtained for the electromagnetic field in Region 1. The approximations $r_1\sim\rho$, $r_2\sim\rho$ are made in amplitudes in the second form given for each component.

$$\begin{aligned}E_{1\rho}(\rho,\phi,z)=&-\frac{\omega\mu_0}{2\pi k_1^2}\cos\phi\\&\times\Bigg\{k_2 g(\rho;k_1,k_2)e^{ik_2\rho}e^{ik_1(z+d)}-\left(\frac{k_1}{r_1^2}+\frac{i}{r_1^3}\right)e^{ik_1r_1}\\&+\left(\frac{z+d}{\rho}\right)\left(\frac{ik_1}{\rho^2}-\frac{3}{2\rho^3}\right)e^{ik_1r_2}\\&-\frac{1}{2}\Bigg[\left(\frac{z-d}{r_1}\right)^2\left(\frac{ik_1^2}{r_1}-\frac{3k_1}{r_1^2}-\frac{3i}{r_1^3}\right)e^{ik_1r_1}\\&+\left(\frac{z+d}{r_2}\right)^2\left(\frac{ik_1^2}{r_2}-\frac{3k_1}{r_2^2}-\frac{3i}{r_2^3}\right)e^{ik_1r_2}\Bigg]\Bigg\}\end{aligned}\tag{5.5.37a}$$

$$\begin{aligned}=&-\frac{\omega\mu_0}{2\pi k_1^2}\cos\phi\\&\times\Bigg[k_2 g(\rho;k_1,k_2)e^{ik_2\rho}e^{ik_1(z+d)}-\left(\frac{k_1}{\rho^2}+\frac{i}{\rho^3}\right)e^{ik_1r_1}\\&+\left(\frac{z+d}{\rho}\right)\left(\frac{ik_1}{\rho^2}-\frac{3}{2\rho^3}\right)e^{ik_1r_2}\Bigg],\end{aligned}\tag{5.5.37b}$$

$$E_{1\phi}(\rho,\phi,z)=\frac{\omega\mu_0}{2\pi k_1^2}\sin\phi$$

$$\times \left\{ k_2 h(\rho; k_1, k_2) e^{ik_2\rho} e^{ik_1(z+d)} \right.$$
$$+ \frac{1}{2}\left[\left(\frac{ik_1^2}{r_2} - \frac{3k_1}{r_2^2} - \frac{3i}{r_2^3}\right)e^{ik_1r_2}\right.$$
$$\left. - \left(\frac{ik_1^2}{r_1} - \frac{k_1}{r_1^2} - \frac{i}{r_1^3}\right)e^{ik_1r_1}\right]$$
$$+ i\left(\frac{z+d}{r_2}\right)\left(\frac{ik_1^2}{\rho} - \frac{3k_1}{2\rho^2} - \frac{5i}{8\rho^3}\right)e^{ik_1r_2}$$
$$\left. - \left(\frac{z+d}{r_2}\right)^2\left(\frac{ik_1^2}{r_2} - \frac{3k_1}{r_2^2} - \frac{3i}{r_2^3}\right)e^{ik_1r_2} \right\} \tag{5.5.38a}$$

$$= \frac{\omega\mu_0}{2\pi k_1^2}\sin\phi$$
$$\times \left\{ k_2 h(\rho; k_1, k_2) e^{ik_2\rho} e^{ik_1(z+d)} \right.$$
$$+ \frac{1}{2}\left[\left(\frac{ik_1^2}{\rho} - \frac{3k_1}{\rho^2} - \frac{3i}{\rho^3}\right)e^{ik_1r_2}\right.$$
$$\left. - \left(\frac{ik_1^2}{\rho} - \frac{k_1}{\rho^2} - \frac{i}{\rho^3}\right)e^{ik_1r_1}\right]$$
$$\left. + i\left(\frac{z+d}{\rho}\right)\left(\frac{ik_1^2}{\rho} - \frac{3k_1}{2\rho^2} - \frac{5i}{8\rho^3}\right)e^{ik_1r_2} \right\}, \tag{5.5.38b}$$

$$E_{1z}(\rho, \phi, z)$$
$$= \frac{\omega\mu_0}{2\pi k_1^2}\cos\phi$$
$$\times \left\{ \frac{k_2^2}{k_1}\left[f(\rho; k_1, k_2)e^{ik_2\rho}e^{ik_1(z+d)} - \left(\frac{i}{\rho^2} - \frac{3}{2k_1\rho^3}\right)e^{ik_1r_2}\right]\right.$$
$$- \frac{1}{2}\left[\left(\frac{\rho}{r_1}\right)\left(\frac{z-d}{r_1}\right)\left(\frac{ik_1^2}{r_1} - \frac{3k_1}{r_1^2} - \frac{3i}{r_1^3}\right)e^{ik_1r_1}\right.$$
$$\left.\left. + \left(\frac{\rho}{r_2}\right)\left(\frac{z+d}{r_2}\right)\left(\frac{ik_1^2}{r_2} - \frac{3k_1}{r_2^2} - \frac{3i}{r_2^3}\right)e^{ik_1r_2}\right]\right\} \tag{5.5.39a}$$

$$= \frac{\omega\mu_0}{2\pi k_1^2}\cos\phi$$
$$\times \left\{ \frac{k_2^2}{k_1}\left[f(\rho; k_1, k_2)e^{ik_2\rho}e^{ik_1(z+d)} - \left(\frac{i}{\rho^2} - \frac{3}{2k_1\rho^3}\right)e^{ik_1r_2}\right]\right.$$
$$\left. - \frac{1}{2}\left(\frac{z-d}{\rho}\,e^{ik_1r_1} + \frac{z+d}{\rho}\,e^{ik_1r_2}\right)\left(\frac{ik_1^2}{\rho} - \frac{3k_1}{\rho^2} - \frac{3i}{\rho^3}\right)\right\}, \tag{5.5.39b}$$

$$\begin{aligned}
&B_{1\rho}(\rho, \phi, z)\\
&= -\frac{\mu_0}{2\pi k_1}\sin\phi\\
&\quad\times\Bigg\{k_2 h(\rho; k_1, k_2)e^{ik_2\rho}e^{ik_1(z+d)}\\
&\qquad+\left[\frac{ik_1}{\rho^2} - \frac{7}{2\rho^3} + \left(\frac{z+d}{r_2}\right)\left(\frac{ik_1^2}{r_2} - \frac{5k_1}{r_2^2} - \frac{12i}{r_2^3}\right)\right]e^{ik_1r_2}\\
&\qquad-\frac{1}{2}\Bigg[\left(\frac{z-d}{r_1}\right)\left(\frac{ik_1^2}{r_1} - \frac{k_1}{r_1^2}\right)e^{ik_1r_1}\\
&\qquad\qquad+\left(\frac{z+d}{r_2}\right)\left(\frac{ik_1^2}{r_2} - \frac{k_1}{r_2^2}\right)e^{ik_1r_2}\Bigg]\Bigg\}
\end{aligned} \tag{5.5.40a}$$

$$\begin{aligned}
&= -\frac{\mu_0}{2\pi k_1}\sin\phi\\
&\quad\times\Bigg\{k_2 h(\rho; k_1, k_2)e^{ik_2\rho}e^{ik_1(z+d)}\\
&\qquad+\left[\frac{ik_1}{\rho^2} - \frac{7}{2\rho^3} + \left(\frac{z+d}{\rho}\right)\left(\frac{ik_1^2}{\rho} - \frac{5k_1}{\rho^2} - \frac{12i}{\rho^3}\right)\right]e^{ik_1r_2}\\
&\qquad-\frac{1}{2}\left(\frac{z-d}{\rho}e^{ik_1r_1} + \frac{z+d}{\rho}e^{ik_1r_2}\right)\left(\frac{ik_1^2}{\rho} - \frac{k_1}{\rho^2}\right)\Bigg\},
\end{aligned} \tag{5.5.40b}$$

$$\begin{aligned}
&B_{1\phi}(\rho, \phi, z)\\
&= -\frac{\mu_0}{2\pi k_1}\cos\phi\\
&\quad\times\Bigg\{k_2 g(\rho; k_1, k_2)e^{ik_2\rho}e^{ik_1(z+d)}\\
&\qquad+\frac{1}{2}\Bigg[\left(\frac{2}{\rho^3} + \frac{3i}{k_1\rho^4}\right)e^{ik_1r_2} - \left(\frac{z-d}{r_1}\right)\left(\frac{ik_1^2}{r_1} - \frac{k_1}{r_1^2}\right)e^{ik_1r_1}\\
&\qquad\qquad-\left(\frac{z+d}{r_2}\right)\left(\frac{ik_1^2}{r_2} - \frac{3k_1}{r_2^2} - \frac{6i}{r_2^3}\right)e^{ik_1r_2}\Bigg]\Bigg\}
\end{aligned} \tag{5.5.41a}$$

$$\begin{aligned}
&= -\frac{\mu_0}{2\pi k_1}\cos\phi\\
&\quad\times\Bigg\{k_2 g(\rho; k_1, k_2)e^{ik_2\rho}e^{ik_1(z+d)}\\
&\qquad+\frac{1}{2}\Bigg[\left(\frac{2}{\rho^3} + \frac{3i}{k_1\rho^4}\right)e^{ik_1r_2} - \left(\frac{z-d}{\rho}\right)\left(\frac{ik_1^2}{\rho} - \frac{k_1}{\rho^2}\right)e^{ik_1r_1}\\
&\qquad\qquad-\left(\frac{z+d}{\rho}\right)\left(\frac{ik_1^2}{\rho} - \frac{3k_1}{\rho^2} - \frac{6i}{\rho^3}\right)e^{ik_1r_2}\Bigg]\Bigg\},
\end{aligned} \tag{5.5.41b}$$

$$
\begin{aligned}
B_{1z}(\rho,\phi,z) = \frac{\mu_0}{2\pi}\sin\phi
&\times\Bigg\{\frac{k_2^2}{k_1^2}\left(\frac{1}{\rho^2}+\frac{3i}{k_2\rho^3}-\frac{3}{k_2^2\rho^4}\right)e^{ik_2\rho}e^{ik_1(z+d)}\\
&-\left(\frac{\rho}{r_2}\right)\Bigg[\frac{1}{r_2^2}+\frac{3i}{k_1r_2^3}-\frac{3}{k_1^2r_2^4}\\
&\qquad+\left(\frac{z+d}{\rho}\right)\left(\frac{k_1}{\rho}+\frac{3i}{2\rho^2}-\frac{9}{8k_1\rho^3}\right)\\
&\qquad+\left(\frac{z+d}{r_2}\right)^2\left(\frac{ik_1}{r_2}-\frac{6}{r_2^2}-\frac{15i}{k_1r_2^3}\right)\Bigg]e^{ik_1r_2}\\
&-\frac{1}{2}\Bigg[\left(\frac{\rho}{r_1}\right)\left(\frac{ik_1}{r_1}-\frac{1}{r_1^2}\right)e^{ik_1r_1}\\
&\qquad-\left(\frac{\rho}{r_2}\right)\left(\frac{ik_1}{r_2}-\frac{1}{r_2^2}\right)e^{ik_1r_2}\Bigg]\Bigg\}
\end{aligned}
\tag{5.5.42a}
$$

$$
\begin{aligned}
= \frac{\mu_0}{2\pi}\sin\phi
&\times\Bigg\{\frac{k_2^2}{k_1^2}\left(\frac{1}{\rho^2}+\frac{3i}{k_2\rho^3}-\frac{3}{k_2^2\rho^4}\right)e^{ik_2\rho}e^{ik_1(z+d)}\\
&-\Bigg[\frac{1}{\rho^2}+\frac{3i}{k_1\rho^3}-\frac{3}{k_1^2\rho^4}\\
&\qquad+\left(\frac{z+d}{\rho}\right)\left(\frac{k_1}{\rho}+\frac{3i}{2\rho^2}-\frac{9}{8k_1\rho^3}\right)\Bigg]e^{ik_1r_2}\\
&-\frac{1}{2}\left(e^{ik_1r_1}-e^{ik_1r_2}\right)\left(\frac{ik_1}{\rho}-\frac{1}{\rho^2}\right)\Bigg\}.
\end{aligned}
\tag{5.5.42b}
$$

In these formulas,

$$f(\rho;k_1,k_2) = \frac{ik_2}{\rho}-\frac{1}{\rho^2}-\frac{k_2^3}{k_1}\left(\frac{\pi}{k_2\rho}\right)^{1/2}e^{-ip}\mathcal{F}(p), \tag{5.5.43}$$

$$g(\rho;k_1,k_2) = \frac{ik_2}{\rho}-\frac{1}{\rho^2}-\frac{i}{k_2\rho^3}-\frac{k_2^3}{k_1}\left(\frac{\pi}{k_2\rho}\right)^{1/2}e^{-ip}\mathcal{F}(p), \tag{5.5.44}$$

$$h(\rho;k_1,k_2) = \frac{2}{\rho^2}+\frac{2i}{k_2\rho^3}+\frac{ik_2^2}{k_1\rho}\left(\frac{\pi}{k_2\rho}\right)^{1/2}e^{-ip}\mathcal{F}(p), \tag{5.5.45}$$

where $\mathcal{F}(p)$ and p are defined in (5.5.35). For large values of $|k_2\rho|$,

$$f(\rho;k_1,k_2)\sim g(\rho;k_1,k_2)\sim -\frac{k_1^2}{k_2^2\rho^2};\qquad |k_2\rho|\geq 8\left|\frac{k_1^2}{k_2^2}\right|. \tag{5.5.46}$$

This approximation is discussed in Chapter 3, Section 3.4.

The direct field in these expressions is given by all terms multiplied by $\exp(ik_1r_1)$, the complete reflected field includes all terms with the factor

$\exp(ik_1r_2)$, and the lateral-wave field consists of the terms multiplied by $\exp(ik_2\rho)\exp[ik_1(z+d)]$. The quantities $\exp(ik_1r_1)$ and $\exp(ik_1r_2)$ represent spherical waves traveling radially outward from the dipole at $(0,0,d)$ or the image dipole at $(0,0,-d)$. The quantity $\exp[ik_1(z+d)]\exp(ik_2\rho)$ suggests a plane wave $\exp(ik_1d)$ that travels upward in Region 1 from the source at $(0,0,d)$ to the boundary at $(0,0,0)$, then radially outward in Region 2 as a cylindrical wave $\exp(ik_2\rho)$ to $(\rho,0,0)$, and finally vertically downward from the boundary in Region 1 as a plane wave $\exp(ik_1z)$ to the point of observation at $(\rho,0,z)$.

The corresponding formulas for the electromagnetic field in Region 2 as obtained from (5.7.14), (5.7.15), (5.7.25), (5.7.45), (5.7.52), and (5.7.57) are given below. They are evaluated in Section 5.7. (A generalization of these formulas to eliminate the restriction $\rho \geq 5|z|$ is in Section 5.10.)

$$\begin{aligned}
E_{2\rho}(\rho,\phi,z) &= -\frac{\omega\mu_0}{2\pi k_1^2}\cos\phi \\
&\times\Bigg(e^{ik_1d}e^{ik_2r_0}\Bigg\{k_2\left[\frac{ik_2}{r_0}-\frac{1}{r_0^2}-\frac{i}{k_2r_0^3}-\frac{k_2^3}{k_1}\left(\frac{\pi}{k_2\rho}\right)^{1/2}e^{-iP}\mathcal{F}(P)\right] \\
&\qquad + k_1\left(\frac{z}{r_0}\right)\left(\frac{ik_2}{r_0}-\frac{1}{r_0^2}\right)\Bigg\} \\
&\quad - e^{k_1z}e^{ik_1r_d}k_1\left[\frac{1}{r_d^2}+\frac{i}{k_1r_d^3}-\left(\frac{d}{\rho}\right)\left(\frac{i}{\rho^2}-\frac{3}{2k_1\rho^3}\right)\right. \\
&\qquad \left.+\left(\frac{d}{r_d}\right)^2\left(\frac{ik_1}{r_d}-\frac{3}{r_d^2}-\frac{3i}{k_1r_d^3}\right)\right]\Bigg); \quad z\leq 0,
\end{aligned} \tag{5.5.47}$$

$$\begin{aligned}
E_{2\phi}(\rho,\phi,z) &= \frac{\omega\mu_0}{2\pi k_1^2}\sin\phi \\
&\times\Bigg(e^{ik_1d}e^{ik_2r_0}\Bigg\{k_2\left[\frac{2}{r_0^2}+\frac{2i}{k_2r_0^3}+\frac{ik_2^2}{k_1\rho}\left(\frac{\pi}{k_2\rho}\right)^{1/2}e^{-iP}\mathcal{F}(P)\right. \\
&\qquad \left.+\left(\frac{z}{r_0}\right)^2\left(\frac{ik_2}{r_0}-\frac{3}{r_0^2}-\frac{3i}{k_2r_0^3}\right)\right] \\
&\qquad + k_1\left(\frac{z}{r_0}\right)\left(\frac{ik_2}{r_0}-\frac{1}{r_0^2}\right)\Bigg\} \\
&\quad - e^{k_1z}e^{ik_1r_d}k_1\left[\frac{1}{r_d^2}+\frac{i}{k_1r_d^3}-i\left(\frac{d}{r_d}\right)\left(\frac{ik_1}{\rho}-\frac{3}{2\rho^2}-\frac{5i}{8k_1\rho^3}\right)\right. \\
&\qquad \left.+\left(\frac{d}{r_d}\right)^2\left(\frac{ik_1}{r_d}-\frac{3}{r_d^2}-\frac{3i}{k_1r_d^3}\right)\right]\Bigg); \quad z\leq 0,
\end{aligned} \tag{5.5.48}$$

$$E_{2z}(\rho,\phi,z) = \frac{\omega\mu_0}{2\pi k_1}\cos\phi\left\{e^{ik_1d}e^{ik_2r_0}\left[\left(\frac{\rho}{r_0}\right)\left(\frac{ik_2}{r_0}-\frac{1}{r_0^2}\right)\right.\right.$$

$$- \frac{k_2^3}{k_1}\left(\frac{\pi}{k_2\rho}\right)^{1/2} e^{-iP}\mathcal{F}(P)\Bigg]$$

$$- e^{k_1 z} e^{ik_1 r_d}\left(\frac{i}{\rho^2} - \frac{3}{2k_1\rho^3}\right)\Bigg\}; \quad z \le 0, \quad (5.5.49)$$

$$\begin{aligned}
&B_{2\rho}(\rho,\phi,z)\\
&= -\frac{\mu_0}{2\pi k_1}\sin\phi\\
&\quad\times\Bigg\{e^{ik_1 d}e^{ik_2 r_0}k_2\Bigg[\frac{2}{r_0^2} + \frac{2i}{k_2 r_0^3} + \frac{ik_2^2}{k_1\rho}\left(\frac{\pi}{k_2\rho}\right)^{1/2} e^{-iP}\mathcal{F}(P)\\
&\qquad + \left(\frac{z^2}{r_0^2}\right)\left(\frac{ik_2}{r_0} - \frac{3}{r_0^2} - \frac{3i}{k_2 r_0^3}\right)\\
&\qquad + \left(\frac{k_2 z}{k_1 r_0}\right)\left(\frac{4}{r_0^2} - \frac{1}{\rho^2} + \frac{12i}{k_2 r_0^3}\right)\Bigg]\\
&\quad + e^{k_1 z}e^{ik_1 r_d}k_1\Bigg[\frac{i}{\rho^2} - \frac{7}{2k_1\rho^3}\\
&\qquad + \left(\frac{d}{r_d}\right)\left(\frac{ik_1}{r_d} - \frac{5}{r_d^2} - \frac{12i}{k_1 r_d^3}\right)\Bigg]\Bigg\}; \quad z \le 0,
\end{aligned} \quad (5.5.50a)$$

$$\begin{aligned}
&= -\frac{\mu_0}{2\pi k_1}\sin\phi\\
&\quad\times\Bigg\{e^{ik_1 d}e^{ik_2 r_0}k_2\Bigg[\frac{ik_2 z^2}{r_0^3} + \frac{2\rho^2 - z^2}{r_0^2}\left(\frac{1}{r_0^2} + \frac{i}{k_2 r_0^3}\right)\\
&\qquad + \frac{ik_2^2}{k_1\rho}\left(\frac{\pi}{k_2\rho}\right)^{1/2} e^{-iP}\mathcal{F}(P)\\
&\qquad + \left(\frac{k_2 z}{k_1 r_0}\right)\left(\frac{4}{r_0^2} - \frac{1}{\rho^2} + \frac{12i}{k_2 r_0^3}\right)\Bigg]\\
&\quad + e^{k_1 z}e^{ik_1 r_d}k_1\Bigg[\frac{i}{\rho^2} - \frac{7}{2k_1\rho^3}\\
&\qquad + \left(\frac{d}{r_d}\right)\left(\frac{ik_1}{r_d} - \frac{5}{r_d^2} - \frac{12i}{k_1 r_d^3}\right)\Bigg]\Bigg\}; \quad z \le 0,
\end{aligned} \quad (5.5.50b)$$

$$\begin{aligned}
B_{2\phi}(\rho,\phi,z) &= -\frac{\mu_0}{2\pi k_1}\cos\phi\\
&\quad\times\Bigg\{e^{ik_1 d}e^{ik_2 r_0}k_2\Bigg[\frac{ik_2}{r_0} - \frac{1}{r_0^2} - \frac{i}{k_2 r_0^3}
\end{aligned}$$

$$
\left. - \frac{k_2^3}{k_1}\left(\frac{\pi}{k_2\rho}\right)^{1/2} e^{-iP}\mathcal{F}(P)\right]
$$

$$
+ e^{k_1 z} e^{ik_1 r_d} k_1 \left[\frac{1}{k_1\rho^3} + \frac{3i}{2k_1^2\rho^4} + \left(\frac{d}{r_d}\right)\left(\frac{1}{r_d^2} + \frac{3i}{k_1 r_d^3}\right)\right]\Bigg\}; \quad z \le 0, \tag{5.5.51}
$$

$$
B_{2z}(\rho,\phi,z) = \frac{\mu_0 k_2}{2\pi k_1}\sin\phi
$$

$$
\times \Bigg(e^{ik_1 d} e^{ik_2 r_0} \Bigg\{ \left(\frac{\rho z}{r_0^2}\right)\left(\frac{ik_2}{r_0} - \frac{3}{r_0^2} - \frac{3i}{k_2 r_0^3}\right)
$$

$$
+ \left(\frac{k_2\rho}{k_1 r_0}\right)\left[\frac{1}{r_0^2} + \frac{3i}{k_2 r_0^3} - \frac{3}{k_2^2 r_0^4} + \left(\frac{z^2}{r_0^2}\right)\left(\frac{k_2}{r_0} + \frac{6i}{r_0^2} - \frac{15}{k_2 r_0^3}\right)\right]\Bigg\}
$$

$$
- e^{k_1 z} e^{ik_1 r_d}\left(\frac{k_1\rho}{k_2 r_d}\right)\left[\frac{1}{r_d^2} + \frac{3i}{k_1 r_d^3} - \frac{3}{k_1^2 r_d^4} + \left(\frac{d}{\rho}\right)\left(\frac{k_1}{\rho} + \frac{3i}{2\rho^2} - \frac{9}{8k_1\rho^3}\right) + \left(\frac{d}{r_d}\right)^2\left(\frac{ik_1}{r_d} - \frac{6}{r_d^2} - \frac{15i}{k_1 r_d^3}\right)\right]\Bigg);
$$

$$
z \le 0. \tag{5.5.52}
$$

In these formulas

$$
P = \frac{(R-Z)^2}{R}; \qquad R = \frac{k_2^3\rho}{2k_1^2}; \qquad Z = \frac{k_2^2 z}{2k_1} \le 0. \tag{5.5.53}
$$

When $Z = 0,\ P = R = p$. Also,

$$
\mathcal{F}(P) = \tfrac{1}{2}(1+i) - C_2(P) - iS_2(P), \tag{5.5.54}
$$

$$
C_2(P) + iS_2(P) = \int_0^P \frac{e^{it}}{(2\pi t)^{1/2}}\,dt, \tag{5.5.55}
$$

$$
r_0 = (\rho^2 + z^2)^{1/2}; \qquad r_d = (\rho^2 + d^2)^{1/2}. \tag{5.5.56}
$$

With the boundary conditions, the formulas for the field in Region 2 can be compared with those of the field in Region 1 along the boundary $z = 0$. The formulas at $z = 0$ with the dipole at the depth d in Region 1 are

$$
E_{2\rho}(\rho,\phi,0) = E_{1\rho}(\rho,\phi,0)
$$

$$
= -\frac{\omega\mu_0}{2\pi k_1^2}\cos\phi
$$

$$\times \Bigg\{ k_2 g(\rho; k_1, k_2) e^{ik_2\rho} e^{ik_1 d} - \left[\frac{k_1}{r_d^2} + \frac{i}{r_d^3} - \left(\frac{d}{\rho}\right)\left(\frac{ik_1}{\rho^2} - \frac{3}{2\rho^3}\right) + \left(\frac{d}{\rho}\right)^2 \left(\frac{ik_1^2}{r_d} - \frac{3k_1}{r_d^2} - \frac{3i}{r_d^3}\right)\right] e^{ik_1 r_d} \Bigg\}, \tag{5.5.57}$$

$$E_{2\phi}(\rho,\phi,0) = E_{1\phi}(\rho,\phi,0) = \frac{\omega\mu_0}{2\pi k_1^2}\sin\phi \times \Bigg\{ k_2 h(\rho; k_1, k_2) e^{ik_2\rho} e^{ik_1 d} - \left[\frac{k_1}{r_d^2} + \frac{i}{r_d^3} - i\left(\frac{d}{r_d}\right)\left(\frac{ik_1^2}{\rho} - \frac{3k_1}{2\rho^2} - \frac{5i}{8\rho^3}\right) + \left(\frac{d}{r_d}\right)^2 \left(\frac{ik_1^2}{r_d} - \frac{3k_1}{r_d^2} - \frac{3i}{r_d^3}\right)\right] e^{ik_1 r_d} \Bigg\}, \tag{5.5.58}$$

$$E_{2z}(\rho,\phi,0) = \frac{k_1^2}{k_2^2} E_{1z}(\rho,\phi,0) = \frac{\omega\mu_0}{2\pi k_1}\cos\phi \left[f(\rho; k_1, k_2) e^{ik_2\rho} e^{ik_1 d} - \left(\frac{i}{\rho^2} - \frac{3}{2k_1\rho^3}\right) e^{ik_1 r_d} \right], \tag{5.5.59}$$

$$B_{2\rho}(\rho,\phi,0) = B_{1\rho}(\rho,\phi,0) = -\frac{\mu_0}{2\pi k_1}\sin\phi \times \Bigg\{ k_2 h(\rho; k_1, k_2) e^{ik_2\rho} e^{ik_1 d} + \left[\frac{ik_1}{\rho^2} - \frac{7}{2\rho^3} + \left(\frac{d}{r_d}\right)\left(\frac{ik_1^2}{r_d} - \frac{5k_1}{r_d^2} - \frac{12i}{r_d^3}\right)\right] e^{ik_1 r_d} \Bigg\}, \tag{5.5.60}$$

$$B_{2\phi}(\rho,\phi,0) = B_{1\phi}(\rho,\phi,0) = -\frac{\mu_0}{2\pi k_1}\cos\phi \times \Bigg\{ k_2 g(\rho; k_1, k_2) e^{ik_2\rho} e^{ik_1 d} + \left[\frac{1}{\rho^3} + \frac{3i}{2k_1\rho^4} + \left(\frac{d}{r_d}\right)\left(\frac{k_1}{r_d^2} + \frac{3i}{r_d^3}\right)\right] e^{ik_1 r_d} \Bigg\}, \tag{5.5.61}$$

$$
\begin{aligned}
B_{2z}(\rho,\phi,0) &= B_{1z}(\rho,\phi,0)\\
&= \frac{\mu_0}{2\pi k_1^2}\sin\phi\\
&\quad\times\left\{\left(\frac{k_2^2}{\rho^2}+\frac{3ik_2}{\rho^3}-\frac{3}{\rho^4}\right)e^{ik_2\rho}e^{ik_1 d}\right.\\
&\qquad-\left[\frac{k_1^2}{r_d^2}+\frac{3ik_1}{r_d^3}-\frac{3}{r_d^4}+\left(\frac{d}{\rho}\right)\left(\frac{k_1^3}{\rho}+\frac{3ik_1^2}{2\rho^2}-\frac{9k_1}{8\rho^3}\right)\right.\\
&\qquad\left.\left.+\left(\frac{d}{r_d}\right)^2\left(\frac{ik_1^3}{r_d}-\frac{6k_1^2}{r_d^2}-\frac{15k_1}{r_d^3}\right)\right]\left(\frac{\rho}{r_d}\right)e^{ik_1 r_d}\right\}.
\end{aligned}
\tag{5.5.62}
$$

A graphical representation of the three components of the electric field given in (5.5.57)–(5.5.59) is in Fig. 5.5.1 at $f = 600$ MHz when Region 1 is sea water and Region 2 is air. The complete field including the direct-field terms is shown and also the field given by the lateral-wave terms alone. These are seen to represent the entire field when $k_2\rho > 1$. Also shown for comparison are the two components of the electric field for the vertical dipole on the boundary in air. The components of the magnetic field are in Fig. 5.5.2. A discussion of the individual curves in these figures follows.

Curve 3 + 3D in Fig. 5.5.1 represents $|E_{1\rho}(\rho,0,0)|_{1h} = |E_{2\rho}(\rho,0,0)|_{1h}$, i.e., the radial electric field at the boundary due to a horizontal electric dipole in Region 1 at $(0,0,0)$. It includes the contribution by the direct-field terms. It is reproduced in Fig. 5.5.3 together with its constituent parts. These include the direct field CP_1 which decreases exponentially with distance; the terms CP_2 which decrease as $1/\rho^3$ for small values of ρ and as $1/\rho$ for the larger values of ρ; and the Fresnel-integral term CF which decreases as $1/\rho^{1/2}$ when ρ is not too large and then, for large values of ρ, as $(ik_2/\rho) + (k_1^2/k_2^2\rho^2)$. When $k_2\rho < 1$, the direct-field term is significant and combines with the surface-wave term in an interference pattern. In general, $|E_{1\rho}(\rho,0,0)|_{1h}$ decreases as $1/\rho^3$ in the near field, as $1/\rho$ in the intermediate range, and as $1/\rho^2$ in the asymptotic range. Curve 4 + 4D is the magnitude of the complete component $|E_{1z}(\rho,0,0)|_{1h}$ (including the direct field) of the horizontal dipole source; curve 4 + 4L is the lateral-wave part.

Curve 5 + 5D in Fig. 5.5.1 is for $|E_{1\phi}(\rho,\pi/2,0)|_{1h}$, i.e., for the transverse component of the electric field in Region 1 due to a horizontal dipole also in Region 1 at the surface. It decreases as $1/\rho^3$ for small values of ρ, and as $1/\rho^2$ in both intermediate and asymptotic ranges so that, while initially larger than $|E_{1\rho}(\rho,0,0)|_{1h}$ by a factor 2, it decreases rapidly to much smaller values as ρ increases. Curve 5 + 5L is the lateral-wave part of $|E_{1\phi}(\rho,\pi/2,0)|_{1h}$, i.e., the field without the contributions by the direct-field terms.

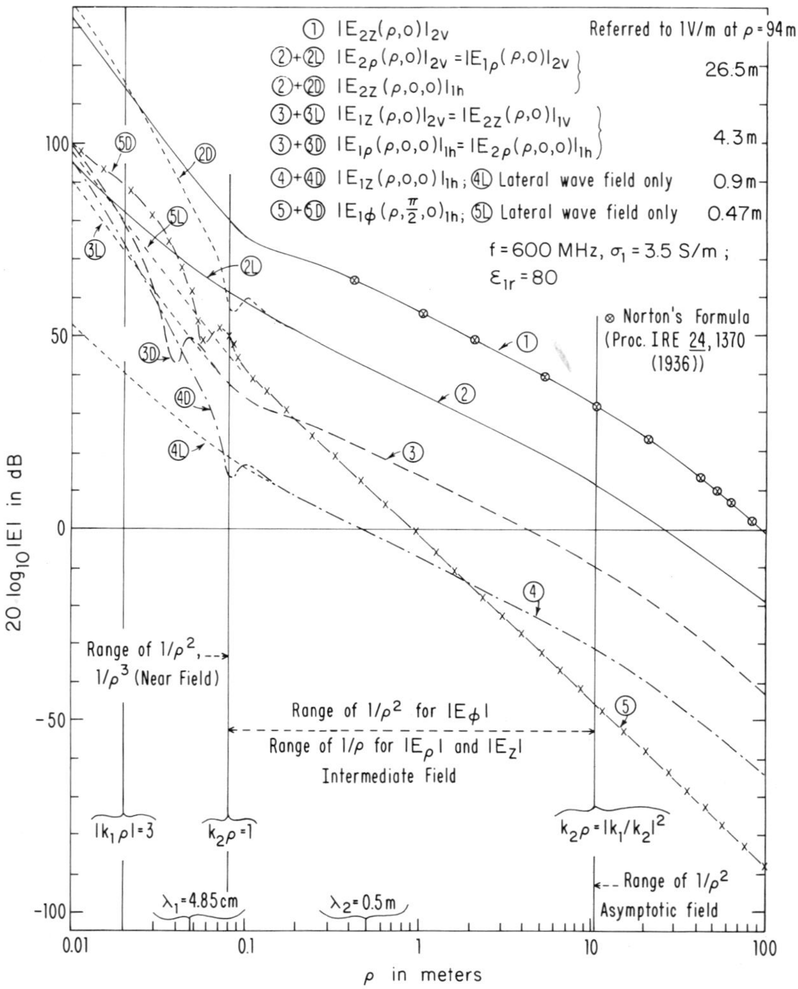

Figure 5.5.1. Components of electric field $\mathbf{E}(\rho, z)$ of unit vertical electric dipole ($I\Delta l = 1$ A m) in air (Region 2, $z \leq 0$), and $\mathbf{E}(\rho, \phi, z)$ of unit horizontal electric dipole in sea water (Region 1, $z \geq 0$) near surface ($z \sim 0$). (Subscript notation: first, field in Region 1 or 2; second, component of field; third, source dipole in Region 1 or 2; fourth, vertical or horizontal dipole.)

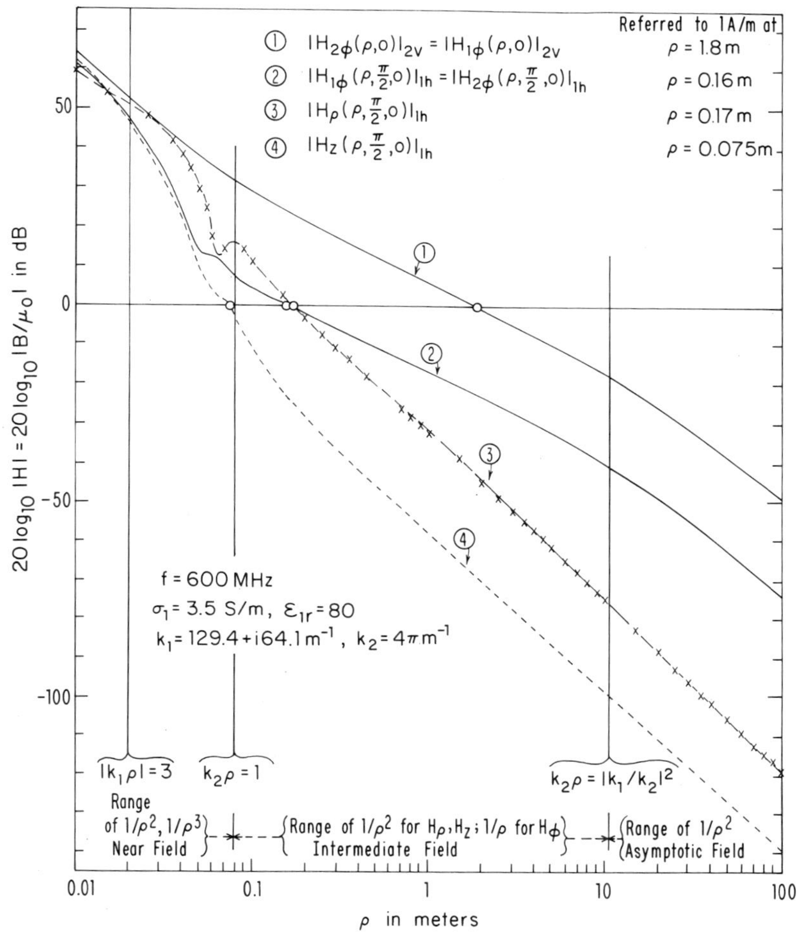

Figure 5.5.2. Components of magnetic field $\mathbf{H}(\rho, z)$ of unit vertical electric dipole in air and $\mathbf{H}(\rho, \phi, z)$ of unit horizontal electric dipole in sea water.

In Fig. 5.5.2 for the magnetic field, curve 2 is for the complete component $H_\phi = B_\phi/\mu_0$ for the horizontal dipole including the direct-field terms. B_ϕ is associated with E_z and E_ρ in a field of electric type. Curves 3 and 4 are for $H_\rho = B_\rho/\mu_0$ and $H_z = B_z/\mu_0$. They are associated with E_ϕ in a field of magnetic type. Note that E_ϕ, B_ρ, and B_z have the directional factor $\sin\phi$; E_ρ, E_z, and B_ϕ have the factor $\cos\phi$.

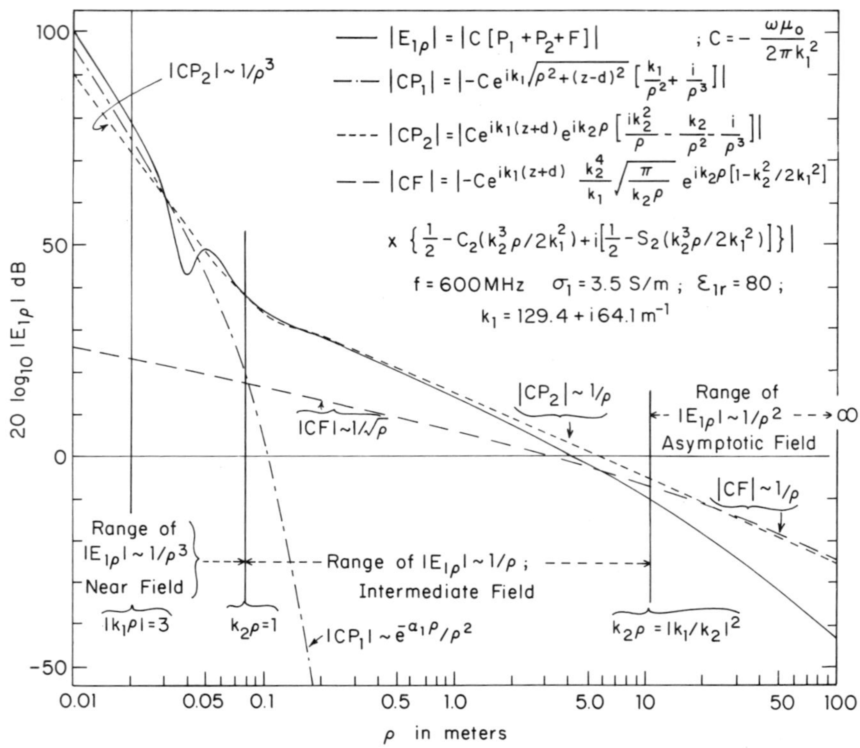

Figure 5.5.3. Radial electric field $E_\rho(\rho, \phi, z)$ near surface ($z \sim 0$, $\phi = 0$) of unit ($I\Delta l = 1$ A m) horizontal electric dipole at depth $d \sim 0$ in sea water. (Referred to 1 V/m at $\rho = 4.1$ m.)

5.6 Evaluation of the Integrals for the Electromagnetic Field in Region 1 of the Horizontal Electric Dipole at the Depth d in Region 1

The integral $F_{\rho 1}(\rho, z+d)$ defined in (5.5.4) consists of two parts, viz.,

$$F_{\rho 1}(\rho, z+d) = F_{\rho 2}(\rho, z+d) + F_{\rho 3}(\rho, z+d), \tag{5.6.1}$$

where

$$F_{\rho 2}(\rho, z+d) = \int_0^\infty \frac{\gamma_1}{2}\,(Q-1)[J_0(\lambda\rho) - J_2(\lambda\rho)]e^{i\gamma_1(z+d)}\,\lambda\,d\lambda, \tag{5.6.2}$$

$$F_{\rho 3}(\rho, z+d) = -k_1^2 \int_0^\infty \frac{1}{2\gamma_1}(P+1)[J_0(\lambda\rho) + J_2(\lambda\rho)]e^{i\gamma_1(z+d)}\,\lambda\,d\lambda, \tag{5.6.3}$$

and

$$Q - 1 = -\frac{2k_2^2\gamma_1}{k_1^2\gamma_2 + k_2^2\gamma_1}; \qquad P + 1 = \frac{2\gamma_2}{\gamma_2 + \gamma_1}. \tag{5.6.4}$$

With these values, the integrands contain the terms

$$\begin{aligned}\frac{\gamma_1^2}{k_1^2\gamma_2 + k_2^2\gamma_1} &= \gamma_1^2\left(\frac{1}{k_1^2\gamma_2 + k_2^2\gamma_1} - \frac{1}{k_1^2\gamma_2}\right) + \frac{(\gamma_1^2 - \gamma_2^2) + \gamma_2^2}{k_1^2\gamma_2} \\ &= \gamma_1^2\left(\frac{1}{k_1^2\gamma_2 + k_2^2\gamma_1} - \frac{1}{k_1^2\gamma_2}\right) + \frac{k_1^2 - k_2^2}{k_1^2\gamma_2} + \frac{\gamma_2}{k_1^2} \\ &\sim \gamma_1^2\left(\frac{1}{k_1^2\gamma_2 + k_2^2\gamma_1} - \frac{1}{k_1^2\gamma_2}\right) + \frac{1}{\gamma_2} + \frac{\gamma_2}{k_1^2},\end{aligned} \tag{5.6.5}$$

and

$$\begin{aligned}\frac{\gamma_2}{\gamma_1(\gamma_1 + \gamma_2)} &= \frac{\gamma_2(\gamma_1 - \gamma_2)}{\gamma_1(\gamma_1^2 - \gamma_2^2)} = \frac{\gamma_1^2 - \gamma_2^2 - \gamma_1^2 - \gamma_1\gamma_2}{\gamma_1(\gamma_1^2 - \gamma_2^2)} \\ &= \frac{1}{\gamma_1} - \frac{\gamma_1}{k_1^2 - k_2^2} + \frac{\gamma_2}{k_1^2 - k_2^2} \\ &\sim \frac{1}{\gamma_1} - \frac{\gamma_1}{k_1^2} + \frac{\gamma_2}{k_1^2},\end{aligned} \tag{5.6.6}$$

where use has been made of $\gamma_j^2 = k_j^2 - \lambda^2$ and $|k_1^2| \gg |k_2^2|$. When (5.6.4) with (5.6.5) and (5.6.6) are used in (5.6.2) and (5.6.3), these become

$$\begin{aligned}F_{\rho 1}(\rho, z+d) &= H_{\rho 1}(\rho, z+d) \\ &\quad - k_2^2 \int_0^\infty \gamma_2^{-1}[J_0(\lambda\rho) - J_2(\lambda\rho)]e^{i\gamma_1(z+d)}\lambda\,d\lambda \\ &\quad - \frac{k_2^2}{k_1^2}\int_0^\infty \gamma_2[J_0(\lambda\rho) - J_2(\lambda\rho)]e^{i\gamma_1(z+d)}\lambda\,d\lambda \\ &\quad - k_1^2 \int_0^\infty \gamma_1^{-1}[J_0(\lambda\rho) + J_2(\lambda\rho)]e^{i\gamma_1(z+d)}\lambda\,d\lambda \\ &\quad + \int_0^\infty \gamma_1[J_0(\lambda\rho) + J_2(\lambda\rho)]e^{i\gamma_1(z+d)}\lambda\,d\lambda \\ &\quad - \int_0^\infty \gamma_2[J_0(\lambda\rho) + J_2(\lambda\rho)]e^{i\gamma_1(z+d)}\lambda\,d\lambda,\end{aligned} \tag{5.6.7}$$

where

$$H_{\rho 1}(\rho, z+d) = -k_2^2 \int_0^\infty \gamma_1^2\left(\frac{1}{k_1^2\gamma_2 + k_2^2\gamma_1} - \frac{1}{k_1^2\gamma_2}\right)$$

$$\times [J_0(\lambda\rho) - J_2(\lambda\rho)] e^{i\gamma_1(z+d)} \lambda \, d\lambda. \tag{5.6.8}$$

The second integral in (5.6.7) is multiplied by the small factor k_2^2/k_1^2 so that it can be neglected. In the integral in (5.6.8), significant contributions to the integrand occur only when $\lambda \sim k_2$ when $\gamma_2 = (k_2^2 - \lambda^2)^{1/2}$ becomes small. It follows that $\gamma_1 = (k_1^2 - \lambda^2)^{1/2}$ is well approximated by $\gamma_1 \sim k_1$ so that the exponential factor can be replaced by

$$e^{i\gamma_1(z+d)} \sim e^{ik_1(z+d)}. \tag{5.6.9}$$

The first and last integrals in (5.6.7) involve both γ_1 and γ_2 which makes them more difficult to integrate. They are evaluated, respectively, as $I_K(\rho, z)$ in (D30) and $I_L(\rho, z)$ in (D31) of Appendix D. The third and fourth integrals in (5.6.7) are given in Appendix A. With these several values, (5.6.7) can be written as follows:

$$F_{\rho 1}(\rho, z+d) = e^{ik_1(z+d)} H_{\rho 1}(\rho, 0) - k_2^2 I_K(\rho, z+d) - k_1^2 I_3(k_1; \rho, z+d) + I_{11}(k_1; \rho, z+d) - I_L(\rho, z+d). \tag{5.6.10}$$

The values of the several integrals in Appendices A and D can be inserted to give for (5.6.10),

$$\begin{aligned} F_{\rho 1}(\rho, z+d) &\sim e^{ik_1(z+d)} H_{\rho 1}(\rho, 0) \\ &- 2k_2^2 \left\{ \left(\frac{z+d}{\rho}\right) \left[\frac{e^{ik_1(z+d)}}{k_1 \rho^2} - \left(\frac{1}{\rho} + \frac{i}{2k_1\rho^2} + \frac{3}{8k_1^2\rho^3}\right) e^{ik_1 r_2} \right] \right. \\ &\qquad \left. - \left[\frac{e^{ik_1(z+d)}}{k_2\rho^2} + \left(\frac{i}{\rho} - \frac{1}{k_2\rho^2}\right) e^{ik_2\rho} e^{ik_1(z+d)} \right] \right\} \\ &- \frac{2k_1}{\rho^2} \left(e^{ik_1(z+d)} - e^{ik_1 r_2}\right) \\ &+ 2\left[\frac{k_1}{\rho^2} e^{ik_1(z+d)} + \frac{i e^{ik_1 r_2}}{r_2^3} - \left(\frac{z}{r_2}\right)^2 \frac{k_1}{\rho^2} e^{ik_1 r_2} \right] \\ &- 2\left[\left(\frac{k_2}{\rho^2} + \frac{i e^{ik_2\rho}}{\rho^3}\right) e^{ik_1(z+d)} - \left(\frac{z+d}{\rho}\right)\left(\frac{ik_1}{\rho^2} - \frac{3}{2\rho^3}\right) e^{ik_1 r_2} \right]. \end{aligned} \tag{5.6.11}$$

When small terms of the order k_2^2/k_1^2 compared with 1 are omitted, this reduces to

$$\begin{aligned} &F_{\rho 1}(\rho, z+d) \\ &\quad = e^{ik_1(z+d)} \left[H_{\rho 1}(\rho, 0) + 2k_2 e^{ik_2\rho} \left(\frac{ik_2}{\rho} - \frac{1}{\rho^2} - \frac{i}{k_2\rho^3} \right) \right] \\ &\qquad + 2k_1 e^{ik_1 r_2} \left[\frac{1}{r_2^2} + \frac{i}{k_1 r_2^3} + \left(\frac{z+d}{\rho}\right)\left(\frac{i}{\rho^2} - \frac{3}{2k_1\rho^3}\right) \right]. \end{aligned} \tag{5.6.12}$$

The integral $H_{\rho 1}(\rho, 0)$ is evaluated in Appendix E; the result is given in (E63). With it,

$$\begin{aligned} F_{\rho 1}(\rho, z+d) &= 2k_2 e^{ik_1(z+d)} e^{ik_2\rho} \\ &\quad \times \left[\frac{ik_2}{\rho} - \frac{1}{\rho^2} - \frac{i}{k_2\rho^3} - \frac{k_2^3}{k_1}\left(\frac{\pi}{k_2\rho}\right)^{1/2} e^{-ip}\mathcal{F}(p) \right] \\ &\quad + 2k_1 e^{ik_1r_2}\left[\frac{1}{r_2^2} + \frac{i}{k_1r_2^3} \right. \\ &\qquad \left. + \left(\frac{z+d}{\rho}\right)\left(\frac{i}{\rho^2} - \frac{3}{2k_1\rho^3}\right)\right]. \end{aligned} \tag{5.6.13}$$

The integral $F_{\phi 1}(\rho, z+d)$ defined in (5.5.8) is

$$F_{\phi 1}(\rho, z+d) = F_{\phi 2}(\rho, z+d) + F_{\phi 3}(\rho, z+d), \tag{5.6.14}$$

where

$$F_{\phi 2}(\rho, z+d) = \int_0^\infty \frac{\gamma_1}{2}(Q-1)[J_0(\lambda\rho) + J_2(\lambda\rho)]e^{i\gamma_1(z+d)}\lambda\, d\lambda, \tag{5.6.15}$$

$$F_{\phi 3}(\rho, z+d) = -k_1^2 \int_0^\infty \frac{1}{2\gamma_1}(P+1)[J_0(\lambda\rho) - J_2(\lambda\rho)]e^{i\gamma_1(z+d)}\lambda\, d\lambda, \tag{5.6.16}$$

where $Q-1$ and $P+1$ are defined in (5.6.4). These integrals are like those in (5.6.2) and (5.6.3) except for the sign of $J_2(\lambda\rho)$. It follows that (5.6.7) and (5.6.8) apply if the sign of $J_2(\lambda\rho)$ is changed. The integrated results are then like (5.6.10) with the subscripts K and M, L and N, 3 and 4, and 11 and 12, respectively, interchanged. Thus,

$$\begin{aligned} F_{\phi 1}(\rho, z+d) &= e^{ik_1(z+d)} H_{\phi 1}(\rho, 0) - k_2^2 I_M(\rho, z+d) \\ &\quad - k_1^2 I_4(k_1; \rho, z+d) + I_{12}(k_1; \rho, z+d) - I_N(\rho, z+d), \end{aligned} \tag{5.6.17}$$

where

$$\begin{aligned} H_{\phi 1}(\rho, z+d) &= -k_2^2 \int_0^\infty \gamma_1^2 \left(\frac{1}{k_1^2\gamma_2 + k_2^2\gamma_1} - \frac{1}{k_1^2\gamma_2} \right) \\ &\qquad \times [J_0(\lambda\rho) + J_2(\lambda\rho)]e^{i\gamma_1(z+d)}\lambda\, d\lambda \\ &\sim e^{ik_1(z+d)} H_{\phi 1}(\rho, 0). \end{aligned} \tag{5.6.18}$$

The several integrals I are tabulated in Appendix A and evaluated in Appendix B or Appendix D. With them,

$$\begin{aligned} F_{\phi 1}(\rho, z+d) &= e^{ik_1(z+d)} H_{\phi 1}(\rho, 0) \\ &\quad - 2k_2^2 \left[\frac{e^{ik_1(z+d)}}{k_2\rho^2}(1 - e^{ik_2\rho}) + \left(\frac{z+d}{\rho}\right)\frac{ie^{ik_1r_2}}{k_1\rho^2} \right] \end{aligned}$$

$$
\begin{aligned}
&+ 2k_1 \left[\frac{1}{\rho^2} e^{ik_1(z+d)} + \left(\frac{ik_1}{r_2} - \frac{1}{\rho^2} \right) e^{ik_1 r_2} \right] \\
&- 2k_1 \left[\frac{1}{\rho^2} e^{ik_1(z+d)} + \left(\frac{1}{r_2^2} + \frac{2i}{k_1 r_2^3} \right) e^{ik_1 r_2} \right. \\
&\qquad \left. + \left(\frac{z+d}{r_2} \right)^2 \left(\frac{ik_1}{r_2} - \frac{3}{r_2^2} - \frac{3i}{k_1 r_2^3} - \frac{1}{\rho^2} \right) e^{ik_1 r_2} \right] \\
&+ 2ik_1 \left(\frac{z+d}{r_2} \right) \left(\frac{ik_1}{\rho} - \frac{3}{2\rho^2} - \frac{5i}{8k_1\rho^3} \right) e^{ik_1 r_2} \\
&+ 2k_2 \left[\frac{1}{\rho^2} + \left(\frac{1}{\rho^2} + \frac{2i}{k_2\rho^3} \right) e^{ik_2\rho} \right] e^{ik_1(z+d)}.
\end{aligned} \tag{5.6.19}
$$

After all cancellations have been made and one term of order k_2^2/k_1^2 has been neglected, the result is

$$
\begin{aligned}
F_{\phi 1}(\rho, z+d) &= e^{ik_1(z+d)} \left[H_{\phi 1}(\rho, 0) + 4k_2 \left(\frac{1}{\rho^2} + \frac{i}{k_2\rho^3} \right) e^{ik_2\rho} \right] \\
&\quad + 4k_1 \left(\frac{ik_1}{2r_2} - \frac{1}{r_2^2} - \frac{i}{k_1 r_2^3} \right) e^{ik_1 r_2} \\
&\quad + 2ik_1 \left(\frac{z+d}{r_2} \right) \left(\frac{ik_1}{\rho} - \frac{3}{2\rho^2} - \frac{5i}{8k_1\rho^3} \right) e^{ik_1 r_2} \\
&\quad - 2k_1 \left(\frac{z+d}{r_2} \right)^2 \left(\frac{ik_1}{r_2} - \frac{3}{r_2^2} - \frac{3i}{k_1 r_2^3} \right) e^{ik_1 r_2}.
\end{aligned} \tag{5.6.20}
$$

The integral $H_{\phi 1}(\rho, 0)$ is evaluated in (E64) of Appendix E. With it,

$$
\begin{aligned}
F_{\phi 1}(\rho, z+d) &= 4k_2 e^{ik_1(z+d)} e^{ik_2\rho} \\
&\quad \times \left[\frac{1}{\rho^2} + \frac{i}{k_2\rho^3} + \frac{ik_2^2}{2k_1\rho} \left(\frac{\pi}{k_2\rho} \right)^{1/2} e^{-ip} \mathcal{F}(p) \right] \\
&\quad + 4k_1 e^{ik_1 r_2} \left(\frac{ik_1}{2r_2} - \frac{1}{r_2^2} - \frac{i}{k_1 r_2^3} \right) \\
&\quad + 2ik_1 e^{ik_1 r_2} \left(\frac{z+d}{r_2} \right) \left(\frac{ik_1}{\rho} - \frac{3}{2\rho^2} - \frac{5i}{8k_1\rho^3} \right) \\
&\quad - 2k_1 e^{ik_1 r_2} \left(\frac{z+d}{r_2} \right)^2 \left(\frac{ik_1}{r_2} - \frac{3}{r_2^2} - \frac{3i}{k_1 r_2^3} \right).
\end{aligned} \tag{5.6.21}
$$

The integral $F_{z1}(\rho, z+d)$ is

$$
F_{z1}(\rho, z+d) = \int_0^\infty (Q-1) J_1(\lambda\rho) e^{i\gamma_1(z+d)} \lambda^2 \, d\lambda, \tag{5.6.22}
$$

where $Q-1$ is defined in (5.6.4). With it, (5.6.22) becomes

$$
F_{z1}(\rho, z+d) = -2k_2^2 \int_0^\infty \frac{\gamma_1}{k_1^2\gamma_2 + k_2^2\gamma_1} J_1(\lambda\rho) e^{i\gamma_1(z+d)} \lambda^2 \, d\lambda. \tag{5.6.23}
$$

Here the integrand can be expanded to give

$$F_{z1}(\rho, z+d) = H_{z1}(\rho, z+d) - \frac{2k_2^2}{k_1^2} \int_0^\infty \frac{\gamma_1}{\gamma_2} J_1(\lambda\rho) e^{i\gamma_1(z+d)} \lambda^2 \, d\lambda, \quad (5.6.24)$$

where

$$H_{z1}(\rho, z+d) = -2k_2^2 \int_0^\infty \gamma_1 \left(\frac{1}{k_1^2\gamma_2 + k_2^2\gamma_1} - \frac{1}{k_1^2\gamma_2} \right) \times J_1(\lambda\rho) e^{i\gamma_1(z+d)} \lambda^2 \, d\lambda. \quad (5.6.25)$$

The integral on the right in (5.6.24) is evaluated in (D1) and (D22) of Appendix D. Thus,

$$F_{z1}(\rho, z+d) = H_{z1}(\rho, z+d) - \frac{2k_2^2}{k_1^2} \frac{\rho}{2} I_A(\rho, z+d). \quad (5.6.26)$$

With (D22) this becomes

$$F_{z1}(\rho, z+d) = H_{z1}(\rho, z+d) + \frac{2k_2^2}{k_1} \left[e^{ik_1(z+d)} e^{ik_2\rho} \left(\frac{k_2}{\rho} + \frac{i}{\rho^2} \right) - e^{ik_1 r_2} \left(\frac{1}{\rho^2} + \frac{3i}{2k_1\rho^3} \right) \right]. \quad (5.6.27)$$

The integral $H_{z1}(\rho, z+d) \sim e^{ik_1(z+d)} H_{z1}(\rho, 0)$ is evaluated in (E65) of Appendix E. With it,

$$F_{z1}(\rho, z+d) = \frac{2k_2^2}{k_1} \Bigg\{ e^{ik_1(z+d)} e^{ik_2\rho} \times \left[\frac{k_2}{\rho} + \frac{i}{\rho^2} + \frac{ik_2^3}{k_1} \left(\frac{\pi}{k_2\rho} \right)^{1/2} e^{-ip} \mathcal{F}(p) \right] - e^{ik_1 r_2} \left(\frac{1}{\rho^2} + \frac{3i}{2k_1\rho^3} \right) \Bigg\}. \quad (5.6.28)$$

The integrals $G_{\rho 1}(\rho, z+d)$ and $G_{\phi 1}(\rho, z+d)$ defined in (5.5.17) can be separated as follows:

$$G_{\rho 1}(\rho, z+d) = G_{\rho 2}(\rho, z+d) + G_{\rho 3}(\rho, z+d),$$
$$G_{\phi 1}(\rho, z+d) = G_{\phi 2}(\rho, z+d) + G_{\phi 3}(\rho, z+d), \quad (5.6.29)$$

where

$$\left. \begin{matrix} G_{\rho 2}(\rho, z+d) \\ G_{\phi 2}(\rho, z+d) \end{matrix} \right\} = \int_0^\infty \frac{1}{2} (Q-1)[J_0(\lambda\rho) \pm J_2(\lambda\rho)] e^{i\gamma_1(z+d)} \lambda \, d\lambda, \quad (5.6.30)$$

$$\left. \begin{matrix} G_{\rho 3}(\rho, z+d) \\ G_{\phi 3}(\rho, z+d) \end{matrix} \right\} = -\int_0^\infty \frac{1}{2} (P+1)[J_0(\lambda\rho) \mp J_2(\lambda\rho)] e^{i\gamma_1(z+d)} \lambda \, d\lambda. \quad (5.6.31)$$

With (5.6.4),

$$\left.\begin{array}{l} G_{\rho 2}(\rho, z+d) \\ G_{\phi 2}(\rho, z+d) \end{array}\right\} = \left\{\begin{array}{l} H_{B\rho 1}(\rho, z+d) \\ H_{B\phi 1}(\rho, z+d) \end{array}\right\} - \frac{k_2^2}{k_1^2} \int_0^\infty \frac{\gamma_1}{\gamma_2} [J_0(\lambda\rho) \pm J_2(\lambda\rho)] e^{i\gamma_1(z+d)} \lambda \, d\lambda, \tag{5.6.32}$$

where

$$\left.\begin{array}{l} H_{B\rho 1}(\rho, z+d) \\ H_{B\phi 1}(\rho, z+d) \end{array}\right\} \sim -k_2^2 e^{ik_1(z+d)} \int_0^\infty \gamma_1 \left(\frac{1}{k_1^2\gamma_2 + k_2^2\gamma_1} - \frac{1}{k_1^2\gamma_2} \right) \times [J_0(\lambda\rho) \pm J_2(\lambda\rho)] \lambda \, d\lambda. \tag{5.6.33}$$

With (5.6.4),

$$\tfrac{1}{2}(P+1) = \frac{\gamma_2}{\gamma_1 + \gamma_2} = \frac{\gamma_1\gamma_2 - \gamma_2^2}{\gamma_1^2 - \gamma_2^2} = \frac{\gamma_1\gamma_2 - k_2^2 + \lambda^2}{k_1^2 - k_2^2} \sim \frac{\gamma_1\gamma_2}{k_1^2} - \frac{k_2^2}{k_1^2} + \frac{\lambda^2}{k_1^2}. \tag{5.6.34}$$

Hence, (5.6.31) can be written

$$\left.\begin{array}{l} G_{\rho 3}(\rho, z+d) \\ G_{\phi 3}(\rho, z+d) \end{array}\right\} \sim -k_1^{-2} \int_0^\infty \gamma_1\gamma_2 [J_0(\lambda\rho) \mp J_2(\lambda\rho)] e^{i\gamma_1(z+d)} \lambda \, d\lambda + k_1^{-2}k_2^2 \int_0^\infty [J_0(\lambda\rho) \mp J_2(\lambda\rho)] e^{i\gamma_1(z+d)} \lambda \, d\lambda - k_1^{-2} \int_0^\infty [J_0(\lambda\rho) \mp J_2(\lambda\rho)] e^{i\gamma_1(z+d)} \lambda^3 \, d\lambda. \tag{5.6.35}$$

The integrals on the right in (5.6.32) are evaluated in (D25) and (D23) with (D4), (D2), and (D20). With these,

$$G_{\rho 2}(\rho, z+d) = e^{ik_1(z+d)} H_{B\rho 1}(\rho, 0) - \frac{k_2^2}{k_1^2} \left[\left(\frac{i}{k_1\rho^3} - \frac{z+d}{\rho^3} + \frac{2k_1}{k_2\rho^2} \right) e^{ik_1(z+d)} + \frac{2e^{ik_1r_2}}{k_1\rho^3} - \frac{2k_1}{k_2\rho^2} e^{ik_1(z+d)} e^{ik_2\rho} \right], \tag{5.6.36a}$$

$$G_{\phi 2}(\rho, z+d) = e^{ik_1(z+d)} H_{B\phi 1}(\rho, 0) + \frac{2k_2^2}{k_1^2} \left[\left(\frac{k_1}{k_2\rho^2} - \frac{z+d}{\rho^3} + \frac{i}{k_1\rho^3} \right) e^{ik_1(z+d)} - i \left(\frac{1}{\rho^2} + \frac{i}{2k_1\rho^3} + \frac{3}{8k_1^2\rho^4} \right) e^{ik_1r_2} + \frac{k_1}{k_2} \left(\frac{ik_2}{\rho} - \frac{1}{\rho^2} \right) e^{ik_2\rho} e^{ik_1(z+d)} \right]. \tag{5.6.36b}$$

The first integrals in (5.6.35) are evaluated in (D29) and (D24) with (D8), (D3), and (D21) in Appendix D. The second and third integrals are $I_{19}(\rho, z)$ and $I_{18}(\rho, z)$ in (A19a) and (A18a) and $I_{23}(\rho, z)$ and $I_{22}(\rho, z)$ in (A23a) and (A22a) of Appendix A. These are evaluated in Appendix C. Thus,

$$\begin{aligned}
G_{\rho 3}(\rho, z+d) &= -k_1^{-2} I_H(\rho, z+d) + \frac{k_2^2}{k_1^2} I_{19}(\rho, z+d) - k_1^{-2} I_{23}(\rho, z+d) \\
&= k_1^{-2} \Bigg\{ 2k_1 \left[ik_1 \left(\frac{1}{\rho^2} + \frac{7i}{2k_1\rho^3} \right) e^{ik_1 r_2} + \frac{k_2}{\rho^2} e^{ik_1(z+d)} \right. \\
&\qquad \left. + k_2 \left(\frac{1}{\rho^2} + \frac{2i}{k_2\rho^3} \right) e^{ik_2\rho} e^{ik_1(z+d)} \right] \\
&\qquad - 2k_2^2 \left[e^{ik_1 r_2} \left(\frac{z+d}{r_2} \right) \left(\frac{ik_1}{r_2} - \frac{1}{r_2^2} - \frac{1}{\rho^2} \right) \right. \\
&\qquad \left. + \frac{e^{ik_1(z+d)}}{\rho^2} \right] \\
&\qquad - 2k_1 e^{ik_1 r_2} \left(\frac{z+d}{r_2} \right) \left(-\frac{ik_1^2}{r_2} + \frac{5k_1}{r_2^2} + \frac{12i}{r_2^3} \right) \Bigg\}.
\end{aligned} \tag{5.6.37a}$$

After the small term with the factor k_2^2/k_1^2 is neglected, the result is

$$\begin{aligned}
G_{\rho 3}(\rho, z+d) = 2 \Bigg\{ & e^{ik_1 r_2} \left[\frac{i}{\rho^2} - \frac{7}{2k_1\rho^3} \right. \\
& \left. - \left(\frac{z+d}{r_2} \right) \left(-\frac{ik_1}{r_2} + \frac{5}{r_2^2} + \frac{12i}{k_1 r_2^3} \right) \right] \\
& + \frac{k_2}{k_1} e^{ik_1(z+d)} \left[\frac{1}{\rho^2} + e^{ik_2\rho} \left(\frac{1}{\rho^2} + \frac{2i}{k_2\rho^3} \right) \right] \Bigg\}.
\end{aligned} \tag{5.6.37b}$$

The corresponding expression for $G_{\phi 3}(\rho, z+d)$ is

$$\begin{aligned}
G_{\phi 3}(\rho, z+d) &= -k_1^{-2} I_C(\rho, z+d) + \frac{k_2^2}{k_1^2} I_{18}(\rho, z+d) - k_1^{-2} I_{22}(\rho, z+d) \\
&= 2k_1^{-1} \left[\left(\frac{1}{\rho^3} + \frac{3i}{2k_1\rho^4} \right) e^{ik_1 r_2} \right. \\
&\qquad \left. - k_2 \left(\frac{1}{\rho^2} + \frac{i}{k_2\rho^3} e^{ik_2\rho} \right) e^{ik_1(z+d)} \right] \\
&\quad + \frac{2k_2^2}{k_1^2} \left(\frac{1}{\rho^2} e^{ik_1(z+d)} - \frac{z+d}{r_2} \frac{e^{ik_1 r_2}}{\rho^2} \right) \\
&\quad + 2e^{ik_1 r_2} \left(\frac{1}{r_2^2} + \frac{3i}{k_1 r_2^3} \right) \left(\frac{z+d}{r_2} \right).
\end{aligned} \tag{5.6.38}$$

When terms of order k_2^2/k_1^2 have been neglected,

$$G_{\rho 1}(\rho, z+d) = e^{ik_1(z+d)} H_{B\rho 1}(\rho, 0) + \frac{2k_2}{k_1} e^{ik_1(z+d)} e^{ik_2\rho}\left(\frac{2}{\rho^2} + \frac{2i}{k_2\rho^3}\right)$$
$$+ 2e^{ik_1 r_2}\left[\frac{i}{\rho^2} - \frac{7}{2k_1\rho^3} + \left(\frac{z+d}{r_2}\right)\left(\frac{ik_1}{r_2} - \frac{5}{r_2^2} - \frac{12i}{k_1 r_2^3}\right)\right], \tag{5.6.39}$$

$$G_{\phi 1}(\rho, z+d) = \frac{1}{k_1}\left\{e^{ik_1(z+d)}\left[k_1 H_{B\phi 1}(\rho, 0) + 2k_2 e^{ik_2\rho}\left(\frac{ik_2}{\rho} - \frac{1}{\rho^2} - \frac{i}{k_2\rho^3}\right)\right]\right.$$
$$+ 2e^{ik_1 r_2}\left(\frac{1}{\rho^3} + \frac{3i}{2k_1\rho^4}\right)$$
$$\left. + 2k_1 e^{ik_1 r_2}\left(\frac{z+d}{r_2}\right)\left(\frac{1}{r_2^2} + \frac{3i}{k_1 r_2^3}\right)\right\}. \tag{5.6.40}$$

The integrals $H_{B\rho 1}(\rho, 0)$ and $H_{B\phi 1}(\rho, 0)$ are given in (E66) and (E67) of Appendix E. With them,

$$G_{\rho 1}(\rho, z+d) = \frac{2k_2}{k_1} e^{ik_1(z+d)} e^{ik_2\rho}$$
$$\times \left[\frac{2}{\rho^2} + \frac{2i}{k_2\rho^3} + \frac{ik_2^2}{k_1\rho}\left(\frac{\pi}{k_2\rho}\right)^{1/2} e^{-ip}\mathcal{F}(p)\right]$$
$$+ 2e^{ik_1 r_2}\left[\frac{i}{\rho^2} - \frac{7}{2k_1\rho^3} + \left(\frac{z+d}{r_2}\right)\left(\frac{ik_1}{r_2} - \frac{5}{r_2^2} - \frac{12i}{k_1 r_2^3}\right)\right], \tag{5.6.41}$$

$$G_{\phi 1}(\rho, z+d) = \frac{2k_2}{k_1} e^{ik_1(z+d)} e^{ik_2\rho}$$
$$\times \left[\frac{ik_2}{\rho} - \frac{1}{\rho^2} - \frac{i}{k_2\rho^3} - \frac{k_2^3}{k_1}\left(\frac{\pi}{k_2\rho}\right)^{1/2} e^{-ip}\mathcal{F}(p)\right]$$
$$+ 2e^{ik_1 r_2}\left[\frac{1}{k_1\rho^3} + \frac{3i}{2k_1^2\rho^4} + \left(\frac{z+d}{r_2}\right)\left(\frac{1}{r_2^2} + \frac{3i}{k_1 r_2^3}\right)\right]. \tag{5.6.42}$$

The integral $G_{z1}(\rho, z+d)$ given in (5.5.21) is

$$G_{z1}(\rho, z+d) = -\int_0^\infty (P-1)\gamma_1^{-1} J_1(\lambda\rho) e^{i\gamma_1(z+d)} \lambda^2 \, d\lambda, \tag{5.6.43}$$

where $-(P-1) = 2\gamma_1/(\gamma_1+\gamma_2)$ so that $-(P-1)\gamma_1^{-1} = 2/(\gamma_1+\gamma_2)$. This can be multiplied above and below by $\gamma_1 - \gamma_2$ to give $2(\gamma_1-\gamma_2)/(k_1^2-k_2^2)$. Hence,

$$\begin{aligned} G_{z1}(\rho, z+d) = \frac{2}{k_1^2-k_2^2}\bigg(\int_0^\infty & \gamma_1 J_1(\lambda\rho)e^{i\gamma_1(z+d)}\lambda^2\,d\lambda \\ & - i\int_0^\infty J_1(\lambda\rho)e^{i\gamma_1(z+d)}\lambda^3\,d\lambda \\ & - e^{ik_1(z+d)}\int_0^\infty \gamma_2 J_1(\lambda\rho)\lambda^2\,d\lambda \\ & + ie^{ik_1(z+d)}\int_0^\infty J_1(\lambda\rho)\lambda^3\,d\lambda\bigg). \end{aligned} \tag{5.6.44}$$

These integrals are in Appendix A, so that

$$\begin{aligned} & G_{z1}(\rho, z+d) \\ &= \frac{\rho}{k_1^2-k_2^2}\,[I_{15}(k_1;\rho,z+d) - iI_{24}(k_1;\rho,z+d) \\ &\qquad - e^{ik_1(z+d)}I_{15}(k_2;\rho,0) + ie^{ik_1(z+d)}I_{24}(\rho,0)] \\ &\sim 2i\bigg\{\left(\frac{\rho}{r_2}\right)\bigg[\frac{1}{r_2^2} + \frac{3i}{k_1 r_2^3} - \frac{3}{k_1^2 r_2^4} \\ &\qquad + \left(\frac{z+d}{\rho}\right)\left(\frac{k_1}{\rho} + \frac{3i}{2\rho^2} - \frac{9}{8k_1\rho^3}\right) \\ &\qquad + \left(\frac{z+d}{r_2}\right)^2\left(\frac{ik_1}{r_2} - \frac{6}{r_2^2} - \frac{15i}{k_1 r_2^3}\right)\bigg]e^{ik_1 r_2} \\ &\quad - \frac{k_2^2}{k_1^2}\left(\frac{1}{\rho^2} + \frac{3i}{k_2\rho^3} - \frac{3}{k_2^2\rho^4}\right)e^{ik_1(z+d)}e^{ik_2\rho}\bigg\}. \end{aligned} \tag{5.6.45}$$

5.7 Evaluation of the Integrals of the Field in Region 2 of the Horizontal Electric Dipole at the Depth d in Region 1

The six components of the electromagnetic field in Region 2 ($z \le 0$) due to a unit horizontal electric dipole at the depth d in Region 1 ($z \ge 0$) are given in integral form in (5.4.35)–(5.4.40). These can be evaluated in the same manner as the integrals for the field in Region 1, but the procedure is more complicated because the exponential factors involve both $\gamma_1 = (k_1^2-\lambda^2)^{1/2}$ and $\gamma_2 = (k_2^2-\lambda^2)^{1/2}$. In the evaluation, terms of the order of magnitude of

k_2^2/k_1^2, z^2/ρ^2, d^2/ρ^2, and $k_2|z|/k_1\rho$ are neglected since the field is evaluated subject to the conditions

$$|k_1|^2 \gg |k_2|^2; \qquad \rho \geq 5d; \qquad \rho \geq 5z. \tag{5.7.1}$$

These conditions are well satisfied in the applications of the horizontal dipole treated in this book. They considerably simplify the evaluation of the surface-wave terms.

In the evaluation of $E_{2\rho}(\rho, \phi, z)$ and $E_{2\phi}(\rho, \phi, z)$ as defined in (5.4.35) and (5.4.36), a rearrangement into the form of $E_{1\rho}(\rho, \phi, z)$ and $E_{1\phi}(\rho, \phi, z)$ as given in (5.4.29) and (5.4.30) is necessary so that the Fresnel-integral terms that appear in (5.5.43)–(5.5.45) can be derived. The desired equivalent forms of (5.4.35) and (5.4.36) are

$$\begin{aligned}
\left.\begin{matrix} E_{2\rho}(\rho,\phi,z) \\ E_{2\phi}(\rho,\phi,z) \end{matrix}\right\} &= \mp\frac{\omega\mu_0}{4\pi k_1^2}\left\{\begin{matrix}\cos\phi \\ \sin\phi\end{matrix}\right\} \\
&\quad\times\left[\int_0^\infty 2\left(\frac{\gamma_1}{2}\,[J_0(\lambda\rho) \mp J_2(\lambda\rho)]\right.\right. \\
&\qquad\left. + \frac{k_1^2}{2\gamma_1}\,[J_0(\lambda\rho) \pm J_2(\lambda\rho)]\right)e^{i\gamma_1 d}e^{-i\gamma_2 z}\lambda\,d\lambda \\
&\quad + \int_0^\infty\left(\frac{\gamma_1}{2}\,(Q-1)[J_0(\lambda\rho) \mp J_2(\lambda\rho)]\right. \\
&\qquad\left. - \frac{k_1^2}{2\gamma_1}\,(P+1)[J_0(\lambda\rho) \pm J_2(\lambda\rho)]\right) \\
&\qquad\left.\times\, e^{i\gamma_1 d}e^{-i\gamma_2 z}\lambda\,d\lambda\right]; \qquad z \leq 0,
\end{aligned} \tag{5.7.2}$$

where, as given in (5.6.4),

$$Q - 1 = -\frac{2k_2^2\gamma_1}{N}; \qquad P + 1 = \frac{2\gamma_2}{M}, \tag{5.7.3a}$$

with

$$N = k_1^2\gamma_2 + k_2^2\gamma_1; \qquad M = \gamma_1 + \gamma_2. \tag{5.7.3b}$$

It is readily verified that (5.7.2) is equivalent to (5.4.35) and (5.4.36). The formulas (5.7.2) are like the last two terms in (5.5.1) with (5.5.3) and (5.5.4) and in (5.5.5) with (5.5.7) and (5.5.8) except that the exponential factors $\exp[i\gamma_1(z+d)]$ are replaced by $\exp[i(\gamma_2|z| + \gamma_1 d)]$. Specifically,

$$E_{2\rho}(\rho, \phi, z) = -\frac{\omega\mu_0}{4\pi k_1^2}\cos\phi\;[2F_{\rho 0}(\rho, -z, d) + F_{\rho 1}(\rho, -z, d)], \tag{5.7.4a}$$

$$E_{2\phi}(\rho, \phi, z) = \frac{\omega\mu_0}{4\pi k_1^2}\sin\phi\;[2F_{\phi 0}(\rho, -z, d) + F_{\phi 1}(\rho, -z, d)], \tag{5.7.4b}$$

where $F_0(\rho, -z, d)$ is the first integral in (5.7.2) and $F_1(\rho, -z, d)$ is the second integral. The first integral can be evaluated with the approximate procedure described in Appendix D in the form

$$f(k_1,k_2;\lambda) \sim f_1(k_1,k_2;\lambda) + f_2(k_1,k_2;\lambda) - f_3(k_1,k_2;\lambda), \tag{5.7.5a}$$

where

$$\begin{aligned} f_1(k_1,k_2;\lambda) &= f(k_1,k_2;\lambda \sim k_1),\\ f_2(k_1,k_2;\lambda) &= f(k_1,k_2;\lambda \sim k_2), \end{aligned} \tag{5.7.5b}$$

and

$$f_3(k_1,k_2;\lambda) = f_1(k_1,k_2;\lambda \sim k_2) = f_2(k_1,k_2;\lambda \sim k_1). \tag{5.7.5c}$$

When $\lambda \sim k_1$, $\gamma_2 = (k_2^2 - \lambda^2)^{1/2} \sim i\lambda \sim ik_1$; when $\lambda \sim k_2$, $\gamma_1 \sim k_1$. When applied to the first integral in (5.7.2), the result is

$$\begin{aligned} \left.\begin{matrix} F_{\rho 0}(\rho,-z,d) \\ F_{\phi 0}(\rho,-z,d) \end{matrix}\right\} \sim \frac{e^{k_1 z}}{2} \int_0^\infty \Big(& \gamma_1[J_0(\lambda\rho) \mp J_2(\lambda\rho)] \\ &+ \frac{k_1^2}{\gamma_1}\left[J_0(\lambda\rho) \pm J_2(\lambda\rho)\right]\Big) e^{i\gamma_1 d}\lambda\, d\lambda \\ &+ k_1 e^{ik_1 d} \int_0^\infty J_0(\lambda\rho) e^{-i\gamma_2 z}\lambda\, d\lambda \\ &- k_1 e^{ik_1 d} e^{k_1 z} \int_0^\infty J_0(\lambda\rho)\lambda\, d\lambda. \end{aligned} \tag{5.7.6a}$$

With Appendix A, (5.7.6a) is equivalent to

$$\begin{aligned} \left.\begin{matrix} F_{\rho 0}(\rho,-z,d) \\ F_{\phi 0}(\rho,-z,d) \end{matrix}\right\} \sim \frac{e^{k_1 z}}{2} &\left(\begin{bmatrix} I_{12}(k_1;\rho,d) \\ I_{11}(k_1;\rho,d) \end{bmatrix} + k_1^2 \begin{bmatrix} I_3(k_1;\rho,d) \\ I_4(k_1;\rho,d) \end{bmatrix} \right) \\ &+ k_1 e^{ik_1 d} I_{26}(k_2;\rho,-z) - k_1 e^{ik_1 d} e^{k_1 z} I_{26}(\rho,0). \end{aligned} \tag{5.7.6b}$$

It follows that

$$\begin{aligned} &F_{\rho 0}(\rho,-z,d) \\ &= e^{k_1 z}\Bigg\{ -k_1 \Bigg[\frac{e^{ik_1 d}}{\rho^2} + \left(\frac{1}{r_d^2} + \frac{2i}{k_1 r_d^3}\right) e^{ik_1 r_d} \\ &\qquad + \left(\frac{d}{r_d}\right)^2 \left(\frac{ik_1}{r_d} - \frac{3}{r_d^2} - \frac{3i}{k_1 r_d^3} - \frac{1}{\rho^2}\right) e^{ik_1 r_d} \Bigg] \\ &\qquad + \frac{k_1}{\rho^2}\left(e^{ik_1 d} - e^{ik_1 r_d}\right)\Bigg\} + k_1 e^{ik_1 d}\left(\frac{z}{r_0}\right)\left(\frac{ik_2}{r_0} - \frac{1}{r_0^2}\right) e^{ik_2 r_0} \\ &= k_1 \Bigg\{ e^{ik_1 d} e^{ik_2 r_0} \left(\frac{z}{r_0}\right)\left(\frac{ik_2}{r_0} - \frac{1}{r_0^2}\right) \\ &\qquad - e^{k_1 z} e^{ik_1 r_d} \left[\frac{2}{r_d^2} + \frac{2i}{k_1 r_d^3} + \left(\frac{d}{r_d}\right)^2 \left(\frac{ik_1}{r_d} - \frac{3}{r_d^2} - \frac{3i}{k_1 r_d^3}\right) \right] \Bigg\}, \end{aligned} \tag{5.7.7a}$$

$$
\begin{aligned}
F_{\phi 0}(\rho, -z, d)& \\
&= e^{k_1 z}\left[k_1\left(\frac{e^{ik_1 d}}{\rho^2} + \frac{ie^{ik_1 r_d}}{k_1 r_d^3}\right) - \frac{k_1}{\rho^2}\left(e^{ik_1 d} - e^{ik_1 r_d}\right) - \frac{ik_1^2 e^{ik_1 r_d}}{r_d}\right] \\
&\quad + k_1 e^{ik_1 d}\left(\frac{z}{r_0}\right)\left(\frac{ik_2}{r_0} - \frac{1}{r_0^2}\right)e^{ik_2 r_0} \\
&= k_1\left[e^{ik_1 d}e^{ik_2 r_0}\left(\frac{z}{r_0}\right)\left(\frac{ik_2}{r_0} - \frac{1}{r_0^2}\right) - e^{k_1 z}e^{ik_1 r_d}\left(\frac{ik_1}{r_d} - \frac{1}{\rho^2} - \frac{i}{k_1 r_d^3}\right)\right],
\end{aligned}
\tag{5.7.7b}
$$

where $r_0 = (\rho^2 + z^2)^{1/2}$ and $r_d = (\rho^2 + d^2)^{1/2}$.

The second integral pair in (5.7.2) is the same as (5.6.7) and (5.6.19) except for the exponential factors which are $\exp(i\gamma_1 d)\exp(-i\gamma_2 z)$ instead of $\exp[i\gamma_1(z+d)]$. Specifically, and with terms of the order of k_2^2/k_1^2 omitted,

$$
\begin{aligned}
\left.\begin{matrix} F_{\rho 1}(\rho, -z, d) \\ F_{\phi 1}(\rho, -z, d) \end{matrix}\right\} &= \left\{\begin{matrix} H_{\rho 1}(\rho, -z, d) \\ H_{\phi 1}(\rho, -z, d) \end{matrix}\right\} \\
&\quad - k_2^2\int_0^\infty \gamma_2^{-1}[J_0(\lambda\rho) \mp J_2(\lambda\rho)]e^{i\gamma_1 d}e^{-i\gamma_2 z}\lambda\, d\lambda \\
&\quad - k_1^2\int_0^\infty \gamma_1^{-1}[J_0(\lambda\rho) \pm J_2(\lambda\rho)]e^{i\gamma_1 d}e^{-i\gamma_2 z}\lambda\, d\lambda \\
&\quad + \int_0^\infty \gamma_1[J_0(\lambda\rho) \pm J_2(\lambda\rho)]e^{i\gamma_1 d}e^{-i\gamma_2 z}\lambda\, d\lambda \\
&\quad - \int_0^\infty \gamma_2[J_0(\lambda\rho) \pm J_2(\lambda\rho)]e^{i\gamma_1 d}e^{-i\gamma_2 z}\lambda\, d\lambda,
\end{aligned}
\tag{5.7.8a}
$$

where

$$
\begin{aligned}
\left.\begin{matrix} H_{\rho 1}(\rho, -z, d) \\ H_{\phi 1}(\rho, -z, d) \end{matrix}\right\} &= -k_2^2\int_0^\infty \gamma_1^2\left(\frac{1}{N} - \frac{1}{N_0}\right)[J_0(\lambda\rho) \mp J_2(\lambda\rho)] \\
&\qquad \times e^{i\gamma_1 d}e^{-i\gamma_2 z}\lambda\, d\lambda.
\end{aligned}
\tag{5.7.8b}
$$

In (5.7.8b), contributions by the integrand are significant only in the range $\lambda \sim k_2$ where $N_0 = k_1^2\gamma_2 = k_1^2(k_2^2 - \lambda^2)^{1/2}$ becomes small. The other integrals in (5.7.8a) can be evaluated with the approximate procedure in (5.7.5a–c). Thus,

$$
\begin{aligned}
&\left.\begin{matrix} F_{\rho 1}(\rho, -z, d) \\ F_{\phi 1}(\rho, -z, d) \end{matrix}\right\} \\
&\quad \sim e^{ik_1 d}\left[-k_2^2 k_1^2\int_0^\infty \left(\frac{1}{N} - \frac{1}{N_0}\right)[J_0(\lambda\rho) \mp J_2(\lambda\rho)]e^{-i\gamma_2 z}\lambda\, d\lambda\right. \\
&\qquad - k_2^2\int_0^\infty \gamma_2^{-1}[J_0(\lambda\rho) \mp J_2(\lambda\rho)]e^{-i\gamma_2 z}\lambda\, d\lambda
\end{aligned}
$$

$$
\begin{aligned}
&- k_1 \int_0^\infty [J_0(\lambda\rho) \pm J_2(\lambda\rho)] e^{-i\gamma_2 z} \lambda\, d\lambda \\
&+ k_1 \int_0^\infty [J_0(\lambda\rho) \pm J_2(\lambda\rho)] e^{-i\gamma_2 z} \lambda\, d\lambda \\
&- \int_0^\infty \gamma_2 [J_0(\lambda\rho) \pm J_2(\lambda\rho)] e^{-i\gamma_2 z} \lambda\, d\lambda \Big] \\
+ e^{k_1 z} \Big(&ik_2^2 \int_0^\infty [J_0(\lambda\rho) \mp J_2(\lambda\rho)] e^{i\gamma_1 d}\, d\lambda \\
&- k_1^2 \int_0^\infty \gamma_1^{-1} [J_0(\lambda\rho) \pm J_2(\lambda\rho)] e^{i\gamma_1 d} \lambda\, d\lambda \\
&+ \int_0^\infty \gamma_1 [J_0(\lambda\rho) \pm J_2(\lambda\rho)] e^{i\gamma_1 d} \lambda\, d\lambda \\
&- i \int_0^\infty [J_0(\lambda\rho) \pm J_2(\lambda\rho)] e^{i\gamma_1 d} \lambda^2\, d\lambda \Big) \\
- e^{k_1 z} e^{ik_1 d} \Big(&ik_2^2 \int_0^\infty [J_0(\lambda\rho) \mp J_2(\lambda\rho)]\, d\lambda \\
&- k_1 \int_0^\infty [J_0(\lambda\rho) \pm J_2(\lambda\rho)] \lambda\, d\lambda \\
&+ k_1 \int_0^\infty [J_0(\lambda\rho) \pm J_2(\lambda\rho)] \lambda\, d\lambda \\
&- i \int_0^\infty [J_0(\lambda\rho) \pm J_2(\lambda\rho)] \lambda^2\, d\lambda \Big).
\end{aligned} \tag{5.7.9}
$$

The first integral in (5.7.9) is of the type evaluated in Appendix E for large values of the arguments of the Bessel functions. From (E13)–(E15) with (E16) and (E44), it follows that

$$
\begin{aligned}
&\int_0^\infty \left(\frac{1}{N} - \frac{1}{N_0}\right) [J_0(\lambda\rho) \mp J_2(\lambda\rho)] e^{-i\gamma_2 z} \lambda\, d\lambda \\
&\quad \sim 2 \int_0^\infty \left(\frac{1}{N} - \frac{1}{N_0}\right) \begin{Bmatrix} J_0(\lambda\rho) \\ (\lambda\rho)^{-1} J_1(\lambda\rho) \end{Bmatrix} e^{-i\gamma_2 z} \lambda\, d\lambda \\
&\quad \sim \begin{Bmatrix} 2k_2^{-2} G_{2z}(\rho, z) \\ 2k_2^{-2} \rho^{-1} G_{2\phi}(\rho, z) \end{Bmatrix}.
\end{aligned} \tag{5.7.10}
$$

These are like (E6) and (E5) except for the exponent on λ. However, in the evaluation the approximation (E44), i.e., $\lambda \sim k_2$, is made. With this, the simple relations with (E5) and (E6) are established as shown on the right. The remaining integrals in (5.7.9) are listed in Appendix A. Thus,

$$
\begin{aligned}
&F_{\rho 1}(\rho, -z, d) \\
&\quad \sim e^{ik_1 d} [-2k_1^2 G_{2z}(\rho, z) - k_2^2 I_4(k_2; \rho, -z) - I_{11}(k_2; \rho, -z)] \\
&\qquad + e^{k_1 z} [ik_2^2 I_{17}(k_1; \rho, d) - k_1^2 I_3(k_1; \rho, d)
\end{aligned}
$$

$$
\begin{aligned}
&+ I_{11}(k_1;\rho,d) - iI_{20}(k_1;\rho,d)] \\
&- e^{ik_1 d}e^{k_1 z}[ik_2^2 I_{17}(\rho,0) - iI_{20}(\rho,0)] \\
\sim\ & 2e^{ik_1 d}\left[-\frac{k_2^4}{k_1}\left(\frac{\pi}{k_2\rho}\right)^{1/2} e^{ik_2 r_0}e^{-iP}\mathcal{F}(P)\right. \\
&+ k_2^2\left(\frac{ie^{ik_2 r_0}}{r_0} + \frac{1}{k_2\rho^2}\left(e^{-ik_2 z} - e^{ik_2 r_0}\right)\right) \\
&\left.- k_2\,\frac{e^{-ik_2 z}}{\rho^2} - \frac{ie^{ik_2 r_0}}{r_0^3} + \frac{z^2}{r_0^2}\frac{k_2}{\rho^2}\,e^{ik_2 r_0}\right] \\
&+ 2e^{k_1 z}\left\{-\frac{k_2^2}{k_1^2}\left(\frac{d}{\rho}\right)\left[\frac{k_1}{\rho^2}\,e^{ik_1 d} - \left(\frac{k_1^2}{\rho} + \frac{ik_1}{2\rho^2} + \frac{3}{8\rho^3}\right)e^{ik_1 r_d}\right]\right. \\
&- k_1\left(\frac{e^{ik_1 d}}{\rho^2} - \frac{e^{ik_1 r_d}}{\rho^2}\right) + k_1\,\frac{e^{ik_1 d}}{\rho^2} + \frac{ie^{ik_1 r_d}}{r_d^3} \\
&\left.- \frac{z^2}{r_d^2}\frac{k_1}{\rho^2}\,e^{ik_1 r_d} - \frac{ie^{ik_1 d}}{\rho^3} + e^{ik_1 r_d}\left(\frac{d}{\rho}\right)\left(\frac{ik_1}{\rho^2} - \frac{3}{2\rho^3}\right)\right\} \\
&+ 2ie^{ik_1 d}\,\frac{e^{k_1 z}}{\rho^3} \\
\sim\ & 2e^{ik_1 d}e^{ik_2 r_0}k_2\left[\frac{ik_2}{r_0} - \frac{1}{r_0^2} - \frac{i}{k_2 r_0^3} - \frac{k_2^3}{k_1}\left(\frac{\pi}{k_2\rho}\right)^{1/2} e^{-iP}\mathcal{F}(P)\right] \\
&+ 2e^{k_1 z}e^{ik_1 r_d}k_1\left[\frac{1}{r_d^2} + \frac{i}{k_1 r_d^3} + \left(\frac{d}{\rho}\right)\left(\frac{i}{\rho^2} - \frac{3}{2k_1\rho^3}\right)\right], \qquad (5.7.11)
\end{aligned}
$$

where $P = (R-Z)^2/R$, $R = k_2^3\rho/2k_1^2$, and $Z = k_2^2 z/2k_1$. Note that the approximation $k_2\rho + Z^2/R = k_2(\rho + z^2/2\rho) \sim k_2 r_0$ when $\rho^2 \gg z^2$ has been made in the exponent of the Fresnel-integral term.

Similarly,

$$
\begin{aligned}
&F_{\phi 1}(\rho,-z,d) \\
&= e^{ik_1 d}\left(-\frac{2k_1^2}{\rho}\,G_{2\phi}(\rho,z) - k_2^2 I_3(k_2;\rho,-z) - I_{12}(k_2;\rho,-z)\right) \\
&\quad + e^{k_1 z}[ik_2^2 I_{16}(k_1;\rho,d) - k_1^2 I_4(k_1;\rho,d) \\
&\qquad + I_{12}(k_1;\rho,d) - iI_{21}(k_1;\rho,d)] \\
&\quad - e^{ik_1 d}e^{k_1 z}[ik_2^2 I_{16}(\rho,0) - iI_{21}(\rho,0)] \\
&= 2e^{ik_1 d}\left\{\frac{ik_2^3}{k_1\rho}\left(\frac{\pi}{k_2\rho}\right)^{1/2} e^{ik_2 r_0}e^{-iP}\mathcal{F}(P) - k_2\left(\frac{e^{-ik_2 z}}{\rho^2} - \frac{e^{ik_2 r_0}}{\rho^2}\right)\right. \\
&\quad + k_2\left[\frac{e^{-ik_2 z}}{\rho^2} + \left(\frac{1}{r_0^2} + \frac{2i}{k_2 r_0^3}\right)e^{ik_2 r_0}\right. \\
&\qquad \left.\left. + \left(\frac{z}{r_0}\right)^2\left(\frac{ik_2}{r_0} - \frac{3}{r_0^2} - \frac{3i}{k_2 r_0^3} - \frac{1}{\rho^2}\right)e^{ik_2 r_0}\right]\right\}
\end{aligned}
$$

$$
\begin{aligned}
&+ 2e^{k_1 z}\Bigg\{ ik_2^2 \left[e^{ik_1 d}\left(\frac{1}{\rho} - \frac{id}{2k_1\rho^3}\right) - \left(\frac{d}{\rho}\right)\frac{e^{ik_1 r_d}}{k_1\rho^2}\right] \\
&\qquad + k_1^2 \left[\frac{ie^{ik_1 r_d}}{r_d} + \frac{1}{k_1\rho^2}\left(e^{ik_1 d} - e^{ik_1 r_d}\right)\right] \\
&\qquad - k_1 \left[\frac{e^{ik_1 d}}{\rho^2} + \left(\frac{1}{r_d^2} + \frac{2i}{k_1 r_d^3}\right)e^{ik_1 r_d} \right. \\
&\qquad\qquad \left. + \left(\frac{d}{r_d}\right)^2\left(\frac{ik_1}{r_d} - \frac{3}{r_d^2} - \frac{3i}{k_1 r_d^3} - \frac{1}{\rho^2}\right)e^{ik_1 r_d}\right] \\
&\qquad + \frac{2ie^{ik_1 d}}{\rho^3} + ik_1\left(\frac{d}{r_d}\right)\left(\frac{ik_1}{\rho} - \frac{3}{2\rho^2} - \frac{5i}{8k_1\rho^3}\right)e^{ik_1 r_d}\Bigg\} \\
&- 2e^{ik_1 d}e^{k_1 z}\left(\frac{ik_2^2}{\rho} + \frac{2i}{\rho^3}\right).
\end{aligned} \tag{5.7.12}
$$

After one term of order k_2^2/k_1^2 has been neglected and cancellations made, the result is

$$
\begin{aligned}
&F_{\phi 1}(\rho, -z, d) \\
&\quad \sim 2e^{ik_1 d}e^{ik_2 r_0}k_2\left[\frac{2}{r_0^2} + \frac{2i}{k_2 r_0^3} + \left(\frac{z}{r_0}\right)^2\left(\frac{ik_2}{r_0} - \frac{3}{r_0^2} - \frac{3i}{k_2 r_0^3}\right) \right. \\
&\qquad\qquad \left. + \frac{ik_2^2}{k_1\rho}\left(\frac{\pi}{k_2\rho}\right)^{1/2} e^{-iP}\mathcal{F}(P)\right] \\
&\quad + 2e^{k_1 z}e^{ik_1 r_d}k_1\left[\frac{ik_1}{r_d} - \frac{2}{r_d^2} - \frac{2i}{k_1 r_d^3} \right. \\
&\qquad\qquad + i\left(\frac{d}{r_d}\right)\left(\frac{ik_1}{\rho} - \frac{3}{2\rho^2} - \frac{5i}{8k_1\rho^3}\right) \\
&\qquad\qquad \left. - \left(\frac{d}{r_d}\right)^2\left(\frac{ik_1}{r_d} - \frac{3}{r_d^2} - \frac{3i}{k_1 r_d^3}\right)\right].
\end{aligned} \tag{5.7.13}
$$

The formulas for F_0 and F_1 can now be combined to form (5.7.4a,b). This gives the final formulas for $E_{2\rho}(\rho, \phi, z)$ and $E_{2\phi}(\rho, \phi, z)$. They are

$$
\begin{aligned}
&E_{2\rho}(\rho, \phi, z) = -\frac{\omega\mu_0}{2\pi k_1^2}\cos\phi \\
&\times\Bigg(e^{ik_1 d}e^{ik_2 r_0}\Bigg\{k_2\left[\frac{ik_2}{r_0} - \frac{1}{r_0^2} - \frac{i}{k_2 r_0^3} - \frac{k_2^3}{k_1}\left(\frac{\pi}{k_2\rho}\right)^{1/2} e^{-iP}\mathcal{F}(P)\right] \\
&\qquad + k_1\left(\frac{z}{r_0}\right)\left(\frac{ik_2}{r_0} - \frac{1}{r_0^2}\right)\Bigg\} \\
&\quad - e^{k_1 z}e^{ik_1 r_d}k_1\left[\frac{1}{r_d^2} + \frac{i}{k_1 r_d^3} - \left(\frac{d}{\rho}\right)\left(\frac{i}{\rho^2} - \frac{3}{2k_1\rho^3}\right)\right.
\end{aligned}
$$

$$+\left(\frac{d}{r_d}\right)^2\left(\frac{ik_1}{r_d}-\frac{3}{r_d^2}-\frac{3i}{k_1r_d^3}\right)\Bigg]\Bigg); \qquad z\le 0, \tag{5.7.14}$$

$$\begin{aligned}E_{2\phi}(\rho,\phi,z) &= \frac{\omega\mu_0}{2\pi k_1^2}\sin\phi\\
&\times\Bigg(e^{ik_1d}e^{ik_2r_0}\Bigg\{k_2\Bigg[\frac{2}{r_0^2}+\frac{2i}{k_2r_0^3}+\left(\frac{z}{r_0}\right)^2\left(\frac{ik_2}{r_0}-\frac{3}{r_0^2}-\frac{3i}{k_2r_0^3}\right)\\
&\qquad+\frac{ik_2^2}{k_1\rho}\left(\frac{\pi}{k_2\rho}\right)^{1/2}e^{-iP}\mathcal{F}(P)\Bigg]\\
&\qquad+k_1\left(\frac{z}{r_0}\right)\left(\frac{ik_2}{r_0}-\frac{1}{r_0^2}\right)\Bigg\}\\
&\qquad-e^{k_1z}e^{ik_1r_d}k_1\Bigg[\frac{1}{r_d^2}+\frac{i}{k_1r_d^3}-i\left(\frac{d}{r_d}\right)\left(\frac{ik_1}{\rho}-\frac{3}{2\rho^2}-\frac{5i}{8k_1\rho^3}\right)\\
&\qquad+\left(\frac{d}{r_d}\right)^2\left(\frac{ik_1}{r_d}-\frac{3}{r_d^2}-\frac{3i}{k_1r_d^3}\right)\Bigg]\Bigg); \qquad z\le 0.\end{aligned} \tag{5.7.15}$$

At the boundary $z=0$, these reduce to

$$\begin{aligned}E_{2\rho}(\rho,\phi,0) &= -\frac{\omega\mu_0}{2\pi k_1^2}\cos\phi\\
&\times\Bigg\{e^{ik_1d}e^{ik_2\rho}k_2g(\rho;k_1,k_2)\\
&\qquad-k_1e^{ik_1r_d}\Bigg[\frac{1}{r_d^2}+\frac{i}{k_1r_d^3}-\left(\frac{d}{\rho}\right)\left(\frac{i}{\rho^2}-\frac{3}{2k_1\rho^3}\right)\\
&\qquad+\left(\frac{d}{r_d}\right)^2\left(\frac{ik_1}{r_d}-\frac{3}{r_d^2}-\frac{3i}{k_1r_d^3}\right)\Bigg]\Bigg\},\end{aligned} \tag{5.7.16}$$

$$\begin{aligned}E_{2\phi}(\rho,\phi,0) &= \frac{\omega\mu_0}{2\pi k_1^2}\sin\phi\\
&\times\Bigg\{e^{ik_1d}e^{ik_2\rho}k_2h(\rho;k_1,k_2)\\
&\qquad-k_1e^{ik_1r_d}\Bigg[\frac{1}{r_d^2}+\frac{i}{k_1r_d^3}\\
&\qquad-i\left(\frac{d}{r_d}\right)\left(\frac{ik_1}{\rho}-\frac{3}{2\rho^2}-\frac{5i}{8k_1\rho^3}\right)\\
&\qquad+\left(\frac{d}{r_d}\right)^2\left(\frac{ik_1}{r_d}-\frac{3}{r_d^2}-\frac{3i}{k_1r_d^3}\right)\Bigg]\Bigg\},\end{aligned} \tag{5.7.17}$$

where $g(\rho;k_1,k_2)$ and $h(\rho;k_1,k_2)$ are defined in (5.5.44) and (5.5.45). These satisfy the boundary conditions

$$E_{2\rho}(\rho,\phi,0)=E_{1\rho}(\rho,\phi,0); \qquad E_{2\phi}(\rho,\phi,0)=E_{1\phi}(\rho,\phi,0), \tag{5.7.18}$$

where $E_{1\rho}(\rho, \phi, 0)$ and $E_{1\phi}(\rho, \phi, 0)$ are given by (5.5.37a) and (5.5.38a) with $z = 0$.

The formula (5.4.37) for $E_{2z}(\rho, \phi, z)$ is

$$E_{2z}(\rho, \phi, z) = -\frac{i\omega\mu_0}{2\pi}\cos\phi \int_0^\infty \frac{\gamma_1}{N} J_1(\lambda\rho)e^{i\gamma_1 d}e^{-i\gamma_2 z}\lambda^2\, d\lambda. \tag{5.7.19}$$

With (5.7.1) and (5.7.3b), the integrand can be expanded into

$$\frac{\gamma_1}{N} = \gamma_1\left(\frac{1}{N} - \frac{1}{N_0}\right) + \frac{\gamma_1}{N_0} = \gamma_1\left(\frac{1}{N} - \frac{1}{N_0}\right) + \frac{\gamma_1}{k_1^2\gamma_2}, \tag{5.7.20}$$

so that

$$\begin{aligned}&\int_0^\infty \frac{\gamma_1}{N} J_1(\lambda\rho)e^{i\gamma_1 d}e^{-i\gamma_2 z}\lambda^2\, d\lambda \\ &\quad = k_1^{-2}\int_0^\infty \frac{\gamma_1}{\gamma_2} J_1(\lambda\rho)e^{i\gamma_1 d}e^{-i\gamma_2 z}\lambda^2\, d\lambda + H_z(\rho, -z, d),\end{aligned} \tag{5.7.21}$$

where

$$\begin{aligned}H_z(\rho, -z, d) &= \int_0^\infty \gamma_1\left(\frac{1}{N} - \frac{1}{N_0}\right)J_1(\lambda\rho)e^{i\gamma_1 d}e^{-i\gamma_2 z}\lambda^2\, d\lambda \\ &\sim k_1 e^{ik_1 d}\int_0^\infty \left(\frac{1}{N} - \frac{1}{N_0}\right)J_1(\lambda\rho)e^{-i\gamma_2 z}\lambda^2\, d\lambda\end{aligned} \tag{5.7.22}$$

since significant contributions come only when $\lambda \sim k_2$ so that $\gamma_1 \sim k_1$. With (E5) and $k_2\rho + Z^2/R = k_2(\rho + z^2/2\rho) \sim k_2 r_0$,

$$\begin{aligned}H_z(\rho, -z, d) &= k_1 e^{ik_1 d}G_{2\phi}(\rho, z) \\ &= -\frac{ik_2^3}{k_1^2}\left(\frac{\pi}{k_2\rho}\right)^{1/2} e^{ik_1 d}e^{ik_2 r_0}e^{-iP}\mathcal{F}(P),\end{aligned} \tag{5.7.23}$$

where $P = (R - Z)^2/R,\ Z \leq 0$. The approximate procedure in (5.7.5a–c) can be applied to the integral in (5.7.21) to give

$$\begin{aligned}&k_1^{-2}\int_0^\infty \frac{\gamma_1}{\gamma_2} J_1(\lambda\rho)e^{i\gamma_1 d}e^{-i\gamma_2 z}\lambda^2\, d\lambda \\ &\sim k_1^{-2}\bigg(k_1 e^{ik_1 d}\int_0^\infty \gamma_2^{-1}J_1(\lambda\rho)e^{-i\gamma_2 z}\lambda^2\, d\lambda \\ &\qquad - ie^{k_1 z}\int_0^\infty \gamma_1 J_1(\lambda\rho)e^{i\gamma_1 d}\lambda\, d\lambda \\ &\qquad + ik_1 e^{ik_1 d}e^{k_1 z}\int_0^\infty J_1(\lambda\rho)\lambda\, d\lambda\bigg) \\ &= \frac{k_1^{-2}\rho}{2}\left[k_1 e^{ik_1 d}I_7(k_2;\rho,-z) - ie^{k_1 z}I_{13}(k_1;\rho,d) + ik_1 e^{ik_1 d}e^{k_1 z}I_{20}(\rho,0)\right]\end{aligned}$$

$$
\begin{aligned}
&= k_1^{-2}\rho\Big\{k_1 e^{ik_1 d}(-k_2)\left(\frac{1}{r_0^2}+\frac{i}{k_2 r_0^3}\right)e^{ik_2 r_0} \\
&\qquad - ik_1 e^{k_1 z}\left[\frac{1}{\rho^3}\,e^{ik_1 d} + ie^{ik_1 r_d}\left(\frac{1}{\rho^3}+\frac{3i}{2k_1\rho^4}\right)\right] \\
&\qquad + ik_1 e^{ik_1 d}e^{k_1 z}\,\frac{1}{\rho^3}\Big\} \\
&= k_1^{-2}\Big\{-k_1 k_2 e^{ik_1 d}\left(\frac{\rho}{r_0^2}+\frac{i\rho}{k_2 r_0^3}\right)e^{ik_2 r_0} \\
&\qquad - ik_1 e^{k_1 z}\left[\frac{e^{ik_1 d}}{\rho^2} + e^{ik_1 r_d}\left(\frac{i}{\rho^2}-\frac{3}{2k_1\rho^3}\right)\right] + \frac{ik_1}{\rho^2}\,e^{ik_1 d}e^{k_1 z}\Big\} \\
&= -\left[\frac{k_2}{k_1}\,e^{ik_1 d}e^{ik_2 r_0}\left(\frac{\rho}{r_0^2}+\frac{i\rho}{k_2 r_0^3}\right) + \frac{i}{k_1}\,e^{k_1 z}e^{ik_1 r_d}\left(\frac{i}{\rho^2}-\frac{3}{2k_1\rho^3}\right)\right] \\
&= -\frac{1}{k_1}\left[e^{ik_1 d}e^{ik_2 r_0}\left(\frac{\rho}{r_0}\right)\left(\frac{k_2}{r_0}+\frac{i}{r_0^2}\right) + ie^{k_1 z}e^{ik_1 r_d}\left(\frac{i}{\rho^2}-\frac{3}{2k_1\rho^3}\right)\right].
\end{aligned}
\tag{5.7.24}
$$

When (5.7.23) and (5.7.24) are combined in (5.7.21) and this is substituted in (5.7.19), the final formula for $E_{2z}(\rho,\phi,z)$ is

$$
\begin{aligned}
E_{2z}(\rho,\phi,z) = \frac{\omega\mu_0}{2\pi k_1}\cos\phi\,\Big\{e^{ik_1 d}e^{ik_2 r_0}\Big[\left(\frac{\rho}{r_0}\right)\left(\frac{ik_2}{r_0}-\frac{1}{r_0^2}\right) \\
- \frac{k_2^3}{k_1}\left(\frac{\pi}{k_2\rho}\right)^{1/2} e^{-iP}\mathcal{F}(P)\Big] \\
- e^{k_1 z}e^{ik_1 r_d}\left(\frac{i}{\rho^2}-\frac{3}{2k_1\rho^3}\right)\Big\}; \quad z\le 0.
\end{aligned}
\tag{5.7.25}
$$

This satisfies the boundary condition $k_1^2 E_{1z}(\rho,\phi,0) = k_2^2 E_{2z}(\rho,\phi,0)$ at $z = 0$, where $E_{1z}(\rho,\phi,0)$ is obtained from (5.5.39) with $z = 0$.

The evaluation of the integrals (5.4.38) and (5.4.39) can be carried out together since they differ only in the sign of $J_2(\lambda\rho)$. The integrals are

$$
\begin{aligned}
\left.\begin{matrix} B_{2\phi}(\rho,\phi,z) \\ B_{2\rho}(\rho,\phi,z)\end{matrix}\right\} = \frac{\mu_0}{4\pi}\begin{Bmatrix}\cos\phi \\ \sin\phi\end{Bmatrix}\int_0^\infty\Big(\frac{\gamma_2}{M}\,[J_0(\lambda\rho)\pm J_2(\lambda\rho)] \\
+ \frac{k_2^2\gamma_1}{N}\,[J_0(\lambda\rho)\mp J_2(\lambda\rho)]\Big) \\
\times e^{i\gamma_1 d}e^{-i\gamma_2 z}\lambda\,d\lambda.
\end{aligned}
\tag{5.7.26}
$$

The integrands can be expanded as follows with $|k_2|^2 \ll |k_1|^2$:

$$
\frac{1}{M} = \frac{1}{\gamma_1+\gamma_2} = \frac{\gamma_1-\gamma_2}{\gamma_1^2-\gamma_2^2} = \frac{\gamma_1-\gamma_2}{k_1^2-k_2^2} \sim k_1^{-2}(\gamma_1-\gamma_2), \tag{5.7.27}
$$

$$\frac{1}{N} = \left(\frac{1}{N} - \frac{1}{N_0}\right) + \frac{1}{N_0}; \qquad N_0 = k_1^2\gamma_2, \tag{5.7.28a}$$

$$\frac{k_2^2\gamma_1}{N} = k_2^2\gamma_1\left(\frac{1}{N} - \frac{1}{N_0}\right) + \frac{k_2^2\gamma_1}{k_1^2\gamma_2}. \tag{5.7.28b}$$

It follows that

$$\begin{aligned}
&\left.\begin{matrix} B_{2\phi}(\rho,\phi,z) \\ B_{2\rho}(\rho,\phi,z) \end{matrix}\right\} \\
&\quad = \frac{\mu_0}{4\pi}\left\{\begin{matrix}\cos\phi \\ \sin\phi\end{matrix}\right\}\Bigg(k_1^{-2}\int_0^\infty \Big[(\gamma_1\gamma_2 - \gamma_2^2)[J_0(\lambda\rho) \pm J_2(\lambda\rho)] \\
&\qquad\qquad + k_2^2\left(\frac{\gamma_1}{\gamma_2}\right)[J_0(\lambda\rho) \mp J_2(\lambda\rho)]\Big] \\
&\qquad\qquad \times e^{i\gamma_1 d}e^{-i\gamma_2 z}\lambda\, d\lambda \\
&\qquad\qquad + \left\{\begin{matrix} H_\phi(\rho,-z,d) \\ H_\rho(\rho,-z,d)\end{matrix}\right\}\Bigg),
\end{aligned} \tag{5.7.29}$$

where

$$\begin{aligned}
&\left.\begin{matrix} H_\phi(\rho,-z,d) \\ H_\rho(\rho,-z,d) \end{matrix}\right\} \\
&\quad = k_2^2\int_0^\infty \gamma_1\left(\frac{1}{N} - \frac{1}{N_0}\right)[J_0(\lambda\rho) \mp J_2(\lambda\rho)]e^{i\gamma_1 d}e^{-i\gamma_2 z}\lambda\, d\lambda \\
&\quad \sim k_1k_2^2e^{ik_1d}\int_0^\infty \left(\frac{1}{N} - \frac{1}{N_0}\right)[J_0(\lambda\rho) \mp J_2(\lambda\rho)]e^{-i\gamma_2 z}\lambda\, d\lambda.
\end{aligned} \tag{5.7.30}$$

With Appendix E,

$$\begin{aligned}
H_\phi(\rho,-z,d) &\sim 2k_1k_2^2e^{ik_1d}\int_0^\infty \left(\frac{1}{N} - \frac{1}{N_0}\right)J_0(\lambda\rho)e^{-i\gamma_2 z}\lambda\, d\lambda \\
&\sim 2k_1k_2^2e^{ik_1d}[k_2^{-2}G_{2z}(\rho,z)],
\end{aligned} \tag{5.7.31}$$

$$\begin{aligned}
H_\rho(\rho,-z,d) &\sim \frac{2k_1k_2^2}{\rho}\, e^{ik_1d}\int_0^\infty \left(\frac{1}{N} - \frac{1}{N_0}\right)J_1(\lambda\rho)e^{-i\gamma_2 z}\, d\lambda \\
&\sim \frac{2k_1k_2^2}{\rho}\, e^{ik_1d}[k_2^{-2}G_{2\phi}(\rho,z)].
\end{aligned} \tag{5.7.32}$$

The integrals in (5.7.29) are generalizations of the integrals evaluated in Appendix D. They can be expanded by the same approximate procedure, as follows:

$$\begin{aligned}
\vartheta_1 &= k_1^{-2}\int_0^\infty \gamma_1\gamma_2[J_0(\lambda\rho) \pm J_2(\lambda\rho)]e^{i\gamma_1 d}e^{-i\gamma_2 z}\lambda\, d\lambda \\
&= k_1^{-2}\Bigg(k_1e^{ik_1d}\int_0^\infty \gamma_2[J_0(\lambda\rho) \pm J_2(\lambda\rho)]e^{-i\gamma_2 z}\lambda\, d\lambda
\end{aligned}$$

$$+ ie^{k_1 z}\int_0^\infty \gamma_1[J_0(\lambda\rho) \pm J_2(\lambda\rho)]e^{i\gamma_1 d}\lambda^2\,d\lambda$$

$$- ik_1 e^{k_1 z}e^{ik_1 d}\int_0^\infty [J_0(\lambda\rho) \pm J_2(\lambda\rho)]\lambda^2\,d\lambda\bigg), \tag{5.7.33}$$

$$\begin{aligned}
\vartheta_2 &= k_1^{-2}\int_0^\infty (-k_2^2+\lambda^2)[J_0(\lambda\rho) \pm J_2(\lambda\rho)]e^{i\gamma_1 d}e^{-i\gamma_2 z}\lambda\,d\lambda \\
&= k_1^{-2}\bigg(-k_2^2\int_0^\infty [J_0(\lambda\rho) \pm J_2(\lambda\rho)]e^{i\gamma_1 d}e^{-i\gamma_2 z}\lambda\,d\lambda \\
&\qquad + \int_0^\infty [J_0(\lambda\rho) \pm J_2(\lambda\rho)]e^{i\gamma_1 d}e^{-i\gamma_2 z}\lambda^3\,d\lambda\bigg) \\
&\sim k_1^{-2}\bigg(-k_2^2 e^{ik_1 d}\int_0^\infty [J_0(\lambda\rho) \pm J_2(\lambda\rho)]e^{-i\gamma_2 z}\lambda\,d\lambda \\
&\qquad - k_2^2 e^{k_1 z}\int_0^\infty [J_0(\lambda\rho) \pm J_2(\lambda\rho)]e^{i\gamma_1 d}\lambda\,d\lambda \\
&\qquad + k_2^2 e^{ik_1 d}e^{k_1 z}\int_0^\infty [J_0(\lambda\rho) \pm J_2(\lambda\rho)]\lambda\,d\lambda \\
&\qquad + e^{ik_1 d}\int_0^\infty [J_0(\lambda\rho) \pm J_2(\lambda\rho)]e^{-i\gamma_2 z}\lambda^3\,d\lambda \\
&\qquad + e^{k_1 z}\int_0^\infty [J_0(\lambda\rho) \pm J_2(\lambda\rho)]e^{i\gamma_1 d}\lambda^3\,d\lambda \\
&\qquad - e^{k_1 z}e^{ik_1 d}\int_0^\infty [J_0(\lambda\rho) \pm J_2(\lambda\rho)]\lambda^3\,d\lambda\bigg),
\end{aligned} \tag{5.7.34}$$

$$\begin{aligned}
\vartheta_3 &= k_1^{-2}k_2^2\int_0^\infty \frac{\gamma_1}{\gamma_2}[J_0(\lambda\rho) \mp J_2(\lambda\rho)]e^{i\gamma_1 d}e^{-i\gamma_2 z}\lambda\,d\lambda \\
&\sim k_1^{-2}k_2^2\bigg(k_1 e^{ik_1 d}\int_0^\infty \gamma_2^{-1}[J_0(\lambda\rho) \mp J_2(\lambda\rho)]e^{-i\gamma_2 z}\lambda\,d\lambda \\
&\qquad - ie^{k_1 z}\int_0^\infty \gamma_1[J_0(\lambda\rho) \mp J_2(\lambda\rho)]e^{i\gamma_1 d}\,d\lambda \\
&\qquad + ik_1 e^{ik_1 d}e^{k_1 z}\int_0^\infty [J_0(\lambda\rho) \mp J_2(\lambda\rho)]\,d\lambda\bigg).
\end{aligned} \tag{5.7.35}$$

With Appendix A, the following listed integrals are involved:

$$\begin{aligned}
\vartheta_1 &= k_1^{-2}\bigg(k_1 e^{ik_1 d}\begin{bmatrix} I_{11}(k_2;\rho,-z) \\ I_{12}(k_2;\rho,-z)\end{bmatrix} + ie^{k_1 z}\begin{bmatrix} I_{13}(k_1;\rho,d) \\ I_{14}(k_1;\rho,d)\end{bmatrix} \\
&\qquad - ik_1 e^{ik_1 d}e^{k_1 z}\begin{bmatrix} I_{20}(\rho,0) \\ I_{21}(\rho,0)\end{bmatrix}\bigg),
\end{aligned} \tag{5.7.36}$$

$$\vartheta_2 = k_1^{-2}\bigg(-k_2^2 e^{ik_1 d}\begin{bmatrix} I_{18}(k_2;\rho,-z) \\ I_{19}(k_2;\rho,-z)\end{bmatrix} - k_2^2 e^{k_1 z}\begin{bmatrix} I_{18}(k_1;\rho,d) \\ I_{19}(k_1;\rho,d)\end{bmatrix}$$

$$+ k_2^2 e^{k_1 z} e^{ik_1 d} \begin{bmatrix} I_{18}(\rho, 0) \\ I_{19}(\rho, 0) \end{bmatrix} + e^{ik_1 d} \begin{bmatrix} I_{22}(k_2; \rho, -z) \\ I_{23}(k_2; \rho, -z) \end{bmatrix}$$

$$+ e^{k_1 z} \begin{bmatrix} I_{22}(k_1; \rho, d) \\ I_{23}(k_1; \rho, d) \end{bmatrix} - e^{k_1 z} e^{ik_1 d} \begin{bmatrix} I_{22}(\rho, 0) \\ I_{23}(\rho, 0) \end{bmatrix} \Bigg), \tag{5.7.37}$$

$$\vartheta_3 = k_1^{-2} k_2^2 \Bigg(k_1 e^{ik_1 d} \begin{bmatrix} I_4(k_2; \rho, -z) \\ I_3(k_2; \rho, -z) \end{bmatrix} - i e^{k_1 z} \begin{bmatrix} I_{10}(k_1; \rho, d) \\ I_9(k_1; \rho, d) \end{bmatrix}$$

$$+ i k_1 e^{ik_1 d} e^{k_1 z} \begin{bmatrix} I_{17}(\rho, 0) \\ I_{16}(\rho, 0) \end{bmatrix} \Bigg). \tag{5.7.38}$$

With (5.7.31) and (E83),

$$\begin{aligned} H_\phi(\rho, -z, d) &= 2k_1 k_2^2 e^{ik_1 d} [k_2^{-2} G_{2z}(\rho, z)] \\ &= 2k_1 k_2^2 e^{ik_1 d} \left[\frac{k_2^2}{k_1^3} \left(\frac{\pi}{k_2 \rho} \right)^{1/2} e^{ik_2 \rho} e^{iZ^2/R} e^{-iP} \mathcal{F}(P) \right] \\ &= \frac{2k_2^4}{k_1^2} \left(\frac{\pi}{k_2 \rho} \right)^{1/2} e^{ik_1 d} e^{ik_2 r_0} e^{-iP} \mathcal{F}(P). \end{aligned} \tag{5.7.39}$$

With (5.7.32) and (E82),

$$\begin{aligned} H_\rho(\rho, -z, d) &= \frac{2k_1 k_2^2}{\rho} e^{ik_1 d} [k_2^{-2} G_{2\phi}(\rho, z)] \\ &= \frac{2k_1 k_2^2}{\rho} e^{ik_1 d} \left[-\frac{ik_2}{k_1^3} \left(\frac{\pi}{k_2 \rho} \right)^{1/2} e^{ik_2 \rho} e^{iZ^2/R} e^{-iP} \mathcal{F}(P) \right] \\ &= -\frac{2ik_2^3}{k_1^2 \rho} \left(\frac{\pi}{k_2 \rho} \right)^{1/2} e^{ik_1 d} e^{ik_2 r_0} e^{-iP} \mathcal{F}(P). \end{aligned} \tag{5.7.40}$$

Thus, with the upper sign understood in ϑ_1, ϑ_2, and ϑ_3,

$$B_{2\phi}(\rho, \phi, z) = \frac{\mu_0}{4\pi} \cos\phi \, [H_\phi(\rho, -z, d) + \vartheta_1 + \vartheta_2 + \vartheta_3], \tag{5.7.41}$$

where, with Appendix A, the following results are obtained:

$$\begin{aligned} &B_{2\phi}(\rho, \phi, z) \\ &\quad = \frac{\mu_0}{4\pi} \cos\phi \, \{H_\phi(\rho, -z, d) \\ &\qquad + k_1^{-2} [k_1 e^{ik_1 d} I_{11}(k_2; \rho, -z) + i e^{k_1 z} I_{13}(k_1; \rho, d) \\ &\qquad - i k_1 e^{ik_1 d} e^{k_1 z} I_{20}(\rho, 0) - k_2^2 e^{ik_1 d} I_{18}(k_2; \rho, -z) \\ &\qquad - k_2^2 e^{k_1 z} I_{18}(k_1; \rho, d) + k_2^2 e^{ik_1 d} e^{k_1 z} I_{18}(\rho, 0) \\ &\qquad + e^{ik_1 d} I_{22}(k_2; \rho, -z) + e^{k_1 z} I_{22}(k_1; \rho, d) \\ &\qquad - e^{k_1 z} e^{ik_1 d} I_{22}(\rho, 0) + k_2^2 k_1 e^{ik_1 d} I_4(k_2; \rho, -z) \end{aligned}$$

$$- ik_2^2 e^{k_1 z} I_{10}(k_1; \rho, d) + ik_2^2 k_1 e^{ik_1 d} e^{k_1 z} I_{17}(\rho, 0)]\} \tag{5.7.42}$$

$$\begin{aligned}
= \frac{\mu_0}{4\pi} \cos\phi \Big(H_\phi(\rho, -z, d) \\
& + k_1^{-2}\{e^{ik_1 d}[k_1 I_{11}(k_2; \rho, -z) - k_2^2 I_{18}(k_2; \rho, -z) \\
& \qquad + I_{22}(k_2; \rho, -z) + k_2^2 k_1 I_4(k_2; \rho, -z)] \\
& + e^{k_1 z}[iI_{13}(k_1; \rho, d) - k_2^2 I_{18}(k_1; \rho, d) \\
& \qquad + I_{22}(k_1; \rho, d) - ik_2^2 I_{10}(k_1; \rho, d)] \\
& + e^{ik_1 d} e^{k_1 z}[-ik_1 I_{20}(\rho, 0) + k_2^2 I_{18}(\rho, 0) \\
& \qquad - I_{22}(\rho, 0) + ik_2^2 k_1 I_{17}(\rho, 0)]\}\Big)
\end{aligned} \tag{5.7.43}$$

$$\begin{aligned}
= \frac{\mu_0}{4\pi k_1^2} \cos\phi \Bigg(k_1^2 H_\phi(\rho, -z, d) \\
& + e^{ik_1 d}\Bigg\{ 2k_1 \left[\frac{k_2 e^{-ik_2 z}}{\rho^2} + \frac{ie^{ik_2 r_0}}{r_0^3} - \left(\frac{z}{r_0}\right)^2 \frac{k_2}{\rho^2}\, e^{ik_2 r_0} \right] \\
& \quad - 2k_2^2 \left(\frac{e^{-ik_2 z}}{\rho^2} + \frac{z}{r_0}\, \frac{e^{ik_2 r_0}}{\rho^2} \right) \\
& \quad - 2k_2 e^{ik_2 r_0} \left(-\frac{z}{r_0} \right) \left(\frac{k_2}{r_0^2} + \frac{3i}{r_0^3} - \frac{3}{k_2 r_0^4} \right) \\
& \quad - 2k_2^2 k_1 \left[\frac{ie^{ik_2 r_0}}{r_0} + \frac{1}{k_2 \rho^2}\, (e^{-ik_2 z} - e^{ik_2 r_0}) \right] \Bigg\} \\
& + e^{k_1 z}\Bigg\{ \frac{2ik_1}{\rho^3} \left[e^{ik_1 d} + ie^{ik_1 r_d} \left(1 + \frac{3i}{2k_1 \rho} \right) \right] \\
& \quad - 2k_2^2 \left(\frac{e^{ik_1 d}}{\rho^2} - \frac{d}{r_d}\, \frac{e^{ik_1 r_d}}{\rho^2} \right) \\
& \quad - 2k_1 e^{ik_1 r_d} \left(\frac{d}{r_d} \right) \left(\frac{k_1}{r_d^2} + \frac{3i}{r_d^3} - \frac{3}{k_1 r_d^4} \right) \\
& \quad - 2ik_2^2 \Bigg[\left(\frac{id}{\rho^3} + \frac{1}{k_1 \rho^3} \right) e^{ik_1 d} \\
& \qquad - \left(\frac{1}{\rho^2} + \frac{i}{2k_1 \rho^3} + \frac{3}{8k_1^2 \rho^4} \right) e^{ik_1 r_d} \Bigg] \Bigg\} \\
& + e^{ik_1 d} e^{k_1 z} \left(-ik_1\, \frac{2}{\rho^3} + k_2^2\, \frac{2}{\rho^2} \right) \Bigg).
\end{aligned} \tag{5.7.44}$$

With (5.7.39) in (5.7.44) and the omission of small terms of the order of k_2^2/k_1^2 and $k_2 z/k_1 \rho$, the final formula for $B_{2\phi}(\rho, \phi, z)$ is

$$B_{2\phi}(\rho,\phi,z) = -\frac{\mu_0}{2\pi k_1}\cos\phi$$
$$\times\left\{e^{ik_2r_0}e^{ik_1d}k_2\left[\frac{ik_2}{r_0}-\frac{1}{r_0^2}-\frac{i}{k_2r_0^3}-\frac{k_2^3}{k_1}\left(\frac{\pi}{k_2\rho}\right)^{1/2}e^{-iP}\mathcal{F}(P)\right]\right.$$
$$\left.+\,e^{k_1z}e^{ik_1r_d}k_1\left[\frac{1}{k_1\rho^3}+\frac{3i}{2k_1^2\rho^4}+\left(\frac{d}{r_d}\right)\left(\frac{1}{r_d^2}+\frac{3i}{k_1r_d^3}\right)\right]\right\};$$
$$z\le 0. \qquad (5.7.45)$$

Note that this formula for $B_{2\phi}(\rho,\phi,z)$ satisfies the boundary condition $B_{2\phi}(\rho,\phi,0)=B_{1\phi}(\rho,\phi,0)$ as is readily verified from a comparison of (5.7.45) with (5.5.41) when $z=0$.

The lower signs in (5.7.29) and the relevant integrals give

$$B_{2\rho}(\rho,\phi,z) = \frac{\mu_0}{4\pi}\sin\phi\,[H_\rho(\rho,-z,d)+\vartheta_1+\vartheta_2+\vartheta_3], \qquad (5.7.46)$$

from which, with Appendix A, the following formula is obtained:

$$B_{2\rho}(\rho,\phi,z)$$
$$\begin{aligned}=\frac{\mu_0}{4\pi}\sin\phi\,\{&H_\rho(\rho,-z,d)\\
&+k_1^{-2}[k_1e^{ik_1d}I_{12}(k_2;\rho,-z)+ie^{k_1z}I_{14}(k_1;\rho,d)\\
&\quad-ik_1e^{ik_1d}e^{k_1z}I_{21}(\rho,0)-k_2^2e^{ik_1d}I_{19}(k_2;\rho,-z)\\
&\quad-k_2^2e^{k_1z}I_{19}(k_1;\rho,d)+k_2^2e^{ik_1d}e^{k_1z}I_{19}(\rho,0)\\
&\quad+e^{ik_1d}I_{23}(k_2;\rho,-z)+e^{k_1z}I_{23}(k_1;\rho,d)\\
&\quad-e^{ik_1d}e^{k_1z}I_{23}(\rho,0)+k_2^2k_1e^{ik_1d}I_3(k_2;\rho,-z)\\
&\quad-ik_2^2e^{k_1z}I_9(k_1;\rho,d)+ik_2^2k_1e^{ik_1d}e^{k_1z}I_{16}(\rho,0)]\}\end{aligned} \qquad (5.7.47)$$

$$\begin{aligned}=\frac{\mu_0}{4\pi k_1^2}\sin\phi\,\{&k_1^2H_\rho(\rho,-z,d)\\
&+e^{ik_1d}[k_1I_{12}(k_2;\rho,-z)-k_2^2I_{19}(k_2;\rho,-z)\\
&\qquad+I_{23}(k_2;\rho,-z)+k_2^2k_1I_3(k_2;\rho,-z)]\\
&+e^{k_1z}[iI_{14}(k_1;\rho,d)-k_2^2I_{19}(k_1;\rho,d)\\
&\qquad+I_{23}(k_1;\rho,d)-ik_2^2I_9(k_1;\rho,d)]\\
&+e^{ik_1d}e^{k_1z}[-ik_1I_{21}(\rho,0)+k_2^2I_{19}(\rho,0)\\
&\qquad-I_{23}(\rho,0)+ik_2^2k_1I_{16}(\rho,0)]\}\end{aligned} \qquad (5.7.48)$$

$$
\begin{aligned}
&= \frac{\mu_0}{4\pi k_1^2}\sin\phi \\
&\quad\times\Bigg[k_1^2 H_\rho(\rho,-z,d) \\
&\qquad + e^{ik_1 d}\Bigg(-2k_1k_2\Bigg\{\frac{e^{-ik_2z}}{\rho^2}+\Bigg[\frac{1}{r_0^2}+\frac{2i}{k_2r_0^3} \\
&\qquad\qquad + \left(\frac{z^2}{r_0^2}\right)\left(\frac{ik_2}{r_0}-\frac{3}{r_0^2}-\frac{3i}{k_2r_0^3}-\frac{1}{\rho^2}\right)\Bigg]e^{ik_2r_0}\Bigg\} \\
&\qquad\qquad - 2k_2^2 e^{ik_2r_0}\left(\frac{z}{r_0}\right)\left(\frac{ik_2}{r_0}-\frac{1}{r_0^2}-\frac{1}{\rho^2}\right)+2k_2^2\,\frac{e^{-ik_2z}}{\rho^2} \\
&\qquad\qquad + 2k_2e^{ik_2r_0}\left(-\frac{z}{r_0}\right)\left(-\frac{ik_2^2}{r_0}+\frac{5k_2}{r_0^2}+\frac{12i}{r_0^3}\right) \\
&\qquad\qquad + 2k_2k_1\left(\frac{e^{-ik_2z}}{\rho^2}-\frac{e^{ik_2r_0}}{\rho^2}\right)\Bigg) \\
&\qquad + e^{k_1z}\Bigg\{-\frac{4ik_1}{\rho^3}\,e^{ik_1d}-2ik_1^2\left(\frac{1}{\rho^2}+\frac{7i}{2k_1\rho^3}\right)e^{ik_1r_d} \\
&\qquad\qquad + 2k_2^2\left[e^{ik_1r_d}\left(\frac{d}{r_d}\right)\left(\frac{ik_1}{r_d}-\frac{1}{r_d^2}-\frac{1}{\rho^2}\right)+\frac{e^{ik_1d}}{\rho^2}\right] \\
&\qquad\qquad + 2k_1e^{ik_1r_d}\left(\frac{d}{r_d}\right)\left(-\frac{ik_1^2}{r_d}+\frac{5k_1}{r_d^2}+\frac{12i}{r_d^3}\right) \\
&\qquad\qquad - 2ik_2^2\left[\left(\frac{k_1}{\rho}-\frac{1}{2k_1\rho^3}-\frac{id}{2\rho^3}\right)e^{ik_1d}+\frac{ie^{ik_1r_d}}{k_1\rho^3}\right]\Bigg\} \\
&\qquad + e^{ik_1d}e^{k_1z}\left(\frac{4ik_1}{\rho^3}-\frac{2k_2^2}{\rho^2}+\frac{2ik_2^2k_1}{\rho}\right)\Bigg].
\end{aligned}
\tag{5.7.49}
$$

After small terms of the order of k_2^2/k_1^2 have been neglected, the result is

$$
\begin{aligned}
B_{2\rho}(\rho,\phi,z) &= \frac{\mu_0}{4\pi k_1^2}\sin\phi \\
&\quad\times\Bigg\{k_1^2H_\rho(\rho,-z,d) \\
&\qquad - 2k_1k_2\left[\frac{2}{r_0^2}+\frac{2i}{k_2r_0^3}+\left(\frac{z^2}{r_0^2}\right)\left(\frac{ik_2}{r_0}-\frac{3}{r_0^2}-\frac{3i}{k_2r_0^3}\right)\right]e^{ik_2r_0}e^{ik_1d} \\
&\qquad - 2k_2^2\left(\frac{z}{r_0}\right)\left(\frac{4}{r_0^2}-\frac{1}{\rho^2}+\frac{12i}{k_2r_0^3}\right)e^{ik_2r_0}e^{ik_1d} \\
&\qquad - 2k_1^2\left[\frac{i}{\rho^2}-\frac{7}{2k_1\rho^3}\right.
\end{aligned}
$$

$$+\left(\frac{d}{r_d}\right)\left(\frac{ik_1}{r_d}-\frac{5}{r_d^2}-\frac{12i}{k_1r_d^3}\right)\Bigg]e^{k_1z}e^{ik_1r_d}\Bigg\}. \tag{5.7.50}$$

The formula for $H_\rho(\rho,-z,d)$ is in (5.7.40). With it, the final formula for $B_{2\rho}(\rho,\phi,z)$ becomes

$$\begin{aligned}B_{2\rho}(\rho,\phi,z) &= -\frac{\mu_0}{2\pi k_1}\sin\phi\\ &\times\Bigg\{e^{ik_2r_0}e^{ik_1d}k_2\Bigg[\frac{2}{r_0^2}+\frac{2i}{k_2r_0^3}+\left(\frac{z^2}{r_0^2}\right)\left(\frac{ik_2}{r_0}-\frac{3}{r_0^2}-\frac{3i}{k_2r_0^3}\right)\\ &\qquad+\frac{ik_2^2}{k_1\rho}\left(\frac{\pi}{k_2\rho}\right)^{1/2}e^{-iP}\mathcal{F}(P)\\ &\qquad+\left(\frac{k_2z}{k_1r_0}\right)\left(\frac{4}{r_0^2}-\frac{1}{\rho^2}+\frac{12i}{k_2r_0^3}\right)\Bigg]\\ &+e^{k_1z}e^{ik_1r_d}k_1\left[\frac{i}{\rho^2}-\frac{7}{2k_1\rho^3}+\left(\frac{d}{r_d}\right)\left(\frac{ik_1}{r_d}-\frac{5}{r_d^2}-\frac{12i}{k_1r_d^3}\right)\right]\Bigg\}.\end{aligned} \tag{5.7.51}$$

The terms can be rearranged as follows:

$$\begin{aligned}B_{2\rho}(\rho,\phi,z) &= -\frac{\mu_0}{2\pi k_1}\sin\phi\\ &\quad\times\Bigg\{e^{ik_1d}e^{ik_2r_0}k_2\Bigg[\frac{ik_2z^2}{r_0^3}+\frac{2\rho^2-z^2}{r_0^2}\left(\frac{1}{r_0^2}+\frac{i}{k_2r_0^3}\right)\\ &\qquad+\frac{ik_2^2}{k_1\rho}\left(\frac{\pi}{k_2\rho}\right)^{1/2}e^{-iP}\mathcal{F}(P)\\ &\qquad+\left(\frac{k_2z}{k_1r_0}\right)\left(\frac{4}{r_0^2}-\frac{1}{\rho^2}+\frac{12i}{k_2r_0^3}\right)\Bigg]\\ &\quad+e^{k_1z}e^{ik_1r_d}\Bigg[\frac{ik_1}{\rho^2}-\frac{7}{2\rho^3}\\ &\qquad+\left(\frac{d}{r_d}\right)\left(\frac{ik_1^2}{r_d}-\frac{5k_1}{r_d^2}-\frac{12i}{r_d^3}\right)\Bigg]\Bigg\};\\ &\qquad z\le 0.\end{aligned} \tag{5.7.52}$$

The integral for $B_{2z}(\rho,\phi,z)$ is given by (5.4.40). With $M^{-1}\sim k_1^{-2}(\gamma_1-\gamma_2)$, it becomes

$$B_{2z}(\rho,\phi,z)=\frac{i\mu_0}{2\pi k_1^2}\sin\phi\int_0^\infty(\gamma_1-\gamma_2)J_1(\lambda\rho)e^{i\gamma_1d}e^{-i\gamma_2z}\lambda^2\,d\lambda. \tag{5.7.53}$$

The two integrals involved are readily expanded with the approximate procedure described in Appendix D in the form

$$f(k_1,k_2;\lambda)=f_1(k_1,k_2;\lambda)+f_2(k_1,k_2;\lambda)-f_3(k_1,k_2;\lambda), \tag{5.7.54}$$

where $f_1(k_1, k_2; \lambda)$, $f_2(k_1; k_2; \lambda)$, and $f_3(k_1, k_2; \lambda)$ are given in (5.7.5b,c). Specifically, when applied to (5.7.53), the result is

$$\begin{aligned} I &= \int_0^\infty (\gamma_1 - \gamma_2) J_1(\lambda\rho) e^{i\gamma_1 d} e^{-i\gamma_2 z} \lambda^2 \, d\lambda \\ &\sim e^{k_1 z} \int_0^\infty (\gamma_1 - i\lambda) J_1(\lambda\rho) e^{i\gamma_1 d} \lambda^2 \, d\lambda \\ &\quad + e^{ik_1 d} \int_0^\infty (k_1 - \gamma_2) J_1(\lambda\rho) e^{-i\gamma_2 z} \lambda^2 \, d\lambda \\ &\quad - e^{k_1 z} e^{ik_1 d} \int_0^\infty (k_1 - i\lambda) J_1(\lambda\rho) \lambda^2 \, d\lambda. \end{aligned} \tag{5.7.55}$$

With Appendix A, this is equivalent to

$$\begin{aligned} I &\sim \frac{\rho e^{k_1 z}}{2} \left[I_{15}(k_1; \rho, d) - i I_{24}(k_1; \rho, d) \right] \\ &\quad + \frac{\rho e^{ik_1 d}}{2} \left[k_1 I_{22}(k_2; \rho, -z) - I_{15}(k_2; \rho, -z) \right] \\ &\quad - \frac{\rho e^{k_1 z}}{2} e^{ik_1 d} \left[k_1 I_{22}(\rho, 0) - i I_{24}(\rho, 0) \right] \\ &\sim \frac{\rho e^{k_1 z}}{2} \left[2 e^{ik_1 r_d} \left(\frac{ik_1^2}{r_d^3} - \frac{3k_1}{r_d^4} - \frac{3i}{r_d^5} \right) \right. \\ &\qquad + 2ik_1^2 \left(\frac{d}{r_d} \right) \left(\frac{k_1}{\rho^2} + \frac{3i}{2\rho^3} - \frac{9}{8k_1\rho^4} \right) e^{ik_1 r_d} \\ &\qquad \left. - 2k_1^2 \left(\frac{d}{r_d} \right)^2 \left(\frac{k_1}{r_d^2} + \frac{6i}{r_d^3} - \frac{15}{k_1 r_d^4} \right) e^{ik_1 r_d} \right] \\ &\quad + \frac{\rho e^{ik_1 d}}{2} \left[2k_1 k_2 e^{ik_2 r_0} \left(\frac{z}{r_0} \right) \left(\frac{k_2}{r_0^2} + \frac{3i}{r_0^3} - \frac{3}{k_2 r_0^4} \right) \right. \\ &\qquad - 2 e^{ik_2 r_0} \left(\frac{ik_2^2}{r_0^3} - \frac{3k_2}{r_0^4} - \frac{3i}{r_0^5} \right) \\ &\qquad \left. + 2k_2^2 e^{ik_2 r_0} \left(\frac{z}{r_0} \right)^2 \left(\frac{k_2}{r_0^2} + \frac{6i}{r_0^3} - \frac{15}{k_2 r_0^4} \right) \right] \\ &= -e^{ik_1 d} e^{ik_2 r_0} k_2 \left(\frac{\rho}{r_0} \right) \left[\frac{ik_2}{r_0^2} - \frac{3}{r_0^3} - \frac{3i}{k_2 r_0^4} \right. \\ &\qquad - k_1 \left(\frac{z}{r_0} \right) \left(\frac{k_2}{r_0} + \frac{3i}{r_0^2} - \frac{3}{k_2 r_0^3} \right) \\ &\qquad \left. - k_2 \left(\frac{z}{r_0} \right)^2 \left(\frac{k_2}{r_0} + \frac{6i}{r_0^2} - \frac{15}{k_2 r_0^3} \right) \right] \\ &\quad + ie^{k_1 z} e^{ik_1 r_d} k_1^2 \left(\frac{\rho}{r_d} \right) \left[\frac{1}{r_d^2} + \frac{3i}{k_1 r_d^3} - \frac{3}{k_1^2 r_d^4} \right. \end{aligned}$$

$$+\left(\frac{d}{\rho}\right)\left(\frac{k_1}{\rho}+\frac{3i}{2\rho^2}-\frac{9}{8k_1\rho^3}\right)$$
$$+\left(\frac{d}{r_d}\right)^2\left(\frac{ik_1}{r_d}-\frac{6}{r_d^2}-\frac{15i}{k_1r_d^3}\right)\Bigg]. \qquad (5.7.56)$$

It follows with (5.7.53) that the final formula for $B_{2z}(\rho,\phi,z)$ is

$$\begin{aligned}
B_{2z}(\rho,\phi,z) &= \frac{\mu_0 k_2}{2\pi k_1}\sin\phi \\
&\times\Bigg(e^{ik_1d}e^{ik_2r_0}\Bigg\{\left(\frac{\rho z}{r_0^2}\right)\left(\frac{ik_2}{r_0}-\frac{3}{r_0^2}-\frac{3i}{k_2r_0^3}\right) \\
&\quad+\left(\frac{k_2\rho}{k_1r_0}\right)\Bigg[\frac{1}{r_0^2}+\frac{3i}{k_2r_0^3}-\frac{3}{k_2^2r_0^4} \\
&\qquad+\left(\frac{z^2}{r_0^2}\right)\left(\frac{k_2}{r_0}+\frac{6i}{r_0^2}-\frac{15}{k_2r_0^3}\right)\Bigg]\Bigg\} \\
&\quad-e^{k_1z}e^{ik_1r_d}\left(\frac{k_1\rho}{k_2r_d}\right)\Bigg[\frac{1}{r_d^2}+\frac{3i}{k_1r_d^3}-\frac{3}{k_1^2r_d^4} \\
&\qquad+\left(\frac{d}{\rho}\right)\left(\frac{k_1}{\rho}+\frac{3i}{2\rho^2}-\frac{9}{8k_1\rho^3}\right) \\
&\qquad+\left(\frac{d}{r_d}\right)^2\left(\frac{ik_1}{r_d}-\frac{6}{r_d^2}-\frac{15i}{k_1r_d^3}\right)\Bigg]\Bigg);
\end{aligned}$$
$$z \le 0. \qquad (5.7.57)$$

5.8 Alternative Derivation of the Field in Region 2 of the Horizontal Electric Dipole at the Depth d in Region 1

The six components of the electromagnetic field in Region 2 due to a unit horizontal electric dipole at depth d in Region 1 are given in (5.5.47)–(5.5.52) as evaluated directly from the integrals (5.4.35)–(5.4.40). It is also possible to determine the field in Region 2 when the dipole is close to the boundary surface by an alternative approximate method that gives additional physical insight. Since it has been assumed that $|k_1| \ge 3|k_2|$, it follows that total internal reflection occurs at the boundary at angles that exceed the critical angle. This means that the field entering Region 2 is localized in a moderately small area below the dipole.

The field incident on the boundary from the dipole can be obtained directly from the well-known field of a horizontal electric dipole. For an x-directed unit dipole, the components of the field are

$$E_x = \frac{i\omega\mu_0}{4\pi k_1^2} e^{ik_1r_0}\left[\frac{k_1^2}{r_0} + \frac{ik_1}{r_0^2} - \frac{1}{r_0^3} - \frac{x^2}{r_0^2}\left(\frac{k_1^2}{r_0} + \frac{3ik_1}{r_0^2} - \frac{3}{r_0^3}\right)\right], \tag{5.8.1}$$

$$E_y = -\frac{i\omega\mu_0}{4\pi k_1^2} e^{ik_1r_0}\frac{xy}{r_0^2}\left(\frac{k_1^2}{r_0} + \frac{3ik_1}{r_0^2} - \frac{3}{r_0^3}\right), \tag{5.8.2}$$

$$E_z = -\frac{i\omega\mu_0}{4\pi k_1^2} e^{ik_1r_0}\frac{xz}{r_0^2}\left(\frac{k_1^2}{r_0} + \frac{3ik_1}{r_0^2} - \frac{3}{r_0^3}\right), \tag{5.8.3}$$

$$\mathbf{B} = \frac{\mu_0}{4\pi r_0} e^{ik_1r_0}(\hat{\mathbf{y}}z - \hat{\mathbf{z}}y)\left(\frac{ik_1}{r_0} - \frac{1}{r_0^2}\right), \tag{5.8.4}$$

where $r_0 = (\rho^2+z^2)^{1/2}$. The field on the boundary with Region 2 is given by (5.8.1)–(5.8.4) with $z = d$. However, this is not the complete field because of reflection at the boundary.

The plane-wave reflection coefficient R for normal incidence as seen from Region 1 (where the dipole is located) is defined by the components of the field tangent to the boundary, i.e.,

$$\begin{aligned} B_{\text{tang}} &= B_{\text{tang}}^{\text{inc}}(e^{-ik_1z} + Re^{ik_1z}), \\ E_{\text{tang}} &= E_{\text{tang}}^{\text{inc}}(e^{-ik_1z} - Re^{ik_1z}) \end{aligned} \tag{5.8.5}$$

at $z = 0$. In (5.8.5),

$$R = \frac{k_2 - k_1}{k_2 + k_1}. \tag{5.8.6}$$

Since $|k_2|^2 \ll |k_1|^2$, it follows that

$$R \sim -1 + \frac{2k_2}{k_1}. \tag{5.8.7}$$

On the boundary $z = 0$,

$$B_{\text{tang}} \sim \frac{2k_2}{k_1} B_{\text{tang}}^{\text{inc}}; \qquad E_{\text{tang}} \sim 2E_{\text{tang}}^{\text{inc}}\left(1 - \frac{k_2}{k_1}\right). \tag{5.8.8}$$

As a first approximation, let $B_{\text{tang}} \sim 0$ and $E_{\text{tang}} \sim 2E_{\text{tang}}^{\text{inc}}$. This is equivalent to an unbounded Region 1 with a source dipole at $z = d$ and an image dipole at $z = -d$ with equal and codirectional electric moment. Thus, the tangential electric field is double the incident electric field in Region 1 at the boundary plane $z = 0$. It is

$$E_{1x} \sim \frac{i\omega\mu_0}{2\pi k_1^2} e^{ik_1r_d}\left[\frac{k_1^2}{r_d} + \frac{ik_1}{r_d^2} - \frac{1}{r_d^3} - \frac{x^2}{r_d^2}\left(\frac{k_1^2}{r_d} + \frac{3ik_1}{r_d^2} - \frac{3}{r_d^3}\right)\right], \tag{5.8.9}$$

$$E_{1y} \sim -\frac{i\omega\mu_0}{2\pi k_1^2} e^{ik_1r_d}\frac{xy}{r_d^2}\left(\frac{k_1^2}{r_d} + \frac{3ik_1}{r_d^2} - \frac{3}{r_d^3}\right), \tag{5.8.10}$$

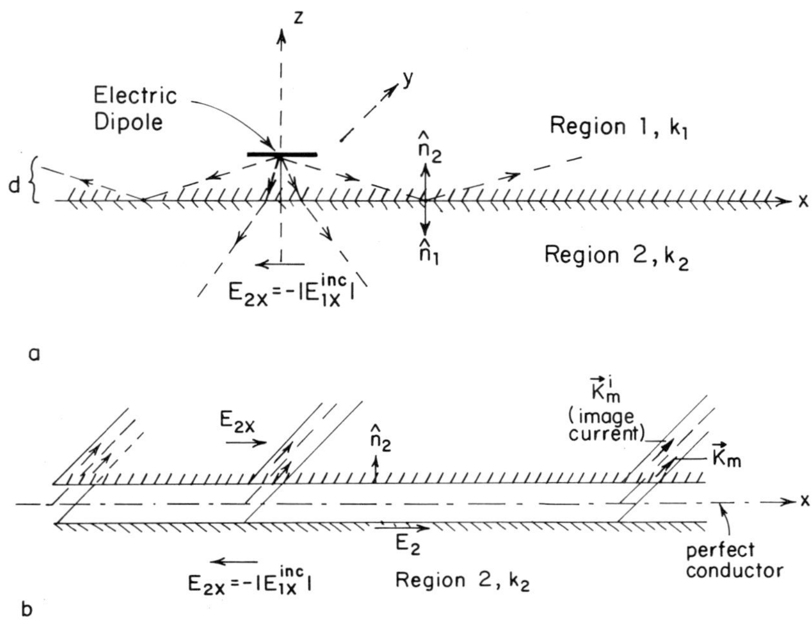

Figure 5.8.1. (a) Unit x-directed electric dipole in Region 1 near boundary with Region 2 ($|k_2|^2 \ll |k_1|^2$). (b) Layer of y-directed magnetic current $2\mathbf{K}_m = \mathbf{K}_m + \mathbf{K}_m^i$ which maintains same field in Region 2 as electric dipole.

where $r_d = (\rho^2 + d^2)^{1/2}$. The unit x-directed dipole in Region 1, $z \geq 0$, at a height d over the boundary, $z = 0$, with Region 2, $z \leq 0$, is shown in Fig. 5.8.1a. As shown schematically, the field generated by the dipole is transmitted across the boundary into Region 2 when the angle of incidence is smaller than the critical angle. At larger angles, the field is almost completely reflected. It follows that the primary field is significant on the boundary surface as a source that generates the field in Region 2 only within the cone bounded by the critical angle.

The primary field tangent to the boundary can be represented by an equivalent sheet of fictitious magnetic surface current $\mathbf{K}_m$ that generates the same field in Region 2 as the transmitted primary field. This current is

$$\begin{aligned}\mathbf{K}_m = -\hat{\mathbf{n}}_2 \times \mathbf{E}_2 = -\hat{\mathbf{n}}_2 \times \mathbf{E}_1^{\text{inc}} &= -\hat{\mathbf{z}} \times (\hat{\mathbf{x}} E_{1x}^{\text{inc}} + \hat{\mathbf{y}} E_{1y}^{\text{inc}}) \\ &= -\hat{\mathbf{y}} E_{1x}^{\text{inc}} + \hat{\mathbf{x}} E_{1y}^{\text{inc}}. \qquad (5.8.11)\end{aligned}$$

In order to calculate the field in Region 2 from this sheet of current on its boundary, it is necessary initially to take note of the fact that $|k_1|^2 \gg |k_2|^2$ and approximate the boundary $z = 0$ on which the current $\mathbf{K}_m$ is located by a perfect conductor where $\mathbf{E}_{\text{tang}} = 0$. This approximation will be corrected later. The field in Region 2 ($z \leq 0$) is then determined by the current $\mathbf{K}_m$ and by an image current $\mathbf{K}_m^i = \mathbf{K}_m$ located adjacent to $\mathbf{K}_m$, as shown

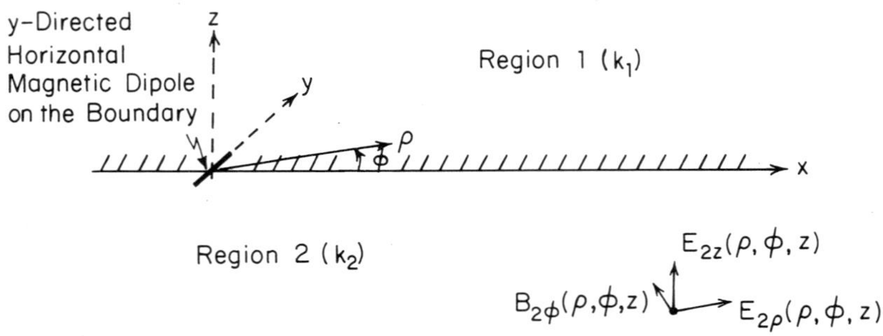

Figure 5.8.2. Equivalent magnetic dipole (including image) on boundary for determining field in Region 2.

in Fig. 5.8.1b. It is the sheet of current $2\mathbf{K}_m$ that generates the field in Region 2.

Since the current $2\mathbf{K}_m$ is significant on a relatively small area near $z = 0$, its effect in generating the field in Region 2 can be approximated by a magnetic dipole (including image) located at $x = y = 0$ on the boundary $z = 0$ with the total magnetic current moment $\mathbf{V}l$ or the combined magnetic moment $\mathbf{m} = \hat{\mathbf{x}}m_x + \hat{\mathbf{y}}m_y$ given by

$$V_y l_y = i\omega\mu_0 m_y = 2\int_{-\infty}^{\infty}\int_{-\infty}^{\infty} K_{my}\,dx\,dy$$

$$= -2\int_{-\infty}^{\infty}\int_{-\infty}^{\infty} E_{1x}\,dx\,dy = \frac{2\omega\mu_0}{k_1}\,e^{ik_1 d}, \tag{5.8.12}$$

$$V_x l_x = i\omega\mu_0 m_x = 2\int_{-\infty}^{\infty}\int_{-\infty}^{\infty} K_{mx}\,dx\,dy$$

$$= 2\int_{-\infty}^{\infty}\int_{-\infty}^{\infty} E_{1y}\,dx\,dy = 0, \tag{5.8.13}$$

where E_{1x} and E_{1y} are the components (5.8.9) and (5.8.10) of the primary field.

Evidently, the field in Region 2 ($z \le 0$) not too close to the origin or the boundary surface is the field of a y-directed magnetic dipole with the moment given in (5.8.12) and located at the origin. This is illustrated in Fig. 5.8.2. In cylindrical coordinates ρ, ϕ, z, where ϕ is measured from the positive x-axis and the dipole lies along the y-axis, the components of the field are

$$E_{2\rho}(\rho,\phi,z) = -\frac{\omega\mu_0}{2\pi k_1}\,e^{ik_1 d}\cos\phi\left(\frac{z}{r_0}\right)\left(\frac{ik_2}{r_0} - \frac{1}{r_0^2}\right)e^{ik_2 r_0}, \tag{5.8.14}$$

$$E_{2\phi}(\rho,\phi,z) = \frac{\omega\mu_0}{2\pi k_1}\,e^{ik_1 d}\sin\phi\left(\frac{z}{r_0}\right)\left(\frac{ik_2}{r_0} - \frac{1}{r_0^2}\right)e^{ik_2 r_0}, \tag{5.8.15}$$

$$E_{2z}(\rho, \phi, z) = \frac{\omega\mu_0}{2\pi k_1}\, e^{ik_1 d} \cos\phi \left(\frac{\rho}{r_0}\right)\left(\frac{ik_2}{r_0} - \frac{1}{r_0^2}\right) e^{ik_2 r_0}, \tag{5.8.16}$$

$$B_{2\rho}(\rho, \phi, z) = -\frac{\mu_0 k_2}{2\pi k_1}\, e^{ik_1 d} \sin\phi \times \left[\frac{ik_2 z^2}{r_0^3} + \frac{2\rho^2 - z^2}{r_0^2}\left(\frac{1}{r_0^2} + \frac{i}{k_2 r_0^3}\right)\right] e^{ik_2 r_0}, \tag{5.8.17}$$

$$B_{2\phi}(\rho, \phi, z) = -\frac{\mu_0 k_2}{2\pi k_1}\, e^{ik_1 d} \cos\phi \left(\frac{ik_2}{r_0} - \frac{1}{r_0^2} - \frac{i}{k_2 r_0^3}\right) e^{ik_2 r_0}, \tag{5.8.18}$$

$$B_{2z}(\rho, \phi, z) = \frac{\mu_0 k_2}{2\pi k_1}\, e^{ik_1 d} \sin\phi \left(\frac{\rho z}{r_0^2}\right)\left(\frac{ik_2}{r_0} - \frac{3}{r_0^2} - \frac{3i}{k_2 r_0^3}\right) e^{ik_2 r_0}, \tag{5.8.19}$$

where $r_0 = (\rho^2 + z^2)^{1/2}$. This field is consistent with the boundary condition $E_{2\,\text{tang}} \sim 0$ for a source in Region 2. This is an approximation of the actual boundary condition,

$$\begin{aligned}\mathbf{E}_{1\,\text{tang}}(\rho, \phi, 0) &= \mathbf{E}_{2\,\text{tang}}(\rho, \phi, 0) \\ &= -(\omega/k_1)\hat{\mathbf{z}} \times \mathbf{B}_{2\,\text{tang}}(\rho, \phi, 0),\end{aligned} \tag{5.8.20}$$

where $z \leq 0$ in Region 2. Use can be made of this accurate boundary condition to extend the range of (5.8.14)–(5.8.19) to the boundary. The condition

$$|n^2| \equiv \left|\frac{k_2}{k_1}\right|^2 \ll 1 \tag{5.8.21}$$

is assumed.

The formulation begins with Maxwell's equations in cylindrical coordinates, viz.,

$$\frac{1}{\rho}\frac{\partial B_z}{\partial\phi} - \frac{\partial B_\phi}{\partial z} = -\frac{ik^2}{\omega} E_\rho, \tag{5.8.22}$$

$$\frac{\partial B_\rho}{\partial z} - \frac{\partial B_z}{\partial\rho} = -\frac{ik^2}{\omega} E_\phi, \tag{5.8.23}$$

$$\frac{1}{\rho}\left(\frac{\partial}{\partial\rho}(\rho B_\phi) - \frac{\partial B_\rho}{\partial\phi}\right) = -\frac{ik^2}{\omega} E_z, \tag{5.8.24}$$

$$\frac{1}{\rho}\frac{\partial E_z}{\partial\phi} - \frac{\partial E_\phi}{\partial z} = i\omega B_\rho, \tag{5.8.25}$$

$$\frac{\partial E_\rho}{\partial z} - \frac{\partial E_z}{\partial\rho} = i\omega B_\phi, \tag{5.8.26}$$

$$\frac{1}{\rho}\left(\frac{\partial}{\partial\rho}(\rho E_\phi) - \frac{\partial E_\rho}{\partial\phi}\right) = i\omega B_z. \tag{5.8.27}$$

These are to be applied to the field (5.8.14)–(5.8.19) after the variables ρ and z have been changed according to

$$R = \frac{k_2^3 \rho}{2k_1^2} = \frac{n^2 k_2 \rho}{2}; \qquad Z = \frac{k_2^2 z}{2k_1} = \frac{n k_2 z}{2} \tag{5.8.28}$$

or

$$k_2 \rho = \frac{2R}{n^2}; \qquad k_2 z = \frac{2Z}{n}. \tag{5.8.29}$$

Note that R is the "numerical distance" introduced by Sommerfeld. Since the correction to the formulas (5.8.14)–(5.8.19) of the field in Region 2 is to apply only near the boundary $z = 0$ where $\rho^2 \gg z^2$, it is adequate to set $r_0 = (\rho^2 + z^2)^{1/2} \sim \rho$ in amplitudes and $r_0 \sim \rho + z^2/2\rho$ in phases. To simplify the evaluation, the correction will be derived only for the leading terms; the omitted other terms can be replaced later.

When expressed in terms of the variables (5.8.28) and with

$$A \equiv \frac{i\omega\mu_0 k_2^2 n^2}{4\pi k_1}\, e^{ik_1 d}, \tag{5.8.30}$$

the leading term in each of the cylindrical components (5.8.14)–(5.8.19) is

$$E_{2\rho} \sim A \cos\phi\, e^{i2R/n^2} n\tilde{E}_{2\rho}; \qquad \tilde{E}_{2\rho} = -\frac{Z}{R^2}\, e^{iZ^2/R}, \tag{5.8.31}$$

$$E_{2\phi} \sim A \sin\phi\, e^{i2R/n^2} n\tilde{E}_{2\phi}; \qquad \tilde{E}_{2\phi} = \frac{Z}{R^2}\, e^{iZ^2/R}, \tag{5.8.32}$$

$$E_{2z} \sim A \cos\phi\, e^{i2R/n^2} \tilde{E}_{2z}; \qquad \tilde{E}_{2z} = \frac{1}{R}\, e^{iZ^2/R}, \tag{5.8.33}$$

$$B_{2\rho} \sim A\, \frac{k_2}{\omega} \sin\phi\, e^{i2R/n^2} n^2 \tilde{B}_{2\rho}; \qquad \tilde{B}_{2\rho} = \left(\frac{1}{R^2} - \frac{Z^2}{R^3}\right) e^{iZ^2/R}, \tag{5.8.34}$$

$$B_{2\phi} \sim A\, \frac{k_2}{\omega} \cos\phi\, e^{i2R/n^2} \tilde{B}_{2\phi}; \qquad \tilde{B}_{2\phi} = -\frac{1}{R}\, e^{iZ^2/R}, \tag{5.8.35}$$

$$B_{2z} \sim A\, \frac{k_2}{\omega} \sin\phi\, e^{i2R/n^2} n\tilde{B}_{2z}; \qquad \tilde{B}_{2z} = \frac{Z}{R^2}\, e^{iZ^2/R}. \tag{5.8.36}$$

The next step is to insert (5.8.31)–(5.8.36) into the Maxwell equations (5.8.22)–(5.8.27) and arrange the terms in the order of powers of the small quantity $n = k_2/k_1$. The following first-order relations are obtained. [Note that $\partial E/\partial\rho = ik_2 E + (n^2 k_2/2)(\partial E/\partial R)$.]

$$\frac{\partial \tilde{B}_{2\phi}}{\partial Z} = 2i\tilde{E}_{2\rho}; \qquad \tilde{B}_{2z} = \tilde{E}_{2\phi}, \tag{5.8.37}$$

$$\frac{\partial \tilde{E}_{2\phi}}{\partial Z} + \frac{\tilde{E}_{2z}}{R} = -2i\tilde{B}_{2\rho}; \qquad \tilde{E}_{2z} = -\tilde{B}_{2\phi}. \tag{5.8.38}$$

With appropriate combination, the following second-order equations are obtained:

$$\left(\frac{1}{R}\frac{\partial}{\partial R} R + \frac{\partial}{\partial R}\right)\tilde{B}_{2\phi} = -\frac{\partial \tilde{E}_{2\rho}}{\partial Z}, \tag{5.8.39}$$

$$\frac{1}{R}\frac{\partial}{\partial R}(R\tilde{E}_{2\phi}) + \frac{\partial \tilde{B}_{2z}}{\partial R} - \frac{\partial \tilde{B}_{2\rho}}{\partial Z} = -\frac{\tilde{E}_{2\rho}}{R}. \tag{5.8.40}$$

When (5.8.37) and (5.8.38) are combined with (5.8.39) and (5.8.40), two equations remain to be solved. The first is

$$\left(\frac{1}{R} + 2\frac{\partial}{\partial R}\right)\tilde{B}_{2\phi} - \frac{i}{2}\frac{\partial^2 \tilde{B}_{2\phi}}{\partial Z^2} = 0, \tag{5.8.41}$$

subject to the boundary condition

$$\tilde{E}_{2\rho} = \tilde{B}_{2\phi} \quad \text{at } Z = 0, \tag{5.8.42}$$

and, from (5.8.31) and (5.8.35),

$$\left.\begin{aligned} \tilde{E}_{2\rho} &\to -\frac{Z}{R^2}\, e^{iZ^2/R} \\ \tilde{B}_{2\phi} &\to -\frac{1}{R}\, e^{iZ^2/R} \end{aligned}\right\} \quad \text{far from the boundary at } Z = 0. \tag{5.8.43}$$

The second equation is

$$\left(\frac{1}{R} + 2\frac{\partial}{\partial R}\right)\tilde{E}_{2\phi} - \frac{\partial \tilde{B}_{2\rho}}{\partial Z} = -\frac{\tilde{E}_{2\rho}}{R}, \tag{5.8.44}$$

subject to the boundary condition

$$\tilde{E}_{2\phi} \sim 0 \quad \text{at } Z = 0, \tag{5.8.45}$$

and, from (5.8.32),

$$\tilde{E}_{2\phi} \to \frac{Z}{R^2}\, e^{iZ^2/R} \quad \text{far from the boundary.} \tag{5.8.46}$$

The solution of (5.8.41) is carried out in Chapter 3 with the following results:

$$-\tilde{E}_{2z} = \tilde{B}_{2\phi} = -\frac{e^{iZ^2/R}}{R}(1 + \mathcal{G}), \tag{5.8.47}$$

$$\tilde{E}_{2\rho} = -\frac{e^{iZ^2/R}}{R}\left(\frac{Z}{R} + 1 + \mathcal{G}\right), \tag{5.8.48}$$

where

$$\mathcal{G} \equiv i(2\pi R)^{1/2} e^{-i(R-Z)^2/R}\, \mathcal{F}[(R-Z)^2/R]; \qquad Z \le 0, \tag{5.8.49}$$

and

$$\mathcal{F}(u) = \tfrac{1}{2}(1+i) - C_2(u) - iS_2(u) = \int_u^\infty \frac{e^{it}}{(2\pi t)^{1/2}}\, dt, \tag{5.8.50}$$

$$C_2(u) + iS_2(u) = \int_0^u \frac{e^{it}}{(2\pi t)^{1/2}}\, dt. \tag{5.8.51}$$

With (5.8.31), (5.8.33), and (5.8.35), the leading terms are

$$E_{2\rho} \sim -A\cos\phi\, e^{i2R/n^2} e^{iZ^2/R}\, \frac{n}{R}\left(\frac{Z}{R} + 1 + \mathcal{G}\right), \tag{5.8.52}$$

$$E_{2z} \sim A\cos\phi\, e^{i2R/n^2}\, \frac{e^{iZ^2/R}}{R}\,(1+\mathcal{G}), \tag{5.8.53}$$

$$B_{2\phi} \sim -A\,\frac{k_2}{\omega}\cos\phi\, e^{i2R/n^2}\, \frac{e^{iZ^2/R}}{R}\,(1+\mathcal{G}). \tag{5.8.54}$$

The second equation that must be solved is (5.8.44), viz.,

$$\left(\frac{1}{R} + 2\,\frac{\partial}{\partial R}\right)\tilde{E}_{2\phi} - \frac{\partial \tilde{B}_{2\rho}}{\partial Z} = -\frac{\tilde{E}_{2\rho}}{R}. \tag{5.8.55}$$

When the two equations in (5.8.38) are combined, the result is

$$\tilde{B}_{2\rho} = \frac{i}{2}\left(\frac{\partial \tilde{E}_{2\phi}}{\partial Z} - \frac{\tilde{B}_{2\phi}}{R}\right), \tag{5.8.56}$$

so that, with (5.8.37),

$$\frac{\partial \tilde{B}_{2\rho}}{\partial Z} = \frac{i}{2}\left(\frac{\partial^2 \tilde{E}_{2\phi}}{\partial Z^2} - \frac{1}{R}\,\frac{\partial \tilde{B}_{2\phi}}{\partial Z}\right) = \frac{i}{2}\,\frac{\partial^2 \tilde{E}_{2\phi}}{\partial Z^2} + \frac{1}{R}\,\tilde{E}_{2\rho}. \tag{5.8.57}$$

When this is substituted in (5.8.55), this reduces to

$$\left(\frac{1}{R} + 2\,\frac{\partial}{\partial R} - \frac{i}{2}\,\frac{\partial^2}{\partial Z^2}\right)\tilde{E}_{2\phi} = 0, \tag{5.8.58}$$

which has the solution

$$\tilde{E}_{2\phi} = \frac{Z}{R^2}\, e^{iZ^2/R}. \tag{5.8.59}$$

This satisfies the conditions (5.8.45) and (5.8.46). When it is substituted in (5.8.32), it follows that

$$E_{2\phi} \sim A\sin\phi\,\frac{nZ}{R^2}\, e^{i2R/n^2} e^{iZ^2/R}. \tag{5.8.60}$$

With (5.8.37) and (5.8.36),

$$B_{2z} \sim A\,\frac{k_2}{\omega}\sin\phi\,\frac{nZ}{R^2}\, e^{i2R/n^2} e^{iZ^2/R}. \tag{5.8.61}$$

The substitution of (5.8.59) and (5.8.47) in (5.8.56) gives

$$\tilde{B}_{2\rho} = \left(\frac{i}{R^2}\,(1 + \tfrac{1}{2}\mathcal{G}) - \frac{Z^2}{R^3}\right) e^{iZ^2/R}, \tag{5.8.62}$$

so that (5.8.34) leads to

$$B_{2\rho} \sim A\,\frac{k_2}{\omega}\sin\phi\, e^{i2R/n^2} e^{iZ^2/R} n^2 \left(\frac{i}{R^2}\,(1 + \tfrac{1}{2}\mathcal{G}) - \frac{Z^2}{R^3}\right). \tag{5.8.63}$$

The leading terms of $E_{2\rho}$, E_{2z}, and $B_{2\phi}$ are corrected for use near the boundary in (5.8.52)–(5.8.54). The same is true of $E_{2\phi}$, B_{2z}, and $B_{2\rho}$ in (5.8.60), (5.8.61), and (5.8.63). When the variables are changed back to ρ and z with (5.8.28), and the higher-order terms (which were omitted in the long derivation of the correction term) and $r_0 = (\rho^2 + z^2)^{1/2}$ are reintroduced, the formulas for the electromagnetic field in Region 2 are

$$E_{2\rho}(\rho, \phi, z) \sim -\frac{\omega\mu_0}{2\pi k_1} e^{ik_1 d} \cos\phi \times \left[\left(\frac{ik_2}{r_0} - \frac{1}{r_0^2}\right)\left(\frac{z}{r_0} + \frac{k_2}{k_1}\right) + \frac{ik_2^2}{k_1 r_0} \mathcal{G}\right] e^{ik_2 r_0}, \tag{5.8.64}$$

$$E_{2\phi}(\rho, \phi, z) \sim \frac{\omega\mu_0}{2\pi k_1} e^{ik_1 d} \sin\phi \left(\frac{z}{r_0}\right)\left(\frac{ik_2}{r_0} - \frac{1}{r_0^2}\right) e^{ik_2 r_0}, \tag{5.8.65}$$

$$E_{2z}(\rho, \phi, z) \sim \frac{\omega\mu_0}{2\pi k_1} e^{ik_1 d} \cos\phi \left[\left(\frac{\rho}{r_0}\right)\left(\frac{ik_2}{r_0} - \frac{1}{r_0^2}\right) + \frac{ik_2}{r_0} \mathcal{G}\right] e^{ik_2 r_0}, \tag{5.8.66}$$

$$B_{2\rho}(\rho, \phi, z) \sim -\frac{\mu_0 k_2}{2\pi k_1} e^{ik_1 d} \sin\phi \times \left[\frac{ik_2 z^2}{r_0^3} + \frac{2\rho^2 - z^2}{r_0^2}\left(\frac{1}{r_0^2} + \frac{i}{k_2 r_0^3}\right) + \frac{1}{r_0^2} \mathcal{G}\right] e^{ik_2 r_0}, \tag{5.8.67}$$

$$B_{2\phi}(\rho, \phi, z) \sim -\frac{\mu_0 k_2}{2\pi k_1} e^{ik_1 d} \cos\phi \left(\frac{ik_2}{r_0} - \frac{1}{r_0^2} - \frac{i}{k_2 r_0^3} + \frac{ik_2}{r_0} \mathcal{G}\right) e^{ik_2 r_0}, \tag{5.8.68}$$

$$B_{2z}(\rho, \phi, z) \sim \frac{\mu_0 k_2}{2\pi k_1} e^{ik_1 d} \sin\phi \left(\frac{\rho z}{r_0^2}\right)\left(\frac{ik_2}{r_0} - \frac{3}{r_0^2} - \frac{3i}{k_2 r_0^3}\right) e^{ik_2 r_0}. \tag{5.8.69}$$

A comparison of these formulas with those obtained directly from the integrals shows that they are good approximations everywhere in Region 2 except near the source, i.e., in the range $|k_1 r_0| \gg 1$. $\mathcal{G}$ in (5.8.64)–(5.8.68) is defined in (5.8.49) in terms of the variables $R = k_2^3 \rho / 2k_1^2$ and $Z = k_2^2 z / 2k_1 \leq 0$. Their derivation in terms of a magnetic dipole with the moment $(2\omega\mu_0/k_1)e^{ik_1 d}$ explains the appearance of this factor in the approximate evaluation of the integrals carried out in Section 5.7.

5.9 The Locus of the Poynting Vector and the Depth of Penetration of the Lateral Wave in Region 2

The general expressions (5.5.47)–(5.5.52) for the complete electromagnetic field in Region 2 include the lateral wave that characterizes the field along the boundary. In order to obtain insight into the path taken by the lateral wave in Region 2 and its depth of penetration, it is useful to study the locus of the Poynting vector. Because their properties are different and not very closely coupled, it is convenient to separate the components of the electromagnetic field in Region 2 into those of electric type and those of magnetic type. The two groups are $E_{2\rho}$, E_{2z}, and $B_{2\phi}$ of electric type and $B_{2\rho}$, B_{2z}, and $E_{2\phi}$ of magnetic type. The associated parts of the Poynting vector are

$$\mathbf{S}_{2E} = \frac{1}{2\mu_0}(\mathbf{E}_{2E} \times \mathbf{B}^*_{2E}) = \frac{1}{2\mu_0}(\hat{\mathbf{z}}E_{2\rho}B^*_{2\phi} - \hat{\boldsymbol{\rho}}E_{2z}B^*_{2\phi}), \tag{5.9.1}$$

$$\mathbf{S}_{2M} = \frac{1}{2\mu_0}(\mathbf{E}_{2M} \times \mathbf{B}^*_{2M}) = \frac{1}{2\mu_0}(-\hat{\mathbf{z}}E_{2\phi}B^*_{2\rho} + \hat{\boldsymbol{\rho}}E_{2\phi}B^*_{2z}). \tag{5.9.2}$$

These are coupled by a small mutual term, viz.,

$$\mathbf{S}_{2EM} = \frac{\hat{\boldsymbol{\phi}}}{2\mu_0}(E_{2z}B^*_{2\rho} - E_{2\rho}B^*_{2z}), \tag{5.9.3}$$

but this involves only a circulation and does not contribute to the outward flow of energy.

The locus of the Poynting vector in Region 2 is defined by

$$\frac{dz}{d\rho} = -\frac{dz'}{d\rho} = \frac{\operatorname{Re} S_{2z}}{\operatorname{Re} S_{2\rho}} = -\frac{\operatorname{Re} E_{2\rho}B^*_{2\phi}}{\operatorname{Re} E_{2z}B^*_{2\phi}} \tag{5.9.4}$$

for the electric type. For the field of magnetic type it is defined by

$$\frac{dz}{d\rho} = -\frac{dz'}{d\rho} = \frac{\operatorname{Re} S_{2z}}{\operatorname{Re} S_{2\rho}} = -\frac{\operatorname{Re} E_{2\phi}B^*_{2\rho}}{\operatorname{Re} E_{2\phi}B^*_{2z}}. \tag{5.9.5}$$

Since z is negative in Region 2, it is convenient to introduce $z' = -z$. In the range $1 \leq |k_2\rho| \leq 4|k_1/k_2|^2$, which excludes the far field, the contribution from the Fresnel-integral terms is small and the entire intermediate range is included. This is of primary interest in important applications.

5.9.1 The Locus of the Poynting Vector for the Field of Electric Type

The components S_{2z} and $S_{2\rho}$ are readily obtained with (5.5.47), (5.5.49), and (5.5.51). Let the formulas for $E_{2\rho}(\rho,\phi,z)$ and $E_{2z}(\rho,\phi,z)$ be multiplied by k_2 outside the brackets and divided by k_2 inside the brackets. The products $E_{2\rho}B^*_{2\phi}$ and $E_{2z}B^*_{2\phi}$ then have the common factor $(\omega\mu_0^2 k_2 k_2^*/4\pi^2 k_1 k_1^*)\cos^2\phi$, which is real and can be canceled in the quotient $\operatorname{Re} S_{2z}/\operatorname{Re} S_{2\rho}$. Since the far-field terms and Fresnel-integral terms are negligible in the near and intermediate ranges, they can be omitted in determining the locus of the Poynting vector in that range. The resulting formula is

$$\frac{dz}{d\rho} = -\frac{\operatorname{Re} E_{2\rho} B_{2\phi}^*}{\operatorname{Re} E_{2z} B_{2\phi}^*} = \frac{\operatorname{Re}\left[\left(\frac{z}{r_0} + \frac{k_2}{k_1}\right)\left(\frac{1}{k_2 r_0^2}\right)\left(\frac{1}{r_0^2} - \frac{i}{k_2^* r_0^3}\right)\right]}{\operatorname{Re}\left[\left(\frac{\rho}{r_0}\right)\left(\frac{1}{k_2 r_0^2}\right)\left(\frac{1}{r_0^2} - \frac{i}{k_2^* r_0^3}\right)\right]}$$

$$= \frac{\operatorname{Re}\left(\frac{z}{r_0} + \frac{k_2}{k_1}\right)(k_2^* r_0 - i)}{\operatorname{Re}\left(\frac{\rho}{r_0}\right)(k_2^* r_0 - i)}, \tag{5.9.6}$$

where $r_0 = (\rho^2 + z^2)^{1/2}$. In general,

$$\frac{k_2}{k_1} = \frac{\beta_2 + i\alpha_2}{\beta_1 + i\alpha_1} = \frac{\beta_1\beta_2 + \alpha_1\alpha_2}{\beta_1^2 + \alpha_1^2} - i\,\frac{\alpha_1\beta_2 - \alpha_2\beta_1}{\beta_1^2 + \alpha_1^2} \equiv \delta_0 - ic_0. \tag{5.9.7}$$

It follows that

$$\frac{dz}{d\rho} = \frac{\operatorname{Re}\left(\frac{z}{r_0} + \delta_0 - ic_0\right)[(\beta_2 - i\alpha_2)r_0 - i]}{\operatorname{Re}\left(\frac{\rho}{r_0}\right)[(\beta_2 - i\alpha_2)r_0 - i]}$$

$$= \frac{\operatorname{Re}\left(\frac{z}{r_0} + \delta_0 - ic_0\right)[\beta_2 r_0 - i(1 + \alpha_2 r_0)]}{\beta_2\rho}$$

$$= \operatorname{Re}\left(\frac{z}{r_0} + \delta_0 - ic_0\right)\left[\frac{r_0}{\rho} - i\left(\frac{1}{\beta_2\rho} + \frac{\alpha_2 r_0}{\beta_2\rho}\right)\right]$$

$$\sim \frac{z}{\rho} + \delta_0 - \frac{c_0}{\beta_2\rho} - c_0\,\frac{\alpha_2}{\beta_2}. \tag{5.9.8}$$

Since both δ_0 and c_0 are small, it is possible to let $r_0 = (\rho^2 + z^2)^{1/2} \sim \rho$ since z^2/ρ^2 is small on all parts of the locus except near the origin $z = 0$ where none of the formulas is accurate. With this approximation,

$$\frac{dz'}{d\rho} = \frac{z' + c}{\rho} - \delta, \tag{5.9.9}$$

where

$$c = \frac{c_0}{\beta_2} = \frac{\alpha_1\beta_2 - \alpha_2\beta_1}{\beta_2(\alpha_1^2 + \beta_1^2)}, \tag{5.9.10a}$$

$$\delta = \delta_0 - c_0\,\frac{\alpha_2}{\beta_2} = \frac{\beta_1(\beta_2^2 + \alpha_2^2)}{\beta_2(\beta_1^2 + \alpha_1^2)}, \tag{5.9.10b}$$

and where $z' = -z$ is positive in Region 2.

Two special cases are of interest: (a) When Region 2 is air with $\alpha_2 = 0$, $\beta_2 = \omega/c$, and Region 1 is sea water with $\alpha_1 = \beta_1 = (\omega\mu_0\sigma_1/2)^{1/2}$ so that

$$\delta = \frac{\beta_1}{\beta_2} \cdot \frac{\beta_2^2}{2\beta_1^2} = \frac{\beta_2}{2\beta_1} = \left(\frac{\omega\epsilon_0}{2\sigma_1}\right)^{1/2}; \qquad c = \frac{\alpha_1}{\alpha_1^2 + \beta_1^2} = \frac{1}{2\beta_1}. \tag{5.9.11}$$

(b) When Region 2 is the oceanic crust, Region 1 the sea, and the frequency is sufficiently low so that $\alpha_1 \sim \beta_1 = (\omega\mu_0\sigma_1/2)^{1/2}$, $\alpha_2 \sim \beta_2 = (\omega\mu_0\sigma_2/2)^{1/2}$, and

$$\delta = \frac{\beta_1}{\beta_2} \cdot \frac{2\beta_2^2}{2\beta_1^2} = \frac{\beta_2}{\beta_1} = \left(\frac{\sigma_2}{\sigma_1}\right)^{1/2}; \qquad c = 0. \tag{5.9.12}$$

In this case, δ is independent of the frequency.

Equation (5.9.9) is readily solved with the substitutions $(z'+c) = s\rho$ and $(dz'/d\rho) = s + \rho(ds/d\rho)$, which lead to the following differential equation and its solution:

$$ds = -\delta\,\frac{d\rho}{\rho}; \qquad s = -\delta\,\ln\left(\frac{\rho}{\rho_0}\right), \tag{5.9.13}$$

where ρ_0 is the constant of integration. With $s = (z'+c)/\rho$, the equation for the locus is

$$z' = -c - \delta\rho\,\ln\left(\frac{\rho}{\rho_0}\right). \tag{5.9.14}$$

Here c is a small distance which ranges from zero for the locus in the oceanic crust below the sea floor to $1/2\beta_1$ for the locus in air above the sea. At $\rho = 0$, $z' = -c$; $z' = 0$ at ρ_1 given by

$$\rho_1\,\ln\left(\frac{\rho_1}{\rho_0}\right) \sim \rho_1\left(\frac{\rho_1}{\rho_0} - 1\right) = -\frac{c}{\delta}$$

or

$$\rho_1^2 - \rho_1\rho_0 + \frac{\rho_0 c}{\delta} = 0 \tag{5.9.15a}$$

so that

$$\rho_1 \sim \frac{1}{2}\left[\rho_0 \pm \left(\rho_0^2 - \frac{4\rho_0 c}{\delta}\right)^{1/2}\right] \sim \rho_0 - \frac{c}{\delta}. \tag{5.9.15b}$$

The approximation in (5.9.15a) implies that $(\rho_1/\rho_0)^2 < 1$. With $c = 1/2\beta_1$ and $\delta = \beta_2/2\beta_1$,

$$\rho_1 = \rho_0 - \frac{1}{\beta_2} \sim \rho_0. \tag{5.9.16}$$

Thus, when $\rho_0 \gg 1/\beta_2$, the locus defined by

$$z' = -\delta\rho\,\ln\left(\frac{\rho}{\rho_0}\right) \tag{5.9.17}$$

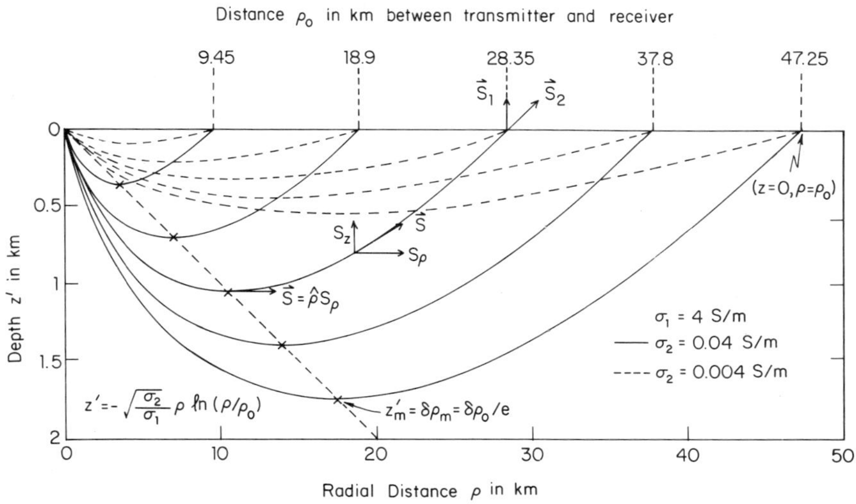

Figure 5.9.1. Locus of Poynting vector for field of electric type in half-space model of lithosphere; $k_1 \sim (i\omega\mu_0\sigma_1)^{1/2}$, $k_2 \sim (i\omega\mu_0\sigma_2)^{1/2}$; $f = 1$ Hz. At $z = 0$, $S_{1\rho} = (\sigma_2/\sigma_1)S_{2\rho}$ and $S_{1z} = S_{2z}$.

is approximate near the dipole at $z' = 0$ and actually returns to the surface $z' = 0$ at a radius ρ_1 very slightly less than ρ_0. For all practical purposes, (5.9.17) is adequate.

The depth of penetration of the lateral wave in Region 2 depends on the radial distance ρ_0 where the field is observed. The maximum depth is at (ρ_m, z'_m), where

$$\rho_m = \frac{\rho_0}{e}; \qquad z'_m = \frac{\delta\rho_0}{e}, \tag{5.9.18}$$

and $e = 2.718$ is the base of the natural logarithm.

On the conical surface defined by (5.9.18), the Poynting vector is purely radial, i.e.,

$$\mathbf{S}_2 = \hat{\boldsymbol{\rho}}S_{2\rho}; \qquad S_{2z} = 0. \tag{5.9.19}$$

The apex of the half-angle Θ'_m of the cone is defined by

$$\Theta'_m = \frac{\pi}{2} - \tan^{-1}\delta \sim \frac{\pi}{2} - \delta, \tag{5.9.20}$$

where the expression on the right follows since δ is small. Graphs of the locus defined by (5.9.17) are shown in Fig. 5.9.1 specifically for a geophysical application that is described in Chapter 8. If note is taken of the factor 10 in the vertical and horizontal scales in Fig. 5.9.1, the depth of penetration of the lateral wave into Region 2 (in this case the sea floor) is seen to be a very small fraction of the radial range.

5.9.2 The Locus of the Poynting Vector for the Field of Magnetic Type

Since in all three components of the field of magnetic type the $1/r_0$ (far-field) terms vanish along the boundary $z = 0$, they can contribute only to the spherical wave and not to the lateral wave. Accordingly, they may be omitted in the determination of the locus of the Poynting vector of the lateral wave. Terms multiplied by $\exp(k_1 z) = \exp(-k_1 z')$, where $z' = -z$ is positive in Region 2, exist only very close to the boundary in terms of the wavelength $\lambda_2 = 2\pi/\beta_2$ since $\lambda_1 = 2\pi/\beta_1$ and $k_1 = \beta_1 + i\alpha_1$ is large compared to $k_2 = \beta_2 + i\alpha_2$. These terms have no factor $\exp(ik_2 r_0)$ so that they are not part of a wave propagating in Region 2 and hence play no part in the lateral wave. The Fresnel-integral terms contribute significantly only to the far field and can also be omitted. The remaining terms in (5.5.48), (5.5.50b), and (5.5.52) give

$$E_{2\phi} = \frac{\omega\mu_0 k_2}{2\pi k_1} \sin\phi \; e^{ik_1 d} e^{ik_2 r_0} \left(\frac{2}{k_1} + iz\right)\left(\frac{1}{r_0^2} + \frac{i}{k_2 r_0^3}\right), \tag{5.9.21}$$

$$\begin{aligned} B_{2\rho} = {} & -\frac{\mu_0 k_2}{2\pi k_1} \sin\phi \; e^{ik_1 d} e^{ik_2 r_0} \\ & \times \left[\frac{ik_2}{r_0}\left(\frac{z^2}{r_0^2}\right) + \frac{2\rho^2 - z^2}{r_0^2}\left(\frac{1}{r_0^2} + \frac{i}{k_2 r_0^3}\right) \right. \\ & \left. + \frac{k_2 z}{k_1 \rho}\left(\frac{ik_2}{r_0} - \frac{3}{r_0^2} - \frac{6i}{k_2 r_0^3}\right)\right], \end{aligned} \tag{5.9.22}$$

$$\begin{aligned} B_{2z} = {} & \frac{\mu_0 k_2}{2\pi k_1} \sin\phi \; e^{ik_1 d} e^{ik_2 r_0} \\ & \times \left[\left(\frac{\rho}{r_0}\right)\left(\frac{ik_2}{r_0} - \frac{3}{r_0^2} - \frac{3i}{k_2 r_0^3}\right)\left(\frac{z}{r_0} - \frac{i}{k_1 r_0}\right)\right]. \end{aligned} \tag{5.9.23}$$

As explained above, the far-field terms can also be omitted. They include two terms in $B_{2\rho}$ and one term in B_{2z}. The remaining terms are

$$\begin{aligned} B_{2\rho} = {} & -\frac{\mu_0 k_2}{2\pi k_1} \sin\phi \; e^{ik_1 d} e^{ik_2 r_0} \\ & \times \left[\frac{2\rho^2 - z^2}{r_0^2}\left(\frac{1}{r_0^2} + \frac{i}{k_2 r_0^3}\right) - \frac{3k_2 z}{k_1 r_0}\left(\frac{1}{r_0^2} + \frac{2i}{k_2 r_0^3}\right)\right], \end{aligned} \tag{5.9.24}$$

$$B_{2z} = -\frac{\mu_0 k_2}{2\pi k_1} \sin\phi \; e^{ik_1 d} e^{ik_2 r_0} \left[\left(\frac{3\rho}{r_0^2}\right)\left(\frac{1}{r_0^2} + \frac{i}{k_2 r_0^3}\right)\left(z - \frac{i}{k_1}\right)\right]. \tag{5.9.25}$$

When these formulas are substituted in (5.9.5), the following equation is obtained. The real common factor $(\omega\mu_0^2 k_2 k_2^* / 4\pi^2 k_1 k_1^*) \sin^2\phi$ has been canceled.

$$\frac{dz}{d\rho} = -\frac{\operatorname{Re} E_{2\phi} B_{2\rho}^*}{\operatorname{Re} E_{2\phi} B_{2z}^*}$$

$$= -\frac{\mathrm{Re}\left(\dfrac{2}{k_1} + iz\right)\left(\dfrac{1}{r_0^2} + \dfrac{i}{k_2 r_0^3}\right)\left(\dfrac{2\rho^2 - z^2}{r_0^2} - \dfrac{3k_2^* z}{k_1^* r_0}\right)\left(\dfrac{1}{r_0^2} - \dfrac{i}{k_2^* r_0^3}\right)}{\mathrm{Re}\left(\dfrac{2}{k_1} + iz\right)\left(\dfrac{1}{r_0^2} + \dfrac{i}{k_2 r_0^3}\right)\left(\dfrac{3\rho}{r_0^2}\right)\left(\dfrac{1}{r_0^2} - \dfrac{i}{k_2^* r_0^3}\right)\left(z + \dfrac{i}{k_1^*}\right)}. \tag{5.9.26}$$

The term $(2i/k_2 r_0^3)$ has been approximated by $(i/k_2 r_0^3)$ in the last and very small term in $B_{2\rho}$. The factor

$$\left(\frac{1}{r_0^2} + \frac{i}{k_2 r_0^3}\right)\left(\frac{1}{r_0^2} - \frac{i}{k_2^* r_0^3}\right)$$

has a negligibly small imaginary part and can be canceled to leave

$$-\frac{dz}{d\rho} = \frac{\mathrm{Re}\left(\dfrac{2}{k_1} + iz\right)\left(2\rho^2 - z^2 - 3\,\dfrac{k_2^* z}{k_1^* \rho}\,r_0^2\right)}{\mathrm{Re}\,3\rho\left(\dfrac{2}{k_1} + iz\right)\left(z + \dfrac{i}{k_1^*}\right)}$$

$$= \frac{\mathrm{Re}\left(\dfrac{2(\beta_1 - i\alpha_1)}{k_1 k_1^*} + iz\right)\left(2\rho^2 - z^2 - 3zr_0\,\dfrac{(\beta_2 - i\alpha_2)(\beta_1 + i\alpha_1)}{k_1 k_1^*}\right)}{\mathrm{Re}\,3\rho\left(\dfrac{2z(\beta_1 - i\alpha_1)}{k_1 k_1^*} + \dfrac{2i}{k_1 k_1^*} + iz^2 - \dfrac{z(\beta_1 + i\alpha_1)}{k_1 k_1^*}\right)}, \tag{5.9.27}$$

so that

$$-\frac{dz}{d\rho} = \frac{\mathrm{Re}\left[\dfrac{2\beta_1}{k_1 k_1^*} + i\left(z - \dfrac{2\alpha_1}{k_1 k_1^*}\right)\right][2\rho^2 - z^2 - 3zr_0(\delta_r + i\delta_i)]}{3\rho\,\dfrac{\beta_1 z}{k_1 k_1^*}}$$

$$= \frac{\mathrm{Re}\left[2 + i\left(\dfrac{k_1 k_1^* z}{\beta_1} - \dfrac{2\alpha_1}{\beta_1}\right)\right][2\rho^2 - z^2 - 3zr_0(\delta_r + i\delta_i)]}{3\rho z}, \tag{5.9.28}$$

where

$$\delta_r = \frac{\beta_1\beta_2 + \alpha_1\alpha_2}{\beta_1^2 + \alpha_1^2}; \qquad \delta_i = \frac{\alpha_1\beta_2 - \alpha_2\beta_1}{\beta_1^2 + \alpha_1^2}. \tag{5.9.29}$$

It now follows that

$$-\frac{dz}{d\rho} = \frac{2(2\rho^2 - z^2 - 3zr_0\delta_r) + 3zr_0\delta_i\left(\dfrac{k_1 k_1^* z}{\beta_1} - \dfrac{2\alpha_1}{\beta_1}\right)}{3\rho z}$$

$$= \frac{2(2\rho^2 - z^2 - 3zr_0\delta) + 3z^2 r_0 c}{3\rho z}, \tag{5.9.30}$$

where the following symbolism has been introduced:

$$\delta = \delta_r + \frac{\alpha_1 \delta_i}{\beta_1} = \frac{\beta_2}{\beta_1}. \tag{5.9.31a}$$

Also,

$$c = \frac{k_1 k_1^*}{\beta_1}\,\delta_i = \frac{\alpha_1\beta_2 - \beta_1\alpha_2}{\beta_1} = \frac{\beta_2\alpha_1}{\beta_1} - \alpha_2. \tag{5.9.31b}$$

In applications on the sea floor or the air–earth or air–sea boundary at frequencies that are not too high, $\alpha_1 \sim \beta_1$, so that $c = \beta_2 - \alpha_2$. When Region 2 is air, $\alpha_2 = 0$ and $c = \beta_2$; when Region 2 is the oceanic crust at very low frequencies, $\alpha_2 \sim \beta_2$ and $c = 0$. Actually the surface wave of magnetic type plays no significant role in communication over the earth or sea, and a determination of the locus of the part of the Poynting vector associated with it is of little interest. It is of interest in the geophysical exploration of the oceanic crust at very low frequencies. Therefore, the solution of (5.9.30) is restricted to this simpler problem in which $c = 0$ and $\delta = \beta_2/\beta_1$ is small. The equation to be solved is

$$-\frac{dz}{d\rho} = 2\left[\frac{2\rho^2 - z^2}{3\rho z} - \delta\left(1 + \frac{z^2}{\rho^2}\right)^{1/2}\right] \sim 2\left(\frac{2\rho^2 - z^2}{3\rho z} - \delta\right). \tag{5.9.32}$$

Since the range of z/ρ is from zero to about 1, $(1 + z^2/\rho^2)^{1/2}$ ranges from 1 to $\sqrt{2}$. When multiplied by the small quantity $\delta = \beta_2/\beta_1 = (\sigma_2/\sigma_1)^{1/2}$, the approximation $(1 + z^2/\rho^2)^{1/2} \sim 1$ is adequate, as indicated on the right. Since $-(dz/d\rho) = dz'/d\rho$ is the slope of the locus, δ represents a small constant change in slope. The basic nature of the contour can be determined with $\delta = 0$ and the small effect of δ assessed later. Thus, with $z' = -z$, the equation to be solved is

$$-\frac{dz'}{d\rho} = \frac{2(2\rho^2 - z'^2)}{3\rho z'}. \tag{5.9.33}$$

This is readily accomplished with the substitutions $z' = s\rho$ and $dz'/d\rho = s + \rho(ds/d\rho)$. Thus,

$$-\left(s + \rho\frac{ds}{d\rho}\right) = \frac{2(2 - s^2)}{3s}$$

or

$$-\rho\frac{ds}{d\rho} = \frac{4 + s^2}{3s}$$

or

$$\frac{2s\,ds}{s^2 + 4} = -\frac{2\,d\rho}{3\rho}. \tag{5.9.34}$$

The integrals of the two sides are

$$\ln(s^2+4) = \ln\left(\frac{\rho_1}{\rho}\right)^{2/3}, \tag{5.9.35}$$

where $\ln \rho_1$ is a constant of integration. The solution for z' is

$$s = \frac{z'}{\rho} = \left[\left(\frac{\rho_1}{\rho}\right)^{2/3} - 4\right]^{1/2} \tag{5.9.36}$$

so that

$$z' = \rho^{2/3}(\rho_1^{2/3} - 4\rho^{2/3})^{1/2}. \tag{5.9.37}$$

The constant ρ_1 can now be expressed in terms of the radial distance ρ_0 where the locus reaches the surface $z' = 0$. Thus,

$$\rho_1^{2/3} = 4\rho_0^{2/3} \quad \text{or} \quad \rho_1 = 4^{3/2}\rho_0$$

so that

$$z' = 2\rho^{2/3}(\rho_0^{2/3} - \rho^{2/3})^{1/2}. \tag{5.9.38}$$

This equation defines the locus of the part of the Poynting vector associated with the surface wave of magnetic type. The maximum depth of penetration into Region 2 occurs at (ρ_m, z'_m) where $dz'/d\rho = 0$. With (5.9.32) and $\delta = 0$,

$$z'_m = \rho_m\sqrt{2}. \tag{5.9.39}$$

When this is substituted in (5.9.38) with $z' = z'_m$ and $\rho = \rho_m$, the result is

$$\rho_m = \left(\frac{2}{3}\right)^{3/2}\rho_0 = 0.544\rho_0; \qquad z'_m = \rho_m\sqrt{2} = 0.77\rho_0. \tag{5.9.40}$$

On the conical surface defined by (ρ_m, z'_m), the Poynting vector is radial. The half-angle Θ'_m of the cone is

$$\begin{aligned}\Theta'_m &= \frac{\pi}{2} - \tan^{-1}\left(\frac{z'_m}{\rho_m}\right) = \frac{\pi}{2} - \tan^{-1}\sqrt{2} \\ &= 0.615 \quad \text{or} \quad 35^\circ.3.\end{aligned} \tag{5.9.41}$$

It is independent of the properties of the two media so long as $|k_1/k_2| \geq 3$. The effect of the small terms involving $\delta = \beta_2/\beta_1$ is usually negligible so that the loci of the Poynting vector of magnetic type are actually independent of both k_1 and k_2. Graphs of (5.9.38) are shown in Fig. 5.9.2 with $\sigma_1 = 4$ S/m and $f = 1$ Hz for a specific application on the sea floor that is described in Chapter 8.

When the small term δ is left in (5.9.32),

$$2\rho_m^2 - z_m'^2 - 3\rho_m z'_m \delta = 0$$

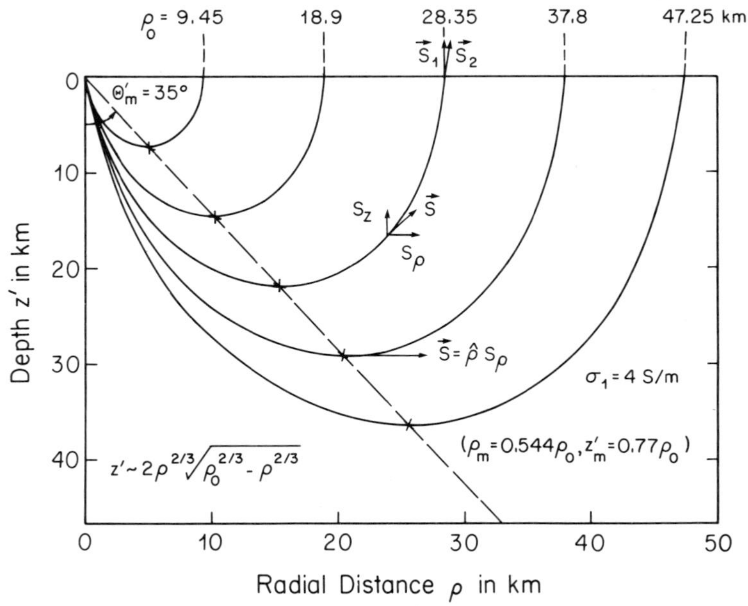

Figure 5.9.2. Locus of Poynting vector for field of magnetic type in half-space model of lithosphere; $k_1 \sim (i\omega\mu_0\sigma_1)^{1/2}$, $k_2 \sim (i\omega\mu_0\sigma_2)^{1/2}$; $f = 1$ Hz.

or

$$z'_m = \frac{-3\rho_m\delta \pm (9\rho_m^2\delta^2 + 8\rho_m^2)^{1/2}}{2} \sim (\sqrt{2} - 1.5\delta)\rho_m \tag{5.9.42}$$

and

$$\rho_m = \left[\frac{3}{2}\left(1 + \frac{\delta}{2}\right)\right]^{-3/2} \rho_0. \tag{5.9.43}$$

With $\delta = \beta_2/\beta_1 = (\sigma_2/\sigma_1)^{1/2} = [(1.6 \times 10^{-3})/4]^{1/2} = 4 \times 10^{-2}$,

$$\rho_m = \tfrac{3}{2}\,(1.02)^{-3/2}\rho_0 = 0.528\rho_0, \tag{5.9.44}$$

$$z'_m = (\sqrt{2} - 1.5 \times 4 \times 10^{-2})\rho_m = 1.35\rho_m = 0.71\rho_0. \tag{5.9.45}$$

Thus, the effect of the small term δ is to slightly distort the locus so that the maximum depth of penetration is slightly smaller and a little nearer the source.

It is significant to note the very large differences between the loci of the Poynting vectors of the lateral waves of electric and magnetic types. The field of electric type ($E_{2\rho}$, E_{2z}, $B_{2\phi}$) depends critically on the ratio k_2/k_1 and the depth of penetration of the lateral wave into Region 2 is very small compared to the radial range. The lateral-wave field is bound close to the surface. The lateral-wave field of magnetic type ($E_{2\phi}$, $B_{2\rho}$, B_{2z}) is independent of the ratio k_2/k_1 so long as $|k_1/k_2| \geq 3$. The depth of penetration is more than half the radial range. For example, on the sea floor at $f = 1$ Hz, the radial distance $\rho_0 = 28.25$ km involves a maximum depth of penetration into the oceanic crust when this has a conductivity $\sigma_2 = 0.04$ S/m of only $z'_m = 1$ km for the field of electric type and of $z'_m = 21.8$ km for the field of magnetic type.

5.9.3 The Spherical Wave

The general expressions (5.5.47)–(5.5.52) give the complete electromagnetic field in Region 2 due to an x-directed horizontal electric dipole at $(0, 0, d)$ in Region 1. Its principal part can be expressed in terms of the field of a y-directed magnetic dipole in Region 2 at $(0, 0, 0)$ on the boundary. The cylindrical coordinates (ρ, ϕ, z) with the z-axis perpendicular to the boundary are used because they are most convenient for the representation of the lateral waves that travel along the boundary surface $z = 0$. The field in Region 2 not too close to the boundary is an outward-traveling spherical wave given by (5.8.14)–(5.8.19) in cylindrical coordinates. This is easily converted to the spherical coordinates (r_0, Θ, Φ), where Θ is measured from the positive z-axis which points into Region 1. The transformation relations are

$$E_r(r_0, \Theta, \Phi) = E_\rho(\rho, \phi, z)\sin\Theta + E_z(\rho, \phi, z)\cos\Theta, \tag{5.9.46}$$

$$E_\Theta(r_0, \Theta, \Phi) = E_\rho(\rho, \phi, z)\cos\Theta - E_z(\rho, \phi, z)\sin\Theta, \tag{5.9.47}$$

$$E_\Phi(r_0, \Theta, \Phi) = E_\phi(\rho, \phi, z), \tag{5.9.48}$$

and similar ones for the magnetic field. With these formulas the spherical wave in Region 2 has three components of electric type, viz.,

$$E_{2r}(r_0, \Theta, \Phi) = 0, \tag{5.9.49}$$

$$E_{2\Theta}(r_0, \Theta, \Phi) = -\frac{\omega\mu_0}{2\pi k_1}\, e^{ik_1 d} e^{ik_2 r_0}\left(\frac{ik_2}{r_0} - \frac{1}{r_0^2}\right)\cos\Phi, \tag{5.9.50}$$

$$B_{2\Phi}(r_0, \Theta, \Phi) = -\frac{\mu_0 k_2}{2\pi k_1}\, e^{ik_1 d} e^{ik_2 r_0}\left(\frac{ik_2}{r_0} - \frac{1}{r_0^2} - \frac{i}{k_2 r_0^3}\right)\cos\Phi, \tag{5.9.51}$$

and three components of magnetic type given by

$$E_{2\Phi}(r_0, \Theta, \Phi) = \frac{\omega\mu_0}{2\pi k_1}\, e^{ik_1 d} e^{ik_2 r_0}\left(\frac{ik_2}{r_0} - \frac{1}{r_0^2}\right)\sin\Phi\cos\Theta, \tag{5.9.52}$$

$$B_{2r}(r_0, \Theta, \Phi) = -\frac{\mu_0 k_2}{2\pi k_1}\, e^{ik_1 d} e^{ik_2 r_0}\left(\frac{2}{r_0^2} + \frac{2i}{k_2 r_0^3}\right)\sin\Phi\sin\Theta, \tag{5.9.53}$$

$$B_{2\Theta}(r_0, \Theta, \Phi) = -\frac{\mu_0 k_2}{2\pi k_1}\, e^{ik_1 d} e^{ik_2 r_0}\left(\frac{ik_2}{r_0} - \frac{1}{r_0^2} - \frac{i}{k_2 r_0^3}\right)\sin\Phi\cos\Theta. \tag{5.9.54}$$

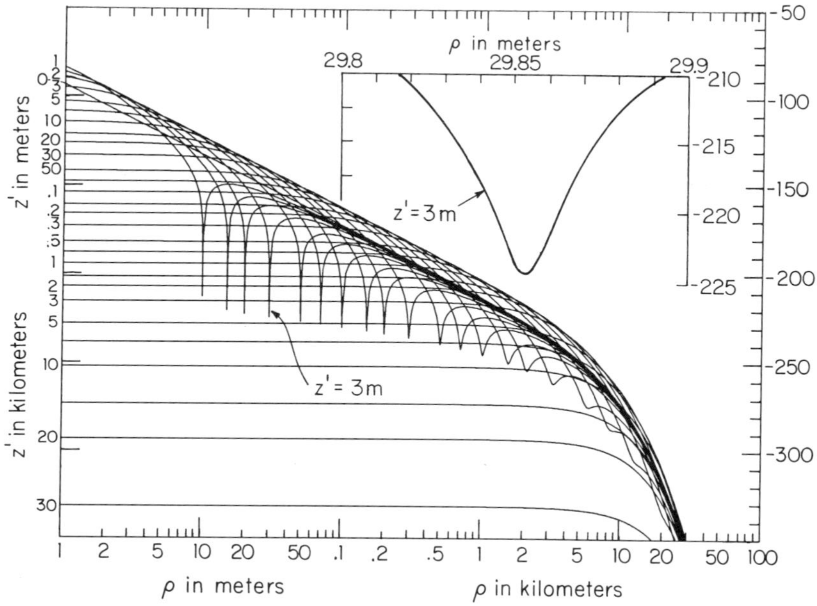

Figure 5.9.3. Radial electric field in V/m in the oceanic crust ($\sigma_2 = 0.04$ S/m, $\epsilon_{2r} = 10$) due to a unit horizontal electric dipole in the ocean ($\sigma_1 = 4$ S/m, $\epsilon_{1r} = 80$) at distance $d = 1$ m from sea floor; $f = 1$ Hz.

When $|k_2 r_0| > 1$, the $1/r_0$ terms dominate and the outward-traveling spherical wave has the form $\exp(ik_2 r_0)/r_0 = \exp(-\alpha_2 r_0)\exp(i\beta_2 r_0)/r_0$. It consists of a transverse electric-type wave ($E_{2r} = 0$, $E_{2\Theta}$, $B_{2\Phi}$) and a magnetic-type wave ($B_{2r} \sim$ small, $B_{2\Theta}$, $E_{2\Phi}$). When referred to the negative z-axis, $z' = -z$, which points down into Region 2, $\Theta' = \pi - \Theta$.

5.9.4 *The Complete Electromagnetic Field in Region 2*

Graphical representations of the complete electromagnetic field in Region 2 ($z \leq 0$) as given by (5.5.47)–(5.5.52) are shown in Figs. 5.9.3–5.9.8 specifically for the parameters that characterize Fig. 5.9.1. These correspond to the field generated by a unit horizontal electric dipole at $f = 1$ Hz located at $d = 1$ m above the sea floor when the oceanic crust is well approximated by a homogeneous, isotropic half-space of rock with $\sigma_2 = 0.04$ S/m (Figs. 5.9.3–5.9.5) and $\sigma_2 = 0.004$ S/m (Figs. 5.9.6–5.9.8). In Figs. 5.9.3 and 5.9.6, the magnitude of the radial electric field in the form $20 \log_{10} |E_{2\rho}(\rho, 0, z)|$ with $E_{2\rho}(\rho, 0, z)$ in V/m is shown against the radial distance ρ with the depth of the point of observation $z' = -z$, $z \leq 0$, as the parameter. At the boundary surface $z' = 0$, the field decreases uniformly with increasing

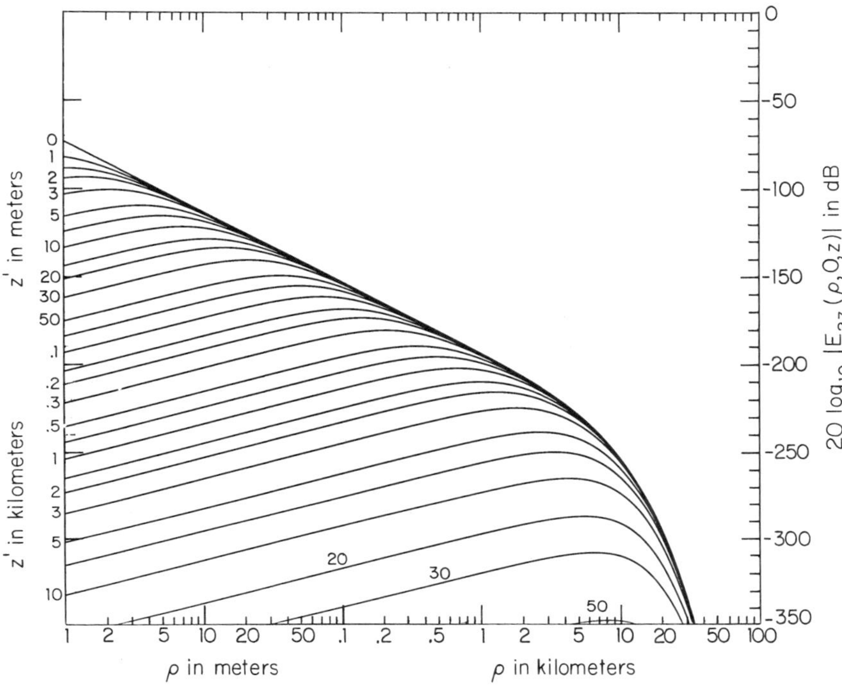

Figure 5.9.4. Vertical electric field in V/m in the oceanic crust ($\sigma_2 = 0.04$ S/m, $\epsilon_{2r} = 10$) due to a unit horizontal electric dipole in the ocean ($\sigma_1 = 4$ S/m, $\epsilon_{1r} = 80$) at distance $d = 1$ m from sea floor; $f = 1$ Hz.

ρ since $E_{2\rho}(\rho, \phi, 0) = E_{1\rho}(\rho, \phi, 0)$. When $z' > 0$, the radial electric field has a deep minimum near $z' = (k_2/k_1)\rho = \rho(\sigma_2/\sigma_1)^{1/2}$. This minimum lies on the line z'_m (shown, for example, in Fig. 5.9.1 for $\sigma_2 = 0.04$ S/m), so long as the Fresnel-integral term contributes negligibly. (This term is omitted in the calculation of the graphs in Fig. 5.9.1.) This is true for the entire range shown in Fig. 5.9.1 for $\sigma_2 = 0.004$ S/m. With $\sigma_2 = 0.04$ S/m, the Fresnel-integral term contributes increasingly as ρ increases. The effect is to cause the line $z'_m = \delta\rho_m$ in Fig. 5.9.1 to bend upward gradually so that it ends at $z' = 2$ km, $\rho = 28$ km instead of $z' = 2$ km, $\rho = 20$ km. The locus of the minima shown in Figs. 5.9.3 and 5.9.6 corresponds to the locus defined by the equation $|E^L_{2\rho}| = |E^S_{2\rho}|$, that is, the contour along which the magnitude of the lateral wave is equal to the magnitude of the spherical wave.

The vertical components $E_{2z}(\rho, 0, z)$ that correspond to the radial components in Figs. 5.9.3 and 5.9.6 are represented in Figs. 5.9.4 and 5.9.7. These graphs show that $|E_{2z}(\rho, 0, z)|$ grows in magnitude with increasing radial distance until it reaches a maximum after which it decreases rapidly. The radial distance to this maximum is larger for greater depths z'. This increase is due to the factor ρ in the formula (5.5.49). It reaches a maximum

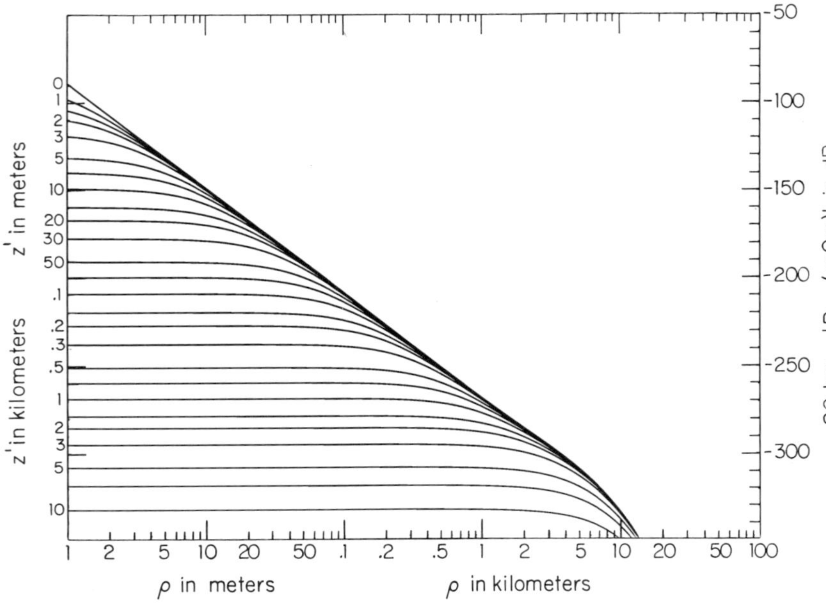

Figure 5.9.5. Magnetic field in V sec/m^2 in the oceanic crust ($\sigma_2 = 0.04$ S/m, $\epsilon_{2r} = 10$) due to a unit horizontal electric dipole in the ocean ($\sigma_1 = 4$ S/m, $\epsilon_{1r} = 80$) at distance $d = 1$ m from sea floor; $f = 1$ Hz.

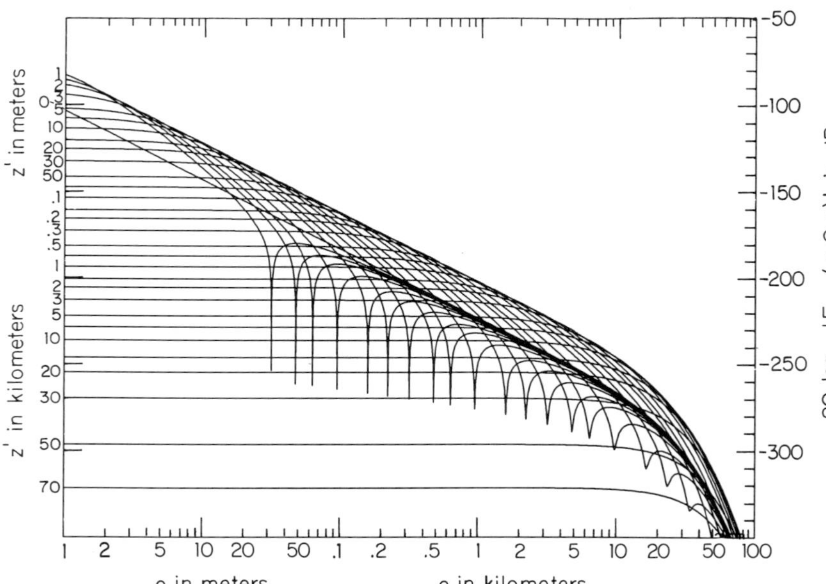

Figure 5.9.6. Radial electric field in V/m in the oceanic crust ($\sigma_2 = 0.004$ S/m, $\epsilon_{2r} = 10$) due to a unit horizontal electric dipole in the ocean ($\sigma_1 = 4$ S/m, $\epsilon_{1r} = 80$) at distance $d = 1$ m from sea floor; $f = 1$ Hz.

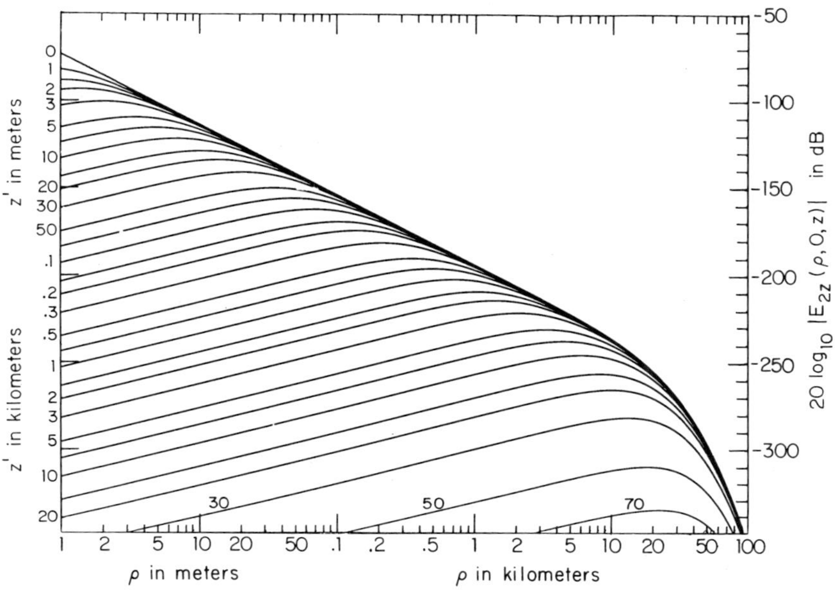

Figure 5.9.7. Vertical electric field in V/m in the oceanic crust ($\sigma_2 = 0.004$ S/m, $\epsilon_{2r} = 10$) due to a unit horizontal electric dipole in the ocean ($\sigma_1 = 4$ S/m, $\epsilon_{1r} = 80$) at distance $d = 1$ m from sea floor; $f = 1$ Hz.

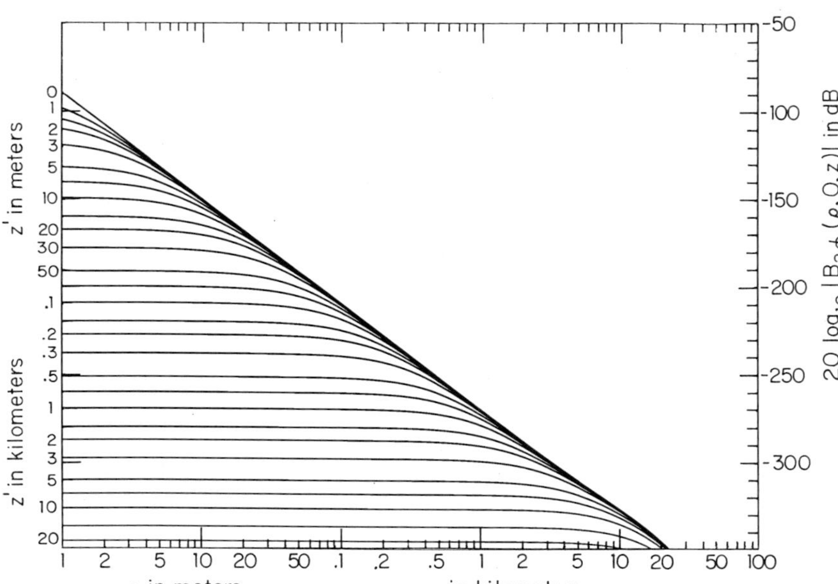

Figure 5.9.8. Magnetic field in V sec/m^2 in the oceanic crust ($\sigma_2 = 0.004$ S/m, $\epsilon_{2r} = 10$) due to a unit horizontal electric dipole in the ocean ($\sigma_1 = 4$ S/m, $\epsilon_{1r} = 80$) at distance $d = 1$ m from sea floor; $f = 1$ Hz.

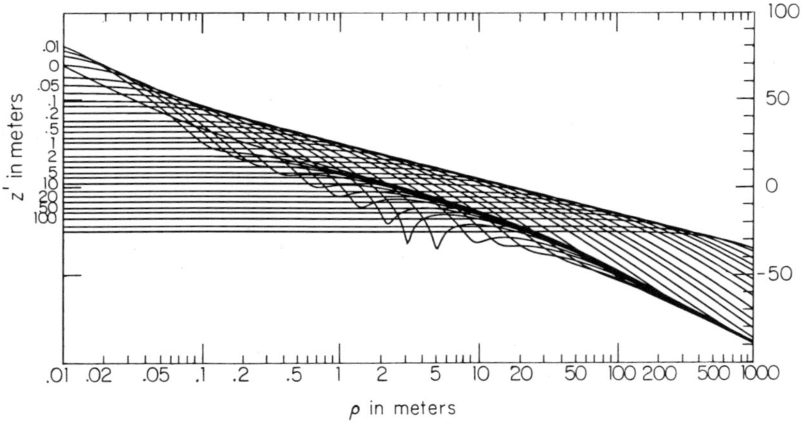

Figure 5.9.9. Radial electric field in V/m in air due to a unit horizontal electric dipole in salt water ($\sigma_1 = 3.5$ S/m, $\epsilon_{1r} = 80$) at depth $d = 7$ mm; $f = 600$ MHz.

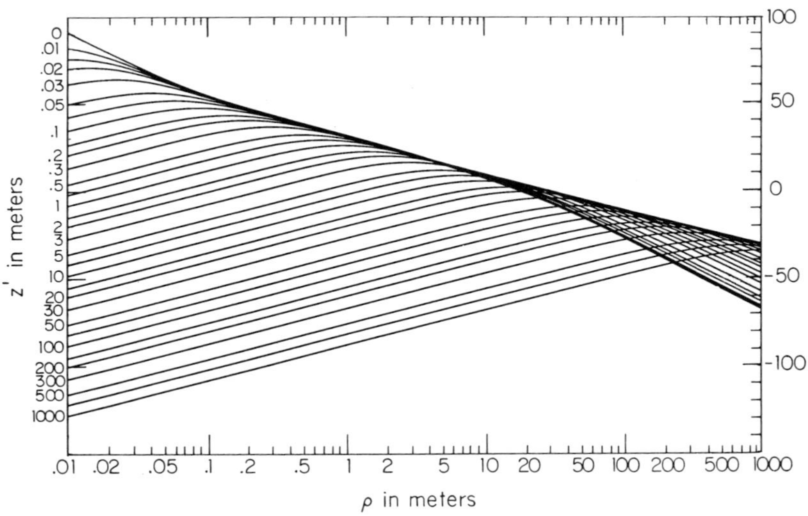

Figure 5.9.10. Vertical electric field in V/m in air due to a unit horizontal electric dipole in salt water ($\sigma_1 = 3.5$ S/m, $\epsilon_{1r} = 80$) at depth $d = 7$ mm; $f = 600$ MHz.

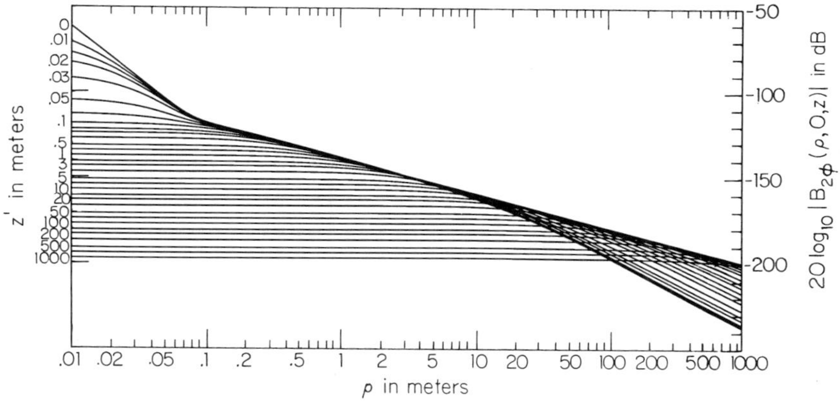

Figure 5.9.11. Magnetic field in V sec/m^2 in air due to a unit horizontal electric dipole in salt water ($\sigma_1 = 3.5$ S/m, $\epsilon_{1r} = 80$) at depth $d = 7$ mm; $f = 600$ MHz.

approximately with ρ/r_0^2, i.e., near $\rho = z'$. The magnetic field $B_{2\phi}(\rho, 0, z)$ associated with $E_{2\rho}(\rho, 0, z)$ and $E_{2z}(\rho, 0, z)$ is shown in Figs. 5.9.5 and 5.9.8. It remains almost constant as the radial distance ρ is increased so long as $z^2 > \rho^2$ in $r_0 = (\rho^2 + z^2)^{1/2}$.

Graphs for $E_{2\rho}(\rho, 0, z)$, $E_{2z}(\rho, 0, z)$, and $B_{2\phi}(\rho, 0, z)$ when Region 2 is air above Region 1, sea water, at $f = 600$ MHz are shown in Figs. 5.9.9–5.9.11.

5.10 Generalization of the Field in Region 2

For possible applications that require a knowledge of the field throughout Region 2, the condition $\rho \geq 5|z|$ (which restricts the formulas to a small angle above the boundary) can be removed. Actually, this condition has been used only in the derivation of the Fresnel-integral terms. A more general treatment—necessary in Section 7.10—can easily be applied. With it, the restriction $\rho \geq 5|z|$ is eliminated to obtain formulas valid everywhere in Region 2. The necessary changes to the formulas (5.5.47)–(5.5.53) are as follows.

In formulas (5.5.47), (5.5.49), and (5.5.51), respectively, for $E_{2\rho}(\rho, \phi, z)$, $E_{2z}(\rho, \phi, z)$, and $B_{2\phi}(\rho, \phi, z)$, replace the term

$$-\frac{k_2^3}{k_1}\left(\frac{\pi}{k_2\rho}\right)^{1/2} e^{-iP}\mathcal{F}(P) \tag{5.10.1a}$$

by

$$-\frac{k_2^3}{k_1}\left(\frac{r_2}{\rho}\right)\left(\frac{\pi}{k_2 r_2}\right)^{1/2} e^{-iP_2}\mathcal{F}(P_2). \tag{5.10.1b}$$

In formulas (5.5.48) and (5.5.50a,b), respectively, for $E_{2\phi}(\rho, \phi, z)$ and $B_{2\rho}(\rho, \phi, z)$, replace the term

$$\frac{ik_2^2}{k_1\rho}\left(\frac{\pi}{k_2\rho}\right)^{1/2} e^{-iP}\mathcal{F}(P) \tag{5.10.2a}$$

by

$$\frac{ik_2^2}{k_1\rho}\left(\frac{r_2^2}{\rho^2}\right)\left(\frac{\pi}{k_2 r_2}\right)^{1/2} e^{-iP_2}\mathcal{F}(P_2), \tag{5.10.2b}$$

where

$$P_2 = S_2\left(\frac{S_2 - Z}{R}\right)^2, \tag{5.10.3}$$

with

$$S_2 = \frac{k_2^3 r_2}{2k_1^2}; \qquad R = \frac{k_2^3\rho}{2k_1^2}; \qquad Z = \frac{k_2^2 z}{2k_1}; \quad z \le 0. \tag{5.10.4}$$

With these changes, the formulas are valid for all values of ρ and z in Region 2. Note that when $\rho^2 \gg z^2$ and $r_2 \sim \rho$, the generalized formulas reduce exactly to (5.5.47)–(5.5.51).

6
Interference Patterns; Comparison of Approximate Formulas with General Integrals and Measurements

6.1 Introduction

In Chapters 3 and 5 the exact general integrals for the electromagnetic field of vertical and horizontal electric dipoles near the plane boundary between electrically different half-spaces are expressed in terms of integrated formulas that are very good approximations within wide ranges of the variables and parameters. In order to determine the accuracy of the representations in a direct quantitative manner, it is appropriate to compare fields calculated from these formulas with corresponding ones obtained by the direct numerical evaluation of the integrals. Such comparisons are best made under critical conditions when even small differences are clearly delineated. Since $E_{1\rho}(\rho, \phi, z)$ is the most important and generally useful component, an extensive study of its properties over wide ranges of distances, frequencies, and material parameters is suggested. Less extensive comparisons for the other five components are also indicated.

6.2 Radial Electric Field and Interference Patterns

The general integral for the radial electric field at (ρ, ϕ, z) in Region 1 ($z \geq 0$, k_1) due to a unit horizontal electric dipole at $(0, 0, d)$ also in Region 1 is given in (5.4.29). The equivalent approximate integrated formula is in (5.5.37). It has the form

$$E_{1\rho}(\rho,\phi,z) = -\frac{\omega\mu_0}{2\pi k_1^2}\cos\phi\left[k_2 g(\rho;k_1,k_2)e^{ik_2\rho}e^{ik_1(z+d)} - \left(\frac{k_1}{\rho^2}+\frac{i}{\rho^3}\right)e^{ik_1r_1} + \left(\frac{z+d}{\rho}\right)\left(\frac{ik_1}{\rho^2}-\frac{3}{2\rho^3}\right)e^{ik_1r_2}\right], \tag{6.2.1}$$

where

$$g(\rho;k_1,k_2) = \frac{ik_2}{\rho} - \frac{1}{\rho^2} - \frac{i}{k_2\rho^3} - \frac{k_2^3}{k_1}\left(\frac{\pi}{k_2\rho}\right)^{1/2} e^{-ip}\mathcal{F}(p). \tag{6.2.2}$$

Here, $p = k_2\rho(k_2^2/2k_1^2)$ and $\mathcal{F}(p)$ is the Fresnel integral. For large arguments,

$$g(\rho;k_1,k_2) \sim -\frac{k_1^2}{k_2^2\rho^2}; \qquad |k_2\rho| \geq 8\left|\frac{k_1^2}{k_2^2}\right|. \tag{6.2.3}$$

As a first range of comparison, the field along an air–sea boundary is useful. This is shown in Fig. 6.2.1 in the form of a graph of $20\log_{10}|E_{1\rho}(\rho,0,z)|$ at $f = 600$ MHz in the wide radial range from $\rho = 1$ cm to 100 km along the air–ocean boundary. The numerical evaluation of the general integral is shown in solid line; points calculated from (6.2.1) with (6.2.2) are shown as small circles. They are seen to fall precisely on the solid-line graph everywhere except in the range $\rho < 1$ m, where they deviate slightly. Also shown on the graph in dashed lines are calculations from the formulas of Baños, (1.4.11), (1.4.8), and (1.4.5), for the near, intermediate, and asymptotic ranges, respectively. These approximate the correct values over important parts of the three ranges, but with intervening gaps. Owing to the high attenuation in sea water, the contribution to $E_{1\rho}(\rho,\phi,z)$ by the direct-field terms in (6.2.1) is negligible over most of the range represented in Fig. 6.2.1 and possible interference patterns can occur only very close to the source.

In order to obtain significant contributions from the direct-field terms, it is necessary to select a much less absorbing medium than sea water. Lake water is a good possibility and—although not practical in the sense of being available in the field—distilled water is a superior medium for carrying out a comparison of interference patterns purely as a check on the accuracy of the integrated formulas. Graphs of $20\log_{10}|E_{1\rho}(\rho,0,z)|$ at $f = 10$ MHz with $d = z = 15$ cm and ρ ranging from 0 to 200 m are shown in Fig. 6.2.2 for $\epsilon_{1r} = 80$ and five values of σ_1 ranging from 4×10^{-6} S/m to 4 S/m. The top and bottom graphs with $\sigma_1 = 4\times10^{-6}$ S/m are primarily of theoretical interest since they apply to distilled water; α_1 is so small that the direct wave is attenuated only slowly. As a consequence, the interference pattern persists for a large radial distance. The curves in solid

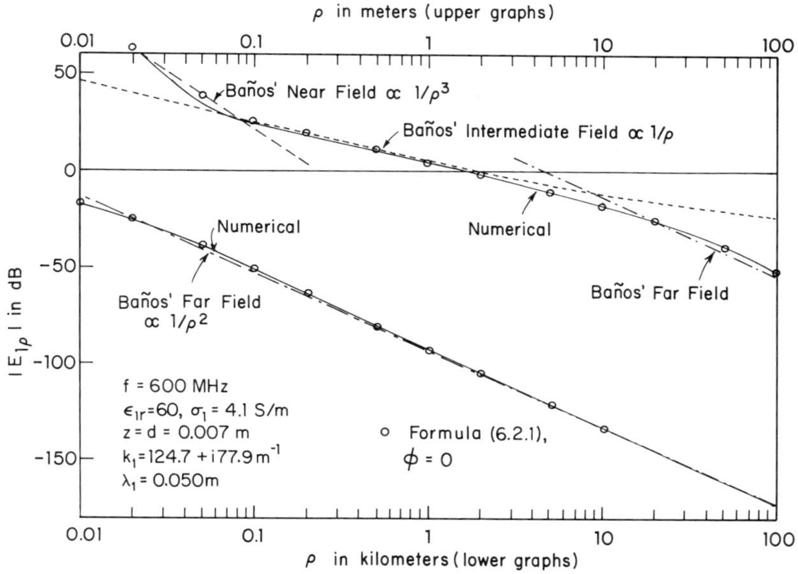

Figure 6.2.1. Theoretical radial electric field $|E_{1\rho}(\rho,0,z)|$ in V/m at depth z of unit horizontal electric dipole at depth d in sea water below air; $f = 600$ MHz, $z = d = 0.007$ m.

line are calculated from (6.2.1), the dotted curve is evaluated numerically from the general integral (5.4.29), both with $\phi = 0$. The central section of the graph for $|E_{1\rho}(\rho,0,z)|$ and the associated real and imaginary parts, $E_{1\rho} = E_{1\rho R} + iE_{1\rho I} = |E_{1\rho}| \exp \theta_{1\rho}$, are shown in Fig. 6.2.3 on a larger scale. The agreement with the numerical calculations is seen to be excellent.

The graph in Fig. 6.2.2 for $\sigma_1 = 4 \times 10^{-3}$ S/m characterizes $|E_{1\rho}|$ in lake water. Owing to the much greater value of the attenuation constant α_1 in $k_1 = \beta_1 + i\alpha_1$, the amplitude of the direct wave decreases quite rapidly so that the interference pattern diminishes to a negligible amplitude in about 50 m. In less pure lake water with $\sigma_1 = 0.04$ S/m, the interference pattern persists only out to about $\rho = 8$ m. In ocean sediment with $\sigma_1 = 0.4$ S/m and in sea water with $\sigma_1 = 4$ S/m, the direct wave has a negligible amplitude even at $\rho = 1$ m. Note that the graphs calculated from formula (6.2.1) are all in excellent agreement with the corresponding numerical values calculated from the general integrals.

The interference patterns at $f = 1$ GHz and 100 MHz in lake water with $\sigma_1 = 4 \times 10^{-3}$ S/m are shown in Fig. 6.2.4. Since the wavelengths in the water are much shorter than with $f = 10$ MHz, the deepest minima occur near $\rho = 0.57$ m with $f = 1$ GHz and near $\rho = 4$ m with $f = 100$ MHz, instead of near $\rho = 58$ m with $f = 10$ MHz. However, since α_1 is substantially smaller at the higher frequencies, the amplitude of the

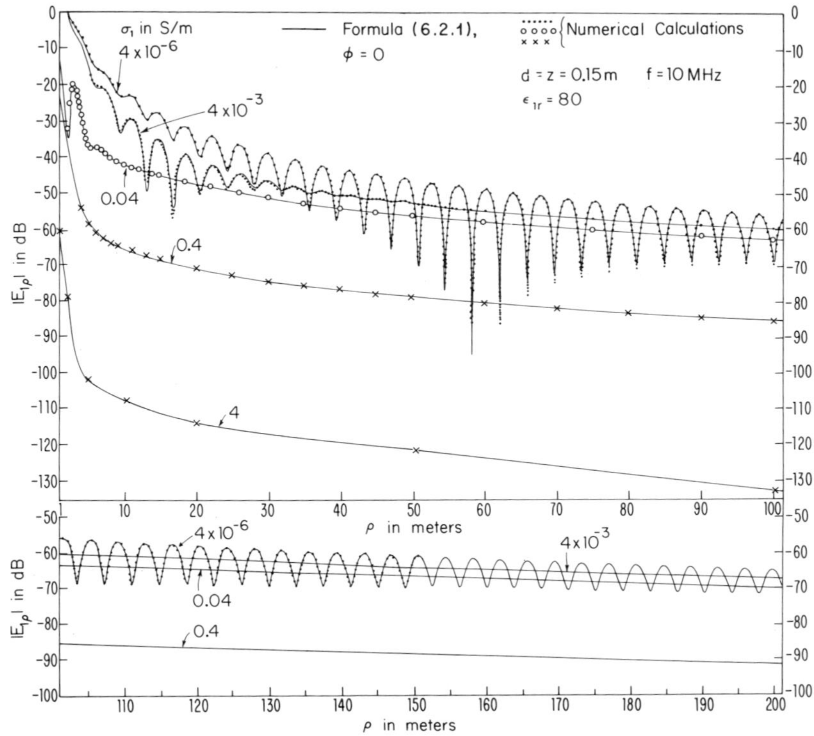

Figure 6.2.2. Theoretical radial electric field $|E_{1\rho}(\rho, 0, z)|$ in V/m at depth z of unit horizontal electric dipole at depth d in water below air; $f = 10$ MHz, $z = d = 0.15$ m.

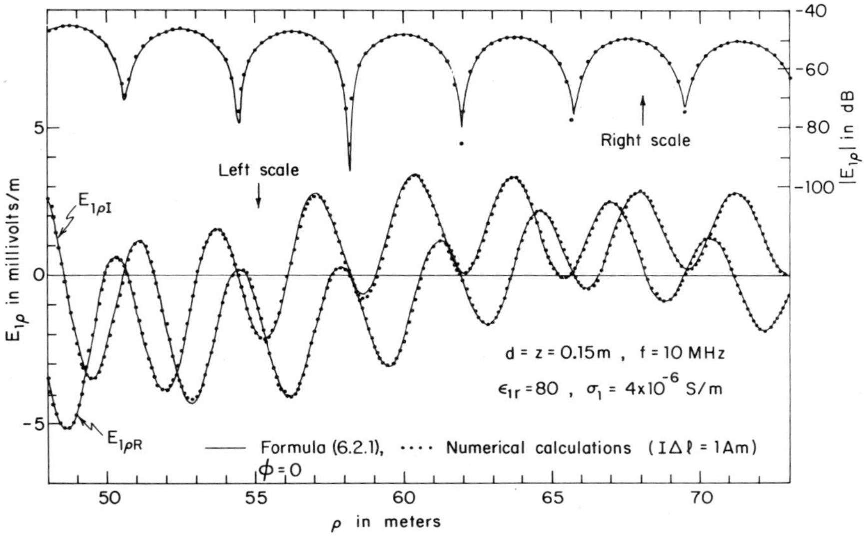

Figure 6.2.3. Theoretical real and imaginary parts and magnitude of radial electric field, $E_\rho = E_{\rho R} + iE_{\rho I} = |E_\rho| \exp \theta_\rho$, in critical range; $\sigma_1 = 4\times10^{-6}$ S/m, $f = 10$ MHz, $z = d = 0.15$ m.

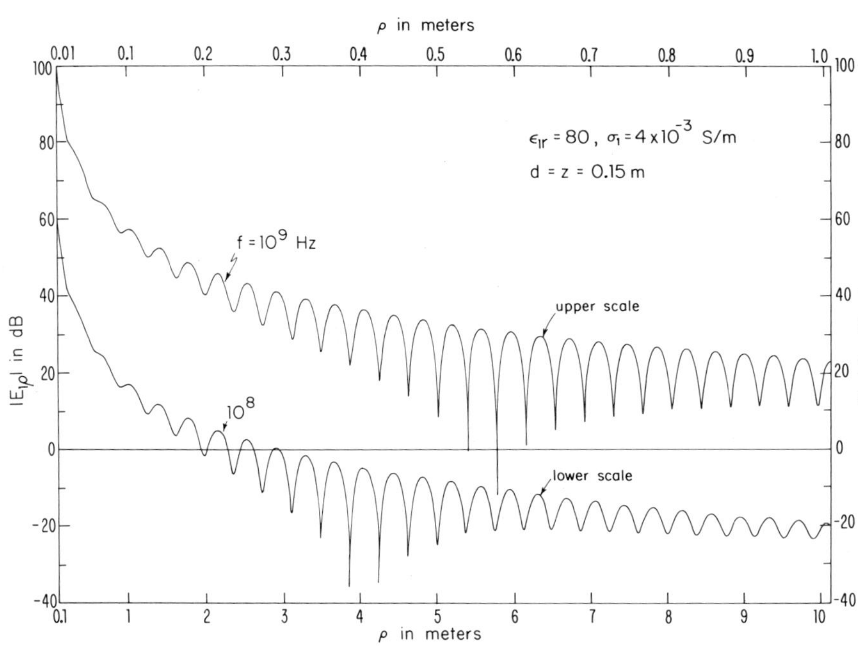

Figure 6.2.4. Theoretical radial electric field $|E_{1\rho}(\rho, 0, z)|$ in V/m at depth z of unit horizontal electric dipole at depth d in lake water below air; $\sigma_1 = 4 \times 10^{-3}$ S/m, $z = d = 0.15$ m; $f = 10^8$ and 10^9 Hz.

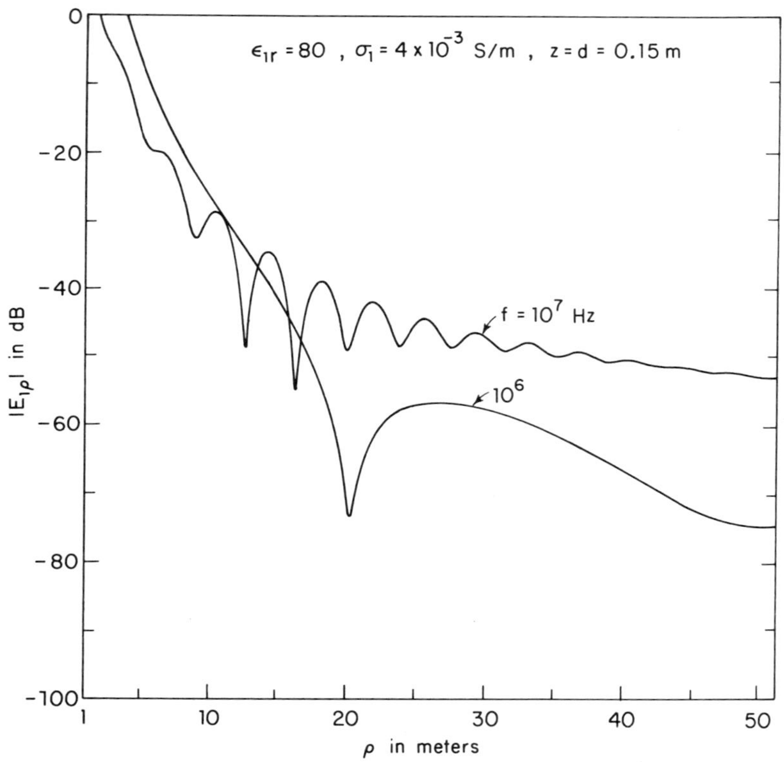

Figure 6.2.5. Theoretical radial electric field $|E_{1\rho}(\rho, 0, z)|$ in V/m at depth z of unit horizontal electric dipole at depth d in lake water below air; $\sigma_1 = 4 \times 10^{-3}$ S/m, $z = d = 0.15$ m; $f = 10^6$ and 10^7 Hz.

interference patterns decreases only slowly with radial distance. Graphs of $20 \log_{10} |E_{1\rho}(\rho, 0, z)|$ for lake water over a wide range of frequencies are shown in Figs. 6.2.5 and 6.2.6 to different scales. A significant interference pattern occurs when $f = 1$ MHz. Graphs of $20 \log_{10} |E_{1\rho}(\rho, 0, z)|$ in sea water ($\sigma_1 = 4$ S/m) at the high frequencies $f = 100$ MHz and $f = 1$ GHz are shown in Fig. 6.2.7 at very close range. It is seen that interference patterns exist out 30 to 40 cm from the source.

Graphs like those in Figs. 6.2.4 through 6.2.7 for water with $\epsilon_{1r} = 80$ are shown in Figs. 6.2.8 through 6.2.11, respectively, at $f = 1$ GHz, 100 MHz, 10 MHz, and 1 MHz for a material medium with $\epsilon_{1r} = 20$ and a range of values of σ_1. The interference patterns generated by the superposition of the direct and lateral waves are similar to those in water with the patterns dominated by the wavelength characteristic of the medium at each frequency. A comparison of the calculations from (6.2.1) with those from the exact integrals is shown in Fig. 6.2.10. The agreement is very good.

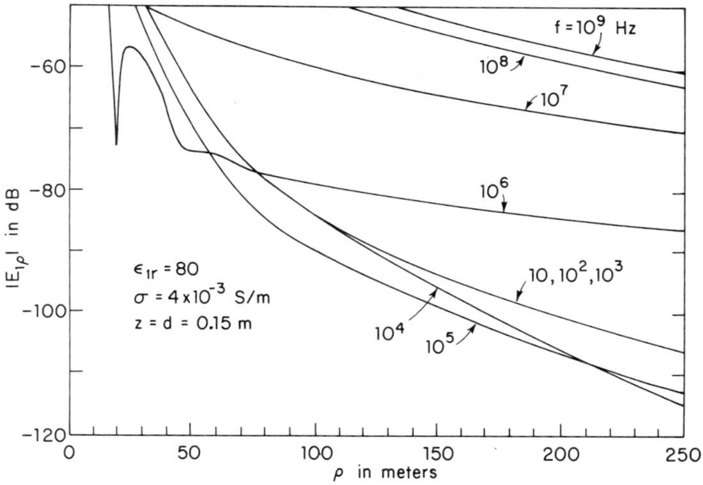

Figure 6.2.6. Theoretical radial electric field $|E_{1\rho}(\rho, 0, z)|$ in V/m at depth z of unit horizontal electric dipole at depth d in lake water below air; $\sigma_1 = 4 \times 10^{-3}$ S/m, $z = d = 0.15$ m; $f = 10$ to 10^9 Hz.

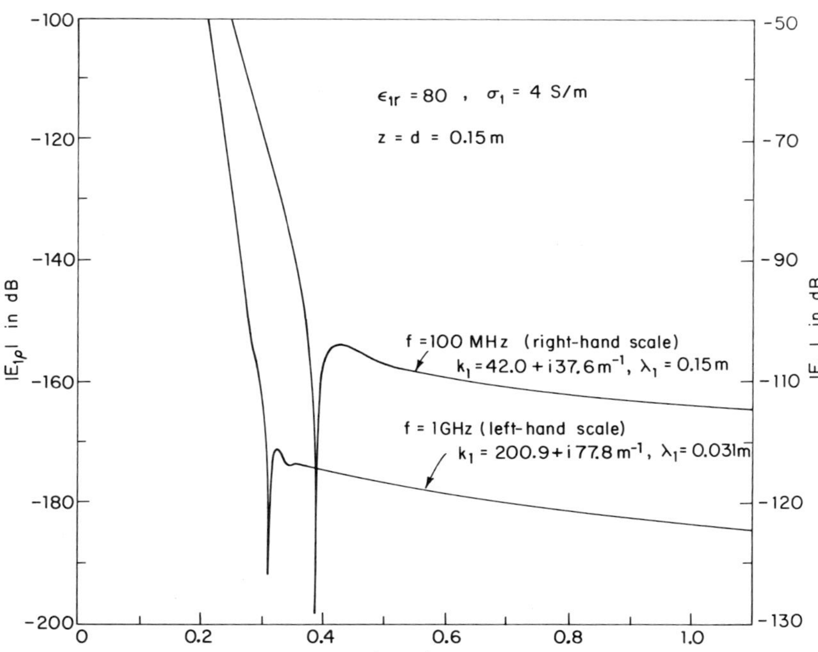

Figure 6.2.7. Theoretical radial electric field $|E_{1\rho}(\rho, 0, z)|$ in V/m at depth z of unit horizontal electric dipole at depth d in sea water below air; $\sigma_1 = 4$ S/m, $z = d = 0.15$ m; $f = 10^8$ and 10^9 Hz.

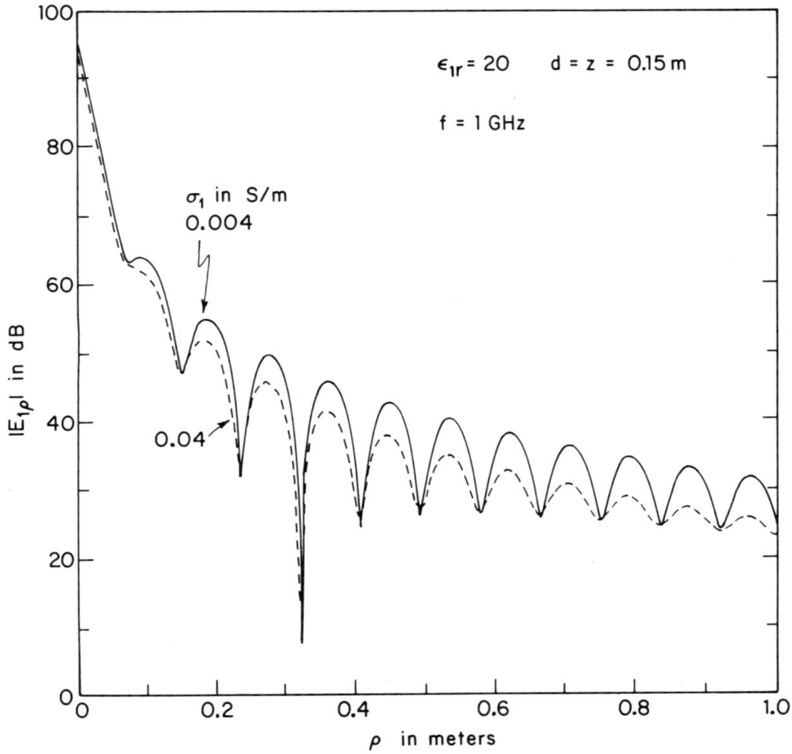

Figure 6.2.8. Theoretical radial electric field $|E_{1\rho}(\rho, 0, z)|$ in V/m at depth z of unit horizontal electric dipole at depth d in Region 1 below air; $\epsilon_{1r} = 20$, $f = 1$ GHz, $z = d = 0.15$ m.

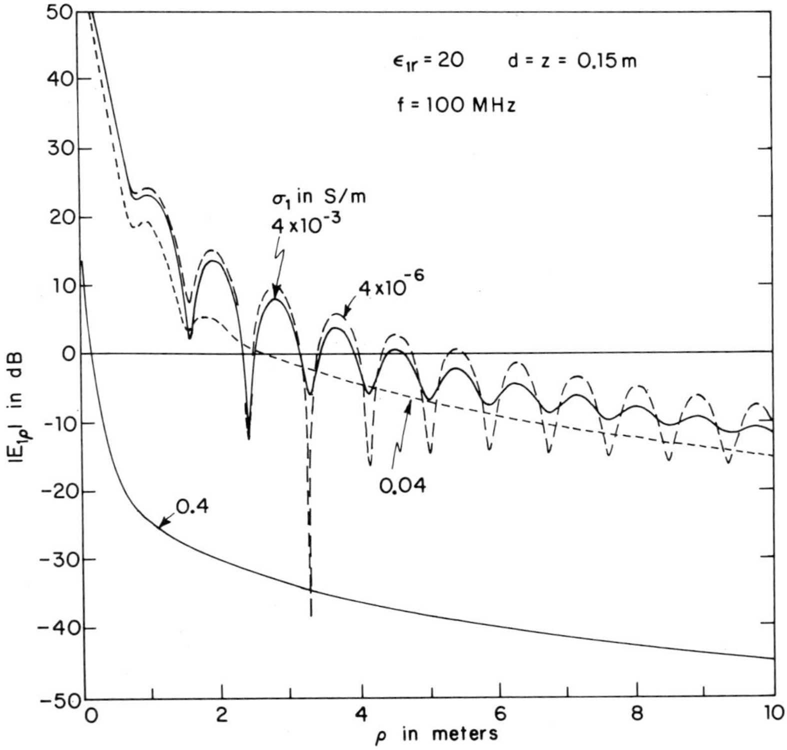

Figure 6.2.9. Theoretical radial electric field $|E_{1\rho}(\rho, 0, z)|$ in V/m at depth z of unit horizontal electric dipole at depth d in Region 1 below air; $\epsilon_{1r} = 20$, $f = 100$ MHz, $z = d = 0.15$ m.

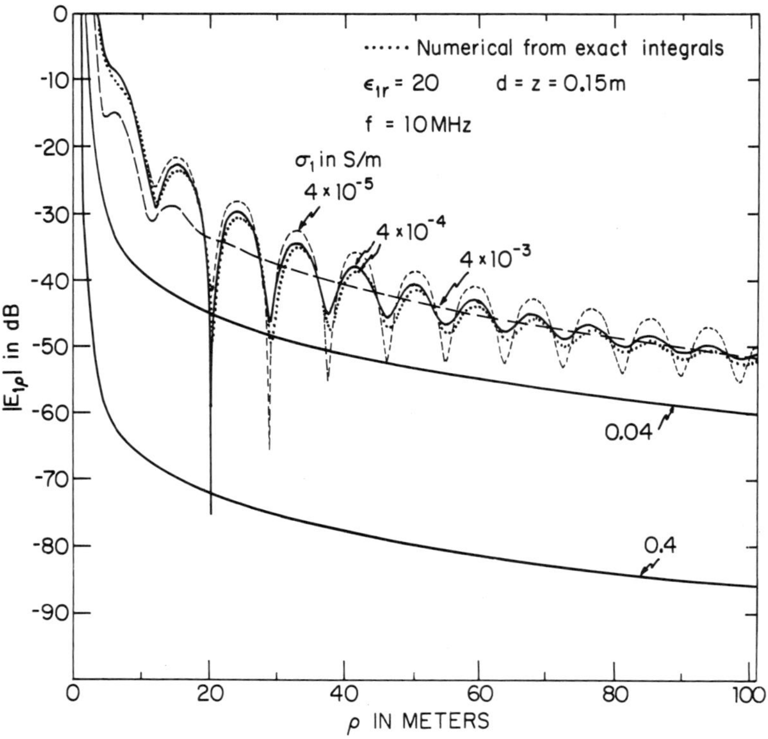

Figure 6.2.10. Theoretical radial electric field $|E_{1\rho}(\rho, 0, z)|$ in V/m at depth z of unit horizontal electric dipole at depth d in Region 1 below air; $\epsilon_{1r} = 20,\ f = 10$ MHz, $z = d = 0.15$ m.

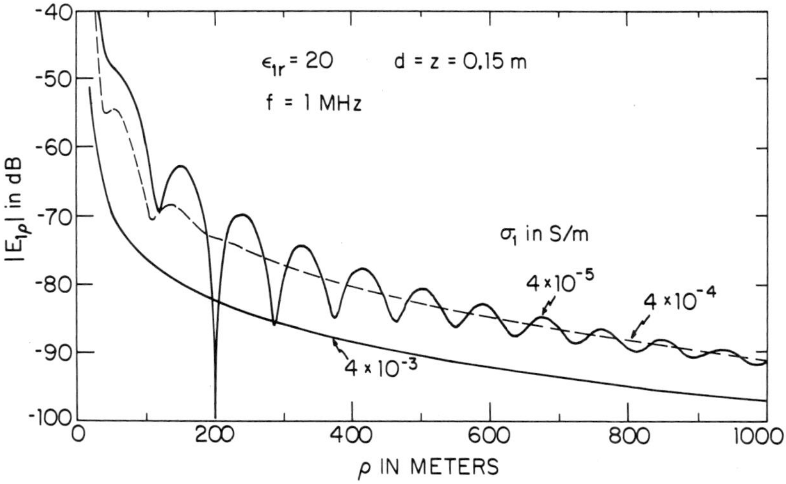

Figure 6.2.11. Theoretical radial electric field $|E_{1\rho}(\rho, 0, z)|$ in V/m at depth z of unit horizontal electric dipole at depth d in Region 1 below air; $\epsilon_{1r} = 20$, $f = 1$ MHz, $z = d = 0.15$ m.

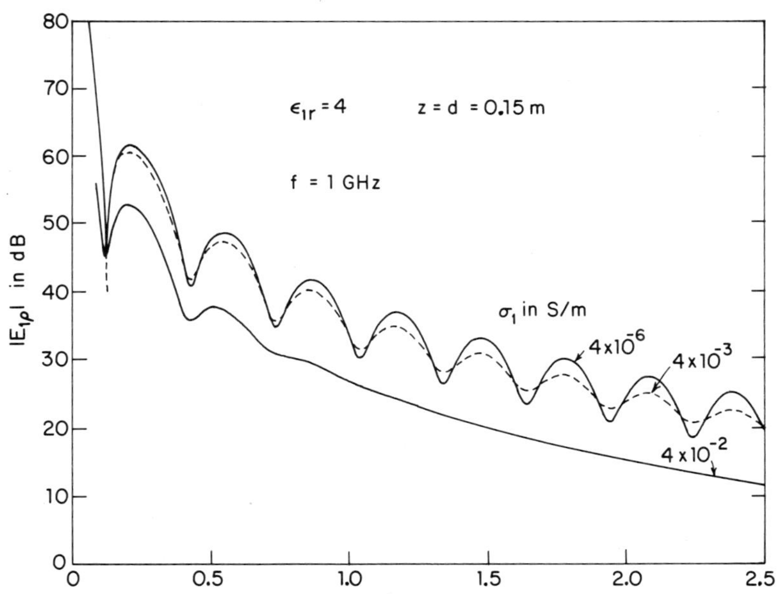

Figure 6.2.12. Theoretical radial electric field $|E_{1\rho}(\rho, 0, z)|$ in V/m at depth z of unit horizontal electric dipole at depth d in Region 1 below air; $\epsilon_{1r} = 4$, $f = 1$ GHz, $z = d = 0.15$ m.

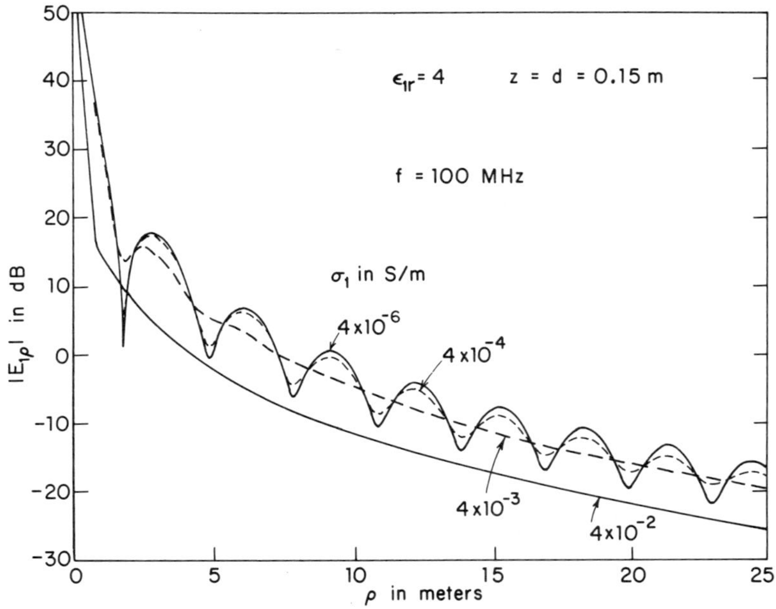

Figure 6.2.13. Theoretical radial electric field $|E_{1\rho}(\rho, 0, z)|$ in V/m at depth z of unit horizontal electric dipole at depth d in Region 1 below air; $\epsilon_{1r} = 4$, $f = 100$ MHz, $z = d = 0.15$ m.

Graphs for dry earth with $\epsilon_{1r} = 4$ and several values of σ_1 are in Figs. 6.2.12 through 6.2.14, respectively, at $f = 1$ GHz, 100 MHz, and 10 MHz. As seen from Fig. 6.2.14, the agreement between the values calculated from (6.2.1) and from the exact integrals is not as good with $\epsilon_{1r} = 4$ as with $\epsilon_{1r} = 20$ and 80. This is to be expected since the condition $|k_1| \geq 3|k_2|$ is not satisfied. However, whereas the amplitude of the interference pattern is not quite accurate, the agreement in the actual location of the maxima and minima is excellent.

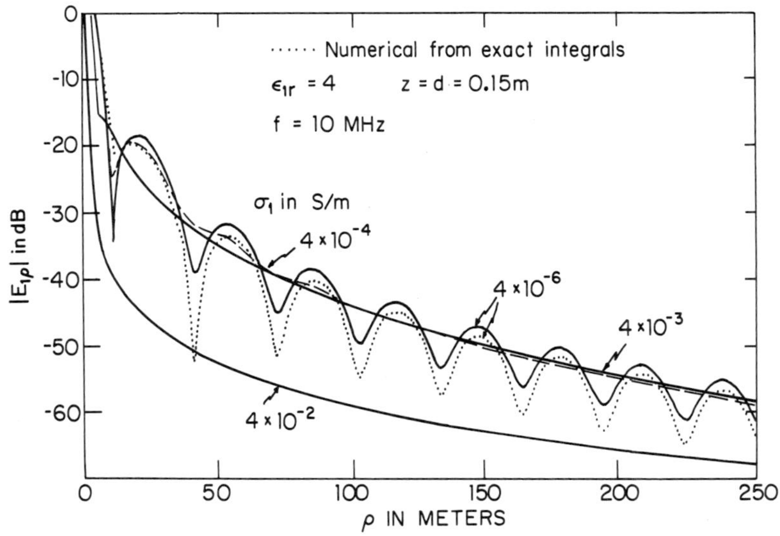

Figure 6.2.14. Theoretical radial electric field $|E_{1\rho}(\rho, 0, z)|$ in V/m at depth z of unit horizontal electric dipole at depth d in Region 1 below air; $\epsilon_{1r} = 4$, $f = 10$ MHz, $z = d = 0.15$ m.

6.3 The Complete Electric and Magnetic Fields

In order to check the accuracy of the components $E_{1\phi}(\rho, \phi, z)$, $E_{1z}(\rho, \phi, z)$, $B_{1\rho}(\rho, \phi, z)$, $B_{1\phi}(\rho, \phi, z)$, and $B_{1z}(\rho, \phi, z)$ as given by (5.5.38)–(5.5.42), these have been evaluated at $f = 600$ MHz for an air–sea water boundary. The exact integrals for these components have been calculated for comparison. To provide the means for intercomparing the several formulas, $E_{1\rho}(\rho, \phi, z)$ is also included. The electric-type components $E_{1\rho}(\rho, \phi, z)$, $E_{1z}(\rho, \phi, z)$, and $H_{1\phi}(\rho, \phi, z) = B_{1\phi}(\rho, \phi, z)/\mu_0$ are evaluated at $\phi = 0$; the magnetic-type components $E_{1\phi}(\rho, \phi, z)$, $H_{1\rho}(\rho, \phi, z) = B_{1\rho}(\rho, \phi, z)/\mu_0$, and $H_{1z}(\rho, \phi, z) = B_{1z}(\rho, \phi, z)/\mu_0$ are evaluated at $\phi = \pi/2$. Region 1 is salt water with $\sigma_1 = 3.5$ S/m and $\epsilon_{1r} = 45$; Region 2 is air. Since agreement at very large distances is no problem, special attention is directed to the range within 20 wavelengths in air of the source. This is of importance since the approximate integrated formulas are good approximations only when $\rho \geq 5d$ and $\rho \geq 5z$.

Graphs showing $20 \log_{10} |E_{1\rho}|$, $20 \log_{10} |E_{1\phi}|$, and $20 \log_{10} |E_{1z}|$ are in Fig. 6.3.1. Graphs of $20 \log_{10} |H_{1\rho}|$, $20 \log_{10} |H_{1\phi}|$, and $20 \log_{10} |H_{1z}|$ are

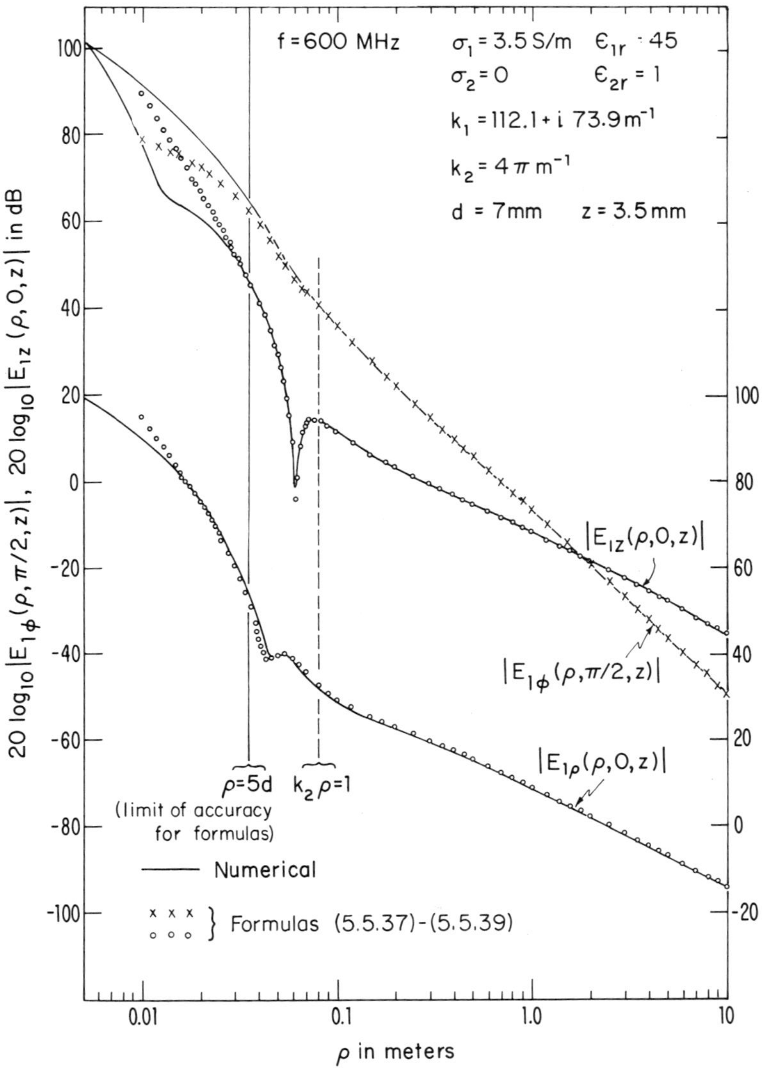

Figure 6.3.1. Theoretical electric field $\mathbf{E}_1(\rho, \phi, z)$ in V/m at depth z of unit horizontal electric dipole at depth d in sea water below air; $f = 600$ MHz, $\sigma_1 = 3.5$ S/m, $\epsilon_{1r} = 45$, $z = 0.0035$ m, $d = 0.007$ m.

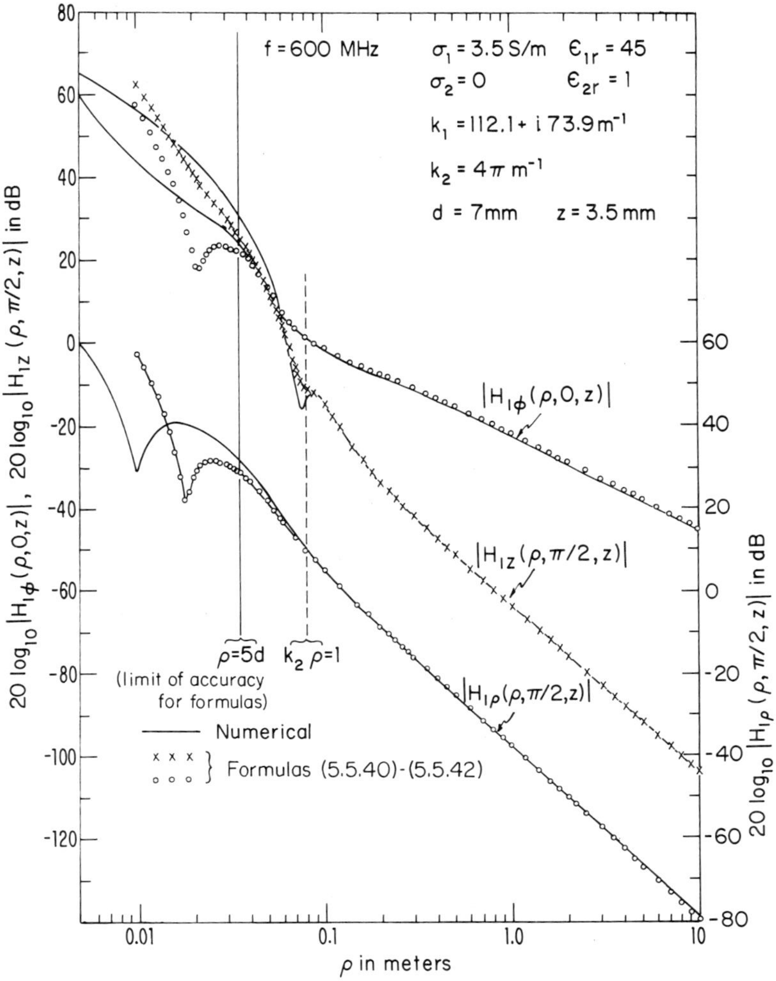

Figure 6.3.2. Theoretical magnetic field $\mathbf{H}_1(\rho, \phi, z) = \mathbf{B}_1(\rho, \phi, z)/\mu_0$ in A/m at depth z of unit horizontal electric dipole at depth d in sea water below air; $f = 600$ MHz, $\sigma_1 = 3.5$ S/m, $\epsilon_{1r} = 45$, $z = 0.0035$ m, $d = 0.007$ m.

in Fig. 6.3.2. The radial range is $0.01 \leq \rho \leq 10$ m with $z = 0.0035$ m, $d = 0.007$ m. The agreement between the integrated formulas and the numerically evaluated general integrals is very satisfactory for all six components in the specified range: $\rho \geq 5d$, $\rho \geq 5z$.

6.4 Comparison of Theory with Measurements

Since the general integrals for the six components of the electromagnetic field generated by a unit horizontal electric dipole at $(0, 0, d)$ in Region 1 $(z \geq 0)$ are exact solutions of Maxwell's equations and these are known to be correct, they require no experimental verification. The approximations involved in the representation of these integrals by integrated formulas are quantitatively clear and, in the range specified for them, corroborating measurements are not essential since they have been compared directly with the numerically evaluated integrals. However, both the integrals and their approximate representations are idealizations in the sense that they apply to two homogeneous, isotropic and electrically different half-spaces with a smooth plane boundary between them. In order to have them as useful tools in the practical experimental study of the propagation of electromagnetic waves along the boundaries between finite regions that may have irregularities, obstacles, and discontinuities, it is desirable to investigate experimentally the degree in which conditions can be provided in the laboratory that approximate the ideal. Accordingly, a series of experiments has been carried out (Brown, King, and Wu 1982, 1984; King and Brown, 1984) to observe the radial electric field due to a horizontal electric monopole (with image in a vertical metal ground plane) just below the surface of salt water in an indoor tank. These are described in Chapter 10. Results relevant to the basic theory are the measurement in amplitude and phase of the radial component $E_{1\rho}(\rho, \phi, z)$ due to a horizontal electric dipole at the depth d in salt water. The aspects of interest are the distribution of the field along the radial line $\phi = 0$ [which should have the approximate form $\exp(ik_2\rho)/\rho$ in the intermediate zone defined roughly by $1 \leq k_2\rho \leq |k_1^2/k_2^2|$]; the field pattern as a function of ϕ (which should have the form $\cos\phi$ for a short bare dipole); and the amplitude and phase as functions of the combined depths of the source at $(0, 0, d)$ and the receiving dipole at $(\rho, 0, z)$ {which should behave like $\exp[-\alpha_1(z + d)] \exp[i\beta_1(z + d)] \exp(ik_2\rho)$}.

Measured graphs of the amplitude and phase along the radial line $\phi = 0$ are shown in Fig. 6.4.1, the measured field pattern is in Fig. 6.4.2, and the decrease in amplitude and change in phase with depth are shown in Fig. 6.4.3. All of the measured graphs are in good agreement with the theory. A small standing wave due to both the finite size of the tank and reflections from the ceiling is the principal difference between the measurements and the theory. Also of interest and shown in Fig. 6.4.2 is the measured directional field pattern of a two-element array of horizontal traveling-wave insulated monopoles backed by a vertical ground plane.

The actual generation and measurement in the laboratory of lateral electromagnetic waves serves an additional useful purpose. The fact that lateral electromagnetic waves are the only means of communication along the surface of the earth or sea is not always recognized. Similarly, the importance

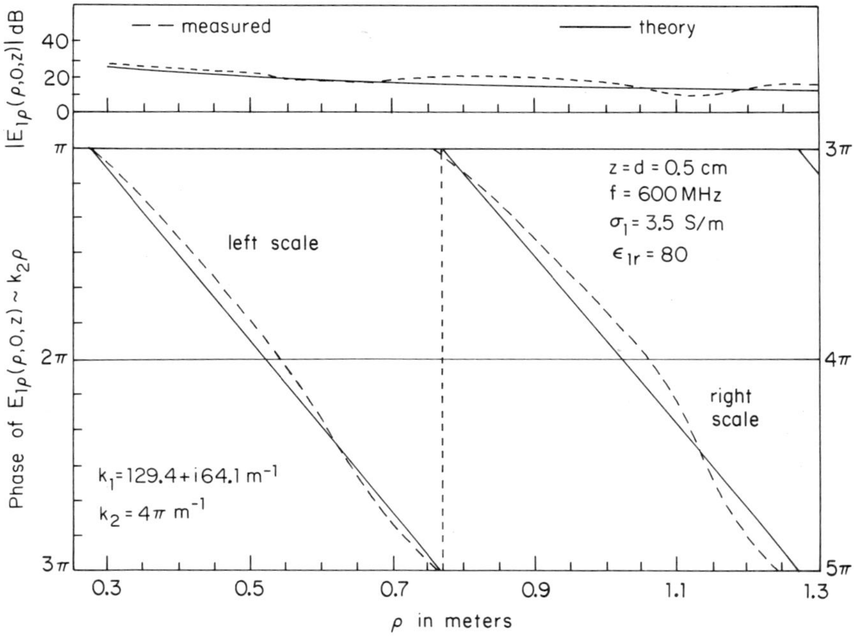

Figure 6.4.1. Measured magnitude (in V/m) and phase (in radians) of radial electric field $E_{1\rho}(\rho, 0, z) \sim \exp(ik_2\rho)/\rho$ (intermediate range); $f = 600$ MHz, $\sigma_1 = 3.5$ S/m, $\epsilon_{1r} = 80$, $z = d = 0.005$ m. (Measured by M. F. Brown.)

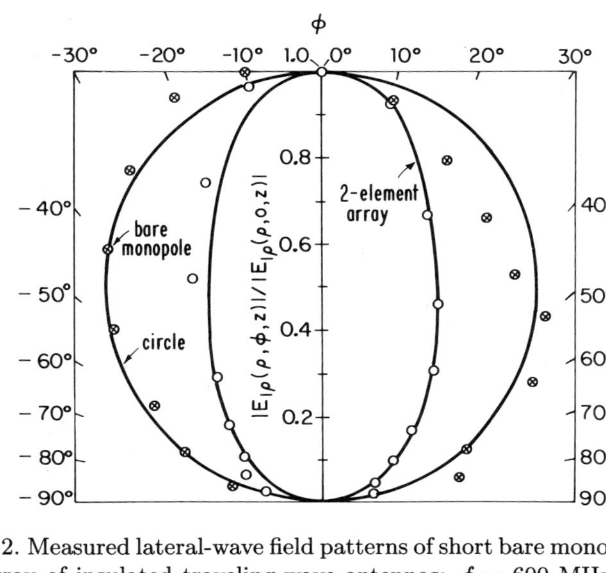

Figure 6.4.2. Measured lateral-wave field patterns of short bare monopole and two-element array of insulated traveling-wave antennas; $f = 600$ MHz, $\rho = 0.8$ m, $z = d = 0.005$ m. (Measured by M. F. Brown.)

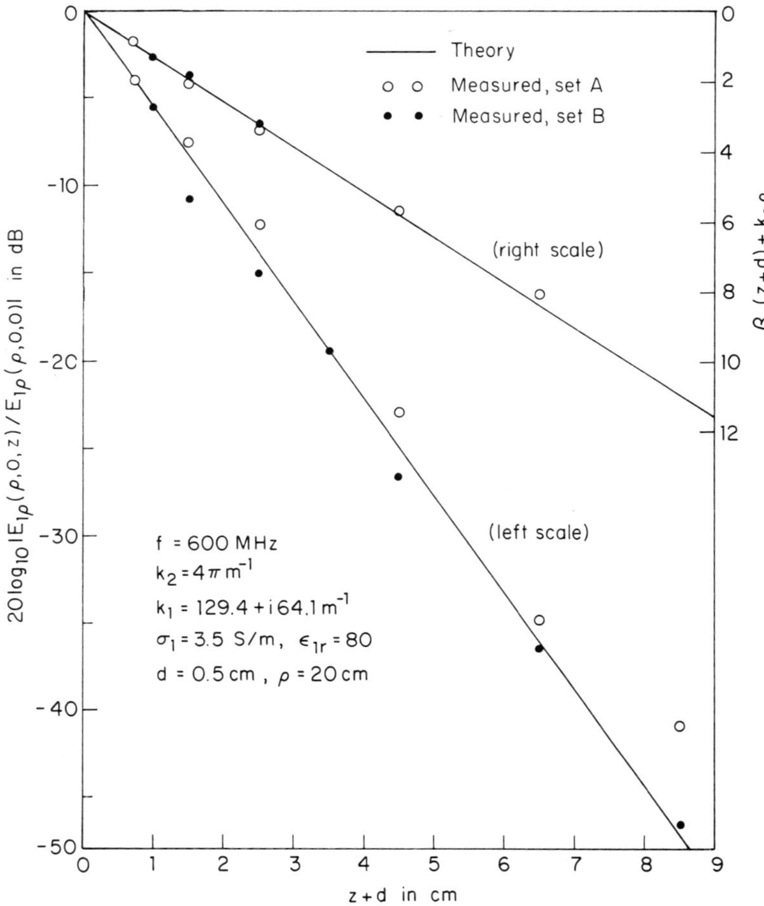

Figure 6.4.3. Experimental test of $E_{1\rho}(\rho, 0, z) \sim \exp[-\alpha_1(z+d)]\exp\{i[\beta_1(z+d)+k_2\rho]\}$; $f = 600$ MHz, $\rho = 0.2$ m, $z = d = 0.005$ m. (Measured by M. F. Brown.)

of lateral electromagnetic waves in all measurements on the sea floor when the source is a horizontal electric dipole also on the sea floor has not been generally appreciated by the geophysical community. The actual measurement of the properties of lateral waves in the laboratory demonstrates that these waves are real and can be readily produced and studied.

7 Applications of the Theory of the Horizontal Dipole Near the Boundary Between Air and Earth or Sea

7.1 Comparison with the Vertical Dipole

In Chapter 4, applications of the vertical electric dipole to communication with submarines and to geophysical measurements on the sea floor are described. The horizontal electric dipole can be used for similar purposes. It has also been used for remote sensing from the surface of the earth. For these applications the radial component of the electric field is generally the most useful and its properties in the direction $\phi = 0$ of its maximum are essentially the same for the horizontal as for the vertical dipole. However, the relative magnitudes of the field for unit dipoles are quite different. In the intermediate $1/\rho$ range, the ratios of the fields for dipoles at the boundary in Regions 1 and 2 (wave numbers k_1 and k_2 with $|k_1| \geq 3|k_2|$) are

$$[\text{Vertical in } 2 : \text{Horizontal in } 1 \text{ or } 2 : \text{Vertical in } 1] = \left[1 : \frac{k_2}{k_1} : \frac{k_2^2}{k_1^2}\right]. \tag{7.1.1}$$

For communication with submarines, the unit vertical dipole on the boundary in air (Region 2) is to be preferred; for transmission along the sea floor with the source in the sea (Region 1), the unit horizontal dipole gives the larger field. But actual antennas are not unit dipoles. At low frequencies—which must usually be used for these applications—electrically long antennas are possible with horizontal elements and generally physically impractical with vertical antennas.

7.2 Lateral-Wave Propagation over the Surface of the Earth and into the Ocean—Unit Horizontal Dipoles

When Region 2 is air, the radial component of the electric field at the point (ρ, ϕ, z) in the earth or sea (Region 1) due to a unit horizontal electric dipole at the point $(0, 0, d)$ also in the earth or sea is given by (5.5.37). The lateral-wave part of this can be written in the form

$$E_{1\rho}^{L}(\rho, \phi, z) = E_{1\rho}^{L}(\rho, 0, 0) \cos\phi \; e^{ik_1(z+d)}, \tag{7.2.1}$$

where $E_{1\rho}^{L}(\rho, 0, 0)$ is the field at the surface $(z = 0)$ in the direction $\phi = 0$ due to a horizontal dipole also at the surface $(d = 0)$ in either medium since $E_{1\rho}(\rho, \phi, 0) = E_{2\rho}(\rho, \phi, 0)$. It is given by

$$E_{1\rho}^{L}(\rho, 0, 0) = -\frac{\omega\mu_0 k_2}{2\pi k_1^2}\, g(\rho; k_1, k_2) e^{ik_2\rho}, \tag{7.2.2}$$

where $g(\rho; k_1, k_2)$ is defined in (5.5.44). In the present case, k_2 is real and $\omega\mu_0/2\pi k_2 = \zeta_0/2\pi = 60\ \Omega$ so that

$$|E_{1\rho}^{L}(\rho, 0, 0)| = 60 \left|\frac{k_2^2}{k_1^2}\right| |g(\rho; k_1, k_2)|. \tag{7.2.3}$$

When Region 1 is the sea, the quantity $20 \log_{10} |E_{1\rho}^{L}(\rho, 0, 0)|$ has the form shown graphically in Fig. 7.2.1 as a function of ρ with the frequency as the parameter. On the graph for each frequency, a short vertical line above the graph indicates the radial distance at which $k_2\rho = 1$, which is the boundary between the near field where the amplitude decreases as $1/\rho^3$ and the intermediate field where the amplitude decreases as $1/\rho$. Similarly, a vertical arrow below each graph locates the radial distance at which $k_2\rho = |k_1/k_2|^2$, which is the approximate boundary between the intermediate field and the asymptotic field. In the latter, the amplitude decreases as $1/\rho^2$. It is seen from Fig. 7.2.1 that, for a range of 1000 km, frequencies of $f = 10^4$ Hz or higher give the greatest radial field at the surface of the sea. In order to penetrate into the ocean, lower frequencies are required. As shown in Section 4.4 for the field generated by a vertical dipole, transmission into the ocean for communication with submerged submarines requires lower frequencies. The optimum frequencies for specified depths of penetration into the sea are readily evaluated for radial distances in the intermediate range. In this range and in the direction of maximum, $\phi = 0$, the magnitude of (7.2.1) is

$$|E_{1\rho}^{L}(\rho, 0, z)| = \frac{60 k_2^3}{\rho |k_1^2|}\, e^{-\alpha_1(z+d)}. \tag{7.2.4}$$

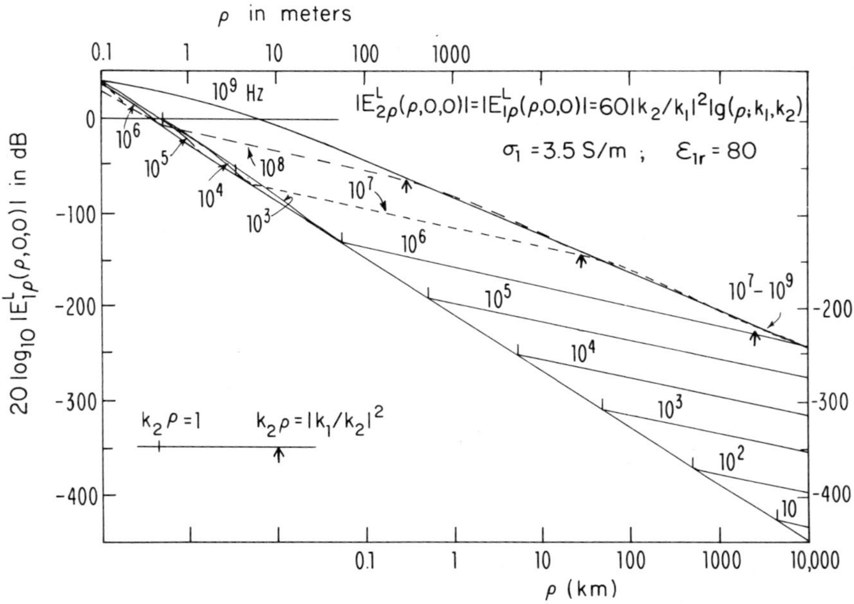

Figure 7.2.1. Magnitude of $E^L_{1\rho}(\rho,0,0)$ in V/m due to horizontal unit electric dipole in salt water just below the surface with air.

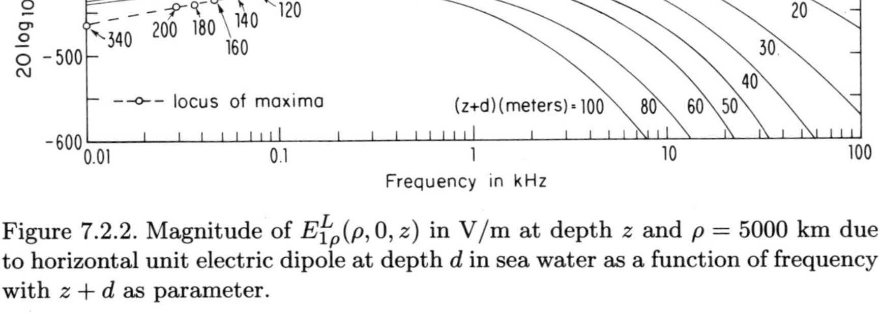

Figure 7.2.2. Magnitude of $E^L_{1\rho}(\rho,0,z)$ in V/m at depth z and $\rho = 5000$ km due to horizontal unit electric dipole at depth d in sea water as a function of frequency with $z+d$ as parameter.

For a dipole on the surface, $d = 0$. The magnitude of (7.2.4) is readily maximized with respect to frequency with $\partial |E^L_{1\rho}(\rho, 0, z)|/\partial f = 0$, which gives $\alpha_1(z+d) = 4$, so that the maximizing frequency satisfies the condition: $fz^2 = 16/\pi\mu_0\sigma_1$. Graphs of $20 \log_{10} |E^L_{1\rho}(\rho, 0, z)|$ are shown in Fig. 7.2.2 as a function of the frequency with the combined depths of the point of observation and of the dipole, $z + d$, as the parameter and with $\rho = 5000$ km. These graphs are very similar to those in Fig. 4.4.3 for the vertical dipole on the surface in air but are smaller by the factor $k_2/|k_1|$. It is seen that the maximizing frequency for a combined depth $z + d = 50$ m is 0.46 kHz, but that any frequency between 0.1 and 3 kHz is acceptable. The associated maximum value of $20 \log_{10} |E^L_{1\rho}(\rho, 0, z)|$ is -396 dB. This is 81 dB below the field of a unit vertical dipole at the same frequency and depth. A very substantial gain over that of the unit dipole must be achieved by any antenna that is to provide a useful field. This is discussed in the following sections.

7.3 The Wave Antenna for Communication over the Earth

A useful antenna for transmission or reception between points on or near the surface of the earth or sea is the wave antenna first described by Beverage, Rice, and Kellog (1923). It is the subject of Chapter 18. It consists of a long wire close to and parallel to the earth or sea. The wire is terminated at each end by connections to the earth or sea so that it carries a traveling wave of current. A schematic diagram of the Beverage antenna for transmission is shown in Fig. 7.3.1a. For reception, the generator is replaced by a receiver.

Closely related to the Beverage antenna is the horizontal-wire antenna shown in Fig. 7.3.1b. It differs from the Beverage antenna in that the long horizontal wire has no connection to the earth but is terminated with suitable impedances in series with resonant open ends. A traveling wave of current is maintained on the long wire between the terminating impedances. The current on the open ends is a standing wave.

In the applications of the wave antenna of either type, interest is in the *field along the surface of the earth or sea.* This is the field involved in the transmission from a driven wave antenna to a receiving antenna—which may be a second wave antenna or any vertical receiving antenna. It is the scattered field received by a wave antenna from an obstacle near the horizon when this has been illuminated by a radar beam from the antenna in its transmitting mode. It is also the field at the surface that determines the field received by a submarine.

The conventional Beverage antenna consists of a long horizontal conductor that extends from $x = 0$ to $x = h$ in Region 2 at the constant

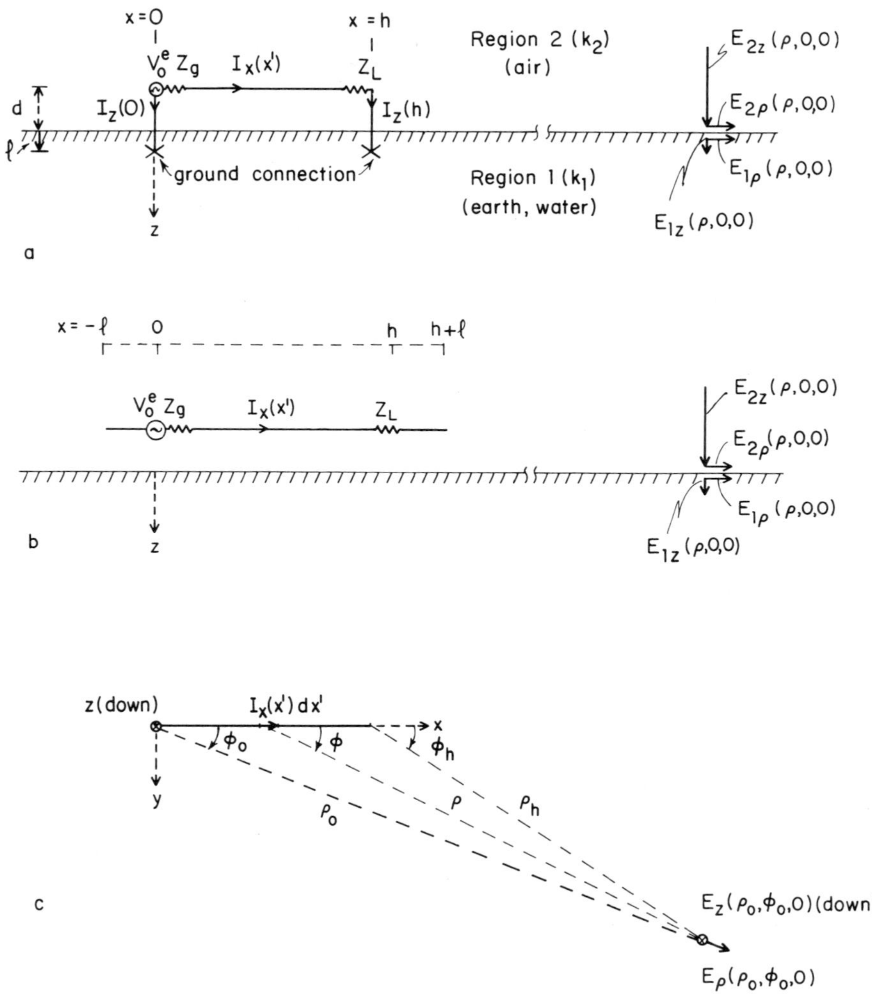

Figure 7.3.1. Wave antennas. (a) Conventional Beverage antenna; (b) horizontal-wire antenna; (c) coordinates.

height d above the x-axis and two vertical conductors in series with suitable impedances that assure a traveling wave of current along the horizontal wire. The vertical conductors enter the ground or sea at $z = 0$ and continue down in Region 1 a distance l where they end in radial ground networks. The antenna is driven at $x = 0,\ z = -d$. The radius of each conductor is a. The horizontal-wire antenna consists of the same horizontal wire between $x = 0$ and $x = h$ at the height d as the Beverage antenna. It is driven at $x = 0,\ z = -d$, by a generator with emf V_0^e and impedance Z_g; the horizontal wire extends beyond the generator to $x = -l$ as a monopole. The antenna is terminated at $x = h$ in the impedance Z_L; the wire extends horizontally beyond $x = h$ to $x = h + l$.

When close to the surface of the earth, the horizontal wire in both forms of the wave antenna is so closely coupled to the earth that it does not behave like an isolated antenna carrying a current with the wave number k_2 of air, but like an eccentrically insulated antenna lying on the surface of the earth. The current in such an antenna is much like that in a transmission line with the wave number $k_L = \beta_L + i\alpha_L$ and the characteristic impedance Z_c that depend on the inductance, capacitance, and resistance per unit length of the wire. For an antenna isolated in the air, radiation is a property of the antenna as a whole that cannot be assigned per unit length. For the antenna close to the earth, radiation is a property per unit length. Its properties are discussed in King and Smith (1981, Sec. 1.9) and analyzed in King, Wu, and Shen (1974). The formula for k_L is

$$k_L = k_2\left\{1 + \frac{2}{\ln(2d/a)}\left[\frac{1}{(2k_1d)^2} - \frac{K_1(2k_1d)}{2k_1d} + \frac{i\pi I_1(2k_1d)}{4k_1d} - i\left(\frac{2k_1d}{3} + \frac{(2k_1d)^3}{45} + \frac{(2k_1d)^5}{1575} + \cdots\right)\right]\right\}^{1/2}, \tag{7.3.1}$$

where $K_1(x)$ and $I_1(x)$ are the modified Bessel functions. When $|k_1d|$ is small, i.e., $0 < |k_1d| \leq 0.8$,

$$k_L \sim k_2\left(1 - \frac{\ln(k_1d) + \gamma - \frac{1}{2} - i(\frac{1}{2}\pi - \frac{4}{3}k_1d)}{\ln(2d/a)}\right)^{1/2}, \tag{7.3.2}$$

with $\gamma = 0.5772$. This formula is obtained with

$$K_1(x) \sim \frac{1}{x} + I_1(x)\left[\ln\left(\frac{x}{2}\right) + \gamma - \frac{1}{2}\right]; \qquad I_1(x) \sim \frac{x}{2}.$$

The formula for Z_c is

$$Z_c = \frac{\omega\mu_0 k_L}{2\pi k_2^2}\ln\left(\frac{2d}{a}\right) = \frac{60k_L}{k_2}\ln\left(\frac{2d}{a}\right)\ \Omega. \tag{7.3.3}$$

The expression on the right in (7.3.3) applies when Region 2 is air with $k_2 = \omega(\mu_0\epsilon_0)^{1/2}$ and $(\mu_0/\epsilon_0)^{1/2} = \zeta_0 = 120\pi\ \Omega$. The current on the conductor has the general transmission-line form:

$$I_x(x) = -\frac{iV_0^e}{Z_c} \frac{\sin[k_L(h-x)+i\theta_h]}{\cos(k_L h + i\theta_h)}, \tag{7.3.4}$$

where $\theta_h = \coth^{-1}(Z_h/Z_c)$ and Z_h is the terminating impedance at $x = h$. When $Z_h = Z_c$, $\theta_h = \infty$ and $I_x(x)$ is a traveling wave given by

$$I_x(x) = \frac{V_0^e}{Z_c} e^{ik_L x}. \tag{7.3.5}$$

The impedance of the antenna at the driving point $x = 0$ is $Z_{\text{in}} = Z + Z_g + Z_0$, where $Z = iZ_c \cot(k_L h + i\theta_h)$, Z_g is the impedance of the generator, and Z_0 is the impedance of the terminating section at $x = 0$. For traveling-wave operation, the effective impedance $Z_h = Z_L + Z_0$ of the terminating section at $x = h$ is made equal to the characteristic impedance Z_c so that $Z = Z_h = Z_c$. With the Beverage antenna, Z_0 is the impedance of each of the grounded vertical conductors. With the horizontal-wire antenna, Z_0 is the impedance of the open-ended section of the eccentrically insulated wire of length l, i.e., with $\theta_h = -i\pi$ for an ideal open end, $Z_0 = iZ_c \cot k_L l$.

In Chapter 18, the electromagnetic field generated by the currents in the several parts of the Beverage and horizontal-wire antennas is determined. It is shown that the contributions by the currents in the terminations are of minor significance when $|k_1 d|$ is small and the length h is adjusted for near optimum. The properties of both antennas are comparable and the field in Region 2 at radial distances $\rho_0 \gg h$ has the following components:

$$E_{2\rho}(\rho_0, \phi_0, 0) = I_x(0) h_e(\phi_0) [E_{2\rho}(\rho_0, \phi_0, 0)]_h, \tag{7.3.6}$$

$$E_{2z}(\rho_0, \phi_0, 0) = I_x(0) h_e(\phi_0) [E_{2z}(\rho_0, \phi_0, 0)]_h, \tag{7.3.7}$$

$$B_{2\phi}(\rho_0, \phi_0, 0) = \frac{k_1}{\omega} E_{2\rho}(\rho_0, \phi_0, 0), \tag{7.3.8}$$

where

$$h_e(\phi_0) = \frac{i(1 - e^{i(k_L - k_2 \cos\phi_0)h})}{k_L - k_2 \cos\phi_0} \tag{7.3.9}$$

is the complex effective length and $[E_{2\rho}(\rho_0, \phi_0, 0)]_h$ and $[E_{2z}(\rho_0, \phi_0, 0)]_h$ are the components of a horizontal electric dipole with unit electric moment located at $(0, 0, 0)$. The radial distance ρ_0 and the angle ϕ_0 are shown in Fig. 7.3.1c. The magnitude of (7.3.9) in the direction $\phi_0 = 0$ is

$$|h_e(0)| = \frac{[1 - 2e^{-\alpha_L h}\cos(\beta_L - k_2)h + e^{-2\alpha_L h}]^{1/2}}{[(\beta_L - k_2)^2 + \alpha_L^2]^{1/2}}. \tag{7.3.10}$$

The maximizing condition, $d|h_e(0)|/dh = 0$, yields the equation

$$(\beta_L - k_2)h_m = \pi - \tan^{-1}\left[\frac{\alpha_L}{\beta_L - k_2}\left(1 - \frac{e^{-\alpha_L h_m}}{\cos(\beta_L - k_2)h_m}\right)\right]. \tag{7.3.11}$$

Table 7.3.1. Properties of the Wave Antenna; $a = 1$ mm

d/a	d/λ_2	$\beta_L + i\alpha_L$ m^{-1}	$\beta_L - k_2$ m^{-1}	$(\beta_L - k_2)h_m$	h_m m	h_e m	$e^{-\alpha_L h_m}$
$f = 300$ MHz, $\lambda_2 = 1$ m, $k_2 = 2\pi$ m^{-1}; Over salt water: $\sigma_1 = 3.9$ S/m, $\epsilon_{1r} = 80$, $k_1 = 80.3 + i57.1$ m^{-1}							
10	0.01	$6.981 + i0.528$	0.698	2.49	3.57	1.28	0.15
20	0.02	$6.641 + i0.295$	0.358	2.45	6.84	2.38	0.13
50	0.05	$6.428 + i0.126$	0.145	2.43	16.76	5.69	0.12
100	0.10	$6.353 + i0.044$	0.070	2.42	48.46	13.22	0.12
$f = 300$ MHz, $\lambda_2 = 1$ m, $k_2 = 2\pi$ m^{-1}; Over fresh water: $\sigma_1 = 0.092$ S/m, $\epsilon_{1r} = 80$, $k_1 = 56.6 + i1.92$ m^{-1}							
10	0.01	$7.188 + i1.043$	0.905	2.28	2.52	0.76	0.072
20	0.02	$6.604 + i0.609$	0.321	2.06	6.40	1.47	0.020
50	0.05	$6.390 + i0.258$	0.107	1.96	19.2	3.59	0.007
100	0.10	$6.327 + i0.107$	0.038	1.91	50.6	8.81	0.004
$f = 30$ MHz, $\lambda_2 = 10$ m, $k_2 = 2\pi \times 10^{-1}$ m^{-1}; Over earth: $\sigma_1 = 0.001$ S/m, $\epsilon_{1r} = 8$, $k_1 = 5.62 + i0.021$ m^{-1}							
100	0.01	$6.63 + i0.046$	0.035	2.22	63.4	17.9	0.054
$f = 3$ MHz, $\lambda_2 = 100$ m, $k_2 = 2\pi \times 10^{-2}$ m^{-1}; Over earth: $\sigma_1 = 0.001$ S/m, $\epsilon_{1r} = 8$, $k_1 = 0.56 + i0.021$ m^{-1}							
10^3	0.01	$(6.53 + i0.31) \times 10^{-2}$	0.25×10^{-2}	2.25	900	261	0.061
$f = 300$ kHz, $\lambda_2 = 1$ km, $k_2 = 2\pi \times 10^{-3}$ m^{-1}; Over earth: $\sigma_1 = 0.001$ S/m, $\epsilon_{1r} = 8$, $k_1 = (3.67 + i3.21) \times 10^{-2}$ m^{-1}							
10^4	0.01	$(6.72 + i1.07) \times 10^{-3}$	0.45×10^{-3}	1.97	4380	850	0.05

For a typical wave antenna, $2d/a \gg 1$ and $\ln(1/k_1 d)$ is significantly smaller than $\ln(2d/a)$. In this case, $(\beta_L - k_2)$ is quite small and often comparable to α_L, as in the example evaluated in Chapter 18. This means that $\alpha_L h_m$ is of the order of π and $\exp(-\alpha_L h_m)$ is quite small. When this is true, (7.3.11) reduces to

$$(\beta_L - k_2)h_m \sim \pi - \tan^{-1}\left(\frac{\alpha_L}{\beta_L - k_2}\right), \tag{7.3.12}$$

which is readily solved for h_m. Solutions of (7.3.12) for the maximizing values of h_m and the associated effective length $h_e \equiv h_e(0)$ are listed in Table 7.3.1 under a variety of conditions. These include the wave antenna over salt and fresh water with a range of different heights at $f = 300$ MHz with $\lambda_2 = 1$ m and the wave antenna over earth at $f = 30$ MHz, 3 MHz, and 300 kHz with $\lambda_2 = 10$ m, 100 m, and 1 km, respectively. It is seen from Table 7.3.1 that in each case the maximizing length is several wavelengths in air and the effective length is of the order of one-quarter to one-third the actual length. The total electric moment in each case is $I_x(0)h_e$.

For use in over-the-horizon radar in the frequency range from 3 to 30 MHz, the optimum lengths range from 900 m to 63.4 m, and the electric moments from $261I_x(0)$ to $17.9I_x(0)$. The electromagnetic field is given by (7.3.6)–(7.3.8). The gain can be increased by arranging many identical elements parallel to one another in an upward- and downward-directed broadside array.

The wave antenna is not practical for communicating with submarines at frequencies as low as 1 kHz. The required optimum lengths of the antenna are excessive.

7.4 Communicating with Submarines with Horizontal Antennas

An alternative to the wave antenna at a small height d *in the air* above the earth or sea (in effect an eccentrically insulated antenna) is the symmetrically insulated traveling-wave antenna located at a small depth d *in the earth or sea.* It is described and analyzed in Chapter 17. It consists of a conductor (radius a, length h) enclosed in an insulating sheath (outer radius b) and is driven at $x = 0$ against a bare monopole with a length $l = \lambda_1/4$ (where λ_1 is the wavelength in the earth or sea) and an impedance Z_0. Because of the rapid exponential decrease of the current along a bare conductor in salt water, longer antennas offer no advantage and they have essentially the same impedance. The insulated conductor is terminated at $x = h$ in a second bare monopole in series with an impedance Z which is chosen so that $Z+Z_0 = Z_c$, where Z_c is the characteristic impedance of the insulated antenna. With this termination, the current along the insulated horizontal wire is a traveling wave given by

$$I_x(x) = I_x(0)e^{ik_L x}; \qquad 0 \le x \le h, \tag{7.4.1}$$

with

$$I_x(0) = \frac{V_0^e}{Z_0 + Z_c}. \tag{7.4.2}$$

The properties of the insulated antenna are analyzed in King and Smith (1981, Chap. 8). In particular, the wave number $k_L = \beta_L + i\alpha_L$ is determined in Section 8.3, eq. (3.4). It has the following quite simple form when $f \ge 1$ kHz and ohmic losses are negligible:

$$k_L = k_2\left(1 + \frac{H_0^{(1)}(k_1 b)}{k_1 b \ln(b/a) H_1^{(1)}(k_1 b)}\right)^{1/2}. \tag{7.4.3}$$

The associated characteristic impedance is

$$Z_c = 60\,\frac{k_L}{k_2}\ln\left(\frac{b}{a}\right)\ \Omega. \tag{7.4.4}$$

At radial distances ρ_0 that are large compared to the length h of the antenna, i.e., $\rho_0^2 \gg h^2$, a typical component of the electromagnetic field is well approximated by

$$E_{1\rho}(\rho_0, \phi_0, z) = I_x(0) h_e(\phi_0) [E_{1\rho}(\rho_0, \phi_0, z)]_h, \tag{7.4.5}$$

where $[E_{1\rho}(\rho_0, \phi_0, z)]_h$ is the radial electric field of a unit horizontal electric dipole at $(0,0,0)$. By (7.4.1), the effective electric moment is

$$\begin{aligned} h_e(\phi_0) &= \frac{1}{I_x(0)}\int_{-l}^{h+l} I_x(x') e^{-ik_2 x' \cos\phi_0}\,dx' \\ &\sim \frac{1}{I_x(0)}\int_0^h I_x(x') e^{-ik_2 x' \cos\phi_0}\,dx' \\ &\sim \int_0^h e^{i(k_L - k_2\cos\phi_0)x'}\,dx', \end{aligned} \tag{7.4.6}$$

so that

$$h_e(\phi_0) = \frac{i(1 - e^{i(k_L - k_2\cos\phi_0)h})}{k_L - k_2\cos\phi_0}. \tag{7.4.7}$$

This is formally the same as (7.3.9) for the wave antenna, but k_L is quite different. The maximizing length h_m in the direction $\phi_0 = 0$ is given by (7.3.11), but for the insulated antenna $(\beta_L - k_2)$ is usually much larger than α_L so that $\alpha_L h_m$ is quite small. Thus, $\cos(\beta_L - k_2)h_m \sim -1$ and $\exp(-\alpha_L h_m) \sim 1$, so that (7.3.11) reduces to

$$(\beta_L - k_2)h_m \sim \pi - \tan^{-1}\frac{2\alpha_L}{\beta_L - k_2} \sim \pi - \frac{2\alpha_L}{\beta_L - k_2}. \tag{7.4.8}$$

The maximum effective length is

$$|h_{em}(0)| = \left|\frac{1 - e^{i(k_L - k_2)h_m}}{k_L - k_2}\right| \sim \frac{1 + e^{-\alpha_L h_m}}{[(\beta_L - k_2)^2 + \alpha_L^2]^{1/2}}. \tag{7.4.9}$$

For communicating with submarines to a depth of 50 m in the ocean at radial distances up to $\rho_0 = 5000$ km, a frequency near $f = 1$ kHz is preferred. Consider, as an example, an insulated antenna with a conductor with radius $a = 0.735$ cm and a styrofoam insulating sheath with outer radius $b = 4.41$ cm. The wave numbers are $k_2 = 2.09 \times 10^{-5}$ m^{-1}, $k_1 = 0.178e^{i\pi/4} = 0.126(1+i)$ m^{-1}, $k_1 b = 0.0078e^{i\pi/4}$, $H_0^{(1)}(k_1 b) \sim 0.5 - i3.16 = 3.20e^{-i1.41}$, $k_1 b H_1^{(1)}(k_1 b) \sim -i0.636$ so that

$$k_L = \beta_L + i\alpha_L = (4.07 + i0.24) \times 10^{-5} \text{ m}^{-1}, \tag{7.4.10}$$

$$Z_c = 209.7 + i12.5\ \Omega. \tag{7.4.11}$$

With (7.4.10), (7.4.8) gives

$$h_m = \left(\frac{\pi}{1.98} - \frac{0.48}{(1.98)^2}\right) \times 10^5 = 1.464 \times 10^5 \text{ m} = 146.4 \text{ km}. \tag{7.4.12}$$

When this value is substituted in (7.4.9), the result is

$$|h_{em}(0)| \sim 0.85 \times 10^5 \text{ m} = 85 \text{ km}. \tag{7.4.13}$$

It is seen from Fig. 7.2.2 that the field at $\rho_0 = 5000$ km and the depth $z = 20$ m due to a unit horizontal electric dipole at the surface is -368 dB referred to 1 V/m. At the radial distance $\rho_0 = 1000$ km, it is -354 dB. If a single insulated traveling-wave antenna with the optimum length of 146.4 km is substituted for the unit dipole and it carries a current $I_x(0) = 100$ A, the effective moment is

$$I_x(0)|h_{em}(0)| = 100 \times 85 \times 10^3 = 8.5 \times 10^6 \text{ A m}. \tag{7.4.14}$$

This corresponds to a gain of $20 \log_{10} |I_x(0)h_e(0)| = 139$ dB and raises the level of the field at the depth $z = 20$ m to -229 dB at $\rho_0 = 5000$ km and to -215 dB at $\rho_0 = 1000$ km. The corresponding values at the depths $z = 30$, 40, and 50 m are -239, -249, and -259 dB at $\rho_0 = 5000$ km and -225, -235, and -245 dB at $\rho_0 = 1000$ km. This field can be increased further if the single element is replaced by an array of 100 parallel identical elements spaced 25 m apart with each carrying 100 A. This provides an additional gain of 40 dB so that the fields at the depths 20, 30, 40, and 50 m become -189, -199, -209, and -219 dB at $\rho_0 = 5000$ km. At $\rho_0 = 1000$ km, they become -175, -185, -195, and -205 dB. These values are all greater than a suggested minimum detectable value of -220 dB, which corresponds to a field of order 10^{-11} V/m. With a receiving antenna with effective length 10^3 m, a receiver with a sensitivity of 0.01 μV is required.

With the adjacent elements of the array separated by the distance $s =$ 25 m, the coupling between the elements in sea water is negligible because $\exp(-\alpha_1 s) = \exp(-25 \times 0.126) = 0.04$. It follows that the driving-point impedance of each element is the same as when alone, i.e., $Z_{\text{in}} = Z_c$ when the antenna is matched. With (7.4.11) and (7.4.14), the power to each element is

$$P_1 = I_x^2(0)R_c = 100^2 \times 209.7 = 2.1 \text{ MW}. \tag{7.4.15}$$

This is the same as the power supplied to the Cutler, Maine antenna in the frequency range from 14 to 28.5 kHz. The total power to the 100-element submerged array at 1 kHz is

$$P_{100} = 210 \text{ MW}. \tag{7.4.16}$$

This very large amount of power is required for penetrating 50 m of highly attenuating sea water at radial distances up to 5000 km from the source.

7.5 Lateral Waves from a Horizontal Antenna for Remote Sensing in the Earth or Water

Buried or submerged objects, inhomogeneities in the earth, and regions with electrical properties that differ significantly from the surrounding earth can be detected and localized from measurements of the scattered field made in the air. In one arrangement, a horizontal dipole is located in the air on the surface of the earth or water and measurements are made with a receiving antenna that is moved about on the surface of the earth in an area over a possible buried object. This is illustrated in Fig. 7.5.1 specifically for an experimental model for which the transmitting dipole is at the small height d_t over the point $x_0 = y_0 = 0$ on the surface of salt water. The submerged object to be located is taken to be an insulated metal rod centered at either of two locations given by ($x_0 = \rho_c$, $y_0 = 0$, $z = d$) or ($x_0 = 0$, $y_0 = \rho_c$, $z = d$). The electromagnetic field is to be investigated in square areas designated as Areas #I and #II on the surface over the submerged insulated conductor. The electric fields incident on the center of the conductor in each of the two locations are the lateral-wave components $E_{1\rho}^{\text{inc}}(\rho_c, 0, d)$ and $E_{1\phi}^{\text{inc}}(\rho_c, \pi/2, d)$. These are given by

$$E_{1\rho}^{\text{inc}}(\rho, \phi, z) = -Gg(\rho; k_1, k_2)e^{ik_2\rho}e^{ik_1 z}\cos\phi = \frac{\omega}{k_1} B_{1\phi}^{\text{inc}}(\rho, \phi, z), \tag{7.5.1}$$

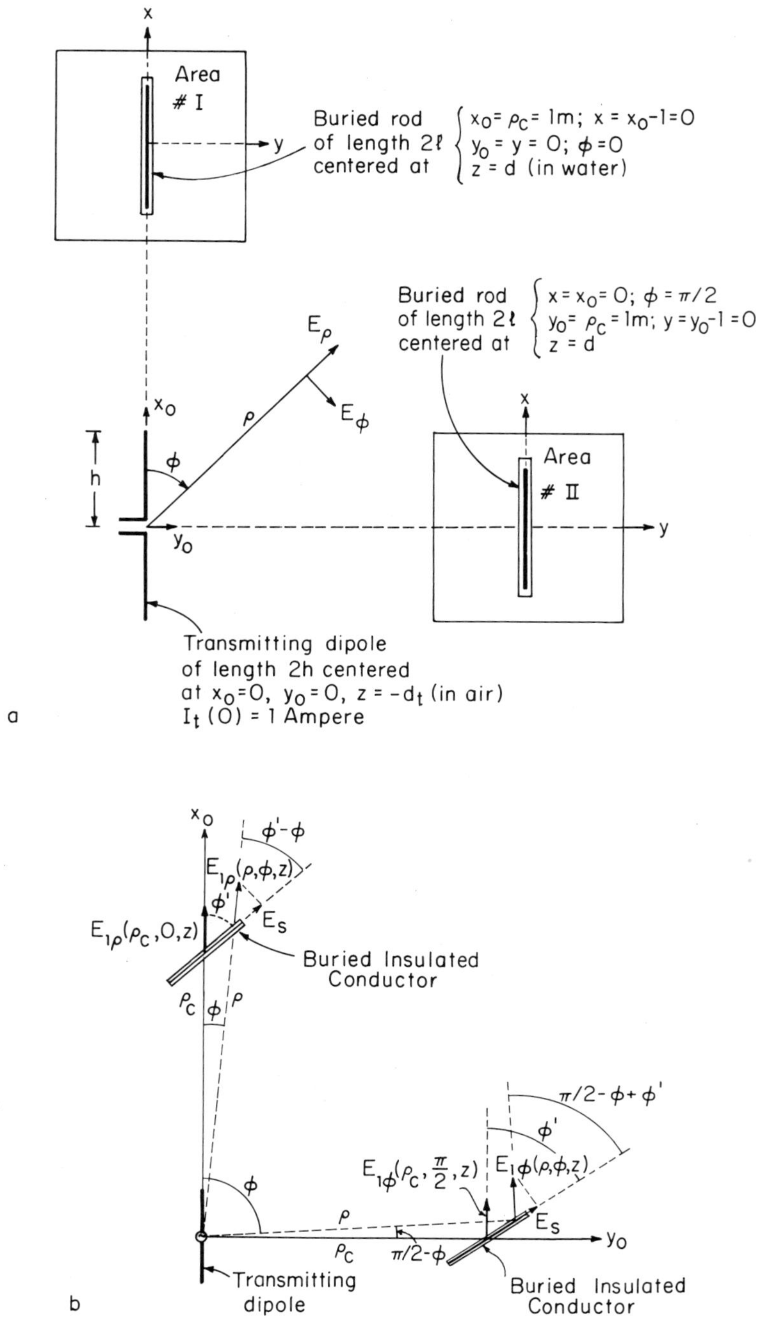

Figure 7.5.1. Two locations of a buried insulated conductor relative to a transmitting dipole on the surface of the earth. (a) Orientations of scattering antenna in areas of interest. (b) General orientations of scattering antenna.

$$E_{1\phi}^{\text{inc}}(\rho,\phi,z) = 2Gh(\rho;k_1,k_2)e^{ik_2\rho}e^{ik_1z}\sin\phi = -\frac{\omega}{k_1}B_{1\rho}^{\text{inc}}(\rho,\phi,z), \tag{7.5.2}$$

where $g(\rho;k_1,k_2)$ and $h(\rho;k_1,k_2)$ are defined respectively in (5.5.44) and (5.5.45), and where

$$G \equiv \frac{\omega\mu_0 k_2}{2\pi k_1^2} I_x(0)M_0(\phi). \tag{7.5.3}$$

The effective length of the center-driven transmitting antenna is

$$M_0(\phi) = \frac{2k_L}{\sin k_L h}\left(\frac{\cos(k_2 h\cos\phi) - \cos k_L h}{k_L^2 - k_2^2\cos^2\phi}\right). \tag{7.5.4}$$

In (7.5.4), k_L is defined in (7.3.1) and $M_0(\phi)$ is obtained from

$$I_x(0)M_0(\phi) = \int_{-h}^{h} I_x(x')e^{-ik_2x'\cos\phi}\,dx', \tag{7.5.5}$$

with

$$I_x(x') = I_x(0)\,\frac{\sin k_L(h-|x'|)}{\sin k_L h}, \tag{7.5.6}$$

for a dipole with half-length h and open ends. From (7.3.4) with $\theta_h = 0$,

$$I_x(0) = -\frac{iV_0^e}{Z_c}\tan k_L h, \tag{7.5.7}$$

with Z_c given by (7.3.3).

Consider first the current induced in a buried conductor in Area #I in Fig. 7.5.1a. The insulated conductor with length $2l$ is located at depth d in the earth with its center at $x_0 = \rho = \rho_c$, $y_0 = 0$, as shown in Fig. 7.5.1b. The subtended angle $\pm\phi$ and with it $\sin\phi$ are small so that the component $E_{1\phi}^{\text{inc}}(\rho,\phi,d) = E_{1\phi}^{\text{inc}}(\rho,0,d)\sin\phi$ is small compared to $E_{1\rho}^{\text{inc}}(\rho,\phi,d) = E_{1\rho}^{\text{inc}}(\rho,0,d)\cos\phi \sim E_{1\rho}^{\text{inc}}(\rho,0,d)$. It follows that the induced currents are due primarily to $E_{1\rho}^{\text{inc}}(\rho,\phi,d)$. The distance along the conductor from its center is s' with $-l \le s' \le l$. The radial incident field at s' along the conductor is

$$E_{1\rho}^{\text{inc}}(\rho,\phi,d) = -Gg(\rho;k_1,k_2)e^{ik_2\rho}e^{ik_1d}\cos\phi, \tag{7.5.8}$$

where, with the angle ϕ' shown in Fig. 7.5.1b,

$$\rho = (\rho_c^2 + s'^2 + 2\rho_c s'\cos\phi')^{1/2} \sim \rho_c + s'\cos\phi'. \tag{7.5.9}$$

The component of the incident radial electric field along the submerged conductor is

$$E_s(s') = E_{1\rho}^{\text{inc}}(\rho,\phi,d)\cos(\phi'-\phi) \sim E_{1\rho}^{\text{inc}}(\rho_c,0,d)\cos\phi'\,e^{ik_2s'\cos\phi'}. \tag{7.5.10}$$

It is this field that induces the current in the conductor.

An insulated conductor in Region 1 with $|k_1| \gg |k_i|$ (where k_i is the wave number of the insulating sheath) has the properties of a transmission line with the wave number k_L and the characteristic impedance Z_c. Since it is at the distance d from the air boundary, it is, in effect, coupled to an image antenna at the distance $2d$. Because the reflection coefficient of a wave that is normally incident on the boundary from Region 1 is

$$f_{mr} = \frac{k_1 - k_2}{k_1 + k_2} \sim 1, \tag{7.5.11}$$

the current in the image antenna is in phase with that in the insulated wire. Hence, the wave number and characteristic impedance are those of two identical parallel conductors excited in phase. Coupled insulated antennas are analyzed in King and Smith (1981, Sec. 1.8), where the formulas for the wave number and characteristic impedance of two antennas driven in phase are eqs. (8.7) and (8.8) with $m = 0$. These are

$$k_L' = k_i \left(1 + \frac{H_0^{(1)}(k_1 b) + H_0^{(1)}(2k_1 d)}{k_1 b \ln(b/a) H_1^{(1)}(k_1 b)} \right)^{1/2}, \tag{7.5.12}$$

$$Z_c' = \frac{\omega \mu_0 k_L'}{2\pi k_i^2} \ln\left(\frac{b}{a}\right). \tag{7.5.13}$$

In these formulas a is the radius of the conductor and b the outer radius of the insulating sheath.

The current induced by the electric field along the entire length of the conductor is

$$I_s = -\frac{i}{Z_c' \sin 2k_L' l} \left(\sin k_L'(l-s) \int_{-l}^{s} E_s(s') \sin k_L'(l+s')\, ds' + \sin k_L'(l+s) \int_{s}^{l} E_s(s') \sin k_L'(l-s')\, ds' \right). \tag{7.5.14}$$

With (7.5.10), this leads to

$$I_s = \frac{ik_L' E_{1\rho}^{\text{inc}}(\rho_c, 0, d) \cos\phi'}{Z_c'(k_L'^2 - k_2^2 \cos^2\phi')} \left(e^{ik_2 s \cos\phi'} - \frac{\sin(k_L' + k_2\cos\phi')l}{\sin 2k_L' l}\, e^{ik_L' s} - \frac{\sin(k_L' - k_2\cos\phi')l}{\sin 2k_L' l}\, e^{-ik_L' s} \right). \tag{7.5.15}$$

Of particular interest is the direction $\phi' = 0$ of maximum $E_{1\rho}^{\text{inc}}(\rho_c, \phi, d)$. For this, $s = x$ and (7.5.15) reduces to

$$I_x = \frac{ik_L' E_{1\rho}^{\text{inc}}(\rho_c, 0, d)}{Z_c'(k_L'^2 - k_2^2)} \left(e^{ik_2 x} - \frac{\sin(k_L' + k_2)l}{\sin 2k_L' l}\, e^{ik_L' x} - \frac{\sin(k_L' - k_2)l}{\sin 2k_L' l}\, e^{-ik_L' x} \right). \tag{7.5.16}$$

When the buried conductor is in Area #II with its center at $x_0 = 0$, $y_0 = \rho = \rho_c$, as shown in Fig. 7.5.1b, the induced current is readily shown to be

$$I_s = \frac{ik'_L E^{\text{inc}}_{1\phi}(\rho_c, \pi/2, d)\cos\phi'}{Z'_c(k'^2_L - k_2^2\sin^2\phi')}\left(e^{ik_2 s\sin\phi'} - \frac{\sin(k'_L + k_2\sin\phi')l}{\sin 2k'_L l}\, e^{ik'_L s} - \frac{\sin(k'_L - k_2\sin\phi')l}{\sin 2k'_L l}\, e^{-ik'_L s}\right), \tag{7.5.17}$$

which reduces to

$$I_x = -\frac{iE^{\text{inc}}_{1\phi}(\rho_c, \pi/2, d)}{Z'_c k'_L}\,\frac{\cos k'_L x - \cos k'_L l}{\cos k'_L l} \tag{7.5.18}$$

when the conductor is parallel to the x-axis with $\phi' = 0$.

The currents in (7.5.15) through (7.5.18) generate the scattered field that travels out from the submerged insulated conductor. Of interest is the field on the surface ($z = 0-$) in air in an area above the conductor, but it is first necessary to determine the field on the surface ($z = 0+$) in the earth or water. For simplicity, let the conductor—when in each area—be parallel to the x-axis so that the respective induced currents have the simpler forms (7.5.16) and (7.5.18).

The field radiated by these currents is given by

$$E^{\text{scat}}_{1x}(x, y, 0) = \frac{i\omega\mu_0}{4\pi}\int_{-l}^{l} I_x(x')\left(\frac{e^{ik_1 r}}{r} + \frac{1}{k_1^2}\frac{\partial^2}{\partial x^2}\frac{e^{ik_1 r}}{r}\right)dx', \tag{7.5.19}$$

$$E^{\text{scat}}_{1y}(x, y, 0) = -\frac{i\omega\mu_0}{4\pi k_1^2}\int_{-l}^{l} I_x(x')\,\frac{\partial^2}{\partial x\partial y}\frac{e^{ik_1 r}}{r}\,dx', \tag{7.5.20}$$

where $I_x(x')$ in (7.5.19) is given by (7.5.16) and $I_x(x')$ in (7.5.20) by (7.5.18), with x replaced by x' in both cases. The evaluation of these integrals is long and leads to generalized exponential integrals. The complete expressions for (7.5.19) and (7.5.20) involve the following notation:

$$\delta_1 = \frac{\alpha_1}{\beta_1}; \qquad \delta_{12} = \frac{\alpha_{12}}{\beta_{12}}; \qquad \delta_{1L} = \frac{\alpha_{1L}}{\beta_{1L}}, \tag{7.5.21}$$

$$A_1 = \beta_1\sqrt{y^2 + d^2}; \quad A_{12} = \beta_{12}\sqrt{y^2 + d^2}; \quad A_{1L} = \beta_{1L}\sqrt{y^2 + d^2}, \tag{7.5.22}$$

$$U_1 = \beta_1(l - x); \qquad U_{12} = \beta_{12}(l - x); \qquad U_{1L} = \beta_{1L}(l - x), \tag{7.5.23}$$

$$U_2 = \beta_1(l + x); \qquad U_{22} = \beta_{12}(l + x); \qquad U_{2L} = \beta_{1L}(l + x), \tag{7.5.24}$$

$$k_1 = \beta_1 + i\alpha_1; \qquad k_{12} = \sqrt{k_1^2 - k_2^2} = \beta_{12} + i\alpha_{12}; \qquad k_{1L} = \sqrt{k_1^2 - k'^2_L} = \beta_{1L} + i\alpha_{1L}. \tag{7.5.25}$$

The complete expressions for the scattered field just below the air surface are

$$\begin{aligned}
&E_{1x}^{\text{scat}}(x,y,0)\\
&\quad = E_{1\rho}^{\text{inc}}(\rho_c,0,d)D_1\\
&\qquad \times \left\{e^{ik_2x}[\bar{E}(\delta_{12},A_{12},U_{12})+\bar{E}(\delta_{12},A_{12},U_{22})]\right.\\
&\qquad\quad -(Be^{ik'_Lx}+Ce^{-ik'_Lx})[\bar{E}(\delta_{1L},A_{1L},U_{1L})+\bar{E}(\delta_{1L},A_{1L},U_{2L})]\\
&\qquad\quad \left.+C_{1x}(l)\right\}; \qquad x\le l
\end{aligned} \tag{7.5.26}$$

for Area #I, and

$$\begin{aligned}
E_{1x}^{\text{scat}}(x,y,0) = E_{1\phi}^{\text{inc}}(\rho_c,\pi/2,d)D_2&\\
\times\Bigg([\bar{E}(\delta_1,A_1,U_1)&+\bar{E}(\delta_1,A_1,U_2)]\\
&-\frac{\cos k'_Lx}{\cos k'_Ll}[\bar{E}(\delta_{1L},A_{1L},U_{1L})+\bar{E}(\delta_{1L},A_{1L},U_{2L})]\\
&+C_{2x}(l)\Bigg); \qquad x\le l
\end{aligned} \tag{7.5.27}$$

for Area #II. Formulas for the range $|x|>l$ are like (7.5.26) and (7.5.27) with x where it occurs explicitly in the three exponentials in (7.5.26) and in the cosine in (7.5.27) replaced by l. In (7.5.26) and (7.5.27), the generalized exponential integral $\bar{E}(\delta,A,U)$ is defined as

$$\bar{E}(\delta,A,U)\equiv\int_0^U\frac{e^{-\delta W}e^{iW}}{W}\,dU; \qquad W=\sqrt{U^2+A^2}, \tag{7.5.28}$$

and

$$D_1=-\frac{\omega\mu_0k'_L}{4\pi Z'_c(k_L'^2-k_2^2)}; \qquad D_2=-\frac{\omega\mu_0}{4\pi Z'_ck'_L}, \tag{7.5.29}$$

$$B=\frac{\sin(k'_L+k_2)l}{\sin 2k'_Ll}; \qquad C=\frac{\sin(k'_L-k_2)l}{\sin 2k'_Ll}, \tag{7.5.30}$$

$$\begin{aligned}
C_{1x}(l)=-\frac{i}{k_1^2}\Bigg(&[k_2e^{ik_2l}-k'_L(Be^{ik'_Ll}-Ce^{-ik'_Ll})]\frac{e^{ik_1r_{1l}}}{r_{1l}}\\
&-[k_2e^{-ik_2l}-k'_L(Be^{-ik'_Ll}-Ce^{ik'_Ll})]\frac{e^{ik_1r_{2l}}}{r_{2l}}\Bigg),
\end{aligned} \tag{7.5.31}$$

$$C_{2x}(l)=-\frac{k'_L}{k_1^2}\left(\frac{e^{ik_1r_{1l}}}{r_{1l}}+\frac{e^{ik_1r_{2l}}}{r_{2l}}\right)\tan k'_Ll, \tag{7.5.32}$$

$$r_{1l}=[(l-x)^2+y^2+d^2]^{1/2},$$

$$r_{2l} = [(l+x)^2 + y^2 + d^2]^{1/2}. \tag{7.5.33}$$

In the integration it is assumed that the buried wire is long compared to its depth, i.e., $2l \gg d$.

In order to obtain the desired scattered fields in the air just above the boundary, it is necessary to multiply (7.5.26) and (7.5.27) by the transmission coefficient when the electric field is perpendicular to the plane of incidence; that is,

$$\begin{aligned} f &= \frac{2\cos\theta}{\cos\theta + [(k_2/k_1)^2 - \sin^2\theta]^{1/2}} \\ &= \frac{2d}{d + [(k_2/k_1)^2 r^2 - (x'-x)^2 - y^2]^{1/2}} \end{aligned} \tag{7.5.34}$$

since $\cos\theta = d/r$ and $\sin\theta = [(x'-x)^2 + y^2]^{1/2}/r$. Because $(k_2/k_1)^2$ is small and $r^2 \sim (x'-x)^2 + y^2$, it follows that

$$\begin{aligned} f &\sim \frac{2d}{d + i[(x'-x)^2 + y^2]^{1/2}} \\ &\sim \begin{cases} \dfrac{2d}{d+iy}, & x \le l \\ \dfrac{2d}{d + i[(l-x)^2 + y^2]^{1/2}}, & x > l \end{cases}. \end{aligned} \tag{7.5.35}$$

The approximations on the right assume that $l^2 \gg d^2$. Since f in the form (7.5.35) is independent of the integration with respect to x', the scattered field in Region 2 just above the boundary is

$$E_{2x}^{\text{scat}}(x,y,0) = f E_{1x}^{\text{scat}}(x,y,0), \tag{7.5.36}$$

where f is given by (7.5.35) and $E_{1x}^{\text{scat}}(x,y,0)$ by (7.5.26) for Area #I and by (7.5.27) for Area #II.

The total field in the air just above the surface is the sum of the incident field and the scattered field. The incident field is obtained from (7.5.1) and (7.5.2) with the boundary conditions $E_{2\rho}(\rho,\phi,0) = E_{1\rho}(\rho,\phi,0)$ and $E_{2\phi}(\rho,\phi,0) = E_{1\phi}(\rho,\phi,0)$. The x-component is

$$\begin{aligned} E_{2x}^{\text{inc}}(\rho,\phi,0) &= E_{2\rho}^{\text{inc}}(\rho,\phi,0)\cos\phi + E_{2\phi}^{\text{inc}}(\rho,\phi,0)\sin\phi \\ &= -Gg(\rho;k_1,k_2)e^{ik_2\rho}\cos^2\phi + 2Gh(\rho;k_1,k_2)e^{ik_2\rho}\sin^2\phi. \end{aligned} \tag{7.5.37}$$

In the Cartesian coordinates x_0, y_0 with the origin at the center of the transmitting antenna,

$$\rho = \sqrt{x_0^2 + y_0^2}; \qquad \cos\phi = \frac{x_0}{\sqrt{x_0^2 + y_0^2}}; \qquad \sin\phi = \frac{y_0}{\sqrt{x_0^2 + y_0^2}}.$$

Hence,

$$E_{2x}^{\mathrm{inc}}(x_0, y_0, 0) = -\frac{Ge^{ik_2\rho}}{x_0^2 + y_0^2}\left[x_0^2 g(\sqrt{x_0^2 + y_0^2}; k_1, k_2)\right.$$
$$\left. - 2y_0^2 h(\sqrt{x_0^2 + y_0^2}; k_1, k_2)\right], \tag{7.5.38}$$

where G is defined in (7.5.3). This expression can be referred to the coordinates x, y with $x_0 = \rho_c + x$, $y_0 = y$ for Area #I and with $x_0 = x$, $y_0 = \rho_c + y$ for Area #II.

The total field in either area has the form

$$E_{2x}^t(x, y, 0) = E_{2x}^{\mathrm{inc}}(x, y, 0) + E_{2x}^{\mathrm{scat}}(x, y, 0), \tag{7.5.39}$$

where $E_{2x}^{\mathrm{inc}}(x, y, 0)$ is obtained from (7.5.38) and $E_{2x}^{\mathrm{scat}}(x, y, 0)$ from (7.5.36) with the appropriate specialization to Area #I or Area #II. The magnitude is

$$|E_{2x}^t(x, y, 0)| = |E_{2x}^{\mathrm{inc}}(x, y, 0)[1 + Ke^{i\psi}e^{ik_1 d}]|, \tag{7.5.40}$$

where

$$Ke^{i\psi} = \frac{E_{2x}^{\mathrm{scat}}(x, y, 0)}{E_{2x}^{\mathrm{inc}}(x, y, 0)},$$

and $\exp(ik_1 d) = \exp(-\alpha_1 d)\exp(i\beta_1 d)$. It follows that

$$\left|\frac{E_{2x}^t(x, y, 0)}{E_{2x}^{\mathrm{inc}}(x, y, 0)}\right| = [1 + 2Ke^{-\alpha_1 d}\cos(\psi + \beta_1 d) + K^2 e^{-2\alpha_1 d}]^{1/2}. \tag{7.5.41}$$

Evidently the total field on the surface varies periodically with the electrical depth $\beta_1 d$ of the buried conductor. There are maxima near $(\psi + \beta_1 d) = n\pi$ with n even, viz.,

$$\left|\frac{E_{2x}^t(x, y, 0)}{E_{2x}^{\mathrm{inc}}(x, y, 0)}\right|_{\max} \sim 1 + Ke^{-\alpha_1 d}, \tag{7.5.42}$$

and minima near $(\psi + \beta_1 d) = n\pi$ with n odd, viz.,

$$\left|\frac{E_{2x}^t(x, y, 0)}{E_{2x}^{\mathrm{inc}}(x, y, 0)}\right|_{\min} \sim 1 - Ke^{-\alpha_1 d}. \tag{7.5.43}$$

The amplitude of the oscillation decreases exponentially with depth, i.e., as $\exp(-\alpha_1 d) = \exp(-d/d_s)$ where d_s is the skin depth.

Of interest in the air above a buried or submerged object are the incident lateral-wave field, the scattered field from the object, and the total field—which is the only one actually measured and used in the sensing of buried objects. Calculations have been made of these fields in Area #I at $f = 300$ MHz for a horizontal half-wave transmitting dipole and an insulated metal scattering rod of length $2l = 33.4$ cm and radius $a = 0.159$ cm, encased in a styrofoam sheath with outer radius $b = 0.413$ cm. Graphs of β_L' and α_L'

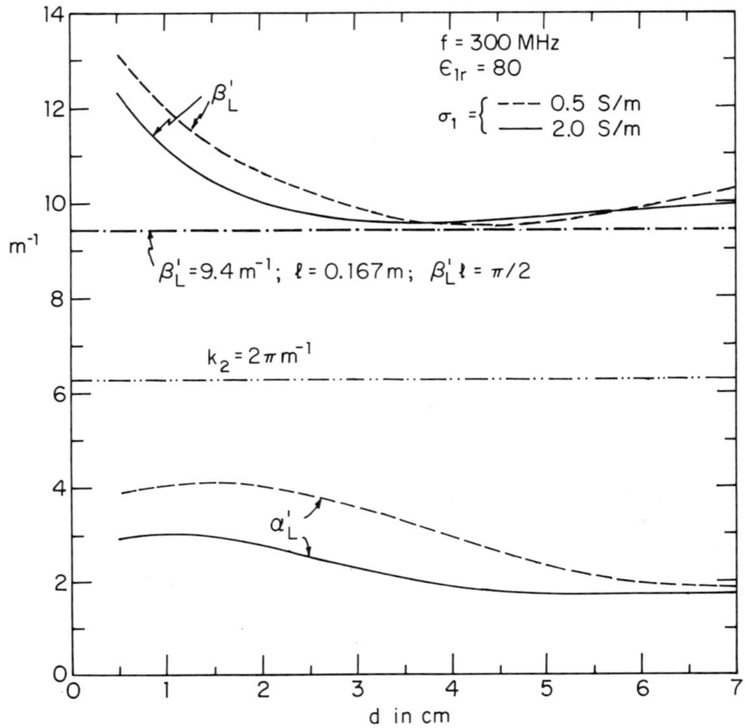

Figure 7.5.2. Wave number $k_L' = \beta_L' + i\alpha_L'$ for induced current on insulated scattering rod; Area #I.

for the wave number $k_L' = \beta_L' + i\alpha_L'$ of the current in the scattering rod as a function of its depth d are shown in Fig. 7.5.2. Since k_L' varies with the depth, the length $2l = 33.4$ cm is chosen so that β_L' is near its minimum and $\beta_L' l \geq \pi/2$ for all values of d in the range $0 < d \leq 8$ cm. The ambient medium is salt water with $\epsilon_{1r} = 80$ and $\sigma_1 = 2$ S/m for one set of data and $\sigma_1 = 0.5$ S/m for a second set of data.

Consider first the field at the point $x = y = 0$ directly above the center of the scattering rod as a function of the depth d. This is shown in Fig. 7.5.3. The incident lateral-wave field is independent of the depth but is greater when the conductivity is smaller due to the factor k_1^2 in the denominator of the amplitude. The magnitude of the scattered field decreases with depth but is also modified by the coupling of the insulated rod to its image, especially at the lower conductivity. The magnitude of the total field oscillates about the incident field.

In Fig. 7.5.4 the total electric field along the x-axis directly over the scattering rod is shown for $\sigma_1 = 2$ S/m on the right and for $\sigma_1 = 0.5$ S/m on the left. Contour diagrams of the magnitude of the scattered field in Area #I with $d = 1.5$ cm and $d = 4.5$ cm for $\sigma_1 = 2$ S/m are shown in

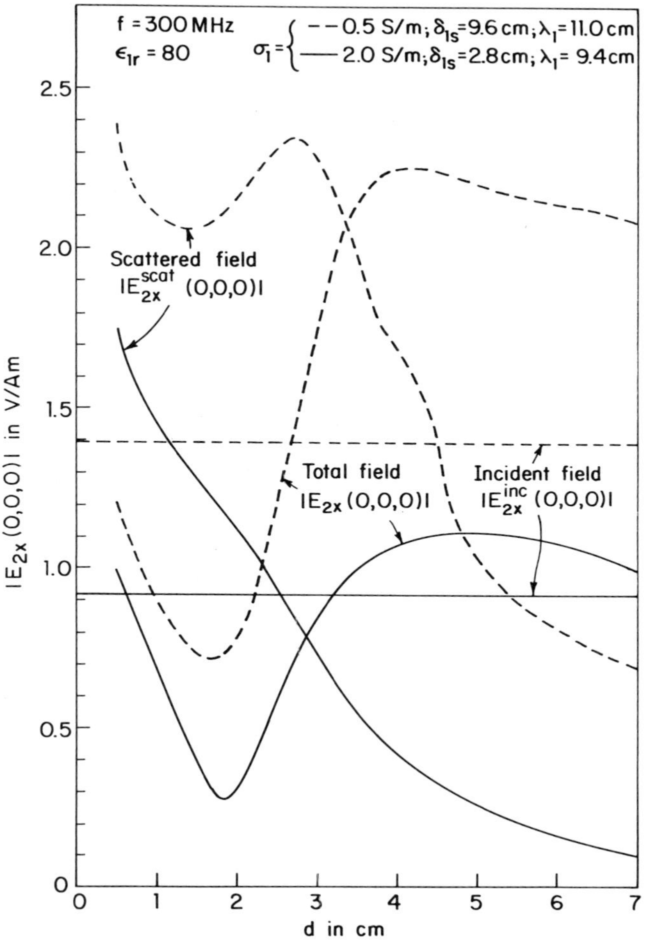

Figure 7.5.3. Electric field in air directly over center of insulated scattering rod as a function of its depth d in salt water; Area #I.

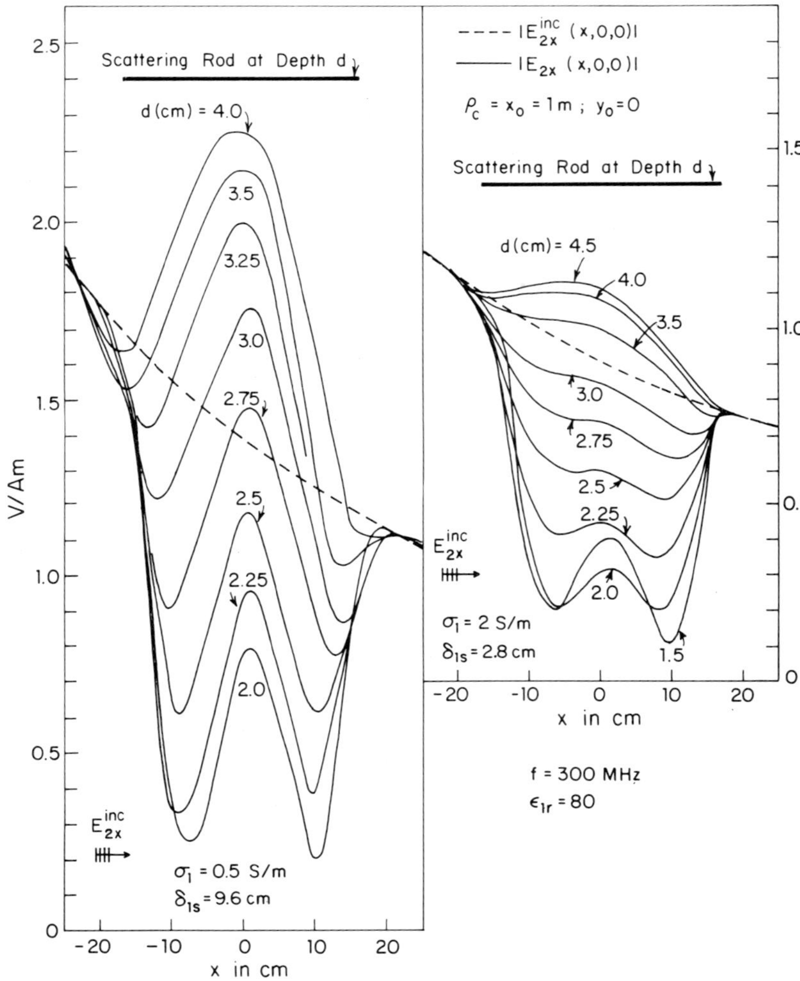

Figure 7.5.4. Electric field along x-axis directly over insulated scattering rod at depth d in salt water; Area #I.

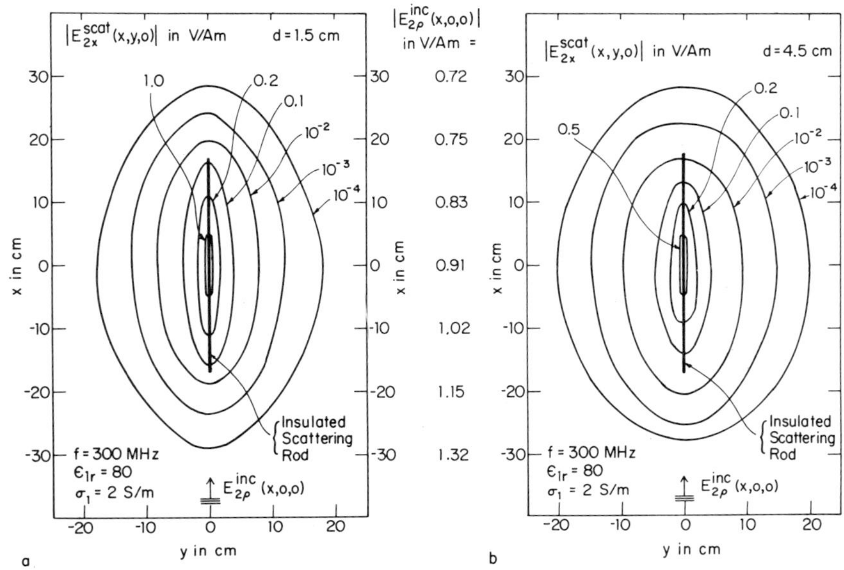

Figure 7.5.5. Contour diagrams of scattered electric field on surface above insulated scatterer of length $2l$ = 33.4 cm at depth d in salt water (σ_1 = 2 S/m); Area #I. (a) $d = 1.5$ cm. (b) $d = 4.5$ cm.

Fig. 7.5.5. The associated total field is shown in Fig. 7.5.6. It is evident from these diagrams that measurements of $|E_{2x}(x, y, 0)|$ with a receiving dipole moved along the surface of the water clearly indicate the location and general shape of the submerged object. Alternatively, a receiving loop can be used to measure $|B_{2y}(x, y, 0)|$ with substantially the same results.

Diagrams of the electric field in Area #II corresponding to those in Figs. 7.5.3 through 7.5.6 are shown in Figs. 7.5.7 and 7.5.8. The amplitude of the incident field is somewhat smaller but, for distances ρ_c not too far from the source, useful measurements can be made in either area. For the larger distances an area like Area #I off the end of the dipole provides substantially greater incident fields because $E_{2\rho}(\rho, 0, 0)$ and $B_{2\phi}(\rho, 0, 0)$ are involved and these are greater than $E_{2\phi}(\rho, \pi/2, 0)$ and $B_{2\rho}(\rho, \pi/2, 0)$.

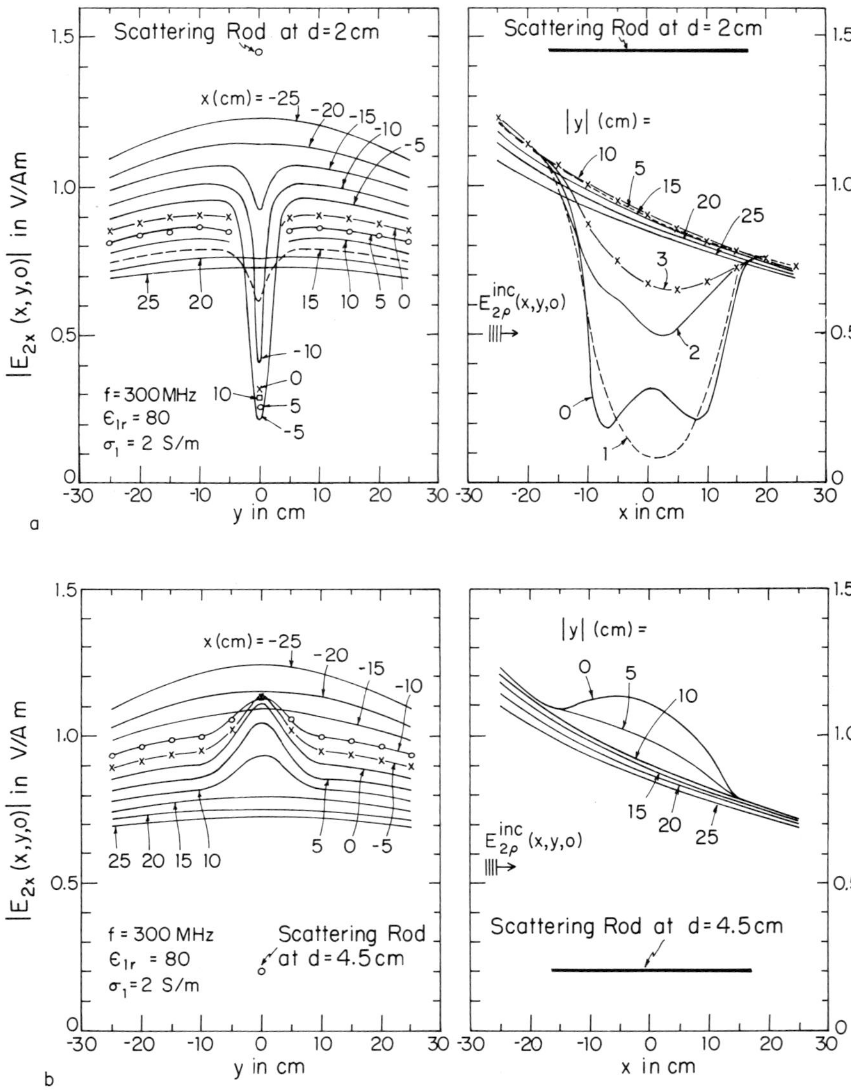

Figure 7.5.6. Electric field in air above insulated scattering rod of length $2l = 33.4$ cm at depth d in salt water ($\sigma_1 = 2$ S/m); Area #I. (a) $d = 2$ cm. (b) $d = 4.5$ cm.

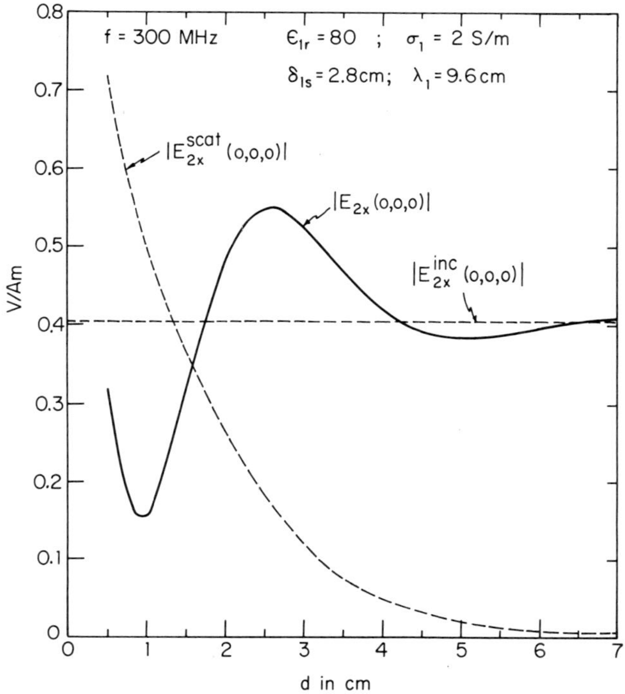

Figure 7.5.7. Electric field in air directly over center of insulated scattering rod at depth d in salt water ($\sigma_1 = 2$ S/m); Area #II.

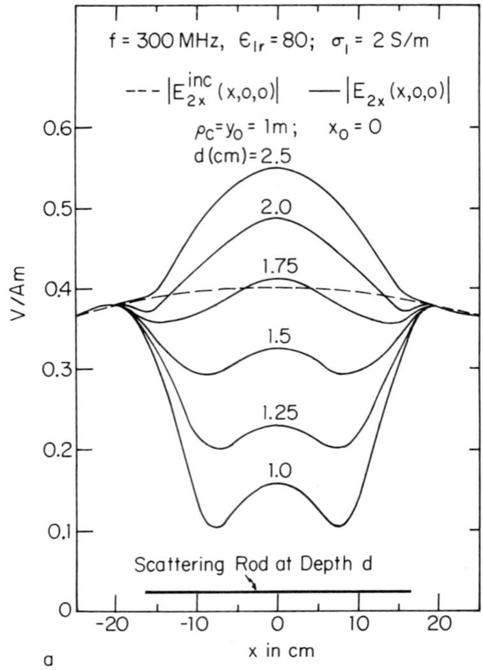

Figure 7.5.8a. Electric field along x-axis in air directly over insulated scattering rod at depth d in salt water ($\sigma_1 = 2$ S/m); Area #II.

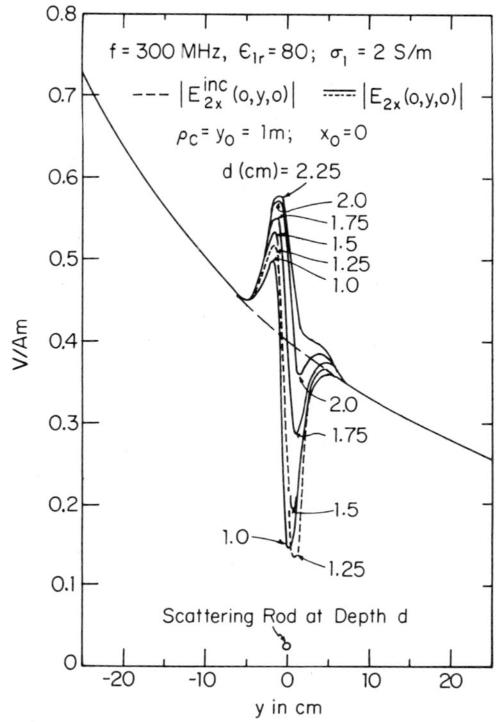

b

Figure 7.5.8b. Electric field along y-axis in air directly over insulated scattering rod at depth d in salt water ($\sigma_1 = 2$ S/m); Area #II.

7.6 Lateral Electromagnetic Waves from a Horizontal Antenna for Remote Sensing in the Ocean

The method described in Section 7.5 for finding a buried metal cylinder in the earth can be adapted to the search for cylindrical metal objects like mines and submarines in the ocean or lying on the sea floor. Because salt water has quite different properties from earth, both the theoretical adaptation and the practical application of the method involve significant modifications that are scientifically interesting. These include the following. (a) The fluid nature of water makes it possible to locate and move both the transmitting and receiving antennas, not in the air, but in the same medium as the cylinder to be located. This greatly increases the magnitude of the receivable scattered signal and significantly reduces the possible interference by reflections from the ionosphere because the transmission coefficient of a plane wave normally incident on the surface of the sea from the air is $f_t = 2k_0/(k_1 + k_0)$ which is very small at the frequencies involved. (b) The much higher conductivity of sea water ($\sigma_1 \sim 4$ S/m) compared to earth leads to a much greater complex wave number for electromagnetic waves. This is given by

$$k_1 = \beta_1 + i\alpha_1 \sim (i\omega\mu_0\sigma_1)^{1/2} = (1+i)\left(\frac{\omega\mu_0\sigma_1}{2}\right)^{1/2}. \tag{7.6.1}$$

The associated wavelength $\lambda_1 = 2\pi/\beta_1$ is much shorter than that in the earth at the same frequency. It follows that a much lower frequency must be used in order to have the scattering cylinder with length $2h$ near resonance, that is, in the range

$$1 \leq \beta_1 h \leq \frac{\pi}{2}. \tag{7.6.2}$$

For example, if the object to be detected is a metal cylinder with a length $2h \sim 5$ m, the useful frequencies are in the range

$$10 \leq f \leq 25 \text{ kHz}. \tag{7.6.3}$$

If the object is a submarine with a length $2h \sim 100$ m, the appropriate range is

$$25 \leq f \leq 62.5 \text{ Hz}. \tag{7.6.4}$$

Associated with the use of a lower frequency is a helpful decrease in the exponential attenuation $\exp(-\alpha_1 r)$ of a traveling wave and a disadvantageous shift in the properties of lateral waves from those of the intermediate field with a radial decrease with distance of ρ^{-1} to the near field with a radial decrease of ρ^{-3}. Also, the properties of an insulated conductor change significantly as the frequency is reduced below $f \sim 0.01/a^2$ Hz, where a is the radius of the conductor in meters. When this is true, the internal resistance per unit length, $r^i \sim r_0 = 1/\pi a^2\sigma_c$, of the conductor becomes dominant and the radiation resistance per unit length negligibly small. This means that the insulated conductor is no longer an antenna, but a lossy transmission line. Account is readily taken in the analysis of these important facts. A further complication, that introduced by the increased importance of noise as the frequency is reduced, is beyond the scope of the present study and will not be treated. It is nonetheless an essential consideration in any practical application.

In its application in the ocean, the modified method of remote sensing described in Section 7.5 involves a horizontal transmitting antenna a short distance d below the surface of the sea and a receiving antenna a similar distance $z \sim d$ also in the sea. The transmitting antenna must be moved to successive locations around each of which the receiving antenna measures the electric (or magnetic) field in order to detect the weak field scattered by the submerged cylindrical object. The scattered field is detectable only when a significant axial current is induced in the cylinder. It follows that the field generated by the transmitting antenna must have an adequate magnitude along the axis of a submerged cylinder *regardless of its orientation.*

An antenna that consists of a suitably terminated insulated x-directed conductor generates a lateral wave with the components

$$E_{1\rho}(\rho,\phi,z) = E_{1\rho}(\rho,0,z)\cos\phi, \tag{7.6.5a}$$

$$E_{1\phi}(\rho,\phi,z) = E_{1\phi}(\rho,\pi/2,z)\sin\phi, \tag{7.6.5b}$$

where ϕ is measured from the positive x-axis. Similarly, an antenna that consists of an identical y-directed conductor generates the lateral-wave field

$$\begin{aligned} E'_{1\rho}(\rho,\phi+\pi/2,z) &= E_{1\rho}(\rho,0,z)\cos(\phi+\pi/2) \\ &= -E_{1\rho}(\rho,0,z)\sin\phi, \end{aligned} \tag{7.6.6a}$$

$$\begin{aligned} E'_{1\phi}(\rho,\phi+\pi/2,z) &= E_{1\phi}(\rho,\pi/2,z)\sin(\phi+\pi/2) \\ &= E_{1\phi}(\rho,\pi/2,z)\cos\phi. \end{aligned} \tag{7.6.6b}$$

If these two mutually perpendicular antennas are operated together with the current in the y-directed element driven 90° out of phase with respect to that in the x-directed antenna, the combined field has the components:

$$\begin{aligned} E_{1\rho}(\rho,\phi,z) + iE'_{1\rho}(\rho,\phi+\pi/2,z) &= E_{1\rho}(\rho,0,z)(\cos\phi - i\sin\phi) \\ &= E_{1\rho}(\rho,0,z)e^{-i\phi}, \end{aligned} \tag{7.6.7a}$$

$$\begin{aligned} E_{1\phi}(\rho,\phi,z) + iE'_{1\phi}(\rho,\phi+\pi/2,z) &= E_{1\phi}(\rho,\pi/2,z)(\sin\phi + i\cos\phi) \\ &= iE_{1\phi}(\rho,\pi/2,z)e^{-i\phi}. \end{aligned} \tag{7.6.7b}$$

Thus, the magnitudes of the ρ- and ϕ-components of the electric field of the crossed antennas are equal to their respective maxima, $E_{1\rho}(\rho,0,z)$ and $E_{1\phi}(\rho,\pi/2,z)$, in *all directions*, i.e., *independent of* ϕ. Hence, a submerged cylinder located at any point (ρ_c,ϕ_c,z_c) in the sea is illuminated by an incident electric field that has the same maximum value independent of the orientation of the cylinder in any horizontal plane.

The receiving antenna must be mobile and designed to respond to the weak field radiated by the currents induced in the submerged cylinder in the presence of a much stronger signal from the source. It must be moved horizontally through the water or lowered into it to successive points of observation.

Since the field to be detected is linearly polarized in the unknown axial direction of the submerged cylinder, the receiving antenna should have an omnidirectional response. This is true of a pair of center-loaded crossed dipoles (or crossed loops). The size of these must be sufficiently small for easy towing by boat while submerged or for easy raising and lowering if transported by helicopter.

7.7 On the Detection of Submerged Submarines with Lateral Electromagnetic Waves

Consider the specific problem of locating a submarine equivalent to a horizontal cylinder with the length $2h = 100$ m and the diameter $2a = 12.5$ m, so that $h/a = 8$. At the lowest frequency $f = 25$ Hz in the range given by (7.6.4), the complex wave number for sea water as defined in (7.6.1) is

$$k_1 = \beta_1 + i\alpha_1 = 0.02(1+i)\ \mathrm{m}^{-1} \tag{7.7.1}$$

and the associated wavelength is $\lambda_1 = 2\pi/\beta_1 = 314$ m. By contrast, the wave number in the air is $k_0 = \omega/c = 0.52 \times 10^{-6}\ \mathrm{m}^{-1}$ and the associated wavelength is $\lambda_0 = 2\pi/k_0 = 1.2 \times 10^4$ km. It follows that $k_0\rho < 1$ when $\rho < k_0^{-1} = 2 \times 10^3$ km. In this range, the near-field formulas for the lateral-wave part of the field generated by the transmitting antenna are good approximations. If this consists of two crossed infinitesimal dipoles with the electric moments $2h_eI_x(0) = 1$ A m and $2h_eI_y(0) = 1$ A m, the lateral-wave electric field at (ρ, ϕ, z) is obtained from (7.6.7a,b) to be

$$\begin{aligned} E_{1\rho}^L(\rho,\phi,z) &\sim \frac{i\omega\mu_0}{2\pi k_1^2\rho^3}\, e^{-i\phi}e^{ik_0\rho}e^{ik_1(z+d)} \\ &= \frac{1}{2\pi\sigma_1\rho^3}\, e^{-i\phi}e^{ik_0\rho}e^{ik_1(z+d)}, \end{aligned} \tag{7.7.2}$$

$$\begin{aligned} E_{1\phi}^L(\rho,\phi,z) &\sim -\frac{\omega\mu_0}{\pi k_1^2\rho^3}\, e^{-i\phi}e^{ik_0\rho}e^{ik_1(z+d)} \\ &= \frac{i}{\pi\sigma_1\rho^3}\, e^{-i\phi}e^{ik_0\rho}e^{ik_1(z+d)}. \end{aligned} \tag{7.7.3}$$

The associated direct field through the sea floor is

$$E_{1\rho}^d(\rho,\phi,z) = \frac{\omega\mu_0}{2\pi k_1^2}\left(\frac{k_1}{\rho^2} + \frac{i}{\rho^3}\right)e^{-i\phi}e^{ik_1r_0}, \tag{7.7.4}$$

$$E_{1\phi}^d(\rho,\phi,z) = -\frac{i\omega\mu_0}{2\pi k_1^2}\left(\frac{k_1}{\rho^2} + \frac{i}{\rho^3}\right)e^{-i\phi}e^{ik_1r_0}, \tag{7.7.5}$$

where $r_0 = (\rho^2 + z^2)^{1/2}$. At radial distances $\rho \geq 0.3$ km, $k_1\rho \geq 6(1+i)$ so that, with $z = 0$, $|(k_1/\rho^2)e^{ik_1r_0}| \leq 0.028(9 \times 10^4)^{-1}e^{-6} \leq 7.7 \times 10^{-10}$ m^{-3}. On the other hand, $1/\rho^3 \leq 3.7 \times 10^{-8}\ \mathrm{m}^{-3}$. It follows that the direct field is negligible compared with the lateral field when $\rho \geq 0.3$ km where the entire significant field is given by (7.7.2) and (7.7.3).

The field given by (7.7.2) and (7.7.3) is rotationally symmetric in its magnitude. Owing to the rapid decrease with the radial distance ρ of the near-field formulas (7.7.2) and (7.7.3), it is necessary to limit the scanning area to moderate values of ρ. Specifically, let the area of search be defined by (see Fig. 7.7.1)

$$1 \leq \rho_c \leq 3\ \text{km}. \tag{7.7.6}$$

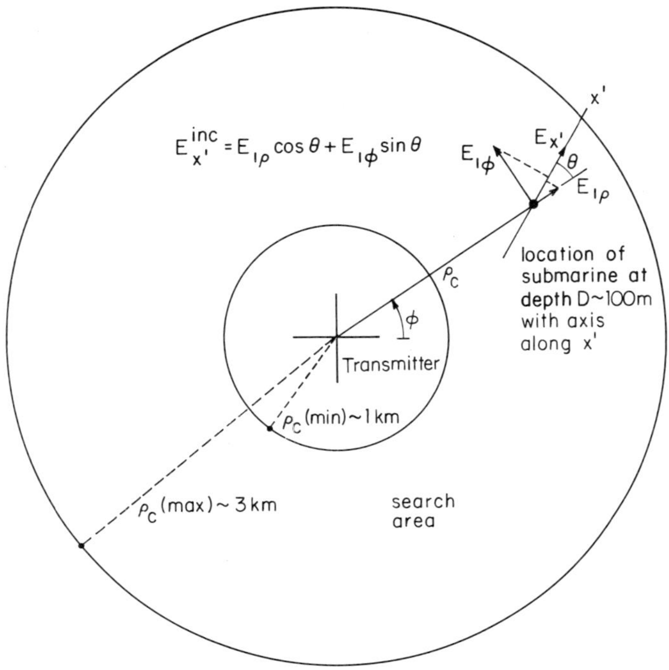

Figure 7.7.1. Transmitting antenna, area of search, and representative location and orientation of submarine.

At the largest distance $\rho_c = 3$ km, with the depths of the transmitting and receiving antennas set at $z = d = 1$ m and since $\exp(ik_0\rho_c) = \exp(i0.0016) \sim 1$,

$$E_{1\rho}(3 \times 10^3 \text{ m}, \phi, 1 \text{ m}) = \frac{e^{-i\phi}}{2\pi \times 4 \times 27 \times 10^9} e^{-0.04(1-i)}$$

$$= 1.41 \times 10^{-12} e^{i0.04} e^{-i\phi} \text{ V/m}, \tag{7.7.7a}$$

$$E_{1\phi}(3 \times 10^3 \text{ m}, \phi, 1 \text{ m}) = i2.82 \times 10^{-12} e^{i0.04} e^{-i\phi} \text{ V/m}. \tag{7.7.7b}$$

This is the horizontal electric field at the location of the receiving antenna, $(\rho_c, \phi, z) = (3 \times 10^3 \text{ m}, \phi, 1 \text{ m})$, due to a pair of crossed unit dipoles at $(0, 0, 1 \text{ m})$.

The electric field incident on a submarine at a depth $D = 100$ m directly below the point $(\rho_c, \phi, 0)$ has the following horizontal components (see Fig. 7.7.1):

$$E_{1\rho}(3 \times 10^3 \text{ m}, \phi, 100 \text{ m}) = \frac{e^{-i\phi}}{2\pi \times 4 \times 27 \times 10^9} e^{-2.02(1-i)}$$

$$= 1.95 \times 10^{-13} e^{i2.02} e^{-i\phi} \text{ V/m}, \tag{7.7.8a}$$

$$E_{1\phi}(3 \times 10^3 \text{ m}, \phi, 100 \text{ m}) = 2iE_{1\rho}(3 \times 10^3 \text{ m}, \phi, 100 \text{ m}). \tag{7.7.8b}$$

Let the center of the submarine be at the origin of the coordinates (x', y', z') with its axis along the horizontal x'-axis and z' vertical, as shown in Fig. 7.7.1. Let the angle between x' and ρ be θ. Then the incident field along the submarine is

$$\begin{aligned} E_{x'}^{\text{inc}} &= E_{1\rho}\cos\theta + E_{1\phi}\sin\theta \\ &= E_{1\rho}(\cos\theta + 2i\sin\theta) \\ &= E_{1\rho}(\cos^2\theta + 4\sin^2\theta)^{1/2}\exp[i\tan^{-1}(2\tan\theta)]. \end{aligned} \tag{7.7.9}$$

The magnitude of this field differs from $E_{1\rho}$ by at most a factor 2. Let it be assumed for simplicity that the angle $\theta = 0$ so that

$$\begin{aligned} E_{x'}^{\text{inc}} &= E_{1\rho}(3\times 10^3 \text{ m}, \phi, 100 \text{ m}) \\ &= 1.95\times 10^{-13} e^{i2.02} e^{-i\phi} \text{ V/m}. \end{aligned} \tag{7.7.10}$$

The axial current induced in the submarine by the incident field $E_{x'}^{\text{inc}}$ is approximated by

$$I_{x'}(x') = -\frac{i4\pi E_{x'}^{\text{inc}}}{\omega\mu_0}\left(\frac{\cos k_1 x' - \cos k_1 h}{\Psi\cos k_1 h - \Psi(h)}\right), \tag{7.7.11a}$$

where the parameters Ψ and $\Psi(h)$ are given by

$$\Psi = 2\ \ln(2h/a) - 3 + \frac{5}{12}\,k_1^2h^2 + i\,\frac{2}{9}\,k_1^3h^3, \tag{7.7.11b}$$

$$\Psi(h) = k_1^2h^2 + i\,\frac{2}{3}\,k_1^3h^3. \tag{7.7.11c}$$

When $|k_1^2h^2| \ll 1$, the induced current is approximately frequency-independent since, with $k_1^2 = i\omega\mu_0\sigma_1$,

$$\begin{aligned} I_{x'}(x') &\sim -\frac{i2\pi E_{x'}^{\text{inc}}}{\omega\mu_0}\left(\frac{k_1^2h^2(1 - x'^2/h^2)}{\Psi(1 - \frac{1}{2}k_1^2h^2) - \Psi(h)}\right) \\ &\sim \frac{2\pi\sigma_1 E_{x'}^{\text{inc}}(h^2 - x'^2)}{2\ \ln(2h/a) - 3}. \end{aligned} \tag{7.7.12}$$

With $k_1 = \beta_1\sqrt{2}\,e^{i\pi/4}$, $k_1^2 = 2i\beta_1^2$, and specifically $\beta_1 h = 1$, $f = 25$ Hz, and $h/a = 8$,

$$\Psi = 2\ \ln(16) - \frac{31}{9} + i\,\frac{7}{18} = 2.10 + i0.39, \tag{7.7.13a}$$

$$\Psi(h) = i\,\frac{2}{3} - \frac{4}{3} = -1.33 + i0.67, \tag{7.7.13b}$$

$$1 - \frac{k_1^2h^2}{2} = 1 - i\beta_1^2h^2 = 1 - i, \tag{7.7.13c}$$

and

$$I_{x'}(x') = -\frac{i2\pi E_{x'}^{\text{inc}}\times 2i(1 - x'^2/h^2)}{50\pi\times 4\pi\times 10^{-7}[(2.10 + i0.39)(1 - i) + 1.33 - i0.67]}$$

$$
\begin{aligned}
&= \frac{E_{x'}^{\mathrm{inc}}(1 - x'^2/h^2)}{50\pi \times 10^{-7}(2.10 - i2.10 + 0.39 + i0.39 + 1.33 - i0.67)} \\
&= \frac{E_{x'}^{\mathrm{inc}} \times 10^7 (1 - x'^2/h^2)}{600 - i374} \\
&= 1.41 \times 10^4 e^{i0.56} E_{x'}^{\mathrm{inc}} (1 - x'^2/h^2).
\end{aligned} \tag{7.7.14}
$$

Use can now be made of (7.7.10) to give

$$
\begin{aligned}
I_{x'}(0) &= 1.41 \times 10^4 e^{i0.56} \times 1.95 \times 10^{-13} e^{i2.02} e^{-i\phi} \\
&= 2.73 \times 10^{-9} e^{i2.58} e^{-i\phi} \text{ A}.
\end{aligned} \tag{7.7.15}
$$

The x'-directed electric field of a dipole with its axis along the x'-axis and an electric moment $2h_e I_{x'}(0)$ is

$$
\begin{aligned}
E_{x'} = \frac{i\omega\mu_0 \times 2h_e I_{x'}(0)}{4\pi k_1^2} \Bigg[\frac{k_1^2}{r} + \frac{ik_1}{r^2} - \frac{1}{r^3} \\
- \frac{x'^2}{r^2} \left(\frac{k_1^2}{r} + \frac{3ik_1}{r^2} - \frac{3}{r^3} \right) \Bigg] e^{ik_1 r},
\end{aligned} \tag{7.7.16}
$$

where $r = (x'^2 + y'^2 + z'^2)^{1/2}$ is the distance from the center of the dipole and $2h_e$ is its effective length. The induced current makes the submarine a radiating dipole that generates the field in (7.7.16). Directly above the submarine at $x' = y' = 0,\ z' = D - z$,

$$
E_{x'} = \frac{i\omega\mu_0 h_e I_{x'}(0)}{2\pi(D - z)} \left(1 + \frac{i}{k_1(D - z)} - \frac{1}{k_1^2(D - z)^2} \right) e^{ik_1(D - z)}. \tag{7.7.17}
$$

With $D - z = 99$ m, $I_{x'}(0)$ as given in (7.7.15), and the effective half-length

$$
h_e = \int_0^h \left(1 - \frac{x'^2}{h^2} \right) dx' = \frac{2h}{3} = \frac{100}{3} = 33.3 \text{ m}, \tag{7.7.18}
$$

the scattered field (7.7.17) is

$$
\begin{aligned}
E_{x'} &= \frac{i\omega\mu_0 \times 33.3 \times 2.73 \times 10^{-9} e^{i2.58} e^{-i\phi}}{2\pi \times 99} \\
&\quad \times \left(1 + \frac{ie^{-i\pi/4}}{0.028 \times 99} + \frac{i}{(0.028 \times 99)^2} \right) e^{-0.02\times 99} e^{i0.02\times 99} \\
&= 5.23 \times 10^{-15} e^{i0.15} e^{-i\phi} \text{ V/m},
\end{aligned} \tag{7.7.19a}
$$

or

$$
20 \log_{10} |E_{x'}| = -286 \text{ dB}. \tag{7.7.19b}
$$

This is the scattered field when the transmitting antenna is a pair of crossed dipoles each with an electric moment of 1 A m. The scattered field of a pair of crossed transmitting antennas, each with the electric moment $2h_{et}I_t(0)$, is

$$
E_{x'} = 5.23 \times 10^{-15} e^{i0.15} e^{-i\phi} \times 2h_{et} I_t(0). \tag{7.7.20}
$$

The crossed transmitting antenna is located at the small depth $d = 1$ m in the ocean. Each element is a center-driven dipole with an insulated center section of length $2h_i$ terminated at each end in a bare section with the length l. As shown in Chapter 16, the optimum length l of a bare monopole in sea water and operated at a low frequency is $l \geq 0.15\lambda_1 = 0.15 \times 314 = 47$ m. Since the internal resistance per unit length dominates in the insulated section when the frequency satisfies the condition $f \leq 0.01/a^2$, where a is the radius of the conductor in meters, it serves only as a lossy transmission line that feeds the bare ends. Its wave number and characteristic impedance are

$$k_L = (1+i)\left(\frac{\omega c_L r_0}{2}\right)^{1/2}; \qquad Z_c = (1+i)\left(\frac{r_0}{2\omega c_L}\right)^{1/2}, \tag{7.7.21}$$

where

$$r_0 = \frac{1}{\pi a^2 \sigma_c}; \qquad c_L = \frac{2\pi\epsilon_d}{\ln(b/a)}. \tag{7.7.22}$$

If the radius of the conductor in both bare and insulated parts is $a = 0.52$ cm, the outer radius of the insulation is $b = 0.80$ cm, and its permittivity is that of neoprene, $\epsilon_d = 6.6\epsilon_0$, and if $\sigma_c = 5.65 \times 10^7$ S/m is the conductivity of copper, it follows that $r_0 = 2.08 \times 10^{-4}$ Ω/m, $c_L = 0.85 \times 10^{-9}$ F/m, and

$$k_L = 0.37 \times 10^{-5}(1+i)\ \text{m}^{-1}; \qquad Z_c = 27.9(1+i)\ \Omega. \tag{7.7.23}$$

Since k_L is very small, any practical antenna is electrically short with $|k_L h| < 1$. When this is true, it follows from Chapter 17 that the input impedance is

$$Z_{\text{in}} \sim -iZ_c k_L h_i = r_0 h_i + 2Z_l, \tag{7.7.24}$$

where Z_l is the terminal impedance of a bare monopole in sea water. It is given by (see Chapter 16)

$$2Z_l = \frac{\Psi}{2\pi\sigma_1 l} + \frac{2}{3} l r_0 - \frac{i\omega\mu_0\Psi l}{6\pi}, \tag{7.7.25}$$

where $\Psi = 2\ln(l/a) - 2$. It follows that

$$Z_{\text{in}} = \frac{\Psi}{2\pi\sigma_1 l} + \left(h_i + \frac{2}{3}l\right) r_0 - \frac{i\omega\mu_0\Psi l}{6\pi}, \tag{7.7.26}$$

where the first term is the external resistance and the second term the internal resistance. With $l = 47$ m, $a = 0.52$ cm, and $\Psi = 16.2$,

$$2Z_l = \frac{16.2}{2\pi \times 4 \times 47} + \frac{2}{3} \times 47 \times 2.08 \times 10^{-4} - \frac{i50\pi \times 4\pi \times 10^{-7} \times 16.2 \times 47}{6\pi}$$

$$= 0.0137 + 0.0065 - i0.008 = 0.0202 - i0.008\ \Omega, \tag{7.7.27}$$

$$Z_{\text{in}} = 2.08 \times 10^{-4} h_i + 0.0202 - i0.008\ \Omega. \tag{7.7.28}$$

Note that $R^e = 0.0137\ \Omega$ and $R^i = 0.0065 + 2.08 \times 10^{-4} h_i\ \Omega$. Here, $h_{ei} = h_i$ is the effective half-length of the insulated part. For an electrically short antenna terminated in a low impedance, the effective length $2h_{ei}$ is equal to the actual length $2h_i$ since the current is uniform.

The effective length of the bare monopoles is

$$l_e \sim \frac{l}{2} \quad \text{or} \quad 2l_e = l. \tag{7.7.29}$$

Thus, the effective length of the entire antenna is

$$2h_{et} = 2(l_e + h_i) = l + 2h_i, \tag{7.7.30}$$

and the total scattered field, from (7.7.20), is

$$E_{x'} = 5.23 \times 10^{-15} e^{i0.15} e^{-i\phi} \times (l + 2h_i) I_t(0). \tag{7.7.31}$$

An appropriate design for the receiving antenna is a pair of crossed bare metal dipoles with half-lengths $h_r = 2$ m, radius $a = 0.5$ cm, and each element center-loaded with an impedance Z_L. When located a small distance $z = 1$ m below the surface of the ocean, the impedance of each electrically short dipole in the presence of its image in the boundary is

$$Z_e = \frac{\Psi_e}{\pi \sigma_1} k_1 \cot k_1 h_r \sim \frac{\Psi_e}{\pi \sigma_1 h_r}, \tag{7.7.32}$$

where $\Psi_e = 2 \ln(h_r / a_e) - 2$. The effective radius is $a_e = (2az)^{1/2} = 0.1$ m so that $\Psi_e = 2 \ln(20) - 2 = 4.0$ and

$$Z_e = \frac{4}{\pi \times 4 \times 2} = 0.159\ \Omega. \tag{7.7.33}$$

The effective half-length of each of the electrically short crossed antennas is

$$h_{re} \sim \frac{h_r}{2} = 1 \text{ m}. \tag{7.7.34}$$

If the orientation of the x'-axis through the scattering cylinder (submarine) makes an angle θ with respect to one of the crossed elements of the receiving antenna, the voltages across the two loads are

$$V_{L1} = 2h_{re} E_{x'} \cos\theta \left(\frac{Z_L}{Z_L + Z_e} \right) \sim 2h_{re} E_{x'} \cos\theta, \tag{7.7.35a}$$

$$V_{L2} = 2h_{re} E_{x'} \sin\theta \left(\frac{Z_L}{Z_L + Z_e} \right) \sim 2h_{re} E_{x'} \sin\theta. \tag{7.7.35b}$$

From these relations,

$$V_L = (V_{L1}^2 + V_{L2}^2)^{1/2} = 2h_{re} E_{x'} = 2E_{x'} \quad \text{when } h_{re} = 1 \text{ m}, \tag{7.7.36}$$

and

$$\theta = \tan^{-1}\left(\frac{V_{L2}}{V_{L1}}\right). \tag{7.7.37}$$

In order to detect the presence of the submarine, the received voltage $V_L = 2E_{x'}$ must be large enough to be observed by the receiver. If the sensitivity of the receiver is 0.001 μV, it is necessary that $E_{x'}$ as given by (7.7.31) satisfy the following inequality:

$$|E_{x'}| = 5.23 \times 10^{-15}(l + 2h_i)I_t(0) \geq 0.5 \times 10^{-9},$$

$$(l + 2h_i)I_t(0) \geq \frac{10^6}{10.46} = 0.96 \times 10^5 \text{ A m}. \tag{7.7.38}$$

This condition can be satisfied with a range of possible combinations limited, on the one hand, by the maximum acceptable current and, on the other hand, by the maximum acceptable length of the crossed elements of the transmitting antenna. The required power is reduced when the antenna is lengthened and the current decreased. Alternatively, the efficiency of power transfer to the ocean medium is increased when the insulated section of the antenna is shortened and the current increased. Consider the following numerical examples.

If the maximum current-carrying capacity of the insulated conductor when immersed in the sea is 500 A, the shortest possible effective length is

$$2h_{et} = 47 + 2h_i = \frac{0.96 \times 10^5}{500} = 192; \qquad h_i = 72.5 \text{ m}. \tag{7.7.39}$$

The actual overall length is $2(l + h_i) = 239$ m and the input impedance is

$$\begin{aligned} Z_{\text{in}} &= 2.08 \times 10^{-4} \times 72.5 + 0.0202 - i0.008 \\ &= 0.0151 + 0.0202 - i0.008 = 0.0353 - i0.008 \ \Omega. \end{aligned} \tag{7.7.40}$$

The total input power (P_t), the power to maintain the external field in the ocean (P^e), and the power dissipated in heating the conductors (P^i) are

$$P_t = 8.83 \text{ kW}; \qquad P^e = 3.43 \text{ kW}; \qquad P^i = 5.40 \text{ kW}. \tag{7.7.41}$$

The efficiency is $P^e/P_t = 33.8\%$. The overall length of 239 m for each of the crossed arms of the transmitting antenna is quite long for moving in the ocean. The towing can be simplified by arranging the transverse arms of the cross to fold onto the longitudinal section while in motion and to fold out for transmission.

If the maximum acceptable length of the antenna can be extended to 1 km,

$$2(47 + h_i) = 1000; \qquad h_i = 453 \text{ m}. \tag{7.7.42}$$

In this case the current is

$$I_t(0) = \frac{0.96 \times 10^5}{953} = 100.7 \text{ A}. \tag{7.7.43}$$

The input impedance is

$$Z_{\text{in}} = 2.08 \times 10^{-4} \times 453 + 0.0202 - i0.008$$
$$= 0.1144 - i0.008 \ \Omega. \tag{7.7.44}$$

The total, external, and internal powers are

$$P_t = 1.16 \text{ kW}; \qquad P^e = 0.14 \text{ kW}; \qquad P^i = 1.02 \text{ kW}. \tag{7.7.45}$$

The efficiency is 12%. The required power is reduced with increasing length of the antenna.

The field scattered by the submarine generates the small voltage $V_L^s \sim 0.001\ \mu\text{V}$ in the receiver. Simultaneously, the lateral-wave field from the transmitter generates the field (7.7.7a) multiplied by (7.7.38) at the receiver. That is,

$$V_L = 2E_{1\rho}(3 \times 10^3 \text{ m}, \phi, 1 \text{ m}) \times 0.96 \times 10^5$$
$$= 2 \times 1.41 \times 10^{-12} e^{i0.04} e^{-i\phi} \times 0.96 \times 10^5$$
$$= 0.271 \times 10^{-6} e^{i0.04} e^{-i\phi} \text{ V}. \tag{7.7.46}$$

The voltage $|V_L| = 0.27\ \mu\text{V}$ is the background voltage that must be kept balanced out in order to detect the field scattered by the submarine which produces the voltage $V_L^s \sim 0.001\ \mu\text{V}$ across the load.

If the depth of the sea below the submarine is small enough so that reflections from the sea floor are significant at the point of observation, these are added to the background field. For a sea floor with the conductivity $\sigma_2 = 4 \times 10^{-2}$ S/m, the reflection coefficient is

$$f_r = \frac{k_1 - k_2}{k_1 + k_2} = \frac{\sqrt{\sigma_1} - \sqrt{\sigma_2}}{\sqrt{\sigma_1} + \sqrt{\sigma_2}} = \frac{2 - 0.2}{2 + 0.2} = 0.82. \tag{7.7.47}$$

For an ocean depth of $s = 200$ m, the reflected field at $z = 1$ m from the surface is

$$E_{1\rho}(3 \times 10^3 \text{ m}, \phi, 1 \text{ m}) \times 0.82 e^{i2k_1(s-z)}$$
$$= E_{1\rho}(3 \times 10^3 \text{ m}, \phi, 1 \text{ m}) \times 2.86 \times 10^{-4} e^{i1.68}, \tag{7.7.48}$$

and its contribution to the voltage is

$$V_L^r = 7.74 \times 10^{-11} e^{i1.72} e^{-i\phi} \text{ V}. \tag{7.7.49}$$

This is somewhat smaller than the voltage due to the scattered field, $V_L^s = 10^{-9}$ V. If the sea floor is reasonably level, the voltage due to the reflected waves merely adds a constant to the background field from the transmitter.

An accurate treatment of possible reflections from the ionosphere requires the analysis of the entire problem with consideration of an additional conducting layer above the sea and air. At $f = 25$ Hz, the wavelength in air is so large, $\lambda_0 = 1.2 \times 10^4$ km, that the thickness of the air is necessarily electrically small since the height of the ionosphere layer is under 100 km. Also, the reflection coefficient of the ionosphere is near 1 as is that of the sea so that the air between them is a waveguide and the electromagnetic field in it is not simple. The transmission coefficient of the surface of the sea for a normally incident plane wave from above is

$$f_t = \frac{2k_0}{k_1 + k_0}$$
$$\sim \frac{2k_0}{k_1} = \frac{2 \times 0.52 \times 10^{-6}}{2.8 \times 10^{-2} e^{i\pi/4}} = 0.372 \times 10^{-4} e^{-i\pi/4}. \tag{7.7.50}$$

This is a measure of the reduction in the field for a receiving antenna below the surface as compared to one above the surface. Whatever the actual field is that reaches the receiving antenna at depth $z = 1$ m in the sea from ionospheric reflection, it is added to the background field which consists of the lateral wave directly from the transmitter and possible reflections from the sea floor and the ionosphere. The scattered field from the submerged submarine—which is significant only in a small region directly above the submarine—must be detected above this background. This should be possible provided the background is constant.

7.8 Detection of Magnetic Field Instead of Electric Field

If it is desired to observe the magnetic field corresponding to the electric field (7.7.31), this is simply

$$B_{\phi'} = \frac{k_1}{\omega} E_{x'} = \frac{5.23k_1}{\omega} \times 10^{-15} e^{i0.15} e^{-i\phi} \times 2h_{et} I_t(0). \tag{7.8.1}$$

With $f = 25$ Hz and $k_1 = 0.02(1+i)$ m$^{-1} = 0.02\sqrt{2}\, e^{i\pi/4}$,

$$\frac{k_1}{\omega} = \frac{0.028 e^{i\pi/4}}{2\pi \times 25} = 1.8 \times 10^{-4} e^{i\pi/4} \text{ sec/m}, \tag{7.8.2}$$

so that

$$B_{\phi'} = 0.941 \times 10^{-18} e^{i0.935} e^{-i\phi} \times 2h_{et} I_t(0) \text{ V sec/m}^2 \tag{7.8.3}$$

or

$$H_{\phi'} = \frac{B_{\phi'}}{\mu_0} = 0.749 \times 10^{-12} e^{-i\phi} \times 2h_{et} I_t(0) \text{ A/m}. \tag{7.8.4}$$

With $2h_{et} = 192$ m and $I_t(0) = 500$ A,

$$B_{\phi'} = 0.9 \times 10^{-13} e^{i0.935} e^{-i\phi} \text{ V sec/m}^2. \tag{7.8.5}$$

The associated H-field is

$$H_{\phi'} = \frac{B_{\phi'}}{\mu_0} = 0.7 \times 10^{-7} e^{i0.935} e^{-i\phi} \text{ A/m.} \tag{7.8.6}$$

This field can be measured with a magnetometer at the low frequency of 25 Hz.

7.9 Determination of the Permittivity and Conductivity of the Earth from Measurement of the Admittance of a Horizontal Antenna

When a horizontal antenna is located close to the boundary between two electrically different regions, its distribution of current and driving-point admittance are functions of the electrical properties of both regions. This suggests that the permittivity and conductivity of one of the regions could be determined from measurements of the admittance of the antenna when it is located in the other region. Two examples are of interest: (a) The determination of the permittivity and conductivity of the earth when the dipole is in the air close to the surface and (b) the determination of the permittivity and conductivity of the sea floor when the antenna is in the sea close to the boundary. Since these applications do not directly involve the lateral waves generated by the dipole, they will be treated only briefly.

(a) Consider first a center-driven horizontal dipole antenna with half-length h and radius a at a height d in the air over the surface of the earth. This is the antenna used in Section 7.5. It is an eccentrically insulated dipole lying on the surface of the earth. The effective thickness of the air insulation is d. The admittance is given by (7.5.7), viz.,

$$Y = G - iB = \frac{I_z(0)}{V_0^e} = -\frac{i}{Z_c} \tan k_L h, \tag{7.9.1}$$

where k_L is given by (7.3.1) or (7.3.2) and Z_c by (7.3.3). It is assumed that $|k_1| \geq 3k_2$, $a^2 \ll d^2$, and $k_1 d < 1$. Since k_L and Z_c are explicit functions of the known quantities h, a, d, and k_2 (the real wave number of the air) and of the unknown quantity $k_1 = \beta_1 + i\alpha_1$ (the complex wave number of the adjacent earth), it is possible to determine this latter from measured values of G and B. The permittivity ϵ_1 and conductivity σ_1 of the earth are related to α_1 and β_1 by the equation

$$k_1 = \beta_1 + i\alpha_1 = \omega \left[\mu_0 \left(\epsilon_1 + \frac{i\sigma_1}{\omega} \right) \right]^{1/2}. \tag{7.9.2}$$

Although explicit formulas for α_1 and β_1 in terms of G and B are not conveniently written down, they can be determined by interpolation from suitable sets of graphs or by computer.

If the earth consists of a layer of thickness l characterized by the wave number k_3 and a half-space below this characterized by the wave number k_4, the theory of the two-layered eccentrically insulated antenna (King 1989) can be applied to obtain the admittance (7.9.1) with the generalized wave number k_L and characteristic impedance Z_c given by

$$k_L = k_2\left(\frac{\Omega_{ac} + \Delta(A)}{\Omega_{ab} + (k_2/k_3)\Omega_{bc}}\right)^{1/2}, \tag{7.9.3}$$

and

$$Z_c = \frac{\omega\mu_0}{2\pi k_2}\left[\Omega_{ac} + \left(\frac{k_2}{k_3}\right)^2 \Omega_{bc}\right]^{1/2} [\Omega_{ac} + \Delta(A)]^{1/2}. \tag{7.9.4}$$

Here

$$\Delta(A) = 2\left[\frac{1}{A^2} - \frac{K_1(A)}{A} + \frac{i\pi I_1(A)}{2A} - i\left(\frac{A}{3} + \frac{A^3}{45} + \frac{A^5}{1575} + \cdots\right)\right]^{1/2}, \tag{7.9.5}$$

with $A = k_4[d + (d^2 - a^2)^{1/2}]$. The height of the axis of the antenna over the air–earth boundary is $d-l$ and the height over the lower half-space is d. I_1 and K_1 are the modified Bessel functions. Also,

$$\Omega_{ac} = \cosh^{-1}\frac{d}{a}, \tag{7.9.6}$$

$$\Omega_{bc} = \cosh^{-1}\left(\frac{d^2 - a^2 + l^2}{d^2 - a^2 - l^2}\right), \tag{7.9.7}$$

$$\Omega_{ab} = \cosh^{-1}\left(\frac{d(d^2 - a^2 + l^2) - 2l(d^2 - a^2)}{a(d^2 - a^2 - l^2)}\right). \tag{7.9.8}$$

Formulas (7.3.1), (7.3.3), (7.9.3), and (7.9.4) assume the antenna to be perfectly conducting. When this is not a good approximation at very low frequencies, terms that take account of the internal impedance of the antenna are readily added. Note that (7.9.3) and (7.9.4) reduce to (7.3.1) and (7.3.2) when $l = 0$ and $a^2 \ll d^2$.

In the case at hand, there are five unknowns: α_3, β_3, α_4, β_4, and l. The determination of combinations of these that yield a measured pair of values of G and B requires a fairly elaborate computer program. Which of the possible combinations is the correct one is not readily determined without additional measurements. Since, in general, G and B are not very sensitive to changes in any of the five unknowns, this is not a particularly good method for determining the electrical properties of a two-layered region.

(b) When an antenna is located parallel and close to the sea floor, the antenna is in sea water with a known value of $k_1 = \beta_1 + i\alpha_1$ that is much larger in magnitude than the wave number $k_2 = \beta_2 + i\alpha_2$ of the oceanic crust or of an intervening layer of sediment. As a consequence, the reflection coefficient is $(k_1 - k_2)/(k_1 + k_2) \sim 1$ so that the contribution to the admittance of the antenna by the sea floor when this is at a small distance d from the antenna is well approximated by an image antenna with an equal and *in-phase* current. The admittance of the bare antenna is determined in Chapter 16 and that of the insulated antenna in Chapter 17. In the approximation described above, the admittance is independent of the properties of the oceanic crust so long as $|k_1|^2 \gg |k_2|^2$. This indicates that very high precision is required in both the measurement of and a theoretical formula for the admittance if account is to be taken of the small departure from the image theory. Clearly, this is not a sensitive method for the accurate determination of the complex wave number of the oceanic crust.

It is perhaps well to add that the accurate measurement and theoretical calculation of the admittance of a center-driven dipole are complicated because the admittance depends on the nature of the driving terminals. If a two-wire or shielded-pair transmission line is used, the finite separation of the two conductors introduces a negative capacitive junction effect and transmission-line end effect that depend on the spacing of the transmission line. [See, for example, Secs. II.33–II.36 of King (1956).] With a bare antenna, this can significantly modify the susceptance, so that only the conductance is readily determined. With the insulated antenna, both the susceptance and the conductance are affected by the line spacing. Meaningful results that depend on admittance or impedance measurements and calculations with center-driven antennas cannot be expected unless careful account is taken of the terminal and junction effects.

7.10 Field of a Horizontal Electric Dipole in the Air over Earth or Sea; Application to FM Communication and Television

The analysis in Chapter 5 and the applications in this chapter are concerned with horizontal electric dipoles and insulated traveling-wave antennas located in the earth or sea near or at the air boundary or near the sea floor. Included is the horizontal-wire or Beverage antenna that is located in the air so close to the earth that it can be treated as an eccentrically insulated antenna on the surface. The analysis includes the determination of the complete electromagnetic field in the earth or sea and in the air. However, in the calculation of the field in the air, the surface-wave term is evaluated subject to the condition $\rho^2 \gg z^2$ so that it applies only to low-angle radiation.

Although most radio communication over the surface of the earth or sea makes use of vertical antennas and a vertically polarized electric field, there are important exceptions. These include FM and television transmission which make use of horizontal transmitting and receiving dipoles that are located in the air well above the earth's surface. Radio communication between two horizontal-wire or Beverage antennas involves only the lateral wave very close to the surface of the earth. This is also true of such antennas when used for over-the-horizon radar. Actually, what is called over-the-horizon radar makes use of ionospheric reflections and not the surface wave. This involves quite-high-angle radiation to the ionosphere that is several hundred kilometers high. For these applications it is necessary to generalize the theory to include horizontal dipoles at any height d' in the air and the electromagnetic field also at any height z'. This requires the determination of the direct and reflected fields and a generalization of the surface-wave field that eliminates the condition $\rho^2 \gg z'^2$. For this purpose use can be made of the integrals (5.5.1)–(5.5.21) with the subscript 1 changed to 2. Since the positive z-direction is into Region 1, the corresponding direction when the subscripts are changed is $z' = -z$ into Region 2 with $E_{z'} = -E_z$ and $B_{z'} = -B_z$. The associated angle is $\phi' = -\phi$ with $E_{\phi'} = -E_\phi$ and $B_{\phi'} = -B_\phi$. The dipole is at the height d'. The desired formulas involve a negative instead of positive image and the consequent use of the coefficients $Q+1$ and $P-1$ instead of $Q-1$ and $P+1$. They are

$$E_{2\rho} = -\frac{\omega\mu_0}{4\pi k_2^2}\cos\phi'\,[F_{\rho 0}(\rho, z'-d') - F_{\rho 0}(\rho, z'+d') + F_{\rho 1}(\rho, z'+d')], \tag{7.10.1}$$

where

$$F_{\rho 0}(\rho, z'-d') = \int_0^\infty \left(\frac{\gamma_2}{2}[J_0(\lambda\rho) - J_2(\lambda\rho)] + \frac{k_2^2}{2\gamma_2}[J_0(\lambda\rho) + J_2(\lambda\rho)]\right) e^{i\gamma_2|z'-d'|}\lambda\,d\lambda, \tag{7.10.2}$$

$$F_{\rho 0}(\rho, z'+d') = \int_0^\infty \left(\frac{\gamma_2}{2}[J_0(\lambda\rho) - J_2(\lambda\rho)] + \frac{k_2^2}{2\gamma_2}[J_0(\lambda\rho) + J_2(\lambda\rho)]\right) e^{i\gamma_2(z'+d')}\lambda\,d\lambda, \tag{7.10.3}$$

$$F_{\rho 1}(\rho, z'+d') = \int_0^\infty \left(\frac{\gamma_2}{2}(Q+1)[J_0(\lambda\rho) - J_2(\lambda\rho)] - \frac{k_2^2}{2\gamma_2}(P-1)[J_0(\lambda\rho) + J_2(\lambda\rho)]\right) \times e^{i\gamma_2(z'+d')}\lambda\,d\lambda; \tag{7.10.4}$$

$$E_{2\phi'} = \frac{\omega\mu_0}{4\pi k_2^2}\sin\phi'\ [F_{\phi'0}(\rho, z'-d') - F_{\phi'0}(\rho, z'+d') + F_{\phi'1}(\rho, z'+d')], \tag{7.10.5}$$

where

$$F_{\phi'0}(\rho, z'-d') = \int_0^\infty \left(\frac{\gamma_2}{2}[J_0(\lambda\rho) + J_2(\lambda\rho)] + \frac{k_2^2}{2\gamma_2}[J_0(\lambda\rho) - J_2(\lambda\rho)]\right) e^{i\gamma_2|z'-d'|}\lambda\, d\lambda, \tag{7.10.6}$$

$$F_{\phi'0}(\rho, z'+d') = \int_0^\infty \left(\frac{\gamma_2}{2}[J_0(\lambda\rho) + J_2(\lambda\rho)] + \frac{k_2^2}{2\gamma_2}[J_0(\lambda\rho) - J_2(\lambda\rho)]\right) e^{i\gamma_2(z'+d')}\lambda\, d\lambda, \tag{7.10.7}$$

$$F_{\phi'1}(\rho, z'+d') = \int_0^\infty \left(\frac{\gamma_2}{2}(Q+1)[J_0(\lambda\rho) + J_2(\lambda\rho)] - \frac{k_2^2}{2\gamma_2}(P-1)[J_0(\lambda\rho) - J_2(\lambda\rho)]\right) \times e^{i\gamma_2(z'+d')}\lambda\, d\lambda; \tag{7.10.8}$$

$$E_{2z'} = \frac{i\omega\mu_0}{4\pi k_2^2}\cos\phi'\ [F_{z'0}(\rho, z'-d') - F_{z'0}(\rho, z'+d') + F_{z'1}(\rho, z'+d')], \tag{7.10.9}$$

where

$$F_{z'0}(\rho, z'-d') = \pm\int_0^\infty J_1(\lambda\rho)e^{i\gamma_2|z'-d'|}\lambda^2\, d\lambda; \qquad \begin{cases} z' > d' \\ 0 \le z' \le d' \end{cases}, \tag{7.10.10}$$

$$F_{z'0}(\rho, z'+d') = \int_0^\infty J_1(\lambda\rho)e^{i\gamma_2(z'+d')}\lambda^2\, d\lambda, \tag{7.10.11}$$

$$F_{z'1}(\rho, z'+d') = \int_0^\infty (Q+1)J_1(\lambda\rho)e^{i\gamma_2(z'+d')}\lambda^2\, d\lambda; \tag{7.10.12}$$

$$B_{2\rho} = -\frac{\mu_0}{4\pi}\sin\phi'\ [G_{\rho 0}(\rho, z'-d') - G_{\rho 0}(\rho, z'+d') + G_{\rho 1}(\rho, z'+d')], \tag{7.10.13}$$

$$B_{2\phi'} = -\frac{\mu_0}{4\pi}\cos\phi'\ [G_{\phi'0}(\rho, z'-d') - G_{\phi'0}(\rho, z'+d') + G_{\phi'1}(\rho, z'+d')], \tag{7.10.14}$$

where

$$G_{\rho0}(\rho, z'-d') = G_{\phi'0}(\rho, z'-d') = \pm\int_0^\infty J_0(\lambda\rho)e^{i\gamma_2|z'-d'|}\lambda\, d\lambda; \qquad \begin{cases} z' > d' \\ 0 \le z' \le d' \end{cases}, \tag{7.10.15}$$

$$G_{\rho0}(\rho, z'+d') = G_{\phi'0}(\rho, z'+d') = \int_0^\infty J_0(\lambda\rho)e^{i\gamma_2(z'+d')}\lambda\, d\lambda, \tag{7.10.16}$$

$$\left.\begin{matrix} G_{\rho1}(\rho, z'+d') \\ G_{\phi'1}(\rho, z'+d') \end{matrix}\right\} = \int_0^\infty \Big(\frac{1}{2}(Q+1)[J_0(\lambda\rho) \pm J_2(\lambda\rho)] - \frac{1}{2}(P-1)[J_0(\lambda\rho) \mp J_2(\lambda\rho)]\Big) \times e^{i\gamma_2(z'+d')}\lambda\, d\lambda; \tag{7.10.17}$$

$$B_{2z'} = \frac{i\mu_0}{4\pi}\sin\phi'\ [G_{z'0}(\rho, z'-d') - G_{z'0}(\rho, z'+d') + G_{z'1}(\rho, z'+d')], \tag{7.10.18}$$

where

$$G_{z'0}(\rho, z'-d') = \int_0^\infty \gamma_2^{-1}J_1(\lambda\rho)e^{i\gamma_2|z'-d'|}\lambda^2\, d\lambda, \tag{7.10.19}$$

$$G_{z'0}(\rho, z'+d') = \int_0^\infty \gamma_2^{-1}J_1(\lambda\rho)e^{i\gamma_2(z'+d')}\lambda^2\, d\lambda, \tag{7.10.20}$$

$$G_{z'1}(\rho, z'+d') = -\int_0^\infty (P-1)\gamma_2^{-1}J_1(\lambda\rho)e^{i\gamma_2(z'+d')}\lambda^2\, d\lambda. \tag{7.10.21}$$

The integrated exact formulas for the contributions $F_{m0}(\rho, z'-d')$ and $G_{m0}(\rho, z'-d')$ by the direct field and $F_{m0}(\rho, z'+d')$ and $G_{m0}(\rho, z'+d')$ by the ideal negative-image field ($m = \rho,\ \phi',\ z'$) are given by (5.5.23)–(5.5.27) with z and d replaced by z' and d', and k_1 by k_2. The parameters P and Q are given by (5.4.28) with the subscripts 1 and 2 interchanged. With them,

$$Q+1 = \frac{2k_2^2\gamma_1}{k_1^2\gamma_2 + k_2^2\gamma_1}; \qquad P-1 = -\frac{2\gamma_2}{\gamma_2+\gamma_1}. \tag{7.10.22}$$

The integrals that remain to be evaluated are

$$F_{\rho1}(\rho, z'+d') = F_{\rho2}(\rho, z'+d') + F_{\rho3}(\rho, z'+d'), \tag{7.10.23}$$

$$F_{\phi'1}(\rho, z'+d') = F_{\phi'2}(\rho, z'+d') + F_{\phi'3}(\rho, z'+d'), \tag{7.10.24}$$

with

$$\left.\begin{array}{l} F_{\rho 2}(\rho, z'+d') \\ F_{\phi' 2}(\rho, z'+d') \end{array}\right\} = \frac{k_2^2}{k_1^2} \int_0^\infty \frac{\gamma_1\gamma_2}{\gamma_2 + (k_2^2/k_1^2)\gamma_1} \times [J_0(\lambda\rho) \mp J_2(\lambda\rho)] e^{i\gamma_2(z'+d')} \lambda \, d\lambda, \quad (7.10.25a)$$

$$\left.\begin{array}{l} F_{\rho 3}(\rho, z'+d') \\ F_{\phi' 3}(\rho, z'+d') \end{array}\right\} = k_2^2 \int_0^\infty \frac{1}{\gamma_2 + \gamma_1} [J_0(\lambda\rho) \pm J_2(\lambda\rho)] e^{i\gamma_2(z'+d')} \lambda \, d\lambda, \quad (7.10.25b)$$

$$F_{z'1}(\rho, z'+d') = \frac{2k_2^2}{k_1^2} \int_0^\infty \frac{\gamma_1}{\gamma_2 + (k_2^2/k_1^2)\gamma_1} J_1(\lambda\rho) e^{i\gamma_2(z'+d')} \lambda^2 \, d\lambda, \quad (7.10.26)$$

$$\left.\begin{array}{l} G_{\rho 2}(\rho, z'+d') \\ G_{\phi' 2}(\rho, z'+d') \end{array}\right\} = \frac{k_2^2}{k_1^2} \int_0^\infty \frac{\gamma_1}{\gamma_2 + (k_2^2/k_1^2)\gamma_1} \times [J_0(\lambda\rho) \pm J_2(\lambda\rho)] e^{i\gamma_2(z'+d')} \lambda \, d\lambda, \quad (7.10.27a)$$

$$\left.\begin{array}{l} G_{\rho 3}(\rho, z'+d') \\ G_{\phi' 3}(\rho, z'+d') \end{array}\right\} = \int_0^\infty \frac{\gamma_2}{\gamma_1 + \gamma_2} [J_0(\lambda\rho) \mp J_2(\lambda\rho)] e^{i\gamma_2(z'+d')} \lambda \, d\lambda, \quad (7.10.27b)$$

$$G_{z'1}(\rho, z'+d') = 2 \int_0^\infty \frac{1}{\gamma_1 + \gamma_2} J_1(\lambda\rho) e^{i\gamma_2(z'+d')} \lambda^2 \, d\lambda. \quad (7.10.28)$$

The integrals can be evaluated with the help of the approximate procedure described in Appendix D. With the condition

$$|k_1|^2 \gg k_2^2 \quad \text{or} \quad |k_1| \geq 3k_2, \quad (7.10.29)$$

$$f(k_1, k_2; \lambda) \sim f_1(k_1, k_2; \lambda) + f_2(k_1, k_2; \lambda) - f_3(k_1, k_2; \lambda), \quad (7.10.30a)$$

where

$$f_1(k_1, k_2; \lambda) = f(k_1, k_2; \lambda \sim k_1),$$
$$f_2(k_1, k_2; \lambda) = f(k_1, k_2; \lambda \sim k_2), \quad (7.10.30b)$$

and

$$f_3(k_1, k_2; \lambda) = f_1(k_1, k_2; \lambda \sim k_2) = f_2(k_1, k_2; \lambda \sim k_1). \quad (7.10.30c)$$

With (7.10.25a),

$$f(k_1, k_2; \lambda) = \frac{\gamma_1\gamma_2}{\gamma_2 + (k_2^2/k_1^2)\gamma_1} e^{i\gamma_2(z'+d')}, \quad (7.10.31)$$

$$f_1(k_1, k_2; \lambda) = \frac{i\lambda\gamma_1}{i\lambda + (k_2^2/k_1^2)\gamma_1} e^{-k_1(z'+d')}, \quad (7.10.32a)$$

$$f_2(k_1, k_2; \lambda) = \frac{k_1\gamma_2}{\gamma_2 + k_2^2/k_1} e^{i\gamma_2(z'+d')}, \quad (7.10.32b)$$

$$f_3(k_1, k_2; \lambda) = \frac{i\lambda k_1}{i\lambda + k_2^2/k_1} e^{-k_1(z'+d')}. \tag{7.10.32c}$$

It follows that, with

$$\epsilon = \frac{k_2}{k_1}, \qquad \epsilon' = k_2\epsilon = \frac{k_2^2}{k_1}, \tag{7.10.33}$$

$$\begin{aligned} &\left.\begin{array}{l} F_{\rho 2}(\rho, z'+d') \\ F_{\phi' 2}(\rho, z'+d') \end{array}\right\} \\ &\quad = \epsilon' \int_0^\infty \frac{\gamma_2}{\gamma_2 + \epsilon'} [J_0(\lambda\rho) \mp J_2(\lambda\rho)] e^{i\gamma_2(z'+d')} \lambda \, d\lambda \\ &\qquad + i\epsilon' e^{-k_1(z'+d')} \int_0^\infty \frac{\gamma_1}{i\lambda k_1 + \gamma_1\epsilon'} [J_0(\lambda\rho) \mp J_2(\lambda\rho)] \lambda^2 \, d\lambda \\ &\qquad - i\epsilon' e^{-k_1(z'+d')} \int_0^\infty \frac{1}{i\lambda + \epsilon'} [J_0(\lambda\rho) \mp J_2(\lambda\rho)] \lambda^2 \, d\lambda \\ &\quad \sim \epsilon' \int_0^\infty \frac{\gamma_2}{\gamma_2 + \epsilon'} [J_0(\lambda\rho) \mp J_2(\lambda\rho)] e^{i\gamma_2(z'+d')} \lambda \, d\lambda. \end{aligned} \tag{7.10.34}$$

Since the second and third terms both have the large exponential attenuation factor $\exp[-k_1(z'+d')]$, they can be omitted. They contribute nothing significant to the propagating field.

In (7.10.25b), note that

$$\frac{1}{\gamma_1 + \gamma_2} = \frac{\gamma_1 - \gamma_2}{\gamma_1^2 - \gamma_2^2} = \frac{\gamma_1 - \gamma_2}{k_1^2 - k_2^2}, \tag{7.10.35}$$

so that

$$\begin{aligned} &\left.\begin{array}{l} F_{\rho 3}(\rho, z'+d') \\ F_{\phi' 3}(\rho, z'+d') \end{array}\right\} \\ &\quad = \frac{k_2^2}{k_1^2 - k_2^2} \int_0^\infty (\gamma_1 - \gamma_2)[J_0(\lambda\rho) \pm J_2(\lambda\rho)] e^{i\gamma_2(z'+d')} \lambda \, d\lambda. \end{aligned} \tag{7.10.36}$$

The integral $F_{z'1}(\rho, z'+d')$ in (7.10.26) can be treated like that in (7.10.25a). That is,

$$F_{z'1}(\rho, z'+d') \sim 2\epsilon' \int_0^\infty \frac{1}{\gamma_2 + \epsilon'} J_1(\lambda\rho) e^{i\gamma_2(z'+d')} \lambda^2 \, d\lambda. \tag{7.10.37}$$

Similarly,

$$\left.\begin{array}{l} G_{\rho 2}(\rho, z'+d') \\ G_{\phi' 2}(\rho, z'+d') \end{array}\right\} \sim \epsilon' \int_0^\infty \frac{1}{\gamma_2 + \epsilon'} [J_0(\lambda\rho) \pm J_2(\lambda\rho)] e^{i\gamma_2(z'+d')} \lambda \, d\lambda, \tag{7.10.38}$$

$$\left.\begin{matrix} G_{\rho 3}(\rho, z'+d') \\ G_{\phi' 3}(\rho, z'+d') \end{matrix}\right\}$$

$$= \frac{1}{k_1^2 - k_2^2} \int_0^\infty (\gamma_1\gamma_2 - \gamma_2^2)[J_0(\lambda\rho) \mp J_2(\lambda\rho)] e^{i\gamma_2(z'+d')} \lambda \, d\lambda, \tag{7.10.39}$$

$$G_{z'1}(\rho, z'+d') = \frac{2}{k_1^2 - k_2^2} \int_0^\infty (\gamma_1 - \gamma_2) J_1(\lambda\rho) e^{i\gamma_2(z'+d')} \lambda^2 \, d\lambda. \tag{7.10.40}$$

The evaluation of the integrals in (7.10.34), (7.10.37), and (7.10.38) can be carried out. With $z'_d = z' + d'$, and $C(\lambda\rho)$ standing for $[J_0(\lambda\rho) \pm J_2(\lambda\rho)]$, $[J_0(\lambda\rho) \mp J_2(\lambda\rho)]$, or $\lambda J_1(\lambda\rho)$, let

$$K(\rho, z'_d) = \int_0^\infty \frac{\gamma_2}{\gamma_2 + \epsilon'} C(\lambda\rho) e^{i\gamma_2 z'_d} \lambda \, d\lambda \tag{7.10.41}$$

and

$$J(\rho, z'_d) = \int_0^\infty \frac{1}{\gamma_2 + \epsilon'} C(\lambda\rho) e^{i\gamma_2 z'_d} \lambda \, d\lambda. \tag{7.10.42}$$

Then

$$\frac{dJ(\rho, z'_d)}{dz'_d} = iK(\rho, z'_d). \tag{7.10.43}$$

Also,

$$\frac{\gamma_2}{\gamma_2 + \epsilon'} - 1 = -\frac{\epsilon'}{\gamma_2 + \epsilon'} \tag{7.10.44}$$

so that

$$K(\rho, z'_d) = K_0(\rho, z'_d) - \epsilon' J(\rho, z'_d), \tag{7.10.45}$$

where

$$K_0(\rho, z'_d) = \int_0^\infty C(\lambda\rho) e^{i\gamma_2 z'_d} \lambda \, d\lambda. \tag{7.10.46}$$

Hence,

$$-i \frac{dJ(\rho, z'_d)}{dz'_d} = K_0(\rho, z'_d) - \epsilon' J(\rho, z'_d) \tag{7.10.47}$$

or

$$\frac{dJ(\rho, z'_d)}{dz'_d} + i\epsilon' J(\rho, z'_d) = iK_0(\rho, z'_d). \tag{7.10.48}$$

This is an ordinary differential equation with the solution

$$J(\rho, z'_d) = -i \int_{z'_d}^\infty e^{i\epsilon'(z - z'_d)} K_0(\rho, z) \, dz. \tag{7.10.49}$$

Let

$$K_1(\rho, z'_d) = \int_0^\infty \frac{1}{\gamma_2} C(\lambda\rho) e^{i\gamma_2 z'_d} \lambda \, d\lambda. \tag{7.10.50}$$

Then

$$\frac{dK_1(\rho, z'_d)}{dz'_d} = iK_0(\rho, z'_d) \tag{7.10.51}$$

and

$$J(\rho, z'_d) = -\int_{z'_d}^{\infty} e^{i\epsilon'(z-z'_d)} \frac{dK_1(\rho, z)}{dz}\, dz. \tag{7.10.52}$$

This is readily integrated by parts to give

$$J(\rho, z'_d) = K_1(\rho, z'_d) - \epsilon' K_2(\rho, z'_d), \tag{7.10.53a}$$

where

$$K_2(\rho, z'_d) = -i\int_{z'_d}^{\infty} K_1(\rho, z) e^{i\epsilon'(z-z'_d)}\, dz. \tag{7.10.53b}$$

It follows that

$$K(\rho, z'_d) = K_0(\rho, z'_d) - \epsilon' K_1(\rho, z'_d) + \epsilon'^2 K_2(\rho, z'_d). \tag{7.10.54}$$

With (7.10.41) and (7.10.42), it follows that

$$\begin{aligned}
\left.\begin{matrix} F_{\rho 2}(\rho, z'+d') \\ F_{\phi' 2}(\rho, z'+d') \end{matrix}\right\} &= \epsilon'\left(\begin{Bmatrix} K_{0\rho}(\rho, z'+d') \\ K_{0\phi'}(\rho, z'+d') \end{Bmatrix} - \epsilon'\begin{Bmatrix} K_{1\rho}(\rho, z'+d') \\ K_{1\phi'}(\rho, z'+d') \end{Bmatrix}\right. \\
&\qquad \left. + \epsilon'^2\begin{Bmatrix} K_{2\rho}(\rho, z'+d') \\ K_{2\phi'}(\rho, z'+d') \end{Bmatrix}\right) \\
&= \epsilon'\left(\int_0^{\infty} [J_0(\lambda\rho) \mp J_2(\lambda\rho)] e^{i\gamma_2(z'+d')}\lambda\, d\lambda\right. \\
&\qquad - \epsilon'\int_0^{\infty} \frac{1}{\gamma_2}[J_0(\lambda\rho) \mp J_2(\lambda\rho)] e^{i\gamma_2(z'+d')}\lambda\, d\lambda \\
&\qquad \left. + \epsilon'^2\begin{Bmatrix} K_{2\rho}(\rho, z'+d') \\ K_{2\phi'}(\rho, z'+d') \end{Bmatrix}\right) \\
&= \epsilon'\begin{Bmatrix} I_{19}(\rho, z'+d') \\ I_{18}(\rho, z'+d') \end{Bmatrix} - \epsilon'^2\begin{Bmatrix} I_4(\rho, z'+d') \\ I_3(\rho, z'+d') \end{Bmatrix} \\
&\qquad + \epsilon'^3\begin{Bmatrix} K_{2\rho}(\rho, z'+d') \\ K_{2\phi'}(\rho, z'+d') \end{Bmatrix},
\end{aligned} \tag{7.10.55}$$

$$\begin{aligned}
F_{z'1}(\rho, z'+d') &= 2\epsilon'[K_{1z'}(\rho, z'+d') - \epsilon' K_{2z'}(\rho, z'+d')] \\
&= 2\epsilon'\left(\int_0^{\infty} \frac{1}{\gamma_2} J_1(\lambda\rho) e^{i\gamma_2(z'+d')}\lambda^2\, d\lambda\right. \\
&\qquad \left. - \epsilon' K_{2z'}(\rho, z'+d')\right) \\
&= 2\epsilon'\left(\frac{\rho}{2} I_7(\rho, z'+d') - \epsilon' K_{2z'}(\rho, z'+d')\right),
\end{aligned} \tag{7.10.56}$$

$$
\begin{aligned}
\left.\begin{matrix} G_{\rho 2}(\rho, z'+d') \\ G_{\phi' 2}(\rho, z'+d') \end{matrix}\right\} &= \epsilon' \left(\left\{ \begin{matrix} K_{1m\rho}(\rho, z'+d') \\ K_{1m\phi'}(\rho, z'+d') \end{matrix} \right\} \right. \\
&\qquad \left. - \epsilon' \left\{ \begin{matrix} K_{2m\rho}(\rho, z'+d') \\ K_{2m\phi'}(\rho, z'+d') \end{matrix} \right\} \right) \\
&= \epsilon' \left(\int_0^\infty \frac{1}{\gamma_2} [J_0(\lambda\rho) \pm J_2(\lambda\rho)] e^{i\gamma_2(z'+d')} \lambda \, d\lambda \right. \\
&\qquad \left. - \epsilon' \left\{ \begin{matrix} K_{2m\rho}(\rho, z'+d') \\ K_{2m\phi'}(\rho, z'+d') \end{matrix} \right\} \right) \\
&= \epsilon' \left\{ \begin{matrix} I_3(\rho, z'+d') \\ I_4(\rho, z'+d') \end{matrix} \right\} - \epsilon'^2 \left\{ \begin{matrix} K_{2m\rho}(\rho, z'+d') \\ K_{2m\phi'}(\rho, z'+d') \end{matrix} \right\}.
\end{aligned} \tag{7.10.57}
$$

The integrals $K_2(\rho, z'+d')$ appearing at the ends of these expressions are defined in general by

$$
K_2(\rho, z'+d') = -i \int_{z'+d'}^\infty K_1(\rho, z) e^{i\epsilon'(z-z'-d')} \, dz, \tag{7.10.58}
$$

where $K_1(\rho, z'+d')$ is the function appropriate for each component.

The integrals $K_0(\rho, z'+d')$ and $K_1(\rho, z'+d')$ can be evaluated with the help of Appendix A:

$$
\begin{aligned}
F_{\rho 2}(\rho, z'+d') = &-2\epsilon' \left[\frac{e^{ik_2(z'+d')}}{\rho^2} + \left(\frac{z'+d'}{r_2} \right) \left(\frac{ik_2}{r_2} - \frac{1}{r_2^2} - \frac{1}{\rho^2} \right) e^{ik_2 r_2} \right] \\
&+ \frac{2\epsilon'^2}{k_2} \left[\frac{e^{ik_2(z'+d')}}{\rho^2} + \left(\frac{ik_2}{r_2} - \frac{1}{\rho^2} \right) e^{ik_2 r_2} \right] \\
&+ \epsilon'^3 K_{2\rho}(\rho, z'+d'),
\end{aligned} \tag{7.10.59}
$$

$$
\begin{aligned}
F_{\phi' 2}(\rho, z'+d') = &\, 2\epsilon' \left[\frac{e^{ik_2(z'+d')}}{\rho^2} - \left(\frac{z'+d'}{r_2} \right) \frac{e^{ik_2 r_2}}{\rho^2} \right] \\
&- \frac{2\epsilon'^2}{k_2} \left(\frac{e^{ik_2(z'+d')}}{\rho^2} - \frac{e^{ik_2 r_2}}{\rho^2} \right) + \epsilon'^3 K_{2\phi'}(\rho, z'+d'),
\end{aligned} \tag{7.10.60}
$$

$$
\begin{aligned}
F_{z'1}(\rho, z'+d') = &-2\epsilon' k_2 \left(\frac{\rho}{r_2} \right) \left(\frac{1}{r_2} + \frac{i}{k_2 r_2^2} \right) e^{ik_2 r_2} \\
&- 2\epsilon'^2 K_{2z'}(\rho, z'+d'),
\end{aligned} \tag{7.10.61}
$$

$$
G_{\rho 2}(\rho, z'+d') = \frac{2\epsilon'}{k_2} \left(\frac{e^{ik_2(z'+d')}}{\rho^2} - \frac{e^{ik_2 r_2}}{\rho^2} \right) - \epsilon'^2 K_{2m\rho}(\rho, z'+d'), \tag{7.10.62}
$$

$$G_{\phi'2}(\rho, z'+d') = -\frac{2\epsilon'}{k_2}\left[\frac{e^{ik_2(z'+d')}}{\rho^2} + \left(\frac{ik_2}{r_2} - \frac{1}{\rho^2}\right)e^{ik_2r_2}\right] - \epsilon'^2 K_{2m\phi'}(\rho, z'+d'). \tag{7.10.63}$$

The functions $F_3(\rho, z'+d')$, $G_3(\rho, z'+d')$, and $G_{z'1}(\rho, z'+d')$ are readily evaluated with Appendix A. Thus,

$$\left.\begin{matrix} F_{\rho 3}(\rho, z'+d') \\ F_{\phi' 3}(\rho, z'+d') \end{matrix}\right\} \sim \frac{k_2^2}{k_1^2 - k_2^2}\left(k_1 \int_0^\infty [J_0(\lambda\rho) \pm J_2(\lambda\rho)]e^{i\gamma_2(z'+d')}\lambda\, d\lambda - \int_0^\infty \gamma_2[J_0(\lambda\rho) \pm J_2(\lambda\rho)]e^{i\gamma_2(z'+d')}\lambda\, d\lambda\right)$$

$$\sim \frac{k_2^2}{k_1^2}\left(k_1\left\{\begin{matrix} I_{18}(\rho, z'+d') \\ I_{19}(\rho, z'+d') \end{matrix}\right\} - \left\{\begin{matrix} I_{11}(\rho, z'+d') \\ I_{12}(\rho, z'+d') \end{matrix}\right\}\right), \tag{7.10.64}$$

so that

$$F_{\rho 3}(\rho, z'+d') = \frac{2k_2^2}{k_1^2}\left(k_1\left[\frac{e^{ik_2(z'+d')}}{\rho^2} - \left(\frac{z'+d'}{r_2}\right)\frac{e^{ik_2r_2}}{\rho^2}\right] - k_2\left\{\frac{e^{ik_2(z'+d')}}{\rho^2} + \left[\frac{i}{k_2r_2^3} - \left(\frac{z'+d'}{r_2}\right)^2\frac{1}{\rho^2}\right]e^{ik_2r_2}\right\}\right), \tag{7.10.65}$$

$$F_{\phi' 3}(\rho, z'+d') = \frac{2k_2^2}{k_1^2}\left(-k_1\left[\frac{e^{ik_2(z'+d')}}{\rho^2} + \left(\frac{z'+d'}{r_2}\right)\left(\frac{ik_2}{r_2} - \frac{1}{r_2^2} - \frac{1}{\rho^2}\right)e^{ik_2r_2}\right] + k_2\left\{\frac{e^{ik_2(z'+d')}}{\rho^2} + \left[\frac{1}{r_2^2} + \frac{2i}{k_2r_2^3} + \left(\frac{z'+d'}{r_2}\right)^2\left(\frac{ik_2}{r_2} - \frac{3}{r_2^2} - \frac{3i}{k_2r_2^3} - \frac{1}{\rho^2}\right)\right]e^{ik_2r_2}\right\}\right); \tag{7.10.66}$$

$$\left.\begin{matrix} G_{\rho 3}(\rho, z'+d') \\ G_{\phi' 3}(\rho, z'+d') \end{matrix}\right\} \sim \frac{1}{k_1^2 - k_2^2}\left(k_1 \int_0^\infty \gamma_2[J_0(\lambda\rho) \mp J_2(\lambda\rho)]e^{i\gamma_2(z'+d')}\lambda\, d\lambda - k_2^2 \int_0^\infty [J_0(\lambda\rho) \mp J_2(\lambda\rho)]e^{i\gamma_2(z'+d')}\lambda\, d\lambda\right.$$

$$+ \int_0^\infty [J_0(\lambda\rho) \mp J_2(\lambda\rho)] e^{i\gamma_2(z'+d')} \lambda^3 \, d\lambda \Big)$$

$$\sim \frac{1}{k_1} \begin{Bmatrix} I_{12}(\rho, z'+d') \\ I_{11}(\rho, z'+d') \end{Bmatrix} - \frac{k_2^2}{k_1^2} \begin{Bmatrix} I_{19}(\rho, z'+d') \\ I_{18}(\rho, z'+d') \end{Bmatrix}$$

$$+ \frac{1}{k_1^2} \begin{Bmatrix} I_{23}(\rho, z'+d') \\ I_{22}(\rho, z'+d') \end{Bmatrix}, \tag{7.10.67}$$

so that

$$G_{\rho 3}(\rho, z'+d')$$

$$= -\frac{2k_2}{k_1} \left\{ \frac{e^{ik_2(z'+d')}}{\rho^2} + \left[\frac{1}{r_2^2} + \frac{2i}{k_2 r_2^3} \right.\right.$$

$$\left.\left. + \left(\frac{z'+d'}{r_2} \right)^2 \left(\frac{ik_2}{r_2} - \frac{3}{r_2^2} - \frac{3i}{k_2 r_2^3} - \frac{1}{\rho^2} \right) \right] e^{ik_2 r_2} \right\}$$

$$+ \frac{2k_2^2}{k_1^2} \left[\frac{e^{ik_2(z'+d')}}{\rho^2} + \left(\frac{z'+d'}{r_2} \right) \left(\frac{ik_2}{r_2} - \frac{1}{r_2^2} - \frac{1}{\rho^2} \right) e^{ik_2 r_2} \right]$$

$$+ \frac{2k_2^2}{k_1^2} \left(\frac{z'+d'}{r_2} \right) \left(-\frac{ik_2}{r_2} + \frac{5}{r_2^2} + \frac{12i}{k_2 r_2^3} \right) e^{ik_2 r_2}, \tag{7.10.68}$$

$$G_{\phi' 3}(\rho, z'+d')$$

$$= \frac{2k_2}{k_1} \left\{ \frac{e^{ik_2(z'+d')}}{\rho^2} + \left[\frac{i}{k_2 r_2^3} - \left(\frac{z'+d'}{r_2} \right)^2 \frac{1}{\rho^2} \right] e^{ik_2 r_2} \right\}$$

$$- \frac{2k_2^2}{k_1^2} \left[\frac{e^{ik_2(z'+d')}}{\rho^2} - \left(\frac{z'+d'}{r_2} \right) \frac{e^{ik_2 r_2}}{\rho^2} \right]$$

$$- \frac{2k_2^2}{k_1^2} \left(\frac{z'+d'}{r_2} \right) \left(\frac{1}{r_2^2} + \frac{3i}{k_2 r_2^3} - \frac{3}{k_2^2 r_2^4} \right) e^{ik_2 r_2}; \tag{7.10.69}$$

and

$$G_{z'1}(\rho, z'+d')$$

$$\sim \frac{2}{k_1^2 - k_2^2} \left(k_1 \int_0^\infty J_1(\lambda\rho) e^{i\gamma_2(z'+d')} \lambda^2 \, d\lambda \right.$$

$$\left. - \int_0^\infty \gamma_2 J_1(\lambda\rho) e^{i\gamma_2(z'+d')} \lambda^2 \, d\lambda \right)$$

$$\sim \frac{2}{k_1^2} \left(\frac{k_1 \rho}{2} I_{22}(\rho, z'+d') - \frac{\rho}{2} I_{15}(\rho, z'+d') \right)$$

$$= \frac{2}{k_1^2} \left\{ -k_1 k_2 \left(\frac{z'+d'}{r_2} \right) \left(\frac{\rho}{r_2} \right) \left(\frac{k_2}{r_2} + \frac{3i}{r_2^2} - \frac{3}{k_2 r_2^3} \right) e^{ik_2 r_2} \right.$$

$$- \left(\frac{\rho}{r_2} \right) \left[\frac{ik_2^2}{r_2^2} - \frac{3k_2}{r_2^3} - \frac{3i}{r_2^4} \right.$$

$$-\left(\frac{z'+d'}{r_2}\right)^2\left(\frac{k_2^3}{r_2}+\frac{6ik_2^2}{r_2^2}-\frac{15k_2}{r_2^3}-\frac{15i}{r_2^4}\right)\Bigg]e^{ik_2r_2}\Bigg\}$$

$$\begin{aligned}=-\frac{2e^{ik_2r_2}}{k_1^2}\left(\frac{\rho}{r_2}\right)\Bigg[&\frac{ik_2^2}{r_2^2}-\frac{3k_2}{r_2^3}-\frac{3i}{r_2^4}\\&+k_1k_2\left(\frac{z'+d'}{r_2}\right)\left(\frac{k_2}{r_2}+\frac{3i}{r_2^2}-\frac{3}{k_2r_2^3}\right)\\&-\left(\frac{z'+d'}{r_2}\right)^2\left(\frac{k_2^3}{r_2}+\frac{6ik_2^2}{r_2^2}-\frac{15k_2}{r_2^3}\right)\Bigg].\end{aligned}\tag{7.10.70}$$

With (7.10.59) and (7.10.65), it follows that

$$\begin{aligned}&F_{\rho1}(\rho,z'+d')=F_{\rho2}(\rho,z'+d')+F_{\rho3}(\rho,z'+d')\\&=-2\epsilon'\left[\frac{e^{ik_2(z'+d')}}{\rho^2}+\left(\frac{z'+d'}{r_2}\right)\left(\frac{ik_2}{r_2}-\frac{1}{r_2^2}-\frac{1}{\rho^2}\right)e^{ik_2r_2}\right]\\&\quad+\frac{2\epsilon'^2}{k_2}\left[\frac{e^{ik_2(z'+d')}}{\rho^2}+\left(\frac{ik_2}{r_2}-\frac{1}{\rho^2}\right)e^{ik_2r_2}\right]+\epsilon'^3K_{2\rho}(\rho,z'+d')\\&\quad+2\epsilon'\left[\frac{e^{ik_2(z'+d')}}{\rho^2}-\left(\frac{z'+d'}{r_2}\right)\frac{e^{ik_2r_2}}{\rho^2}\right]\\&\quad-\frac{2\epsilon'^2}{k_2}\left\{\frac{e^{ik_2(z'+d')}}{\rho^2}+\left[\frac{i}{k_2r_2^3}-\left(\frac{z'+d'}{r_2}\right)^2\frac{1}{\rho^2}\right]e^{ik_2r_2}\right\}\\&=e^{ik_2r_2}\Bigg[-2\epsilon'\left(\frac{z'+d'}{r_2}\right)\left(\frac{ik_2}{r_2}-\frac{1}{r_2^2}\right)\\&\qquad+\frac{2\epsilon'^2}{k_2}\left(\frac{ik_2}{r_2}-\frac{1}{r_2^2}-\frac{i}{k_2r_2^3}\right)\Bigg]+\epsilon'^3K_{2\rho}(\rho,z'+d').\end{aligned}\tag{7.10.71}$$

As discussed in Appendix F, the evaluation of the integrals $K_2(\rho,z'+d')$ requires only the leading term in $1/r_2$ in $K_1(\rho,z'+d')$. From (7.10.55) for $K_{1\rho}(\rho,z'+d')$, this is

$$K_{1\rho}(\rho,z'+d')\sim-\frac{2i}{r_2}\,e^{ik_2r_2}=-\frac{2ik_2}{\bar r_2}\,e^{i\bar r_2},\tag{7.10.72}$$

where $\bar r_2=k_2r_2=(\bar\rho^2+\bar z_d'^2)^{1/2}$ and $\bar\rho=k_2\rho,\ \bar z_d'=k_2z_d'=k_2(z'+d')$. With (7.10.58) and $\bar z=k_2z$,

$$K_{2\rho}(\rho,z'+d')=-i\int_{z'+d'}^{\infty}K_{1\rho}(\rho,z)e^{i\epsilon'(z-z'-d')}\,dz,\tag{7.10.73a}$$

$$K_{2\rho}(\rho,\bar z_d')\sim2i\int_{\bar z_d'}^{\infty}\frac{i}{\bar r_2}\,e^{i\bar r_2}e^{i\epsilon(\bar z-\bar z_d')}\,d\bar z.\tag{7.10.73b}$$

This is like (F38) multiplied by $(2i\bar r_2/\bar\rho)$. Hence, with (F42),

$$K_{2\rho}(\rho,\bar z_d')\sim-2\left(\frac{\bar r_2}{\bar\rho}\right)\left(\frac{\pi}{\bar r_2}\right)^{1/2}e^{i\bar r_2}e^{-iP_2}\mathcal{F}(P_2),$$

$$K_{2\rho}(\rho, z'+d') \sim -2\left(\frac{r_2}{\rho}\right)\left(\frac{\pi}{k_2 r_2}\right)^{1/2} e^{ik_2 r_2} e^{-iP_2}\mathcal{F}(P_2), \tag{7.10.73c}$$

where

$$P_2 = S_2\left(\frac{S_2 + Z' + D'}{R}\right)^2, \tag{7.10.74a}$$

$$S_2 = \frac{k_2^3 r_2}{2k_1^2}; \qquad R = \frac{k_2^3\rho}{2k_1^2}; \qquad Z' = \frac{k_2^2 z'}{2k_1}; \qquad D' = \frac{k_2^2 d'}{2k_1}. \tag{7.10.74b}$$

The complete expression for $E_{2\rho}$ is obtained from (7.10.1). With terms of the order k_2^2/k_1^2 neglected, the formula is

$$\begin{aligned}
E_{2\rho}(\rho, \phi', z') &= \frac{\omega\mu_0}{4\pi k_2}\cos\phi' \\
\times \Bigg(& e^{ik_2 r_1}\left[\frac{2}{r_1^2} + \frac{2i}{k_2 r_1^3} + \left(\frac{z'-d'}{r_1}\right)^2\left(\frac{ik_2}{r_1} - \frac{3}{r_1^2} - \frac{3i}{k_2 r_1^3}\right)\right] \\
& - e^{ik_2 r_2}\Bigg\{\frac{2}{r_2^2} + \frac{2i}{k_2 r_2^3} + \left(\frac{z'+d'}{r_2}\right)^2\left(\frac{ik_2}{r_2} - \frac{3}{r_2^2} - \frac{3i}{k_2 r_2^3}\right) \\
& \qquad - \frac{2k_2}{k_1}\left(\frac{z'+d'}{r_2}\right)\left(\frac{ik_2}{r_2} - \frac{1}{r_2^2}\right) \\
& \qquad + \frac{2k_2^2}{k_1^2}\Bigg[\frac{ik_2}{r_2} - \frac{1}{r_2^2} - \frac{i}{k_2 r_2^3} \\
& \qquad\qquad - \frac{k_2^3}{k_1}\left(\frac{r_2}{\rho}\right)\left(\frac{\pi}{k_2 r_2}\right)^{1/2} e^{-iP_2}\mathcal{F}(P_2)\Bigg]\Bigg\}\Bigg).
\end{aligned} \tag{7.10.75}$$

When $z' = d' = 0$, this reduces to

$$\begin{aligned}
E_{2\rho}(\rho, \phi', 0) = -\frac{\omega\mu_0 k_2}{2\pi k_1^2}\cos\phi' \Bigg[&\frac{ik_2}{\rho} - \frac{1}{\rho^2} - \frac{i}{k_2\rho^3} \\
&- \frac{k_2^3}{k_1}\left(\frac{\pi}{k_2\rho}\right)^{1/2} e^{-ip}\mathcal{F}(p)\Bigg] e^{ik_2\rho},
\end{aligned} \tag{7.10.76}$$

where $p = R = k_2^3\rho/2k_1^2$.

The formula for $F_{\phi'1}(\rho, z'+d')$ is, from (7.10.60) and (7.10.66),

$$\begin{aligned}
F_{\phi'1}(\rho, z'+d') &= F_{\phi'2}(\rho, z'+d') + F_{\phi'3}(\rho, z'+d') \\
&= 2\epsilon'\left[\frac{e^{ik_2(z'+d')}}{\rho^2} - \left(\frac{z'+d'}{r_2}\right)\frac{e^{ik_2 r_2}}{\rho^2}\right] \\
&\quad - \frac{2\epsilon'^2}{k_2}\left(\frac{e^{ik_2(z'+d')}}{\rho^2} - \frac{e^{ik_2 r_2}}{\rho^2}\right) + \epsilon'^3 K_{2\phi'}(\rho, z'+d')
\end{aligned}$$

$$
\begin{aligned}
&- 2\epsilon'\left[\frac{e^{ik_2(z'+d')}}{\rho^2} + \left(\frac{z'+d'}{r_2}\right)\left(\frac{ik_2}{r_2} - \frac{1}{r_2^2} - \frac{1}{\rho^2}\right)e^{ik_2r_2}\right] \\
&+ \frac{2\epsilon'^2}{k_2}\left\{\frac{e^{ik_2(z'+d')}}{\rho^2} + \left[\frac{1}{r_2^2} + \frac{2i}{k_2r_2^3}\right.\right. \\
&\qquad \left.\left. + \left(\frac{z'+d'}{r_2}\right)^2\left(\frac{ik_2}{r_2} - \frac{3}{r_2^2} - \frac{3i}{k_2r_2^3} - \frac{1}{\rho^2}\right)\right]e^{ik_2r_2}\right\} \\
= &\left\{-2\epsilon'\left(\frac{z'+d'}{r_2}\right)\left(\frac{ik_2}{r_2} - \frac{1}{r_2^2}\right)\right. \\
&\left. + \frac{2\epsilon'^2}{k_2}\left[\frac{2}{r_2^2} + \frac{2i}{k_2r_2^3} + \left(\frac{z'+d'}{r_2}\right)^2\left(\frac{ik_2}{r_2} - \frac{3}{r_2^2} - \frac{3i}{k_2r_2^3}\right)\right]\right\}e^{ik_2r_2} \\
&+ \epsilon'^3 K_{2\phi'}(\rho, z'+d').
\end{aligned} \tag{7.10.77}
$$

As seen with (7.10.55), there is no $1/r_2$ term in $K_{1\phi'}(\rho, z'+d')$. The leading $1/r_2^2$ term is

$$
K_{1\phi'}(\rho, z'+d') = -\frac{2}{k_2\rho^2}\, e^{ik_2r_2}. \tag{7.10.78}
$$

This differs from $K_{1\rho}(\rho, z'+d')$ by the factor $(-ir_2/k_2\rho^2)$. Hence,

$$
K_{2\phi'}(\rho, z'+d') = \frac{2i}{k_2\rho}\,\frac{r_2^2}{\rho^2}\left(\frac{\pi}{k_2r_2}\right)^{1/2} e^{ik_2r_2}e^{-iP_2}\mathcal{F}(P_2). \tag{7.10.79}
$$

It follows with (7.10.5) that

$$
\begin{aligned}
E_{2\phi'}(\rho, \phi', z') = &-\frac{\omega\mu_0}{4\pi k_2}\sin\phi' \\
\times &\left(e^{ik_2r_1}\left(\frac{ik_2}{r_1} - \frac{1}{r_1^2} - \frac{i}{k_2r_1^3}\right) - e^{ik_2r_2}\left(\frac{ik_2}{r_2} - \frac{1}{r_2^2} - \frac{i}{k_2r_2^3}\right)\right. \\
&- e^{ik_2r_2}\left\{-\frac{2k_2}{k_1}\left(\frac{z'+d'}{r_2}\right)\left(\frac{ik_2}{r_2} - \frac{1}{r_2^2}\right)\right. \\
&\quad + \frac{2k_2^2}{k_1^2}\left[\frac{2}{r_2^2} + \frac{2i}{k_2r_2^3}\right. \\
&\qquad \left. + \left(\frac{z'+d'}{r_2}\right)^2\left(\frac{ik_2}{r_2} - \frac{3}{r_2^2} - \frac{3i}{k_2r_2^3}\right)\right] \\
&\quad \left.\left. + \frac{2ik_2^4}{k_1^3\rho}\left(\frac{r_2^2}{\rho^2}\right)\left(\frac{\pi}{k_2r_2}\right)^{1/2} e^{-iP_2}\mathcal{F}(P_2)\right\}\right).
\end{aligned} \tag{7.10.80}
$$

With $z' = d' = 0$, this reduces to

$$
E_{2\phi'}(\rho, \phi', 0) = \frac{\omega\mu_0 k_2}{2\pi k_1^2}\sin\phi'
$$

$$\times \left[\frac{2}{\rho^2} + \frac{2i}{k_2\rho^3} + \frac{ik_2^2}{k_1\rho}\left(\frac{\pi}{k_2\rho}\right)^{1/2} e^{-ip}\mathcal{F}(p)\right] e^{ik_2\rho}. \tag{7.10.81}$$

The leading term in $K_{1z'}(\rho, z'+d')$ is, from (7.10.56),

$$K_{1z'}(\rho, z'+d') = -k_2\,\frac{\rho}{r_2^2}\,e^{ik_2r_2}. \tag{7.10.82}$$

This differs from (7.10.72) by the factor $(-ik_2\rho/2r_2)$, so that with (7.10.73c),

$$K_{2z'}(\rho, z'+d') = ik_2\left(\frac{\pi}{k_2r_2}\right)^{1/2} e^{ik_2r_2}e^{-iP_2}\mathcal{F}(P_2). \tag{7.10.83}$$

With (7.10.9) and (7.10.61), the complete formula for $E_{2z'}$ is

$$\begin{aligned} E_{2z'}(\rho, \phi', z') = {}& \frac{\omega\mu_0}{4\pi k_2}\cos\phi' \\ & \times \left\{ -e^{ik_2r_1}\left(\frac{\rho}{r_1}\right)\left(\frac{z'-d'}{r_1}\right)\left(\frac{ik_2}{r_1} - \frac{3}{r_1^2} - \frac{3i}{k_2r_1^3}\right) \right. \\ & \quad + e^{ik_2r_2}\left(\frac{\rho}{r_2}\right)\left(\frac{z'+d'}{r_2}\right)\left(\frac{ik_2}{r_2} - \frac{3}{r_2^2} - \frac{3i}{k_2r_2^3}\right) \\ & \quad - \frac{2k_2}{k_1}\,e^{ik_2r_2}\left[\left(\frac{\rho}{r_2}\right)\left(\frac{ik_2}{r_2} - \frac{1}{r_2^2}\right)\right. \\ & \quad \left.\left. - \frac{k_2^3}{k_1}\left(\frac{\pi}{k_2r_2}\right)^{1/2} e^{-iP_2}\mathcal{F}(P_2)\right]\right\}. \end{aligned} \tag{7.10.84}$$

With $z' = d' = 0$, this becomes

$$\begin{aligned} E_{2z'}(\rho, \phi', 0) = {}& -\frac{\omega\mu_0}{2\pi k_1}\cos\phi' \\ & \times \left[\frac{ik_2}{\rho} - \frac{1}{\rho^2} - \frac{k_2^3}{k_1}\left(\frac{\pi}{k_2\rho}\right)^{1/2} e^{-ip}\mathcal{F}(p)\right] e^{ik_2\rho}. \end{aligned} \tag{7.10.85}$$

The formula for $G_{\rho1}(\rho, z'+d')$ is obtained with (7.10.62) and (7.10.68). With terms of the order k_2^2/k_1^2 neglected, it is

$$\begin{aligned} G_{\rho1}(\rho, z'+d') &= G_{\rho2}(\rho, z'+d') + G_{\rho3}(\rho, z'+d') \\ &= \frac{2\epsilon'}{k_2}\left(\frac{e^{ik_2(z'+d')}}{\rho^2} - \frac{e^{ik_2r_2}}{\rho^2}\right) - \epsilon'^2 K_{2m\rho}(\rho, z'+d') \\ &\quad - \frac{2\epsilon'}{k_2}\left\{\frac{e^{ik_2(z'+d')}}{\rho^2} + \left[\frac{1}{r_2^2} + \frac{2i}{k_2r_2^3}\right.\right. \\ &\quad \left.\left. + \left(\frac{z'+d'}{r_2}\right)^2\left(\frac{ik_2}{r_2} - \frac{3}{r_2^2} - \frac{3i}{k_2r_2^3} - \frac{1}{\rho^2}\right)\right] e^{ik_2r_2}\right\}. \end{aligned} \tag{7.10.86}$$

From (7.10.57), the leading term in $K_{1m\rho}(\rho, z'+d')$ is

$$K_{1m\rho}(\rho, z'+d') = -\frac{2}{k_2}\frac{e^{ik_2r_2}}{\rho^2}. \tag{7.10.87a}$$

This differs from $K_{1\rho}(\rho, z'+d')$ by the factor $(-ir_2/k_2\rho^2)$ so that

$$K_{2m\rho}(\rho, z'+d') = \frac{2ir_2^2}{k_2\rho^3}\left(\frac{\pi}{k_2r_2}\right)^{1/2} e^{ik_2r_2}e^{-iP_2}\mathcal{F}(P_2). \tag{7.10.87b}$$

It follows that

$$\begin{aligned} &G_{\rho 1}(\rho, z'+d') \\ &= -\frac{2k_2}{k_1}e^{ik_2r_2}\left[\frac{2}{r_2^2} + \frac{2i}{k_2r_2^3} + \left(\frac{z'+d'}{r_2}\right)^2\left(\frac{ik_2}{r_2} - \frac{3}{r_2^2} - \frac{3i}{k_2r_2^3}\right)\right. \\ &\qquad\left. + \frac{ik_2^2}{k_1\rho}\left(\frac{r_2^2}{\rho^2}\right)\left(\frac{\pi}{k_2r_2}\right)^{1/2} e^{-iP_2}\mathcal{F}(P_2)\right]. \end{aligned} \tag{7.10.88}$$

Similarly, the formula for $G_{\phi'1}(\rho, z'+d')$ is obtained from (7.10.63) and (7.10.69). It is

$$\begin{aligned} &G_{\phi'1}(\rho, z'+d') = G_{\phi'2}(\rho, z'+d') + G_{\phi'3}(\rho, z'+d') \\ &= -\frac{2\epsilon'}{k_2}\left[\frac{e^{ik_2(z'+d')}}{\rho^2} + \left(\frac{ik_2}{r_2} - \frac{1}{\rho^2}\right)e^{ik_2r_2}\right] - \epsilon'^2 K_{2m\phi'}(\rho, z'+d') \\ &\quad + \frac{2\epsilon'}{k_2}\left\{\frac{e^{ik_2(z'+d')}}{\rho^2} + \left[\frac{i}{k_2r_2^3} - \left(\frac{z'+d'}{r_2}\right)^2\frac{1}{\rho^2}\right]e^{ik_2r_2}\right\}. \end{aligned} \tag{7.10.89}$$

Here,

$$K_{1m\phi'}(\rho, z'+d') = -\frac{2i}{r_2}e^{ik_2r_2}. \tag{7.10.90a}$$

This is like $K_{1\rho}(\rho, z'+d')$ so that

$$K_{2m\phi'}(\rho, z'+d') = -2\left(\frac{r_2}{\rho}\right)\left(\frac{\pi}{k_2r_2}\right)^{1/2} e^{ik_2r_2}e^{-iP_2}\mathcal{F}(P_2) \tag{7.10.90b}$$

and

$$\begin{aligned} G_{\phi'1}(\rho, z'+d') = -\frac{2k_2}{k_1}e^{ik_2r_2}&\left[\frac{ik_2}{r_2} - \frac{1}{r_2^2} - \frac{i}{k_2r_2^3}\right. \\ &\left. - \frac{k_2^3}{k_1}\left(\frac{r_2}{\rho}\right)\left(\frac{\pi}{k_2r_2}\right)^{1/2} e^{-iP_2}\mathcal{F}(P_2)\right]. \end{aligned} \tag{7.10.91}$$

When the functions $G_{\rho 1}(\rho, z'+d')$, $G_{\phi' 1}(\rho, z'+d')$, and $G_{z' 1}(\rho, z'+d')$ are combined in (7.10.13), (7.10.14), and (7.10.18), the results are

$$\begin{aligned} B_{2\rho}(\rho, \phi', z') &= \frac{\mu_0}{4\pi} \sin\phi' \\ &\times \left\{ e^{ik_2 r_1} \left(\frac{z'-d'}{r_1} \right) \left(\frac{ik_2}{r_1} - \frac{1}{r_1^2} \right) - e^{ik_2 r_2} \left(\frac{z'+d'}{r_2} \right) \left(\frac{ik_2}{r_2} - \frac{1}{r_2^2} \right) \right. \\ &\quad + \frac{2k_2}{k_1} e^{ik_2 r_2} \left[\frac{2}{r_2^2} + \frac{2i}{k_2 r_2^3} + \frac{ik_2^2}{k_1 \rho} \left(\frac{r_2^2}{\rho^2} \right) \left(\frac{\pi}{k_2 r_2} \right)^{1/2} e^{-iP_2} \mathcal{F}(P_2) \right. \\ &\quad \left. \left. + \left(\frac{z'+d'}{r_2} \right)^2 \left(\frac{ik_2}{r_2} - \frac{3}{r_2^2} - \frac{3i}{k_2 r_2^3} \right) \right] \right\}, \end{aligned} \tag{7.10.92}$$

$$\begin{aligned} B_{2\phi'}(\rho, \phi', z') &= \frac{\mu_0}{4\pi} \cos\phi' \\ &\times \left\{ e^{ik_2 r_1} \left(\frac{z'-d'}{r_1} \right) \left(\frac{ik_2}{r_1} - \frac{1}{r_1^2} \right) - e^{ik_2 r_2} \left(\frac{z'+d'}{r_2} \right) \left(\frac{ik_2}{r_2} - \frac{1}{r_2^2} \right) \right. \\ &\quad + \frac{2k_2}{k_1} e^{ik_2 r_2} \left[\frac{ik_2}{r_2} - \frac{1}{r_2^2} - \frac{i}{k_2 r_2^3} \right. \\ &\quad \left. \left. - \frac{k_2^3}{k_1} \left(\frac{r_2}{\rho} \right) \left(\frac{\pi}{k_2 r_2} \right)^{1/2} e^{-iP_2} \mathcal{F}(P_2) \right] \right\}, \end{aligned} \tag{7.10.93}$$

and

$$\begin{aligned} B_{2z'}(\rho, \phi', z') &= -\frac{\mu_0}{4\pi} \sin\phi' \\ &\times \left(e^{ik_2 r_1} \left(\frac{\rho}{r_1} \right) \left(\frac{ik_2}{r_1} - \frac{1}{r_1^2} \right) - e^{ik_2 r_2} \left(\frac{\rho}{r_2} \right) \left(\frac{ik_2}{r_2} - \frac{1}{r_2^2} \right) \right. \\ &\quad + 2e^{ik_2 r_2} \left(\frac{\rho}{r_2} \right) \left\{ \frac{k_2}{k_1} \left(\frac{z'+d'}{r_2} \right) \left(\frac{ik_2}{r_2} - \frac{3}{r_2^2} - \frac{3i}{k_2 r_2^3} \right) \right. \\ &\quad - \frac{k_2^2}{k_1^2} \left[\frac{1}{r_2^2} + \frac{3i}{k_2 r_2^3} - \frac{3}{k_2^2 r_2^4} \right. \\ &\quad \left. \left. \left. + \left(\frac{z'+d'}{r_2} \right)^2 \left(\frac{ik_2}{r_2} - \frac{6}{r_2^2} - \frac{15i}{k_2 r_2^3} \right) \right] \right\} \right). \end{aligned} \tag{7.10.94}$$

With $z' = d' = 0$, these become

$$\begin{aligned} B_{2\rho}(\rho, \phi', 0) &= \frac{\mu_0 k_2}{2\pi k_1} \sin\phi' \left[\frac{2}{\rho^2} + \frac{2i}{k_2 \rho^3} \right. \\ &\quad \left. + \frac{ik_2^2}{k_1 \rho} \left(\frac{\pi}{k_2 \rho} \right)^{1/2} e^{-ip} \mathcal{F}(p) \right] e^{ik_2 \rho}, \end{aligned} \tag{7.10.95}$$

$$B_{2\phi'}(\rho,\phi',0) = \frac{\mu_0 k_2}{2\pi k_1}\cos\phi'\left[\frac{ik_2}{\rho} - \frac{1}{\rho^2} - \frac{i}{k_2\rho^3} - \frac{k_2^3}{k_1}\left(\frac{\pi}{k_2\rho}\right)^{1/2} e^{-ip}\mathcal{F}(p)\right]e^{ik_2\rho}, \tag{7.10.96}$$

$$B_{2z'}(\rho,\phi',0) = \frac{\mu_0}{2\pi k_1^2}\sin\phi'\left(\frac{k_2^2}{\rho^2} + \frac{3ik_2}{\rho^3} - \frac{3}{\rho^4}\right)e^{ik_2\rho}. \tag{7.10.97}$$

The formulas (7.10.76), (7.10.81), (7.10.85), and (7.10.95)–(7.10.97) agree with the corresponding formulas (5.5.37)–(5.5.42) when $z + d = 0$ to satisfy the boundary conditions

$$E_{1\rho}(\rho,\phi,0) = E_{2\rho}(\rho,-\phi',0); \qquad E_{1\phi}(\rho,\phi,0) = -E_{2\phi'}(\rho,-\phi',0),$$

$$E_{1z}(\rho,\phi,0) = -\frac{k_2^2}{k_1^2}E_{2z'}(\rho,-\phi',0); \qquad \mathbf{B}_1(\rho,\phi,0) = \mathbf{B}_2(\rho,-\phi',0),$$

except for certain terms in $\mathbf{E}_1$ and $\mathbf{B}_1$ multiplied by $\exp(ik_1\rho)$. The corresponding terms in $\mathbf{E}_2$ and $\mathbf{B}_2$ were omitted in the evaluation of the integrals because they are multiplied by the factor $\exp[-k_1(z'+d')]$ which makes them insignificant in Region 2 except when $z'+d'$ is equal to or near zero. If the relevant integrals had been retained and evaluated, they would contribute the required terms. Note that such terms also occur in $\mathbf{E}_2$ and $\mathbf{B}_2$ when the dipole source is in Region 1. In this case, they have been evaluated and are contained in (5.7.14)–(5.7.17), (5.7.25), (5.7.45), (5.7.52), and (5.7.57).

In television and FM radio transmission and reception, both the transmitting and receiving elements are at heights of several wavelengths so that the lateral-wave terms are usually negligible. The far-field $(1/r_1)$ and $(1/r_2)$ terms are

$$E_{2\rho}^r(\rho,\phi',z') = \frac{i\omega\mu_0}{4\pi}\cos\phi'\left[\left(\frac{z'-d'}{r_1}\right)^2\frac{e^{ik_2r_1}}{r_1} - \left(\frac{z'+d'}{r_2}\right)^2\frac{e^{ik_2r_2}}{r_2}\right], \tag{7.10.98}$$

$$E_{2\phi'}^r(\rho,\phi',z') = -\frac{i\omega\mu_0}{4\pi}\sin\phi'\left(\frac{e^{ik_2r_1}}{r_1} - \frac{e^{ik_2r_2}}{r_2}\right), \tag{7.10.99}$$

$$E_{2z'}^r(\rho,\phi',z') = -\frac{i\omega\mu_0}{4\pi}\cos\phi'\left[\left(\frac{\rho}{r_1}\right)\left(\frac{z'-d'}{r_1}\right)\frac{e^{ik_2r_1}}{r_1} - \left(\frac{\rho}{r_2}\right)\left(\frac{z'+d'}{r_2}\right)\frac{e^{ik_2r_2}}{r_2}\right], \tag{7.10.100}$$

$$B_{2\rho}^r(\rho,\phi',z') = \frac{i\mu_0k_2}{4\pi}\sin\phi'\left[\left(\frac{z'-d'}{r_1}\right)\frac{e^{ik_2r_1}}{r_1} - \left(\frac{z'+d'}{r_2}\right)\frac{e^{ik_2r_2}}{r_2}\right], \tag{7.10.101}$$

$$B_{2\phi'}^r(\rho,\phi',z') = \frac{i\mu_0 k_2}{4\pi}\cos\phi'\left[\left(\frac{z'-d'}{r_1}\right)\frac{e^{ik_2r_1}}{r_1} - \left(\frac{z'+d'}{r_2}\right)\frac{e^{ik_2r_2}}{r_2}\right], \tag{7.10.102}$$

$$B_{2z'}^r(\rho,\phi',z') = -\frac{i\mu_0 k_2}{4\pi}\sin\phi'\left[\left(\frac{\rho}{r_1}\right)\frac{e^{ik_2r_1}}{r_1} - \left(\frac{\rho}{r_2}\right)\frac{e^{ik_2r_2}}{r_2}\right]. \tag{7.10.103}$$

At sufficiently great distances, i.e., $r_0 = (\rho^2 + z'^2)^{1/2} \gg d'$,

$$r_1 \sim r_0 - d'\cos\Theta; \qquad r_2 \sim r_0 + d'\cos\Theta,$$

where $\cos\Theta = z'/r_0$ and Θ is measured from the positive z'-axis. Also, in amplitudes $r_1 \sim r_2 \sim r_0$. With these values,

$$e^{ik_2r_1} \sim e^{ik_2r_0}e^{-ik_2d'\cos\Theta}; \qquad e^{ik_2r_2} \sim e^{ik_2r_0}e^{ik_2d'\cos\Theta}, \tag{7.10.104a}$$

$$\begin{aligned}\frac{e^{ik_2r_2}}{r_2} - \frac{e^{ik_2r_1}}{r_1} &\sim \frac{e^{ik_2r_0}}{r_0}\left(e^{ik_2d'\cos\Theta} - e^{-ik_2d'\cos\Theta}\right)\\ &= \frac{2ie^{ik_2r_0}}{r_0}\sin(k_2d'\cos\Theta),\end{aligned} \tag{7.10.104b}$$

and (7.10.98)–(7.10.103) become

$$E_{2\rho}^r(r_0,\Theta,\Phi') = \frac{\omega\mu_0}{2\pi}\frac{e^{ik_2r_0}}{r_0}\cos\Phi'\cos^2\Theta\sin(k_2d'\cos\Theta), \tag{7.10.105}$$

$$E_{2\phi'}^r(r_0,\Theta,\Phi') = -\frac{\omega\mu_0}{2\pi}\frac{e^{ik_2r_0}}{r_0}\sin\Phi'\sin(k_2d'\cos\Theta), \tag{7.10.106}$$

$$E_{2z'}^r(r_0,\Theta,\Phi') = -\frac{\omega\mu_0}{2\pi}\frac{e^{ik_2r_0}}{r_0}\cos\Phi'\sin\Theta\cos\Theta\sin(k_2d'\cos\Theta), \tag{7.10.107}$$

$$B_{2\rho}^r(r_0,\Theta,\Phi') = \frac{\mu_0 k_2}{2\pi}\frac{e^{ik_2r_0}}{r_0}\sin\Phi'\cos\Theta\sin(k_2d'\cos\Theta), \tag{7.10.108}$$

$$B_{2\phi'}^r(r_0,\Theta,\Phi') = \frac{\mu_0 k_2}{2\pi}\frac{e^{ik_2r_0}}{r_0}\cos\Phi'\cos\Theta\sin(k_2d'\cos\Theta), \tag{7.10.109}$$

$$B_{2z'}^r(r_0,\Theta,\Phi') = -\frac{\mu_0 k_2}{2\pi}\frac{e^{ik_2r_0}}{r_0}\sin\Phi'\sin\Theta\sin(k_2d'\cos\Theta), \tag{7.10.110}$$

where $\Phi' = \phi'$.

Since $\cos\Theta = z'/r_0$ and $\sin\Theta = \rho/r_0$, $\cos\Theta$ is small and $\sin\Theta \sim 1$ when $r_0 \gg z'$. This means that $E_{2\phi'}^r(r_0,\Theta,\Phi')$ and $B_{2z'}^r(r_0,\Theta,\Phi')$ are the largest components of the electromagnetic field. These are given by

$$E_{2\phi'}^r(r_0,\Theta,\Phi')\sin\Theta = \frac{\omega}{k_2}B_{2z'}^r(r_0,\Theta,\Phi')$$

$$= -\frac{\omega\mu_0}{2\pi}\frac{e^{ik_2r_0}}{r_0}\sin\Phi'\sin\Theta\,\sin(k_2d'\cos\Theta). \quad (7.10.111)$$

Reception of this field makes use of a horizontal electric dipole—in practice a folded dipole in conjunction with directors and a reflector in a properly designed Yagi-Uda antenna. Alternatively, a loop antenna can be used. For maximum signal, the plane of the loop must be perpendicular to $B^r_{2z'}(r_0, \Theta, \Phi')$ and, therefore, parallel to the surface of the earth.

Application of the generalized formulas for the field of a horizontal electric dipole to the horizontal-wire antenna and over-the-horizon radar is carried out in Chapter 18.

8
The Measurement of the Conductivity of the Oceanic Lithosphere with a Horizontal Antenna as the Source

8.1 The Homogeneous Isotropic Half-Space Model of the Oceanic Lithosphere—Theory

An important application of the horizontal electric dipole near the boundary between two electrically different regions is to the measurement of the conductivity of the oceanic crust. For simplicity it is advantageous to consider first an idealized sea floor that consists of a smooth plane boundary between sea water and rock or sediment. The effect of a boundary that is not smooth is considered in Chapter 10. The depths of the sea water and of the rock or sediment are required to be sufficient so that each is well approximated by a half-space. This excludes very shallow water and applies to a layer of sediment only when this is either electrically so thin that it can be ignored or so thick that reflections from the rock beyond it are negligible. The intermediate three-layer problem—which applies specifically to a layer of sediment with arbitrary thickness between sea water and rock—is treated in Chapter 12.

For applications on the sea bottom where Region 1 is salt water ($z \geq 0$) and Region 2 is rock or sediment ($z \leq 0$), the source and the points of observation of the electromagnetic field are advantageously kept in the sea water if only for practical reasons of deployment and mobility. The field generated by an x-directed horizontal unit dipole located close to the boundary ($d \sim 0$) consists of the six components given by (5.5.37)–(5.5.42) in Region 1, the sea. When the field is observed on or very near the sea floor, it is appropriate to set $z \sim 0$. At all frequencies useful for measurements on the sea floor, sea water is highly attenuating so that, except very close to the source, $\exp(-\alpha_1 \rho)$ is small and all terms with the factor $\exp(ik_1\rho) = \exp(-\alpha_1\rho)\exp(i\beta_1\rho)$ are negligible compared with the lateral-wave terms. Hence, only the lateral-wave terms in (5.5.37)–(5.5.42) are significant and the entire field near the boundary in Region 1 is given by

$$E_{1\rho}(\rho, \phi, 0) = -\frac{\omega\mu_0 k_2}{2\pi k_1^2} \cos\phi \; g(\rho; k_1, k_2) e^{ik_2\rho}$$

$$= \frac{\omega}{k_1} B_{1\phi}(\rho, \phi, 0), \tag{8.1.1}$$

$$E_{1z}(\rho, \phi, 0) = \frac{\omega\mu_0 k_2^2}{2\pi k_1^3} \cos\phi \; f(\rho; k_1, k_2) e^{ik_2\rho}, \tag{8.1.2}$$

$$E_{1\phi}(\rho, \phi, 0) = \frac{\omega\mu_0 k_2}{2\pi k_1^2} \sin\phi \; h(\rho; k_1, k_2) e^{ik_2\rho}$$

$$= -\frac{\omega}{k_1} B_{1\rho}(\rho, \phi, 0), \tag{8.1.3}$$

$$B_{1z}(\rho, \phi, 0) = \frac{\mu_0}{2\pi k_1^2} \sin\phi \left(\frac{k_2^2}{\rho^2} + \frac{3ik_2}{\rho^3} - \frac{3}{\rho^4} \right) e^{ik_2\rho}, \tag{8.1.4}$$

where the functions $f(\rho; k_1, k_2)$, $g(\rho; k_1, k_2)$, and $h(\rho; k_1, k_2)$ are defined in (5.5.43)–(5.5.45). Since $k_2 = \beta_2 + i\alpha_2$ is the wave number for the rock or sediment, α_2 is very significant and not zero as in air. This means that measurements of the field given by (8.1.1)–(8.1.4) cannot be made at very large radial distances but are generally restricted to the near and intermediate ranges. That is,

$$|k_2\rho| \leq 4 \left| \frac{k_1^2}{k_2^2} \right|. \tag{8.1.5}$$

Subject to (8.1.5), the Fresnel-integral terms in the three radial functions $f(\rho; k_1, k_2)$, $g(\rho; k_1, k_2)$, and $h(\rho; k_1, k_2)$ are negligible and these reduce to

$$f(\rho; k_1, k_2) = f(\rho, k_2) = \frac{ik_2}{\rho} - \frac{1}{\rho^2}, \tag{8.1.6}$$

$$g(\rho; k_1, k_2) = g(\rho, k_2) = \frac{ik_2}{\rho} - \frac{1}{\rho^2} - \frac{i}{k_2\rho^3}, \tag{8.1.7}$$

$$h(\rho; k_1, k_2) = h(\rho, k_2) = \frac{2}{\rho^2} + \frac{2i}{k_2\rho^3}. \tag{8.1.8}$$

The formulas (8.1.1)–(8.1.4) suggest that the lateral wave travels down from the source at $z = d \sim 0$ in the ocean across the boundary into the oceanic crust, then along the sea floor in the rock or sediment a radial distance ρ, and finally upward across the boundary into the ocean to the point of observation at $z \sim 0$. This description is superficial because it ignores the phase changes that occur in the transitions from the near field to the intermediate field and finally to the far field. The actual path is described in Chapter 5, Section 5.9. At frequencies for which

$$\omega\epsilon_1 \ll \sigma_1; \qquad \omega\epsilon_2 \ll \sigma_2 \tag{8.1.9a}$$

or

$$k_1 \sim (i\omega\mu_0\sigma_1)^{1/2}; \qquad k_2 \sim (i\omega\mu_0\sigma_2)^{1/2}, \tag{8.1.9b}$$

the locus of the Poynting vector is defined by (5.9.14) or (5.9.17) for the three components of electric type, $E_{2\rho}$, E_{2z}, and $B_{2\phi}$. For them, the maximum depth of penetration into the rock or sediment is

$$z'_m = \frac{\delta\rho_0}{e} \quad \text{at } \rho = \rho_m = \frac{\rho_0}{e}, \tag{8.1.10}$$

where $e = 2.718$ and

$$\begin{aligned} \delta &= \frac{\beta_1(\beta_2^2 + \alpha_2^2)}{\beta_2(\beta_1^2 + \alpha_1^2)} \\ &= \begin{cases} (\beta_2^2 + \alpha_2^2)/2\beta_2\alpha_1, & \text{when } \omega\epsilon_1 \ll \sigma_1, \\ \alpha_2/\alpha_1 = (\sigma_2/\sigma_1)^{1/2}, & \text{when } \omega\epsilon_1 \ll \sigma_1,\ \omega\epsilon_2 \ll \sigma_2. \end{cases} \end{aligned} \tag{8.1.11}$$

Similarly, for the components $E_{2\phi}$, $B_{2\rho}$, and B_{2z} of magnetic type, the locus of the Poynting vector is defined by (5.9.38). For them, the maximum depth of penetration into the rock or sediment is

$$z'_m = 0.77\rho_0 \quad \text{at } \rho = \rho_m = 0.544\rho_0. \tag{8.1.12}$$

In (8.1.10) and (8.1.12), ρ_0 is the radial distance from the source to the point of observation—both on the sea floor.

It is important to remember that the components of electric and magnetic type are not independent and that only a part of the field is associated with the surface wave. There is also a spherical wave that travels into Region 2. The somewhat arbitrary separation into fields of electric and magnetic type is possible because the coupling between the two is loose. It is convenient as a means for clarifying important properties of the field that in some respects behave as quasi-independent. For example, the two different depths of penetration for the components of electric and magnetic type associated with a Poynting vector that returns to the surface indicate that the structure of the nonspherical part of the complete field changes significantly as the depth into Region 2 increases.

The significant observation regarding the components of the field defined in (8.1.1)–(8.1.4) is that they provide the means for determining the wave number $k_2 = \beta_2 + i\alpha_2$ for the rock or sediment from measurements made only in the adjacent accessible sea water. Once α_2 and β_2 are known, ϵ_2 and σ_2 are readily calculated since $(\beta_2 + i\alpha_2)^2 = \omega^2\mu_0\epsilon_2 + i\omega\mu_0\sigma_2$ so that

$$\epsilon_2 = \frac{\beta_2^2 - \alpha_2^2}{\omega^2\mu_0}; \qquad \sigma_2 = \frac{2\alpha_2\beta_2}{\omega\mu_0}. \tag{8.1.13}$$

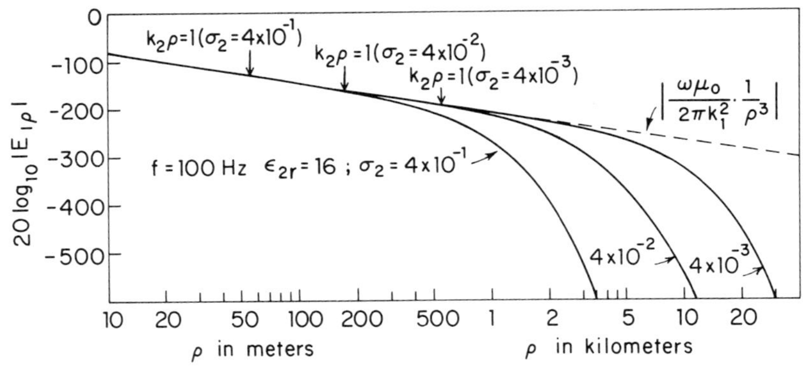

Figure 8.1.1. Radial electric field $|E_{1\rho}(\rho, 0, z)|$ in V/m at $z = 15$ cm in sea water ($\sigma_1 = 3.5$ S/m, $\epsilon_{1r} = 80$) above rock or deep sediment (σ_2, $\epsilon_{2r} = 16$) as function of radial distance ρ from unit horizontal electric dipole at height $d = 15$ cm above sea bottom; $f = 100$ Hz.

When $\sigma_2 \gg \omega\epsilon_2$, as at low frequencies and high conductivities, $\beta_2 \sim \alpha_2$, ϵ_2 is irrelevant, and

$$\sigma_2 = \frac{2\alpha_2^2}{\omega\mu_0}. \tag{8.1.14}$$

The permittivity and conductivity determined from measured values of α_2 and β_2 apply to the region actually traversed by the electric field observed at $(\rho = \rho_0, \phi, 0)$. When $E_{1\rho}(\rho_0, \phi, 0)$ or $B_{1\phi}(\rho_0, \phi, 0)$ is used to determine α_2 and β_2, these are characteristic of the rock or sediment above the depth z'_m given by (8.1.10). When $E_{1\phi}(\rho_0, \phi, 0)$ or $B_{1\rho}(\rho_0, \phi, 0)$ is used to determine α_2 and β_2, these values are characteristic of the rock or sediment above the depth z'_m given in (8.1.12). This is much greater than that for $E_{1\rho}(\rho_0, \phi, 0)$ or $B_{1\phi}(\rho_0, \phi, 0)$.

The graphs in Figs. 8.1.1 and 8.1.2 show $|E_{1\rho}(\rho, 0, z)|$ as a function of the radial distance ρ at the frequencies $f = 0.1$, 1, 10, and 100 kHz. They include ranges with a significant exponential decrease in amplitude given by $|\exp(ik_2\rho)| = \exp(-\alpha_2\rho)$. An alternative representation, shown in Fig. 8.1.3, is $|E_{1\rho}(\rho, 0, z)|$ as a function of the frequency at selected radial distances but with $(z + d)$ sufficiently small so that $|\exp[ik_1(z + d)]| \sim 1$. In both cases, σ_2 is the parameter. At the lower frequencies and higher conductivities, the graphs in Fig. 8.1.3 are similar to those in Fig. 8.1.2. Since α_2 is not linear in the frequency and decreases to small values when σ_2 is reduced, $\exp(-\alpha_2\rho)$ with ρ fixed has a rapid decrease only when $\alpha_2 \sim (\omega\mu_0\sigma_2/2)^{1/2}$ and approaches unity when $\sigma_2 \ll \omega\epsilon_2$. This behavior is illustrated in Fig. 8.1.4, which shows α_2/β_2 as a function of the frequency. A rapid exponential decay occurs only in ranges where α_2/β_2 is not much smaller than one. When $\sigma_2 \gg \omega\epsilon_2$, the factor k_2^2/k_1^2—which occurs in the expression (8.1.1) for $E_{1\rho}(\rho, \phi, 0)$ in the intermediate zone—reduces to

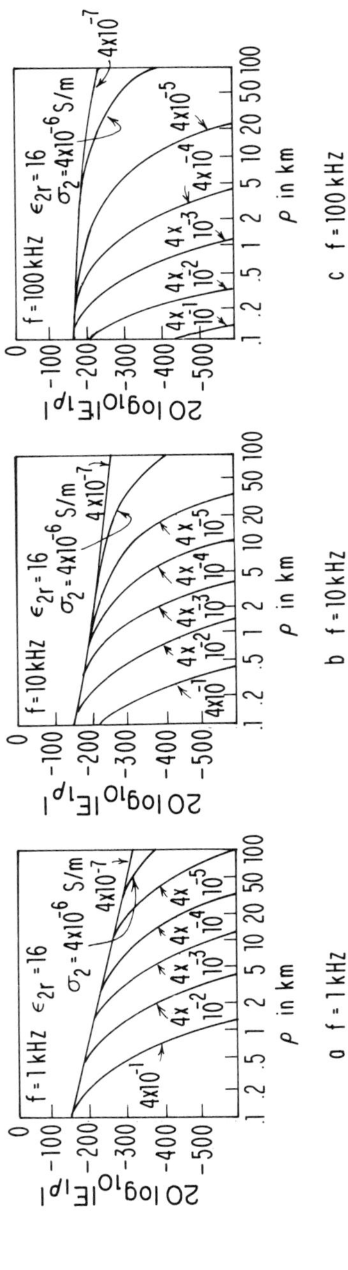

Figure 8.1.2. Radial electric field $|E_{1\rho}(\rho, 0, z)|$ in V/m at $z = 15$ cm in sea water ($\sigma_1 = 4$ S/m, $\epsilon_{1r} = 80$) above rock or deep sediment (σ_2, $\epsilon_{2r} = 16$) as function of radial distance ρ from unit horizontal dipole at height $d = 15$ cm above sea bottom.

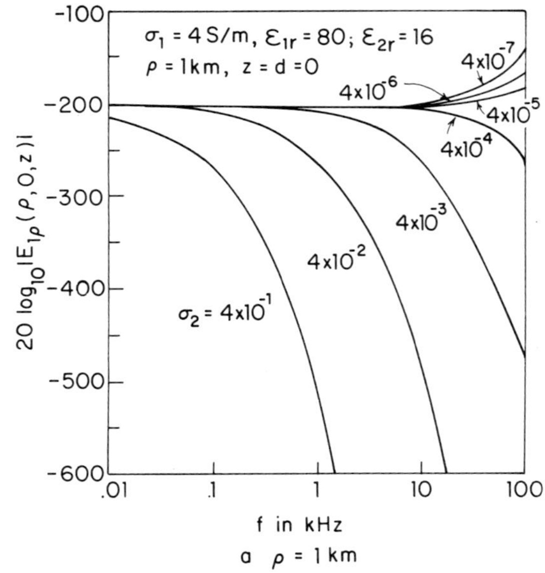

Figure 8.1.3a. Radial electric field $|E_{1\rho}(\rho, 0, z)|$ in V/m at $z = 0$ in sea water ($\sigma_1 = 4$ S/m, $\epsilon_{1r} = 80$) above rock or deep sediment (σ_2, $\epsilon_{2r} = 16$) due to unit horizontal dipole on sea bottom ($d = 0$) as function of frequency; $\rho = 1$ km.

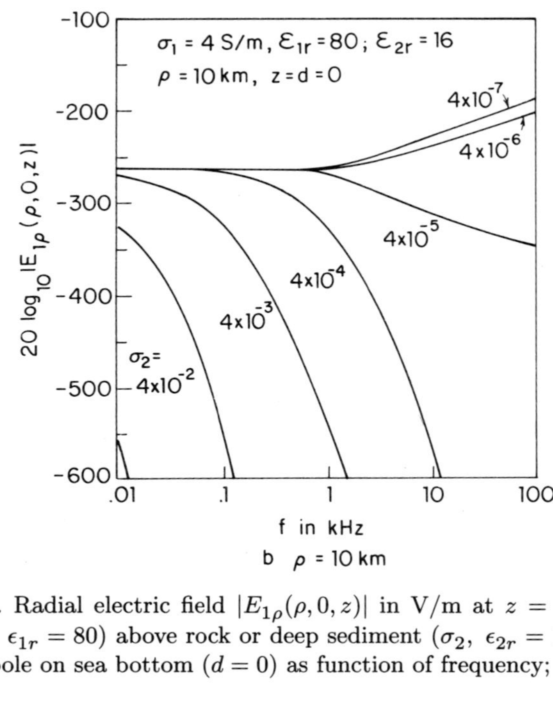

Figure 8.1.3b. Radial electric field $|E_{1\rho}(\rho, 0, z)|$ in V/m at $z = 0$ in sea water ($\sigma_1 = 4$ S/m, $\epsilon_{1r} = 80$) above rock or deep sediment (σ_2, $\epsilon_{2r} = 16$) due to unit horizontal dipole on sea bottom ($d = 0$) as function of frequency; $\rho = 10$ km.

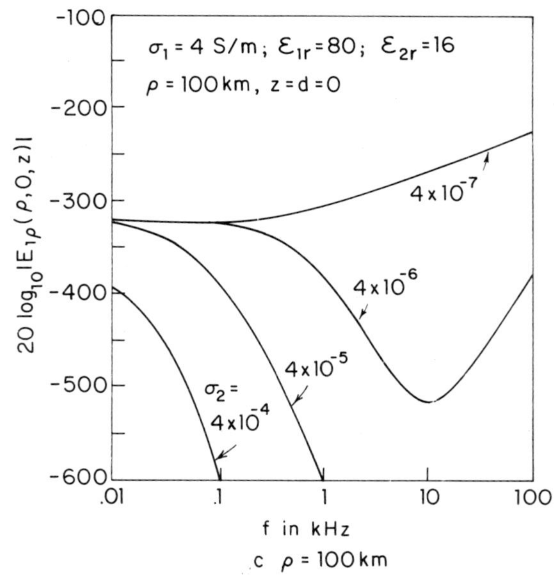

Figure 8.1.3c. Radial electric field $|E_{1\rho}(\rho, 0, z)|$ in V/m at $z = 0$ in sea water ($\sigma_1 = 4$ S/m, $\epsilon_{1r} = 80$) above rock or deep sediment (σ_2, $\epsilon_{2r} = 16$) due to unit horizontal dipole on sea bottom ($d = 0$) as function of frequency; $\rho = 100$ km.

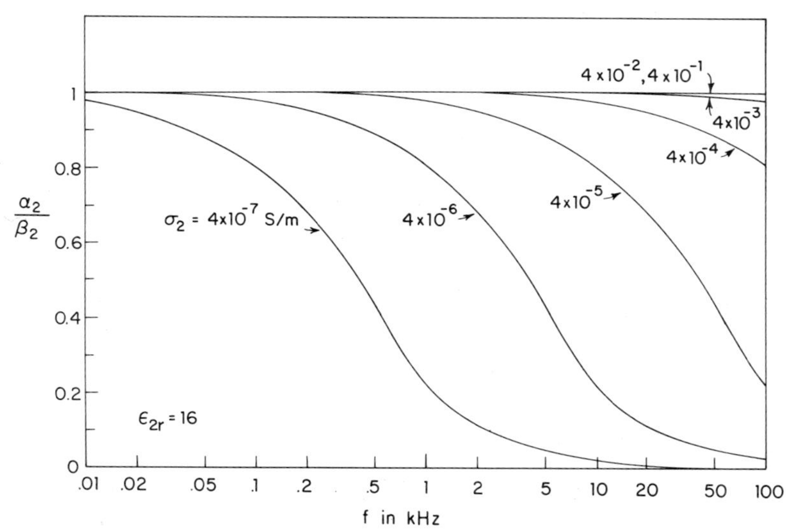

Figure 8.1.4. Ratio of attenuation to phase constant in $k_2 = \beta_2 + i\alpha_2$ for $E_{1\rho}$ in sea water bounded by rock or sediment.

σ_2/σ_1 which is *independent of the frequency* so that when $\alpha_1(z+d) \ll 1$,

$$E_{1\rho}(\rho,0,0) \sim -\frac{i\omega\mu_0\sigma_2}{2\pi\sigma_1\rho}\, e^{-\alpha_2\rho}e^{i\beta_2\rho}. \tag{8.1.15}$$

This is essentially what is represented in Fig. 8.1.3 when σ_2 is not too small and f is not too high so that a rapid exponential decrease in amplitude with frequency occurs. Where this is true, the graphs of Fig. 8.1.3 are as useful as those of Figs. 8.1.1 and 8.1.2 for determining σ_2.

8.2 The Measurement of α_2

A schematic diagram of an apparatus for carrying out measurements from which α_2 of the sea floor can be determined is shown in Fig. 8.2.1. It consists of a transmitting antenna in the form of an insulated conductor lying on the bottom of the ocean with its axis in the direction $\phi = 0$ along which observations are to be made. Power to the antenna is supplied by a cable to the surface where a boat or float is located. The antenna generates an electromagnetic field that is complicated near the antenna but is limited to a surface wave beyond a short distance. Its dominant components are $E_{1\rho}(\rho,\phi,z)$ and $B_{1\phi}(\rho,\phi,z)$, as given by (8.1.1). These have their maxima in the direction $\phi = 0$ along the axis of the antenna. In order to determine α_2 in $k_2 = \beta_2 + i\alpha_2$, measurements are needed that permit the construction of graphs like those in Figs. 8.1.1 and 8.1.2 or like those in Fig. 8.1.3. This can be accomplished by lowering a receiving dipole to the bottom of the ocean at a succession of known radial distances (or by dragging a suitably protected antenna along the bottom) and observing $|E_{1\rho}(\rho,0,z)|$ at each location at selected frequencies. From these data, graphs can be constructed of the measured relative values of $20\log_{10}|E_{1\rho}(\rho,0,z)|$ both as functions of the radial distance at each frequency and as functions of the frequency at each radial distance. The particular curve that includes a rapid decrease with increasing radial distance or increasing frequency can then be used to determine α_2. The calculation of α_2 from curves like those in Figs. 8.1.1 and 8.1.2 in conjunction with (8.1.1) is simpler since, when ρ is the variable at a fixed frequency, k_1 is a constant. On the other hand, when f is the variable, k_1 also varies, but it is known quite accurately for salt water at each frequency. Note that in (8.1.15), which is adequate in the useful range of frequency, the frequency dependence in k_1 is absent because $(z+d)$ is sufficiently small.

Alternatively, and especially in the ranges $0.1 \le f \le 100$ kHz, $4 \times 10^{-5} \le \sigma_2 \le 0.4$ S/m, the graphs in Figs. 8.1.1 and 8.1.2 can be combined and extended into the single set shown in Fig. 8.2.2. Here each curve can represent up to six combinations of frequency and conductivity and σ_2 can be determined by interpolation between adjacent measured curves.

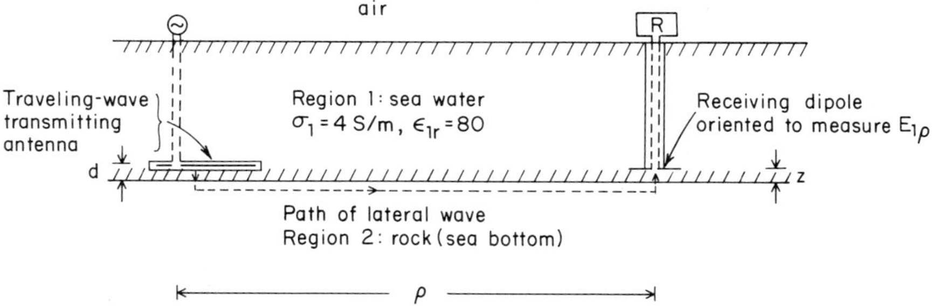

Figure 8.2.1. Schematic diagram of directional transmitter and receiver near bottom of sea.

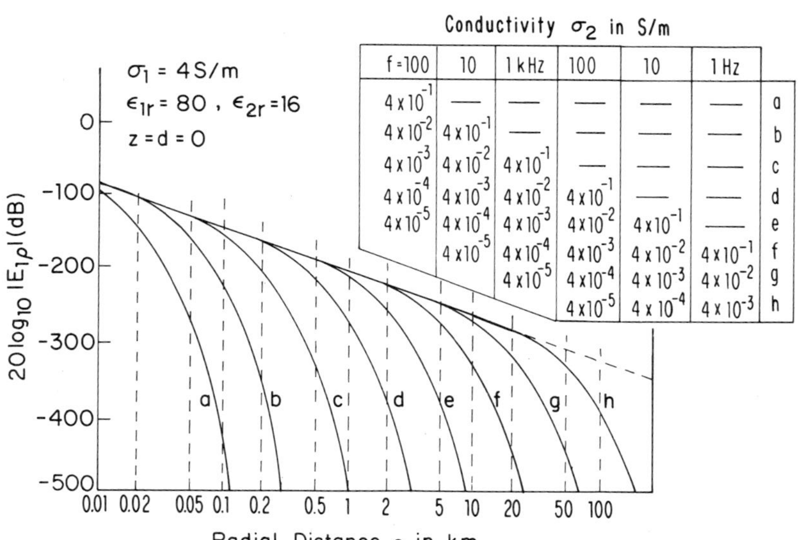

Figure 8.2.2. Radial electric field $|E_{1\rho}(\rho, 0, 0)|$ in V/m of antenna on bottom of sea ($\sigma_1 = 4$ S/m, $\epsilon_{1r} = 80$) above rock or deep sediment (σ_2, $\epsilon_{2r} = 16$). Electric moment is 1 A m.

8.3 The Measurement of β_2

At the higher frequencies and over rock with sufficiently low conductivity so that α_2 is significantly smaller than β_2 (specifically, $\alpha_2 \leq 0.1\beta_2$), the phase constant β_2 along the ocean bottom can be determined in a manner similar to that used for α_2, but with measurements carried out on a standing wave instead of a traveling wave. Two identical transmitting antennas are located on the sea floor separated by the distance $2\rho_0$. They point toward each other and are driven 180° out of phase with equal amplitudes. The lateral waves that travel radially outward from the two antennas combine to form a lateral standing wave. When referred to the midpoint between the two transmitting antennas, the resultant field in the intermediate zone is

$$E_{1\rho}(\rho,0,0) = -\frac{\omega\mu_0 k_2}{2\pi k_1^2}\left(\frac{e^{ik_2(\rho_0+x)}}{\rho_0+x} - \frac{e^{ik_2(\rho_0-x)}}{\rho_0-x}\right) \tag{8.3.1}$$

when $\exp[ik_1(z+d)] \sim 1$. Here x is the distance from the point $\rho = \rho_0$ halfway between the two transmitters with positive values toward antenna 2. When x is kept sufficiently small so that $\alpha_2 x \ll 1$ and $\exp(-\alpha_2 x) \sim \exp(\alpha_2 x) \sim 1$,

$$E_{1\rho}(\rho,0,0) \sim \frac{2ie^{ik_2\rho_0}}{\rho_0(1-x^2/\rho_0^2)}\left(\sin\beta_2 x + i\,\frac{x}{\rho_0}\cos\beta_2 x\right) \tag{8.3.2}$$

and

$$|E_{1\rho}(\rho,0,0)| \sim \frac{2e^{-\alpha_2\rho_0}}{\rho_0(1-x^2/\rho_0^2)}\left(\sin^2\beta_2 x + \frac{x^2}{\rho_0^2}\cos^2\beta_2 x\right)^{1/2}. \tag{8.3.3}$$

This has the following maxima:

$$|E_{1\rho}(\rho,0,0)|_{\max} \sim \frac{2e^{-\alpha_2\rho_0}}{\rho_0(1-x^2/\rho_0^2)}; \qquad \beta_2 x = \frac{(2n+1)\pi}{2}, \quad n = 0,\,1,\,2,\,\ldots, \tag{8.3.4}$$

and deep minima:

$$|E_{1\rho}(\rho,0,0)|_{\min} \sim \frac{2|x|e^{-\alpha_2\rho_0}}{\rho_0^2(1-x^2/\rho_0^2)}; \qquad \beta_2 x = n\pi\left(\frac{\beta_2^2\rho_0^2}{\beta_2^2\rho_0^2-1}\right) \sim n\pi. \tag{8.3.5}$$

This last step is valid when $\beta_2^2\rho_0^2 \gg 1$.

The procedure for measuring β_2 is first to locate the point $\rho = \rho_0$ halfway between the antennas where $|E_{1\rho}(\rho,0,0)| = 0$ and where the receiver on the sea floor should record a null. The receiver is now moved toward transmitter 2 until a minimum is located at $\beta_2 x = \pi$. From the measured distance x between the null and the first minimum, β_2 is determined from

$$\beta_2 = \frac{\pi}{x}. \tag{8.3.6}$$

From the measured values of α_2 and β_2, σ_2 and ϵ_2 can be evaluated from (8.1.13). Since these values are obtained from measurements of $|E_{1\rho}(\rho, 0, 0)|$, they apply to the rock or sediment at depths up to z'_m as given by (8.1.10). With reference to the conductivity range in Fig. 8.2.2, $(\sigma_2/\sigma_1)^{1/2}$ ranges from 0.316 to 0.00316 and z'_m from $0.116\rho_0$ to $0.00116\rho_0$. For a radial distance $\rho_0 = 20$ km, z'_m ranges from 2.3 km to 23 m when σ_2 ranges from 0.4 S/m to 4×10^{-5} S/m.

8.4 Comparison with Measurements

An interesting set of measurements of the electric field on the sea floor generated by a horizontal antenna also on the ocean floor has been reported (Young and Cox 1981). The location was near the East Pacific Rise off the coast of Mexico. At the upper left in Fig. 8.4.1 is a diagram showing the relative locations and orientations of the insulated transmitting antenna T and the crossed bare dipoles of the receiver R. The radial distance between the centers of the fixed transmitting and receiving antennas was $\rho = 18.9$ km. The ocean floor in the area between the antennas consisted of rock with very little sediment. The sea water had a conductivity $\sigma_1 = 3.2$ S/m. The orientation of the crossed dipoles of the receiver was along arbitrary X and Y axes as indicated in the diagram. The positive X-axis was at a small but unspecified angle ψ from the positive ϕ-direction; the positive Y-axis was at the same angle ψ with respect to the negative ρ-direction. In terms of the basic components $E_{1\rho}(\rho, \phi, z)$ and $E_{1\phi}(\rho, \phi, z)$ of the field generated by the transmitter, the components actually measured were

$$E_X = E_{1\rho}(\rho, \phi, 0) \sin\psi + E_{1\phi}(\rho, \phi, 0) \cos\psi, \tag{8.4.1}$$

$$E_Y = -E_{1\rho}(\rho, \phi, 0) \cos\psi + E_{1\phi}(\rho, \phi, 0) \sin\psi. \tag{8.4.2}$$

Since ψ is small, E_X is dominated by $E_{1\phi}(\rho, \phi, 0)$ and E_Y by $-E_{1\rho}(\rho, \phi, 0)$. The angles ψ and ϕ were not specified but can be estimated from the diagram. The components E_X and E_Y were measured in the frequency range from 0.25 to 2.25 Hz.

The measured data are shown by the boxes in the lower part of Fig. 8.4.1. $|E_X|$ on the lower left is seen to decrease rapidly and smoothly as the frequency is increased. On the other hand, $|E_Y|$ on the lower right increases to a maximum between $f = 1$ and 1.5 Hz and then decreases. With the measured data are shown numerically evaluated graphs (Young and Cox 1981) of $|E_X|$ and $|E_Y|$ for a lithosphere represented successively by (a) a homogeneous isotropic half-space with $\sigma_2 = 0.004$ S/m (dotted lines) and (b) three different stacks of eight horizontal layers in a total depth of 33 km (solid, short- and long-dashed lines). The four models are

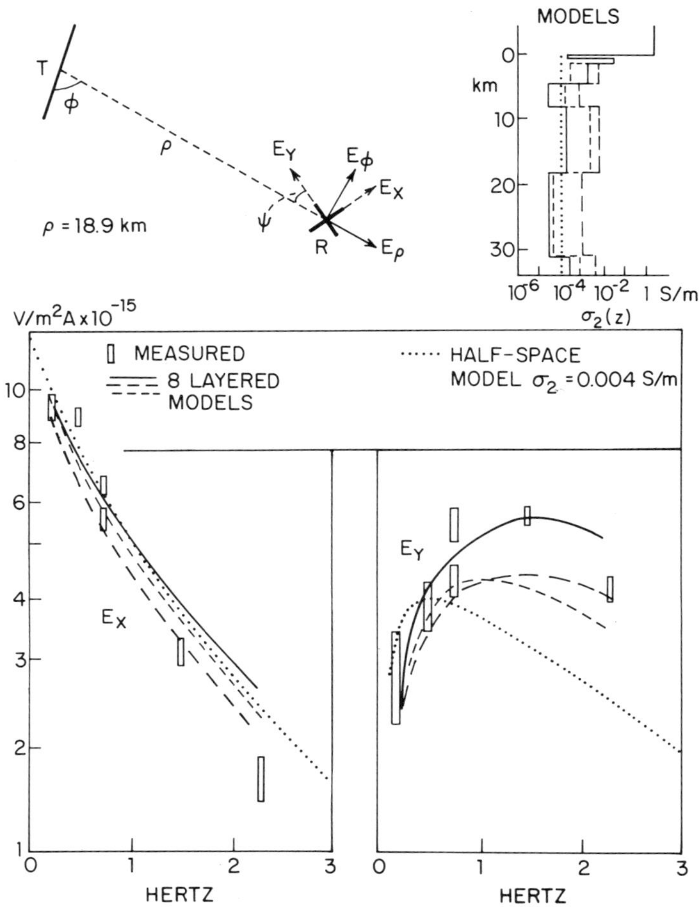

Figure 8.4.1. Data of Young and Cox (1981).

sketched on the upper right in Fig. 8.4.1 . All four graphs of $|E_X|$, although calculated for very different models, differ negligibly from one another and all agree quite well with the measured values. The corresponding graphs of $|E_Y|$ differ significantly from one another and only one, the "best-fit" curve in solid line, agrees reasonably well with the measured data. All have maxima but these occur at different frequencies; that for the half-space model with $\sigma_2 = 0.004$ S/m is near $f = 0.5$ Hz; that for the "best-fit" eight-layer 33-km-deep model is near $f = 1.5$ Hz.

The comparison of the numerically calculated fields for the four models with the measured data and the selection of a "best-fit model" raise interesting questions. First, why is the frequency dependence of $|E_X|$ different from that of $|E_Y|$? Second, how can $|E_Y|$ be sensitive to all eight layers, whereas $|E_X|$ is not? $|E_X|$ as calculated for the half-space model is in good agreement with the measurements and the presence of the hypothetical eight layers with any one of three different combinations of conductivity to a depth of 33 km has virtually no effect on the field. If only $|E_X|$ had been measured, the obvious conclusion would have been that the lithosphere is a homogeneous half-space with $\sigma_2 = 0.004$ S/m, at least down to a depth of 33 km. For $|E_Y|$, the situation is very different. The field of the half-space model with $\sigma_2 = 0.004$ S/m is in poor agreement with measurements, its maximum is at the wrong frequency, and the three different layered models all lead to different fields—all in better agreement with the measurements than the half-space model. It is from $|E_Y|$ that the "best-fit model" was selected. But, how can $|E_X|$ be almost totally insensitive to the presence of the eight layers with their widely different conductivity distributions, while at the same time $|E_Y|$ is apparently significantly reflected from all of the laterally homogeneous layers?

The field observed on the sea floor due to multiple reflections from layers deep in the earth cannot be determined from a model of the oceanic lithosphere as a homogeneous isotropic half-space—which is the subject of this chapter. However, much insight can be gained from a detailed investigation of the nature of the field in such a model. The complete general formulas for such a study are available and relatively simple in form. Indeed, for the range $0.25 \le f \le 3$ Hz with $\sigma_1 = 3.2$ S/m, the specified radial distance $\rho = 18.9$ km between the transmitting and receiving antennas, and $z = d = 1$ m, significant further simplification is possible. Specifically, all contributions from direct and reflected fields are negligible; the same is true of the Fresnel-integral terms of the lateral-wave field. Also, $\exp(ik_1 d) \sim 1$ and $\exp(ik_1 z) \sim 1$. It follows that the complete, accurate expressions for the components (5.5.37) and (5.5.38) are

$$E_{1\rho}(\rho,\phi,0) = \frac{\cos\phi}{2\pi\sigma_1}\, e^{ik_2\rho}\left(\frac{1}{\rho^3} - \frac{ik_2}{\rho^2} - \frac{k_2^2}{\rho}\right) \tag{8.4.3a}$$

$$= -\frac{i\cos\phi}{2\pi\sigma_1}\,\frac{e^{-\alpha_2\rho}e^{i\alpha_2\rho}}{\rho^3}\left[\alpha_2\rho + 2\alpha_2^2\rho^2 + i(1+\alpha_2\rho)\right], \tag{8.4.3b}$$

$$E_{1\phi}(\rho,\phi,0) = \frac{\sin\phi}{\pi\sigma_1}\, e^{ik_2\rho}\left(\frac{1}{\rho^3} - \frac{ik_2}{\rho^2}\right) \tag{8.4.4a}$$

$$= -\frac{i\sin\phi}{\pi\sigma_1}\,\frac{e^{-\alpha_2\rho}e^{i\alpha_2\rho}}{\rho^3}\left[\alpha_2\rho + i(1+\alpha_2\rho)\right], \tag{8.4.4b}$$

where $k_2 = (1+i)\alpha_2$, $k_2^2 = 2i\alpha_2^2$, and $\alpha_2 = (\omega\mu_0\sigma_2/2)^{1/2}$. $|E_{1\rho}(\rho,\phi,0)|$ has a maximum at $\alpha_2\rho = (1+\sqrt{5})/2 = 1.618$. With $\rho = 18.9$ km, this gives $f_m\sigma_2 = 1.84 \times 10^{-3}$ so that, with $\sigma_2 = 0.004$ S/m, $f_m = 0.46$ Hz in precise agreement with the curve in dotted line on the lower right in Fig. 8.4.1, which was evaluated numerically by Young and Cox (1981). The maximizing frequency for $\sigma_2 = 0.002$ S/m is 0.92 Hz.

The angles ψ and ϕ are not given by Young and Cox. With the choice $\phi = 50°$, $\psi = 10°$, and with $\rho = 18.9$ km, graphs of $|E_{1\rho}(\rho,\phi,0)|$ and

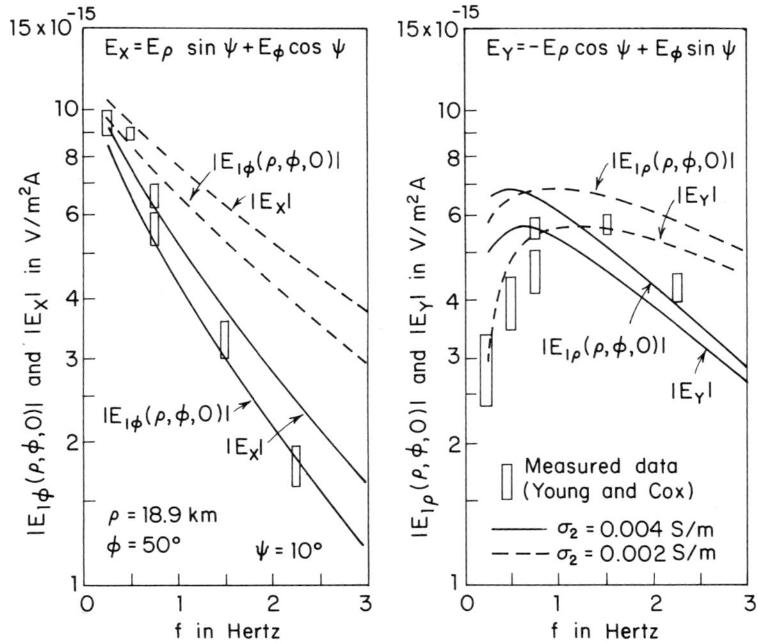

Figure 8.4.2. $|E_{1\rho}(\rho,\phi,0)|$ and $|E_{1\phi}(\rho,\phi,0)|$ calculated from (8.4.3) and (8.4.4), $|E_X|$ and $|E_Y|$ from (8.4.1) and (8.4.2), with measured data of Young and Cox (1981).

$|E_{1\phi}(\rho,\phi,0)|$ are shown in Fig. 8.4.2 for both $\sigma_2 = 0.004$ S/m and $\sigma_2 = 0.002$ S/m. Also shown are the curves for $|E_X|$ and $|E_Y|$ for both $\sigma_2 = 0.004$ S/m and $\sigma_2 = 0.002$ S/m, as calculated from (8.4.1) and (8.4.2). It is evident from these graphs that, in order to have a maximum of $|E_Y|$ near $f = 1$ Hz to agree with the measured data, it is necessary to choose $\sigma_2 = 0.002$ S/m. On the other hand, in order to have $|E_X|$ agree with the measured data, it is necessary that $\sigma_2 = 0.004$ S/m. If measurements had been restricted to E_Y [or better, to $E_{1\rho}(\rho,0,0)$], the good agreement with the measured points would suggest that the *half-space model* with $\sigma_2 = 0.002$ S/m is correct. On the other hand, if only E_X [or better, $E_{1\phi}(\rho,\pi/2,0)$] had been measured, the equally good agreement with the measured points would indicate a *half-space model* with $\sigma_2 = 0.004$ S/m. No reflecting layers are indicated in either case. When both E_X and E_Y are measured, these two conclusions are inconsistent and both are rejected by Young and Cox in favor of an eight-layered region with a depth of 33 km which is selected for best agreement with measured values of E_Y. This leaves unresolved the fact that none of the three eight-layered models leads to a theoretical curve for E_X that differs significantly from E_X as determined from the half-space model with $\sigma_2 = 0.004$ S/m.

In the half-space model, the entire field at the point of observation at $(\rho_0, \phi, 0)$ on the sea floor is due to the lateral wave. For the components of electric type, viz., $E_{1\rho}(\rho,\phi,0)$, $E_{1z}(\rho,\phi,0)$, and $B_{1\phi}(\rho,\phi,0)$, the locus of the Poynting vector is shown in Fig. 5.9.1 in dashed lines for $\sigma_1 = 4$ S/m and $\sigma_2 = 0.004$ S/m. The curve ending at $\rho_0 = 18.9$ km applies specifically to the case at hand. It is independent of the frequency and is the same for all frequencies in the range $0.25 \leq f \leq 3$ Hz, which are involved in the measured data shown in Fig. 8.4.1. The maximum depth of penetration is

$$z'_m = \left(\frac{\sigma_2}{\sigma_1}\right)^{1/2} \frac{\rho_0}{2.718} = \left(\frac{4 \times 10^{-3}}{4}\right)^{1/2} \frac{18.9 \times 10^3}{2.718} = 220 \text{ m}.$$

It occurs at the radial distance $\rho_m = \rho_0/2.718 = 6.95$ km. The field of electric type varies as $\cos\phi$.

For the components of magnetic type, viz., $E_{1\phi}(\rho,\phi,0)$, $B_{1\rho}(\rho,\phi,0)$, and $B_{1z}(\rho,\phi,0)$, the locus of the Poynting vector is shown in Fig. 5.9.2. The curve ending at $\rho_0 = 18.9$ km applies to the case at hand. It is independent of the frequency and the values of σ_1 and σ_2 so long as $|k_1| \geq 3|k_2|$. The maximum depth of penetration is

$$z'_m = 0.77\rho_0 = 14.6 \text{ km}.$$

It occurs at $\rho_m = 0.544\rho_0 = 10.3$ km. The field of magnetic type varies as $\sin\phi$.

The depths of penetration for the components of electric type are very different from those of magnetic type. The former are part of a sharply defined surface wave that travels very close to the boundary, the latter penetrate quite deeply into the rock before returning to the surface. Measurements of $E_{1\rho}(\rho,\phi,0)$ at $\rho_0 = 18.9$ km involve a wave that has traveled only in a thin layer 220 m deep; measurements of $E_{1\phi}(\rho,\phi,0)$ are for a wave that has penetrated a region 14.6 km deep.

8.5 Reflections from an Ideal Reflecting Layer

The field in the half-space of rock (Region 2, $z \leq 0$) includes not only the surface wave but also a spherical wave that travels downward. The complete spherical part of the electromagnetic field is given by (5.9.49)–(5.9.54) in the range $\Theta \geq \pi/2$ or the range $\Theta' \leq \pi/2$, where $\Theta' = \pi - \Theta$ is referred to the negative z-axis so that $z' = -z$ is positive downward into the rock—Region 2. It is the downward-traveling spherical wave that is partly reflected upward and partly transmitted downward when incident on a surface of discontinuity in σ_2. However, since it is not a plane wave, is not normally incident, and is not independent of the transverse coordinates

ρ and ϕ, it is not readily determined except when the reflection coefficient for plane waves is ± 1. That is, when there is a perfect electric or perfect magnetic conductor at depth D. In this case, the problem becomes a three-layer one which is discussed in Chapter 11. However, the magnitude of the once-reflected wave can be obtained directly since it is the field of an image dipole with equal and codirectional or equal and opposite magnetic moment. Furthermore, in a highly attenuating medium [such as rock at low frequencies with $\alpha_2 = \beta_2 = (\omega\mu_0\sigma_2/2)^{1/2}$], the contributions from second-, third-, and higher-order reflections decrease rapidly. It follows that the magnitude of the first reflection is a good estimate of the entire reflected field at the point of observation on the sea floor.

If a perfectly conducting boundary is at the depth $z' = D$, the image dipole is at depth $z' = 2D$ and the once-reflected field at $(\rho_0, \phi, 0)$ just below the sea floor in the rock is given by (5.9.49)–(5.9.54) with $r_0 = (\rho^2 + z^2)^{1/2}$ replaced by $2r_0 = (\rho^2 + 4D^2)^{1/2}$; also $\cos\Theta' = -\cos\Theta = D/r_0$ and $\sin\Theta' = \sin\Theta = \rho/2r_0$. The path of the reflected field and of the locus of the lateral wave as defined by the Poynting vector are shown schematically in Fig. 8.5.1 in the xz-plane ($\phi = 0$) for the components E_ρ, E_z, and B_ϕ of electric type and in Fig. 8.5.2 in the yz-plane ($\phi = \pi/2$) for the components E_ϕ, B_ρ, and B_z of magnetic type.

When $|k_2 r_0| > 1$, the reflected components $E^S_{2\rho}$ and $E^S_{2\phi}$ in Region 2 (rock) just below the point of observation in Region 1 (sea) are

$$E^S_{2\rho} = E^S_{2\Theta}\cos\Theta = \frac{i\omega\mu_0 k_2}{2\pi k_1}\, e^{ik_1 d}\, \frac{e^{i2k_2 r_0}}{2r_0}\, \frac{D}{r_0}\cos\phi, \tag{8.5.1}$$

$$E^S_{2\phi} = -\frac{i\omega\mu_0 k_2}{2\pi k_1}\, e^{ik_1 d}\, \frac{e^{i2k_2 r_0}}{2r_0}\, \frac{D}{r_0}\sin\phi. \tag{8.5.2}$$

The plane-wave transmission coefficient for the electric field of electric type (i.e., $\mathbf{E}_2$ in the plane of incidence) for transmission from Region 2 to Region 1, i.e., from the rock into the sea, is

$$T_e = \frac{k_2}{k_1}\left(\frac{2k_1\cos\Theta'}{k_1\cos\Theta' + k_2[1 - (k_2^2/k_1^2)\sin^2\Theta']^{1/2}}\right) \sim \frac{2k_2}{k_1}. \tag{8.5.3}$$

For the electric field of magnetic type (i.e., $\mathbf{E}_2$ perpendicular to the plane of incidence), the transmission coefficient is

$$\begin{aligned} T_m &= \frac{2k_2\cos\Theta'}{k_2\cos\Theta' + k_1[1 - (k_2^2/k_1^2)\sin^2\Theta']^{1/2}} \\ &\sim \frac{2k_2\cos\Theta'}{k_1} = \frac{2k_2 D}{k_1 r_0}. \end{aligned} \tag{8.5.4}$$

The approximations on the right in (8.5.3) and (8.5.4) depend on the requirement that $|k_2|$ be small compared with $|k_1|$. The once-reflected field

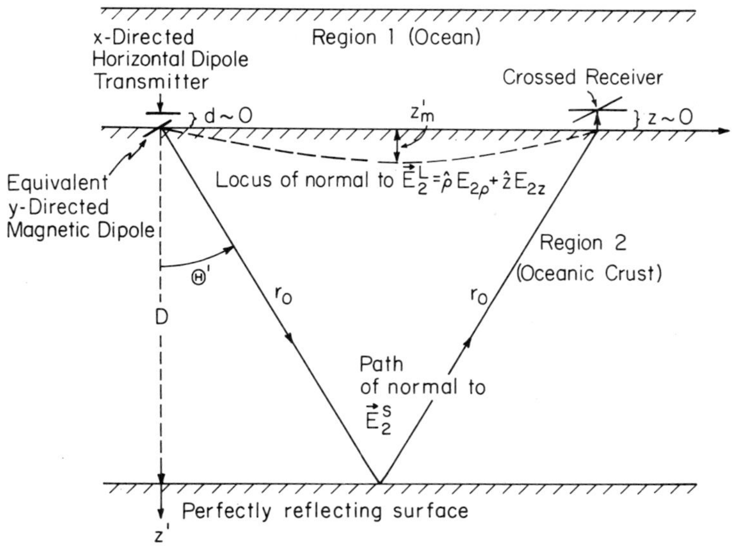

Figure 8.5.1. Schematic diagram to show paths in xz-plane of lateral wave and reflected spherical wave of electric type (E_ρ, E_z, B_ϕ) from equivalent source at $(0,0,0)$ in Region 2 to point of observation in Region 1.

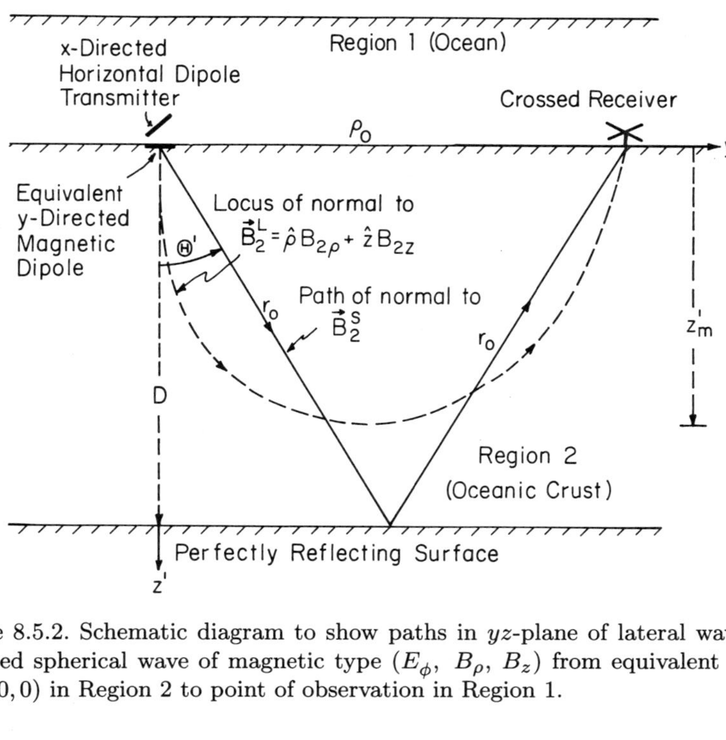

Figure 8.5.2. Schematic diagram to show paths in yz-plane of lateral wave and reflected spherical wave of magnetic type (E_ϕ, B_ρ, B_z) from equivalent source at $(0,0,0)$ in Region 2 to point of observation in Region 1.

in the ocean (Region 1) just across the boundary from the rock (Region 2) at $z = 0$ is approximated by

$$E_{1\rho}^{S}(\rho,\phi,0) = T_e E_{2\rho}^{S} = \frac{i\omega\mu_0 k_2^2}{2\pi k_1^2}\, e^{ik_1 d}\, \frac{D e^{i2k_2 r_0}}{r_0^2} \cos\phi, \tag{8.5.5}$$

$$E_{1\phi}^{S}(\rho,\phi,0) = T_m E_{2\phi}^{S} = -\frac{i\omega\mu_0 k_2^2}{2\pi k_1^2}\, e^{ik_1 d}\, \frac{D^2 e^{i2k_2 r_0}}{r_0^3} \sin\phi. \tag{8.5.6}$$

The magnitudes of these fields are

$$|E_{1\rho}^{S}(\rho,\phi,0)| = \frac{\omega\mu_0}{2\pi}\left|\frac{k_2^2}{k_1^2}\right| e^{-\alpha_1 d}\, \frac{D}{r_0}\, \frac{e^{-2\alpha_2 r_0}}{r_0}\, |\cos\phi|, \tag{8.5.7}$$

$$|E_{1\phi}^{S}(\rho,\phi,0)| = \frac{\omega\mu_0}{2\pi}\left|\frac{k_2^2}{k_1^2}\right| e^{-\alpha_1 d}\, \frac{D^2}{r_0^2}\, \frac{e^{-2\alpha_2 r_0}}{r_0}\, |\sin\phi|. \tag{8.5.8}$$

The magnitudes of the lateral-wave components at the same location in the range $1 \le |k_2\rho| \le 4|k_1^2/k_2^2|$ are

$$|E_{1\rho}^{L}(\rho,\phi,0)| = \frac{\omega\mu_0}{2\pi}\left|\frac{k_2^2}{k_1^2}\right| e^{-\alpha_1 d}\, \frac{e^{-\alpha_2\rho}}{\rho}\, |\cos\phi|, \tag{8.5.9}$$

$$|E_{1\phi}^{L}(\rho,\phi,0)| = \frac{\omega\mu_0}{\pi}\left|\frac{k_2^2}{k_1^2}\right| e^{-\alpha_1 d}\, \frac{e^{-\alpha_2\rho}}{|k_2|\rho^2}\, |\sin\phi|. \tag{8.5.10}$$

The ratios of the once-reflected components to the lateral-wave components at $(\rho_0,\phi,0)$ are

$$\mathcal{R}_\rho = \frac{|E_{1\rho}^{S}(\rho_0,\phi,0)|}{|E_{1\rho}^{L}(\rho_0,\phi,0)|} = \frac{\rho_0 D}{r_0^2}\, e^{-\alpha_2(2r_0-\rho_0)}, \tag{8.5.11}$$

$$\mathcal{R}_\phi = \frac{|E_{1\phi}^{S}(\rho_0,\phi,0)|}{|E_{1\phi}^{L}(\rho_0,\phi,0)|} = \frac{|k_2|\rho_0^2 D^2}{2r_0^3}\, e^{-\alpha_2(2r_0-\rho_0)}. \tag{8.5.12}$$

Graphs of $\mathcal{R}_\rho$ for a range of values of σ_2 and at two frequencies are in Fig. 8.5.3 together with the associated depths of penetration of the lateral wave with $\rho_0 = 20$ km. Corresponding graphs of $\mathcal{R}_\phi$ are in Fig. 8.5.4. In this case, the depth of penetration of the magnetic-type wave is essentially independent of both σ_2 and the frequency.

Also of interest are the ratios $\mathcal{R}_\rho^L$ and $\mathcal{R}_\rho^S$, $\mathcal{R}_\phi^L$ and $\mathcal{R}_\phi^S$ defined by

$$\mathcal{R}_j^L = (1+\mathcal{R}_j)^{-1} = \frac{|E_j^L|}{|E_j^L| + |E_j^S|}, \tag{8.5.13}$$

$$\mathcal{R}_j^S = (1+\mathcal{R}_j^{-1})^{-1} = \frac{|E_j^S|}{|E_j^L| + |E_j^S|}, \tag{8.5.14}$$

where $j = \rho$ or ϕ, $\mathcal{R}_\rho$ is defined by (8.5.11), and $\mathcal{R}_\phi$ by (8.5.12). Graphs of $\mathcal{R}_\rho^L$ and $\mathcal{R}_\rho^S$ are shown in Fig. 8.5.5 and those of $\mathcal{R}_\phi^L$ and $\mathcal{R}_\phi^S$ in Fig. 8.5.6.

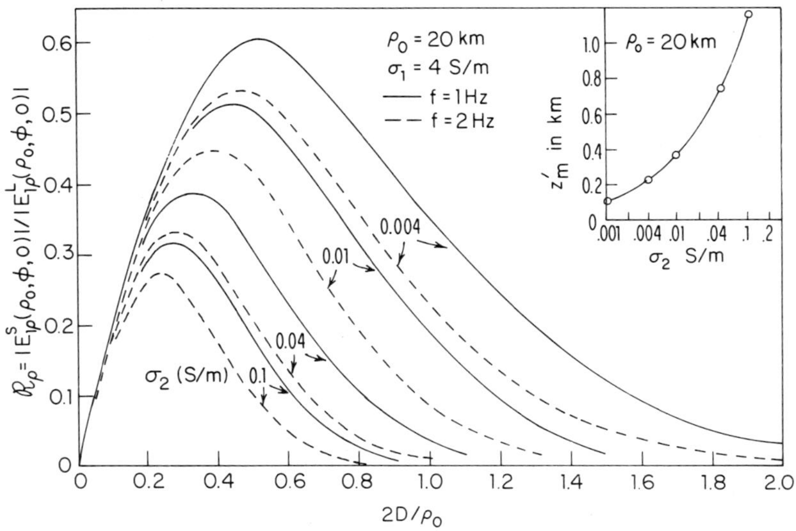

Figure 8.5.3. Ratio $\mathcal{R}_\rho$ of once-reflected spherical wave $|E^S_{1\rho}(\rho_0, \phi, 0)|$ to lateral wave $|E^L_{1\rho}(\rho_0, \phi, 0)|$ on sea floor; z'_m is depth of penetration of lateral wave.

These curves show that the once-reflected spherical wave first rises to a maximum as the depth D of the reflecting layer is increased and then decays. The initial rise is due to the increase in $\cos\Theta'$ as the depth D of the perfectly reflecting surface becomes larger; the decay is caused by the exponential factor which decreases rapidly as D and, with it, $2r_0 = (\rho_0^2 + 4D^2)^{1/2}$ are increased.

Even at its greatest value with $\sigma_2 = 0.004$ S/m, $D \sim 0.25\rho_0 = 5$ km and $f = 1$ Hz, the contribution by the once-reflected spherical wave $E^S_{1\rho}$ to the total field $E_{1\rho}(\rho_0, \phi, 0)$ observed on the sea floor (Figs. 8.5.3 and 8.5.5) is less than 40% with the lateral wave contributing over 60%. For other depths and conductivities, the lateral wave may contribute as much as 90% or more to the total field observed at $\rho_0 = 20$ km on the sea floor. Since the lateral wave of electric type penetrates only a short distance into the oceanic crust, as shown in the insert in Fig. 8.5.3 and schematically in Fig. 8.5.1, great care must be exercised in interpreting measured values of the total field.

The graphs in Figs. 8.5.4 and 8.5.6 show that the relative magnitude of the once-reflected wave for the field of magnetic type is comparable to that for the field of electric type. At the particular depth D of the perfectly reflecting layer which gives a maximum for $\mathcal{R}^S_\phi$, the once-reflected wave $E^S_{1\phi}$ is less than 35% of the total field $E_{1\phi}(\rho_0, \phi, 0)$. A principal and important difference between the components of electric and magnetic types is that the depth of penetration of the latter is very much greater. It is essentially

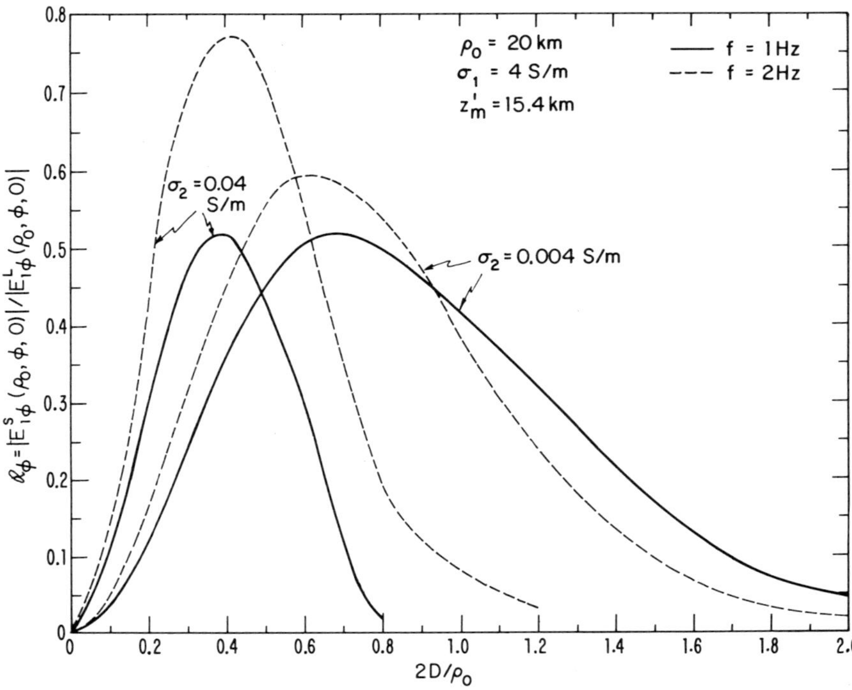

Figure 8.5.4. Ratio $\mathcal{R}_\phi$ of once-reflected spherical wave $|E^S_{1\phi}(\rho_0, \phi, 0)|$ to lateral wave $|E^L_{1\phi}(\rho_0, \phi, 0)|$ on sea floor; z'_m is depth of penetration of lateral wave.

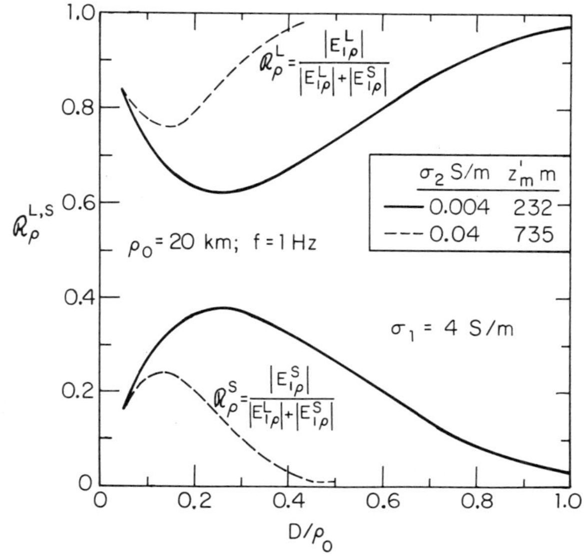

Figure 8.5.5. Ratio $\mathcal{R}_\rho^L$ of lateral wave $|E_{1\rho}^L|$ to $|E_{1\rho}^L| + |E_{1\rho}^S|$ and ratio $\mathcal{R}_\rho^S$ of once-reflected spherical wave $|E_{1\rho}^S|$ to $|E_{1\rho}^L| + |E_{1\rho}^S|$ at point of observation.

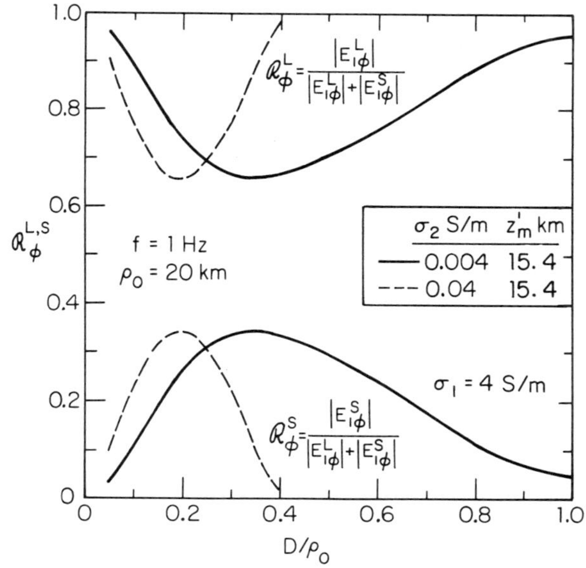

Figure 8.5.6. Ratio $\mathcal{R}_\phi^L$ of lateral wave $|E_{1\phi}^L|$ to $|E_{1\phi}^L| + |E_{1\phi}^S|$ and ratio $\mathcal{R}_\phi^S$ of once-reflected spherical wave $|E_{1\phi}^S|$ to $|E_{1\phi}^L| + |E_{1\phi|}^S$ at point of observation.

$z'_m = 0.77\rho_0$ which, as indicated in Fig. 8.5.2, means that for depths $D$of the reflecting layer less than z'_m the lateral wave becomes a part of the reflected wave. It follows that the measured values of $E_{1\phi}(\rho_0, \phi, 0)$ must be interpreted quite differently from measured values of $E_{1\rho}(\rho_0, \phi, 0)$.

8.6 Interpretation of Measured Data

It is instructive to examine the measured data in Fig. 8.4.1 and the eight-layered "best-fit" model that is constructed to represent the physical reality of the oceanic crust. The measurements give $E_X \sim E_{1\phi}(\rho_0, \phi, 0)$ and $E_Y \sim -E_{1\rho}(\rho_0, \phi, 0)$ since ψ is small. It has been shown that the depth of penetration of the magnetic-type field represented by $E_{2\phi}(\rho, \phi, z)$ is 14.6 km when $\rho_0 = 18.9$ km. As seen in the diagram at the upper right in Fig. 8.4.1, this is deeper than at least four of the eight layers. It follows that the lateral-wave field cannot be separated from the spherical wave insofar as reflections from the layers is concerned. The entire magnetic-type field penetrates down to all layers that can contribute a significant reflected field at the point of observation. It is seen from Fig. 8.5.6 that layers deeper than the depth of penetration of the lateral wave contribute negligibly. The fact that the three quite different combinations of conductivity in these layers yield virtually the same $E_{1\phi}(\rho_0, \phi, 0)$ for all frequencies as the half-space model with uniform $\sigma_2 = 0.004$ S/m requires an explanation.

The depth of penetration of the electric-type field is only $z'_m = 220$ m when $\rho_0 = 18.9$ km and it is seen from Fig. 8.5.5 that reflection of the entire spherical wave from a perfectly reflecting boundary at the optimum depth contributes only a small fraction to $E_{1\rho}(\rho_0, \phi, 0)$. The component $E_{1\rho}(\rho_0, \phi, 0)$ measured at $\rho_0 = 18.9$ km consists predominantly of the lateral wave that has penetrated only 220 m into the rock. In the interpretation of the results from the eight-layer model, the presence of the lateral wave is not mentioned. This is reasonable for $E_{1\phi}(\rho_0, \phi, 0)$, but would seem to be difficult to justify for $E_{1\rho}(\rho_0, \phi, 0)$.

An alternative, physically more acceptable interpretation of the measured data is indicated in Fig. 8.4.2. Specifically, the observation that E_X is in good agreement with measurements when $\sigma_2 = 0.004$ S/m and simultaneously E_Y is in good agreement with measurements when $\sigma_2 = 0.002$ S/m suggests that the rock may be anisotropic in conductivity in a layer at least 220 m thick. It seems probable that all measured data would be well satisfied by a half-space model that is homogeneous but one-dimensionally anisotropic in conductivity with $\sigma_{2x} = \sigma_{2y} \sim 0.004$ S/m and $\sigma_{2z} \sim 0.002$ S/m. This model is consistent with the assumption of approximately horizontal layering but the layers could consist of a stack of quite thin horizontal slabs with conductivities that alternate between $\sigma_{2x} = \sigma_{2y} = 0.004$ S/m and a lower value such that the effective conductivity $\sigma_{2z} = 0.002$ S/m. Such a model requires no reflections from deeper, widely spaced horizontal layers to explain the data. Also, a half-space is not required—only a thickness that is larger than the depth of penetration $z'_m = 220$ m. An analytical study of the one-dimensionally anisotropic half-space is carried out in Chapter 9.

9
Lateral Waves in a One-Dimensionally Anisotropic Half-Space

9.1 Introduction

Anisotropy in conductivity is found in various stratified media including alternating layers of dense rock with low conductivity and less dense rock with higher conductivity shown schematically in Fig. 9.1.1. In such media, the conductivity transverse to the bedding surfaces is always smaller than that along these surfaces (Parkhomenko 1967). For horizontal surfaces, this means that σ_{2z} is smaller than $\sigma_{2x} = \sigma_{2y}$, e.g., $\sigma_{2z} \sim 0.002$ S/m and $\sigma_{2x} = \sigma_{2y} \sim 0.004$ S/m. Anisotropy in conductivity is also found in only slightly stratified hard clay (for which the conductivity ratio of 2 is appropriate) and in rocks with different conductivities along the principal crystallographic axes. It may be added that a slightly inclined stratification must also exhibit different conductivities to the components of current parallel to the bedding surfaces and those perpendicular to them. In this case, σ_{2x} and σ_{2y} may differ somewhat from each other and more substantially from σ_{2z}.

Since the properties of the measured components $E_X \sim E_{1\phi}(\rho, \phi, z)$ and $E_Y \sim -E_{1\rho}(\rho, \phi, z)$ of the electric field excited on the sea floor by a horizontal electric dipole also on the sea floor seemed to suggest in Chapter 8 that $\sigma_{2z} \sim 0.002$ S/m and $\sigma_{2x} = \sigma_{2y} \sim 0.004$ S/m, it is necessary to reanalyze the problem treated in Chapter 5, generalized so that in Region 2 (the oceanic crust) the volume density of current $\mathbf{J}_2 = \hat{\mathbf{x}}J_{2x} + \hat{\mathbf{y}}J_{2y} + \hat{\mathbf{z}}J_{2z}$ with $J_{2x} = \sigma_T E_{2x}$, $J_{2y} = \sigma_T E_{2y}$, and $J_{2z} = \sigma_L E_{2z}$ (Pan 1985). For greater simplicity, it is convenient to assume the permittivity to be similarly

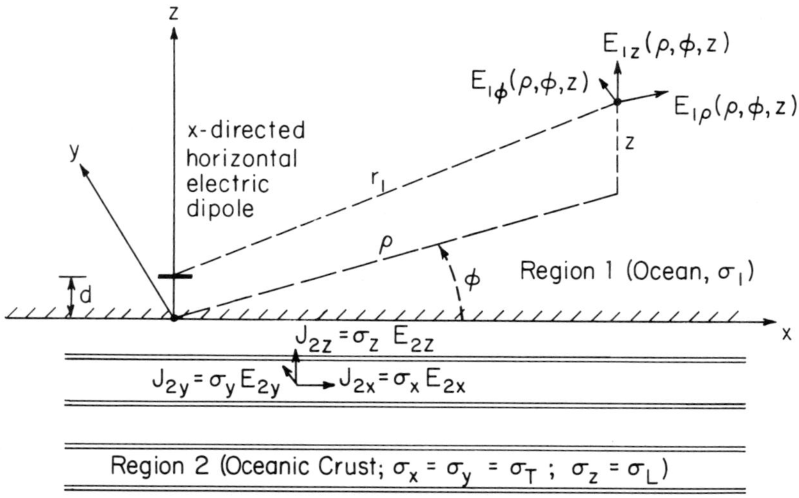

Figure 9.1.1. Components of electric field at (ρ, ϕ, z) due to an x-directed unit horizontal electric dipole ($I\Delta l = 1$ A m) located at $(0, 0, d)$ in homogeneous isotropic Region 1 ($z \geq 0$) over homogeneous anisotropic Region 2 ($z \leq 0$). Horizontal layering of rock or sand gives $\sigma_x = \sigma_y > \sigma_z$.

stratified with $\epsilon_{2x} = \epsilon_{2y} = \epsilon_T$ and $\epsilon_{2z} = \epsilon_L$. This leads to the following representation for the complex permittivity $\tilde{\epsilon}$,

$$\tilde{\epsilon}_1 = \begin{bmatrix} \epsilon_1 + i\sigma_1/\omega & 0 & 0 \\ 0 & \epsilon_1 + i\sigma_1/\omega & 0 \\ 0 & 0 & \epsilon_1 + i\sigma_1/\omega \end{bmatrix}, \tag{9.1.1}$$

$$\tilde{\epsilon}_2 = \begin{bmatrix} \epsilon_T + i\sigma_T/\omega & 0 & 0 \\ 0 & \epsilon_T + i\sigma_T/\omega & 0 \\ 0 & 0 & \epsilon_L + i\sigma_L/\omega \end{bmatrix}, \tag{9.1.2}$$

in Maxwell's equations

$$\nabla \times \mathbf{E}_j = i\omega \mathbf{B}_j, \tag{9.1.3}$$

$$\nabla \times \mathbf{B}_j = \mu_0(-i\omega\tilde{\epsilon}_j\,\mathbf{E}_j + \hat{\mathbf{x}}J_x^e), \tag{9.1.4}$$

where

$$J_x^e = \delta(x)\delta(y)\delta(z-d) \tag{9.1.5}$$

is the externally maintained current in the active unit dipole.

9.2 Transform of Maxwell's Equations; Equations for $\tilde{E}_x$ and $\tilde{B}_x$

Fourier transforms of the form

$$\mathbf{E}(x,y,z) = \frac{1}{(2\pi)^2} \int_{-\infty}^{\infty} d\xi \int_{-\infty}^{\infty} d\eta \, e^{i(\xi x + \eta y)} \tilde{\mathbf{E}}(\xi,\eta,z), \tag{9.2.1}$$

and similar transforms for $\mathbf{B}(x,y,z)$ and $J_x(x,y,z)$, are applied to Maxwell's equations. Thus,

$$\tilde{J}_x^e(\xi,\eta,z) = \delta(z-d). \tag{9.2.2}$$

The transformed Maxwell equations in Cartesian coordinates are given by (5.2.8)–(5.2.13) modified for Region 2 in accordance with (9.1.2). For Region 1,

$$i\eta\tilde{E}_{1z} - \frac{\partial}{\partial z}\tilde{E}_{1y} = i\omega\tilde{B}_{1x}, \tag{9.2.3}$$

$$\frac{\partial}{\partial z}\tilde{E}_{1x} - i\xi\tilde{E}_{1z} = i\omega\tilde{B}_{1y}, \tag{9.2.4}$$

$$i\xi\tilde{E}_{1y} - i\eta\tilde{E}_{1x} = i\omega\tilde{B}_{1z}, \tag{9.2.5}$$

$$i\eta\tilde{B}_{1z} - \frac{\partial}{\partial z}\tilde{B}_{1y} = -\frac{ik_1^2}{\omega}\tilde{E}_{1x} + \mu_0\delta(z-d), \tag{9.2.6}$$

$$\frac{\partial}{\partial z}\tilde{B}_{1x} - i\xi\tilde{B}_{1z} = -\frac{ik_1^2}{\omega}\tilde{E}_{1y}, \tag{9.2.7}$$

$$i\xi\tilde{B}_{1y} - i\eta\tilde{B}_{1x} = -\frac{ik_1^2}{\omega}\tilde{E}_{1z}. \tag{9.2.8}$$

For Region 2,

$$i\eta\tilde{E}_{2z} - \frac{\partial}{\partial z}\tilde{E}_{2y} = i\omega\tilde{B}_{2x}, \tag{9.2.9}$$

$$\frac{\partial}{\partial z}\tilde{E}_{2x} - i\xi\tilde{E}_{2z} = i\omega\tilde{B}_{2y}, \tag{9.2.10}$$

$$i\xi\tilde{E}_{2y} - i\eta\tilde{E}_{2x} = i\omega\tilde{B}_{2z}, \tag{9.2.11}$$

$$i\eta\tilde{B}_{2z} - \frac{\partial}{\partial z}\tilde{B}_{2y} = -\frac{ik_T^2}{\omega}\tilde{E}_{2x}, \tag{9.2.12}$$

$$\frac{\partial}{\partial z}\tilde{B}_{2x} - i\xi\tilde{B}_{2z} = -\frac{ik_T^2}{\omega}\tilde{E}_{2y}, \tag{9.2.13}$$

$$i\xi\tilde{B}_{2y} - i\eta\tilde{B}_{2x} = -\frac{ik_L^2}{\omega}\tilde{E}_{2z}. \tag{9.2.14}$$

In these equations

$$k_j^2 = \omega^2\mu_0\tilde{\epsilon}_j; \qquad j = 1,\, L,\, T. \tag{9.2.15}$$

Equations (9.2.3)–(9.2.14) can be used to solve for $\tilde{E}_y$, $\tilde{E}_z$, $\tilde{B}_y$, and $\tilde{B}_z$ in terms of $\tilde{E}_x$ and $\tilde{B}_x$. The results for Region 1 are

$$\tilde{E}_{1y} = \frac{1}{k_1^2 - \xi^2}\left(-\xi\eta\tilde{E}_{1x} + i\omega\frac{\partial}{\partial z}\tilde{B}_{1x}\right), \tag{9.2.16}$$

$$\tilde{E}_{1z} = \frac{1}{k_1^2 - \xi^2}\left(i\xi\frac{\partial}{\partial z}\tilde{E}_{1x} + \omega\eta\tilde{B}_{1x}\right), \tag{9.2.17}$$

$$\tilde{B}_{1y} = \frac{1}{k_1^2 - \xi^2}\left(-\frac{ik_1^2}{\omega}\frac{\partial}{\partial z}\tilde{E}_{1x} - \xi\eta\tilde{B}_{1x}\right), \tag{9.2.18}$$

$$\tilde{B}_{1z} = \frac{1}{k_1^2 - \xi^2}\left(-\frac{k_1^2}{\omega}\eta\tilde{E}_{1x} + i\xi\frac{\partial}{\partial z}\tilde{B}_{1x}\right). \tag{9.2.19}$$

For Region 2,

$$\tilde{E}_{2y} = \frac{1}{k_T^2 - \xi^2}\left(-\xi\eta\tilde{E}_{2x} + i\omega\frac{\partial}{\partial z}\tilde{B}_{2x}\right), \tag{9.2.20}$$

$$\tilde{E}_{2z} = \frac{1}{k_L^2 - \xi^2}\left(i\xi\frac{\partial}{\partial z}\tilde{E}_{2x} + \omega\eta\tilde{B}_{2x}\right), \tag{9.2.21}$$

$$\tilde{B}_{2y} = \frac{1}{k_L^2 - \xi^2}\left(-\frac{ik_L^2}{\omega}\frac{\partial}{\partial z}\tilde{E}_{2x} - \xi\eta\tilde{B}_{2x}\right), \tag{9.2.22}$$

$$\tilde{B}_{2z} = \frac{1}{k_T^2 - \xi^2}\left(-\frac{k_T^2}{\omega}\eta\tilde{E}_{2x} + i\xi\frac{\partial}{\partial z}\tilde{B}_{2x}\right). \tag{9.2.23}$$

The substitution of (9.2.16)–(9.2.19) into (9.2.3) and (9.2.6) leads to the following ordinary differential equations for $\tilde{E}_{1x}$ and $\tilde{B}_{1x}$:

$$\frac{\partial^2}{\partial z^2}\tilde{B}_{1x} + \gamma_1^2\tilde{B}_{1x} = 0, \tag{9.2.24}$$

$$\frac{\partial^2}{\partial z^2}\tilde{E}_{1x} + \gamma_1^2\tilde{E}_{1x} = \frac{k_1^2 - \xi^2}{i\omega\tilde{\epsilon}_1}\delta(z-d), \tag{9.2.25}$$

where

$$\gamma_1 = (k_1^2 - \xi^2 - \eta^2)^{1/2}. \tag{9.2.26}$$

In (9.2.26) the imaginary part of the square root is positive.

The substitution of (9.2.20)–(9.2.23) into (9.2.9) and (9.2.12) yields the corresponding equations for $\tilde{E}_{2x}$ and $\tilde{B}_{2x}$. They are

$$\frac{1}{k_T^2 - \xi^2}\frac{\partial^2}{\partial z^2}\tilde{B}_{2x} + \left(1 - \frac{\eta^2}{k_L^2 - \xi^2}\right)\tilde{B}_{2x} + \frac{i\xi\eta}{\omega}\left(\frac{1}{k_T^2 - \xi^2} - \frac{1}{k_L^2 - \xi^2}\right)\frac{\partial}{\partial z}\tilde{E}_{2x} = 0, \tag{9.2.27}$$

$$\frac{k_L^2}{k_L^2 - \xi^2}\frac{\partial^2}{\partial z^2}\tilde{E}_{2x} + k_T^2\left(1 - \frac{\eta^2}{k_T^2 - \xi^2}\right)\tilde{E}_{2x} + i\omega\xi\eta\left(\frac{1}{k_T^2 - \xi^2} - \frac{1}{k_L^2 - \xi^2}\right)\frac{\partial}{\partial z}\tilde{B}_{2x} = 0. \tag{9.2.28}$$

9.3 Solutions of the Differential Equations for $\tilde{E}_x$ and $\tilde{B}_x$

With the application of the radiation condition for $z \to \infty$, the solutions of (9.2.24) and (9.2.25) for $\tilde{E}_{1x}$ and $\tilde{B}_{1x}$ are

$$\tilde{E}_{1x} = C_2 e^{i\gamma_1 z} - \frac{k_1^2 - \xi^2}{i\omega\tilde{\epsilon}_1\gamma_1} e^{i\gamma_1 d} \sin\gamma_1 z; \qquad 0 \le z \le d, \tag{9.3.1}$$

$$\tilde{E}_{1x} = \left(C_2 - \frac{k_1^2 - \xi^2}{i\omega\tilde{\epsilon}_1\gamma_1} \sin\gamma_1 d\right) e^{i\gamma_1 z}; \qquad d \le z, \tag{9.3.2}$$

$$\tilde{B}_{1x} = \mu_0 C_1 e^{i\gamma_1 z}; \qquad 0 \le z, \tag{9.3.3}$$

where C_1 and C_2 are constants of integration.

The solutions for $\tilde{E}_{2x}$ and $\tilde{B}_{2x}$ in Region 2 from (9.2.27) and (9.2.28) are obtained with the substitutions:

$$\tilde{B}_{2x} = \mu_0 A e^{-i\gamma z}; \qquad \tilde{E}_{2x} = B e^{-i\gamma z}. \tag{9.3.4}$$

When these expressions are substituted in (9.2.27) and (9.2.28), the following algebraic equations are obtained:

$$\left(1 - \frac{\gamma^2}{k_T^2 - \xi^2} - \frac{\eta^2}{k_L^2 - \xi^2}\right)\mu_0 A + \frac{\xi\eta\gamma}{\omega}\left(\frac{1}{k_T^2 - \xi^2} - \frac{1}{k_L^2 - \xi^2}\right) B = 0, \tag{9.3.5}$$

$$-\omega\xi\eta\gamma\left(\frac{1}{k_L^2 - \xi^2} - \frac{1}{k_T^2 - \xi^2}\right)\mu_0 A + \left(k_T^2 - \frac{k_T^2\eta^2}{k_T^2 - \xi^2} - \frac{k_L^2\gamma^2}{k_L^2 - \xi^2}\right) B = 0. \tag{9.3.6}$$

The nonzero solutions of these equations must satisfy the following relations:

$$\begin{aligned} &\gamma^4 k_L^2 + \gamma^2[(k_L^2 + k_T^2)(\xi^2 + \eta^2) - 2k_L^2 k_T^2] \\ &\quad + k_T^2(k_L^2 - \xi^2 - \eta^2)(k_T^2 - \xi^2 - \eta^2) = 0. \end{aligned} \tag{9.3.7}$$

The solutions are

$$\begin{aligned} \gamma_T &= (k_T^2 - \lambda^2)^{1/2}, \\ \gamma_e &= \left(k_T^2 - \frac{k_T^2}{k_L^2}\lambda^2\right)^{1/2} = \frac{k_T}{k_L}\gamma_L = \frac{k_T}{k_L}(k_L^2 - \lambda^2)^{1/2}, \end{aligned} \tag{9.3.8}$$

where

$$\lambda^2 = \xi^2 + \eta^2. \tag{9.3.9}$$

The imaginary parts of γ_T and γ_e must be positive when λ is real. The coefficients A and B in (9.3.4) to (9.3.6) must satisfy the equations

$$\frac{B_T}{A_T} = -\frac{\omega\mu_0\eta}{\xi\gamma_T}; \qquad \frac{B_e}{A_e} = \frac{\omega\mu_0\xi\gamma_e}{k_T^2\eta}. \tag{9.3.10}$$

It follows that

$$\tilde{B}_{2x} = \mu_0\left(A_T e^{-i\gamma_T z} + A_e e^{-i\gamma_e z}\right); \qquad z \le 0, \tag{9.3.11}$$

$$\tilde{E}_{2x} = -\frac{\omega\mu_0\eta}{\xi\gamma_T}\, A_T e^{-i\gamma_T z} + \frac{\omega\mu_0\xi\gamma_e}{k_T^2\eta}\, A_e e^{-i\gamma_e z}; \qquad z \le 0. \tag{9.3.12}$$

The other four components of the electromagnetic field in Region 2 are obtained from (9.2.20)–(9.2.23) with (9.3.11) and (9.3.12). They are

$$\tilde{B}_{2y} = \mu_0\left(\frac{\eta}{\xi}\, A_T e^{-i\gamma_T z} - \frac{\xi}{\eta}\, A_e e^{-i\gamma_e z}\right); \qquad z \le 0, \tag{9.3.13}$$

$$\tilde{B}_{2z} = \frac{\mu_0\lambda^2}{\xi\gamma_T}\, A_T e^{-i\gamma_T z}; \qquad z \le 0, \tag{9.3.14}$$

$$\tilde{E}_{2y} = \omega\mu_0\left(\frac{A_T}{\gamma_T}\, e^{-i\gamma_T z} + \frac{A_e\gamma_e}{k_T^2}\, e^{-i\gamma_e z}\right); \qquad z \le 0, \tag{9.3.15}$$

$$\tilde{E}_{2z} = \frac{\omega\mu_0\lambda^2}{k_L^2\eta}\, A_e e^{-i\gamma_e z}; \qquad z \le 0. \tag{9.3.16}$$

On the boundary defined by $z = 0$, the conditions $\tilde{B}_{1z} = \tilde{B}_{2z}$ and $k_1^2\tilde{E}_{1z} = k_L^2\tilde{E}_{2z}$ obtain. With them, the coefficients A_T and A_e can be expressed in terms of C_1 and C_2. Thus,

$$A_T = -\frac{\xi\gamma_T}{\lambda^2(k_1^2 - \xi^2)}\,(\xi\gamma_1 C_1 + \omega\tilde{\epsilon}_1\eta C_2), \tag{9.3.17}$$

$$A_e = \frac{k_1^2\eta}{\omega\mu_0\lambda^2}\left(\frac{\omega\mu_0\eta C_1 - \xi\gamma_1 C_2}{k_1^2 - \xi^2} - \frac{\xi}{\omega\tilde{\epsilon}_1}\, e^{i\gamma_1 d}\right). \tag{9.3.18}$$

Similarly, on the boundary $z = 0$, $\tilde{E}_{2y} = \tilde{E}_{1y}$ and $\tilde{B}_{2y} = \tilde{B}_{1y}$. With them, the following equations are obtained:

$$-\frac{\omega\mu_0\gamma_1}{k_1^2 - \xi^2}\, C_1 - \frac{\xi\eta}{k_1^2 - \xi^2}\, C_2 - \frac{\omega\mu_0}{\gamma_T}\, A_T - \frac{\omega\mu_0\gamma_e}{k_T^2}\, A_e = 0, \tag{9.3.19}$$

$$-\frac{\xi\eta}{k_1^2 - \xi^2}\, C_1 + \frac{\omega\tilde{\epsilon}_1\gamma_1}{k_1^2 - \xi^2}\, C_2 - \frac{\eta}{\xi}\, A_T + \frac{\xi}{\eta}\, A_e = -e^{i\gamma_1 d}. \tag{9.3.20}$$

These equations with (9.3.17) and (9.3.18) lead to

$$C_1 = \frac{\xi\eta e^{i\gamma_1 d}}{\lambda^2 M_T N_e}\,(k_1^2\gamma_T\gamma_e - k_T^2\gamma_1^2), \tag{9.3.21}$$

$$C_2 = -\frac{\omega\mu_0 e^{i\gamma_1 d}}{\lambda^2 M_T N_e}\,\{\lambda^2 N_e + \xi^2[\gamma_1(\gamma_T\gamma_e - k_T^2) - \lambda^2\gamma_e]\}, \tag{9.3.22}$$

where

$$M_T = \gamma_1 + \gamma_T; \qquad N_e = k_T^2\gamma_1 + k_1^2\gamma_e = \frac{k_T}{k_L}\,(k_L k_T \gamma_1 + k_1^2\gamma_L). \tag{9.3.23}$$

With (9.3.17), (9.3.18), (9.3.21), (9.3.22), and the Fourier integrals like (9.2.1), the complete solution for all six components of the electromagnetic field in Region 1 in the range $0 \le z \le d$ is

$$\begin{aligned} E_{1x} &= \frac{1}{(2\pi)^2}\int_{-\infty}^{\infty} d\xi \int_{-\infty}^{\infty} d\eta\, e^{i(\xi x+\eta y)} \\ &\quad \times \left(\frac{-\omega\mu_0 e^{i\gamma_1(z+d)}}{\lambda^2 M_T N_e} \{\lambda^2 N_e + \xi^2[\gamma_1(\gamma_T\gamma_e - k_T^2) - \lambda^2\gamma_e]\} \right. \\ &\qquad \left. - \frac{e^{i\gamma_1 d}}{i\omega\tilde{\epsilon}_1\gamma_1}\,(k_1^2 - \xi^2)\sin\gamma_1 z \right), \end{aligned} \tag{9.3.24}$$

$$\begin{aligned} E_{1y} &= \frac{1}{(2\pi)^2}\int_{-\infty}^{\infty} d\xi \int_{-\infty}^{\infty} d\eta\, e^{i(\xi x+\eta y)}\xi\eta \\ &\quad \times \left(\frac{\omega\mu_0 e^{i\gamma_1(z+d)}}{\lambda^2 M_T N_e}\,(k_T^2\gamma_1 + \lambda^2\gamma_e - \gamma_1\gamma_T\gamma_e) + \frac{e^{i\gamma_1 d}}{i\omega\tilde{\epsilon}_1\gamma_1}\sin\gamma_1 z \right), \end{aligned} \tag{9.3.25}$$

$$\begin{aligned} E_{1z} &= \frac{1}{(2\pi)^2}\int_{-\infty}^{\infty} d\xi \int_{-\infty}^{\infty} d\eta\, e^{i(\xi x+\eta y)}\xi \\ &\quad \times \left(\frac{\omega\mu_0\gamma_e e^{i\gamma_1(z+d)}}{N_e} - \frac{e^{i\gamma_1 d}}{\omega\tilde{\epsilon}_1}\cos\gamma_1 z \right), \end{aligned} \tag{9.3.26}$$

$$B_{1x} = \frac{\mu_0}{(2\pi)^2}\int_{-\infty}^{\infty} d\xi \int_{-\infty}^{\infty} d\eta\, e^{i(\xi x+\eta y)}\xi\eta\, \frac{e^{i\gamma_1(z+d)}}{\lambda^2 M_T N_e}\,(k_1^2\gamma_T\gamma_e - k_T^2\gamma_1^2), \tag{9.3.27}$$

$$\begin{aligned} B_{1y} &= -\frac{\mu_0}{(2\pi)^2}\int_{-\infty}^{\infty} d\xi \int_{-\infty}^{\infty} d\eta\, e^{i(\xi x+\eta y)} \\ &\quad \times \left(\frac{e^{i\gamma_1(z+d)}}{\lambda^2 M_T N_e}\,[\lambda^2\gamma_1 N_e + \xi^2(k_1^2\gamma_T\gamma_e - k_T^2\gamma_1^2)] - e^{i\gamma_1 d}\cos\gamma_1 z \right), \end{aligned} \tag{9.3.28}$$

$$B_{1z} = \frac{\mu_0}{(2\pi)^2}\int_{-\infty}^{\infty} d\xi \int_{-\infty}^{\infty} d\eta\, e^{i(\xi x+\eta y)}\eta \left(\frac{e^{i\gamma_1(z+d)}}{M_T} - \frac{ie^{i\gamma_1 d}}{\gamma_1}\sin\gamma_1 z \right). \tag{9.3.29}$$

The corresponding formulas for the range $d < z$ could readily be written down but are not needed.

The expressions (9.3.24)–(9.3.29) can now be expressed in the cylindrical coordinates ρ, ϕ, z with the relations

$$x = \rho\cos\phi; \quad y = \rho\sin\phi; \qquad \xi = \lambda\cos\phi'; \quad \eta = \lambda\sin\phi', \tag{9.3.30}$$

and the integral representation of the Bessel functions, viz.,

$$J_n(\lambda\rho) = \frac{i^{-n}}{2\pi}\int_0^{2\pi} e^{i\lambda\rho\cos\theta}e^{in\theta}\,d\theta. \tag{9.3.31}$$

The final formulas are

$$\begin{aligned} E_{1\rho}(\rho,\phi,z) = {} & -\frac{\omega\mu_0}{4\pi k_1^2}\cos\phi \\ & \times\left[\int_0^\infty\left(k_1^2 J_0(\lambda\rho) - \frac{\lambda^2}{2}\left[J_0(\lambda\rho) - J_2(\lambda\rho)\right]\right)\frac{e^{i\gamma_1|z-d|}}{\gamma_1}\lambda\,d\lambda\right. \\ & + \int_0^\infty\left(\frac{\gamma_1 Q_e}{2}\left[J_0(\lambda\rho) - J_2(\lambda\rho)\right]\right. \\ & \left.\left. - \frac{k_1^2 P_T}{2\gamma_1}\left[J_0(\lambda\rho) + J_2(\lambda\rho)\right]\right)e^{i\gamma_1(z+d)}\lambda\,d\lambda\right], \end{aligned} \tag{9.3.32}$$

$$\begin{aligned} E_{1\phi}(\rho,\phi,z) = {} & \frac{\omega\mu_0}{4\pi k_1^2}\sin\phi \\ & \times\left[\int_0^\infty\left(k_1^2 J_0(\lambda\rho) - \frac{\lambda^2}{2}\left[J_0(\lambda\rho) + J_2(\lambda\rho)\right]\right)\frac{e^{i\gamma_1|z-d|}}{\gamma_1}\lambda\,d\lambda\right. \\ & + \int_0^\infty\left(\frac{\gamma_1 Q_e}{2}\left[J_0(\lambda\rho) + J_2(\lambda\rho)\right]\right. \\ & \left.\left. - \frac{k_1^2 P_T}{2\gamma_1}\left[J_0(\lambda\rho) - J_2(\lambda\rho)\right]\right)e^{i\gamma_1(z+d)}\lambda\,d\lambda\right], \end{aligned} \tag{9.3.33}$$

$$E_{1z}(\rho,\phi,z) = \frac{i\omega\mu_0}{4\pi k_1^2}\cos\phi\int_0^\infty\left(\pm e^{i\gamma_1|z-d|} + Q_e e^{i\gamma_1(z+d)}\right)J_1(\lambda\rho)\lambda^2\,d\lambda, \tag{9.3.34}$$

$$\begin{aligned} B_{1\rho}(\rho,\phi,z) = {} & -\frac{\mu_0}{4\pi}\sin\phi \\ & \times\left[\pm\int_0^\infty J_0(\lambda\rho)e^{i\gamma_1|z-d|}\lambda\,d\lambda\right. \\ & + \int_0^\infty\left(\frac{Q_e}{2}\left[J_0(\lambda\rho) + J_2(\lambda\rho)\right]\right. \\ & \left.\left. - \frac{P_T}{2}\left[J_0(\lambda\rho) - J_2(\lambda\rho)\right]\right)e^{i\gamma_1(z+d)}\lambda\,d\lambda\right], \end{aligned} \tag{9.3.35}$$

$$B_{1\phi}(\rho,\phi,z) = -\frac{\mu_0}{4\pi}\cos\phi \times \left[\pm\int_0^\infty J_0(\lambda\rho)e^{i\gamma_1|z-d|}\lambda\,d\lambda + \int_0^\infty \left(\frac{Q_e}{2}\left[J_0(\lambda\rho) - J_2(\lambda\rho)\right] - \frac{P_T}{2}\left[J_0(\lambda\rho) + J_2(\lambda\rho)\right]\right)e^{i\gamma_1(z+d)}\lambda\,d\lambda\right], \tag{9.3.36}$$

$$B_{1z}(\rho,\phi,z) = \frac{i\mu_0}{4\pi}\sin\phi\int_0^\infty (e^{i\gamma_1|z-d|} - P_T e^{i\gamma_1(z+d)})\gamma_1^{-1}J_1(\lambda\rho)\lambda^2\,d\lambda, \tag{9.3.37}$$

where

$$Q_e = \frac{k_1^2\gamma_e - k_T^2\gamma_1}{k_1^2\gamma_e + k_T^2\gamma_1} = \frac{k_1^2\gamma_L - k_L k_T\gamma_1}{k_1^2\gamma_L + k_L k_T\gamma_1}; \qquad P_T = \frac{\gamma_T - \gamma_1}{\gamma_T + \gamma_1}. \tag{9.3.38}$$

The lower sign applies to the range $0 \leq z \leq d$; the upper sign is added for the range $d < z$. The formulas (9.3.32)–(9.3.37) are identical with those in (5.4.29)–(5.4.34) for an isotropic Region 2 except for the appearance of Q_e instead of Q and P_T instead of P. If Region 2 is made isotropic by setting $k_L = k_T = k_2$, then $\gamma_e = \gamma_T = \gamma_2$, $Q_e \to Q$, and $P_T \to P$.

9.4 Approximate Integrated Formulas for the Complete Field in Region 1

The general and exact integrals (9.3.32)–(9.3.37) for the electromagnetic field at (ρ,ϕ,z) in Region 1 ($z \geq 0$) due to a horizontal unit electric dipole also in Region 1 at $(0,0,d)$—as shown in Fig. 9.1.1—can be evaluated to obtain relatively simple, very good approximations by the same procedure used for the corresponding integrals when Region 2 is isotropic. The evaluation depends on the conditions

$$|k_1| \geq 3|k_L|; \qquad |k_1| \geq 3|k_T|; \qquad \rho \geq 5|z|; \qquad \rho \geq 5d. \tag{9.4.1}$$

It is assumed that these conditions are adequate to permit the neglect of $|k_L^2/k_1^2|$, $|k_T^2/k_1^2|$, and $|k_L k_T/k_1^2|$ compared to one.

Since the integrals for the direct and ideally reflected fields are the same for the isotropic and anisotropic cases, and their values are known exactly, it is not necessary to reevaluate them. Actually, they are usually negligible in measurements on the sea floor. The evaluation of the integrals that lead to the complete field is carried out in Section 9.6.

The final formulas for the lateral-wave components are

$$E^L_{1\rho}(\rho,\phi,z) = -\frac{\omega\mu_0 k_T}{2\pi k_1^2}\cos\phi\; e^{ik_1(z+d)} \times \left\{ e^{ik_L\rho}\left[\frac{ik_L}{\rho} - \frac{1}{\rho^2} - \frac{k_L^2 k_T}{k_1}\left(\frac{\pi}{k_L\rho}\right)^{1/2} e^{-ip_e}\,\mathcal{F}(p_e)\right] - \frac{i}{k_T\rho^3}\,e^{ik_T\rho}\right\}, \tag{9.4.2}$$

$$E^L_{1\phi}(\rho,\phi,z) = \frac{\omega\mu_0 k_T}{2\pi k_1^2}\sin\phi\; e^{ik_1(z+d)} \times \left\{ e^{ik_L\rho}\left[\frac{1}{\rho^2} + \frac{ik_L k_T}{k_1\rho}\left(\frac{\pi}{k_L\rho}\right)^{1/2} e^{-ip_e}\mathcal{F}(p_e)\right] + e^{ik_T\rho}\left(\frac{1}{\rho^2} + \frac{2i}{k_T\rho^3}\right)\right\}, \tag{9.4.3}$$

$$E^L_{1z}(\rho,\phi,z) = \frac{\omega\mu_0 k_L k_T}{2\pi k_1^3}\cos\phi\; e^{ik_1(z+d)}e^{ik_L\rho} \times \left[\frac{ik_L}{\rho} - \frac{1}{\rho^2} - \frac{k_L^2 k_T}{k_1}\left(\frac{\pi}{k_L\rho}\right)^{1/2} e^{-ip_e}\mathcal{F}(p_e)\right], \tag{9.4.4}$$

$$B^L_{1\rho}(\rho,\phi,z) = -\frac{k_1}{\omega}\,E^L_{1\phi}(\rho,\phi,z), \tag{9.4.5}$$

$$B^L_{1\phi}(\rho,\phi,z) = \frac{k_1}{\omega}\,E^L_{1\rho}(\rho,\phi,z), \tag{9.4.6}$$

$$B^L_{1z}(\rho,\phi,z) = \frac{\mu_0 k_T^2}{2\pi k_1^2}\sin\phi\; e^{ik_1(z+d)}e^{ik_T\rho}\left(\frac{1}{\rho^2} + \frac{3i}{k_T\rho^3} - \frac{3}{k_T^2\rho^4}\right). \tag{9.4.7}$$

In these formulas,

$$p_e = k_L\rho\left(\frac{k_T^2}{2k_1^2}\right) \tag{9.4.8}$$

and

$$\mathcal{F}(p_e) = \tfrac{1}{2}(1+i) - C_2(p_e) - iS_2(p_e), \tag{9.4.9}$$

$$C_2(p_e) + iS_2(p_e) = \int_0^{p_e} \frac{e^{it}}{(2\pi t)^{1/2}}\,dt. \tag{9.4.10}$$

Subject to the condition for the asymptotic or far-field range, viz.,

$$|k_L\rho| \gg \left|\frac{2k_1^2}{k_T^2}\right|, \tag{9.4.11}$$

the Fresnel-integral term can be approximated by

$$e^{-ip_e}\mathcal{F}(p_e) \sim \frac{ik_1}{k_T}\left(\frac{1}{\pi k_L\rho}\right)^{1/2}\left(1-\frac{ik_1^2}{k_T^2 k_L\rho}\right), \tag{9.4.12}$$

and the components of the field in the asymptotic range are

$$E_{1\rho}^L(\rho,\phi,z) \sim \frac{\omega\mu_0}{2\pi k_T}\cos\phi\; e^{ik_1(z+d)}\,\frac{e^{ik_L\rho}}{\rho^2}, \tag{9.4.13}$$

$$E_{1\phi}^L(\rho,\phi,z) \sim \frac{\omega\mu_0}{2\pi k_T}\sin\phi\; e^{ik_1(z+d)} \times\left(\frac{i}{k_L\rho^3}\,e^{ik_L\rho}+\frac{k_T^2}{k_1^2\rho^2}\,e^{ik_T\rho}\right), \tag{9.4.14}$$

$$E_{1z}^L(\rho,\phi,z) \sim -\frac{\omega\mu_0 k_L}{2\pi k_1 k_T}\cos\phi\; e^{ik_1(z+d)}\,\frac{e^{ik_L\rho}}{\rho^2}, \tag{9.4.15}$$

$$B_{1\rho}^L(\rho,\phi,z) \sim -\frac{k_1}{\omega}\,E_{1\phi}^L(\rho,\phi,z), \tag{9.4.16}$$

$$B_{1\phi}^L(\rho,\phi,z) \sim \frac{k_1}{\omega}\,E_{1\rho}^L(\rho,\phi,z), \tag{9.4.17}$$

$$B_{1z}^L(\rho,\phi,z) \sim \frac{\mu_0 k_T^2}{2\pi k_1^2}\sin\phi\; e^{ik_1(z+d)}\,\frac{e^{ik_T\rho}}{\rho^2}. \tag{9.4.18}$$

The $1/\rho^3$ term is retained in (9.4.14) because the $1/\rho^2$ term is multiplied by the small factor k_T^2/k_1^2 and by the exponential $\exp(ik_T\rho) = \exp(-\alpha_T\rho)\exp(i\beta_T\rho)$ which is usually smaller than $\exp(ik_L\rho) = \exp(-\alpha_L\rho) \times \exp(i\beta_L\rho)$. It is evident that the field of magnetic type with the components $E_{1\phi}^L(\rho,\phi,z)$, $B_{1\rho}^L(\rho,\phi,z)$, and $B_{1z}^L(\rho,\phi,z)$ is very small in the asymptotic range compared to the field of electric type with the components $E_{1\rho}^L(\rho,\phi,z)$, $E_{1z}^L(\rho,\phi,z)$, and $B_{1\phi}^L(\rho,\phi,z)$. Except for the appearance of k_T in the amplitude factor, the asymptotic field of electric type depends only on k_L and hence on σ_L. Its properties are like those in Region 1 (the ocean) when Region 2 (the oceanic crust) is isotropic with the conductivity $\sigma_2 = \sigma_L$.

Actually, the asymptotic field is not generally useful for measurements on the sea floor owing to the exponential attenuation in $\exp(ik_L\rho) = \exp(-\alpha_L\rho)\exp(i\beta_L\rho)$. The near and intermediate ranges provide much greater amplitudes. If the asymptotic range is excluded with the condition

$$|k_L\rho| \le \left|\frac{2k_1^2}{k_T^2}\right|, \tag{9.4.19}$$

the Fresnel-integral terms have a negligible effect on the nature of the field.

Without them, the components have the following simple forms:

$$E_{1\rho}^{L}(\rho,\phi,z) = -\frac{\omega\mu_0 k_T}{2\pi k_1^2}\cos\phi\; e^{ik_1(z+d)}\left[\left(\frac{ik_L}{\rho}-\frac{1}{\rho^2}\right)e^{ik_L\rho}-\frac{i}{k_T\rho^3}\,e^{ik_T\rho}\right], \tag{9.4.20}$$

$$E_{1\phi}^{L}(\rho,\phi,z) = \frac{\omega\mu_0 k_T}{2\pi k_1^2}\sin\phi\; e^{ik_1(z+d)}\left[\frac{1}{\rho^2}\,e^{ik_L\rho}+\left(\frac{1}{\rho^2}+\frac{2i}{k_T\rho^3}\right)e^{ik_T\rho}\right], \tag{9.4.21}$$

$$E_{1z}^{L}(\rho,\phi,z) = \frac{\omega\mu_0 k_L k_T}{2\pi k_1^3}\cos\phi\; e^{ik_1(z+d)}\left(\frac{ik_L}{\rho}-\frac{1}{\rho^2}\right)e^{ik_L\rho}, \tag{9.4.22}$$

$$B_{1\rho}^{L}(\rho,\phi,z) = -\frac{k_1}{\omega}\,E_{1\phi}^{L}(\rho,\phi,z), \tag{9.4.23}$$

$$B_{1\phi}^{L}(\rho,\phi,z) = \frac{k_1}{\omega}\,E_{1\rho}^{L}(\rho,\phi,z), \tag{9.4.24}$$

$$B_{1z}^{L}(\rho,\phi,z) = \frac{\mu_0 k_T^2}{2\pi k_1^2}\sin\phi\; e^{ik_1(z+d)}\left(\frac{1}{\rho^2}+\frac{3i}{k_T\rho^3}-\frac{3}{k_T^2\rho^4}\right)e^{ik_T\rho}. \tag{9.4.25}$$

It is seen that, except in the amplitude factor, $E_{1\rho}^{L}(\rho,\phi,z)$ and $B_{1\phi}^{L}(\rho,\phi,z)$ involve k_T only in the $1/\rho^3$ term, that is, in the quasi-static range, and that $E_{1z}^{L}(\rho,\phi,z)$ is independent of k_T. This means that in the intermediate range defined by

$$1 \le |k_L\rho| \le \left|\frac{2k_1^2}{k_T^2}\right|, \tag{9.4.26}$$

the entire field of electric type has the same properties it would have if Region 2 (the oceanic crust) were isotropic with $\sigma_2 = \sigma_L$.

The quasi-static $1/\rho^3$ terms in the components of magnetic type also depend on k_T. On the other hand, in the intermediate range defined by (9.4.26), $E_{1\phi}^{L}(\rho,\phi,z)$ and $B_{1\rho}^{L}(\rho,\phi,z)$ are symmetric in k_L and k_T with terms of the type

$$\frac{1}{\rho^2}\,(e^{ik_L\rho}+e^{ik_T\rho}). \tag{9.4.27}$$

This means that interchanging k_L and k_T (σ_L and σ_T) has no effect.

Graphs of $|E_{1\rho}(\rho,0,1\text{ m})|$ as a function of the radial distance from the source at $(0,0,1\text{ m})$ are shown in Fig. 9.4.1 for the frequency $f = 1$ Hz and in Fig. 9.4.2 for $f = 1$ kHz with $\sigma_1 = 3.6$ S/m. The solid lines are for Region 2 anisotropic with $\sigma_x = \sigma_y = \sigma_T = 0.002$ S/m and $\sigma_z = \sigma_L = 0.004$ S/m; the dashed lines are with $\sigma_x = \sigma_y = \sigma_T = 0.004$ S/m and $\sigma_z = \sigma_L = 0.002$ S/m. Only the latter case is physically meaningful on the sea floor (Parkhomenko 1967) but the interchange is provided for theoretical interest.

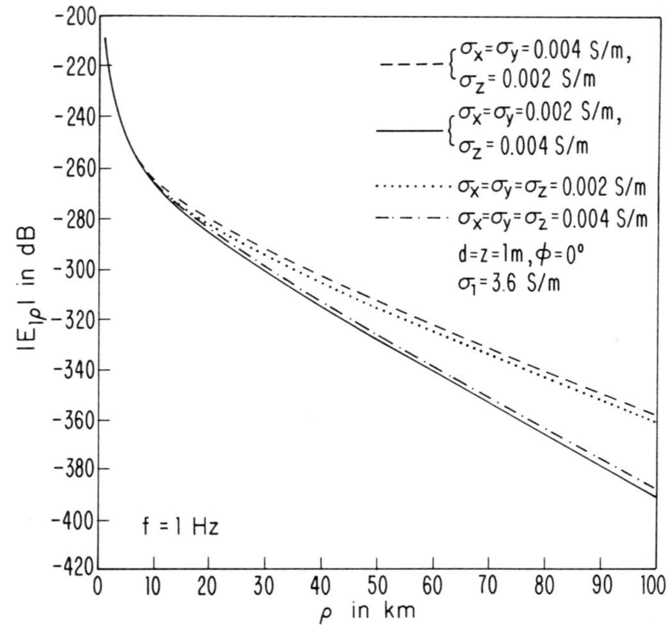

Figure 9.4.1. $|E_{1\rho}(\rho, 0, 1 \text{ m})|$ in V/m due to unit horizontal electric dipole at $(0, 0, 1 \text{ m})$ in sea water (Region 1, $\sigma_1 = 3.6$ S/m) over one-dimensionally anisotropic rock (Region 2, $\sigma_x = \sigma_y = \sigma_T$, $\sigma_z = \sigma_L$); $f = 1$ Hz.

The dotted and dot-dashed graphs are for Region 2 isotropic with $\sigma_L = \sigma_T = \sigma_2 = 0.002$ S/m and $\sigma_L = \sigma_T = \sigma_2 = 0.004$ S/m, respectively. The corresponding curves for $|E_{1\phi}(\rho, \pi/2, 1 \text{ m})|$ are in Fig. 9.4.3 for $f = 1$ Hz and Fig. 9.4.4 for $f = 1$ kHz.

In Fig. 9.4.1 for $|E_{1\rho}(\rho, 0, 1 \text{ m})|$ at $f = 1$ Hz, it is seen that the solid and dot-dashed lines are very close together. This means that $|E_{1\rho}(\rho, 0, 1 \text{ m})|$ is virtually the same when Region 2 is anisotropic with $\sigma_x = \sigma_y = 0.002$ S/m and $\sigma_z = 0.004$ S/m, or isotropic with $\sigma_x = \sigma_y = \sigma_z = 0.004$ S/m. The same conclusion follows from the proximity of the dashed and dotted curves, which show that $|E_{1\rho}(\rho, 0, 1 \text{ m})|$ is virtually the same when Region 2 is anisotropic with $\sigma_x = \sigma_y = 0.004$ S/m and $\sigma_z = 0.002$ S/m, or isotropic with $\sigma_x = \sigma_y = \sigma_z = 0.002$ S/m. Clearly, $|E_{1\rho}(\rho, 0, 1 \text{ m})|$ is determined by σ_z independent of $\sigma_x = \sigma_y$.

When the frequency is increased to $f = 1$ kHz in Fig. 9.4.2, the dashed and dotted lines are again very close together indicating that the term in $\exp(ik_T\rho)$ in (9.4.20) is negligible. Note that in this case $\sigma_L = \sigma_z < \sigma_T = \sigma_x = \sigma_y$ so that $k_L = \alpha_L(1 + i) < k_T = \alpha_T(1 + i)$. It follows that $\exp(ik_L\rho)$ decays more slowly than $\exp(ik_T\rho)$. On the other hand, the solid and dot-dashed lines remain together only for $\rho \leq 6$ km. For $\rho > 6$ km, the two lines diverge with the solid line (for $\sigma_T = \sigma_x = \sigma_y = 0.002$ S/m

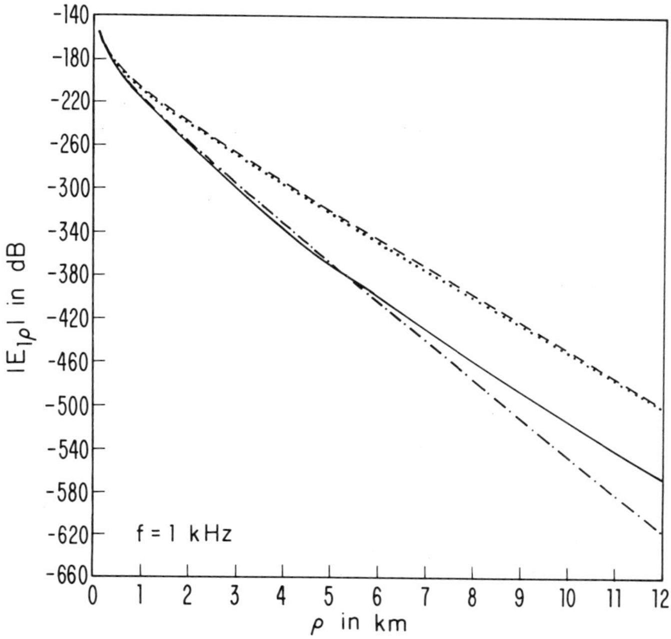

Figure 9.4.2. $|E_{1\rho}(\rho, 0, 1 \text{ m})|$ in V/m due to unit horizontal electric dipole at $(0, 0, 1 \text{ m})$ in sea water (Region 1, $\sigma_1 = 3.6$ S/m) over one-dimensionally anisotropic rock (Region 2, $\sigma_x = \sigma_y = \sigma_T$, $\sigma_z = \sigma_L$); $f = 1$ kHz.

and $\sigma_L = \sigma_z = 0.004$ S/m) showing a slower rate of decrease than the dot-dashed line (for $\sigma_x = \sigma_y = \sigma_z = 0.004$ S/m). In this case, $k_L > k_T$, so that in (9.4.20) the term in $\exp(ik_T\rho)$ dominates when ρ becomes large enough (~ 6 km) and the rate of decrease is less than for $k_2 = k_L$ for the isotropic Region 2.

In Figs. 9.4.3 and 9.4.4 for $|E_{1\phi}(\rho, \pi/2, 1 \text{ m})|$ at $f = 1$ Hz and $f = 1$ kHz, the solid and dashed lines coincide and lie between the dotted and dot-dashed curves. This shows that when Region 2 is anisotropic, the dependence of $|E_{1\phi}(\rho, \phi, z)|$ is symmetrical with respect to k_T and k_L so that an interchange of σ_L with σ_T leaves $|E_{1\phi}(\rho, \phi, z)|$ virtually unchanged. With either choice, the term that depends on the smaller conductivity (i.e., 0.002 S/m) dominates. The graphs for the anisotropic case are closer to the dotted curve for the isotropic Region 2 because this is for $\sigma_2 = 0.002$ S/m and involves smaller exponential attenuation than $\sigma_2 = 0.004$ S/m for the dot-dashed curve.

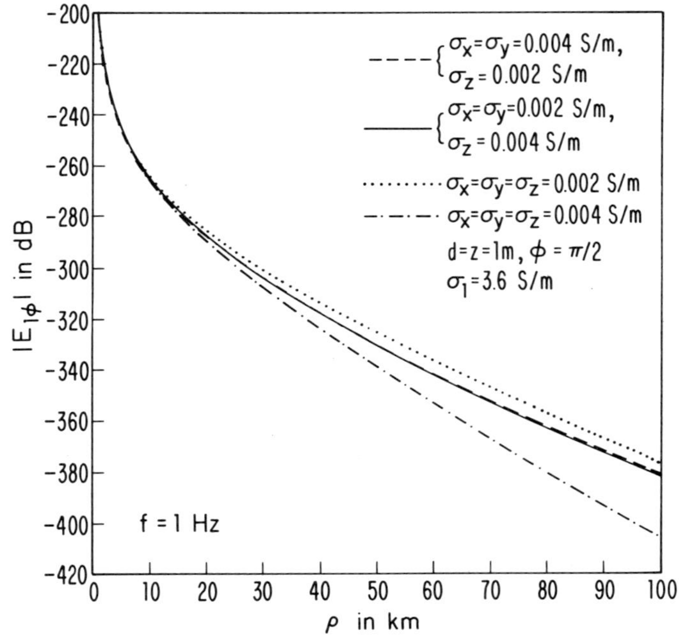

Figure 9.4.3. $|E_{1\phi}(\rho, \pi/2, 1 \text{ m})|$ in V/m due to unit horizontal electric dipole at $(0, 0, 1 \text{ m})$ in sea water (Region 1, $\sigma_1 = 3.6$ S/m) over one-dimensionally anisotropic rock (Region 2, $\sigma_x = \sigma_y = \sigma_T$, $\sigma_z = \sigma_L$); $f = 1$ Hz.

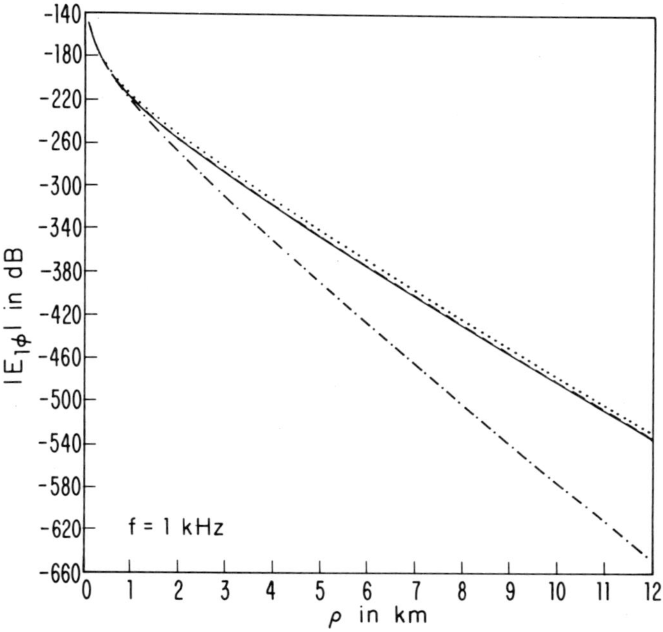

Figure 9.4.4. $|E_{1\phi}(\rho, \pi/2, 1 \text{ m})|$ in V/m due to unit horizontal electric dipole at $(0, 0, 1 \text{ m})$ in sea water (Region 1, $\sigma_1 = 3.6$ S/m) over one-dimensionally anisotropic rock (Region 2, $\sigma_x = \sigma_y = \sigma_T$, $\sigma_z = \sigma_L$); $f = 1$ kHz.

9.5 Comparison with Measurements

The attempt to interpret the measured data of Young and Cox (1981) for

$$E_X = E_{1\rho}(\rho, \phi, z) \sin\psi + E_{1\phi}(\rho, \phi, z) \cos\psi, \tag{9.5.1}$$

$$E_Y = -E_{1\rho}(\rho, \phi, z) \cos\psi + E_{1\phi}(\rho, \phi, z) \sin\psi \tag{9.5.2}$$

on the sea floor in the frequency range $0.25 \leq f \leq 2.25$ Hz with $\rho = \rho_0 = 18.9$ km led to the suggestion in Chapter 8 that the oceanic crust at the site of the measurements might be anisotropic in the conductivity. With the complete theory available for Region 2 an anisotropic half-space with $\sigma_T = \sigma_x = \sigma_y \neq \sigma_L = \sigma_z$, it is appropriate to apply it to the conditions of the measurements. The experimental layout is shown at the top in Fig. 9.5.1; the measured values of E_X and E_Y are represented by the boxes in the lower part of the figure. The angles ψ and ϕ are not given, but can be estimated from the diagram. It is important to note that ψ is small so that $E_X \sim E_{1\phi}(\rho, \phi, z)$ and $E_Y \sim -E_{1\rho}(\rho, \phi, z)$. It would have been better if the receiving antenna had been oriented so that $\psi = 0$.

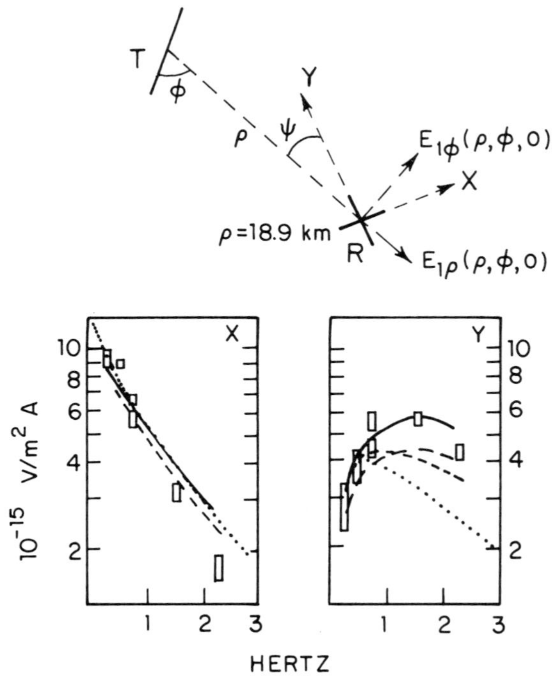

Figure 9.5.1. Experimental layout and graphs traced from Figs. 2 and 3 of Young and Cox (1981).

Computations with $\phi = 60°$, ψ from 5° to 10°, and σ_L and σ_T from 0.0025 to 0.005 S/m are shown in Figs. 9.5.2 through 9.5.5. In these figures, the solid lines and dot-dashed lines represent the one-dimensionally anisotropic rock; the solid lines are for the physically realistic $\sigma_L < \sigma_T$ and the dot-dashed lines for the physically unavailable but theoretically interesting $\sigma_L > \sigma_T$. The dotted lines and dashed lines represent isotropic rock. From these figures it is seen that for E_Y only the solid lines ($\sigma_L < \sigma_T$) agree well with the measured data. In contrast, for E_X the solid lines and dot-dashed lines are nearly coincident and both agree well with the measured data. The reason why E_X is not sensitive to an interchange of the two anisotropic models ($\sigma_L > \sigma_T$, $\sigma_L < \sigma_T$) is because E_X is nearly equal to $E_{1\phi}(\rho, \phi, z)$ and for this the coefficients of the terms multiplied by $\exp(ik_L\rho)$ are approximately equal to those of the terms multiplied by $\exp(ik_T\rho)$ in a range of ρ in which the exponentials do not dominate.

A final computation with the values suggested in Chapter 8, viz., $\psi = 10°$, $\phi = 50°$, $\sigma_{2z} = \sigma_L = 0.002$ S/m, and $\sigma_{2x} = \sigma_{2y} = \sigma_T = 0.004$ S/m, is represented in Fig. 9.5.6. Graphs of $|E_{1\rho}(\rho, \phi, 0)|$ and $|E_{1\phi}(\rho, \phi, 0)|$ are shown together with $|E_X|$ and $|E_Y|$ as calculated from (9.5.1) and (9.5.2).

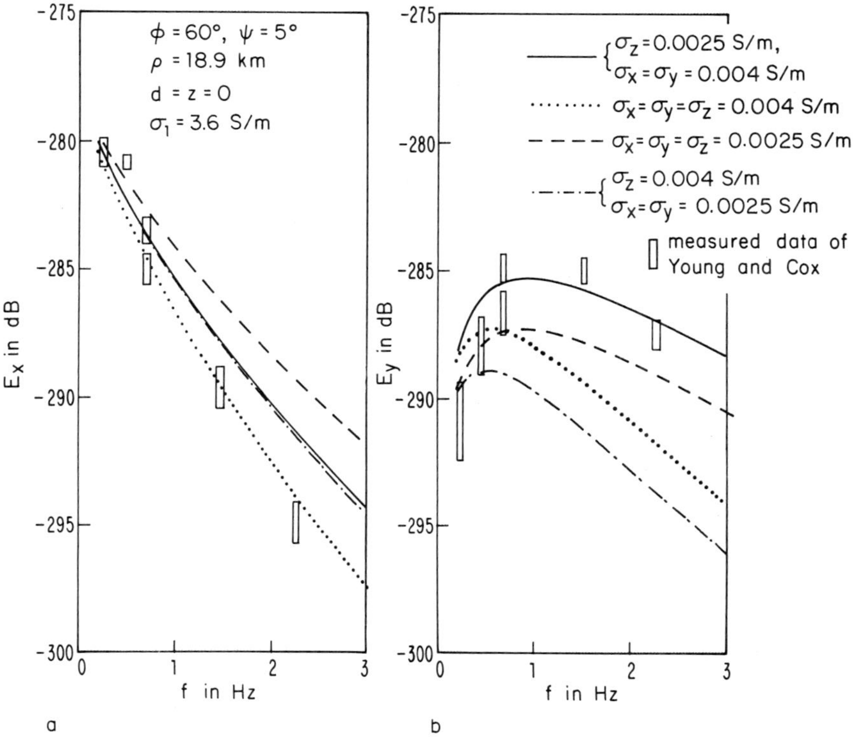

Figure 9.5.2. Comparison of surface-wave field components calculated for a one-dimensionally anisotropic rock (Region 2, $\sigma_x = \sigma_y = \sigma_T$, $\sigma_z = \sigma_L$) with sea-floor measurements by Young and Cox (1981). (a) $E_X = E_{1\rho} \sin\psi + E_{1\phi} \cos\psi$. (b) $E_Y = -E_{1\rho} \cos\psi + E_{1\phi} \sin\psi$.

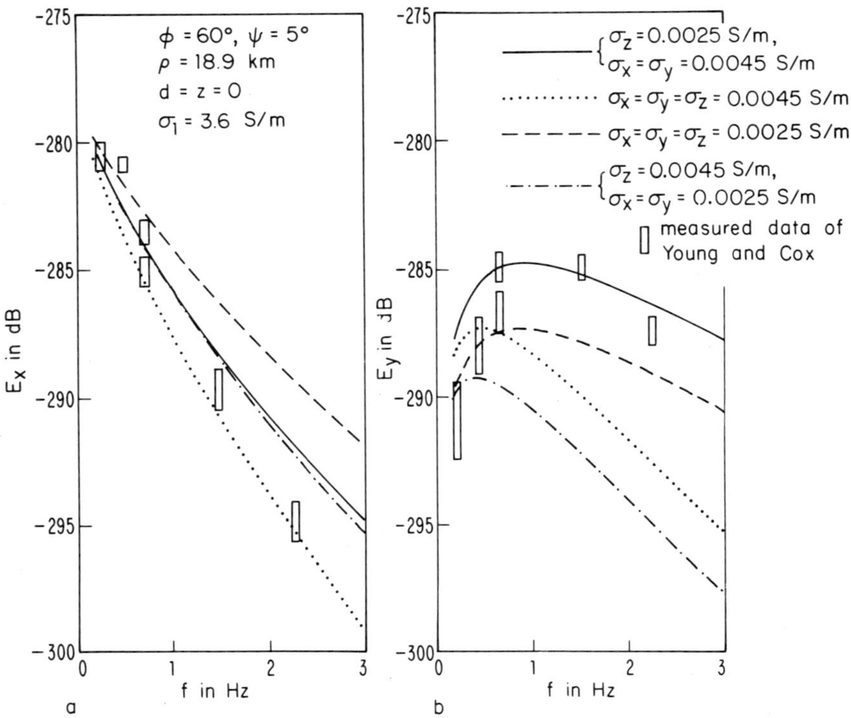

Figure 9.5.3. Comparison of surface-wave field components calculated for a one-dimensionally anisotropic rock (Region 2, $\sigma_x = \sigma_y = \sigma_T$, $\sigma_z = \sigma_L$) with sea-floor measurements by Young and Cox (1981). (a) $E_X = E_{1\rho} \sin\psi + E_{1\phi} \cos\psi$. (b) $E_Y = -E_{1\rho} \cos\psi + E_{1\phi} \sin\psi$.

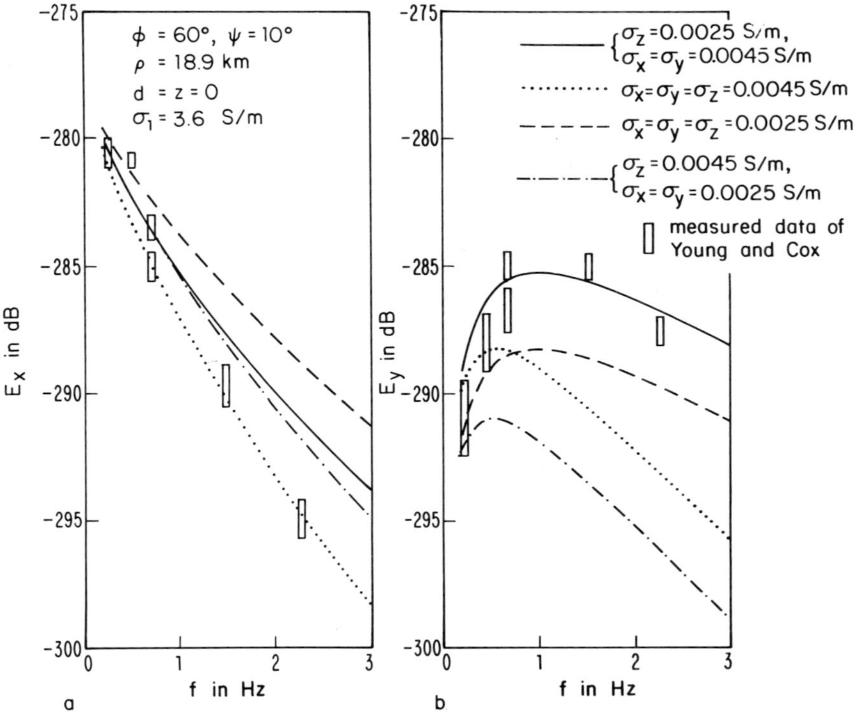

Figure 9.5.4. Comparison of surface-wave field components calculated for a one-dimensionally anisotropic rock (Region 2, $\sigma_x = \sigma_y = \sigma_T$, $\sigma_z = \sigma_L$) with sea-floor measurements by Young and Cox (1981). (a) $E_X = E_{1\rho} \sin\psi + E_{1\phi} \cos\psi$. (b) $E_Y = -E_{1\rho} \cos\psi + E_{1\phi} \sin\psi$.

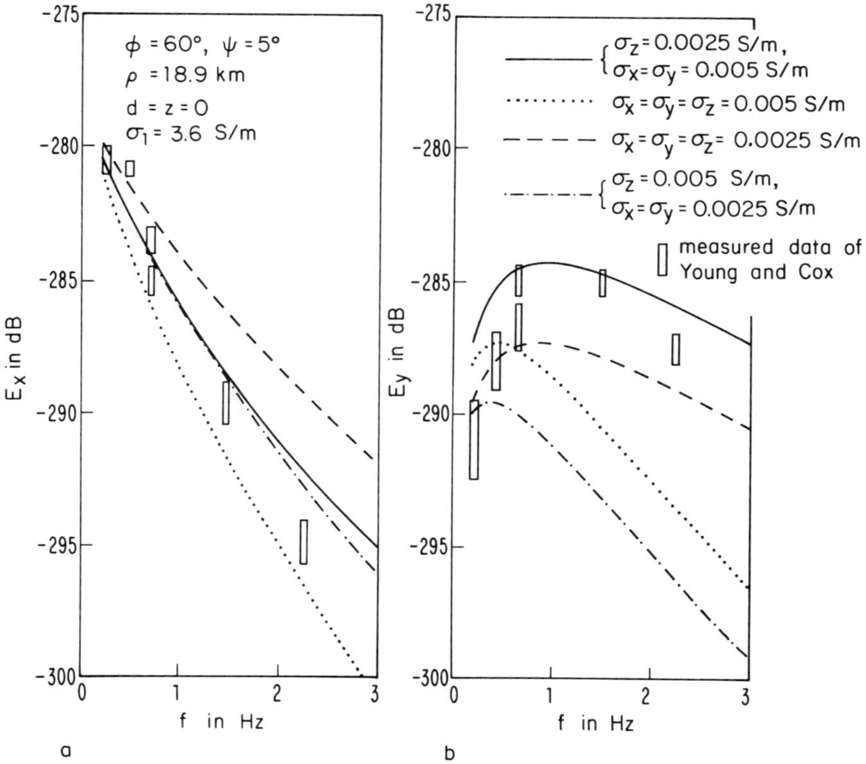

Figure 9.5.5. Comparison of surface-wave field components calculated for a one-dimensionally anisotropic rock (Region 2, $\sigma_x = \sigma_y = \sigma_T$, $\sigma_z = \sigma_L$) with sea-floor measurements by Young and Cox (1981). (a) $E_X = E_{1\rho} \sin\psi + E_{1\phi} \cos\psi$. (b) $E_Y = -E_{1\rho} \cos\psi + E_{1\phi} \sin\psi$.

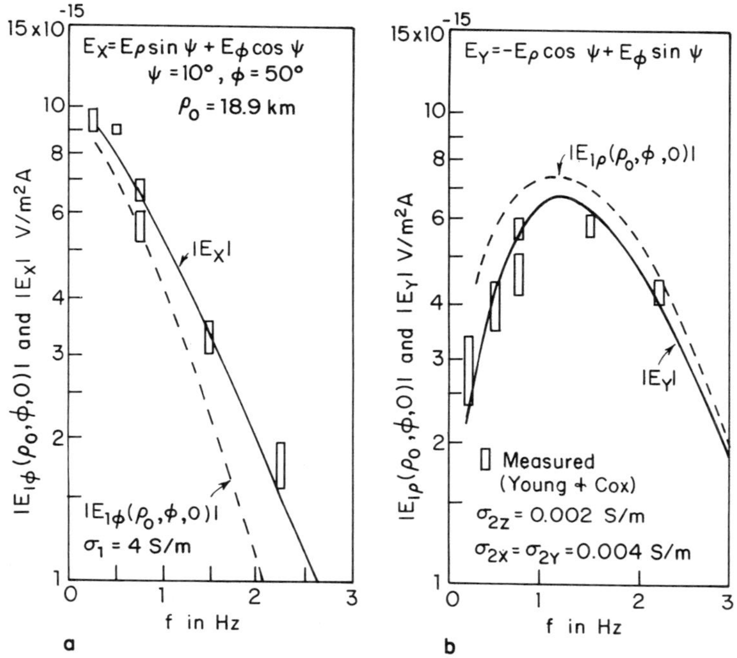

Figure 9.5.6. Comparison of surface-wave field components calculated for a one-dimensionally anisotropic rock (Region 2, $\sigma_x = \sigma_y = 0.004$ S/m, $\sigma_z = 0.002$ S/m) with sea-floor measurements by Young and Cox (1981). (a) $E_X = E_{1\rho} \sin\psi + E_{1\phi}\cos\psi$. (b) $E_Y = -E_{1\rho}\cos\psi + E_{1\phi}\sin\psi$.

The experimental data of Young and Cox (1981) are included for direct comparison. The agreement between the measured components and the theoretical ones is excellent over the entire range for both E_X and E_Y. The close relationship $E_X \sim E_{1\phi}(\rho,\phi,0)$ and $E_Y \sim -E_{1\rho}(\rho,\phi,0)$ is also demonstrated.

From a study of the several figures it may be concluded that the measured data are well represented by a one-dimensionally anisotropic half-space model with $\sigma_{2z} = \sigma_L$ smaller than $\sigma_{2x} = \sigma_{2y} = \sigma_T$. Reasonable values are $\sigma_L \sim 0.002$ to 0.0025 S/m and $\sigma_T \sim 0.004$ to 0.005 S/m.

The good agreement of the anisotropic half-space model with the measured data suggests that it may be an alternative to the so-called "best fit" eight-layer model of Young and Cox discussed in Chapter 8. It has the advantage that it is consistent with the requirements of the theory. Specifically, it is in accord with the theoretical requirement that from 60% to over 90% of the measured component $E_Y \sim -E_{1\rho}(\rho,\phi,0)$ is due to the lateral wave that penetrates only a few hundred meters into the oceanic crust. The anisotropic model also provides a clear explanation for the differences between $E_X \sim E_{1\phi}(\rho,\phi,0)$ and $E_Y \sim -E_{1\rho}(\rho,\phi,0)$. The insensitivity of $E_X \sim E_{1\phi}(\rho,\phi,0)$ to the different properties of the three eight-layer models and the half-space model remains unexplained. Theory shows that $E_{1\phi}(\rho,\phi,0)$ penetrates quite deeply into the crust and that it is reflected from a perfectly reflecting boundary and transmitted into the salt water even more strongly than $E_{1\rho}(\rho,\phi,0)$.

9.6 Evaluation of the Integrals for the Field in Region 1 over a Homogeneous Anisotropic Region 2

The integral expression (9.3.32) for $E_{1\rho}(\rho,\phi,z)$ can be rewritten in the form

$$E_{1\rho}(\rho,\phi,z) = -\frac{\omega\mu_0}{4\pi k_1^2}\cos\phi\,[F_{\rho 0}(\rho,z-d) + F_{\rho 0}(\rho,z+d) + F_{\rho 1}(\rho,z+d)], \tag{9.6.1}$$

where $F_{\rho 0}(\rho,z-d)$, $F_{\rho 0}(\rho,z+d)$, and $F_{\rho 1}(\rho,z+d)$ represent the direct wave, reflected wave, and lateral wave, respectively. The direct and reflected terms are identical to the corresponding integrals for the isotropic case. The well-known expressions for these fields are given in (5.5.23).

The lateral-wave term can be separated into two parts,

$$F_{\rho 1}(\rho,z+d) = F_{\rho 2}(\rho,z+d) + F_{\rho 3}(\rho,z+d), \tag{9.6.2}$$

where, with (9.3.8) and $\gamma_L = (k_L^2-\lambda^2)^{1/2}$,

$$F_{\rho 2}(\rho,z+d) = -k_L k_T\int_0^\infty \frac{\gamma_1^2}{k_1^2\gamma_L + k_L k_T\gamma_1}[J_0(\lambda\rho) - J_2(\lambda\rho)] \times e^{i\gamma_1(z+d)}\lambda\,d\lambda. \tag{9.6.3}$$

Similarly, with $\gamma_T = (k_T^2-\lambda^2)^{1/2}$,

$$F_{\rho 3}(\rho,z+d) = -k_1^2\int_0^\infty \frac{\gamma_T}{\gamma_1(\gamma_T+\gamma_1)}[J_0(\lambda\rho) + J_2(\lambda\rho)] \times e^{i\gamma_1(z+d)}\lambda\,d\lambda. \tag{9.6.4}$$

These formulas are like (5.6.2) and (5.6.3) with $k_2^2\gamma_1^2/(k_1^2\gamma_2+k_2^2\gamma_1)$ in (5.6.4) replaced by $k_L k_T\gamma_1^2/(k_1^2\gamma_L + k_L k_T\gamma_1)$ to obtain (9.6.3) and $\gamma_2/(\gamma_2+\gamma_1)$ in (5.6.4) replaced by $\gamma_T/(\gamma_T+\gamma_1)$ to obtain (9.6.4). It follows, as in (5.6.5), that

$$\frac{\gamma_1^2}{k_1^2\gamma_L + k_L k_T\gamma_1} = \gamma_1^2\left(\frac{1}{k_1^2\gamma_L + k_L k_T\gamma_1} - \frac{1}{k_1^2\gamma_L}\right) + \frac{(\gamma_1^2-\gamma_L^2)+\gamma_L^2}{k_1^2\gamma_L}$$

$$= \gamma_1^2\left(\frac{1}{k_1^2\gamma_L + k_L k_T\gamma_1} - \frac{1}{k_1^2\gamma_L}\right) + \frac{k_1^2 - k_L^2}{k_1^2\gamma_L} + \frac{\gamma_L}{k_1^2}$$

$$\sim \gamma_1^2\left(\frac{1}{k_1^2\gamma_L + k_L k_T\gamma_1} - \frac{1}{k_1^2\gamma_L}\right) + \frac{1}{\gamma_L} + \frac{\gamma_L}{k_1^2}. \tag{9.6.5}$$

Similarly, as in (5.6.6), it follows that

$$\frac{\gamma_T}{\gamma_1(\gamma_T + \gamma_1)} \sim \frac{1}{\gamma_1} - \frac{\gamma_1}{k_1^2} + \frac{\gamma_T}{k_1^2}. \tag{9.6.6}$$

When these formulas are substituted in (9.6.3) and (9.6.4) to obtain (9.6.2), the result parallels (5.6.7) with suitable changes in notation. Thus,

$$\begin{aligned} F_{\rho 1}(\rho, z+d) = H_{\rho 1}(\rho, z+d) & \\ & - k_L k_T \int_0^\infty \gamma_L^{-1}[J_0(\lambda\rho) - J_2(\lambda\rho)]e^{i\gamma_1(z+d)}\lambda\, d\lambda \\ & - \frac{k_L k_T}{k_1^2}\int_0^\infty \gamma_L[J_0(\lambda\rho) - J_2(\lambda\rho)]e^{i\gamma_1(z+d)}\lambda\, d\lambda \\ & - k_1^2\int_0^\infty \gamma_1^{-1}[J_0(\lambda\rho) + J_2(\lambda\rho)]e^{i\gamma_1(z+d)}\lambda\, d\lambda \\ & + \int_0^\infty \gamma_1[J_0(\lambda\rho) + J_2(\lambda\rho)]e^{i\gamma_1(z+d)}\lambda\, d\lambda \\ & - \int_0^\infty \gamma_T[J_0(\lambda\rho) + J_2(\lambda\rho)]e^{i\gamma_1(z+d)}\lambda\, d\lambda, \end{aligned} \tag{9.6.7}$$

where

$$\begin{aligned} H_{\rho 1}(\rho, z+d) &= -k_L k_T\int_0^\infty \gamma_1^2\left(\frac{1}{k_1^2\gamma_L + k_L k_T\gamma_1} - \frac{1}{k_1^2\gamma_L}\right) \\ &\qquad\qquad \times [J_0(\lambda\rho) - J_2(\lambda\rho)]e^{i\gamma_1(z+d)}\lambda\, d\lambda \\ &\sim e^{ik_1(z+d)}H_{\rho 1}(\rho, 0). \end{aligned} \tag{9.6.8}$$

The integrals in (9.6.7) are the same as those in (5.6.7) with the substitution of k_T or k_L for k_2. The integrated formula corresponding to (5.6.11) is

$$\begin{aligned} F_{\rho 1}(\rho, z+d) &= e^{ik_1(z+d)}H_{\rho 1}(\rho, 0) \\ &- 2k_L k_T\left\{\left(\frac{z+d}{\rho}\right)\left[\frac{e^{ik_1(z+d)}}{k_1\rho^2} - \left(\frac{1}{\rho} + \frac{i}{2k_1\rho^2} + \frac{3}{8k_1^2\rho^3}\right)e^{ik_1 r_2}\right]\right. \\ &\qquad \left. - e^{ik_1(z+d)}\left[\frac{1}{k_L\rho^2} + \left(\frac{i}{\rho} - \frac{1}{k_L\rho^2}\right)e^{ik_L\rho}\right]\right\} \\ &+ \frac{2k_L k_T}{k_1^2}\left\{ik_1\left(\frac{z+d}{\rho}\right)\left(\frac{ik_1}{\rho} - \frac{3}{2\rho^2} - \frac{5i}{8k_1\rho^3}\right)e^{ik_1 r_2}\right. \\ &\qquad \left. + k_L\left[\frac{e^{ik_1(z+d)}}{\rho^2} + \left(\frac{1}{\rho^2} + \frac{2i}{k_L\rho^3}\right)e^{ik_L\rho}e^{ik_1(z+d)}\right]\right\} \end{aligned}$$

$$
\begin{aligned}
&- \frac{2k_1}{\rho^2}\left[e^{ik_1(z+d)} - e^{ik_1 r_2}\right] + 2k_1\left(\frac{e^{ik_1(z+d)}}{\rho^2} + \frac{i}{k_1\rho^3}\, e^{ik_1 r_2}\right) \\
&- 2k_T\left(\frac{1}{\rho^2} + \frac{i}{k_T\rho^3}\, e^{ik_T\rho}\right)e^{ik_1(z+d)} \\
&+ 2k_1\left(\frac{z+d}{\rho}\right)\left(\frac{i}{\rho^2} - \frac{3}{2k_1\rho^3}\right)e^{ik_1 r_2}.
\end{aligned} \tag{9.6.9}
$$

The integral $H_{\rho 1}(\rho, 0)$ is evaluated in Section 9.7. It is

$$
H_{\rho 1}(\rho, 0) = -\frac{2k_L^2 k_T^2}{k_1}\left(\frac{\pi}{k_L\rho}\right)^{1/2} e^{ik_L\rho}\, e^{-ip_e}\mathcal{F}(p_e), \tag{9.6.10}
$$

where p_e and $\mathcal{F}(p_e)$ are defined, respectively, in (9.4.8) and (9.4.9).

The use of (5.5.23) and (9.6.9) with (9.6.10) in (9.6.1), with the omission of small terms of the order $k_L k_T/k_1^2$ compared with one, leads to the following formula for $E_{1\rho}$:

$$
\begin{aligned}
E_{1\rho}(\rho, \phi, z) = &-\frac{\omega\mu_0}{2\pi k_1^2}\cos\phi \\
&\times \Bigg(e^{ik_1(z+d)}\Bigg\{e^{ik_L\rho}\Bigg[\frac{ik_L k_T}{\rho} - \frac{k_T}{\rho^2} \\
&\qquad - \frac{k_L^2 k_T^2}{k_1}\left(\frac{\pi}{k_L\rho}\right)^{1/2} e^{-ip_e}\,\mathcal{F}(p_e)\Bigg] - \frac{ie^{ik_T\rho}}{\rho^3}\Bigg\} \\
&- e^{ik_1 r_1}\left(\frac{k_1}{\rho^2} + \frac{i}{\rho^3}\right) + e^{ik_1 r_2}\left(\frac{z+d}{\rho}\right)\left(\frac{ik_1}{\rho^2} - \frac{3}{2\rho^3}\right)\Bigg).
\end{aligned} \tag{9.6.11}
$$

The integral expression (9.3.33) for $E_{1\phi}(\rho, \phi, z)$ can be rewritten in the form

$$
E_{1\phi}(\rho, \phi, z) = \frac{\omega\mu_0}{4\pi k_1^2}\sin\phi\,\left[F_{\phi 0}(\rho, z-d) + F_{\phi 0}(\rho, z+d) + F_{\phi 1}(\rho, z+d)\right], \tag{9.6.12}
$$

where $F_{\phi 0}(\rho, z-d)$, $F_{\phi 0}(\rho, z+d)$, and $F_{\phi 1}(\rho, z+d)$ are the direct, reflected, and lateral waves, respectively. The functions $F_{\phi 0}(\rho, z-d)$ and $F_{\phi 0}(\rho, z+d)$ coincide with the corresponding formulas (5.5.24) for an isotropic Region 2. The lateral-wave term is

$$
F_{\phi 1}(\rho, z+d) = F_{\phi 2}(\rho, z+d) + F_{\phi 3}(\rho, z+d), \tag{9.6.13}
$$

with

$$
F_{\phi 2}(\rho, z+d) = -k_L k_T \int_0^\infty \frac{\gamma_1^2}{k_1^2\gamma_L + k_L k_T\gamma_1}\left[J_0(\lambda\rho) + J_2(\lambda\rho)\right]
$$

$$\times e^{i\gamma_1(z+d)}\lambda\, d\lambda, \tag{9.6.14}$$

$$F_{\phi 3}(\rho, z+d) = -k_1^2 \int_0^\infty \frac{\gamma_T}{\gamma_1(\gamma_T+\gamma_1)} [J_0(\lambda\rho) - J_2(\lambda\rho)] \times e^{i\gamma_1(z+d)}\lambda\, d\lambda. \tag{9.6.15}$$

These integrals are like (9.6.3) and (9.6.4) with the sign of $J_2(\lambda\rho)$ reversed. It follows that (9.6.7) applies if the sign of $J_2(\lambda\rho)$ is changed. Thus,

$$\begin{aligned} F_{\phi 1}(\rho, z+d) &= H_{\phi 1}(\rho, z+d) \\ &\quad - k_L k_T \int_0^\infty \gamma_L^{-1}[J_0(\lambda\rho) + J_2(\lambda\rho)]e^{i\gamma_1(z+d)}\lambda\, d\lambda \\ &\quad - \frac{k_L k_T}{k_1^2}\int_0^\infty \gamma_L[J_0(\lambda\rho) + J_2(\lambda\rho)]e^{i\gamma_1(z+d)}\lambda\, d\lambda \\ &\quad - k_1^2 \int_0^\infty \gamma_1^{-1}[J_0(\lambda\rho) - J_2(\lambda\rho)]e^{i\gamma_1(z+d)}\lambda\, d\lambda \\ &\quad + \int_0^\infty \gamma_1[J_0(\lambda\rho) - J_2(\lambda\rho)]e^{i\gamma_1(z+d)}\lambda\, d\lambda \\ &\quad - \int_0^\infty \gamma_T[J_0(\lambda\rho) - J_2(\lambda\rho)]e^{i\gamma_1(z+d)}\lambda\, d\lambda, \end{aligned} \tag{9.6.16}$$

where

$$\begin{aligned} H_{\phi 1}(\rho, z+d) &= -k_L k_T \int_0^\infty \gamma_1^2 \left(\frac{1}{k_1^2\gamma_L + k_L k_T \gamma_1} - \frac{1}{k_1^2\gamma_L}\right) \\ &\qquad \times [J_0(\lambda\rho) + J_2(\lambda\rho)]e^{i\gamma_1(z+d)}\lambda\, d\lambda \\ &\sim e^{ik_1(z+d)} H_{\phi 1}(\rho, 0). \end{aligned} \tag{9.6.17}$$

The integrals in (9.6.16) are the same as those in (5.6.17). The integrated formula corresponding to (5.6.19) is

$$\begin{aligned} F_{\phi 1}(\rho, z+d) &= e^{ik_1(z+d)} H_{\phi 1}(\rho, 0) \\ &\quad - 2k_L k_T \left[\frac{e^{ik_1(z+d)}}{k_L\rho^2}(1 - e^{ik_L\rho}) + \left(\frac{z+d}{\rho}\right)\frac{ie^{ik_1 r_2}}{k_1\rho^2}\right] \\ &\quad - \frac{2k_L^2 k_T}{k_1^2}\left(\frac{1}{\rho^2} + \frac{i}{k_L\rho^3}e^{ik_L\rho}\right)e^{ik_1(z+d)} \\ &\quad + \frac{2k_L k_T}{k_1}\left(\frac{z+d}{\rho}\right)\left(\frac{i}{\rho^2} - \frac{3}{2k_1\rho^3}\right)e^{ik_1 r_2} \\ &\quad + 2k_1\left[\frac{e^{ik_1(z+d)}}{\rho^2} + \left(\frac{ik_1}{\rho} - \frac{1}{\rho^2}\right)e^{ik_1 r_2}\right] \\ &\quad - 2k_1\left[\frac{e^{ik_1(z+d)}}{\rho^2} + \left(\frac{1}{\rho^2} + \frac{2i}{k_1\rho^3}\right)e^{ik_1 r_2}\right] \end{aligned}$$

$$+ 2ik_1\left(\frac{z+d}{\rho}\right)\left(\frac{ik_1}{\rho} - \frac{3}{2\rho^2} - \frac{5i}{8k_1\rho^3}\right)e^{ik_1r_2}$$

$$+ 2k_T\left[\frac{e^{ik_1(z+d)}}{\rho^2} + \left(\frac{1}{\rho^2} + \frac{2i}{k_T\rho^3}\right)e^{ik_T\rho}e^{ik_1(z+d)}\right], \quad (9.6.18)$$

where, from Section 9.7,

$$H_{\phi 1}(\rho, 0) = \frac{2ik_L^2k_T^2}{k_1}\frac{\pi^{1/2}}{(k_L\rho)^{3/2}}\,e^{ik_L\rho}\,e^{-ip_e}\mathcal{F}(p_e). \quad (9.6.19)$$

After terms of order k_Lk_T/k_1^2 have been neglected in (9.6.18), the result is

$$F_{\phi 1}(\rho, z+d) = 2k_Te^{ik_1(z+d)}\left\{e^{ik_L\rho}\left[\frac{1}{\rho^2} + \frac{ik_Lk_T}{k_1\rho}\left(\frac{\pi}{k_L\rho}\right)^{1/2} e^{-ip_e}\mathcal{F}(p_e)\right]\right.$$
$$\left. + e^{ik_T\rho}\left(\frac{1}{\rho^2} + \frac{2i}{k_T\rho^3}\right)\right\}$$
$$+ 4e^{ik_1r_2}\left(\frac{ik_1^2}{2\rho} - \frac{k_1}{\rho^2} - \frac{i}{\rho^3}\right)$$
$$+ 2ik_1e^{ik_1r_2}\left(\frac{z+d}{\rho}\right)\left(\frac{ik_1}{\rho} - \frac{3}{2\rho^2} - \frac{5i}{8k_1\rho^3}\right). \quad (9.6.20)$$

When this is substituted in (9.6.12) with (5.5.24), the result is

$$E_{1\phi}(\rho, \phi, z) = \frac{\omega\mu_0}{2\pi k_1^2}\sin\phi$$
$$\times \left(k_Te^{ik_1(z+d)}\left\{e^{ik_L\rho}\left[\frac{1}{\rho^2} + \frac{ik_Lk_T}{k_1\rho}\left(\frac{\pi}{k_L\rho}\right)^{1/2} e^{-ip_e}\mathcal{F}(p_e)\right]\right.\right.$$
$$\left. + e^{ik_T\rho}\left(\frac{1}{\rho^2} + \frac{2i}{k_T\rho^3}\right)\right\}$$
$$+ \frac{k_1}{2}\left[e^{ik_1r_2}\left(\frac{ik_1}{\rho} - \frac{3}{\rho^2} - \frac{3i}{k_1\rho^3}\right) - e^{ik_1r_1}\left(\frac{ik_1}{\rho} - \frac{1}{\rho^2} - \frac{i}{k_1\rho^3}\right)\right]$$
$$\left. + ik_1e^{ik_1r_2}\left(\frac{z+d}{\rho}\right)\left(\frac{ik_1}{\rho} - \frac{3}{2\rho^2} - \frac{5i}{8k_1\rho^3}\right)\right). \quad (9.6.21)$$

The integral expression (9.3.34) for $E_{1z}(\rho, \phi, z)$ can be written as follows:

$$E_{1z}(\rho, \phi, z) = \frac{i\omega\mu_0}{4\pi k_1^2}\cos\phi\,[F_{z0}(\rho, z-d) + F_{z0}(\rho, z+d) + F_{z1}(\rho, z+d)]. \quad (9.6.22)$$

The direct-wave term $F_{z0}(\rho, z-d)$ and the reflected-wave term $F_{z0}(\rho, z+d)$ are identical to the formulas (5.5.25) for an isotropic medium. The lateral-wave term is

$$F_{z1}(\rho, z+d) = \int_0^\infty e^{i\gamma_1(z+d)}(Q_e - 1)J_1(\lambda\rho)\lambda^2\,d\lambda$$

$$= -2k_L k_T \int_0^\infty e^{i\gamma_1(z+d)} \frac{\gamma_1}{k_1^2\gamma_L + k_L k_T \gamma_1} J_1(\lambda\rho)\lambda^2\, d\lambda. \tag{9.6.23}$$

This integral can be expanded to give

$$F_{z1}(\rho, z+d) = H_{z1}(\rho, z+d) - \frac{2k_L k_T}{k_1^2} \int_0^\infty \frac{\gamma_1}{\gamma_L} J_1(\lambda\rho) e^{i\gamma_1(z+d)}\lambda^2\, d\lambda, \tag{9.6.24}$$

where

$$\begin{aligned} H_{z1}(\rho, z+d) &= -2k_L k_T \int_0^\infty \gamma_1 \left(\frac{1}{k_1^2\gamma_L + k_L k_T\gamma_1} - \frac{1}{k_1^2\gamma_L}\right) \\ &\qquad \times J_1(\lambda\rho) e^{i\gamma_1(z+d)}\lambda^2\, d\lambda \\ &\sim e^{ik_1(z+d)} H_{z1}(\rho, 0). \end{aligned} \tag{9.6.25}$$

The integral on the right in (9.6.24) is the same as the corresponding integral in (5.6.24). With (5.6.27), the result is

$$\begin{aligned} F_{z1}(\rho, z+d) = H_{z1}(\rho, z+d) + \frac{2k_L k_T}{k_1}\Bigg[& e^{ik_1(z+d)} e^{ik_L\rho}\left(\frac{k_L}{\rho} + \frac{i}{\rho^2}\right) \\ & - e^{ik_1 r_2}\left(\frac{1}{\rho^2} + \frac{3i}{2k_1\rho^3}\right)\Bigg]. \end{aligned} \tag{9.6.26}$$

The integral $H_{z1}(\rho, 0)$ is evaluated in Section 9.7 and given in (9.7.13). With it and (5.6.28), the formula for $F_{z1}(\rho, z+d)$ is

$$\begin{aligned} & F_{z1}(\rho, z+d) \\ & = \frac{2k_L k_T}{k_1}\Bigg\{ e^{ik_1(z+d)} e^{ik_L\rho}\left[\frac{k_L}{\rho} + \frac{i}{\rho^2} + \frac{ik_T k_L^2}{k_1}\left(\frac{\pi}{k_L\rho}\right)^{1/2} e^{-ip_e}\mathcal{F}(p_e)\right] \\ & \qquad - e^{ik_1 r_2}\left(\frac{1}{\rho^2} + \frac{3i}{2k_1\rho^3}\right)\Bigg\}. \end{aligned} \tag{9.6.27}$$

The substitution of (9.6.27) and (5.5.25) in (9.6.22) gives the following final formula for $E_{1z}(\rho, \phi, z)$:

$$\begin{aligned} E_{1z}(\rho, \phi, z) &= \frac{\omega\mu_0}{2\pi k_1^2}\cos\phi \\ &\times \Bigg(\frac{k_L k_T}{k_1}\Bigg\{ e^{ik_1(z+d)} e^{ik_L\rho}\left[\frac{ik_L}{\rho} - \frac{1}{\rho^2} - \frac{k_L^2 k_T}{k_1}\left(\frac{\pi}{k_L\rho}\right)^{1/2} e^{-ip_e}\mathcal{F}(p_e)\right] \\ &\qquad - \left(\frac{i}{\rho^2} - \frac{3}{2k_1\rho^3}\right) e^{ik_1 r_2}\Bigg\} \\ &\quad - \frac{1}{2}\left(\frac{z-d}{\rho} e^{ik_1 r_1} + \frac{z+d}{\rho} e^{ik_1 r_2}\right)\left(\frac{ik_1^2}{\rho} - \frac{3k_1}{\rho^2} - \frac{3i}{\rho^3}\right)\Bigg). \end{aligned} \tag{9.6.28}$$

In a similar manner, the integral expressions (9.3.35) for $B_{1\rho}(\rho, \phi, z)$ and (9.3.36) for $B_{1\phi}(\rho, \phi, z)$ may be rewritten as follows:

$$B_{1\rho}(\rho, \phi, z) = -\frac{\mu_0}{4\pi} \sin\phi \, [G_{\rho 0}(\rho, z-d) + G_{\rho 0}(\rho, z+d) + G_{\rho 1}(\rho, z+d)], \tag{9.6.29}$$

$$B_{1\phi}(\rho, \phi, z) = -\frac{\mu_0}{4\pi} \cos\phi \, [G_{\phi 0}(\rho, z-d) + G_{\phi 0}(\rho, z+d) + G_{\phi 1}(\rho, z+d)]. \tag{9.6.30}$$

The direct-wave terms $G_{\rho 0}(\rho, z-d)$ and $G_{\phi 0}(\rho, z-d)$ and the reflected-wave terms $G_{\rho 0}(\rho, z+d)$ and $G_{\phi 0}(\rho, z+d)$ are like the corresponding formulas in (5.5.26) for an isotropic medium. The lateral-wave terms $G_{\rho 1}(\rho, z+d)$ and $G_{\phi 1}(\rho, z+d)$ are

$$G_{\rho 1}(\rho, z+d) = G_{\rho 2}(\rho, z+d) + G_{\rho 3}(\rho, z+d), \tag{9.6.31}$$
$$G_{\phi 1}(\rho, z+d) = G_{\phi 2}(\rho, z+d) + G_{\phi 3}(\rho, z+d), \tag{9.6.32}$$

where

$$\left.\begin{matrix} G_{\rho 2}(\rho, z+d) \\ G_{\phi 2}(\rho, z+d) \end{matrix}\right\} = -k_L k_T \int_0^\infty \frac{\gamma_1}{k_1^2 \gamma_L + k_L k_T \gamma_1} \times [J_0(\lambda\rho) \pm J_2(\lambda\rho)] e^{i\gamma_1(z+d)} \lambda \, d\lambda$$
$$= \left\{\begin{matrix} H_{B\rho 1}(\rho, z+d) \\ H_{B\phi 1}(\rho, z+d) \end{matrix}\right\} - \frac{k_L k_T}{k_1^2} \int_0^\infty \frac{\gamma_1}{\gamma_L} [J_0(\lambda\rho) \pm J_2(\lambda\rho)] e^{i\gamma_1(z+d)} \lambda \, d\lambda, \tag{9.6.33}$$

with

$$\left.\begin{matrix} H_{B\rho 1}(\rho, z+d) \\ H_{B\phi 1}(\rho, z+d) \end{matrix}\right\} = -k_L k_T \int_0^\infty \gamma_1 \left(\frac{1}{k_1^2 \gamma_L + k_L k_T \gamma_1} - \frac{1}{k_1^2 \gamma_L} \right) \times [J_0(\lambda\rho) \pm J_2(\lambda\rho)] e^{i\gamma_1(z+d)} \lambda \, d\lambda$$
$$\sim e^{ik_1(z+d)} \left\{\begin{matrix} H_{B\rho 1}(\rho, 0) \\ H_{B\phi 1}(\rho, 0) \end{matrix}\right\}, \tag{9.6.34}$$

and

$$\left.\begin{matrix} G_{\rho 3}(\rho, z+d) \\ G_{\phi 3}(\rho, z+d) \end{matrix}\right\} = -\int_0^\infty \frac{\gamma_T}{\gamma_T + \gamma_1} [J_0(\lambda\rho) \mp J_2(\lambda\rho)] e^{i\gamma_1(z+d)} \lambda \, d\lambda. \tag{9.6.35}$$

The integrals on the right in (9.6.33) are like those in (5.6.32) with k_L substituted for k_2; the integrals in (9.6.35) are like those in (5.6.31) with k_T

substituted for k_2. The integrals for $H_{B\rho 1}(\rho, z+d)$ and $H_{B\phi 1}(\rho, z+d)$ differ, respectively, from $H_{\phi 1}(\rho, z+d)$ in (9.6.17) and $H_{\rho 1}(\rho, z+d)$ in (9.6.8) only by the appearance of γ_1 instead of γ_1^2 in the integrand. Since the evaluations of $H_{\phi 1}(\rho, z+d) \sim e^{ik_1(z+d)} H_{\phi 1}(\rho, 0)$ and $H_{\rho 1}(\rho, z+d) \sim e^{ik_1(z+d)} H_{\rho 1}(\rho, 0)$ in Section 9.7 make the approximation $\gamma_1 \sim k_1$, it follows that

$$\left. \begin{array}{l} H_{B\rho 1}(\rho, z+d) \\ H_{B\phi 1}(\rho, z+d) \end{array} \right\} = k_1^{-1} \left\{ \begin{array}{l} H_{\phi 1}(\rho, z+d) \\ H_{\rho 1}(\rho, z+d) \end{array} \right\}$$

$$= k_1^{-1} e^{ik_1(z+d)} \left\{ \begin{array}{l} H_{\phi 1}(\rho, 0) \\ H_{\rho 1}(\rho, 0) \end{array} \right\}, \tag{9.6.36}$$

where $H_{\phi 1}(\rho, 0)$ is given in (9.6.19) and $H_{\rho 1}(\rho, 0)$ in (9.6.10). It follows that

$$\begin{aligned} G_{\rho 2}(\rho, z+d) = e^{ik_1(z+d)} H_{B\rho 1}(\rho, 0) \\ - \frac{k_L k_T}{k_1^2} \bigg[\left(\frac{i}{k_1 \rho^3} - \frac{z+d}{\rho^3} + \frac{2k_1}{k_L \rho^2} \right) e^{ik_1(z+d)} \\ + \frac{2e^{ik_1 r_2}}{k_1 \rho^3} - \frac{2k_1}{k_L \rho^2} e^{ik_1(z+d)} e^{ik_L \rho} \bigg], \end{aligned} \tag{9.6.37}$$

$$\begin{aligned} G_{\phi 2}(\rho, z+d) = e^{ik_1(z+d)} H_{B\phi 1}(\rho, 0) \\ + \frac{2k_L k_T}{k_1^2} \bigg[\left(\frac{k_1}{k_L \rho^2} - \frac{z+d}{\rho^3} + \frac{i}{k_1 \rho^3} \right) e^{ik_1(z+d)} \\ - i \left(\frac{1}{\rho^2} + \frac{i}{2k_1 \rho^3} + \frac{3}{8k_1^2 \rho^4} \right) e^{ik_1 r_2} \\ + \frac{k_1}{k_L} \left(\frac{ik_L}{\rho} - \frac{1}{\rho^2} \right) e^{ik_L \rho} e^{ik_1(z+d)} \bigg] \end{aligned} \tag{9.6.38}$$

and

$$\begin{aligned} G_{\rho 3}(\rho, z+d) = 2 \bigg\{ e^{ik_1 r_2} \left[\frac{i}{\rho^2} - \frac{7}{2k_1 \rho^3} + \left(\frac{z+d}{\rho} \right) \left(\frac{ik_1}{\rho} - \frac{5}{\rho^2} - \frac{12i}{k_1 \rho^3} \right) \right] \\ + \frac{k_T}{k_1} e^{ik_1(z+d)} \left[\frac{1}{\rho^2} + e^{ik_T \rho} \left(\frac{1}{\rho^2} + \frac{2i}{k_T \rho^3} \right) \right] \bigg\}, \end{aligned} \tag{9.6.39}$$

$$\begin{aligned} G_{\phi 3}(\rho, z+d) = 2k_1^{-1} \bigg[e^{ik_1 r_2} \left(\frac{1}{\rho^3} + \frac{3i}{2k_1 \rho^4} \right) \\ - k_T \left(\frac{1}{\rho^2} + \frac{i}{k_T \rho^3} e^{ik_T \rho} \right) e^{ik_1(z+d)} \bigg] \\ + \frac{2k_T^2}{k_1^2} \left(\frac{1}{\rho^2} e^{ik_1(z+d)} - \frac{z+d}{\rho^3} e^{ik_1 r_2} \right) \\ + 2e^{ik_1 r_2} \left(\frac{1}{\rho^2} + \frac{3i}{k_1 \rho^3} \right) \left(\frac{z+d}{\rho} \right). \end{aligned} \tag{9.6.40}$$

When terms of order $k_L k_T/k_1^2$, k_T^2/k_1^2, and k_L^2/k_1^2 have been neglected and (9.6.37) and (9.6.39) have been combined in (9.6.31), the result is

$$\begin{aligned} &G_{\rho 1}(\rho, z+d) \\ &\quad = e^{ik_1(z+d)}\left\{H_{B\rho 1}(\rho,0) + \frac{2k_T}{k_1}\left[\left(\frac{1}{\rho^2} + \frac{2i}{k_T\rho^3}\right)e^{ik_T\rho} + \frac{1}{\rho^2}\,e^{ik_L\rho}\right]\right\} \\ &\qquad + 2e^{ik_1r_2}\left[\frac{i}{\rho^2} - \frac{7}{2k_1\rho^3} + \left(\frac{z+d}{\rho}\right)\left(\frac{ik_1}{\rho} - \frac{5}{\rho^2} - \frac{12i}{k_1\rho^3}\right)\right]. \end{aligned} \tag{9.6.41}$$

With (9.6.36) and (9.6.19),

$$\begin{aligned} &G_{\rho 1}(\rho, z+d) \\ &\quad = \frac{2k_T}{k_1}\,e^{ik_1(z+d)}\left\{\left[\frac{1}{\rho^2} + \frac{ik_Lk_T}{k_1\rho}\left(\frac{\pi}{k_L\rho}\right)^{1/2} e^{-ip_e}\mathcal{F}(p_e)\right]e^{ik_L\rho}\right. \\ &\qquad\qquad \left. + \left(\frac{1}{\rho^2} + \frac{2i}{k_T\rho^3}\right)e^{ik_T\rho}\right\} \\ &\qquad + 2e^{ik_1r_2}\left[\frac{i}{\rho^2} - \frac{7}{2k_1\rho^3} + \left(\frac{z+d}{\rho}\right)\left(\frac{ik_1}{\rho} - \frac{5}{\rho^2} - \frac{12i}{k_1\rho^3}\right)\right]. \end{aligned} \tag{9.6.42}$$

With (9.6.42) and (5.5.26), (9.6.29) becomes

$$\begin{aligned} B_{1\rho}(\rho,\phi,z) &= -\frac{\mu_0}{2\pi k_1}\sin\phi \\ &\times\left(e^{ik_1(z+d)}\left\{e^{ik_L\rho}\left[\frac{k_T}{\rho^2} + \frac{ik_Lk_T^2}{k_1\rho}\left(\frac{\pi}{k_L\rho}\right)^{1/2} e^{-ip_e}\mathcal{F}(p_e)\right]\right.\right. \\ &\qquad \left. + e^{ik_T\rho}\left(\frac{k_T}{\rho^2} + \frac{2i}{\rho^3}\right)\right\} \\ &\quad + e^{ik_1r_2}\left[\frac{ik_1}{\rho^2} - \frac{7}{2\rho^3} + \left(\frac{z+d}{\rho}\right)\left(\frac{ik_1^2}{\rho} - \frac{5k_1}{\rho^2} - \frac{12i}{\rho^3}\right)\right] \\ &\quad \left. - \frac{1}{2}\left(\frac{z-d}{\rho}\,e^{ik_1r_1} + \frac{z+d}{\rho}\,e^{ik_1r_2}\right)\left(\frac{ik_1^2}{\rho} - \frac{k_1}{\rho^2}\right)\right). \end{aligned} \tag{9.6.43}$$

Similarly, the combination of (9.6.38) and (9.6.40) in (9.6.32) and the omission of small terms of the order k_Lk_T/k_1^2, k_T^2/k_1^2, and k_L^2/k_1^2 yield the following expression for $G_{\phi 1}(\rho, z+d)$:

$$\begin{aligned} &G_{\phi 1}(\rho, z+d) \\ &\quad = e^{ik_1(z+d)}\left[H_{B\phi 1}(\rho,0) + \frac{2k_T}{k_1}\left(\frac{ik_L}{\rho} - \frac{1}{\rho^2}\right)e^{ik_L\rho} - \frac{2i}{k_1\rho^3}\,e^{ik_T\rho}\right] \\ &\qquad + \frac{2}{k_1}\,e^{ik_1r_2}\left(\frac{1}{\rho^3} + \frac{3i}{2k_1\rho^4}\right) + 2e^{ik_1r_2}\left(\frac{1}{\rho^2} + \frac{3i}{k_1\rho^3}\right)\left(\frac{z+d}{\rho}\right). \end{aligned} \tag{9.6.44}$$

With (9.6.36) and (9.6.10),

$$G_{\phi 1}(\rho, z+d) = 2e^{ik_1(z+d)}\left[\frac{k_T}{k_1}\left(\frac{ik_L}{\rho} - \frac{1}{\rho^2}\right)e^{ik_L\rho} - \frac{i}{k_T\rho^3}\, e^{ik_T\rho} \right.$$
$$\left. - \frac{k_L^2 k_T^2}{k_1^2}\left(\frac{\pi}{k_L\rho}\right)^{1/2} e^{ik_L\rho}e^{-ip_e}\mathcal{F}(p_e)\right]$$
$$+ 2e^{ik_1r_2}\left[\frac{1}{k_1\rho^3} + \frac{3i}{2k_1^2\rho^4} + \left(\frac{z+d}{\rho}\right)\left(\frac{1}{\rho^2} + \frac{3i}{k_1\rho^3}\right)\right]. \tag{9.6.45}$$

With (9.6.45) and (5.5.26), (9.6.30) becomes

$$B_{1\phi}(\rho,\phi,z) = -\frac{\mu_0}{2\pi k_1}\cos\phi$$
$$\times \left(e^{ik_1(z+d)}\left\{e^{ik_L\rho}\left[\frac{ik_Lk_T}{\rho} - \frac{k_T}{\rho^2} - \frac{k_L^2k_T^2}{k_1}\left(\frac{\pi}{k_L\rho}\right)^{1/2} e^{-ip_e}\mathcal{F}(p_e)\right]\right.\right.$$
$$\left. - \frac{i}{\rho^3}\, e^{ik_T\rho}\right\}$$
$$+ \frac{1}{2}\, e^{ik_1r_2}\left(\frac{2}{\rho^3} + \frac{3i}{k_1\rho^4}\right) - \frac{1}{2}\, e^{ik_1r_1}\left(\frac{z-d}{\rho}\right)\left(\frac{ik_1^2}{\rho} - \frac{k_1}{\rho^2}\right)$$
$$\left. - \frac{1}{2}\, e^{ik_1r_2}\left(\frac{z+d}{\rho}\right)\left(\frac{ik_1^2}{\rho} - \frac{3k_1}{\rho^2} - \frac{6i}{\rho^3}\right)\right). \tag{9.6.46}$$

The integral expression (9.3.37) for $B_{1z}(\rho,\phi,z)$ is identical to the corresponding formula (5.4.34) for an isotropic medium except that the factor γ_2 in P has been changed to γ_T in P_T. It follows that the final formula is

$$B_{1z}(\rho,\phi,z) = \frac{\mu_0}{2\pi}\sin\phi$$
$$\times\left\{\frac{k_T^2}{k_1^2}\left(\frac{1}{\rho^2} + \frac{3i}{k_T\rho^3} - \frac{3}{k_T^2\rho^4}\right)e^{ik_1(z+d)}e^{ik_T\rho}\right.$$
$$- e^{ik_1r_2}\left[\frac{1}{\rho^2} + \frac{3i}{k_1\rho^3} - \frac{3}{k_1^2\rho^4}\right.$$
$$\left. + \left(\frac{z+d}{\rho}\right)\left(\frac{k_1}{\rho} + \frac{3i}{2\rho^2} - \frac{9}{8k_1\rho^3}\right)\right]$$
$$\left. - \frac{1}{2}\left(e^{ik_1r_1} - e^{ik_1r_2}\right)\left(\frac{ik_1}{\rho} - \frac{1}{\rho^2}\right)\right\}. \tag{9.6.47}$$

9.7 Evaluation of $H_{\rho 1}(\rho, 0)$, $H_{\phi 1}(\rho, 0)$, and $H_{z1}(\rho, 0)$

The evaluation of the integrals in this section follows the work of Pan (1985). The function $H_{\rho 1}(\rho, 0)$ is defined in (9.6.8) as follows:

$$\begin{aligned} H_{\rho 1}(\rho, 0) &= -k_L k_T \int_0^\infty \gamma_1^2 \left(\frac{1}{k_1^2 \gamma_L + k_L k_T \gamma_1} - \frac{1}{k_1^2 \gamma_L} \right) \\ &\qquad \times [J_0(\lambda\rho) - J_2(\lambda\rho)] \lambda \, d\lambda \\ &= -\frac{k_L k_T}{2} \int_0^\infty \gamma_1^2 \left(\frac{1}{k_1^2 \gamma_L + k_L k_T \gamma_1} - \frac{1}{k_1^2 \gamma_L} \right) \\ &\qquad \times [H_0^{(1)}(\lambda\rho) - H_2^{(1)}(\lambda\rho)] \lambda \, d\lambda. \end{aligned} \tag{9.7.1}$$

The integrand of (9.7.1) has two pairs of branch points at $\lambda = \pm k_1$ and $\lambda = \pm k_L$ so that the integral contour may be changed to be along the branch cut in the upper λ-plane. Because $\operatorname{Im} k_1 \gg \operatorname{Im} k_L$, only one branch cut from k_L to infinity is necessary. When $|k_L \rho| \geq 3$, the Hankel functions may be approximated as follows:

$$H_n^{(1)}(\lambda\rho) = \left(\frac{2}{\pi\lambda\rho} \right)^{1/2} e^{i(\lambda\rho - \pi/4 - n\pi/2)}, \tag{9.7.2a}$$

$$H_0^{(1)}(\lambda\rho) - H_2^{(1)}(\lambda\rho) = 2 \left(\frac{2}{\pi\lambda\rho} \right)^{1/2} e^{i(\lambda\rho - \pi/4)}. \tag{9.7.2b}$$

Now let

$$\lambda = k_L(1 + i\tau); \qquad d\lambda = i k_L \, d\tau. \tag{9.7.3}$$

The most important contribution to the integral is from the neighborhood of the branch point $\lambda = k_L$ so that

$$\gamma_1 \sim k_1, \tag{9.7.4}$$

$$\gamma_L \sim \pm e^{-i\pi/4} k_L \sqrt{2\tau}. \tag{9.7.5}$$

The plus sign is for the left side of the branch cut and the minus sign for the right side. With (9.7.2)–(9.7.5), (9.7.1) may be rewritten as

$$\begin{aligned} H_{\rho 1}(\rho, 0) &= -2ik_T k_L^2 \left(\frac{2}{\pi k_L \rho} \right)^{1/2} e^{ik_L\rho} \\ &\qquad \times \int_0^\infty \left(\frac{1}{\sqrt{2\tau}} - \frac{\sqrt{2\tau}}{2(\tau - ik_T^2/2k_1^2)} \right) e^{-k_L\rho\tau} \, d\tau \\ &= -\frac{2k_T^3 k_L^2}{k_1^2} \left(\frac{1}{\pi k_L \rho} \right)^{1/2} e^{ik_L\rho} \int_0^\infty \frac{e^{-k_L\rho t^2}}{t^2 - ik_T^2/2k_1^2} \, dt \\ &= -e^{i\pi/4} \frac{k_T^2 k_L^2}{k_1} \left(\frac{2\pi}{k_L \rho} \right)^{1/2} e^{ik_L\rho} \, e^{-ip_e} \operatorname{erfc}[(-ip_e)^{1/2}] \end{aligned}$$

$$= -\frac{2k_T^2 k_L^2}{k_1}\left(\frac{\pi}{k_L\rho}\right)^{1/2} e^{ik_L\rho}\, e^{-ip_e}\, \mathcal{F}(p_e), \tag{9.7.6}$$

where

$$p_e = k_L\rho\left(\frac{k_T^2}{2k_1^2}\right) \tag{9.7.7}$$

and $\mathcal{F}(p_e)$ is defined in (9.4.9) in terms of the Fresnel integral $C_2(p_e) + iS_2(p_e)$ given in (9.4.10).

The function $H_{\phi 1}(\rho,0)$ is defined in (9.6.17) as follows:

$$\begin{aligned} H_{\phi 1}(\rho,0) &= -k_L k_T \int_0^\infty \gamma_1^2\left(\frac{1}{k_1^2\gamma_L + k_L k_T \gamma_1} - \frac{1}{k_1^2\gamma_L}\right) \\ &\qquad \times [J_0(\lambda\rho) + J_2(\lambda\rho)]\lambda\, d\lambda \\ &= -\frac{k_L k_T}{2}\int_0^\infty \gamma_1^2\left(\frac{1}{k_1^2\gamma_L + k_L k_T \gamma_1} - \frac{1}{k_1^2\gamma_L}\right) \\ &\qquad \times [H_0^{(1)}(\lambda\rho) + H_2^{(1)}(\lambda\rho)]\lambda\, d\lambda. \end{aligned} \tag{9.7.8}$$

With a standard formula and the large-argument approximation, the following relations apply:

$$\begin{aligned} H_0^{(1)}(\lambda\rho) + H_2^{(1)}(\lambda\rho) &= \frac{2}{\lambda\rho} H_1^{(1)}(\lambda\rho) \\ &\sim \frac{2}{\lambda\rho}\left(\frac{2}{\pi\lambda\rho}\right)^{1/2} e^{i(\lambda\rho - 3\pi/4)}. \end{aligned} \tag{9.7.9}$$

When (9.7.9) is compared with (9.7.2), it is seen that, with $\lambda \sim k_L$, $H_{\phi 1}(\rho,0)$ can be related to $H_{\rho 1}(\rho,0)$ in (9.7.6) as follows:

$$\begin{aligned} H_{\phi 1}(\rho,0) &= -\frac{i}{k_L\rho} H_{\rho 1}(\rho,0) \\ &= \frac{2ik_L^2 k_T^2}{k_1}\frac{\pi^{1/2}}{(k_L\rho)^{3/2}} e^{ik_L\rho}\, e^{-ip_e}\, \mathcal{F}(p_e). \end{aligned} \tag{9.7.10}$$

The function $H_{z1}(\rho,0)$ is defined in (9.6.25) as follows:

$$H_{z1}(\rho,0) = -2k_L k_T \int_0^\infty \gamma_1\left(\frac{1}{k_1^2\gamma_L + k_L k_T\gamma_1} - \frac{1}{k_1^2\gamma_L}\right) J_1(\lambda\rho)\lambda^2\, d\lambda. \tag{9.7.11}$$

With $J_1(\lambda\rho) = (\lambda\rho/2)[J_0(\lambda\rho) + J_2(\lambda\rho)]$,

$$\begin{aligned} H_{z1}(\rho,0) &= -k_L k_T \rho \int_0^\infty \gamma_1\left(\frac{1}{k_1^2\gamma_L + k_L k_T\gamma_1} - \frac{1}{k_1^2\gamma_L}\right) \\ &\qquad \times [J_0(\lambda\rho) + J_2(\lambda\rho)]\lambda^3\, d\lambda. \end{aligned} \tag{9.7.12}$$

This differs from $H_{\phi 1}(\rho, 0)$ in (9.7.8) by the factor $\rho\lambda^2/\gamma_1$. In the evaluation of $H_{\phi 1}(\rho, 0)$, the approximations $\lambda \sim k_L$ and $\gamma_1 \sim k_1$ are made. Hence, with (9.7.10),

$$\begin{aligned} H_{z1}(\rho, 0) &= \frac{\rho k_L^2}{k_1} H_{\phi 1}(\rho, 0) \\ &= \frac{2ik_L^3 k_T^2}{k_1^2} \left(\frac{\pi}{k_L \rho} \right)^{1/2} e^{ik_L \rho} e^{-ip_e} \mathcal{F}(p_e). \end{aligned} \tag{9.7.13}$$

9.8 Three-Dimensionally Anisotropic Region 2

The case when all three rectangular components of the conductivity of the oceanic crust are different so that $\sigma_x \neq \sigma_y \neq \sigma_z$ is also of importance. Another interesting possibility is that of inclined or not-quite-horizontal layering. These are special cases of a constant tensor conductivity. The solution of these problems involves the asymptotic evaluation of a double integral which, in principle, can also be carried out. However, this has not actually been accomplished.

10 The Propagation of Lateral Electromagnetic Waves in Air over Vertical Discontinuities

10.1 Review of Formulas for the Field in Homogeneous Isotropic Half-Spaces

When a horizontal electric dipole is located at a depth d below the surface of the earth or sea, the electromagnetic field along the boundary ($z = 0$) is dominated—at all but very short distances from the source—by the lateral-wave terms. This is a consequence of the fact that the direct field decays rapidly with distance in lossy media. The lateral-wave field of electric type has the following components in the earth or sea—Region 1 ($z \geq 0,\ k_1$)—when $|k_1| \gg |k_2|$ (see Chapter 5):

$$E_{1\rho}^L(\rho, \phi, z) = A(\phi)\,\frac{k_2}{k_1}\,e^{ik_1 z}e^{ik_2\rho}g(\rho; k_1, k_2), \tag{10.1.1}$$

$$E_{1z}^L(\rho, \phi, z) = -A(\phi)\,\frac{k_2^2}{k_1^2}\,e^{ik_1 z}e^{ik_2\rho}f(\rho; k_1, k_2), \tag{10.1.2}$$

$$B_{1\phi}^L(\rho, \phi, z) = \frac{k_1}{\omega}\,E_{1\rho}^L(\rho, \phi, z), \tag{10.1.3}$$

where

$$A(\phi) \equiv -\frac{\omega\mu_0}{2\pi k_1}\,e^{ik_1 d}\cos\phi \tag{10.1.4}$$

and

$$g(\rho;k_1,k_2) = f(\rho;k_1,k_2) - \frac{i}{k_2\rho^3}$$
$$= \frac{ik_2}{\rho} - \frac{1}{\rho^2} - \frac{i}{k_2\rho^3} - \frac{k_2^3}{k_1}\left(\frac{\pi}{k_2\rho}\right)^{1/2} e^{-ip_1}\mathcal{F}(p_1). \tag{10.1.5}$$

In (10.1.5),

$$p_1 \equiv k_2\rho\left(\frac{k_2^2}{2k_1^2}\right), \tag{10.1.6}$$

$$\mathcal{F}(p_1) = \tfrac{1}{2}(1+i) - C_2(p_1) - iS_2(p_1), \tag{10.1.7a}$$

with

$$C_2(p_1) + iS_2(p_1) = \int_0^{p_1} \frac{e^{it}}{(2\pi t)^{1/2}}\, dt. \tag{10.1.7b}$$

The field along both sides of the boundary is obtained from (10.1.1)–(10.1.3) with $z = 0$ and the boundary conditions. It is

$$E_{2\rho}^L(\rho,\phi,0) = E_{1\rho}^L(\rho,\phi,0);$$
$$k_2^2 E_{2z}^L(\rho,\phi,0) = k_1^2 E_{1z}^L(\rho,\phi,0), \tag{10.1.8}$$
$$B_{2\phi}^L(\rho,\phi,0) = B_{1\phi}^L(\rho,\phi,0). \tag{10.1.9}$$

For large arguments where

$$|p_1| \geq 4 \quad \text{or} \quad |k_2\rho| \geq 8\left|\frac{k_1^2}{k_2^2}\right|, \tag{10.1.10}$$

(10.1.5) reduces to

$$g(\rho;k_1,k_2) \sim f(\rho;k_1,k_2) \sim -\frac{k_1^2}{k_2^2\rho^2}. \tag{10.1.11}$$

The first three terms on the right in (10.1.5) are simple powers of the inverse radial distance; taken together they will be called the *radial or ρ term.* The remaining fourth term is called the *Fresnel term.* A comparison of the Fresnel and radial terms of $E_{1\rho}^L(\rho,0,z)$ with the direct field is given in Fig. 10.1.1 when Region 1 is sea water, Region 2 is air, $d = 7$ mm, and $z = 3.5$ mm at $f = 600$ MHz. The rapid decay of the direct field is clearly evident. It is negligible only a small fraction of the wavelength in air from the source. Beyond this short distance, the radial term of the lateral wave is greatest for about 40 air wavelengths when the Fresnel term becomes comparable and even slightly larger. The relative magnitudes of the lateral-wave terms and the direct wave for $E_{1z}(\rho,0,z)$ and $B_{1\phi}(\rho,0,z)$ are shown in Fig. 10.1.2.

Associated with the field (10.1.1)–(10.1.3) in Region 1 ($z \geq 0,\ k_1$) is the field in Region 2 ($z \leq 0,\ k_2$). It is

$$E_{2\rho}^L(\rho,\phi,z) = A(\phi)e^{ik_2r_0}\left[\left(\frac{z}{r_0}\right)\left(\frac{ik_2}{r_0} - \frac{1}{r_0^2}\right) + \frac{k_2}{k_1}\, g(r_0;k_1,k_2)\right], \tag{10.1.12}$$

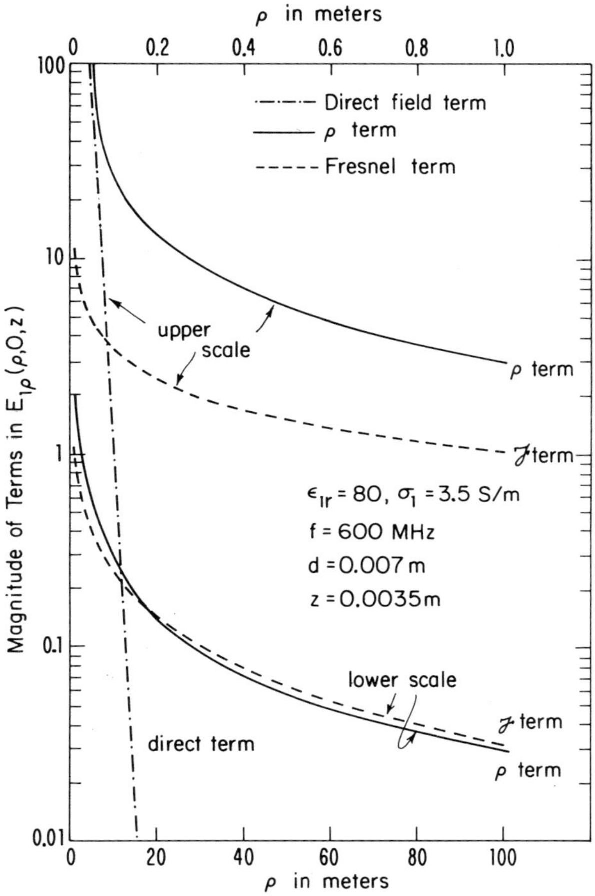

Figure 10.1.1. Magnitudes of the direct-field term and the lateral-wave terms of $E_{1\rho}(\rho, 0, z)$ in dB referred to 1 V/m.

$$E^L_{2z}(\rho, \phi, z) = -A(\phi)e^{ik_2r_0}\left(\frac{\rho}{r_0}\right) f(r_0; k_1, k_2), \tag{10.1.13}$$

$$B^L_{2\phi}(\rho, \phi, z) = A(\phi)\left(\frac{k_2}{\omega}\right)e^{ik_2r_0}g(r_0; k_1, k_2), \tag{10.1.14}$$

where

$$r_0 = (\rho^2 + z^2)^{1/2} \tag{10.1.15}$$

and

$$\begin{aligned} g(r_0; k_1, k_2) &= f(r_0; k_1, k_2) - \frac{i}{k_2 r_0^3} \\ &= \frac{ik_2}{r_0} - \frac{1}{r_0^2} - \frac{i}{k_2 r_0^3} - \frac{k_2}{r_0}(2\pi R_1)^{1/2}e^{-iP_1}\mathcal{F}(P_1). \end{aligned} \tag{10.1.16}$$

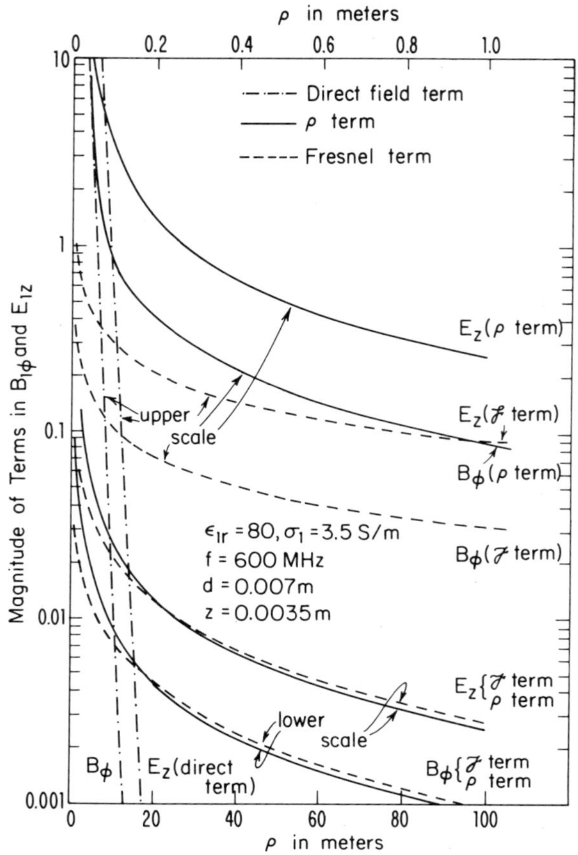

Figure 10.1.2. Magnitudes of the direct-field term and the lateral-wave terms of $E_{1z}(\rho, 0, z)$ and $B_{1\phi}(\rho, 0, z)$ in dB referred to 1 V/m.

In (10.1.16),

$$P_1 = \frac{(R_1 - Z_1)^2}{R_1}, \tag{10.1.17}$$

$$R_1 = p_1 = k_2\rho\left(\frac{k_2^2}{2k_1^2}\right); \qquad Z_1 = k_2 z\left(\frac{k_2}{2k_1}\right). \tag{10.1.18}$$

These formulas are good approximations at all points that are not too close to the source. That is, it is required that

$$k_2 r_0 \geq 4, \tag{10.1.19}$$

in which case the $1/r_0^3$ term is negligible in $g(r_0; k_1, k_2)$, so that

$$g(r_0; k_1, k_2) \sim f(r_0; k_1, k_2). \tag{10.1.20}$$

Note that when $Z_1 \to 0$, it follows that $r_0 \to \rho$, $P_1 \to R_1 = p_1$, and $\mathcal{F}(P_1) \to \mathcal{F}(p_1)$. Subject to (10.1.19), formulas (10.1.12)–(10.1.14) reduce precisely to (10.1.1)–(10.1.3) at $z = 0$.

10.2 Formulation of the Problem When the Properties of Region 1 Are Discontinuous

The purpose in this chapter is to study the propagation of lateral waves across a vertical boundary in Region 1. This is conveniently carried out with the horizontal dipole source located at the depth d at the center of a circular idealized "island" with the radius ρ_b. The island ($\rho \leq \rho_b$) is designated Region 1 with $|k_1| \geq 3|k_2|$. It is surrounded by an infinite "ocean" ($\rho \geq \rho_b$), Region 3, with $|k_3| \geq 3|k_2|$. Over both regions is air (Region 2, k_2) and the boundary between Regions 1 and 3 and Region 2 is the smooth plane, $z = 0$. It is assumed that $|k_2\rho_b| > 4$ so that the distance $\rho = \rho_b$ is sufficiently great to make the field at $\rho = \rho_b$ the lateral-wave field with no significant contributions from direct-field terms.

A rigorous determination of the field at all points in Regions 1, 2, and 3 is difficult because of the presence of a sharp corner. However, the incident lateral wave in the air has very unusual properties that suggest a simple approximate solution. This is directly evident from (10.1.12)–(10.1.14) for the lateral-wave field in air over a half-space of earth or sea. Thus, in the range $k_2\rho \leq 2|k_1^2/k_2^2|$, where the Fresnel-integral term is negligible, $E_{2z}^L(\rho, \phi, z)$ and $B_{2\phi}^L(\rho, \phi, z)$ are completely independent of the properties of the region over which the wave travels. The wave number k_1 appears only in the initial amplitude and the vertical travel in Region 1 to the boundary in $A(\phi)$. A change in the properties of Region 1 between the source and a point of observation in the near or intermediate range can have no effect on $E_{2z}^L(\rho, \phi, z)$ or $B_{2\phi}^L(\rho, \phi, z)$. Except near the boundary where z is small, $E_{2\rho}^L(\rho, \phi, z)$ is also independent of k_1. Sufficiently near to and at the boundary, the small term in (10.1.12) with the factor k_2/k_1 becomes significant and this depends on k_1. This evidently means that when a lateral wave in the air passes over a change in the properties of the earth—as from over an island to over the sea—only $E_{2\rho}^L(\rho, \phi, z)$ quite close to the surface is affected.

With reference to the circular "island" model, the incident field over Region 1, $\rho \leq \rho_b$, unperturbed by the boundary at $\rho = \rho_b$, is given by (10.1.12)–(10.1.14). With (10.1.19) and (10.1.20), the components are

$$[E_{2\rho}^i(\rho, \phi, z)]_1 = A(\phi)e^{ik_2r_0}\left[\left(\frac{z}{r_0}\right)\left(\frac{ik_2}{r_0} - \frac{1}{r_0^2}\right)\right.$$

$$+ \frac{k_2}{k_1} f(r_0; k_1, k_2) \Bigg]; \quad \rho \leq \rho_b, \tag{10.2.1}$$

$$[E^i_{2z}(\rho, \phi, z)]_1 = -A(\phi) e^{ik_2 r_0} \left(\frac{\rho}{r_0} \right) f(r_0; k_1, k_2); \quad \rho \leq \rho_b, \tag{10.2.2}$$

$$[B^i_{2\phi}(\rho, \phi, z)]_1 = A(\phi) \left(\frac{k_2}{\omega} \right) e^{ik_2 r_0} f(r_0; k_1, k_2); \quad \rho \leq \rho_b. \tag{10.2.3}$$

The subscript 1 on the square brackets indicates over Region 1. The corresponding field over Region 3, $\rho \geq \rho_b$, is

$$[E^i_{2\rho}(\rho, \phi, z)]_3 = A(\phi) e^{ik_2 r_0} \Bigg[\left(\frac{z}{r_0} \right) \left(\frac{ik_2}{r_0} - \frac{1}{r_0^2} \right)$$

$$+ \frac{k_2}{k_3} \mathcal{R}_{13}(r_b) f(r_0; k_3, k_2) \Bigg]; \quad \rho \geq \rho_b, \tag{10.2.4}$$

$$[E^i_{2z}(\rho, \phi, z)]_3 = -A(\phi) e^{ik_2 r_0} \left(\frac{\rho}{r_0} \right) \mathcal{R}_{13}(r_b) f(r_0; k_3, k_2); \quad \rho \geq \rho_b, \tag{10.2.5}$$

$$[B^i_{2\phi}(\rho, \phi, z)]_3 = A(\phi) \left(\frac{k_2}{\omega} \right) e^{ik_2 r_0} \mathcal{R}_{13}(r_b) f(r_0; k_3, k_2); \quad \rho \geq \rho_b, \tag{10.2.6}$$

where

$$\mathcal{R}_{13}(r_b) \equiv \frac{f(r_b; k_1, k_2)}{f(r_b; k_3, k_2)}$$

$$= \frac{\dfrac{ik_2}{r_b} - \dfrac{1}{r_b^2} - \dfrac{k_2}{r_b} (2\pi R_{1b})^{1/2} e^{-iP_{1b}} \mathcal{F}(P_{1b})}{\dfrac{ik_2}{r_b} - \dfrac{1}{r_b^2} - \dfrac{k_2}{r_b} (2\pi R_{3b})^{1/2} e^{-iP_{3b}} \mathcal{F}(P_{3b})} \tag{10.2.7}$$

with

$$r_b = (\rho_b^2 + z^2)^{1/2} \tag{10.2.8}$$

and

$$R_{1b} = k_2 \rho_b \left(\frac{k_2^2}{2k_1^2} \right); \qquad R_{3b} = k_2 \rho_b \left(\frac{k_2^2}{2k_3^2} \right), \tag{10.2.9}$$

$$P_{1b} = \frac{(R_{1b} - Z_1)^2}{R_{1b}}; \qquad P_{3b} = \frac{(R_{3b} - Z_3)^2}{R_{3b}}, \tag{10.2.10}$$

$$Z_1 = k_2 z \left(\frac{k_2}{2k_1} \right); \qquad Z_3 = k_2 z \left(\frac{k_2}{2k_3} \right). \tag{10.2.11}$$

The factor $\mathcal{R}_{13}(r_b)$ is introduced to make the field continuous at $\rho = \rho_b$. Note that when $|R_{1b}| \leq 1$, $|R_{3b}| \leq 1$, the Fresnel-integral terms are small and

$$|\mathcal{R}_{13}(r_b)| \sim 1; \quad \mathcal{R}_{13}(r_b) f(r_0; k_3, k_2) \sim f(r_0; k_1, k_2) \sim f(r_0, k_2). \tag{10.2.12}$$

Also, when $|R_{1b}| \geq 4$, $|R_{3b}| \geq 4$,

$$\mathcal{R}_{13}(r_b) \sim \frac{k_1^2}{k_3^2}; \quad \mathcal{R}_{13}(r_b) f(r_0; k_3, k_2) \sim f(r_0; k_1, k_2) = -\frac{k_1^2}{k_2^2 r_0^2}. \tag{10.2.13}$$

Clearly, it is only in the range

$$1 \leq |R_{1b}| \leq 4; \qquad 1 \leq |R_{3b}| \leq 4, \tag{10.2.14}$$

that the wave number k_3 appears in $\mathcal{R}_{13}(r_b) f(r_0; k_3, k_2)$. Outside this range, the components $[E^i_{2z}(\rho, \phi, z)]_3$ and $[B^i_{2\phi}(\rho, \phi, z)]_3$ are independent of k_3. The same is true of $[E^i_{2\rho}(\rho, \phi, z)]_3$ except close to the boundary where $(z/r_0) \lesssim |k_2/k_3|$. Here, the radial fields given by (10.2.1) and (10.2.4) are discontinuous at $\rho = \rho_b$ and therefore incomplete. There must be added terms due to the field scattered from the boundary between Regions 1 and 3 that combine with $[E^i_{2\rho}(\rho, \phi, z)]_1$ and $[E^i_{2\rho}(\rho, \phi, z)]_3$ to yield a continuous $E_{2\rho}(\rho, \phi, z)$ at $\rho = \rho_b$.

The field in Regions 1 and 3 generated by the incident field follows directly from that in Region 2 with the appropriate boundary condition at $z = 0$. Thus,

$$E^i_{1\rho}(\rho, \phi, z) = [E^i_{2\rho}(\rho, \phi, 0)]_1 e^{ik_1 z} = A(\phi) \frac{k_2}{k_1} e^{ik_2\rho} e^{ik_1 z} f(\rho; k_1, k_2), \tag{10.2.15}$$

$$E^i_{1z}(\rho, \phi, z) = \frac{k_2^2}{k_1^2} [E^i_{2z}(\rho, \phi, 0)]_1 e^{ik_1 z} = -A(\phi) \frac{k_2^2}{k_1^2} e^{ik_2\rho} e^{ik_1 z} f(\rho; k_1, k_2), \tag{10.2.16}$$

$$B^i_{1\phi}(\rho, \phi, z) = [B^i_{2\phi}(\rho, \phi, 0)]_1 e^{ik_1 z} = \frac{k_1}{\omega} E^i_{1\rho}(\rho, \phi, z), \tag{10.2.17}$$

$$E^i_{3\rho}(\rho, \phi, z) = [E^i_{2\rho}(\rho, \phi, 0)]_3 e^{ik_3 z} = A(\phi) \frac{k_2}{k_3} e^{ik_2\rho} e^{ik_3 z} \mathcal{R}_{13}(\rho_b) f(\rho; k_3, k_2), \tag{10.2.18}$$

$$E^i_{3z}(\rho, \phi, z) = \frac{k_2^2}{k_3^2} [E^i_{2z}(\rho, \phi, 0)]_3 e^{ik_3 z} = -A(\phi) \frac{k_2^2}{k_3^2} e^{ik_2\rho} e^{ik_3 z} \mathcal{R}_{13}(\rho_b) f(\rho; k_3, k_2), \tag{10.2.19}$$

$$B^i_{3\phi}(\rho, \phi, z) = [B^i_{2\phi}(\rho, \phi, 0)]_3 e^{ik_3 z} = \frac{k_3}{\omega} E^i_{3\rho}(\rho, \phi, z). \tag{10.2.20}$$

The general boundary conditions for the total field—incident plus scattered fields—at $\rho = \rho_b$, $z \geq 0$, between Regions 1 and 3 are

$$k_1^2 E^t_{1\rho}(\rho_b, \phi, z) - k_3^2 E^t_{3\rho}(\rho_b, \phi, z) = 0, \tag{10.2.21}$$

$$E^t_{1z}(\rho_b, \phi, z) - E^t_{3z}(\rho_b, \phi, z) = 0, \tag{10.2.22}$$

$$B^t_{1\phi}(\rho_b, \phi, z) - B^t_{3\phi}(\rho_b, \phi, z) = 0. \tag{10.2.23}$$

The equivalent relations for the incident field alone, viz., (10.2.15)–(10.2.20), are

$$k_1^2 E^i_{1\rho}(\rho_b, \phi, z) - k_3^2 E^i_{3\rho}(\rho_b, \phi, z)$$
$$= A(\phi) k_2 e^{ik_2\rho_b} f(\rho_b; k_1, k_2)(k_1 e^{ik_1 z} - k_3 e^{ik_3 z}), \tag{10.2.24}$$
$$E^i_{1z}(\rho_b, \phi, z) - E^i_{3z}(\rho_b, \phi, z)$$
$$= -A(\phi) k_2^2 e^{ik_2\rho_b} f(\rho_b; k_1, k_2)\left(\frac{e^{ik_1 z}}{k_1^2} - \frac{e^{ik_3 z}}{k_3^2}\right), \tag{10.2.25}$$
$$B^i_{1\phi}(\rho_b, \phi, z) - B^i_{3\phi}(\rho_b, \phi, z)$$
$$= A(\phi) \frac{k_2}{\omega} e^{ik_2\rho_b} f(\rho_b; k_1, k_2)(e^{ik_1 z} - e^{ik_3 z}). \tag{10.2.26}$$

These are physically correct boundary conditions for the two large components E^i_ρ and B^i_ϕ if a fictitious surface density of current, $K_z(\rho_b, \phi, z)$, is located on the boundary surface. That is,

$$B^i_{1\phi}(\rho_b, \phi, z) - B^i_{3\phi}(\rho_b, \phi, z) = -\mu_0 K_z(\rho_b, \phi, z) \tag{10.2.27}$$

and

$$\tilde{\epsilon}_1 E^i_{1\rho}(\rho_b, \phi, z) - \tilde{\epsilon}_3 E^i_{3\rho}(\rho_b, \phi, z) = \frac{i}{\omega}\frac{\partial}{\partial z} K_z(\rho_b, \phi, z) \tag{10.2.28}$$

or

$$k_1^2 E^i_{1\rho}(\rho_b, \phi, z) - k_3^2 E^i_{3\rho}(\rho_b, \phi, z) = i\omega\mu_0 \frac{\partial}{\partial z} K_z(\rho_b, \phi, z). \tag{10.2.29}$$

The last equation is obtained with $k_j^2 = \omega^2\mu_0\tilde{\epsilon}_j$, $j = 1, 3$. With (10.2.26), (10.2.27) gives

$$K_z(\rho_b, \phi, z) = -A(\phi) \frac{k_2}{\omega\mu_0} e^{ik_2\rho_b} f(\rho_b; k_1, k_2)(e^{ik_1 z} - e^{ik_3 z}). \tag{10.2.30}$$

When (10.2.30) is substituted in (10.2.29), this becomes

$$i\omega\mu_0 \frac{\partial}{\partial z} K_z(\rho_b, \phi, z)$$
$$= A(\phi) k_2 e^{ik_2\rho_b} f(\rho_b; k_1, k_2)(k_1 e^{ik_1 z} - k_3 e^{ik_3 z}), \tag{10.2.31}$$

in complete agreement with (10.2.24).

The z-directed currents $K_z(\rho_b, \phi, z)$ given in (10.2.30) generate a scattered field that travels radially in both directions from the boundary $\rho = \rho_b$. It consists of a reflected field that travels radially inward in Region 1 and the transmitted field that travels radially outward in Region 3. It includes the radiated direct field that is by far the largest part of E_z on and near the boundary. However, this is attenuated rapidly with radial distance so that the radiated lateral-wave terms dominate when $k_2(\rho - \rho_b) > 1$ in Region 3 and $k_2(\rho_b - \rho) > 1$ in Region 1.

10.3 The Scattered Field in Regions 1 and 3

The field scattered at the boundary, $\rho = \rho_b,\ z \geq 0$, in the form of a reflected field in Region 1 ($\rho \leq \rho_b,\ z \geq 0$) and a transmitted field in Region 3 ($\rho \geq \rho_b$, $z \geq 0$) is generated by the currents $K_z(\rho_b, \phi, z)$ on the boundary surface. Since these currents are z-directed, they can be represented by a suitable distribution of vertical electric dipoles over the entire boundary surface, $\rho = \rho_b,\ z \geq 0$. The lateral-wave field of a unit vertical electric dipole at (ρ_b, ϕ', z') on the boundary has the following three components at the radial distance R and depth z. In Region 1, $\rho < \rho_b$, they are

$$E_{1R}^L(R, z) = -\frac{\omega\mu_0 k_2^2}{2\pi k_1^3}\, e^{ik_1(z+z')} e^{ik_2R} f(R; k_1, k_2), \tag{10.3.1}$$

$$E_{1z}^L(R, z) = \frac{\omega\mu_0 k_2^3}{2\pi k_1^4}\, e^{ik_1(z+z')} e^{ik_2R} g(R; k_1, k_2), \tag{10.3.2}$$

$$B_{1\phi}^L(R, z) = \frac{k_1}{\omega}\, E_{1\rho}^L(R, z). \tag{10.3.3}$$

At distances $k_2R \geq 1$, $g(R; k_1, k_2) \sim f(R; k_1, k_2)$. In Region 3, $\rho > \rho_b$, the three components are like (10.3.1)–(10.3.3) with k_1 replaced by k_3. The functions $f(R; k_1, k_2)$ and $g(R; k_1, k_2)$ are defined in (10.1.5) and (10.1.6) with ρ replaced by R, where

$$R = [\rho_b^2 + \rho^2 - 2\rho\rho_b \cos(\phi - \phi')]^{1/2}. \tag{10.3.4}$$

The distance from the dipole at (ρ_b, ϕ', z') to the point of observation at (ρ, ϕ, z) is $[R^2 + (z - z')^2]^{1/2}$. The dipole also generates a direct field, but it is assumed that the point of observation is sufficiently far from the boundary so that this is negligible.

The electric moment due to the element of current $dK_z(\rho_b, \phi', z') = K_z(\rho_b, \phi', z')\rho_b\, d\phi'\, dz'$ is given by

$$dK_z(\rho_b, \phi', z') = -\mu_0^{-1} B_{2\phi}(\rho_b, 0, 0)(e^{ik_1z'} - e^{ik_3z'})\rho_b \cos\phi'\, d\phi'\, dz', \tag{10.3.5}$$

where $B_{2\phi}(\rho_b, 0, 0)$ is given by (10.2.3) or (10.2.6) with $\rho = \rho_b$, $\phi = 0$, and $z = 0$. With (10.2.17) and (10.2.20),

$$B_{2\phi}(\rho_b, 0, 0) = \frac{k_1}{\omega}\, E_{1\rho}^i(\rho_b, 0, 0) = \frac{k_3}{\omega}\, E_{3\rho}^i(\rho_b, 0, 0). \tag{10.3.6}$$

The scattered radial electric field in the two regions due to a vertical line source that extends from $z' = 0$ to $z' = \infty$ at ρ_b, ϕ' and that has the width $\rho_b\, d\phi'$ is

$$dE_{1R}^s(R, z) = -\frac{k_1}{\omega\mu_0}\, E_{1\rho}^i(\rho_b, 0, 0) \int_0^\infty E_{1R}^L(R, z)(e^{ik_1z'} - e^{ik_3z'})$$

$$\times \rho_b \cos\phi'\, d\phi'\, dz'; \qquad \rho \le \rho_b, \tag{10.3.7}$$

$$dE^s_{3R}(R,z) = -\frac{k_3}{\omega\mu_0} E^i_{3\rho}(\rho_b,0,0) \int_0^\infty E^L_{3R}(R,z)(e^{ik_1z'} - e^{ik_3z'}) \times \rho_b \cos\phi'\, d\phi'\, dz'; \qquad \rho \ge \rho_b. \tag{10.3.8}$$

Here $E^L_{1R}(R,z)$ is given by (10.3.1) and $E^L_{3R}(R,z)$ by (10.3.1) with k_1 replaced by k_3. The integrations with respect to z' are elementary and the incremental fields are

$$dE^s_{1R}(R,z) = \frac{ik_2^2(k_3-k_1)}{4\pi k_1^3(k_3+k_1)} E^i_{1\rho}(\rho_b,0,0) \times e^{ik_1z}e^{ik_2R} f(R;k_1,k_2)\rho_b\cos\phi'\,d\phi'; \qquad \rho \le \rho_b, \tag{10.3.9}$$

$$dE^s_{3R}(R,z) = \frac{ik_2^2(k_3-k_1)}{4\pi k_3^3(k_3+k_1)} E^i_{3\rho}(\rho_b,0,0) \times e^{ik_3z}e^{ik_2R} f(R;k_3,k_2)\rho_b\cos\phi'\,d\phi'; \qquad \rho \ge \rho_b. \tag{10.3.10}$$

In order to simplify the determination of the complete field due to all of the line sources around the cylindrical boundary $\rho = \rho_b$, let the field be determined specifically on the radial line $\phi = 0$ along which $E^i_{1\rho}(\rho,\phi,z)$ due to the horizontal electric dipole at $(0,0,d)$ has its maximum. Along this line, the resultant electric field due to pairs of line sources at ϕ' and $-\phi'$ is directed radially inward toward the origin in Region 1, radially outward in Region 3, as shown in Fig. 10.3.1. The actual evaluation of the scattered radial electric fields in Regions 1 and 3 is somewhat involved. The details are available in (King 1984). The final formulas for the normalized scattered electric field are

$$\begin{aligned} e_{1\rho}(\rho,0,z) &= \frac{E^s_{1\rho}(\rho,0,z)}{E^i_{1\rho}(\rho_b,0,z)} \\ &= \frac{ik_2^2(k_3-k_1)}{2\pi k_1^3(k_3+k_1)} \\ &\quad \times \int_0^\pi e^{ik_2A}\Bigg\{\left(\frac{ik_2}{R} - \frac{1}{R^2}\right)\cos k_2B \\ &\qquad - \frac{k_2^3}{k_1}\left(\frac{\pi}{k_2R}\right)^{1/2} e^{-ik_2A(k_2^2/2k_1^2)}\mathcal{F}(p_{1R}) \\ &\qquad \times \cos\left[k_2B\left(1-\frac{k_2^2}{2k_1^2}\right)\right]\Bigg\} \\ &\quad \times \rho_b\cos\psi'\cos\phi'\,d\phi'; \qquad \rho \le \rho_b, \end{aligned} \tag{10.3.11}$$

$$e_{3\rho}(\rho,0,z) = \frac{E^s_{3\rho}(\rho,0,z)}{E^i_{3\rho}(\rho_b,0,z)}$$

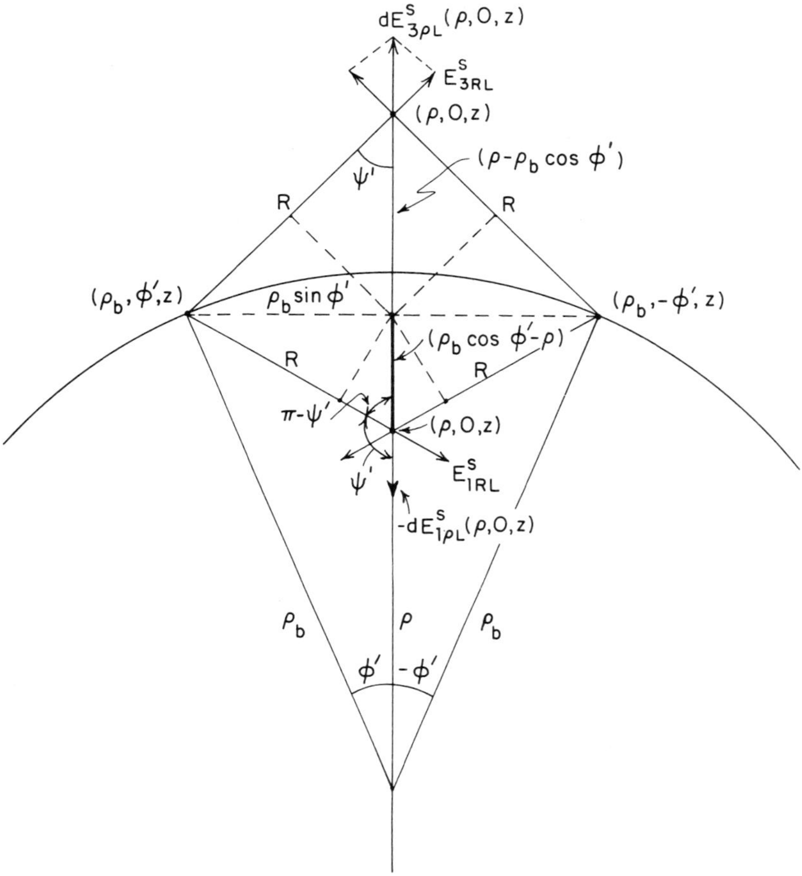

Figure 10.3.1. Notation with respect to two points of observation at $(\rho, 0, z)$ with $\rho < \rho_b$ and $\rho > \rho_b$.

$$\begin{aligned}
&= \frac{ik_2^2(k_3 - k_1)}{2\pi k_3^3(k_3 + k_1)} \\
&\quad \times \int_0^\pi e^{ik_2 A}\left\{\left(\frac{ik_2}{R} - \frac{1}{R^2}\right)\cos k_2 B \right. \\
&\qquad\qquad - \frac{k_2^3}{k_3}\left(\frac{\pi}{k_2 R}\right)^{1/2} e^{-ik_2 A(k_2^2/2k_3^2)}\mathcal{F}(p_{3R}) \\
&\qquad\qquad \left. \times \cos\left[k_2 B\left(1 - \frac{k_2^2}{2k_3^2}\right)\right]\right\} \\
&\qquad \times \rho_b \cos\psi' \cos\phi' \, d\phi'; \qquad \rho \geq \rho_b,
\end{aligned} \tag{10.3.12}$$

where

$$R = (\rho_b^2 + \rho^2 - 2\rho\rho_b \cos\phi')^{1/2}, \tag{10.3.13a}$$

$$p_{1R} = k_2 R\left(\frac{k_2^2}{2k_1^2}\right); \qquad p_{3R} = k_2 R\left(\frac{k_2^2}{2k_3^2}\right), \tag{10.3.13b}$$

and

$$A = (\rho - \rho_b \cos\phi')\cos\psi' = \frac{(\rho - \rho_b\cos\phi')^2}{R}, \tag{10.3.14a}$$

$$B = (\rho_b \sin\phi')\sin\psi' = \frac{(\rho_b \sin\phi')^2}{R}. \tag{10.3.14b}$$

The angle ψ' in Region 1 ($\rho < \rho_b$) and Region 3 ($\rho > \rho_b$) is shown in Fig. 10.3.1.

In these formulas,

$$E_{1\rho}^i(\rho_b, 0, z) = E_{1\rho}^i(\rho_b, 0, 0)e^{ik_1 z} \tag{10.3.15a}$$

and

$$E_{3\rho}^i(\rho_b, 0, z) = E_{3\rho}^i(\rho_b, 0, 0)e^{ik_3 z} \tag{10.3.15b}$$

are the incident radial lateral-wave electric fields that travel down from Region 2 into Regions 1 and 3 along their common boundary $\rho = \rho_b$. Specifically, from (10.2.15) and (10.2.18),

$$E_{1\rho}^i(\rho_b, 0, 0) = A(0)\,\frac{k_2}{k_1}\,e^{ik_2\rho_b} f(\rho_b; k_1, k_2), \tag{10.3.16}$$

$$E_{3\rho}^i(\rho_b, 0, 0) = A(0)\,\frac{k_2}{k_3}\,e^{ik_2\rho_b} f(\rho_b; k_1, k_2). \tag{10.3.17}$$

The total lateral-wave radial electric field along the central radius $\phi = 0$ in each region is the sum of the relevant incident and scattered fields. Thus,

$$\begin{aligned} E_{1\rho}^t(\rho, 0, z) &= E_{1\rho}^i(\rho, 0, z) + E_{1\rho}^s(\rho, 0, z) \\ &= E_{1\rho}^i(\rho, 0, z) + E_{1\rho}^i(\rho_b, 0, z)e_{1\rho}(\rho, 0, z), \end{aligned} \tag{10.3.18}$$

where $E_{1\rho}^i(\rho, 0, z) = E_{1\rho}^i(\rho, 0, 0)\exp(ik_1 z) = [E_{2\rho}^i(\rho, 0, 0)]_1 \exp(ik_1 z)$ and $[E_{2\rho}^i(\rho, 0, 0)]_1$ is given by (10.2.1) with $z = 0$ and $\phi = 0$; $e_{1\rho}(\rho, 0, z)$ is in (10.3.11). Similarly,

$$\begin{aligned} E_{3\rho}^t(\rho, 0, z) &= E_{3\rho}^i(\rho, 0, z) + E_{3\rho}^s(\rho, 0, z) \\ &= E_{3\rho}^i(\rho, 0, z) + E_{3\rho}^i(\rho_b, 0, z)e_{3\rho}(\rho, 0, z), \end{aligned} \tag{10.3.19}$$

where $E_{3\rho}^i(\rho, 0, z) = E_{3\rho}^i(\rho, 0, 0)\exp(ik_3 z) = [E_{2\rho}^i(\rho, 0, 0)]_3 \exp(ik_3 z)$ and $[E_{2\rho}^i(\rho, 0, 0)]_3$ is given by (10.2.18) with $z = 0$ and $\phi = 0$; $e_{3\rho}(\rho, 0, z)$ is in (10.3.12). In expanded form,

$$E_{1\rho}^t(\rho, 0, z) = A(0)e^{ik_1 z}\,\frac{k_2}{k_1}$$

$$\times\,[e^{ik_2\rho}f(\rho;k_1,k_2)+e^{ik_2\rho_b}f(\rho_b;k_1,k_2)e_{1\rho}(\rho,0,z)];$$

$$(\rho_b-\rho)\geq 5z,\quad k_2(\rho_b-\rho)\geq 1, \quad (10.3.20)$$

$$E^t_{3\rho}(\rho,0,z)=A(0)e^{ik_3z}\,\frac{k_2}{k_3}\,\mathcal{R}_{13}(\rho_b)$$

$$\times\,[e^{ik_2\rho}f(\rho;k_3,k_2)+e^{ik_2\rho_b}f(\rho_b;k_3,k_2)e_{3\rho}(\rho,0,z)];$$

$$(\rho-\rho_b)\geq 5z,\quad k_2(\rho-\rho_b)\geq 1. \quad (10.3.21)$$

Note that $\mathcal{R}_{13}(\rho_b)f(\rho_b;k_3,k_2)=f(\rho_b;k_1,k_2)$. In (10.3.20) and (10.3.21),

$$A(0)=-\frac{\omega\mu_0}{2\pi k_1}\,e^{ik_1d}. \quad (10.3.22)$$

10.4 Numerical Evaluation of the Scattered Field; The Total Field

The radial component, $E^i_{1\rho}(\rho,0,z)$, of the incident lateral-wave electric field in Region 1 due to a horizontal electric dipole with unit electric moment at depth $d=0.15$ m has been evaluated also at the depth $z=0.15$ m from (10.2.15) with $\phi=0$ at $f=600$ MHz. Region 1 has the properties of salt water with $\sigma_1=3.5$ S/m and $\epsilon_{1r}=80$ so that $k_1=\beta_1+i\alpha_1=129.4+i64.1$ m^{-1}. It extends from $\rho=0$ to $\rho_b=1.86$ m where it joins Region 3 defined by $\rho\geq\rho_b=1.86$ m. Region 2 is air with $k_2=4\pi$ m^{-1}. The magnitude of $E^i_{1\rho}(\rho,0,z)$ is shown at the top of Fig. 10.4.1 in the range $0.5\leq\rho\leq 1.86$ m. For comparison, the quantity $|E^i_{1\rho}(1\text{ m},0,z)|/\rho$ is also shown. Evidently, the range $0.5\leq\rho\leq 1.86$ m is in the intermediate zone defined by $1\leq k_2\rho\leq|2k_1^2/k_2^2|$ or $0.08\leq\rho\leq 264$ m. In this range, $|E^i_{1\rho}(\rho,0,z)|$ varies as $1/\rho$.

The normalized scattered radial electric field $|e_{1\rho}(\rho,0,z)|=|E^s_{1\rho}(\rho,0,z)/E^i_{1\rho}(\rho_b,0,z)|$ is shown at the bottom in Fig. 10.4.1 with $\rho_b=1.86$ m, $z=0.15$ m, when Region 1 is salt water with $\sigma_1=3.5$ S/m and $\epsilon_{1r}=80$, and with three different Regions 3, viz., lake water with $\sigma_3=0.0035$ S/m and $\epsilon_{3r}=80$, water with $\sigma_3=0.35$ S/m and $\epsilon_{3r}=80$, and metal with $\sigma_3\sim\infty$. The small oscillations are a consequence of the lack of rotational symmetry. The formula (10.3.11) for $e_{1\rho}(\rho,0,z)$ is not valid near the boundary $\rho=\rho_b=1.86$ m because the direct-field terms are not included. The valid range is $0.1\leq\rho\leq 1.76$ m. However, when Region 3 is metal with $\sigma_3\sim\infty$, the reflected field at the boundary $\rho=\rho_b$ is known to be equal in magnitude to the incident field. That is, $|E^s_{1\rho}(\rho_b,0,z)/E^i_{1\rho}(\rho_b,0,z)|=|e_{1\rho}(\rho_b,0,z)|=1$.

What is primarily significant about the scattered field calculated in Fig. 10.4.1 is the rapidity of its decrease with radial distance from the boundary

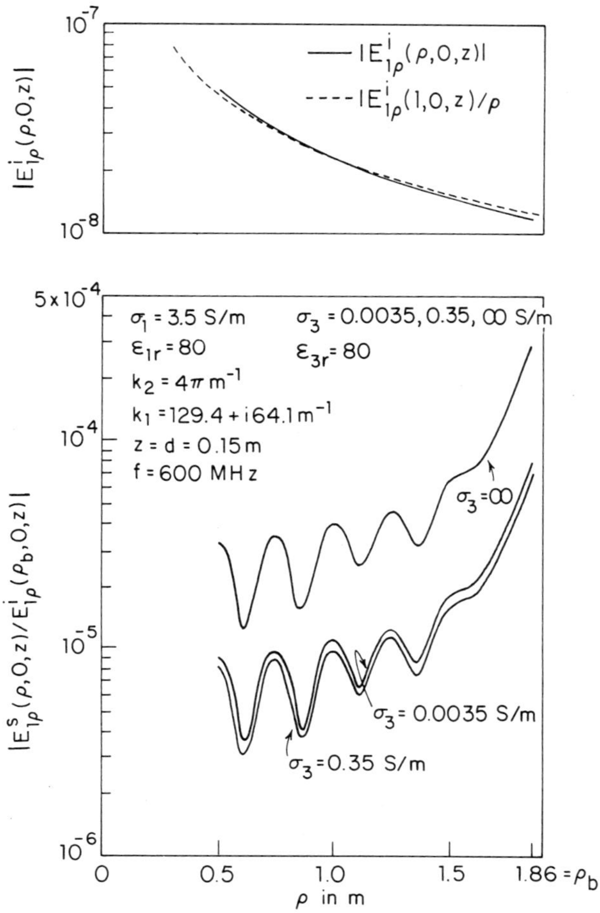

Figure 10.4.1. Incident and scattered fields in Region 1 ($\rho \leq \rho_b$, $z \geq 0$).

to very small values. Even when Region 3 is a perfect conductor, the relative average amplitude of the scattered field when compared to the incident field decreases from 1 at $\rho = \rho_b = 1.86$ m to 10^{-4} at $\rho = 1.76$ m. This value is so small that the resultant total radial electric field differs negligibly from the incident field. The conclusion follows that the reflection of lateral waves from discontinuities in Region 1 or the properties of the half-space $z \geq 0$ is insignificant so long as the ratio $|k_2^2/k_1^2|$ remains small and the scattering discontinuity includes only vertical boundaries within the relatively short distances from the surface $z = 0$ for which $|\exp(ik_1 z)| = \exp(-\alpha_1 z)$ is significant. Since the scattered field transmitted into Region 3 ($\rho \geq \rho_b$) has a magnitude that is comparable to that of the reflected field in Region 1, it is also insignificant at short distances from the boundary $\rho = \rho_b$.

The rapid decay of the scattered field to negligibly small values at quite short radial distances from the boundary indicates that the field given in (10.2.15)–(10.2.20) is a good approximation in Regions 1 and 3 except very close to the boundary $\rho = \rho_b$. Because of the boundary conditions at $z = 0$, it also indicates that the field in (10.2.1)–(10.2.6) is a good approximation in Region 2 except close to the junction point $\rho = \rho_b$, $z = 0$, of Regions 1, 2, and 3.

The rapid decay of the field scattered from the boundary $\rho = \rho_b$ clarifies and justifies the approximation made in its evaluation. This consists primarily of the substitution of the boundary conditions (10.2.24)–(10.2.26) for the actual ones, (10.2.21)–(10.2.23). Evidently, the fictitious sheet of current $K_z(\rho_b, \phi, z)$ does not actually reside on the boundary but is included in the distribution of $J_{1z}(\rho, \phi, z)$, $\rho \leq \rho_b$, and $J_{3z}(\rho, \phi, z)$, $\rho \geq \rho_b$, within short radial distances. With such localized volume distributions of current, $B_{1\phi}(\rho, \phi, z)$ and $B_{3\phi}(\rho, \phi, z)$ are continuous across the boundary $\rho = \rho_b$ but vary rapidly near it. Since $E_z(\rho, \phi, z)$ is directly related to $J_z(\rho, \phi, z)$ by the relations $E_{1z}(\rho, \phi, z) = J_{1z}(\rho, \phi, z)/\sigma_1$ and $E_{3z}(\rho, \phi, z) = J_{3z}(\rho, \phi, z)/\sigma_3$, the actual distribution of $J_{1z}(\rho, \phi, z)$ and $J_{3z}(\rho, \phi, z)$ near $\rho = \rho_b$ will be such that $E_{1z}(\rho_b, \phi, z) = E_{3z}(\rho_b, \phi, z)$ and, therefore, $J_{1z}(\rho_b, \phi, z)/\sigma_1 = J_{3z}(\rho_b, \phi, z)/\sigma_3$. The substitution of localized modifications in the distributions of $J_{1z}(\rho, \phi, z)$ and $J_{3z}(\rho, \phi, z)$ near $\rho = \rho_b$ for $K_z(\rho_b, \phi, z)$ involves a corresponding substitution of a localized modification in $\partial J_{1z}(\rho, \phi, z)/\partial z$ and $\partial J_{3z}(\rho, \phi, z)/\partial z$ for $\partial K_z(\rho_b, \phi, z)/\partial z$ which corresponds to the addition of the scattered-field components $E_{1\rho}^s(\rho, \phi, z)$ and $E_{3\rho}^s(\rho, \phi, z)$ to the incident fields. As a consequence, $k_1^2 E_{1\rho}^t(\rho, \phi, z)$ and $k_3^2 E_{3\rho}^t(\rho, \phi, z)$ are continuous across $\rho = \rho_b$, as required by (10.2.21). This means that $E_{1\rho}(\rho_b, \phi, 0) = E_{3\rho}(\rho_b, \phi, 0) = E_{2\rho}(\rho_b, \phi, 0) = 0$ at the junction $\rho = \rho_b$, $z = 0$, of Regions 1, 2, and 3. Also, $E_{2\rho}(\rho, \phi, z)$ is everywhere continuous but undergoes fairly rapid change near $\rho = \rho_b$, $z = 0$.

With this explanation of the conditions near $\rho = \rho_b$, $z \geq 0$, approximated by the fictitious current $K_z(\rho_b, \phi, z)$ on the boundary between Regions 1 and 3, and the demonstrated very rapid decrease with radial distance of the field scattered at the boundary $\rho = \rho_b$, $z \geq 0$, it can be understood why the incident fields given by (10.2.15)–(10.2.20) are good approximations of the total field in Regions 1 and 3 except in small ranges on each side of the boundary $\rho = \rho_b$ and the incident fields (10.2.1)–(10.2.6) are good approximations of the field in Region 2 except quite near $\rho = \rho_b$, $z = 0$.

The electric field in Region 2 (air) just above the boundary $z = 0$ with Region 1 (salt water) due to a unit horizontal electric dipole in Region 1 at $\rho = 0,\ z \sim 0$, is shown in solid lines in Fig. 10.4.2. In the notation $[E_{2z}(\rho, 0, 0)]_{1h}$ and $[E_{2\rho}(\rho, 0, 0)]_{1h}$, the subscripts $2z$ and 2ρ denote the z and ρ components in Region 2; the subscripts $1h$ stand for a horizontal dipole source in Region 1. The argument $(\rho, 0, 0)$ locates the point of observation at $\rho,\ \phi = 0,\ z = 0$. Note that $E_{2\rho}$ is smaller than E_{2z} by a factor k_2/k_1. Also shown in dashed lines in Fig. 10.4.2 is the electric field in Region 2 (air) just above the boundary $z = 0$ with Region 1 (aluminum) due to a unit vertical electric dipole in Region 2 at $\rho = 0,\ z = 0$. In $[E_{2z}(\rho, 0)]_{2v}$ and $[E_{2\rho}(\rho, 0)]_{2v}$, the subscripts $2v$ denote a vertical dipole source in Region 2. The field of the vertical dipole over aluminum in the range shown is the $1/\rho$ far field. The component $[E_{2\rho}(\rho, 0)]_{2v}$ tangent to the metal surface is, of course, very much smaller (by the factor k_2/k_1) than the component $[E_{2z}(\rho, 0)]_{2v}$ perpendicular to the metal.

When the horizontal dipole is located in Region 1 (salt water) at $(0, 0, 0)$ and Region 1 becomes Region 3 (aluminum) when $\rho \geq \rho_b = 1.86$ m, the simple theory outlined above gives the component $[E_{2z}(\rho, 0, 0)]_{1h}$ shown by the curve on the upper left in Fig. 10.4.3 and, in an enlarged section, by the upper curve in Fig. 10.4.4. The field over salt water is in solid line, that over metal in dashed line. The associated radial component $[E_{2\rho}(\rho, 0, 0)]_{1h}$ is discontinuous at $\rho = \rho_b$ in the simple theory, but is readily approximated near $\rho = \rho_b$ where it must vanish as shown by the curve on the lower left in Fig. 10.4.3.

The inverse situation with a vertical electric dipole on an aluminum Region 1 ($\rho \leq \rho_b = 1.86$ m) and Region 3 (salt water) extending beyond ρ_b ($\rho \geq \rho_b = 1.86$ m) is illustrated on the right in Fig. 10.4.3 and by the lower curve in Fig. 10.4.4.

Sets of theoretical curves similar to those in Figs. 10.4.2–10.4.4 are shown in Fig. 10.4.5 at the much higher frequency $f = 4.765$ GHz. Part (a) in Fig. 10.4.5 shows $|[E_{2z}(\rho, 0)]_{2v}|$ due to a unit vertical dipole on metal and $|[E_{2z}(\rho, 0, 0)]_{1h}|$ due to a unit horizontal dipole in distilled water just below the air surface. In (b) and (c) in the same figure, the graphs for E_{2z} over metal and water are combined, with water (Region 1, $\rho \leq 0.4$ m) and metal (Region 3, $\rho \geq 0.4$ m) in (b), and metal (Region 1, $\rho \leq 0.4$ m) and water (Region 3, $\rho \geq 0.4$ m) in (c). Sections of these theoretical graphs are combined in Fig. 10.4.6 with measured data of King, Cho, and Jaggard (1973) as reported in King and Wait (1976). The measurements were made with the open end of a C-band waveguide serving as both the source of the forward-traveling wave and the receiver of a backward-traveling wave

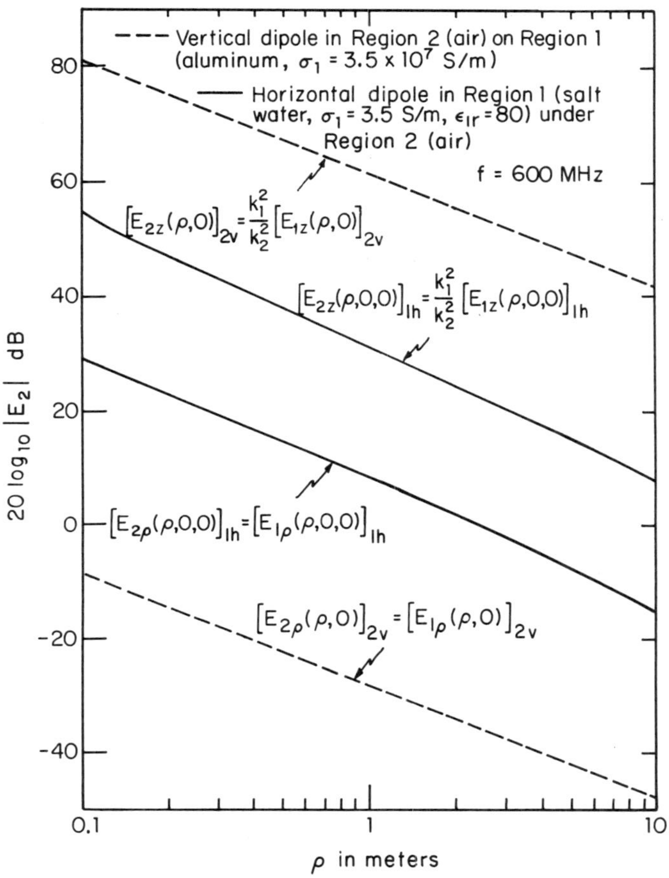

Figure 10.4.2. Electric field in Region 2 (air) due to vertical dipole in air over aluminum (dashed curves) and horizontal dipole in salt water under air (solid curves).

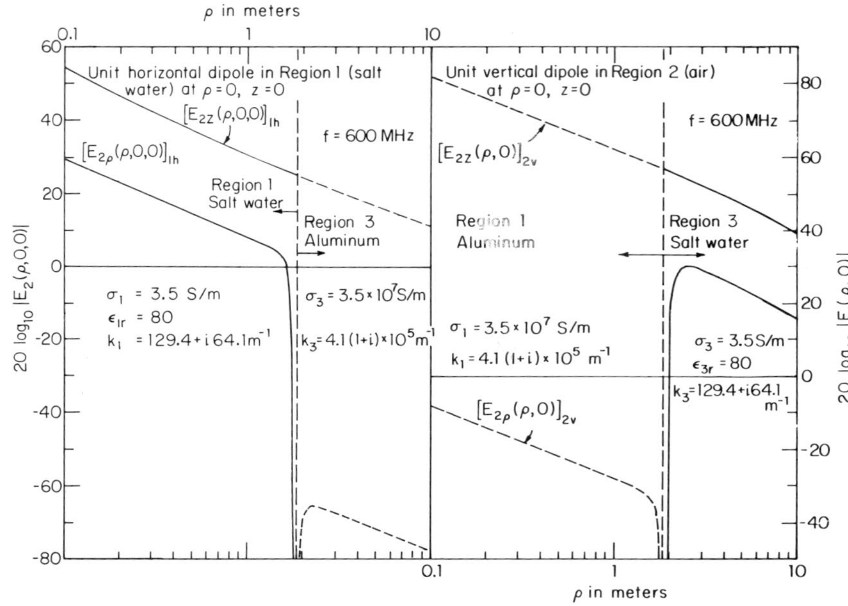

Figure 10.4.3. Electric field in dB in Region 2 (air) over: salt water (Region 1, $\rho \leq 1.86$ m) and aluminum (Region 3, $\rho \geq 1.86$ m) (left); and aluminum (Region 1, $\rho \leq 1.86$ m) and salt water (Region 3, $\rho \geq 1.86$ m) (right). Source at $\rho = 0$, $z = 0$, is: unit horizontal dipole in Region 1, salt water (left); and unit vertical dipole in Region 2, air (right).

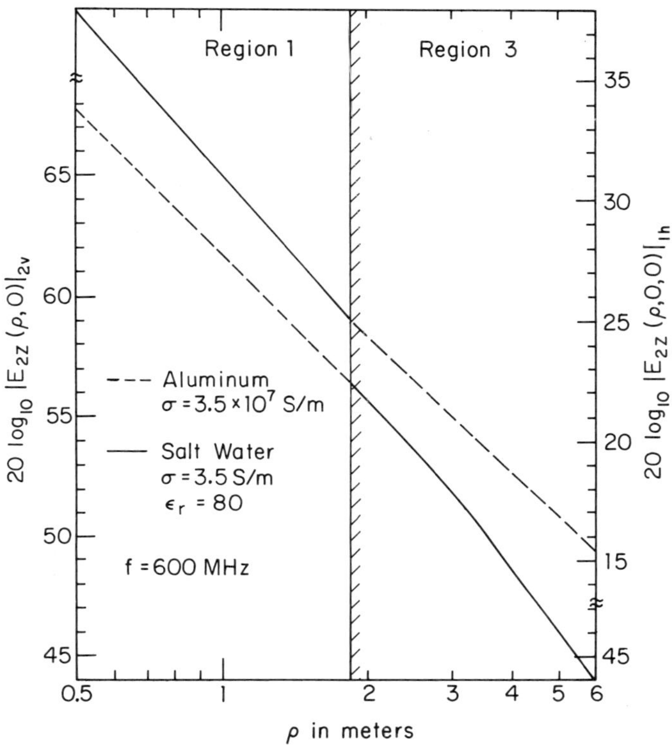

Figure 10.4.4. Vertical electric field in dB in Region 2 (air) over Region 1 ($\rho \leq$ 1.86 m) and Region 3 ($\rho \geq 1.86$ m). (Enlarged section of Fig. 10.4.3.)

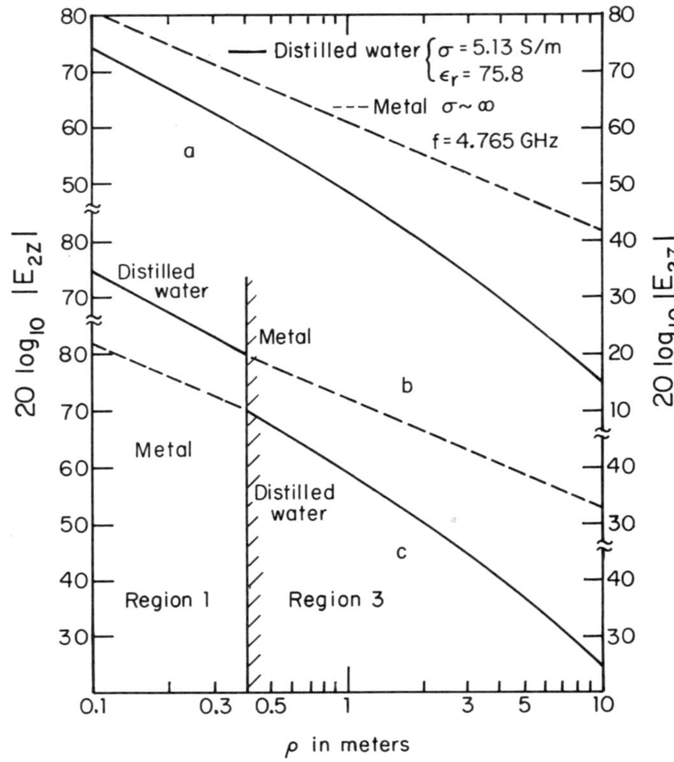

Figure 10.4.5. Vertical electric field in dB in Region 2 (air) over (a) metal alone (vertical dipole in air) and distilled water alone (horizontal dipole in water); (b) distilled water ($\rho \leq 0.4$ m) and metal ($\rho \geq 0.4$ m); and (c) metal ($\rho \leq 0.4$ m) and distilled water ($\rho \geq 0.4$ m).

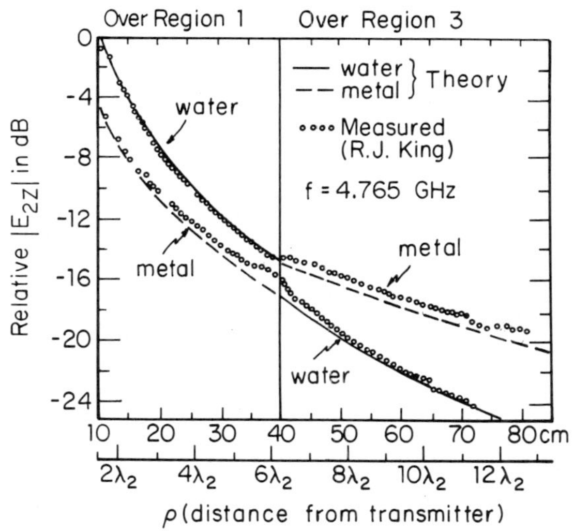

Figure 10.4.6. Theoretical data of Fig. 10.4.5 with measurements of King et al. (1973) in King and Wait (1976).

modulated by and reflected from a diode with short leads end-loaded with brass disks. Its overall length was of the order of $0.1\lambda_2$ to $0.2\lambda_2$ and the disks provided significant coupling to the adjacent metal or water surface. The agreement between the simple theory in (10.2.15)–(10.2.20) and the measured values is quite good except close to the boundary. In the transition from water to metal, there is a local flattening of the experimental curve within a few centimeters of the boundary which leads to an upward displacement of the measured curve from the theoretical curve over the metal but the two curves continue parallel with the required $1/\rho$ decrease. In the transition from metal to water, there is a localized rise in amplitude near the boundary which leaves the measured curve very slightly higher over the water, but again close to and parallel to the theoretical curve.

10.5 Propagation in the Presence of a Metal Wall in the Air; Induced Current and Scattered Field

A second type of vertical discontinuity is presented by a vertical metal wall extending upward from the boundary $z = 0$ between Region 1 ($z \geq 0$) and Region 2 ($z \leq 0$). With a horizontal dipole source at the depth d in Region 1, the component $E_{1z}(\rho, \phi, z)$ is quite small compared with $E_{1\rho}(\rho, \phi, z)$. The reverse is true in Region 2 (air) because of the boundary condition

$$k_1^2 E_{1z}(\rho, \phi, 0) = k_2^2 E_{2z}(\rho, \phi, 0) \tag{10.5.1}$$

and the assumed inequality $|k_1|^2 \gg |k_2|^2$. This suggests that a vertical metal reflecting wall in Region 2 should have a much more significant effect than a similar wall in Region 1. This is readily investigated with a cylindrical metal wall of height L located at $\rho = \rho_b$, $-L \leq z \leq 0$, in Region 2 ($z \leq 0$) over Region 1 ($z \geq 0$).

The field incident on the cylinder is given by (10.1.12)–(10.1.14). When the height L of the cylinder is electrically small so that

$$k_2 L \leq 1.5, \tag{10.5.2}$$

this field reduces to

$$E_{2\rho}(\rho, \phi, 0) = A(\phi)\, \frac{k_2}{k_1}\, e^{ik_2\rho} g(\rho; k_1, k_2), \tag{10.5.3}$$

$$E_{2z}(\rho, \phi, 0) = -A(\phi) e^{ik_2\rho} f(\rho; k_1, k_2), \tag{10.5.4}$$

$$B_{2\phi}(\rho, \phi, 0) = A(\phi)\, \frac{k_2}{\omega}\, e^{ik_2\rho} g(\rho; k_1, k_2), \tag{10.5.5}$$

where $f(\rho; k_1, k_2)$ and $g(\rho; k_1, k_2)$ are given by (10.1.5). This incident field induces a current in the metal cylinder which generates the scattered field.

In order to determine the current induced in the tubular cylinder on salt water, use is made of an analysis by Myers (1962) which determines the current in and the scattered field of an infinitely wide strip of height $2L$ in a normally incident plane electromagnetic field with either E- or H-polarization. Since salt water is a good reflector, the current density induced in the metal cylinder of height L over salt water when $\rho_b \gg L$, $k_2\rho_b \gg 1$, is well approximated by the current induced in the upper half of an isolated infinitely wide strip of height $2L$ in an H-polarized field. Myers reduced the determination of the current density to a differential equation that he solved numerically. The surface density of current on the strip can be expressed in a notation appropriate for the metal tube with the electrically large radius ρ_b and a small height L in the form

$$K_z(\rho_b, 0, z) = K_z(\rho_b, 0, 0) g(z)$$

$$= -\mu_0^{-1} B_{2\phi}^i(\rho_b, 0, 0) K_N(z) e^{i\theta_N(z)}, \tag{10.5.6}$$

where $K_z(\rho_b, 0, 0)$ is the current density at the center of the strip or at the base of the cylinder, $z = 0$. It is tabulated in the work of Myers (1962) in the normalized form

$$K_N(0) e^{i\theta_N(0)} = -\frac{\mu_0 K_z(\rho_b, 0, 0)}{B_{2\phi}^i(\rho_b, 0, 0)} = \frac{K_z(\rho_b, 0, 0)}{H_{2\phi}^i(\rho_b, 0, 0)}. \tag{10.5.7}$$

The numerical values of the amplitude $K_N(0)$ and the argument $\theta_N(0)$ of the normalized current density are shown graphically in Fig. 10.5.1 as a function of L. It is seen that a maximum current is obtained near $k_2 L = 1.5$ or $L \sim \lambda_2/4$. The transverse distribution of current on the strip is $g(z)$. When $k_2 L$ is sufficiently small,

$$g(z) \sim \left(1 - \frac{z^2}{L^2}\right)^{1/2}. \tag{10.5.8}$$

The actual distributions $K_N(z)$ as determined by Myers (1962) for $k_2 L = 1.0$ and 1.5 are shown in Fig. 10.5.2 together with the quantity $K_N(0)(1 - z^2/L^2)^{1/2}$. It is seen that even when $k_2 L$ is as large as 1.5, the approximation is acceptable. The argument $\theta_N(z)$ is not shown, but Myers' results indicate that $\theta_N(z) \sim \theta_N(0)$ within a few percent when $k_2 L \leq 1.5$. It follows that, for $k_2 L \leq 1.5$, the current density on the tubular cylinder is well approximated by

$$K_z(\rho_b, \phi', z') = -\mu_0^{-1} B_{2\phi}^i(\rho_b, 0, 0) K_N(0) e^{i\theta_N(0)} \left(1 - \frac{z'^2}{L^2}\right)^{1/2} \cos\phi', \tag{10.5.9}$$

where $K_N(0)$ and $\theta_N(0)$ are given in Fig. 10.5.1.

The element of surface current on the metal cylinder at (ρ_b, ϕ', z') in Region 2 with $-L \leq z' \leq 0$ is

$$dK_z(\rho_b, \phi', z') = -\mu_0^{-1} B_{2\phi}^i(\rho_b, 0, 0) K_N(0) e^{i\theta_N(0)} \left(1 - \frac{z'^2}{L^2}\right)^{1/2} \times \cos\phi' \, \rho_b \, d\phi' \, dz'. \tag{10.5.10}$$

The scattered radial lateral-wave electric field at the depth z in Region 1 due to a vertical line source in Region 2 extending from $z' = -L$ to $z' = 0$ and of width $\rho_b \, d\phi'$ is

$$[dE_{1RL}^s(R, z)]_2 = -\mu_0^{-1} B_{2\phi}^i(\rho_b, 0, 0) \times \int_{-L}^{0} dz' \, [E_{1RL}^V(R, z)]_2 K_N(0) e^{i\theta_N(0)} \left(1 - \frac{z'^2}{L^2}\right)^{1/2} \times \rho_b \cos\phi' \, d\phi'; \qquad \rho \leq \rho_b, \tag{10.5.11}$$

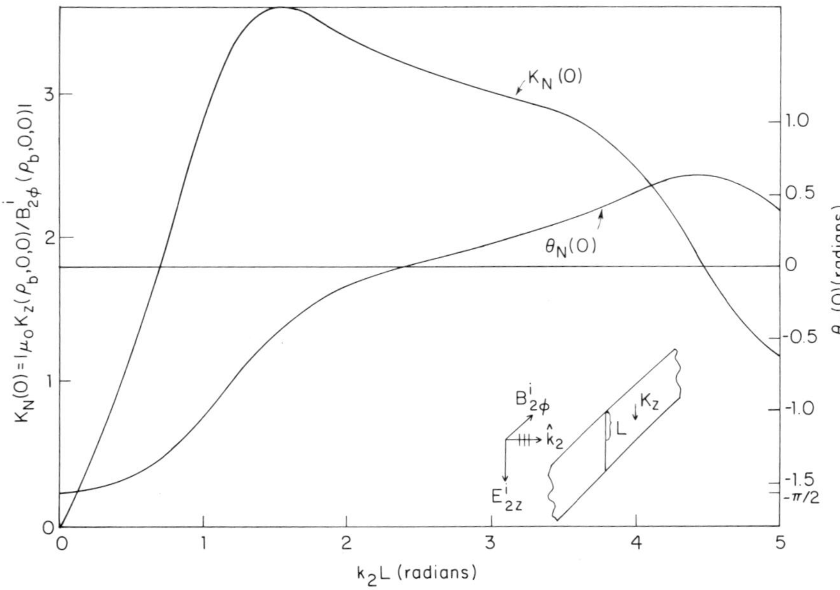

Figure 10.5.1. Current density on infinite strip of height $2L$ in H-polarized field. Data of Myers (1962).

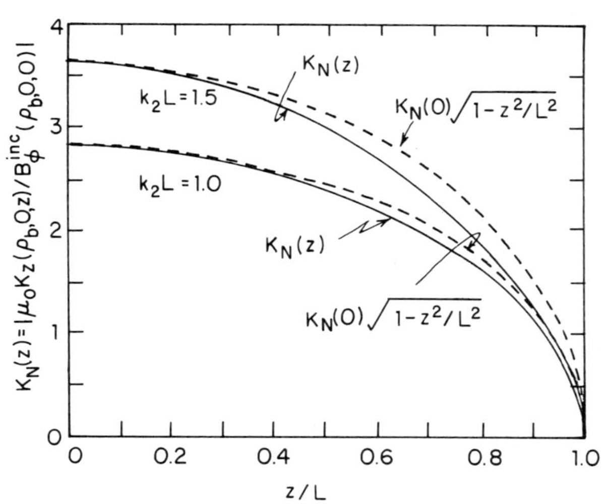

Figure 10.5.2. Actual and approximate normalized distributions of current on strip in H-polarized, normally incident field. Data of Myers (1962).

where $[E^V_{1RL}(R,z)]_2$ is the radial electric field in Region 1 due to a unit vertical electric dipole at $\rho_b, \phi', 0$ in Region 2 (air). It is given by (10.3.1) with $z' = 0$ and multiplied by the factor (k_1^2/k_2^2). With $B^i_{2\phi}(\rho_b, 0, 0) = (k_1/\omega)E^i_{2\rho}(\rho_b, 0, 0)$ and

$$\int_{-L}^{0} \left(1 - \frac{z'^2}{L^2}\right)^{1/2} dz' = -\frac{\pi L}{4}, \tag{10.5.12}$$

(10.5.11) becomes

$$[dE^s_{1RL}(R,z)]_2 = -\frac{L}{8} K_N(0) e^{i\theta_N(0)} E^i_{2\rho}(\rho_b, 0, 0) e^{ik_1 z} e^{ik_2 R} \times f(R; k_1, k_2) \rho_b \cos\phi' \, d\phi'. \tag{10.5.13}$$

This formula can be compared with (10.3.9) with $k_3 \to \infty$, viz.,

$$dE^s_{1R}(R,z) = \frac{ik_2^2}{4\pi k_1^3} E^i_{2\rho}(\rho_b, 0, 0) e^{ik_1 z} e^{ik_2 R} f(R; k_1, k_2) \times \rho_b \cos\phi' \, d\phi'. \tag{10.5.14}$$

The ratio of (10.5.13) to (10.5.14) is

$$\mathcal{R} = \frac{\pi k_2 L}{2} \frac{k_1^3}{k_2^3} K_N(0) e^{i[\theta_N(0) + \pi/2]}. \tag{10.5.15}$$

Since this expression is independent of ϕ', the integral of (10.5.13) for the scattered field in Region 1 due to the sheet of current induced in the metal tube in Region 2 is the same as the integral of (10.5.14) over the cylindrical metal surface bounding Region 1 multiplied by the ratio $\mathcal{R}$. The total radial lateral-wave field along the central radius $\phi = 0$ at (ρ, z) in Region 1 is the sum of the incident and scattered fields. It is

$$E^t_{1\rho}(\rho, 0, z) = E^i_{1\rho}(\rho, 0, z)[1 + \mathcal{R} e_{1\rho}(\rho, 0, z)], \tag{10.5.16}$$

where $e_{1\rho}(\rho, 0, z)$ is given in (10.3.11). It is evaluated for $k_3 = \infty$ in the curve marked $\sigma_3 = \infty$ in Fig. 10.4.1 at $f = 600$ MHz with $\sigma_1 = 3.5$ S/m, $\epsilon_{1r} = 80$, at $z = 15$ cm and $d = 15$ cm for $\rho_b = 1.86$ m. Calculations for the total field given by (10.5.16) have been made with $k_2 L = 0.65$ for which $K_N(0) = 1.63$, $\theta_N(0) = -1.37$, and $\mathcal{R} = 2.5 \times 10^3 \exp(i1.581)$, and with $k_2 L = 1.5$ for which $K_N(0) = 3.60$, $\theta_N(0) = -0.4685$, and $\mathcal{R} = 12.9 \times 10^3 \exp(i2.482)$.

Graphs of the magnitudes of the incident, scattered, and total fields are shown in Figs. 10.5.3 and 10.5.4, respectively, for the two values of $k_2 L$. At $f = 600$ MHz, $L = 5.2$ cm and $L/\lambda_2 \sim 0.1$ in Fig. 10.5.3; $L = 11.9$ cm and $L/\lambda_2 \sim 0.25$ in Fig. 10.5.4. It is seen that at the lower height, the reflected field and the amplitude of the standing waves are quite small. At the greater height (Fig. 10.5.4), the reflected field is much larger and

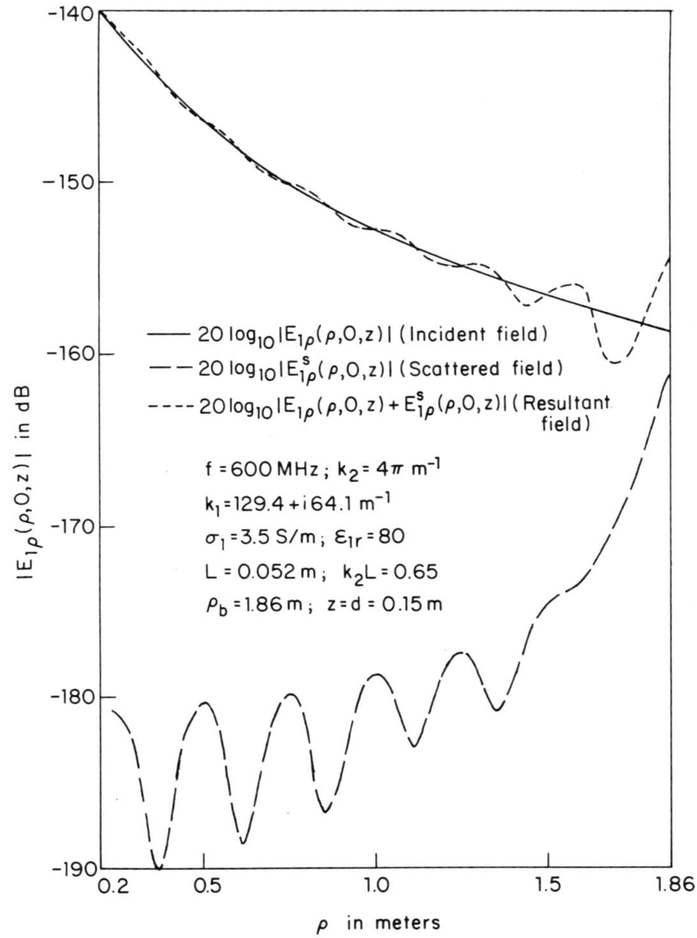

Figure 10.5.3. Standing waves due to reflection from metal wall of height $L = 5.2$ cm $= 0.10\lambda_2$ at $\rho = \rho_b = 1.86$ m.

significant standing waves are produced, especially near the reflecting wall at $\rho = \rho_b = 1.86$ m. This evidently means that the field transmitted across the metal wall is only slightly affected when the wall is no higher than $0.1\lambda_2$, but substantially reduced when the height is $\lambda_2/4$ or greater.

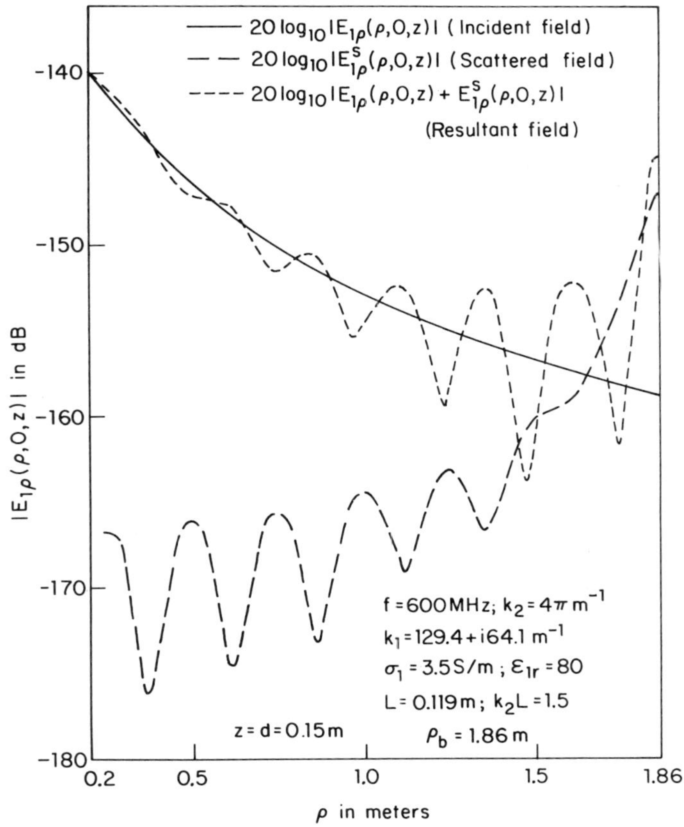

Figure 10.5.4. Standing waves due to reflection from metal wall of height $L = 11.9$ cm $= 0.24\lambda_2$ at $\rho = \rho_b = 1.86$ m.

10.6 Experiments on the Reflection of Lateral Electromagnetic Waves

Because the reflection of lateral waves from any but the simplest obstacles and discontinuities is analytically difficult, it is advantageous to develop supplementary experimental methods to illuminate their observable nature. One method is to select a frequency or range of frequencies that is sufficiently high so that meaningful measurements can be made in a relatively small tank of water with variable salinity (Brown, King, and Wu 1982, 1984). In order to avoid reflections from side walls, a circular tank

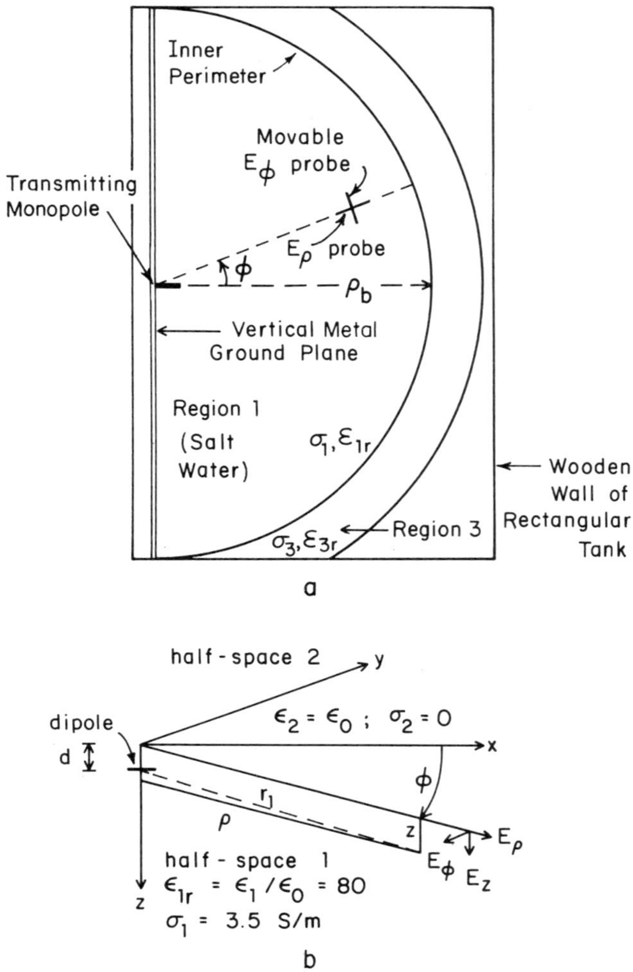

Figure 10.6.1. (a) Schematic diagram of semicircular tank. (b) Field components of Hertzian dipole near planar interface; $I_x \Delta x = 1$ A m.

is convenient with the source at the center. Alternatively, a semicircular tank bounded along the diameter by a large vertical ground plane of metal can be used. Such a tank is illustrated schematically in Fig. 10.6.1. The source is a horizontal monopole that projects from the ground plane into salt water at the small depth d below the air surface. The depth s of the water must be sufficient so that $\exp(-\alpha_1 s) \ll 1$, where α_1 is the imaginary part of the complex wave number $k_1 = \beta_1 + i\alpha_1$ of the water. (For most measurements, the salinity was chosen to make the conductivity $\sigma_1 \sim 3.5$ S/m.) This makes reflections from the floor of the tank negligible. In order to be feasible in the laboratory, the radius of the tank was quite small, viz.,

1.86 m. At the frequency $f = 600$ MHz or $\lambda_2 = 50$ cm, this radial distance is over 3 wavelengths in air; at $f = 1.5$ GHz or $\lambda_2 = 20$ cm, it is over 9 wavelengths. The ceiling of reinforced concrete over the tank was at a height of $D = 2.625$ m or over 5 wavelengths at $f = 600$ MHz and over 13 wavelengths at $f = 1.5$ GHz.

Measurements were made of $E_{1\rho}(\rho, \phi, z)$ with $\phi = 0$ and $z = 5$ mm by means of a movable dipole with a vertical support that rested on the bottom of the tank. The output cable extended vertically down to the floor of the tank and then along this to the edge of the tank and out to a receiver. A typical measured graph of $20 \log_{10} |E_{1\rho}(\rho, 0, z)|$ at $f = 1.5$ GHz is shown later (in Fig. 10.6.5) together with the corresponding theoretical curve. Note that $k_2\rho = 1$ occurs at $\rho = 10$ cm and $k_2\rho = |k_1/k_2|^2$ at $\rho = 2.88$ m so that the entire graph in Fig. 10.6.5 is in the intermediate range where the field decreases with distance as $1/\rho$. The measured curve is quite smooth with a small standing wave appearing beyond $\rho \sim 80$ cm.

With reflections from side walls and the bottom of the tank eliminated, there remain possible reflections from the radial boundary at $\rho = \rho_b = 1.86$ m and from the ceiling at $z = -D = -2.625$ m. In order to visualize the nature of such reflections, approximate analyses can be carried out. Along the radial line $\phi = 0$ (where measurements are made), the maximum effect of reflections at $\rho = \rho_b$ can be evaluated with a vertical metal plane at $\rho = \rho_b$. The standing wave induced by the interaction of the reflected wave and the incident lateral wave is shown at the top in Fig. 10.6.2 at $f = 600$ MHz and at the top in Fig. 10.6.3 at $f = 1.5$ GHz. Similarly, the maximum effect of the reflections from the ceiling can be evaluated with a horizontal metal plane at $|z| = D$. The superposition of the lateral wave and the once-reflected wave is shown in Fig. 10.6.2, curve 3, at $f = 600$ MHz. An alternative approximation of a reflection from the ceiling is a plane wave of suitable amplitude incident on the water surface from above. The interaction of such a wave with the lateral wave is shown in Fig. 10.6.2, curve 4, and in the center of Fig. 10.6.3. The interference pattern resulting from this interaction is seen to be quite different from the standing-wave pattern due to reflections along the radial boundary. The simultaneous interaction of the incident lateral wave with reflections from both the radial boundary and the ceiling is shown at the bottom in Fig. 10.6.3 for $f = 1.5$ GHz. The superposition of an incident lateral wave, reflections from a circular wall of height $L = 0.052$ m at $\rho = \rho_b = 1.86$ m, and a plane wave traveling down from the ceiling for $f = 600$ MHz is shown in Fig. 10.6.4. Figures 10.6.2 through 10.6.4 show the possible standing-wave and interference patterns that may be encountered in the small indoor tank. Because metal reflectors are assumed, the amplitudes of the standing waves are larger than what is to be expected.

Actual measurements of the radial electric field $E_{1\rho}(\rho, 0, z)$ made in the tank represented in Fig. 10.6.1 are shown in Fig. 10.6.5 at $f = 1.5$ GHz. Region 1 ($\rho \leq 1.86$ m) is salt water; Region 3 ($\rho \geq 1.86$ m) is aluminum. The

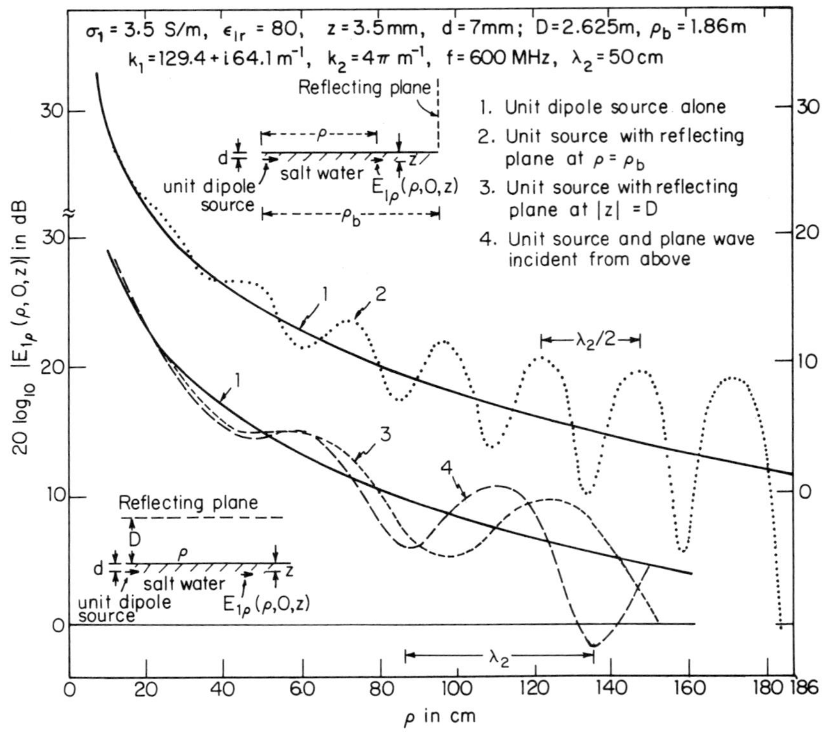

Figure 10.6.2. Lateral-wave field $E_{1\rho}(\rho, \phi, z)$ due to unit horizontal electric dipole at $(0, 0, z)$ in salt water: (1) alone; (2) in presence of perfectly reflecting vertical plane at $\rho = \rho_b$; (3) in presence of perfectly reflecting horizontal plane at $|z| = D$; and (4) in combination with plane wave incident from above.

dotted curves are the theoretical unperturbed incident field. The solid-line curve (a) is the observed field when the water surface goes over smoothly into the metal surface at $\rho = 1.86$ m. The small standing wave is evidently due to reflections from the ceiling since the distance between maxima is λ_2 as in the middle curve in Fig. 10.6.3. The agreement with the unperturbed theoretical curve is very good. For the measurements in curve (b), a circular metal wall with height $L = 15$ cm was located at $\rho = \rho_b$. The resulting standing wave is a superposition of reflections from the ceiling and from the metal wall at $\rho = \rho_b = 1.86$ m. Comparable measurements at $f = 644$ MHz are in Fig. 10.6.6. Here the solid-line curve involves no reflecting wall at $\rho = \rho_b$ and the standing waves are clearly a consequence of ceiling reflections. The long-dashed curve is the field when a circular metal wall of height $L = 5$ cm is placed at $\rho = \rho_b$. Since the distribution remains substantially the same with a full wavelength between successive maxima, as in curve 4 of Fig. 10.6.2, and little indication of a superimposed distribution like that in the upper curve in Fig. 10.6.2 for a reflecting wall

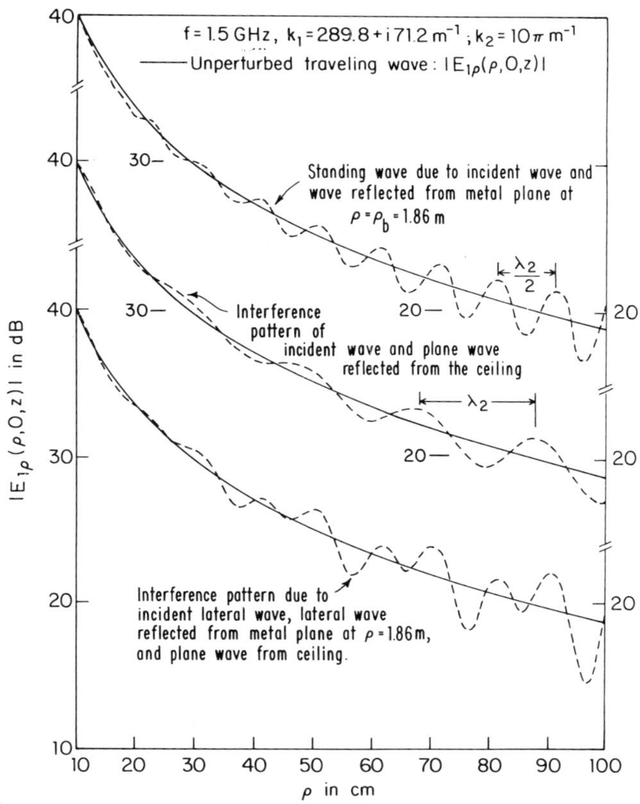

Figure 10.6.3. Standing-wave and interference patterns due to reflections from a metal plate at $\rho = 1.86$ m and from the ceiling; $f = 1.5$ GHz.

at $\rho = \rho_b$, it may be concluded that a reflecting wall only 5 cm or $0.11\lambda_2$ in height has no significant effect on the field. This agrees with the prediction based on theory at the end of Section 10.5. When the height of the reflecting wall is raised to $L = 15$ cm, the measured field shown in the lower curve in Fig. 10.6.7 is obtained. This is clearly dominated by the type of distribution characteristic of reflection from a wall at $\rho = \rho_b$ and illustrated by the upper curve in Fig. 10.6.2. The effect of the reflected wave from the ceiling is clearly evident in the alternately larger and smaller maxima. It is evident that a wall of height $L = 15$ cm or $L/\lambda_2 = 0.32$ is an effective reflector. The upper curve in Fig. 10.6.7 is the measured field when $L = 95$ cm or $L/\lambda_2 = 2.0$. The reflecting wall is now so high that it acts virtually like an infinitely high wall. The large standing wave due to reflections from the wall at $\rho = \rho_b$ is hardly altered by the relatively small superimposed reflections from the ceiling.

It is seen from Figs. 10.6.2 through 10.6.7 that useful measurements of lateral-wave fields can be made in a small indoor tank in spite of not

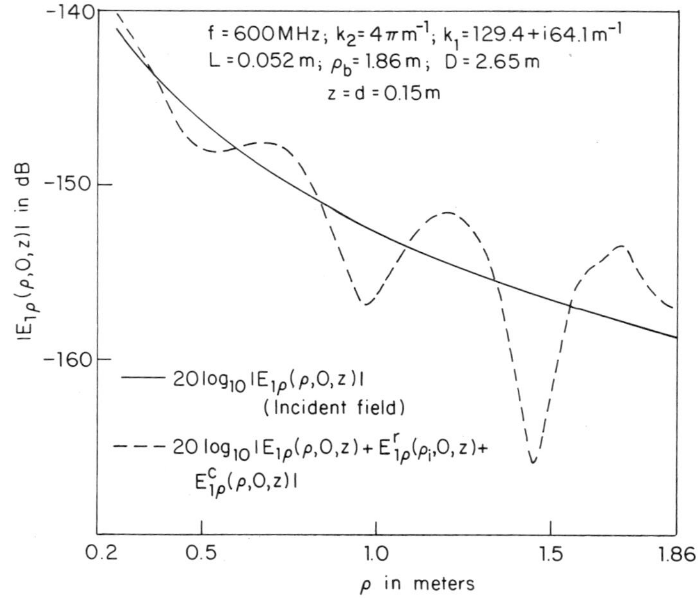

Figure 10.6.4. Superposition of incident lateral wave, lateral wave reflected from vertical metal wall of height L, and plane wave reflected from ceiling; $f = 600$ MHz.

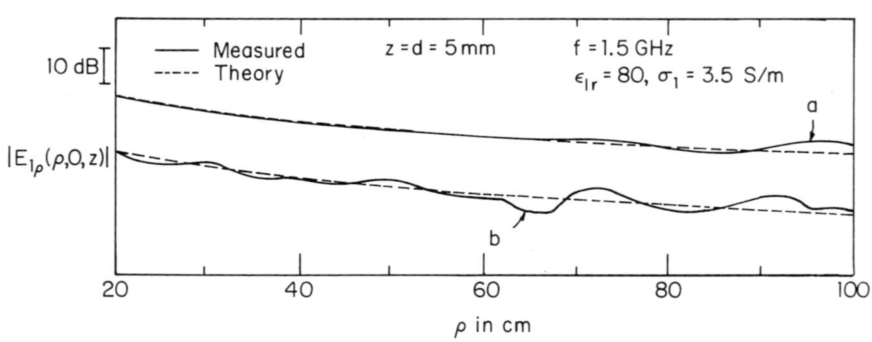

Figure 10.6.5. 20 $\log_{10} |E_{1\rho}(\rho, 0, 0)|$ in Region 1 ($\rho \leq 186$ cm, salt water) when (a) Region 3 ($\rho \geq 186$ cm) is aluminum; (b) like (a) but with vertical circular metal wall of height $L = 15$ cm in air at $\rho = 186$ cm. (Measured by M. F. Brown.)

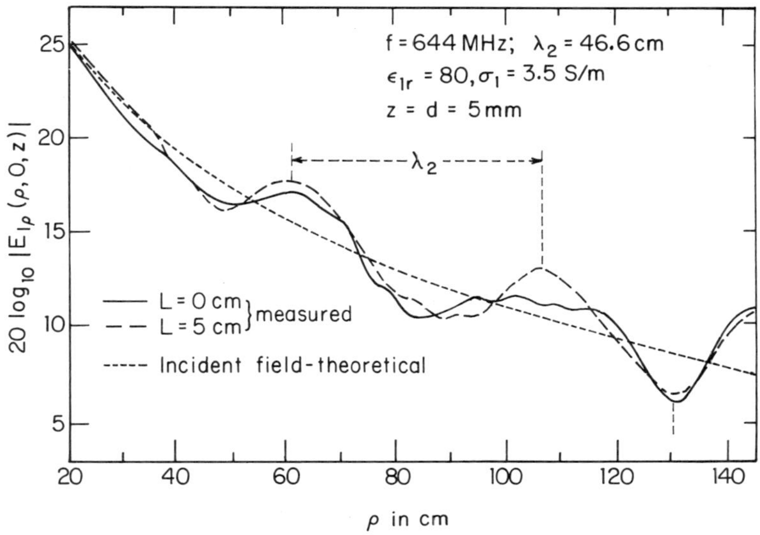

Figure 10.6.6. Radial electric field in dB in Region 1 (salt water) with vertical metal wall in Region 1 and of height L in Region 2 at $\rho = \rho_b = 186$ cm. (Measured by M. F. Brown.)

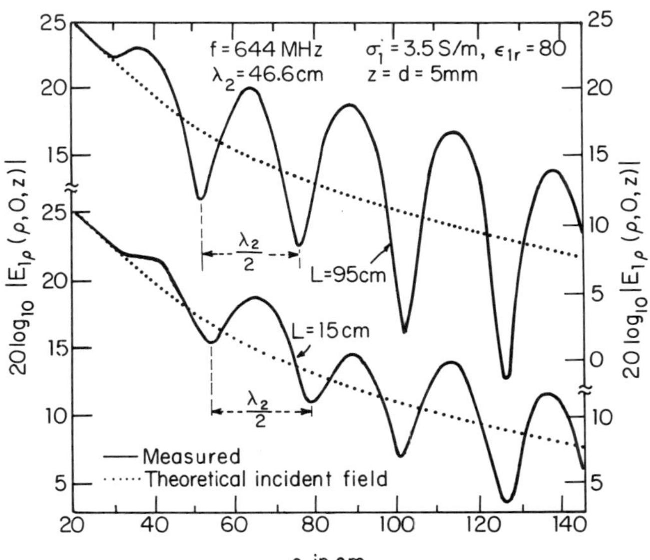

Figure 10.6.7. Measured radial electric field in dB in salt water with circular metal wall of height L at $\rho = \rho_b = 186$ cm. (Measured by M. F. Brown.)

insignificant reflections from the ceiling. A prerequisite to the correct interpretation of measured data is a comprehensive understanding of the nature of lateral-wave fields and their reflections in vertical and horizontal boundaries.

11 The Horizontally Layered Half-Space

11.1 Description and Notation; Reflection Coefficients

The theoretical and experimental discussions in the preceding chapters are concerned with the electromagnetic fields generated by dipoles near the boundary between two homogeneous but electrically different half-spaces. Both half-spaces are also assumed to be isotropic except in Chapter 9 in which one of the half-spaces is anisotropic in the conductivity with the horizontal values, $\sigma_x = \sigma_y$, different from the vertical value, σ_z. A more general form of horizontal layering is when one of the half-spaces is composed of a succession of n horizontally bounded layers, each with arbitrary thickness l_j, $j = 1, 2, \ldots, n$, and arbitrary wave number k_j, as illustrated in Fig. 11.1.1. This shows a homogeneous, isotropic half-space (Region 0, $z \geq 0$) above a half-space with n horizontal layers (Regions 1 to n, $-\sum_{j=1}^{n} l_j \leq z \leq 0$) of which each is homogeneous and isotropic but the wave number is discontinuous across the boundaries. Below the nth layer (Region $n+1$, $z \leq -\sum_{j=1}^{n} l_j$), the rest of the half-space is homogeneous and isotropic to infinity. This layered model is used by geophysicists in the study of the oceanic crust. Specifically, the measurements by Young and Cox (1981) described in Chapters 8 and 9 were interpreted by them in terms of an eight-layer model with the eight layers occupying the lower half-space to a depth of some 30 km beyond which the earth was assumed to be homogeneous and isotropic.

The analytical formulation of the electromagnetic field above a layered half-space proceeds from the general integrals for the field of a vertical or

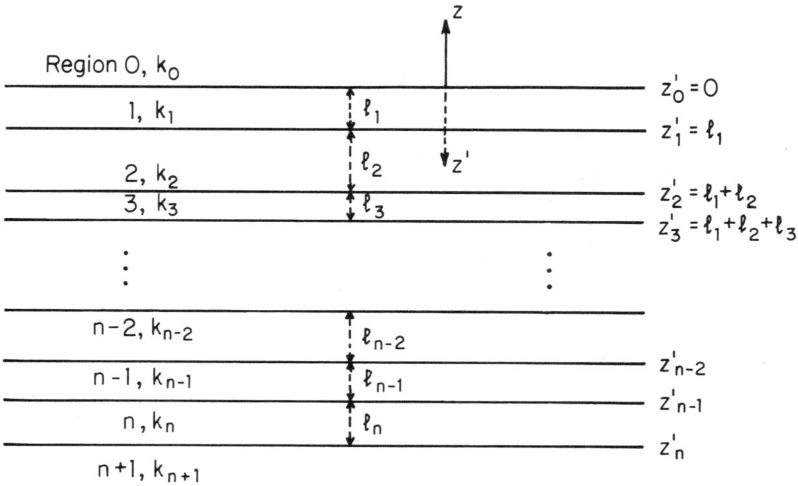

Figure 11.1.1. Horizontally layered half-space $z \leq 0$ below isotropic half-space $z \geq 0$.

horizontal dipole in Region 0 near the boundary $z = 0$ with Region 1, the first of the n layers. Typical integrals are those for $E_{0\rho}$. For the vertical dipole, this is given in (3.3.10). This formula can be rearranged by combining the last two terms. With the region designations changed from 1 to 0 and from 2 to 1 and with the suggested rearrangement,

$$E_{0\rho}(\rho, z) = \frac{i\omega\mu_0}{4\pi k_0^2}\int_0^\infty (\pm e^{i\gamma_0|z-d|} - Qe^{i\gamma_0(z+d)})J_1(\lambda\rho)\lambda^2\,d\lambda; \qquad \begin{cases} d < z \\ 0 \leq z \leq d \end{cases}, \tag{11.1.1}$$

where

$$Q = \frac{k_0^2\gamma_1 - k_1^2\gamma_0}{k_0^2\gamma_1 + k_1^2\gamma_0} = \frac{k_0^2(k_1^2-\lambda^2)^{1/2} - k_1^2(k_0^2-\lambda^2)^{1/2}}{k_0^2(k_1^2-\lambda^2)^{1/2} + k_1^2(k_0^2-\lambda^2)^{1/2}}. \tag{11.1.2}$$

The corresponding formula for the horizontal dipole is (5.4.29). With subscripts $1 \to 0$ and $2 \to 1$, it is

$$\begin{aligned} E_{0\rho}(\rho, \phi, z) = {} & -\frac{\omega\mu_0}{4\pi k_0^2}\cos\phi \\ & \times \left[\int_0^\infty \left(k_0^2 J_0(\lambda\rho) - \frac{\lambda^2}{2}\left[J_0(\lambda\rho) - J_2(\lambda\rho)\right]\right)\gamma_0^{-1} e^{i\gamma_0|z-d|}\lambda\,d\lambda \right. \\ & + \int_0^\infty \left(\frac{\gamma_0 Q}{2}\left[J_0(\lambda\rho) - J_2(\lambda\rho)\right]\right. \\ & \left.\left. - \frac{k_0^2 P}{2\gamma_0}\left[J_0(\lambda\rho) + J_2(\lambda\rho)\right]\right)e^{i\gamma_0(z+d)}\lambda\,d\lambda\right], \end{aligned} \tag{11.1.3}$$

where Q is defined in (11.1.2) and

$$P = \frac{\gamma_1 - \gamma_0}{\gamma_1 + \gamma_0} = \frac{(k_1^2 - \lambda^2)^{1/2} - (k_0^2 - \lambda^2)^{1/2}}{(k_1^2 - \lambda^2)^{1/2} + (k_0^2 - \lambda^2)^{1/2}}. \tag{11.1.4}$$

In both expressions, $\gamma_j = (k_j^2 - \lambda^2)^{1/2}$, $j = 0, 1$.

It follows from Chapter 2 that the reflection coefficients for a plane wave incident from Region 0 with the vector wave number

$$\mathbf{k}_0 = \hat{\mathbf{x}} k_{0x} + \hat{\mathbf{z}}' k_{0z'} = \hat{\mathbf{x}} k_0 \sin \Theta_0' + \hat{\mathbf{z}}' k_0 \cos \Theta_0', \tag{11.1.5}$$

and transmitted into Region 1 with the vector wave number

$$\mathbf{k}_1 = \hat{\mathbf{x}} k_{1x} + \hat{\mathbf{z}}' k_{1z'} = \hat{\mathbf{x}} k_1 \sin \Theta_1' + \hat{\mathbf{z}}' k_1 \cos \Theta_1' \tag{11.1.6}$$

are

$$f_{er} = \frac{k_1^2 k_{0z'} - k_0^2 k_{1z'}}{k_1^2 k_{0z'} + k_0^2 k_{1z'}} = \frac{k_0^{-1} \cos \Theta_0' - k_1^{-1} \cos \Theta_1'}{k_0^{-1} \cos \Theta_0' + k_1^{-1} \cos \Theta_1'} \tag{11.1.7}$$

when the electric field is in the plane of polarization and

$$f_{mr} = \frac{k_{0z'} - k_{1z'}}{k_{0z'} + k_{1z'}} = \frac{k_0 \cos \Theta_0' - k_1 \cos \Theta_1'}{k_0 \cos \Theta_0' + k_1 \cos \Theta_1'} \tag{11.1.8}$$

when the magnetic field is in the plane of polarization. For convenience in the description of waves propagating from Region 0 into Region 1 and continuing into Regions 2, 3, ..., $n+1$, the coordinate $z' = -z$ is introduced. It increases downward. Since $k^2 = k_x^2 + k_{z'}^2$, it follows that

$$k_{z'} = (k^2 - k_x^2)^{1/2} \quad \text{and} \quad \cos \Theta' = (1 - \sin^2 \Theta')^{1/2}. \tag{11.1.9}$$

Hence,

$$\begin{aligned} f_{er} &= \frac{k_1^2 (k_0^2 - k_{0x}^2)^{1/2} - k_0^2 (k_1^2 - k_{1x}^2)^{1/2}}{k_1^2 (k_0^2 - k_{0x}^2)^{1/2} + k_0^2 (k_1^2 - k_{1x}^2)^{1/2}} \\ &= \frac{k_1^2 (k_0^2 - k_x^2)^{1/2} - k_0^2 (k_1^2 - k_x^2)^{1/2}}{k_1^2 (k_0^2 - k_x^2)^{1/2} + k_0^2 (k_1^2 - k_x^2)^{1/2}}, \end{aligned} \tag{11.1.10}$$

$$\begin{aligned} f_{mr} &= \frac{(k_0^2 - k_{0x}^2)^{1/2} - (k_1^2 - k_{1x}^2)^{1/2}}{(k_0^2 - k_{0x}^2)^{1/2} + (k_1^2 - k_{1x}^2)^{1/2}} \\ &= \frac{(k_0^2 - k_x^2)^{1/2} - (k_1^2 - k_x^2)^{1/2}}{(k_0^2 - k_x^2)^{1/2} + (k_1^2 - k_x^2)^{1/2}}. \end{aligned} \tag{11.1.11}$$

The last steps in (11.1.10) and (11.1.11) make use of Snell's law,

$$k_{0x} = k_{1x} = k_x \quad \text{or} \quad k_0 \sin \Theta_0' = k_1 \sin \Theta_1'. \tag{11.1.12}$$

A comparison of (11.1.10) with (11.1.2) and (11.1.11) with (11.1.4) shows that

$$f_{er} = -Q; \qquad f_{mr} = -P \tag{11.1.13}$$

if λ is identified with k_x. It is evident from (11.1.1) and (11.1.3) that, since λ is in fact the radial wave number for an incremental cylindrical wave, $-Q$ and $-P$ are the reflection coefficients for locally plane waves for which $k_x = k_\rho = \lambda$.

11.2 The n-Layered Half-Space

When the half-space $z' \geq 0$ (designated above as Region 1) is replaced by n electrically different layers, all with horizontal boundaries at $z' = l_1,\ l_1 + l_2,\ \ldots,\ \sum_{j=1}^{n} l_j$, Snell's law applies at each boundary so that

$$k_{1x} = k_{2x} = \cdots = k_{jx} = \cdots = k_{n-1,x} = k_{nx} = k_x. \tag{11.2.1}$$

This means that the electromagnetic fields in the n layers depend only on k_x and $k_{1z'},\ k_{2z'},\ \ldots,\ k_{nz'},\ k_{n+1,z'}$. That is, the properties of the fields are effectively one-dimensional insofar as the effects of the n boundaries are concerned. The field that emerges from the surface $z' = 0$ as a result of multiple partial transmissions and reflections at the n boundaries plays the same role in determining the complete field in Region 0 as the once-reflected field given by the reflection coefficient or response function. If, in (11.1.1) and (11.1.3), the coefficients Q and P are replaced by the negative of the reflection coefficients, respectively of electric and magnetic types, of the n layers as observed in Region 0, the field given by the integrals will be the complete field of the dipole over the layered region. It is, therefore, necessary to determine these reflection coefficients. This is accomplished quite simply in terms of the surface impedance for the electric type and the surface admittance for the magnetic type. These quantities are defined in (2.5.50) and (2.5.51) when the plane wave is incident from Region 2 on the boundary of Region 1. In the numbering in this section, the subscripts 2 and 1 must be replaced, respectively, by 1 and 0. Instead of the argument Θ_2, referring to the angle of incidence, it is more convenient to use the vertical coordinate z'_j that locates the boundary of interest. The surface impedance of the boundary z'_j, $j = 0,\ 1,\ \ldots,\ n$, is

$$Z_{sj}(z'_j) = \frac{\mu_0 E_{jx}(0, z'_j)}{B_{jy}(0, z'_j)} = -\frac{i\omega\mu_0[\partial B_{jy}(x, z')/\partial z']_{z'=z'_j}}{k_j^2 B_{jy}(0, z'_j)}, \tag{11.2.2}$$

$$Y_{sj}(z'_j) = -\frac{B_{jx}(0, z'_j)}{\mu_0 E_{jy}(0, z'_j)} = \frac{[\partial E_{jy}(x, z')/\partial z']_{z'=z'_j}}{i\omega\mu_0 E_{jy}(0, z'_j)}. \tag{11.2.3}$$

The associated reflection coefficients at $z'_j = z'_0 = 0$ are

$$f_{er}(0) = \frac{B_{0y}^{\mathrm{ref}}(0,0)}{B_{0y}^{\mathrm{inc}}(0,0)} = \frac{\cos\Theta'_0 - k_0 Z_{s0}(0)/\omega\mu_0}{\cos\Theta'_0 + k_0 Z_{s0}(0)/\omega\mu_0}$$

$$= \frac{k_{0z'} - (k_0^2/\omega\mu_0)Z_{s0}(0)}{k_{0z'} + (k_0^2/\omega\mu_0)Z_{s0}(0)}$$

$$= \frac{\gamma_0 - (k_0^2/\omega\mu_0)Z_{s0}(0)}{\gamma_0 + (k_0^2/\omega\mu_0)Z_{s0}(0)}, \tag{11.2.4}$$

$$f_{mr}(0) = \frac{E_{0y}^{\mathrm{ref}}(0,0)}{E_{0y}^{\mathrm{inc}}(0,0)} = \frac{\cos\Theta_0' - \omega\mu_0 k_0^{-1} Y_{s0}(0)}{\cos\Theta_0' + \omega\mu_0 k_0^{-1} Y_{s0}(0)}$$

$$= \frac{\gamma_0 - \omega\mu_0 Y_{s0}(0)}{\gamma_0 + \omega\mu_0 Y_{s0}(0)}, \tag{11.2.5}$$

where $k_{0z'} = \gamma_0 = k_0 \cos\Theta_0'$. Note that the surface impedance and admittance are continuous across all boundaries because the tangential components of E and B are continuous. That is, $Z_{s,j-1}(z_j') = Z_{sj}(z_j')$ and $Y_{s,j-1}(z_j') = Y_{sj}(z_j')$. The reflection coefficients at the several boundaries are not continuous.

11.3 The Surface Impedance and Reflection Coefficient for an n-Layered Region: Electric Type

In order to determine the surface impedance and reflection coefficient of a plane wave incident from Region 0 on the boundary at $z' = z_0' = 0$, it is necessary to begin at the boundary $z' = z_n'$ between the last of the n layers and the infinite half-space beyond this. The general expressions for the components B_y and E_x are obtained from (2.5.32) and (2.5.33). They are

$$B_{jy}(x,z') = B_{jy}(0,z_j')e^{ik_{jx}x}(A_j e^{ik_{jz'}z'} + B_j e^{-ik_{jz'}z'}), \tag{11.3.1}$$

$$E_{jx}(x,z') = \frac{\omega k_{jz'}}{k_j^2} B_{jy}(0,z_j')e^{ik_{jx}x}(A_j e^{ik_{jz'}z'} - B_j e^{-ik_{jz'}z'});$$

$$z_{j-1}' \le z' \le z_j'. \tag{11.3.2}$$

When $j = n+1$, $B_{n+1} = 0$ so that with

$$\gamma_j = k_{jz'} \tag{11.3.3}$$

the surface impedance at z_n' is, from (11.2.2),

$$Z_{s,n+1}(z_n') = \frac{\omega\mu_0\gamma_{n+1}}{k_{n+1}^2}. \tag{11.3.4}$$

The quantity $Z_{sn}(z')$ with $z_{n-1}' \le z' \le z_n'$ is

$$Z_{sn}(z') = \frac{\omega\mu_0\gamma_n}{k_n^2}\left(\frac{A_n e^{i\gamma_n z'} - B_n e^{-i\gamma_n z'}}{A_n e^{i\gamma_n z'} + B_n e^{-i\gamma_n z'}}\right)$$

$$
\begin{aligned}
&= \frac{\omega\mu_0\gamma_n}{k_n^2}\left(\frac{(A_n/B_n)^{1/2}e^{i\gamma_n z'} - (A_n/B_n)^{-1/2}e^{-i\gamma_n z'}}{(A_n/B_n)^{1/2}e^{i\gamma_n z'} + (A_n/B_n)^{-1/2}e^{-i\gamma_n z'}}\right) \\
&= \frac{\omega\mu_0\gamma_n}{k_n^2}\left(\frac{e^{\ln(A_n/B_n)^{1/2}}e^{i\gamma_n z'} - e^{-\ln(A_n/B_n)^{1/2}}e^{-i\gamma_n z'}}{e^{\ln(A_n/B_n)^{1/2}}e^{i\gamma_n z'} + e^{-\ln(A_n/B_n)^{1/2}}e^{-i\gamma_n z'}}\right) \\
&= \frac{\omega\mu_0\gamma_n}{k_n^2}\tanh\left[\ln\left(\frac{A_n}{B_n}\right)^{1/2} + i\gamma_n z'\right]; \qquad z'_{n-1} \le z' \le z'_n.
\end{aligned}
\tag{11.3.5}
$$

Since continuity of the surface impedance requires that

$$
Z_{sn}(z'_n) = Z_{s,n+1}(z'_n), \tag{11.3.6}
$$

it follows that, with (11.3.4) and (11.3.5) at $z' = z'_n$,

$$
\frac{\gamma_n}{k_n^2}\tanh\left[\ln\left(\frac{A_n}{B_n}\right)^{1/2} + i\gamma_n z'_n\right] = \frac{\gamma_{n+1}}{k_{n+1}^2}
$$

and

$$
\ln\left(\frac{A_n}{B_n}\right)^{1/2} = \tanh^{-1}\left(\frac{\gamma_{n+1}k_n^2}{\gamma_n k_{n+1}^2}\right) - i\gamma_n z'_n. \tag{11.3.7}
$$

The surface impedance at $z' = z'_{n-1}$ is obtained from (11.3.5) with $z' = z'_{n-1}$ and (11.3.7) substituted for $\ln(A_n/B_n)^{1/2}$. It is

$$
Z_{sn}(z'_{n-1}) = \frac{\omega\mu_0\gamma_n}{k_n^2}\tanh\left[-i\gamma_n l_n + \tanh^{-1}\left(\frac{\gamma_{n+1}k_n^2}{\gamma_n k_{n+1}^2}\right)\right], \tag{11.3.8}
$$

where

$$
l_n = z'_n - z'_{n-1} \tag{11.3.9}
$$

is the thickness of the nth layer. With the identity

$$
\tanh(a + \tanh^{-1} b) = \coth(a + \coth^{-1} b), \tag{11.3.10}
$$

(11.3.8) can be expressed in the following alternative form:

$$
Z_{sn}(z'_{n-1}) = \frac{\omega\mu_0\gamma_n}{k_n^2}\coth\left[-i\gamma_n l_n + \coth^{-1}\left(\frac{\gamma_{n+1}k_n^2}{\gamma_n k_{n+1}^2}\right)\right]. \tag{11.3.11}
$$

The subtraction of 1 from the index n in (11.3.5) gives, for the $(n-1)$st layer,

$$
Z_{s,n-1}(z') = \frac{\omega\mu_0\gamma_{n-1}}{k_{n-1}^2}\tanh\left[\ln\left(\frac{A_{n-1}}{B_{n-1}}\right)^{1/2} + i\gamma_{n-1}z'\right]; \qquad z'_{n-2} \le z' \le z'_{n-1}. \tag{11.3.12}
$$

With the required continuity relation

$$
Z_{s,n-1}(z'_{n-1}) = Z_{sn}(z'_{n-1}), \tag{11.3.13}
$$

it follows that

$$\ln\left(\frac{A_{n-1}}{B_{n-1}}\right)^{1/2} = \tanh^{-1}\left\{\left(\frac{\gamma_n k_{n-1}^2}{\gamma_{n-1}k_n^2}\right)\tanh\left[-i\gamma_n l_n + \tanh^{-1}\left(\frac{\gamma_{n+1}k_n^2}{\gamma_n k_{n+1}^2}\right)\right]\right\} - i\gamma_{n-1}z'_{n-1} \tag{11.3.14}$$

and

$$Z_{s,n-1}(z'_{n-2}) = \frac{\omega\mu_0\gamma_{n-1}}{k_{n-1}^2}\tanh\left\{-i\gamma_{n-1}l_{n-1} + \tanh^{-1}\left[\left(\frac{\gamma_n k_{n-1}^2}{\gamma_{n-1}k_n^2}\right)\tanh\left[-i\gamma_n l_n + \tanh^{-1}\left(\frac{\gamma_{n+1}k_n^2}{\gamma_n k_{n+1}^2}\right)\right]\right]\right\}, \tag{11.3.15}$$

where

$$l_{n-1} = z'_{n-1} - z'_{n-2} \tag{11.3.16}$$

is the thickness of the $(n-1)$st layer.

If this procedure is continued to the boundary at $z' = z'_0 = 0$ between Regions 0 and 1, the desired surface impedance is obtained. It is

$$Z_{s0}(z'_0) = \frac{\omega\mu_0\gamma_0}{k_0^2}\tanh\left\{\tanh^{-1}\left[\left(\frac{\gamma_1 k_0^2}{\gamma_0 k_1^2}\right)\tanh\left[-i\gamma_1 l_1 + \tanh^{-1}\left[\left(\frac{\gamma_2 k_1^2}{\gamma_1 k_2^2}\right)\tanh\left[-i\gamma_2 l_2 + \tanh^{-1}\left[\left(\frac{\gamma_3 k_2^2}{\gamma_2 k_3^2}\right)\tanh\left[-i\gamma_3 l_3 + \cdots + \tanh^{-1}\left[\left(\frac{\gamma_n k_{n-1}^2}{\gamma_{n-1}k_n^2}\right)\tanh\left[-i\gamma_n l_n + \tanh^{-1}\left(\frac{\gamma_{n+1}k_n^2}{\gamma_n k_{n+1}^2}\right)\right]\right]\cdots\right]\right]\right]\right]\right]\right\}. \tag{11.3.17}$$

Note that when $l_1 = l_2 = \cdots = l_n = 0$, this reduces to

$$Z_{s0}(z'_0) = \frac{\omega\mu_0\gamma_0}{k_0^2}\frac{\gamma_{n+1}k_0^2}{\gamma_0 k_{n+1}^2} = \frac{\omega\mu_0\gamma_{n+1}}{k_{n+1}^2} = \frac{\omega\mu_0\gamma_1}{k_1^2}. \tag{11.3.18}$$

The last step follows since, when $l_1 = l_2 = \cdots = l_n = 0$, there are only two regions so that $n = 0$. Clearly, (11.3.17) reduces to

$$Z_{s0}(z'_0) = Z_{s1}(z'_0) = \frac{\omega\mu_0\gamma_1}{k_1^2}\tanh\left\{-i\gamma_1 l_1 + \tanh^{-1}\left[\left(\frac{\gamma_2 k_1^2}{\gamma_1 k_2^2}\right)\tanh\left[-i\gamma_2 l_2\right.\right.\right.$$

$$+\tanh^{-1}\left[\left(\frac{\gamma_3 k_2^2}{\gamma_2 k_3^2}\right)\tanh\left[-i\gamma_3 l_3 + \cdots\right.\right.$$

$$+\tanh^{-1}\left[\left(\frac{\gamma_n k_{n-1}^2}{\gamma_{n-1} k_n^2}\right)\tanh\left[-i\gamma_n l_n\right.\right.$$

$$\left.\left.\left.\left.+\tanh^{-1}\left(\frac{\gamma_{n+1} k_n^2}{\gamma_n k_{n+1}^2}\right)\right]\right]\cdots\right]\right]\right]\right\}. \quad (11.3.19)$$

Since (11.3.19) does not involve Region 0 in any way, it cannot be used with $l_1 = l_2 = \cdots = l_n = 0$ to obtain (11.3.18). With the identity (11.3.10), formula (11.3.17) can also be written with coth and $\coth^{-1}$ substituted for tanh and $\tanh^{-1}$. With this change and a modified notation, (11.3.17) is equivalent to the recurrence relation (2.85) of Rokityansky (1982, p. 47).

The substitution of $Z_{s0}(0)$ as given by (11.3.17) or (11.3.19) into (11.2.4) with $k_{jz'} = \gamma_j$ gives

$$f_{er}(0) = \frac{\gamma_0 - (k_0^2/\omega\mu_0)Z_{s0}(0)}{\gamma_0 + (k_0^2/\omega\mu_0)Z_{s0}(0)} = -Q_n. \quad (11.3.20)$$

With $n = 0$ and $l_1 = l_2 = \cdots = l_n = 0$, (11.3.17) gives (11.3.18) and with it

$$f_{er}(0) = \frac{\gamma_0 - k_0^2\gamma_1/k_1^2}{\gamma_0 + k_0^2\gamma_1/k_1^2} = \frac{k_1^2\gamma_0 - k_0^2\gamma_1}{k_1^2\gamma_0 + k_0^2\gamma_1} = -Q. \quad (11.3.21)$$

11.4 The Surface Admittance and Reflection Coefficient for an n-Layered Region: Magnetic Type

The general expressions for the components B_x and E_y for the electromagnetic field of magnetic type are

$$E_{jy}(x, z') = E_{jy}(0, z_j')e^{ik_{jx}x}(A_j e^{ik_{jz'}z'} + B_j e^{-ik_{jz'}z'}), \quad (11.4.1)$$

$$B_{jx}(x, z') = -\frac{k_{jz'}}{\omega} E_{jy}(0, z_j')e^{ik_{jx}x}(A_j e^{ik_{jz'}z'} - B_j e^{-ik_{jz'}z'});$$
$$z_{j-1}' \leq z' \leq z_j'. \quad (11.4.2)$$

When $j = n + 1$, $B_{n+1} = 0$ at $z' = z_n'$ so that with (11.2.3) and $\gamma_j = k_{jz'}$,

$$Y_{s,n+1}(z_n') = -\frac{B_{n+1,x}(0, z_n')}{\mu_0 E_{n+1,y}(0, z_n')} = \frac{\gamma_{n+1}}{\omega\mu_0}. \quad (11.4.3)$$

The general quantity $Y_{sn}(z')$ with $z'_{n-1} \leq z' \leq z'_n$ is

$$\begin{aligned} Y_{sn}(z') &= \frac{\gamma_n}{\omega\mu_0}\left(\frac{A_n e^{i\gamma_n z'} - B_n e^{-i\gamma_n z'}}{A_n e^{i\gamma_n z'} + B_n e^{-i\gamma_n z'}}\right) \\ &= \frac{\gamma_n}{\omega\mu_0}\tanh\left[\ln\left(\frac{A_n}{B_n}\right)^{1/2} + i\gamma_n z'\right]; \qquad z'_{n-1} \leq z' \leq z'_n. \end{aligned} \tag{11.4.4}$$

The steps in (11.4.4) are the same as those in (11.3.5). With

$$Y_{sn}(z'_n) = Y_{s,n+1}(z'_n), \tag{11.4.5}$$

(11.4.3) and (11.4.4) with $z' = z'_n$ give

$$\ln\left(\frac{A_n}{B_n}\right)^{1/2} = \tanh^{-1}\left(\frac{\gamma_{n+1}}{\gamma_n}\right) - i\gamma_n z'_n. \tag{11.4.6}$$

This parallels (11.3.7). Following the same steps as in Section 11.3, it follows directly that

$$\begin{aligned} Y_{s0}(z'_0) = \frac{\gamma_0}{\omega\mu_0}\tanh\Bigg\{ &\tanh^{-1}\Bigg[\left(\frac{\gamma_1}{\gamma_0}\right)\tanh\Bigg[-i\gamma_1 l_1 \\ &+ \tanh^{-1}\Bigg[\left(\frac{\gamma_2}{\gamma_1}\right)\tanh\Bigg[-i\gamma_2 l_2 \\ &+ \tanh^{-1}\Bigg[\left(\frac{\gamma_3}{\gamma_2}\right)\tanh\Bigg[-i\gamma_3 l_3 + \cdots \\ &+ \tanh^{-1}\Bigg[\left(\frac{\gamma_n}{\gamma_{n-1}}\right)\tanh\Bigg[-i\gamma_n l_n \\ &+ \tanh^{-1}\left(\frac{\gamma_{n+1}}{\gamma_n}\right)\Bigg]\Bigg]\cdots\Bigg]\Bigg]\Bigg]\Bigg]\Bigg]\Bigg]\Bigg\}. \end{aligned} \tag{11.4.7}$$

When $l_1 = l_2 = \cdots = l_n = 0$, this reduces to (11.4.3) with $n = 0$. Thus,

$$Y_{s0}(z'_0) = \frac{\gamma_0}{\omega\mu_0}\frac{\gamma_{n+1}}{\gamma_0} = \frac{\gamma_{n+1}}{\omega\mu_0} = \frac{\gamma_1}{\omega\mu_0}. \tag{11.4.8}$$

The alternative form of (11.4.7) is

$$\begin{aligned} Y_{s0}(z'_0) &= Y_{s1}(z'_0) \\ &= \frac{\gamma_1}{\omega\mu_0}\tanh\Bigg\{-i\gamma_1 l_1 + \tanh^{-1}\Bigg[\left(\frac{\gamma_2}{\gamma_1}\right)\tanh\Bigg[-i\gamma_2 l_2 \\ &\qquad + \tanh^{-1}\Bigg[\left(\frac{\gamma_3}{\gamma_2}\right)\tanh\Bigg[-i\gamma_3 l_3 + \cdots \\ &\qquad + \tanh^{-1}\Bigg[\left(\frac{\gamma_n}{\gamma_{n-1}}\right)\tanh\Bigg[-i\gamma_n l_n \\ &\qquad + \tanh^{-1}\left(\frac{\gamma_{n+1}}{\gamma_n}\right)\Bigg]\Bigg]\cdots\Bigg]\Bigg]\Bigg]\Bigg]\Bigg\}. \end{aligned} \tag{11.4.9}$$

The substitution of $Y_{s0}(z_0')$ as given by (11.4.7) or (11.4.9) into (11.2.5) gives the reflection coefficient of magnetic type. It is

$$f_{mr}(0) = \frac{\gamma_0 - \omega\mu_0 Y_{s0}(0)}{\gamma_0 + \omega\mu_0 Y_{s0}(0)} = -P_n. \tag{11.4.10}$$

When $l_1 = l_2 = \cdots = l_n = 0$, this gives

$$f_{mr}(0) = \frac{\gamma_0 - \gamma_1}{\gamma_0 + \gamma_1} = -P. \tag{11.4.11}$$

11.5 The Electromagnetic Field in Region 1 over an n-Layered Half-Space

The radial electric field $E_{0\rho}(\rho, z)$ in Region 0 ($z \geq 0$) due to a vertical electric dipole at the height d in Region 0 when there are n layers between $z = 0$ and $z = z_n$ is given by (11.1.1) with Q replaced by Q_n, as given in (11.3.20). The other components, viz., $E_{0z}(\rho, z)$ and $B_{0\phi}(\rho, z)$, are given by (3.3.9) and (3.3.11) with the subscripts $1 \to 0$ and $2 \to 1$ and with the last two terms in each combined. The resulting expressions for the complete field are

$$B_{0\phi}(\rho, z) = \frac{i\mu_0}{4\pi}\int_0^\infty \gamma_0^{-1}(e^{i\gamma_0|z-d|} - Q_n e^{i\gamma_0(z+d)})J_1(\lambda\rho)\lambda^2\,d\lambda, \tag{11.5.1}$$

$$E_{0\rho}(\rho, z) = \frac{i\omega\mu_0}{4\pi k_0^2}\int_0^\infty (\pm e^{i\gamma_0|z-d|} - Q_n e^{i\gamma_0(z+d)})J_1(\lambda\rho)\lambda^2\,d\lambda; \quad \begin{cases} d < z \\ 0 \leq z \leq d \end{cases}, \tag{11.5.2}$$

$$E_{0z}(\rho, z) = -\frac{\omega\mu_0}{4\pi k_0^2}\int_0^\infty \gamma_0^{-1}(e^{i\gamma_0|z-d|} - Q_n e^{i\gamma_0(z+d)})J_0(\lambda\rho)\lambda^3\,d\lambda. \tag{11.5.3}$$

With a suitable change in notation, these formulas are like those given by Chave and Cox [1982, eqs. (27)–(29), p. 5330] except that their formula for B_ϕ incorrectly contains $J_0(\lambda\rho)$ instead of $J_1(\lambda\rho)$. Also their formulas are restricted to $k_0 = (i\omega\mu_0\sigma_0)^{1/2}$ and $k_1 = (i\omega\mu_0\sigma_1)^{1/2}$.

The corresponding expressions for the electromagnetic field of a horizontal dipole are given by (5.4.29)–(5.4.34) with Q_n substituted for Q and P_n for P. They are

$$E_{0\rho}(\rho, \phi, z) = \frac{\omega\mu_0}{4\pi k_0^2}\cos\phi$$
$$\times\left[\int_0^\infty \left(k_0^2 J_0(\lambda\rho) - \frac{\lambda^2}{2}\left[J_0(\lambda\rho) - J_2(\lambda\rho)\right]\right)\gamma_0^{-1}e^{i\gamma_0|z-d|}\lambda\,d\lambda\right.$$

$$+\int_0^\infty\left(\frac{\gamma_0 Q_n}{2}\,[J_0(\lambda\rho)-J_2(\lambda\rho)]\right.$$
$$\left.\left.-\frac{k_0^2 P_n}{2\gamma_0}\,[J_0(\lambda\rho)+J_2(\lambda\rho)]\right)e^{i\gamma_0(z+d)}\lambda\,d\lambda\right], \tag{11.5.4}$$

$$E_{0\phi}(\rho,\phi,z)=\frac{\omega\mu_0}{4\pi k_0^2}\sin\phi$$
$$\times\left[\int_0^\infty\left(k_0^2 J_0(\lambda\rho)-\frac{\lambda^2}{2}\,[J_0(\lambda\rho)+J_2(\lambda\rho)]\right)\gamma_0^{-1}e^{i\gamma_0|z-d|}\lambda\,d\lambda\right.$$
$$+\int_0^\infty\left(\frac{\gamma_0 Q_n}{2}\,[J_0(\lambda\rho)+J_2(\lambda\rho)]\right.$$
$$\left.\left.-\frac{k_0^2 P_n}{2\gamma_0}\,[J_0(\lambda\rho)-J_2(\lambda\rho)]\right)e^{i\gamma_0(z+d)}\lambda\,d\lambda\right], \tag{11.5.5}$$

$$E_{0z}(\rho,\phi,z)=\frac{i\omega\mu_0}{4\pi k_0^2}\cos\phi\int_0^\infty\left(\pm e^{i\gamma_0|z-d|}+Q_n e^{i\gamma_0(z+d)}\right)J_1(\lambda\rho)\lambda^2\,d\lambda, \tag{11.5.6}$$

$$B_{0\rho}(\rho,\phi,z)$$
$$=-\frac{\mu_0}{4\pi}\sin\phi\left[\pm\int_0^\infty J_0(\lambda\rho)e^{i\gamma_0|z-d|}\lambda\,d\lambda\right.$$
$$+\int_0^\infty\left(\frac{Q_n}{2}\,[J_0(\lambda\rho)+J_2(\lambda\rho)]\right.$$
$$\left.\left.-\frac{P_n}{2}\,[J_0(\lambda\rho)-J_2(\lambda\rho)]\right)e^{i\gamma_0(z+d)}\lambda\,d\lambda\right], \tag{11.5.7}$$

$$B_{0\phi}(\rho,\phi,z)$$
$$=-\frac{\mu_0}{4\pi}\cos\phi\left[\pm\int_0^\infty J_0(\lambda\rho)e^{i\gamma_0|z-d|}\lambda\,d\lambda\right.$$
$$+\int_0^\infty\left(\frac{Q_n}{2}\,[J_0(\lambda\rho)-J_2(\lambda\rho)]\right.$$
$$\left.\left.-\frac{P_n}{2}\,[J_0(\lambda\rho)+J_2(\lambda\rho)]\right)e^{i\gamma_0(z+d)}\lambda\,d\lambda\right], \tag{11.5.8}$$

$$B_{0z}(\rho,\phi,z)=\frac{i\mu_0}{4\pi}\sin\phi\int_0^\infty\left(e^{i\gamma_0|z-d|}-P_n e^{i\gamma_0(z+d)}\right)\gamma_0^{-1}J_1(\lambda\rho)\lambda^2\,d\lambda. \tag{11.5.9}$$

Where two signs appear, the upper one is for $z>d$ and the lower one for $0\le z\le d$.

With $Q_n=R_{TM}$ and $P_n=R_{TE}$, these formulas correspond to those given by Chave and Cox [1982, eqs. (30)–(35)]. They are, however, not restricted to $\omega\epsilon_j\ll\sigma_j$, $j=1,\,2,\,\ldots,\,n+1$, as are those of Chave and Cox.

11.6 The Three-Layered Region: General Formulas

Three-layered regions are of considerable importance since they include the oceanic crust when this is separated from the sea by a layer of sediment, the Arctic ice between air and sea, and the air–dielectric–conductor environment for conductors located on the air–dielectric boundary in microstrip antennas and circuits. The general formulas for the surface impedance and admittance for the three-layered region are obtained from (11.3.19) and (11.4.9). They are

$$Z_{s0}(0) = \frac{\omega\mu_0\gamma_1}{k_1^2}\tanh\left[-i\gamma_1 l_1 + \tanh^{-1}\left(\frac{\gamma_2 k_1^2}{\gamma_1 k_2^2}\right)\right], \tag{11.6.1}$$

$$Y_{s0}(0) = \frac{\gamma_1}{\omega\mu_0}\tanh\left[-i\gamma_1 l_1 + \tanh^{-1}\left(\frac{\gamma_2}{\gamma_1}\right)\right]. \tag{11.6.2}$$

With the identity

$$\tanh(x+y) = \frac{\tanh x + \tanh y}{1 + \tanh x \tanh y} \tag{11.6.3}$$

and $\tanh ix = i\tan x$, these become

$$Z_{s0}(0) = \left(\frac{\omega\mu_0\gamma_1}{k_1^2}\right)\left(\frac{-i\tan\gamma_1 l_1 + (\gamma_2 k_1^2/\gamma_1 k_2^2)}{1 - i(\gamma_2 k_1^2/\gamma_1 k_2^2)\tan\gamma_1 l_1}\right), \tag{11.6.4}$$

$$Y_{s0}(0) = \left(\frac{\gamma_1}{\omega\mu_0}\right)\left(\frac{-i\tan\gamma_1 l_1 + (\gamma_2/\gamma_1)}{1 - i(\gamma_2/\gamma_1)\tan\gamma_1 l_1}\right). \tag{11.6.5}$$

The associated reflection coefficients are

$$f_{er}(0) = \frac{\gamma_0 - (k_0^2/\omega\mu_0)Z_{s0}(0)}{\gamma_0 + (k_0^2/\omega\mu_0)Z_{s0}(0)}$$

$$= \frac{\gamma_0 k_1^2\left(1 - \dfrac{i\gamma_2 k_1^2}{\gamma_1 k_2^2}\tan\gamma_1 l_1\right) - \gamma_1 k_0^2\left(-i\tan\gamma_1 l_1 + \dfrac{\gamma_2 k_1^2}{\gamma_1 k_2^2}\right)}{\gamma_0 k_1^2\left(1 - \dfrac{i\gamma_2 k_1^2}{\gamma_1 k_2^2}\tan\gamma_1 l_1\right) + \gamma_1 k_0^2\left(-i\tan\gamma_1 l_1 + \dfrac{\gamma_2 k_1^2}{\gamma_1 k_2^2}\right)}$$

$$= \frac{\gamma_0 k_2^2 - \gamma_2 k_0^2 - i\left(\dfrac{k_1^2\gamma_0\gamma_2}{\gamma_1} - \dfrac{\gamma_1 k_0^2 k_2^2}{k_1^2}\right)\tan\gamma_1 l_1}{\gamma_0 k_2^2 + \gamma_2 k_0^2 - i\left(\dfrac{k_1^2\gamma_0\gamma_2}{\gamma_1} + \dfrac{\gamma_1 k_0^2 k_2^2}{k_1^2}\right)\tan\gamma_1 l_1} = -Q_3, \tag{11.6.6}$$

$$f_{mr}(0) = \frac{\gamma_0 - \omega\mu_0 Y_{s0}(0)}{\gamma_0 + \omega\mu_0 Y_{s0}(0)}$$

$$= \frac{\gamma_0 - (i\gamma_0\gamma_2/\gamma_1)\tan\gamma_1 l_1 + i\gamma_1 \tan\gamma_1 l_1 - \gamma_2}{\gamma_0 - (i\gamma_0\gamma_2/\gamma_1)\tan\gamma_1 l_1 - i\gamma_1 \tan\gamma_1 l_1 + \gamma_2}$$

$$= \frac{\gamma_0 - \gamma_2 - i\left(\dfrac{\gamma_0\gamma_2}{\gamma_1} - \gamma_1\right)\tan\gamma_1 l_1}{\gamma_0 + \gamma_2 - i\left(\dfrac{\gamma_0\gamma_2}{\gamma_1} + \gamma_1\right)\tan\gamma_1 l_1} = -P_3. \tag{11.6.7}$$

It is seen that, when $l_1 = 0$, (11.6.6) and (11.6.7) reduce to the plane-wave reflection coefficients for the boundary between Regions 0 and 2. Alternatively, when $l_1 \to \infty$, $\tan\gamma_1 l_1 \to i$ and

$$f_{er}(0) \to \frac{\gamma_0 k_1^2\left(\dfrac{\gamma_2}{\gamma_1}\right) - \gamma_1 k_0^2\left(\dfrac{k_2^2}{k_1^2}\right) + \gamma_0 k_1^2\left(\dfrac{k_2^2}{k_1^2}\right) - \gamma_1 k_0^2\left(\dfrac{\gamma_2}{\gamma_1}\right)}{\gamma_0 k_1^2\left(\dfrac{\gamma_2}{\gamma_1}\right) + \gamma_1 k_0^2\left(\dfrac{k_2^2}{k_1^2}\right) + \gamma_0 k_1^2\left(\dfrac{k_2^2}{k_1^2}\right) + \gamma_1 k_0^2\left(\dfrac{\gamma_2}{\gamma_1}\right)}$$

$$= \frac{(\gamma_0 k_1^2 - \gamma_1 k_0^2)\left(\dfrac{\gamma_2}{\gamma_1} + \dfrac{k_2^2}{k_1^2}\right)}{(\gamma_0 k_1^2 + \gamma_1 k_0^2)\left(\dfrac{\gamma_2}{\gamma_1} + \dfrac{k_2^2}{k_1^2}\right)} = \frac{\gamma_0 k_1^2 - \gamma_1 k_0^2}{\gamma_0 k_1^2 + \gamma_1 k_0^2}. \tag{11.6.8}$$

Similarly,

$$f_{mr}(0) \to \frac{\gamma_0 - \gamma_1}{\gamma_0 + \gamma_1}. \tag{11.6.9}$$

These are the reflection coefficients for the boundary between Regions 0 and 1 when the latter occupies the entire half-space.

The general formulas (11.6.6) and (11.6.7) for the reflection coefficients of the three-layered region can be simplified in the principal applications. When Region 0 is the sea and it contains the antenna, Region 1 is sediment, and Region 2 is rock, the inequalities $|k_0|^2 \gg |k_1|^2 \gg |k_2|^2$ are valid. On the other hand, when Region 0 is air, Region 1 ice, and Region 2 sea, the reverse inequalities apply. That is, $|k_0|^2 \ll |k_1|^2 \ll |k_2|^2$. The latter inequalities also apply to the conductor on the air–dielectric boundary when the dielectric is backed by a conducting plane. These are studied separately in Chapters 12 and 15.

12
The Three-Layer Problem for Sediment on the Oceanic Crust

12.1 Introduction

The propagation of electromagnetic waves along the boundary between two different media like sea water and air or sea water and rock when the exciting source is a horizontal antenna in the sea close to that boundary involves a surface wave of the lateral-wave type. When the boundary also includes a layer of a third material with intermediate properties such as ice between the air and the sea in the Arctic or sediment on the sea floor, the electromagnetic waves excited by the same horizontal antenna—now in the sea just below the ice or just above the sediment—are in general much more complicated. If the intermediate layer has only small losses as in the case of fairly pure ice, so-called "trapped" waves of both TE and TM types can propagate in it. The number of possible propagating modes depends on the electrical thickness of the ice. In addition, a lateral wave will travel in the ice if this is sufficiently thick. Alternatively, if it is electrically relatively thin, the lateral wave may traverse the ice almost vertically and propagate in the air along the air–ice boundary. If the layer is quite lossy and electrically not too thick, as in the case of sediment on the sea floor, propagating "trapped" modes are rapidly attenuated. In this case, the field beyond their range is dominated by the lateral wave that travels through the sediment, then along the sediment–rock boundary in the rock, and finally back through the sediment to a point of observation in the sea close to the sea floor. It is the determination of the properties of this lateral wave that is the primary concern of this chapter. Explicit formulas for it are derived but no attempt is made to determine the associated highly attenuated "trapped" modes in the sediment or the complete direct field that travels in the sea water. The formulas obtained are primarily useful at radial distances from the source to the point of observation—both in the sea—that are large enough so that waves that travel in the sea or in the sediment have negligible amplitudes compared with the lateral waves that travel in the rock.

The sea floor in some locations is simply the boundary between sea water and rock. More often, there is an intervening layer of sediment with a thickness l that varies from near zero to several kilometers. Let the sea be Region 1 ($z \geq 0$) with the wave number k_1, the sediment Region 2 ($0 \geq z \geq -l$) with the wave number k_2, and the rock Region 3 ($-l \geq z > -\infty$) with the wave number k_3. The sediment usually has a conductivity σ_2 that is between σ_1 for sea water and σ_3 for rock. When this is true, the following inequalities are usually good approximations:

$$|k_1|^2 \gg |k_2|^2 \gg |k_3|^2 \quad \text{or} \quad |k_1| \geq 3|k_2| \geq 9|k_3|. \tag{12.1.1}$$

At low frequencies when $k_j \sim (i\omega\mu_0\sigma_j)^{1/2}$, $j = 1, 2, 3$, these relations are equivalent to

$$\sigma_1 \geq 9\sigma_2 \geq 81\sigma_3. \tag{12.1.2}$$

With $\sigma_1 \sim 4$ S/m, this requires that $\sigma_2 \leq 4/9 = 0.44$ S/m. If the conductivity of rock satisfies the condition $\sigma_3 \leq 0.01$ S/m, then $\sigma_2 \geq 0.09$ S/m. Thus, the conductivity of the sediment should satisfy the following physically reasonable conditions:

$$0.09 \leq \sigma_2 \leq 0.44 \text{ S/m}. \tag{12.1.3}$$

It is assumed in the following analysis that (12.1.1) is satisfied. It is evident from (12.1.3) that this can be expected to be true for most sediments.

The plane-wave reflection coefficients of the sea–sediment surface for a wave incident from the sea are given by (11.6.6) and (11.6.7) with the subscript indices advanced by 1. They are

$$-f_{er} = Q_3 = \frac{\gamma_3 k_1^2 - \gamma_1 k_3^2 - i(\gamma_2 k_1^2 k_3^2/k_2^2 - k_2^2\gamma_1\gamma_3/\gamma_2)\tan\gamma_2 l}{\gamma_3 k_1^2 + \gamma_1 k_3^2 - i(\gamma_2 k_1^2 k_3^2/k_2^2 + k_2^2\gamma_1\gamma_3/\gamma_2)\tan\gamma_2 l}, \tag{12.1.4}$$

$$-f_{mr} = P_3 = \frac{\gamma_3 - \gamma_1 - i(\gamma_2 - \gamma_1\gamma_3/\gamma_2)\tan\gamma_2 l}{\gamma_3 + \gamma_1 - i(\gamma_2 + \gamma_1\gamma_3/\gamma_2)\tan\gamma_2 l}. \tag{12.1.5}$$

Because the general integrals of the type (11.5.4)–(11.5.9) can be evaluated exactly when $Q_n = P_n = 1$, it is advantageous to formulate the quantities

$$Q_3 - 1 = \frac{-2[\gamma_1 k_3^2 - i(k_2^2\gamma_1\gamma_3/\gamma_2)\tan\gamma_2 l]}{\gamma_3 k_1^2 + \gamma_1 k_3^2 - i(\gamma_2 k_1^2 k_3^2/k_2^2 + k_2^2\gamma_1\gamma_3/\gamma_2)\tan\gamma_2 l} \tag{12.1.6}$$

and

$$P_3 + 1 = \frac{2(\gamma_3 - i\gamma_2\tan\gamma_2 l)}{\gamma_3 + \gamma_1 - i(\gamma_2 + \gamma_1\gamma_3/\gamma_2)\tan\gamma_2 l}, \tag{12.1.7}$$

just as in the two-layer case.

At the low frequencies of interest in geophysical exploration of the sea floor, $k_2 = \beta_2 + i\alpha_2 \sim (1+i)(\omega\mu_0\sigma_2/2)^{1/2}$ so that $\alpha_2 \sim \beta_2$. This precludes waveguide-like propagation in the sediment layer (which is possible when $\sigma_2 \ll \omega\epsilon_2$). Since it is assumed in (12.1.1) that $\alpha_3 < \alpha_2$, it is to be expected that the principal lateral-wave propagation is in Region 3 provided Region 2 is not too thick. Clearly, there is no lateral wave in Region 3 when the thickness l of Region 2 and the radial distance from the source to the point of observation in Region 1 are such that l greatly exceeds the maximum depth of penetration z'_m of the lateral wave into Region 2. In this case, the field observed at (ρ, ϕ, z) in Region 1 has traveled entirely in Region 2. Evidently, a lateral wave that travels primarily in Region 3 requires the thickness l of Region 2 to be substantially smaller than z'_m. An examination of Fig. 5.9.1 suggests that for the components $E_{1\rho}(\rho, \phi, z)$, $E_{1z}(\rho, \phi, z)$, and $B_{1\phi}(\rho, \phi, z)$ of electric type generated by a horizontal electric dipole, l should satisfy the condition

$$l_E \le 0.5z'_m = \frac{0.5\delta\rho}{2.718}, \tag{12.1.8}$$

where ρ is the radial distance to the point of observation and

$$\delta = \frac{\beta_1(\beta_2^2 + \alpha_2^2)}{\beta_2(\beta_1^2 + \alpha_1^2)}. \tag{12.1.9}$$

In the case of primary interest here, $\alpha_1 \sim \beta_1$ and $\alpha_2 \sim \beta_2$ so that

$$\delta = \frac{\beta_2}{\beta_1}. \tag{12.1.10}$$

The corresponding condition for the components of magnetic type, viz., $E_{1\phi}(\rho, \phi, z)$, $B_{1\rho}(\rho, \phi, z)$, and $B_{1z}(\rho, \phi, z)$, is less severe. It is

$$l_M \le 0.5z'_m = 0.385\rho. \tag{12.1.11}$$

If the electric dipole is vertical, the condition on l is

$$l_V \le 0.5z'_m = 0.075\delta\beta_2\rho^2. \tag{12.1.12}$$

At sufficiently low frequencies—which are of primary interest in measurements on the sea floor—$\alpha_j = \beta_j$, $j = 1, 2, 3$, so that the three conditions (12.1.8), (12.1.11), and (12.1.12) become

$$l_E \le \frac{0.5(\sigma_2/\sigma_1)^{1/2}\rho}{2.718}, \tag{12.1.13a}$$

$$l_M \le 0.385\rho, \tag{12.1.13b}$$

$$l_V \le 0.075\left(\frac{\sigma_2}{\sigma_1}\right)^{1/2}\left(\frac{\omega\mu_0\sigma_2}{2}\right)^{1/2}\rho^2. \tag{12.1.13c}$$

12.2 Integrals for the Components of the Field of a Horizontal Electric Dipole

The general integrals for the six components of the electromagnetic field in Region 1 (sea water, $z \geq 0$) due to a unit horizontal electric dipole also in Region 1 at the height d above the boundary at $z = 0$ between Region 1 and Region 2 (sediment, $0 \geq z \geq -l$) are given by (5.4.29)–(5.4.34) with Q_3 and P_3 as given by (12.1.4) and (12.1.5) substituted for Q and P as given by (5.4.28). The evaluation of these integrals with the new, much more complicated values of Q and P can be approached in the same general manner as in Chapter 5. Specifically, for the components of the electric field,

$$E_{1\rho}(\rho, \phi, z) = -\frac{\omega\mu_0}{4\pi k_1^2} \cos\phi \, [F_{\rho 0}(\rho, z-d) + F_{\rho 0}(\rho, z+d) + F_{\rho 1}(\rho, z+d)], \tag{12.2.1}$$

$$E_{1\phi}(\rho, \phi, z) = \frac{\omega\mu_0}{4\pi k_1^2} \sin\phi \, [F_{\phi 0}(\rho, z-d) + F_{\phi 0}(\rho, z+d) + F_{\phi 1}(\rho, z+d)], \tag{12.2.2}$$

$$E_{1z}(\rho, \phi, z) = \frac{i\omega\mu_0}{4\pi k_1^2} \cos\phi \, [F_{z0}(\rho, z-d) + F_{z0}(\rho, z+d) + F_{z1}(\rho, z+d)], \tag{12.2.3}$$

where the functions F are defined in (5.5.2)–(5.5.4), (5.5.6)–(5.5.8), and (5.5.10)–(5.5.12). Since only the three functions $F_1(\rho, z+d)$ involve Q and P, the other six are unchanged when Q_3 is substituted for Q and P_3 for P. These latter are the same as given in (5.5.23)–(5.5.25). Similarly, the three components of the magnetic field are given by (5.5.13), (5.5.14), and (5.5.18). They are

$$B_{1\rho}(\rho, \phi, z) = -\frac{\mu_0}{4\pi} \sin\phi \, [G_{\rho 0}(\rho, z-d) + G_{\rho 0}(\rho, z+d) + G_{\rho 1}(\rho, z+d)], \tag{12.2.4}$$

$$B_{1\phi}(\rho, \phi, z) = -\frac{\mu_0}{4\pi} \cos\phi \, [G_{\phi 0}(\rho, z-d) + G_{\phi 0}(\rho, z+d) + G_{\phi 1}(\rho, z+d)], \tag{12.2.5}$$

$$B_{1z}(\rho, \phi, z) = \frac{i\mu_0}{4\pi} \sin\phi \, [G_{z0}(\rho, z-d) - G_{z0}(\rho, z+d) + G_{z1}(\rho, z+d)]. \tag{12.2.6}$$

The functions G are defined in (5.5.15)–(5.5.17) and (5.5.19)–(5.5.21). Since $G_{\rho 0}(\rho, z \pm d)$, $G_{\phi 0}(\rho, z \pm d)$, and $G_{z0}(\rho, z \pm d)$ do not involve Q and P, they are the same as given by (5.5.26) and (5.5.27). Specifically,

$$F_{\rho 0}(\rho, z \pm d) = -e^{ik_1 r}\left[\frac{2k_1}{r^2} + \frac{2i}{r^3} + \frac{(z \pm d)^2}{r^2}\left(\frac{ik_1^2}{r} - \frac{3k_1}{r^2} - \frac{3i}{r^3}\right)\right], \tag{12.2.7}$$

$$F_{\phi 0}(\rho, z \pm d) = -e^{ik_1 r}\left(\frac{ik_1^2}{r} - \frac{k_1}{r^2} - \frac{i}{r^3}\right), \tag{12.2.8}$$

$$F_{z0}(\rho, z \pm d) = -e^{ik_1 r}\left(\frac{\rho}{r}\right)\left(\frac{z \pm d}{r}\right)\left(\frac{k_1^2}{r} + \frac{3ik_1}{r^2} - \frac{3}{r^3}\right), \tag{12.2.9}$$

$$G_{\rho 0}(\rho, z \pm d) = G_{\phi 0}(\rho, z \pm d) = -e^{ik_1 r}\left(\frac{z \pm d}{r}\right)\left(\frac{ik_1}{r} - \frac{1}{r^2}\right), \tag{12.2.10}$$

$$G_{z0}(\rho, z \pm d) = -e^{ik_1 r}\left(\frac{\rho}{r}\right)\left(\frac{k_1}{r} + \frac{i}{r^2}\right). \tag{12.2.11}$$

In these formulas, $r = r_2 = [\rho^2 + (z + d)^2]^{1/2}$ for the upper sign; $r = r_1 = [\rho^2 + (z - d)^2]^{1/2}$ for the lower sign. It remains to evaluate the following integrals:

$$F_{\rho 1}(\rho, z + d) = \int_0^\infty \left(\frac{\gamma_1}{2}(Q_3 - 1)[J_0(\lambda\rho) - J_2(\lambda\rho)] - \frac{k_1^2}{2\gamma_1}(P_3 + 1)[J_0(\lambda\rho) + J_2(\lambda\rho)]\right)e^{i\gamma_1(z+d)}\lambda\, d\lambda, \tag{12.2.12}$$

$$F_{\phi 1}(\rho, z + d) = \int_0^\infty \left(\frac{\gamma_1}{2}(Q_3 - 1)[J_0(\lambda\rho) + J_2(\lambda\rho)] - \frac{k_1^2}{2\gamma_1}(P_3 + 1)[J_0(\lambda\rho) - J_2(\lambda\rho)]\right)e^{i\gamma_1(z+d)}\lambda\, d\lambda, \tag{12.2.13}$$

$$F_{z1}(\rho, z + d) = \int_0^\infty (Q_3 - 1)J_1(\lambda\rho)e^{i\gamma_1(z+d)}\lambda^2\, d\lambda, \tag{12.2.14}$$

$$\left.\begin{matrix} G_{\rho 1}(\rho, z + d) \\ G_{\phi 1}(\rho, z + d) \end{matrix}\right\} = \int_0^\infty \left(\frac{1}{2}(Q_3 - 1)[J_0(\lambda\rho) \pm J_2(\lambda\rho)] - \frac{1}{2}(P_3 + 1)[J_0(\lambda\rho) \mp J_2(\lambda\rho)]\right)e^{i\gamma_1(z+d)}\lambda\, d\lambda, \tag{12.2.15}$$

$$G_{z1}(\rho, z + d) = -\int_0^\infty (P_3 - 1)\gamma_1^{-1}J_1(\lambda\rho)e^{i\gamma_1(z+d)}\lambda^2\, d\lambda. \tag{12.2.16}$$

12.3 Approximate Formulas for the Quantity $(\gamma_1/2)(Q_3 - 1)$

In order to simplify the complicated factor $(Q_3 - 1)$ that occurs in some of the general integrals for the components of the electromagnetic field in Region 1 near its boundary with Region 2, use is made of the approximate procedure described in Appendix D and applied in Chapter 5 to the simpler two-layer problem. For use in the three-layered region, the method must be generalized from two to three parameters k_j, $j = 1, 2, 3$, with the relations $|k_1|^2 \gg |k_2|^2 \gg |k_3|^2$. In effect, the procedure replaces a single integral containing γ_1, γ_2, and γ_3, where $\gamma_j = (k_j^2 - \lambda^2)^{1/2}$, by a set of five integrals such that there is one integral that contains γ_1, a second that contains γ_2, a third that contains γ_3, and two that contain no γ's. If available, the integrated formulas that would represent the five integrals would include terms which characterize radially propagating waves, as well as other radially nonpropagating terms that would cancel to zero when all of the terms from all of the integrals were added. That this is true is shown in Chapter 5 for the two-layer region. Since in the approximate analysis of the three-layer region in this chapter attention is directed only to the lateral wave that propagates in Region 3 (rock), and the integrals for the field in Region 2 (sediment) are not evaluated, some of the radially nonpropagating terms obtained are not canceled. Since, in a complete evaluation of all integrals, all such terms would be canceled, it is appropriate simply to omit those that are obtained as soon as they are identified.

The generalized form of the approximate procedure described in Appendix D is carried out in terms of a function $f(\lambda; k_1, k_2, k_3)$. This is approximated by a sum of five terms defined as follows:

$$f_1 = f(\lambda \sim k_1; k_1, k_2, k_3), \tag{12.3.1a}$$

$$f_2 = f(\lambda \sim k_2; k_1, k_2, k_3), \tag{12.3.1b}$$

$$f_3 = f(\lambda \sim k_3; k_1, k_2, k_3), \tag{12.3.1c}$$

$$f_4 = f_1(\lambda \sim k_2); \qquad f_5 = f_2(\lambda \sim k_3). \tag{12.3.2}$$

Then,

$$f(\lambda; k_1, k_2, k_3) \sim f_1 + f_2 + f_3 - f_4 - f_5. \tag{12.3.3}$$

When applied to

$$f \equiv f(\lambda; k_1, k_2, k_3) = \frac{\gamma_1}{2}(Q_3 - 1)e^{i\gamma_1(z+d)}, \tag{12.3.4}$$

subject to the conditions $|k_1|^2 \gg |k_2|^2 \gg |k_3|^2$ and with $(Q_3 - 1)$ given by (12.1.6), the result is

$$f = -\frac{\gamma_1^2 k_3^2[1 - i(k_2^2\gamma_3/k_3^2\gamma_2)\tan\gamma_2 l]e^{i\gamma_1(z+d)}}{\gamma_3 k_1^2 + \gamma_1 k_3^2 - i(\gamma_2 k_1^2 k_3^2/k_2^2 + k_2^2\gamma_1\gamma_3/\gamma_2)\tan\gamma_2 l}, \tag{12.3.5}$$

$$f_1 = -\frac{\gamma_1^2(k_3^2 - ik_2^2 \tan i\lambda l)e^{i\gamma_1(z+d)}}{i\lambda k_1^2 + \gamma_1 k_3^2 - i(i\lambda k_1^2 k_3^2/k_2^2 + k_2^2\gamma_1)\tan i\lambda l}, \tag{12.3.6}$$

$$f_2 = -\frac{k_1^2 k_3^2[1 - i(i\lambda k_2^2/k_3^2\gamma_2)\tan\gamma_2 l]e^{ik_1(z+d)}}{i\lambda k_1^2 + k_1 k_3^2 - i(\gamma_2 k_1^2 k_3^2/k_2^2 + k_2^2 k_1 i\lambda/\gamma_2)\tan\gamma_2 l}, \tag{12.3.7}$$

$$f_3 = -\frac{k_1^2 k_3^2[1 - i(k_2\gamma_3/k_3^2)\tan k_2 l]e^{ik_1(z+d)}}{\gamma_3 k_1^2 + k_1 k_3^2 - i(k_1^2 k_3^2/k_2 + k_1 k_2\gamma_3)\tan k_2 l}, \tag{12.3.8}$$

$$f_4 = -\frac{k_1^2(k_3^2 - ik_2^2\tan i\lambda l)e^{ik_1(z+d)}}{i\lambda k_1^2 + k_1 k_3^2 - i(i\lambda k_1^2 k_3^2/k_2^2 + k_1 k_2^2)\tan i\lambda l}, \tag{12.3.9}$$

$$f_5 = -\frac{k_1^2 k_3^2[1 + (\lambda k_2/k_3^2)\tan k_2 l]e^{ik_1(z+d)}}{i\lambda k_1^2 + k_1 k_3^2 - i(k_1^2 k_3^2/k_2 + i\lambda k_1 k_2)\tan k_2 l}. \tag{12.3.10}$$

An examination of these five functions reveals that f_1 involves only $\gamma_1 = (k_1^2 - \lambda^2)^{1/2}$ so that in the integration over λ with Bessel functions with argument $\lambda\rho$, only exponential terms of the type $\exp(ik_1\rho)$ can occur. This means that f_1 describes propagation in Region 1. More precisely, f_1 has branch points at $\lambda = \pm k_1$; it is the contribution from the vicinity of these branch points that leads to the exponential $\exp(ik_1\rho)$. Similarly, f_3 involves only $\gamma_3 = (k_3^2 - \lambda^2)^{1/2}$ and has branch points at $\lambda = \pm k_3$ so that it describes propagation of the type $\exp(ik_3\rho)$ in Region 3. On the other hand, f_2 involves only $\gamma_2 = (k_2^2 - \lambda^2)^{1/2}$ but has no branch points at $\lambda = \pm k_2$. This can be seen readily by noticing that each term in f_2 is an even function of γ_2. The absence of these branch points in f_2 means that there are no terms of the type $\exp(ik_2\rho)$. Instead, the exponents are related to the zeros of the denominator of f_2. Physically, this is readily understood: Region 2 acts as a parallel-plate waveguide, with one magnetic boundary and one electric boundary, approximately. The propagation in such a waveguide is controlled by the waveguide modes, not the free-space wave number. Therefore, f_2 needs to be treated in a way quite different from f_1 and f_3. This treatment is being studied, but is beyond the scope and the specific needs of this chapter. It may be noted that when $|k_2 l| \ll 1$, there are no radial propagating modes in Region 2.

The simplification of the five functions $f_1, \ldots, f_5$, so that the integrals in which they occur can be evaluated, is to be carried out for f_1 and $(f_3 - f_5)$. It will not be attempted for $(f_2 - f_4)$ for the reason given above. This means that the field calculated at the point of observation will include explicit formulas only for the direct field in Region 1 and for the lateral wave that travels in Region 3. It is an adequate representation of the complete field only when the "trapped" waves traveling in Region 2 are attenuated so as to make them negligible compared with the lateral waves traveling in Region 3. This is generally true at sufficient radial distances from the source when Region 2 is a layer of sediment on the sea floor with significant attenuation and Region 3 is the rock of the oceanic crust with relatively low attenuation. Since Region 1 is sea water with very high attenuation, the direct field arriving at the point of observation will be undetectably small compared to the lateral wave. These observations suggest that only the terms $(f_3 - f_5)$ need to be evaluated to obtain the desired field since all others will contribute only negligibly. Actually, it is of interest also to evaluate the field that travels in the sea water, Region 1, in order to compare this with the corresponding field for the two-layer case described in Chapter 5. It is to be expected that it must be largely independent of the presence or absence of the layer of sediment so long as $|k_1|^2 \gg |k_2|^2 \gg |k_3|^2$.

In the following rather long calculations, all contributions to the fields that propagate in Region 1 and Region 3 will be evaluated. No attempt will be made to evaluate the field in Region 2.

The principal contributions to the field in Region 1 come from the integrals with $(k_1^2/2\gamma_1)(P_3+1)$ in the integrand. These are an order of magnitude greater than contributions from integrals with $(\gamma_1/2)(Q_3-1)$ in the integrand. Only in components that involve only integrals with (Q_3-1) in the integrand is it necessary to evaluate f_1 (or its counterpart) even though it is very small. For this purpose, only the leading term in f_1 is required. This is obtained from (12.3.6) by retaining only the term $i\lambda k_1^2$ in the denominator. This approximation leads to

$$f_1 \sim \frac{i\gamma_1^2}{\lambda k_1^2}(k_3^2 - ik_2^2 \tan i\lambda l)e^{i\gamma_1(z+d)}. \tag{12.3.11}$$

The function f_3 can be simplified with the following rearrangement:

$$f_3 = -\frac{k_1^2(k_3^2 - ik_2\gamma_3 \tan k_2 l)e^{ik_1(z+d)}}{k_1^2\gamma_3[1 - i(k_2/k_1)\tan k_2 l] + k_3^2 k_1[1 - i(k_1/k_2)\tan k_2 l]}. \tag{12.3.12}$$

Let

$$b \equiv \frac{1 - i(k_1/k_2)\tan k_2 l}{1 - i(k_2/k_1)\tan k_2 l}. \tag{12.3.13}$$

Then,

$$f_3 = -\frac{k_1^2(k_3^2 - ik_2\gamma_3 \tan k_2 l)e^{ik_1(z+d)}}{(k_1^2\gamma_3 + bk_1k_3^2)[1 - i(k_2/k_1)\tan k_2 l]}.$$

This can be expanded into

$$\begin{aligned} f_3 &= -\frac{e^{ik_1(z+d)}}{1 - i(k_2/k_1)\tan k_2 l} \\ &\quad \times \left(\frac{k_1^2 k_3^2 - ik_2 \tan k_2 l\,[k_1^2\gamma_3 - (k_1^2\gamma_3 + bk_1k_3^2) + (k_1^2\gamma_3 + bk_1k_3^2)]}{k_1^2\gamma_3 + bk_1k_3^2}\right) \\ &= -\frac{e^{ik_1(z+d)}}{1 - i(k_2/k_1)\tan k_2 l}\left(\frac{k_1k_3^2(k_1 + ibk_2\tan k_2 l)}{k_1^2\gamma_3 + bk_1k_3^2} - ik_2\tan k_2 l\right). \end{aligned} \tag{12.3.14}$$

The numerator in (12.3.14) can be expanded with (12.3.13) to give

$$\begin{aligned} k_1^2 k_3^2 &\left[1 + \frac{ik_2}{k_1} \tan k_2 l \left(\frac{1 - i(k_1/k_2)\tan k_2 l}{1 - i(k_2/k_1)\tan k_2 l}\right)\right] \\ &= \frac{k_1^2 k_3^2 [1 - i(k_2/k_1)\tan k_2 l + i(k_2/k_1)\tan k_2 l + \tan^2 k_2 l]}{1 - i(k_2/k_1)\tan k_2 l} \\ &= \frac{k_1^2 k_3^2 (1 + \tan^2 k_2 l)}{1 - i(k_2/k_1)\tan k_2 l}. \end{aligned}$$

When this is used in (12.3.14), this becomes

$$f_3 = -k_1^2 k_3^2 A e^{ik_1(z+d)} \left(\frac{1}{k_1^2 \gamma_3 + b k_3^2 k_1}\right) + \frac{ik_2 \tan k_2 l \, e^{ik_1(z+d)}}{1 - i(k_2/k_1)\tan k_2 l}, \tag{12.3.15}$$

where

$$A \equiv \left(\cos k_2 l - \frac{ik_2}{k_1}\sin k_2 l\right)^{-2}. \tag{12.3.16}$$

Since f_5 differs from f_3 only in the appearance of $i\lambda$ for γ_3, it can be rearranged in the same manner as f_3 to give

$$f_5 = -k_1^2 k_3^2 A e^{ik_1(z+d)} \left(\frac{1}{i\lambda k_1^2 + b k_3^2 k_1}\right) + \frac{ik_2 \tan k_2 l \, e^{ik_1(z+d)}}{1 - i(k_2/k_1)\tan k_2 l}. \tag{12.3.17}$$

It follows that

$$f_3 - f_5 = -k_1^2 k_3^2 A e^{ik_1(z+d)} \left(\frac{1}{\gamma_3 k_1^2 + b k_3^2 k_1} - \frac{1}{i\lambda k_1^2 + b k_3^2 k_1}\right). \tag{12.3.18}$$

Following the method used in Chapter 5, the first term in (12.3.18) can be expanded as follows:

$$\frac{1}{\gamma_3 k_1^2 + b k_3^2 k_1} = \left(\frac{1}{\gamma_3 k_1^2 + b k_3^2 k_1} - \frac{1}{\gamma_3 k_1^2}\right) + \frac{1}{\gamma_3 k_1^2}. \tag{12.3.19a}$$

The last term in (12.3.18) can be rewritten to give

$$\frac{1}{i\lambda k_1^2 + b k_3^2 k_1} = -\frac{i}{k_1^2}\left(\frac{1}{\lambda - ik_b}\right); \qquad k_b \equiv \frac{b k_3^2}{k_1}. \tag{12.3.19b}$$

Hence,

$$\begin{aligned} f_3 - f_5 = &-k_1^2 k_3^2 A e^{ik_1(z+d)} \left(\frac{1}{\gamma_3 k_1^2 + b k_3^2 k_1} - \frac{1}{\gamma_3 k_1^2}\right) \\ &- \frac{k_3^2 A e^{ik_1(z+d)}}{\gamma_3} - \frac{ik_3^2 A e^{ik_1(z+d)}}{\lambda - ik_b}. \end{aligned} \tag{12.3.20}$$

12.4 Approximate Formulas for the Quantity $(k_1^2/2\gamma_1)(P_3+1)$

The factor (P_3+1) which occurs in the general integrals for the components of the field can be treated in a manner similar to that used for (Q_3-1) in Section 12.3. In this case, the general function is

$$\begin{aligned} g(\lambda; k_1, k_2, k_3) &= \frac{k_1^2}{2\gamma_1}(P_3+1)e^{i\gamma_1(z+d)} \\ &= \frac{k_1^2}{\gamma_1}\left(\frac{(\gamma_3 - i\gamma_2\tan\gamma_2 l)e^{i\gamma_1(z+d)}}{\gamma_3[1-i(\gamma_1/\gamma_2)\tan\gamma_2 l] + \gamma_1[1-i(\gamma_2/\gamma_1)\tan\gamma_2 l]}\right) \\ &\sim g_1 + g_2 + g_3 - g_4 - g_5, \end{aligned} \tag{12.4.1}$$

where

$$g_1 = \frac{k_1^2}{\gamma_1}\left(\frac{i\lambda(1-i\tan i\lambda l)e^{i\gamma_1(z+d)}}{i\lambda[1-i(\gamma_1/i\lambda)\tan i\lambda l] + \gamma_1[1-i(i\lambda/\gamma_1)\tan i\lambda l]}\right), \tag{12.4.2}$$

$$g_2 = k_1\left(\frac{(i\lambda - i\gamma_2\tan\gamma_2 l)e^{ik_1(z+d)}}{i\lambda[1-i(k_1/\gamma_2)\tan\gamma_2 l] + k_1[1-i(\gamma_2/k_1)\tan\gamma_2 l]}\right), \tag{12.4.3}$$

$$g_3 = k_1\left(\frac{(\gamma_3 - ik_2\tan k_2 l)e^{ik_1(z+d)}}{\gamma_3[1-i(k_1/k_2)\tan k_2 l] + k_1[1-i(k_2/k_1)\tan k_2 l]}\right), \tag{12.4.4}$$

$$g_4 = k_1\left(\frac{i\lambda(1-i\tan i\lambda l)e^{ik_1(z+d)}}{i\lambda[1-i(k_1/i\lambda)\tan i\lambda l] + k_1[1-i(i\lambda/k_1)\tan i\lambda l]}\right), \tag{12.4.5}$$

$$g_5 = k_1\left(\frac{(i\lambda - ik_2\tan k_2 l)e^{ik_1(z+d)}}{i\lambda[1-i(k_1/k_2)\tan k_2 l] + k_1[1-i(k_2/k_1)\tan k_2 l]}\right). \tag{12.4.6}$$

The functions g_1 and $(g_3 - g_5)$ can also be rearranged and simplified. As with $(f_2 - f_4)$, the terms $(g_2 - g_4)$ are omitted. Specifically,

$$\begin{aligned} g_1 &= \frac{k_1^2}{\gamma_1}\left(\frac{i\lambda(1-i\tan i\lambda l)e^{i\gamma_1(z+d)}}{i\lambda(1-i\tan i\lambda l) + \gamma_1(1-i\tan i\lambda l)}\right) \\ &= \frac{ik_1^2\lambda e^{i\gamma_1(z+d)}}{\gamma_1(i\lambda+\gamma_1)} = \frac{ik_1^2\lambda e^{i\gamma_1(z+d)}}{\gamma_1}\,\frac{\gamma_1 - i\lambda}{k_1^2 - \lambda^2 + \lambda^2} \\ &= e^{i\gamma_1(z+d)}\left(i\lambda + \frac{\lambda^2}{\gamma_1}\right), \end{aligned} \tag{12.4.7}$$

$$g_3 = k_1\left(\frac{(\gamma_3 - ik_2\tan k_2 l)e^{ik_1(z+d)}}{[1-i(k_1/k_2)\tan k_2 l](\gamma_3 + k_{12})}\right), \tag{12.4.8}$$

and

$$g_5 = k_1\left(\frac{(i\lambda - ik_2\tan k_2 l)e^{ik_1(z+d)}}{[1-i(k_1/k_2)\tan k_2 l](i\lambda + k_{12})}\right), \tag{12.4.9}$$

where

$$k_{12} = \frac{k_1}{b} = k_1\left(\frac{1 - i(k_2/k_1)\tan k_2 l}{1 - i(k_1/k_2)\tan k_2 l}\right). \tag{12.4.10}$$

It follows that

$$\begin{aligned}
g_3 - g_5 &= \frac{k_1 e^{ik_1(z+d)}}{1 - i(k_1/k_2)\tan k_2 l}\left(\frac{\gamma_3 - ik_2 \tan k_2 l}{\gamma_3 + k_{12}} - \frac{i\lambda - ik_2\tan k_2 l}{i\lambda + k_{12}}\right) \\
&= \frac{k_1 e^{ik_1(z+d)}}{1 - i(k_1/k_2)\tan k_2 l}\left[\frac{\gamma_3}{\gamma_3 + k_{12}} - \frac{ik_2 \tan k_2 l}{k_{12}}\left(1 - \frac{\gamma_3}{\gamma_3 + k_{12}}\right)\right. \\
&\qquad \left. - \frac{i\lambda}{i\lambda + k_{12}} + \frac{ik_2\tan k_2 l}{k_{12}}\left(1 - \frac{i\lambda}{i\lambda + k_{12}}\right)\right] \\
&= \frac{k_1 e^{ik_1(z+d)}}{1 - i(k_1/k_2)\tan k_2 l}\left[\frac{\gamma_3}{\gamma_3 + k_{12}}\left(1 + \frac{ik_2}{k_{12}}\tan k_2 l\right)\right. \\
&\qquad \left. - \frac{i\lambda}{i\lambda + k_{12}}\left(1 + \frac{ik_2}{k_{12}}\tan k_2 l\right)\right] \\
&= \frac{k_1[1 + i(k_2/k_{12})\tan k_2 l]}{1 - i(k_1/k_2)\tan k_2 l}\left(\frac{\gamma_3}{\gamma_3 + k_{12}} - \frac{i\lambda}{i\lambda + k_{12}}\right)e^{ik_1(z+d)}.
\end{aligned} \tag{12.4.11}$$

However, with (12.4.10),

$$\begin{aligned}
1 + \frac{ik_2}{k_{12}}\tan k_2 l &= 1 + \frac{ik_2 \tan k_2 l\,[1 - i(k_1/k_2)\tan k_2 l]}{k_1[1 - i(k_2/k_1)\tan k_2 l]} \\
&= \frac{k_1 + k_1\tan^2 k_2 l}{k_1[1 - i(k_2/k_1)\tan k_2 l]} \\
&= \frac{\sec^2 k_2 l}{1 - i(k_2/k_1)\tan k_2 l},
\end{aligned} \tag{12.4.12}$$

so that (12.4.11) becomes

$$\begin{aligned}
g_3 - g_5 = &\frac{k_1 \sec^2 k_2 l\, e^{ik_1(z+d)}}{[1 - i(k_1/k_2)\tan k_2 l][1 - i(k_2/k_1)\tan k_2 l]} \\
&\times \left(\frac{\gamma_3}{\gamma_3 + k_{12}} - \frac{i\lambda}{i\lambda + k_{12}}\right).
\end{aligned}$$

With (12.3.13) and (12.3.16), this is

$$g_3 - g_5 = \frac{k_1 A}{b}\left(\frac{\gamma_3}{\gamma_3 + k_{12}} - \frac{i\lambda}{i\lambda + k_{12}}\right)e^{ik_1(z+d)}. \tag{12.4.13}$$

When Region 2 is a layer of sediment on the sea floor and the very low frequencies involved in geophysical exploration are used, k_2 is complex with values that keep $\tan k_2 l$ of the order of magnitude of 1. When this is true, the wave number k_{12} defined in (12.4.10) has the order of magnitude of k_2 so that

$$|k_{12}|^2 \gg |k_3|^2. \tag{12.4.14}$$

The quantity (12.4.13)—which occurs in the integrands of the integrals for the components of the electromagnetic field in the sea near the sea floor—can be expressed in simpler form. This is accomplished with the expansion of the quantity $\gamma_3/\gamma_{12}(\gamma_3+\gamma_{12})$, where $\gamma_{12} = (k_{12}^2 - \lambda^2)^{1/2}$. With (12.4.14), the approximate procedure for the two parameters k_{12} and k_3 gives

$$\frac{\gamma_3}{\gamma_{12}(\gamma_{12}+\gamma_3)} \sim \frac{i\lambda}{\gamma_{12}(\gamma_{12}+i\lambda)} + \frac{\gamma_3}{k_{12}(k_{12}+\gamma_3)} - \frac{i\lambda}{k_{12}(k_{12}+i\lambda)}. \tag{12.4.15}$$

It follows that (12.4.13) can be written in the alternative form

$$\begin{aligned} g_3 - g_5 &\sim \frac{k_1 k_{12} A}{b}\left(\frac{\gamma_3}{\gamma_{12}(\gamma_{12}+\gamma_3)} - \frac{i\lambda}{\gamma_{12}(\gamma_{12}+i\lambda)}\right) e^{ik_1(z+d)} \\ &= \frac{k_1 k_{12} A}{b}\left(\frac{\gamma_3(\gamma_{12}-\gamma_3)}{\gamma_{12}(k_{12}^2 - k_3^2)} - \frac{i\lambda(\gamma_{12}-i\lambda)}{\gamma_{12}k_{12}^2}\right) e^{ik_1(z+d)} \\ &= \frac{k_1 k_{12} A}{b}\left(\frac{\gamma_{12}^2 - \gamma_3^2 - \gamma_{12}^2 + \gamma_3\gamma_{12}}{\gamma_{12}(k_{12}^2 - k_3^2)} - \frac{i\lambda}{k_{12}^2} - \frac{\lambda^2}{k_{12}^2\gamma_{12}}\right) e^{ik_1(z+d)} \\ &\sim \frac{k_1 k_{12} A}{b}\left(\frac{1}{\gamma_{12}} - \frac{\gamma_{12}}{k_{12}^2} + \frac{\gamma_3}{k_{12}^2} - \frac{i\lambda}{k_{12}^2} - \frac{\lambda^2}{k_{12}^2\gamma_{12}}\right) e^{ik_1(z+d)} \\ &= \frac{k_1 k_{12} A}{b}\left[\frac{1}{\gamma_{12}} + \frac{1}{k_{12}^2}\left(\gamma_3 - \gamma_{12} - i\lambda - \frac{\lambda^2}{\gamma_{12}}\right)\right] e^{ik_1(z+d)}. \end{aligned} \tag{12.4.16}$$

This completes the simplification of $(\gamma_1/2)(Q_3 - 1)$ and $(k_1^2/2\gamma_1)(P_3+1)$.

12.5 Evaluation of $F_{\rho 1}(\rho, z+d)$ and $F_{\phi 1}(\rho, z+d)$

The formulas for $E_{1\rho}(\rho, \phi, z)$ and $E_{1\phi}(\rho, \phi, z)$ are

$$E_{1\rho}(\rho,\phi,z) = -\frac{\omega\mu_0}{4\pi k_1^2}\cos\phi\,[F_{\rho 0}(\rho, z-d) + F_{\rho 0}(\rho, z+d) + F_{\rho 1}(\rho, z+d)], \tag{12.5.1a}$$

$$E_{1\phi}(\rho,\phi,z) = \frac{\omega\mu_0}{4\pi k_1^2}\sin\phi\,[F_{\phi 0}(\rho, z-d) + F_{\phi 0}(\rho, z+d) + F_{\phi 1}(\rho, z+d)], \tag{12.5.1b}$$

where $F_{\rho 0}(\rho, z \pm d)$ and $F_{\phi 0}(\rho, z \pm d)$ are given in (12.2.7) and (12.2.8) and, from (12.2.12) and (12.2.13),

$$\left.\begin{matrix} F_{\rho 1}(\rho, z+d) \\ F_{\phi 1}(\rho, z+d) \end{matrix}\right\} = \int_0^\infty \left(\frac{\gamma_1}{2}(Q_3 - 1)[J_0(\lambda\rho) \mp J_2(\lambda\rho)] - \frac{k_1^2}{2\gamma_1}(P_3 + 1)[J_0(\lambda\rho) \pm J_2(\lambda\rho)] \right) e^{i\gamma_1(z+d)} \lambda \, d\lambda. \tag{12.5.2}$$

Here,

$$\frac{\gamma_1}{2}(Q_3 - 1)e^{i\gamma_1(z+d)} = f_1 + (f_2 - f_4) + (f_3 - f_5), \tag{12.5.3}$$

$$\frac{k_1^2}{2\gamma_1}(P_3 + 1)e^{i\gamma_1(z+d)} = g_1 + (g_2 - g_4) + (g_3 - g_5), \tag{12.5.4}$$

where the f_j are defined in (12.3.6)–(12.3.10) and the g_j in (12.4.2)–(12.4.6). Let the corresponding integrals be defined as follows:

$$L_j = \int_0^\infty f_j[J_0(\lambda\rho) \mp J_2(\lambda\rho)]\lambda \, d\lambda; \qquad j = 1, 2, 3, 4, 5, \tag{12.5.5}$$

$$M_j = \int_0^\infty g_j[J_0(\lambda\rho) \pm J_2(\lambda\rho)]\lambda \, d\lambda; \qquad j = 1, 2, 3, 4, 5. \tag{12.5.6}$$

Then,

$$F_{\rho 1}(\rho, z+d) = F_{\rho 2}(\rho, z+d) + F_{\rho 3}(\rho, z+d),$$
$$F_{\phi 1}(\rho, z+d) = F_{\phi 2}(\rho, z+d) + F_{\phi 3}(\rho, z+d),$$

where

$$\left.\begin{matrix} F_{\rho 2}(\rho, z+d) \\ F_{\phi 2}(\rho, z+d) \end{matrix}\right\} = L_1 + (L_2 - L_4) + (L_3 - L_5), \tag{12.5.7a}$$

$$\left.\begin{matrix} F_{\rho 3}(\rho, z+d) \\ F_{\phi 3}(\rho, z+d) \end{matrix}\right\} = -[M_1 + (M_2 - M_4) + (M_3 - M_5)]. \tag{12.5.7b}$$

Hence,

$$\left.\begin{matrix} F_{\rho 1}(\rho, z+d) \\ F_{\phi 1}(\rho, z+d) \end{matrix}\right\} = (L_1 - M_1) + (L_2 - L_4 - M_2 + M_4) + (L_3 - L_5 - M_3 + M_5). \tag{12.5.8}$$

In (12.5.8), the term $(L_1 - M_1)$ gives the entire contribution from $F_{\rho 1}(\rho, z+d)$ and $F_{\phi 1}(\rho, z+d)$ to the parts of the electric-field components $E_{1\rho}(\rho, \phi, z)$ and $E_{1\phi}(\rho, \phi, z)$ that propagate in Region 1 with the wave number k_1 and the factor $\exp(ik_1\rho)$. A comparison of the integrals in L_1 with those in M_1 shows that the latter are much greater so that L_1 is negligible compared with M_1. This conclusion is confirmed by the parallel two-layer cases with $l \to 0$ and $l \to \infty$. In both of these, the contributions to the field by the integrals with $(\gamma_1/2)(Q-1)$ in the integrand are negligible compared with the contributions from the integrals with $(k_1^2/2\gamma_1)(P+1)$ in the integrand. This means that $(L_1 - M_1) \sim -M_1$.

The integrals $(L_2 - L_4 - M_2 + M_4)$ represent the parts of the field that propagate in Region 2. For the special situation investigated in this chapter, Region 2 is a layer of sediment on the oceanic crust. It is highly attenuating compared with the rock in Region 3 so that, at sufficiently great radial distances from the source, the contributions to the field in Region 1 by waves in Region 2 are negligible compared with those from the lateral waves that travel in Region 3. At such distances, the contributions from the field that propagates in Region 1—the sea—are even smaller so that the entire significant field is the lateral wave. This is represented by $(L_3 - L_5 - M_3 + M_5)$.

Even though the field that propagates in the sea water (Region 1) is negligible at the points of observation, it is of interest to evaluate it for comparison with the corresponding field in the two-layer region, i.e., when $l \to 0$. It is to be expected that the field that travels in the sea water is largely independent of the presence of the layer of sediment.

With (12.3.20),

$$L_3 - L_5 = Ae^{ik_1(z+d)}\left(-k_3^2 \int_0^\infty \gamma_3^{-1}[J_0(\lambda\rho) \mp J_2(\lambda\rho)]\lambda\, d\lambda + \begin{Bmatrix} S_\rho(b;\rho,0) \\ S_\phi(b;\rho,0) \end{Bmatrix} + \begin{Bmatrix} H_{\rho 1}(b;\rho,0) \\ H_{\phi 1}(b;\rho,0) \end{Bmatrix}\right), \tag{12.5.9}$$

where

$$\begin{rcases} S_\rho(b;\rho,0) \\ S_\phi(b;\rho,0) \end{rcases} = -ik_3^2 \int_0^\infty \frac{1}{\lambda - ik_b}\,[J_0(\lambda\rho) \mp J_2(\lambda\rho)]\lambda\, d\lambda, \tag{12.5.10a}$$

$$\begin{rcases} H_{\rho 1}(b;\rho,0) \\ H_{\phi 1}(b;\rho,0) \end{rcases} = -k_3^2 k_1^2 \int_0^\infty \left(\frac{1}{k_1^2\gamma_3 + bk_3^2 k_1} - \frac{1}{k_1^2\gamma_3}\right) \times [J_0(\lambda\rho) \mp J_2(\lambda\rho)]\lambda\, d\lambda. \tag{12.5.10b}$$

With Appendix A, the first integral in (12.5.9) is

$$-k_3^2 \int_0^\infty \gamma_3^{-1}[J_0(\lambda\rho) - J_2(\lambda\rho)]\lambda\, d\lambda = -k_3^2 I_4(k_3;\rho,0) = 2k_3\left[\frac{1}{\rho^2} + \left(\frac{ik_3}{\rho} - \frac{1}{\rho^2}\right)e^{ik_3\rho}\right], \tag{12.5.11a}$$

$$-k_3^2 \int_0^\infty \gamma_3^{-1}[J_0(\lambda\rho) + J_2(\lambda\rho)]\lambda\, d\lambda = -k_3^2 I_3(k_3;\rho,0) = -\frac{2k_3}{\rho^2}\,(1 - e^{ik_3\rho}). \tag{12.5.11b}$$

The second integral in (12.5.9) is readily shown to contribute no radially propagating terms so that it must ultimately be canceled. The derivation follows.

$$\left.\begin{matrix} S_\rho \\ S_\phi \end{matrix}\right\} = -ik_3^2 \int_0^\infty \frac{\lambda}{\lambda + c} \left[J_0(\lambda\rho) \mp J_2(\lambda\rho)\right] d\lambda; \qquad c = -ik_b, \quad k_b = \frac{bk_3^2}{k_1}.$$

This can be rearranged with the two relations

$$\frac{\lambda}{\lambda + c} = 1 - \frac{c}{\lambda + c}; \qquad \frac{\lambda}{\lambda + c} = \frac{1}{c}\left(\lambda - \frac{\lambda^2}{\lambda + c}\right).$$

With these,

$$\left.\begin{matrix} S_\rho \\ S_\phi \end{matrix}\right\} = -ik_3^2 \left[\int_0^\infty \left(1 - \frac{c}{\lambda + c}\right) J_0(\lambda\rho)\, d\lambda \right.$$
$$\left. \mp \frac{1}{c} \int_0^\infty \left(\lambda - \frac{\lambda^2}{\lambda + c}\right) J_2(\lambda\rho)\, d\lambda \right].$$

Two of the integrals are simple Bessel functions. Thus,

$$\int_0^\infty J_0(\lambda\rho)\, d\lambda = \frac{1}{\rho}; \qquad \int_0^\infty J_2(\lambda\rho)\lambda\, d\lambda = \frac{2}{\rho^2}.$$

The other two integrals are evaluated with formula (1) of Watson (1958, p. 431). The results are

$$\int_0^\infty \frac{J_0(\lambda\rho)}{\lambda + c}\, d\lambda = \frac{\pi}{2}\left[H_0(c\rho) - Y_0(c\rho)\right]$$
$$= \frac{\pi}{2}\left[H_0(-ik_b\rho) - Y_0(-ik_b\rho)\right],$$
$$\int_0^\infty \frac{\lambda^2 J_2(\lambda\rho)}{\lambda + c}\, d\lambda = \frac{\pi c^2}{2}\left[H_{-2}(-ik_b\rho) - Y_{-2}(-ik_b\rho)\right].$$

Here, $H(-ik_b\rho)$ is the Struve function and $Y(-ik_b\rho)$ is the Bessel function of the second kind. With these values,

$$\left.\begin{matrix} S_\rho \\ S_\phi \end{matrix}\right\} = -ik_3^2 \left(\frac{1}{\rho} + \frac{ik_b\pi}{2}\left[H_0(-ik_b\rho) - Y_0(-ik_b\rho)\right] \right.$$
$$\left. \mp \frac{2i}{k_b\rho^2} \mp \frac{ik_b\pi}{2}\left[H_{-2}(-ik_b\rho) - Y_{-2}(-ik_b\rho)\right]\right)$$
$$= -\frac{ik_3^2}{\rho} \mp \frac{2k_3^2}{k_b\rho^2} + \frac{k_3^2 k_b\pi}{2}\{\left[H_0(-ik_b\rho) - Y_0(-ik_b\rho)\right]$$
$$\mp \left[H_{-2}(-ik_b\rho) - Y_{-2}(-ik_b\rho)\right]\}.$$

Since none of these terms represents a radially propagating wave, all must be canceled in the complete formula. They will be omitted.

The integrals $H_{\rho 1}(b; \rho, 0)$ and $H_{\phi 1}(b; \rho, 0)$ involve the new factor b when compared with the corresponding integrals in Chapter 5. They are readily evaluated in the manner carried out with $b = 1$ in Appendix E. For all but small arguments,

$$J_0(\lambda\rho) - J_2(\lambda\rho) \sim \left(\frac{2}{\pi\lambda\rho}\right)^{1/2} e^{i(\lambda\rho - \pi/4)}, \tag{12.5.12a}$$

$$J_0(\lambda\rho) + J_2(\lambda\rho) = \frac{2}{\lambda\rho} J_1(\lambda\rho) \sim -\frac{i}{\lambda\rho}\left(\frac{2}{\pi\lambda\rho}\right)^{1/2} e^{i(\lambda\rho - \pi/4)}. \tag{12.5.12b}$$

With the approximation (12.5.12a), (12.5.10b) with the upper sign becomes

$$H_{\rho 1}(b; \rho, 0) = -k_3^2 k_1^2 \int_0^\infty \left(\frac{1}{k_1^2\gamma_3 + bk_3^2 k_1} - \frac{1}{k_1^2\gamma_3}\right) \times \left(\frac{2}{\pi\lambda\rho}\right)^{1/2} e^{i(\lambda\rho - \pi/4)} \lambda\, d\lambda. \tag{12.5.13}$$

The variable of integration is now changed from λ to τ with the substitution

$$\lambda = k_3(1 + m\tau); \qquad d\lambda = k_3 m\, d\tau; \qquad m = \frac{k_3^2}{2k_1^2}. \tag{12.5.14}$$

It follows that

$$\lambda^2 \sim k_3^2(1 + 2m\tau); \qquad \gamma_3 = (k_3^2 - \lambda^2)^{1/2} \sim (-2m\tau)^{1/2} k_3. \tag{12.5.15}$$

In phases, the substitution (12.5.14) is made; in amplitudes, $\lambda \sim k_3$. Thus, with

$$R = mk_3\rho = \frac{k_3^3\rho}{2k_1^2}, \tag{12.5.16}$$

(12.5.13) becomes

$$H_{\rho 1}(b; \rho, 0) = -k_1 k_3 \int_{-\infty}^\infty \left(\frac{1}{k_1(-2m\tau)^{1/2} + bk_3} - \frac{1}{k_1(-2m\tau)^{1/2}}\right) \times \left(\frac{2}{\pi k_3\rho}\right)^{1/2} e^{i(k_3\rho - \pi/4)} e^{iR\tau} m k_3^2\, d\tau. \tag{12.5.17}$$

The lower limit $\tau = -1/m = -2k_1^2/k_3^2$ is approximated by $-\infty$ since $|k_1^2/k_3^2| \gg 1$. This expression can be written as follows:

$$H_{\rho 1}(b; \rho, 0) = -mk_1 k_3^2 \left(\frac{2}{\pi k_3\rho}\right)^{1/2} e^{i(k_3\rho - \pi/4)} H(R), \tag{12.5.18}$$

where

$$H(R) = \int_{-\infty}^\infty \left(\frac{1}{\sqrt{-\tau} + b} - \frac{1}{\sqrt{-\tau}}\right) e^{iR\tau}\, d\tau. \tag{12.5.19}$$

With $t = \tau/b^2$ and $\sqrt{-1} = -i$, (12.5.19) becomes

$$H(R) = b \int_{-\infty}^{\infty} \frac{1}{\sqrt{t}\,(\sqrt{t}+i)}\, e^{iRb^2 t}\, dt. \tag{12.5.20}$$

This integral is the same as (E46) in Appendix E for $I_{\phi 1}$ but with Rb^2 appearing in place of R. The integrated value is given in (E58). With it, the integrated value of (12.5.20) is

$$H(R) = 4\pi b \left(\frac{i}{2}\right)^{1/2} e^{-iRb^2} \mathcal{F}(Rb^2), \tag{12.5.21}$$

where

$$\begin{aligned} \mathcal{F}(Rb^2) &= \int_{Rb^2}^{\infty} \frac{e^{it}}{(2\pi t)^{1/2}}\, dt \\ &= \tfrac{1}{2}(1+i) - C_2(Rb^2) - iS_2(Rb^2) \end{aligned} \tag{12.5.22}$$

and

$$C_2(Rb^2) + iS_2(Rb^2) = \int_0^{Rb^2} \frac{e^{it}}{(2\pi t)^{1/2}}\, dt \tag{12.5.23}$$

is the Fresnel integral. With (12.5.21), (12.5.18) becomes

$$H_{\rho 1}(b;\rho,0) = -2b\,\frac{k_3^4}{k_1}\left(\frac{\pi}{k_3\rho}\right)^{1/2} e^{ik_3\rho} e^{-iRb^2} \mathcal{F}(Rb^2). \tag{12.5.24}$$

Here

$$b = \frac{1 - i(k_1/k_2)\tan k_2 l}{1 - i(k_2/k_1)\tan k_2 l}. \tag{12.5.25}$$

It is readily verified that with (12.5.12b) replacing (12.5.12a),

$$\begin{aligned} H_{\phi 1}(b;\rho,0) &= -\frac{i}{k_3\rho}\, H_{\rho 1}(b;\rho,0) \\ &= 2ib\,\frac{k_3^3}{k_1\rho}\left(\frac{\pi}{k_3\rho}\right)^{1/2} e^{ik_3\rho} e^{-iRb^2} \mathcal{F}(Rb^2). \end{aligned} \tag{12.5.26}$$

With (12.5.24), (12.5.26), and (12.5.11a,b), (12.5.9) becomes

$$\begin{aligned} (L_3 - L_5)_\rho &= 2Ae^{ik_1(z+d)} \\ &\times \left\{ \frac{k_3}{\rho^2} + \left[\frac{ik_3^2}{\rho} - \frac{k_3}{\rho^2} - \frac{bk_3^4}{k_1}\left(\frac{\pi}{k_3\rho}\right)^{1/2} e^{-iRb^2}\mathcal{F}(Rb^2)\right] e^{ik_3\rho} \right\}, \end{aligned} \tag{12.5.27a}$$

$$\begin{aligned} (L_3 - L_5)_\phi &= 2Ae^{ik_1(z+d)} \\ &\times \left\{ -\frac{k_3}{\rho^2} + \left[\frac{k_3}{\rho^2} + \frac{ibk_3^3}{k_1\rho}\left(\frac{\pi}{k_3\rho}\right)^{1/2} e^{-iRb^2}\mathcal{F}(Rb^2)\right] e^{ik_3\rho} \right\}. \end{aligned} \tag{12.5.27b}$$

The radially propagating fields in these expressions have the factor $\exp(ik_3\rho)$. Those without this factor are terms that must be canceled by comparable terms in the other integrals. Since these are not all evaluated, the terms not associated with radially propagating waves are omitted. The desired components are

$$\left.\begin{array}{l} F_{\rho 2}(\rho, z+d) \\ F_{\phi 2}(\rho, z+d) \end{array}\right\} = L_1 + (L_2 - L_4) + (L_3 - L_5)$$
$$\sim (L_2 - L_4) + (L_3 - L_5)$$
$$\sim (L_2 - L_4) + (L_3 - L_5)(e^{ik_3\rho})$$

so that

$$F_{\rho 2}(\rho, z+d) \sim (L_2 - L_4)_\rho + 2Ae^{ik_1(z+d)}e^{ik_3\rho}\left[\frac{ik_3^2}{\rho} - \frac{k_3}{\rho^2} - \frac{bk_3^4}{k_1}\left(\frac{\pi}{k_3\rho}\right)^{1/2} e^{-iRb^2}\mathcal{F}(Rb^2)\right], \tag{12.5.28a}$$

$$F_{\phi 2}(\rho, z+d) \sim (L_2 - L_4)_\phi + 2Ae^{ik_1(z+d)}e^{ik_3\rho}\left[\frac{k_3}{\rho^2} + \frac{ibk_3^3}{k_1\rho}\left(\frac{\pi}{k_3\rho}\right)^{1/2} e^{-iRb^2}\mathcal{F}(Rb^2)\right]. \tag{12.5.28b}$$

The integrals M_j that involve $(k_1^2/2\gamma_1)(P_3+1)$ are evaluated in a similar manner. Thus, with (12.4.7) substituted in (12.5.6),

$$M_1 = i\int_0^\infty [J_0(\lambda\rho) \pm J_2(\lambda\rho)]e^{i\gamma_1(z+d)}\lambda^2\,d\lambda + \int_0^\infty \gamma_1^{-1}[J_0(\lambda\rho) \pm J_2(\lambda\rho)]e^{i\gamma_1(z+d)}\lambda^3\,d\lambda$$
$$= i\left\{\begin{array}{l} I_{20}(k_1; \rho, z+d) \\ I_{21}(k_1; \rho, z+d) \end{array}\right\} + \left\{\begin{array}{l} I_7(k_1; \rho, z+d) \\ I_8(k_1; \rho, z+d) \end{array}\right\}. \tag{12.5.29}$$

With Appendix A,

$$M_{1\rho} = i\left[\frac{2}{\rho^3}\,e^{ik_1(z+d)} - 2\left(\frac{z+d}{r_2}\right)\left(\frac{k_1}{\rho^2} + \frac{3i}{2\rho^3}\right)e^{ik_1r_2}\right] - 2\left(\frac{k_1}{r_2^2} + \frac{i}{r_2^3}\right)e^{ik_1r_2}, \tag{12.5.30a}$$

$$M_{1\phi} = i\left[-\frac{4}{\rho^3}\,e^{ik_1(z+d)} - 2\left(\frac{z+d}{\rho}\right)\left(\frac{ik_1^2}{\rho} - \frac{3k_1}{2\rho^2} - \frac{5i}{8\rho^3}\right)e^{ik_1r_2}\right]$$

$$- 2\left[\frac{ik_1^2}{r_2} - \frac{2k_1}{r_2^2} - \frac{2i}{r_2^3} - \left(\frac{z+d}{r_2}\right)^2\left(\frac{ik_1^2}{r_2} - \frac{3k_1}{r_2^2} - \frac{3i}{r_2^3}\right)\right]e^{ik_1r_2}. \tag{12.5.30b}$$

With (12.4.16) in (12.5.6),

$$\begin{aligned} M_3 - M_5 = \frac{k_1 A}{b}\, e^{ik_1(z+d)} &\times \left(k_{12}\int_0^\infty \gamma_{12}^{-1}[J_0(\lambda\rho) \pm J_2(\lambda\rho)]\lambda\, d\lambda\right. \\ &+ \frac{1}{k_{12}}\int_0^\infty \gamma_3[J_0(\lambda\rho) \pm J_2(\lambda\rho)]\lambda\, d\lambda \\ &- \frac{1}{k_{12}}\int_0^\infty \gamma_{12}[J_0(\lambda\rho) \pm J_2(\lambda\rho)]\lambda\, d\lambda \\ &- \frac{i}{k_{12}}\int_0^\infty [J_0(\lambda\rho) \pm J_2(\lambda\rho)]\lambda^2\, d\lambda \\ &\left. - \frac{1}{k_{12}}\int_0^\infty \gamma_{12}^{-1}[J_0(\lambda\rho) \pm J_2(\lambda\rho)]\lambda^3\, d\lambda\right). \end{aligned} \tag{12.5.31}$$

The several integrals are listed in Appendix A.

$$\int_0^\infty \gamma_{12}^{-1}[J_0(\lambda\rho) + J_2(\lambda\rho)]\lambda\, d\lambda = I_3(k_{12};\rho,0) = \frac{2}{k_{12}\rho^2}\,(1 - e^{ik_{12}\rho}), \tag{12.5.32a}$$

$$\int_0^\infty \gamma_{12}^{-1}[J_0(\lambda\rho) - J_2(\lambda\rho)]\lambda\, d\lambda = I_4(k_{12};\rho,0) = -\frac{2}{k_{12}}\left[\frac{1}{\rho^2} + \left(\frac{ik_{12}}{\rho} - \frac{1}{\rho^2}\right)e^{ik_{12}\rho}\right], \tag{12.5.32b}$$

$$\int_0^\infty \gamma_3[J_0(\lambda\rho) + J_2(\lambda\rho)]\lambda\, d\lambda = I_{11}(k_3;\rho,0) = 2k_3\left(\frac{1}{\rho^2} + \frac{i}{k_3\rho^3}\, e^{ik_3\rho}\right), \tag{12.5.33a}$$

$$\int_0^\infty \gamma_3[J_0(\lambda\rho) - J_2(\lambda\rho)]\lambda\, d\lambda = I_{12}(k_3;\rho,0) = -2k_3\left[\frac{1}{\rho^2} + \left(\frac{1}{\rho^2} + \frac{2i}{k_3\rho^3}\right)e^{ik_3\rho}\right], \tag{12.5.33b}$$

$$\int_0^\infty \gamma_{12}[J_0(\lambda\rho) + J_2(\lambda\rho)]\lambda\, d\lambda = I_{11}(k_{12};\rho,0) = 2k_{12}\left(\frac{1}{\rho^2} + \frac{i}{k_{12}\rho^3}\, e^{ik_{12}\rho}\right), \tag{12.5.34a}$$

$$\int_0^\infty \gamma_{12}[J_0(\lambda\rho) - J_2(\lambda\rho)]\lambda\, d\lambda$$

$$= I_{12}(k_{12};\rho,0) = -2k_{12}\left[\frac{1}{\rho^2} + \left(\frac{1}{\rho^2} + \frac{2i}{k_{12}\rho^3}\right)e^{ik_{12}\rho}\right], \qquad (12.5.34b)$$

$$\int_0^\infty [J_0(\lambda\rho) + J_2(\lambda\rho)]\lambda^2\, d\lambda = I_{20}(\rho,0) = \frac{2}{\rho^3}, \qquad (12.5.35a)$$

$$\int_0^\infty [J_0(\lambda\rho) - J_2(\lambda\rho)]\lambda^2\, d\lambda = I_{21}(\rho,0) = -\frac{4}{\rho^3}, \qquad (12.5.35b)$$

$$\int_0^\infty \gamma_{12}^{-1}[J_0(\lambda\rho) + J_2(\lambda\rho)]\lambda^3\, d\lambda$$

$$= I_7(k_{12};\rho,0) = -2k_{12}\left(\frac{1}{\rho^2} + \frac{i}{k_{12}\rho^3}\right)e^{ik_{12}\rho}, \qquad (12.5.36a)$$

$$\int_0^\infty \gamma_{12}^{-1}[J_0(\lambda\rho) - J_2(\lambda\rho)]\lambda^3\, d\lambda$$

$$= I_8(k_{12};\rho,0) = -2k_{12}\left(\frac{ik_{12}}{\rho} - \frac{2}{\rho^2} - \frac{2i}{k_{12}\rho^3}\right)e^{ik_{12}\rho}. \qquad (12.5.36b)$$

It follows that

$$M_3 - M_5 = \frac{k_1 A}{k_{12} b}\, e^{ik_1(z+d)}\left[k_{12}^2\begin{Bmatrix} I_3(k_{12};\rho,0) \\ I_4(k_{12};\rho,0)\end{Bmatrix} + \begin{Bmatrix} I_{11}(k_3;\rho,0) \\ I_{12}(k_3;\rho,0)\end{Bmatrix} - \begin{Bmatrix} I_{11}(k_{12};\rho,0) \\ I_{12}(k_{12};\rho,0)\end{Bmatrix} - i\begin{Bmatrix} I_{20}(\rho,0) \\ I_{21}(\rho,0)\end{Bmatrix} - \begin{Bmatrix} I_7(k_{12};\rho,0) \\ I_8(k_{12};\rho,0)\end{Bmatrix}\right]. \qquad (12.5.37)$$

With (12.4.10), $k_{12}b = k_1$ so that

$$(M_3 - M_5)_\rho = 2Ae^{ik_1(z+d)}\left(\frac{k_3}{\rho^2} - \frac{i}{\rho^3} + \frac{i}{\rho^3}\, e^{ik_3\rho}\right). \qquad (12.5.38a)$$

Similarly,

$$(M_3 - M_5)_\phi = 2Ae^{ik_1(z+d)}\left[-\frac{k_3}{\rho^2} + \frac{4i}{\rho^3} - \left(\frac{k_3}{\rho^2} + \frac{2i}{\rho^3}\right)e^{ik_3\rho}\right]. \qquad (12.5.38b)$$

With (12.5.30a,b) and (12.5.38a,b) in (12.5.7b), it follows that

$$F_{\rho 3}(\rho, z+d) = -(M_2 - M_4)_\rho - \frac{2i}{\rho^3}\, e^{ik_1(z+d)} + 2i\left(\frac{z+d}{\rho}\right)\left(\frac{k_1}{\rho^2} + \frac{3i}{2\rho^3}\right)e^{ik_1 r_2} + 2\left(\frac{k_1}{r_2^2} + \frac{i}{r_2^3}\right)e^{ik_1 r_2} - 2Ae^{ik_1(z+d)}\left(\frac{k_3}{\rho^2} - \frac{i}{\rho^3} + \frac{i}{\rho^3}\, e^{ik_3\rho}\right), \qquad (12.5.39)$$

$$F_{\phi 3}(\rho, z+d) = -(M_2 - M_4)_\phi + \frac{4i}{\rho^3} e^{ik_1(z+d)} + \left\{ 2i\left(\frac{z+d}{r_2}\right)\left(\frac{ik_1^2}{\rho} - \frac{3k_1}{2\rho^2} - \frac{5i}{8\rho^3}\right) + 2\left[\frac{ik_1^2}{r_2} - \frac{2k_1}{r_2^2} - \frac{2i}{r_2^3} - \left(\frac{z+d}{r_2}\right)^2 \left(\frac{ik_1^2}{r_2} - \frac{3k_1}{r_2^2} - \frac{3i}{r_2^3}\right)\right]\right\} e^{ik_1 r_2} - 2Ae^{ik_1(z+d)}\left[-\frac{k_3}{\rho^2} + \frac{4i}{\rho^3} - \left(\frac{k_3}{\rho^2} + \frac{2i}{\rho^3}\right)e^{ik_3\rho}\right]. \quad (12.5.40)$$

When only the radially propagating fields are retained, the results are

$$\left.\begin{array}{l} F_{\rho 3}(\rho, z+d) \\ F_{\phi 3}(\rho, z+d) \end{array}\right\} = -[M_1 + (M_2 - M_4) + (M_3 - M_5)] \sim -[M_1(e^{ik_1 r_2}) + (M_2 - M_4) + (M_3 - M_5)(e^{ik_3\rho})] \quad (12.5.41a)$$

so that

$$F_{\rho 3}(\rho, z+d) = -(M_2 - M_4)_\rho + 2k_1 e^{ik_1 r_2}\left[\frac{1}{r_2^2} + \frac{i}{k_1 r_2^3} + \left(\frac{z+d}{\rho}\right)\left(\frac{i}{\rho^2} - \frac{3}{2k_1\rho^3}\right)\right] - 2Ae^{ik_1(z+d)}e^{ik_3\rho}\left(\frac{i}{\rho^3}\right), \quad (12.5.41b)$$

$$F_{\phi 3}(\rho, z+d) = -(M_2 - M_4)_\phi + 2e^{ik_1 r_2}\left[\frac{ik_1^2}{r_2} - \frac{2k_1}{r_2^2} - \frac{2i}{r_2^3} - \left(\frac{z+d}{r_2}\right)^2\left(\frac{ik_1^2}{r_2} - \frac{3k_1}{r_2^2} - \frac{3i}{r_2^3}\right) + i\left(\frac{z+d}{r_2}\right)\left(\frac{ik_1^2}{\rho} - \frac{3k_1}{2\rho^2} - \frac{5i}{8\rho^3}\right)\right] + 2Ae^{ik_1(z+d)}e^{ik_3\rho}\left(\frac{k_3}{\rho^2} + \frac{2i}{\rho^3}\right). \quad (12.5.41c)$$

When the $F_{m1}(\rho, z+d) = F_{m2}(\rho, z+d) + F_{m3}(\rho, z+d)$, with $m = \rho$ and ϕ, are formed with (12.5.28a,b) and (12.5.41b,c), the results are

$$F_{\rho 1}(\rho, z+d) = (L_2 - L_4 - M_2 + M_4)_\rho + 2k_1 e^{ik_1 r_2}\left[\frac{1}{r_2^2} + \frac{i}{k_1 r_2^3} + \left(\frac{z+d}{\rho}\right)\left(\frac{i}{\rho^2} - \frac{3}{2k_1\rho^3}\right)\right]$$

$$+ 2k_3 A e^{ik_1(z+d)} e^{ik_3\rho}\left[\frac{ik_3}{\rho} - \frac{1}{\rho^2} - \frac{i}{k_3\rho^3} - \frac{bk_3^3}{k_1}\left(\frac{\pi}{k_3\rho}\right)^{1/2} e^{-iRb^2}\mathcal{F}(Rb^2)\right], \tag{12.5.42}$$

$$\begin{aligned} F_{\phi 1}(\rho, z+d) &= (L_2 - L_4 - M_2 + M_4)_\phi \\ &+ 2e^{ik_1r_2}\left[\frac{ik_1^2}{r_2} - \frac{2k_1}{r_2^2} - \frac{2i}{r_2^3} - \left(\frac{z+d}{r_2}\right)^2\left(\frac{ik_1^2}{r_2} - \frac{3k_1}{r_2^2} - \frac{3i}{r_2^3}\right)\right] \\ &+ 2ik_1 e^{ik_1r_2}\left(\frac{z+d}{r_2}\right)\left(\frac{ik_1}{\rho} - \frac{3}{2\rho^2} - \frac{5i}{8k_1\rho^3}\right) \\ &+ 4k_3 A e^{ik_1(z+d)} e^{ik_3\rho}\left[\frac{1}{\rho^2} + \frac{i}{k_3\rho^3} + \frac{ibk_3^2}{2k_1\rho}\left(\frac{\pi}{k_3\rho}\right)^{1/2} e^{-iRb^2}\mathcal{F}(Rb^2)\right], \end{aligned} \tag{12.5.43}$$

where

$$b = \frac{1 - i(k_1/k_2)\tan k_2 l}{1 - i(k_2/k_1)\tan k_2 l}; \qquad r_2 = [\rho^2 + (z+d)^2]^{1/2}, \tag{12.5.44}$$

$$A = \left(\cos k_2 l - \frac{ik_2}{k_1}\sin k_2 l\right)^{-2}; \qquad R = \frac{k_3^3\rho}{2k_1^2}, \tag{12.5.45}$$

$$\mathcal{F}(Rb^2) = \tfrac{1}{2}(1+i) - C_2(Rb^2) - iS_2(Rb^2). \tag{12.5.46}$$

When $l = 0$, these formulas reduce exactly to the corresponding ones in Chapter 5.

12.6 Evaluation of $F_{z1}(\rho, z+d)$; Formulas for the Electric Field

The integral for $F_{z1}(\rho, z+d)$ is (12.2.14), viz.,

$$F_{z1}(\rho, z+d) = \int_0^\infty (Q_3 - 1)e^{i\gamma_1(z+d)} J_1(\lambda\rho)\lambda^2\, d\lambda, \tag{12.6.1}$$

where $Q_3 - 1$ is given by (12.1.6). The quantity $(Q_3 - 1)e^{i\gamma_1(z+d)}$ can be expanded in the same manner as $(\gamma_1/2)(Q_3 - 1)e^{i\gamma_1(z+d)}$. With (12.3.4)–(12.3.10), it follows that the quantity of interest is

$$(Q_3 - 1)e^{i\gamma_1(z+d)} = \frac{2f}{\gamma_1} \sim 2\left(\frac{f_1}{\gamma_1} + \frac{(f_2 - f_4)}{k_1} + \frac{(f_3 - f_5)}{k_1}\right). \tag{12.6.2}$$

As before, the integrals $2(f_2 - f_4)/k_1$ are not involved in the lateral-wave transmission of interest here and are not evaluated. In the evaluation of f for use in $F_{\rho 1}(\rho, z+d)$ and $F_{\phi 1}(\rho, z+d)$, the contribution by f_1 was negligible since there were much greater contributions from g_1. In (12.6.1), there are no contributions from other terms so that $2f_1/\gamma_1$ is the entire field that propagates in Region 1 in the form $\exp(ik_1 r_1)$ or $\exp(ik_1 r_2)$. The wave traveling in Region 3 is given by $2(f_3 - f_5)/k_1$.

The leading terms in $2f_1/\gamma_1$, as obtained from (12.3.11), are

$$\frac{2f_1}{\gamma_1} \sim \frac{2i\gamma_1(k_3^2 - ik_2^2 \tan i\lambda l)e^{i\gamma_1(z+d)}}{\lambda k_1^2} \tag{12.6.3}$$

so that

$$\begin{aligned} L_{1z} &= 2\int_0^\infty \frac{f_1}{\gamma_1} J_1(\lambda\rho)\lambda^2 \, d\lambda \\ &= \frac{2ik_3^2}{k_1^2}\int_0^\infty \gamma_1 e^{i\gamma_1(z+d)} J_1(\lambda\rho)\lambda \, d\lambda \\ &\quad + \frac{2k_2^2}{k_1^2}\int_0^\infty \gamma_1 e^{i\gamma_1(z+d)} \tan i\lambda l \, J_1(\lambda\rho)\lambda \, d\lambda \\ &= \frac{2ik_3^2}{k_1^2}\frac{\rho}{2} I_{13}(k_1; \rho, z+d) + \frac{2k_2^2}{k_1^2} T_z(k_2 l, e^{ik_1\rho}) \\ &= \frac{2ik_3^2}{k_1}\left[\frac{1}{\rho^2} e^{ik_1(z+d)} + ie^{ik_1 r_2}\left(\frac{1}{\rho^2} + \frac{3i}{2k_1\rho^3}\right)\right] \\ &\quad + \frac{2k_2^2}{k_1^2} T_z(k_2 l, e^{ik_1\rho}), \end{aligned} \tag{12.6.4}$$

where the unevaluated integral T_z involves propagation in Region 1 that depends on the thickness l of the sediment. It is of order k_2^2/k_1^2 so that it is very small.

With (12.3.20),

$$\begin{aligned} \frac{2(f_3 - f_5)}{k_1} &= -2k_1 k_3^2 A e^{ik_1(z+d)}\left(\frac{1}{\gamma_3 k_1^2 + bk_3^2 k_1} - \frac{1}{\gamma_3 k_1^2}\right) \\ &\quad - \frac{2k_3^2 A e^{ik_1(z+d)}}{k_1\gamma_3} - \frac{ik_3^2 A e^{ik_1(z+d)}}{k_1(\lambda - ik_b)}. \end{aligned} \tag{12.6.5}$$

The associated integrals are

$$\begin{aligned} (L_3 - L_5)_z &= \frac{2}{k_1}\int_0^\infty (f_3 - f_5) J_1(\lambda\rho)\lambda^2 \, d\lambda \\ &= AH_{z1}(b; \rho, z) - \frac{2k_3^2}{k_1} A e^{ik_1(z+d)} \int_0^\infty \frac{1}{\gamma_3} J_1(\lambda\rho)\lambda^2 \, d\lambda \\ &\quad - \frac{ik_3^2}{k_1} A e^{ik_1(z+d)} \int_0^\infty \frac{J_1(\lambda\rho)\lambda^2}{\lambda - ik_b} \, d\lambda, \end{aligned} \tag{12.6.6}$$

where

$$H_{z1}(b;\rho,z) = H_{z1}(b;\rho,0)e^{ik_1(z+d)}$$
$$= -2k_1k_3^2e^{ik_1(z+d)}\int_0^\infty\left(\frac{1}{\gamma_3k_1^2+bk_1k_3^2}-\frac{1}{\gamma_3k_1^2}\right)$$
$$\times J_1(\lambda\rho)\lambda^2\,d\lambda. \tag{12.6.7}$$

Since, with (12.5.12b),

$$2J_1(\lambda\rho)\lambda^2 = [J_0(\lambda\rho)+J_2(\lambda\rho)]\lambda^3\rho \tag{12.6.8}$$

and since in the evaluation of the integral $\lambda \sim k_3$, it follows that

$$H_{z1}(b;\rho,0) = \frac{k_3^2\rho}{k_1}H_{\phi1}(b;\rho,0)$$
$$= 2ib\,\frac{k_3^5}{k_1^2}\left(\frac{\pi}{k_3\rho}\right)^{1/2}e^{ik_3\rho}e^{-iRb^2}\mathcal{F}(Rb^2). \tag{12.6.9}$$

The second integral in (12.6.6) is

$$\int_0^\infty\frac{1}{\gamma_3}J_1(\lambda\rho)\lambda^2\,d\lambda = \frac{\rho}{2}I_7(k_3;\rho,0) = -\left(\frac{k_3}{\rho}+\frac{i}{\rho^2}\right)e^{ik_3\rho}. \tag{12.6.10}$$

The last integral in (12.6.6) can be expressed in Struve and Bessel functions. It contains only radially nonpropagating terms that must cancel so that they are omitted. The result is

$$(L_3-L_5)_z = \frac{2k_3^2}{k_1}Ae^{ik_1(z+d)}e^{ik_3\rho}$$
$$\times\left[\frac{k_3}{\rho}+\frac{i}{\rho^2}+\frac{ibk_3^3}{k_1}\left(\frac{\pi}{k_3\rho}\right)^{1/2}e^{-iRb^2}\mathcal{F}(Rb^2)\right]. \tag{12.6.11}$$

When the several terms are combined to form (12.6.1) and all radially nonpropagating terms are omitted, the results are

$$F_{z1}(\rho,z+d) \sim L_{1z}(e^{ik_1r_2}) + (L_2-L_4)_z + (L_3-L_5)_z(e^{ik_3\rho})$$
$$= (L_2-L_4)_z$$
$$+\frac{2k_3^2}{k_1}Ae^{ik_1(z+d)}e^{ik_3\rho}$$
$$\times\left[\frac{k_3}{\rho}+\frac{i}{\rho^2}+\frac{ibk_3^3}{k_1}\left(\frac{\pi}{k_3\rho}\right)^{1/2}e^{-iRb^2}\mathcal{F}(Rb^2)\right]$$
$$-\frac{2k_3^2}{k_1}e^{ik_1r_2}\left(\frac{1}{\rho^2}+\frac{3i}{2k_1\rho^3}\right)+\frac{2k_2^2}{k_1^2}T_z(k_2l,e^{ik_1\rho}). \tag{12.6.12}$$

This formula reduces to (5.6.28) for the two-layered region when $l \to 0$.

When the formulas for $F_{\rho 1}(\rho, z+d)$, $F_{\phi 1}(\rho, z+d)$, and $F_{z1}(\rho, z+d)$ are combined with the formulas for $F_{\rho 0}(\rho, z\pm d)$, $F_{\phi 0}(\rho, z\pm d)$, and $F_{z0}(\rho, z\pm d)$—which are identical with those for the two-layer region in Chapter 5—the results are

$$\begin{aligned}
E_{1\rho}(\rho,\phi,z) = &-\frac{\omega\mu_0}{2\pi k_1^2}\cos\phi \\
&\times\left\{Ak_3 e^{ik_1(z+d)}e^{ik_3\rho}g(\nu,\rho,k_3) - \left(\frac{k_1}{r_1^2}+\frac{i}{r_1^3}\right)e^{ik_1r_1}\right. \\
&\quad + \left(\frac{z+d}{\rho}\right)\left(\frac{ik_1}{\rho^2}-\frac{3}{2\rho^3}\right)e^{ik_1r_2} \\
&\quad - \frac{1}{2}\left[\left(\frac{z-d}{r_1}\right)^2\left(\frac{ik_1^2}{r_1}-\frac{3k_1}{r_1^2}-\frac{3i}{r_1^3}\right)e^{ik_1r_1}\right. \\
&\qquad \left.\left. + \left(\frac{z+d}{r_2}\right)^2\left(\frac{ik_1^2}{r_2}-\frac{3k_1}{r_2^2}-\frac{3i}{r_2^3}\right)e^{ik_1r_2}\right]\right\} \\
&+ E_{1\rho}^{(2)}(\rho,\phi,z),
\end{aligned} \tag{12.6.13}$$

$$\begin{aligned}
E_{1\phi}(\rho,\phi,z) = &\frac{\omega\mu_0}{2\pi k_1^2}\sin\phi \\
&\times\left\{Ak_3 e^{ik_1(z+d)}e^{ik_3\rho}h(\nu,\rho,k_3)\right. \\
&\quad + \frac{1}{2}\left[e^{ik_1r_2}\left(\frac{ik_1^2}{r_2}-\frac{3k_1}{r_2^2}-\frac{3i}{r_2^3}\right)\right. \\
&\qquad \left. - e^{ik_1r_1}\left(\frac{ik_1^2}{r_1}-\frac{k_1}{r_1^2}-\frac{i}{r_1^3}\right)\right] \\
&\quad + ie^{ik_1r_2}\left(\frac{z+d}{r_2}\right)\left(\frac{ik_1^2}{\rho}-\frac{3k_1}{2\rho^2}-\frac{5i}{8\rho^3}\right) \\
&\quad \left. - e^{ik_1r_2}\left(\frac{z+d}{r_2}\right)^2\left(\frac{ik_1^2}{r_2}-\frac{3k_1}{r_2^2}-\frac{3i}{r_2^3}\right)\right\} \\
&+ E_{1\phi}^{(2)}(\rho,\phi,z),
\end{aligned} \tag{12.6.14}$$

$$\begin{aligned}
E_{1z}(\rho,\phi,z) = &\frac{\omega\mu_0}{2\pi k_1^2}\cos\phi \\
&\times\left\{\frac{k_3^2}{k_1}\left[Ae^{ik_1(z+d)}e^{ik_3\rho}f(\nu,\rho,k_3)\right.\right. \\
&\qquad \left. - e^{ik_1r_2}\left(\frac{i}{\rho^2}-\frac{3}{2k_1\rho^3}\right)\right] + \frac{ik_2^2}{k_1^2}T_z(k_2l, e^{ik_1\rho}) \\
&\quad - \frac{1}{2}\left[e^{ik_1r_1}\left(\frac{\rho}{r_1}\right)\left(\frac{z-d}{r_1}\right)\left(\frac{ik_1^2}{r_1}-\frac{3k_1}{r_1^2}-\frac{3i}{r_1^3}\right)\right.
\end{aligned}$$

$$+ e^{ik_1 r_2}\left(\frac{\rho}{r_2}\right)\left(\frac{z+d}{r_2}\right)\left(\frac{ik_1^2}{r_2} - \frac{3k_1}{r_2^2} - \frac{3i}{r_2^3}\right)\Bigg]\Bigg\}$$

$$+ E_{1z}^{(2)}(\rho, \phi, z). \tag{12.6.15}$$

The functions $E^{(2)}(\rho, \phi, z)$ represent the unevaluated parts of the field that involve propagation in Region 2. The term $(k_2^2/k_1^2)T_z(k_2 l, e^{ik_1\rho})$ is negligible when $(k_2^2/k_1^2) \ll (z+d)/r_2$. All terms multiplied by $\exp(ik_1 r_1)$ or $\exp(ik_1 r_2)$ are negligible when Region 1 is sea water and $|k_1\rho| > 3$. The formulas (12.6.13)–(12.6.15) reduce to the corresponding formulas for the two-layer region derived in Chapter 5 when $l \to 0$. In these formulas,

$$A = \left(\cos k_2 l - \frac{ik_2}{k_1}\sin k_2 l\right)^{-2}, \tag{12.6.16}$$

$$f(\nu, \rho, k_3) = \frac{ik_3}{\rho} - \frac{1}{\rho^2} - \nu k_3^2\left(\frac{\pi}{k_3\rho}\right)^{1/2} e^{-ik_3\rho\nu^2/2}\mathcal{F}\left(\frac{k_3\rho\nu^2}{2}\right), \tag{12.6.17}$$

$$g(\nu, \rho, k_3) = \frac{ik_3}{\rho} - \frac{1}{\rho^2} - \frac{i}{k_3\rho^3} - \nu k_3^2\left(\frac{\pi}{k_3\rho}\right)^{1/2} e^{-ik_3\rho\nu^2/2}\mathcal{F}\left(\frac{k_3\rho\nu^2}{2}\right), \tag{12.6.18}$$

$$h(\nu, \rho, k_3) = \frac{2}{\rho^2} + \frac{2i}{k_3\rho^3} + \frac{i\nu k_3}{\rho}\left(\frac{\pi}{k_3\rho}\right)^{1/2} e^{-ik_3\rho\nu^2/2}\mathcal{F}\left(\frac{k_3\rho\nu^2}{2}\right). \tag{12.6.19}$$

In order to relate directly to available graphical representations, the parameter ν is introduced to replace b. The relation is simple, viz.,

$$\nu = \frac{k_3 b}{k_1}; \qquad b = \frac{1 - i(k_1/k_2)\tan k_2 l}{1 - i(k_2/k_1)\tan k_2 l}. \tag{12.6.20}$$

It follows that

$$Rb^2 = \frac{k_3^3\rho}{2k_1^2}\frac{k_1^2\nu^2}{k_3^2} = \frac{k_3\rho\nu^2}{2}, \tag{12.6.21}$$

$$\mathcal{F}\left(\frac{k_3\rho\nu^2}{2}\right) = \tfrac{1}{2}(1+i) - C_2\left(\frac{k_3\rho\nu^2}{2}\right) - iS_2\left(\frac{k_3\rho\nu^2}{2}\right), \tag{12.6.22}$$

$$\begin{aligned} C_2&\left(\frac{k_3\rho\nu^2}{2}\right) + iS_2\left(\frac{k_3\rho\nu^2}{2}\right) \\ &= \int_0^{k_3\rho\nu^2/2} \frac{e^{it}}{(2\pi t)^{1/2}}\,dt \\ &= \int_0^{k_3\rho\nu^2/2} \frac{\cos t}{(2\pi t)^{1/2}}\,dt + i\int_0^{k_3\rho\nu^2/2} \frac{\sin t}{(2\pi t)^{1/2}}\,dt, \end{aligned} \tag{12.6.23}$$

$$r_1 = [\rho^2 + (z-d)^2]^{1/2}; \qquad r_2 = [\rho^2 + (z+d)^2]^{1/2}. \tag{12.6.24}$$

12.7 Evaluation of $G_{\rho 1}(\rho, z+d)$ and $G_{\phi 1}(\rho, z+d)$

The formulas (12.2.4) and (12.2.5) for the magnetic-field components $B_{1\rho}(\rho, \phi, z)$ and $B_{1\phi}(\rho, \phi, z)$ include the integrated terms $G_0(\rho, z-d)$ and $G_0(\rho, z+d)$ given in (12.2.10) and the terms

$$G_{\rho 1}(\rho, z+d) = G_{\rho 2}(\rho, z+d) + G_{\rho 3}(\rho, z+d), \tag{12.7.1}$$

$$G_{\phi 1}(\rho, z+d) = G_{\phi 2}(\rho, z+d) + G_{\phi 3}(\rho, z+d), \tag{12.7.2}$$

where

$$\left.\begin{matrix} G_{\rho 2}(\rho, z+d) \\ G_{\phi 2}(\rho, z+d) \end{matrix}\right\} = \int_0^\infty \tfrac{1}{2}(Q_3 - 1)[J_0(\lambda\rho) \pm J_2(\lambda\rho)]e^{i\gamma_1(z+d)}\lambda\, d\lambda, \tag{12.7.3}$$

$$\left.\begin{matrix} G_{\rho 3}(\rho, z+d) \\ G_{\phi 3}(\rho, z+d) \end{matrix}\right\} = -\int_0^\infty \tfrac{1}{2}(P_3 + 1)[J_0(\lambda\rho) \mp J_2(\lambda\rho)]e^{i\gamma_1(z+d)}\lambda\, d\lambda. \tag{12.7.4}$$

The quantities $\frac{1}{2}(Q_3-1)$ and $\frac{1}{2}(P_3+1)$ differ from the quantities $(\gamma_1/2)(Q_3-1)$ and $(k_1^2/2\gamma_1)(P_3+1)$ only in the factors γ_1 and k_1^2/γ_1. It follows that the integrands in (12.7.3) and (12.7.4) differ from those in (12.5.5) and (12.5.6) only in these factors. That is,

$$L_{jm} = \frac{L_j}{\gamma_1} = \int_0^\infty \frac{f_j}{\gamma_1}[J_0(\lambda\rho) \mp J_2(\lambda\rho)]\lambda\, d\lambda, \tag{12.7.5}$$

$$M_{jm} = \frac{\gamma_1 M_j}{k_1^2} = k_1^{-2}\int_0^\infty \gamma_1 g_j[J_0(\lambda\rho) \pm J_2(\lambda\rho)]\lambda\, d\lambda, \tag{12.7.6}$$

where the f_j are defined in (12.3.6)–(12.3.10) and the g_j in (12.4.2)–(12.4.6). In this case, however, the upper sign in L_{jm} and M_{jm} applies to the ϕ-component and the lower sign to the ρ-component. It follows that

$$\begin{aligned} \left.\begin{matrix} G_{\phi 2}(\rho, z+d) \\ G_{\rho 2}(\rho, z+d) \end{matrix}\right\} &= L_{1m} + (L_{2m} - L_{4m}) + (L_{3m} - L_{5m}) \\ &= L_{1m} + \frac{L_2 - L_3}{k_1} + \frac{L_3 - L_5}{k_1}, \end{aligned} \tag{12.7.7a}$$

$$\begin{aligned} \left.\begin{matrix} G_{\phi 3}(\rho, z+d) \\ G_{\rho 3}(\rho, z+d) \end{matrix}\right\} &= -[M_{1m} + (M_{2m} - M_{4m}) + (M_{3m} - M_{5m})] \\ &= -\left(M_{1m} + \frac{M_2 - M_4}{k_1} + \frac{M_3 - M_5}{k_1}\right). \end{aligned} \tag{12.7.7b}$$

As with the electric field, no attempt will be made to evaluate $(L_{2m} - L_{4m})$ or $(M_{2m} - M_{4m})$ which would contribute the magnetic field in the layer, i.e., in Region 2. Also, as with $F_{\rho 1}(\rho, z+d)$ and $F_{\phi 1}(\rho, z+d)$, the contribution by L_{1m} to the field propagating in Region 1 is negligible. It follows that the contribution of interest and significance to $G_{\phi 2}(\rho, z+d)$ and $G_{\rho 2}(\rho, z+d)$ comes from the same terms as in the case of $F_{\rho 2}(\rho, z+d)$ and $F_{\phi 2}(\rho, z+d)$ and they differ only by the factor k_1^{-1}. Thus, $G_{\phi 2}(\rho, z+d) = k_1^{-1}F_{\rho 2}(\rho, z+d)$ and $G_{\rho 2}(\rho, z+d) = k_1^{-1}F_{\phi 2}(\rho, z+d)$. With (12.5.28a,b), these are

$$\begin{aligned} G_{\phi 2}(\rho, z+d) \sim (L_{2m} - L_{4m})_\phi &+ 2Ae^{ik_1(z+d)}e^{ik_3\rho}k_1^{-1} \\ &\times \left[\frac{ik_3^2}{\rho} - \frac{k_3}{\rho^2} - \frac{bk_3^4}{k_1}\left(\frac{\pi}{k_3\rho}\right)^{1/2} e^{-iRb^2}\mathcal{F}(Rb^2)\right], \end{aligned} \tag{12.7.8}$$

$$\begin{aligned} G_{\rho 2}(\rho, z+d) \sim (L_{2m} - L_{4m})_\rho &+ 2Ae^{ik_1(z+d)}e^{ik_3\rho}k_1^{-1} \\ &\times \left[\frac{k_3}{\rho^2} + \frac{ibk_3^3}{k_1\rho}\left(\frac{\pi}{k_3\rho}\right)^{1/2} e^{-iRb^2}\mathcal{F}(Rb^2)\right]. \end{aligned} \tag{12.7.9}$$

With g_1 given in (12.4.7), the M_{1m} integrals are

$$\begin{aligned} \left.\begin{matrix} M_{1m\phi} \\ M_{1m\rho} \end{matrix}\right\} &= ik_1^{-2}\int_0^\infty \gamma_1[J_0(\lambda\rho) \pm J_2(\lambda\rho)]e^{i\gamma_1(z+d)}\lambda^2\,d\lambda \\ &\quad + k_1^{-2}\int_0^\infty [J_0(\lambda\rho) \pm J_2(\lambda\rho)]e^{i\gamma_1(z+d)}\lambda^3\,d\lambda \\ &= ik_1^{-2}\left\{\begin{matrix} I_{13}(k_1;\rho, z+d) \\ I_{14}(k_1;\rho, z+d) \end{matrix}\right\} + k_1^{-2}\left\{\begin{matrix} I_{22}(k_1;\rho, z+d) \\ I_{23}(k_1;\rho, z+d) \end{matrix}\right\}. \end{aligned} \tag{12.7.10}$$

With Appendix A,

$$\begin{aligned} M_{1m\phi} &= \frac{2i}{k_1}\left[\frac{e^{ik_1(z+d)}}{\rho^3} + ie^{ik_1r_2}\left(\frac{1}{\rho^3} + \frac{3i}{2k_1\rho^4}\right)\right] \\ &\quad - \frac{2}{k_1}\left(\frac{z+d}{r_2}\right)\left(\frac{k_1}{r_2^2} + \frac{3i}{r_2^3}\right)e^{ik_1r_2}, \end{aligned} \tag{12.7.11}$$

$$\begin{aligned} M_{1m\rho} &= -\frac{4i}{k_1\rho^3}e^{ik_1(z+d)} - 2i\left(\frac{1}{\rho^2} + \frac{7i}{2k_1\rho^3}\right)e^{ik_1r_2} \\ &\quad + \frac{2}{k_1}\left(\frac{z+d}{r_2}\right)\left(-\frac{ik_1^2}{r_2} + \frac{5k_1}{r_2^2} + \frac{12i}{r_2^3}\right)e^{ik_1r_2}. \end{aligned} \tag{12.7.12}$$

Similarly, $(M_{3m} - M_{5m})$ can be obtained from $(M_3 - M_5)$ as given in (12.5.31) and in final form in (12.5.38a,b). Thus,

$$\begin{aligned} (M_{3m} - M_{5m})_\phi &= k_1^{-1}(M_3 - M_5)_\rho \\ &= 2Ae^{ik_1(z+d)}k_1^{-1}\left(\frac{k_3}{\rho^2} - \frac{i}{\rho^3} + \frac{ie^{ik_3\rho}}{\rho^3}\right), \end{aligned} \tag{12.7.13}$$

$$(M_{3m} - M_{5m})_\rho = k_1^{-1}(M_3 - M_5)_\phi$$

$$= 2Ae^{ik_1(z+d)}k_1^{-1}\left[-\frac{k_3}{\rho} + \frac{4i}{\rho^3} - \left(\frac{k_3}{\rho^2} + \frac{2i}{\rho^3}\right)e^{ik_3\rho}\right]. \tag{12.7.14}$$

With these values, the radially propagating terms are

$$\begin{aligned}
&G_{\rho 3}(\rho, z+d) \\
&\quad \sim -[M_{1m\rho}(e^{ik_1 r_2}) + (M_{2m} - M_{4m})_\rho + (M_{3m} - M_{5m})_\rho(e^{ik_3\rho})] \\
&\quad = -(M_{2m} - M_{4m})_\rho \\
&\qquad + 2\Bigg\{e^{ik_1 r_2}\left[\frac{i}{\rho^2} - \frac{7}{2k_1\rho^3} - \left(\frac{z+d}{r_2}\right)\left(-\frac{ik_1}{r_2} + \frac{5}{r_2^2} + \frac{12i}{k_1 r_2^3}\right)\right] \\
&\qquad\qquad + Ae^{ik_1(z+d)}\,\frac{k_3}{k_1}\,e^{ik_3\rho}\left(\frac{1}{\rho^2} + \frac{2i}{k_3\rho^3}\right)\Bigg\},
\end{aligned} \tag{12.7.15}$$

$$\begin{aligned}
&G_{\phi 3}(\rho, z+d) \\
&\quad \sim -[M_{1m\phi}(e^{ik_1 r_2}) + (M_{2m} - M_{4m})_\phi + (M_{3m} - M_{5m})_\phi(e^{ik_3\rho})] \\
&\quad = -(M_{2m} - M_{4m})_\phi \\
&\qquad + 2\Bigg\{e^{ik_1 r_2}\left[\frac{1}{k_1\rho^3} + \frac{3i}{2k_1^2\rho^4} + \left(\frac{z+d}{r_2}\right)\left(\frac{1}{r_2^2} + \frac{3i}{k_1 r_2^3}\right)\right] \\
&\qquad\qquad - Ae^{ik_1(z+d)}\,\frac{ik_3}{k_1}\,\frac{e^{ik_3\rho}}{k_3\rho^3}\Bigg\}.
\end{aligned} \tag{12.7.16}$$

It now follows that

$$\begin{aligned}
&G_{\rho 1}(\rho, z+d) = G_{\rho 2}(\rho, z+d) + G_{\rho 3}(\rho, z+d) \\
&\quad = (L_{2m} - L_{4m} - M_{2m} + M_{4m})_\rho \\
&\qquad + 2Ae^{ik_1(z+d)}e^{ik_3\rho}\,\frac{k_3}{k_1}\left[\frac{2}{\rho^2} + \frac{2i}{k_3\rho^3} + \frac{ibk_3^2}{k_1\rho}\left(\frac{\pi}{k_3\rho}\right)^{1/2} e^{-iRb^2}\mathcal{F}(Rb^2)\right] \\
&\qquad + 2e^{ik_1 r_2}\left[\frac{i}{\rho^2} - \frac{7}{2k_1\rho^3} + \left(\frac{z+d}{r_2}\right)\left(\frac{ik_1}{r_2} - \frac{5}{r_2^2} - \frac{12i}{k_1 r_2^3}\right)\right],
\end{aligned} \tag{12.7.17}$$

$$\begin{aligned}
&G_{\phi 1}(\rho, z+d) = G_{\phi 2}(\rho, z+d) + G_{\phi 3}(\rho, z+d) \\
&\quad = (L_{2m} - L_{4m} - M_{2m} + M_{4m})_\phi \\
&\qquad + 2Ae^{ik_1(z+d)}e^{ik_3\rho}\,\frac{k_3}{k_1}\Bigg[\frac{ik_3}{\rho} - \frac{1}{\rho^2} - \frac{i}{k_3\rho^3} \\
&\qquad\qquad\qquad - \frac{bk_3^3}{k_1}\left(\frac{\pi}{k_3\rho}\right)^{1/2} e^{-iRb^2}\mathcal{F}(Rb^2)\Bigg] \\
&\qquad + 2e^{ik_1 r_2}\left[\frac{1}{k_1\rho^3} + \frac{3i}{2k_1^2\rho^4} + \left(\frac{z+d}{r_2}\right)\left(\frac{1}{r_2^2} + \frac{3i}{k_1 r_2^3}\right)\right].
\end{aligned} \tag{12.7.18}$$

When $l \to 0$, these values agree with (5.6.41) and (5.6.42).

12.8 Evaluation of $G_{z1}(\rho, z + d)$; Formulas for the Magnetic Field

In formula (12.2.6) for $B_{1z}(\rho, \phi, z)$, viz.,

$$B_{1z}(\rho, \phi, z) = \frac{i\mu_0}{4\pi} \sin\phi \, [G_{z0}(\rho, z-d) - G_{z0}(\rho, z+d) + G_{z1}(\rho, z+d)], \tag{12.8.1}$$

the direct- and image-field terms are given in (12.2.11). The remaining term is the integral

$$G_{z1}(\rho, z+d) = -\int_0^\infty (P_3 - 1)\gamma_1^{-1} e^{i\gamma_1(z+d)} J_1(\lambda\rho)\lambda^2 \, d\lambda. \tag{12.8.2}$$

The term $-(P_3 - 1)\gamma_1^{-1}$ in the integrand is readily obtained from the corresponding term $(k_1^2/2\gamma_1)(P_3+1)$ which occurs in the integrals for $E_{1\rho}(\rho, \phi, z)$ and $E_{1\phi}(\rho, \phi, z)$. Specifically, with

$$g = \frac{k_1^2}{2\gamma_1}(P_3 + 1)e^{i\gamma_1(z+d)}, \tag{12.8.3a}$$

let

$$h \equiv -\frac{(P_3 - 1)e^{i\gamma_1(z+d)}}{\gamma_1} = -2\left(\frac{g}{k_1^2} - \frac{1}{\gamma_1} e^{i\gamma_1(z+d)}\right). \tag{12.8.3b}$$

With (12.4.1),

$$h = \frac{2[1 - i(\gamma_3/\gamma_2)\tan\gamma_2 l]e^{i\gamma_1(z+d)}}{\gamma_3[1 - i(\gamma_1/\gamma_2)\tan\gamma_2 l] + \gamma_1[1 - i(\gamma_2/\gamma_1)\tan\gamma_2 l]}. \tag{12.8.4}$$

Let

$$h = h_1 + (h_2 - h_4) + (h_3 - h_5), \tag{12.8.5}$$

where the h_j are defined in the manner of the g_j in (12.4.1). Specifically,

$$\begin{aligned} h_1 &= \frac{2(1 - i\tan i\lambda l)e^{i\gamma_1(z+d)}}{i\lambda[1 - i(\gamma_1/i\lambda)\tan i\lambda l] + \gamma_1[1 - i(i\lambda/\gamma_1)\tan i\lambda l]} \\ &= \frac{2}{\gamma_1 + i\lambda} e^{i\gamma_1(z+d)} = \frac{2(\gamma_1 - i\lambda)}{k_1^2} e^{i\gamma_1(z+d)}. \end{aligned} \tag{12.8.6}$$

Note that the intermediate Region 2 contributes nothing to k_1. In $h_2, \ldots, h_5$, $\gamma_1 \sim k_1$ so that

$$h_2 - h_4 = -\frac{2}{k_1^2}[(g_2 - g_4) - k_1 e^{ik_1(z+d)}], \tag{12.8.7}$$

$$h_3 - h_5 = -\frac{2}{k_1^2}\left[(g_3 - g_5) - k_1 e^{ik_1(z+d)}\right]. \tag{12.8.8}$$

The integrals involved with (12.8.6)–(12.8.8) in (12.8.2) are

$$G_{z1}(\rho, z+d) = \int_0^\infty [h_1 + (h_2 - h_4) + (h_3 - h_5)] J_1(\lambda\rho)\lambda^2\, d\lambda. \tag{12.8.9}$$

The evaluations follow. With (12.8.6),

$$\begin{aligned} H_{1mz} &= \int_0^\infty h_1 J_1(\lambda\rho)\lambda^2\, d\lambda \\ &= \frac{2}{k_1^2}\int_0^\infty \gamma_1 e^{i\gamma_1(z+d)} J_1(\lambda\rho)\lambda^2\, d\lambda \\ &\quad - \frac{2i}{k_1^2}\int_0^\infty e^{i\gamma_1(z+d)} J_1(\lambda\rho)\lambda^3\, d\lambda \\ &= \frac{\rho}{k_1^2}\left[I_{15}(k_1;\rho, z+d) - iI_{24}(k_1;\rho, z+d)\right] \\ &= 2ie^{ik_1r_2}\left[\left(\frac{\rho}{r_2}\right)\left(\frac{1}{r_2^2} + \frac{3i}{k_1r_2^3} - \frac{3}{k_1^2r_2^4}\right)\right. \\ &\qquad \left. - \left(\frac{z+d}{r_2}\right)^2\left(-\frac{ik_1}{r_2} + \frac{6}{r_2^2} + \frac{15i}{k_1r_2^3}\right)\right] \\ &\quad + 2ie^{ik_1r_2}\left(\frac{z+d}{r_2}\right)\left(\frac{k_1}{\rho} + \frac{3i}{2\rho^2} - \frac{9}{8k_1\rho^3}\right). \end{aligned} \tag{12.8.10}$$

The field in Region 2 is not evaluated. It is represented by

$$(H_{2m} - H_{4m})_z = \int_0^\infty (h_2 - h_4) J_1(\lambda\rho)\lambda^2\, d\lambda. \tag{12.8.11}$$

With (12.8.8),

$$\begin{aligned} (H_{3m} - H_{5m})_z &= \int_0^\infty (h_3 - h_5) J_1(\lambda\rho)\lambda^2\, d\lambda \\ &= -\frac{2}{k_1^2}\int_0^\infty \left[(g_3 - g_5) - k_1 e^{ik_1(z+d)}\right] \\ &\qquad \times J_1(\lambda\rho)\lambda^2\, d\lambda. \end{aligned} \tag{12.8.12}$$

From (A22b) in Appendix A, the second integral in (12.8.12) is equal to zero. With (12.4.14) and (12.4.16), (12.8.12) becomes

$$\begin{aligned} (H_{3m} - H_{5m})_z &= -\frac{2k_{12}A}{bk_1} e^{ik_1(z+d)} \\ &\quad \times \int_0^\infty \left[\frac{1}{\gamma_{12}} + \frac{1}{k_{12}^2}\left(\gamma_3 - \gamma_{12} - i\lambda - \frac{\lambda^2}{\gamma_{12}}\right)\right] \\ &\qquad \times J_1(\lambda\rho)\lambda^2\, d\lambda, \end{aligned} \tag{12.8.13}$$

where $\gamma_{12} = (k_{12}^2 - \lambda^2)^{1/2}$ and $k_{12} = k_1/b$ with b given in (12.6.20). The several integrals in (12.8.13) are

$$\int_0^\infty \frac{1}{\gamma_{12}} J_1(\lambda\rho)\lambda^2\,d\lambda$$

$$= \frac{\rho}{2} I_7(k_{12};\rho,0) = -\left(\frac{k_{12}}{\rho} + \frac{i}{\rho^2}\right)e^{ik_{12}\rho}, \tag{12.8.14a}$$

$$\frac{1}{k_{12}^2}\int_0^\infty \gamma_3 J_1(\lambda\rho)\lambda^2\,d\lambda$$

$$= \frac{\rho}{2k_{12}^2} I_{15}(k_3;\rho,0) = \frac{ik_3^2}{k_{12}^2}\left(\frac{1}{\rho^2} + \frac{3i}{k_3\rho^3} - \frac{3}{k_3^2\rho^4}\right)e^{ik_3\rho}, \tag{12.8.14b}$$

$$-\frac{1}{k_{12}^2}\int_0^\infty \gamma_{12} J_1(\lambda\rho)\lambda^2\,d\lambda$$

$$= -\frac{\rho}{2k_{12}^2} I_{15}(k_{12};\rho,0) = -i\left(\frac{1}{\rho^2} + \frac{3i}{k_{12}\rho^3}\right)e^{ik_{12}\rho}, \tag{12.8.14c}$$

$$-\frac{i}{k_{12}^2}\int_0^\infty J_1(\lambda\rho)\lambda^3\,d\lambda = -\frac{i\rho}{2k_{12}^2} I_{24}(\rho,0) = 0, \tag{12.8.14d}$$

$$-\frac{1}{k_{12}^2}\int_0^\infty \frac{1}{\gamma_{12}} J_1(\lambda\rho)\lambda^4\,d\lambda$$

$$= -\int_0^\infty \frac{1}{\gamma_{12}} J_1(\lambda\rho)\lambda^2\,d\lambda + \frac{1}{k_{12}^2}\int_0^\infty \gamma_{12}J_1(\lambda\rho)\lambda^2\,d\lambda$$

$$= -\frac{\rho}{2} I_7(k_{12};\rho,0) + \frac{1}{k_{12}^2}\frac{\rho}{2} I_{15}(k_{12};\rho,0)$$

$$= \left(\frac{k_{12}}{\rho} + \frac{2i}{\rho^2} - \frac{3}{k_{12}\rho^3}\right)e^{ik_{12}\rho}. \tag{12.8.14e}$$

The evaluation of (12.8.14e) makes use of (B52) in Appendix B, viz.,

$$\lambda^2(k^2-\lambda^2)^{-1/2} = k^2(k^2-\lambda^2)^{-1/2} - (k^2-\lambda^2)^{1/2}.$$

The substitution of (12.8.14a–e) in (12.8.13) gives

$$(H_{3m} - H_{5m})_z = -2iA\,\frac{k_3^2}{k_1^2}\,e^{ik_3\rho}e^{ik_1(z+d)}\left(\frac{1}{\rho^2} + \frac{3i}{k_3\rho^3} - \frac{3}{k_3^2\rho^4}\right). \tag{12.8.15}$$

When (12.8.10), (12.8.11), and (12.8.15) are combined to form (12.8.9), the result is

$$\begin{aligned}G_{z1}(\rho, z+d) = (H_{2m} &- H_{4m})_z \\ &- 2iA\,\frac{k_3^2}{k_1^2}\left(\frac{1}{\rho^2} + \frac{3i}{k_3\rho^3} - \frac{3}{k_3^2\rho^4}\right)e^{ik_3\rho}e^{ik_1(z+d)} \\ &+ 2i\left(\frac{\rho}{r_2}\right)\left[\frac{1}{r_2^2} + \frac{3i}{k_1r_2^3} - \frac{3}{k_1^2r_2^4}\right.\end{aligned}$$

$$+\left(\frac{z+d}{r_2}\right)^2\left(\frac{ik_1}{r_2}-\frac{6}{r_2^2}-\frac{15i}{k_1r_2^3}\right)\Bigg]e^{ik_1r_2}$$

$$+2i\left(\frac{z+d}{r_2}\right)\left(\frac{k_1}{\rho}+\frac{3i}{2\rho^2}-\frac{9}{8k_1\rho^3}\right)e^{ik_1r_2}. \tag{12.8.16}$$

This formula agrees with (5.6.45) for the two-layer region when $l \to 0$ and $A = b = 1$.

With the notation in (12.6.16)–(12.6.19), the final formulas for the magnetic field are

$$\begin{aligned}B_{1\rho}(\rho,\phi,z) = &-\frac{\mu_0}{2\pi k_1}\sin\phi\\ &\times\Bigg\{Ae^{ik_1(z+d)}e^{ik_3\rho}k_3h(\nu,\rho,k_3)\\ &+e^{ik_1r_2}\Bigg[\frac{ik_1}{\rho^2}-\frac{7}{2\rho^3}\\ &\qquad+\left(\frac{z+d}{r_2}\right)\left(\frac{ik_1^2}{r_2}-\frac{5k_1}{r_2^2}-\frac{12i}{r_2^3}\right)\Bigg]\\ &-\frac{1}{2}\Bigg[e^{ik_1r_1}\left(\frac{z-d}{r_1}\right)\left(\frac{ik_1^2}{r_1}-\frac{k_1}{r_1^2}\right)\\ &\qquad+e^{ik_1r_2}\left(\frac{z+d}{r_2}\right)\left(\frac{ik_1^2}{r_2}-\frac{k_1}{r_2^2}\right)\Bigg]\Bigg\}\\ &+B_{1\rho}^{(2)}(\rho,\phi,z),\end{aligned} \tag{12.8.17}$$

$$\begin{aligned}B_{1\phi}(\rho,\phi,z) = &-\frac{\mu_0}{2\pi k_1}\cos\phi\\ &\times\Bigg\{Ae^{ik_1(z+d)}e^{ik_3\rho}k_3g(\nu,\rho,k_3)\\ &+e^{ik_1r_2}\Bigg[\frac{1}{\rho^3}+\frac{3i}{2k_1\rho^4}\\ &\qquad+\left(\frac{z+d}{r_2}\right)\left(\frac{k_1}{r_2^2}+\frac{3i}{r_2^3}-\frac{3}{k_1r_2^4}\right)\Bigg]\\ &-\frac{1}{2}\Bigg[e^{ik_1r_1}\left(\frac{z-d}{r_1}\right)\left(\frac{ik_1^2}{r_1}-\frac{k_1}{r_1^2}\right)\\ &\qquad+e^{ik_1r_2}\left(\frac{z+d}{r_2}\right)\left(\frac{ik_1^2}{r_2}-\frac{k_1}{r_2^2}\right)\Bigg]\Bigg\}\\ &+B_{1\phi}^{(2)}(\rho,\phi,z),\end{aligned} \tag{12.8.18}$$

$$\begin{aligned}B_{1z}(\rho,\phi,z) = &\frac{\mu_0}{2\pi}\sin\phi\\ &\times\Bigg\{Ae^{ik_1(z+d)}e^{ik_3\rho}\frac{k_3^2}{k_1^2}\left(\frac{1}{\rho^2}+\frac{3i}{k_3\rho^3}-\frac{3}{k_3^2\rho^4}\right)\end{aligned}$$

$$
\begin{aligned}
&- e^{ik_1 r_2}\left(\frac{\rho}{r_2}\right)\left[\frac{1}{r_2^2} + \frac{3i}{k_1 r_2^3} - \frac{3}{k_1^2 r_2^4}\right. \\
&\qquad + \left(\frac{z+d}{r_2}\right)^2\left(\frac{ik_1}{r_2} - \frac{6}{r_2^2} - \frac{15i}{k_1 r_2^3}\right) \\
&\qquad \left. + \left(\frac{z+d}{\rho}\right)\left(\frac{k_1}{\rho} + \frac{3i}{2\rho^2} - \frac{9}{8k_1\rho^3}\right)\right] \\
&- \frac{1}{2}\left[e^{ik_1 r_1}\left(\frac{\rho}{r_1}\right)\left(\frac{ik_1}{r_1} - \frac{1}{r_1^2}\right)\right. \\
&\qquad \left.\left. - e^{ik_1 r_2}\left(\frac{\rho}{r_2}\right)\left(\frac{ik_1}{r_2} - \frac{1}{r_2^2}\right)\right]\right\} \\
&+ B_{1z}^{(2)}(\rho,\phi,z). \qquad (12.8.19)
\end{aligned}
$$

The functions $B^{(2)}(\rho,\phi,z)$ indicate the unevaluated terms that involve propagation in Region 2, the layer of sediment between the sea floor and the rock of the oceanic crust.

A comparison of the formulas (12.6.13)–(12.6.15) and (12.8.17)–(12.8.19) for the six components of the electromagnetic field for the three-layer case with the corresponding formulas (5.5.37a)–(5.5.42a) shows that the direct field that propagates in the sea water (Region 1) is essentially the same. The presence of the layer of sediment has no effect on the larger terms representing the field propagating in the form $\exp(ik_1 r_1)$ or $\exp(ik_1 r_2)$, where $r_1 = [\rho^2 + (z-d)^2]^{1/2}$ and $r_2 = [\rho^2 + (z+d)^2]^{1/2}$. Only certain very small terms of the order of magnitude k_2^2/k_1^2 and represented by the unintegrated terms $T(k_2 l, e^{ik_1\rho})$ are dependent on the properties of the sedimentary layer (Region 2). Such terms are negligible in all components except $E_{1z}(\rho,\phi,z)$ where they are the principal part of the direct field when the source and the point of observation are both very close to the boundary, i.e., when $z+d$ is very small. When this is true, the direct field has the form $(k_2^2/k_1^2)\exp(ik_1\rho)$, the lateral-wave field $(k_3^2/k_1^2)\exp(ik_3\rho)$. Since $(k_2^2/k_1^2) \gg (k_3^2/k_1^2)$, the contribution by the direct field to $E_{1z}(\rho,\phi,z)$ is negligible only when $|(k_3^2/k_2^2)\exp(ik_3\rho)| \gg |\exp(ik_1\rho)|$. Actually, the component $E_{1z}(\rho,\phi,z)$ is of little interest in measurements on the sea floor that are concerned primarily with $E_{1\rho}(\rho,\phi,z)$, $E_{1\phi}(\rho,\phi,z)$, and $B_{1\phi}(\rho,\phi,z)$ for which this problem does not arise.

12.9 The Lateral-Wave Field; Comparison of Theoretical and Numerical Results

The complete lateral-wave field at the point of observation (ρ,ϕ,z) in Region 1 (sea water) due to a horizontal dipole source at $\rho = 0$, $z = d$, also in the sea water has the following components:

$$E^L_{1\rho}(\rho, \phi, z) = -\frac{\omega\mu_0}{2\pi k_1^2} \cos\phi\, A k_3 g(\nu, \rho, k_3) e^{ik_3\rho} e^{ik_1(z+d)}, \tag{12.9.1}$$

$$E^L_{1\phi}(\rho, \phi, z) = \frac{\omega\mu_0}{2\pi k_1^2} \sin\phi\, A k_3 h(\nu, \rho, k_3) e^{ik_3\rho} e^{ik_1(z+d)}, \tag{12.9.2}$$

$$E^L_{1z}(\rho, \phi, z) = \frac{\omega\mu_0}{2\pi k_1^2} \cos\phi\, A \frac{k_3^2}{k_1} f(\nu, \rho, k_3) e^{ik_3\rho} e^{ik_1(z+d)}, \tag{12.9.3}$$

$$B^L_{1\rho}(\rho, \phi, z) = -\frac{\mu_0}{2\pi k_1} \sin\phi\, A k_3 h(\nu, \rho, k_3) e^{ik_3\rho} e^{ik_1(z+d)}, \tag{12.9.4}$$

$$B^L_{1\phi}(\rho, \phi, z) = -\frac{\mu_0}{2\pi k_1} \cos\phi\, A k_3 g(\nu, \rho, k_3) e^{ik_3\rho} e^{ik_1(z+d)}, \tag{12.9.5}$$

$$B^L_{1z}(\rho, \phi, z) = \frac{\mu_0}{2\pi k_1^2} \sin\phi \times A k_3^2 \left(\frac{1}{\rho^2} + \frac{3i}{k_3\rho^3} - \frac{3}{k_3^2\rho^4} \right) e^{ik_3\rho} e^{ik_1(z+d)}. \tag{12.9.6}$$

The amplitude A, the three functions $f(\nu, \rho, k_3)$, $g(\nu, \rho, k_3)$, and $h(\nu, \rho, k_3)$, and the parameter ν are defined in (12.6.16)–(12.6.20).

In view of the numerous approximations made in the evaluation of the formulas for the components of the electromagnetic field in Region 1 (sea water) when in the presence of a layer of Region 2 (sediment) over Region 3 (rock), it is desirable to make an explicit comparison of the field specified by these approximate formulas with the same field calculated numerically directly from their integral forms. For most purposes, the component $E_{1\rho}(\rho, \phi, z)$ as given in (12.9.1) is the most important because it is the largest at significant distances from the source. At all but quite small distances, the lateral-wave terms dominate because the high exponential attenuation of both the sea water and the sediment makes the direct- and image-field terms and the field propagating in the sediment negligible.

In order to understand the properties of $E^L_{1\rho}(\rho, \phi, z)$ in their dependence on the intermediate layer, it is instructive to study the behavior of the parameters ν and A as functions of the thickness l. (Note that when $l = 0$, $\nu \to k_3/k_1$ and $A \to 1$.) This is carried out here at the high frequency $f = 600$ MHz with Region 3 taken to be air. These values have no direct application to the geophysical problem on the sea floor, but provide a critical test of the formulation under conditions that also lend themselves to experimental checks in the laboratory. If the formulas can be shown to be valid in one case that involves high attenuation in the sea and sediment, they are certainly true at extremely low frequencies when attenuation in the sea and sediment is even greater. The parameters used are for Region 1, salt water, $\sigma_1 = 3.5$ S/m and $\epsilon_{1r} = 80$; for Region 2, sediment, $\sigma_2 = 0.1$ S/m and $\epsilon_{2r} = 10$; and for Region 3, air, $\sigma_3 = 0$ and $\epsilon_{3r} = 1$, or rock, $\sigma_3 = 0.01$

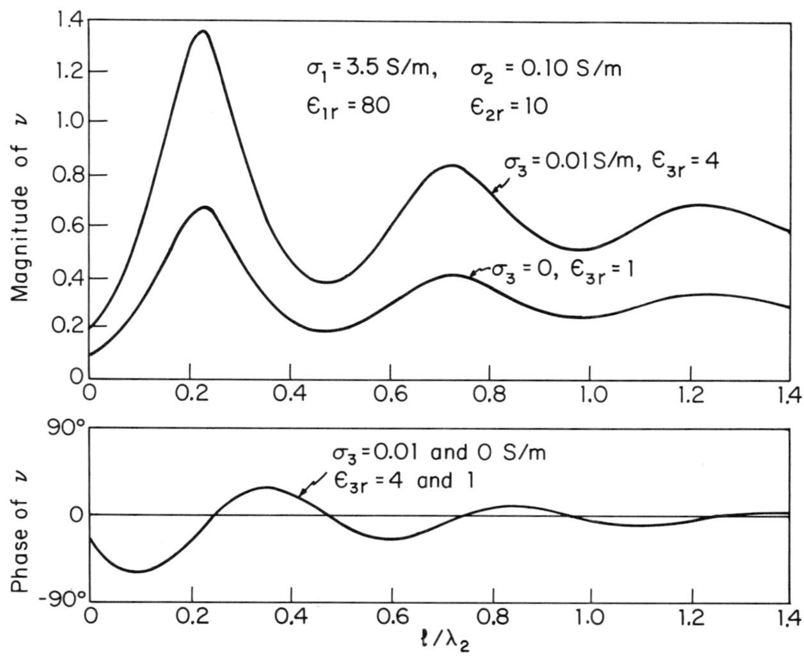

Figure 12.9.1. The parameter $\nu = (k_3/k_2)(k_2 - ik_1 \tan k_2 l)(k_1 - ik_2 \tan k_2 l)^{-1}$ for a three-layered region at $f = 600$ MHz.

S/m and $\epsilon_{3r} = 4$. The source dipole is located at $d = 3.5$ mm, the point of observation in the plane $z = 7$ mm in Region 1. The wavelengths in the three regions are $\lambda_1 = 4.86$ cm (sea water), $\lambda_2 = 15.6$ cm (sediment), and $\lambda_3 = 50$ cm (air) or $\lambda_3 = 25$ cm (rock). The distances in which the amplitude decreases from 1 to $1/e$ (decay lengths) are $\alpha_1^{-1} = 1.56$ cm (sea water), $\alpha_2^{-1} = 17.0$ cm (sediment), and $\alpha_3^{-1} = \infty$ (air) or $\alpha_3^{-1} = 1.06$ m (rock). Observations made at distances $\rho \geq 25$ cm involve only the lateral wave that travels in the air. The "trapped" waves in Region 2 and the direct waves in Region 1 are negligible.

In Fig. 12.9.1 are shown the magnitude and phase of the parameter ν as a function of $l/\lambda_2 = k_2 l/2\pi$. In Fig. 12.9.2 are the magnitude and phase of the amplitude A also as a function of l/λ_2. The range is $0 \leq l/\lambda_2 \leq 1.4$. It is seen in Fig. 12.9.1 that there are resonances near $l/\lambda_2 = 0.25$, 0.75, and 1.25, but these are highly attenuated so that $|\nu|$ is never large and $|\nu^2|$ is small compared to unity as assumed in the approximations. There are similar resonances in the amplitude A, but the attenuation is so severe with increasing values of l that only the one at $l/\lambda_2 = 0.25$ is significant.

The quantity $20 \log_{10} |E_{1\rho}(\rho, 0, z)|$ as evaluated from (12.9.1) with (12.6.18) is shown in Figs. 12.9.3a–c as a function of the radial distance ρ with the thickness l as the parameter. Values of l range from 0 (the

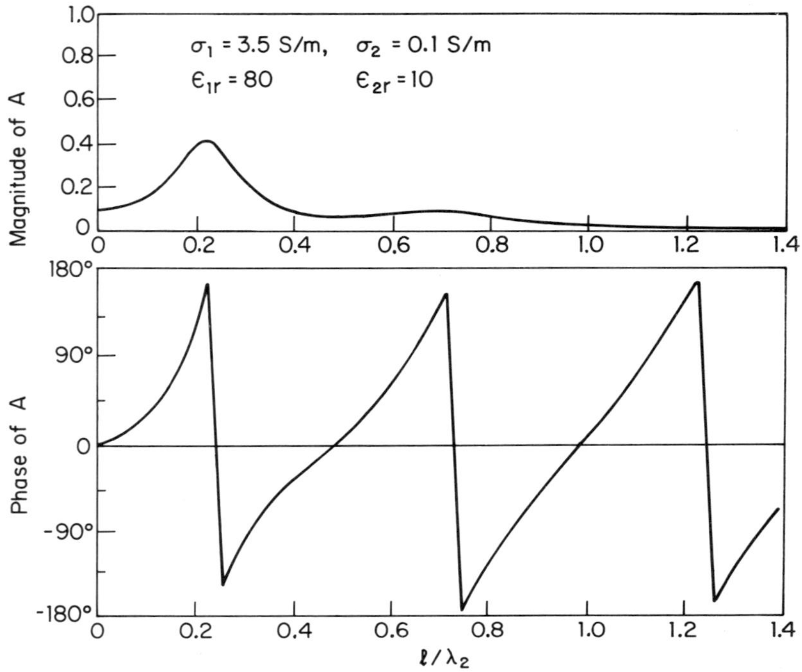

Figure 12.9.2. The amplitude $A = [\cos k_2 l - i(k_2/k_1)\sin k_2 l]^{-2}$ at $f = 600$ MHz.

two-layer case) to $l = 1.0\lambda_2$. Also shown by crosses are numerical values calculated directly from the integral formula without approximations. The agreement is seen to be very satisfactory throughout, so that it may be assumed that (12.9.1) is quite accurate. This confirms the earlier work with a different approach by Dunn (1984, 1986).

It is interesting to compare the radial electric field in Region 1 when there is no layer of Region 2 between it and Region 3 with the field when there is a thin layer present. This is carried out graphically in Fig. 12.9.4 which shows 20 $\log_{10}|E_{1\rho}(\rho, 0, z)|$ for the two-layer region consisting of Regions 1 and 3 and for the three layers when Region 2 with thickness $l = 0.05\lambda_2$ is between them. It is seen that, in the intermediate range $1 \le k_3\rho \le |k_1^2/k_3^2|$, the field for the three-layered region differs from that of the two-layered region only by the effect of the amplitude factor A, the magnitude of which is slightly greater than one. In the far-field range $k_3\rho > |k_1^2/k_3^2|$, the effect of the Fresnel-integral term becomes dominant and the argument $k_3\rho\nu^2/2$ for the three-layered region is significantly greater than the argument $k_3^3\rho/2k_1^2$ for the two-layered region. The value of ν^2 for $l/\lambda_2 = 0.05$ is of the order $\nu^2 \sim 0.04$, whereas $|k_3^2/k_1^2| \sim 0.007$. This greatly reduces the amplitude of the field in Region 1 when the layer of Region 2 is present. Note that in the intermediate range $E_{1\rho}$ decreases as $1/\rho$ and in the far-field range as

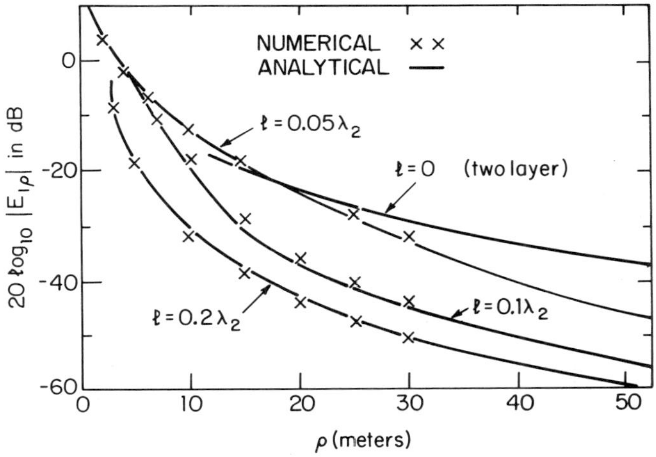

Figure 12.9.3a. The radial electric field $|E_{1\rho}(\rho, 0, z)|$ in V/m in Region 1 with $\sigma_1 = 3.5$ S/m, $\epsilon_{1r} = 80$; $\sigma_2 = 0.1$ S/m, $\epsilon_{2r} = 10$; $\sigma_3 = 0$, $\epsilon_{3r} = 1$; $f = 600$ MHz, $z = 7$ mm, $d = 3.5$ mm; $l = 0,\ 0.05\lambda_2$, and $0.2\lambda_2$. [Calculations of Dunn (1984).]

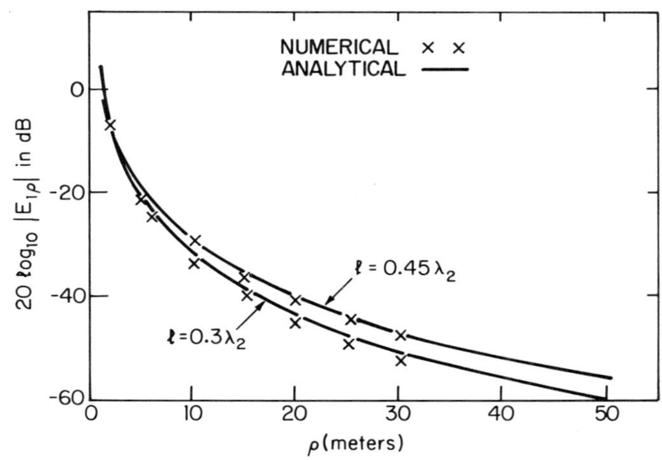

Figure 12.9.3b. The radial electric field $|E_{1\rho}(\rho, 0, z)|$ in V/m in Region 1 with $\sigma_1 = 3.5$ S/m, $\epsilon_{1r} = 80$; $\sigma_2 = 0.1$ S/m, $\epsilon_{2r} = 10$; $\sigma_3 = 0$, $\epsilon_{3r} = 1$; $f = 600$ MHz, $z = 7$ mm, $d = 3.5$ mm; $l = 0.3\lambda_2$ and $0.45\lambda_2$. [Calculations of Dunn (1984).]

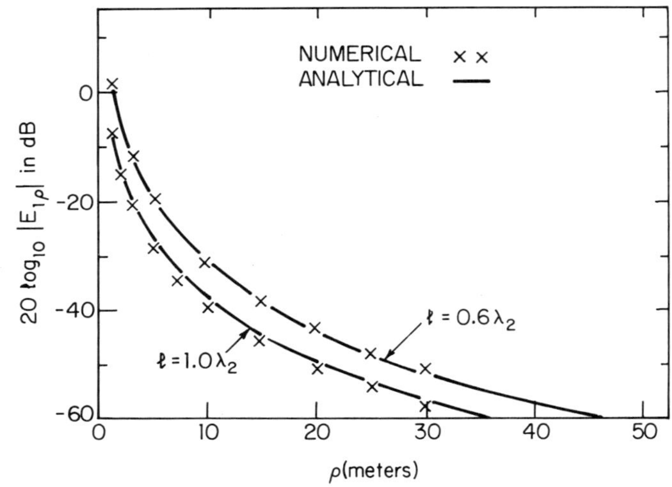

Figure 12.9.3c. The radial electric field $|E_{1\rho}(\rho, 0, z)|$ in V/m in Region 1 with $\sigma_1 = 3.5$ S/m, $\epsilon_{1r} = 80$; $\sigma_2 = 0.1$ S/m, $\epsilon_{2r} = 10$; $\sigma_3 = 0$, $\epsilon_{3r} = 1$; $f = 600$ MHz, $z = 7$ mm, $d = 3.5$ mm; $l = 0.6\lambda_2$ and $1.0\lambda_2$. [Calculations of Dunn (1984).]

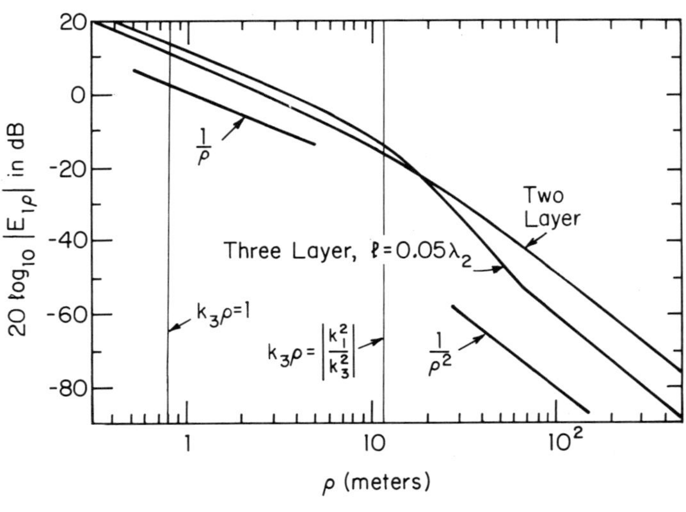

Figure 12.9.4. Comparison of radial field $|E_{1\rho}(\rho, 0, z)|$ in V/m in Region 1 when a layer of Region 2 is and is not present; $z = 7$ mm, $d = 3.5$ mm. [Calculations of Dunn (1984).]

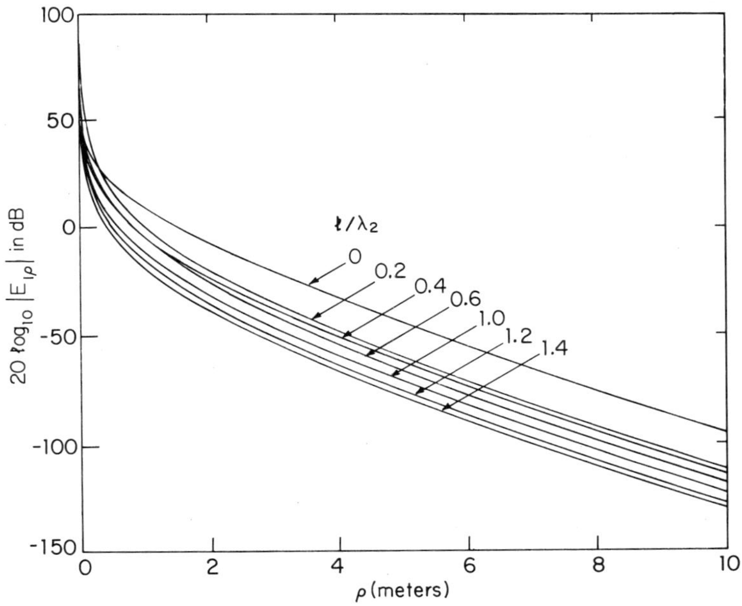

Figure 12.9.5. The radial electric field $|E_{1\rho}(\rho, 0, z)|$ in V/m in Region 1 with $\sigma_1 = 3.5$ S/m, $\epsilon_{1r} = 80$; $\sigma_2 = 0.1$ S/m, $\epsilon_{2r} = 10$; $\sigma_3 = 0.01$ S/m, $\epsilon_{3r} = 4$; $f = 600$ MHz, $z = 7$ mm, $d = 3.5$ mm.

$1/\rho^2$, both in the presence of the layer of Region 2 and in its absence.

Graphs of $|E_{1\rho}(\rho, 0, z)|$ corresponding to those in Figs. 12.9.3a–c but with Region 3 composed of rock with $\sigma_3 = 0.01$ S/m and $\epsilon_{3r} = 4$ instead of air are shown in Fig. 12.9.5. Note that at $f = 600$ MHz the wave number in the rock is $k_3 = \beta_3 + i\alpha_3 = 25.16 + i0.94$ m^{-1} and that in air is $k_3 = 12.56$ m^{-1}. In the rock, there is exponential attenuation as the lateral wave travels in Region 3 since $\exp(ik_3\rho) = \exp(-\alpha_3\rho)\exp(i\beta_3\rho)$ and the wavelength is reduced from $\lambda_3 = 50$ cm in air to $\lambda_3 = 25$ cm in the rock. The decay length is $\alpha_3^{-1} = 1.06$ m.

It is necessary to determine the range of ρ for each value of l in which the condition (12.1.8) is satisfied. Specifically,

$$\rho \geq \frac{5.436\beta_1 l}{\beta_2}. \tag{12.9.7}$$

With $\sigma_1 = 3.5$ S/m, $\epsilon_{1r} = 80$, $\sigma_2 = 0.1$ S/m, $\epsilon_{2r} = 10$, and $f = 600$ MHz, it follows that $k_1 = \beta_1 + i\alpha_1 = 129.3 + i64.0$ m^{-1} and $k_2 = \beta_2 + i\alpha_2 = 40.2 + i5.9$ m^{-1}. Hence, $\beta_1/\beta_2 = 3.22$; also, $\lambda_2 = 2\pi/\beta_2 = 0.156$ m. Thus,

$$\rho \geq 17.50l = \frac{2.73l}{\lambda_2}. \tag{12.9.8}$$

With this formula, Table 12.9.1 can be constructed relating l/λ_2 to the

Table 12.9.1. Minimum Value of ρ Satisfying Condition (12.1.8)

$l/\lambda_2 =$	0.05	0.1	0.2	0.3	0.45	0.6	1.0
ρ (m) $\geq$	0.137	0.273	0.546	0.819	1.229	1.638	2.73

minimum acceptable value of ρ. It is seen that all of the curves shown in Figs. 12.9.3a–c are almost entirely in the valid ranges.

12.10 Application of the Three-Layer Formula to the Sea Floor

If a horizontal dipole is located on the sea floor as a controlled source in the frequency range of 0.1 to 3.0 Hz, as in the measurements of Young and Cox (1981), and the three-layer formulas derived in this chapter are to be applied, the condition (12.9.7) must be satisfied. In the very-low-frequency range, it becomes

$$\rho \geq 5.436l\left(\frac{\sigma_1}{\sigma_2}\right)^{1/2}. \tag{12.10.1}$$

With the conductivity of sea water $\sigma_1 = 4$ S/m and the conductivity of a layer of sediment in the range $\sigma_2 = 0.6$ to 0.2 S/m,

$$\rho \geq 14.0l \quad \text{to} \quad 24.3l. \tag{12.10.2}$$

This means that, with a sediment layer of thickness $l \leq 0.75$ km, the point of observation on the sea floor must be at radial distances of at least $\rho \geq 10.5$ to 18.2 km. Since the measurements by Young and Cox (1981) were made at the radial distance of 18.9 km, their results could be interpreted with the three-layer formulas developed in this chapter. Actually, the layer of sediment involved at the location of their experiments was known to be of negligible thickness.

It is nevertheless of interest to apply the three-layer formulas for $E_{1\rho}(\rho, \phi, z)$ and $E_{1\phi}(\rho, \phi, z)$ to the conditions of the Young and Cox experiment with an added hypothetical layer of sediment. Let the conductivity of the sea be $\sigma_1 = 3.5$ S/m. Let the layer of sediment have a thickness l with $0 \leq l \leq 1.5$ km and a conductivity σ_2 in the range $0.03 \leq \sigma_2 \leq 0.3$ S/m. The oceanic crust is a half-space of rock with conductivity in the range $0.002 \leq \sigma_3 \leq 0.004$ S/m.

Graphs of $E_{1\rho}(\rho, \phi, z)$ with $\rho = 18.9$ km, $\phi = 0$, and $z = 7$ mm and of $E_{1\phi}(\rho, \phi, z)$ with $\rho = 18.9$ km, $\phi = \pi/2$, and $z = 7$ mm are shown in Figs. 12.10.1 through 12.10.4 in the frequency range $0.1 \leq f \leq 3$ Hz. In

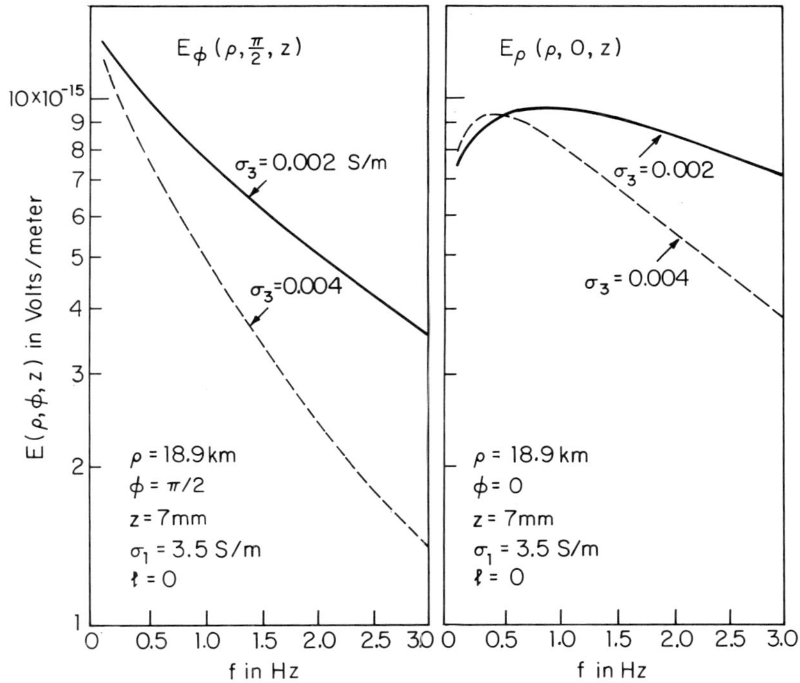

Figure 12.10.1. The components $E_{1\phi}(\rho, \pi/2, z)$ and $E_{1\rho}(\rho, 0, z)$ for the two-layered region of Young and Cox; horizontal dipole at $d = 3.5$ mm above the sea floor; $z = 7$ mm.

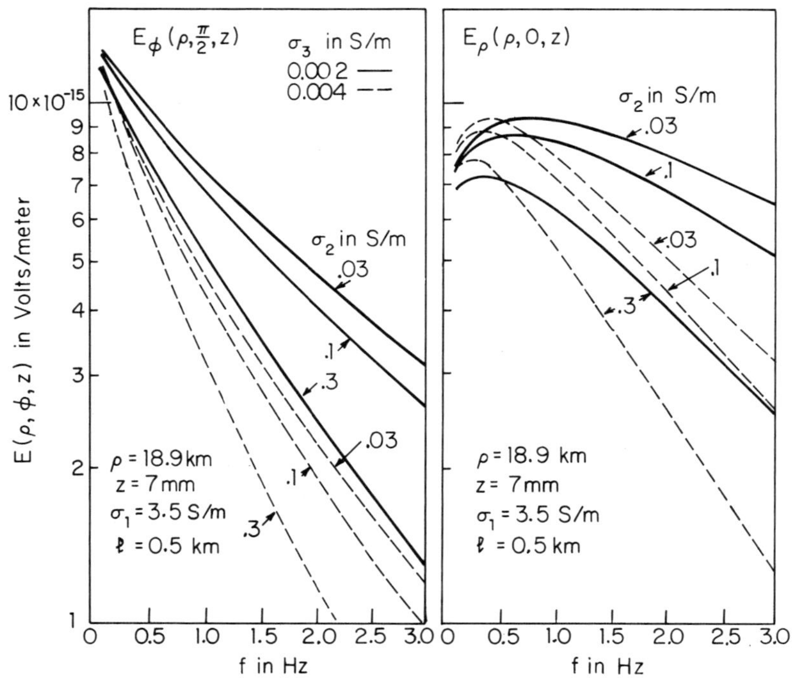

Figure 12.10.2. The components $E_{1\phi}(\rho, \pi/2, z)$ and $E_{1\rho}(\rho, 0, z)$ for the region of Fig. 12.10.1 with an added layer of thickness $l = 0.5$ km and a range of conductivities σ_2; $z = 7$ mm, $d = 3.5$ mm.

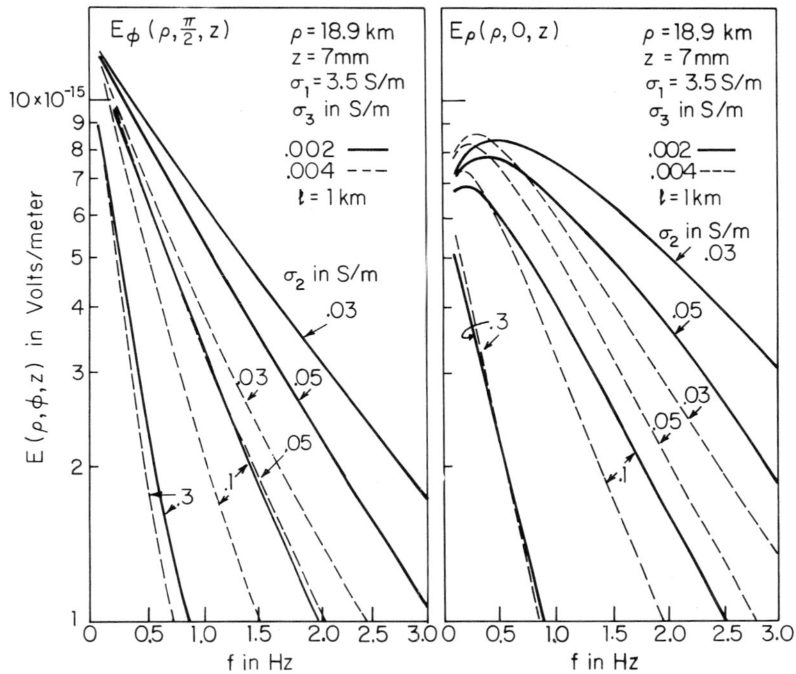

Figure 12.10.3. The components $E_{1\phi}(\rho, \pi/2, z)$ and $E_{1\rho}(\rho, 0, z)$ for the region of Fig. 12.10.1 with an added layer of thickness $l = 1$ km and a range of conductivities σ_2; $z = 7$ mm, $d = 3.5$ mm.

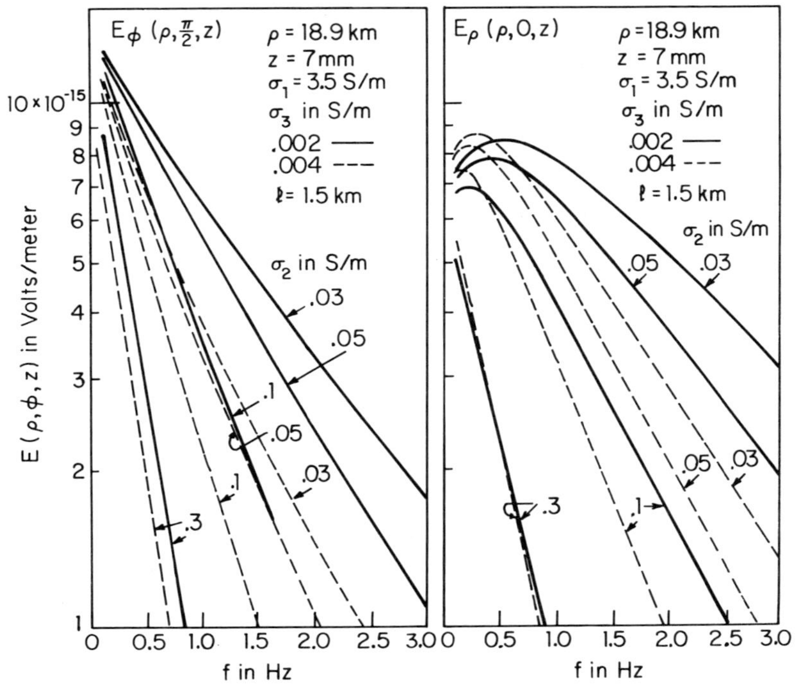

Figure 12.10.4. The components $E_{1\phi}(\rho, \pi/2, z)$ and $E_{1\rho}(\rho, 0, z)$ for the region of Fig. 12.10.1 with an added layer of thickness $l = 1.5$ km and a range of conductivities σ_2; $z = 7$ mm, $d = 3.5$ mm.

Fig. 12.10.1, $l = 0$ and σ_3 is the parameter at the values 0.002 and 0.004 S/m. This is the two-layered region of salt water over rock. $E_{1\phi}(\rho, \pi/2, z)$ is seen to decrease almost linearly with increasing frequency with a much weaker field when $\sigma_3 = 0.004$ S/m than when $\sigma_3 = 0.002$ S/m. This is to be expected since the lateral wave travels in the rock and the exponential attenuation $\exp(-\alpha_3\rho)$ is smaller when $\sigma_3 = 0.002$ S/m than when $\sigma_3 = 0.004$ S/m. The behavior of $E_{1\rho}(\rho, 0, z)$ is quite different. It has a maximum at $\alpha_3\rho = (1+\sqrt{5})/\sqrt{2}$. At frequencies below the maximizing value, the exponential attenuation is not dominant in determining the magnitude of the field and this is greater for $\sigma_3 = 0.004$ S/m than for $\sigma_3 = 0.002$ S/m. At frequencies above the maximizing value, the exponential attenuation becomes dominant and the field is much smaller for the higher conductivity $\sigma_3 = 0.004$ S/m than for the lower conductivity $\sigma_3 = 0.002$ S/m.

The effect of a layer of sediment with the thickness l and conductivity σ_2 between the salt water and rock is simply to add an additional exponential attenuation experienced by the lateral wave as it traverses the layer, travels along the boundary in the rock, and again traverses the layer to reach the point of observation in the sea water. This is evident in Figs. 12.10.2 to 12.10.4 that apply to layers with thicknesses 0.5, 1.0, and 1.5 km. For each thickness, curves for several conductivities are shown. The curves are moved downward, indicating a field of smaller magnitude, as the conductivity of the layer is increased and when it is made thicker. These observations apply equally to the component $E_{1\rho}(\rho, 0, z)$ of electric type and the component $E_{1\phi}(\rho, \pi/2, z)$ of magnetic type. This observation is at variance with the corresponding fields calculated by Young and Cox (1981) for an eight-layered model of the oceanic crust. In their graphs, $E_{1X} \sim E_{1\phi}$ is virtually independent of the properties of the layers and is accurately given by the two-layer model with $\sigma_3 = 0.004$ S/m for all four eight-layer models. On the other hand, $E_{1Y} \sim E_{1\rho}$ behaves very much like $E_{1\rho}$ in Figs. 12.10.2 through 12.10.4 in its sensitivity to the properties of the layers. No explanation for this anomalous behavior of $E_{1X} \sim E_{1\phi}$ is available.

Extensive calculations have been made of the components

$$E_{1X} = E_{1\rho}(\rho, \phi, z)\sin\psi + E_{1\phi}(\rho, \phi, z)\cos\psi, \tag{12.10.3}$$

$$E_{1Y} = -E_{1\rho}(\rho, \phi, z)\cos\psi + E_{1\phi}(\rho, \phi, z)\sin\psi, \tag{12.10.4}$$

calculated by Young and Cox in order to determine whether both components can be fitted to the measured data. The following ranges of the several variables and parameters were used:

$$45^\circ \le \phi \le 60^\circ; \qquad 5^\circ \le \psi \le 15^\circ, \tag{12.10.5}$$

$$0 \le l \le 1.5 \text{ km}; \qquad 0.03 \le \sigma_2 \le 0.3 \text{ S/m}, \tag{12.10.6}$$

$$\sigma_3 = 0.002 \quad \text{and} \quad 0.004 \text{ S/m}. \tag{12.10.7}$$

A number of combinations were found that provided good approximations of the experimentally obtained data for each component separately. But parameters that gave good approximations of E_{1X} did not also provide good approximations of E_{1Y} and vice versa. If E_{1X} had been insensitive to the parameters l and σ_2 (it was insensitive to the thicknesses l_j and conductivities σ_j of eight layers in the calculations of Young and Cox), it would have been a simple matter to choose σ_3 to provide a good fit for E_{1X} and adjust σ_2 and l to provide a good approximation for E_{1Y}. An important alternative to the multilayered region is given in Chapter 9, where the Young and Cox data are shown to be in good agreement with a Region 2 that is anisotropic in conductivity.

12.11 Integrals for the Components of the Field of a Vertical Electric Dipole

The integrals for the components of the electromagnetic field in Region 1 (sea water) in the presence of a layer of Region 2 (sediment) over Region 3 (rock) due to a vertical dipole at $z = d$ in Region 1 are

$$E_{1\rho}(\rho, z) = \frac{i\omega\mu_0}{4\pi k_1^2} \int_0^\infty [\pm e^{i\gamma_1|z-d|} - e^{i\gamma_1(z+d)} - (Q_3 - 1)e^{i\gamma_1(z+d)}]J_1(\lambda\rho)\lambda^2\, d\lambda; \quad \begin{cases} z > d \\ 0 \le z \le d \end{cases}, \tag{12.11.1}$$

$$E_{1z}(\rho, z) = -\frac{\omega\mu_0}{4\pi k_1^2} \int_0^\infty \gamma_1^{-1}[e^{i\gamma_1|z-d|} - e^{i\gamma_1(z+d)} - (Q_3 - 1)e^{i\gamma_1(z+d)}]J_0(\lambda\rho)\lambda^3\, d\lambda, \tag{12.11.2}$$

$$B_{1\phi}(\rho, z) = \frac{i\mu_0}{4\pi} \int_0^\infty \gamma_1^{-1}[e^{i\gamma_1|z-d|} - e^{i\gamma_1(z+d)} - (Q_3 - 1)e^{i\gamma_1(z+d)}]J_1(\lambda\rho)\lambda^2\, d\lambda. \tag{12.11.3}$$

The evaluation of these integrals parallels that for the horizontal dipole. Specifically, in (12.11.1)–(12.11.3) the first two integrals are the same as for the two-layer region so that the integrals that remain to be evaluated are

$$F_{1\rho}(\rho, z + d) = -\int_0^\infty (Q_3 - 1)e^{i\gamma_1(z+d)} J_1(\lambda\rho)\lambda^2\, d\lambda, \tag{12.11.4}$$

$$F_{1z}(\rho, z + d) = -\int_0^\infty (Q_3 - 1)e^{i\gamma_1(z+d)}\gamma_1^{-1} J_0(\lambda\rho)\lambda^3\, d\lambda, \tag{12.11.5}$$

$$F_{1\phi}(\rho, z + d) = -\int_0^\infty (Q_3 - 1)e^{i\gamma_1(z+d)}\gamma_1^{-1} J_1(\lambda\rho)\lambda^2\, d\lambda. \tag{12.11.6}$$

A comparison of (12.6.1) with (12.11.4) shows them to differ only by a factor -1. Hence, it follows with (12.6.12) that

$$\begin{aligned} F_{1\rho}(\rho, z+d) &= (L_2 - L_4)^v_\rho \\ &\quad - \frac{2k_3^2}{k_1} A e^{ik_1(z+d)} e^{ik_3\rho} \\ &\quad \times \left[\frac{k_3}{\rho} + \frac{i}{\rho^2} + \frac{ibk_3^3}{k_1} \left(\frac{\pi}{k_3\rho} \right)^{1/2} e^{-iRb^2} \mathcal{F}(Rb^2) \right] \\ &\quad + \frac{2k_3^2}{k_1} e^{ik_1r_2} \left(\frac{1}{\rho^2} + \frac{3i}{2k_1\rho^3} \right) - \frac{2k_2^2}{k_1^2} T^v_\rho(k_2l, e^{ik_1\rho}). \end{aligned} \tag{12.11.7}$$

The integrand in (12.11.5) differs from that in (12.11.4) by the factor γ_1^{-1}, as well as the appearance of $J_0(\lambda\rho)\lambda^3$ in place of $J_1(\lambda\rho)\lambda^2$. It follows with (12.6.2) that it can be expanded in the form

$$-\frac{(Q_3 - 1)}{\gamma_1} e^{i\gamma_1(z+d)} = -\frac{2f}{\gamma_1^2} \sim -2\left(\frac{f_1}{\gamma_1^2} + \frac{(f_2 - f_4)}{k_1^2} + \frac{(f_3 - f_5)}{k_1^2} \right). \tag{12.11.8}$$

The leading term in $-2f_1/\gamma_1^2$ is obtained from (12.6.3) to be

$$-\frac{2f_1}{\gamma_1^2} \sim -\frac{2i(k_3^2 - ik_2^2 \tan i\lambda l) e^{i\gamma_1(z+d)}}{\lambda k_1^2} \tag{12.11.9}$$

so that

$$\begin{aligned} L^v_{1z} &= -2 \int_0^\infty \frac{f_1}{\gamma_1^2} J_0(\lambda\rho)\lambda^3 \, d\lambda \\ &\sim -\frac{2ik_3^2}{k_1^2} \int_0^\infty e^{i\gamma_1(z+d)} J_0(\lambda\rho)\lambda^2 \, d\lambda \\ &\quad - \frac{2k_2^2}{k_1^2} \int_0^\infty e^{i\gamma_1(z+d)} \tan i\lambda l \, J_0(\lambda\rho)\lambda^2 \, d\lambda \\ &= -\frac{2ik_3^2}{k_1^2} \frac{1}{2} [I_{20}(k_1; \rho, z+d) + I_{21}(k_1; \rho, z+d)] \\ &\quad - \frac{2k_2^2}{k_1^2} T^v_z(k_2l, e^{ik_1\rho}). \end{aligned} \tag{12.11.10}$$

With Appendix A,

$$\begin{aligned} L^v_{1z} &= -\frac{2ik_3^2}{k_1^2} \frac{1}{2} \left[\frac{2}{\rho^3} e^{ik_1(z+d)} - 2\left(\frac{z+d}{\rho} \right) \left(\frac{k_1}{\rho^2} + \frac{3i}{2\rho^3} \right) e^{ik_1r_2} \right. \\ &\qquad \left. - \frac{4}{\rho^3} e^{ik_1(z+d)} - 2\left(\frac{z+d}{r_2} \right) \left(\frac{ik_1^2}{\rho} - \frac{3k_1}{2\rho^2} - \frac{5i}{8\rho^3} \right) e^{ik_1r_2} \right] \\ &\quad - \frac{2k_2^2}{k_1^2} T^v_z(k_2l, e^{ik_1\rho}) \\ &\sim \frac{2ik_3^2}{k_1^2} \left[\frac{1}{\rho^3} e^{ik_1(z+d)} + \left(\frac{z+d}{\rho} \right) \left(\frac{ik_1^2}{\rho} - \frac{k_1}{2\rho^2} + \frac{7i}{8\rho^3} \right) e^{ik_1r_2} \right] \end{aligned}$$

$$-\frac{2k_2^2}{k_1^2}\,T_z^v(k_2 l, e^{ik_1\rho}). \tag{12.11.11}$$

The term $-2(f_2 - f_4)/k_1^2$ is not evaluated since it contributes nothing to the lateral wave observed at a point in the ocean near the sea floor. The term $-2(f_3 - f_5)/k_1^2$ is the same as (12.6.5) with an extra factor $-k_1^{-1}$. That is,

$$\begin{aligned}(L_3 - L_5)_z^v &= -\frac{2}{k_1^2}\int_0^\infty (f_3 - f_5)J_0(\lambda\rho)\lambda^3\,d\lambda \\ &= Ae^{ik_1(z+d)}\bar{H}_{z1}(b;\rho,0) \\ &\quad + \frac{2k_3^2}{k_1^2}\,Ae^{ik_1(z+d)}\int_0^\infty \frac{1}{\gamma_3}\,J_0(\lambda\rho)\lambda^3\,d\lambda \\ &\quad + \frac{ik_3^2}{k_1^2}\,Ae^{ik_1(z+d)}\int_0^\infty \frac{J_0(\lambda\rho)\lambda^3}{\lambda - ik_b}\,d\lambda.\end{aligned} \tag{12.11.12}$$

The function $\bar{H}_{z1}(b;\rho,0)$ differs from $H_{z1}(b;\rho,0)$ not only in the factor $-k_1^{-1}$ but also in the factors $J_0(\lambda\rho)\lambda^3$ in place of $J_1(\lambda\rho)\lambda^2$. In the evaluation of $H_{z1}(b;\rho,0)$, the large-argument form of the Bessel function is used. As seen from (E13) and (E14) in Appendix E, $J_0(\lambda\rho) \sim e^{i\pi/2}J_1(\lambda\rho) = iJ_1(\lambda\rho)$. Also, λ is replaced by k_3 in the integration. Hence, with (12.6.9),

$$\begin{aligned}\bar{H}_{z1}(b;\rho,0) &= -\frac{ik_3}{k_1}\,H_{z1}(b;\rho,0) \\ &= 2b\,\frac{k_3^6}{k_1^3}\left(\frac{\pi}{k_3\rho}\right)^{1/2} e^{ik_3\rho}e^{-iRb^2}\mathcal{F}(Rb^2).\end{aligned} \tag{12.11.13}$$

The second integral in (12.11.12) is

$$\begin{aligned}\int_0^\infty \frac{1}{\gamma_3}\,J_0(\lambda\rho)\lambda^3\,d\lambda &= \tfrac{1}{2}[I_7(k_3;\rho,0) + I_8(k_3;\rho,0)] \\ &= -k_3\left(\frac{1}{\rho^2} + \frac{i}{k_3\rho^3} + \frac{ik_3}{\rho} - \frac{2}{\rho^2} - \frac{2i}{k_3\rho^3}\right)e^{ik_3\rho} \\ &= -k_3\left(\frac{ik_3}{\rho} - \frac{1}{\rho^2} - \frac{i}{k_3\rho^3}\right)e^{ik_3\rho}.\end{aligned} \tag{12.11.14}$$

The last integral in (12.11.12) leads to Struve functions and nonpropagating terms that must cancel in the complete expression for the field. It will be omitted.

With nonpropagating terms omitted, the complete expression for $(L_3 - L_5)_z^v$ is

$$(L_3 - L_5)_z^v = -\frac{2k_3^2}{k_1^2}\,Ae^{ik_1(z+d)}e^{ik_3\rho}$$

$$\times \left[\frac{ik_3^2}{\rho} - \frac{k_3}{\rho^2} - \frac{i}{\rho^3} - \frac{bk_3^4}{k_1} \left(\frac{\pi}{k_3 \rho} \right)^{1/2} e^{-iRb^2} \mathcal{F}(Rb^2) \right]. \tag{12.11.15}$$

It now follows that

$$\begin{aligned} &F_{1z}(\rho, z+d) \\ &\quad \sim L_{1z}^v(e^{ik_1 r_2}) + (L_2 - L_4)_z^v + (L_3 - L_5)_z^v(e^{ik_3\rho}) \\ &\quad = (L_2 - L_4)_z^v \\ &\qquad - \frac{2k_3^2}{k_1^2} \left\{ Ae^{ik_1(z+d)} e^{ik_3\rho} \right. \\ &\qquad\qquad \times \left[\frac{ik_3^2}{\rho} - \frac{k_3}{\rho^2} - \frac{i}{\rho^3} - \frac{bk_3^4}{k_1} \left(\frac{\pi}{k_3\rho} \right)^{1/2} e^{-iRb^2} \mathcal{F}(Rb^2) \right] \\ &\qquad\qquad \left. - i \left(\frac{z+d}{\rho} \right) \left(\frac{ik_1^2}{\rho} - \frac{k_1}{2\rho^2} + \frac{7i}{8\rho^3} \right) e^{ik_1 r_2} \right\} \\ &\qquad - \frac{2k_2^2}{k_1^2} T_z^v(k_2 l, e^{ik_1\rho}). \end{aligned} \tag{12.11.16}$$

The integrand in (12.11.6) is like that in (12.11.5) but with $J_1(\lambda\rho)\lambda^2$ in place of $J_0(\lambda\rho)\lambda^3$. This means that (12.11.8) again applies. Paralleling Section 12.6 as before,

$$\begin{aligned} L_{1\phi}^v &= -2 \int_0^\infty \frac{f_1}{\gamma_1^2} J_1(\lambda\rho)\lambda^2 \, d\lambda \\ &= -\frac{2ik_3^2}{k_1^2} \int_0^\infty e^{i\gamma_1(z+d)} J_1(\lambda\rho)\lambda \, d\lambda \\ &\quad - \frac{2k_2^2}{k_1^2} \int_0^\infty e^{i\gamma_1(z+d)} \tan i\lambda l \, J_1(\lambda\rho)\lambda \, d\lambda \\ &= -\frac{2ik_3^2}{k_1^2} \frac{\rho}{2} I_{20}(k_1; \rho, z+d) - \frac{2k_2^2}{k_1^2} T_\phi^v(k_2 l, e^{ik_1\rho}) \\ &= -\frac{2ik_3^2}{k_1^2} \left[\frac{e^{ik_1(z+d)}}{\rho^2} - \left(\frac{z+d}{\rho} \right) \left(\frac{k_1}{\rho} + \frac{3i}{2\rho^2} \right) e^{ik_1 r_2} \right] \\ &\quad - \frac{2k_2^2}{k_1^2} T_\phi^v(k_2 l, e^{ik_1\rho}) \end{aligned} \tag{12.11.17}$$

and

$$\begin{aligned} (L_3 - L_5)_\phi^v &= -\frac{2k_3^2}{k_1^2} Ae^{ik_1(z+d)} e^{ik_3\rho} \\ &\quad \times \left[\frac{k_3}{\rho} + \frac{i}{\rho^2} + \frac{ibk_3^3}{k_1} \left(\frac{\pi}{k_3\rho} \right)^{1/2} e^{-iRb^2} \mathcal{F}(Rb^2) \right]. \end{aligned} \tag{12.11.18}$$

Finally, the propagating terms in $F_{1\phi}(\rho, z+d)$ are

$$\begin{aligned}
F_{1\phi}(\rho, z+d) &\sim L_{1\phi}^{v}(e^{ik_1r_2}) + (L_2 - L_4)_{\phi}^{v} + (L_3 - L_5)_{\phi}^{v}(e^{ik_3\rho}) \\
&= (L_2 - L_4)_{\phi}^{v} \\
&\quad - \frac{2k_3^2}{k_1^2}\Bigg\{Ae^{ik_1(z+d)}e^{ik_3\rho} \\
&\qquad \times \left[\frac{k_3}{\rho} + \frac{i}{\rho^2} + \frac{ibk_3^3}{k_1}\left(\frac{\pi}{k_3\rho}\right)^{1/2} e^{-iRb^2}\mathcal{F}(Rb^2)\right] \\
&\qquad - ie^{ik_1r_2}\left(\frac{z+d}{\rho}\right)\left(\frac{k_1}{\rho} + \frac{3i}{2\rho^2}\right)\Bigg\} \\
&\quad - \frac{2k_2^2}{k_1^2}T_{\phi}^{v}(k_2 l, e^{ik_1\rho}).
\end{aligned} \tag{12.11.19}$$

The complete formulas for the integrals in (12.11.1)–(12.11.3) are obtained by combining the values of the first two integrals in each formula, as obtained from the two-layer case, with (12.11.7), (12.11.16), and (12.11.19). The former are obtained from (3.4.3)–(3.4.8). The complete results are

$$\begin{aligned}
E_{1\rho}(\rho, z) = -\frac{\omega\mu_0}{2\pi k_1^2}&\Bigg\{\frac{k_3^2}{k_1}\Bigg[Af(\nu, \rho, k_3)e^{ik_3\rho}e^{ik_1(z+d)} \\
&\quad - ie^{ik_1r_2}\left(\frac{1}{\rho^2} + \frac{3i}{2k_1\rho^3}\right)\Bigg] + \frac{ik_2^2}{k_1^2}T_{\rho}^{v}(k_2 l, e^{ik_1\rho}) \\
&\quad + \frac{e^{ik_1r_1}}{2}\left(\frac{ik_1^2}{r_1} - \frac{3k_1}{r_1^2} - \frac{3i}{r_1^3}\right)\left(\frac{\rho}{r_1}\right)\left(\frac{z-d}{r_1}\right) \\
&\quad - \frac{e^{ik_1r_2}}{2}\left(\frac{ik_1^2}{r_2} - \frac{3k_1}{r_2^2} - \frac{3i}{r_2^3}\right)\left(\frac{\rho}{r_2}\right)\left(\frac{z+d}{r_2}\right)\Bigg\} \\
+ E_{1\rho}^{(2)}(\rho, z),&
\end{aligned} \tag{12.11.20}$$

$$\begin{aligned}
E_{1z}(\rho, z) = \frac{\omega\mu_0}{2\pi k_1^2}&\Bigg\{\frac{k_3^2}{k_1^2}\Bigg[Ak_3 g(\nu, \rho, k_3)e^{ik_3\rho}e^{ik_1(z+d)} \\
&\quad - ie^{ik_1r_2}\left(\frac{z+d}{\rho}\right)\left(\frac{ik_1^2}{\rho} - \frac{k_1}{2\rho^2} + \frac{7i}{8\rho^3}\right)\Bigg] \\
&\quad + \frac{k_2^2}{k_1^2}T_{z}^{v}(k_2 l, e^{ik_1\rho}) \\
&\quad + \frac{e^{ik_1r_1}}{2}\Bigg[\frac{ik_1^2}{r_1} - \frac{k_1}{r_1^2} - \frac{i}{r_1^3} \\
&\qquad - \left(\frac{z-d}{r_1}\right)^2\left(\frac{ik_1^2}{r_1} - \frac{3k_1}{r_1^2} - \frac{3i}{r_1^3}\right)\Bigg] \\
&\quad - \frac{e^{ik_1r_2}}{2}\Bigg[\frac{ik_1^2}{r_2} - \frac{k_1}{r_2^2} - \frac{i}{r_2^3}
\end{aligned}$$

$$- \left(\frac{z+d}{r_2}\right)^2 \left(\frac{ik_1^2}{r_2} - \frac{3k_1}{r_2^2} - \frac{3i}{r_2^3}\right)\Bigg]\Bigg\}$$

$$+ E_{1z}^{(2)}(\rho, z), \tag{12.11.21}$$

$$B_{1\phi}(\rho, z) = -\frac{\mu_0}{2\pi}\Bigg\{\frac{k_3^2}{k_1^2}\Bigg[Af(\nu, \rho, k_3)e^{ik_3\rho}e^{ik_1(z+d)} - ie^{ik_1r_2}\left(\frac{z+d}{\rho}\right)\left(\frac{ik_1}{\rho} - \frac{3}{2\rho^2}\right)\Bigg] + \frac{ik_2^2}{k_1^2}T_\phi^v(k_2l, e^{ik_1\rho}) + \frac{e^{ik_1r_1}}{2}\left(\frac{ik_1}{r_1} - \frac{1}{r_1^2}\right)\left(\frac{\rho}{r_1}\right) - \frac{e^{ik_1r_2}}{2}\left(\frac{ik_1}{r_2} - \frac{1}{r_2^2}\right)\left(\frac{\rho}{r_2}\right)\Bigg\} + B_{1\phi}^{(2)}(\rho, z). \tag{12.11.22}$$

The functions $E_{1\rho}^{(2)}(\rho, z)$, $E_{1z}^{(2)}(\rho, z)$, and $B_{1\phi}^{(2)}(\rho, z)$ represent the unevaluated terms that involve propagation in Region 2. The quantities A, $f(\nu, \rho, k_3)$, and $g(\nu, \rho, k_3)$ are defined in (12.6.16)–(12.6.18), ν and b are in (12.6.20), and $\mathcal{F}(Rb^2)$ in (12.6.22). Note that $Rb^2 = R\nu^2k_1^2/k_3^2 = k_3\rho\nu^2/2$. When $l = 0$, $A = b = 1$ and $\nu = k_3/k_1$ so that these formulas reduce to those for the two-layer region in Chapter 3, viz., (3.6.2)–(3.6.4). This completes the derivation of the electromagnetic field in Region 1 (sea) in the presence of Region 2 (sediment) over Region 3 (rock). These formulas presuppose (12.1.1) and (12.1.12).

When applied to the field generated by a vertical electric dipole on the sea floor at the low frequencies of interest to geophysicists, the condition (12.1.12) simplifies to (12.1.13c), i.e., to

$$\rho \geq 3.65\left(\frac{\sigma_1}{\sigma_2}\right)^{1/4}\left(\frac{2}{\omega\mu_0\sigma_2}\right)^{1/4} l^{1/2}. \tag{12.11.23}$$

For the conditions that obtained in the measurements of Edwards et al. (1985) described in Chapter 4, the following numerical values apply: $f = 0.125$ Hz, $l \sim 640$ m, $\sigma_1 = 2.85$ S/m, $\sigma_2 \sim 0.4$ S/m, so that

$$\begin{aligned} \rho &\geq 3.65\left(\frac{2.85}{0.4}\right)^{1/4}\left(\frac{2}{2\pi \times 0.125 \times 4\pi \times 10^{-7} \times 0.4}\right)^{1/4}(640)^{1/2} \\ &\geq 3.65 \times 1.634 \times (0.0507 \times 10^8)^{1/4} \times 25.3 \\ &\geq 3.65 \times 1.634 \times 47.4 \times 25.3 \\ &\geq 7.15 \text{ km}. \end{aligned} \tag{12.11.24}$$

Since the measurements of Edwards et al. (1985) were all made in the range $0.1 \leq \rho \leq 2$ km, the three-layer formulas developed in this section have no application. This is as would be expected since, as shown in Section 4.7, the depth of penetration of the lateral waves into the sediment is much smaller than the thickness of the sediment so that the lateral waves travel primarily in Region 2, not largely in Region 3 as required by the formulas in this section. It is because of this that the two-layer formulas supplemented with a single reflection as carried out in Section 4.7 are adequate.

13
Exact Formulas for the Lateral Electromagnetic Pulses Generated by Vertical Dipoles

13.1 Introduction

Instead of exciting a dipole near the plane boundary between two electrically different half-spaces with a continuous wave source at an angular frequency ω so that its electric moment is

$$I(t)h_e = \operatorname{Re} I(0)h_e e^{-i\omega t}, \tag{13.1.1}$$

a single pulse (or sequence of such pulses) can be applied. If the pulse is a δ-function, the electric moment is

$$I(t)h_e = I(0)h_e\delta(t), \tag{13.1.2}$$

where the δ-function can be expressed in terms of the Fourier integral as follows:

$$\delta(t) = \frac{1}{2\pi}\int_{-\infty}^{\infty} e^{-i\omega t}\, d\omega. \tag{13.1.3}$$

Before pulse excitation is studied for a dipole on the plane boundary between two electrically different half-spaces, it is of interest to consider the isolated dipole in an infinite homogeneous dielectric with relative permittivity $\epsilon_r = \epsilon$ independent of frequency. This problem is most easily solved by treating directly the time-dependent Maxwell's equations. However, it is difficult to generalize such an approach to the case with a planar boundary between two electrically different half-spaces. Therefore, in this section the case of the infinite homogeneous dielectric is treated by applying the Fourier transform to the time-independent problem. The time-independent expression for the component E_z in the equatorial plane, $z = 0$, is

$$\begin{aligned}\tilde{E}_z(\rho, \omega) &= -\frac{\omega\mu_0}{4\pi}\int_0^\infty \frac{J_0(\lambda\rho)\lambda^3\,d\lambda}{k^2(k^2-\lambda^2)^{1/2}}\\ &= \frac{\omega\mu_0}{4\pi}\,e^{ik\rho}\left(\frac{i}{\rho}-\frac{1}{k\rho^2}-\frac{i}{k^2\rho^3}\right),\end{aligned}\tag{13.1.4}$$

where $k^2 = k_0^2\epsilon = \omega^2\epsilon/c^2$. With $\lambda' = c\lambda$ and $\rho' = \rho/c$, the integral form of (13.1.4) becomes

$$\tilde{E}_z(\rho', \omega) = -\frac{\omega\mu_0}{4\pi c\epsilon}\int_0^\infty \frac{J_0(\lambda'\rho')\lambda'^3\,d\lambda'}{\omega^2(\omega^2\epsilon-\lambda'^2)^{1/2}},\tag{13.1.5}$$

with the branch cut and contour of integration shown in Fig. 13.1.1. With a δ-function current with unit moment, i.e., $I(t)h_e = I(0)h_e\delta(t)$ with $I(0)h_e = 1$ A m, the time-dependent field is

$$\begin{aligned}E_z(\rho', t) &= \frac{1}{\pi}\,\mathrm{Re}\int_0^\infty d\omega\, e^{-i\omega t}\tilde{E}_z(\rho', \omega)\\ &= -\frac{\mu_0}{4\pi^2 c\epsilon}\,\mathrm{Re}\int_0^\infty d\omega\, e^{-i\omega t}\omega^2\int_0^\infty \frac{J_0(\omega\xi\rho')\xi^3\,d\xi}{(\epsilon-\xi^2)^{1/2}},\end{aligned}\tag{13.1.6}$$

where the change in variable $\xi = \lambda'/\omega$ has been made. This is equivalent to

$$E_z(\rho', t) = \frac{\mu_0}{4\pi^2 c\epsilon}\frac{\partial^2}{\partial t^2}\,\mathrm{Re}\int_0^\infty d\xi\,\frac{\xi^3}{(\epsilon-\xi^2)^{1/2}}\int_0^\infty J_0(\omega\xi\rho')e^{-i\omega t}\,d\omega.\tag{13.1.7}$$

The integral with respect to ω can be carried out with formula (6.611-1) of Gradshteyn and Ryzhik (1980, p. 707). For $t > \xi\rho'$, this gives

$$\int_0^\infty J_0(\omega\xi\rho')e^{-i\omega t}\,d\omega = -i(t^2-\xi^2\rho'^2)^{-1/2}\tag{13.1.8}$$

so that

$$E_z(\rho', t) = \frac{\mu_0}{4\pi^2 c\rho'\epsilon}\frac{\partial^2}{\partial t^2}\,\mathrm{Im}\int_0^\infty \frac{\xi^3\,d\xi}{(\epsilon-\xi^2)^{1/2}(t^2/\rho'^2-\xi^2)^{1/2}}.\tag{13.1.9}$$

Now let the contour of integration be chosen as shown in Fig. 13.1.2. When $t/\rho' < \epsilon^{1/2}$, the integrand in (13.1.9) is real so that there is no contribution to the relevant imaginary part of the integral. When $t/\rho' > \epsilon^{1/2}$, the integral becomes

$$E_z(\rho', t) = \frac{\mu_0}{4\pi^2 c\rho'\epsilon}\frac{\partial^2}{\partial t^2}\,\mathrm{Im}\int_0^\infty \frac{\xi^3\,d\xi}{i(\xi^2-\epsilon)^{1/2}(t^2/\rho'^2-\xi^2)^{1/2}}.\tag{13.1.10}$$

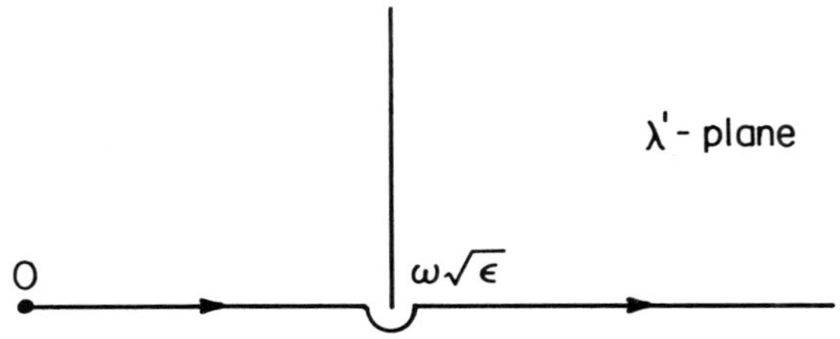

Figure 13.1.1. Branch cut and contour of integration in the λ'-plane.

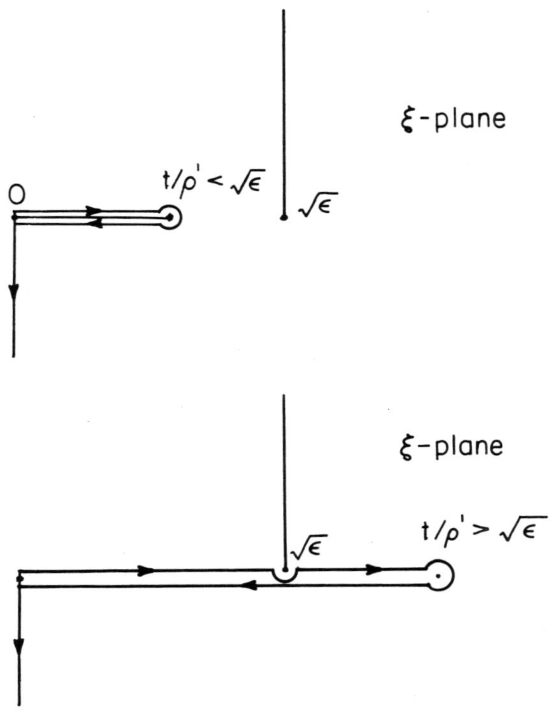

Figure 13.1.2. Branch-cut structure and contour of integration in the ξ-plane.

Along the indicated contours, the only contribution is

$$E_z(\rho',t) = -\frac{\mu_0}{4\pi^2 c\rho'\epsilon}\frac{\partial^2}{\partial t^2}\left(\int_{\epsilon^{1/2}}^{t/\rho'} \frac{\xi^3\,d\xi}{(\xi^2-\epsilon)^{1/2}(t^2/\rho'^2-\xi^2)^{1/2}} - \int_{\epsilon^{1/2}}^{t/\rho'} \frac{\xi^3\,d\xi}{(\xi^2-\epsilon)^{1/2}[-(t^2/\rho'^2-\xi^2)^{1/2}]}\right)$$

$$= -\frac{\mu_0}{2\pi^2 c\rho'\epsilon}\frac{\partial^2}{\partial t^2}\int_{\epsilon^{1/2}}^{t/\rho'} \frac{\xi^3\,d\xi}{(\xi^2-\epsilon)^{1/2}(t^2/\rho'^2-\xi^2)^{1/2}}. \tag{13.1.11}$$

The integral is elementary. Let $x=\xi^2$. Then,

$$E_z(\rho',t) = -\frac{\mu_0}{4\pi^2 c\rho'\epsilon}\frac{\partial^2}{\partial t^2} f(t/\rho'), \tag{13.1.12}$$

where

$$f(t/\rho') = \int_{\epsilon}^{t^2/\rho'^2} \frac{x\,dx}{X^{1/2}}$$

$$= -X^{1/2}\Big|_{\epsilon}^{t^2/\rho'^2} + \frac{1}{2}\left(\epsilon+\frac{t^2}{\rho'^2}\right)\int_{\epsilon}^{t^2/\rho'^2}\frac{dx}{X^{1/2}}, \tag{13.1.13}$$

where $X=(x-\epsilon)(t^2/\rho'^2-x) = -\epsilon t^2/\rho'^2+(\epsilon+t^2/\rho'^2)x-x^2 = a+bx+cx^2$ with $a=-\epsilon t^2/\rho'^2$, $b=(\epsilon+t^2/\rho'^2)$, and $c=-1$. The second integral gives

$$\int \frac{dx}{X^{1/2}} = -\frac{1}{(-c)^{1/2}}\sin^{-1}\frac{2cx+b}{(-q)^{1/2}}, \tag{13.1.14}$$

where $q=4ac-b^2=-(t^2/\rho'^2-\epsilon)^2$. The substitution of the limits shows that the first term on the right in (13.1.13) is zero and the arcsines have the arguments -1 and $+1$ so that

$$f(t/\rho') = \frac{\pi}{2}\left(\frac{t^2}{\rho'^2}+\epsilon\right). \tag{13.1.15}$$

The differentiation of this function must be carried out with care since it and its derivatives have discontinuous behaviors at $t/\rho'=\epsilon^{1/2}$. Thus, since $E_z(\rho',t)$ and $f(t/\rho')$ are zero when $t/\rho'<\epsilon^{1/2}$ so that $f(\epsilon^{1/2}-)=0$, whereas $f(\epsilon^{1/2}+)=\pi\epsilon$, there is a step increase in $f(t/\rho')$ as t/ρ' increases across $t/\rho'=\epsilon^{1/2}$. When $t/\rho'>\epsilon^{1/2}$, $f(t/\rho')=\frac{1}{2}\pi(t^2/\rho'^2+\epsilon)$. Now let $f'(t/\rho')=\partial f(t/\rho')/\partial t$. When $t/\rho'<\epsilon^{1/2}$, $f'(t/\rho')=0$; when $t/\rho'>\epsilon^{1/2}$, $f'(t/\rho')=\pi t/\rho'^2$. Thus,

$$f'(t/\rho') = \pi\epsilon\delta(t-\epsilon^{1/2}\rho') + \begin{cases} 0, & t/\rho'<\epsilon^{1/2} \\ \pi t/\rho'^2, & t/\rho'>\epsilon^{1/2} \end{cases}. \tag{13.1.16}$$

Evidently, there is a step increase by $(\pi t/\rho'^2)$ at $t/\rho'=\epsilon^{1/2}$. Let $f''(t/\rho') = \partial f'(t/\rho')/\partial t$. When $t/\rho'<\epsilon^{1/2}$, $f''(t/\rho')=0$; when $t/\rho'>\epsilon^{1/2}$, $f''(t/\rho') = \pi/\rho'^2$. Hence,

$$f''(t/\rho') = \pi\epsilon\delta'(t-\epsilon^{1/2}\rho') + \frac{\pi t}{\rho'^2}\,\delta(t-\epsilon^{1/2}\rho')$$

$$+\begin{cases} 0, & t/\rho' < \epsilon^{1/2} \\ \pi/\rho'^2, & t/\rho' > \epsilon^{1/2} \end{cases}. \tag{13.1.17}$$

However, $(t/\rho')\delta(t - \epsilon^{1/2}\rho') = \epsilon^{1/2}\delta(t - \epsilon^{1/2}\rho')$. Hence, with $\rho' = \rho/c$ and $c^2 = (\mu_0\epsilon_0)^{-1}$,

$$E_z(\rho, t) = -\frac{\mu_0}{4\pi\rho}\,\delta'\left(t - \frac{\epsilon^{1/2}\rho}{c}\right) - \frac{\zeta_e}{4\pi\rho^2}\,\delta\left(t - \frac{\epsilon^{1/2}\rho}{c}\right)$$
$$- \frac{1}{4\pi\epsilon\epsilon_0\rho^3}\begin{Bmatrix} 0 \\ 1 \end{Bmatrix}; \quad \begin{cases} ct/\rho < \epsilon^{1/2} \\ ct/\rho > \epsilon^{1/2} \end{cases}, \tag{13.1.18}$$

where $\zeta_e = (\mu_0/\epsilon\epsilon_0)^{1/2}$.

Formula (13.1.18) indicates that, in the equatorial plane of the dipole, pulses of the form $\delta'(t - \epsilon^{1/2}\rho/c)$ and $\delta(t - \epsilon^{1/2}\rho/c)$ travel radially outward with the velocity $v = c/\epsilon^{1/2}$ with the amplitude factors $\mu_0/4\pi\rho$ and $\zeta_e/4\pi\rho^2$, respectively. The δ' pulse corresponds to the far-field term $(i\omega\mu_0/4\pi\rho)\exp[i(k\rho - \omega t)]$ and the δ pulse corresponds to the term $(-\zeta_e/4\pi\rho^2)\exp[i(k\rho - \omega t)]$ in the steady-state field (13.1.4). After the pulses have passed any radial distance ρ, a static electric field with magnitude $1/4\pi\epsilon\epsilon_0\rho^3$ remains. This is due to the charge that remains separated on the dipole after the current pulses have occurred. The corresponding term in the steady-state formula (13.1.4) is the quasi-static term $(-i\omega\mu_0/4\pi k^2\rho^3)\exp[i(k\rho - \omega t)] \equiv (-i/4\pi\omega\epsilon\epsilon_0\rho^3)\exp[i(k\rho - \omega t)]$.

13.2 The Time-Independent Field of a Unit Vertical Electric Dipole on the Boundary Between Electrically Different Half-Spaces

Consider next a vertical dipole in Region 1 (lake water) on the boundary, i.e., at $d = 0$, between Region 1 and Region 2 (air). The wave numbers of these regions are $k_1 = \epsilon_{1r}^{1/2}\omega/c$ and $k_2 = \omega/c$, where $\epsilon_1 = \epsilon_{1r}\epsilon_0$ is the real permittivity of Region 1. The components of the electromagnetic field on the boundary $z = 0$ in Region 2 are given by (3.3.12)–(3.3.14) with $z = d = 0$. They are

$$\tilde{B}_\phi(\rho, \omega) = \frac{i\mu_0 k_2^2}{2\pi}\int_0^\infty N^{-1}J_1(\lambda\rho)\lambda^2\,d\lambda, \tag{13.2.1}$$

$$\tilde{E}_\rho(\rho, \omega) = -\frac{i\omega\mu_0}{2\pi}\int_0^\infty \gamma_2 N^{-1}J_1(\lambda\rho)\lambda^2\,d\lambda, \tag{13.2.2}$$

$$\frac{\epsilon_1}{\epsilon_0}\,\tilde{E}_{1z}(\rho, \omega) = \tilde{E}_{2z}(\rho, \omega) = -\frac{\omega\mu_0}{2\pi}\int_0^\infty N^{-1}J_0(\lambda\rho)\lambda^3\,d\lambda, \tag{13.2.3}$$

where

$$N = k_1^2\gamma_2 + k_2^2\gamma_1; \qquad \gamma_j = (k_j^2 - \lambda^2)^{1/2}, \quad j = 1,\, 2. \tag{13.2.4}$$

These are the time-independent field components generated by a current varying periodically at the angular frequency ω and with a unit electric moment, i.e., $I(t)h_e = I(0)h_e \exp(-i\omega t)$ with $I(0)h_e = 1$ A m. They are the Fourier transforms of the time-dependent components $B_\phi(\rho,t)$, $E_\rho(\rho,t)$, and $E_z(\rho,t)$, which are to be determined.

13.3 The Time-Dependent Component $E_{2z}(\rho,t)$ with a δ-Function Excitation

The integral for $\tilde{E}_{2z}(\rho,\omega)$ in explicit form is

$$\tilde{E}_{2z}(\rho,\omega) = -\frac{\omega\mu_0}{2\pi}\int_0^\infty \frac{\lambda^3 J_0(\lambda\rho)}{k_1^2(k_2^2-\lambda^2)^{1/2} + k_2^2(k_1^2-\lambda^2)^{1/2}}\,d\lambda. \tag{13.3.1}$$

It is convenient to express this in terms of ω instead of the wave numbers. Thus, with $k_1^2 = \omega^2\epsilon/c^2$ and $k_2^2 = \omega^2/c^2$, where $c = (\mu_0\epsilon_0)^{-1/2}$ is the velocity of light and $\epsilon = \epsilon_{1r} = \epsilon_1/\epsilon_0$,

$$\tilde{E}_{2z}(\rho',\omega) = -\frac{\omega\mu_0}{2\pi c}\int_0^\infty \frac{\lambda'^3 J_0(\lambda'\rho')}{\omega^2\epsilon(\omega^2-\lambda'^2)^{1/2} + \omega^2(\omega^2\epsilon-\lambda'^2)^{1/2}}\,d\lambda'. \tag{13.3.2}$$

In this formula, $\lambda' = c\lambda$ and $\rho' = \rho/c$. The path of integration in the λ'-plane is shown in Fig. 13.3.1. The integral in (13.3.2) is analytic in the upper half of the ω-plane. This follows since, in moving from the positive real axis to the upper half-plane as shown in Fig. 13.3.2, there is no need at all to deform the contour. Furthermore, since $\tilde{E}_{2z}(\rho',\omega)$ as defined in (13.3.2) is the Fourier transform of the *real* time-dependent component $E_{2z}(\rho',t)$, it follows that

$$\tilde{E}_{2z}(\rho',\omega) = \tilde{E}^*_{2z}(\rho',\omega), \tag{13.3.3}$$

where the asterisk denotes the complex conjugate.

Let the vertical dipole be excited by a unit moment that is a δ-function pulse in time, i.e., $I(t)h_e = I(0)h_e\delta(t)$ with $I(0)h_e = 1$ A m. In order to carry out the Fourier transform explicitly in a way similar to that of Section 13.1, the assumption or approximation is made that the relative dielectric constant ϵ_r is independent of frequency. The electric field, which is of course real, is then given by the Fourier transform

$$E_{2z}(\rho',t) = \frac{1}{\pi}\,\text{Re}\int_0^\infty e^{-i\omega t}\tilde{E}_{2z}(\rho',\omega)\,d\omega. \tag{13.3.4}$$

Figure 13.3.1. The λ'-plane and the path of integration along the real axis.

With (13.3.2), this becomes

$$E_{2z}(\rho', t) = -\frac{\mu_0}{2\pi^2 c} \operatorname{Re} \int_0^\infty d\omega\, e^{-i\omega t} \omega \times \int_0^\infty d\lambda' \frac{\lambda'^3 J_0(\lambda'\rho')}{\omega^2[\epsilon(\omega^2 - \lambda'^2)^{1/2} + (\omega^2\epsilon - \lambda'^2)^{1/2}]}. \tag{13.3.5}$$

Now let the variable be changed to ξ with $\lambda' = \omega\xi$ and $d\lambda' = \omega\, d\xi$. The result is

$$E_{2z}(\rho', t) = -\frac{\mu_0}{2\pi^2 c} \operatorname{Re} \int_0^\infty d\omega\, e^{-i\omega t} \omega \times \int_0^\infty d\xi\, \omega \frac{\xi^3 J_0(\omega\xi\rho')}{\epsilon(1 - \xi^2)^{1/2} + (\epsilon - \xi^2)^{1/2}}. \tag{13.3.6}$$

This integral is equal to the following:

$$E_{2z}(\rho', t) = \frac{\mu_0}{2\pi^2 c} \frac{\partial^2}{\partial t^2} \operatorname{Re} \int_0^\infty d\xi \frac{\xi^3}{\epsilon(1 - \xi^2)^{1/2} + (\epsilon - \xi^2)^{1/2}} \times \int_0^\infty d\omega\, e^{-i\omega t} J_0(\omega\xi\rho'). \tag{13.3.7}$$

The integration with respect to ω can be carried out with formula (6.611-1) of Gradshteyn and Ryzhik (1980, p. 707). With $t > 0$, the integral is

$$\int_0^\infty d\omega\, e^{-i\omega t} J_0(\omega\xi\rho') = \begin{cases} -i(t^2 - \xi^2\rho'^2)^{-1/2}, & \xi\rho' < t \\ (\xi^2\rho'^2 - t^2)^{-1/2}, & \xi\rho' > t \end{cases}, \tag{13.3.8}$$

so that, with $t > \xi\rho'$,

$$E_{2z}(\rho', t) = \frac{\mu_0}{2\pi^2 \rho' c} \times \frac{\partial^2}{\partial t^2} \operatorname{Im} \int_0^\infty d\xi \frac{\xi^3}{[\epsilon(1 - \xi^2)^{1/2} + (\epsilon - \xi^2)^{1/2}](t^2/\rho'^2 - \xi^2)^{1/2}}, \tag{13.3.9}$$

with the branch-cut structure shown in Fig. 13.3.3. In (13.3.9), ξ has been taken to be real. It is now convenient to let ξ be complex and to deform the contour of integration from the real axis to that shown in Fig. 13.3.4. Since the integrand is real on the vertical piece of the contour, integration

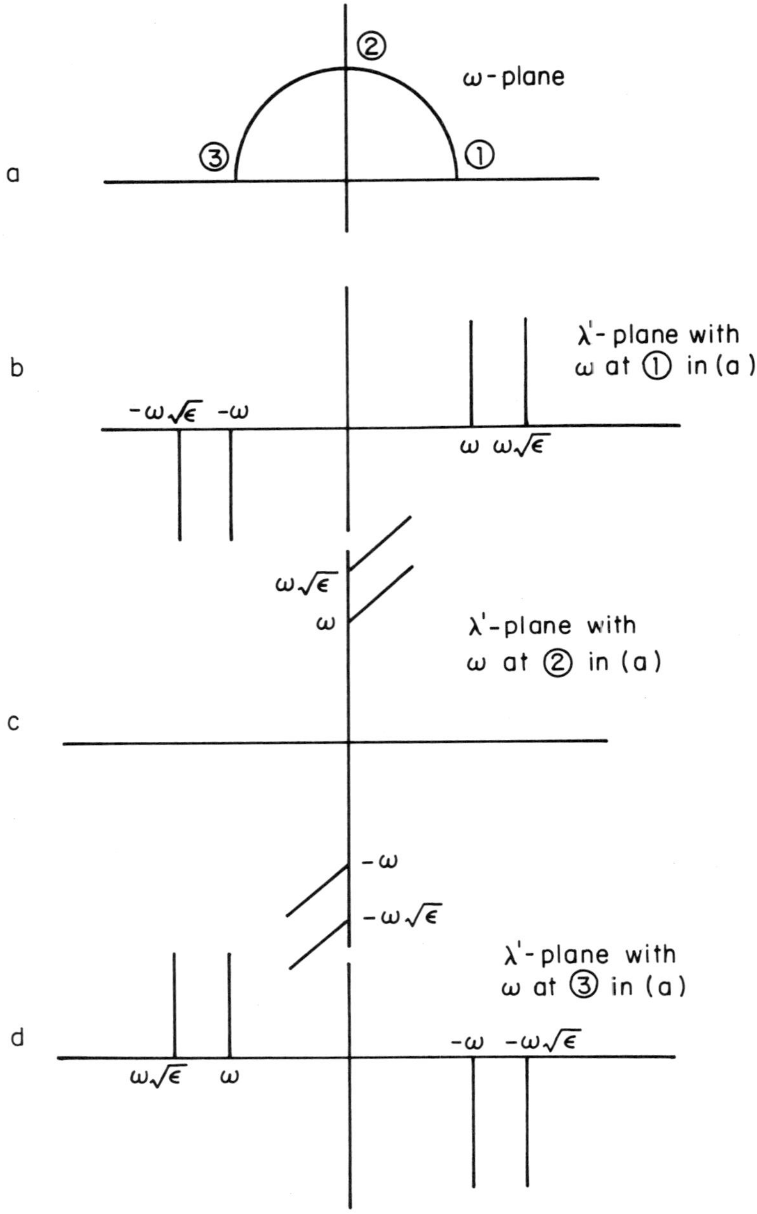

Figure 13.3.2. The ω-plane and the λ'-plane for three locations of ω.

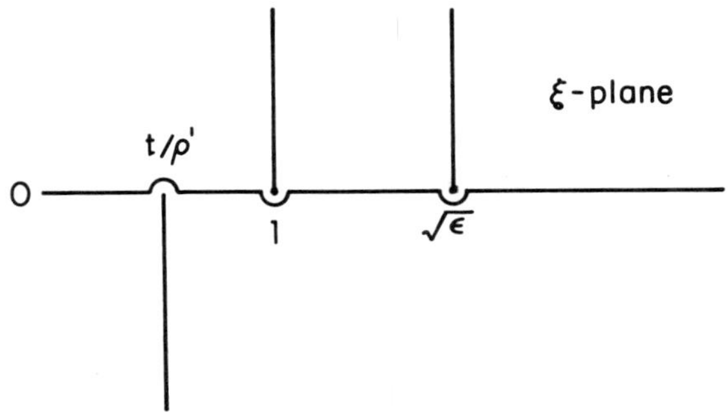

Figure 13.3.3. Branch-cut structure for the integral in (13.3.9).

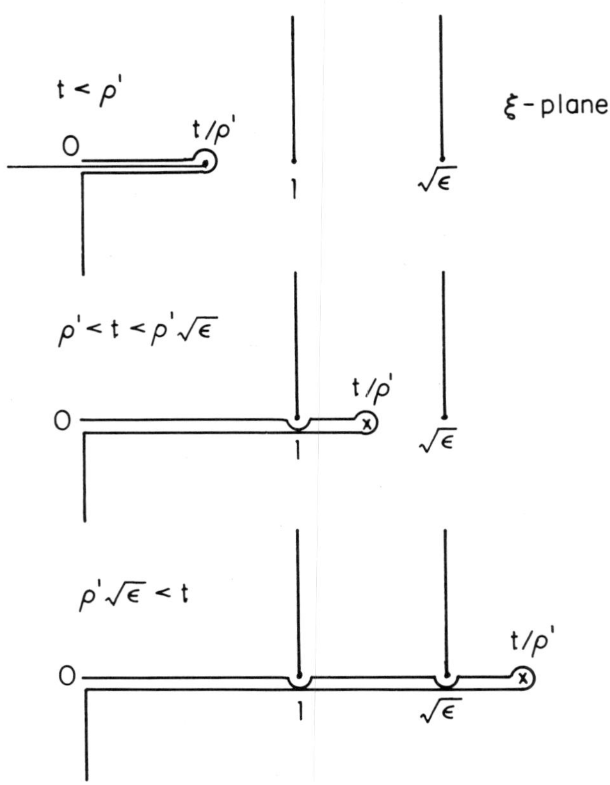

Figure 13.3.4. Contours of integration for (13.3.9).

along it yields zero. For $t < \rho'$, the integrand also remains real along the rest of the contour. Hence,

$$E_{2z}(\rho', t) = 0; \qquad t/\rho' < 1. \tag{13.3.10}$$

When $1 < t/\rho' < \epsilon^{1/2}$, there is a contribution to the integral. Thus,

$$\begin{aligned} E_{2z}(\rho', t) = {} & \frac{\mu_0}{2\pi^2\rho' c}\frac{\partial^2}{\partial t^2}\,\mathrm{Im}\bigg[\int_0^1 d\xi\,\xi^3 \\ & \times\bigg(\frac{1}{[\epsilon(1-\xi^2)^{1/2}+(\epsilon-\xi^2)^{1/2}](t^2/\rho'^2-\xi^2)^{1/2}} \\ & \quad - \frac{1}{[\epsilon(1-\xi^2)^{1/2}+(\epsilon-\xi^2)^{1/2}][-(t^2/\rho'^2-\xi^2)^{1/2}]}\bigg) \\ & + \int_1^{t/\rho'} d\xi\,\xi^3\bigg(\frac{1}{[i\epsilon(\xi^2-1)^{1/2}+(\epsilon-\xi^2)^{1/2}](t^2/\rho'^2-\xi^2)^{1/2}} \\ & \quad - \frac{1}{[i\epsilon(\xi^2-1)^{1/2}+(\epsilon-\xi^2)^{1/2}][-(t^2/\rho'^2-\xi^2)^{1/2}]}\bigg)\bigg]. \end{aligned} \tag{13.3.11}$$

Here the integrand of the first integral is real so that there is no contribution to the imaginary part. The integrand in the second integral is complex so that

$$E_{2z}(\rho', t) = \frac{\mu_0}{\pi^2\rho' c} \times \frac{\partial^2}{\partial t^2}\,\mathrm{Im}\int_1^{t/\rho'} d\xi\,\xi^3\,\frac{1}{[i\epsilon(\xi^2-1)^{1/2}+(\epsilon-\xi^2)^{1/2}](t^2/\rho'^2-\xi^2)^{1/2}}. \tag{13.3.12}$$

The real and imaginary parts are readily separated and only the relevant imaginary part is retained. This leaves

$$E_{2z}(\rho', t) = \frac{\mu_0}{\pi^2\rho' c}\frac{\partial^2}{\partial t^2}\int_1^{t/\rho'} d\xi\,\xi^3\,\frac{-\epsilon(\xi^2-1)^{1/2}}{(\epsilon-1)[(\epsilon+1)\xi^2-\epsilon](t^2/\rho'^2-\xi^2)^{1/2}}. \tag{13.3.13}$$

It is now convenient to change the variable to $\zeta = \xi^2$, $d\zeta = 2\xi\,d\xi$, and obtain

$$E_{2z}(\rho', t) = -\frac{\epsilon}{\epsilon-1}\frac{\mu_0}{2\pi^2\rho' c}\frac{\partial^2}{\partial t^2}\int_1^{t^2/\rho'^2}\frac{(\zeta-1)^{1/2}\zeta\,d\zeta}{[(\epsilon+1)\zeta-\epsilon](t^2/\rho'^2-\zeta)^{1/2}}; \qquad 1 < t/\rho' < \epsilon^{1/2}. \tag{13.3.14}$$

The expression corresponding to (13.3.12) for the range $t/\rho' > \epsilon^{1/2}$ is

$$E_{2z}(\rho', t) = \frac{\mu_0}{\pi^2\rho' c}$$

$$\times \frac{\partial^2}{\partial t^2} \operatorname{Im}\left(\int_1^{\epsilon^{1/2}} \frac{\xi^3 \, d\xi}{[i\epsilon(\xi^2-1)^{1/2} + (\epsilon - \xi^2)^{1/2}](t^2/\rho'^2 - \xi^2)^{1/2}} + \int_{\epsilon^{1/2}}^{t/\rho'} \frac{\xi^3 \, d\xi}{[i\epsilon(\xi^2-1)^{1/2} + i(\xi^2 - \epsilon)^{1/2}](t^2/\rho'^2 - \xi^2)^{1/2}} \right). \tag{13.3.15}$$

The imaginary part is

$$E_{2z}(\rho', t) = \frac{\mu_0}{\pi^2 \rho' c} \times \frac{\partial^2}{\partial t^2} \left(\int_1^{\epsilon^{1/2}} \frac{-\epsilon(\xi^2-1)^{1/2}\xi^3 \, d\xi}{(\epsilon-1)[(\epsilon+1)\xi^2 - \epsilon](t^2/\rho'^2 - \xi^2)^{1/2}} + \int_{\epsilon^{1/2}}^{t/\rho'} \frac{[-\epsilon(\xi^2-1)^{1/2} + (\xi^2-\epsilon)^{1/2}]\xi^3 \, d\xi}{(\epsilon-1)[(\epsilon+1)\xi^2 - \epsilon](t^2/\rho'^2 - \xi^2)^{1/2}} \right). \tag{13.3.16}$$

The first part of the second integral can be combined with the first integral to obtain

$$E_{2z}(\rho', t) = -\frac{1}{\epsilon - 1} \frac{\mu_0}{\pi^2 \rho' c} \times \frac{\partial^2}{\partial t^2} \left(\epsilon \int_1^{t/\rho'} \frac{(\xi^2-1)^{1/2}\xi^3 \, d\xi}{[(\epsilon+1)\xi^2 - \epsilon](t^2/\rho'^2 - \xi^2)^{1/2}} - \int_{\epsilon^{1/2}}^{t/\rho'} \frac{(\xi^2-\epsilon)^{1/2}\xi^3 \, d\xi}{[(\epsilon+1)\xi^2 - \epsilon](t^2/\rho'^2 - \xi^2)^{1/2}} \right). \tag{13.3.17}$$

In terms of the variable $\zeta = \xi^2$, this becomes

$$E_{2z}(\rho', t) = -\frac{1}{\epsilon - 1} \frac{\mu_0}{2\pi^2 \rho' c} \times \frac{\partial^2}{\partial t^2} \left(\epsilon \int_1^{t^2/\rho'^2} \frac{(\zeta-1)^{1/2}\zeta \, d\zeta}{[(\epsilon+1)\zeta - \epsilon](t^2/\rho'^2 - \zeta)^{1/2}} - \int_{\epsilon}^{t^2/\rho'^2} \frac{(\zeta-\epsilon)^{1/2}\zeta \, d\zeta}{[(\epsilon+1)\zeta - \epsilon](t^2/\rho'^2 - \zeta)^{1/2}} \right); \quad t/\rho' > \epsilon^{1/2}. \tag{13.3.18}$$

13.4 Evaluation of the Integrals for $E_{2z}(\rho, t)$

The expressions (13.3.14) and (13.3.18) are more convenient for integration with the notation

$$T_0 = \frac{t^2}{\rho'^2} - 1; \qquad T_\epsilon = \frac{t^2}{\rho'^2} - \epsilon, \tag{13.4.1}$$

and the change in variables $x' = \zeta - 1$ and $y' = \zeta - \epsilon$. Thus, using ϑ_0 and ϑ_ϵ to denote the two integrals that appear in (13.3.14) and (13.3.18),

$$\begin{aligned}\vartheta_0 &= \int_1^{t^2/\rho'^2} \frac{(\zeta-1)^{1/2}\zeta\, d\zeta}{[(\epsilon+1)\zeta-\epsilon](t^2/\rho'^2-\zeta)^{1/2}} \\ &= \int_1^{T_0+1} \frac{(\zeta-1)^{1/2}\zeta\, d\zeta}{[(\epsilon+1)\zeta-\epsilon][T_0-(\zeta-1)]^{1/2}} \\ &= \int_0^{T_0} \frac{x'(x'+1)\, dx'}{[(\epsilon+1)(x'+1)-\epsilon][(T_0-x')x']^{1/2}},\end{aligned} \tag{13.4.2}$$

$$\begin{aligned}\vartheta_\epsilon &= \int_\epsilon^{t^2/\rho'^2} \frac{(\zeta-\epsilon)^{1/2}\zeta\, d\zeta}{[(\epsilon+1)\zeta-\epsilon](t^2/\rho'^2-\zeta)^{1/2}} \\ &= \int_\epsilon^{T_\epsilon+\epsilon} \frac{(\zeta-\epsilon)^{1/2}\zeta\, d\zeta}{[(\epsilon+1)\zeta-\epsilon][T_\epsilon-(\zeta-\epsilon)]^{1/2}} \\ &= \int_0^{T_\epsilon} \frac{y'(y'+\epsilon)\, dy'}{[(\epsilon+1)(y'+\epsilon)-\epsilon][(T_\epsilon-y')y']^{1/2}}.\end{aligned} \tag{13.4.3}$$

As a temporary shorthand, let $E_0 = 1/(\epsilon+1)$ and $E_\epsilon = \epsilon^2/(\epsilon+1)$; also, introduce $x = x' + E_0$ and $y = y' + E_\epsilon$. The integrals now become

$$\vartheta_0 = \frac{1}{\epsilon+1}\int_{E_0}^{T_0+E_0} \frac{(x+1-E_0)(x-E_0)\, dx}{x[(T_0+E_0-x)(x-E_0)]^{1/2}}, \tag{13.4.4}$$

$$\vartheta_\epsilon = \frac{1}{\epsilon+1}\int_{E_\epsilon}^{T_\epsilon+E_\epsilon} \frac{(y+\epsilon-E_\epsilon)(y-E_\epsilon)\, dy}{y[(T_\epsilon+E_\epsilon-y)(y-E_\epsilon)]^{1/2}}. \tag{13.4.5}$$

These can be reduced to standard form. Let

$$\begin{aligned}X &= (T+E-x)(x-E) \\ &= -E(T+E) + (T+2E)x - x^2 = a + bx + cx^2,\end{aligned} \tag{13.4.6}$$

where T stands for T_0 or T_ϵ, E for E_0 or E_ϵ, and x for x or y. Note that $q = 4ac - b^2 = -T^2$ so that $(-q)^{1/2} = T$. With this notation

$$\vartheta = \frac{1}{\epsilon+1}\left(\int_E^{T+E} \frac{x\, dx}{X^{1/2}} + (1-2E)\int_E^{T+E} \frac{dx}{X^{1/2}}\right.$$

$$- E(1-E)\int_E^{T+E} \frac{dx}{xX^{1/2}}\Bigg). \tag{13.4.7}$$

These are tabulated integrals. When the limits are inserted in the integrated expressions, the first integral vanishes and the second and third integrals yield arcsines with arguments ± 1. The final integrated formula is

$$\vartheta = \frac{\pi}{2(\epsilon+1)}\left(T - 2(E-1) + \frac{2(E-1)E}{[E(T+E)]^{1/2}}\right). \tag{13.4.8}$$

The substitutions T_0 and T_ϵ for T, E_0 and E_ϵ for E, and 1 and ϵ for 1 in $(E-1)$ can now be made, followed by the substitutions $T_0 = (t^2/\rho'^2) - 1$, $T_\epsilon = (t^2/\rho'^2) - \epsilon$, $E_0 = (\epsilon+1)^{-1}$, and $E_\epsilon = \epsilon^2(\epsilon+1)^{-1}$. The resulting formulas are

$$\vartheta_0 = \frac{\pi}{2(\epsilon+1)}\left[\frac{t^2}{\rho'^2} - 1 + \frac{2\epsilon}{\epsilon+1} - \frac{2\epsilon}{\epsilon+1}\left((\epsilon+1)\frac{t^2}{\rho'^2} - \epsilon\right)^{-1/2}\right], \tag{13.4.9}$$

$$\vartheta_\epsilon = \frac{\pi}{2(\epsilon+1)}\left[\frac{t^2}{\rho'^2} - \epsilon + \frac{2\epsilon}{\epsilon+1} - \frac{2\epsilon^2}{\epsilon+1}\left((\epsilon+1)\frac{t^2}{\rho'^2} - \epsilon\right)^{-1/2}\right]. \tag{13.4.10}$$

13.5 Evaluation of $E_{2z}(\rho, t)$

With the integrated values (13.4.9) and (13.4.10), the expressions (13.3.14) and (13.3.18) for $E_{2z}(\rho', t)$ become

$$E_{2z}(\rho', t) = -\frac{\epsilon}{\epsilon^2-1}\frac{\mu_0}{4\pi\rho' c}\frac{\partial^2}{\partial t^2}\left[\frac{t^2}{\rho'^2} - 1 + \frac{2\epsilon}{\epsilon+1} - \frac{2\epsilon}{\epsilon+1}\left((\epsilon+1)\frac{t^2}{\rho'^2} - \epsilon\right)^{-1/2}\right]; \quad 1 < t/\rho' < \epsilon^{1/2}, \tag{13.5.1}$$

$$\begin{aligned} E_{2z}(\rho', t) = &-\frac{\epsilon}{\epsilon^2-1}\frac{\mu_0}{4\pi\rho' c}\frac{\partial^2}{\partial t^2}\left[\frac{t^2}{\rho'^2} - 1 + \frac{2\epsilon}{\epsilon+1} - \frac{2\epsilon}{\epsilon+1}\left((\epsilon+1)\frac{t^2}{\rho'^2} - \epsilon\right)^{-1/2}\right] \\ &+ \frac{1}{\epsilon^2-1}\frac{\mu_0}{4\pi\rho' c}\frac{\partial^2}{\partial t^2}\left[\frac{t^2}{\rho'^2} - \epsilon + \frac{2\epsilon}{\epsilon+1} - \frac{2\epsilon^2}{\epsilon+1}\left((\epsilon+1)\frac{t^2}{\rho'^2} - \epsilon\right)^{-1/2}\right]; \end{aligned} \quad t/\rho' > \epsilon^{1/2}. \tag{13.5.2}$$

The second formula reduces to

$$E_{2z}(\rho',t) = -\frac{1}{\epsilon+1}\frac{\mu_0}{4\pi\rho' c}\frac{\partial^2}{\partial t^2}\left(\frac{t^2}{\rho'^2}+\frac{2\epsilon}{\epsilon+1}\right); \qquad t/\rho' > \epsilon^{1/2}. \tag{13.5.3}$$

Similar to the case of the infinite homogeneous dielectric of Section 13.1, the differentiation of (13.5.1) and (13.5.3) must be carried out with care, owing to the presence of discontinuities at $t/\rho' = 1$ and $\epsilon^{1/2}$. Thus, with

$$f(t/\rho') \equiv \begin{cases} 0, & t/\rho' < 1, \quad (13.5.4)\\ \dfrac{\epsilon}{\epsilon^2-1} \\ \times\left[\dfrac{t^2}{\rho'^2}-1+\dfrac{2\epsilon}{\epsilon+1}\right. \\ \qquad \left. -\dfrac{2\epsilon}{\epsilon+1}\left((\epsilon+1)\dfrac{t^2}{\rho'^2}-\epsilon\right)^{-1/2}\right], & 1 < t/\rho' < \epsilon^{1/2}, \quad (13.5.5)\\ \dfrac{1}{\epsilon+1}\left(\dfrac{t^2}{\rho'^2}+\dfrac{2\epsilon}{\epsilon+1}\right), & t/\rho' > \epsilon^{1/2}, \quad (13.5.6) \end{cases}$$

it follows that

$$E_{2z}(\rho',t) = -\frac{\mu_0}{4\pi\rho' c}\frac{\partial^2}{\partial t^2} f(t/\rho'). \tag{13.5.7}$$

Consider first the continuity of $f(t/\rho')$ at $t/\rho' = 1$ and $\epsilon^{1/2}$. From (13.5.4) and (13.5.5),

$$f(1-) = f(1+) = 0. \tag{13.5.8}$$

Similarly, from (13.5.5) and (13.5.6),

$$f(\epsilon^{1/2}-) = f(\epsilon^{1/2}+) = \frac{\epsilon(\epsilon+3)}{(\epsilon+1)^2}. \tag{13.5.9}$$

It follows that $f(t/\rho')$ is everywhere continuous. Consider next the behavior of $\partial f(t/\rho')/\partial t$. Let $f'(t/\rho') = \partial f(t/\rho')/\partial t$. Then

$$f'(t/\rho') = \begin{cases} 0, & t/\rho' < 1, \quad (13.5.10)\\ \dfrac{\epsilon}{\epsilon^2-1}\dfrac{2t}{\rho'^2} \\ \times\left[1+\epsilon\left((\epsilon+1)\dfrac{t^2}{\rho'^2}-\epsilon\right)^{-3/2}\right], & 1 < t/\rho' < \epsilon^{1/2}, \quad (13.5.11)\\ \dfrac{1}{\epsilon+1}\dfrac{2t}{\rho'^2}, & t/\rho' > \epsilon^{1/2}. \quad (13.5.12) \end{cases}$$

At $t/\rho' = 1$ and $\epsilon^{1/2}$, $f'(t/\rho')$ has the following behavior:

$$f'(1-) = 0; \qquad f'(1+) = \frac{\epsilon}{\epsilon-1}\frac{2}{\rho'}. \tag{13.5.13}$$

Hence, there is a step discontinuity of $2\epsilon/(\epsilon-1)\rho'$ in $f'(t/\rho')$ at $t/\rho'=1$. Similarly,

$$f'(\epsilon^{1/2}-)=\frac{\epsilon}{\epsilon^2-1}\frac{2\epsilon^{1/2}}{\rho'}(1+\epsilon^{-2})=\frac{\epsilon^2+1}{\epsilon^{1/2}(\epsilon^2-1)}\frac{2}{\rho'}, \tag{13.5.14}$$

$$f'(\epsilon^{1/2}+)=\frac{\epsilon^{1/2}}{\epsilon+1}\frac{2}{\rho'}. \tag{13.5.15}$$

There is a step discontinuity of

$$\left(\frac{\epsilon^{1/2}}{\epsilon+1}-\frac{\epsilon^2+1}{\epsilon^{1/2}(\epsilon^2-1)}\right)\frac{2}{\rho'}=-\frac{1}{\epsilon^{1/2}(\epsilon-1)}\frac{2}{\rho'}$$

at $t/\rho'=\epsilon^{1/2}$. Now, let $f''(t/\rho')=\partial f'(t/\rho')/\partial t$. Its behavior is

$$f''(t/\rho')=\begin{cases}0, & t/\rho'<1, \quad (13.5.16)\\[2mm] \dfrac{\epsilon}{\epsilon^2-1}\dfrac{2}{\rho'^2} & \\ \quad\times\left[1-\dfrac{\epsilon}{(\epsilon+1)^{3/2}}\left(\dfrac{2t^2}{\rho'^2}+\dfrac{\epsilon}{\epsilon+1}\right)\right. & \\ \qquad\left.\times\left(\dfrac{t^2}{\rho'^2}-\dfrac{\epsilon}{\epsilon+1}\right)^{-5/2}\right], & 1<t/\rho'<\epsilon^{1/2}, \quad (13.5.17)\\[2mm] \dfrac{1}{\epsilon+1}\dfrac{2}{\rho'^2}, & t/\rho'>\epsilon^{1/2}. \quad (13.5.18)\end{cases}$$

In view of the discontinuities of $f''(t/\rho')$ at 1 and $\epsilon^{1/2}$, it follows that, with $\rho'=\rho/c$ and $\mu_0c^2=\epsilon_0^{-1}$, the complete final expression for $E_{2z}(\rho,t)=-(\mu_0/4\pi\rho)f''(t/\rho)$ is

$$E_{2z}(\rho,t)=-\frac{1}{2\pi\epsilon_0\rho^2c(\epsilon-1)}\left[\epsilon\delta\left(t-\frac{\rho}{c}\right)-\epsilon^{-1/2}\delta\left(t-\frac{\epsilon^{1/2}\rho}{c}\right)\right]$$
$$-\frac{1}{2\pi\epsilon_0\rho^3(\epsilon+1)}\left\{\begin{array}{l}0\\ \dfrac{\epsilon}{\epsilon-1}\left[1-\dfrac{\epsilon}{(\epsilon+1)^{3/2}}\left(\dfrac{2c^2t^2}{\rho^2}+\dfrac{\epsilon}{\epsilon+1}\right)\right.\\ \qquad\qquad\left.\times\left(\dfrac{c^2t^2}{\rho^2}-\dfrac{\epsilon}{\epsilon+1}\right)^{-5/2}\right]\\ 1\end{array}\right\};$$
$$\begin{cases}ct/\rho<1\\ 1<ct/\rho<\epsilon^{1/2}\\ ct/\rho>\epsilon^{1/2}\end{cases}. \tag{13.5.19}$$

This is the vertical electric field in Region 2 just above the boundary. The field in Region 1 just below the boundary is obtained from the boundary condition. It is

$$E_{1z}(\rho, t) = \epsilon^{-1} E_{2z}(\rho, t), \tag{13.5.20}$$

where $\epsilon \equiv \epsilon_{1r}$. Note that the dipole source is located on the boundary in Region 1. At any radial distance ρ, the electric field is initially zero until the instant $t = \rho/c$ when a downward-directed electric field arrives in the form of a δ-function pulse. That is, the magnitude of the field increases momentarily to infinity and decreases to the value

$$E_{2z}(\rho, \rho/c) = -\frac{1}{2\pi\epsilon_0\rho^3} \frac{\epsilon(1 - 3\epsilon^2 - 2\epsilon)}{\epsilon^2 - 1}. \tag{13.5.21}$$

This pulse has traveled along the boundary in Region 2 (air) with the velocity c. The field then rises gradually with time according to

$$E_{2z}(\rho, t) = -\frac{1}{2\pi\epsilon_0\rho^3} \frac{\epsilon}{\epsilon^2 - 1} \left[1 - \frac{\epsilon}{(\epsilon+1)^{3/2}} \left(\frac{2c^2t^2}{\rho^2} + \frac{\epsilon}{\epsilon+1} \right) \times \left(\frac{c^2t^2}{\rho^2} - \frac{\epsilon}{\epsilon+1} \right)^{-5/2} \right] \tag{13.5.22}$$

until $t = \rho\epsilon^{1/2}/c$ when it has the value

$$E_{2z}(\rho, \rho\epsilon^{1/2}/c) = -\frac{1}{2\pi\epsilon_0\rho^3} \frac{\epsilon^3 - 2\epsilon - 3}{\epsilon^2(\epsilon^2 - 1)}. \tag{13.5.23}$$

At this instant an upward-directed δ-function rises momentarily in magnitude to infinity and decreases to

$$E_{2z}(\rho, t) = -\frac{1}{2\pi\epsilon_0\rho^3(\epsilon+1)}. \tag{13.5.24}$$

The second pulse that arrives at $t = \rho\epsilon^{1/2}/c$ has traveled along the boundary in Region 1 (the dielectric) with the velocity $c/\epsilon^{1/2}$. When it has passed, the field drops to the static value due to the charged infinitesimal dipole. Note that the electric field (13.5.19) is always directed downward except during the instant that the second δ-function pulse passes, when it is directed upward.

It is of interest to note that for the first pulse, traveling in air, the adjacent boundary with the dielectric acts approximately like an electric wall so that the vertically directed field is approximately doubled in amplitude along the boundary surface. On the other hand, for the second pulse, traveling in the dielectric, the adjacent boundary with air acts approximately like a magnetic wall so that the vertically directed reflected field is almost canceled by the direct field along the boundary. In the two cases, the doubling is exact and the cancellation complete when the permittivity of the dielectric half-space is infinite, i.e., $\epsilon \to \infty$. Specifically, when $\epsilon < \infty$, the amplitude factor for the pulse in air is not exactly 2 but is given by $2\epsilon/(\epsilon-1) = 1+(\epsilon+1)/(\epsilon-1) \to 2$ when $\epsilon \to \infty$; similarly, the cancellation in the dielectric is not complete and the difference between the direct and reflected fields is the quantity $-1/(\epsilon-1) = 1-\epsilon/(\epsilon-1) \to 0$ when $\epsilon \to \infty$. Evidently, when ϵ is large but finite, the positive contribution to the field in the air from the imperfect image is slightly greater than the direct field so that the total field is positive and slightly more than doubled; similarly, the negative contribution to the field in the dielectric by the imperfect image is slightly greater than the direct field so that the total field in this case is small and negative.

It is significant to note that the amplitude of the δ-function pulses in (13.5.19) includes the factor $1/\rho^2$. This is also the radial dependence *in the far field* of the frequency-dependent field, (3.5.7) with (3.4.35), and consists entirely of the lateral wave. Since the spectrum of the δ-function is dominated by high frequencies, this behavior is readily understood. Note that the pulse of the dipole in a homogeneous single medium also follows the far-field behavior of the steady state but this is the direct field with $1/\rho$ decrease. When the dipole is on the boundary, the far field is entirely due to lateral waves or a lateral pulse with $1/\rho^2$ decrease with distance.

13.6 The Time-Dependent Component $B_{2\phi}(\rho, t)$ with a δ-Function Excitation

The integral for $\tilde{B}_{2\phi}(\rho, \omega)$ in explicit form is

$$\tilde{B}_{2\phi}(\rho', \omega) = \frac{i\mu_0\omega^2}{2\pi c^2}\int_0^\infty \frac{J_1(\lambda'\rho')\lambda'^2\,d\lambda'}{\omega^2\epsilon(\omega^2-\lambda'^2)^{1/2}+\omega^2(\omega^2\epsilon-\lambda'^2)^{1/2}}, \tag{13.6.1}$$

where $\lambda' = c\lambda$ and $\rho' = \rho/c$. The corresponding time-dependent component for the excitation (13.1.2) is

$$\begin{aligned} B_{2\phi}(\rho', t) &= \frac{1}{\pi}\,\mathrm{Re}\int_0^\infty e^{-i\omega t}\tilde{B}_{2\phi}(\rho', \omega)\,d\omega \\ &= \mathrm{Re}\,\frac{i\mu_0}{2\pi^2c^2}\int_0^\infty d\omega\, e^{-i\omega t}\omega^2 \\ &\quad\times\int_0^\infty \frac{J_1(\lambda'\rho')\lambda'^2\,d\lambda'}{\omega^2[\epsilon(\omega^2-\lambda'^2)^{1/2}+(\omega^2\epsilon-\lambda'^2)^{1/2}]}. \end{aligned} \tag{13.6.2}$$

Now let the variable be changed to ξ with the substitutions $\lambda' = \omega\xi$ and $d\lambda' = \omega\,d\xi$. The result is

$$B_{2\phi}(\rho', t) = \frac{\mu_0}{2\pi^2c^2}\int_0^\infty d\xi\int_0^\infty d\omega\;\mathrm{Re}\,i\omega^2e^{-i\omega t}\,\frac{\xi^2J_1(\omega\xi\rho')}{\epsilon(1-\xi^2)^{1/2}+(\epsilon-\xi^2)^{1/2}}$$

$$= \frac{\mu_0}{2\pi^2 c^2} \frac{\partial^2}{\partial t^2} \int_0^\infty d\xi \ \mathrm{Im} \frac{\xi^2}{\epsilon(1-\xi^2)^{1/2} + (\epsilon - \xi^2)^{1/2}} \times \int_0^\infty d\omega \, e^{-i\omega t} J_1(\omega \xi \rho'). \tag{13.6.3}$$

The integration with respect to ω can be carried out with formula (6.611-1) of Gradshteyn and Ryzhik (1980, p. 707). The result is

$$B_{2\phi}(\rho', t) = \frac{\mu_0}{2\pi^2 c^2} \frac{\partial^2}{\partial t^2} \mathrm{Im} \int_0^\infty \frac{\xi^2}{\epsilon(1-\xi^2)^{1/2} + (\epsilon - \xi^2)^{1/2}} \times \left(1 - \frac{t}{(t^2 - \xi^2 \rho'^2)^{1/2}}\right) \frac{d\xi}{\xi \rho'}. \tag{13.6.4}$$

This integral can be separated into two parts as follows:

$$B_{2\phi}(\rho', t) = \frac{\mu_0}{2\pi^2 c^2 \rho'} \times \frac{\partial^2}{\partial t^2} \mathrm{Im} \left(\int_0^\infty \frac{\xi \, d\xi}{\epsilon(1-\xi^2)^{1/2} + (\epsilon - \xi^2)^{1/2}} - \frac{t}{\rho'} \int_0^\infty \frac{\xi \, d\xi}{[\epsilon(1-\xi^2)^{1/2} + (\epsilon - \xi^2)^{1/2}](t^2/\rho'^2 - \xi^2)^{1/2}} \right). \tag{13.6.5}$$

With the branch-cut structure in Fig. 13.3.3, it follows that

$$B_{2\phi}(\rho', t) = 0; \qquad t/\rho' < 1, \tag{13.6.6}$$

$$B_{2\phi}(\rho', t) = \frac{\mu_0}{2\pi^2 c^2 \rho'} \times \frac{\partial^2}{\partial t^2} \mathrm{Im} \left(\int_0^\infty \frac{\xi \, d\xi}{i\epsilon(\xi^2 - 1)^{1/2} + (\epsilon - \xi^2)^{1/2}} - \frac{t}{\rho'} \int_0^\infty \frac{\xi \, d\xi}{[i\epsilon(\xi^2-1)^{1/2} + (\epsilon - \xi^2)^{1/2}](t^2/\rho'^2 - \xi^2)^{1/2}} \right); \quad 1 < t/\rho' < \epsilon^{1/2}. \tag{13.6.7}$$

The real and imaginary parts are readily separated and only the imaginary part is retained. The result is

$$B_{2\phi}(\rho', t) = -\frac{\epsilon}{\epsilon - 1} \frac{\mu_0}{2\pi^2 c^2 \rho'} \times \frac{\partial^2}{\partial t^2} \left(\int_0^\infty \frac{(\xi^2 - 1)^{1/2} \xi \, d\xi}{(\epsilon + 1)\xi^2 - \epsilon} - \frac{t}{\rho'} \int_0^\infty \frac{(\xi^2 - 1)^{1/2} \xi \, d\xi}{[(\epsilon+1)\xi^2 - \epsilon](t^2/\rho'^2 - \xi^2)^{1/2}} \right). \tag{13.6.8}$$

With the contour in Fig. 13.3.4, this becomes

$$
\begin{aligned}
B_{2\phi}(\rho', t) &= -\frac{\epsilon}{\epsilon - 1} \frac{\mu_0}{2\pi^2 c^2 \rho'} \\
&\quad \times \frac{\partial^2}{\partial t^2}\Bigg(\int_1^{t/\rho'} \frac{(\xi^2 - 1)^{1/2}\xi \, d\xi}{(\epsilon + 1)\xi^2 - \epsilon} \\
&\qquad - \frac{t}{\rho'} \int_1^{t/\rho'} \frac{(\xi^2 - 1)^{1/2}\xi \, d\xi}{[(\epsilon + 1)\xi^2 - \epsilon](t^2/\rho'^2 - \xi^2)^{1/2}} \\
&\qquad - \int_1^{t/\rho'} \frac{(\xi^2 - 1)^{1/2}\xi \, d\xi}{(\epsilon + 1)\xi^2 - \epsilon} \\
&\qquad + \frac{t}{\rho'} \int_1^{t/\rho'} \frac{(\xi^2 - 1)^{1/2}\xi \, d\xi}{[(\epsilon + 1)\xi^2 - \epsilon][-(t^2/\rho'^2 - \xi^2)^{1/2}]}\Bigg) \\
&= \frac{\epsilon}{\epsilon - 1} \frac{\mu_0}{\pi^2 c^2 \rho'} \\
&\quad \times \frac{\partial^2}{\partial t^2}\left(\frac{t}{\rho'} \int_1^{t/\rho'} \frac{(\xi^2 - 1)^{1/2}\xi \, d\xi}{[(\epsilon + 1)\xi^2 - \epsilon](t^2/\rho'^2 - \xi^2)^{1/2}}\right).
\end{aligned}
\tag{13.6.9}
$$

With $\zeta = \xi^2$,

$$
\begin{aligned}
B_{2\phi}(\rho', t) &= \frac{\epsilon}{\epsilon - 1} \frac{\mu_0}{2\pi^2 c^2 \rho'} \\
&\quad \times \frac{\partial^2}{\partial t^2}\left(\frac{t}{\rho'} \int_1^{t^2/\rho'^2} \frac{(\zeta - 1)^{1/2} \, d\zeta}{[(\epsilon + 1)\zeta - \epsilon](t^2/\rho'^2 - \zeta)^{1/2}}\right); \\
&\qquad\qquad 1 < t/\rho' < \epsilon^{1/2}.
\end{aligned}
\tag{13.6.10}
$$

Let $T_0 = t^2/\rho'^2 - 1$ and $x' = \zeta - 1$. Then

$$
\begin{aligned}
B_{2\phi}(\rho', t) &= \frac{\epsilon}{\epsilon - 1} \frac{\mu_0}{2\pi^2 c^2 \rho'} \\
&\quad \times \frac{\partial^2}{\partial t^2}\left(\frac{t}{\rho'} \int_0^{T_0} \frac{x' \, dx'}{[(\epsilon + 1)(x' + 1) - \epsilon][(T_0 - x')x']^{1/2}}\right).
\end{aligned}
\tag{13.6.11}
$$

Finally, with $x = x' + E_0$ and $E_0 = 1/(\epsilon + 1)$,

$$
\begin{aligned}
B_{2\phi}(\rho', t) &= \frac{\epsilon}{\epsilon^2 - 1} \frac{\mu_0}{2\pi^2 c^2 \rho'} \\
&\quad \times \frac{\partial^2}{\partial t^2}\left[\frac{t}{\rho'}\left(\int_{E_0}^{T_0 + E_0} \frac{dx}{X_0^{1/2}} - E_0 \int_{E_0}^{T_0 + E_0} \frac{dx}{x X_0^{1/2}}\right)\right],
\end{aligned}
\tag{13.6.12}
$$

where $X_0 = a_0 + b_0 x + c_0 x^2$ with $a_0 = -E_0(T_0 + E_0)$, $b_0 = T_0 + 2E_0$, and $c_0 = -1$. These integrals are like those in (13.4.7). They give

$$
B_{2\phi}(\rho', t) = \frac{\epsilon}{\epsilon^2 - 1} \frac{\mu_0}{2\pi c^2 \rho'} \frac{\partial^2}{\partial t^2}\left\{\frac{t}{\rho'}\left[1 - \left((\epsilon + 1)\frac{t^2}{\rho'^2} - \epsilon\right)^{-1/2}\right]\right\};
$$

$$1 < t/\rho' < \epsilon^{1/2}. \quad (13.6.13)$$

In the remaining range, $t/\rho' > \epsilon^{1/2}$,

$$\begin{aligned} B_{2\phi}(\rho', t) = &\frac{\mu_0}{2\pi^2 c^2 \rho'} \\ &\times \frac{\partial^2}{\partial t^2} \operatorname{Im}\biggl(\int_0^{\epsilon^{1/2}} \frac{\xi\, d\xi}{i\epsilon(\xi^2-1)^{1/2} + (\epsilon-\xi^2)^{1/2}} \\ &\quad - \frac{t}{\rho'} \int_0^{\epsilon^{1/2}} \frac{\xi\, d\xi}{[i\epsilon(\xi^2-1)^{1/2} + (\epsilon-\xi^2)^{1/2}](t^2/\rho'^2 - \xi^2)^{1/2}} \\ &\quad + \int_{\epsilon^{1/2}}^{\infty} \frac{\xi\, d\xi}{i\epsilon(\xi^2-1)^{1/2} + i(\xi^2-\epsilon)^{1/2}} \\ &\quad - \frac{t}{\rho'} \int_{\epsilon^{1/2}}^{\infty} \frac{\xi\, d\xi}{[i\epsilon(\xi^2-1)^{1/2} + i(\xi^2-\epsilon)^{1/2}](t^2/\rho'^2 - \xi^2)^{1/2}}\biggr). \end{aligned} \quad (13.6.14)$$

The contributing imaginary part of the integrals is

$$\begin{aligned} B_{2\phi}(\rho', t) = &-\frac{1}{\epsilon-1}\frac{\mu_0}{2\pi^2 c^2 \rho'} \\ &\times \frac{\partial^2}{\partial t^2}\biggl(\epsilon \int_0^{\epsilon^{1/2}} \frac{(\xi^2-1)^{1/2}\xi\, d\xi}{(\epsilon+1)\xi^2 - \epsilon} \\ &\quad - \frac{\epsilon t}{\rho'} \int_0^{\epsilon^{1/2}} \frac{(\xi^2-1)^{1/2}\xi\, d\xi}{[(\epsilon+1)\xi^2 - \epsilon](t^2/\rho'^2 - \xi^2)^{1/2}} \\ &\quad - \int_{\epsilon^{1/2}}^{\infty} \frac{[-\epsilon(\xi^2-1)^{1/2} + (\xi^2-\epsilon)^{1/2}]\xi\, d\xi}{(\epsilon+1)\xi^2 - \epsilon} \\ &\quad + \frac{t}{\rho'} \int_{\epsilon^{1/2}}^{\infty} \frac{[-\epsilon(\xi^2-1)^{1/2} + (\xi^2-\epsilon)^{1/2}]\xi\, d\xi}{[(\epsilon+1)\xi^2 - \epsilon](t^2/\rho'^2 - \xi^2)^{1/2}}\biggr). \end{aligned} \quad (13.6.15)$$

Here the integrands of the first terms in the third and fourth integrals are, respectively, the same as the integrands in the first and second integrals so that they can be combined to give

$$\begin{aligned} B_{2\phi}(\rho', t) = &-\frac{1}{\epsilon-1}\frac{\mu_0}{2\pi^2 c^2 \rho'} \\ &\times \frac{\partial^2}{\partial t^2}\biggl(\epsilon \int_0^{\infty} \frac{(\xi^2-1)^{1/2}\xi\, d\xi}{(\epsilon+1)\xi^2 - \epsilon} \\ &\quad - \frac{\epsilon t}{\rho'} \int_0^{\infty} \frac{(\xi^2-1)^{1/2}\xi\, d\xi}{[(\epsilon+1)\xi^2 - \epsilon](t^2/\rho'^2 - \xi^2)^{1/2}} \\ &\quad - \int_{\epsilon^{1/2}}^{\infty} \frac{(\xi^2-\epsilon)^{1/2}\xi\, d\xi}{(\epsilon+1)\xi^2 - \epsilon} \end{aligned}$$

$$+\frac{t}{\rho'}\int_{\epsilon^{1/2}}^{\infty}\frac{(\xi^2-\epsilon)^{1/2}\xi\,d\xi}{[(\epsilon+1)\xi^2-\epsilon](t^2/\rho'^2-\xi^2)^{1/2}}\Bigg). \tag{13.6.16}$$

When the contour of integration is changed to that in Fig. 13.3.4 and the variable to $\zeta = \xi^2$, the contributing integrals are

$$B_{2\phi}(\rho', t) = -\frac{1}{\epsilon - 1}\frac{\mu_0}{2\pi^2 c^2\rho'}$$
$$\times\frac{\partial^2}{\partial t^2}\left[\frac{t}{\rho'}\left(-\epsilon\int_1^{t^2/\rho'^2}\frac{(\zeta-1)^{1/2}\,d\zeta}{[(\epsilon+1)\zeta-\epsilon](t^2/\rho'^2-\zeta)^{1/2}}\right.\right.$$
$$\left.\left.+\int_\epsilon^{t^2/\rho'^2}\frac{(\zeta-\epsilon)^{1/2}\,d\zeta}{[(\epsilon+1)\zeta-\epsilon](t^2/\rho'^2-\zeta)^{1/2}}\right)\right]. \tag{13.6.17}$$

Here the first integral is the same as the integral in (13.6.10). The second integral can be treated in the same manner with the substitutions $x' = \zeta - \epsilon$ and $T_\epsilon = t^2/\rho'^2 - \epsilon$. These lead to the integral

$$\mathcal{I}_\epsilon = \int_0^{T_\epsilon}\frac{x'\,dx'}{[(\epsilon+1)x'+\epsilon^2][(T_\epsilon - x')x']^{1/2}}, \tag{13.6.18}$$

in which the substitution $x = x' + E_\epsilon$ with $E_\epsilon = \epsilon^2/(\epsilon+1)$ gives

$$\mathcal{I}_\epsilon = \frac{1}{\epsilon+1}\left(\int_{E_\epsilon}^{T_\epsilon+E_\epsilon}\frac{dx}{X_\epsilon^{1/2}} - E_\epsilon\int_{E_\epsilon}^{T_\epsilon+E_\epsilon}\frac{dx}{xX_\epsilon^{1/2}}\right), \tag{13.6.19}$$

where $X_\epsilon = a_\epsilon + b_\epsilon x + c_\epsilon x^2$ with $a_\epsilon = -E_\epsilon(T_\epsilon + E_\epsilon)$, $b_\epsilon = T_\epsilon + 2E_\epsilon$, and $c_\epsilon = -1$. The integrals in (13.6.19) integrate into arcsines and the substitution of the limits reduces their arguments to ± 1. Hence,

$$\mathcal{I}_\epsilon = \frac{\pi}{\epsilon+1}\left[1 - \epsilon\left((\epsilon+1)\frac{t^2}{\rho'^2} - \epsilon\right)^{-1/2}\right]. \tag{13.6.20}$$

When this is substituted for the second integral in (13.6.17) and the value of the first integral is obtained from (13.6.13), the result is

$$B_{2\phi}(\rho', t) = \frac{1}{\epsilon^2 - 1}\frac{\mu_0}{2\pi c^2\rho'}$$
$$\times\frac{\partial^2}{\partial t^2}\left(\frac{t}{\rho'}\left\{\epsilon\left[1 - \left((\epsilon+1)\frac{t^2}{\rho'^2} - \epsilon\right)^{-1/2}\right]\right.\right.$$
$$\left.\left.-\left[1 - \epsilon\left((\epsilon+1)\frac{t^2}{\rho'^2} - \epsilon\right)^{-1/2}\right]\right\}\right)$$
$$= \frac{1}{\epsilon+1}\frac{\mu_0}{2\pi c^2\rho'}\frac{\partial^2}{\partial t^2}\frac{t}{\rho'}; \qquad t/\rho' > \epsilon^{1/2}. \tag{13.6.21}$$

The magnetic field in the three ranges (13.6.6), (13.6.13), and (13.6.21) is summarized as follows. Let

$$f(t/\rho') = \begin{cases} 0, & t/\rho' < 1, \\ \dfrac{\epsilon}{\epsilon^2-1}\dfrac{t}{\rho'} \\ \quad \times\left[1-\left((\epsilon+1)\dfrac{t^2}{\rho'^2}-\epsilon\right)^{-1/2}\right], & 1 < t/\rho' < \epsilon^{1/2}, \\ \dfrac{1}{\epsilon+1}\dfrac{t}{\rho'}, & t/\rho' > \epsilon^{1/2}. \end{cases} \tag{13.6.22}$$

Then

$$B_{2\phi}(\rho', t) = \frac{\mu_0}{2\pi c^2 \rho'}\frac{\partial^2}{\partial t^2} f(t/\rho'). \tag{13.6.23}$$

Since $f(1-) = f(1+) = 0$ and $f(\epsilon^{1/2}-) = f(\epsilon^{1/2}+) = \epsilon^{1/2}/(\epsilon+1)$, $f(t/\rho')$ is everywhere continuous. But, with $f'(t/\rho') = \partial f(t/\rho')/\partial t$,

$$f'(t/\rho') = \begin{cases} 0, & t/\rho' < 1, \\ \dfrac{\epsilon}{\epsilon^2-1}\dfrac{1}{\rho'} \\ \quad \times\left[1-\left((\epsilon+1)\dfrac{t^2}{\rho'^2}-\epsilon\right)^{-1/2}\right. \\ \quad\quad \left. +\dfrac{t^2}{\rho'^2}(\epsilon+1)\left((\epsilon+1)\dfrac{t^2}{\rho'^2}-\epsilon\right)^{-3/2}\right], & 1 < t/\rho' < \epsilon^{1/2}, \\ \dfrac{1}{\epsilon+1}\dfrac{1}{\rho'}, & t/\rho' > \epsilon^{1/2}, \end{cases} \tag{13.6.24}$$

it follows that $f'(1-) = 0$ and $f'(1+) = \epsilon/(\epsilon-1)\rho'$. Thus, $f'(t/\rho')$ has a step discontinuity of $\epsilon/(\epsilon-1)\rho'$ at $t/\rho' = 1$. Similarly, $f'(\epsilon^{1/2}-) = (\epsilon^2 + 1)/\epsilon(\epsilon^2-1)\rho'$ and $f'(\epsilon^{1/2}+) = 1/(\epsilon+1)\rho'$. This is a step of $-1/\epsilon(\epsilon-1)\rho'$.

With $f''(t/\rho') = \partial f'(t/\rho')/\partial t$,

$$f''(t/\rho') = \begin{cases} 0, & t/\rho' < 1, \\ -\dfrac{3\epsilon^2}{(\epsilon-1)}\dfrac{1}{\rho'^2}\dfrac{t}{\rho'} \\ \quad \times\left((\epsilon+1)\dfrac{t^2}{\rho'^2}-\epsilon\right)^{-5/2}, & 1 < t/\rho' < \epsilon^{1/2}, \\ 0, & t/\rho' > \epsilon^{1/2}. \end{cases} \tag{13.6.25}$$

When these results are combined with (13.6.23) and $\rho' = \rho/c$, the final formula for the magnetic field is

$$B_{1\phi}(\rho,t) = B_{2\phi}(\rho,t)$$
$$= \frac{\mu_0}{2\pi\rho^2}\frac{1}{\epsilon-1}\left[\epsilon\delta\left(t-\frac{\rho}{c}\right) - \epsilon^{-1}\delta\left(t-\frac{\epsilon^{1/2}\rho}{c}\right)\right]$$
$$-\frac{\mu_0 c}{2\pi\rho^3}\frac{\epsilon^2}{\epsilon-1}\begin{Bmatrix} 0 \\ \dfrac{3ct}{\rho}\left((\epsilon+1)\dfrac{c^2t^2}{\rho^2}-\epsilon\right)^{-5/2} \\ 0 \end{Bmatrix};$$
$$\begin{cases} ct/\rho < 1 \\ 1 < ct/\rho < \epsilon^{1/2} \\ ct/\rho > \epsilon^{1/2} \end{cases}. \quad (13.6.26)$$

13.7 The Time-Dependent Component $E_{2\rho}(\rho,t)$ with a δ-Function Excitation

The integral for $\tilde{E}_{2\rho}(\rho,\omega)$ in explicit form is

$$\tilde{E}_{2\rho}(\rho',\omega) = -\frac{i\omega\mu_0}{2\pi c}\int_0^\infty \frac{(\omega^2-\lambda'^2)^{1/2}\lambda'^2 J_1(\lambda'\rho')\,d\lambda'}{\omega^2\epsilon(\omega^2-\lambda'^2)^{1/2}+\omega^2(\omega^2\epsilon-\lambda'^2)^{1/2}}, \quad (13.7.1)$$

where $\lambda' = c\lambda$ and $\rho' = \rho/c$. The time-dependent component is

$$E_{2\rho}(\rho',t) = \frac{1}{\pi}\,\mathrm{Re}\int_0^\infty e^{-i\omega t}\tilde{E}_{2\rho}(\rho',\omega)\,d\omega$$
$$= -\,\mathrm{Re}\,\frac{i\mu_0}{2\pi^2 c}\int_0^\infty d\omega\, e^{-i\omega t}\omega$$
$$\times\int_0^\infty \frac{(\omega^2-\lambda'^2)^{1/2}J_1(\lambda'\rho')\lambda'^2\,d\lambda'}{\omega^2[\epsilon(\omega^2-\lambda'^2)^{1/2}+(\omega^2\epsilon-\lambda'^2)^{1/2}]}. \quad (13.7.2)$$

With the change in variable $\lambda' = \omega\xi$,

$$E_{2\rho}(\rho',t) = -\frac{\mu_0}{2\pi^2 c}\int_0^\infty d\xi\int_0^\infty d\omega\;\mathrm{Re}\, i\omega^2\,\frac{(1-\xi^2)^{1/2}\xi^2 J_1(\omega\xi\rho')}{\epsilon(1-\xi^2)^{1/2}+(\epsilon-\xi^2)^{1/2}}$$
$$= -\frac{\mu_0}{2\pi^2 c}\frac{\partial^2}{\partial t^2}\int_0^\infty d\xi\;\mathrm{Im}\,\frac{(1-\xi^2)^{1/2}\xi^2}{\epsilon(1-\xi^2)^{1/2}+(\epsilon-\xi^2)^{1/2}}$$

$$\times \int_0^\infty d\omega\, e^{-i\omega t} J_1(\omega\xi\rho'). \tag{13.7.3}$$

The integration with respect to ω is the same as in (13.6.3). The result is

$$E_{2\rho}(\rho', t) = -\frac{\mu_0}{2\pi^2 c}\frac{\partial^2}{\partial t^2}\,\mathrm{Im}\int_0^\infty \frac{\xi^2(1-\xi^2)^{1/2}}{\epsilon(1-\xi^2)^{1/2}+(\epsilon-\xi^2)^{1/2}} \times\left(1-\frac{t}{(t^2-\xi^2\rho'^2)^{1/2}}\right)\frac{d\xi}{\xi\rho'}. \tag{13.7.4}$$

This gives the following two integrals:

$$E_{2\rho}(\rho', t) = -\frac{\mu_0}{2\pi^2 c\rho'} \times\frac{\partial^2}{\partial t^2}\,\mathrm{Im}\left(\int_0^\infty \frac{\xi(1-\xi^2)^{1/2}\,d\xi}{\epsilon(1-\xi^2)^{1/2}+(\epsilon-\xi^2)^{1/2}} - \frac{t}{\rho'}\int_0^\infty \frac{\xi(1-\xi^2)^{1/2}\,d\xi}{[\epsilon(1-\xi^2)^{1/2}+(\epsilon-\xi^2)^{1/2}](t^2/\rho'^2-\xi^2)^{1/2}}\right). \tag{13.7.5}$$

They lead to some elliptic integrals and hence are not expressible in terms of elementary functions.

13.8 The Hertz Potential with δ-Function Excitation

The formulation of Sommerfeld (1909, 1926, 1935) and many other authors begins with the Hertz potential for a vertical electric dipole on the boundary $z = 0$ between two half-spaces. The transient calculation of Van der Pol (1956) also evaluates the Hertz potential. It is, therefore, of interest to apply the methods of this chapter to the evaluation of the Hertz potential in the time domain. As defined by Sommerfeld (1935), the Hertz potential in the frequency domain for Region 2, $z < 0$, with the vertical electric dipole in Region 1, $z > 0$, on the boundary with Region 2 is

$$\tilde{\Pi}_{2z}(\rho, \omega) = \frac{p}{2\pi\epsilon_0}\int_0^\infty \frac{ik_2^2 J_0(\lambda\rho)e^{i\gamma_2 z}}{k_1^2(k_2^2-\lambda^2)^{1/2}+k_2^2(k_1^2-\lambda^2)^{1/2}}\,\lambda\, d\lambda, \tag{13.8.1}$$

where

$$p = \frac{iIh_e}{\omega} = \frac{i}{\omega} \quad \text{with } Ih_e = 1 \text{ A m}. \tag{13.8.2}$$

With $k_1^2 = \omega^2\epsilon/c^2$, $k_2^2 = \omega^2/c^2$, $\lambda' = c\lambda$, $\rho' = \rho/c$, and $z = 0$, the Hertz potential on the boundary in Region 2 is

$$\tilde{\Pi}_{2z}(\rho',\omega) = -\frac{\omega}{2\pi\epsilon_0 c}\int_0^\infty \frac{\lambda' J_0(\lambda'\rho')}{\omega^2\epsilon(\omega^2-\lambda'^2)^{1/2}+\omega^2(\omega^2\epsilon-\lambda'^2)^{1/2}}\,d\lambda'. \tag{13.8.3}$$

Except for the factor in front of the integral sign and the appearance of λ' instead of λ'^3 in the numerator of the integrand, (13.8.3) is like (13.3.2). If the same steps are taken, the time-domain expression corresponding to (13.3.9) is

$$\Pi_{2z}(\rho',t) = -\frac{1}{2\pi^2\epsilon_0\rho' c} \times \operatorname{Im}\int_0^\infty d\xi\,\frac{\xi}{[\epsilon(1-\xi^2)^{1/2}+(\epsilon-\xi^2)^{1/2}](t^2/\rho'^2-\xi^2)^{1/2}}. \tag{13.8.4}$$

The further manipulation leading to (13.3.10), (13.3.14), and (13.3.18) gives

$$\Pi_{2z}(\rho',t) = 0; \qquad t/\rho' < 1, \tag{13.8.5}$$

$$\Pi_{2z}(\rho',t) = \frac{\epsilon}{\epsilon-1}\,\frac{\vartheta_0}{2\pi^2\epsilon_0\rho' c}; \qquad 1 < t/\rho' < \epsilon^{1/2}, \tag{13.8.6}$$

$$\Pi_{2z}(\rho',t) = \frac{1}{\epsilon-1}\,\frac{1}{2\pi^2\epsilon_0\rho' c}\,(\epsilon\vartheta_0-\vartheta_\epsilon); \qquad t/\rho' > \epsilon^{1/2}, \tag{13.8.7}$$

where, with (13.4.2) through (13.4.7),

$$\vartheta_0 = \int_1^{t^2/\rho'^2} \frac{(\zeta-1)^{1/2}\,d\zeta}{[(\epsilon+1)\zeta-\epsilon](t^2/\rho'^2-\zeta)^{1/2}} = \frac{1}{\epsilon+1}\left(\int_{E_0}^{T_0+E_0}\frac{dx}{X^{1/2}} - E_0\int_{E_0}^{T_0+E_0}\frac{dx}{xX^{1/2}}\right), \tag{13.8.8}$$

$$\vartheta_\epsilon = \int_\epsilon^{t^2/\rho'^2} \frac{(\zeta-\epsilon)^{1/2}\,d\zeta}{[(\epsilon+1)\zeta-\epsilon](t^2/\rho'^2-\zeta)^{1/2}} = \frac{1}{\epsilon+1}\left(\int_{E_\epsilon}^{T_\epsilon+E_\epsilon}\frac{dx}{X^{1/2}} - E_\epsilon\int_{E_\epsilon}^{T_\epsilon+E_\epsilon}\frac{dx}{xX^{1/2}}\right). \tag{13.8.9}$$

The integrated expressions corresponding to (13.4.9) and (13.4.10) are

$$\vartheta_0 = \frac{\pi}{\epsilon+1}\left[1-\left((\epsilon+1)\frac{t^2}{\rho'^2}-\epsilon\right)^{-1/2}\right], \tag{13.8.10}$$

$$\vartheta_\epsilon = \frac{\pi}{\epsilon+1}\left[1-\epsilon\left((\epsilon+1)\frac{t^2}{\rho'^2}-\epsilon\right)^{-1/2}\right]. \tag{13.8.11}$$

When these values are substituted in (13.8.6) and (13.8.7), with $\rho = c\rho'$, the results are

$$\Pi_{2z}(\rho,t) = \frac{1}{2\pi\epsilon_0\rho(\epsilon+1)}\left\{\begin{array}{l} 0 \\ \dfrac{\epsilon}{\epsilon-1}\left[1-\left((\epsilon+1)\dfrac{c^2t^2}{\rho^2}-\epsilon\right)^{-1/2}\right] \\ 1 \end{array}\right\};$$

$$\left\{\begin{array}{l} ct/\rho < 1 \\ 1 < ct/\rho < \epsilon^{1/2} \\ ct/\rho > \epsilon^{1/2} \end{array}\right. . \qquad (13.8.12)$$

In terms of the step function $U(t)$ [which is defined as $U(t) = 0,\ t < 0$; $U(t) = 1,\ t \geq 0$], this can be written as

$$\begin{aligned}\Pi_{2z}(\rho,t) = {} & \frac{1}{2\pi\epsilon_0\rho}\,\frac{\epsilon}{\epsilon^2-1} \\ & \times\left[U\left(t-\frac{\rho}{c}\right)-\epsilon^{-1}U\left(t-\frac{\rho\epsilon^{1/2}}{c}\right)\right. \\ & \left.-\left((\epsilon+1)\frac{c^2t^2}{\rho^2}-\epsilon\right)^{-1/2}U\left(t-\frac{\rho}{c}\right)U\left(\frac{\rho\epsilon^{1/2}}{c}-t\right)\right].\end{aligned} \qquad (13.8.13)$$

If this is multiplied on the right by $2\pi\epsilon_0(\epsilon+1)$ to obtain the same normalization introduced by Van der Pol (1956), (13.8.13) is the same as equation (17) in that work. Van der Pol did not derive the formulas for the components of the electromagnetic field from the following expressions:

$$E_{2z} = k_2^2\Pi_{2z} + \frac{\partial^2\Pi_{2z}}{\partial z^2}; \qquad E_{2\rho} = \frac{\partial^2\Pi_{2z}}{\partial\rho\,\partial z}, \qquad (13.8.14)$$

$$B_{2\phi} = -\frac{1}{c^2}\frac{\partial^2\Pi_{2z}}{\partial\rho\,\partial t}. \qquad (13.8.15)$$

Since (13.8.12) or (13.8.13) is valid only along the boundary $z = 0$, it cannot be used to evaluate the electric field from (13.8.14). On the other hand, $B_{2\phi}(\rho,t)$ as given in (13.6.26) is readily evaluated from (13.8.15) with (13.8.12) or (13.8.13) to obtain (13.6.26).

14 Approximate Formulas for Lateral Electromagnetic Pulses Generated by Vertical and Horizontal Electric Dipoles

14.1 Integrals for the Approximate Time-Dependent Field of a Vertical Dipole; δ-Function Excitation

In the preceding chapter, exact formulas are derived for the principal components E_z and B_ϕ of the transient electromagnetic field generated by a vertical electric dipole on the boundary between two perfect dielectric half-spaces when the exciting current is a δ-function current with unit amplitude. This was accomplished by the evaluation of the general Sommerfeld integrals. The corresponding calculation of the field generated by the same dipole when excited by a periodically varying current of unit amplitude is carried out in Chapter 3. However, it is approximate in that the final formulas include terms that are good approximations only in the range specified by the conditions

$$|k_1^2| \geq 9|k_2^2|; \qquad |k_1\rho| \geq 3|k_2\rho| \geq 3. \tag{14.1.1}$$

Here the second condition is a consequence of the use of large-argument approximations of the Bessel functions in the derivation of the Fresnel-integral terms and in the evaluation of certain integrals. For perfect dielectrics $k_1^2 = (\omega^2/c^2)\epsilon$ and $k_2^2 = \omega^2/c^2$, where $\epsilon \equiv \epsilon_{1r}$ is the relative permittivity of Region 1 and Region 2 is taken to be air; $c = 3 \times 10^8$ m/sec is the velocity of light. It follows that (14.1.1) is equivalent to

$$\epsilon \geq 9; \qquad \omega\rho' \geq 1, \tag{14.1.2}$$

where

$$\rho' = \frac{\rho}{c}. \tag{14.1.3}$$

For the dielectric half-spaces, the complete electromagnetic field on the boundary, $z = 0$, due to a unit vertical electric dipole on the boundary at $d = 0$ in Region 1 is given by

$$\tilde{B}_{2\phi}(\rho', \omega) = \tilde{B}_{1\phi}(\rho', \omega)$$

$$= -\frac{\mu_0}{2\pi\epsilon c^2} e^{i\omega\rho'} \left[\frac{i\omega}{\rho'} - \frac{1}{\rho'^2} - \omega^{3/2}\left(\frac{\pi}{\epsilon\rho'}\right)^{1/2} e^{-ip}\mathcal{F}(p)\right], \tag{14.1.4}$$

$$\tilde{E}_{2\rho}(\rho', \omega) = \tilde{E}_{1\rho}(\rho', \omega)$$

$$= -\frac{\mu_0}{2\pi\epsilon^{3/2}c}\left\{e^{i\omega\rho'}\left[\frac{i\omega}{\rho'} - \frac{1}{\rho'^2} - \omega^{3/2}\left(\frac{\pi}{\epsilon\rho'}\right)^{1/2} e^{-ip}\mathcal{F}(p)\right] - \frac{ie^{i\omega\epsilon^{1/2}\rho'}}{\rho'^2}\right\}, \tag{14.1.5}$$

$$\tilde{E}_{2z}(\rho', \omega) = \epsilon\tilde{E}_{1z}(\rho', \omega)$$

$$= \frac{\mu_0}{2\pi\epsilon c} e^{i\omega\rho'}\left[\frac{i\omega}{\rho'} - \frac{1}{\rho'^2} - \frac{i}{\omega\rho'^3} - \omega^{3/2}\left(\frac{\pi}{\epsilon\rho'}\right)^{1/2} e^{-ip}\mathcal{F}(p)\right], \tag{14.1.6}$$

where

$$p = \frac{k_2^3\rho}{2k_1^2} = \frac{\omega\rho'}{2\epsilon}, \tag{14.1.7}$$

$$\mathcal{F}(p) = \tfrac{1}{2}(1+i) - \int_0^p \frac{e^{i\tau}}{(2\pi\tau)^{1/2}}\, d\tau = \int_p^\infty \frac{e^{i\tau}}{(2\pi\tau)^{1/2}}\, d\tau. \tag{14.1.8}$$

The real time-dependent field generated by a δ-function current in the dipole as defined in (13.1.2) and (13.1.3) is

$$B_{2\phi}(\rho', t) = \frac{1}{\pi}\int_0^\infty \mathrm{Re}\, e^{-i\omega t}\tilde{B}_{2\phi}(\rho', \omega)\, d\omega, \tag{14.1.9}$$

$$E_{2\rho}(\rho', t) = \frac{1}{\pi}\int_0^\infty \mathrm{Re}\, e^{-i\omega t}\tilde{E}_{2\rho}(\rho', \omega)\, d\omega, \tag{14.1.10}$$

$$E_{2z}(\rho', t) = \frac{1}{\pi}\int_0^\infty \mathrm{Re}\, e^{-i\omega t}\tilde{E}_{2z}(\rho', \omega)\, d\omega. \tag{14.1.11}$$

With (14.1.4) through (14.1.6), the integrals in (14.1.9) through (14.1.11) can be expressed as follows:

$$B_{2\phi}(\rho', t) = -\frac{\mu_0}{2\pi\epsilon c^2}\left(\frac{I_1}{\rho'} - \frac{I_2}{\rho'^2} + \frac{I_4}{(\pi\epsilon\rho')^{1/2}}\right), \tag{14.1.12}$$

$$E_{2\rho}(\rho', t) = -\frac{\mu_0}{2\pi\epsilon^{3/2}c}\left(\frac{I_1}{\rho'} - \frac{I_2}{\rho'^2} + \frac{I_4}{(\pi\epsilon\rho')^{1/2}}\right), \tag{14.1.13}$$

$$E_{2z}(\rho', t) = \frac{\mu_0}{2\pi\epsilon c}\left(\frac{I_1}{\rho'} - \frac{I_2}{\rho'^2} - \frac{I_3}{\rho'^3} + \frac{I_4}{(\pi\epsilon\rho')^{1/2}}\right), \tag{14.1.14}$$

where

$$I_1 = \operatorname{Re} \frac{1}{\pi}\int_0^\infty i\omega e^{-i\omega(t-\rho')}\, d\omega, \tag{14.1.15}$$

$$I_2 = \operatorname{Re} \frac{1}{\pi}\int_0^\infty e^{-i\omega(t-\rho')}\, d\omega;$$

$$I_{2\epsilon} = \operatorname{Re} \frac{1}{\pi}\int_0^\infty e^{-i\omega(t-\epsilon^{1/2}\rho')}\, d\omega, \tag{14.1.16}$$

$$I_3 = \operatorname{Re} \frac{1}{\pi}\int_0^\infty \frac{i}{\omega} e^{-i\omega(t-\rho')}\, d\omega;$$

$$I_{3\epsilon} = \operatorname{Re} \frac{1}{\pi}\int_0^\infty \frac{i}{\omega} e^{-i\omega(t-\epsilon^{1/2}\rho')}\, d\omega, \tag{14.1.17}$$

$$I_4 = -\operatorname{Re}\int_0^\infty \omega^{3/2} e^{-i\omega(t-\rho'+\rho'/2\epsilon)} \times \left(\tfrac{1}{2}(1+i) - \int_0^{\omega\rho'/2\epsilon} \frac{e^{i\tau}}{(2\pi\tau)^{1/2}}\, d\tau\right) d\omega. \tag{14.1.18}$$

[The integrals $I_{2\epsilon}$ and $I_{3\epsilon}$ in (14.1.16) and (14.1.17) do not occur in (14.1.12) through (14.1.14). They are included here for later reference.]

Note that the single term in $\exp(ik_1\rho) = \exp(i\omega\epsilon^{1/2}\rho')$ in $\tilde{E}_{2\rho}(\rho', \omega)$ leads to a pure imaginary so that it contributes nothing to the real part in (14.1.13).

14.2 Evaluation of Integrals

The five integrals (14.1.15) through (14.1.17) can be evaluated exactly. (The variants $I_{2\epsilon}$ and $I_{3\epsilon}$ are included for later application.) The integrated values are

$$I_1 = -\frac{\partial}{\partial t}\left(\frac{1}{\pi}\int_0^\infty e^{-i\omega(t-\rho')}\, d\omega\right) = -\frac{\partial}{\partial t}\delta(t-\rho') = -\delta'(t-\rho'), \tag{14.2.1}$$

$$I_2 = \delta(t-\rho'); \qquad I_{2\epsilon} = \delta(t-\epsilon^{1/2}\rho'), \tag{14.2.2}$$

$$I_3 = \begin{cases} 0, & t < \rho' \\ 1, & t > \rho' \end{cases}; \qquad I_{3\epsilon} = \begin{cases} 0, & t < \epsilon^{1/2}\rho' \\ 1, & t > \epsilon^{1/2}\rho' \end{cases}, \tag{14.2.3}$$

$$I_4 = \text{Re}\,\frac{\partial^2}{\partial t^2}\biggl(\tfrac{1}{2}(1+i)\int_0^\infty e^{-i\omega(t-\bar{\alpha})}\,\frac{d\omega}{\omega^{1/2}} - \frac{1}{(2\pi)^{1/2}}\int_0^\infty e^{-i\omega(t-\bar{\alpha})}\,\frac{d\omega}{\omega^{1/2}}\int_0^p \frac{e^{i\tau}}{\tau^{1/2}}\,d\tau\biggr), \tag{14.2.4}$$

where

$$\bar{\alpha} = \rho'\left(1 - \frac{1}{2\epsilon}\right) = \rho' - \beta; \qquad \beta = \frac{\rho'}{2\epsilon}. \tag{14.2.5}$$

Let

$$\alpha = \frac{t - \bar{\alpha}}{\beta}. \tag{14.2.6}$$

Then, with (14.1.7),

$$\omega = \frac{p}{\beta}. \tag{14.2.7}$$

With (14.2.5) through (14.2.7) in (14.2.4),

$$I_4 = \text{Re}\,\frac{\partial^2}{\partial t^2}\left[\left(\frac{1+i}{2\beta^{1/2}}\right)\int_0^\infty e^{-ip\alpha}\,\frac{dp}{p^{1/2}} - \frac{1}{(2\pi\beta)^{1/2}}\int_0^\infty e^{-ip\alpha}\,\frac{dp}{p^{1/2}}\int_0^p \frac{e^{i\tau}}{\tau^{1/2}}\,d\tau\right]. \tag{14.2.8}$$

Here the first integral is

$$\vartheta_1 = \int_0^\infty e^{-ip\alpha}\,\frac{dp}{p^{1/2}} = \int_0^\infty \frac{\cos\alpha p}{p^{1/2}}\,dp - i\int_0^\infty \frac{\sin\alpha p}{p^{1/2}}\,dp. \tag{14.2.9}$$

With $x = \alpha p$,

$$\vartheta_1 = \frac{1}{\alpha^{1/2}}\int_0^\infty \frac{\cos x}{x^{1/2}}\,dx - \frac{i}{\alpha^{1/2}}\int_0^\infty \frac{\sin x}{x^{1/2}}\,dx = \left(\frac{\pi}{2\alpha}\right)^{1/2}(1-i). \tag{14.2.10}$$

With $p = \xi^2/\alpha$ and $\tau = \zeta^2$, the second integral in (14.2.8) becomes

$$\vartheta_2 = \frac{4}{\alpha^{1/2}}\int_0^\infty e^{-i\xi^2}\,d\xi\int_0^{\xi/\alpha^{1/2}} e^{i\zeta^2}\,d\zeta. \tag{14.2.11}$$

Cylindrical coordinates can now be introduced with

$$\xi = \rho\cosh\theta, \qquad \zeta = \rho\sinh\theta, \qquad d\xi\,d\zeta = \rho\,d\rho\,d\theta. \tag{14.2.12}$$

With (14.2.12), (14.2.11) becomes

$$\vartheta_2 = \frac{4}{\alpha^{1/2}}\int_0^\infty e^{-i\rho^2}\,\rho\,d\rho\int_0^{\tanh^{-1}(1/\alpha^{1/2})} d\theta$$

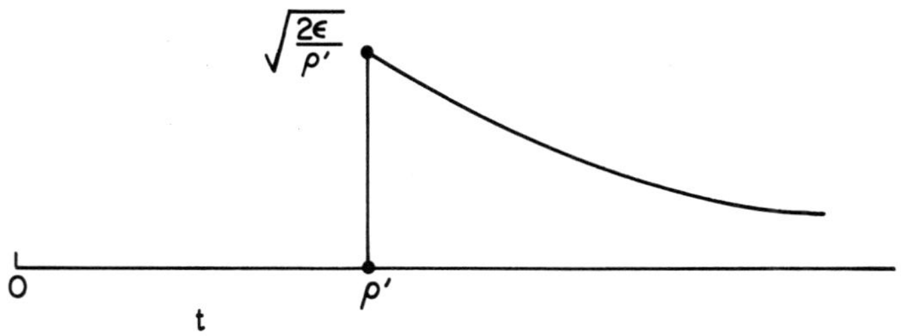

Figure 14.2.1. The function $(t - \rho' + \rho'/2\epsilon)^{-1/2}$.

$$= \frac{4}{\alpha^{1/2}} \tanh^{-1} \frac{1}{\alpha^{1/2}} \int_0^\infty e^{-i\rho^2} \rho \, d\rho. \tag{14.2.13}$$

With $u = \rho^2$ and $du = 2\rho \, d\rho$, this becomes

$$\vartheta_2 = \frac{2}{\alpha^{1/2}} \tanh^{-1} \frac{1}{\alpha^{1/2}} \int_0^\infty e^{-iu} \, du = -\frac{2i}{\alpha^{1/2}} \tanh^{-1} \frac{1}{\alpha^{1/2}}. \tag{14.2.14}$$

With (14.2.14) and (14.2.10) in (14.2.8), this becomes

$$I_4 = \text{Re} \frac{\partial^2}{\partial t^2} \left[\left(\frac{1+i}{2\beta^{1/2}} \right) \left(\frac{\pi}{2\alpha} \right)^{1/2} (1-i) + \frac{2i}{(2\pi\alpha\beta)^{1/2}} \tanh^{-1} \frac{1}{\alpha^{1/2}} \right]. \tag{14.2.15}$$

Here the second term is a pure imaginary so that it contributes nothing to I_4. Thus,

$$I_4 = \frac{\partial^2}{\partial t^2} \left(\frac{\pi}{2\alpha\beta} \right)^{1/2} = \frac{\partial^2}{\partial t^2} \left(\frac{\pi}{2} \right)^{1/2} \frac{1}{(t - \bar{\alpha})^{1/2}}. \tag{14.2.16}$$

Since this must be real, it is necessary that $t > \bar{\alpha}$ and

$$I_4 = \begin{cases} 0, & t < \bar{\alpha}, \\ \dfrac{\partial^2}{\partial t^2} \left(\dfrac{\pi}{2} \right)^{1/2} \dfrac{1}{(t - \bar{\alpha})^{1/2}} & \\ \quad = \left(\dfrac{\pi}{2} \right)^{1/2} \dfrac{\partial^2}{\partial t^2} \dfrac{1}{(t - \rho' + \rho'/2\epsilon)^{1/2}}, & t > \bar{\alpha}. \end{cases} \tag{14.2.17}$$

Actually, since $\rho' = \rho/c$ and since the signal cannot travel with a velocity greater than c, it follows that $I_4 = 0$ for $t < \rho' = \rho/c$. Thus, I_4 has the form shown in Fig. 14.2.1. It is given by

$$I_4 = \left(\frac{\pi}{2} \right)^{1/2} \frac{\partial^2}{\partial t^2} \frac{1}{(t - \rho' + \rho'/2\epsilon)^{1/2}} U(t - \rho'), \tag{14.2.18}$$

where $U(t - \rho')$ is the Heaviside function or step function.

The differentiation in (14.2.18) can now be carried out. For later use, it is convenient to introduce a separate notation for the first time derivative. Thus,

$$I_5 = -\int_0^\infty \operatorname{Re} e^{-i\omega(t-\rho')} i\omega^{1/2} e^{-ip}\mathcal{F}(p)\,d\omega$$
$$= \left(\frac{\pi}{2}\right)^{1/2}\left[\left(\frac{2\epsilon}{\rho'}\right)^{1/2}\delta(t-\rho') - \frac{1}{2}\left(t-\rho'+\frac{\rho'}{2\epsilon}\right)^{-3/2}U(t-\rho')\right], \tag{14.2.19a}$$

and

$$I_4 = \frac{\partial}{\partial t}I_5 = \left(\frac{\pi}{2}\right)^{1/2}\left[\left(\frac{2\epsilon}{\rho'}\right)^{1/2}\delta'(t-\rho') - \frac{1}{2}\left(\frac{2\epsilon}{\rho'}\right)^{3/2}\delta(t-\rho') + \frac{3}{4}\left(t-\rho'+\frac{\rho'}{2\epsilon}\right)^{-5/2}U(t-\rho')\right]. \tag{14.2.19b}$$

14.3 The Approximate Time-Dependent Electromagnetic Field of the Vertical Dipole with δ-Function Excitation; Comparison with Exact Field

When the four integrals (14.2.1), (14.2.2), (14.2.3), and (14.2.19b) are substituted into (14.1.12) through (14.1.14), these become

$$B_{2\phi}(\rho',t) = -\frac{\mu_0}{2\pi\epsilon c^2}\left\{-\frac{\delta'(t-\rho')}{\rho'} - \frac{\delta(t-\rho')}{\rho'^2} + \frac{1}{(2\epsilon\rho')^{1/2}}\left[\left(\frac{2\epsilon}{\rho'}\right)^{1/2}\delta'(t-\rho') - \frac{1}{2}\left(\frac{2\epsilon}{\rho'}\right)^{3/2}\delta(t-\rho') + \frac{3}{4}\left(t-\rho'+\frac{\rho'}{2\epsilon}\right)^{-5/2}U(t-\rho')\right]\right\}. \tag{14.3.1}$$

With $\rho' = \rho/c$, this reduces to

$$B_{2\phi}(\rho,t) = \frac{\mu_0}{2\pi\rho^2}\left(\frac{\epsilon+1}{\epsilon}\right)\delta\left(t-\frac{\rho}{c}\right) - \frac{\mu_0 c}{2\pi\rho^3}\left\{\begin{matrix} 0 \\ 3\epsilon\left[2\epsilon\left(\dfrac{ct}{\rho}-1\right)+1\right]^{-5/2} \end{matrix}\right\};\quad \left\{\begin{matrix} t<\rho/c \\ t>\rho/c \end{matrix}\right.. \tag{14.3.2}$$

The expression for the time-dependent radial electric field follows directly from (14.1.13) with (14.2.1), (14.2.2), and (14.2.19b). It is

$$E_{2\rho}(\rho,t) = \frac{1}{2\pi\epsilon_0 c\rho^2}\left(\frac{\epsilon+1}{\epsilon^{3/2}}\right)\delta\left(t-\frac{\rho}{c}\right) - \frac{1}{2\pi\epsilon_0\rho^3}\left\{\begin{matrix} 0 \\ 3\epsilon^{1/2}\left[2\epsilon\left(\dfrac{ct}{\rho}-1\right)+1\right]^{-5/2}\end{matrix}\right\}; \quad \begin{cases} t<\rho/c \\ t>\rho/c \end{cases}. \tag{14.3.3}$$

The formula for $E_{2z}(\rho,t)$ is obtained from (14.1.14). It is

$$E_{2z}(\rho,t) = -\frac{1}{2\pi\epsilon_0 c\rho^2}\left(\frac{\epsilon+1}{\epsilon}\right)\delta\left(t-\frac{\rho}{c}\right) - \frac{1}{2\pi\epsilon_0\epsilon\rho^3}\left\{\begin{matrix} 0 \\ 1-3\epsilon^2\left[2\epsilon\left(\dfrac{ct}{\rho}-1\right)+1\right]^{-5/2}\end{matrix}\right\}; \quad \begin{cases} t<\rho/c \\ t>\rho/c \end{cases}. \tag{14.3.4}$$

Since (14.3.2) through (14.3.4) are Fourier transforms of formulas for the steady-state components that are good approximations in a wide but nonetheless restricted range of frequency, it is important to compare them with the corresponding exact formulas derived in Chapter 13. These latter are available only for $B_{2\phi}(\rho,t)$ and $E_{2z}(\rho,t)$. Let the exact formulas be denoted with a subscript e. For greater formal unity, let the δ-functions be denoted as follows:

$$\delta\left(t-\frac{\rho}{c}\right) = \frac{c}{\rho}\,\delta\left(\frac{ct}{\rho}-1\right) = \frac{c}{\rho}\,\delta(\tau-1), \tag{14.3.5}$$

where

$$\tau = \frac{ct}{\rho}. \tag{14.3.6}$$

The formulas are

$$B_{2\phi}(\rho,\tau) = \left\{\begin{matrix} 0 \\ \dfrac{\mu_0 c}{2\pi\rho^3}\left(\dfrac{\epsilon+1}{\epsilon}\,\delta(\tau-1) - 3\epsilon[2\epsilon(\tau-1)+1]^{-5/2}\right)\end{matrix}\right\}; \quad \begin{cases} \tau<1 \\ \tau>1 \end{cases}, \tag{14.3.7}$$

$$[B_{2\phi}(\rho,\tau)]_e = \begin{cases} 0 \\ \dfrac{\mu_0 c}{2\pi\rho^3}\left(\dfrac{\epsilon}{\epsilon-1}\,\delta(\tau-1) - \dfrac{1}{\epsilon(\epsilon-1)}\,\delta(\tau-\epsilon^{1/2}) \right. \\ \qquad \left. - \dfrac{3\epsilon^2\tau}{\epsilon-1}\,[(\epsilon+1)\tau^2-\epsilon]^{-5/2}\right) \\ 0 \end{cases}; \quad \begin{cases} \tau < 1 \\ 1 < \tau < \epsilon^{1/2} \\ \tau > \epsilon^{1/2} \end{cases}, \tag{14.3.8}$$

$$E_{2z}(\rho,\tau) = \begin{cases} 0 \\ -\dfrac{1}{2\pi\epsilon_0\epsilon\rho^3}\,\{(\epsilon+1)\delta(\tau-1) \\ \qquad + 1 - 3\epsilon^2[2\epsilon(\tau-1)+1]^{-5/2}\} \end{cases}; \quad \begin{cases} \tau < 1 \\ \tau > 1 \end{cases}, \tag{14.3.9}$$

$$[E_{2z}(\rho,\tau)]_e = \begin{cases} 0 \\ -\dfrac{1}{2\pi\epsilon_0\epsilon\rho^3}\left[\dfrac{\epsilon^2}{\epsilon-1}\,\delta(\tau-1) - \dfrac{\epsilon^{1/2}}{\epsilon-1}\,\delta(\tau-\epsilon^{1/2}) \right. \\ \qquad \left. + \dfrac{\epsilon^2}{\epsilon^2-1}\left(1 - \dfrac{\epsilon[2(\epsilon+1)\tau^2+\epsilon]}{[(\epsilon+1)\tau^2-\epsilon]^{5/2}}\right)\right] \\ -\dfrac{1}{2\pi\epsilon_0(\epsilon+1)\rho^3} \end{cases}; \quad \begin{cases} \tau < 1 \\ 1 < \tau < \epsilon^{1/2} \\ \tau > \epsilon^{1/2} \end{cases}. \tag{14.3.10}$$

A comparison of the approximate with the exact formulas shows that the coefficients of the pulses at $\tau = ct/\rho = 1$ are essentially the same since, with the assumed condition $\epsilon^2 \gg 1$, it follows that $\epsilon/(\epsilon-1) = \epsilon(\epsilon+1)/(\epsilon^2-1) \sim (\epsilon+1)/\epsilon$. The pulse which occurs at $\tau = \epsilon^{1/2}$ in the exact formulas is absent in the approximate formulas. Note, however, that the coefficient of this pulse is very much smaller than that of the pulse at $\tau = 1$ by the factors ϵ^{-2} and $\epsilon^{-3/2}$. The approximate and exact formulas are not readily compared in a quantitative sense in the range $1 < \tau < \epsilon^{1/2}$ between the two pulses except by direct numerical evaluation for specific values of ϵ. Such an evaluation has been made for $\epsilon = 80$ and 10. The results are listed in Table 14.3.1. The quite remarkable agreement between the two

Table 14.3.1. Comparison of Exact and Approximate Formulas for $B_{2\phi}(\rho,\tau)$ and $E_{2z}(\rho,\tau)$ with δ-Function Excitation; $\tau = ct/\rho \geq 1$

	$(-2\pi\rho^3/\mu_0 c)\, B_{2\phi}(\rho,\tau)$			
	Approximate		Exact	
	$\dfrac{3\epsilon}{[2\epsilon(\tau-1)+1]^{5/2}}$		$\dfrac{3\epsilon^2\tau}{(\epsilon-1)[(\epsilon+1)\tau^2-\epsilon]^{5/2}}$	
τ	$\epsilon = 80$	$\epsilon = 10$	$\epsilon = 80$	$\epsilon = 10$
1.0	240	30.0	243	33.3
1.01	22.0	19.0	21.9	20.4
1.05	0.987	5.30	0.967	5.30
1.1	0.201	1.92	0.194	1.84
1.2	0.0384	0.536	0.0359	0.485
1.5	0.406×10^{-2}	0.0748	0.345×10^{-2}	0.0598
2.0	0.73×10^{-3}	0.0148	0.52×10^{-3}	0.0099
3.0	0.13×10^{-3}	0.00279	0.068×10^{-3}	0.00134
$\epsilon^{1/2}$	0.416×10^{-5}	0.00230	0.066×10^{-5}	0.00105

	$-2\pi\epsilon_0\epsilon\rho^3\, E_{2z}(\rho,\tau)$			
	Approximate		Exact	
	$1 - \dfrac{3\epsilon^2}{[2\epsilon(\tau-1)+1]^{5/2}}$		$\dfrac{\epsilon^2}{\epsilon^2-1}\left(1 - \dfrac{\epsilon[2(\epsilon+1)\tau^2+\epsilon]}{[(\epsilon+1)\tau^2-\epsilon]^{5/2}}\right)$	
τ	$\epsilon = 80$	$\epsilon = 10$	$\epsilon = 80$	$\epsilon = 10$
1.0	$-19,200$	-299	$-19,400$	-322
1.01	$-1,760$	-189	$-1,750$	-198
1.05	-78.0	-52.0	-77.4	-51.4
1.1	-15.1	-18.2	-15.0	-17.5
1.2	-2.07	-4.37	-2.08	-4.10
1.3	-0.142	-1.31	-0.159	-1.19
1.32	0.025	-1.01	0.008	-0.91
2.0	0.942	0.85	0.937	0.86
3.0	0.9896	0.97	0.9887	0.98
$\epsilon^{1/2}$	0.99967	0.977	0.99984	0.987

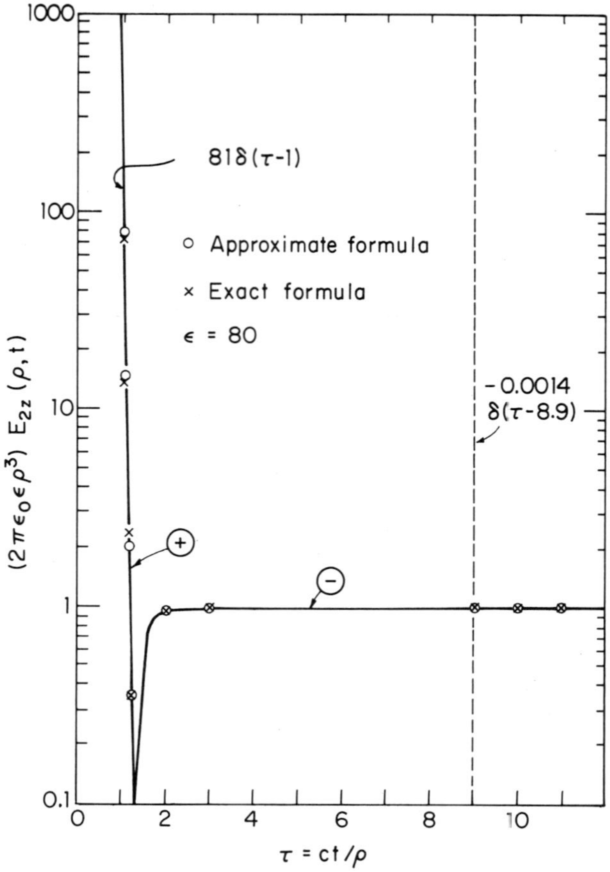

Figure 14.3.1. Comparison of the approximate and exact formulas for $E_{2z}(\rho, t)$ with $\epsilon = 80$; δ-function excitation. Graphs of Table 14.3.1, columns for $\epsilon = 80$ and $\tau > 1$.

superficially quite different formulas is best seen from a comparison of the corresponding numerical values over their very wide range of magnitudes. A graph for $E_{2z}(\rho, t)$ with $\epsilon = 80$ is shown in Fig. 14.3.1. If desired, the approximate formula can be improved arbitrarily by the addition of the pulse at $\tau = \epsilon^{1/2}$.

14.4 Generalization to the Field When the Source and Point of Observation Are in the Dielectric Region 1

Formulas (14.1.4) through (14.1.6) apply when the source is on the boundary in Region 1 and the point of observation is on the boundary in Region 2 so that $d = z = 0$. This special case was considered in order to permit a direct comparison of the approximate formulas with the exact ones derived in Chapter 13. The more general case when the source is at the depth d and the point of observation at the depth z in the dielectric Region 1 is readily derived. Note, however, that the conditions $\rho^2 \gg z^2$ and $\rho^2 \gg d^2$ are assumed in the frequency domain.

The largest component and the one of general interest is $E_{1\rho}(\rho, z; t)$. The general frequency-dependent formula is

$$\begin{aligned} E_{1\rho}(\rho, z) &= -\frac{\omega\mu_0}{2\pi k_1^2}\left(\frac{k_2^2}{k_1}\left\{e^{ik_1(z+d)}e^{ik_2\rho}\right.\right. \\ &\qquad \times\left[\frac{ik_2}{\rho} - \frac{1}{\rho^2} - \frac{k_2^3}{k_1}\left(\frac{\pi}{k_2\rho}\right)^{1/2} e^{-ip}\mathcal{F}(p)\right] - \left.\frac{ie^{ik_1r_2}}{\rho^2}\right\} \\ &\qquad + \left.\frac{1}{2}\left(\frac{z-d}{\rho}e^{ik_1r_1} - \frac{z+d}{\rho}e^{ik_1r_2}\right)\left(\frac{ik_1^2}{\rho} - \frac{3k_1}{\rho^2} - \frac{3i}{\rho^3}\right)\right). \end{aligned} \tag{14.4.1}$$

With $k_2 = \omega/c$, $k_1 = \omega\epsilon^{1/2}/c$, and

$$s' = \rho' + \epsilon^{1/2}(z' + d') = \frac{\rho + \epsilon^{1/2}(z+d)}{c} = \frac{s}{c}, \tag{14.4.2}$$

the formula for $E_{1\rho}(\rho, z)$ becomes

$$\begin{aligned} \tilde{E}_{1\rho}(\rho', z'; \omega) &= -\frac{\mu_0}{2\pi c\epsilon}\left(\epsilon^{-1/2}\left\{e^{i\omega s'}\left[\frac{i\omega}{\rho'} - \frac{1}{\rho'^2} - \omega^{3/2}\left(\frac{\pi}{\epsilon\rho'}\right)^{1/2} e^{-ip}\mathcal{F}(p)\right]\right.\right. \end{aligned}$$

$$- \frac{ie^{i\omega\epsilon^{1/2}r_2'}}{\rho'^2}\bigg\}$$
$$+ \frac{1}{2}\left(\frac{z'-d'}{\rho'}\, e^{i\omega\epsilon^{1/2}r_1'} - \frac{z'+d'}{\rho'}\, e^{i\omega\epsilon^{1/2}r_2'}\right)$$
$$\times \left(\frac{i\omega\epsilon}{\rho'} - \frac{3\epsilon^{1/2}}{\rho'^2} - \frac{3i}{\omega\rho'^3}\right)\Bigg). \tag{14.4.3}$$

The notation $r_1' = r_1/c = [\rho^2 + (z-d)^2]^{1/2}/c$ and $r_2' = r_2/c = [\rho^2 + (z+d)^2]^{1/2}/c$ is used; p and $\mathcal{F}(p)$ are defined in (14.1.7) and (14.1.8).

The time-dependent form of (14.4.3) is obtained with (14.1.10). With (14.2.1)–(14.2.3) and (14.2.19b), the following integrals are obtained:

$$I_1 = \text{Re}\, \frac{1}{\pi}\int_0^\infty i\omega e^{-i\omega(t-s')}\, d\omega = -\delta'(t-s'), \tag{14.4.4}$$
$$I_{1\epsilon} = \text{Re}\, \frac{1}{\pi}\int_0^\infty i\omega e^{-i\omega(t-\epsilon^{1/2}r')}\, d\omega = -\delta'(t-\epsilon^{1/2}r'), \tag{14.4.5}$$
$$I_2 = \text{Re}\, \frac{1}{\pi}\int_0^\infty e^{-i\omega(t-s')}\, d\omega = \delta(t-s'), \tag{14.4.6}$$
$$I_{2\epsilon} = \text{Re}\, \frac{1}{\pi}\int_0^\infty e^{-i\omega(t-\epsilon^{1/2}r')}\, d\omega = \delta(t-\epsilon^{1/2}r'), \tag{14.4.7}$$
$$I_3 = \text{Re}\, \frac{1}{\pi}\int_0^\infty \frac{i}{\omega}\, e^{-i\omega(t-s')}\, d\omega = \begin{cases} 0, & t < s' \\ 1, & t > s' \end{cases}, \tag{14.4.8}$$
$$I_{3\epsilon} = \text{Re}\, \frac{1}{\pi}\int_0^\infty \frac{i}{\omega}\, e^{-i\omega(t-\epsilon^{1/2}r')}\, d\omega = \begin{cases} 0, & t < \epsilon^{1/2}r' \\ 1, & t > \epsilon^{1/2}r' \end{cases}, \tag{14.4.9}$$
$$I_4 = -\,\text{Re}\int_0^\infty \omega^{3/2}\, e^{-i\omega(t-s'+\rho'/2\epsilon)}$$
$$\times \left(\tfrac{1}{2}(1+i) - \int_0^p \frac{e^{i\tau}}{(2\pi\tau)^{1/2}}\, d\tau\right) d\omega$$
$$= \left(\frac{\pi}{2}\right)^{1/2}\left[\left(\frac{2\epsilon}{\rho'}\right)^{1/2}\delta'(t-s') - \frac{1}{2}\left(\frac{2\epsilon}{\rho'}\right)^{3/2}\delta(t-s')\right.$$
$$\left. + \frac{3}{4}\left(t - s' + \frac{\rho'}{2\epsilon}\right)^{-5/2} U(t-s')\right]. \tag{14.4.10}$$

With these integrals, the time-dependent form of (14.4.3) is

$$E_{1\rho}(\rho', z'; t) = -\frac{\mu_0}{2\pi c\epsilon^{3/2}}$$
$$\times \left(-\frac{1}{\rho'}\,\delta'(t-s') - \frac{1}{\rho'^2}\,\delta(t-s') + \frac{1}{\rho'}\,\delta'(t-s')\right.$$
$$- \frac{\epsilon}{\rho'^2}\,\delta(t-s') + \frac{3}{4(2\epsilon\rho')^{1/2}}\left(t - s' + \frac{\rho'}{2\epsilon}\right)^{-5/2} U(t-s')$$

$$
\begin{aligned}
&+\frac{\epsilon^{1/2}}{2}\Bigg[\left(\frac{z'-d'}{\rho'}\right)\Bigg(-\frac{\epsilon}{\rho'}\,\delta'(t-\epsilon^{1/2}r_1')-\frac{3\epsilon^{1/2}}{\rho'^2}\,\delta(t-\epsilon^{1/2}r_1') \\
&\qquad\qquad -\frac{3}{\rho'^3}\begin{Bmatrix}0, & t<\epsilon^{1/2}r_1' \\ 1, & t>\epsilon^{1/2}r_1'\end{Bmatrix}\Bigg) \\
&\quad -\left(\frac{z'+d'}{\rho'}\right)\Bigg(-\frac{\epsilon}{\rho'}\,\delta'(t-\epsilon^{1/2}r_2')-\frac{3\epsilon^{1/2}}{\rho'^2}\,\delta(t-\epsilon^{1/2}r_2') \\
&\qquad\qquad -\frac{3}{\rho'^3}\begin{Bmatrix}0, & t<\epsilon^{1/2}r_2' \\ 1, & t>\epsilon^{1/2}r_2'\end{Bmatrix}\Bigg)\Bigg]\Bigg). \qquad (14.4.11)
\end{aligned}
$$

With $\rho' = \rho/c$, this becomes

$$
\begin{aligned}
E_{1\rho}(\rho,z;t) &= \frac{1}{2\pi\epsilon_0 c\rho^2}\left(\frac{\epsilon+1}{\epsilon^{3/2}}\right)\delta\left(t-\frac{s}{c}\right) \\
&\quad -\frac{3\epsilon^{1/2}}{2\pi\epsilon_0\rho^3}\left[2\epsilon\left(\frac{ct-s}{\rho}\right)+1\right]^{-5/2}\begin{Bmatrix}0, & ct<s \\ 1, & ct>s\end{Bmatrix} \\
&\quad +\frac{1}{4\pi\epsilon_0\epsilon c^2}\Bigg(\left(\frac{z-d}{\rho}\right)\Bigg[\frac{\epsilon}{\rho}\,\delta'\left(t-\frac{\epsilon^{1/2}r_1}{c}\right) \\
&\qquad\qquad +\frac{3\epsilon^{1/2}c}{\rho^2}\,\delta\left(t-\frac{\epsilon^{1/2}r_1}{c}\right) \\
&\qquad\qquad +\frac{3c^2}{\rho^3}\begin{Bmatrix}0, & ct<\epsilon^{1/2}r_1 \\ 1, & ct>\epsilon^{1/2}r_1\end{Bmatrix}\Bigg] \\
&\qquad -\left(\frac{z+d}{\rho}\right)\Bigg[\frac{\epsilon}{\rho}\,\delta'\left(t-\frac{\epsilon^{1/2}r_2}{c}\right) \\
&\qquad\qquad +\frac{3\epsilon^{1/2}c}{\rho^2}\,\delta\left(t-\frac{\epsilon^{1/2}r_2}{c}\right) \\
&\qquad\qquad +\frac{3c^2}{\rho^3}\begin{Bmatrix}0, & ct<\epsilon^{1/2}r_2 \\ 1, & ct>\epsilon^{1/2}r_2\end{Bmatrix}\Bigg]\Bigg),
\end{aligned}
\qquad (14.4.12)
$$

where $s = \rho + \epsilon^{1/2}(z+d)$. This is a complicated field that includes the initial lateral pulse that arrives at $ct = \rho + \epsilon^{1/2}(z+d)$, travels vertically downward in the dielectric, and is followed by a tail that decreases rapidly to zero. The direct pulses arrive at $ct = \epsilon^{1/2}r_1$ and travel radially outward from the source. They consist of both δ' and δ pulses and are followed by the static value $3/4\pi\epsilon_0\epsilon\rho^3$. The reflected (or image) pulses arrive at $ct = \epsilon^{1/2}r_2$, consist of δ' and δ pulses that travel radially outward from the image source, and are followed by the static value $-3/4\pi\epsilon_0\epsilon\rho^3$, which exactly cancels the static value generated by the passing direct pulse. Thus, the final radial electric field is zero.

The corresponding expression for the vertical component of the electric field is readily obtained from

$$\begin{aligned}\tilde{E}_{1z}(\rho', z'; \omega) \\ = \frac{\mu_0}{2\pi c\epsilon}\bigg\{&\epsilon^{-1}e^{i\omega s'}\left[\frac{i\omega}{\rho'} - \frac{1}{\rho'^2} - \frac{i}{\omega\rho'^3} - \omega^{3/2}\left(\frac{\pi}{\epsilon\rho'}\right)^{1/2} e^{-ip}\mathcal{F}(p)\right] \\ &+ \frac{1}{2}\left(e^{i\omega\epsilon^{1/2}r_1'} - e^{i\omega\epsilon^{1/2}r_2'}\right)\left(\frac{i\omega\epsilon}{\rho'} - \frac{\epsilon^{1/2}}{\rho'^2} - \frac{i}{\omega\rho'^3}\right)\bigg\}. \end{aligned} \tag{14.4.13}$$

This differs from (14.4.3) only in factors independent of ω so that the individual Fourier transforms in the evaluation of the time-dependent form are the same. Thus,

$$\begin{aligned} E_{1z}(\rho, z; t) \\ = &-\frac{1}{2\pi\epsilon_0 c\rho^2}\left(\frac{\epsilon+1}{\epsilon^2}\right)\delta\left(t - \frac{s}{c}\right) \\ &- \frac{1}{2\pi\epsilon_0\epsilon^2\rho^3}\left\{1 - 3\epsilon^2\left[2\epsilon\left(\frac{ct-s}{\rho}\right) + 1\right]^{-5/2}\right\}\begin{Bmatrix} 0, & ct < s \\ 1, & ct > s \end{Bmatrix} \\ &- \frac{1}{4\pi\epsilon_0\epsilon c^2}\left[\frac{\epsilon}{\rho}\delta'\left(t - \frac{\epsilon^{1/2}r_1}{c}\right) + \frac{\epsilon^{1/2}c}{\rho^2}\delta\left(t - \frac{\epsilon^{1/2}r_1}{c}\right)\right. \\ &\quad + \frac{c^2}{\rho^3}\begin{Bmatrix} 0, & ct < \epsilon^{1/2}r_1 \\ 1, & ct > \epsilon^{1/2}r_1 \end{Bmatrix} - \frac{\epsilon}{\rho}\delta'\left(t - \frac{\epsilon^{1/2}r_2}{c}\right) \\ &\quad \left. - \frac{\epsilon^{1/2}c}{\rho^2}\delta\left(t - \frac{\epsilon^{1/2}r_2}{c}\right) - \frac{c^2}{\rho^3}\begin{Bmatrix} 0, & ct < \epsilon^{1/2}r_2 \\ 1, & ct > \epsilon^{1/2}r_2 \end{Bmatrix}\right], \end{aligned} \tag{14.4.14}$$

where $s = \rho + \epsilon^{1/2}(z+d)$, $r_1 = [\rho^2 + (z-d)^2]^{1/2}$, and $r_2 = [\rho^2 + (z+d)^2]^{1/2}$.

Like $E_{1\rho}(\rho, z; t)$, this is a complicated field. It includes the initial lateral pulse that arrives at $ct = \rho + \epsilon^{1/2}(z + d)$, the direct pulses that arrive at $ct = \epsilon^{1/2}r_1$, and the reflected (image) pulses that arrive at $ct = \epsilon^{1/2}r_2$. The first pulse has a tail that decreases rapidly to the static value $-1/2\pi\epsilon_0\epsilon^2\rho^3$; the second and third pulses are each followed by static values that are equal and opposite and cancel. The final field is the static value left by the passing lateral pulse on the boundary surface, $z = 0$, viz.,

$$E_{1z}(\rho, 0; t) = -\frac{1}{2\pi\epsilon_0\epsilon^2\rho^3}. \tag{14.4.15}$$

The corresponding field in the air is given by (14.3.4). It is

$$E_{2z}(\rho, 0; t) = -\frac{1}{2\pi\epsilon_0\epsilon\rho^3} \tag{14.4.16}$$

and satisfies the boundary condition $\epsilon E_{1z}(\rho, 0; t) = E_{2z}(\rho, 0; t)$.

14.5 Approximate Time-Domain Formulas for the Field of a Horizontal Electric Dipole; δ-Function Excitation

The good agreement between the exact and approximate formulas for $E_{2z}(\rho, 0; t)$ and $B_{2\phi}(\rho, 0; t)$ for the vertical dipole excited by a δ-function pulse suggests that a similar evaluation of the complete field of the horizontal electric dipole be done. Actually, the approximate formulas for $E_\rho(\rho, 0, z)$ and $E_z(\rho, 0, z)$ for the horizontal dipole are very similar, respectively, to $E_z(\rho, z)$ and $E_\rho(\rho, z)$ for the vertical dipole. The three components of the electric field of the horizontal dipole are obtained from (5.5.47)–(5.5.49). With $z = d = 0$, $k_1 = \omega\epsilon^{1/2}/c$, $k_2 = \omega/c$, and $\rho' = \rho/c$, they are

$$\tilde{E}_{2\rho}(\rho', 0; \omega)$$

$$= -\frac{\mu_0}{2\pi\epsilon c}\left\{e^{i\omega\rho'}\left[\frac{i\omega}{\rho'} - \frac{1}{\rho'^2} - \frac{i}{\omega\rho'^3} - \omega^{3/2}\left(\frac{\pi}{\epsilon\rho'}\right)^{1/2} e^{-ip}\mathcal{F}(p)\right] - e^{i\omega\epsilon^{1/2}\rho'}\left(\frac{\epsilon^{1/2}}{\rho'^2} + \frac{i}{\omega\rho'^3}\right)\right\}, \tag{14.5.1}$$

$$\tilde{E}_{2\phi}(\rho', \pi/2; \omega) = \frac{\mu_0}{2\pi\epsilon c}\left\{e^{i\omega\rho'}\left[\frac{2}{\rho'^2} + \frac{2i}{\omega\rho'^3} + \frac{i\omega^{1/2}}{\rho'}\left(\frac{\pi}{\epsilon\rho'}\right)^{1/2} e^{-ip}\mathcal{F}(p)\right] - e^{i\omega\epsilon^{1/2}\rho'}\left(\frac{\epsilon^{1/2}}{\rho'^2} + \frac{i}{\omega\rho'^3}\right)\right\}, \tag{14.5.2}$$

$$\tilde{E}_{2z}(\rho', 0; \omega) = \frac{\mu_0}{2\pi\epsilon^{1/2}c}\left\{e^{i\omega\rho'}\left[\frac{i\omega}{\rho'} - \frac{1}{\rho'^2} - \omega^{3/2}\left(\frac{\pi}{\epsilon\rho'}\right)^{1/2} e^{-ip}\mathcal{F}(p)\right] - \frac{ie^{i\omega\epsilon^{1/2}\rho'}}{\rho'^2}\right\}, \tag{14.5.3}$$

where p and $\mathcal{F}(p)$ are defined in (14.1.7) and (14.1.8).

The corresponding formulas in the time domain are obtained in the same manner as (14.1.9)–(14.1.11). With the integrals defined in (14.1.15)–(14.1.18) and (14.2.19a,b), and by comparison with (14.3.3) and (14.3.4), the results are

$$E_{2\rho}(\rho', 0; t) = E_{1\rho}(\rho', 0; t)$$

$$= \frac{1}{\pi}\int_0^\infty \operatorname{Re} e^{-i\omega t}\tilde{E}_{2\rho}(\rho', 0; \omega)\, d\omega$$

$$= -\frac{\mu_0}{2\pi\epsilon c}\left(\frac{I_1}{\rho'} - \frac{I_2}{\rho'^2} - \frac{I_3}{\rho'^3} + \frac{I_4}{(\pi\epsilon\rho')^{1/2}} - \frac{\epsilon^{1/2}I_{2\epsilon}}{\rho'^2} - \frac{I_{3\epsilon}}{\rho'^3}\right), \tag{14.5.4}$$

so that

$$E_{2\rho}(\rho,0;t)=E_{1\rho}(\rho,0;t)$$
$$=\frac{1}{2\pi\epsilon_0 c\rho^2}\left[\left(\frac{\epsilon+1}{\epsilon}\right)\delta\left(t-\frac{\rho}{c}\right)+\epsilon^{-1/2}\delta\left(t-\frac{\epsilon^{1/2}\rho}{c}\right)\right]$$
$$+\frac{1}{2\pi\epsilon_0\epsilon\rho^3}\left\{\begin{array}{l}0\\ 1-3\epsilon^2\left[2\epsilon\left(\dfrac{ct}{\rho}-1\right)+1\right]^{-5/2}\\ 2-3\epsilon^2\left[2\epsilon\left(\dfrac{ct}{\rho}-1\right)+1\right]^{-5/2}\end{array}\right\};$$
$$\begin{cases}ct/\rho<1\\ 1<ct/\rho<\epsilon^{1/2}\\ ct/\rho>\epsilon^{1/2}\end{cases};\quad(14.5.5)$$

$$E_{2\phi}(\rho',\pi/2;t)=E_{1\phi}(\rho',\pi/2;t)$$
$$=\frac{1}{\pi}\int_0^\infty \mathrm{Re}\,e^{-i\omega t}\tilde{E}_{2\phi}(\rho',\pi/2;\omega)\,d\omega$$
$$=\frac{\mu_0}{2\pi\epsilon c}\left(\frac{2I_2}{\rho'^2}+\frac{2I_3}{\rho'^3}-\frac{I_5}{\rho'(\pi\epsilon\rho')^{1/2}}-\frac{\epsilon^{1/2}I_{2\epsilon}}{\rho'^2}-\frac{I_{3\epsilon}}{\rho'^3}\right),\quad(14.5.6)$$

so that

$$E_{2\phi}(\rho,\pi/2;t)=E_{1\phi}(\rho,\pi/2;t)$$
$$=\frac{1}{2\pi\epsilon_0\epsilon c\rho^2}\left[\delta\left(t-\frac{\rho}{c}\right)-\epsilon^{1/2}\delta\left(t-\frac{\epsilon^{1/2}\rho}{c}\right)\right]$$
$$+\frac{1}{2\pi\epsilon_0\epsilon\rho^3}\left\{\begin{array}{l}0\\ 2+\epsilon\left[2\epsilon\left(\dfrac{ct}{\rho}-1\right)+1\right]^{-3/2}\\ 1+\epsilon\left[2\epsilon\left(\dfrac{ct}{\rho}-1\right)+1\right]^{-3/2}\end{array}\right\};$$
$$\begin{cases}ct/\rho<1\\ 1<ct/\rho<\epsilon^{1/2}\\ ct/\rho>\epsilon^{1/2}\end{cases};\quad(14.5.7)$$

$$E_{2z}(\rho',0;t)=\epsilon E_{1z}(\rho',0;t)$$
$$=\frac{1}{\pi}\int_0^\infty \mathrm{Re}\,e^{-i\omega t}\tilde{E}_{2z}(\rho',0;\omega)\,d\omega$$

$$= \frac{\mu_0}{2\pi\epsilon^{1/2}c}\left(\frac{I_1}{\rho'} - \frac{I_2}{\rho'^2} + \frac{I_4}{(\pi\epsilon\rho')^{1/2}}\right), \tag{14.5.8}$$

so that

$$\begin{aligned} E_{2z}(\rho, 0; t) &= \epsilon E_{1z}(\rho, 0; t) \\ &= -\frac{1}{2\pi\epsilon_0\epsilon^{1/2}c\rho^2}(\epsilon+1)\delta\left(t - \frac{\rho}{c}\right) \\ &\quad + \frac{1}{2\pi\epsilon_0\epsilon^{1/2}\rho^3}\left\{\begin{matrix} 0 \\ 3\epsilon^2\left[2\epsilon\left(\frac{ct}{\rho} - 1\right) + 1\right]^{-5/2} \end{matrix}\right\}; \\ &\qquad\qquad \left\{\begin{matrix} ct/\rho < 1 \\ ct/\rho > 1 \end{matrix}\right. . \end{aligned} \tag{14.5.9}$$

It is to be noted that the electric field at all points on the surface is obtained from (14.5.5), (14.5.7), and (14.5.9) with the following simple formulas:

$$E_{2\rho}(\rho, \phi; t) = E_{2\rho}(\rho, 0; t)\cos\phi, \tag{14.5.10}$$

$$E_{2\phi}(\rho, \phi; t) = E_{2\phi}(\rho, \pi/2; t)\sin\phi, \tag{14.5.11}$$

$$E_{2z}(\rho, \phi; t) = E_{2z}(\rho, 0; t)\cos\phi. \tag{14.5.12}$$

The sequence of pulses and the values of $E_{2\rho}(\rho, 0; t)$ and $E_{2\phi}(\rho, \pi/2; t)$ between and after them, as given by (14.5.5) and (14.5.7), are interesting. $E_{2\rho}$ is seen to approach $1/2\pi\epsilon_0\epsilon\rho^3$ after the first pulse has passed at $ct/\rho = 1$, and then to reach the final static field, $2/2\pi\epsilon_0\epsilon\rho^3$, after the second pulse has passed at $ct/\rho = \epsilon^{1/2}$. On the other hand, $E_{2\phi}$ approaches $2/2\pi\epsilon_0\epsilon\rho^3$ after the first pulse and then the final static field, $1/2\pi\epsilon_0\epsilon\rho^3$, after the second pulse has passed.

Formulas for the three components of the magnetic field in the time domain are not evaluated from the approximate expressions in the frequency domain because they are less accurate. This is a consequence of two important facts. In the first place, the formulas in the frequency domain are good approximations only subject to (14.1.2), notably $\omega\rho' \geq 1$. This means that they are not good approximations at very low frequencies. In spite of this, the formulas for the electric field are correct at zero frequency and their Fourier transforms to the time domain give the correct static field after the δ-function pulses have passed any point. This is not true of the formulas for the magnetic field primarily because they are a power lower in ω. This is seen directly from Maxwell's equations in the frequency domain, e.g.,

$$B_\phi = \frac{i}{\omega}\left(\frac{\partial E_z}{\partial\rho} - \frac{\partial E_\rho}{\partial z}\right). \tag{14.5.13}$$

In the relatively simple field of the vertical dipole, the time-domain formula for $B_\phi(\rho, t)$ is a good approximation of the exact value. In the more complicated field of the horizontal dipole, the time-domain formulas for $B_\rho(\rho, \pi/2; t)$, $B_\phi(\rho, 0; t)$, and $B_z(\rho, \pi/2; t)$ obtained by Fourier transformation from the approximate formulas are not adequate.

It is possible to obtain generalized time-domain expressions for the components of the electric field at the depth z in the dielectric Region 1 when the dipole source is at the depth d in the same manner as for the vertical dipole in Section 14.4. Subject to $\rho^2 \gg z^2$ and $\rho^2 \gg d^2$, the formula for $E_{1\rho}(\rho, 0, z; t)$ is

$$E_{1\rho}(\rho, 0, z; t) = \frac{1}{2\pi\epsilon_0 c\rho^2}\left[\left(\frac{\epsilon+1}{\epsilon}\right)\delta\left(t - \frac{s}{c}\right) + \epsilon^{-1/2}\delta\left(t - \frac{\epsilon^{1/2}r_1}{c}\right)\right] + \frac{1}{2\pi\epsilon_0\epsilon\rho^3}\begin{cases} 0 \\ 1 - 3\epsilon^2\left[2\epsilon\left(\dfrac{ct-s}{\rho}\right)+1\right]^{-5/2} \\ 2 - 3\epsilon^2\left[2\epsilon\left(\dfrac{ct-s}{\rho}\right)+1\right]^{-5/2} \end{cases}; \quad \begin{cases} ct < s \\ s < ct < \epsilon^{1/2}r_1 \\ ct > \epsilon^{1/2}r_1 \end{cases}, \tag{14.5.14}$$

where

$$s = \rho + \epsilon^{1/2}(z+d); \qquad r_1 = [\rho^2 + (z-d)^2]^{1/2},$$
$$\frac{ct-s}{\rho} = \frac{ct}{\rho} - 1 - \frac{\epsilon^{1/2}(z+d)}{\rho}. \tag{14.5.15}$$

14.6 The Approximate Transient Field of a Vertical Dipole Excited by a Gaussian Pulse

A useful alternative to the δ-function pulse is the Gaussian pulse. With its amplitude normalized so that it reduces to the δ-function when its width is reduced to zero, it is defined by

$$f(t) = \frac{e^{-t^2/t_1^2}}{\sqrt{\pi}\,t_1}, \tag{14.6.1}$$

where t_1 is the half-width or time in which the pulse is reduced to $\exp(-1) = 0.367$ of its maximum of unity at $t = 0$. The Fourier transform of the

Gaussian pulse is

$$\tilde{f}(\omega) = \int_{-\infty}^{\infty} \frac{e^{-t^2/t_1^2}}{\sqrt{\pi}\, t_1} e^{i\omega t}\, dt = e^{-\omega^2 t_1^2/4}. \tag{14.6.2}$$

Note that when $t_1 \to 0$, then $f(t) \to \delta(t)$, the δ-function, and $\tilde{f}(\omega) \to 1$, the transform of the δ-function. The function (14.6.2) can be used with (14.1.4)–(14.1.6) to obtain the time-dependent forms. For $E_{2z}(\rho, t)$, the time-dependent form is

$$E_{2z}(\rho', t) = \frac{1}{2\pi} \operatorname{Re} \int_{-\infty}^{\infty} d\omega\, e^{-i\omega t} \tilde{f}(\omega) \tilde{E}_{2z}(\rho', \omega). \tag{14.6.3}$$

With (14.1.6) and (14.6.2), this becomes

$$\begin{aligned} E_{2z}(\rho', t) = \frac{\mu_0}{4\pi^2 \epsilon c} \operatorname{Re} \int_{-\infty}^{\infty} d\omega\, e^{-i\omega t}\, e^{-\omega^2 t_1^2/4}\, e^{i\omega\rho'} \\ \times \left[\frac{i\omega}{\rho'} - \frac{1}{\rho'^2} - \frac{i}{\omega \rho'^3} - \omega^{3/2} \left(\frac{\pi}{\epsilon \rho'} \right)^{1/2} e^{-ip} \mathcal{F}(p) \right]. \end{aligned} \tag{14.6.4}$$

This involves four integrals, viz.,

$$I_1 = \operatorname{Re} \frac{1}{2\pi} \int_{-\infty}^{\infty} i\omega e^{-\omega^2 t_1^2/4} e^{-i\omega(t-\rho')}\, d\omega, \tag{14.6.5}$$

$$I_2 = \operatorname{Re} \frac{1}{2\pi} \int_{-\infty}^{\infty} e^{-\omega^2 t_1^2/4} e^{-i\omega(t-\rho')}\, d\omega, \tag{14.6.6}$$

$$I_3 = \operatorname{Re} \frac{1}{2\pi} \int_{-\infty}^{\infty} \frac{i}{\omega} e^{-\omega^2 t_1^2/4} e^{-i\omega(t-\rho')}\, d\omega, \tag{14.6.7}$$

$$I_4 = -\operatorname{Re} \frac{1}{2} \int_{-\infty}^{\infty} \omega^{3/2} e^{-\omega^2 t_1^2/4} e^{-i\omega(t-\rho')} e^{-ip} \mathcal{F}(p)\, d\omega. \tag{14.6.8}$$

With them,

$$E_{2z}(\rho', t) = \frac{\mu_0}{2\pi\epsilon c} \left(\frac{I_1}{\rho'} - \frac{I_2}{\rho'^2} - \frac{I_3}{\rho'^3} + \frac{I_4}{(\pi\epsilon\rho')^{1/2}} \right). \tag{14.6.9}$$

Similarly, with (14.1.4) and (14.1.5),

$$B_{2\phi}(\rho', t) = -\frac{\mu_0}{2\pi\epsilon c^2} \left(\frac{I_1}{\rho'} - \frac{I_2}{\rho'^2} + \frac{I_4}{(\pi\epsilon\rho')^{1/2}} \right), \tag{14.6.10}$$

$$E_{2\rho}(\rho', t) = -\frac{\mu_0}{2\pi\epsilon^{3/2} c} \left(\frac{I_1}{\rho'} - \frac{I_2}{\rho'^2} + \frac{I_4}{(\pi\epsilon\rho')^{1/2}} \right). \tag{14.6.11}$$

The first three integrals (14.6.5)–(14.6.7) can be evaluated together since

$$I_1 = -\frac{\partial I_2}{\partial t}, \tag{14.6.12}$$

$$\frac{\partial I_3}{\partial t} = I_2. \tag{14.6.13}$$

Consider first I_2. With the notation

$$\tau \equiv t - \rho' = t - \frac{\rho}{c}, \tag{14.6.14}$$

I_2 can be written as

$$\begin{aligned} I_2 &= \mathrm{Re}\,\frac{1}{2\pi}\left(\int_0^\infty e^{-\omega^2 t_1^2/4}\,e^{-i\omega\tau}\,d\omega - \int_{-\infty}^0 e^{-\omega^2 t_1^2/4}\,e^{-i\omega\tau}\,d\omega\right) \\ &= \frac{1}{\pi}\int_0^\infty e^{-\omega^2 t_1^2/4}\cos\omega\tau\,d\omega. \end{aligned} \tag{14.6.15}$$

With formula 3.896-4 of Gradshteyn and Ryzhik (1980, p. 480),

$$I_2 = \frac{1}{\sqrt{\pi}\,t_1}\,e^{-\tau^2/t_1^2}. \tag{14.6.16}$$

Since $\partial/\partial t = \partial/\partial\tau$, it follows that

$$I_1 = -\frac{\partial I_2}{\partial\tau} = \frac{2\tau}{\sqrt{\pi}\,t_1^3}\,e^{-\tau^2/t_1^2}. \tag{14.6.17}$$

Also, with (14.6.13),

$$\frac{\partial I_3}{\partial\tau} = I_2 = \frac{1}{\sqrt{\pi}\,t_1}\,e^{-\tau^2/t_1^2}, \tag{14.6.18}$$

$$\begin{aligned} I_3 &= \frac{1}{\sqrt{\pi}\,t_1}\int_{-\infty}^{\tau} e^{-\tau^2/t_1^2}\,d\tau \\ &= \frac{1}{\sqrt{\pi}}\left(\int_0^{\tau/t_1} e^{-u^2}\,du + \int_0^\infty e^{-u^2}\,du\right), \end{aligned} \tag{14.6.19}$$

where $u = \tau/t_1$ and $du = d\tau/t_1$. The definition of the error function is

$$\mathrm{erf}\,z = \frac{2}{\sqrt{\pi}}\int_0^z e^{-u^2}\,du; \qquad \frac{2}{\sqrt{\pi}}\int_0^\infty e^{-u^2}\,du = 1. \tag{14.6.20}$$

Hence,

$$I_3 = \tfrac{1}{2}[1 + \mathrm{erf}(\tau/t_1)]. \tag{14.6.21}$$

The fourth integral (14.6.8) is more difficult. It is

$$\begin{aligned} I_4 = -\,\mathrm{Re}\,\frac{1}{2}\int_{-\infty}^{\infty} d\omega\;\omega^{3/2}e^{-\omega^2 t_1^2/4}e^{-i\omega(t-\rho'+\rho'/2\epsilon)} \\ \times\left(\tfrac{1}{2}(1+i) - \int_0^p \frac{e^{i\tau}}{(2\pi\tau)^{1/2}}\,d\tau\right). \end{aligned} \tag{14.6.22}$$

This integral is like (14.1.18) for δ-function excitation except for the additional factor $\exp(-\omega^2 t_1^2/4)$, which is real. Hence, the double integral is like the corresponding integral in (14.2.8) but with the added factor

$\exp(-\omega^2 t_1^2/4) = \exp(-p^2 t_1^2/4\beta^2)$. With it, the integral in (14.2.11) becomes

$$\vartheta_2 = \frac{4}{\alpha^{1/2}} \int_0^\infty e^{-i\rho^2}\, \rho\, d\rho \int_0^{\tanh^{-1}(1/\alpha^{1/2})} e^{-(\rho^4 t_1^2/4\alpha^2\beta^2)\cosh^4\theta}\, d\theta. \quad (14.6.23)$$

This is not readily integrated. However, the range of θ is small since $\alpha = 2\epsilon[(ct/\rho) - 1] + 1$ and $\epsilon \geq 9$. With $\tanh^{-1}(1/\alpha^{1/2}) \sim 1/\alpha^{1/2}$, $0 \leq \theta \leq 1/\alpha^{1/2}$; also with $\cosh^4\theta \sim 1 + 2\theta^2$, the condition $2\theta^2 \ll 1$ is adequately satisfied when $\theta \leq 0.3$. Hence, $\alpha^{-1} \leq 0.09$ or $\alpha \geq 11$. This is true when $2\epsilon[(ct/\rho) - 1] + 1 \geq 11$ or $ct/\rho \geq 1 + (5/\epsilon) \geq 1.56$ for $\epsilon = 9$. Since $ct/\rho = 1$ is the center of the pulse, this is not a severe restriction on the useful range. With it, the integral ϑ_2 is the same as (14.2.11) and yields a pure imaginary that contributes nothing to I_4 in (14.6.22). Hence, with

$$T \equiv t - \rho' + \frac{\rho'}{2\epsilon}, \quad (14.6.24)$$

$$I_4 = \frac{\partial^2}{\partial t^2} \operatorname{Re} \tfrac{1}{4}(1+i) \int_{-\infty}^\infty \omega^{-1/2} e^{-\omega^2 t_1^2/4} e^{-i\omega T}\, d\omega. \quad (14.6.25)$$

Let $\omega = \sqrt{2}\tau/t_1$ and $z = \sqrt{2}T/t_1$. Then

$$I_4 = \operatorname{Re} \tfrac{1}{4}(1+i)\, 2^{1/4} t_1^{-1/2} \frac{\partial^2}{\partial t^2} \vartheta_0 = \operatorname{Re} \frac{e^{i\pi/4}\, 2^{1/4}}{2\sqrt{2}\, t_1^{1/2}} \frac{\partial^2}{\partial t^2} \vartheta_0, \quad (14.6.26)$$

where

$$\begin{aligned} \vartheta_0 &= \int_{-\infty}^\infty \tau^{-1/2} e^{-\tau^2/2 - iz\tau}\, d\tau \\ &= \int_0^\infty \tau^{-1/2} e^{-\tau^2/2 - iz\tau}\, d\tau - i \int_0^\infty \tau^{-1/2} e^{-\tau^2/2 + iz\tau}\, d\tau. \end{aligned} \quad (14.6.27)$$

The two integrals on the right are related to the parabolic cylinder functions $U(0, x) \equiv D_{-1/2}(x)$ which have the following integral representation and relation to the Bessel functions [see formula 3 of Bateman (1953, Vol. 2, p. 19) and formulas 19.5.3 and 19.15.9 of Abramowitz and Stegun (1972, pp. 687 and 692)]:

$$\begin{aligned} D_{-1/2}(x) \equiv U(0, x) &= \pi^{-1/2} (\tfrac{1}{2}x)^{1/2} K_{1/4}(\tfrac{1}{4}x^2) \\ &= \pi^{-1/2} e^{-x^2/4} \int_0^\infty e^{-xt - t^2/2}\, t^{-1/2}\, dt. \end{aligned} \quad (14.6.28)$$

[The notation $U(0, x)$—which is not related to the step function $U(x)$—is that used by Abramowitz and Stegun.] With $y = U(0, x)$, this function satisfies the following equation [see formula 19.1.2 of Abramowitz and Stegun (1972, p. 686)]:

$$\frac{d^2 y}{dx^2} - \tfrac{1}{4} x^2 y = 0. \quad (14.6.29)$$

With $x = \pm iz = z\exp(\pm i\pi/2)$, the two integrals in (14.6.27) become

$$\int_0^\infty \tau^{-1/2} e^{-\tau^2/2\mp iz\tau}\, d\tau = \left(\tfrac{1}{2} z e^{\pm i\pi/2}\right)^{1/2} e^{-z^2/4} K_{1/4}\left(\tfrac{1}{4} z^2 e^{\pm i\pi}\right). \tag{14.6.30}$$

With formula 45 of Bateman (1953, Vol. 2, p. 80), the two Bessel functions with imaginary argument can be expanded as follows:

$$K_{1/4}(xe^{\pm i\pi}) = e^{\mp i\pi/4} K_{1/4}(x) \mp i\pi I_{1/4}(x). \tag{14.6.31}$$

With (14.6.31) and $x = \frac{1}{4}z^2$, (14.6.30) becomes

$$\int_0^\infty \tau^{-1/2} e^{-\tau^2/2\mp iz\tau}\, d\tau$$
$$= \left(\tfrac{1}{2}z\right)^{1/2} e^{-z^2/4} \left[K_{1/4}\left(\tfrac{1}{4}z^2\right) \mp i\pi e^{\pm i\pi/4} I_{1/4}\left(\tfrac{1}{4}z^2\right)\right]. \tag{14.6.32}$$

These expressions can now be substituted in (14.6.27) to give

$$\vartheta_0 = \left(\tfrac{1}{2}z\right)^{1/2} e^{-z^2/4} e^{-i\pi/4} f\left(\tfrac{1}{4}z^2\right),$$
$$f\left(\tfrac{1}{4}z^2\right) \equiv \sqrt{2} K_{1/4}\left(\tfrac{1}{4}z^2\right) + 2\pi I_{1/4}\left(\tfrac{1}{4}z^2\right). \tag{14.6.33}$$

The next step is to evaluate

$$\frac{\partial^2 \vartheta_0}{\partial t^2} = \frac{\partial^2 \vartheta_0}{\partial T^2} = \frac{2}{t_1^2}\frac{\partial^2 \vartheta_0}{\partial z^2}. \tag{14.6.34}$$

It is readily verified with (14.6.29) that

$$\frac{\partial^2 \vartheta_0}{\partial t^2} = -\frac{2e^{-i\pi/4}}{t_1^2}\left(\tfrac{1}{2}(1-z^2)y + z\frac{dy}{dz}\right)e^{-z^2/4}. \tag{14.6.35}$$

With the tabulated parabolic cylinder functions $U(a,z)$ and $V(a,z)$ defined on p. 687 and formulas 19.15.1, 19.15.2, 19.15.13, and 19.15.19 in Abramowitz and Stegun (1972),

$$K_{1/4}\left(\tfrac{1}{4}z^2\right) = \pi^{1/2}\left(\tfrac{1}{2}z\right)^{-1/2} U(0,z), \tag{14.6.36}$$
$$I_{1/4}\left(\tfrac{1}{4}z^2\right) = z^{-1/2} V(0,z) - (\pi z)^{-1/2} U(0,z), \tag{14.6.37}$$

and

$$y = \left(\tfrac{1}{2}z\right)^{1/2} f\left(\tfrac{1}{4}z^2\right) = \frac{2\pi}{\sqrt{2}} V(0,z), \tag{14.6.38}$$
$$\frac{dy}{dz} = \frac{2\pi}{\sqrt{2}} V'(0,z) = \frac{2\pi}{\sqrt{2}}\left[\tfrac{1}{2}zV(0,z) - \tfrac{1}{2}V(-1,z)\right], \tag{14.6.39}$$

where $V'(0,z)$ is obtained with formula 19.6.5 of Abramowitz and Stegun (1972, p. 688) with $a = 0$. When y and dy/dz are substituted in (14.6.35), the result is

$$\frac{\partial^2 \vartheta_0}{\partial t^2} = -\frac{2\pi}{\sqrt{2}}\frac{e^{-i\pi/4}e^{-z^2/4}}{t_1^2}\left[V(0,z) - zV(-1,z)\right]; \quad z \geq 0. \tag{14.6.40}$$

For negative values of z, it is convenient to use formula 19.4.2 of Abramowitz and Stegun (1972, p. 687) with $a = 0$ and $a = -1$. This gives

$$V(0,z) = \frac{\Gamma(\frac{1}{2})}{\pi}\,U(0,-z); \qquad V(-1,z) = \frac{\Gamma(-\frac{1}{2})}{\pi}\,U(-1,-z), \tag{14.6.41}$$

where $U(0,x)$ is given by (14.6.28), $\Gamma(\frac{1}{2}) = \pi^{1/2}$, and $\Gamma(-\frac{1}{2}) = -2\pi^{1/2}$. With these values,

$$\frac{\partial^2\vartheta_0}{\partial t^2} = -\frac{2\pi}{\sqrt{2}}\,\frac{e^{-i\pi/4}e^{-z^2/4}}{t_1^2}\,[\pi^{-1/2}\,U(0,-z) + 2\pi^{-1/2}z\,U(-1,-z)];$$
$$z \le 0. \tag{14.6.42}$$

With (14.6.40) and (14.6.42) substituted in (14.6.26),

$$I_4 = -\frac{\pi}{2}\,\frac{2^{1/4}}{t_1^{5/2}}\,e^{-z^2/4}\begin{Bmatrix} V(0,z) - zV(-1,z) \\ \pi^{-1/2}\,U(0,-z) + 2\pi^{-1/2}z\,U(-1,-z)\end{Bmatrix};$$
$$\begin{cases} z \ge 0 \\ z \le 0 \end{cases}. \tag{14.6.43}$$

With $z = \sqrt{2}T/t_1$, where $T = t - \rho' + \rho'/2\epsilon$, this becomes

$$I_4 = -\frac{\pi}{2}\,\frac{2^{1/4}}{t_1^{5/2}}\,e^{-T^2/2t_1^2}$$
$$\times\begin{Bmatrix} V\left(0,\dfrac{\sqrt{2}T}{t_1}\right) - \dfrac{\sqrt{2}T}{t_1}\,V\left(-1,\dfrac{\sqrt{2}T}{t_1}\right) \\ \pi^{-1/2}\left[U\left(0,-\dfrac{\sqrt{2}T}{t_1}\right) + \dfrac{2\sqrt{2}T}{t_1}\,U\left(-1,-\dfrac{\sqrt{2}T}{t_1}\right)\right]\end{Bmatrix};$$
$$\begin{cases} T/t_1 \ge 0 \\ T/t_1 \le 0 \end{cases}, \tag{14.6.44}$$

where $V(0,x)$, $V(-1,x)$ and $U(0,x)$, $U(-1,x)$ are parabolic cylinder functions (Abramowitz and Stegun 1972) and

$$T \equiv t - \rho' + \frac{\rho'}{2\epsilon}; \qquad \rho' = \frac{\rho}{c}. \tag{14.6.45}$$

When (14.6.16), (14.6.17), (14.6.21), and (14.6.44) are substituted in (14.6.9)–(14.6.11) with the following new notation introducing dimensionless variables,

$$\rho_1 \equiv \frac{\rho'}{t_1} = \frac{\rho}{ct_1}; \qquad t' \equiv \frac{t}{t_1}; \qquad \tau' \equiv \frac{\tau}{t_1} = t' - \rho_1, \tag{14.6.46}$$

$$T' \equiv \frac{T}{t_1} = t' - \rho_1\left(1 - \frac{1}{2\epsilon}\right), \tag{14.6.47}$$

the results are

$$B_{2\phi}(\rho,t) = \frac{\mu_0 c}{2\pi\epsilon}\frac{\pi^{-1/2}}{c^3t_1^3}\left[\left(-\frac{2\tau'}{\rho_1}+\frac{1}{\rho_1^2}\right)e^{-\tau'^2} + \frac{\pi}{2}\frac{2^{1/4}}{(\epsilon\rho_1)^{1/2}}e^{-T'^2/2} \times \begin{Bmatrix} V(0,\sqrt{2}\,T') - \sqrt{2}\,T'\,V(-1,\sqrt{2}\,T') \\ \pi^{-1/2}[U(0,-\sqrt{2}\,T') \\ \qquad + 2\sqrt{2}\,T'\,U(-1,-\sqrt{2}\,T')] \end{Bmatrix}\right]; \quad \begin{cases} T' \geq 0 \\ T' \leq 0 \end{cases}, \tag{14.6.48}$$

$$E_{2\rho}(\rho,t) = \frac{c}{\epsilon^{1/2}}B_{2\phi}(\rho,t), \tag{14.6.49}$$

$$E_{2z}(\rho,t) = -\frac{1}{2\pi\epsilon_0\epsilon}\frac{\pi^{-1/2}}{c^3t_1^3}\left[\left(-\frac{2\tau'}{\rho_1}+\frac{1}{\rho_1^2}\right)e^{-\tau'^2} + \frac{\pi^{1/2}}{2\rho_1^3}(\operatorname{erf}\tau'+1) + \frac{\pi}{2}\frac{2^{1/4}}{(\epsilon\rho_1)^{1/2}}e^{-T'^2/2} \times \begin{Bmatrix} V(0,\sqrt{2}\,T') - \sqrt{2}\,T'\,V(-1,\sqrt{2}\,T') \\ \pi^{-1/2}[U(0,-\sqrt{2}\,T') \\ \qquad + 2\sqrt{2}\,T'\,U(-1,-\sqrt{2}\,T')] \end{Bmatrix}\right]; \quad \begin{cases} T' \geq 0 \\ T' \leq 0 \end{cases}. \tag{14.6.50}$$

As with the δ-function pulse, the omission of small terms with magnitudes smaller than ϵ^{-1} in the frequency-domain formulas must lead to the nonappearance of a second pulse of the form $\exp(-\tau_\epsilon'^2)/\rho_1^2$ where $\tau_\epsilon' = t' - \rho_1\epsilon^{1/2}$. Its amplitude is very much smaller than that of the pulse $\exp(-\tau'^2)/\rho_1^2$ because it is multiplied by a factor of the order ϵ^{-2}; it travels with the phase velocity $c/\epsilon^{1/2}$ instead of $c = 3\times 10^8$ m/sec; and it arrives at any radial distance $\rho = \rho_1 ct_1$ at a time $\Delta t = (\rho/c)(\epsilon^{1/2}-1)$ after the first pulse. Formulas (14.6.48)–(14.6.50) can be expected to be quite accurate before the second pulse arrives, i.e., when $t' < \rho_1\epsilon^{1/2}$ or $T' < \rho_1(\epsilon^{1/2} - 1 + 1/2\epsilon)$; they also give the correct static values as $\tau' \to \infty$, viz., $B_{2\phi}(\rho,t) = 0$ and, since $\operatorname{erf}\infty = 1$ and $ct_1\rho_1 = \rho$, $E_{2z}(\rho,t) = -1/2\pi\epsilon_0\epsilon\rho^3$. Although there is no assurance that the form of (14.6.48)–(14.6.50) is correct when $t > \rho\epsilon^{1/2}/c$ or $T > (\epsilon^{1/2} - 1 + 1/2\epsilon)\rho/c$, the exact evaluation of $E_{2z}(\rho,t)$ in Section 14.7 indicates that, except for the absence of the second pulse, the formulas give correct fields for all values of ρ. As was to be expected from the steady-state behavior, there is no $\rho^{-1/2}$ dependence when $t > \rho\epsilon^{1/2}/c$.

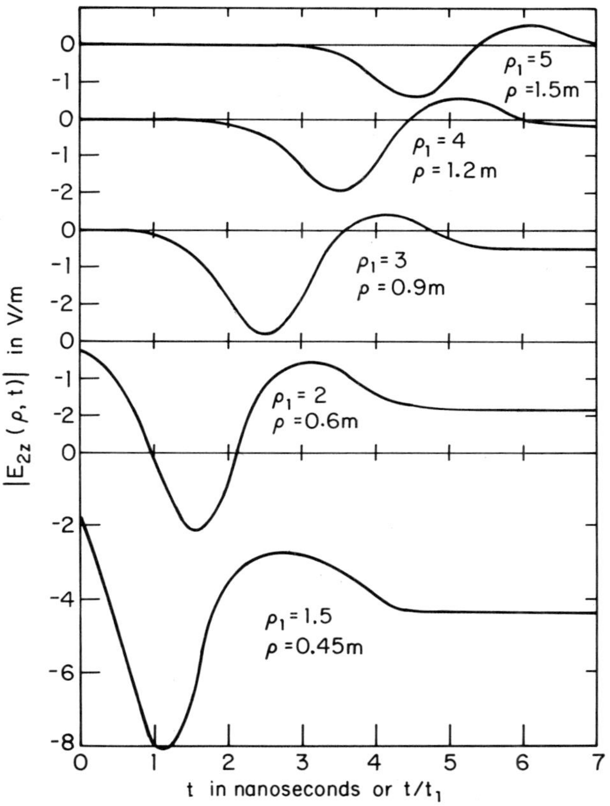

Figure 14.6.1. Approximate vertical electric field on the boundary in air due to a vertical electric dipole on the boundary in a perfect dielectric with relative permittivity $\epsilon = 80$. The dipole is excited by a Gaussian pulse with $t_1 = 1$ nsec.

A graphical representation of the expression in large square brackets in (14.6.50) is shown in Fig. 14.6.1 when the exciting Gaussian pulse has a half-width $t_1 = 1$ nsec, as shown in Fig. 14.6.2. Note that $ct_1 = 0.3$ so that the actual radial distance is $\rho = \rho_1 ct_1 = 0.3\rho_1$. The several curves in Fig. 14.6.1 show the vertical electric field as a function of the time at five radial distances from $\rho = 0.45$ m to $\rho = 1.5$ m. Similar graphs with a different amplitude scale are in Fig. 14.6.3 for $\rho = 4.8$ m and $\rho = 7.5$ m. The shape of the pulse as a function of the time at any given radial distance ρ is determined by the relative contributions from the four terms in (14.6.50) with their very different dependences on both the radial distance and the time.

The third term represents the rapid increase in the vertical electric field as the pulse passes and leaves the final electrostatic field of the charged infinitesimal dipole. This part of the field decreases rapidly as $1/\rho^3$ with distance. The second term is a true Gaussian pulse like that in Fig. 14.6.2.

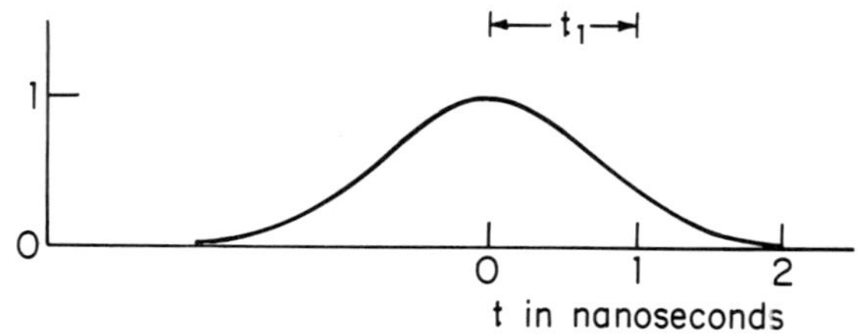

Figure 14.6.2. Gaussian pulse with unit amplitude and half-width $t_1 = 1$ nsec.

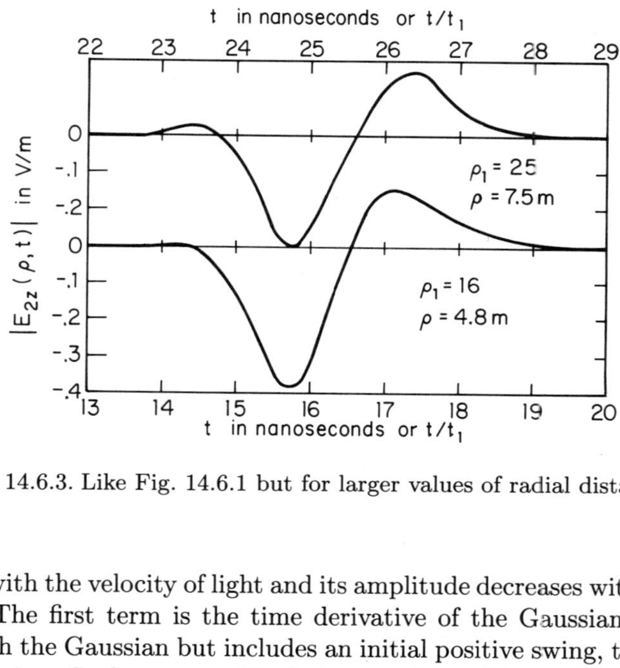

Figure 14.6.3. Like Fig. 14.6.1 but for larger values of radial distance ρ.

It travels with the velocity of light and its amplitude decreases with distance as $1/\rho^2$. The first term is the time derivative of the Gaussian pulse. It travels with the Gaussian but includes an initial positive swing, then a final negative swing. It decreases with distance only as $1/\rho$ so that it persists for much greater distances than the $1/\rho^2$ term. Unlike the field with a δ-function pulse or the steady-state field, the fourth term due to the Fresnel integral does not completely cancel the $1/\rho$ term. It behaves similar to a negative derivative of the Gaussian pulse and cancels a significant part of the first integral but not all of it. However, the Fresnel-integral term decreases only as $\rho^{-1/2}$ so that at sufficiently great distances it is the entire field within the range $t < \rho\epsilon^{1/2}/c$. When $t > \rho\epsilon^{1/2}/c$, the field is presumably not characterized by $\rho^{-1/2}$.

The graphs in Figs. 14.6.1 and 14.6.3 indicate that the time-derivative terms dominate both near and far so that there is little change in the shape of the outward-traveling pulses in the range $t < \rho\epsilon^{1/2}/c$. This is evident especially from Fig. 14.6.3. The amplitudes of the first and fourth terms in (14.6.50) are approximately equal when $\rho = 4.8$ m; the fourth term dominates when ρ is as large as 7.5 m.

In Figs. 14.6.1 and 14.6.3, the second pulse should arrive at $t_2 = \epsilon^{1/2}\rho/c = 9\rho/(3\times 10^8) = 30\rho$ nsec when ρ is in meters. At $\rho = 0.45$ m, $t_2 = 13.5$ nsec; at $\rho = 7.5$ m, $t_2 = 225$ nsec. These times are off the scales in Figs. 14.6.1 and 14.6.3.

14.7 The Exact Transient Field of a Vertical Dipole Excited by a Gaussian Pulse

In the preceding section, the steady-state formulas for the electromagnetic field generated by a unit vertical electric dipole are Fourier transformed to obtain the corresponding field when the dipole is excited by a Gaussian pulse. Since the steady-state formulas are approximate in the sense that small terms of the order $k_2^2/k_1^2 = \epsilon_2/\epsilon_1$ are neglected, the transient formulas are also approximate. In particular, the second pulse due to propagation in the dielectric Region 1 is not represented.

An alternative more accurate approach is to make use of the available exact formulas for the transient field due to a δ-function pulse in order to obtain the transient field due to a Gaussian pulse. This is readily done for E_z. Specifically, for the δ-function pulse, (13.5.19) is

$$[E_{2z}(\rho,t)]_\delta = -\frac{1}{2\pi\epsilon_0\rho^2 c(\epsilon-1)}\left[\epsilon\delta\left(t-\frac{\rho}{c}\right) - \epsilon^{-1/2}\delta\left(t-\frac{\rho\epsilon^{1/2}}{c}\right)\right] - \frac{1}{2\pi\epsilon_0\rho^3(\epsilon+1)}\begin{cases} 0 & ct/\rho < 1 \\ \dfrac{\epsilon}{\epsilon-1}\left[1-\dfrac{\epsilon}{(\epsilon+1)^{3/2}}\left(\dfrac{2c^2t^2}{\rho^2}+\dfrac{\epsilon}{\epsilon+1}\right)\times\left(\dfrac{c^2t^2}{\rho^2}-\dfrac{\epsilon}{\epsilon+1}\right)^{-5/2}\right] & 1 < ct/\rho < \epsilon^{1/2} \\ 1 & ct/\rho > \epsilon^{1/2} \end{cases}; \tag{14.7.1}$$

The exact field with Gaussian excitation is

$$E_{2z}(\rho,t) = \frac{1}{t_1\sqrt{\pi}}\int_{-\infty}^{\infty}[E_{2z}(\rho,t-\zeta)]_\delta e^{-\zeta^2/t_1^2}\,d\zeta, \tag{14.7.2}$$

where t_1 is the half-width of the Gaussian pulse defined by

$$f(t) = \frac{e^{-t^2/t_1^2}}{t_1\sqrt{\pi}}. \tag{14.7.3}$$

This field is conveniently written as the sum of several integrals. Thus,

$$t_1\sqrt{\pi}\,E_{2z}(\rho,t) = -\frac{1}{2\pi\epsilon_0\rho^2 c(\epsilon-1)}\left(\epsilon I_1 - \epsilon^{-1/2} I_2\right)$$
$$-\frac{1}{2\pi\epsilon_0\rho^3(\epsilon+1)}\left\{\begin{matrix} 0 \\ \dfrac{\epsilon}{\epsilon-1}I_3 - \dfrac{\epsilon^2}{(\epsilon-1)(\epsilon+1)^{3/2}}I_4 \\ I_5 \end{matrix}\right\};$$
$$\left\{\begin{matrix} c(t-\zeta)/\rho < 1 \\ 1 < c(t-\zeta)/\rho < \epsilon^{1/2} \\ c(t-\zeta)/\rho > \epsilon^{1/2} \end{matrix}\right. . \tag{14.7.4}$$

Note that the range of ζ from $-\infty$ to ∞ is subdivided into three parts. For the integrals I_3, I_4, and I_5, the two relevant parts are

$$1 < \frac{c(t-\zeta)}{\rho} < \epsilon^{1/2} \quad \text{or} \quad t - \frac{\rho}{c} > \zeta > t - \frac{\rho\epsilon^{1/2}}{c}, \tag{14.7.5a}$$

$$\epsilon^{1/2} < \frac{c(t-\zeta)}{\rho} < \infty \quad \text{or} \quad t - \frac{\rho\epsilon^{1/2}}{c} > \zeta > -\infty. \tag{14.7.5b}$$

The five integrals are

$$I_1 = \int_{-\infty}^{\infty} \delta\left(t - \frac{\rho}{c} - \zeta\right) e^{-\zeta^2/t_1^2}\,d\zeta$$
$$= e^{-(t-\rho/c)^2/t_1^2} = e^{-(t'-\rho_1)^2}, \tag{14.7.6}$$

$$I_2 = \int_{-\infty}^{\infty} \delta\left(t - \frac{\rho\epsilon^{1/2}}{c} - \zeta\right) e^{-\zeta^2/t_1^2}\,d\zeta$$
$$= e^{-(t-\rho\epsilon^{1/2}/c)^2/t_1^2} = e^{-(t'-\rho_1\epsilon^{1/2})^2}, \tag{14.7.7}$$

$$I_3 = \int_{t-\rho\epsilon^{1/2}/c}^{t-\rho/c} e^{-\zeta^2/t_1^2}\,d\zeta$$
$$= t_1\left(\int_0^{(t-\rho/c)/t_1} e^{-x^2}\,dx - \int_0^{(t-\rho\epsilon^{1/2}/c)/t_1} e^{-x^2}\,dx\right)$$
$$= \frac{t_1\sqrt{\pi}}{2}\left[\operatorname{erf}(t'-\rho_1) - \operatorname{erf}(t'-\rho_1\epsilon^{1/2})\right], \tag{14.7.8}$$

$$I_4 = \int_{t-\rho\epsilon^{1/2}/c}^{t-\rho/c} \left(\frac{2c^2(t-\zeta)^2}{\rho^2} + \frac{\epsilon}{\epsilon+1}\right)$$

$$\times \left(\frac{c^2(t-\zeta)^2}{\rho^2} - \frac{\epsilon}{\epsilon+1} \right)^{-5/2} e^{-\zeta^2/t_1^2}\, d\zeta, \tag{14.7.9}$$

$$\begin{aligned} I_5 &= \int_{-\infty}^{t-\rho\epsilon^{1/2}/c} e^{-\zeta^2/t_1^2}\, d\zeta \\ &= t_1 \left(\int_0^{\infty} e^{-x^2}\, dx - \int_0^{-(t-\rho\epsilon^{1/2}/c)/t_1} e^{-x^2}\, dx \right) \\ &= \frac{t_1\sqrt{\pi}}{2}\, [1 + \operatorname{erf}(t' - \rho_1\epsilon^{1/2})], \end{aligned} \tag{14.7.10}$$

where

$$t' = \frac{t}{t_1}; \qquad \rho_1 = \frac{\rho}{ct_1}. \tag{14.7.11a}$$

I_4 cannot be evaluated analytically but is readily evaluated numerically. For this purpose, it is convenient to introduce the additional notation:

$$\zeta' = \frac{\zeta}{t_1}; \qquad a^2 = \frac{\epsilon}{\epsilon+1}. \tag{14.7.11b}$$

With (14.7.11a,b),

$$I_4 = t_1 \int_{t'-\rho_1\epsilon^{1/2}}^{t'-\rho_1} \left(\frac{2(t'-\zeta')^2}{\rho_1^2} + a^2 \right) \left(\frac{(t'-\zeta')^2}{\rho_1^2} - a^2 \right)^{-5/2} e^{-\zeta'^2}\, d\zeta'. \tag{14.7.12}$$

An alternative is with $x = (t' - \zeta')/\rho_1$, $dx = -d\zeta'/\rho_1$, and $\zeta' = t' - \rho_1 x$. The integral is

$$I_4 = \rho_1 t_1 \int_1^{\epsilon^{1/2}} \frac{2x^2 + a^2}{(x^2 - a^2)^{5/2}}\, e^{-(t'-\rho_1 x)^2}\, dx. \tag{14.7.13}$$

This integral has been evaluated numerically and combined with the four integrated formulas to give $E_{2z}(\rho, t) \times t_1\sqrt{\pi}$. The results are shown in Fig. 14.7.1 in a form that parallels the approximate field shown in Fig. 14.6.3. The agreement is seen to be very good. The only difference is a somewhat more rounded minimum and a somewhat sharper maximum in the exact pulse.

The question about the second pulse that travels in the dielectric and arrives at a later time is completely resolved in the exact solution. In Fig. 14.7.2, the time is extended far enough for the second pulse to appear in three of the radial distances. It arrives at the predicted time and has a Gaussian shape with an amplitude that is very much smaller than the first pulse, as expected. The relative amplitude would, of course, be relatively greater for values of permittivity less than $\epsilon = 80$.

The exact solution indicates that the approximate form is useful just as predicted in the range prior to the time of arrival of the second pulse. It

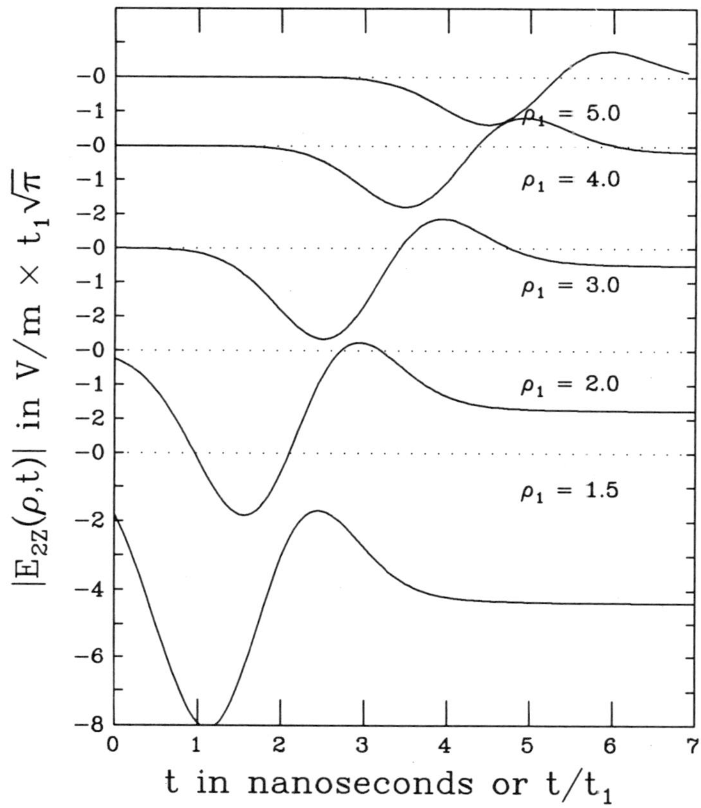

Figure 14.7.1. Exact vertical electric field on the boundary in air due to a vertical electric dipole on the boundary in a perfect dielectric with relative permittivity $\epsilon = 80$. The dipole is excited by a Gaussian pulse with $t_1 = 1$ nsec; $\rho_1 = \rho/ct_1$.

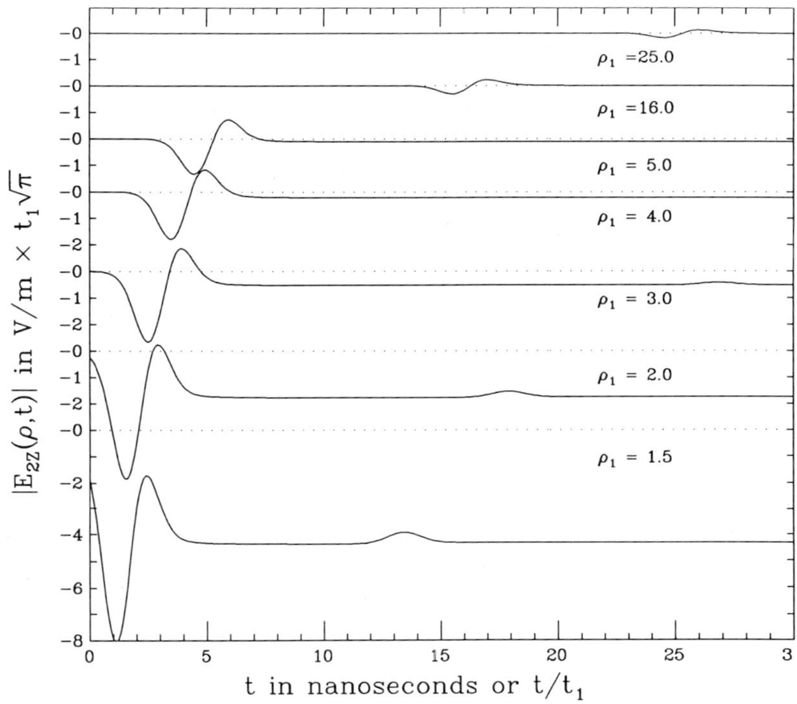

Figure 14.7.2. Like Fig. 14.7.1 but with an extended time scale to show the second pulse on the curves with $\rho_1 = 1.5$, 2.0, and 3.0. It is off the scale for the curves with $\rho_1 = 4.0$ to 25.0.

is now seen that, except for the nonappearance of the second pulse, the approximate form is valid for all times and distances. This is of importance since exact formulas are available only for E_z and B_ϕ so that E_ρ can be evaluated only from the approximate steady-state formula.

14.8 The Approximate Transient Field of a Horizontal Dipole Excited by a Gaussian Pulse

The vertical and radial components of the electric field of a horizontal dipole on the boundary between air (Region 2) and a perfect dielectric (Region 1) with relative permittivity $\epsilon_{1r} = \epsilon$ are readily obtained from the corresponding components of the vertical dipole when the exciting pulse has Gaussian form. With the respective dipoles at $d = 0$ in Region 1, the following formulas apply in the frequency domain:

Vertical Dipole:

$$\begin{aligned}[\tilde{E}_{2\rho}(\rho',\omega)]_v &= [\tilde{E}_{1\rho}(\rho',\omega)]_v \\ &= -\frac{\mu_0}{2\pi\epsilon^{3/2}c}\left\{e^{i\omega\rho'}\left[\frac{i\omega}{\rho'} - \frac{1}{\rho'^2} - \omega^{3/2}\left(\frac{\pi}{\epsilon\rho'}\right)^{1/2} e^{-ip}\mathcal{F}(p)\right]\right. \\ &\qquad \left. - \frac{ie^{i\omega\rho'\epsilon^{1/2}}}{\rho'^2}\right\}, \end{aligned} \tag{14.8.1}$$

$$\begin{aligned}[\tilde{E}_{2z}(\rho',\omega)]_v &= \epsilon[\tilde{E}_{1z}(\rho',\omega)]_v \\ &= \frac{\mu_0}{2\pi\epsilon c}e^{i\omega\rho'}\left[\frac{i\omega}{\rho'} - \frac{1}{\rho'^2} - \frac{i}{\omega\rho'^3}\right. \\ &\qquad \left. - \omega^{3/2}\left(\frac{\pi}{\epsilon\rho'}\right)^{1/2} e^{-ip}\mathcal{F}(p)\right]. \end{aligned} \tag{14.8.2}$$

Horizontal Dipole:

$$\begin{aligned}[\tilde{E}_{2\rho}(\rho',0;\omega)]_h &= [\tilde{E}_{1\rho}(\rho',0;\omega)]_h \\ &= -\frac{\mu_0}{2\pi\epsilon c}\left\{e^{i\omega\rho'}\left[\frac{i\omega}{\rho'} - \frac{1}{\rho'^2} - \frac{i}{\omega\rho'^3}\right.\right. \\ &\qquad \left. - \omega^{3/2}\left(\frac{\pi}{\epsilon\rho'}\right)^{1/2} e^{-ip}\mathcal{F}(p)\right] \\ &\qquad \left. - e^{i\omega\rho'\epsilon^{1/2}}\left(\frac{\epsilon^{1/2}}{\rho'^2} + \frac{i}{\omega\rho'^3}\right)\right\}, \end{aligned} \tag{14.8.3}$$

$$[\tilde{E}_{2z}(\rho', 0; \omega)]_h = \epsilon[\tilde{E}_{1z}(\rho', 0; \omega)]_h$$
$$= \frac{\mu_0}{2\pi\epsilon^{1/2}c}\left\{e^{i\omega\rho'}\left[\frac{i\omega}{\rho'} - \frac{1}{\rho'^2} - \omega^{3/2}\left(\frac{\pi}{\epsilon\rho'}\right)^{1/2} e^{-ip}\mathcal{F}(p)\right] - \frac{ie^{i\omega\rho'\epsilon^{1/2}}}{\rho'^2}\right\}. \tag{14.8.4}$$

In the above formulas

$$\rho' = \frac{\rho}{c}; \qquad p = \frac{\omega\rho'}{2\epsilon}; \qquad \mathcal{F}(p) = \int_p^\infty \frac{e^{i\tau}}{(2\pi\tau)^{1/2}}\,d\tau. \tag{14.8.5}$$

It is seen that

$$\epsilon^{-1}[\tilde{E}_{2z}(\rho', 0; \omega)]_h = [\tilde{E}_{1z}(\rho', 0; \omega)]_h$$
$$= -[\tilde{E}_{1\rho}(\rho', \omega)]_v = -[\tilde{E}_{2\rho}(\rho', \omega)]_v, \tag{14.8.6}$$

$$[\tilde{E}_{2\rho}(\rho', 0; \omega)]_h = [\tilde{E}_{1\rho}(\rho', 0; \omega)]_h$$
$$= -[\tilde{E}_{2z}(\rho', \omega)]_v + \frac{\mu_0\epsilon^{1/2}}{2\pi c}\,e^{i\omega\rho'_\epsilon}\left(\frac{1}{\rho'^2_\epsilon} + \frac{i}{\omega\rho'^3_\epsilon}\right), \tag{14.8.7}$$

where

$$\rho'_\epsilon = \rho'\epsilon^{1/2} = \frac{\rho\epsilon^{1/2}}{c} = \frac{\rho}{v}. \tag{14.8.8}$$

In (14.8.6) and (14.8.7), the subscripts v and h have been added to denote, respectively, the fields due to the vertical and horizontal dipoles. In (14.8.8), $v = c\epsilon^{-1/2}$ is the velocity of propagation in the dielectric Region 1.

Since the transient components E_ρ and E_z with Gaussian pulse excitation are given in (14.6.49) and (14.6.50) for the vertical dipole, the corresponding components for the horizontal dipole are readily obtained from (14.8.6) and (14.8.7). Specifically for the horizontal dipole,

$$\epsilon^{-1}[E_{2z}(\rho, 0; t)]_h$$
$$= [E_{1z}(\rho, 0; t)]_h = -[E_{2\rho}(\rho, t)]_v$$
$$= -\frac{1}{2\pi\epsilon_0\epsilon^{3/2}}\frac{\pi^{-1/2}}{c^3t_1^3}\left[\left(-\frac{2\tau'}{\rho_1} + \frac{1}{\rho_1^2}\right)e^{-\tau'^2} + \frac{\pi}{2}\frac{2^{1/4}}{(\epsilon\rho_1)^{1/2}}e^{-T'^2/2} \times \begin{Bmatrix} V(0, \sqrt{2}\,T') - \sqrt{2}\,T'\,V(-1, \sqrt{2}\,T') \\ \pi^{-1/2}[U(0, -\sqrt{2}\,T') \\ \quad + 2\sqrt{2}\,T'\,U(-1, -\sqrt{2}\,T')] \end{Bmatrix}\right]; \quad \begin{cases} T' \geq 0 \\ T' \leq 0 \end{cases}, \tag{14.8.9}$$

where

$$\rho_1 = \frac{\rho'}{t_1} = \frac{\rho}{ct_1}; \qquad \tau' = \frac{t-\rho/c}{t_1}; \qquad T' = \frac{t - \dfrac{\rho}{c}\left(1-\dfrac{1}{2\varepsilon}\right)}{t_1}, \tag{14.8.10}$$

and V and U are the parabolic cylinder functions introduced in (14.6.36) and (14.6.37).

Similarly, with (14.6.16) and (14.6.21) with ρ'_ϵ substituted for ρ' in (14.6.14), it follows that

$$\begin{aligned}[E_{2\rho}(\rho',0;t)]_h &= [E_{1\rho}(\rho',0;t)]_h \\ &= -[E_{2z}(\rho',t)]_v + \frac{\mu_0\epsilon^{1/2}}{2\pi c}\left(\frac{I_{2\epsilon}}{\rho_\epsilon'^2} + \frac{I_{3\epsilon}}{\rho_\epsilon'^3}\right),\end{aligned} \tag{14.8.11}$$

where

$$I_{2\epsilon} = \frac{1}{t_1\sqrt{\pi}}e^{-\tau_\epsilon'^2}; \qquad I_{3\epsilon} = \tfrac{1}{2}(1+\operatorname{erf}\tau'_\epsilon), \tag{14.8.12}$$

and $\tau'_\epsilon = (t-\rho'_\epsilon)/t_1$ with $\rho'_\epsilon = \rho\epsilon^{1/2}/c$. With $\mu_0 = 1/c^2\epsilon_0$ and (14.6.50), (14.8.11) becomes

$$\begin{aligned}[E_{2\rho}(\rho,0;t)]_h &= [E_{1\rho}(\rho,0;t)]_h \\ &= \frac{1}{2\pi\epsilon_0\epsilon}\frac{\pi^{-1/2}}{c^3t_1^3} \\ &\quad\times\Bigg[\left(-\frac{2\tau'}{\rho_1}+\frac{1}{\rho_1^2}\right)e^{-\tau'^2} + \frac{\pi^{1/2}}{2\rho_1^3}(\operatorname{erf}\tau'+1) \\ &\quad + \epsilon^{3/2}\left(\frac{e^{-\tau_\epsilon'^2}}{\rho_{1\epsilon}^2} + \frac{\pi^{1/2}}{2\rho_{1\epsilon}^3}(\operatorname{erf}\tau'_\epsilon+1)\right) \\ &\quad + \frac{\pi}{2}\frac{2^{1/4}}{(\epsilon\rho_1)^{1/2}}e^{-T'^2/2} \\ &\quad \times\left\{\begin{array}{l} V(0,\sqrt{2}T') - \sqrt{2}T'V(-1,\sqrt{2}T') \\ \pi^{-1/2}[U(0,-\sqrt{2}T') \\ \qquad + 2\sqrt{2}T'U(-1,-\sqrt{2}T')] \end{array}\right\}\Bigg]; \\ &\qquad\qquad \left\{\begin{array}{l} T'\geq 0 \\ T'\leq 0 \end{array}\right., \end{aligned} \tag{14.8.13}$$

where

$$\rho_1 = \frac{\rho}{ct_1}; \quad \rho_{1\epsilon} = \frac{\rho\epsilon^{1/2}}{ct_1}; \qquad \tau' = \frac{t-\rho/c}{t_1}; \quad \tau'_\epsilon = \frac{t-\rho\epsilon^{1/2}/c}{t_1}. \tag{14.8.14}$$

Note that $[E_{1z}(\rho,0;t)]_h$ consists of a pulse that travels in the air and is characterized by $\rho_1 = \rho/ct_1$ and $\tau' = (t-\rho/c)/t_1$, whereas $[E_{1\rho}(\rho,0;t)]_h$

consists of both a pulse that travels in the air and a later pulse that travels in the dielectric characterized by $\rho_{1\epsilon} = \rho\epsilon^{1/2}/ct_1$ and $\tau'_\epsilon = (t - \rho\epsilon^{1/2}/c)/t_1$. In both cases, the pulse in the air includes as the dominant term the derivative $[-2\tau' \exp(-\tau'^2)]$ of the Gaussian pulse $\exp(-\tau'^2)$. The derivative term has the radial amplitude $1/\rho$ which makes it the leading term except close to the source where the Gaussian pulse with its $1/\rho^2$ amplitude is greatest. Note that the pulse in the dielectric has no derivative term with a $1/\rho$ amplitude. Its leading term is the Gaussian pulse $\exp(-\tau_\epsilon'^2)$ with a $1/\rho^2$ amplitude. After passing a given point, both pulses contribute to the static field in the error-function terms with the amplitude factor $1/\rho^3$.

The Fresnel-integral term in the steady-state field contributes the terms that involve the parabolic cylinder functions U and V to the transient field. Although this has the amplitude factor $1/\rho^{1/2}$, this combines with the $1/\rho$ term to leave only a negligibly small contribution to the field at significant distances ρ.

The graphs in Figs. 14.6.1 and 14.6.3 for $[E_{2z}(\rho,t)]_v = [E_{1z}(\rho,t)]_v$ also apply to the lateral-wave part of $[E_{1\rho}(\rho,0;t)]_h$; the same is true of Figs. 14.7.1 and 14.7.2. The Gaussian pulse in the dielectric is larger than the pulse in the air near the source but it decreases more rapidly with distance so that the pulse in the air with its derivative term dominates the transient field except quite close to the source.

The leading terms of the transient component $[E_{2\phi}(\rho,\pi/2;t)]_h = [E_{1\phi}(\rho,\pi/2;t)]_h$ are readily obtained from (14.5.2). Thus,

$$E_{2\phi}(\rho',\pi/2;t) = \frac{\mu_0}{2\pi\epsilon c}\left[\frac{2I_2}{\rho'^2} + \frac{2I_3}{\rho'^3} - \epsilon^{3/2}\left(\frac{I_{2\epsilon}}{\rho_\epsilon'^2} + \frac{I_{3\epsilon}}{\rho_\epsilon'^3}\right)\right], \tag{14.8.15}$$

where the integrals I are given in (14.6.16), (14.6.21), and (14.8.12). With them,

$$E_{2\phi}(\rho,\pi/2;t) = \frac{1}{2\pi\epsilon_0\epsilon}\frac{\pi^{-1/2}}{c^3t_1^3}\left[\frac{2e^{-\tau'^2}}{\rho_1^2} + \frac{\pi^{1/2}}{\rho_1^3}(\operatorname{erf}\tau' + 1) - \epsilon^{3/2}\left(\frac{e^{-\tau_\epsilon'^2}}{\rho_{1\epsilon}^2} + \frac{\pi^{1/2}}{\rho_{1\epsilon}^3}(\operatorname{erf}\tau'_\epsilon + 1)\right)\right]. \tag{14.8.16}$$

The contribution to the transient field in (14.8.16) by the Fresnel-integral term in the steady-state formula has been omitted since it is small.

15 The Propagation of Signals Along a Three-Layered Region: Open Microstrip

15.1 Introduction

A useful type of electric circuit consists of conducting paths in the form of thin copper strips on a dielectric layer or substrate that is bounded on one side by a half-space of air and on the other side by a conducting sheet that is thick compared to the skin depth. The substrate is bonded to the conducting sheet and the combination is generally known as microstrip. Microstrip circuits, including especially transmission lines and antennas, are printed or etched on the substrate and, at appropriate points, connected through it to the conducting sheet. They are of primary importance in the frequency band from 0.1 to 100 GHz.

Microstrip transmission lines have been a subject of theoretical and experimental investigation for more than a quarter century. It is well known that they do not support a TEM mode (Wu 1957; Shafer 1957) as do the familiar coaxial and open-wire lines; even TE and TM modes of conventional type are not possible. As a consequence, the accurate analysis of microstrip transmission lines is difficult. A principal approach was introduced by Wu (1957) in the form of a Fourier transform. A generalization of this method to include the application of various integral transforms, such as the Fourier and Hankel transforms, to boundary-value problems like microstrip is commonly referred to as the "spectral-domain approach" (Jansen 1985). There is a very extensive literature on this and related subjects (Denlinger 1971; Fujiki, Hayashi, and Suzuki 1972; Kowalski and

Pregla 1972; Itoh and Mittra 1973; Kuester and Chang 1979; Hashimoto 1985). Unfortunately, the application of an integral transform and the subsequent solution of the closed-form integral equation by numerical methods are generally complicated and time-consuming and not conveniently used with practical circuits. Alternative approximate procedures include, for example, (1) the determination of the effective permittivity of an "equivalent" single-dielectric TEM-mode transmission line with a phase velocity equal to that of the microstrip line (Deibele and Beyer 1987), and (2) the assumption that "the quasi-TEM mode on microstrip is primarily a single longitudinal-section electric mode" combined with the introduction of an appropriate model with a U-shaped cross section that can be analyzed and that has dispersion characteristics resembling those of microstrip (Getsinger 1973). These considerations are no less significant for transient signals and pulses (Veghte and Balanis 1986). Similar problems of dispersion and coupling occur in the analysis of microstrip antennas (Carver and Mink 1981; Mailloux, McIlvenna, and Kernweis 1981).

It is, of course, well known that a dielectric-coated conducting plane supports a surface wave that propagates in both the dielectric and the air (Attwood 1951). Such surface waves are usually derived as possible solutions of Maxwell's equations when the appropriate conditions are imposed. These include especially the requirement that the electromagnetic field in the air be exponentially attenuated in the direction perpendicular to the surface of the dielectric. The phase velocity of propagation along the surface in both air and the dielectric is intermediate to the velocity c in air and $c\epsilon^{-1/2}$ in the dielectric.

The rigorous approach to the analysis of microstrip circuits and the resolution of perplexing problems of dispersion and coupling must necessarily follow the path already taken in the spectral-domain formulation with a complete satisfaction of all boundary conditions. However, there is an alternative to attempting to solve the resulting complicated and physically obscure integrals numerically. It is to seek a fundamental understanding of the nature of electromagnetic-wave propagation in microstrip by obtaining simple analytical approximations of the integrals in a manner similar to that carried out for the problem of two electrically different half-spaces.

Actually microstrip is a special three-layered region and some of its properties can be investigated in terms of the electromagnetic field generated by a unit horizontal electric dipole on the air–dielectric surface. The problem resembles the three-layered problem analyzed in Chapter 12 with application to the sea, sediment, and crustal rock, but the ratios of the wave numbers are very different since in microstrip the dipole is not in the conducting region (sea water) on its boundary with a layer of sediment, but on the boundary between the dielectric layer and the air.

15.2 The Horizontal Electric Dipole on or above the Air–Dielectric Boundary

The circuit to be treated is illustrated in Fig. 15.2.1. It consists of a unit electric dipole [electric moment $I_x(0)h_e = 1$ A m] at a height d' above the air–dielectric boundary with its axis parallel to the x-axis. The air is Region 0 ($z \leq 0$, wave number $k_0 = \omega/c$), the dielectric is Region 1 ($0 \leq z \leq l$, wave number $k_1 = \beta_1 + i\alpha_1$, $\alpha_1 \ll \beta_1$), and the conducting half-space is Region 2 ($l \leq z < \infty$, wave number $k_2 = \beta_2 + i\alpha_2$, $\alpha_2 \sim \beta_2$). In typical applications, $\sigma_2 \sim 1.7 \times 10^6$ S/m and $l \sim 0.1$ cm, $\epsilon_{1r} \sim 2-4$, and $f \sim 0.8$ GHz, or $l \sim 0.66$ mm, $\epsilon_{1r} \sim 9$, and $f \sim 0.5-7$ GHz. In millimeter-wave applications, f can be as high as 100 GHz.

The determination of the electromagnetic field differs from that in Chapter 12 since the relative magnitudes of the three wave numbers, k_0 for air, k_1 for the dielectric substrate, and k_2 for the conducting sheet, are inverted. Specifically,

$$k_0^2 \ll |k_1^2| \ll |k_2^2|, \tag{15.2.1}$$

with the source dipole in the medium with the smallest wave number (air, k_0) instead of the largest (salt water) as in Chapter 12.

Of primary interest in microstrip circuits and antennas is the electric field at the air–dielectric boundary, i.e., at $z = 0$, and in the air above it when the horizontal dipole is on the boundary at $d' = 0$. Note that the current-carrying conductors in printed circuits are located on the air surface of the substrate with one side in contact with the dielectric and the other in contact with the air. Also present are vertical conductors that connect those on the air–dielectric surface to the ground plate. In the following analysis, the unit horizontal dipole is assumed to be in the air at a height d' above the surface of the dielectric. For applications in microstrip, $d' = 0$.

Because the tangential components of the electric field are continuous across the boundary, it follows that in the cylindrical coordinates ρ, ϕ, z, $E_{0\rho}(\rho, \phi, 0) = E_{1\rho}(\rho, \phi, 0)$ and $E_{0\phi}(\rho, \phi, 0) = E_{1\phi}(\rho, \phi, 0)$. However, $k_0^2 E_{0z}(\rho, \phi, 0) = k_1^2 E_{1z}(\rho, \phi, 0)$. These components are readily obtained from (5.5.1)–(5.5.21) with the subscript 1 changed to 0. Since the positive z-direction in these formulas is into Region 1, the corresponding direction when the subscripts are changed is $z' = -z$ into Region 0. The associated angle is $\phi' = -\phi$. The dipole is at the distance d' from the boundary in Region 0. The desired formulas involve a negative instead of positive image and the coefficients $Q_3 + 1$ and $P_3 - 1$ instead of $Q_3 - 1$ and $P_3 + 1$. They are

$$E_{0\rho} = -\frac{\omega\mu_0}{4\pi k_0^2} \cos\phi' \, [F_{\rho 0}(\rho, z' - d') - F_{\rho 0}(\rho, z' + d') + F_{\rho 1}(\rho, z' + d')], \tag{15.2.2}$$

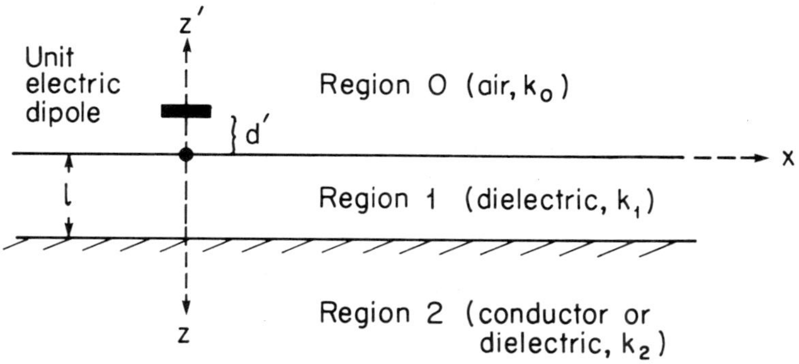

Figure 15.2.1. Unit electric dipole at height d' over plane boundary $z = 0$ between air and a sheet of dielectric with thickness l over a conducting (or dielectric) half-space.

where

$$F_{\rho 0}(\rho, z' - d') = \int_0^\infty \left(\frac{\gamma_0}{2} \left[J_0(\lambda\rho) - J_2(\lambda\rho) \right] + \frac{k_0^2}{2\gamma_0} \left[J_0(\lambda\rho) + J_2(\lambda\rho) \right] \right) e^{i\gamma_0 |z' - d'|} \lambda \, d\lambda, \tag{15.2.3}$$

$$F_{\rho 0}(\rho, z' + d') = \int_0^\infty \left(\frac{\gamma_0}{2} \left[J_0(\lambda\rho) - J_2(\lambda\rho) \right] + \frac{k_0^2}{2\gamma_0} \left[J_0(\lambda\rho) + J_2(\lambda\rho) \right] \right) e^{i\gamma_0 (z' + d')} \lambda \, d\lambda, \tag{15.2.4}$$

$$F_{\rho 1}(\rho, z' + d') = \int_0^\infty \left(\frac{\gamma_0}{2} (Q_3 + 1) \left[J_0(\lambda\rho) - J_2(\lambda\rho) \right] - \frac{k_0^2}{2\gamma_0} (P_3 - 1) \left[J_0(\lambda\rho) + J_2(\lambda\rho) \right] \right) \times e^{i\gamma_0 (z' + d')} \lambda \, d\lambda; \tag{15.2.5}$$

$$E_{0\phi'} = \frac{\omega\mu_0}{4\pi k_0^2} \sin\phi' \left[F_{\phi' 0}(\rho, z' - d') - F_{\phi' 0}(\rho, z' + d') + F_{\phi' 1}(\rho, z' + d') \right], \tag{15.2.6}$$

where

$$F_{\phi' 0}(\rho, z' - d') = \int_0^\infty \left(\frac{\gamma_0}{2} \left[J_0(\lambda\rho) + J_2(\lambda\rho) \right] \right.$$

$$+ \frac{k_0^2}{2\gamma_0}\left[J_0(\lambda\rho) - J_2(\lambda\rho)\right]\Bigg) e^{i\gamma_0|z'-d'|}\lambda\, d\lambda, \tag{15.2.7}$$

$$F_{\phi'0}(\rho, z'+d') = \int_0^\infty \Bigg(\frac{\gamma_0}{2}\left[J_0(\lambda\rho) + J_2(\lambda\rho)\right] + \frac{k_0^2}{2\gamma_0}\left[J_0(\lambda\rho) - J_2(\lambda\rho)\right]\Bigg) e^{i\gamma_0(z'+d')}\lambda\, d\lambda, \tag{15.2.8}$$

$$F_{\phi'1}(\rho, z'+d') = \int_0^\infty \Bigg(\frac{\gamma_0}{2}(Q_3+1)\left[J_0(\lambda\rho) + J_2(\lambda\rho)\right] - \frac{k_0^2}{2\gamma_0}(P_3-1)\left[J_0(\lambda\rho) - J_2(\lambda\rho)\right]\Bigg) \times e^{i\gamma_0(z'+d')}\lambda\, d\lambda; \tag{15.2.9}$$

$$E_{0z'} = \frac{i\omega\mu_0}{4\pi k_0^2}\cos\phi'\ \left[F_{z'0}(\rho, z'-d') - F_{z'0}(\rho, z'+d') + F_{z'1}(\rho, z'+d')\right], \tag{15.2.10}$$

where

$$F_{z'0}(\rho, z'-d') = \pm\int_0^\infty J_1(\lambda\rho) e^{i\gamma_0|z'-d'|}\lambda^2\, d\lambda; \quad \begin{cases} z' > d' \\ 0 \le z' \le d' \end{cases}, \tag{15.2.11}$$

$$F_{z'0}(\rho, z'+d') = \int_0^\infty J_1(\lambda\rho) e^{i\gamma_0(z'+d')}\lambda^2\, d\lambda, \tag{15.2.12}$$

$$F_{z'1}(\rho, z'+d') = \int_0^\infty (Q_3+1) J_1(\lambda\rho) e^{i\gamma_0(z'+d')}\lambda^2\, d\lambda; \tag{15.2.13}$$

$$B_{0\rho} = -\frac{\mu_0}{4\pi}\sin\phi'\ \left[G_{\rho 0}(\rho, z'-d') - G_{\rho 0}(\rho, z'+d') + G_{\rho 1}(\rho, z'+d')\right], \tag{15.2.14}$$

$$B_{0\phi'} = -\frac{\mu_0}{4\pi}\cos\phi'\ \left[G_{\phi'0}(\rho, z'-d') - G_{\phi'0}(\rho, z'+d') + G_{\phi'1}(\rho, z'+d')\right], \tag{15.2.15}$$

where

$$G_{\rho 0}(\rho, z'-d') = G_{\phi'0}(\rho, z'-d') = \pm\int_0^\infty J_0(\lambda\rho) e^{i\gamma_0|z'-d'|}\lambda\, d\lambda; \quad \begin{cases} z' > d' \\ 0 \le z' \le d' \end{cases}, \tag{15.2.16}$$

$$G_{\rho 0}(\rho, z'+d') = G_{\phi' 0}(\rho, z'+d') = \int_0^\infty J_0(\lambda\rho) e^{i\gamma_0(z'+d')}\lambda\, d\lambda, \tag{15.2.17}$$

$$\left.\begin{matrix} G_{\rho 1}(\rho, z'+d') \\ G_{\phi' 1}(\rho, z'+d') \end{matrix}\right\} = \int_0^\infty \{\tfrac{1}{2}(Q_3+1)[J_0(\lambda\rho) \pm J_2(\lambda\rho)] - \tfrac{1}{2}(P_3-1)[J_0(\lambda\rho) \mp J_2(\lambda\rho)]\} e^{i\gamma_0(z'+d')}\lambda\, d\lambda; \tag{15.2.18}$$

$$B_{0z'} = \frac{i\mu_0}{4\pi}\sin\phi'\,[G_{z'0}(\rho, z'-d') - G_{z'0}(\rho, z'+d') + G_{z'1}(\rho, z'+d')], \tag{15.2.19}$$

where

$$G_{z'0}(\rho, z'-d') = \int_0^\infty \gamma_0^{-1} J_1(\lambda\rho) e^{i\gamma_0|z'-d'|}\lambda^2\, d\lambda, \tag{15.2.20}$$

$$G_{z'0}(\rho, z'+d') = \int_0^\infty \gamma_0^{-1} J_1(\lambda\rho) e^{i\gamma_0(z'+d')}\lambda^2\, d\lambda, \tag{15.2.21}$$

$$G_{z'1}(\rho, z'+d') = -\int_0^\infty (P_3-1)\gamma_0^{-1} J_1(\lambda\rho) e^{i\gamma_0(z'+d')}\lambda^2\, d\lambda. \tag{15.2.22}$$

In these formulas, $F_{m0}(\rho, z'-d')$ and $G_{m0}(\rho, z'-d')$, $m = \rho$, ϕ', and z', represent the direct field in Region 0 of the horizontal electric dipole at $(0, 0, d')$, and $F_{m0}(\rho, z'+d')$ and $G_{m0}(\rho, z'+d')$ represent the ideal reflected field or the field of an image dipole at $(0, 0, -d')$. The exact expressions for these fields are given in (5.5.23)–(5.5.27) with the appropriate changes in notation. They are

$$F_{\rho 0}(\rho, z' \pm d') = -e^{ik_0 r}\left[\frac{2k_0}{r^2} + \frac{2i}{r^3} + \left(\frac{z' \pm d'}{r}\right)^2 \left(\frac{ik_0^2}{r} - \frac{3k_0}{r^2} - \frac{3i}{r^3}\right)\right], \tag{15.2.23}$$

$$F_{\phi' 0}(\rho, z' \pm d') = -e^{ik_0 r}\left(\frac{ik_0^2}{r} - \frac{k_0}{r^2} - \frac{i}{r^3}\right), \tag{15.2.24}$$

$$F_{z'0}(\rho, z' \pm d') = -e^{ik_0 r}\left(\frac{\rho}{r}\right)\left(\frac{z' \pm d'}{r}\right)\left(\frac{k_0^2}{r} + \frac{3ik_0}{r^2} - \frac{3}{r^3}\right), \tag{15.2.25}$$

$$G_{\rho 0}(\rho, z' \pm d') = G_{\phi' 0}(\rho, z' \pm d') = -e^{ik_0 r}\left(\frac{z' \pm d'}{r}\right)\left(\frac{ik_0}{r} - \frac{1}{r^2}\right), \tag{15.2.26}$$

$$G_{z'0}(\rho, z' \pm d') = -e^{ik_0 r}\left(\frac{\rho}{r}\right)\left(\frac{k_0}{r} + \frac{i}{r^2}\right). \tag{15.2.27}$$

In these formulas, $r = r_2 = [\rho^2 + (z' + d')^2]^{1/2}$ for the upper signs and $r = r_1 = [\rho^2 + (z' - d')^2]^{1/2}$ for the lower signs.

The formulas for $F_{m1}(\rho, z' + d')$ and $G_{m1}(\rho, z' + d')$, $m = \rho$, ϕ', and z', involve

$$\gamma_0 = (k_0^2 - \lambda^2)^{1/2}, \tag{15.2.28}$$

and the coefficients

$$Q_3 = -f_{er}; \qquad P_3 = -f_{mr}, \tag{15.2.29}$$

where f_{er} and f_{mr} are given by (11.6.6) and (11.6.7). Specifically,

$$f_r = \frac{x_0 - x_2 - i[(x_0 x_2/x_1) - x_1] \tan \gamma_1 l}{x_0 + x_2 - i[(x_0 x_2/x_1) + x_1] \tan \gamma_1 l} \tag{15.2.30}$$

and

$$f_{er} = f_r \quad \text{with } x_j = \frac{\gamma_j}{k_j^2}; \quad j = 0,\, 1,\, 2, \tag{15.2.31}$$

$$f_{mr} = f_r \quad \text{with } x_j = \gamma_j; \quad j = 0,\, 1,\, 2. \tag{15.2.32}$$

When the general formulas for Q_3 and P_3 are used in (15.2.5) and (15.2.9), they lead to integrals for the electromagnetic field that are complicated. They can be greatly simplified if an approximation is made at the outset that does not alter the fundamental properties. The approximation is to assume that the highly conducting base—Region 2—is perfectly conducting, i.e., $k_2 \sim \infty$. This means that for Q_3 and the reflection coefficient f_{er}, $x_2 = \gamma_2/k_2^2 \to 0$, whereas for P_3 and f_{mr}, $x_2 = \gamma_2 = (k_2^2 - \lambda^2)^{1/2} \to \infty$. Thus, with (15.2.29) and (15.2.30)–(15.2.32),

$$Q_3 = -\frac{x_0 + i x_1 \tan \gamma_1 l}{x_0 - i x_1 \tan \gamma_1 l}; \qquad x_j = \frac{\gamma_j}{k_j^2}, \tag{15.2.33}$$

$$P_3 = \frac{x_1 + i x_0 \tan \gamma_1 l}{x_1 - i x_0 \tan \gamma_1 l}; \qquad x_j = \gamma_j. \tag{15.2.34}$$

Note that when $l \to \infty$ with $\alpha_1 > 0$, $\tan \gamma_1 l \to i$ and these reduce to the values for the two-layered region, viz.,

$$Q_3 \to \frac{x_1 - x_0}{x_1 + x_0} = \frac{k_0^2\gamma_1 - k_1^2\gamma_0}{k_0^2\gamma_1 + k_1^2\gamma_0}; \qquad P_3 \to \frac{x_1 - x_0}{x_1 + x_0} = \frac{\gamma_1 - \gamma_0}{\gamma_1 + \gamma_0}. \tag{15.2.35}$$

The quantities $Q_3 + 1$ and $P_3 - 1$ are

$$Q_3 + 1 = -\frac{2 i x_1 \tan \gamma_1 l}{x_0 - i x_1 \tan \gamma_1 l} = \frac{2\gamma_1 k_0^2 K}{k_1^2 \gamma_0 + k_0^2 \gamma_1 K}, \tag{15.2.36}$$

$$P_3 - 1 = \frac{2 i x_0 \tan \gamma_1 l}{x_1 - i x_0 \tan \gamma_1 l} = -\frac{2\gamma_0 K}{\gamma_1 + \gamma_0 K}, \tag{15.2.37}$$

where

$$K = -i \tan \gamma_1 l. \tag{15.2.38}$$

15.3 Evaluation of the Integrals for the Field of the Horizontal Dipole over Microstrip

The integrals to be evaluated are in the functions $F_{m1}(\rho, z'+d')$ and $G_{m1}(\rho, z'+d')$, $m = \rho,\ \phi'$, and z', with the specific values of Q_3+1 and P_3-1 given in (15.2.36) and (15.2.37). They are

$$F_{\rho 1}(\rho, z'+d') = F_{\rho 2}(\rho, z'+d') + F_{\rho 3}(\rho, z'+d'), \tag{15.3.1}$$

$$F_{\phi' 1}(\rho, z'+d') = F_{\phi' 2}(\rho, z'+d') + F_{\phi' 3}(\rho, z'+d'), \tag{15.3.2}$$

with

$$\left.\begin{array}{l} F_{\rho 2}(\rho, z'+d') \\ F_{\phi' 2}(\rho, z'+d') \end{array}\right\} = \frac{k_0^2}{k_1^2}\int_0^\infty \frac{\gamma_0\gamma_1 K}{\gamma_0 + (k_0^2/k_1^2)\gamma_1 K} \times [J_0(\lambda\rho) \mp J_2(\lambda\rho)] e^{i\gamma_0(z'+d')}\lambda\, d\lambda, \tag{15.3.3a}$$

$$\left.\begin{array}{l} F_{\rho 3}(\rho, z'+d') \\ F_{\phi' 3}(\rho, z'+d') \end{array}\right\} = k_0^2\int_0^\infty \frac{K}{\gamma_1 + \gamma_0 K}[J_0(\lambda\rho) \pm J_2(\lambda\rho)] \times e^{i\gamma_0(z'+d')}\lambda\, d\lambda, \tag{15.3.3b}$$

and

$$F_{z'1}(\rho, z'+d') = \frac{2k_0^2}{k_1^2}\int_0^\infty \frac{\gamma_1 K}{\gamma_0 + (k_0^2/k_1^2)\gamma_1 K} J_1(\lambda\rho) \times e^{i\gamma_0(z'+d')}\lambda^2\, d\lambda; \tag{15.3.4}$$

$$G_{\rho 1}(\rho, z'+d') = G_{\rho 2}(\rho, z'+d') + G_{\rho 3}(\rho, z'+d'), \tag{15.3.5}$$

$$G_{\phi' 1}(\rho, z'+d') = G_{\phi' 2}(\rho, z'+d') + G_{\phi' 3}(\rho, z'+d'), \tag{15.3.6}$$

with

$$\left.\begin{array}{l} G_{\rho 2}(\rho, z'+d') \\ G_{\phi' 2}(\rho, z'+d') \end{array}\right\} = \frac{k_0^2}{k_1^2}\int_0^\infty \frac{\gamma_1 K}{\gamma_0 + (k_0^2/k_1^2)\gamma_1 K} \times [J_0(\lambda\rho) \pm J_2(\lambda\rho)] e^{i\gamma_0(z'+d')}\lambda\, d\lambda, \tag{15.3.7a}$$

$$\left.\begin{array}{l} G_{\rho 3}(\rho, z'+d') \\ G_{\phi' 3}(\rho, z'+d') \end{array}\right\} = \int_0^\infty \frac{\gamma_0 K}{\gamma_1 + \gamma_0 K}[J_0(\lambda\rho) \mp J_2(\lambda\rho)] e^{i\gamma_0(z'+d')}\lambda\, d\lambda, \tag{15.3.7b}$$

and

$$G_{z'1}(\rho, z'+d') = 2\int_0^\infty \frac{K}{\gamma_1 + \gamma_0 K} J_1(\lambda\rho) e^{i\gamma_0(z'+d')}\lambda^2\, d\lambda. \tag{15.3.8}$$

In order to carry out the integrations, the coefficients in the integrands can be expanded with the help of the approximate procedure outlined in

Appendix D. That is,

$$f(k_0, k_1; \lambda) \sim f_1(k_0, k_1; \lambda) + f_2(k_0, k_1; \lambda) - f_3(k_0, k_1; \lambda), \tag{15.3.9}$$

where

$$\begin{aligned} f_1(k_0, k_1; \lambda) &= f(k_0, k_1; \lambda \sim k_0), \\ f_2(k_0, k_1; \lambda) &= f(k_0, k_1; \lambda \sim k_1), \end{aligned} \tag{15.3.10a}$$

and

$$f_3(k_0, k_1; \lambda) = f_1(k_0, k_1; \lambda \sim k_1) = f_2(k_0, k_1; \lambda \sim k_0). \tag{15.3.10b}$$

Specifically in (15.3.3a),

$$\begin{aligned} f(k_0, k_1; \lambda) &= \frac{\gamma_0 \gamma_1 K e^{i\gamma_0(z'+d')}}{\gamma_0 + (k_0^2/k_1^2)\gamma_1 K} \\ &= -\frac{i\gamma_0\gamma_1 \tan\gamma_1 l}{\gamma_0 - i(k_0^2/k_1^2)\gamma_1 \tan\gamma_1 l}\, e^{i\gamma_0(z'+d')}, \end{aligned} \tag{15.3.11}$$

$$f_1(k_0, k_1; \lambda) = -\frac{i\gamma_0 k_1 \tan k_1 l}{\gamma_0 - i(k_0^2/k_1)\tan k_1 l}\, e^{i\gamma_0(z'+d')}, \tag{15.3.12}$$

$$f_2(k_0, k_1; \lambda) = \frac{\lambda\gamma_1 \tan\gamma_1 l}{i\lambda - i(k_0^2/k_1^2)\gamma_1 \tan\gamma_1 l}\, e^{-k_1(z'+d')}, \tag{15.3.13}$$

$$f_3(k_0, k_1; \lambda) = \frac{\lambda k_1 \tan k_1 l}{i\lambda - i(k_0^2/k_1)\tan k_1 l}\, e^{-k_1(z'+d')}. \tag{15.3.14}$$

Since $f_2(k_0, k_1; \lambda)$ and $f_3(k_0, k_1; \lambda)$ have the factor $\exp[-k_1(z'+d')]$, they are exponentially highly attenuated upward from the air–dielectric surface and contribute nothing to the field propagating in the radial direction. Since such terms are not of interest, as already seen in Chapter 12, they are hereafter omitted so that

$$f(k_0, k_1; \lambda) \sim f_1(k_0, k_1; \lambda) = \frac{k_1^2}{k_0^2}\, \frac{\gamma_0 \epsilon' e^{i\gamma_0(z'+d')}}{\gamma_0 + \epsilon'}, \tag{15.3.15}$$

where

$$\epsilon' = -\frac{ik_0^2}{k_1}\tan k_1 l \sim -ik_0^2 l = k_0\epsilon. \tag{15.3.16}$$

In (15.3.16), the approximate form assumes that

$$k_1^2 l^2 \ll 1 \tag{15.3.17}$$

so that $\tan k_1 l \sim k_1 l$. Condition (15.3.17) is true in microstrip where the dielectric layer is electrically thin. When (15.3.17) is satisfied, the dimensionless parameter ϵ is small, viz.,

$$|\epsilon| = |-ik_0 l| \ll 1. \tag{15.3.18}$$

Similarly in (15.3.4) and (15.3.7a),

$$\begin{aligned} f(k_0, k_1; \lambda) &= \frac{\gamma_1 K e^{i\gamma_0(z'+d')}}{\gamma_0 + (k_0^2/k_1^2)\gamma_1 K} \\ &= -\frac{i\gamma_1 \tan\gamma_1 l \; e^{i\gamma_0(z'+d')}}{\gamma_0 - i(k_0^2/k_1^2)\gamma_1 \tan\gamma_1 l} \\ &\sim f_1(k_0, k_1; \lambda) = \frac{k_1^2}{k_0^2}\frac{\epsilon'}{\gamma_0 + \epsilon'} e^{i\gamma_0(z'+d')}. \end{aligned} \tag{15.3.19}$$

It follows that

$$\left.\begin{matrix} F_{\rho 2}(\rho, z'+d') \\ F_{\phi' 2}(\rho, z'+d') \end{matrix}\right\} = \epsilon' \int_0^\infty \frac{\gamma_0}{\gamma_0 + \epsilon'} [J_0(\lambda\rho) \mp J_2(\lambda\rho)] e^{i\gamma_0(z'+d')} \lambda \, d\lambda, \tag{15.3.20}$$

$$F_{z'1}(\rho, z'+d') = 2\epsilon' \int_0^\infty \frac{1}{\gamma_0 + \epsilon'} J_1(\lambda\rho) e^{i\gamma_0(z'+d')} \lambda^2 \, d\lambda, \tag{15.3.21}$$

$$\left.\begin{matrix} G_{\rho 2}(\rho, z'+d') \\ G_{\phi' 2}(\rho, z'+d') \end{matrix}\right\} = \epsilon' \int_0^\infty \frac{1}{\gamma_0 + \epsilon'} [J_0(\lambda\rho) \pm J_2(\lambda\rho)] e^{i\gamma_0(z'+d')} \lambda \, d\lambda. \tag{15.3.22}$$

The integrands in (15.3.3b), (15.3.7b), and (15.3.8) are greatly simplified with the condition (15.3.17). Thus,

$$\frac{K}{\gamma_1 + \gamma_0 K} = -\frac{i\tan\gamma_1 l}{\gamma_1 - i\gamma_0 \tan\gamma_1 l} \sim -\frac{i\gamma_1 l}{\gamma_1 - i\gamma_0\gamma_1 l} = -\frac{il}{1 - i\gamma_0 l}. \tag{15.3.23}$$

The denominator in (15.3.23) can be expanded in series as follows:

$$\frac{1}{1 - i\gamma_0 l} = 1 + i\gamma_0 l - \gamma_0^2 l^2 - \cdots. \tag{15.3.24}$$

It follows that in (15.3.3b), (15.3.7b), and (15.3.8),

$$\left.\begin{matrix} F_{\rho 3}(\rho, z'+d') \\ F_{\phi' 3}(\rho, z'+d') \end{matrix}\right\} \sim \epsilon' \int_0^\infty (1 + i\gamma_0 l - \gamma_0^2 l^2)[J_0(\lambda\rho) \pm J_2(\lambda\rho)] \times e^{i\gamma_0(z'+d')} \lambda \, d\lambda, \tag{15.3.25}$$

$$\left.\begin{matrix} G_{\rho 3}(\rho, z'+d') \\ G_{\phi' 3}(\rho, z'+d') \end{matrix}\right\} \sim \frac{\epsilon'}{k_0^2} \int_0^\infty \gamma_0(1 + i\gamma_0 l - \gamma_0^2 l^2)[J_0(\lambda\rho) \mp J_2(\lambda\rho)] \times e^{i\gamma_0(z'+d')} \lambda \, d\lambda, \tag{15.3.26}$$

$$G_{z'1}(\rho, z'+d') \sim \frac{2\epsilon'}{k_0^2} \int_0^\infty (1 + i\gamma_0 l - \gamma_0^2 l^2) J_1(\lambda\rho) e^{i\gamma_0(z'+d')} \lambda^2 \, d\lambda. \tag{15.3.27}$$

The integrals in (15.3.25)–(15.3.27) can be evaluated directly with Appendix A. Thus,

$$\left.\begin{matrix} F_{\rho 3}(\rho, z'+d') \\ F_{\phi' 3}(\rho, z'+d') \end{matrix}\right\} = \epsilon'\left(\left\{\begin{matrix} I_{18}(\rho, z'+d') \\ I_{19}(\rho, z'+d') \end{matrix}\right\} + il\left\{\begin{matrix} I_{11}(\rho, z'+d') \\ I_{12}(\rho, z'+d') \end{matrix}\right\} - k_0^2 l^2\left\{\begin{matrix} I_{18}(\rho, z'+d') \\ I_{19}(\rho, z'+d') \end{matrix}\right\} + l^2\left\{\begin{matrix} I_{22}(\rho, z'+d') \\ I_{23}(\rho, z'+d') \end{matrix}\right\}\right). \tag{15.3.28}$$

Since $k_0 l \ll 1$, the first two terms are adequate. That is,

$$\begin{aligned} F_{\rho 3}(\rho, z'+d') &\sim \epsilon'[I_{18}(\rho, z'+d') + ilI_{11}(\rho, z'+d')] \\ &= 2\epsilon'\left\{\frac{e^{ik_0(z'+d')}}{\rho^2} - \frac{z'+d'}{r_2}\frac{e^{ik_0 r_2}}{\rho^2} + ik_0 l\left[\frac{e^{ik_0(z'+d')}}{\rho^2} + \frac{ie^{ik_0 r_2}}{k_0 r_2^3} - \left(\frac{z'+d'}{\rho}\right)^2 \frac{e^{ik_0 r_2}}{r_2^2}\right]\right\}, \end{aligned} \tag{15.3.29}$$

$$\begin{aligned} F_{\phi' 3}(\rho, z'+d') &\sim \epsilon'[I_{19}(\rho, z'+d') + ilI_{12}(\rho, z'+d')] \\ &= -2\epsilon'\left(\frac{e^{ik_0(z'+d')}}{\rho^2} + \left(\frac{z'+d'}{r_2}\right)\left(\frac{ik_0}{r_2} - \frac{1}{r_2^2} - \frac{1}{\rho^2}\right)e^{ik_0 r_2}\right. \\ &\quad + ik_0 l\left\{\frac{e^{ik_0(z'+d')}}{\rho^2} + \left[\frac{1}{r_2^2} + \frac{2i}{k_0 r_2^3}\right.\right. \\ &\quad \left.\left.\left. + \left(\frac{z'+d'}{r_2}\right)^2\left(\frac{ik_0}{r_2} - \frac{3}{r_2^2} - \frac{3i}{k_0 r_2^3} - \frac{1}{\rho^2}\right)\right]e^{ik_0 r_2}\right\}\right). \end{aligned} \tag{15.3.30}$$

Similarly,

$$\left.\begin{matrix} G_{\rho 3}(\rho, z'+d') \\ G_{\phi' 3}(\rho, z'+d') \end{matrix}\right\} = \frac{\epsilon'}{k_0^2}\left(\left\{\begin{matrix} I_{12}(\rho, z'+d') \\ I_{11}(\rho, z'+d') \end{matrix}\right\} + ilk_0^2\left\{\begin{matrix} I_{19}(\rho, z'+d') \\ I_{18}(\rho, z'+d') \end{matrix}\right\} - il\left\{\begin{matrix} I_{23}(\rho, z'+d') \\ I_{22}(\rho, z'+d') \end{matrix}\right\}\right), \tag{15.3.31}$$

so that

$$\begin{aligned} G_{\rho 3}(\rho, z'+d') &= -\frac{2\epsilon'}{k_0^2}\left(k_0\left\{\frac{e^{ik_0(z'+d')}}{\rho^2} + \left[\frac{1}{r_2^2} + \frac{2i}{k_0 r_2^3}\right.\right.\right. \end{aligned}$$

$$+\left(\frac{z'+d'}{r_2}\right)^2\left(\frac{ik_0}{r_2}-\frac{3}{r_2^2}-\frac{3i}{k_0r_2^3}-\frac{1}{\rho^2}\right)\Bigg]e^{ik_0r_2}\Bigg\}$$

$$-\epsilon'\left[\frac{e^{ik_0(z'+d')}}{\rho^2}+\left(\frac{z'+d'}{r_2}\right)\left(\frac{ik_0}{r_2}-\frac{1}{r_2^2}-\frac{1}{\rho^2}\right)e^{ik_0r_2}\right]$$

$$-\epsilon'\left(\frac{z'+d'}{r_2}\right)\left(-\frac{ik_0}{r_2}+\frac{5}{r_2^2}+\frac{12i}{k_0r_2^3}\right)e^{ik_0r_2}\Bigg), \tag{15.3.32}$$

$$G_{\phi'3}(\rho, z'+d')$$

$$=\frac{2\epsilon'}{k_0^2}\Bigg\{k_0\left[\frac{e^{ik_0(z'+d')}}{\rho^2}+\frac{ie^{ik_0r_2}}{k_0r_2^3}-\left(\frac{z'+d'}{\rho}\right)^2\frac{e^{ik_0r_2}}{r_2^2}\right]$$

$$-\epsilon'\left[\frac{e^{ik_0(z'+d')}}{\rho^2}-\left(\frac{z'+d'}{r_2}\right)\frac{e^{ik_0r_2}}{\rho^2}\right]$$

$$-\epsilon'\left(\frac{z'+d'}{r_2}\right)\left(\frac{1}{r_2^2}+\frac{3i}{k_0r_2^3}-\frac{3}{k_0^2r_2^4}\right)e^{ik_0r_2}\Bigg\}, \tag{15.3.33}$$

and

$$G_{z'1}(\rho, z'+d')$$

$$=\frac{2\epsilon'}{k_0^2}\left(\frac{\rho}{2}\,I_{22}(\rho, z'+d')+il\,\frac{\rho}{2}\,I_{15}(\rho, z'+d')\right)$$

$$=\frac{2\epsilon'}{k_0^2}\Bigg\{-k_0\left(\frac{\rho}{r_2}\right)\left(\frac{z'+d'}{r_2}\right)\left(\frac{k_0}{r_2}+\frac{3i}{r_2^2}-\frac{3}{k_0r_2^3}\right)e^{ik_0r_2}$$

$$+il\left(\frac{\rho}{r_2}\right)\Bigg[\frac{ik_0^2}{r_2^2}-\frac{3k_0}{r_2^3}$$

$$-\left(\frac{z'+d'}{r_2}\right)^2\left(\frac{k_0^3}{r_2}+\frac{6ik_0^2}{r_2^2}-\frac{15k_0}{r_2^3}\right)\Bigg]e^{ik_0r_2}\Bigg\}$$

$$=\frac{2\epsilon'}{k_0}\Bigg\{-\left(\frac{\rho}{r_2}\right)\left(\frac{z'+d'}{r_2}\right)\left(\frac{k_0}{r_2}+\frac{3i}{r_2^2}-\frac{3}{k_0r_2^3}\right)e^{ik_0r_2}$$

$$-\frac{\epsilon'}{k_0}\left(\frac{\rho}{r_2}\right)\Bigg[\frac{i}{r_2^2}-\frac{3}{k_0r_2^3}-\frac{3i}{k_0^2r_2^4}$$

$$-\left(\frac{z'+d'}{r_2}\right)^2\left(\frac{k_0}{r_2}+\frac{6i}{r_2^2}-\frac{15}{k_0r_2^3}\right)\Bigg]e^{ik_0r_2}\Bigg\}. \tag{15.3.34}$$

In order to evaluate the integrals (15.3.20)–(15.3.22), let

$$K(\rho, z'_d)=\int_0^\infty\frac{\gamma_0}{\gamma_0+\epsilon'}\,C(\lambda\rho)e^{i\gamma_0z'_d}\lambda\,d\lambda \tag{15.3.35}$$

and

$$J(\rho, z'_d)=\int_0^\infty\frac{1}{\gamma_0+\epsilon'}\,C(\lambda\rho)e^{i\gamma_0z'_d}\lambda\,d\lambda, \tag{15.3.36}$$

where $z'_d = z' + d'$ and $C(\lambda\rho)$ stands for $[J_0(\lambda\rho) \mp J_2(\lambda\rho)]$, $\lambda J_1(\lambda\rho)$, or $[J_0(\lambda\rho) \pm J_2(\lambda\rho)]$. Then

$$\frac{dJ(\rho, z'_d)}{dz'_d} = iK(\rho, z'_d). \tag{15.3.37}$$

Also,

$$\frac{\gamma_0}{\gamma_0 + \epsilon'} - 1 = -\frac{\epsilon'}{\gamma_0 + \epsilon'} \tag{15.3.38}$$

so that

$$K(\rho, z'_d) = K_0(\rho, z'_d) - \epsilon' J(\rho, z'_d), \tag{15.3.39}$$

where

$$K_0(\rho, z'_d) = \int_0^\infty C(\lambda\rho) e^{i\gamma_0 z'_d} \lambda \, d\lambda. \tag{15.3.40}$$

Hence,

$$-i \frac{dJ(\rho, z'_d)}{dz'_d} = K_0(\rho, z'_d) - \epsilon' J(\rho, z'_d) \tag{15.3.41a}$$

or

$$\frac{dJ(\rho, z'_d)}{dz'_d} + i\epsilon' J(\rho, z'_d) = iK_0(\rho, z'_d). \tag{15.3.41b}$$

This is an ordinary differential equation with the solution

$$J(\rho, z'_d) = -i \int_{z'_d}^\infty e^{i\epsilon'(z - z'_d)} K_0(\rho, z) \, dz. \tag{15.3.42}$$

Let

$$K_1(\rho, z'_d) = \int_0^\infty \frac{1}{\gamma_0} C(\lambda\rho) e^{i\gamma_0 z'_d} \lambda \, d\lambda. \tag{15.3.43}$$

Then

$$\frac{dK_1(\rho, z'_d)}{dz'_d} = iK_0(\rho, z'_d) \tag{15.3.44}$$

and

$$J(\rho, z'_d) = -\int_{z'_d}^\infty e^{i\epsilon'(z - z'_d)} \frac{dK_1(\rho, z)}{dz} \, dz. \tag{15.3.45}$$

This is readily integrated by parts to give

$$J(\rho, z'_d) = K_1(\rho, z'_d) + i\epsilon' \int_{z'_d}^\infty K_1(\rho, z) e^{i\epsilon'(z - z'_d)} \, dz. \tag{15.3.46}$$

Hence,

$$K(\rho, z'_d) = K_0(\rho, z'_d) - \epsilon' K_1(\rho, z'_d) - i\epsilon'^2 \int_{z'_d}^{\infty} K_1(\rho, z) e^{i\epsilon'(z - z'_d)}\, dz. \tag{15.3.47}$$

With

$$K_2(\rho, z'_d) = -i \int_{z'_d}^{\infty} K_1(\rho, z) e^{i\epsilon'(z - z'_d)}\, dz, \tag{15.3.48}$$

it follows that

$$J(\rho, z'_d) = K_1(\rho, z'_d) - \epsilon' K_2(\rho, z'_d) \tag{15.3.49a}$$

and

$$K(\rho, z'_d) = K_0(\rho, z'_d) - \epsilon' K_1(\rho, z'_d) + \epsilon'^2 K_2(\rho, z'_d). \tag{15.3.49b}$$

When (15.3.35) with (15.3.49b) is used in (15.3.20) with $\epsilon' = -ik_0^2 l$, the result is

$$F_2(\rho, z' + d') = \epsilon'[K_0(\rho, z' + d') - \epsilon' K_1(\rho, z' + d') + \epsilon'^2 K_2(\rho, z' + d')] \tag{15.3.50}$$

so that

$$\begin{aligned} F_{\rho 2}(\rho, z' + d') &= \epsilon' \bigg(\int_0^{\infty} [J_0(\lambda\rho) - J_2(\lambda\rho)] e^{i\gamma_0(z'+d')} \lambda\, d\lambda \\ &\qquad - \epsilon' \int_0^{\infty} \frac{1}{\gamma_0} [J_0(\lambda\rho) - J_2(\lambda\rho)] e^{i\gamma_0(z'+d')} \lambda\, d\lambda \\ &\qquad + \epsilon'^2 K_{2\rho}(\rho, z' + d') \bigg) \\ &= \epsilon' I_{19}(\rho, z' + d') - \epsilon'^2 I_4(\rho, z' + d') + \epsilon'^3 K_{2\rho}(\rho, z' + d') \\ &= 2ik_0^2 l \left[\frac{e^{ik_0(z'+d')}}{\rho^2} + \left(\frac{z' + d'}{r_2} \right) \left(\frac{ik_0}{r_2} - \frac{1}{r_2^2} - \frac{1}{\rho^2} \right) e^{ik_0 r_2} \right] \\ &\qquad - 2k_0^3 l^2 \left[\frac{e^{ik_0(z'+d')}}{\rho^2} + \left(\frac{ik_0}{r_2} - \frac{1}{\rho^2} \right) e^{ik_0 r_2} \right] + ik_0^6 l^3 K_{2\rho}(\rho, z' + d'), \end{aligned} \tag{15.3.51}$$

$$\begin{aligned} F_{\phi' 2}(\rho, z' + d') &= \epsilon' \bigg(\int_0^{\infty} [J_0(\lambda\rho) + J_2(\lambda\rho)] e^{i\gamma_0(z'+d')} \lambda\, d\lambda \\ &\qquad - \epsilon' \int_0^{\infty} \frac{1}{\gamma_0} [J_0(\lambda\rho) + J_2(\lambda\rho)] e^{i\gamma_0(z'+d')} \lambda\, d\lambda \end{aligned}$$

$$
\begin{aligned}
&\quad + \epsilon'^2 K_{2\phi'}(\rho, z'+d')\Big) \\
&= \epsilon' I_{18}(\rho, z'+d') - \epsilon'^2 I_3(\rho, z'+d') + \epsilon'^3 K_{2\phi'}(\rho, z'+d') \\
&= -2ik_0^2 l \left(\frac{e^{ik_0(z'+d')}}{\rho^2} - \frac{z'+d'}{r_2}\frac{e^{ik_0 r_2}}{\rho^2} \right) \\
&\quad + 2k_0^3 l^2 \left(\frac{e^{ik_0(z'+d')}}{\rho^2} - \frac{e^{ik_0 r_2}}{\rho^2} \right) + ik_0^6 l^3 K_{2\phi'}(\rho, z'+d'). \qquad (15.3.52)
\end{aligned}
$$

Similarly, with (15.3.36) and (15.3.49a) used in (15.3.21) and (15.3.22),

$$
\begin{aligned}
F_{z'1}(\rho, z'+d') &= 2\epsilon'[K_{1z'}(\rho, z'+d') - \epsilon' K_{2z'}(\rho, z'+d')] \\
&= 2\epsilon' \Big(\int_0^\infty \frac{1}{\gamma_0} J_1(\lambda\rho) e^{i\gamma_0(z'+d')} \lambda^2 \, d\lambda \\
&\qquad - \epsilon' K_{2z'}(\rho, z'+d') \Big) \\
&= 2\epsilon' \left(\frac{\rho}{2} I_7(\rho, z'+d') - \epsilon' K_{2z'}(\rho, z'+d') \right) \\
&= -2\epsilon' \Big[\left(\frac{\rho}{r_2} \right) \left(\frac{k_0}{r_2} + \frac{i}{r_2^2} \right) e^{ik_0 r_2} \\
&\qquad - ik_0^2 l K_{2z'}(\rho, z'+d') \Big], \qquad (15.3.53)
\end{aligned}
$$

and, corresponding to (15.3.50) with the subscript m denoting magnetic-field terms,

$$
G_2(\rho, z'+d') = \epsilon'[K_{1m}(\rho, z'+d') - \epsilon' K_{2m}(\rho, z'+d')] \qquad (15.3.54)
$$

so that

$$
\begin{aligned}
G_{\rho 2}(\rho, z'+d') &= \epsilon' \Big(\int_0^\infty \frac{1}{\gamma_0} [J_0(\lambda\rho) + J_2(\lambda\rho)] e^{i\gamma_0(z'+d')} \lambda \, d\lambda \\
&\qquad - \epsilon' K_{2m\rho}(\rho, z'+d') \Big) \\
&= \epsilon' I_3(\rho, z'+d') - \epsilon'^2 K_{2m\rho}(\rho, z'+d') \\
&= -2ik_0 l \left(\frac{e^{ik_0(z'+d')}}{\rho^2} - \frac{e^{ik_0 r_2}}{\rho^2} \right) \\
&\quad + k_0^4 l^2 K_{2m\rho}(\rho, z'+d'), \qquad (15.3.55) \\
G_{\phi' 2}(\rho, z'+d') &= \epsilon' \Big(\int_0^\infty \frac{1}{\gamma_0} [J_0(\lambda\rho) - J_2(\lambda\rho)] e^{i\gamma_0(z'+d')} \lambda \, d\lambda \\
&\qquad - \epsilon' K_{2m\phi'}(\rho, z'+d') \Big) \\
&= \epsilon' I_4(\rho, z'+d') - \epsilon'^2 K_{2m\phi'}(\rho, z'+d')
\end{aligned}
$$

$$
\begin{aligned}
&= 2ik_0 l\left[\frac{e^{ik_0(z'+d')}}{\rho^2} + \left(\frac{ik_0}{r_2} - \frac{1}{\rho^2}\right)e^{ik_0 r_2}\right] \\
&\quad + k_0^4 l^2 K_{2m\phi'}(\rho, z' + d'). \qquad (15.3.56)
\end{aligned}
$$

The integrals $K_2(\rho, z' + d')$ appearing at the ends of these expressions are defined in general in (15.3.48). That is,

$$
K_2(\rho, z' + d') = -i\int_{z'+d'}^{\infty} K_1(\rho, z)e^{i\epsilon'(z-z'-d')}\,dz, \qquad (15.3.57)
$$

where $K_1(\rho, z' + d')$ is the form appropriate to each component. These are all given in integrated form in (15.3.50)–(15.3.56).

The several integrals $K_2(\rho, z' + d')$ are closely related to the Fresnel integrals evaluated in Appendix F. It is, therefore, convenient to introduce the notation used in Appendix F. This is

$$
\begin{aligned}
&\bar{z} = k_0 z; \qquad \bar{z}' = k_0(z' + d'); \qquad \bar{\rho} = k_0\rho, \\
&\bar{r} = (\bar{\rho}^2 + \bar{z}^2)^{1/2}; \qquad \bar{r}_2 = k_0 r_2 = (\bar{\rho}^2 + \bar{z}'^2)^{1/2}. \qquad (15.3.58)
\end{aligned}
$$

The parameter ϵ' used in this chapter is related to the parameter ϵ used in Appendix F by

$$
\epsilon' = k_0\epsilon, \qquad (15.3.59)
$$

with $-il$ replacing k_1^{-1} and subscripts 2 replaced by 0. In the analysis in Appendix F, it is shown that the integrals involved contribute significantly to the complete field only at large values of r so that only the leading $1/r$ terms need to be retained. Thus, from (15.3.51),

$$
K_{2\rho}(\bar{\rho}, \bar{z}') \sim 2i\int_{\bar{z}'}^{\infty} \frac{i}{\bar{r}}\, e^{i\bar{r}} e^{i\epsilon(\bar{z}-\bar{z}')}\,d\bar{z}. \qquad (15.3.60)
$$

With

$$
\phi(\bar{z}) = \epsilon(\bar{z} - \bar{z}') + \bar{r} - \bar{r}_2, \qquad (15.3.61)
$$

(15.3.60) becomes

$$
K_{2\rho}(\bar{\rho}, \bar{z}') = -2e^{i\bar{r}_2}\int_{\bar{z}'}^{\infty} e^{i\phi(\bar{z})}\,\frac{1}{\bar{r}}\,d\bar{z}. \qquad (15.3.62)
$$

With the Taylor expansion of $\phi(\bar{z})$ in (F37) and $\bar{r}$ approximated by $\bar{r}_2$,

$$
\begin{aligned}
K_{2\rho}(\bar{\rho}, \bar{z}') \sim -\frac{2e^{i\bar{r}_2}}{\bar{r}_2}\int_{\bar{z}'}^{\infty} \exp\Bigg\{i\bigg[&\left(\epsilon + \frac{\bar{z}'}{\bar{r}_2}\right)(\bar{z} - \bar{z}') \\
&+ \frac{\bar{\rho}^2}{2\bar{r}_2^3}(\bar{z} - \bar{z}')^2\bigg]\Bigg\}\,d\bar{z}. \qquad (15.3.63)
\end{aligned}
$$

This is the same as (F38) multiplied by $2i\bar{r}_2/\bar{\rho}$. Hence, the final formula for $K_{2\rho}(\bar{\rho}, \bar{z}')$ is (F42) multiplied by this factor. It is

$$
K_{2\rho}(\bar{\rho}, \bar{z}') \sim -2\left(\frac{\bar{r}_2}{\bar{\rho}}\right)\left(\frac{\pi}{\bar{r}_2}\right)^{1/2} e^{i\bar{r}_2}e^{-iP_2}\mathcal{F}(P_2), \qquad (15.3.64)
$$

where

$$P_2 = \frac{\bar{r}_2}{2}\left(\frac{\epsilon\bar{r}_2 + \bar{z}'}{\bar{\rho}}\right)^2 \tag{15.3.65}$$

and

$$\mathcal{F}(P_2) = \tfrac{1}{2}(1+i) - \int_0^{P_2} \frac{e^{it}}{(2\pi t)^{1/2}}\,dt. \tag{15.3.66}$$

This can be converted to the notation of this chapter as follows:

$$K_{2\rho}(\rho, z'+d') = -2\left(\frac{r_2}{\rho}\right)\left(\frac{\pi}{k_0 r_2}\right)^{1/2} e^{ik_0 r_2} e^{-iP_2}\mathcal{F}(P_2), \tag{15.3.67}$$

$$P_2 = \frac{k_0 r_2}{2}\left(\frac{\epsilon' r_2 + k_0(z'+d')}{k_0\rho}\right)^2 = \frac{k_0 r_2}{2}\left(\frac{-ik_0 l r_2 + z' + d'}{\rho}\right)^2, \tag{15.3.68}$$

$$\epsilon'^3 K_{2\rho}(\rho, z'+d') = -2ik_0^6 l^3\left(\frac{r_2}{\rho}\right)\left(\frac{\pi}{k_0 r_2}\right)^{1/2} e^{ik_0 r_2} e^{-iP_2}\mathcal{F}(P_2). \tag{15.3.69}$$

Since $K_{1\phi'}(\rho, z'+d')$ from (15.3.52) includes no $1/r$ term, the contribution of the integral $K_{2\phi'}(\rho, z'+d')$ to the field is small. It is included for completeness. Thus,

$$\begin{aligned} K_{2\phi'}(\bar{\rho}, \bar{z}') &= -i\int_{\bar{z}'}^{\infty} K_{1\phi'}(\bar{\rho}, \bar{z}) e^{i\epsilon(\bar{z}-\bar{z}')}\,d\bar{z} \\ &= 2i\int_{\bar{z}'}^{\infty} \frac{1}{\bar{\rho}^2} e^{i\bar{r}} e^{i\epsilon(\bar{z}-\bar{z}')}\,d\bar{z}. \end{aligned} \tag{15.3.70}$$

This differs from (15.3.60) by a factor $-i(\bar{r}/\bar{\rho}^2)$ so that the solution in (15.3.64) has to be multiplied by this factor to give

$$K_{2\phi'}(\bar{\rho}, \bar{z}') = 2i\left(\frac{\bar{r}_2^2}{\bar{\rho}^3}\right)\left(\frac{\pi}{\bar{r}_2}\right)^{1/2} e^{i\bar{r}_2} e^{-iP_2}\mathcal{F}(P_2) \tag{15.3.71}$$

so that

$$K_{2\phi'}(\rho, z'+d') = \frac{2i}{k_0}\left(\frac{r_2^2}{\rho^3}\right)\left(\frac{\pi}{k_0 r_2}\right)^{1/2} e^{ik_0 r_2} e^{-iP_2}\mathcal{F}(P_2) \tag{15.3.72}$$

and

$$\epsilon'^3 K_{2\phi'}(\rho, z'+d') = -2k_0^5 l^3\left(\frac{r_2^2}{\rho^3}\right)\left(\frac{\pi}{k_0 r_2}\right)^{1/2} e^{ik_0 r_2} e^{-iP_2}\mathcal{F}(P_2). \tag{15.3.73}$$

From (15.3.57) with (15.3.53),

$$K_{2z'}(\bar{\rho}, \bar{z}') \sim ik_0\int_{\bar{z}'}^{\infty}\left(\frac{\bar{\rho}}{\bar{r}^2}\right) e^{i\bar{r}} e^{i\epsilon(\bar{z}-\bar{z}')}\,d\bar{z}. \tag{15.3.74}$$

This differs from (F33) in Appendix F only in the factor k_0. Hence, the final formula is (F42) multiplied by k_0. That is,

$$
\begin{aligned}
K_{2z'}(\bar{\rho}, \bar{z}') &= ik_0\left(\frac{\pi}{\bar{r}_2}\right)^{1/2} e^{i\bar{r}_2} e^{-iP_2}\mathcal{F}(P_2) \\
&= ik_0\left(\frac{\pi}{k_0 r_2}\right)^{1/2} e^{ik_0 r_2} e^{-iP_2}\mathcal{F}(P_2). \qquad (15.3.75)
\end{aligned}
$$

The formula for $K_{2m\rho}(\rho, z'+d')$ is the same as that for $K_{2\phi'}(\rho, z'+d')$ so that

$$
K_{2m\rho}(\rho, z'+d') = \frac{2i}{k_0}\left(\frac{r_2^2}{\rho^3}\right)\left(\frac{\pi}{k_0 r_2}\right)^{1/2} e^{ik_0 r_2} e^{-iP_2}\mathcal{F}(P_2) \qquad (15.3.76)
$$

and

$$
\epsilon'^2 K_{2m\rho}(\rho, z'+d') = -2i(k_0^3 l^2)\left(\frac{r_2^2}{\rho^3}\right)\left(\frac{\pi}{k_0 r_2}\right)^{1/2} e^{ik_0 r_2} e^{-iP_2}\mathcal{F}(P_2). \qquad (15.3.77)
$$

Since $K_{2m\phi'}(\rho, z'+d')$ has the same integrand as $K_{2\rho}(\rho, z'+d')$, it is given by

$$
K_{2m\phi'}(\rho, z'+d') = -2\left(\frac{r_2}{\rho}\right)\left(\frac{\pi}{k_0 r_2}\right)^{1/2} e^{ik_0 r_2} e^{-iP_2}\mathcal{F}(P_2). \qquad (15.3.78)
$$

The formulas for $K_2(\rho, z'+d')$ can be included in those for $F_2(\rho, z'+d')$ and $G_2(\rho, z'+d')$ and combined with the corresponding formulas for $F_3(\rho, z'+d')$ and $G_3(\rho, z'+d')$ to give $F_1(\rho, z'+d')$ and $G_1(\rho, z'+d')$. Thus, the leading terms are

$$
\begin{aligned}
&F_{\rho 1}(\rho, z'+d') \\
&\quad = F_{\rho 2}(\rho, z'+d') + F_{\rho 3}(\rho, z'+d') \\
&\quad = 2ik_0^2 l\left[\frac{e^{ik_0(z'+d')}}{\rho^2} + \left(\frac{z'+d'}{r_2}\right)\left(\frac{ik_0}{r_2} - \frac{1}{r_2^2} - \frac{1}{\rho^2}\right)e^{ik_0 r_2}\right] \\
&\qquad - 2k_0^3 l^2\left[\frac{e^{ik_0(z'+d')}}{\rho^2} + \left(\frac{ik_0}{r_2} - \frac{1}{\rho^2}\right)e^{ik_0 r_2}\right] \\
&\qquad - 2ik_0^6 l^3\left(\frac{r_2}{\rho}\right)\left(\frac{\pi}{k_0 r_2}\right)^{1/2} e^{ik_0 r_2} e^{-iP_2}\mathcal{F}(P_2) \\
&\qquad - 2ik_0^2 l\left[\frac{e^{ik_0(z'+d')}}{\rho^2} - \left(\frac{z'+d'}{r_2}\right)\frac{e^{ik_0 r_2}}{\rho^2}\right] \\
&\qquad + 2k_0^3 l^2\left[\frac{e^{ik_0(z'+d')}}{\rho^2} + \frac{ie^{ik_0 r_2}}{k_0 r_2^3} - \left(\frac{z'+d'}{\rho}\right)^2\frac{e^{ik_0 r_2}}{r_2^2}\right] \\
&\quad = 2k_0 e^{ik_0 r_2}\left[ik_0 l\left(\frac{z'+d'}{r_2}\right)\left(\frac{ik_0}{r_2} - \frac{1}{r_2^2}\right) - k_0^2 l^2\left(\frac{ik_0}{r_2} - \frac{1}{r_2^2} - \frac{i}{k_0 r_2^3}\right)\right.
\end{aligned}
$$

$$- ik_0^5 l^3 \left(\frac{r_2}{\rho}\right)\left(\frac{\pi}{k_0 r_2}\right)^{1/2} e^{-iP_2}\mathcal{F}(P_2)\bigg], \tag{15.3.79}$$

$$\begin{aligned}
&F_{\phi'1}(\rho, z'+d') \\
&\quad = F_{\phi'2}(\rho, z'+d') + F_{\phi'3}(\rho, z'+d') \\
&\quad = -2ik_0^2 l\left(\frac{e^{ik_0(z'+d')}}{\rho^2} - \frac{z'+d'}{r_2}\frac{e^{ik_0 r_2}}{\rho^2}\right) \\
&\qquad + 2k_0^3 l^2\left(\frac{e^{ik_0(z'+d')}}{\rho^2} - \frac{e^{ik_0 r_2}}{\rho^2}\right) \\
&\qquad + 2ik_0^2 l\left[\frac{e^{ik_0(z'+d')}}{\rho^2} + \left(\frac{z'+d'}{r_2}\right)\left(\frac{ik_0}{r_2} - \frac{1}{r_2^2} - \frac{1}{\rho^2}\right)e^{ik_0 r_2}\right] \\
&\qquad - 2k_0^3 l^2\left\{\frac{e^{ik_0(z'+d')}}{\rho^2} + \left[\frac{1}{r_2^2} + \frac{2i}{k_0 r_2^3}\right.\right. \\
&\qquad\qquad \left.\left. + \left(\frac{z'+d'}{r_2}\right)^2\left(\frac{ik_0}{r_2} - \frac{3}{r_2^2} - \frac{3i}{k_0 r_2^3} - \frac{1}{\rho^2}\right)\right]e^{ik_0 r_2}\right\} \\
&\qquad - 2k_0^5 l^3\left(\frac{r_2^2}{\rho^3}\right)\left(\frac{\pi}{k_0 r_2}\right)^{1/2} e^{ik_0 r_2} e^{-iP_2}\mathcal{F}(P_2) \\
&\quad = 2k_0 e^{ik_0 r_2}\left\{ik_0 l\left(\frac{z'+d'}{r_2}\right)\left(\frac{ik_0}{r_2} - \frac{1}{r_2^2}\right)\right. \\
&\qquad\qquad - k_0^2 l^2\left[\frac{2}{r_2^2} + \frac{2i}{k_0 r_2^3} + \left(\frac{z'+d'}{r_2}\right)^2\left(\frac{ik_0}{r_2} - \frac{3}{r_2^2} - \frac{3i}{k_0 r_2^3}\right)\right] \\
&\qquad\qquad \left. - k_0^4 l^3\left(\frac{r_2^2}{\rho^3}\right)\left(\frac{\pi}{k_0 r_2}\right)^{1/2} e^{-iP_2}\mathcal{F}(P_2)\right\},
\end{aligned} \tag{15.3.80}$$

$$\begin{aligned}
&F_{z'1}(\rho, z'+d') \\
&\quad = 2ik_0^2 l\, e^{ik_0 r_2}\left[\left(\frac{\rho}{r_2}\right)\left(\frac{k_0}{r_2} + \frac{i}{r_2^2}\right) + k_0^3 l\left(\frac{\pi}{k_0 r_2}\right)^{1/2} e^{-iP_2}\mathcal{F}(P_2)\right],
\end{aligned} \tag{15.3.81}$$

$$\begin{aligned}
&G_{\rho 1}(\rho, z'+d') \\
&\quad = G_{\rho 2}(\rho, z'+d') + G_{\rho 3}(\rho, z'+d') \\
&\quad = -2ik_0 l\left(\frac{e^{ik_0(z'+d')}}{\rho^2} - \frac{e^{ik_0 r_2}}{\rho^2}\right) \\
&\qquad + 2ik_0 l\left\{\frac{e^{ik_0(z'+d')}}{\rho^2} + \left[\frac{1}{r_2^2} + \frac{2i}{k_0 r_2^3}\right.\right. \\
&\qquad\qquad \left.\left. + \left(\frac{z'+d'}{r_2}\right)^2\left(\frac{ik_0}{r_2} - \frac{3}{r_2^2} - \frac{3i}{k_0 r_2^3} - \frac{1}{\rho^2}\right)\right]e^{ik_0 r_2}\right\}
\end{aligned}$$

$$- ik_0^5 l^3 \left(\frac{r_2}{\rho}\right)\left(\frac{\pi}{k_0 r_2}\right)^{1/2} e^{-iP_2}\mathcal{F}(P_2)\Bigg], \tag{15.3.79}$$

$$\begin{aligned}
&F_{\phi'1}(\rho, z'+d') \\
&\quad = F_{\phi'2}(\rho, z'+d') + F_{\phi'3}(\rho, z'+d') \\
&\quad = -2ik_0^2 l\left(\frac{e^{ik_0(z'+d')}}{\rho^2} - \frac{z'+d'}{r_2}\frac{e^{ik_0 r_2}}{\rho^2}\right) \\
&\qquad + 2k_0^3 l^2\left(\frac{e^{ik_0(z'+d')}}{\rho^2} - \frac{e^{ik_0 r_2}}{\rho^2}\right) \\
&\qquad + 2ik_0^2 l\left[\frac{e^{ik_0(z'+d')}}{\rho^2} + \left(\frac{z'+d'}{r_2}\right)\left(\frac{ik_0}{r_2} - \frac{1}{r_2^2} - \frac{1}{\rho^2}\right)e^{ik_0 r_2}\right] \\
&\qquad - 2k_0^3 l^2\Bigg\{\frac{e^{ik_0(z'+d')}}{\rho^2} + \Bigg[\frac{1}{r_2^2} + \frac{2i}{k_0 r_2^3} \\
&\qquad\qquad + \left(\frac{z'+d'}{r_2}\right)^2\left(\frac{ik_0}{r_2} - \frac{3}{r_2^2} - \frac{3i}{k_0 r_2^3} - \frac{1}{\rho^2}\right)\Bigg]e^{ik_0 r_2}\Bigg\} \\
&\qquad - 2k_0^5 l^3\left(\frac{r_2^2}{\rho^3}\right)\left(\frac{\pi}{k_0 r_2}\right)^{1/2} e^{ik_0 r_2} e^{-iP_2}\mathcal{F}(P_2) \\
&\quad = 2k_0 e^{ik_0 r_2}\Bigg\{ik_0 l\left(\frac{z'+d'}{r_2}\right)\left(\frac{ik_0}{r_2} - \frac{1}{r_2^2}\right) \\
&\qquad\qquad - k_0^2 l^2\left[\frac{2}{r_2^2} + \frac{2i}{k_0 r_2^3} + \left(\frac{z'+d'}{r_2}\right)^2\left(\frac{ik_0}{r_2} - \frac{3}{r_2^2} - \frac{3i}{k_0 r_2^3}\right)\right] \\
&\qquad\qquad - k_0^4 l^3\left(\frac{r_2^2}{\rho^3}\right)\left(\frac{\pi}{k_0 r_2}\right)^{1/2} e^{-iP_2}\mathcal{F}(P_2)\Bigg\},
\end{aligned} \tag{15.3.80}$$

$$\begin{aligned}
&F_{z'1}(\rho, z'+d') \\
&\quad = 2ik_0^2 l\, e^{ik_0 r_2}\left[\left(\frac{\rho}{r_2}\right)\left(\frac{k_0}{r_2} + \frac{i}{r_2^2}\right) + k_0^3 l\left(\frac{\pi}{k_0 r_2}\right)^{1/2} e^{-iP_2}\mathcal{F}(P_2)\right],
\end{aligned} \tag{15.3.81}$$

$$\begin{aligned}
&G_{\rho 1}(\rho, z'+d') \\
&\quad = G_{\rho 2}(\rho, z'+d') + G_{\rho 3}(\rho, z'+d') \\
&\quad = -2ik_0 l\left(\frac{e^{ik_0(z'+d')}}{\rho^2} - \frac{e^{ik_0 r_2}}{\rho^2}\right) \\
&\qquad + 2ik_0 l\Bigg\{\frac{e^{ik_0(z'+d')}}{\rho^2} + \Bigg[\frac{1}{r_2^2} + \frac{2i}{k_0 r_2^3} \\
&\qquad\qquad + \left(\frac{z'+d'}{r_2}\right)^2\left(\frac{ik_0}{r_2} - \frac{3}{r_2^2} - \frac{3i}{k_0 r_2^3} - \frac{1}{\rho^2}\right)\Bigg]e^{ik_0 r_2}\Bigg\}
\end{aligned}$$

$$
\begin{aligned}
&+ 2ik_0^3 l^2 \left(\frac{r_2^2}{\rho^3}\right)\left(\frac{\pi}{k_0 r_2}\right)^{1/2} e^{ik_0 r_2} e^{-iP_2}\mathcal{F}(P_2) \\
&= 2ik_0 l\, e^{ik_0 r_2}\left[\frac{2}{r_2^2} + \frac{2i}{k_0 r_2^3} + \left(\frac{z'+d'}{r_2}\right)^2 \left(\frac{ik_0}{r_2} - \frac{3}{r_2^2} - \frac{3i}{k_0 r_2^3}\right)\right. \\
&\qquad \left. + k_0^2 l \left(\frac{r_2^2}{\rho^3}\right)\left(\frac{\pi}{k_0 r_2}\right)^{1/2} e^{-iP_2}\mathcal{F}(P_2)\right], \qquad (15.3.82)
\end{aligned}
$$

$$
\begin{aligned}
&G_{\phi'1}(\rho, z'+d') \\
&= G_{\phi'2}(\rho, z'+d') + G_{\phi'3}(\rho, z'+d') \\
&= 2ik_0 l\left[\frac{e^{ik_0(z'+d')}}{\rho^2} + \left(\frac{ik_0}{r_2} - \frac{1}{\rho^2}\right) e^{ik_0 r_2}\right] \\
&\quad - 2k_0^4 l^2 \left(\frac{r_2}{\rho}\right)\left(\frac{\pi}{k_0 r_2}\right)^{1/2} e^{ik_0 r_2} e^{-iP_2}\mathcal{F}(P_2) \\
&\quad - 2ik_0 l\left[\frac{e^{ik_0(z'+d')}}{\rho^2} + \frac{ie^{ik_0 r_2}}{k_0 r_2^3} - \left(\frac{z'+d'}{r_2}\right)^2 \frac{e^{ik_0 r_2}}{\rho^2}\right] \\
&= 2e^{ik_0 r_2}\left[ik_0 l\left(\frac{ik_0}{r_2} - \frac{1}{r_2^2} - \frac{i}{k_0 r_2^3}\right)\right. \\
&\qquad \left. - k_0^4 l^2 \left(\frac{r_2}{\rho}\right)\left(\frac{\pi}{k_0 r_2}\right)^{1/2} e^{-iP_2}\mathcal{F}(P_2)\right], \qquad (15.3.83)
\end{aligned}
$$

$$
\begin{aligned}
&G_{z'1}(\rho, z'+d') \\
&= 2e^{ik_0 r_2}\left\{ik_0 l\left(\frac{\rho}{r_2}\right)\left(\frac{z'+d'}{r_2}\right)\left(\frac{k_0}{r_2} + \frac{3i}{r_2^2} - \frac{3}{k_0 r_2^3}\right)\right. \\
&\qquad + k_0^2 l^2 \left(\frac{\rho}{r_2}\right)\left[\frac{i}{r_2^2} - \frac{3}{k_0 r_2^3} - \frac{3i}{k_0^2 r_2^4}\right. \\
&\qquad\qquad \left.\left. + \left(\frac{z'+d'}{r_2}\right)^2 \left(\frac{k_0}{r_2} + \frac{6i}{r_2^2} - \frac{15}{k_0 r_2^3}\right)\right]\right\}. \\
&\qquad (15.3.84)
\end{aligned}
$$

15.4 Integrated Formulas for the Components of the Electromagnetic Field

The complete expressions for the six cylindrical components of the electromagnetic field of a horizontal electric dipole at the height d' over a dielectric-coated conducting plane are obtained by combining the contributions $F_0(\rho, z'-d')$ or $G_0(\rho, z'-d')$ of the direct field, $F_0(\rho, z'+d')$ or

$G_0(\rho, z'+d')$ of the ideal image field, and $F_1(\rho, z'+d')$ or $G_1(\rho, z'+d')$ of the surface wave. With (15.2.23)–(15.2.27) and (15.3.79)–(15.3.84), the final results are

$$\begin{aligned}
&E_{0\rho}(\rho, \phi', z') \\
&\quad = -\frac{\omega\mu_0}{4\pi k_0^2}\cos\phi'\,[F_{\rho 0}(\rho, z'-d') - F_{\rho 0}(\rho, z'+d') + F_{\rho 1}(\rho, z'+d')] \\
&\quad = \frac{\omega\mu_0}{4\pi k_0}\cos\phi' \\
&\qquad \times \left(e^{ik_0 r_1}\left[\frac{2}{r_1^2} + \frac{2i}{k_0 r_1^3} + \left(\frac{z'-d'}{r_1}\right)^2\left(\frac{ik_0}{r_1} - \frac{3}{r_1^2} - \frac{3i}{k_0 r_1^3}\right)\right]\right. \\
&\qquad\quad - e^{ik_0 r_2}\left[\frac{2}{r_2^2} + \frac{2i}{k_0 r_2^3} + \left(\frac{z'+d'}{r_2}\right)^2\left(\frac{ik_0}{r_2} - \frac{3}{r_2^2} - \frac{3i}{k_0 r_2^3}\right)\right] \\
&\qquad\quad - 2e^{ik_0 r_2}\left\{ik_0 l\left(\frac{z'+d'}{r_2}\right)\left(\frac{ik_0}{r_2} - \frac{1}{r_2^2}\right)\right. \\
&\qquad\qquad - k_0^2 l^2\left[\frac{ik_0}{r_2} - \frac{1}{r_2^2} - \frac{i}{k_0 r_2^3}\right. \\
&\qquad\qquad\quad \left.\left.\left. + ik_0^3 l\left(\frac{r_2}{\rho}\right)\left(\frac{\pi}{k_0 r_2}\right)^{1/2} e^{-iP_2}\mathcal{F}(P_2)\right]\right\}\right),
\end{aligned} \tag{15.4.1}$$

$$\begin{aligned}
&E_{0\phi'}(\rho, \phi', z') \\
&\quad = \frac{\omega\mu_0}{4\pi k_0^2}\sin\phi'\,[F_{\phi' 0}(\rho, z'-d') - F_{\phi' 0}(\rho, z'+d') + F_{\phi' 1}(\rho, z'+d')] \\
&\quad = -\frac{\omega\mu_0}{4\pi k_0}\sin\phi' \\
&\qquad \times \left(e^{ik_0 r_1}\left(\frac{ik_0}{r_1} - \frac{1}{r_1^2} - \frac{i}{k_0 r_1^3}\right) - e^{ik_0 r_2}\left(\frac{ik_0}{r_2} - \frac{1}{r_2^2} - \frac{i}{k_0 r_2^3}\right)\right. \\
&\qquad\quad - 2e^{ik_0 r_2}\left\{ik_0 l\left(\frac{z'+d'}{r_2}\right)\left(\frac{ik_0}{r_2} - \frac{1}{r_2^2}\right)\right. \\
&\qquad\qquad - k_0^2 l^2\left[\frac{2}{r_2^2} + \frac{2i}{k_0 r_2^3}\right. \\
&\qquad\qquad\quad + \left(\frac{z'+d'}{r_2}\right)^2\left(\frac{ik_0}{r_2} - \frac{3}{r_2^2} - \frac{3i}{k_0 r_2^3}\right) \\
&\qquad\qquad\quad \left.\left.\left. + k_0^2 l\left(\frac{r_2^2}{\rho^3}\right)\left(\frac{\pi}{k_0 r_2}\right)^{1/2} e^{-iP_2}\mathcal{F}(P_2)\right]\right\}\right),
\end{aligned} \tag{15.4.2}$$

$E_{0z'}(\rho, \phi', z')$

$$
\begin{aligned}
&= \frac{i\omega\mu_0}{4\pi k_0^2}\cos\phi'\ [F_{z'0}(\rho, z'-d') - F_{z'0}(\rho, z'+d') + F_{z'1}(\rho, z'+d')] \\
&= -\frac{\omega\mu_0}{4\pi k_0}\cos\phi'\Bigg\{ e^{ik_0r_1}\left(\frac{\rho}{r_1}\right)\left(\frac{z'-d'}{r_1}\right)\left(\frac{ik_0}{r_1} - \frac{3}{r_1^2} - \frac{3i}{k_0r_1^3}\right) \\
&\qquad - e^{ik_0r_2}\left(\frac{\rho}{r_2}\right)\left(\frac{z'+d'}{r_2}\right)\left(\frac{ik_0}{r_2} - \frac{3}{r_2^2} - \frac{3i}{k_0r_2^3}\right) \\
&\qquad - 2ik_0l\, e^{ik_0r_2}\Bigg[\left(\frac{\rho}{r_2}\right)\left(\frac{ik_0}{r_2} - \frac{1}{r_2^2}\right) \\
&\qquad\qquad + ik_0^3l\left(\frac{\pi}{k_0r_2}\right)^{1/2} e^{-iP_2}\mathcal{F}(P_2)\Bigg]\Bigg\},
\end{aligned}
\tag{15.4.3}
$$

$$
\begin{aligned}
&B_{0\rho}(\rho, \phi', z') \\
&= -\frac{\mu_0}{4\pi}\sin\phi'\ [G_{\rho 0}(\rho, z'-d') - G_{\rho 0}(\rho, z'+d') + G_{\rho 1}(\rho, z'+d')] \\
&= \frac{\mu_0}{4\pi}\sin\phi'\Bigg\{ e^{ik_0r_1}\left(\frac{z'-d'}{r_1}\right)\left(\frac{ik_0}{r_1} - \frac{1}{r_1^2}\right) \\
&\qquad - e^{ik_0r_2}\left(\frac{z'+d'}{r_2}\right)\left(\frac{ik_0}{r_2} - \frac{1}{r_2^2}\right) \\
&\qquad - 2ik_0l\, e^{ik_0r_2}\Bigg[\frac{2}{r_2^2} + \frac{2i}{k_0r_2^3} \\
&\qquad\qquad + \left(\frac{z'+d'}{r_2}\right)^2\left(\frac{ik_0}{r_2} - \frac{3}{r_2^2} - \frac{3i}{k_0r_2^3}\right) \\
&\qquad\qquad + k_0^2l\left(\frac{r_2^2}{\rho^3}\right)\left(\frac{\pi}{k_0r_2}\right)^{1/2} e^{-iP_2}\mathcal{F}(P_2)\Bigg]\Bigg\},
\end{aligned}
\tag{15.4.4}
$$

$$
\begin{aligned}
&B_{0\phi'}(\rho, \phi', z') \\
&= -\frac{\mu_0}{4\pi}\cos\phi'\ [G_{\phi' 0}(\rho, z'-d') - G_{\phi' 0}(\rho, z'+d') + G_{\phi' 1}(\rho, z'+d')] \\
&= \frac{\mu_0}{4\pi}\cos\phi'\Bigg\{ e^{ik_0r_1}\left(\frac{z'-d'}{r_1}\right)\left(\frac{ik_0}{r_1} - \frac{1}{r_1^2}\right) \\
&\qquad - e^{ik_0r_2}\left(\frac{z'+d'}{r_2}\right)\left(\frac{ik_0}{r_2} - \frac{1}{r_2^2}\right) \\
&\qquad - 2ik_0l\, e^{ik_0r_2}\Bigg[\frac{ik_0}{r_2} - \frac{1}{r_2^2} - \frac{i}{k_0r_2^3} \\
&\qquad\qquad + ik_0^3l\left(\frac{r_2}{\rho}\right)\left(\frac{\pi}{k_0r_2}\right)^{1/2} e^{-iP_2}\mathcal{F}(P_2)\Bigg]\Bigg\},
\end{aligned}
\tag{15.4.5}
$$

$$
\begin{aligned}
&B_{0z'}(\rho,\phi',z')\\
&= \frac{i\mu_0}{4\pi}\sin\phi'\ [G_{z'0}(\rho,z'-d') - G_{z'0}(\rho,z'+d') + G_{z'1}(\rho,z'+d')]\\
&= -\frac{\mu_0}{4\pi}\sin\phi'\\
&\quad\times\Bigg(e^{ik_0r_1}\left(\frac{\rho}{r_1}\right)\left(\frac{ik_0}{r_1}-\frac{1}{r_1^2}\right) - e^{ik_0r_2}\left(\frac{\rho}{r_2}\right)\left(\frac{ik_0}{r_2}-\frac{1}{r_2^2}\right)\\
&\qquad - 2e^{ik_0r_2}\Bigg\{ik_0l\left(\frac{\rho}{r_2}\right)\left(\frac{z'+d'}{r_2}\right)\left(\frac{ik_0}{r_2}-\frac{3}{r_2^2}-\frac{3i}{k_0r_2^3}\right)\\
&\qquad\quad - k_0^2l^2\left(\frac{\rho}{r_2}\right)\Bigg[\frac{1}{r_2^2}+\frac{3i}{k_0r_2^3}-\frac{3}{k_0^2r_2^4}\\
&\qquad\qquad + \left(\frac{z'+d'}{r_2}\right)^2\left(\frac{ik_0}{r_2}-\frac{6}{r_2^2}-\frac{15i}{k_0r_2^3}\right)\Bigg]\Bigg\}\Bigg),
\end{aligned}
\tag{15.4.6}
$$

where P_2 is given in (15.3.68) and $\mathcal{F}(P_2)$ in (15.3.66).

Of particular interest in microstrip circuits—as already stated in Section 15.1—is the field on the dielectric–air surface, $z' = 0$, when the dipole is also on the surface, $d' = 0$. With the boundary conditions, the field at the boundary in the thin substrate is also obtained. The components are

$$
\begin{aligned}
E_{0\rho}(\rho,\phi',0) &= \frac{\omega\mu_0}{2\pi k_0}\cos\phi'\, e^{ik_0\rho}(k_0^2l^2)\\
&\quad\times\left[\frac{ik_0}{\rho}-\frac{1}{\rho^2}-\frac{i}{k_0\rho^3}+ik_0^3l\left(\frac{\pi}{k_0\rho}\right)^{1/2}e^{-ip_2}\mathcal{F}(p_2)\right],
\end{aligned}
\tag{15.4.7}
$$

$$
\begin{aligned}
E_{0\phi'}(\rho,\phi',0) &= -\frac{\omega\mu_0}{2\pi k_0}\sin\phi'\, e^{ik_0\rho}(k_0^2l^2)\\
&\quad\times\left[\frac{2}{\rho^2}+\frac{2i}{k_0\rho^3}+\frac{k_0^2l}{\rho}\left(\frac{\pi}{k_0\rho}\right)^{1/2}e^{-ip_2}\mathcal{F}(p_2)\right],
\end{aligned}
\tag{15.4.8}
$$

$$
\begin{aligned}
E_{0z'}(\rho,\phi',0) &= \frac{i\omega\mu_0}{2\pi k_0}\cos\phi'\, e^{ik_0\rho}(k_0l)\\
&\quad\times\left[\frac{ik_0}{\rho}-\frac{1}{\rho^2}+ik_0^3l\left(\frac{\pi}{k_0\rho}\right)^{1/2}e^{-ip_2}\mathcal{F}(p_2)\right],
\end{aligned}
\tag{15.4.9}
$$

$$
\begin{aligned}
B_{0\rho}(\rho,\phi',0) &= -\frac{i\mu_0}{2\pi}\sin\phi'\, e^{ik_0\rho}(k_0l)\\
&\quad\times\left[\frac{2}{\rho^2}+\frac{2i}{k_0\rho^3}+\frac{k_0^2l}{\rho}\left(\frac{\pi}{k_0\rho}\right)^{1/2}e^{-ip_2}\mathcal{F}(p_2)\right],
\end{aligned}
\tag{15.4.10}
$$

$$B_{0\phi'}(\rho,\phi',0) = -\frac{i\mu_0}{2\pi}\cos\phi'\, e^{ik_0\rho}(k_0 l) \times \left[\frac{ik_0}{\rho} - \frac{1}{\rho^2} - \frac{i}{k_0\rho^3} + ik_0^3 l\left(\frac{\pi}{k_0\rho}\right)^{1/2} e^{-ip_2}\mathcal{F}(p_2)\right], \tag{15.4.11}$$

$$B_{0z'}(\rho,\phi',0) = -\frac{\mu_0}{2\pi}\sin\phi'\, e^{ik_0\rho}(k_0^2 l^2)\left(\frac{1}{\rho^2} + \frac{3i}{k_0\rho^3} - \frac{3}{k_0^2\rho^4}\right), \tag{15.4.12}$$

where, from (15.3.68),

$$p_2 = -\frac{k_0\rho}{2}\,(k_0^2 l^2). \tag{15.4.13}$$

At radial distances that significantly exceed the Sommerfeld numerical distance, the entire Fresnel-integral term has the following form:

$$ik_0^3 l\left(\frac{\pi}{k_0\rho}\right)^{1/2} e^{-ip_2}\mathcal{F}(p_2) \to -\frac{ik_0}{\rho} + \frac{1}{k_0^2 l^2\rho^2} \quad \text{when } k_0\rho \geq \frac{8}{k_0^2 l^2}. \tag{15.4.14}$$

In typical microstrip with $l = 0.1$ cm and $f \sim 0.8$ GHz, $k_0 = 2\pi f/c = 1.6\pi \times 10^9/3 \times 10^8 = 16.75\ \text{m}^{-1}$ and $k_0 l = 0.01675$. The far-field condition is

$$\rho \geq \frac{8}{k_0^3 l^2} = \frac{8}{(0.01675)^2 \times 16.75} = 445.5\ \text{m}. \tag{15.4.15a}$$

Or, with $l = 0.25$ mm and $f = 10$ GHz, $k_0 = 2\pi \times 10^{10}/3 \times 10^8 = 209.4\ \text{m}^{-1}$, $k_0 l = 0.0523$, and the far-field condition is

$$\rho \geq \frac{8}{(0.0523)^2 \times 209.4} = 13.96\ \text{m}. \tag{15.4.15b}$$

These distances are very large for microstrip circuits, but not necessarily for the fields of microstrip antennas for which the finite size of the ground plane becomes important. In general, when

$$k_0\rho \leq \frac{1}{k_0^2 l^2}, \tag{15.4.16}$$

the Fresnel-integral terms contribute negligibly and a good approximation of the field on both sides of the air–dielectric boundary is

$$E_{1\rho}(\rho,\phi',0) = E_{0\rho}(\rho,\phi',0) \sim \frac{\omega\mu_0}{2\pi k_0}\,(k_0^2 l^2)\cos\phi'\, e^{ik_0\rho}\left(\frac{ik_0}{\rho} - \frac{1}{\rho^2} - \frac{i}{k_0\rho^3}\right), \tag{15.4.17}$$

$$E_{1\phi'}(\rho,\phi',0) = E_{0\phi'}(\rho,\phi',0) \sim -\frac{\omega\mu_0}{2\pi k_0}\,(k_0^2 l^2)\sin\phi'\, e^{ik_0\rho}\left(\frac{2}{\rho^2} + \frac{2i}{k_0\rho^3}\right), \tag{15.4.18}$$

$$\frac{k_1^2}{k_0^2} E_{1z'}(\rho, \phi', 0) = E_{0z'}(\rho, \phi', 0)$$

$$\sim \frac{i\omega\mu_0}{2\pi k_0} (k_0 l) \cos\phi' \, e^{ik_0\rho} \left(\frac{ik_0}{\rho} - \frac{1}{\rho^2} \right), \tag{15.4.19}$$

$$B_{1\rho}(\rho, \phi', 0) = B_{0\rho}(\rho, \phi', 0)$$

$$\sim -\frac{i\mu_0}{2\pi} (k_0 l) \sin\phi' \, e^{ik_0\rho} \left(\frac{2}{\rho^2} + \frac{2i}{k_0\rho^3} \right), \tag{15.4.20}$$

$$B_{1\phi'}(\rho, \phi', 0) = B_{0\phi'}(\rho, \phi', 0)$$

$$\sim -\frac{i\mu_0}{2\pi} (k_0 l) \cos\phi' \, e^{ik_0\rho} \left(\frac{ik_0}{\rho} - \frac{1}{\rho^2} - \frac{i}{k_0\rho^3} \right), \tag{15.4.21}$$

$$B_{1z'}(\rho, \phi', 0) = B_{0z'}(\rho, \phi', 0)$$

$$\sim -\frac{\mu_0}{2\pi} (k_0^2 l^2) \sin\phi' \, e^{ik_0\rho} \left(\frac{1}{\rho^2} + \frac{3i}{k_0\rho^3} - \frac{3}{k_0^2\rho^4} \right). \tag{15.4.22}$$

The largest components are $E_{0z'}(\rho, \phi', 0)$ and $B_{0\phi'}(\rho, \phi', 0)$; $E_{1z'}(\rho, \phi', 0)$ in the substrate is smaller by the usually quite small factor $k_0^2/k_1^2 = \epsilon_0/\epsilon_1 = 1/\epsilon_{1r}$.

When the radial distance satisfies the condition (15.4.14), i.e., $k_0\rho \geq 8/k_0^2 l^2$, the far field has the following simple forms in the air on the air–dielectric boundary:

$$E_{0\rho}(\rho, \phi', 0) \sim \frac{\omega\mu_0}{2\pi k_0} \cos\phi' \, \frac{e^{ik_0\rho}}{\rho^2}, \tag{15.4.23}$$

$$E_{0z'}(\rho, \phi', 0) \sim \frac{i\omega\mu_0}{2\pi k_0} \left(\frac{1}{k_0 l} \right) \cos\phi' \, \frac{e^{ik_0\rho}}{\rho^2}, \tag{15.4.24}$$

$$B_{0\phi'}(\rho, \phi', 0) \sim -\frac{i\mu_0}{2\pi} \left(\frac{1}{k_0 l} \right) \cos\phi' \, \frac{e^{ik_0\rho}}{\rho^2}. \tag{15.4.25}$$

In the substrate, $E_{1\rho}(\rho, \phi', 0) = E_{0\rho}(\rho, \phi', 0)$, $B_{1\phi'}(\rho, \phi', 0) = B_{0\phi'}(\rho, \phi', 0)$, and

$$E_{1z'}(\rho, \phi', 0) = \frac{k_0^2}{k_1^2} E_{0z'}(\rho, \phi', 0) \sim \frac{i\omega\mu_0}{2\pi k_1} \left(\frac{1}{k_1 l} \right) \cos\phi' \, \frac{e^{ik_0\rho}}{\rho^2}. \tag{15.4.26}$$

15.5 Comparison of the Electromagnetic Field of the Horizontal Dipole in Three Related Environments

It is instructive to compare the formulas for the components of the electromagnetic field generated by a unit horizontal dipole under the following three conditions: (a) The dipole is on the boundary $z = 0$ between air (Region 0) and a dielectric half-space (Region 1). (b) The dipole is on the boundary $z = 0$ between air (Region 0) and a thin layer of dielectric (Region 1) on a perfectly conducting half-space (Region 2). The thickness of the dielectric is l with $k_1^2 l^2 \ll 1$. (c) The dipole is in the plane $z = 0$ in air (Region 0) at a height l over a perfectly conducting half-space (Region 1). Case (c) is like case (b) with the layer of dielectric replaced by air. The components of the field are to be compared in the plane $z = 0$.

Case (a). The electromagnetic-field components on the boundary $z = 0$ between a half-space of air (Region 0) and a half-space of dielectric (Region 1) are given in Chapter 5. The formulas include terms with the exponential $\exp(ik_1\rho)$ which represent waves propagating in Region 1 and terms with $\exp(ik_0\rho)$ which represent surfaces waves propagating in the air along the boundary surface. Since there are no waves that can propagate with the wave number k_1 in either of the two other cases, these terms will be omitted in the comparisons. Cases (a) and (b) involve a Fresnel-integral term; case (c) does not.

The components of the field on the boundary in air over a dielectric half-space are obtained from (5.7.16), (5.7.17), (5.7.25), (5.7.45), (5.7.51), and (5.7.57) with $z = d = 0$. They are

$$E_{0\rho}(\rho,\phi,0) = -\frac{\omega\mu_0}{2\pi k_1^2}\cos\phi\, k_0\left[\frac{ik_0}{\rho} - \frac{1}{\rho^2} - \frac{i}{k_0\rho^3} - \frac{k_0^3}{k_1}\left(\frac{\pi}{k_0\rho}\right)^{1/2} e^{-ip}\mathcal{F}(p)\right]e^{ik_0\rho}, \tag{15.5.1}$$

$$E_{0\phi}(\rho,\phi,0) = \frac{\omega\mu_0}{2\pi k_1^2}\sin\phi\, k_0\left[\frac{2}{\rho^2} + \frac{2i}{k_0\rho^3} + \frac{ik_0^2}{k_1\rho}\left(\frac{\pi}{k_0\rho}\right)^{1/2} e^{-ip}\mathcal{F}(p)\right]e^{ik_0\rho}, \tag{15.5.2}$$

$$E_{0z}(\rho,\phi,0) = \frac{\omega\mu_0}{2\pi k_1}\cos\phi\left[\frac{ik_0}{\rho} - \frac{1}{\rho^2} - \frac{k_0^3}{k_1}\left(\frac{\pi}{k_0\rho}\right)^{1/2} e^{-ip}\mathcal{F}(p)\right]e^{ik_0\rho}, \tag{15.5.3}$$

$$B_{0\rho}(\rho,\phi,0) = -\frac{\mu_0}{2\pi k_1}\sin\phi\, k_0\left[\frac{2}{\rho^2} + \frac{2i}{k_0\rho^3} + \frac{ik_0^2}{k_1\rho}\left(\frac{\pi}{k_0\rho}\right)^{1/2} e^{-ip}\mathcal{F}(p)\right]e^{ik_0\rho}, \tag{15.5.4}$$

$$B_{0\phi}(\rho,\phi,0) = -\frac{\mu_0}{2\pi k_1}\cos\phi\, k_0\left[\frac{ik_0}{\rho} - \frac{1}{\rho^2} - \frac{i}{k_0\rho^3} - \frac{k_0^3}{k_1}\left(\frac{\pi}{k_0\rho}\right)^{1/2} e^{-ip}\mathcal{F}(p)\right]e^{ik_0\rho}, \tag{15.5.5}$$

$$B_{0z}(\rho,\phi,0) = \frac{\mu_0}{2\pi k_1^2}\sin\phi\, k_0^2\left(\frac{1}{\rho^2} + \frac{3i}{k_0\rho^3} - \frac{3}{k_0^2\rho^4}\right)e^{ik_0\rho}, \tag{15.5.6}$$

where

$$p = \frac{k_0^3\rho}{2k_1^2}. \tag{15.5.7}$$

Case (b). The formulas for the three-layered region that correspond to (15.5.1)–(15.5.6) are (15.4.7)–(15.4.12). Since the amplitude factor of interest is $K \sim -ik_1 l$, this can be introduced explicitly to give

$$\begin{aligned} E_{0\rho}(\rho,\phi,0) = &-\frac{\omega\mu_0}{2\pi k_1^2}(-ik_1 l)^2\cos\phi \\ &\times k_0\left[\frac{ik_0}{\rho} - \frac{1}{\rho^2} - \frac{i}{k_0\rho^3}\right. \\ &\left. - (-ik_1 l)\frac{k_0^3}{k_1}\left(\frac{\pi}{k_0\rho}\right)^{1/2} e^{-ip_2}\mathcal{F}(p_2)\right]e^{ik_0\rho}, \end{aligned} \tag{15.5.8}$$

$$\begin{aligned} E_{0\phi}(\rho,\phi,0) = &\frac{\omega\mu_0}{2\pi k_1^2}(-ik_1 l)^2\sin\phi \\ &\times k_0\left[\frac{2}{\rho^2} + \frac{2i}{k_0\rho^3}\right. \\ &\left. + (-ik_1 l)\frac{ik_0^2}{k_1\rho}\left(\frac{\pi}{k_0\rho}\right)^{1/2} e^{-ip_2}\mathcal{F}(p_2)\right]e^{ik_0\rho}, \end{aligned} \tag{15.5.9}$$

$$\begin{aligned} E_{0z}(\rho,\phi,0) = &-\frac{\omega\mu_0}{2\pi k_1}(-ik_1 l)\cos\phi \\ &\times \left[\frac{ik_0}{\rho} - \frac{1}{\rho^2}\right. \\ &\left. - (-ik_1 l)\frac{k_0^3}{k_1}\left(\frac{\pi}{k_0\rho}\right)^{1/2} e^{-ip_2}\mathcal{F}(p_2)\right]e^{ik_0\rho}, \end{aligned} \tag{15.5.10}$$

$$\begin{aligned} B_{0\rho}(\rho,\phi,0) = &\frac{\mu_0}{2\pi k_1}(-ik_1 l)\sin\phi \\ &\times k_0\left[\frac{2}{\rho^2} + \frac{2i}{k_0\rho^3}\right. \\ &\left. + (-ik_1 l)\frac{ik_0^2}{k_1\rho}\left(\frac{\pi}{k_0\rho}\right)^{1/2} e^{-ip_2}\mathcal{F}(p_2)\right]e^{ik_0\rho}, \end{aligned} \tag{15.5.11}$$

$$\begin{aligned} B_{0\phi}(\rho,\phi,0) = &\frac{\mu_0}{2\pi k_1}(-ik_1 l)\cos\phi \\ &\times k_0\left[\frac{ik_0}{\rho} - \frac{1}{\rho^2} - \frac{i}{k_0\rho^3}\right. \\ &\left. - (-ik_1 l)\frac{k_0^3}{k_1}\left(\frac{\pi}{k_0\rho}\right)^{1/2} e^{-ip_2}\mathcal{F}(p_2)\right]e^{ik_0\rho}, \end{aligned} \tag{15.5.12}$$

$$B_{0z}(\rho, \phi, 0) = \frac{\mu_0}{2\pi k_1^2}(-ik_1 l)^2 \sin\phi\, k_0^2 \left(\frac{1}{\rho^2} + \frac{3i}{k_0\rho^3} - \frac{3}{k_0^2\rho^4}\right)e^{ik_0\rho}, \quad (15.5.13)$$

where

$$p_2 = \frac{k_0^3 \rho}{2k_1^2}(-ik_1 l)^2. \quad (15.5.14)$$

It is of interest at this point to recall that the approximation $\tan k_1 l \sim k_1 l$ was made since the dielectric substrate in microstrip is usually electrically very thin. If $k_1 l$ is replaced by $\tan k_1 l$, it follows that when the dielectric has some loss,

$$-ik_1 l \sim -i \tan k_1 l, \quad (15.5.15a)$$

$$\lim_{l\to\infty} -i \tan k_1 l = 1. \quad (15.5.15b)$$

When $(-ik_1 l)$ is replaced by (15.5.15a,b) in (15.5.8)–(15.5.13), these formulas reduce exactly to (15.5.1)–(15.5.6). This means that the surface wave in the air above the substrate in microstrip is closely related to the surface wave in the air above a dielectric half-space. The amplitude is modified by the factor $(-ik_1 l)$ or multiples thereof, and the Sommerfeld numerical distance is similarly modified by the factor $(-ik_1 l)^2$. The far-field form for the dielectric half-space is

$$-\frac{k_0^3}{k_1}\left(\frac{\pi}{k_0\rho}\right)^{1/2} e^{-ip}\mathcal{F}(p) \to -\frac{ik_0}{\rho} - \frac{k_1^2}{k_0^2\rho^2} \quad (15.5.16a)$$

when $k_0\rho \geq 8|k_1|^2/k_0^2$. The corresponding formula for the field in air over a dielectric-coated conducting half-space is

$$-(-ik_1 l)\frac{k_0^3}{k_1}\left(\frac{\pi}{k_0\rho}\right)^{1/2} e^{-ip_2}\mathcal{F}(p_2) \to -\frac{ik_0}{\rho} - \frac{k_1^2}{k_0^2\rho^2(-ik_1 l)^2} \quad (15.5.16b)$$

when $k_0\rho \geq 8/k_0^2 l^2$.

A comparison of (15.5.8)–(15.5.13) with (15.5.1)–(15.5.6) shows that they are identical except for the amplitude factor $(-ik_1 l)$ or $(-ik_1 l)^2$ and $+$ or $-$ signs due to different positive directions for z and ϕ. The presence of the conducting half-space at the small distance l from the air–dielectric boundary eliminates propagation with the wave number k_1 in the dielectric and provides the amplitude factors in powers of $(-ik_1 l)$. In effect, the surface wave that propagates in the air along the air–dielectric boundary is unaffected except in amplitude by the presence of the conducting half-space at all radial distances from the source smaller than $\rho \sim k_1^2/k_0^2$. The three components of the TM_0 surface-wave mode in the thin dielectric layer differ negligibly from the leading three components of the lateral wave along the surface of the dielectric half-space except in their smaller amplitude by the factors $(-ik_1 l)$ and $(-ik_1 l)^2$.

Case (c). The electromagnetic field of a dipole at the height l above a perfectly conducting half-space is readily calculated from the field of the isolated dipole given in (2.2.17)–(2.2.22). The field in the plane $z = 0$ containing the dipole and at a distance $z = -2l$ from an oppositely directed image dipole is required. For $E_{0\rho}(\rho, \phi, 0)$, it is

$$
\begin{aligned}
E_{0\rho}&(\rho, \phi, 0)\\
&= E_{0\rho}^{d}(\rho, \phi, 0) - E_{0\rho}^{i}(\rho, \phi, z)\\
&= \frac{\omega\mu_0}{4\pi k_0^2}\cos\phi\left\{e^{ik_0\rho}\left(\frac{2k_0}{\rho^2} + \frac{2i}{\rho^3}\right)\right.\\
&\qquad\left. - e^{ik_0r_0}\left[\frac{2k_0}{r_0^2} + \frac{2i}{r_0^3} + \left(\frac{z}{r_0}\right)^2\left(\frac{ik_0^2}{r_0} - \frac{3k_0}{r_0^2} - \frac{3i}{r_0^3}\right)\right]\right\},
\end{aligned}
\tag{15.5.17}
$$

where $z = -2l$ and

$$
r_0 = (\rho^2 + z^2)^{1/2} = (\rho^2 + 4l^2)^{1/2}; \qquad \rho^2 \gg 4l^2; \qquad k_0 l \ll 1. \tag{15.5.18}
$$

In order to obtain the terms of order $k_0^2 l^2$ or $k_0 l$, it is necessary to expand in a power series in l. Let

$$
\begin{aligned}
f(z) &= e^{ik_0\rho}\left(\frac{2k_0}{\rho^2} + \frac{2i}{\rho^3}\right)\\
&\qquad - e^{ik_0r_0}\left[\frac{2k_0}{r_0^2} + \frac{2i}{r_0^3} + \left(\frac{z}{r_0}\right)^2\left(\frac{ik_0^2}{r_0} - \frac{3k_0}{r_0^2} - \frac{3i}{r_0^3}\right)\right]\\
&= f(0) + zf'(0) + \frac{z^2}{2}f''(0) + \cdots,
\end{aligned}
\tag{15.5.19}
$$

$$
f(0) = 0, \tag{15.5.20}
$$

$$
\begin{aligned}
f'(z) &= -ik_0e^{ik_0r_0}\left(\frac{z}{r_0}\right)\left[\frac{2k_0}{r_0^2} + \frac{2i}{r_0^3} + \left(\frac{z}{r_0}\right)^2\left(\frac{ik_0^2}{r_0} - \frac{3k_0}{r_0^2} - \frac{3i}{r_0^3}\right)\right]\\
&\qquad - e^{ik_0r_0}\left[\left(-\frac{4k_0}{r_0^3} - \frac{6i}{r_0^4}\right)\left(\frac{z}{r_0}\right)\right.\\
&\qquad\qquad + \left(\frac{z}{r_0}\right)^2\left(-\frac{ik_0^2}{r_0^2} + \frac{6k_0}{r_0^3} + \frac{9i}{r_0^4}\right)\left(\frac{z}{r_0}\right)\\
&\qquad\qquad + \frac{2z}{r_0^2}\left(\frac{ik_0^2}{r_0} - \frac{3k_0}{r_0^2} - \frac{3i}{r_0^3}\right)\\
&\qquad\qquad \left. - \frac{2z^2}{r_0^3}\left(\frac{ik_0^2}{r_0} - \frac{3k_0}{r_0^2} - \frac{3i}{r_0^3}\right)\left(\frac{z}{r_0}\right)\right],
\end{aligned}
\tag{15.5.21}
$$

$$
f'(0) = 0. \tag{15.5.22}
$$

In the evaluation of $f''(z)$, the only terms that contribute to $f''(0)$ are those not multiplied by z. Therefore, terms multiplied by z need not be evaluated; they are represented by [ ]. Hence, the only terms in $f'(z)$ that need to be differentiated to obtain $f''(z)$ are those multiplied by z in the first power. Thus,

$$f''(z) = -ik_0 e^{ik_0 r_0}\left(\frac{2k_0}{r_0^3} + \frac{2i}{r_0^4}\right) - e^{ik_0 r_0}\left(-\frac{4k_0}{r_0^4} - \frac{6i}{r_0^5} + \frac{2ik_0^2}{r_0^3} - \frac{6k_0}{r_0^4} - \frac{6i}{r_0^5}\right) + \left(\frac{z}{r_0}\right)\Big[\quad\Big], \tag{15.5.23}$$

$$\begin{aligned}\frac{z^2}{2} f''(0) &= e^{ik_0\rho}\left(-\frac{2ik_0^2}{\rho^3} + \frac{2k_0}{\rho^4} + \frac{4k_0}{\rho^4} + \frac{6i}{\rho^5} - \frac{2ik_0^2}{\rho^3} + \frac{6k_0}{\rho^4} + \frac{6i}{\rho^5}\right)\left(\frac{z^2}{2}\right)\\ &= z^2 e^{ik_0\rho}\left(-\frac{2ik_0^2}{\rho^3} + \frac{6k_0}{\rho^4} + \frac{6i}{\rho^5}\right).\end{aligned} \tag{15.5.24}$$

With this value and $z = -2l$, (15.5.17) becomes

$$\begin{aligned}E_{0\rho}(\rho,\phi,0) &= -\frac{\omega\mu_0(-2l)^2}{2\pi k_0^2}\cos\phi\, e^{ik_0\rho}\left(\frac{ik_0^2}{\rho^3} - \frac{3k_0}{\rho^4} - \frac{3i}{\rho^5}\right)\\ &= -\frac{\omega\mu_0 k_0(-2l)^2}{2\pi}\cos\phi\, e^{ik_0\rho}\left(\frac{i}{k_0\rho^3} - \frac{3}{k_0^2\rho^4} - \frac{3i}{k_0^3\rho^5}\right).\end{aligned} \tag{15.5.25}$$

This same procedure can be applied to the other components. Thus, for $E_{0\phi}(\rho,\phi,0)$,

$$\begin{aligned}E_{0\phi}(\rho,\phi,0) &= E_{0\phi}^d(\rho,\phi,0) - E_{0\phi}^i(\rho,\phi,z)\\ &= -\frac{\omega\mu_0}{4\pi k_0^2}\sin\phi\left[e^{ik_0\rho}\left(\frac{ik_0^2}{\rho} - \frac{k_0}{\rho^2} - \frac{i}{\rho^3}\right)\right.\\ &\qquad\left. - e^{ik_0 r_0}\left(\frac{ik_0^2}{r_0} - \frac{k_0}{r_0^2} - \frac{i}{r_0^3}\right)\right].\end{aligned} \tag{15.5.26}$$

Let

$$f(z) = e^{ik_0\rho}\left(\frac{ik_0^2}{\rho} - \frac{k_0}{\rho^2} - \frac{i}{\rho^3}\right) - e^{ik_0 r_0}\left(\frac{ik_0^2}{r_0} - \frac{k_0}{r_0^2} - \frac{i}{r_0^3}\right), \tag{15.5.27}$$

$$f(0) = 0, \tag{15.5.28}$$

$$f'(z) = -ik_0 e^{ik_0 r_0}\left(\frac{z}{r_0}\right)\left(\frac{ik_0^2}{r_0} - \frac{k_0}{r_0^2} - \frac{i}{r_0^3}\right) - e^{ik_0 r_0}\left(-\frac{ik_0^2}{r_0^2} + \frac{2k_0}{r_0^3} + \frac{3i}{r_0^4}\right)\left(\frac{z}{r_0}\right), \tag{15.5.29}$$

$$f'(0) = 0, \tag{15.5.30}$$

$$f''(z) = -ik_0 e^{ik_0 r_0}\left(\frac{ik_0^2}{r_0^2} - \frac{k_0}{r_0^3} - \frac{i}{r_0^4}\right) - e^{ik_0 r_0}\left(-\frac{ik_0^2}{r_0^3} + \frac{2k_0}{r_0^4} + \frac{3i}{r_0^5}\right) + \left(\frac{z}{r_0}\right)\Big[\quad\Big], \tag{15.5.31}$$

$$\frac{z^2}{2}f''(0) = e^{ik_0\rho}\left(\frac{z^2}{2}\right)\left(\frac{k_0^3}{\rho^2} + \frac{ik_0^2}{\rho^3} - \frac{k_0}{\rho^4} + \frac{ik_0^2}{\rho^3} - \frac{2k_0}{\rho^4} - \frac{3i}{\rho^5}\right)$$
$$= e^{ik_0\rho}\left(\frac{z^2}{2}\right)\left(\frac{k_0^3}{\rho^2} + \frac{2ik_0^2}{\rho^3} - \frac{3k_0}{\rho^4} - \frac{3i}{\rho^5}\right). \tag{15.5.32}$$

Hence, with $z = -2l$,

$$E_{0\phi}(\rho, \phi, 0) = -\frac{\omega\mu_0 k_0 (2l)^2}{8\pi}\sin\phi\, e^{ik_0\rho}\left(\frac{1}{\rho^2} + \frac{2i}{k_0\rho^3} - \frac{3}{k_0^2\rho^4} - \frac{3i}{k_0^3\rho^5}\right). \tag{15.5.33}$$

The component $E_{0z}(\rho, \phi, 0)$ is

$$E_{0z}(\rho, \phi, 0) = E_{0z}^d(\rho, \phi, 0) - E_{0z}^i(\rho, \phi, z)$$
$$= \frac{\omega\mu_0}{4\pi k_0^2}\cos\phi\, e^{ik_0 r_0}\left(\frac{\rho z}{r_0^2}\right)\left(\frac{ik_0^2}{r_0} - \frac{3k_0}{r_0^2} - \frac{3i}{r_0^3}\right). \tag{15.5.34}$$

Let

$$f(z) = e^{ik_0 r_0}\left(\frac{\rho z}{r_0^2}\right)\left(\frac{ik_0^2}{r_0} - \frac{3k_0}{r_0^2} - \frac{3i}{r_0^3}\right), \tag{15.5.35}$$
$$f(0) = 0, \tag{15.5.36}$$
$$f'(z) = ik_0 e^{ik_0 r_0}\left(\frac{z^2}{r_0^2}\right)\left(\frac{\rho}{r_0}\right)\left(\frac{ik_0^2}{r_0} - \frac{3k_0}{r_0^2} - \frac{3i}{r_0^3}\right)$$
$$+ e^{ik_0 r_0}\left(\frac{\rho}{r_0^2}\right)\left(\frac{ik_0^2}{r_0} - \frac{3k_0}{r_0^2} - \frac{3i}{r_0^3}\right) + \left(\frac{z}{r_0}\right)\Big[\quad\Big], \tag{15.5.37}$$
$$zf'(0) = ze^{ik_0\rho}\left(\frac{ik_0^2}{\rho^2} - \frac{3k_0}{\rho^3} - \frac{3i}{\rho^4}\right). \tag{15.5.38}$$

Hence, (15.5.34) becomes

$$E_{0z}(\rho, \phi, 0) = -\frac{\omega\mu_0(2l)}{4\pi}\cos\phi\, e^{ik_0\rho}\left(\frac{i}{\rho^2} - \frac{3}{k_0\rho^3} - \frac{3i}{k_0^2\rho^4}\right). \tag{15.5.39}$$

The magnetic-field components are obtained similarly.

$$\left.\begin{matrix} B_{0\rho}(\rho, \phi, 0) \\ B_{0\phi}(\rho, \phi, 0) \end{matrix}\right\} = \left\{\begin{matrix} B_{0\rho}^d(\rho, \phi, 0) \\ B_{0\phi}^d(\rho, \phi, 0) \end{matrix}\right\} - \left\{\begin{matrix} B_{0\rho}^i(\rho, \phi, z) \\ B_{0\phi}^i(\rho, \phi, z) \end{matrix}\right\}$$
$$= -\frac{\mu_0}{4\pi}\left\{\begin{matrix} \sin\phi \\ \cos\phi \end{matrix}\right\} e^{ik_0 r_0}\left(\frac{z}{r_0}\right)\left(\frac{ik_0}{r_0} - \frac{1}{r_0^2}\right). \tag{15.5.40}$$

Let

$$f(z) = e^{ik_0 r_0}\left(\frac{z}{r_0}\right)\left(\frac{ik_0}{r_0} - \frac{1}{r_0^2}\right), \tag{15.5.41}$$
$$f(0) = 0, \tag{15.5.42}$$
$$f'(z) = ik_0 e^{ik_0 r_0}\left(\frac{z^2}{r_0^2}\right)\left(\frac{ik_0}{r_0} - \frac{1}{r_0^2}\right)$$

$$+ e^{ik_0 r_0}\left(\frac{ik_0}{r_0^2} - \frac{1}{r_0^3}\right) + \left(\frac{z}{r_0}\right)\Big[\quad\Big], \tag{15.5.43}$$

$$zf'(0) = ze^{ik_0\rho}\left(\frac{ik_0}{\rho^2} - \frac{1}{\rho^3}\right) \tag{15.5.44}$$

so that

$$B_{0\rho}(\rho, \phi, 0) = -\frac{i\mu_0 k_0(-2l)}{4\pi}\sin\phi\, e^{ik_0\rho}\left(\frac{1}{\rho^2} + \frac{i}{k_0\rho^3}\right) \tag{15.5.45}$$

and

$$B_{0\phi}(\rho, \phi, 0) = -\frac{i\mu_0 k_0(-2l)}{4\pi}\cos\phi\, e^{ik_0\rho}\left(\frac{1}{\rho^2} + \frac{i}{k_0\rho^3}\right). \tag{15.5.46}$$

The final component is

$$B_{0z}(\rho, \phi, 0) = -\frac{\mu_0}{4\pi}\sin\phi\left[e^{ik_0\rho}\left(\frac{ik_0}{\rho} - \frac{1}{\rho^2}\right) - e^{ik_0 r_0}\left(\frac{ik_0}{r_0} - \frac{1}{r_0^2}\right)\left(\frac{\rho}{r_0}\right)\right]. \tag{15.5.47}$$

Let

$$\begin{aligned} f(z) &= e^{ik_0\rho}\left(\frac{ik_0}{\rho} - \frac{1}{\rho^2}\right) - e^{ik_0 r_0}\left(\frac{ik_0}{r_0} - \frac{1}{r_0^2}\right)\left(\frac{\rho}{r_0}\right) \\ &= e^{ik_0\rho}\left(\frac{ik_0}{\rho} - \frac{1}{\rho^2}\right) - \rho e^{ik_0 r_0}\left(\frac{ik_0}{r_0^2} - \frac{1}{r_0^3}\right), \end{aligned} \tag{15.5.48}$$

$$f(0) = 0, \tag{15.5.49}$$

$$\begin{aligned} f'(z) &= -ik_0 e^{ik_0 r_0}\left(\frac{z}{r_0}\right)\left(\frac{ik_0}{r_0} - \frac{1}{r_0^2}\right)\left(\frac{\rho}{r_0}\right) \\ &\quad - \rho e^{ik_0 r_0}\left(-\frac{2ik_0}{r_0^3} + \frac{3}{r_0^4}\right)\left(\frac{z}{r_0}\right), \end{aligned} \tag{15.5.50}$$

$$f'(0) = 0, \tag{15.5.51}$$

$$\begin{aligned} f''(z) &= -ik_0 e^{ik_0 r_0}\left(\frac{ik_0}{r_0^2} - \frac{1}{r_0^3}\right)\left(\frac{\rho}{r_0}\right) \\ &\quad - e^{ik_0 r_0}\left(-\frac{2ik_0}{r_0^3} + \frac{3}{r_0^4}\right)\left(\frac{\rho}{r_0}\right) + \left(\frac{z}{r_0}\right)\Big[\quad\Big], \end{aligned} \tag{15.5.52}$$

$$\begin{aligned} \frac{z^2}{2} f''(0) &= \frac{z^2}{2}\, e^{ik_0\rho}\left(\frac{k_0^2}{\rho^2} + \frac{ik_0}{\rho^3} + \frac{2ik_0}{\rho^3} - \frac{3}{\rho^4}\right) \\ &= \frac{z^2}{2}\, k_0^2 e^{ik_0\rho}\left(\frac{1}{\rho^2} + \frac{3i}{k_0\rho^3} - \frac{3}{k_0^2\rho^4}\right). \end{aligned} \tag{15.5.53}$$

Hence,

$$B_{0z}(\rho, \phi, 0) = -\frac{\mu_0 k_0^2(2l)^2}{8\pi}\sin\phi\, e^{ik_0\rho}\left(\frac{1}{\rho^2} + \frac{3i}{k_0\rho^3} - \frac{3}{k_0^2\rho^4}\right). \tag{15.5.54}$$

In summary, and retaining terms up to order $1/\rho^3$, the field is

$$E_{0\rho}(\rho,\phi,0) = -\frac{\omega\mu_0}{2\pi k_0}\,(2k_0 l)^2\cos\phi\,\frac{ie^{ik_0\rho}}{k_0\rho^3}, \tag{15.5.55}$$

$$E_{0\phi}(\rho,\phi,0) = -\frac{\omega\mu_0}{8\pi k_0}\,(2k_0 l)^2\sin\phi\,e^{ik_0\rho}\left(\frac{1}{\rho^2}+\frac{2i}{k_0\rho^3}\right), \tag{15.5.56}$$

$$E_{0z}(\rho,\phi,0) = -\frac{i\omega\mu_0}{4\pi k_0}\,(2k_0 l)\cos\phi\,e^{ik_0\rho}\left(\frac{1}{\rho^2}+\frac{3i}{k_0\rho^3}\right), \tag{15.5.57}$$

$$B_{0\rho}(\rho,\phi,0) = \frac{i\mu_0}{4\pi}\,(2k_0 l)\sin\phi\,e^{ik_0\rho}\left(\frac{1}{\rho^2}+\frac{i}{k_0\rho^3}\right), \tag{15.5.58}$$

$$B_{0\phi}(\rho,\phi,0) = \frac{i\mu_0}{4\pi}\,(2k_0 l)\cos\phi\,e^{ik_0\rho}\left(\frac{1}{\rho^2}+\frac{i}{k_0\rho^3}\right), \tag{15.5.59}$$

$$B_{0z}(\rho,\phi,0) = -\frac{\mu_0}{8\pi}\,(2k_0 l)^2\sin\phi\,e^{ik_0\rho}\left(\frac{1}{\rho^2}+\frac{3i}{k_0\rho^3}\right). \tag{15.5.60}$$

When these are compared with (15.5.8)–(15.5.13), the major difference is the absence of the $1/\rho$ terms in the formulas for $E_{0\rho}(\rho,\phi,0)$, $E_{0z}(\rho,\phi,0)$, and $B_{0\phi}(\rho,\phi,0)$. It is these components that provide a strong collinear coupling in the presence of the dielectric layer. When they are generated by each element of a microstrip transmission line, the definition of frequency-independent parameters per unit length as required by TEM transmission-line theory becomes impossible.

15.6 The Propagation of Pulses on Microstrip

A simple way in which to show the properties of microstrip under transient conditions is to superimpose a suitable set of frequencies that combine to form a well-shaped pulse (Gulla 1988). This is illustrated in Fig. 15.6.1 which shows a pulse formed by the superposition of the following set of frequencies: 50 MHz to 50 GHz at 50-MHz intervals with Gaussian amplitude weighting. The propagation of this pulse along the boundary between air and a dielectric half-space can be observed by evaluating the field due to the superposition of the same set of frequencies at successive radial distances. The results at four radial distances are shown in Figs. 15.6.2a,b when $\epsilon_{1r} = 11.8$. At the shortest radial distance (top graph in Fig. 15.6.2a), the component pulses due respectively to the superposition of the lateral waves in air and the direct waves in the dielectric are also shown. This separate evaluation is not necessary at larger distances from the source since the two pulses become separated because they have different velocities, viz., the velocity of light $c = 3\times10^8$ m/sec in the air and the velocity $v = c\epsilon_{1r}^{-1/2}$ in the dielectric. Close to the source the lateral-wave

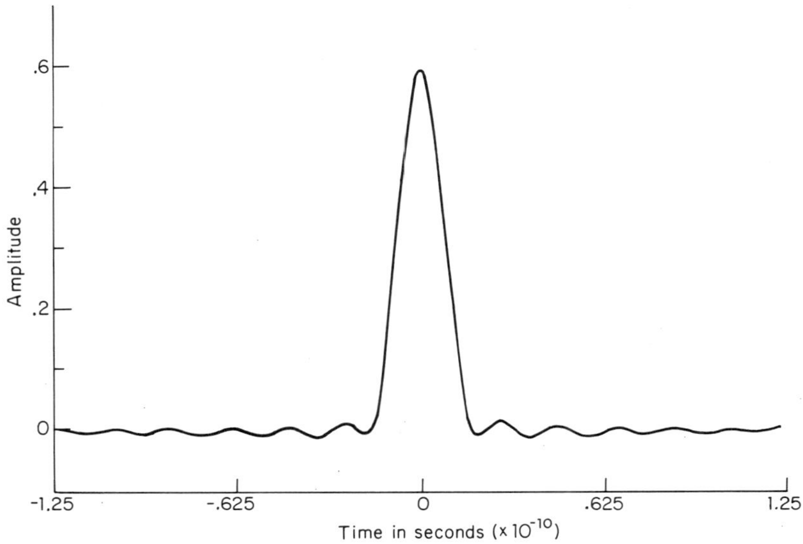

Figure 15.6.1. Input signal: superposition of continuous-wave signals at 50-MHz intervals from 50 MHz to 50 GHz; Gaussian amplitude weighting. Taken from Gulla (1988).

pulse is much smaller than the direct-wave pulse. As the radial distance is increased, the lateral-wave pulse grows relative to the direct-wave pulse. Note that the two shapes are quite different: the direct-wave pulse corresponds closely to the initial pulse in Fig. 15.6.1; the lateral-wave pulse suggests the derivative of the initial pulse. The reason for these differences is explained in Section 15.7.

The same initial pulse can be applied to microstrip by the same procedure, viz., the evaluation of the field at successive distances due to the superposition of the fields generated at the selected set of frequencies. The initial pulse and the propagated pulse for E_{0z} at different radial distances from the source are shown in Figs. 15.6.3a,b for a microstrip with a substrate of thickness $l = 0.25$ mm and relative permittivity $\epsilon_{1r} = 11.8$. (With these values for l and ϵ_{1r}, the condition $k_1^2 l^2 \ll 1$ is satisfied over the entire range of frequencies.) Note that there is no pulse corresponding to the direct wave propagating in the dielectric half-space. The velocity of propagation of the surface wave in the thin dielectric layer differs negligibly from that in the air. The general behavior of the surface-wave pulse is seen to be similar to that of the surface-wave pulse along the two-layered region shown in Figs. 15.6.2a,b except that the shape is different. The surface-wave pulse in the three-layered region has an additional peak due to reflection from the conducting base.

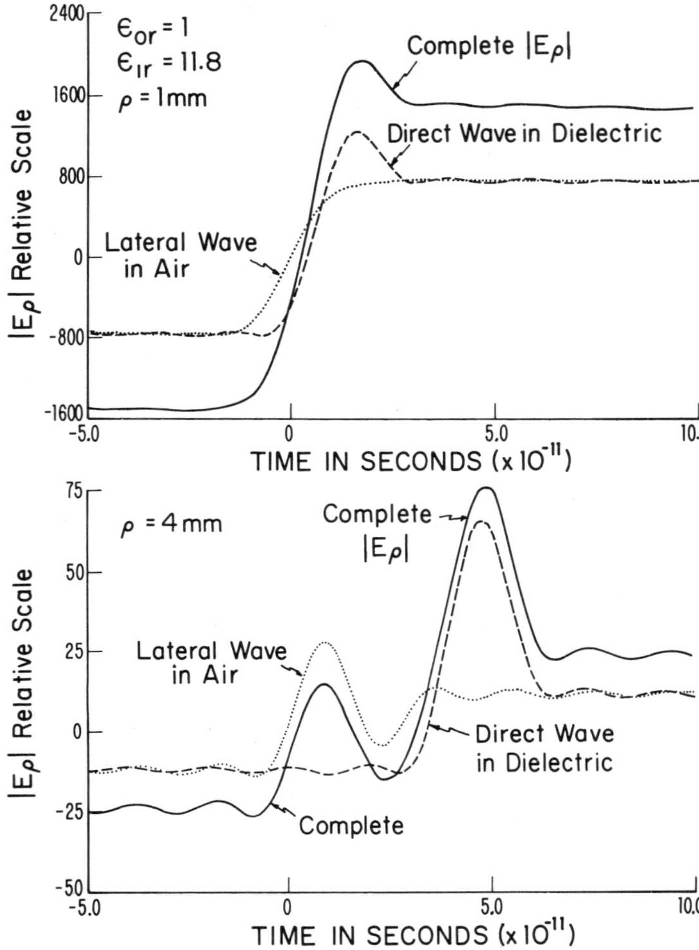

Figure 15.6.2a. Pulse propagating along air–dielectric boundary; taken from Gulla (1988). Top: $\rho = 1$ mm; bottom: $\rho = 4$ mm.

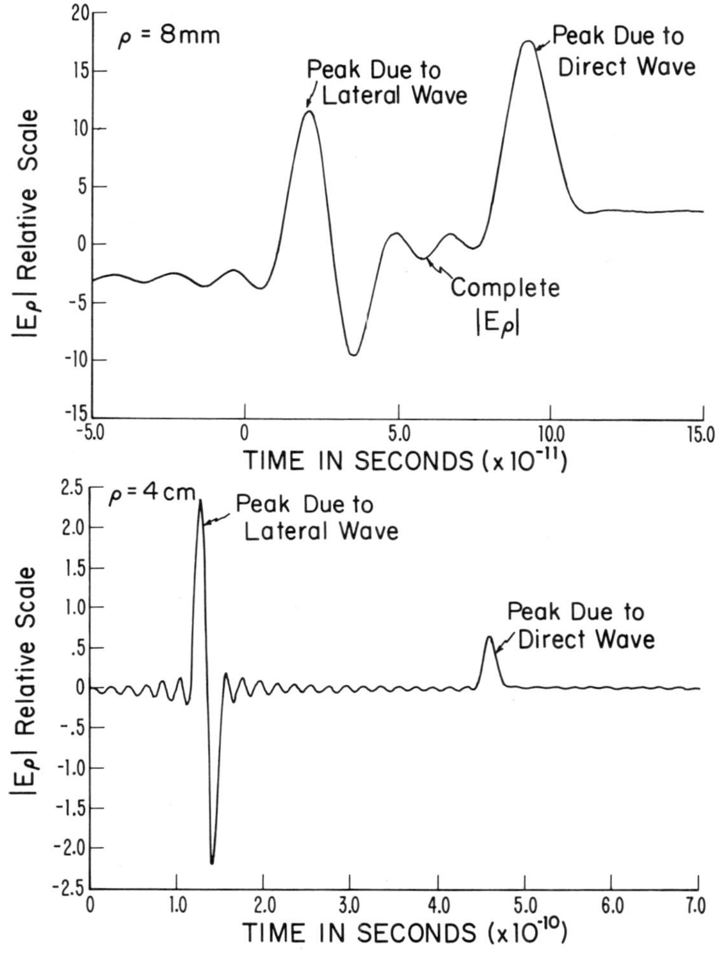

Figure 15.6.2b. Pulse propagating along air–dielectric boundary; taken from Gulla (1988). Top: $\rho = 8$ mm; bottom: $\rho = 4$ cm.

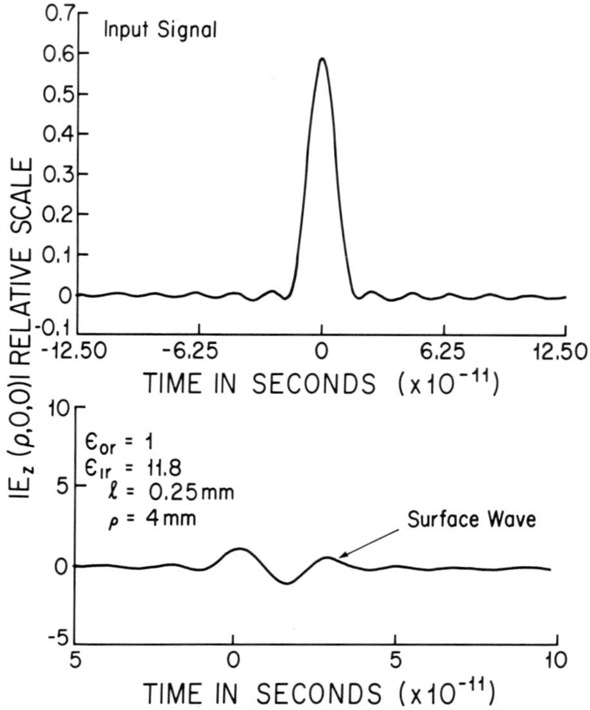

Figure 15.6.3a. Pulse propagating along microstrip; taken from Gulla (1988). Top: input signal; bottom: $\rho = 4$ mm.

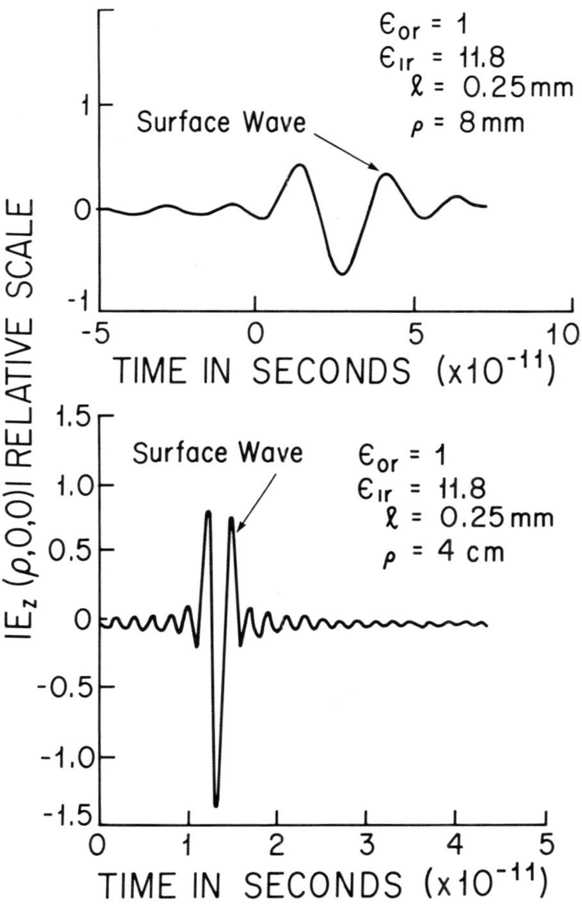

Figure 15.6.3b. Pulse propagating along microstrip; taken from Gulla (1988). Top: $\rho = 8$ mm; bottom: $\rho = 4$ cm.

15.7 The Propagation of a Gaussian Pulse Along the Boundary Between Air and a Dielectric

In the preceding section the propagation of a single pulse along the boundary between air and a dielectric half-space and along the air–substrate surface of open microstrip is examined simply by the superposition of a selection of frequencies that together constitute a single pulse at the source point. A more fundamental alternative involves the derivation of explicit formulas for the propagation of pulses along these boundaries when the exciting source is a horizontal dipole located on the same boundaries. From such formulas, an understanding can be gained of the reasons for the differences in the shapes of the pulses and their rates of decrease in amplitude with distance from the source.

In Chapter 14, the transient electromagnetic field of a vertical electric dipole located on the boundary between two dielectric half-spaces is determined when the dipole is excited by a Gaussian current pulse. In Section 14.8, the corresponding field of a horizontal electric dipole is obtained from that of the vertical dipole. The components of interest here are, from (14.8.9) and (14.8.13),

$$\begin{aligned}
\epsilon^{-1}E_{0z}(\rho,0,0;t) &\\
= E_{1z}(\rho,0,0;t) &\\
= -\frac{1}{2\pi\epsilon_0\epsilon^{3/2}}\,\frac{\pi^{-1/2}}{c^3t_1^3}\Bigg[&\left(-\frac{2\tau'}{\rho_1}+\frac{1}{\rho_1^2}\right)e^{-\tau'^2} \\
&+\frac{\pi}{2}\,\frac{2^{1/4}}{(\epsilon\rho_1)^{1/2}}\,e^{-T'^2/2} \\
&\times\left\{\begin{array}{l} V(0,\sqrt{2}\,T')-\sqrt{2}\,T'\,V(-1,\sqrt{2}\,T') \\ \pi^{-1/2}[U(0,-\sqrt{2}\,T') \\ \qquad +2\sqrt{2}\,T'\,U(-1,-\sqrt{2}\,T')] \end{array}\right\}\Bigg]; \\
&\qquad \left\{\begin{array}{l} T'\ge 0 \\ T'\le 0 \end{array}\right., \qquad (15.7.1)
\end{aligned}$$

$$\begin{aligned}
E_{0\rho}(\rho,0,0;t) &\\
= E_{1\rho}(\rho,0,0;t) &\\
= \frac{1}{2\pi\epsilon_0\epsilon}\,\frac{\pi^{-1/2}}{c^3t_1^3}\Bigg[&\left(-\frac{2\tau'}{\rho_1}+\frac{1}{\rho_1^2}\right)e^{-\tau'^2}+\frac{\pi^{1/2}}{2\rho_1^3}(\operatorname{erf}\tau'+1) \\
&+\epsilon^{3/2}\left(\frac{e^{-\tau_\epsilon'^2}}{\rho_{1\epsilon}^2}+\frac{\pi^{1/2}}{2\rho_{1\epsilon}^3}(\operatorname{erf}\tau_\epsilon'+1)\right) \\
&+\frac{\pi}{2}\,\frac{2^{1/4}}{(\epsilon\rho_1)^{1/2}}\,e^{-T'^2/2}
\end{aligned}$$

$$\times \left\{ \begin{array}{l} V(0,\sqrt{2}\,T') - \sqrt{2}\,T'\,V(-1,\sqrt{2}\,T') \\ \pi^{-1/2}[U(0,-\sqrt{2}\,T') \\ \qquad + 2\sqrt{2}\,T'\,U(-1,-\sqrt{2}\,T')] \end{array} \right\} \Bigg] ;$$

$$\left\{ \begin{array}{l} T' \geq 0 \\ T' \leq 0 \end{array} \right. , \tag{15.7.2}$$

where $\epsilon = \epsilon_{1r}$ is the relative permittivity of Region 1—a perfect dielectric—and Region 0 is air; and where t_1 is the half-width of the normalized Gaussian pulse,

$$f(t) = \frac{e^{-t^2/t_1^2}}{t_1\sqrt{\pi}}, \tag{15.7.3}$$

and

$$\rho_1 = \frac{\rho}{ct_1}, \qquad \rho_{1\epsilon} = \frac{\rho\epsilon^{1/2}}{ct_1}, \tag{15.7.4}$$

$$\tau' = \frac{t - \rho/c}{t_1} = t' - \rho_1, \qquad \tau'_\epsilon = \frac{t - \epsilon^{1/2}\rho/c}{t_1} = t' - \rho_{1\epsilon}, \tag{15.7.5}$$

$$T' = \frac{t - \dfrac{\rho}{c}\left(1 - \dfrac{1}{2\epsilon}\right)}{t_1}. \tag{15.7.6}$$

The functions V and U are the parabolic cylinder functions defined in Abramowitz and Stegun (1972).

The dominant parts of (15.7.1) and (15.7.2) are the same. They occur inside the first pair of large parentheses which contains two terms. The second of these is a Gaussian pulse which decreases with radial distance from the source as $1/\rho^2$. The first term is the time derivative of the Gaussian pulse. It decreases with radial distance as $1/\rho$ so that it dominates except close to the source. This means that what starts as a Gaussian pulse quickly changes to a pulse shaped like the derivative. The two formulas also have in common the term with parabolic cylinder functions and an explicit $1/\rho^{1/2}$ decrease with radial distance. Note, however, that ρ also occurs in the arguments of the parabolic cylinder functions. This entire term is significant only at large distances where it approaches a $1/\rho^2$ dependence on the distance.

The expression (15.7.2) for $E_{1\rho}(\rho, 0, 0; t)$ includes a second Gaussian pulse which decreases as $1/\rho^2\epsilon$ and travels in the dielectric with the velocity $c\epsilon^{-1/2}$. It arrives later than the pulse in air and, at even moderate radial distances, it is much smaller. Also in (15.7.2) are two terms that involve the error function and that decrease with distance as $1/\rho^3$ and $1/\rho^3\epsilon^{3/2}$, respectively. They contribute to the initial rise of the pulse and give the entire static field that persists after both pulses have passed. Although not evident in (15.7.1) because of the approximation involved in the steady-state formula, there is actually also a similar second pulse in the complete transient for $E_{1z}(\rho, 0, 0; t)$.

The transient field given by (15.7.1) and (15.7.2) for the dipole with Gaussian pulse excitation is in agreement with that determined by the superposition of a set of frequencies as detailed in Section 15.6. The constructed pulse shown in Fig. 15.6.1 is not a Gaussian pulse but in its main peak is sufficiently like one so that the field generated by it closely resembles that of a Gaussian pulse. Since explicit formulas are available for the components of the field generated by a Gaussian pulse, these can be used to explain the properties of the field generated by the pulse shown in Fig. 15.6.1. In particular, the fact that the single pulse in Fig. 15.6.1 generates two field pulses of which the first is like the derivative of the generating pulse and the second is like the generating pulse itself is clarified by comparison with the formulas for the Gaussian pulse.

15.8 The Transient Electric Field Generated in Open Microstrip by a Horizontal Dipole Excited by a Gaussian Pulse

The steady-state formulas for the electric field generated by a horizontal electric dipole on the air–dielectric surface of microstrip are given by (15.4.17)–(15.4.19). In the direction $\phi = 0$ along the axis of the dipole, the $E_{0\rho}(\rho, 0, 0)$ and $E_{0z}(\rho, 0, 0)$ components have the following forms with $k_0 = \omega/c$:

$$\tilde{E}_{0\rho}(\rho', 0, 0; \omega) \sim \frac{\mu_0 l^2}{2\pi c^3}\left(\frac{i\omega^3}{\rho'} - \frac{\omega^2}{\rho'^2} - \frac{i\omega}{\rho'^3}\right)e^{i\omega\rho'}, \tag{15.8.1}$$

$$\tilde{E}_{0z}(\rho', 0, 0; \omega) \sim -\frac{\mu_0 l}{2\pi c^2}\left(\frac{\omega^2}{\rho'} + \frac{i\omega}{\rho'^2}\right)e^{i\omega\rho'}, \tag{15.8.2}$$

where

$$\rho' = \frac{\rho}{c} \tag{15.8.3}$$

and l is the thickness of the substrate.

When the dipole is excited by a normalized Gaussian current pulse $f(t)$ with its Fourier transform $f(\omega)$, viz.,

$$f(t) = \frac{e^{-t^2/t_1^2}}{t_1\sqrt{\pi}}; \qquad \tilde{f}(\omega) = e^{-\omega^2 t_1^2/4}, \tag{15.8.4}$$

the transient field is given by

$$\begin{aligned} E_{0\rho}(\rho', 0, 0; t) &= \frac{1}{2\pi} \operatorname{Re} \int_{-\infty}^{\infty} \tilde{E}_{0\rho}(\rho', 0, 0; \omega) e^{-\omega^2 t_1^2/4} e^{-i\omega t}\, d\omega \\ &= \frac{\mu_0 l^2}{2\pi c^3} \left(\frac{I_1''}{\rho'} - \frac{I_1'}{\rho'^2} - \frac{I_1}{\rho'^3} \right). \end{aligned} \tag{15.8.5}$$

Similarly,

$$\begin{aligned} E_{0z}(\rho', 0, 0; t) &= \frac{1}{2\pi} \operatorname{Re} \int_{-\infty}^{\infty} \tilde{E}_{0z}(\rho', 0, 0; \omega) e^{-\omega^2 t_1^2/4} e^{-i\omega t}\, d\omega \\ &= -\frac{\mu_0 l}{2\pi c^2} \left(\frac{I_1'}{\rho'} + \frac{I_1}{\rho'^2} \right), \end{aligned} \tag{15.8.6}$$

where $\tau = t - \rho'$. The requirement that $k_1 l$ be small in the frequency domain implies that in the time domain l be small compared to the half-width of the Gaussian pulse, i.e.,

$$\epsilon_{1r} l^2 \ll (ct_1)^2. \tag{15.8.7}$$

From (14.6.5) and (14.6.17),

$$\begin{aligned} I_1 &= \operatorname{Re} \frac{1}{2\pi} \int_{-\infty}^{\infty} i\omega e^{-\omega^2 t_1^2/4} e^{-i\omega\tau}\, d\omega \\ &= \frac{2\tau}{t_1^3\sqrt{\pi}} e^{-\tau^2/t_1^2}, \end{aligned} \tag{15.8.8}$$

$$\begin{aligned} I_1' &= \operatorname{Re} \frac{1}{2\pi} \int_{-\infty}^{\infty} \omega^2 e^{-\omega^2 t_1^2/4} e^{-i\omega\tau}\, d\omega \\ &= \frac{dI_1}{dt} = \left(\frac{2}{t_1^3\sqrt{\pi}} - \frac{4\tau^2}{t_1^5\sqrt{\pi}} \right) e^{-\tau^2/t_1^2}, \end{aligned} \tag{15.8.9}$$

$$\begin{aligned} I_1'' &= \operatorname{Re} \frac{1}{2\pi} \int_{-\infty}^{\infty} i\omega^3 e^{-\omega^2 t_1^2/4} e^{-i\omega\tau}\, d\omega \\ &= -\frac{dI_1'}{dt} = \left(\frac{4\tau}{t_1^5\sqrt{\pi}} + \frac{8\tau}{t_1^5\sqrt{\pi}} - \frac{8\tau^3}{t_1^7\sqrt{\pi}} \right) e^{-\tau^2/t_1^2} \\ &= \left(\frac{12\tau}{t_1^5\sqrt{\pi}} - \frac{8\tau^3}{t_1^7\sqrt{\pi}} \right) e^{-\tau^2/t_1^2}. \end{aligned} \tag{15.8.10}$$

With these values

$$\begin{aligned} E_{0\rho}(\rho', 0, 0; t) = \frac{\mu_0 l^2}{2\pi c^3} \bigg[&\frac{1}{\rho'} \left(\frac{12\tau}{t_1^5\sqrt{\pi}} - \frac{8\tau^3}{t_1^7\sqrt{\pi}} \right) \\ &- \frac{1}{\rho'^2} \left(\frac{2}{t_1^3\sqrt{\pi}} - \frac{4\tau^2}{t_1^5\sqrt{\pi}} \right) - \frac{1}{\rho'^3} \frac{2\tau}{t_1^3\sqrt{\pi}} \bigg] e^{-\tau^2/t_1^2}, \end{aligned} \tag{15.8.11}$$

$$E_{0z}(\rho',0,0;t) = -\frac{\mu_0 l}{2\pi c^2}\left[\frac{1}{\rho'}\left(\frac{2}{t_1^3\sqrt{\pi}} - \frac{4\tau^2}{t_1^5\sqrt{\pi}}\right) + \frac{1}{\rho'^2}\frac{2\tau}{t_1^3\sqrt{\pi}}\right]e^{-\tau^2/t_1^2}. \tag{15.8.12}$$

With the notation,

$$\tau' = \frac{\tau}{t_1}; \qquad \rho_1 = \frac{\rho'}{t_1} = \frac{\rho}{ct_1}, \tag{15.8.13}$$

and $\mu_0\epsilon_0 = 1/c^2$, these formulas reduce to

$$E_{0\rho}(\rho,0,0;t) = \frac{l^2}{\pi\epsilon_0}\frac{\pi^{-1/2}}{c^5 t_1^5}\left(\frac{2}{\rho_1}(3\tau' - 2\tau'^3) - \frac{1-2\tau'^2}{\rho_1^2} - \frac{\tau'}{\rho_1^3}\right)e^{-\tau'^2}, \tag{15.8.14}$$

$$E_{0z}(\rho,0,0;t) = -\frac{l}{\pi\epsilon_0}\frac{\pi^{-1/2}}{c^4 t_1^4}\left(\frac{1-2\tau'^2}{\rho_1} + \frac{\tau'}{\rho_1^2}\right)e^{-\tau'^2}. \tag{15.8.15}$$

The Gaussian current pulse on the horizontal dipole is shown at the upper left in Fig. 15.8.1. It has the half-width $t_1 = 62.5/3 = 20.8$ psec. Following it is shown the vertical electric-field pulse (E_z) that is generated on the surface of the dielectric at the radial distance $\rho = 6.25$ mm. (Note that the exciting Gaussian current pulse has its central maximum at the reference time $t = 0$.) The pulse arrives at $\rho = 50$ mm at a later time and has the shape shown on the lower left in Fig. 15.8.1. The pulse at 100 mm is shown at the lower right. As soon as the $1/\rho$ term becomes dominant, the amplitude decreases at the $1/\rho$ rate and the shape changes only slightly as the pulse advances with the velocity $c = 3 \times 10^8$ m/sec. Similar graphs for the radial electric-field pulse (E_ρ) are in Fig. 15.8.2. Both the $E_{0\rho}$ and E_{0z} pulses extend into the dielectric substrate as $E_{1\rho}$ and E_{1z}, but the amplitude of the latter is smaller in the dielectric by the factor ϵ_{1r}^{-1}. That is, $E_{1z} = E_{0z}/\epsilon_{1r}$, whereas $E_{1\rho} = E_{0\rho}$. When the dielectric layer is removed so that the dipole is at the height l in air over the highly conducting base, the steady-state formula for $E_{0z}(\rho,0,0;\omega)$ is, from (15.5.39),

$$\tilde{E}_{0z}(\rho,0,0;\omega) = -\frac{\mu_0 l}{2\pi}\left(\frac{i\omega}{\rho^2} - \frac{3c}{\rho^3} - \frac{3ic^2}{\omega\rho^4}\right)e^{i\omega\rho/c}. \tag{15.8.16a}$$

With $\rho' = \rho/c$ and $\mu_0\epsilon_0 = 1/c^2$, this becomes

$$\tilde{E}_{0z}(\rho',0,0;\omega) = -\frac{l}{2\pi\epsilon_0 c^4}\left(\frac{i\omega}{\rho'^2} - \frac{3}{\rho'^3} - \frac{3i}{\omega\rho'^4}\right)e^{i\omega\rho'}. \tag{15.8.16b}$$

Similarly, from (15.5.25),

$$\tilde{E}_{0\rho}(\rho,0,0;\omega) = -\frac{\mu_0(2l)^2}{2\pi}\left(\frac{i\omega}{\rho^3} - \frac{3c}{\rho^4} - \frac{3ic^2}{\omega\rho^5}\right)e^{i\omega\rho/c} \tag{15.8.17a}$$

or

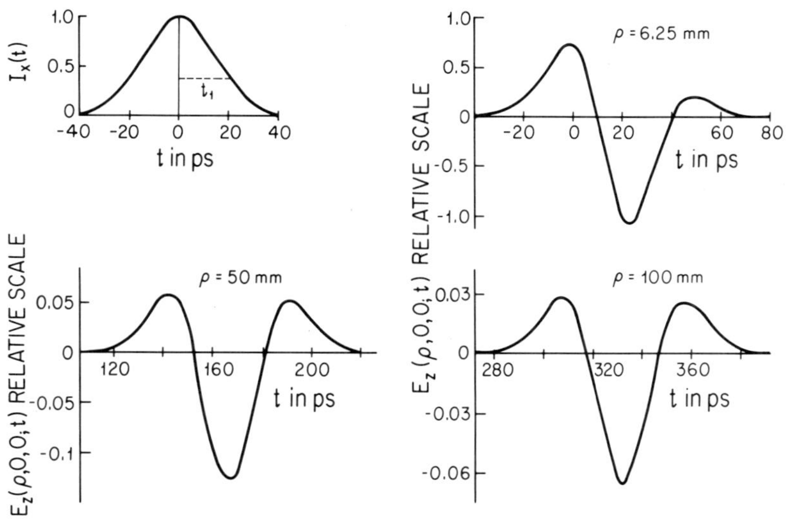

Figure 15.8.1. Surface-wave pulse of $E_z(\rho, 0, 0; t)$ on surface of microstrip with $l =$ 1.5 mm, $\epsilon_{1r} = 13$, due to Gaussian current pulse with half-width $t_1 = 20.8$ psec in horizontal dipole on surface.

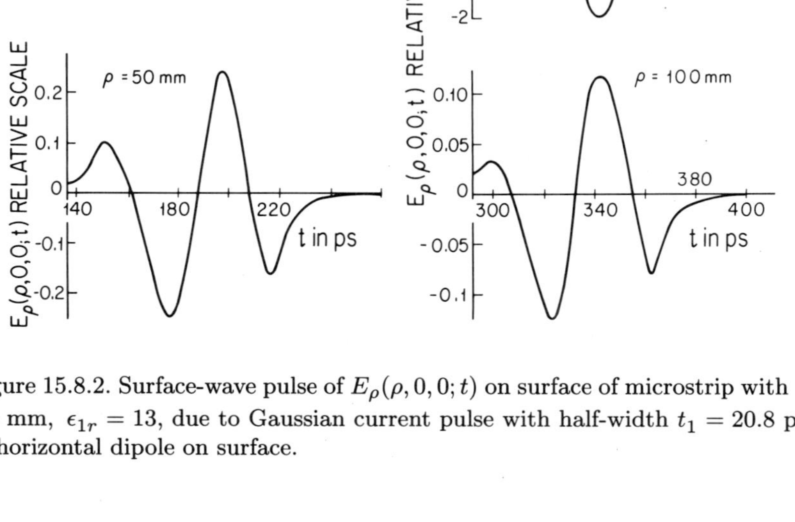

Figure 15.8.2. Surface-wave pulse of $E_\rho(\rho, 0, 0; t)$ on surface of microstrip with $l =$ 1.5 mm, $\epsilon_{1r} = 13$, due to Gaussian current pulse with half-width $t_1 = 20.8$ psec in horizontal dipole on surface.

$$\tilde{E}_{0\rho}(\rho', 0, 0; \omega) = -\frac{(2l)^2}{2\pi\epsilon_0 c^5}\left(\frac{i\omega}{\rho'^3} - \frac{3}{\rho'^4} - \frac{3i}{\omega\rho'^5}\right)e^{i\omega\rho'}. \tag{15.8.17b}$$

When the dipole is excited by the normalized Gaussian pulse defined in (15.8.4), the transient electric field is

$$E_{0z}(\rho', 0, 0; t) = \frac{1}{2\pi}\,\mathrm{Re}\int_{-\infty}^{\infty} \tilde{E}_{0z}(\rho', 0, 0; \omega)e^{-\omega^2 t_1^2/4}e^{-i\omega t}\,d\omega. \tag{15.8.18}$$

With (15.8.16b), this becomes

$$E_{0z}(\rho', 0, 0; t) = -\frac{l}{2\pi\epsilon_0 c^4}\left(\frac{I_1}{\rho'^2} - \frac{3I_2}{\rho'^3} - \frac{3I_3}{\rho'^4}\right), \tag{15.8.19}$$

where, with $\tau = t - \rho'$ and from (14.6.16), (14.6.17), and (14.6.21),

$$\begin{aligned} I_1 &= \mathrm{Re}\,\frac{1}{2\pi}\int_{-\infty}^{\infty} i\omega e^{-\omega^2 t_1^2/4}e^{-i\omega\tau}\,d\omega \\ &= \frac{2\tau}{t_1^3\sqrt{\pi}}\,e^{-\tau^2/t_1^2}, \end{aligned} \tag{15.8.20}$$

$$\begin{aligned} I_2 &= \mathrm{Re}\,\frac{1}{2\pi}\int_{-\infty}^{\infty} e^{-\omega^2 t_1^2/4}e^{-i\omega\tau}\,d\omega \\ &= \frac{1}{t_1\sqrt{\pi}}\,e^{-\tau^2/t_1^2}, \end{aligned} \tag{15.8.21}$$

$$\begin{aligned} I_3 &= \mathrm{Re}\,\frac{1}{2\pi}\int_{-\infty}^{\infty} \frac{i}{\omega}\,e^{-\omega^2 t_1^2/4}e^{-i\omega\tau}\,d\omega \\ &= \tfrac{1}{2}[1 + \mathrm{erf}(\tau/t_1)]. \end{aligned} \tag{15.8.22}$$

With these values,

$$\begin{aligned} E_{0z}(\rho', 0, 0; t) = -\frac{l\pi^{-1/2}}{2\pi\epsilon_0 c}\Bigg[&\left(\frac{1}{c^3 t_1^3}\frac{2\tau}{\rho'^2} - \frac{3}{c^3 t_1\rho'^3}\right)e^{-\tau^2/t_1^2} \\ &- \frac{3\pi^{1/2}}{2\rho'^4}\,[1 + \mathrm{erf}(\tau/t_1)]\Bigg]. \end{aligned} \tag{15.8.23}$$

With $\rho_1 = \rho'/t_1 = \rho/ct_1$ and $\tau' = \tau/t_1$,

$$\begin{aligned} E_{0z}(\rho, 0, 0; t) = -\frac{l\pi^{-1/2}}{2\pi\epsilon_0}\,\frac{1}{c^4 t_1^4}\Bigg[&\left(\frac{2\tau'}{\rho_1^2} - \frac{3}{\rho_1^3}\right)e^{-\tau'^2} \\ &- \frac{3\pi^{1/2}}{2\rho_1^4}\,(1 + \mathrm{erf}\,\tau')\Bigg]. \end{aligned} \tag{15.8.24}$$

Similarly,

$$\begin{aligned} E_{0\rho}(\rho', 0, 0; t) &= \frac{1}{2\pi}\,\mathrm{Re}\int_{-\infty}^{\infty} \tilde{E}_{0\rho}(\rho', 0, 0; \omega)e^{-\omega^2 t_1^2/4}e^{-i\omega t}\,d\omega \\ &= -\frac{(2l)^2}{2\pi\epsilon_0 c^5}\left(\frac{I_1}{\rho'^3} - \frac{3I_2}{\rho'^4} - \frac{3I_3}{\rho'^5}\right) \end{aligned} \tag{15.8.25}$$

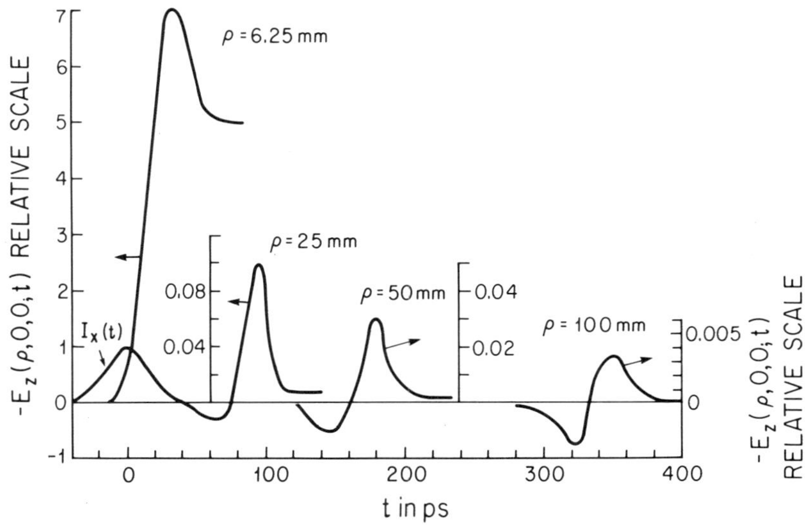

Figure 15.8.3. Pulse of $-E_z(\rho, 0, 0; t)$ at distances ρ from dipole along its axis when dipole is at height $l = 1.5$ mm from perfectly conducting half-space and is excited by Gaussian current pulse of half-width $t_1 = 20.8$ psec.

and

$$E_{0\rho}(\rho, 0, 0; t) = -\frac{(2l)^2\pi^{-1/2}}{2\pi\epsilon_0}\frac{1}{c^5t_1^5}\left[\left(\frac{2\tau'}{\rho_1^3} - \frac{3}{\rho_1^4}\right)e^{-\tau'^2} - \frac{3\pi^{1/2}}{2\rho_1^5}(1 + \operatorname{erf}\tau')\right]. \tag{15.8.26}$$

Note that $E_{0\rho}(\rho, 0, 0; t) = E_{0z}(\rho, 0, 0; t) \times 4l/\rho_1 ct_1 = E_{0z}(\rho, 0, 0; t) \times 4l/\rho$.

Graphs of $-E_{0z}(\rho, 0, 0; t)$ as given in (15.8.24) are shown in Fig. 15.8.3. The exciting Gaussian current pulse is shown at the lower left. Its center is at $t = 0$. The first curve is the pulse as it passes $\rho = 6.25$ mm from the dipole. At this close distance, the $1/\rho^4$ term with the error function dominates. At $\rho = 25$ mm, the amplitude has decreased greatly and the $1/\rho^3$ term with its Gaussian shape is dominant. At $\rho = 50$ mm and $\rho = 100$ mm, the $1/\rho^4$ and $1/\rho^3$ terms contribute little and the derivative of the Gaussian shape with a $1/\rho^2$ decrease in amplitude is the final shape. Note that $-E_{0\rho}(\rho, 0, 0; t)$ is like $-E_{0z}(\rho, 0, 0; t)$ but it decreases in amplitude at an even faster rate with increasing radial distance since it is given by $E_{0\rho}(\rho, 0, 0; t) \sim E_{0z}(\rho, 0, 0; t) \times 4l/\rho$.

When Fig. 15.8.3 is compared with Fig. 15.8.1 for the same vertical field but with the dielectric layer present, very significant differences are obvious. In the absence of the dielectric layer, the decrease in amplitude with radial distance is extremely rapid for $E_z \sim 1/\rho^2$ and even more so for $E_\rho \sim 1/\rho^3$. With the dielectric present and in the intermediate range, both

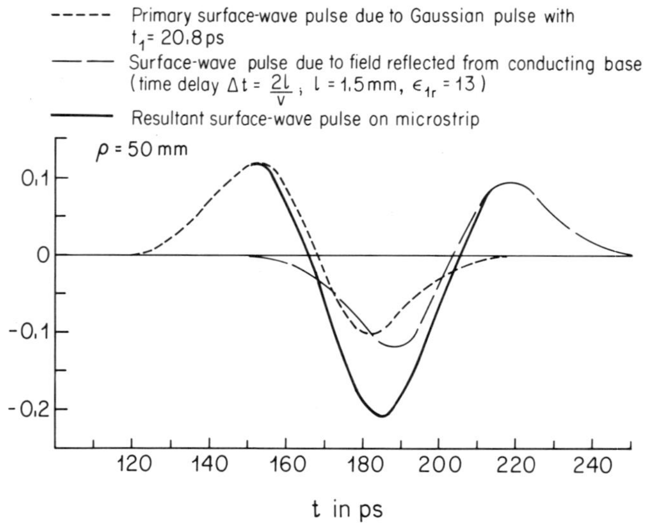

Figure 15.8.4. Surface-wave pulse on microstrip as superposition of pulse due to dipole and pulse due to field reflected from conducting base.

E_z and E_ρ decrease slowly with increasing distance—$\sim 1/\rho$. The shape of the E_z pulse in the absence of the dielectric layer is essentially the superposition of a positive Gaussian pulse and a reflected negative, slightly delayed Gaussian pulse. In the presence of the dielectric layer, the shape of E_z is obtained from the superposition of a pulse with the shape of the derivative of a Gaussian pulse and a similar reflected negative and slightly delayed pulse, as shown in Fig. 15.8.4. This is readily understood if it is recalled that the leading $1/\rho$ term in the surface-wave pulse generated on the boundary between two dielectric half-spaces by a Gaussian current pulse in a horizontal electric dipole [see formula (15.7.1)] is the derivative of a Gaussian pulse. The $1/\rho$ field of the surface wave propagates as the derivative of a Gaussian pulse; the $1/\rho^2$ field of the direct field in air propagates as a Gaussian pulse. In the presence of a conducting half-space, both are reflected and the delayed reflected pulse combines with the direct fields to produce the final pulse shapes.

The structure of microstrip provides the essential properties for the propagation of two different types of surface waves. First, the boundary between air and a dielectric with quite large relative permittivity is the appropriate structure for the propagation of a lateral wave or pulse in the air. Second, the dielectric layer on a highly conducting plane is the essential medium for so-called "trapped" surface waves in the dielectric with an associated evanescent wave in the adjoining air. These were first analyzed by Attwood (1951); they are described in detail by Collin (1960).

A horizontal electric dipole on the air–dielectric surface generates a lateral wave with all six components of the electromagnetic field. These are conveniently grouped as three components of electric (TM) type with the cylindrical components E_ρ, E_z, and B_ϕ, and three components of magnetic (TE) type with the components B_ρ, B_z, and E_ϕ. They have significant values in the air, on the boundary, and in the dielectric into which they travel from the air in a direction almost vertically downward. When a current pulse is impressed on the dipole, a lateral-wave electric-field pulse travels in the air to the point of observation at $\rho = L$. A similar pulse traveling down into the dielectric is reflected at the metal surface with a reversal in direction. It arrives at $\rho = L$ delayed by the extra time of travel over the distance $2l$ with the velocity $c\epsilon_{1r}^{-1/2}$. The two pulses and their superposition are shown in Fig. 15.8.4 with $L = 50$ mm, $\epsilon_{1r} = 13$, and $t_1 = 20.8$ psec. The resultant pulse is seen to approximate the pulse for $\rho = 50$ mm in Fig. 15.8.1.

A horizontal electric dipole on the air–dielectric surface of microstrip also generates "trapped" surface waves in the dielectric. These are of TM and TE types. The former have the components E_x (E_ρ with $\phi = 0$), E_z, and B_y (B_ϕ with $\phi = 0$); the latter have the components B_x, B_z, and E_y. The cutoff frequencies are defined by

$$\frac{2lf_c}{c} = \frac{2l}{\lambda_c} = \frac{n}{2(\epsilon_{1r} - 1)^{1/2}}; \qquad n = 0,\ 1,\ 2,\ \ldots, \tag{15.8.27}$$

where only odd TE modes and even TM modes are possible for a dielectric of thickness l on a conducting plane. Thus, only the TM_0 mode has no cutoff; the lowest possible TE mode, TE_1, has the cutoff frequency

$$f_c = \frac{c}{4l(\epsilon_{1r} - 1)^{1/2}} = 14.4 \text{ GHz} \quad \text{for } l = 1.5 \text{ mm}, \ \epsilon_{1r} = 13. \tag{15.8.28}$$

The three TM components of the lateral-wave field in the air are the same as the three components of the TM_0 mode in the dielectric. At a distance $x = \rho = L$, the field on the surface $z = 0$ consists of a lateral-wave Gaussian pulse that travels the distance L and a TM_0 pulse that traverses the zigzag distance $4[(L/4)^2 + l^2]^{1/2} = [L^2 + (4l)^2]^{1/2}$. With $L = 50$ mm and $l = 1.5$ mm, the zigzag path has a length 50.36 mm. The direction of the electric vector is reversed at each of two reflections at the metal surface and left unaltered by one perfect internal reflection at an angle greater than the critical angle at the air surface. Thus, the multiply-reflected pulse arrives delayed by only 4.3 psec, with somewhat reduced amplitude and the same shape at $\rho = L = 50$ mm, $z = 0$. The superposition of the two similar pulses that arrive almost at the same time yields a very slightly broadened pulse that, in effect, constitutes a single pulse. Thus, along the air–dielectric boundary, the TM_0 "trapped" wave combines with the TM components of the lateral wave to make them indistinguishable. Because of the very small electrical thickness of the dielectric layer, the amplitude of the combined field is determined by that thickness. Its form is essentially like that of the lateral wave along a dielectric half-space.

The formulas for the electromagnetic field generated by a horizontal electric dipole on microstrip are derived from general integrals that include the lateral wave and all "trapped" surface waves in the dielectric. The condition $k_1^2 l^2 \ll 1$ that is imposed in order to obtain simple formulas means that the $\tan k_1 l$ is replaced by $k_1 l$. In effect, this is equivalent to requiring the frequency to be low enough so that all higher "trapped" surface waves in the dielectric do not propagate. Thus, $k_1 l = \omega\epsilon_{1r}^{1/2} l/c < 1$ or $f < c/2\pi\epsilon_{1r}^{1/2} l$ is equivalent to $f < f_c$ where f_c is given by (15.8.28). With $\epsilon_{1r} = 13$ and $l = 1.5$ mm, $f < 8.8$ GHz $< f_c = 14.4$ GHz.

The range of frequencies with significant amplitudes in a Gaussian pulse of half-width t_1 is $\omega t_1 < 2.6$ or $f < 2.6/2\pi t_1$. With $t_1 = 20.8$ psec, $f < 20$ GHz. Since all frequencies in the pulse with large amplitudes are below f_c, none of the higher "trapped" surface-wave modes will propagate and contribute to the field in the dielectric or along its boundary.

15.9 Microstrip Transmission Lines

A microstrip transmission line consists of a long metal strip of width w bonded directly to the surface of the substrate. In its simplest form, its thickness is small compared to the skin depth so that currents are uniformly distributed through it. When the thickness is significantly greater than the skin depth, separate currents can be defined on the top and bottom surfaces of the strip and on its edges. Since each element of current is, in effect, a horizontal electric dipole, the field generated by the current along the entire length of the transmission line must consist of a suitable superposition of the fields of all the dipoles. This suggests that the physical meaning of the numerically determined currents in and electromagnetic field of specific microstrip transmission lines (Hoffmann 1987) should be illuminated by relating them to the fields of their dipole sources.

The complete frequency-dependent electromagnetic field on the surface of the air–substrate boundary and at all points in the air above it due to a horizontal electric dipole is given in Section 15.4. It consists of the components $E_\rho(\rho, \phi, 0)$, $E_\phi(\rho, \phi, 0)$, and $E_z(\rho, \phi, 0)$ and the associated magnetic field components $B_\rho(\rho, \phi, 0)$, $B_\phi(\rho, \phi, 0)$, and $B_z(\rho, \phi, 0)$. Although not independent, the complete field consists of an electric (TM) part including E_ρ, E_z, and B_ϕ and a magnetic part with B_ρ, B_z, and E_ϕ. The field in the air above the boundary surface is closely related to the field in the air over a dielectric half-space as derived in Chapter 5. It consists of a lateral wave that travels in the air near the boundary and a space wave that travels upward into the air. Except quite close to the source, the electric-type components dominate along the air–dielectric surface because they decrease with radial distance as $1/\rho$, whereas the magnetic-type components decrease as $1/\rho^2$.

If a pure TEM mode could propagate in the x-direction along a microstrip transmission line, the electric-field lines would have only E_y and E_z components that form loops that extend from the charges on the surfaces of the strip conductor to charges of opposite sign on the conducting base, as shown in Fig. 15.9.1a. With a traveling wave along the line, there are cross-sectional planes at intervals of a wavelength $\lambda = 2\pi/k$ where the transverse electric field has a positive maximum. Halfway between these are negative maxima, i.e., the direction of the electric vector at each point in the plane is reversed. The electric field is zero at a quarter wavelength from a maximum. The associated magnetic field is also transverse. It forms closed loops around the strip conductor, as shown in Fig. 15.9.1b. For a lossless line, the traveling-wave magnetic field is in phase with the electric field so that the cross sections of maximum magnetic field coincide with the cross sections of maximum electric field.

A graph of the numerically evaluated electric field (Hoffmann 1987) of an actual microstrip transmission line is shown in Fig. 15.9.2a. In the cross sections $x = 0, \lambda/2, \ldots$, where the magnitude of the transverse electric field is maximum, this is essentially like that of the TEM mode in Fig. 15.9.1. In these cross-sectional planes, the field is entirely transverse. However, it does not remain purely transverse in cross-sectional planes between these maxima. A longitudinal component E_x appears. It has its largest value in the planes $x = \lambda/4,\ 3\lambda/4, \ldots$, where $E_y = E_z = 0$. As shown in Fig. 15.9.2a, the complete electric-field lines combine the three components E_x, E_y, and E_z. The loops formed by them include some for which one end terminates on the strip conductor and the other end on the conducting base, and loops that begin and end on the conducting base. The associated magnetic-field lines are shown in Fig. 15.9.2b. In the planes $x = 0,\ \lambda/2$, the field is like that of the TEM line in Fig. 15.9.1b. Between these planes a longitudinal component of B appears and the magnetic lines form axially extended loops that link the currents in the strip twice—once in each direction.

The horizontal component E_x in the air is the lateral-wave component $E_\rho(\rho, 0, z)$ generated by all of the longitudinal currents in the strip. The lateral-wave components $E_z(\rho, 0, z)$ and $B_\phi(\rho, 0, z) = B_y$ combine with the components E_z and B_y of the TEM field. The lateral-wave component $B_\rho(\rho, \pi/2, z) = B_x$ provides the longitudinal magnetic field that occurs in Fig. 15.9.2b. The lateral-wave components $B_z(\rho, \pi/2, z)$ and $E_\phi(\rho, \pi/2, z) = E_y$ combine with the TEM components B_z and E_y. The complete field in the air above a microstrip transmission line can be regarded as a suitably proportioned combination of the TEM field E_y, E_z and B_y, B_z with the lateral-wave field E_x, E_y, E_z and B_x, B_y, B_z. This combination is, of course, complicated by the different phase velocities for the surface wave and the TEM components. The former travels in air with the velocity c; the latter travels with a velocity that involves the permittivity of the substrate. The effect is to make the combination—the quasi-TEM

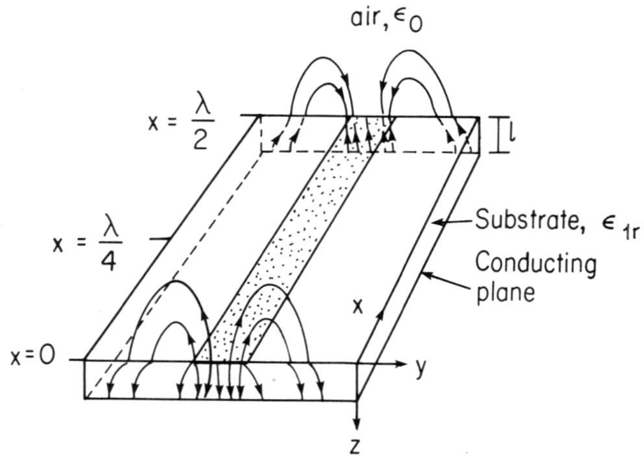

Figure 15.9.1a. Electric-field lines in microstrip transmission line if this could support a pure TEM mode; $\lambda \equiv \lambda_{\text{TEM}}$ is distance between successive maxima.

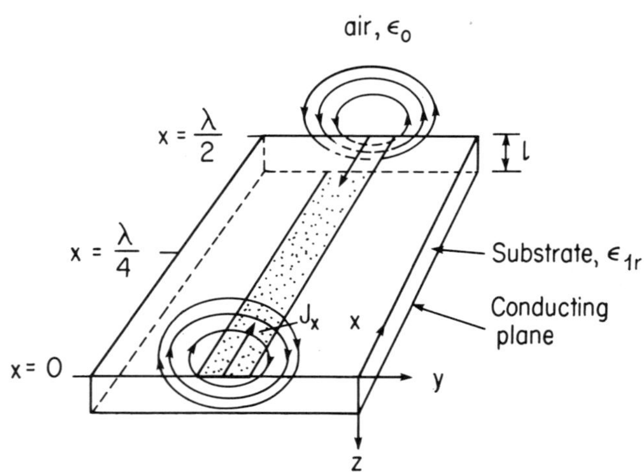

Figure 15.9.1b. Magnetic-field lines associated with electric field in Fig. 15.9.1a.

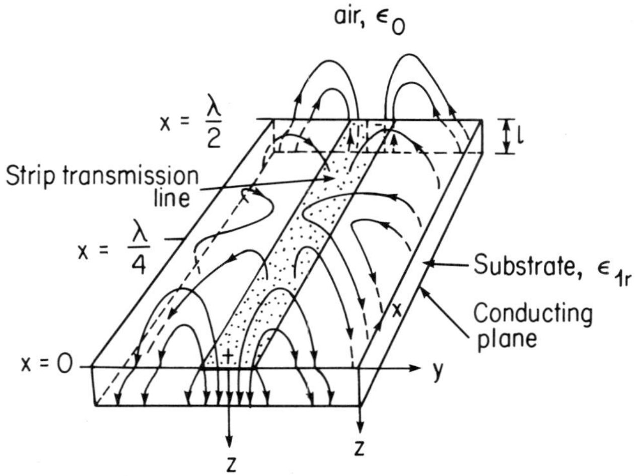

Figure 15.9.2a. Electric-field lines in air and in dielectric substrate generated by currents and charges in microstrip transmission line; λ is distance between successive maxima.

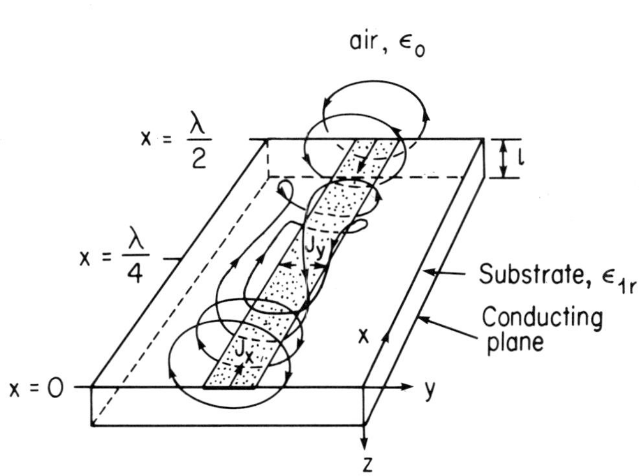

Figure 15.9.2b. Magnetic-field lines associated with electric field in Fig. 15.9.2a.

mode—frequency dependent. The resultant field in the air is due to the superposition of all of the surface-wave and TEM components generated by the current elements in the transmission line.

15.10 Gaussian Pulse Propagation on an Open Microstrip Transmission Line

It is shown in Section 15.8 that a horizontal electric dipole on the surface of microstrip and excited by a Gaussian current pulse generates an electric field outward along the axis of the dipole that consists of a surface-wave pulse that travels with the velocity of light in the air and in the dielectric. The shape of such a pulse is shown in Figs. 15.8.1 and 15.8.2. A Gaussian current pulse applied at one end of a microstrip transmission line generates a field—like that described above—that moves along the strip conductor. The associated radial electric field induces a current in the conductor. It follows that at a given point along the transmission line, the arrival of the Gaussian current pulse is preceded by the superposition of all of the currents induced by the radial electric field that is generated continuously by the approaching pulse. At successive distances this has the forms and relative amplitudes shown in Fig. 15.8.1. The pulse that is actually observed at the given point is the superposition of the currents induced by the approaching and receding pulse and the currents that constitute the original pulse. This composite pulse is spread over a much longer time than the original Gaussian and its shape is determined in no small measure by the distribution in time of the electric field at the point of observation.

Such a pulse has been calculated by Leung and Balanis (1988) by first determining the "effective" wave number $k(\omega) = (\omega/c)\sqrt{\epsilon_{1r\ \text{eff}}(\omega)}$ by the spectral-domain method. The effective permittivity $\epsilon_{1r\ \text{eff}}(\omega)$ as evaluated by them as a function of the frequency is shown in Fig. 15.10.1 for a substrate with the thickness $l = 1.5$ mm and a relative permittivity $\epsilon_{1r} = 13$. A Gaussian voltage pulse $V(t, x = 0)$ with a half-width $t_1 = 20$ psec applied to the microstrip transmission line at $x = 0$ is then substituted in

$$V(t, x = L) = \frac{1}{2\pi} \int_{-\infty}^{\infty} \tilde{V}(\omega, x = 0) e^{j[\omega t - k(\omega)L]}\, d\omega \tag{15.10.1}$$

to determine the voltage pulse at $x = L$. In (15.10.1), $\tilde{V}(\omega, x = 0)$ is the Fourier transform of the Gaussian voltage $V(t, x = 0)$. A graph of the pulse $V(t, x = L)$ with $L = 50$ mm as computed from (15.10.1) by Leung and Balanis (1988) is shown in Fig. 15.10.2. Also shown in dotted line is the pulse that would arrive at $x = L$ if the line supported only the pure TEM mode with the wave number $k = (\omega/c)\sqrt{\epsilon_{1r\ \text{eff static}}}$, where $\epsilon_{1r\ \text{eff static}}$ is the effective static permittivity of the line.

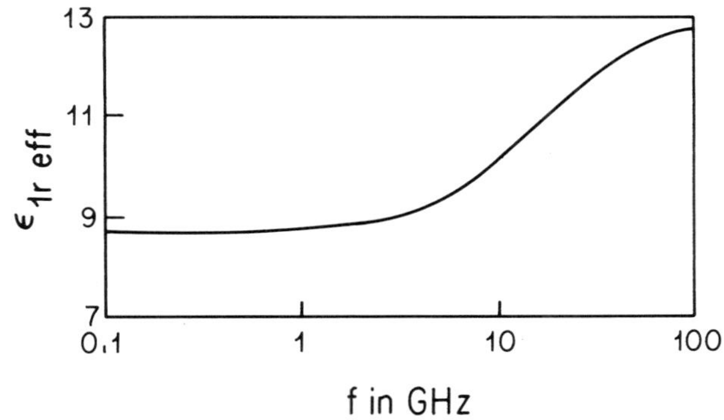

Figure 15.10.1. Effective permittivity of microstrip with $\epsilon_{1_r} = 13$ and $l = 1.5$ mm. Taken from Leung and Balanis (1988).

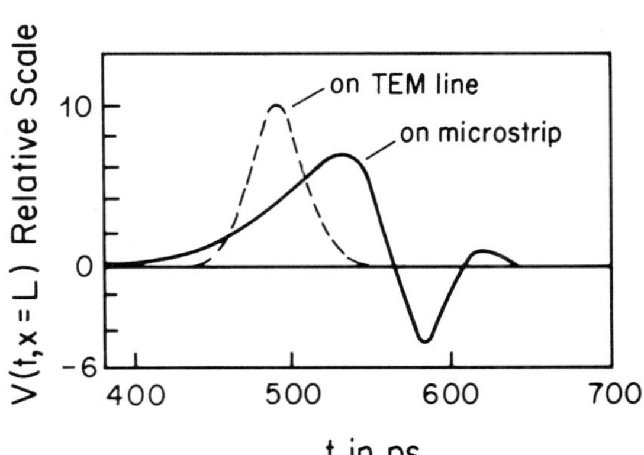

Figure 15.10.2. Pulse at $x = 50$ mm due to Gaussian pulse with half-width $t_1 =$ 20 psec at $x = 0$. Taken from Leung and Balanis (1988).

In Fig. 15.10.2, the center of the Gaussian pulse on the TEM line arrives at $x = 50$ mm after $t = 490$ psec—the correct length of time to traverse 50 mm at the velocity $v = c/\sqrt{\epsilon_{1r\text{ eff static}}}$ where $\epsilon_{1r\text{ eff static}} = 8.65$. It is seen that a significant part of the actual pulse arrives earlier than any part of the Gaussian pulse and a major part later. The actual pulse is spread over a much greater time interval than the Gaussian and its shape is very different. A comparison with Fig. 15.8.1 shows that its shape is actually very much like that of the vertical electric field close to the source dipole, e.g., at $\rho = 6.25$ mm. Clearly, the components of the surface-wave electric field are of major importance in pulse propagation on microstrip transmission lines. They provide the long-range collinear coupling among the current elements that does not exist in a TEM line.

15.11 Radiation from a Unit Dipole on a Dielectric Substrate

The complete electromagnetic field of a unit dipole on the air–dielectric boundary of a conducting plane coated with a layer of dielectric with the thickness l is given by (15.4.1)–(15.4.6) with $d' = 0$ and $r_1 = r_2 = r_0 = (\rho^2 + z'^2)^{1/2}$. Since the Fresnel-integral terms in $E_{0\phi'}(\rho, \phi', 0)$ and $B_{0\rho}(\rho, \phi', 0)$ are of higher order in $(k_0\rho)^{-1}$, they are negligible and are omitted. The components are

$$
\begin{aligned}
E_{0\rho}(\rho, \phi', z') = {} & -\frac{\omega\mu_0}{2\pi k_0} \cos\phi'\, e^{ik_0 r_0} \\
& \times \left\{ ik_0 l \left(\frac{z'}{r_0}\right)\left(\frac{ik_0}{r_0} - \frac{1}{r_0^2}\right) \right. \\
& \quad - k_0^2 l^2 \left[\frac{ik_0}{r_0} - \frac{1}{r_0^2} - \frac{i}{k_0 r_0^3} \right. \\
& \quad \left.\left. + ik_0^3 l \left(\frac{r_0}{\rho}\right)\left(\frac{\pi}{k_0 r_0}\right)^{1/2} e^{-iP_0}\mathcal{F}(P_0) \right] \right\},
\end{aligned} \tag{15.11.1}
$$

$$
\begin{aligned}
E_{0\phi'}(\rho, \phi', z') = {} & \frac{\omega\mu_0}{2\pi k_0} \sin\phi'\, e^{ik_0 r_0} \\
& \times \left\{ ik_0 l \left(\frac{z'}{r_0}\right)\left(\frac{ik_0}{r_0} - \frac{1}{r_0^2}\right) \right. \\
& \quad - k_0^2 l^2 \left[\frac{2}{r_0^2} + \frac{2i}{k_0 r_0^3} \right.
\end{aligned}
$$

$$+\left(\frac{z'}{r_0}\right)^2\left(\frac{ik_0}{r_0}-\frac{3}{r_0^2}-\frac{3i}{k_0r_0^3}\right)\bigg]\bigg\}, \tag{15.11.2}$$

$$\begin{aligned}E_{0z'}(\rho,\phi',z') &= \frac{\omega\mu_0}{2\pi k_0}\cos\phi'\, e^{ik_0r_0}(ik_0l)\\ &\quad\times\left[\left(\frac{\rho}{r_0}\right)\left(\frac{ik_0}{r_0}-\frac{1}{r_0^2}\right)\right.\\ &\quad\left.+\, ik_0^3l\left(\frac{\pi}{k_0r_0}\right)^{1/2}e^{-iP_0}\mathcal{F}(P_0)\right],\end{aligned} \tag{15.11.3}$$

$$\begin{aligned}B_{0\rho}(\rho,\phi',z') &= -\frac{\mu_0}{2\pi}\sin\phi'\, e^{ik_0r_0}(ik_0l)\\ &\quad\times\left[\frac{2}{r_0^2}+\frac{2i}{k_0r_0^3}+\left(\frac{z'}{r_0}\right)^2\left(\frac{ik_0}{r_0}-\frac{3}{r_0^2}-\frac{3i}{k_0r_0^3}\right)\right],\end{aligned} \tag{15.11.4}$$

$$\begin{aligned}B_{0\phi'}(\rho,\phi',z') &= -\frac{\mu_0}{2\pi}\cos\phi'\, e^{ik_0r_0}(ik_0l)\\ &\quad\times\left[\frac{ik_0}{r_0}-\frac{1}{r_0^2}-\frac{i}{k_0r_0^3}\right.\\ &\quad\left.+\, ik_0^3l\left(\frac{r_0}{\rho}\right)\left(\frac{\pi}{k_0r_0}\right)^{1/2}e^{-iP_0}\mathcal{F}(P_0)\right],\end{aligned} \tag{15.11.5}$$

$$\begin{aligned}B_{0z'}(\rho,\phi',z') &= \frac{\mu_0}{2\pi}\sin\phi'\, e^{ik_0r_0}\left(\frac{\rho}{r_0}\right)\\ &\quad\times\left\{ik_0l\left(\frac{z'}{r_0}\right)\left(\frac{ik_0}{r_0}-\frac{3}{r_0^2}-\frac{3i}{k_0r_0^3}\right)\right.\\ &\quad-k_0^2l^2\left[\frac{1}{r_0^2}+\frac{3i}{k_0r_0^3}-\frac{3}{k_0^2r_0^4}\right.\\ &\quad\left.\left.+\left(\frac{z'}{r_0}\right)^2\left(\frac{ik_0}{r_0}-\frac{6}{r_0^2}-\frac{15i}{k_0r_0^3}\right)\right]\right\},\end{aligned} \tag{15.11.6}$$

where, from (15.3.68) with $d'=0$,

$$P_0=\frac{k_0r_0}{2}\left(\frac{-ik_0lr_0+z'}{\rho}\right)^2=-\frac{k_0^3l^2r_0}{2}\left(\frac{r_0+iz'/k_0l}{\rho}\right)^2. \tag{15.11.7}$$

A comparison of the general formula (15.11.1) for $E_{0\rho}(\rho,\phi',z')$ with a numerical evaluation of the exact integrals from which it was derived has been carried out by A. Hoorfar (King 1992) with $\epsilon_{1r}=2.2$ ($k_1l=0.031$), $\epsilon_{1r}=12.9$ ($k_1l=0.075$), and $\epsilon_{1r}=20$ ($k_1l=0.094$). The agreement for $\epsilon_{1r}=12.9$ and 20 is virtually exact since the condition $k_0^2\ll k_1^2$ is well satisfied. The agreement for $\epsilon_{1r}=2.2$ is only fair since the condition $k_0^2\ll k_1^2$ is not satisfied.

The far field is conveniently expressed in spherical coordinates with the angle Θ measured from the vertical z'-axis so that

$$\rho = r_0 \sin\Theta; \qquad z' = r_0 \cos\Theta. \tag{15.11.8}$$

At radial distances r_0 that are sufficiently great so that

$$k_0 r_0 \geq \frac{4}{k_0^2 l^2}, \tag{15.11.9}$$

the entire Fresnel-integral term assumes a simple form. It is

$$\begin{aligned} T &\equiv -ik_0^3 l\left(\frac{\pi}{k_0 r_0}\right)^{1/2} e^{-iP_0}\mathcal{F}(P_0) \\ &\to T^r = -ik_0^3 l\left(\frac{\pi}{k_0 r_0}\right)^{1/2} \frac{1}{(2\pi P_0)^{1/2}}\left(\frac{1}{2P_0} + i\right). \end{aligned} \tag{15.11.10}$$

With (15.11.7), this becomes

$$T^r = \frac{ik_0\rho}{r_0^2[1 + (iz'/k_0 l r_0)]} - \frac{\rho^3}{k_0^2 l^2 r_0^5 [1 + (iz'/k_0 l r_0)]^3}. \tag{15.11.11}$$

In spherical coordinates,

$$T^r = \frac{ik_0 \sin\Theta}{r_0[1 + (i/k_0 l)\cos\Theta]} - \frac{\sin^3\Theta}{k_0^2 l^2 r_0^2 [1 + (i/k_0 l)\cos\Theta]^3}. \tag{15.11.12}$$

The spherical components are obtained from (15.11.1)–(15.11.6) with

$$E_{0r}(r_0, \Theta, \Phi') = E_{0\rho}(\rho, \phi', z')\sin\Theta + E_{0z'}(\rho, \phi', z')\cos\Theta, \tag{15.11.13}$$
$$E_{0\Theta}(r_0, \Theta, \Phi') = E_{0\rho}(\rho, \phi', z')\cos\Theta - E_{0z'}(\rho, \phi', z')\sin\Theta, \tag{15.11.14}$$
$$E_{0\Phi'}(r_0, \Theta, \Phi') = E_{0\phi'}(\rho, \phi', z'), \tag{15.11.15}$$
$$B_{0r}(r_0, \Theta, \Phi') = B_{0\rho}(\rho, \phi', z')\sin\Theta + B_{0z'}(\rho, \phi', z')\cos\Theta, \tag{15.11.16}$$
$$B_{0\Theta}(r_0, \Theta, \Phi') = B_{0\rho}(\rho, \phi', z')\cos\Theta - B_{0z'}(\rho, \phi', z')\sin\Theta, \tag{15.11.17}$$
$$B_{0\Phi'}(r_0, \Theta, \Phi') = B_{0\phi'}(\rho, \phi', z'). \tag{15.11.18}$$

With (15.11.12), the far field obtained from (15.11.13) is

$$\begin{aligned} E_{0r}^r(r_0, \Theta, \Phi') = &-\frac{\omega\mu_0}{2\pi k_0}\cos\Phi'\, e^{ik_0 r_0} \\ &\times\left[\frac{ik_0}{r_0}\left(ik_0 l\cos\Theta - k_0^2 l^2 + \frac{k_0^2 l^2}{1 + (i/k_0 l)\cos\Theta}\right)\sin\Theta\right. \\ &\quad - \frac{ik_0}{r_0}\left(ik_0 l\sin\Theta - \frac{ik_0 l\sin\Theta}{1 + (i/k_0 l)\cos\Theta}\right)\cos\Theta \\ &\quad \left. - \frac{\sin^3\Theta}{k_0^2 l^2 r_0^2[1 + (i/k_0 l)\cos\Theta]^3}\left(k_0^2 l^2 - ik_0 l\cos\Theta\right)\right] \end{aligned}$$

$$= \frac{\omega\mu_0}{2\pi k_0}\cos\Phi'\, e^{ik_0r_0}\left(\frac{\sin^3\Theta\,(k_0^2l^2 - ik_0l\cos\Theta)}{k_0^2l^2r_0^2[1+(i/k_0l)\cos\Theta]^3}\right). \quad (15.11.19)$$

The $E_r(r_0, \Theta, \Phi')$ component has no $1/r_0$ terms. This is necessary since $\nabla\cdot\mathbf{E} = 0$ at all points in space.

When (15.11.1) and (15.11.3) are substituted in (15.11.14), the far-field terms are

$$\begin{aligned} E^r_{0\Theta}(r_0,\Theta,\Phi') &= -\frac{\omega\mu_0}{2\pi k_0}\cos\Phi'\, e^{ik_0r_0} \\ &\quad\times\left[\frac{ik_0}{r_0}\left(ik_0l\cos\Theta - k_0^2l^2 + \frac{k_0^2l^2}{1+(i/k_0l)\cos\Theta}\right)\cos\Theta\right. \\ &\quad + \frac{ik_0}{r_0}\left(ik_0l\sin\Theta - \frac{ik_0l\sin\Theta}{1+(i/k_0l)\cos\Theta}\right)\sin\Theta \\ &\quad\left. + \frac{i\sin^4\Theta}{k_0^2l^2r_0^2[1+(i/k_0l)\cos\Theta]^3}\right] \\ &= -\frac{\omega\mu_0}{2\pi k_0}\cos\Phi'\, e^{ik_0r_0}\left[-\frac{ik_0}{r_0}\left(\frac{k_0l\cos\Theta}{k_0l+i\cos\Theta}\right)\right. \\ &\quad\left. + \frac{i\sin^4\Theta}{k_0^2l^2r_0^2[1+(i/k_0l)\cos\Theta]^3}\right]. \end{aligned} \quad (15.11.20)$$

This includes a $1/r_0$ term and a $1/r_0^2$ term multiplied by the large factor $1/k_0^2l^2$. With (15.11.5) and (15.11.12), (15.11.18) becomes

$$\begin{aligned} &B^r_{0\Phi'}(r_0,\Theta,\Phi') \\ &= -\frac{\mu_0}{2\pi}\cos\Phi'\, e^{ik_0r_0}(ik_0l) \\ &\quad\times\left[\frac{ik_0}{r_0}\left(1 - \frac{1}{1+(i/k_0l)\cos\Theta}\right) + \frac{\sin^2\Theta}{k_0^2l^2r_0^2[1+(i/k_0l)\cos\Theta]^3}\right] \\ &= -\frac{\mu_0}{2\pi}\cos\Phi'\, e^{ik_0r_0} \\ &\quad\times\left[-\frac{ik_0}{r_0}\left(\frac{k_0l\cos\Theta}{k_0l+i\cos\Theta}\right) + \frac{ik_0l\sin^2\Theta}{k_0^2l^2r_0^2[1+(i/k_0l)\cos\Theta]^3}\right]. \end{aligned} \quad (15.11.21)$$

Similarly, with (15.11.2), (15.11.4), and (15.11.6) substituted in (15.11.15)–(15.11.17), the leading $1/r_0$ terms in $E^r_{0\Phi'}$, B^r_{0r}, and $B^r_{0\Theta}$ are

$$E^r_{0\Phi'}(r_0,\Theta,\Phi') = \frac{i\omega\mu_0}{2\pi}\sin\Phi'\,\frac{e^{ik_0r_0}}{r_0}(ik_0l)\cos\Theta, \quad (15.11.22)$$

$$B^r_{0r}(r_0,\Theta,\Phi') = -\frac{\mu_0}{2\pi}\sin\Phi'\, e^{ik_0r_0}\,\frac{ik_0}{r_0}(ik_0l)(\cos^2\Theta\sin\Theta - \cos^2\Theta\sin\Theta)$$

$$\sim 0, \tag{15.11.23}$$

$$B^r_{0\Theta}(r_0, \Theta, \Phi') = -\frac{\mu_0}{2\pi} \sin\Phi' \, e^{ik_0 r_0} \frac{ik_0}{r_0} (ik_0 l)(\cos^3\Theta + \sin^2\Theta \cos\Theta)$$

$$\sim -\frac{i\mu_0 k_0}{2\pi} \sin\Phi' \, \frac{e^{ik_0 r_0}}{r_0} (ik_0 l) \cos\Theta. \tag{15.11.24}$$

In summary, the far-field components of the horizontal dipole located on the surface of the dielectric-coated conducting plane are

$$E^r_{0\Theta}(r_0, \Theta, \Phi') = \frac{\omega\mu_0}{2\pi k_0} \cos\Phi' \, e^{ik_0 r_0} \left[\frac{ik_0}{r_0} \left(\frac{k_0 l \cos\Theta}{k_0 l + i\cos\Theta} \right) - \frac{i \sin^4\Theta}{k_0^2 l^2 r_0^2 [1 + (i/k_0 l)\cos\Theta]^3} \right], \tag{15.11.25}$$

$$B^r_{0\Phi'}(r_0, \Theta, \Phi') = \frac{\mu_0}{2\pi} \cos\Phi' \, e^{ik_0 r_0} \left[\frac{ik_0}{r_0} \left(\frac{k_0 l \cos\Theta}{k_0 l + i\cos\Theta} \right) - \frac{ik_0 l \sin^2\Theta}{k_0^2 l^2 r_0^2 [1 + (i/k_0 l)\cos\Theta]^3} \right], \tag{15.11.26}$$

$$E^r_{0\Phi'}(r_0, \Theta, \Phi') = \frac{i\omega\mu_0}{2\pi} \sin\Phi' \, \frac{e^{ik_0 r_0}}{r_0} (ik_0 l) \cos\Theta, \tag{15.11.27}$$

$$B^r_{0\Theta}(r_0, \Theta, \Phi') = -\frac{i\mu_0 k_0}{2\pi} \sin\Phi' \, \frac{e^{ik_0 r_0}}{r_0} (ik_0 l) \cos\Theta. \tag{15.11.28}$$

These formulas apply when $k_0 r_0 \geq 4/k_0^2 l^2$. Note that the entire field along the surface of the dielectric, $\Theta = \pi/2$, is given by the surface-wave terms in $E^r_{0\Theta}$ and $B^r_{0\Phi'}$. These decrease with radial distance as $1/r_0^2$ but are multiplied by the large factor $1/k_0^2 l^2$. They have their maxima in the directions $\Phi' = 0,\ \pi$, along the axis of the dipole. There is no far field in the directions $\Phi' = \pi/2,\ 3\pi/2$, along the boundary surface.

In the plane $\Phi' = 0$, the magnitude of the $1/r_0$ term in $E^r_{0\Theta}$ is given by

$$\left| \frac{E^r_{0\Theta}(r_0, \Theta, 0)}{E^r_{0\Theta}(r_0, 0, 0)} \right| \sim \frac{|\cos\Theta|}{(k_0^2 l^2 + \cos^2\Theta)^{1/2}}. \tag{15.11.29a}$$

Since $k_0^2 l^2$ is very small compared to 1, the pattern is constant at the value 1 for all angles from $\Theta = 0$ to near $\Theta = \pi/2$ where it drops quickly to zero at $\Theta = \pi/2$. Because of the surface-wave term in (15.11.25), the field at $\Phi' = 0,\ \Theta = \pi/2$, is not zero but equal to the contribution $1/k_0^2 l^2 r_0^2$ at the particular radial distance $r_0 = \rho$. The power pattern is

$$\left| \frac{E^r_{0\Theta}(r_0, \Theta, 0)}{E^r_{0\Theta}(r_0, 0, 0)} \right|^2 \sim \frac{\cos^2\Theta}{k_0^2 l^2 + \cos^2\Theta}. \tag{15.11.29b}$$

The field and power patterns in the plane $\Phi' = \pi/2$ are

$$\left| \frac{E^r_{0\Phi'}(r_0, \Theta, \pi/2)}{E^r_{0\Phi'}(r_0, 0, \pi/2)} \right| \sim |\cos\Theta|, \tag{15.11.30a}$$

$$\left| \frac{E^r_{0\Phi'}(r_0, \Theta, \pi/2)}{E^r_{0\Phi'}(r_0, 0, \pi/2)} \right|^2 \sim \cos^2\Theta. \tag{15.11.30b}$$

These are identically zero at $\Theta = \pi/2$ since $E^r_{0\Phi'}$ has no surface-wave term.

The radial Poynting vector in the far field for the $1/r_0$ terms is

$$\begin{aligned} S_r &= \frac{1}{2\mu_0}\left(E^r_{0\Theta}B^{r\,*}_{0\Phi'} + E^r_{0\Phi'}B^{r\,*}_{0\Theta}\right) \\ &= \frac{\omega\mu_0}{8\pi^2 k_0}\cos^2\Phi' \frac{k_0^2}{r_0^2}\left(\frac{k_0^2 l^2\cos^2\Theta}{k_0^2 l^2 + \cos^2\Theta}\right) \\ &\quad - \frac{\omega\mu_0 k_0}{8\pi^2}\sin^2\Phi' \frac{k_0^2 l^2}{r_0^2}\cos^2\Theta \\ &= \frac{\omega\mu_0 k_0}{8\pi^2}\frac{k_0^2 l^2}{r_0^2}\cos^2\Theta\left(\frac{\cos^2\Phi'}{k_0^2 l^2 + \cos^2\Theta} - \sin^2\Phi'\right). \end{aligned} \tag{15.11.31}$$

The total power radiated is

$$\begin{aligned} P &= \operatorname{Re}\int_0^{\pi/2}\int_0^{2\pi} S_r \cdot r_0^2 \sin\Theta\, d\Theta\, d\Phi' \\ &= \frac{\omega\mu_0 k_0}{8\pi^2}k_0^2 l^2 \int_0^{\pi/2}\int_0^{2\pi}\left(\frac{\cos^2\Theta\cos^2\Phi'}{k_0^2 l^2 + \cos^2\Theta} - \sin^2\Phi'\cos^2\Theta\right) \\ &\qquad \times \sin\Theta\, d\Theta\, d\Phi'. \end{aligned} \tag{15.11.32}$$

The Φ' and Θ integrals are

$$\begin{aligned} \int_0^{2\pi}\cos^2\Phi'\, d\Phi' &= \frac{\Phi'}{2} + \frac{\sin 2\Phi'}{4}\bigg|_0^{2\pi} = \pi, \\ \int_0^{2\pi}\sin^2\Phi'\, d\Phi' &= \frac{\Phi'}{2} - \frac{\sin 2\Phi'}{4}\bigg|_0^{2\pi} = \pi, \\ \int_0^{\pi/2}\cos^2\Theta\sin\Theta\, d\Theta &= -\frac{\cos^3\Theta}{3}\bigg|_0^{\pi/2} = \frac{1}{3}. \end{aligned}$$

Hence,

$$P = \frac{\omega\mu_0 k_0}{8\pi}k_0^2 l^2\left(-\frac{1}{3} + \int_0^{\pi/2}\frac{\cos^2\Theta\sin\Theta}{k_0^2 l^2 + \cos^2\Theta}\, d\Theta\right). \tag{15.11.33}$$

The integral can be separated as follows:

$$\vartheta = \int_0^{\pi/2}\frac{\cos^2\Theta\sin\Theta}{k_0^2 l^2 + \cos^2\Theta}\, d\Theta$$

$$= \int_0^{\pi/2} \sin\Theta \, d\Theta - k_0^2 l^2 \int_0^{\pi/2} \frac{\sin\Theta \, d\Theta}{k_0^2 l^2 + \cos^2\Theta}$$

$$= 1 - k_0^2 l^2 \int_0^1 \frac{du}{k_0^2 l^2 + u^2}. \tag{15.11.34}$$

Since $k_0^2 l^2 \ll 1$,

$$\vartheta \sim 1 - k_0 l \cdot \frac{\pi}{2}. \tag{15.11.35}$$

Hence,

$$P = \frac{\omega\mu_0 k_0}{8\pi} k_0^2 l^2 \left(-\frac{1}{3} + 1 - k_0 l \cdot \frac{\pi}{2} \right)$$

$$= \frac{\omega\mu_0 k_0}{8\pi} k_0^2 l^2 \left(\frac{2}{3} - k_0 l \cdot \frac{\pi}{2} \right)$$

$$\sim \frac{\omega\mu_0 k_0}{12\pi} k_0^2 l^2 = \frac{k_0^2 \zeta_0}{12\pi} k_0^2 l^2 \sim 10 k_0^2 (k_0^2 l^2). \tag{15.11.36}$$

When the electric moment is Ih_e, the power is

$$P \sim 10 k_0^2 h_e^2 I^2 (k_0^2 l^2) \tag{15.11.37}$$

and the radiation resistance is

$$R^e = \frac{2P}{I^2} = 20 k_0^2 h_e^2 (k_0^2 l^2) \ \Omega. \tag{15.11.38}$$

15.12 Microstrip Antennas

Microstrip antennas consist of variously shaped and excited conducting strips or patches on the dielectric substrate (Carver and Mink 1981; Mailloux, McIlvenna, and Kernweis 1981; Weiss 1981). Their primary function is to generate a significant electromagnetic field at distant points. This field is the superposition of contributions from each infinitesimal element of current or unit dipole in the patch. The complete far field of such an x'-directed element is given by (15.11.25)–(15.11.28). The unit dipole has an electric moment $Ih = 1$ A m. When multiplied by $I(x'')\,dx''$, the dipole is an element of a microstrip transmission line or microstrip patch antenna. The complete field of the transmission line or patch antenna is obtained by an integration over the current distribution. In practice, these distributions are obtained by numerical methods for solving the appropriate integral equation or simply by "an educated guess." Since the far field is not sensitive to the details of the current distribution and the accurate analysis of the current in microstrip transmission lines and antennas is beyond the scope of this book, the latter procedure will be followed. This generally provides an adequate picture of the radiating properties including the surface-wave terms which are important along and near the air–dielectric surface.

A typical assumed current density in a conducting rectangle that has a length $2h$ and width $2w$ is (Perlmutter, Shtrikman, and Treves 1985)

$$J_x(x'', y'') = \frac{I_x(0)}{2w} \cos k_L x''; \qquad \begin{cases} -h \le x'' \le h \\ -w \le y'' \le w \end{cases}. \tag{15.12.1}$$

Here $k_L = k_0\sqrt{\epsilon_{1r\,\text{eff}}}$ is the wave number that would characterize the patch as a segment of a microstrip transmission line for which $\epsilon_{1r\,\text{eff}}$ is the relative effective permittivity at the operating frequency, $I_x(0)$ is the total current traversing the center line $x'' = 0$ of the patch, and h is so chosen that $k_L h = \pi/2$. The transverse distribution of the x'-directed current is assumed to be uniform. The actual transverse distribution has large peaks at the edges, $|y''| = w$, and a minimum at the center, $y'' = 0$. In addition, there is a y'-directed surface current. Since this is in opposite directions on the top and bottom surfaces of the patch, its contribution to the far field is negligible. An actual method of exciting such a current is not available. A suggested approximate arrangement is to end-drive the patch at its center, $y'' = 0$, $x'' = -h$, by an x'-directed microstrip transmission line. Since the impedance of the patch is very large when end-driven, the driving-point current is small compared to $I_x(0)$, so that it can be neglected in determining the approximate far field.

Approximate formulas for the relative effective permittivity $\epsilon_{1r\,\text{eff}}$ and characteristic impedance Z_c of a microstrip transmission line are given in Hoffman (1987). Relatively simple examples are in formulas [3.20] to [3.22], viz.,

$$Z_c = \frac{60}{\sqrt{\epsilon_{1r\,\text{eff}}}} \ln\left(\frac{4l}{w} + \frac{w}{2l}\right); \qquad 0 \le \frac{2w}{l} \le 1, \tag{15.12.2}$$

$$Z_c = \frac{120\pi}{\sqrt{\epsilon_{1r\,\text{eff}}}} \left[\frac{2w}{l} + 2.42 - 0.22\,\frac{l}{w} + \left(1 - \frac{l}{2w}\right)^6\right]^{-1}; \qquad \frac{2w}{l} > 1, \tag{15.12.3}$$

$$\epsilon_{1r\,\text{eff}} = \frac{\epsilon_{1r} + 1}{2} + \frac{\epsilon_{1r} - 1}{2}\left(1 + \frac{5l}{w}\right)^{-1/2}. \tag{15.12.4}$$

These apply to a microstrip transmission line of width $2w$ on a dielectric substrate with the thickness l and the relative permittivity ϵ_{1r}. They are the formulas used by Perlmutter, Shtrikman, and Treves (1985).

The electromagnetic far field of the assumed current (15.12.1) in the rectangular patch is obtained by integration from (15.11.25)–(15.11.28). The variables of integration, x'' and y'', occur only in $r = [(x' - x'')^2 + (y' - y'')^2 + z'^2]^{1/2}$. Since the dimensions of the patch are small compared

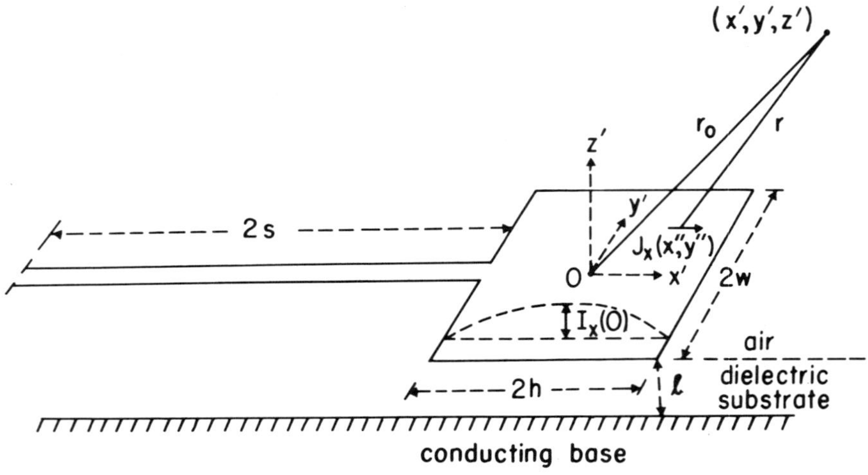

Figure 15.12.1. Microstrip patch antenna end-driven by microstrip transmission line.

to the radial distance to a point in the far field, $r \sim r_0 = (x'^2 + y'^2 + z'^2)^{1/2}$ is adequate in amplitudes. In the phase factor in the exponentials, the more accurate value

$$r \sim (r_0^2 - 2x'x'' - 2y'y'')^{1/2} = r_0\left(1 - 2\,\frac{x'x''}{r_0^2} - 2\,\frac{y'y''}{r_0^2}\right)^{1/2} \tag{15.12.5}$$

must be used. With (15.11.8) and Fig. 15.12.1, it follows that

$$\frac{x'}{r_0} = \frac{x'}{\rho}\cdot\frac{\rho}{r_0} = \cos\Phi'\sin\Theta; \qquad \frac{y'}{r_0} = \frac{y'}{\rho}\cdot\frac{\rho}{r_0} = \sin\Phi'\sin\Theta. \tag{15.12.6}$$

Hence,

$$r \sim r_0 - x''\cos\Phi'\sin\Theta - y''\sin\Phi'\sin\Theta. \tag{15.12.7}$$

The integral to be evaluated in common for all components is simply

$$J = \int_{-h}^{h} dx'' \int_{-w}^{w} dy''\, e^{ik_0 r}\cos k_L x'' \tag{15.12.8}$$

or

$$J = e^{ik_0r_0} J_x\, J_y = e^{ik_0r_0}\int_{-h}^{h}\cos k_L x''\, e^{-ik_0x''\cos\Phi'\sin\Theta}\,dx'' \times \int_{-w}^{w} e^{-ik_0y''\sin\Phi'\sin\Theta}\,dy''. \tag{15.12.9}$$

These are elementary integrals, viz.,

$$J_x = \int e^{au}\cos u\,du = \frac{e^{au}}{a^2+1}\,(a\cos u + \sin u), \tag{15.12.10}$$

$$J_y = \int e^{by}\, dy = \frac{e^{by}}{b}. \tag{15.12.11}$$

With $k_L h = \pi/2$, it follows that

$$J_x = \frac{2k_L \cos(k_0 h \cos\Phi' \sin\Theta)}{k_L^2 - k_0^2 \cos^2\Phi' \sin^2\Theta}, \tag{15.12.12}$$

$$J_y = \frac{2\sin(k_0 w \sin\Phi' \sin\Theta)}{k_0 \sin\Phi' \sin\Theta}, \tag{15.12.13}$$

so that the field factor for the patch is

$$\begin{aligned} P(\Theta, \Phi') &= \frac{I_x(0)}{2w} \cdot J \\ &= I_x(0)\left(\frac{2k_L \cos(k_0 h \cos\Phi' \sin\Theta)}{k_L^2 - k_0^2 \cos^2\Phi' \sin^2\Theta}\right)\left(\frac{\sin(k_0 w \sin\Phi' \sin\Theta)}{k_0 w \sin\Phi' \sin\Theta}\right). \end{aligned} \tag{15.12.14}$$

This factor multiplies each of the components (15.11.25)–(15.11.28) of the infinitesimal-dipole field to obtain the field of the patch antenna with the current distribution (15.12.1). With $\phi' = \Phi'$, the far-field components of the patch antennas are

$$\begin{aligned} [E^r_{0\Theta}(r_0, \Theta, \Phi')]_p = \frac{\omega\mu_0}{2\pi k_0}\, P(\Theta, \Phi') \cos\Phi'\, e^{ik_0 r_0} \\ \times \left[\frac{ik_0}{r_0}\left(\frac{k_0 l \cos\Theta}{k_0 l + i\cos\Theta}\right)\right. \\ \left. - \frac{i\sin^4\Theta}{k_0^2 l^2 r_0^2 [1 + (i/k_0 l)\cos\Theta]^3}\right], \end{aligned} \tag{15.12.15}$$

$$\begin{aligned} [B^r_{0\Phi'}(r_0, \Theta, \Phi')]_p = \frac{\mu_0}{2\pi}\, P(\Theta, \Phi') \cos\Phi'\, e^{ik_0 r_0} \\ \times \left[\frac{ik_0}{r_0}\left(\frac{k_0 l \cos\Theta}{k_0 l + i\cos\Theta}\right)\right. \\ \left. - \frac{ik_0 l \sin^2\Theta}{k_0^2 l^2 r_0^2 [1 + (i/k_0 l)\cos\Theta]^3}\right], \end{aligned} \tag{15.12.16}$$

$$[E^r_{0\Phi'}(r_0, \Theta, \Phi')]_p = \frac{i\omega\mu_0}{2\pi}\, P(\Theta, \Phi') \sin\Phi' \cos\Theta\, (ik_0 l)\, \frac{e^{ik_0 r_0}}{r_0}, \tag{15.12.17}$$

$$[B^r_{0\Theta}(r_0, \Theta, \Phi')]_p = -\frac{i\mu_0 k_0}{2\pi}\, P(\Theta, \Phi') \sin\Phi' \cos\Theta\, (ik_0 l)\, \frac{e^{ik_0 r_0}}{r_0}. \tag{15.12.18}$$

Graphs showing the electric far field given in (15.12.15) and (15.12.17) are shown in Fig. 15.12.2. On the left is $[E^r_{0\Phi'}(r_0, \Theta, \pi/2)]_p$; this vanishes when $\Theta = \pi/2$. In the center is $[E^r_{0\Theta}(r_0, \Theta, 0)]_p$; the space-wave term vanishes when $\Theta = \pi/2$, but the surface wave has its maximum there. At $r_0 = 10$ m and $f = 1$ GHz, the maximum of the surface wave is small

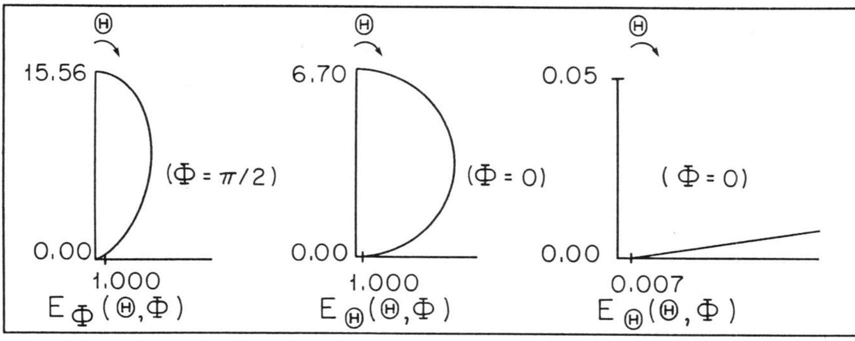

Figure 15.12.2. Electric far field of microstrip patch antenna as given by (15.12.15) and (15.12.17) with $r_0 = 10$ m, $l = 1$ mm, $\epsilon_{1r} = 2$, and $f = 1$ GHz.

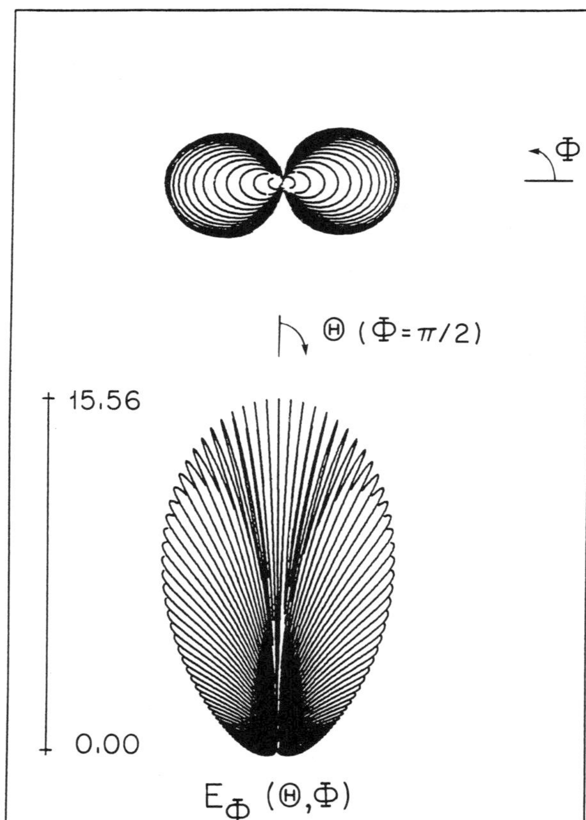

Figure 15.12.3. Three-dimensional diagram of $[E^r_{0\Phi'}(r_0, \Theta, \Phi)]_p$ with $r_0 = 10$ m, $l = 1$ mm, $\epsilon_{1r} = 2$, and $f = 1$ GHz.

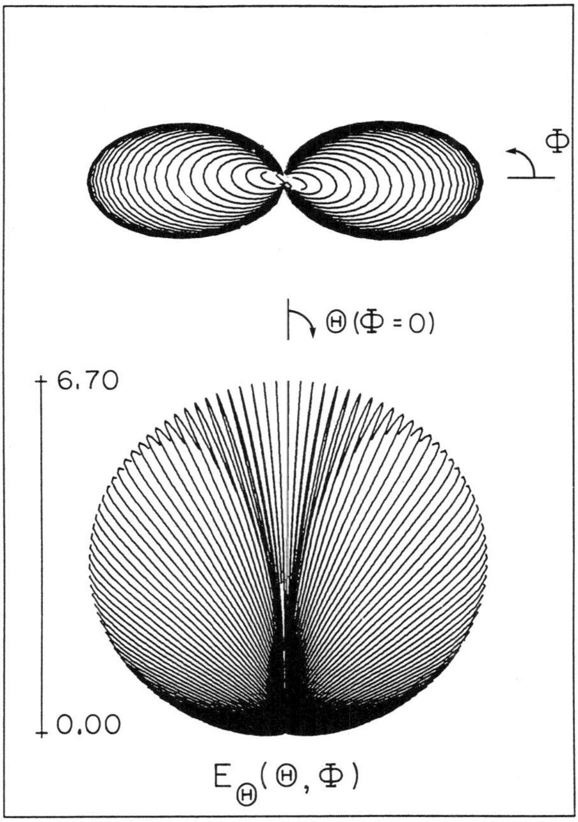

Figure 15.12.4. Three-dimensional diagram of $[E^r_{0\Theta}(r_0, \Theta, \Phi)]_p$ with $r_0 = 10$ m, $l = 1$ mm, $\epsilon_{1r} = 2$, and $f = 1$ GHz.

compared with that of the space wave, but it is the entire field along the surface. This is shown in the greatly enlarged region near $\Theta = \pi/2$ on the right in Fig. 15.12.2. The amplitude of the surface wave at $\Theta = \pi/2$ is 0.007; the maximum of the space wave at $\Theta = 0$ is 6.7. Three-dimensional diagrams of the two components of the electric field are in Figs. 15.12.3 and 15.12.4.

The field generated by the currents in the microstrip transmission line can be determined in the same manner for various types of load, including resonant lines with high- and low-impedance loads and matched lines with traveling-wave currents.

The complete field of small vertical monopoles, including top-loaded monopoles, is readily evaluated with the procedure carried out in Section 4.2. Such monopoles can be driven as end loads on microstrip transmission lines.

15.13 Generalization of the Theory; Application to Surface-Wave Propagation over the Arctic Ice

The analysis of the electromagnetic field generated by a horizontal electric dipole on the surface of the dielectric-coated conductor used in microstrip is simplified at the beginning of this chapter by the replacement of the highly conducting base with one of infinite conductivity. Actually, the analysis can be carried out with little formal change subject to the inequalities

$$k_0^2 \ll |k_1^2| \ll |k_2^2|, \tag{15.13.1}$$

with k_2 relatively large but not infinite. Since this generalization greatly increases the mathematical formalism and the complexity of the integrals that define the surface wave and is unnecessary for microstrip, the simplification $k_2 \to \infty$ is made. However, it is of theoretical interest to determine precisely what is neglected when k_2 is set equal to infinity and, simultaneously, to learn what happens when propagation can take place in three regions. Note that the present situation with the conditions (15.13.1) and the dipole and the point of observation in the air with the smallest wave number differs from that treated in Chapter 12 where the dipole and the point of observation are in the sea with the largest wave number. A useful application of the generalized theory is to surface-wave propagation over the ice-coated sea of the Arctic and possible communication with submarines. The detection of submarines under the ice is also of interest.

The more general analysis makes use of the complete reflection coefficients given by

$$f_r = \frac{x_0 - x_2 - i[(x_0x_2/x_1) - x_1]\tan\gamma_1 l}{x_0 + x_2 - i[(x_0x_2/x_1) + x_1]\tan\gamma_1 l}, \tag{15.13.2a}$$

$$f_{er} = f_r \quad \text{with } x_j = \frac{\gamma_j}{k_j^2}, \quad j = 0, 1, 2; \qquad Q_3 = -f_{er}, \tag{15.13.2b}$$

$$f_{mr} = f_r \quad \text{with } x_j = \gamma_j, \quad j = 0, 1, 2; \qquad P_3 = -f_{mr}. \tag{15.13.2c}$$

With these values,

$$\frac{\gamma_0}{2}(Q_3+1) = \frac{k_0^2\gamma_0}{k_1^2}\left(\frac{k_1^2\gamma_2}{k_2^2} - i\gamma_1\tan\gamma_1 l\right) \times\left(\gamma_0 - \frac{ik_0^2\gamma_1}{k_1^2}\tan\gamma_1 l + \frac{k_0^2\gamma_2}{k_2^2} - \frac{ik_1^2\gamma_0\gamma_2}{k_2^2\gamma_1}\tan\gamma_1 l\right)^{-1}, \tag{15.13.3}$$

$$\frac{k_0^2}{2\gamma_0}(P_3 - 1) = -k_0^2\left(\frac{\gamma_1}{\gamma_2} - i\tan\gamma_1 l\right)$$

$$\times \left(\gamma_1 - i\gamma_0 \tan\gamma_1 l + \frac{\gamma_1}{\gamma_2}(\gamma_0 - i\gamma_1 \tan\gamma_1 l)\right)^{-1} \tag{15.13.4}$$

so that the following integrals are obtained:

$$\begin{aligned}
&\left.\begin{array}{l} F_{\rho 2}(\rho, z'+d') \\ F_{\phi' 2}(\rho, z'+d') \end{array}\right\} \\
&\quad = \frac{k_0^2}{k_1^2}\int_0^\infty \gamma_0\left(\frac{k_1^2\gamma_2}{k_2^2} - i\gamma_1\tan\gamma_1 l\right) \\
&\qquad \times \left(\gamma_0 - \frac{ik_0^2\gamma_1}{k_1^2}\tan\gamma_1 l + \frac{k_0^2\gamma_2}{k_2^2} - \frac{ik_1^2\gamma_0\gamma_2}{k_2^2\gamma_1}\tan\gamma_1 l\right)^{-1} \\
&\qquad \times [J_0(\lambda\rho) \mp J_2(\lambda\rho)]e^{i\gamma_0(z'+d')}\lambda\, d\lambda,
\end{aligned} \tag{15.13.5}$$

$$\begin{aligned}
&\left.\begin{array}{l} F_{\rho 3}(\rho, z'+d') \\ F_{\phi' 3}(\rho, z'+d') \end{array}\right\} \\
&\quad = k_0^2\int_0^\infty \left(\frac{\gamma_1}{\gamma_2} - i\tan\gamma_1 l\right) \\
&\qquad \times \left(\gamma_1 - i\gamma_0\tan\gamma_1 l + \frac{\gamma_1}{\gamma_2}(\gamma_0 - i\gamma_1\tan\gamma_1 l)\right)^{-1} \\
&\qquad \times [J_0(\lambda\rho) \pm J_2(\lambda\rho)]e^{i\gamma_0(z'+d')}\lambda\, d\lambda,
\end{aligned} \tag{15.13.6}$$

$$\begin{aligned}
&F_{z'1}(\rho, z'+d') \\
&\quad = \frac{2k_0^2}{k_1^2}\int_0^\infty \left(\frac{k_1^2\gamma_2}{k_2^2} - i\gamma_1\tan\gamma_1 l\right) \\
&\qquad \times \left(\gamma_0 - \frac{ik_0^2\gamma_1}{k_1^2}\tan\gamma_1 l + \frac{k_0^2\gamma_2}{k_2^2} - \frac{ik_1^2\gamma_0\gamma_2}{k_2^2\gamma_1}\tan\gamma_1 l\right)^{-1} \\
&\qquad \times J_1(\lambda\rho)e^{i\gamma_0(z'+d')}\lambda^2\, d\lambda,
\end{aligned} \tag{15.13.7}$$

$$\begin{aligned}
&\left.\begin{array}{l} G_{\rho 2}(\rho, z'+d') \\ G_{\phi' 2}(\rho, z'+d') \end{array}\right\} \\
&\quad = \frac{k_0^2}{k_1^2}\int_0^\infty \left(\frac{k_1^2\gamma_2}{k_2^2} - i\gamma_1\tan\gamma_1 l\right) \\
&\qquad \times \left(\gamma_0 - \frac{ik_0^2\gamma_1}{k_1^2}\tan\gamma_1 l + \frac{k_0^2\gamma_2}{k_2^2} - \frac{ik_1^2\gamma_0\gamma_2}{k_2^2\gamma_1}\tan\gamma_1 l\right)^{-1} \\
&\qquad \times [J_0(\lambda\rho) \pm J_2(\lambda\rho)]e^{i\gamma_0(z'+d')}\lambda\, d\lambda,
\end{aligned} \tag{15.13.8}$$

$$\begin{aligned}
&\left.\begin{array}{l} G_{\rho 3}(\rho, z'+d') \\ G_{\phi' 3}(\rho, z'+d') \end{array}\right\} \\
&\quad = \int_0^\infty \gamma_0\left(\frac{\gamma_1}{\gamma_2} - i\tan\gamma_1 l\right)
\end{aligned}$$

$$\times\left(\gamma_1 - i\gamma_0 \tan\gamma_1 l + \frac{\gamma_1}{\gamma_2}(\gamma_0 - i\gamma_1 \tan\gamma_1 l)\right)^{-1}$$
$$\times [J_0(\lambda\rho) \mp J_2(\lambda\rho)]e^{i\gamma_0(z'+d')}\lambda\, d\lambda, \tag{15.13.9}$$

$$G_{z'1}(\rho, z'+d')$$
$$= 2\int_0^\infty \left(\frac{\gamma_1}{\gamma_2} - i\tan\gamma_1 l\right)$$
$$\times\left(\gamma_1 - i\gamma_0 \tan\gamma_1 l + \frac{\gamma_1}{\gamma_2}(\gamma_0 - i\gamma_1 \tan\gamma_1 l)\right)^{-1}$$
$$\times J_1(\lambda\rho)e^{i\gamma_0(z'+d')}\lambda^2\, d\lambda. \tag{15.13.10}$$

When these integrals are evaluated with the approximate procedure generalized to include the three parameters k_0, k_1, and k_2 as in Chapter 12, only the integrals in which $\lambda \sim k_0$ yield propagating terms in the air. All others have one of the attenuating exponentials $\exp[-k_1(z'+d')]$ or $\exp[-k_2(z'+d')]$ as a factor. With $\lambda \sim k_0$, $\gamma_1 \sim k_1$, $\gamma_2 \sim k_2$, and $\tan k_1 l \sim k_1 l$, the significant integrals are

$$\left.\begin{matrix} F_{\rho 2}(\rho, z'+d') \\ F_{\phi' 2}(\rho, z'+d') \end{matrix}\right\} \sim k_0^2 \int_0^\infty \frac{\gamma_0(k_2^{-1} - il)}{\gamma_0 - ik_0^2 l + k_0^2/k_2 - i\gamma_0 k_1^2 l/k_2}$$
$$\times [J_0(\lambda\rho) \mp J_2(\lambda\rho)]e^{i\gamma_0(z'+d')}\lambda\, d\lambda$$
$$= \epsilon' \int_0^\infty \frac{\gamma_0}{\gamma_0 + \epsilon'}[J_0(\lambda\rho) \mp J_2(\lambda\rho)]e^{i\gamma_0(z'+d')}\lambda\, d\lambda, \tag{15.13.11}$$

where

$$\epsilon' = \frac{k_0^2(k_2^{-1} - il)}{1 - ik_1^2 l/k_2} \sim k_0\left(\frac{k_0}{k_2} - ik_0 l\right), \tag{15.13.12}$$

$$\left.\begin{matrix} F_{\rho 3}(\rho, z'+d') \\ F_{\phi' 3}(\rho, z'+d') \end{matrix}\right\} \sim k_0^2 \int_0^\infty \frac{(k_2^{-1} - il)}{1 - i\gamma_0 l + \gamma_0/k_2 - ik_1^2 l/k_2}$$
$$\times [J_0(\lambda\rho) \pm J_2(\lambda\rho)]e^{i\gamma_0(z'+d')}\lambda\, d\lambda$$
$$= k_0^2 \int_0^\infty \frac{(k_2^{-1} - il)}{\gamma_0(k_2^{-1} - il) + 1 - ik_1^2 l/k_2}$$
$$\times [J_0(\lambda\rho) \pm J_2(\lambda\rho)]e^{i\gamma_0(z'+d')}\lambda\, d\lambda$$
$$= k_0^2\epsilon' \int_0^\infty \frac{[J_0(\lambda\rho) \pm J_2(\lambda\rho)]e^{i\gamma_0(z'+d')}\lambda\, d\lambda}{k_0^2 + \gamma_0\epsilon'}, \tag{15.13.13}$$

$$F_{z'1}(\rho, z'+d') \sim 2k_0^2 \int_0^\infty \frac{(k_2^{-1} - il)J_1(\lambda\rho)e^{i\gamma_0(z'+d')}\lambda^2\, d\lambda}{\gamma_0 - ik_0^2 l + k_0^2/k_2 - i\gamma_0 k_1^2 l/k_2}$$

$$= 2\epsilon' \int_0^\infty \frac{J_1(\lambda\rho) e^{i\gamma_0(z'+d')} \lambda^2 \, d\lambda}{\gamma_0 + \epsilon'}, \tag{15.13.14}$$

$$\left.\begin{matrix} G_{\rho 2}(\rho, z'+d') \\ G_{\phi' 2}(\rho, z'+d') \end{matrix}\right\} \sim k_0^2 \int_0^\infty \frac{(k_2^{-1} - il)}{\gamma_0 - ik_0^2 l + k_0^2/k_2 - i\gamma_0 k_1^2 l/k_2} \times [J_0(\lambda\rho) \pm J_2(\lambda\rho)] e^{i\gamma_0(z'+d')} \lambda \, d\lambda$$

$$= \epsilon' \int_0^\infty \frac{[J_0(\lambda\rho) \pm J_2(\lambda\rho)] e^{i\gamma_0(z'+d')} \lambda \, d\lambda}{\gamma_0 + \epsilon'}, \tag{15.13.15}$$

$$\left.\begin{matrix} G_{\rho 3}(\rho, z'+d') \\ G_{\phi' 3}(\rho, z'+d') \end{matrix}\right\} \sim \int_0^\infty \frac{\gamma_0(k_2^{-1} - il)}{1 - i\gamma_0 l + \gamma_0/k_2 - ik_1^2 l/k_2} \times [J_0(\lambda\rho) \mp J_2(\lambda\rho)] e^{i\gamma_0(z'+d')} \lambda \, d\lambda$$

$$= \epsilon' \int_0^\infty \frac{\gamma_0 [J_0(\lambda\rho) \mp J_2(\lambda\rho)] e^{i\gamma_0(z'+d')} \lambda \, d\lambda}{k_0^2 + \gamma_0 \epsilon'}, \tag{15.13.16}$$

$$G_{z'1}(\rho, z'+d') \sim 2 \int_0^\infty \frac{(k_2^{-1} - il) J_1(\lambda\rho) e^{i\gamma_0(z'+d')} \lambda^2 \, d\lambda}{1 - i\gamma_0 l + \gamma_0/k_2 - ik_1^2 l/k_2}$$

$$= 2\epsilon' \int_0^\infty \frac{J_1(\lambda\rho) e^{i\gamma_0(z'+d')} \lambda^2 \, d\lambda}{k_0^2 + \gamma_0 \epsilon'}. \tag{15.13.17}$$

An examination of these integrals shows that the thickness l which characterizes the layer of dielectric and the wave number k_2 that characterizes Region 2 always occur in the combination

$$\epsilon' = k_0 \epsilon = \frac{k_0^2}{1 - ik_1^2 l/k_2} (k_2^{-1} - il) \sim k_0 \left(\frac{k_0}{k_2} - ik_0 l \right). \tag{15.13.18}$$

When $k_2 \to \infty$, $\epsilon' = k_0 \epsilon = -ik_0^2 l$ and all integrals reduce to those evaluated in Section 15.3 for microstrip. Since l appears only in the surface-wave integrals contained in $F_{m1}(\rho, z'+d') = F_{m2}(\rho, z'+d') + F_{m3}(\rho, z'+d')$ and $G_{m1}(\rho, z'+d') = G_{m2}(\rho, z'+d') + G_{m3}(\rho, z'+d')$, $m = \rho, \phi', z'$, and not in the direct- and image-field integrals contained in $F_{m0}(\rho, z' \pm d')$ or $G_{m0}(\rho, z' \pm d')$, the complete formulas (15.4.1)–(15.4.6) can be generalized to apply to the three-layered region with k_2 not infinite by the substitution of ϵ for $(-ik_0 l)$. Specifically, with $z' = d' = 0$, (15.4.7)–(15.4.13) become

$$E_{0\rho}(\rho, \phi', 0) = -\frac{\omega\mu_0 \epsilon^2}{2\pi k_0} \cos\phi' \, e^{ik_0\rho} \times \left[\frac{ik_0}{\rho} - \frac{1}{\rho^2} - \frac{i}{k_0 \rho^3} - k_0^2 \epsilon \left(\frac{\pi}{k_0 \rho} \right)^{1/2} e^{-ip_2} \mathcal{F}(p_2) \right], \tag{15.13.19}$$

$$E_{0\phi'}(\rho, \phi', 0) = \frac{\omega\mu_0 \epsilon^2}{2\pi k_0} \sin\phi' \, e^{ik_0\rho}$$

$$\times \left[\frac{2}{\rho^2} + \frac{2i}{k_0\rho^3} + \frac{ik_0\epsilon}{\rho}\left(\frac{\pi}{k_0\rho}\right)^{1/2} e^{-ip_2}\mathcal{F}(p_2)\right], \qquad (15.13.20)$$

$$E_{0z'}(\rho, \phi', 0) = -\frac{\omega\mu_0\epsilon}{2\pi k_0}\cos\phi'\, e^{ik_0\rho}$$

$$\times \left[\frac{ik_0}{\rho} - \frac{1}{\rho^2} - k_0^2\epsilon\left(\frac{\pi}{k_0\rho}\right)^{1/2} e^{-ip_2}\mathcal{F}(p_2)\right], \qquad (15.13.21)$$

$$B_{0\rho}(\rho, \phi', 0) = \frac{\mu_0\epsilon}{2\pi}\sin\phi'\, e^{ik_0\rho}$$

$$\times \left[\frac{2}{\rho^2} + \frac{2i}{k_0\rho^3} + \frac{ik_0\epsilon}{\rho}\left(\frac{\pi}{k_0\rho}\right)^{1/2} e^{-ip_2}\mathcal{F}(p_2)\right], \qquad (15.13.22)$$

$$B_{0\phi'}(\rho, \phi', 0) = \frac{\mu_0\epsilon}{2\pi}\cos\phi'\, e^{ik_0\rho}$$

$$\times \left[\frac{ik_0}{\rho} - \frac{1}{\rho^2} - \frac{i}{k_0\rho^3} - k_0^2\epsilon\left(\frac{\pi}{k_0\rho}\right)^{1/2} e^{-ip_2}\mathcal{F}(p_2)\right], \qquad (15.13.23)$$

$$B_{0z'}(\rho, \phi', 0) = \frac{\mu_0\epsilon^2}{2\pi}\sin\phi'\, e^{ik_0\rho}\left(\frac{1}{\rho^2} + \frac{3i}{k_0\rho^3} - \frac{3}{k_0^2\rho^4}\right), \qquad (15.13.24)$$

with

$$p_2 = \frac{k_0\epsilon^2\rho}{2}; \qquad \epsilon = \frac{k_0(k_2^{-1} - il)}{1 - ik_1^2 l/k_2}. \qquad (15.13.25)$$

When $k_2 \to \infty$, $\epsilon = -ik_0 l$ and these formulas reduce identically to (15.4.7)–(15.4.13). When $l = 0$, a two-layered region remains consisting of air (Region 0) over Region 2 with $|k_2|^2 \gg k_0^2$ and $\epsilon = k_0/k_2$. With this value, the above formulas correspond exactly to the lateral-wave terms in (5.5.57)–(5.5.62) with $d = 0$. (Note that $z' = -z$ and $\phi' = -\phi$.) These latter formulas also include terms that represent propagation in Region 2 with the exponential factor $\exp(ik_2\rho)$. They are included in the complete analysis in Chapter 5 but do not appear in the evaluation in Section 15.3 because there is no propagation in the thin dielectric with the wave number k_1 or in the conducting base when $k_2 = \infty$. They are negligible beyond short distances from the source when Region 2 is a metal or sea water with $\exp(ik_2\rho) = \exp(-\alpha_2\rho)\exp(i\beta_2\rho)$ and $\alpha_2 = \beta_2$. They can be derived if (15.13.3) and (15.13.4) are used instead of (15.2.36) and (15.2.37).

With the generalized formulas available, it is possible to determine the contribution to the field of a dipole on microstrip by the finite conductivity of the base. For example, when this has a conductivity of the order $\sigma_2 \sim 10^7$ S/m, the dielectric has a thickness $l \sim 0.25$ mm, and the frequency is in the range 10–50 GHz, it follows that $|k_2| = (\omega\mu_0\sigma_2)^{1/2} = [2\pi \times (10\text{–}50) \times 10^9 \times 4\pi \times 10^{-7} \times 10^7]^{1/2} = [8\pi^2(1\text{–}5)]^{1/2} \times 10^5 = \pi \times 10^5 (8\text{–}40)^{1/2} = (0.9\text{–}2.0) \times 10^6$

m^{-1}. Hence, $|k_2^{-1}| = (0.5\text{–}1.0) \times 10^{-6}$ m and $l = 2.5 \times 10^{-4}$ m. Clearly, $|k_2^{-1}| \ll l$ so that the finite conductivity contributes nothing significant and the approximation $k_2 = \infty$, $k_2^{-1} = 0$ is a good one.

For application in the Arctic, the horizontal dipole is on the air surface of ice which has a thickness of the order $l = 2.5$ m. As discussed in Section 7.6, the detection of submarines in the ocean in the absence of ice involves a frequency of the order of 25 Hz, so that $k_2 = (1+i)(\omega\mu_0\sigma_2/2)^{1/2} = (1+i)[(2\pi \times 25 \times 4\pi \times 10^{-7} \times 4)/2]^{1/2} = 20\pi\sqrt{10^{-7}}\,(1+i) \sim 0.02(1+i)$ m^{-1}. It follows that $k_2^{-1} = 25(1-i)$ m and

$$k_2^{-1} - il = 25(1-i) - 2.5i = 25 - 27.5i \text{ m}. \tag{15.13.26}$$

Evidently, the amplitude of the surface wave associated with the layer of ice is much smaller than the amplitude of the surface wave associated with the currents in the sea and can be neglected. This means that the field in the sea is essentially the same with the unit dipole on the surface of the sea or on the surface of the ice. In Section 7.6, an insulated antenna with bare ends is located at a depth $d = 1$ m in the sea. In the case at hand, a bare antenna can be laid on the ice and terminated at each end in a vertical element extending down into the sea. In effect, the antenna is eccentrically insulated with a two-layered dielectric. Formulas for the wave number k_L and the characteristic impedance Z_c are not available. Approximate values are k_L and Z_c for the eccentrically insulated antenna with the single dielectric ice. The dielectric between the antenna and the sea is much more important in determining these parameters than the air above the antenna.

The complete analysis of the field in the air over a dielectric-coated region with a large wave number reveals the following interesting properties:

(1) The lateral wave generated in the air along the air–dielectric boundary and the trapped surface wave in the thin dielectric layer on a perfectly conducting plane ($k_2 \to \infty$) combine into a single surface wave when $k_1 l$ is electrically small. This propagates in the air and in the thin dielectric with the wave number k_0. Its amplitude is multiplied by the small quantities $k_0 l$ or $k_0^2 l^2$ so that it vanishes when $l = 0$.

(2) When $k_1 l$ is small and $|k_2|$ is large but finite, a second lateral wave—associated with conduction and polarization currents in Region 2—is generated along the air–dielectric boundary. The complete lateral wave is a superposition of these two waves and involves the combined factor $k_0(k_2^{-1} - il)(1 - ik_1^2 l/k_2)^{-1}$. When $k_2 \to \infty$, this reduces to (1); when $l = 0$, it reduces to k_0/k_2, as already seen for the case of two layers treated in Chapter 5.

16
Antennas in Material Media Near Boundaries: The Bare Metal Dipole

16.1 Introduction

This chapter is concerned with the properties of antennas embedded in a medium like soil, sand, clay, rock, ice, or water near its boundary with air and with similar antennas located in the salt water near the oceanic crust under the sea floor. In all cases and as shown in Fig. 16.1.1, the transmitting and receiving antennas under study are located in Region 1, $z \geq 0$, with the complex wave number $k_1 = \beta_1 + i\alpha_1 = \omega[\mu_0(\epsilon_1 + i\sigma_1/\omega)]^{1/2}$, where β_1 is the phase constant and α_1 is the attenuation constant. The wave number in the adjacent Region 2, $z \leq 0$, is $k_2 = \beta_2 + i\alpha_2 = \omega[\mu_0(\epsilon_2 + i\sigma_2/\omega)]^{1/2}$. In general, it is assumed that $|k_1|^2 \gg |k_2|^2$. The two regions are characterized by their respective permittivities ϵ_j, conductivities σ_j, and permeabilities μ_j, with $j = 1, 2$. For all media considered here, it is assumed that $\mu_1 = \mu_2 = \mu_0$ for air. In either region, the wavelength is $\lambda_j = 2\pi/\beta_j$ and the skin depth $d_{sj} = 1/\alpha_j$. The wave impedance is $\zeta_j = \omega\mu_0/k_j$ and the normalizing factor $\Delta_j = \zeta_0\beta_j/\omega\mu_0$, where $\zeta_0 = (\mu_0/\epsilon_0)^{1/2} \sim 120\pi\ \Omega$. When Region 2 is air, $\sigma_2 = 0$, $\epsilon_2 = \epsilon_0$, and $k_2 = \beta_2 = \omega(\mu_0\epsilon_0)^{1/2}$; $\alpha_2 = 0$. When the frequency is sufficiently low so that $\epsilon_j \ll \sigma_j/\omega$, $k_j \sim (i\omega\mu_0\sigma_j)^{1/2} = (1+i)(\omega\mu_0\sigma_j/2)^{1/2}$ and $\beta_j = \alpha_j$. Numerical values for many of these quantities are listed in Table 16.1.1 for two sets of frequencies, viz., $f = 0.3$, 1, and 3 Hz when Region 1 is sea water with $\sigma_1 = 4$ S/m, $\epsilon_{1r} = 80$; and $f = 0.3 \times 10^5$, 10^5, and 3×10^5 Hz when Region 1 is moist earth with $\sigma_1 = 10^{-3}$ S/m, $\epsilon_{1r} = 8$. Also listed in Table 16.1.1 is the electrical radius $|k_c a|$ of the copper conductor forming the antenna with $|k_c| \sim (\omega\mu_c\sigma_c)^{1/2}$ and $\sigma_c = 5.65 \times 10^7$ S/m.

Table 16.1.1. Quantities Useful for Antennas in a Dissipative Region 1

	Sea Water: $\epsilon_{1r} = 80,\ \sigma_1 = 4$ S/m		
f (Hz)	0.3	1	3
β_1 (m^{-1})	2.18×10^{-3}	3.97×10^{-3}	6.88×10^{-3}
α_1 (m^{-1})	2.18×10^{-3}	3.97×10^{-3}	6.88×10^{-3}
β_1^{-1} (m)	458.7	251.9	145.3
λ_1 (m)	2884	1583	913
a (mm)	5.2	5.2	5.2
a/λ_1	1.8×10^{-6}	3.3×10^{-6}	5.7×10^{-6}
$\lvert k_c a \rvert$	0.06	0.11	0.19
r^i (Ω/m)	2.08×10^{-4}	2.08×10^{-4}	2.08×10^{-4}
$2\pi r^i/\omega\mu_0$	551.7	165.2	55.2
Δ_1	3.46×10^5	1.90×10^5	1.094×10^5

	Moist Earth: $\epsilon_{1r} = 8,\ \sigma_1 = 10^{-3}$ S/m		
f (Hz)	3×10^4	10^5	3×10^5
β_1 (m^{-1})	1.09×10^{-2}	2.02×10^{-2}	3.67×10^{-2}
α_1 (m^{-1})	1.07×10^{-2}	1.94×10^{-2}	3.21×10^{-2}
β_1^{-1} (m)	91.7	49.5	27.2
λ_1 (m)	576	311	171.2
a (mm)	1.04	1.04	1.04
a/λ_1	1.8×10^{-6}	3.3×10^{-6}	6.1×10^{-6}
$\lvert k_c a \rvert$	3.8	6.94	12.02
r^i (Ω/m)	0.73×10^{-2}	1.28×10^{-2}	2.22×10^{-2}
$2\pi r^i/\omega\mu_0$	0.195	0.101	0.059
Δ_1	17.34	9.67	5.84

$$k_1 = \beta_1 + i\alpha_1 = \omega(\mu_0\epsilon_1)^{1/2}\left(1 + \frac{i\sigma_1}{\omega\epsilon_1}\right)^{1/2},$$

$$\lambda_1 = \frac{2\pi}{\beta_1}; \qquad \Delta_1 = \frac{\zeta_0\beta_1}{\omega\mu_0} = \frac{\beta_1}{\beta_0}; \qquad k_c = (i\omega\mu_c\sigma_c)^{1/2},$$

$$z^i = r^i - ix^i = r_0\,\frac{k_c a}{2}\,\frac{J_0(k_c a)}{J_1(k_c a)},$$

$$z^i \sim r_0 = \frac{1}{\pi a^2\sigma_c}, \quad |k_c a| \le 2; \qquad z^i \sim -\frac{ik_c}{2\pi a\sigma_c}, \quad |k_c a| \ge 5$$

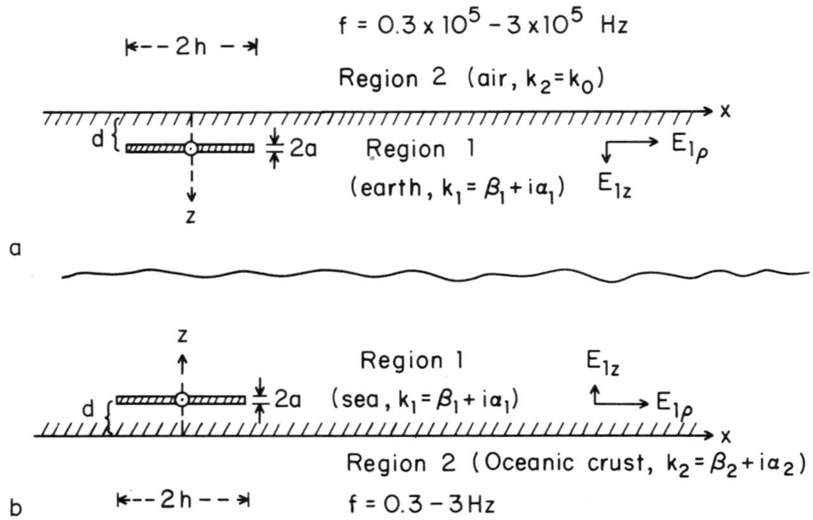

Figure 16.1.1. Bare copper antenna: (a) in earth near air surface; (b) in sea near sea floor.

16.2 The Bare Cylindrical Antenna in a Dissipative Medium

For some applications and as terminations for insulated antennas in the earth or ocean, the bare metal cylinder is useful. Complete general formulas for the current distribution in and the admittance of such an antenna are available (King, Sandler, and Wu 1969) as a function of the half-length h and the radius a when located in an infinite homogeneous and isotropic medium with arbitrary values of permittivity and conductivity. The formulas assume that the very large but finite conductivity of copper ($\sigma_c = 5.65 \times 10^7$ S/m) or other metals is well approximated by $\sigma_c \sim \infty$. This is true at all but the very lowest frequencies. The normalized admittance $Y/\Delta_1 = (G-iB)/\Delta_1$ and impedance $Z\Delta_1 = (R-iX)\Delta_1$ as computed with these formulas for antennas in the two frequency ranges and in the two media listed in Table 16.1.1 are shown in Fig. 16.2.1 as functions of h/λ_1 where $\lambda_1 = 2\pi/\beta_1$. These graphs apply specifically to bare antennas with radius $a = 1.04$ mm (5.2 mm) in moist earth (sea water) at $f = 0.3 \times 10^5$ to 3×10^5 Hz (0.3 to 3 Hz). They show that the conductance rises from zero to a maximum near $h/\lambda_1 \sim 0.15$ or $\beta_1 h \sim 1$ and then decreases to an approximately constant value beyond $h/\lambda_1 \sim 0.25$ or $\beta_1 h \sim \pi/2$. Similarly,

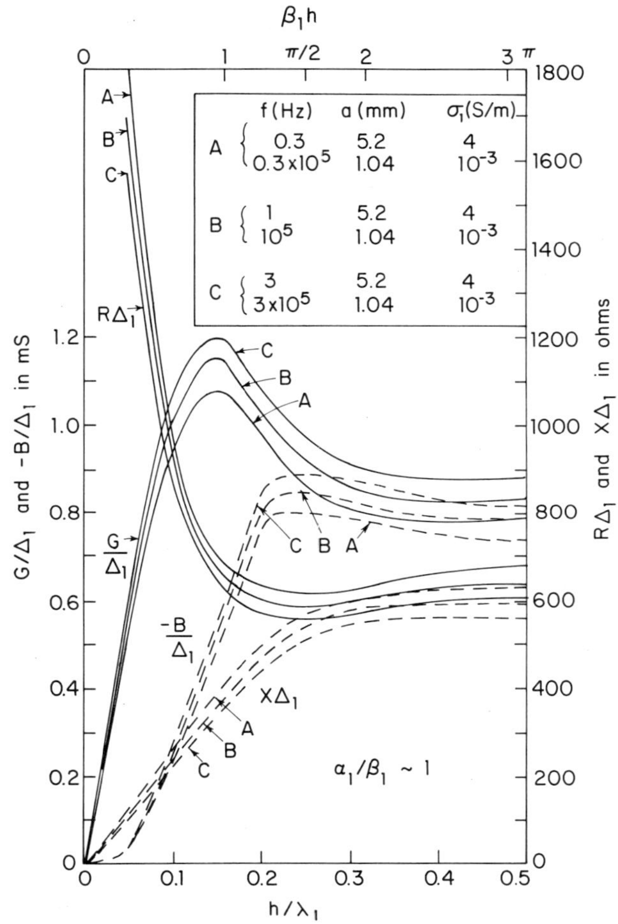

Figure 16.2.1. Normalized admittance $Y/\Delta_1 = (G - iB)/\Delta_1$ and impedance $Z\Delta_1 = (R - iX)\Delta_1$ of bare, perfectly conducting antenna.

the resistance decreases from large values to a minimum near $h/\lambda_1 \sim 0.25$ or $\beta_1 h \sim \pi/2$, beyond which it remains sensibly constant. These values suggest that when $\alpha_1/\beta_1 \sim 1$, the maximum useful length of a bare antenna is $h \sim \lambda_1/4$ or $\beta_1 h \sim \pi/2$. This is a consequence of the fact that the current in the antenna decays to virtually zero beyond $x \sim \lambda_1/4$ when $h \geq \lambda_1/4$. Similar graphs for a range of values of α_1/β_1 are shown in Fig. 16.2.2 for the admittance and in Fig. 16.2.3 for the impedance. It is seen that as α_1/β_1 is reduced, the useful length of the antenna increases.

In order to investigate the effect of the finite conductivity of the metal conductor, it is necessary to reformulate and resolve the integral equation for the current with the actual internal impedance per unit length of the

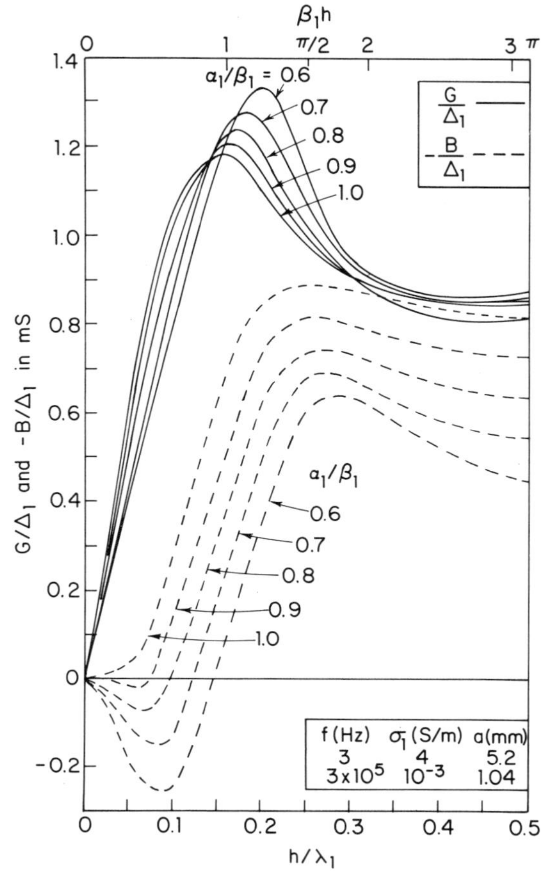

Figure 16.2.2. Normalized admittance $Y/\Delta_1 = (G - iB)/\Delta_1$ of bare, perfectly conducting antenna with α_1/β_1 as parameter.

conductor. This is

$$z^i = r^i - ix^i = r_0 \frac{k_c a}{2} \frac{J_0(k_c a)}{J_1(k_c a)}; \qquad r_0 = (\pi a^2 \sigma_c)^{-1}. \tag{16.2.1}$$

Important limiting forms are

$$|k_c a| \leq 2: \qquad r^i \sim r_0; \quad x^i = \omega l^i \sim \frac{\omega \mu_c}{8\pi}, \tag{16.2.2a}$$

$$|k_c a| \geq 5: \qquad r^i \sim x^i \sim \frac{1}{2\pi a}\left(\frac{\omega \mu_c}{2\sigma_c}\right)^{1/2}. \tag{16.2.2b}$$

This has been carried out (King 1969) and the following expression for the current obtained:

$$I_x(x) = -\frac{i2\pi k_1^2 V_0^e}{\omega \mu_0 k_{\mathrm{I}} \Psi \cos k_{\mathrm{I}} h}\left[\sin k_{\mathrm{I}}(h - |x|) + T_U(\cos k_{\mathrm{I}} x - \cos k_{\mathrm{I}} h)\right.$$

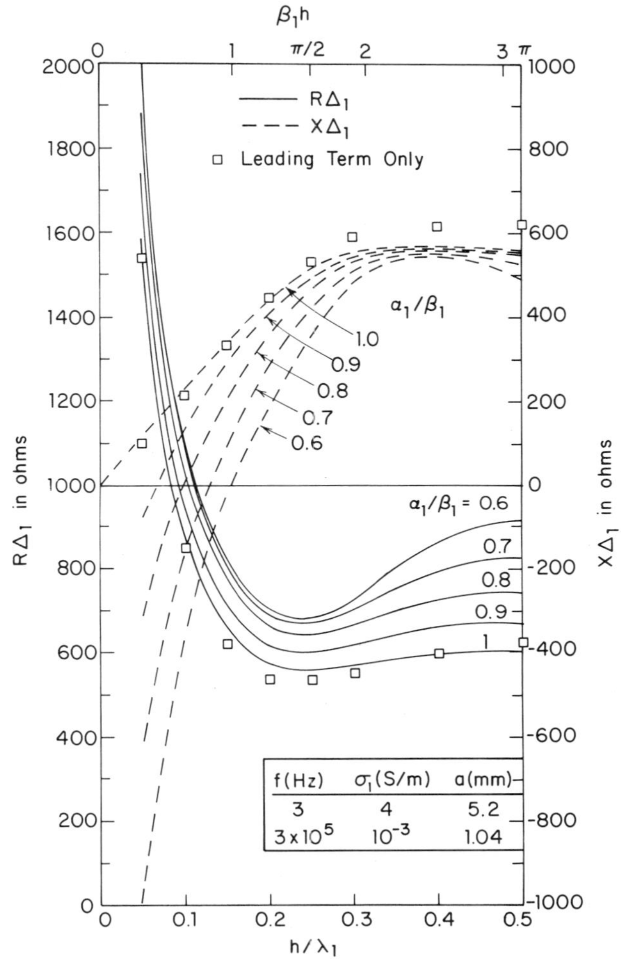

Figure 16.2.3. Normalized impedance $Z\Delta_1 = (R - iX)\Delta_1$ of bare, perfectly conducting antenna with α_1/β_1 as parameter.

$$+ T_D(\cos \tfrac{1}{2}k_1 x - \cos \tfrac{1}{2}k_1 h)], \tag{16.2.3}$$

where

$$k_\mathrm{I} = \beta_\mathrm{I} + i\alpha_\mathrm{I} = k_1\left(1 + \frac{i4\pi z^i}{\omega\mu_0\Psi}\right)^{1/2} \tag{16.2.4}$$

is the complex wave number of the current in the antenna. Formulas for T_U and T_D are given by King (1969). Note that the wave number appearing in the coefficient of T_D is k_1 and not k_I. When $|k_\mathrm{I}|$ is significantly greater than $|k_1|$, the contribution to the current by the third term in (16.2.3) will decay more slowly than the contributions by the other two. This can have only a negligible effect on the current near the driving point. Its contribution to the current elsewhere is also relatively small when $\beta_1 \sim \alpha_1$. In this case, the leading first term in (16.2.3) is a generally adequate approximation. That is,

$$I_x(x) \sim V_0^e Y \frac{\sin k_{\rm I}(h-|x|)}{\sin k_{\rm I} h}; \qquad Y = -\frac{i2\pi k_{\rm I}^2}{\omega\mu_0\Psi}\frac{\tan k_{\rm I}h}{k_{\rm I}}, \tag{16.2.5}$$

where

$$\Psi = 2\ln(h/a) - 2. \tag{16.2.6}$$

It is seen from (16.2.5) that $I_x(x)$ and Y depend only on k_1 and not on $k_{\rm I}$ when $|k_{\rm I}h|^2 \ll 3$. Note that $|k_1 h| < |k_{\rm I}h|$.

Formula (16.2.4) for $k_{\rm I}$ permits the definition of the low-frequency range in which the internal impedance is significant as distinct from higher frequencies for which the copper conductor can be approximated by a perfect conductor. Specifically, let the higher frequencies be defined by

$$(\omega\mu_0)^2 \gg \left(\frac{4\pi r^i}{\Psi}\right)^2 \quad \text{or} \quad \omega\mu_0 \geq 4\pi r^i \tag{16.2.7}$$

since $\Psi \sim 20$. With $r^i \sim 1/\pi a^2\sigma_c$ and $\sigma_c = 5.65 \times 10^7$ S/m,

$$f \geq \frac{r^i \times 10^7}{2\pi} = \frac{10^7}{2\pi^2 a^2\sigma_c} \sim \frac{10^{-2}}{a^2} = \begin{cases} 400 \text{ Hz}, & a = 5.2 \text{ mm} \\ 10^4 \text{ Hz}, & a = 1.04 \text{ mm} \end{cases}. \tag{16.2.8}$$

Thus, the dividing line is between $f = 400$ Hz and 10 kHz depending on the cross-sectional size of the copper conductor. In the following analysis, antennas in the earth will be considered with $f \sim 10^5$ Hz and $a \sim 1$ mm so that the conductor can be treated as if perfect. Also considered are antennas on the sea floor with $a = 5.2$ mm when $f \sim 1$ Hz and (in Chapter 17) $f \sim 1$ kHz. For the latter, the internal impedance is negligibly small; for the former, it is dominant. For earth with $\sigma_1 \sim 10^{-3}$ S/m, $\epsilon_{1r} \sim 8$, and $f \leq 10^6$ Hz and for sea water with $\sigma_1 \sim 4$ S/m, $\epsilon_{1r} \sim 80$, and $f \leq 10$ Hz, $\omega\epsilon_1/\sigma_1$ is sufficiently small to permit the approximation $k_1^2 \sim i\omega\mu_0\sigma_1$. This simplifies (16.2.5) since now

$$Y = G - iB = \frac{2\pi\sigma_1}{\Psi}\frac{\tan k_{\rm I}h}{k_{\rm I}} = Z^{-1}. \tag{16.2.9}$$

The driving-point impedance is

$$Z = R - iX = \frac{\Psi}{2\pi\sigma_1} k_{\rm I}\cot k_{\rm I}h. \tag{16.2.10}$$

In (16.2.9) and (16.2.10),

$$k_{\rm I} = \beta_{\rm I} + i\alpha_{\rm I} = \left(i\omega\mu_0\sigma_1 - \frac{4\pi z^i\sigma_1}{\Psi}\right)^{1/2}. \tag{16.2.11}$$

Explicit formulas for $\beta_{\rm I}$ and $\alpha_{\rm I}$ can be written down, but they are quite long and not actually required since all desired numerical values are easily computed from (16.2.11). With (16.2.10) and the relation

$$\cot k_{\rm I}h = \frac{\sin 2\beta_{\rm I}h - i\sinh 2\alpha_{\rm I}h}{\cosh 2\alpha_{\rm I}h - \cos 2\beta_{\rm I}h}, \tag{16.2.12}$$

it follows that

$$R = \frac{\Psi}{2\pi\sigma_1}\left(\frac{\beta_{\rm I}\sin 2\beta_{\rm I}h + \alpha_{\rm I}\sinh 2\alpha_{\rm I}h}{\cosh 2\alpha_{\rm I}h - \cos 2\beta_{\rm I}h}\right), \tag{16.2.13}$$

$$X = \frac{\Psi}{2\pi\sigma_1}\left(\frac{\beta_{\rm I}\sinh 2\alpha_{\rm I}h - \alpha_{\rm I}\sin 2\beta_{\rm I}h}{\cosh 2\alpha_{\rm I}h - \cos 2\beta_{\rm I}h}\right). \tag{16.2.14}$$

The resistance R in (16.2.13) consists of two parts in the form $R = R^i + R^e$, where R^i is associated with the power dissipated in heating the conductor and R^e with the power transferred to the ambient medium. The former is readily evaluated as follows:

$$I_x(0)I_x^*(0)R^i = r^i\int_{-h}^{h} I_x(x)I_x^*(x)\,dx, \tag{16.2.15}$$

where $I_x(x)$ is given by (16.2.5). The result is

$$R^i = \frac{r^i}{\alpha_{\rm I}\beta_{\rm I}}\left(\frac{\beta_{\rm I}\sinh 2\alpha_{\rm I}h - \alpha_{\rm I}\sin 2\beta_{\rm I}h}{\cosh 2\alpha_{\rm I}h - \cos 2\beta_{\rm I}h}\right). \tag{16.2.16}$$

It follows that

$$R^e = R - R^i. \tag{16.2.17}$$

The corresponding formulas for the electrically short antenna are of particular interest and clarity. Subject to

$$\beta_{\rm I}h < \alpha_{\rm I}h < 1, \tag{16.2.18}$$

the expansion in series of the trigonometric and hyperbolic functions leads to

$$R \sim \frac{\Psi}{2\pi\sigma_1 h}\left(1 + \frac{1}{3}h^2(\alpha_{\rm I}^2 - \beta_{\rm I}^2)\right) = \frac{\Psi}{2\pi\sigma_1 h} + \frac{2}{3}hr^i, \tag{16.2.19}$$

where the last step follows from (16.2.11), which gives $\alpha_{\rm I}^2 - \beta_{\rm I}^2 = 4\pi r^i\sigma_1/\Psi$; and

$$X \sim \frac{\alpha_{\rm I}\beta_{\rm I}h\Psi}{3\pi\sigma_1} = \frac{\omega\mu_0\Psi h}{6\pi}, \tag{16.2.20}$$

where the last step follows from (16.2.11), which gives $2\alpha_{\rm I}\beta_{\rm I} \sim \omega\mu_0\sigma_1$. Also, from (16.2.16) and (16.2.17),

$$R^i \sim \frac{2}{3}hr^i; \qquad R^e = R - R^i \sim \frac{\Psi}{2\pi\sigma_1 h}. \tag{16.2.21}$$

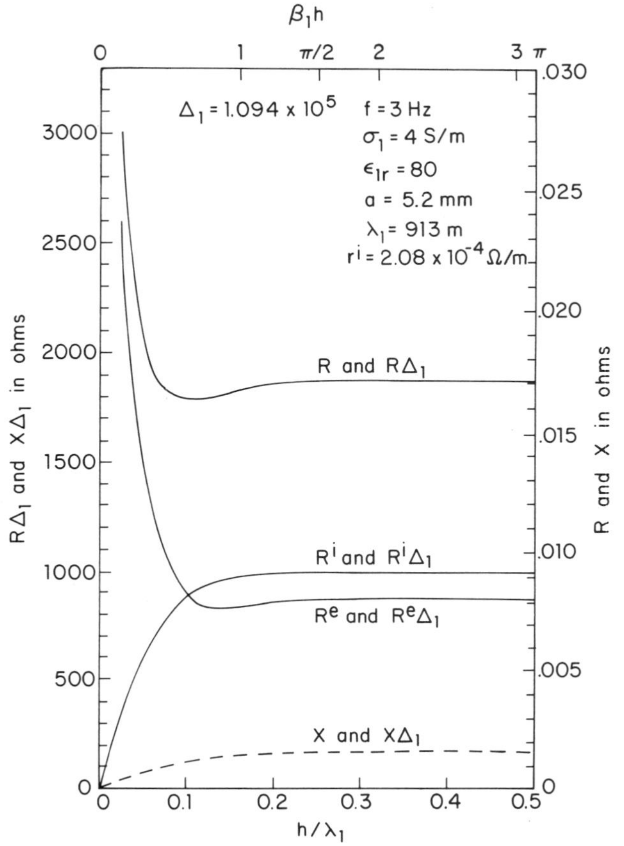

Figure 16.2.4. Impedance $Z = R - iX$ of bare, imperfectly conducting antenna.

The resistances R, R^i, and R^e are shown in Fig. 16.2.4 as a function of h/λ_1 for a copper conductor with radius $a = 5.2$ mm in sea water ($\sigma_1 = 4$ S/m) at $f = 3$ Hz. A scale is also given for the normalized resistances $R\Delta_1$, $R^i\Delta_1$, and $R^e\Delta_1$ to permit direct comparison with Fig. 16.2.3 (curve with $\alpha_1/\beta_1 = 1$), which applies when $r^i = 0$. In this case, $R = R^e$ since $R^i = 0$. Note that the curve for $R = R^e$ in Fig. 16.2.3 is similar to that for R^e in Fig. 16.2.4 but is not the same. This is because the current distribution changes from $\sin k_I(h - |x|)$ when r^i is significant to $\sin k_1(h - |x|)$ when $r^i = 0$. It is seen from Fig. 16.2.4 that the contribution to the total resistance by the internal resistance is insignificant for sufficiently short antennas and becomes comparable and somewhat greater when $h > 0.1\lambda_1$. The constancy of all resistances and the reactance when $h \geq 0.25\lambda_1$ indicates that the current on the conductor is reduced to near zero for $x \geq 0.25\lambda_1$ for all lengths $h > 0.25\lambda_1$.

16.3 The Effective Length and Figure of Merit of the Bare Antenna

The purpose of a transmitting antenna is to generate a useful electromagnetic field at a distance. The magnitude of such a field is determined by the electric moment of the antenna. When the antenna is in Region 1 (with wave number $k_1 = \beta_1 + i\alpha_1$) near and parallel to the boundary with Region 2 ($k_2 = \beta_2 + i\alpha_2$) and $|k_1|^2 \gg |k_2|^2$, the only significant field along the boundary surface in either region is due to the lateral wave which travels along the boundary in Region 2 with the wave number k_2. Thus, the effective half-length for a dipole of length $2h$ as obtained from the leading term given in (16.2.5) is

$$h_e(\phi_0) = \int_0^h \frac{\sin k_I(h-x)}{\sin k_I h} e^{-ik_2 x \cos\phi_0}\, dx, \tag{16.3.1}$$

where ϕ_0 is the angle between the antenna and the direction of observation. The direction of maximum field is $\phi_0 = 0$ and the associated maximum effective length is

$$h_e(0) \equiv h_e = \int_0^h \frac{\sin k_I(h-x)}{\sin k_I h} e^{-ik_2 x}\, dx. \tag{16.3.2a}$$

This integrates into

$$h_e = \frac{k_I}{k_I^2 - k_2^2} \frac{\cos k_2 h - \cos k_I h}{\sin k_I h}. \tag{16.3.2b}$$

For the electrically short antenna with $\beta_I h < \alpha_I h < 1$,

$$|h_e| = \frac{h}{2}. \tag{16.3.3}$$

Since $|k_2|^2 \ll |k_1|^2 \le |k_I|^2$, it is usually adequate to approximate $\cos k_2 h \sim 1$ so that

$$\begin{aligned} |h_e| &\sim \left| \frac{1}{k_I} \tan\left(\frac{k_I h}{2}\right) \right| \\ &= \frac{1}{(\beta_I^2 + \alpha_I^2)^{1/2}} \left(\frac{\cosh \alpha_I h - \cos \beta_I h}{\cosh \alpha_I h + \cos \beta_I h} \right)^{1/2}. \end{aligned} \tag{16.3.4}$$

When $k_I \sim k_1 = \beta_1(1+i)$, the useful range is when $|k_1 h| \le 2$ so that $|h_e| \sim h/2$. Alternatively, at very low frequencies, $\beta_I \ll \alpha_I$ and $|h_e| \sim (1/\alpha_I)\tanh(\alpha_I h/2)$.

The quantity $|h_e|/h$ is shown graphically in Fig. 16.3.1 (in dashed lines) for the bare dipole in the two ranges of interest. These are $f \sim 10^5$ Hz (curve C) with $\sigma_1 = 10^{-3}$ S/m and $a = 1.04$ mm, where the internal

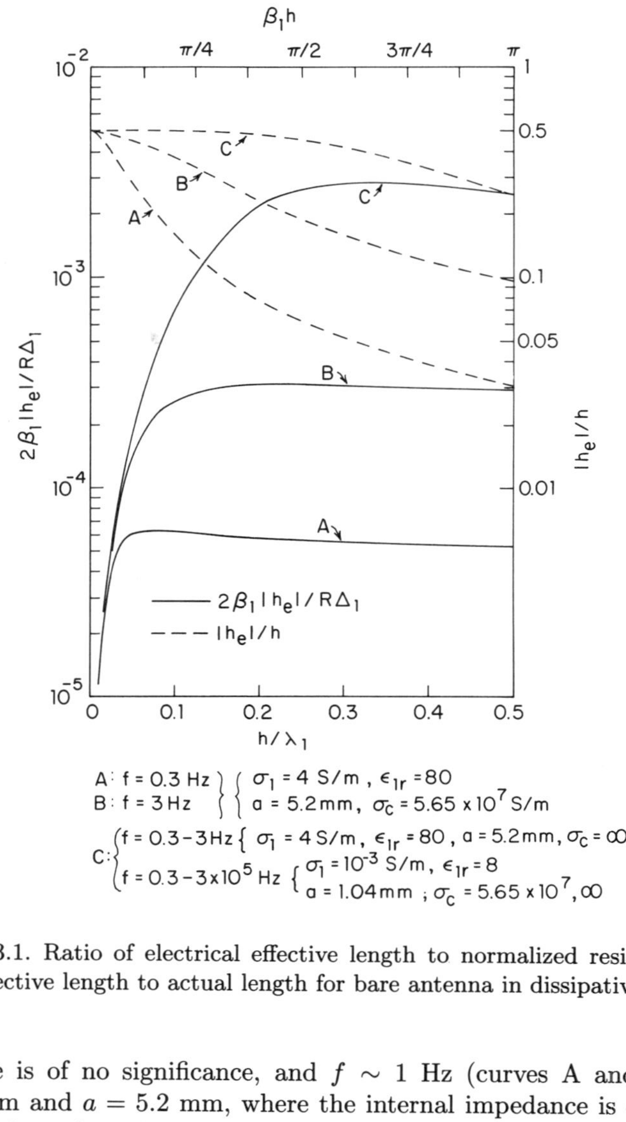

Figure 16.3.1. Ratio of electrical effective length to normalized resistance and ratio of effective length to actual length for bare antenna in dissipative medium.

impedance is of no significance, and $f \sim 1$ Hz (curves A and B) with $\sigma_1 = 4$ S/m and $a = 5.2$ mm, where the internal impedance is dominant. It is seen that when the internal impedance is negligible, $|h_e|/h$ remains close to 0.5 up to $h/\lambda_1 \sim 0.25$, but decreases very rapidly from 0.5 when, at sufficiently low frequencies, r^i is dominant.

In order to generate a large electric moment, it is possible to increase the current to compensate for a small effective length, but only up to the current-carrying capacity of the conductor and the available power. It is evidently desirable to have as large an effective length as possible and use a minimum of power. Accordingly, the ratio $[2|h_e I_x(0)|/|I_x^2(0)|R]$ is to be as large as possible. A related useful figure of merit is defined for unit current in terms of the electrical effective length $\beta_1|h_e|$ in the ambient medium and the normalized resistance $R\Delta_1$, where $\Delta_1 = \beta_1\zeta_0/\omega\mu_0 = \beta_1/\beta_0$. It is

$$F = \frac{2\beta_1|h_e|}{R\Delta_1}. \tag{16.3.5}$$

This quantity is also shown in Fig. 16.3.1 (in solid lines) specifically for antennas in sea water at $f = 0.3$ and 3 Hz for which the internal impedance is important and for antennas in earth at $f = 0.3 \times 10^5$ to 3×10^5 Hz where the internal impedance contributes negligibly and the antennas behave like perfect conductors. It is seen that in this last case (curve C), the ratio has a maximum near $h/\lambda_1 \sim 0.3$ where $|h_e|/h \sim 0.42$. At $f = 3$ Hz (curve B), the maximum of F is at $h/\lambda_1 \sim 0.2$ where $|h_e|/h \sim 0.23$; at $f = 0.3$ Hz (curve A), $F_{\max}$ is at $h/\lambda_1 \sim 0.075$ where $|h_e|/h \sim 0.2$.

16.4 The Bare Cylindrical Antenna in a Dissipative Medium Near a Boundary

In many applications, the antenna is located in a dissipative medium Region 1 ($z \geq 0,\ k_1$; earth, sea water) at a small distance d from its boundary ($z = 0$) with a Region 2 ($z \leq 0,\ k_2$; air, sea floor); $|k_1|^2 \gg |k_2|^2$. Since $|k_1 d| \ll 1$, the antenna in Region 1 is strongly affected by the adjacent, quite different Region 2. Since $|k_1|^2 \gg |k_2|^2$, the reflection coefficient seen by the antenna is $(k_1 - k_2)/(k_1 + k_2) \sim 1$. This means that the reflected electric field is almost equal to and in phase with the incident electric field; the reflected magnetic field is almost equal to and in phase opposition with the incident magnetic field. This condition is approximated by an image antenna at $z = -d$ with current and voltage in phase with those in the actual antenna, as shown in Fig. 16.4.1. Since $|2k_1 d| \ll 1$, the two antennas behave like a single one at $z = 0$ with effective radius $a_e = (2ad)^{1/2}$. Hence, (16.2.6) becomes

$$\Psi_e = 2\ln(h/a_e) - 2. \tag{16.4.1}$$

Since only half the current is in the antenna, this is obtained from (16.2.5) with (16.2.9) and (16.4.1). Thus,

$$I_{1z}(x) = \frac{V_1^e Y_e}{2}\,\frac{\sin k_{\mathrm{I}}(h - |x|)}{\sin k_{\mathrm{I}} h}; \qquad Y_e = \frac{2\pi\sigma_1}{\Psi_e}\,\frac{\tan k_{\mathrm{I}} h}{k_{\mathrm{I}}}, \tag{16.4.2}$$

where Y_e is obtained from (16.2.9). The driving-point admittance Y_1 and impedance Z_1 are

$$Y_1 = \frac{Y_e}{2} = \frac{Y\Psi}{2\Psi_e}; \qquad Z_1 = 2Z_e = \frac{2Z\Psi_e}{\Psi}, \tag{16.4.3}$$

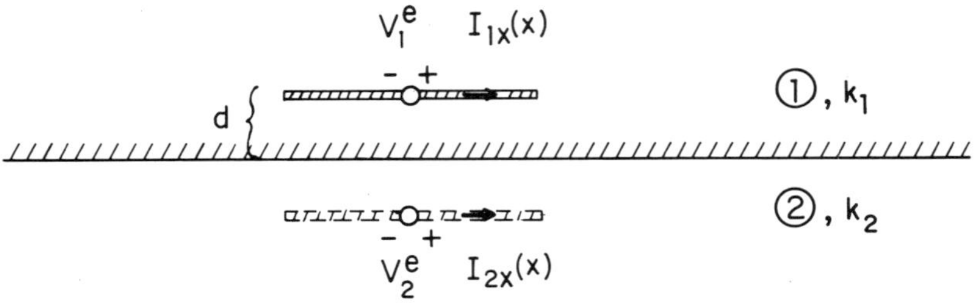

Figure 16.4.1. Horizontal antenna near boundary with image; $V_2^e = V_1^e$, $I_{2x}(x) = I_{1x}(x)$.

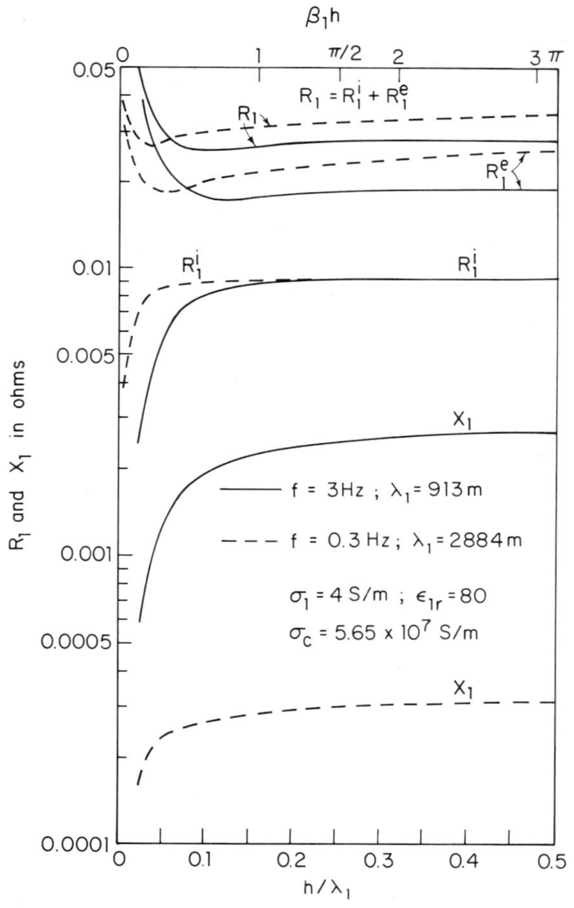

Figure 16.4.2. Impedance $Z_1 = R_1 - iX_1$ of bare copper dipole (radius $a = 5.2$ mm; half-length h) at $d = 10.4$ cm above the sea floor.

where Y and Z are the quantities for the antenna when far from the boundary.

The internal resistance $R_1^i = R^i$ is unaffected by the boundary if this is not so close as to cause a proximity effect. The external resistance R_1^e is approximately double R^e. Graphs of R_1, R_1^e, and R_1^i are shown in Fig. 16.4.2 as functions of h/λ_1 for both $f = 3$ Hz and $f = 0.3$ Hz with $d = 10.4$ cm. For the higher frequencies at which the internal impedance is negligible, the admittance or impedance of an antenna at a small distance d from a boundary is readily obtained from Fig. 16.2.2 or Fig. 16.2.3 with (16.4.3), (16.2.6), and (16.4.1). In Table 16.5.1 (Section 16.5), $d = 50$ cm for $f = 30$ and 300 kHz.

16.5 Summary and Conclusions

The theory of the bare metal dipole antenna has been developed in terms of two specific applications. These are the subsurface antenna for communication between two points below the surface of the earth and the sea-floor antenna for use in studies of the conductivity of the oceanic crust. For the former, the frequency range from 30 to 300 kHz is useful; for the latter, very low frequencies in the range from 0.3 to 3 Hz are in use. In both locations, the transmitting antenna is in a region of large wave number (Region 1), adjacent to a region with a much smaller wave number (Region 2). Furthermore, since the receiving antenna is in a similar location at a more or less distant point along the boundary, the received electromagnetic field is a surface or lateral wave that travels along the boundary in the region of smaller wave number—air for antennas near the surface of the earth, the crustal rock for antennas near the sea floor. Since the lateral wave must travel approximately vertically in Region 1 from the transmitting antenna to the boundary, then radially in Region 2 along the boundary, and finally vertically in Region 1 to the receiving antenna, horizontal antennas parallel to the boundary are superior to vertical ones.

The principal differences between the antennas near the air surface and those near the sea floor are a consequence of the large difference in frequency. At the very low frequencies used on the sea floor, the internal impedance of the copper conductor is important; it is negligible at the higher frequencies used near the air surface. This means that whereas the bare metal antenna is highly efficient in transferring power from the antenna to the exterior media (Regions 1 and 2) at the higher frequencies, it is subject to an unavoidable large internal power loss in heating the antenna at the low frequencies.

The properties of bare copper antennas in the two applications are summarized in Table 16.5.1 where the upper columns apply to a Region 1 of

Table 16.5.1. Effective Lengths, Impedances, and Electric Moments

	Sea Water: $\sigma_1 = 4$ S/m, $a = 5.2$ mm	
f (Hz)	0.3	3
λ_1 (m)	2884	913
h_m/λ_1	0.075	0.2
h_m (m)	216.3	182.6
$\lvert h_e\rvert/h_m$	0.21	0.23
$\lvert h_e\rvert$ (m)	45.4	42.0
α_1/β_1	1	1
Δ_1	3.46×10^5	1.094×10^5
$Z\Delta_1 = R\Delta_1 - iX\Delta_1$	$6193 - i55$	$1860 - i160$
$Z = R - iX$ (Ω)	$(17.9 - i0.16) \times 10^{-3}$	$(1.7 - i0.16) \times 10^{-2}$
d (cm)	10.4	10.4
$Z_1 = R_1 - iX_1$ (Ω)	$(2.84 - i0.025) \times 10^{-2}$	$(2.74 - i0.25) \times 10^{-2}$
$I_{\max}$ (A)	325	325
$2I_{\max}\lvert h_e\rvert$ (A m)	29,510	27,300
$\lvert I_{\max}\rvert^2 R$ (kW)	1.01	1.80
$\lvert I_{\max}\rvert^2 R_1$ (kW)	1.64	2.89

	Moist Earth: $\sigma_1 = 10^{-3}$ S/m; $a = 1.04$ mm	
f (Hz)	0.3×10^5	3×10^5
λ_1 (m)	576	171.2
h_m/λ_1	0.3	0.3
h_m (m)	172.5	51.3
$\lvert h_e\rvert/h_m$	0.42	0.42
$\lvert h_e\rvert$ (m)	72.6	21.6
α_1/β_1	0.98	0.87
Δ_1	17.34	5.84
$Z\Delta_1 = R\Delta_1 - iX\Delta_1$	$620 - i600$	$570 - i550$
$Z = R - iX$ (Ω)	$35.8 - i34.6$	$97.6 - i94.2$
d (cm)	50	50
$Z_1 = R_1 - iX_1$ (Ω)	$49.3 - i47.6$	$126.9 - i122.5$
$I_{\max}$ (A)	20	20
$2I_{\max}\lvert h_e\rvert$ (A m)	2904	864
$\lvert I_{\max}\rvert^2 R$ (kW)	14.3	39.0
$\lvert I_{\max}\rvert^2 R_1$ (kW)	19.7	50.8

sea water over rock, and the lower columns to moist earth under air. The corresponding frequencies are $f = 0.3$ and 3 Hz, and $f = 0.3 \times 10^5$ and 3×10^5 Hz, with copper antennas respectively with radii $a = 5.2$ mm and 1.04 mm. The maximum figure of merit occurs with antennas of half-length h_m obtained from Fig. 16.3.1. Also obtained from this figure is the effective length $|h_e|$ corresponding to this length. The associated driving-point impedances Z for the antenna far from the boundary and Z_1 at the distance d from the boundary are obtained from Figs. 16.2.1, 16.2.3, 16.2.4, and 16.4.2 or the formulas from which these were calculated.

The maximum currents for the copper conductors are those suggested for insulated wire with the specified radii. Bare conductors in contact with an ambient medium like the earth or sea can handle considerably more current without overheating. The dipole moments for the specified maximum currents and effective lengths are seen to be quite large at the low frequencies because the conductor is much thicker and can handle much more current. This is necessary in salt water and for the transmission in the crustal rock or sediment, which at these low frequencies behaves like a conductor with an exponential attenuation with distance. The total power supplied with the specified maximum currents is also listed. Because of the extremely low resistances of the thick low-frequency antennas in the highly conducting salt water, the power supplied at the low frequencies is much less than that at the higher frequencies and about half of this is actually dissipated by the axial current in heating the conductor. The antennas in the relatively poorly conducting earth at the higher frequencies transfer all of the input power to the ambient earth, partly by means of the radial current and partly in radiation by the axial current.

The use of the bare antenna as a termination for an insulated antenna is considered in Chapter 17.

17 Antennas in Material Media Near Boundaries: The Terminated Insulated Antenna

17.1 Circuit Properties of Insulated Antennas

When the ambient medium in which a bare metal antenna is embedded is conducting, radial currents from the antenna into the medium decrease the axial current. This means that often only electrically relatively short antennas are useful with correspondingly small effective lengths. Radial currents can obviously be eliminated by enclosing the metal cylinder (radius a, conductivity σ_c, internal impedance per unit length z^i) in a dielectric sheath (outer radius b, wave number $k_d = \omega[\mu_0(\epsilon_d + i\sigma_d/\omega)]^{1/2}$, $\sigma_d \sim 0$), as shown in Fig. 17.1.1. This greatly changes the properties of the antenna. In particular, the wave number of the current is no longer the wave number $k_\mathrm{I} = k_1[1 + (i4\pi z^i/\omega\mu_0\Psi)]^{1/2}$ as given by (16.2.4)—which reduces simply to the wave number $k_1 = \omega[\mu_0(\epsilon_1 + i\sigma_1/\omega)]^{1/2}$ of the ambient medium except at very low frequencies.

The insulated antenna behaves like a generalized transmission line with the wave number k_L and characteristic impedance Z_c given by the conventional formulas (King and Smith 1981, Chaps. 1 and 8):

$$k_L = (-z_L y_L)^{1/2}; \qquad Z_c = \left(\frac{z_L}{y_L}\right)^{1/2}, \tag{17.1.1}$$

with

$$z_L = z^i - i\omega l + z^e + z_{12}; \qquad y_L = g_L - i\omega c_L. \tag{17.1.2}$$

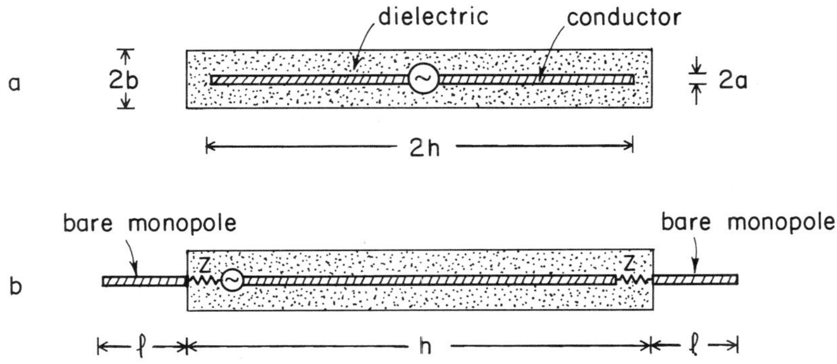

Figure 17.1.1. Insulated antennas. (a) Completely insulated dipole; (b) terminated insulated antenna. Ambient medium: earth, sea water.

Here, z^i is the internal impedance per unit length defined in (16.2.1);

$$l = \frac{\mu_0}{2\pi} \ln \frac{b}{a} \tag{17.1.3}$$

is the series inductance per unit length;

$$z^e = r^e - ix^e = -\frac{i\omega\mu_0}{2\pi}\left(\frac{H_0^{(1)}(k_1 b)}{k_1 b H_1^{(1)}(k_1 b)}\right) \tag{17.1.4}$$

is the series external impedance per unit length; and

$$z_{12} = r_{12} - ix_{12} = -\frac{i\omega\mu_0}{2\pi}\left(\frac{H_0^{(1)}(2k_1 d)}{k_1 b H_1^{(1)}(k_1 b)}\right) \tag{17.1.5}$$

is the series mutual impedance per unit length due to an image with codirectional currents (to take account of the effect of the boundary with Region 2 at the distance d). When the internal impedance is negligible, $z^i \sim 0$; when the antenna is far from the boundary, $z_{12} = 0$. The shunt admittance y_L is defined as follows:

$$y_L = -\frac{i2\pi k_d^2}{\omega\mu_0 \ln(b/a)} = \frac{2\pi\sigma_d}{\ln(b/a)} - i\omega\,\frac{2\pi\epsilon_d}{\ln(b/a)} = g_L - i\omega c_L. \tag{17.1.6}$$

The current in the insulated conductor is distributed like that in a transmission line (King 1965, Chap. 4) but with k_L and Z_c the generalized quantities defined in (17.1.1) with (17.1.2)–(17.1.6). The general formula for a center-driven, end-loaded, insulated antenna with half-length h is

$$\begin{aligned} I_x(x) &= -\frac{iV_0^e}{2Z_c}\,\frac{\sin[k_L(h-|x|)+i\theta_h]}{\cos(k_L h + i\theta_h)} \\ &= V_0^e Y\,\frac{\sin[k_L(h-|x|)+i\theta_h]}{\sin(k_L h + i\theta_h)}, \end{aligned} \tag{17.1.7}$$

where the admittance Y and impedance $Z = 1/Y$ are

$$Y = G - iB = -\frac{i}{2Z_c}\tan(k_L h + i\theta_h), \tag{17.1.8a}$$

$$Z = R - iX = i2Z_c\cot(k_L h + i\theta_h). \tag{17.1.8b}$$

The antenna is terminated at each end in an arbitrary impedance Z_h which is contained in the terminal function θ_h as follows:

$$\theta_h = \rho_h - i\Phi_h = \coth^{-1}\left(\frac{Z_h}{Z_c}\right). \tag{17.1.9}$$

Note that

$$\coth(\rho_h - i\Phi_h) \equiv r_h - ix_h = \mathrm{Re}\left(\frac{Z_h}{Z_c}\right) - i\,\mathrm{Im}\left(\frac{Z_h}{Z_c}\right) \tag{17.1.10a}$$

with

$$r_h = \frac{\sinh 2\rho_h}{\cosh 2\rho_h - \cos 2\Phi_h}; \qquad x_h = -\frac{\sin 2\Phi_h}{\cosh 2\rho_h - \cos 2\Phi_h} \tag{17.1.10b}$$

or

$$\rho_h = \frac{1}{2}\tanh^{-1}\left(\frac{2r_h}{r_h^2 + x_h^2 + 1}\right); \qquad \Phi_h = \frac{1}{2}\tan^{-1}\left(\frac{-2x_h}{r_h^2 + x_h^2 - 1}\right). \tag{17.1.10c}$$

For a low-impedance termination, $|Z_h/Z_c| \ll 1$ and $\rho_h \sim r_h$, $\Phi_h \sim \pi/2 + x_h$; for an insulated end, $|Z_h/Z_c| \to \infty$ and $\rho_h = 0$, $\Phi_h = 0$ or π; for a matched load defined by $Z_h/Z_c = 1$, $r_h = 1$ and $\rho_h = \infty$. For the end-driven insulated antenna with the terminating impedance Z_0 at the generator end and Z_h at the load end, the formulas (17.1.7) and (17.1.8a,b) apply when $2Z_c$ is replaced by $Z_0 + Z_c$.

For applications at frequencies that are not too high and when the cross-sectional size of the antenna is electrically small in the ambient Region 1 so that $|k_1 b| < 1$, the small-argument approximations of the Hankel functions can be used. That is, $H_0^{(1)}(z) \sim 1 + (2i/\pi)[0.5772 + \ln(z/2)]$ and $zH_1^{(1)}(z) \sim -2i/\pi$; also, $k_1 \sim (i\omega\mu_0\sigma_1)^{1/2} = (\omega\mu_0\sigma_1)^{1/2}e^{i\pi/4}$. With these approximations,

$$z^i = r^i - i\omega l^i \sim r_0 - \frac{i\omega\mu_0}{8\pi}; \qquad r_0 = (\pi a^2\sigma_c)^{-1}, \tag{17.1.11}$$

$$z^e = r^e - ix^e \sim \frac{\omega\mu_0}{8} - \frac{i\omega\mu_0}{2\pi}\left(\ln\frac{2}{|k_1 b|} - 0.5772\right), \tag{17.1.12}$$

$$z_{12} = r_{12} - ix_{12} \sim \frac{\omega\mu_0}{8} - \frac{i\omega\mu_0}{2\pi}\left(\ln\frac{1}{|k_1 d|} - 0.5772\right) \tag{17.1.13}$$

so that

$$z_L = r_L - ix_L$$

$$= r_0 + \frac{\omega\mu_0}{4} - \frac{i\omega\mu_0}{2\pi}\left(\ln\frac{b}{a} + \ln\frac{2}{|k_1 b|} + \ln\frac{1}{|k_1 d|} - 0.90\right). \quad (17.1.14)$$

With a good dielectric, $\sigma_d \sim 0$ and

$$y_L \sim -i\omega c_L \quad \text{with} \quad c_L = \frac{2\pi\epsilon_d}{\ln(b/a)}. \quad (17.1.15)$$

In (17.1.14), r_0 is the internal resistance per unit length of the conductor associated with the heat dissipated in the conductor and $\omega\mu_0/4$ is the external resistance per unit length associated with energy transferred to the ambient medium (when the antenna is far from the boundary, this reduces to $\omega\mu_0/8$). With (17.1.14) and (17.1.6), the explicit expressions for the wave number k_L and characteristic impedance Z_c are

$$k_L = k_d\left\{1 + \frac{1}{\ln(b/a)}\left[i\left(\frac{2\pi r_0}{\omega\mu_0} + \frac{\pi}{2}\right) + \ln\frac{2}{|k_1 b|} + \ln\frac{1}{|k_1 d|} - 0.90\right]\right\}^{1/2}, \quad (17.1.16)$$

$$Z_c = \frac{\omega\mu_0 k_L}{2\pi k_d^2}\ln\frac{b}{a}. \quad (17.1.17)$$

At very low frequencies, the leading terms reduce to

$$k_L = (-z_L y_L)^{1/2} \sim (i\omega c_L r_0)^{1/2} = (1+i)\left(\frac{\omega c_L r_0}{2}\right)^{1/2}, \quad (17.1.18)$$

$$Z_c = \left(\frac{z_L}{y_L}\right)^{1/2} \sim \left(\frac{ir_0}{\omega c_L}\right)^{1/2} = (1+i)\left(\frac{r_0}{2\omega c_L}\right)^{1/2}. \quad (17.1.19)$$

These are identically the formulas for a so-called ocean cable, which is defined in conventional coaxial-line theory by $r/\omega l \gg 1$, $g/\omega c \ll 1$. All of the power supplied to the insulated conductor is transferred to the loads at the ends or dissipated as heat in the conductor itself. The structure acts like a lossy transmission line for transferring power to a load, not like an antenna that transfers power directly as radiation.

The impedance of the insulated conductor at very low frequencies is given by (17.1.8b) with (17.1.18) and (17.1.19). With a low-impedance termination, $\theta_h = r_h - i(\pi/2 + x_h)$ and

$$Z = i2Z_c\cot\left(k_L h + ir_h + \frac{\pi}{2} + x_h\right)$$
$$= -i2Z_c\tan(k_L h + ir_h + x_h). \quad (17.1.20)$$

With $|k_L h + ir_h + x_h| < 1$, the impedance of the insulated dipole with half-length h and identical low-impedance terminations at both ends is

$$Z = -i2Z_c(k_L h + ir_h + x_h) = 2(r_0 h + R_h - iX_h). \quad (17.1.21)$$

Note that low-impedance terminations must be used if significant currents are desired. If the ends are insulated ($Z_h = \infty$), $I(0) \sim 0$ and, hence, $I(z) \sim 0$. The terminations as defined in (17.1.10a) are expressed in the following normalized form:

$$r_h - ix_h = \frac{R_h - iX_h}{R_c - iX_c} = \frac{R_hR_c + X_hX_c}{R_c^2 + X_c^2} - i\frac{X_hR_c - X_cR_h}{R_c^2 + X_c^2}. \quad (17.1.22)$$

17.2 The Effective Length of the Insulated Antenna

The insulated antenna has numerous parameters that can be adjusted to change its properties. These include the length h, the terminating impedance Z_h, the ratio b/a of the radius of the insulation to the radius of the conductor, and the electrical radius of the insulation $\beta_1 b$ in $k_1 b$. When $|k_1 b|$ is not too small, the conductor can be located eccentrically in the insulating sleeve to provide a significant transverse directional effect. Insulated antennas with parameters suitably adjusted for many special applications have been described (King and Smith 1981). In this chapter, attention is directed particularly to applications that require frequencies that are in the range from a fraction of a hertz to near a megahertz. These are useful in subsurface communication and geophysical exploration.

One of the most important qualities of the insulated antenna is its electric moment. This depends on the effective length h_e which, in turn, is determined by the current distribution and the actual length h. For the applications to be discussed, a maximum current is maintained along the insulated antenna when it is terminated either in a low impedance or in its characteristic impedance. The relevant current distributions are, for the antiresonant antenna,

$$I_x(x) = I_x(0)\,\frac{\cos[k_L(h-x) + i\theta_h]}{\cos(k_L h + i\theta_h)}, \quad (17.2.1)$$

where $\theta_h = r_h - i(x_h + \pi/2)$ with $r_h \ll 1$ and $|x_h| \ll 1$. In determining the effective length, it is adequate to neglect the effect of the low-impedance termination on the current. For the antenna with matched load ($r_h = 1$) with $\theta_h = \infty$,

$$I_x(x) = I_x(0)e^{ik_L x}. \quad (17.2.2)$$

Since the wave number for these currents is different from k_2 for the lateral electromagnetic waves they generate along the boundary in Region 2, the phase relations among the successive elements of current and the outward-traveling field are important in the definition of the effective electric moment $I_x(0)h_e(\phi_0)$. Note that the effective length $h_e(\phi_0)$ involves the angle ϕ_0 between the direction of propagation along the antenna and the point of observation. The effective length associated with the current (17.2.1) is

$$h_e(\phi_0) = \int_0^h \frac{\cos k_L(h - x')}{\cos k_L h} e^{-ik_2 x' \cos\phi_0} \, dx'. \tag{17.2.3}$$

In the direction $\phi_0 = 0$ of maximum field, this gives

$$h_e(0) \equiv h_e = \frac{-ik_2 e^{-ik_2 h} + ik_2 \cos k_L h + k_L \sin k_L h}{(k_L^2 - k_2^2)\cos k_L h}. \tag{17.2.4}$$

Similarly, with the current (17.2.2),

$$h_e(\phi_0) = \int_0^h e^{ik_L x'} e^{-ik_2 x' \cos\phi_0} \, dx' \tag{17.2.5}$$

and, for $\phi_0 = 0$,

$$h_e(0) \equiv h_e = \frac{i(1 - e^{i(k_L - k_2)h})}{k_L - k_2}. \tag{17.2.6}$$

The maximum magnitude of the effective lengths with respect to the actual lengths h occurs in (17.2.4) near the minimum of the denominator, $\cos k_L h$. This is near

$$\beta_L h = \frac{\pi}{2} \quad \text{or} \quad h = \frac{\pi}{2\beta_L}, \tag{17.2.7}$$

for which $\cos k_L h = -i \sinh \alpha_L h$ and $\sin k_L h = \cosh \alpha_L h$ so that

$$|h_{em}| = \left| \frac{k_2 e^{-ik_2 h} + ik_2 \sinh \alpha_L h + ik_L \cosh \alpha_L h}{(k_L^2 - k_2^2)\sinh \alpha_L h} \right|, \tag{17.2.8}$$

where $k_2 = \beta_2 + i\alpha_2$, $k_L = \beta_L + i\alpha_L$, and h is given by (17.2.7).

The maximum magnitude of h_e in (17.2.6) occurs near

$$(\beta_L - \beta_2)h = \pi \quad \text{or} \quad h = \frac{\pi}{\beta_L - \beta_2}, \tag{17.2.9}$$

where

$$|h_{em}| = \left| \frac{i(1 + e^{-(\alpha_L - \alpha_2)h})}{k_L - k_2} \right| \tag{17.2.10}$$

and h is given by (17.2.9).

When the insulated antenna is terminated either in a low impedance or its characteristic impedance, the terminations often consist of bare monopoles of length l. The effective length of these has been determined in

Chapter 16, but the effective length of the terminated antenna is not just the sum of the effective lengths of the parts since the relative phases make a significant difference. It is possible to evaluate the complete effective length. It has the form

$$h_e = \frac{\tan(k_\mathrm{I} l/2)}{k_\mathrm{I}}(1 + e^{ik_L h}) + \frac{i}{k_L - k_2}(1 - e^{i(k_L - k_2)h}). \tag{17.2.11}$$

At all but the very low frequencies, the internal impedance of the bare sections is negligible and, over the lengths of interest, $k_\mathrm{I}^{-1}\tan(k_\mathrm{I} l/2) \sim l/2$. At the low frequencies where the internal impedance dominates, $k_\mathrm{I}^{-1}\tan(k_\mathrm{I} l/2) \sim \alpha_\mathrm{I}^{-1}\tanh(\alpha_\mathrm{I} l/2)$. Even with these simplifications, the magnitude $|h_e|$ is not a simple formula. For most applications, the contribution to the effective length by the relatively short terminations is relatively small and can be ignored. For any particular case, it can be evaluated from (17.2.11).

17.3 Antennas for Special Purposes: An Underground Antenna

A useful subsurface antenna for communicating from a point $(0, d)$ at depth d below the surface of the earth to a point (ρ, z) below, on, or above the surface is the insulated traveling-wave antenna. As shown in Fig. 17.3.1, this consists of a conductor (radius a, resistance per unit length r^i) enclosed in an insulating sheath (radius b, permittivity ϵ_d) of length h. It is terminated at $x = h$ in a lumped impedance Z in series with a bare metal antenna (length l, input impedance Z_l); the generator (electromotive force V_0^e) is connected in series with the insulated conductor at $x = 0$ and a second bare metal antenna (length l, input impedance Z_l).

The current in the antenna is

$$I_x(x) = I_x(0)e^{ik_L x} = \frac{V_0^e}{Z_\mathrm{in}} e^{ik_L x}; \qquad Z_\mathrm{in} = Z_l + Z_c, \tag{17.3.1}$$

where Z_c is given by (17.1.17) and k_L by (17.1.16). It is required that, at $x = h$, Z is adjusted so that

$$Z_h = Z_l + Z = Z_c. \tag{17.3.2}$$

The input power to the antenna is

$$P_\mathrm{in} = |I_x(0)|^2(R_c + R_l). \tag{17.3.3}$$

The bare metal terminations are monopoles with length l. The normalized input impedance is approximately half the impedance of a dipole, as given in Fig. 16.2.3. For moist earth ($\sigma_1 \sim 10^{-3}$ S/m, $\epsilon_{1r} = 8$) at

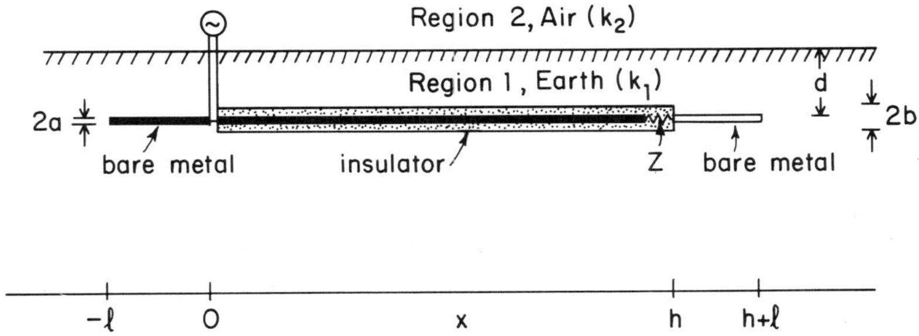

Figure 17.3.1. Terminated insulated antenna at depth d in earth.

$f = 3 \times 10^5$ Hz, the normalizing factor is $\Delta_1 = 5.84$ so that the impedance of a bare monopole with radius $a = 1.04$ mm and length $l = \lambda_1/4 \sim 42$ m, as obtained from Table 16.5.1, is $Z_l = Z_1/2 = 63.9 - i61.0\ \Omega$. This value assumes the antenna is at a depth $d = 50$ cm.

The insulated conductor consists of the copper wire ($\sigma_c = 5.65 \times 10^7$ S/m) with radius $a = 1.04$ mm. In order to have the wave number k_L for the current close to that for the lateral waves in air, the insulation should be styrofoam. Also, as seen from (17.1.16), b/a should be large. Accordingly, the choice $b = 1.04$ cm is made. With data obtained from Tables 16.1.1 and 16.5.1 for $f = 3 \times 10^5$ Hz, it follows from (17.1.16) and (17.1.17) that

$$\begin{aligned} k_L &= k_d(2.36 + i0.141) \\ &= (1.48 + i0.088) \times 10^{-2}\ \text{m}^{-1} = \beta_L + i\alpha_L. \end{aligned} \tag{17.3.4}$$

With styrofoam insulation, $k_d = k_2 = \omega/c = 2\pi \times 10^{-3}\ \text{m}^{-1}$. Also,

$$Z_c = 325 + i19.3\ \Omega. \tag{17.3.5}$$

The terminating impedance is $Z_h = Z_l + Z$, where Z is to be adjusted to make $Z_h = Z_c$. Therefore,

$$Z = Z_c - Z_l = 325 + i19.3 - 63.9 + i61 = 261.1 + i80.3\ \Omega. \tag{17.3.6}$$

The optimum length of the insulated antenna is obtained from (17.2.9) to be

$$h = \frac{\pi}{\beta_L - k_2} \sim 369\ \text{m}. \tag{17.3.7}$$

The associated effective length given by (17.2.10) is

$$|h_e| \sim 201\ \text{m}. \tag{17.3.8}$$

(Note that h and $|h_e|$ can be increased by reducing β_L by increasing b/a.) The input power per unit current is obtained from (17.3.3). It is

$$\frac{P_{\text{in}}}{|I_x(0)|^2} = R_c + R_l = 325 + 63.9 = 389\ \Omega. \tag{17.3.9a}$$

With $|I_x(0)| = 25$ A,

$$P_{\text{in}} = 243 \text{ kW}. \tag{17.3.9b}$$

The radial electric field at the surface at a distance of 500 km is readily evaluated. At this distance, $\rho > 4|k_1^2/k_2^3| = 4|(4.88)^2 \times 10^{-4}/(2\pi)^3 \times 10^{-9}| = 38.4$ km so that the point of observation is in the far field where, with $\exp(-\alpha_1 d) \sim 1$,

$$|E_{1\rho}(\rho, 0, 0)| = |I_x(0)h_e| \frac{\omega\mu_0}{2\pi k_2} \frac{1}{\rho^2}. \tag{17.3.10}$$

With (17.3.10),

$$|E_{1\rho}(500 \text{ km}, 0, 0)| = |I_x(0)| \times 0.48 \times 10^{-7}. \tag{17.3.11}$$

With $|I_x(0)| = 25$ A, the field is

$$|E_{1\rho}(500 \text{ km}, 0, 0)| \sim 1.2 \ \mu\text{V/m}, \tag{17.3.12}$$

which is easily measurable.

17.4 Antennas for Special Purposes: An Antenna for the Sea Floor at $f = 1$ kHz

Terminated insulated antennas can be used on the sea floor to measure the conductivity of the oceanic crust. In one arrangement, the antenna is laid horizontally in the sea water [Region 1, $\sigma_1 = 4$ S/m, $k_1 = (i\omega\mu_0\sigma_1)^{1/2}$] at a small height d over the crust [Region 2, $\sigma_2 = 10^{-4}$ to 10^{-1} S/m, $k_2 = (i\omega\mu_0\sigma_2)^{1/2}$]. The field generated by the antenna is measured at successive radial distances ρ in the sea water just above the crust. Except very close to the antenna, this field is entirely a lateral-wave field if there are no reflecting layers at moderate distances in the crust. But even if such layers are present, the lateral-wave field—which samples only a relatively shallow depth—is from 60% to 90% of the observed field (Chapter 5). In the most useful range of measurement, $1 \leq |k_2\rho| \leq 4|k_1^2/k_2^2|$, the magnitude of the dominant radial component has the following simple form in the direction along the axis of the antenna (Chapter 5):

$$\begin{aligned} |E_{1\rho}(\rho, 0, 0)| &= |I_x(0)h_e| \frac{\omega\mu_0 k_2^2}{2\pi k_1^2} \frac{e^{-\alpha_2\rho}}{\rho} \\ &= |I_x(0)h_e| \frac{\omega\mu_0\sigma_2}{2\pi\sigma_1} \frac{e^{-\alpha_2\rho}}{\rho}, \end{aligned} \tag{17.4.1}$$

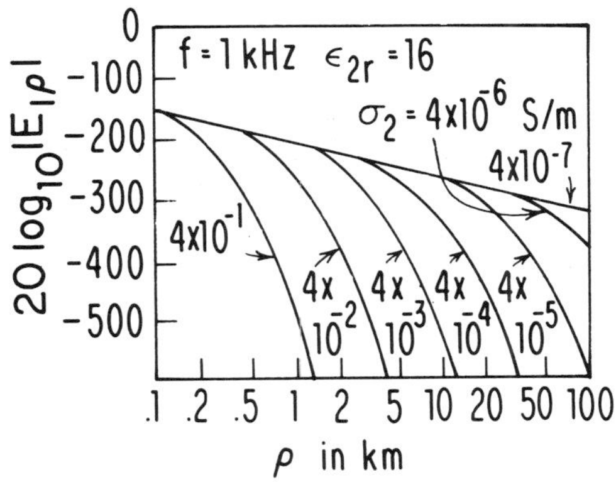

Figure 17.4.1. Radial electric field at $z = 15$ cm in sea water ($\sigma_1 = 4$ S/m, $\epsilon_{1r} = 80$) above rock or deep sediment (σ_2, $\epsilon_{2r} = 16$) as function of radial distance ρ from unit horizontal dipole at height $d = 15$ cm above sea bottom; $f = 1$ kHz.

where $\alpha_2 = (\omega\mu_0\sigma_2/2)^{1/2}$ and $\exp(-\alpha_1 d) \sim 1$. The behavior of this field as a function of the radial distance is shown in Fig. 17.4.1 for σ_2 ranging from 4×10^{-1} to 4×10^{-7} S/m. By measuring the radial field at successive distances, an observed graph of $20 \log_{10} |E_{1\rho}(\rho, 0, 0)|$ can be compared with theoretical ones and σ_2 obtained by interpolation. The rapid exponential decrease at locations that depend on σ_2 assures reasonable accuracy.

Let the transmitting antenna be designed for use at $f = 1$ kHz, as diagramed in Fig. 17.4.2. It consists of an insulated main section (B)—with the conductor radius a and insulation radius b—that extends from $x = 0$ to $x = h$. The conductor with the insulation removed continues from $x = h$ to $x = h+l$ as a bare monopole (A) terminating the insulated section. At the driving point ($x = 0$), the insulated conductor is connected to the inner conductor of the feeding coaxial line (D) from the ship or float. Since this is quite long, efficient power transmission requires the characteristic resistance of the line to be equal to the input resistance of the antenna. As shown in Fig. 17.4.2, the insulated terminated monopole is driven against a bare monopole that consists of a metal sleeve (C) that extends from $x = 0$ to $x = -l$. The sleeve is insulated from the metal walls of the coaxial line except at $x = 0$ where it is conductively connected.

The lengths l of the two bare sections are chosen so that $\beta_1 l > \pi/2$ or $l \sim 12$ m at $f = 1$ kHz. The input impedance of each monopole is of the order $Z_l \sim 0.0135 - i0.0065\ \Omega$. The dimensions of the main insulated section are $a = 0.52$ cm and $b = 0.521$ cm. The insulation is chosen very thin so that k_L is not too different from k_2. It may be neoprene with $\epsilon_{dr} = 6.6$ (or any other dielectric including enamel). At $f = 1$ kHz, $k_d = \omega(\mu_0\epsilon_d)^{1/2} = 5.38 \times 10^{-5}$ m^{-1}. With the antenna at a mean height

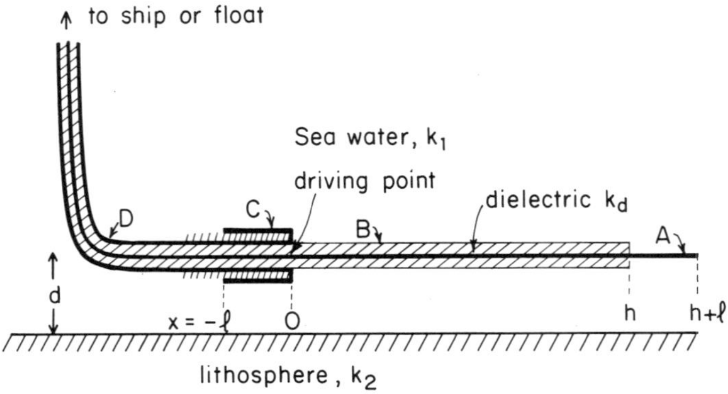

Figure 17.4.2. Terminated insulated antenna for use on sea floor at $f = 1$ kHz. Transverse dimensions are not to scale. A—bare metal monopole of length $l \sim 12$ m; B—insulated antenna of length $h \sim 381$ m; C—sleeve monopole folded back on coaxial line, length $l \sim 12$ m; D—coaxial line to ship, preferably insulated to avoid corrosion.

$d = 5$ cm above the sea bottom, (17.1.16) gives

$$\begin{aligned} k_L &= k_d\left\{1 + 520.5\left[i\left(0.166 + \frac{\pi}{2}\right) + 11.25\right]\right\}^{1/2} \\ &= k_d(76.7 + i5.88) \\ &= (4.12 + i0.32) \times 10^{-3}\ \mathrm{m}^{-1} = \beta_L + i\alpha_L, \end{aligned} \tag{17.4.2}$$

$$Z_c = 3.44 + i0.267\ \Omega. \tag{17.4.3}$$

The normalized resistance and reactance of the terminations are obtained from (17.1.22) to be $r_h = 3.9 \times 10^{-3}\ \Omega$, $x_h = 2.2 \times 10^{-3}\ \Omega$—very small quantities. The length h of the insulated section when this is antiresonant is given by (17.2.7), the associated effective length by (17.2.8). When $\sigma_2 = 10^{-2}$ S/m, they are

$$h = \frac{\pi}{2\beta_L} = 381\ \mathrm{m}; \qquad |h_e| = 9.66\ \mathrm{km}. \tag{17.4.4}$$

The effective length is greater than the actual length because it is referred to the driving-point current $I_x(0)$, which is the minimum current; the current increases to the maximum at $x = h$. Thus,

$$I_x(h) = \frac{I_x(0)}{\sinh(\alpha_L h + r_h)}. \tag{17.4.5}$$

The driving-point impedance of the antenna when $\beta_L h = \pi/2 + x_h \sim \pi/2$ is

$$Z_{\text{in}} = R_{\text{in}} = R_c \coth(\alpha_L h + r_h) = 27.45\ \Omega. \tag{17.4.6}$$

The power input to the antenna is $|I_x(0)|^2 R_{\text{in}}$. The power in the two bare terminations is insignificant as is their contribution to the effective length. The fraction of the power dissipated in the internal resistance of the insulated copper conductor is proportional to the contribution to α_L by the internal impedance. It is seen from (17.4.2) that this is $0.166/(0.166 + \pi/2) = 0.096$—a relatively small amount.

The field to be generated is given by (17.4.1) and represented in Fig. 17.4.1 for unit electric moment. It is seen that measurements over a useful range of σ_2 are possible if they can be made down to $20 \log_{10} |E_{1\rho}(\rho, 0, 0)| = -250$ dB per unit electric moment. With a transmitting antenna that provides an effective electric moment $|I_x(0)h_e| \sim 3 \times 10^5$ A m, the minimum level of -250 dB is raised to -140 dB. The corresponding smallest electric field that must be measured is $|E_{1\rho}(\rho, 0, 0)| \sim 0.1$ μV/m. If the receiving antenna has an effective length of only 10 m, the required sensitivity is 1 μV/m.

With the effective length given in (17.4.4), the required input current is $|I_x(0)| = 31.1$ A. With (17.4.6), the input power is $P_{\text{in}} = (31.1)^2 \times 27.45 = 26.5$ kW.

17.5 Antennas for Special Purposes: An Antenna for the Sea Floor at Very Low Frequencies

Geophysicists use antennas on the sea floor at frequencies so low that the internal resistance of the copper conductor—which plays a negligible role when $f > 100$ Hz—becomes a dominant factor. This is true of both bare and insulated antennas. In the frequency range from 0.3 to 3 Hz, the magnitude of the complete radial electric field in the sea water near the oceanic crust at a radial distance ρ from a unit horizontal dipole is given by

$$|E_{1\rho}(\rho, 0, 0)| = \left| \frac{e^{ik_2\rho}}{2\pi\sigma_1} \left(\frac{1}{\rho^3} - \frac{ik_2}{\rho^2} - \frac{k_2^2}{\rho} \right) \right|$$

$$= \frac{e^{-\alpha_2\rho}}{2\pi\sigma_1\rho^3} (1 + 2\alpha_2\rho + 2\alpha_2^2\rho^2 + 4\alpha_2^3\rho^3 + 4\alpha_2^4\rho^4)^{1/2}. \qquad (17.5.1)$$

This formula assumes the oceanic crust to be a homogeneous isotropic half-space with wave number $k_2 = \beta_2 + i\alpha_2$ and $\beta_2 = \alpha_2 = (\omega\mu_0\sigma_2/2)^{1/2}$. At $f = 3$ Hz and with $\sigma_2 = 4 \times 10^{-3}$ S/m, $\sigma_1 = 4$ S/m, and $\rho = 18.9$ km,

$$20 \log_{10} |E_{1\rho}(\rho, 0, 0)| = -288 \text{ dB}. \qquad (17.5.2)$$

If it is assumed that a receiver is sensitive to 0.1 μV/m and is connected to a receiving antenna with an effective length of the order of 80 m, the transmitting antenna must raise the signal level by 110 dB to −178 dB. This requires an electric moment

$$|I_x(0)h_e| \sim 3.25 \times 10^5 \text{ A m}. \tag{17.5.3}$$

It is seen from Fig. 16.3.1 that a bare termination should be of the order of $0.15\lambda_1$ in length or, at $f = 3$ Hz, $l \sim 0.15 \times 913 = 137$ m for which $l_e = 0.295 \times 137 = 40$ m. From Fig. 16.4.2, the impedance of a bare monopole at a height $d = 10.4$ cm from the sea floor is

$$Z_l = \frac{Z_1}{2} = 0.013 - i0.0011\ \Omega. \tag{17.5.4}$$

The wave number of an insulated section of length h, terminated in Z_l, is obtained from (17.1.18). For a conductor with radius $a = 5.2$ mm and an insulator with radius $b = 8$ mm and relative permittivity $\epsilon_{dr} = 6.6$ (for neoprene), (17.1.18) gives $k_L = 1.29 \times 10^{-6}(1+i)$ m^{-1}. For an electrically short antenna with $|k_L h| \leq 1$,

$$Z_{\text{in}} \sim -iZ_c k_L h = r_0 h + 2Z_l. \tag{17.5.5}$$

The effective length of an electrically short insulated antenna terminated in a very low impedance is the actual length since $I_x(x) \sim I_x(0)$ for all values of x. The maximum current for the insulated conductor is 325 A so that the electric moment is $325|h_e|$ A m. For the required value of $|I_x(0)h_e| = 3.25 \times 10^5$ A m, it is necessary that $|h_e| = h = 1$ km. With the two bare terminations, the total length is

$$h + 2l = 1274 \text{ m}. \tag{17.5.6}$$

The contribution to the effective length by the terminations is small and has not been included.

With this length, the impedance of the antenna is

$$\begin{aligned} Z_{\text{in}} &= 2.08 \times 10^{-4} \times 1000 + 0.026 - i0.002 \\ &= 0.23 - i0.002\ \Omega. \end{aligned} \tag{17.5.7}$$

The input power is

$$P_{\text{in}} = |I_x(0)|^2 R_{\text{in}} = 24.3 \text{ kW}. \tag{17.5.8}$$

The insulated terminated antenna described above is quite similar to an antenna actually used by Young and Cox (1981) for measurements on the sea floor in the frequency range $f = 0.25$ to 2.25 Hz. The insulated part was 800 m long with a conductor radius of $a = 3.5$ mm. It had a resistance $r_0 h = 4.6 \times 10^{-4} \times 800 = 0.36\ \Omega$. Each of the two bare terminations had lengths $l = 15$ m and input resistances of 0.07 Ω when at a height $d \sim 10.4$ cm above the sea floor. Thus, the total resistance was $R_{\text{in}} = r_0 h + 2R_l = 0.5\ \Omega$ and, with $|I_x(0)| = 70$ A, $P_{\text{in}} = 2.45$ kW. Since the bare terminations were short and oriented so that their currents were oppositely directed, their contributions to the field and the effective length canceled. The effective length of the insulated section was $h_e = 800$ m and, with a current of 70 A, the electric moment was 5.6×10^4 A m so that the field at $\rho = 18.9$ km was raised by 95 dB from -288 dB for the unit dipole to -193 dB or $|E_{1\rho}(\rho, 0, 0)| = 2.2 \times 10^{-10}$ V/m. The bare receiving dipole had a length of 9 m and an effective length of 4.5 m so that the voltage across the load was $V_L \sim 10^{-9}$ V. This is so low that noise was a serious problem and required a receiver with a sensitivity higher than 0.001 μV.

17.6 Conclusion and Summary

The insulated conductor with its various possible terminations (including especially bare monopoles) is a highly effective antenna when embedded in a general dielectric or conducting region with a wave number that is much larger in magnitude than the wave number of the insulating sleeve. Depending on the ratio b/a of the radius of the insulating sleeve to that of the conductor, the insulated conductor acts more like an antenna for transferring power to the ambient medium ($b/a \sim 1$) or a transmission line for supplying power to the terminations ($b/a \geq 10$). At frequencies of 1 kHz and higher, the power loss in the internal resistance of the conductor is negligible and all power is either radiated or transferred to the terminations without loss. At frequencies below 100 Hz, the power loss in the internal resistance becomes dominant and the radiated power transferred directly to the ambient medium negligibly small. In this case, the insulated conductor is a highly inefficient transmission line feeding the terminations.

When located in Region 1 (k_1) near a boundary with a half-space (Region 2, k_2) with $|k_1| \geq 3|k_2|$, the insulated conductor is effective as a generator of electromagnetic surface waves that travel close to the boundary in Region 2. They are responsible for the *entire* field in Region 1 at even moderate distances from the source. If Region 2 is not a homogeneous half-space but is horizontally layered, there may be reflections from these layers. When this is true, the reflected field reaching a receiver in Region 1 is necessarily much smaller than the surface-wave field (Chapter 5). The latter penetrates Region 2 only to a depth that is a small fraction of the radial distance from the source to the point of observation in Region 1 near the boundary.

18
The Wave Antenna

18.1 Introduction

The excitation of lateral waves along the boundary between two electrically different media like air and earth or sea or the oceanic crust and the sea is effectively achieved with horizontal antennas in the earth or sea close to the boundary with the air or the sea floor. These are usually insulated terminated antennas, often of the traveling-wave type, such as described in Chapter 17. An alternative to the antenna in the sea or earth is the antenna in the air—it is clearly impractical to insert an antenna in the rock of the sea floor. An example described in Chapter 4 is the vertical monopole erected on the surface of the earth for AM radio broadcasting (Section 4.2) or for communicating with submarines (Section 4.5). Although the vertical dipole is an effective generator of lateral waves, it is difficult at low frequencies to construct one that is not electrically very short and requires very large currents to provide an adequate electric moment. An alternative is the horizontal antenna discussed in Chapter 17 for subsurface use.

There is yet another very useful possibility: the horizontal-wire antenna in the air close to the surface of the earth or sea. This can be a half-wave dipole as described in Section 7.5 for local use in locating buried or submerged objects or, for transmitting and receiving over large distances, the long traveling-wave antenna introduced in Section 7.3. This was called the *wave antenna* in the original paper (Beverage, Rice, and Kellog 1923), but is now often called the Beverage antenna after its inventor. It is an important directional antenna for transmission (communication, radar) between points on or near the surface of the earth. It consists of a long wire parallel to and close to the surface of the earth (sea, lake, ice). It is driven and terminated in a manner to provide a traveling wave of current. The

Beverage antenna has been discussed by numerous authors primarily as a directional receiving antenna (Busch 1923; Ollendorf 1925; Aharoni 1946; Kraus 1950; Belrose et al. 1983). In his book on traveling-wave antennas, Walter (1965) assumes the current on the Beverage antenna to have the form $I_x(x) = I_x(0)\exp(ik_0x)$, where k_0 is the real wave number of air, and determines the far-field pattern of the wire alone. He states that "the presence of the ground complicates the problem" and that "the usual approach is to represent the ground as a complex dielectric and consider the total pattern of the antenna and its properly weighted image." This approach disregards the effect of the earth on the distribution of current in the horizontal wire and assumes that a valid image can be constructed to replace the earth. Furthermore, it fails to provide a means for determining the field along the surface of the earth where a lateral wave dominates.

The current in and driving-point admittance of a horizontal-wire antenna over the earth have been determined (King, Wu, and Shen 1974; Sorbello et al. 1977; King and Smith 1981) for antennas terminated to carry currents with both standing- and traveling-wave distributions. It has been shown that when the horizontal wire is close to the earth, the wave number characteristic of the distribution of current is significantly different from the free-space wave number. Furthermore, a study of the electric field at the surface of the earth near the antenna indicates that the scattered or reflected field cannot be interpreted as due to the current in an image antenna at a specified distance from the boundary.

Of primary interest in the applications of the wave antenna is the field along the surface of the earth. This is the field involved, for example, in the transmission from a driven wave antenna to a second wave or other receiving antenna. It is the reflected field received by a Beverage receiving antenna from an obstacle near the horizon when this is illuminated by a radar beam.

18.2 Description of the Wave Antenna and Its Surroundings

The conventional Beverage antenna shown in Fig. 18.2.1(a) consists of the following: (a) A vertical conductor that extends upward along the negative z-axis from the surface of Region 1 (earth, sea, etc.) at $z = 0$ into Region 2 (air) to $z = -d$; it is driven by a generator with electromotive force (emf) V_0^e and impedance Z_g at $x = 0$, $z = -d$. It is grounded at $z = 0$ to a bare conductor of length l that extends vertically downward along the positive z-axis into Region 1, where it ends in a radial ground network. (b) A horizontal conductor that extends from $x = 0$ to $x = h$ in Region 2 at the constant height d above the x-axis which lies along the boundary

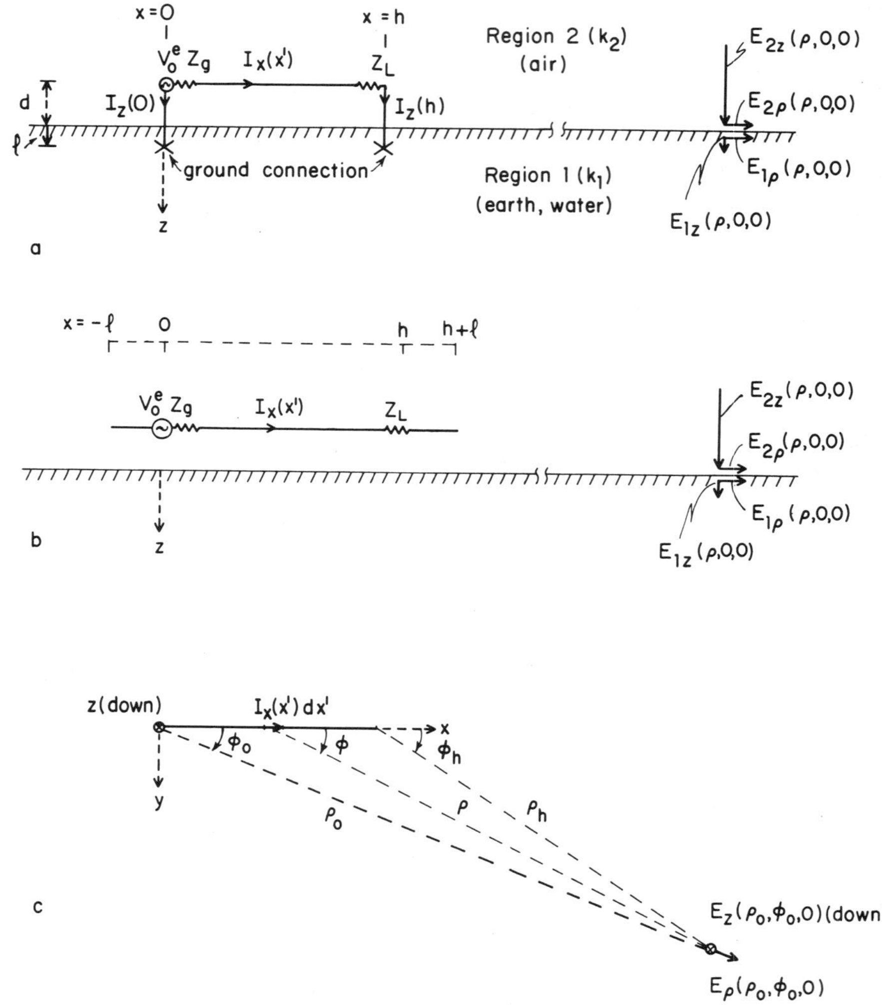

Figure 18.2.1. Wave antennas. (a) Conventional Beverage antenna; (b) horizontal-wire antenna; (c) coordinates.

surface between Regions 2 and 1. (c) A second vertical conductor of length d that extends upward from the ground at $x = h$, $y = z = 0$ to the point $x = h$, $y = 0$, $z = -d$, where it is connected to the horizontal wire through a load impedance Z_L that is selected to assure a traveling wave along the horizontal wire. The vertical conductor enters the ground at $x = h$, $z = 0$ and continues down into Region 1 a distance l, where it ends in a radial ground network. For simplicity, the radius of each conductor is taken to be a.

An alternative form of the wave antenna, shown in Fig. 18.2.1(b), consists of only a horizontal wire at the height d over the earth. It is driven at $x = 0$ at the distance l from the open end at $x = -l$. It is loaded at $x = h$ by a series impedance Z_L and ends at $x = h + l$.

The electrical properties of the two half-spaces, Region 2 (air) at $z < 0$ and Region 1 (earth, sea, lake) at $z > 0$, are characterized by the wave numbers $k_2 = k_0 = \omega(\mu_0\epsilon_0)^{1/2}$ and $k_1 = \beta_1 + i\alpha_1 = \omega(\mu_0\tilde{\epsilon}_1)^{1/2}$, where $\tilde{\epsilon}_1 = \epsilon_1 + i\sigma_1/\omega$. Here, $\epsilon_1 = \epsilon_0\epsilon_{1r}$ and σ_1 are the real effective permittivity and conductivity. Also useful are the wave impedances $\zeta_2 = \zeta_0 = (\mu_0/\epsilon_0)^{1/2} \sim 120\pi\ \Omega$ and $\zeta_1 = (\mu_0/\tilde{\epsilon}_1)^{1/2} = \omega\mu_0/k_1$. The theory to be applied in the analysis of the antennas is accurate subject to the following restriction on the electrical properties of the media and the frequency:

$$|k_1|^2 \gg k_2^2. \tag{18.2.1}$$

For most purposes, this inequality can be interpreted to mean $|k_1|^2 \geq 9k_2^2$ or $|k_1| \geq 3k_2$.

The formulas to be used for the electromagnetic fields of the horizontal and vertical elements are excellent approximations subject to the following restrictions in the cylindrical coordinates (ρ, ϕ, z):

$$\rho^2 = x^2 + y^2 \gg z^2; \qquad |k_1\rho| \geq 3. \tag{18.2.2}$$

(The more general case with no restriction on z is treated in Section 18.11.) The formulas for the horizontal wire require that

$$k_2 d < 1. \tag{18.2.3}$$

When vertical ground connections are used, it is assumed that the length l of the conductor in Region 1 satisfies the inequality

$$|k_1 l| < 1. \tag{18.2.4}$$

A consequence of the conditions (18.2.1) and (18.2.3) is that the fraction of power radiated into Region 2 (air) is negligibly small compared to the fraction transmitted into Region 1 (earth, sea, etc.). The horizontal wire is so closely coupled to the earth that it cannot properly be treated as a wire in air with an added correction to take account of the presence of the earth, as suggested by Walter (1965). The adjacent conducting or dielectric half-space (Region 1) has a dominant effect on the current in the conductor and cannot be disregarded any more than can the sheath of a coaxial line. In effect, the horizontal wire is an eccentrically insulated conductor lying on the surface of the material half-space, Region 1. The thickness of the air insulation is essentially the height d of the wire. When this is sufficiently small, as required by (18.2.3), the current in the wire is distributed like the current in a transmission line with the wave number $k_L = \beta_L + i\alpha_L$ and characteristic impedance Z_c (that depend on the inductance, capacitance, and resistance per unit length of the wire) and not like that of an antenna in air with the wave number k_2 of air. For an antenna in air, radiation is a property of the antenna as a whole; it cannot be assigned per unit length. For the antenna close to the earth, radiation is a property per unit length. The formulas for k_L, Z_c, and the current are given in (7.3.1)–(7.3.4). For convenience they are repeated here:

$$k_L = k_2\left\{1 + \frac{2}{\ln(2d/a)}\left[\frac{1}{(2k_1d)^2} - \frac{K_1(2k_1d)}{2k_1d} + \frac{i\pi I_1(2k_1d)}{4k_1d} - i\left(\frac{2k_1d}{3} + \frac{(2k_1d)^3}{45} + \frac{(2k_1d)^5}{1575} + \cdots\right)\right]\right\}^{1/2} \tag{18.2.5a}$$

$$\sim k_2\left(1 - \frac{\ln(k_1d) + \gamma - \frac{1}{2} - i(\frac{1}{2}\pi - \frac{4}{3}k_1d)}{\ln(2d/a)}\right)^{1/2}; \qquad 0 < |k_1d| \leq 0.8, \tag{18.2.5b}$$

$$Z_c = \frac{\zeta_2 k_L}{2\pi k_2}\ln\left(\frac{2d}{a}\right) = \frac{60k_L}{k_2}\ln\left(\frac{2d}{a}\right)\ \Omega. \tag{18.2.6}$$

Here, $K_1(x)$ and $I_1(x)$ are the modified Bessel functions and $\gamma = 0.5772$. The current on the conductor has the general transmission-line form with an impedanceless generator,

$$I_x(x) = -\frac{iV_0^e}{Z_c}\frac{\sin[k_L(h-x) + i\theta_h]}{\cos(k_Lh + i\theta_h)}, \tag{18.2.7a}$$

where $\theta_h = \coth^{-1}(Z_h/Z_c)$. When $Z_h = Z_c$, $\theta_h \to \infty$ and

$$I_x(x) = \frac{V_0^e}{Z_c}e^{ik_Lx}. \tag{18.2.7b}$$

Note that the wave number for the current is k_L, not k_2.

The impedance of the antenna at the driving point $x = 0$ is

$$Z_{\text{in}} = Z + Z_g + Z_0; \qquad Z = iZ_c\cot(k_Lh + i\theta_h), \tag{18.2.8a}$$

where Z_0 is the impedance of the terminating section at $x = 0$. For traveling-wave operation, the effective impedance $Z_h = Z_L + Z_0$ of the terminating section at $x = h$ is made equal to the characteristic impedance Z_c so that

$$Z = Z_h = Z_c. \tag{18.2.8b}$$

With the Beverage antenna, Z_0 is the impedance of each of the grounded vertical conductors. With the horizontal-wire antenna, Z_0 is the impedance of an open-ended section of the eccentrically insulated wire of length l; that is, with $\theta_h \sim -i\pi$ for an ideal open end,

$$Z_0 = iZ_c\cot k_Ll. \tag{18.2.8c}$$

18.3 The Electric Field Due to Dipoles with Unit Electric Moment

The first step in determining the electromagnetic field generated by the currents in the several parts of the antennas shown in Fig. 18.2.1 is to express the field generated by an element of current $I_z(z')\,dz'$ in the vertical elements and by $I_x(x')\,dx'$ in the horizontal element. Evidently, there are two types of vertical elements, those on the boundary surface in Region 2 ($z < 0$) and those on the boundary surface in Region 1 ($z \geq 0$). These last are bare metal conductors as described in Chapter 16. The eccentrically insulated horizontal element is on the boundary entirely in Region 2.

Simple and accurate formulas for all components of the electromagnetic field in Region 1 and along the boundary in Region 2 are available in Chapters 3 and 5 when the vertical or horizontal electric dipoles are embedded in Region 1 ($z \geq +0$) or are on the surface in either Region 1 ($z = +0$) or Region 2 ($z = -0$). These are the fields of primary interest for transmission to a receiving antenna on the surface in Region 2 or below the surface in Region 1. They include transmission between points on or below the surface of the earth or ocean. In order to be able to distinguish between the fields in Regions 1 and 2 due to vertical and horizontal elements in Regions 1 or 2, the following notation is used: $[E_{1\rho}(\rho,\phi,z)]_{2v}$. This gives the radial component (subscript ρ) of the electric field at the point (ρ,ϕ,z) in Region 1 (subscript 1) due to a vertical dipole in Region 2 (subscripts 2 and v on square bracket). Similarly, $[E_{2z}(\rho,\phi,0)]_{1h}$ is the z-component of the electric field at the point $(\rho,\phi,0)$ (i.e., on the boundary surface) in Region 2 due to a horizontal electric dipole in Region 1.

The formulas in Table 18.3.1 give the radial and vertical components of the electric field (E_ρ and E_z) and the transverse component of the magnetic field (B_ϕ) at distances from the source that are sufficiently great to make contributions from terms that involve $\exp(ik_1\rho) = \exp(-\alpha_1\rho)\exp(i\beta_1\rho)$ negligible. This is only a short distance (much less than a wavelength) when Region 1 is earth or sea water with practically available effective conductivity. In order to conserve space, the following notation is used:

$$F(\rho;k_1,k_2) = \frac{\omega\mu_0}{2\pi}\,[f_2(\rho)e^{ik_2\rho} + f_{21}(\rho)e^{ik_{21}\rho}], \tag{18.3.1}$$

$$G(\rho;k_1,k_2) = \frac{\omega\mu_0}{2\pi}\,[g_2(\rho)e^{ik_2\rho} + f_{21}(\rho)e^{ik_{21}\rho}], \tag{18.3.2}$$

where

$$f_2(\rho) = \frac{ik_2}{\rho} - \frac{1}{\rho^2}; \qquad g_2(\rho) = \frac{ik_2}{\rho} - \frac{1}{\rho^2} - \frac{i}{k_2\rho^3}, \tag{18.3.3}$$

$$f_{21}(\rho) = -\frac{k_2^3}{k_1}\left(\frac{\pi}{k_2\rho}\right)^{1/2}\mathcal{F}\,[k_2\rho(k_2^2/2k_1^2)] \tag{18.3.4}$$

with

$$k_{21} = k_2\left(1 - \frac{k_2^2}{2k_1^2}\right) \tag{18.3.5}$$

and

$$\mathcal{F}[k_2\rho(k_2^2/2k_1^2)] = \tfrac{1}{2}(1+i) - C_2[k_2\rho(k_2^2/2k_1^2)] - iS_2[k_2\rho(k_2^2/2k_1^2)]. \tag{18.3.6}$$

Here, $C_2(u)$ and $S_2(u)$ are Fresnel integrals defined by (Jahnke and Emde 1945; Abramowitz and Stegun 1972)

$$C_2(u) + iS_2(u) = \int_0^u \frac{e^{it}}{(2\pi t)^{1/2}}\,dt. \tag{18.3.7}$$

When $k_2\rho$ is sufficiently large, the Fresnel integrals are well represented by the leading terms in asymptotic formulas. With these,

$$\begin{aligned} F(\rho; k_1, k_2) &\sim G(\rho; k_1, k_2) \\ &\sim -\frac{\omega\mu_0}{2\pi}\frac{k_1^2}{k_2^2}\frac{e^{ik_2\rho}}{\rho^2}; \qquad k_2\rho \geq \frac{8|k_1^2|}{k_2^2}. \end{aligned} \tag{18.3.8}$$

With the above notation, the expressions for the fields of vertical and horizontal electric dipoles with unit electric moments ($I\Delta l = 1$ A m) located on the boundary between Regions 1 and 2 are summarized in Table 18.3.1. Formulas are given for vertical dipoles just above the boundary in Region 2 (air) and just below the boundary in Region 1 (earth, water, etc.). The formulas for the horizontal dipole are the same when the dipole is just above or just below the boundary. The formulas for the components of the electric and magnetic fields are for points on the surface in Region 2 ($z = -0$) and for points on and below the surface in Region 1 ($z \geq +0$). The rotationally symmetric field of the vertical dipole is smaller by the very small factor k_2^2/k_1^2 when the dipole is just below the surface in Region 1 than when it is just above the surface in Region 2. The directional field of the horizontal dipole is smaller by the factor k_2/k_1 than the field of the vertical dipole when this is in Region 2, and larger by the factor k_1/k_2 than the field of the vertical dipole when this is in Region 1. It is, of course, possible to make the electric moment of the horizontal antenna sufficiently great so that its field is greater than that of a vertical antenna in Region 2 with a small electric moment. The fields in Table 18.3.1 are useful for the Beverage antenna, which has both vertical and horizontal conductors in Regions 1 and 2.

Note that the largest component of the electric field in Region 2 is E_{2z}; the largest in Region 1 is $E_{1\rho}$. It is well to add that the electromagnetic field of the horizontal dipole also has the components E_ϕ, B_ρ, and B_z. However, these are not significant in the operation of the Beverage and horizontal-wire antennas. They all have the factor $\sin\phi$ and are, therefore, zero in the direction $\phi = 0$ in which E_ρ, E_z, and B_ϕ have their maxima.

Table 18.3.1. Electromagnetic Field of an Electric Dipole on or below the Surface Between Region 1 (Earth, Water; $z \geq 0$) and Region 2 (Air; $z \leq 0$) $I\Delta l = 1$ A m; $k_1^2 = \omega^2\mu_0(\epsilon_1 + i\sigma_1/\omega)$; $k_2^2 = \omega^2\mu_0\epsilon_0$

Component of the Field	Vertical Electric Dipole on Surface in Air
$E_{2\rho}(\rho, 0)$	$-k_1^{-1}F(\rho; k_1, k_2)$
$E_{1\rho}(\rho, 0)$	$-k_1^{-1}F(\rho; k_1, k_2)\exp(ik_1z)$
$E_{2z}(\rho, 0)$	$k_2^{-1}G(\rho; k_1, k_2)$
$E_{1z}(\rho, 0)$	$k_2k_1^{-2}G(\rho; k_1, k_2)\exp(ik_1z)$
$B_{2\phi}(\rho, 0)$	$-\omega^{-1}F(\rho; k_1, k_2)$
$B_{1\phi}(\rho, 0)$	$-\omega^{-1}F(\rho; k_1, k_2)\exp(ik_1z)$

Component of the Field	Vertical Electric Dipole on Surface in Salt Water
$E_{2\rho}(\rho, 0)$	$-k_2^2k_1^{-3}F(\rho; k_1, k_2)$
$E_{1\rho}(\rho, 0)$	$-k_2^2k_1^{-3}F(\rho; k_1, k_2)\exp(ik_1z)$
$E_{2z}(\rho, 0)$	$k_2k_1^{-2}G(\rho; k_1, k_2)$
$E_{1z}(\rho, 0)$	$k_2^3k_1^{-4}G(\rho; k_1, k_2)\exp(ik_1z)$
$B_{2\phi}(\rho, 0)$	$-\omega^{-1}k_2^2k_1^{-2}F(\rho; k_1, k_2)$
$B_{1\phi}(\rho, 0)$	$-\omega^{-1}k_2^2k_1^{-2}F(\rho; k_1, k_2)\exp(ik_1z)$

Component of the Field	Horizontal Electric Dipole on Surface in Air or Water
$E_{2\rho}(\rho, \phi, 0)$	$-k_2k_1^{-2}G(\rho; k_1, k_2)\cos\phi$
$E_{1\rho}(\rho, \phi, z)$	$-k_2k_1^{-2}G(\rho; k_1, k_2)\cos\phi\,\exp(ik_1z)$
$E_{2z}(\rho, \phi, 0)$	$k_1^{-1}F(\rho; k_1, k_2)\cos\phi$
$E_{1z}(\rho, \phi, z)$	$k_2^2k_1^{-3}F(\rho; k_1, k_2)\cos\phi\,\exp(ik_1z)$
$B_{2\phi}(\rho, \phi, 0)$	$-\omega^{-1}k_2k_1^{-1}G(\rho; k_1, k_2)\cos\phi$
$B_{1\phi}(\rho, \phi, z)$	$-\omega^{-1}k_2k_1^{-1}G(\rho; k_1, k_2)\cos\phi\,\exp(ik_1z)$

18.4 The Electric Field of the Vertical Elements of the Beverage Antenna

The vertical parts of the Beverage antenna shown in Fig. 18.2.1(a) consist of two sections: the one along the z-axis at $x = y = 0$ and the other at $x = h,\ y = 0$. Each is composed of two parts: a length from $z = -d$ to 0 in Region 2 and a length from $z = 0$ to a radial ground system at $z = l$ in Region 1. Since it is assumed that d is electrically short in Region 2 so that $k_2^2d^2 \ll 1$ and l is electrically short in Region 1 so that $|k_1l|^2 \ll 1$, it follows that the currents in these lengths are sensibly constant. In both the lengths d and l, they are given by $I_z(x = 0) \sim -I_x(0)$ at $x = y = 0$ and by $I_z(x = h) \sim I_x(h)$ at $x = h,\ y = 0$. The associated electric moments are $-I_x(0)d$ and $I_x(h)d$ for the lengths in Region 2, $-I_x(0)l$ and $I_x(h)l$ for the lengths in Region 1.

The components of the electric field on the surface $z = 0$ in Region 2 due to vertical elements in Region 2 are taken from Table 18.3.1, multiplied by the respective electric moment. For the elements at $x = 0$ and $x = h$, they are

$$[E_{2z}(\rho_0, \phi_0, 0)]_{2vc} = -k_2^{-1}d[I_x(0)G(\rho_0; k_1, k_2) - I_x(h)G(\rho_h; k_1, k_2)], \tag{18.4.1}$$

$$[E_{2\rho}(\rho_0, \phi_0, 0)]_{2vc} = k_1^{-1}d[I_x(0)F(\rho_0; k_1, k_2) - I_x(h)F(\rho_h; k_1, k_2)], \tag{18.4.2}$$

where ρ_0 is the radial distance from the z-axis and $\rho_h = (\rho_0^2 + h^2 - 2\rho_0 h\cos\phi_0)^{1/2}$. The components of the electric field in Region 2 due to the vertical elements in Region 1 are similarly obtained from Table 18.3.1. They are

$$[E_{2z}(\rho_0, \phi_0, 0)]_{1vc} = -k_2k_1^{-2}l[I_x(0)G(\rho_0; k_1, k_2) - I_x(h)G(\rho_h; k_1, k_2)], \tag{18.4.3}$$

$$[E_{2\rho}(\rho_0, \phi_0, 0)]_{1vc} = k_2^2k_1^{-3}l[I_x(0)F(\rho_0; k_1, k_2) - I_x(h)F(\rho_h; k_1, k_2)], \tag{18.4.4}$$

where $|k_1l|^2 \ll 1$.

It follows from (18.4.1) and (18.4.3) and from (18.4.2) and (18.4.4) that

$$\frac{[E_{2z}(\rho_0, \phi_0, 0)]_{1vc}}{[E_{2z}(\rho_0, \phi_0, 0)]_{2vc}} = \frac{[E_{2\rho}(\rho_0, \phi_0, 0)]_{1vc}}{[E_{2\rho}(\rho_0, \phi_0, 0)]_{2vc}} = \frac{k_2^2 l}{k_1^2 d} \sim \frac{k_2^3}{k_1^3}. \tag{18.4.5}$$

The last step follows since k_2d and $|k_1l|$ are of comparable magnitudes ($k_2^2d^2 \ll 1,\ |k_1l|^2 \ll 1$). It may be concluded that the fields generated by the currents in the two vertical elements of length l in Region 1 are negligible compared to the fields generated by the currents in the vertical elements of length d in Region 2.

The field due to the vertical elements is, with $I_x(h) = I_x(0)\exp(ik_L h)$ and $G(\rho_h; k_1, k_2) \sim G(\rho_0; k_1, k_2)\exp(-ik_2 h\cos\phi_0)$,

$$\begin{aligned}
&[E_{2z}(\rho_0, \phi_0, 0)]_{1vc} + [E_{2z}(\rho_0, \phi_0, 0)]_{2vc} \doteq [E_{2z}(\rho_0, \phi_0, 0)]_{2vc} \\
&\quad = -k_2^{-1} dI_x(0) G(\rho_0; k_1, k_2)(1 - e^{i(k_L - k_2\cos\phi_0)h}), \qquad (18.4.6) \\
&[E_{2\rho}(\rho_0, \phi_0, 0)]_{1vc} + [E_{2\rho}(\rho_0, \phi_0, 0)]_{2vc} \doteq [E_{2\rho}(\rho_0, \phi_0, 0)]_{2vc} \\
&\quad = k_1^{-1} dI_x(0) F(\rho_0; k_1, k_2)(1 - e^{i(k_L - k_2\cos\phi_0)h}). \qquad (18.4.7)
\end{aligned}$$

18.5 The Electric Field of the Horizontal Elements of the Beverage and Horizontal-Wire Antennas

The horizontal conductor in Region 2 (air) at the electrically small height d over Region 1 (earth, water) has been analyzed as an eccentrically insulated antenna on the surface of Region 1. Its wave number k_L is given by (18.2.5a), the characteristic impedance by (18.2.6), and the distribution of current by (18.2.7a). When the terminating impedances are adjusted so that $Z_L + Z_0 = Z_g + Z_0 = Z_c$, the current is a simple traveling wave given by $I_x(x) = I_x(0)\exp(ik_L x)$, where $I_x(0) = V_0^e/2Z_c$.

The field generated by the current in the horizontal wire between $x = 0$ and $x = h$ is the superposition in proper phase and amplitude of the continuity of elements $I_x(x')\,dx'$ along the wire. If the radial distance from the origin at $x = y = z = 0$ to the point of observation at $(\rho_0, \phi_0, 0)$ is ρ_0, the radial distance from an element of current at x', $y = z = 0$ is $\rho_{x'} = (\rho_0^2 + x'^2 - 2\rho_0 x'\cos\phi_0)^{1/2}$. When $\rho_0^2 \gg h^2 \geq x'^2$, $\rho_{x'} \sim \rho_0 - x'\cos\phi_0$ in phases and $\rho_{x'} \sim \rho_0$ in amplitudes. It follows with (18.3.1) and (18.3.2) that

$$\begin{aligned}
&F(\rho_{x'}; k_1, k_2) \\
&\quad \sim \frac{\omega\mu_0}{2\pi}\,[f_2(\rho_0)e^{ik_2(\rho_0 - x'\cos\phi_0)} + f_{21}(\rho_0)e^{ik_{21}(\rho_0 - x'\cos\phi_0)}], \qquad (18.5.1a) \\
&G(\rho_{x'}; k_1, k_2) \\
&\quad \sim \frac{\omega\mu_0}{2\pi}\,[g_2(\rho_0)e^{ik_2(\rho_0 - x'\cos\phi_0)} + f_{21}(\rho_0)e^{ik_{21}(\rho_0 - x'\cos\phi_0)}]. \qquad (18.5.1b)
\end{aligned}$$

The vertical electric field of the entire horizontal wire between $x = 0$ and $x = h$ in both the Beverage and horizontal-wire antennas is given by the following integral:

$$[E_{2z}(\rho_0, \phi_0, 0)]_{hw} = k_1^{-1} I_x(0)\cos\phi_0 \int_0^h F(\rho_{x'}; k_1, k_2)e^{ik_L x'}\,dx'. \qquad (18.5.2)$$

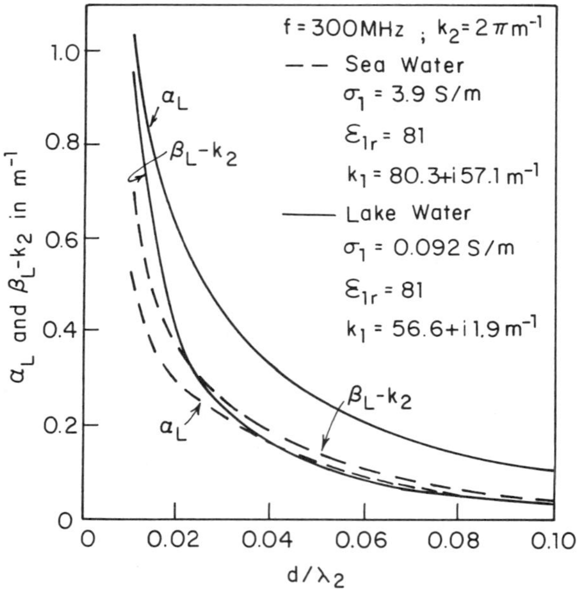

Figure 18.5.1. Attenuation and phase constants of the current on the horizontal wire over sea and lake water.

With (18.5.1a), the integral to be evaluated is

$$\int_0^h e^{i(k_L - k_2 \cos\phi_0)x'}\, dx' = \frac{i(1 - e^{i(k_L - k_2\cos\phi_0)h})}{k_L - k_2 \cos\phi_0}. \tag{18.5.3}$$

There is a similar integral with k_{21} substituted for k_2 in (18.5.3). However, since $k_2^2 \ll |k_1|^2$, it follows that, with $\phi_0 = 0$, $\exp(-ik_{21}h) = \exp[-ik_2 h(1 - k_2^2/2k_1^2)] \sim \exp(-ik_2 h)$ when $|k_2 h(k_2^2/2k_1^2)| \ll 1$, an inequality that is usually satisfied. When it is not valid, the two terms can be treated separately without difficulty. When $\exp(-ik_{21}h) \sim \exp(-ik_2 h)$,

$$[E_{2z}(\rho_0, \phi_0, 0)]_{hw} = I_x(0)\cos\phi_0 \frac{i(1 - e^{i(k_L - k_2\cos\phi_0)h})}{k_1(k_L - k_2\cos\phi_0)} F(\rho_0; k_1, k_2). \tag{18.5.4}$$

Similarly,

$$[E_{2\rho}(\rho_0, \phi_0, 0)]_{hw} = -I_x(0)\cos\phi_0 \frac{ik_2(1 - e^{i(k_L - k_2\cos\phi_0)h})}{k_1^2(k_L - k_2\cos\phi_0)} G(\rho_0; k_1, k_2). \tag{18.5.5}$$

The quantities α_L and $(\beta_L - k_2)$ in the exponents with $\phi_0 = 0$ are shown graphically in Fig. 18.5.1 for sea and lake water at $f = 300$ MHz.

The horizontal-wire antenna has two horizontal terminations that extend a distance l beyond each end of the main section of length h. When $I_x(x') = I_x(0)\exp(ik_L x')$ is the current on the main section, the currents in the end sections are

$$I_x(x') = I_x(0)\,\frac{\sin k_L(l+x')}{\sin k_L l}; \qquad -l \le x' \le 0, \tag{18.5.6}$$

$$I_x(x') = I_x(0)e^{ik_L h}\,\frac{\sin k_L(h+l-x')}{\sin k_L l}; \qquad h \le x' \le h+l. \tag{18.5.7}$$

With these, the vertical component of the electric field due to each is

$$[E_{2z}(\rho_0,\phi_0,0)]_{h0} = \frac{I_x(0)}{k_1\sin k_L l}\cos\phi_0 \times \int_{-l}^{0} F(\rho_{x'};k_1,k_2)\sin k_L(l+x')\,dx', \tag{18.5.8}$$

$$[E_{2z}(\rho_0,\phi_0,0)]_{hh} = \frac{I_x(0)e^{ik_L h}}{k_1\sin k_L l}\cos\phi_0 \times \int_{h}^{h+l} F(\rho_{x'};k_1,k_2)\sin k_L(h+l-x')\,dx'. \tag{18.5.9}$$

With (18.5.1a), the integrals to be evaluated are

$$\int_{-l}^{0}\sin k_L(l+x')e^{-ik_2x'\cos\phi_0}\,dx' = \frac{C(k_2,k_L,\phi_0) - iS(k_2,k_L,\phi_0)}{k_L^2 - k_2^2\cos^2\phi_0}, \tag{18.5.10}$$

$$\int_{h}^{h+l}\sin k_L(h+l-x')e^{-ik_2x'\cos\phi_0}\,dx' = e^{-ik_2h\cos\phi_0}\,\frac{C(k_2,k_L,\phi_0) + iS(k_2,k_L,\phi_0)}{k_L^2 - k_2^2\cos^2\phi_0}, \tag{18.5.11}$$

where

$$C(k_2,k_L,\phi_0) = k_L[\cos(k_2 l\cos\phi_0) - \cos k_L l], \tag{18.5.12}$$

$$S(k_2,k_L,\phi_0) = k_L\sin(k_2 l\cos\phi_0) - k_2\cos\phi_0\,\sin k_L l. \tag{18.5.13}$$

With these expressions, the fields due to the horizontal terminations are

$$[E_{2z}(\rho_0,\phi_0,0)]_{h0} = \frac{I_x(0)F(\rho_0;k_1,k_2)\cos\phi_0}{k_1(k_L^2 - k_2^2\cos^2\phi_0)\sin k_L l} \times [C(k_2,k_L,\phi_0) - iS(k_2,k_L,\phi_0)], \tag{18.5.14}$$

$$[E_{2z}(\rho_0,\phi_0,0)]_{hh} = \frac{I_x(0)F(\rho_0;k_1,k_2)\cos\phi_0\, e^{i(k_L - k_2\cos\phi_0)h}}{k_1(k_L^2 - k_2^2\cos^2\phi_0)\sin k_L l} \times [C(k_2,k_L,\phi_0) + iS(k_2,k_L,\phi_0)]. \tag{18.5.15}$$

The sum gives the field due to both terminations. It is

$$[E_{2z}(\rho_0,\phi_0,0)]_{ht} = \frac{I_x(0)F(\rho_0;k_1,k_2)\cos\phi_0}{k_1(k_L^2 - k_2^2\cos^2\phi_0)\sin k_L l} \times [C(k_2,k_L,\phi_0)(1+e^{i(k_L-k_2\cos\phi_0)h}) - iS(k_2,k_L,\phi_0)(1-e^{i(k_L-k_2\cos\phi_0)h})]. \qquad (18.5.16)$$

Similarly, the radial component of the electric field due to the currents in the terminations is

$$[E_{2\rho}(\rho_0,\phi_0,0)]_{ht} = -\frac{k_2 G(\rho_0;k_1,k_2)}{k_1 F(\rho_0;k_1,k_2)}[E_{2z}(\rho_0,\phi_0,0)]_{ht}. \qquad (18.5.17)$$

The radial ground networks at the ends of the vertical ground connections of length l in the Beverage antenna are constructed of bare horizontal wires. The current on each radial wire has the approximate form $\sin k_1(S-s)$, where S is the radial length of the wire. Since $k_1 = \beta_1 + i\alpha_1$, where α_1 is significant, and the wires are bare, the current on each is quite rapidly attenuated. Consequently, the lengths S can be very small compared to the length h of the wire in air. Furthermore, with a radial distribution of wires, the fields of oppositely directed wires tend to cancel. It follows that the contributions to the field by the currents in the radially arranged horizontal ground wires are negligible.

18.6 The Electric Field of the Beverage Antenna

The field of the Beverage antenna is generated by the currents in both the horizontal and vertical members in Region 2 (air). The contributions from the vertical and horizontal ground wires in Region 1 are negligible. The dominant components of primary interest are the vertical and radial components $E_{2z}(\rho_0,\phi_0,0)$ and $E_{2\rho}(\rho_0,\phi_0,0)$ in the air (Region 2) on the surface of the earth (Region 1) and the radial component $E_{1\rho}(\rho_0,\phi_0,z)$ in Region 1. These are obtained from (18.4.6) with (18.5.4) and (18.4.7) with (18.5.5). They are

$$E_{2z}(\rho_0,\phi_0,0) = \frac{I_x(0)(1-e^{i(k_L-k_2\cos\phi_0)h})}{k_1k_2} \times \left(\frac{ik_2F(\rho_0;k_1,k_2)\cos\phi_0}{k_L - k_2\cos\phi_0} - k_1 dG(\rho_0;k_1,k_2)\right), \qquad (18.6.1)$$

$$E_{2\rho}(\rho_0,\phi_0,0) = -\frac{I_x(0)(1-e^{i(k_L-k_2\cos\phi_0)h})}{k_1^2}$$

$$\times\left(\frac{ik_2G(\rho_0;k_1,k_2)\cos\phi_0}{k_L-k_2\cos\phi_0}-k_1dF(\rho_0;k_1,k_2)\right), \quad (18.6.2a)$$

$$E_{1\rho}(\rho_0,\phi_0,z)=E_{2\rho}(\rho_0,\phi_0,0)e^{ik_1z}. \quad (18.6.2b)$$

The quantity

$$h_e(\phi_0)\equiv\frac{i(1-e^{i(k_L-k_2\cos\phi_0)h})}{k_L-k_2\cos\phi_0}, \quad (18.6.3)$$

which occurs in the first terms in both (18.6.1) and (18.6.2a), is defined to be the complex effective length of the horizontal part of the antenna. Since the contribution from the vertical part is shown to be quite small later in this section, $h_e(\phi_0)$ is a good approximation of the effective length of the entire antenna, as indicated in (7.3.6)–(7.3.9).

For applications in communication and radar, it is desirable to select h, the length of the horizontal conductor and the distance between the two vertical terminations, to make the magnitude of the factor $\{1-\exp[i(k_L-k_2\cos\phi_0)h]\}$ in (18.6.1) and (18.6.2a) as large as possible in the direction $\phi_0=0$. The quantity to be maximized with respect to h is

$$|1-e^{-\alpha_Lh}e^{i(\beta_L-k_2)h}|=[1-2e^{-\alpha_Lh}\cos(\beta_L-k_2)h+e^{-2\alpha_Lh}]^{1/2}. \quad (18.6.4)$$

Differentiation with respect to h leads to the following maximizing equation:

$$\alpha_L\cos(\beta_L-k_2)h+(\beta_L-k_2)\sin(\beta_L-k_2)h=\alpha_Le^{-\alpha_Lh}. \quad (18.6.5)$$

Since α_L and (β_L-k_2) are generally of comparable size and less than 1, the optimum value of h can be expected to be large enough so that $\alpha_L\exp(-\alpha_Lh)$ is quite small. When this is true, the approximate condition for a maximum is that already given in (7.3.12), viz.,

$$(\beta_L-k_2)h_m\sim\pi-\tan^{-1}\left(\frac{\alpha_L}{\beta_L-k_2}\right). \quad (18.6.6)$$

Some solutions of (18.6.6) at $f=300$ MHz are listed in Table 7.3.1 together with the associated small quantity $\exp(-\alpha_Lh_m)$. With such lengths, the factor $\{1-\exp(-\alpha_Lh)\exp[i(\beta_L-k_2)h]\}\sim 1$. The quantity $|1-\exp(-\alpha_Lh)\exp[i(\beta_L-k_2)h]|$ is shown in Fig. 18.6.1 as a function of the length h for $d/\lambda_2=0.02$ for wires over both sea and lake water. It is seen that for lengths h less than half the maximizing value, the amplitude decreases rapidly. Since the exponential term in $|1-\exp(-\alpha_Lh)\exp[i(\beta_L-k_2\cos\phi_0)h]|$ contributes negligibly to the amplitude when $\phi_0=0$ and h is at or near its maximizing value because $\exp(-\alpha_Lh)$ is small, the same is true when $\phi_0\neq 0$. This means that this term contributes negligibly to the directional properties of the Beverage antenna.

Since $G(\rho_0;k_1,k_2)$ and $F(\rho_0;k_1,k_2)$ in (18.6.1) and (18.6.2) differ only in the term $i/k_2\rho^3$, it follows that when $k_2\rho>1$, the quantities $H(\phi_0)=$

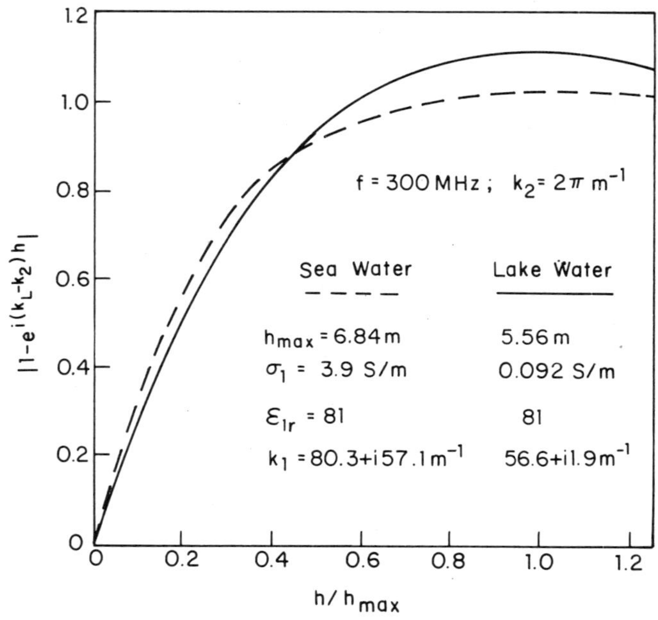

Figure 18.6.1. The magnitude of the factor $\{1 - \exp[i(k_L - k_2)h]\}$ as a function of the length h; $d/\lambda_2 = 0.02$.

$ik_2 \cos\phi_0/(k_L - k_2 \cos\phi_0)$ and $V = -k_1 d$ are measures of the relative importance of the contributions to the field by the horizontal and vertical currents. With $k_L = \beta_L + i\alpha_L$, $k_1 = \beta_1 + i\alpha_1$, and k_2 real,

$$H(\phi_0) = \frac{k_2[\alpha_L + i(\beta_L - k_2\cos\phi_0)]\cos\phi_0}{(\beta_L - k_2\cos\phi_0)^2 + \alpha_L^2},$$
$$V = -(\beta_1 + i\alpha_1)d. \tag{18.6.7}$$

The direction of maximum for $H(\phi_0)$ is $\phi_0 = 0$ for which

$$H(0) = |H|e^{i\theta_H}$$
$$= k_2[(\beta_L - k_2)^2 + \alpha_L^2]^{-1/2} \exp\left[i\tan^{-1}\left(\frac{\beta_L - k_2}{\alpha_L}\right)\right]. \tag{18.6.8}$$

Also,

$$V = |V|e^{i\theta_V} = (\beta_1^2 + \alpha_1^2)^{1/2} d \exp\left\{i\left[\tan^{-1}\left(\frac{\alpha_1}{\beta_1}\right) - 180°\right]\right\}. \tag{18.6.9}$$

Graphs of $|H|$, θ_H and $|V|$, θ_V as functions of d/λ_2 are in Fig. 18.6.2 for sea water with $\sigma_1 = 3.9$ S/m and $\epsilon_{1r} = 81$ and for lake water with $\sigma_1 = 0.092$ S/m and $\epsilon_{1r} = 81$ at $f = 300$ MHz. It is seen that $|H|$ is significantly greater than $|V|$ for all values of d/λ_2 for both lake and sea water. This is

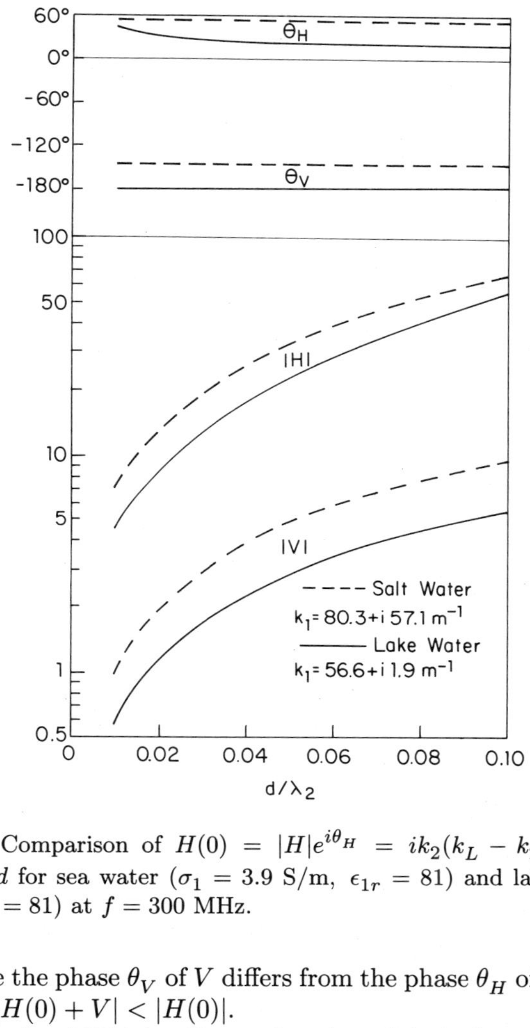

Figure 18.6.2. Comparison of $H(0) = |H|e^{i\theta_H} = ik_2(k_L - k_2)^{-1}$ and $V = |V|e^{i\theta_V} = -k_1 d$ for sea water ($\sigma_1 = 3.9$ S/m, $\epsilon_{1r} = 81$) and lake water ($\sigma_1 = 0.092$ S/m, $\epsilon_{1r} = 81$) at $f = 300$ MHz.

fortunate since the phase θ_V of V differs from the phase θ_H of H by roughly 180°, so that $|H(0) + V| < |H(0)|$.

The magnitude of $H(\phi_0) + V$ as a function of ϕ_0 is shown in Fig. 18.6.3 specifically with $d/\lambda_2 = 0.02$ for both sea water and lake water. It is seen that there is a significant maximum at $\phi_0 = 0$ where $|H(\phi_0)| \gg |V|$. However, $H(\phi_0)$ decreases rapidly when $\phi_0 > 20°$ and, for $\phi_0 > 45°$, $H(\phi_0)$ becomes negligible and V—which is independent of ϕ_0—dominates. Since no other factors in the formulas (18.6.1) and (18.6.2a,b) for $E_{2z}(\rho_0, \phi_0, 0)$, $E_{2\rho}(\rho_0, \phi_0, 0)$, and $E_{1\rho}(\rho_0, \phi_0, z)$ involve ϕ_0, the graphs in Fig. 18.6.3 represent the directional characteristics of the electric field.

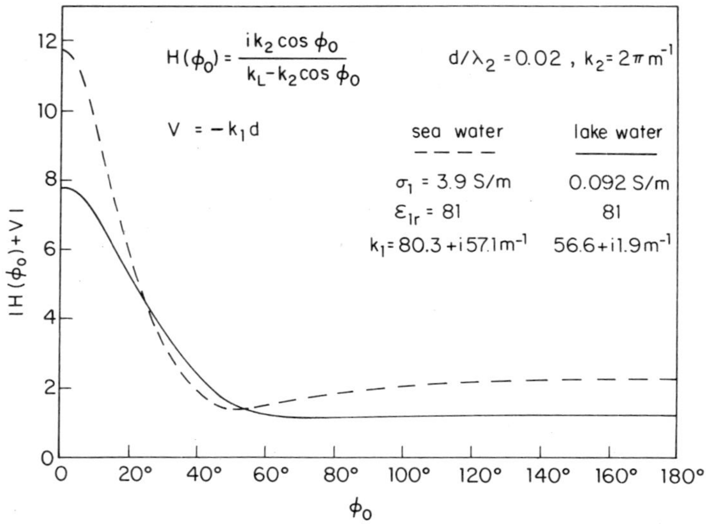

Figure 18.6.3. The magnitude of the factor $H(\phi_0) + V$ as a function of the angle ϕ_0.

18.7 The Electric Field of the Horizontal-Wire Antenna

The electric field of the horizontal-wire antenna includes contributions from the currents in the main section between $x = 0$ and $x = h$ with its traveling wave of current, $I_x(x') = I_x(0) \exp(ik_L x')$, and from the currents in the terminations with their standing waves of current given by (18.5.6) and (18.5.7). The combined field is obtained from (18.5.4) and (18.5.16). It is

$$E_{2z}(\rho_0, \phi_0, 0) = \frac{I_x(0) F(\rho_0; k_1, k_2) \cos\phi_0}{k_1(k_L - k_2 \cos\phi_0)} \times \left[i\left(1 - \frac{S(k_2, k_L, \phi_0)}{(k_L + k_2 \cos\phi_0) \sin k_L l}\right)\left(1 - e^{i(k_L - k_2 \cos\phi_0)h}\right) + \frac{C(k_2, k_L, \phi_0)}{(k_L + k_2 \cos\phi_0) \sin k_L l}\left(1 + e^{i(k_L - k_2 \cos\phi_0)h}\right)\right], \tag{18.7.1}$$

where $C(k_2, k_L, \phi_0)$ is given by (18.5.12) and $S(k_2, k_L, \phi_0)$ by (18.5.13). As with the Beverage antenna, the amplitude is maximized by selecting h as defined in (18.6.6). This is sufficiently great to make $\exp(-\alpha_L h)$ very small so that the exponential terms in (18.7.1) contribute negligibly. Hence,

$$E_{2z}(\rho_0,\phi_0,0) \sim \frac{I_x(0)F(\rho_0;k_1,k_2)\cos\phi_0}{k_1(k_L-k_2\cos\phi_0)} \times \left(i+\frac{C(k_2,k_L,\phi_0)-iS(k_2,k_L,\phi_0)}{(k_L+k_2\cos\phi_0)\sin k_L l}\right). \tag{18.7.2}$$

In order to compare the relative magnitudes of the field due to the currents in the long main section and in the much shorter terminations, let $\phi_0 = 0$ so that

$$C(k_2,k_L,0) = k_L(\cos k_2 l - \cos k_L l), \tag{18.7.3a}$$

$$S(k_2,k_L,0) = k_L \sin k_2 l - k_2 \sin k_L l. \tag{18.7.3b}$$

When $d/\lambda_2 = 0.02$ and Region 1 is sea water, $k_L = \beta_L + i\alpha_L = 6.641 + i0.289\ \mathrm{m}^{-1}$. Let $\beta_L l = \pi/2$ so that $\cos k_L l = -i\sinh\alpha_L l = -i0.070$ and $\sin k_L l = \cosh\alpha_L l = 1.002$; $\cos k_2 l = \cos 2\pi l = 0.084$ and $\sin k_2 l = 0.996$. It follows that $C(k_2,k_L,0) = 0.538 + i0.489$, $S(k_2,k_L,0) = 0.319 + i0.288$, $C(k_2,k_L,0) - iS(k_2,k_L,0) = 0.826 + i0.170$, and $[C(k_2,k_L,0) - iS(k_2,k_L,0)]/(k_L+k_2)\sin k_L l = 0.064 + i0.012$. With this value, the term inside the large parentheses in (18.7.2) with $\phi_0 = 0$ gives 1.014 exp(i1.51) for the combined field, 1.0 exp(i1.57) for the field of the main traveling-wave section alone, and 0.065 exp(i0.18) for the field of the two horizontal terminations alone. Thus, the currents in the horizontal terminations increase the magnitude of the field of the current in the main section by about 1.5%. The corresponding magnitudes for the same antenna with vertical instead of horizontal terminations are 0.86 for the entire antenna, 1.0 for the horizontal part, and 0.14 for the vertical terminations. It is seen that the vertical terminations have a negative effect instead of the small positive effect of the horizontal ones.

18.8 Summary: The Beverage and Horizontal-Wire Antennas for Transmission

The detailed analysis of both the Beverage and horizontal-wire antennas indicates that the contributions to the radiated field by the currents in the terminations are of minor significance when $|k_1 d|$ is small and the length h is adjusted for near optimum. The properties of both antennas are comparable and their field in Region 2 (air) is given by

$$E_{2z}(\rho_0,\phi_0,0) = \frac{iI_x(0)(1-e^{i(k_L-k_2\cos\phi_0)h})}{k_L-k_2\cos\phi_0}[E_{2z}(\rho_0,\phi_0,0)]_h, \tag{18.8.1a}$$

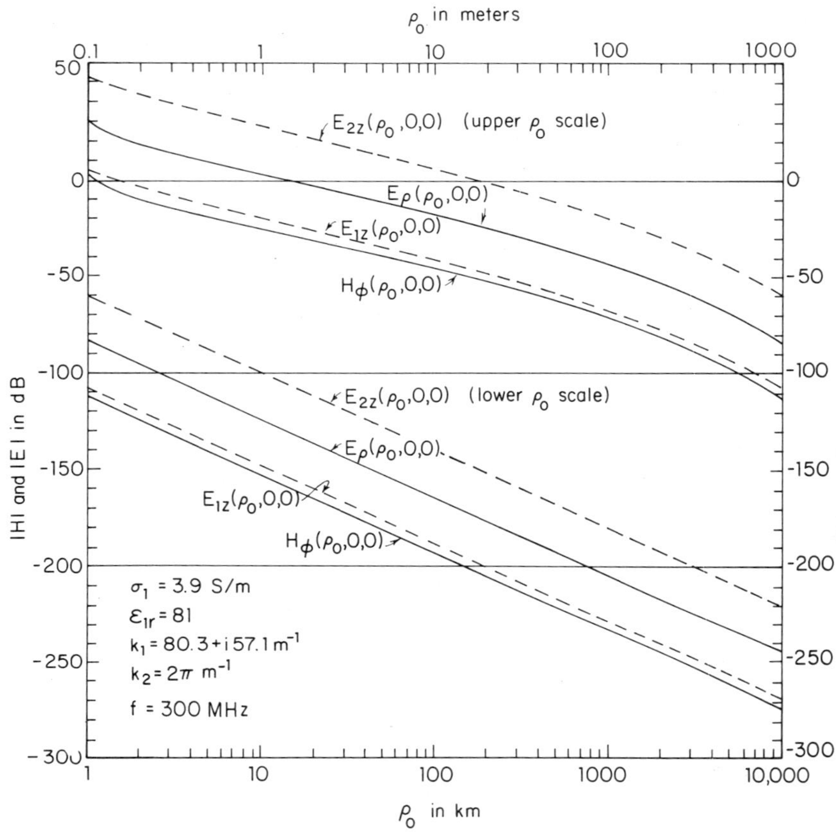

Figure 18.8.1. The components of the lateral-wave electric and magnetic fields on the surface between air and sea water at $f = 300$ MHz due to a horizontal electric dipole with unit electric moment on the surface. ($|E_\rho|$ referred to 1 V/m at $\rho_0 = 1.5$ m; $|H_\phi|$ referred to 1 A/m at $\rho_0 = 0.12$ m; $|E_{1z}|$ referred to 1 V/m at $\rho_0 = 0.15$ m; $|E_{2z}|$ referred to 1 V/m at $\rho_0 = 19$ m.)

$$E_{2\rho}(\rho_0, \phi_0, 0) = \frac{iI_x(0)(1 - e^{i(k_L - k_2 \cos\phi_0)h})}{k_L - k_2 \cos\phi_0} [E_{2\rho}(\rho_0, \phi_0, 0)]_h, \quad (18.8.1b)$$

$$B_{2\phi}(\rho_0, \phi_0, 0) = \frac{k_1}{\omega} E_{2\rho}(\rho_0, \phi_0, 0), \quad (18.8.1c)$$

where $[E_{2z}(\rho_0, \phi_0, 0)]_h$ and $[E_{2\rho}(\rho_0, \phi_0, 0)]_h$ are the components due to a horizontal electric dipole with unit moment. They are given in Table 18.3.1. They are

$$[E_{2z}(\rho_0, \phi_0, 0)]_h = k_1^{-1} F(\rho_0; k_1, k_2) \cos\phi_0, \quad (18.8.2a)$$

$$[E_{2\rho}(\rho_0, \phi_0, 0)]_h = -k_2 k_1^{-2} G(\rho_0; k_1, k_2) \cos\phi_0. \quad (18.8.2b)$$

It also follows that the field of the wave antennas in Region 1 (earth, sea) is

$$E_{1z}(\rho_0, \phi_0, z) = \frac{k_2^2}{k_1^2}\, e^{ik_1 z} E_{2z}(\rho_0, \phi_0, 0), \tag{18.8.3a}$$

$$E_{1\rho}(\rho_0, \phi_0, z) = e^{ik_1 z} E_{2\rho}(\rho_0, \phi_0, 0), \tag{18.8.3b}$$

$$B_{1\phi}(\rho_0, \phi_0, z) = \frac{k_1}{\omega}\, E_{1\rho}(\rho_0, \phi_0, z). \tag{18.8.3c}$$

These formulas are valid when the antennas are terminated so that the current in the long horizontal wire is a pure traveling wave with the form $I_x(x) = I_x(0)\exp(ik_L x)$. In (18.8.2),

$$F(\rho_0; k_1, k_2) = \frac{\omega\mu_0}{2\pi}\, e^{ik_2\rho_0} \left[\frac{ik_2}{\rho_0} - \frac{1}{\rho_0^2} - \frac{k_2^3}{k_1} \left(\frac{\pi}{k_2\rho_0} \right)^{1/2} e^{-ik_2\rho_0(k_2^2/2k_1^2)} \mathcal{F}\left[k_2\rho_0(k_2^2/2k_1^2)\right] \right], \tag{18.8.4a}$$

$$G(\rho_0; k_1, k_2) = \frac{\omega\mu_0}{2\pi}\, e^{ik_2\rho_0} \left[\frac{ik_2}{\rho_0} - \frac{1}{\rho_0^2} - \frac{i}{k_2\rho_0^3} - \frac{k_2^3}{k_1} \left(\frac{\pi}{k_2\rho_0} \right)^{1/2} e^{-ik_2\rho_0(k_2^2/2k_1^2)} \mathcal{F}\left[k_2\rho_0(k_2^2/2k_1^2)\right] \right]. \tag{18.8.4b}$$

When $k_2\rho \geq 8|k_1^2|/k_2^2$,

$$F(\rho_0; k_1, k_2) \sim G(\rho_0; k_1, k_2) \sim -\frac{\omega\mu_0}{2\pi} \frac{k_1^2}{k_2^2} \frac{e^{ik_2\rho_0}}{\rho_0^2}. \tag{18.8.5}$$

Graphs of the magnitudes of $[E_{2z}(\rho_0, 0, 0)]_{2h}$, $[E_{1z}(\rho_0, 0, 0)]_{2h}$, and $[E_\rho(\rho_0, 0, 0)]_{2h}$ and of $[H_\phi(\rho_0, 0, 0)]_{2h} = \mu_0^{-1}[B_\phi(\rho_0, 0, 0)]_{2h}$ are shown in Fig. 18.8.1 at the top with the upper scale in the range 0.1 m $\leq \rho_0 \leq$ 1000 m, and at the bottom with the lower scale in the range 1 km $\leq \rho_0 \leq$ 10,000 km. The graphs are for a unit horizontal electric dipole on the boundary between air ($k_2 = 2\pi$ m^{-1}) and sea water ($\sigma_1 = 3.9$ S/m, $\epsilon_{1r} = 81$, $k_1 = 80.3 + i57.1$ m^{-1}) at $f = 300$ MHz. In the range $0.2 \leq \rho_0 \leq 20$ m, the fields decrease with radial distance approximately as $1/\rho_0$; in the range 200 m $\leq \rho_0 \leq \infty$, the fields decrease with radial distance as $1/\rho_0^2$.

18.9 The Beverage and Horizontal-Wire Antennas for Reception—The Current in the Main Horizontal Wire

The wave antennas shown in Fig. 18.2.1 are converted from transmission to reception by the substitution of a receiver with load Z_r for the generator with impedance Z_g at $x = 0$, $z = -d$, and the adjustment of $Z_0 + Z_r$ to Z_c. This is illustrated in Fig. 18.9.1 for the Beverage antenna where the receiver is represented by the impedance Z_r. In the following, it will be assumed that both impedances Z_r and Z_L in conjunction with their grounding impedances Z_0 are equal to the characteristic impedance Z_c of the eccentrically insulated horizontal wire. That is, $Z_r+Z_0 = Z_L+Z_0 = Z_c$.

The current $I_x(0)$ in the load impedance Z_r can, of course, be obtained by an application of the reciprocal theorem to the transmitting antenna already analyzed. However, the distribution of current along the wire cannot be obtained in this manner. Actually, it is instructive to treat the wave antenna independently for reception.

An important initial question is the description of the incident electromagnetic field that induces currents in the receiving antenna. If the incident field is generated by an aircraft flying at some distance, the incident field is well approximated by a plane wave arriving along a path that subtends an angle ϕ with the x-axis along the horizontal wire and an angle Θ with the vertical and polarized either in the plane of incidence or perpendicular to the plane of incidence. On the other hand, if the source of the incident field is a transmitting antenna—perhaps a Beverage antenna—on the surface of the earth or a missile near the horizon scattering a radar signal, the incident field is a lateral wave traveling along the surface. The vertical and horizontal currents in the source generate the following principal components of the lateral-wave electric field at moderate and large distances:

Field due to vertical currents:

$$E_{2z}^{\text{inc}} = \frac{A}{k_2}\, G(\rho; k_1, k_2); \qquad E_{2\rho}^{\text{inc}} = -\frac{A}{k_1}\, F(\rho; k_1, k_2); \tag{18.9.1a}$$

field due to horizontal currents:

$$E_{2z}^{\text{inc}} = \frac{A}{k_1}\, F(\rho; k_1, k_2)\cos\phi; \qquad E_{2\rho}^{\text{inc}} = -\frac{Ak_2}{k_1^2}\, G(\rho; k_1, k_2)\cos\phi, \tag{18.9.1b}$$

where $F(\rho; k_1, k_2)$ and $G(\rho; k_1, k_2)$ are defined in (18.3.1) and (18.3.2). Here, ρ is the radial distance from a source on or very near the boundary surface between Regions 1 and 2 and A is an amplitude factor determined by the effective length and the magnitude of the currents in the source.

If the incident field is referred to its value at $x = 0$, then

$$E_{2z}^{\text{inc}}(x') = E_{2z}^{\text{inc}}(0)e^{-ik_2x'\cos\phi},$$

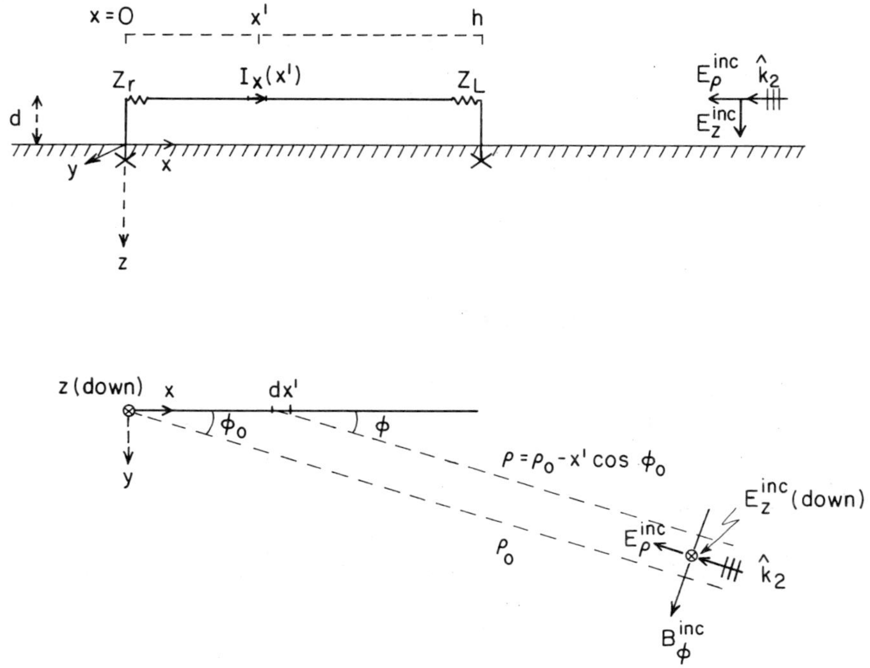

Figure 18.9.1. Beverage receiving antenna in incident lateral-wave field.

$$E_{2x}^{\text{inc}}(x') = -E_{2\rho}^{\text{inc}}(0) \cos \phi \; e^{-ik_2 x' \cos \phi}. \tag{18.9.2}$$

In these expressions, it is assumed that the amplitude of the incident field is sensibly constant over the length h of the horizontal wire and that the phase varies progressively with the wave number k_2. This means that the distances ρ from points on the wire to the source are quite large compared to the length h: $\rho \gg h$.

The current in the horizontal wire shown in Fig. 18.9.1 is induced entirely by $E_{2x}^{\text{inc}}(x)$. It can be determined by treating the incident field as a continuous distribution of generators along the wire. The current due to a generator with the emf $E_{2x}^{\text{inc}}(x')\, dx'$ at x' is

$$dI_x(x) = Z_c^{-1} E_{2x}^{\text{inc}}(x')\, dx' \, e^{ik_L x} e^{-ik_L x'}; \qquad x' \le x \le h, \tag{18.9.3a}$$

$$dI_x(x) = Z_c^{-1} E_{2x}^{\text{inc}}(x')\, dx' \, e^{-ik_L x} e^{ik_L x'}; \qquad 0 \le x \le x'. \tag{18.9.3b}$$

The current due to the entire incident field

$$E_{2x}^{\text{inc}}(x') = -E_{2\rho}^{\text{inc}}(0) \cos \phi \; e^{-ik_2 x' \cos \phi}$$

is given by

$$I_x(x, \phi) = -\frac{E_{2\rho}^{\text{inc}}(0) \cos \phi}{Z_c} \left(e^{ik_L x} \int_0^x e^{-ik_L x'} e^{-ik_2 x' \cos \phi} \, dx' \right.$$

$$+ e^{-ik_L x} \int_x^h e^{ik_L x'} e^{-ik_2 x' \cos\phi}\, dx' \bigg). \tag{18.9.4}$$

This integrates into

$$\begin{aligned} I_x(x,\phi) = &-\frac{iE_{2\rho}^{\text{inc}}(0)\cos\phi}{Z_c(k_L^2 - k_2^2\cos^2\phi)} \\ &\times [2k_L e^{-ik_2 x\cos\phi} - (k_L - k_2\cos\phi)e^{ik_L x} \\ &- (k_L + k_2\cos\phi)e^{i[k_L(h-x) - k_2 h\cos\phi]}]. \end{aligned} \tag{18.9.5}$$

When $\phi = 0$ for maximum received signal,

$$\begin{aligned} I_x(x,0) = -\frac{iE_{2\rho}^{\text{inc}}(0)}{Z_c(k_L^2 - k_2^2)}\, &[2k_L e^{-ik_2 x} - (k_L - k_2)e^{ik_L x} \\ &- (k_L + k_2)e^{i[k_L(h-x) - k_2 h]}]. \end{aligned} \tag{18.9.6}$$

At $x = 0$, this reduces to

$$\begin{aligned} I_x(0,0) &= -\frac{iE_{2\rho}^{\text{inc}}(0)}{Z_c(k_L - k_2)}\,(1 - e^{i(k_L - k_2)h}) \\ &= -\frac{iE_{2\rho}^{\text{inc}}(0)\, e^{-i\tan^{-1}[\alpha_L/(\beta_L - k_2)]}}{Z_c[(\beta_L - k_2)^2 + \alpha_L^2]^{1/2}}\,(1 - e^{-\alpha_L h} e^{i(\beta_L - k_2)h}). \end{aligned} \tag{18.9.7a}$$

At $x = h$,

$$\begin{aligned} I_x(h,0) &= -\frac{iE_{2\rho}^{\text{inc}}(0)}{Z_c(k_L + k_2)}\,(e^{-ik_2 h} - e^{ik_L h}) \\ &= -\frac{iE_{2\rho}^{\text{inc}}(0)\, e^{-i\tan^{-1}[\alpha_L/(\beta_L + k_2)]}}{Z_c[(\beta_L + k_2)^2 + \alpha_L^2]^{1/2}}\,(e^{-ik_2 h} - e^{-\alpha_L h} e^{i\beta_L h}). \end{aligned} \tag{18.9.7b}$$

Equation (18.9.5) can be written in the following manner:

$$I_x(x,\phi) = \frac{E_{2\rho}^{\text{inc}}(0)\cos\phi}{Z_c}\,(Ae^{ia}e^{-ik_2 x\cos\phi} + Be^{ib}e^{ik_L x} + Ce^{ic}e^{ik_L(h-x)}), \tag{18.9.8}$$

where

$$\begin{aligned} A &= \left|\frac{2k_L}{k_L^2 - k_2^2\cos^2\phi}\right| \\ &= 2\left(\frac{\beta_L^2 + \alpha_L^2}{(\beta_L^2 - \alpha_L^2 - k_2^2\cos^2\phi)^2 + 4\alpha_L^2\beta_L^2}\right)^{1/2}, \end{aligned} \tag{18.9.9a}$$

$$a = \tan^{-1}\left(\frac{\alpha_L}{\beta_L}\right) - \tan^{-1}\left(\frac{2\alpha_L\beta_L}{\beta_L^2 - \alpha_L^2 - k_2^2\cos^2\phi}\right) - \frac{\pi}{2}, \tag{18.9.9b}$$

$$B = |k_L + k_2\cos\phi|^{-1} = [(\beta_L + k_2\cos\phi)^2 + \alpha_L^2]^{-1/2}, \tag{18.9.9c}$$

$$b = \frac{\pi}{2} - \tan^{-1}\left(\frac{\alpha_L}{\beta_L + k_2 \cos\phi}\right), \tag{18.9.9d}$$

$$C = |k_L - k_2 \cos\phi|^{-1} = [(\beta_L - k_2 \cos\phi)^2 + \alpha_L^2]^{-1/2}, \tag{18.9.9e}$$

$$c = \frac{\pi}{2} - k_2 h \cos\phi - \tan^{-1}\left(\frac{\alpha_L}{\beta_L - k_2 \cos\phi}\right). \tag{18.9.9f}$$

Also,

$$e^{ik_L x} = e^{-\alpha_L x}\, e^{i\beta_L x}; \qquad e^{ik_L(h-x)} = e^{-\alpha_L(h-x)}\, e^{i\beta_L(h-x)}. \tag{18.9.10}$$

When oriented for maximum received signal with $\phi = 0$, the above expressions are simplified since $\cos\phi = 1$. In (18.9.4)–(18.9.8), $E_{2\rho}^{\text{inc}}(0)$ is given by (18.9.2) with (18.9.1a) or (18.9.1b).

Formula (18.9.5) or (18.9.8) indicates that the current induced on the horizontal wire is the sum of three traveling waves with quite different amplitudes and phases. One of these, represented by $A\exp(ia)\exp(-ik_2x)$ in (18.9.8) with $\phi = 0$, travels with the real wave number k_2 of the incident electric field so that its amplitude remains constant as it progresses from $x = h$ to $x = 0$; the phase is linear and advances by 2π in each wavelength in air ($\lambda_2 = 2\pi/k_2$). A second wave, represented by $B\exp(ib)\exp(ik_Lx)$, travels with the wave number $k_L = \beta_L + i\alpha_L$ in the opposite direction, i.e., from $x = 0$ to $x = h$. Its phase constant β_L is greater than k_2 and the wave is exponentially attenuated with the factor $\exp(-\alpha_Lx)$. The third wave, given by $C\exp(ic)\exp[ik_L(h-x)]$, travels from $x = h$ to $x = 0$ like the first wave but with the wave number k_L instead of k_2. It is exponentially attenuated with the factor $\exp[-\alpha_L(h-x)]$ as $(h-x)$ increases from $x = h$ to $x = 0$. The resultant current is the superposition of the three current waves.

The three current waves and the total current are shown quantitatively in Fig. 18.9.2 for a wire with the optimum length $h = 6.84$ m at a height $d/\lambda_2 = 0.02$ over sea water when $f = 300$ MHz. The first wave has the large constant amplitude A and the linear phase $(a - k_2x)$; the second wave has a very small amplitude $B\exp(-\alpha_Lx)$ that decreases from $x = 0$ to $x = h$ and a linear phase $(b + \beta_Lx)$; the third wave begins with a large amplitude $C\exp[-\alpha_L(h-x)]$ at $x = h$ that decreases exponentially toward $x = 0$. The linear phase $[c + \beta_L(h-x)]$ begins at $x = h$ almost 180° out of phase with the phase $(a - k_2x)$ of the first wave; but since $\beta_L > k_2$, this phase difference diminishes to virtually zero at $x = 0$. The effect of the current associated with this wave is to almost cancel the current of the first wave near $x = h$ where both are nearly equal in amplitude and opposite in phase. The rapid decrease of the third wave as $(h-x)$ increases and the simultaneous decrease in the phase difference allow the resultant current $D(x,0)$ to grow rapidly from $x = h$ to $x = 0$. The second wave is generally unimportant because its amplitude is so small. However, very near $x = 0$ its contribution is to reduce the amplitude of the resultant current. If the

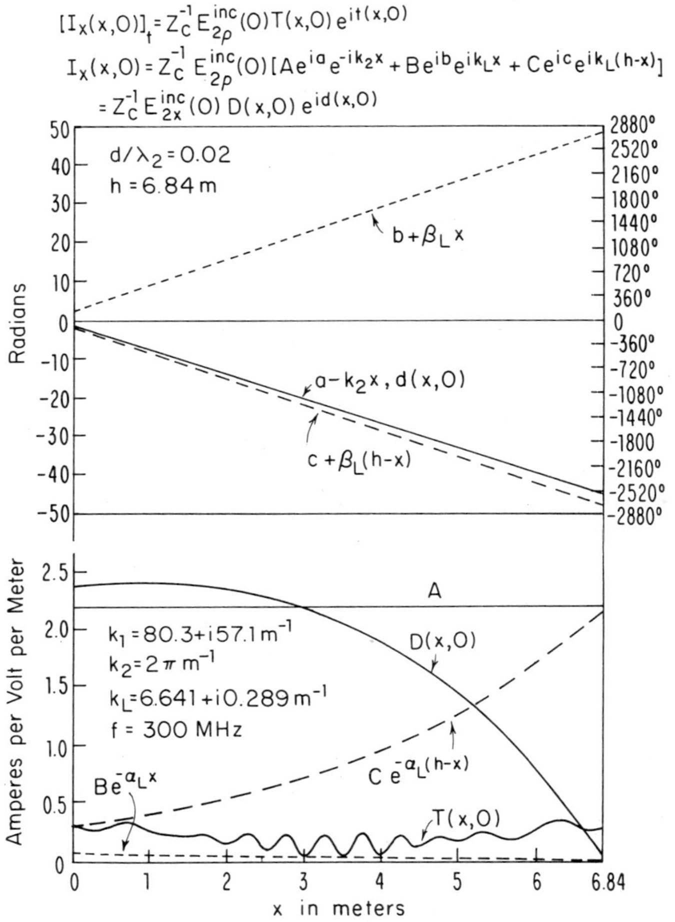

Figure 18.9.2. Components of current on Beverage receiving antenna with optimum length over sea water. $[I_x(x,0)]_t$ is due to voltages induced in terminations; $I_x(x,0)$ is due to voltages induced in main horizontal section.

length of the wire were increased beyond the optimum value, this decrease would become significant. As it is, the current in the load at $x = 0$ is at a maximum.

18.10 The Beverage and Horizontal-Wire Antennas for Reception—The Current Due to the Terminations, in the Load, and in the Entire Antenna

The vertical conductors at $x = 0$ and $x = h$ that connect the horizontal wire to ground in the Beverage antenna are excited by the large z-component E_{2z}^{inc} of the incident field. The electrically short conductor at $x = h$ maintains the voltage $E_{2z}^{\text{inc}}(h)d$ across the end of the eccentrically insulated horizontal wire. With E_{2z} positive down, this is in the positive direction for the current. Similarly, the vertical conductor at $x = 0$ maintains the voltage $E_{2z}^{\text{inc}}(0)d$, but, with E_{2z} down, this is in the negative direction. Since the horizontal wire is terminated in Z_c at each end, the currents along the wire are

$$[I_x(x)]_h = \frac{E_{2z}^{\text{inc}}(h)d}{Z_c} e^{ik_L(h-x)}, \tag{18.10.1a}$$

$$[I_x(x)]_0 = -\frac{E_{2z}^{\text{inc}}(0)d}{Z_c} e^{ik_L x}. \tag{18.10.1b}$$

However, $E_{2z}^{\text{inc}}(h) = E_{2z}^{\text{inc}}(0)\exp(-ik_2 h\cos\phi)$ so that

$$[I_x(x,\phi)]_h = \frac{E_{2z}^{\text{inc}}(0)d}{Z_c} e^{i[k_L(h-x)-k_2 h\cos\phi]} \tag{18.10.2}$$

and the sum of the two currents is

$$[I_x(x,\phi)]_t = -\frac{E_{2z}^{\text{inc}}(0)d}{Z_c}\left(e^{ik_L x} - e^{i[k_L(h-x)-k_2 h\cos\phi]}\right). \tag{18.10.3}$$

This is the superposition of two attenuated traveling waves moving in opposite directions. At $x = 0$,

$$[I_x(0,\phi)]_t = -\frac{E_{2z}^{\text{inc}}(0)d}{Z_c}\left(1 - e^{i(k_L - k_2\cos\phi)h}\right). \tag{18.10.4}$$

At large distances from the source, $F(\rho; k_1, k_2) \sim G(\rho; k_1, k_2)$ so that, from (18.9.1a,b),

$$E_{2z}^{\text{inc}}(0) \sim -\frac{k_1}{k_2} E_{2\rho}^{\text{inc}}(0). \tag{18.10.5}$$

Hence,

$$\begin{aligned}[I_x(x,\phi)]_t &= \frac{k_1}{k_2}\frac{E_{2\rho}^{\text{inc}}(0)d}{Z_c}\left(e^{ik_L x} - e^{i[k_L(h-x)-k_2 h\cos\phi]}\right)\\ &= \frac{E_{2\rho}^{\text{inc}}(0)}{Z_c} T(x,\phi)e^{it(x,\phi)},\end{aligned} \tag{18.10.6}$$

$$[I_x(0,\phi)]_t = \frac{k_1}{k_2}\frac{E_{2\rho}^{\text{inc}}(0)d}{Z_c}\left(1 - e^{i(k_L - k_2\cos\phi)h}\right). \tag{18.10.7}$$

The magnitude of this current with $\phi = 0$ is shown in Fig. 18.9.2. The oppositely directed traveling waves excited by the voltages maintained across the ends of the vertical conductors at $x = 0$ and $x = h$ set up a standing wave. However, the magnitude of the current due to the voltage induced on the vertical elements is quite small compared with that of the currents induced directly in the horizontal wire. Specifically in the load, with $\phi = 0$, the sum of the currents is

$$\begin{aligned} I_x(0,0) &= -\frac{E_{2\rho}^{\text{inc}}(0)}{Z_c k_2}\left(\frac{ik_2}{k_L - k_2} - k_1 d\right)\left(1 - e^{i(k_L - k_2)h}\right) \\ &= -\frac{E_{2\rho}^{\text{inc}}(0)}{Z_c k_2}[H(0) + V]\left(1 - e^{i(k_L - k_2)h}\right), \end{aligned} \tag{18.10.8}$$

where $H(0)$ and V are defined in (18.6.8) and (18.6.9). This current in the load could have been obtained directly from the analysis of the transmitting antenna by means of the reciprocal theorem.

The contribution to the current in the traveling-wave section between $x = 0$ and $x = h$ of a horizontal-wire receiving antenna due to the presence of the horizontal terminations could be evaluated without difficulty. However, it follows from the transmitting case that their contributions are small enough to be negligible so that it may be assumed that the entire current at any point is given by (18.9.5) or (18.9.8) and the current in the load by

$$I_x(0,0) = -\frac{iE_{2\rho}^{\text{inc}}(0)}{Z_c(k_L - k_2)}\left(1 - e^{i(k_L - k_2)h}\right). \tag{18.10.9}$$

18.11 Generalization of the Field of the Wave Antenna for Application to Over-the-Horizon Radar Using Ionospheric Reflections

The electromagnetic field summarized in Section 18.8 includes only the surface-wave terms evaluated subject to the condition $\rho^2 \gg z^2$ which is imposed in (18.2.2). It is entirely adequate for transmission between two wave antennas and for an over-the-horizon radar that makes use of the surface wave in both the transmitting and receiving modes to detect an aircraft or missile near the earth. Actual over-the-horizon radar makes use of reflections from the ionosphere to avoid the exponential attenuation due to radiation by currents following the curved surface of the earth. Transmission to the 300- to 500-km-high ionosphere requires knowledge of the electromagnetic field unrestricted by the condition $\rho^2 \gg z^2$.

Formulas for the complete electromagnetic field at any height z' generated by a horizontal electric dipole at any height d' in the air over the earth or sea are derived in Section 7.10. Of interest here are the components $E_{2\rho}(\rho_0, \phi'_0, z')$, $E_{2z'}(\rho_0, \phi'_0, z')$, and $B_{2\phi'}(\rho_0, \phi'_0, z')$. These are given by (7.10.75), (7.10.84), and (7.10.93). Since the wave antenna is at a very small height d', the direct- and image-field terms practically cancel, leaving the following expressions in the notation of this chapter:

$$[E_{2\rho}(\rho_0, \phi'_0, z')]_h \sim \frac{\omega\mu_0}{2\pi k_1} \cos\phi'_0\, e^{ik_2 r_0} \times \left\{ \left(\frac{z'}{r_0}\right)\left(\frac{ik_2}{r_0} - \frac{1}{r_0^2}\right) - \frac{k_2}{k_1}\left[\frac{ik_2}{r_0} - \frac{1}{r_0^2} - \frac{i}{k_2 r_0^3} - \frac{k_2^3}{k_1}\left(\frac{r_0}{\rho_0}\right)\left(\frac{\pi}{k_2 r_0}\right)^{1/2} e^{-iP_0}\mathcal{F}(P_0)\right]\right\}, \tag{18.11.1}$$

$$[E_{2z'}(\rho_0, \phi'_0, z')]_h \sim -\frac{\omega\mu_0}{2\pi k_1} \cos\phi'_0\, e^{ik_2 r_0} \times \left[\left(\frac{\rho_0}{r_0}\right)\left(\frac{ik_2}{r_0} - \frac{1}{r_0^2}\right) - \frac{k_2^3}{k_1}\left(\frac{\pi}{k_2 r_0}\right)^{1/2} e^{-iP_0}\mathcal{F}(P_0)\right], \tag{18.11.2}$$

$$[B_{2\phi'}(\rho_0, \phi'_0, z')]_h \sim \frac{\mu_0 k_2}{2\pi k_1} \cos\phi'_0\, e^{ik_2 r_0} \times \left[\frac{ik_2}{r_0} - \frac{1}{r_0^2} - \frac{i}{k_2 r_0^3} - \frac{k_2^3}{k_1}\left(\frac{r_0}{\rho_0}\right)\left(\frac{\pi}{k_2 r_0}\right)^{1/2} e^{-iP_0}\mathcal{F}(P_0)\right], \tag{18.11.3}$$

where $z' \geq 0$ and

$$r_0 = (\rho_0^2 + z'^2)^{1/2}; \qquad P_0 = S_0\left(\frac{S_0 + Z'}{R}\right)^2, \tag{18.11.4a}$$

$$S_0 = \frac{k_2^3 r_0}{2k_1^2}; \qquad R = \frac{k_2^3 \rho_0}{2k_1^2}; \qquad Z' = \frac{k_2^2 z'}{2k_1}. \tag{18.11.4b}$$

In practice, it is the far field that is of interest. This consists of the $1/r_0$ terms together with the $1/r_0^2$ terms near the surface of the earth or sea where the $1/r_0$ terms all cancel. The far-field formulas for the Fresnel-integral terms are obtained from the asymptotic formula for

$$T = \frac{k_2^3}{k_1}\left(\frac{\pi}{k_2 r_0}\right)^{1/2} e^{-iP_0}\mathcal{F}(P_0). \tag{18.11.5a}$$

This is derived in Chapter 4 where the coordinates $z' = -z$ and $\phi' = -\phi$ are used. It is

$$T^r = \frac{ik_2 \sin\Theta}{r_0[1 + (k_1/k_2)\cos\Theta]} + \frac{k_1^2 \sin^3\Theta}{k_2^2 r_0^2[1 + (k_1/k_2)\cos\Theta]^3}. \tag{18.11.5b}$$

The spherical coordinate Θ is measured from the positive z'-axis so that

$$\sin\Theta = \frac{\rho_0}{r_0}; \qquad \cos\Theta = \frac{z'}{r_0}. \tag{18.11.6}$$

The spherical coordinate $\Phi_0 = \phi_0'$.

In spherical coordinates, the far-field terms in (18.11.1)–(18.11.3) are

$$\begin{aligned}[E_{2\rho}^r(r_0, \Theta, \Phi_0)]_h = {}& \frac{\omega\mu_0}{2\pi k_1}\cos\Phi_0\, e^{ik_2 r_0} \\ &\times \left\{ \frac{ik_2}{r_0}\left[\cos\Theta - \frac{k_2}{k_1}\left(1 - \frac{1}{1 + (k_1/k_2)\cos\Theta}\right)\right]\right. \\ &\left. + \frac{k_1 \sin^2\Theta}{k_2 r_0^2[1 + (k_1/k_2)\cos\Theta]^3}\right\},\end{aligned} \tag{18.11.7}$$

$$\begin{aligned}[E_{2z'}^r(r_0, \Theta, \Phi_0)]_h = {}& -\frac{\omega\mu_0}{2\pi k_1}\cos\Phi_0\, e^{ik_2 r_0} \\ &\times \left[\frac{ik_2}{r_0}\left(\sin\Theta - \frac{\sin\Theta}{1 + (k_1/k_2)\cos\Theta}\right)\right. \\ &\left. - \frac{k_1^2 \sin^3\Theta}{k_2^2 r_0^2[1 + (k_1/k_2)\cos\Theta]^3}\right],\end{aligned} \tag{18.11.8}$$

$$\begin{aligned}[B_{2\phi'}^r(r_0, \Theta, \Phi_0)]_h = {}& \frac{\mu_0 k_2}{2\pi k_1}\cos\Phi_0\, e^{ik_2 r_0} \\ &\times \left[\frac{ik_2}{r_0}\left(1 - \frac{1}{1 + (k_1/k_2)\cos\Theta}\right)\right. \\ &\left. - \frac{k_1^2 \sin^2\Theta}{k_2^2 r_0^2[1 + (k_1/k_2)\cos\Theta]^3}\right].\end{aligned} \tag{18.11.9}$$

The spherical components are

$$[E_{2r}^r(r_0, \Theta, \Phi_0)]_h = [E_{2\rho}^r(r_0, \Theta, \Phi_0)]_h \sin\Theta + [E_{2z'}^r(r_0, \Theta, \Phi_0)]_h \cos\Theta, \tag{18.11.10}$$

$$[E_{2\Theta}^r(r_0, \Theta, \Phi_0)]_h = [E_{2\rho}^r(r_0, \Theta, \Phi_0)]_h \cos\Theta - [E_{2z'}^r(r_0, \Theta, \Phi_0)]_h \sin\Theta. \tag{18.11.11}$$

With (18.11.7) and (18.11.8), they are

$$
\begin{aligned}
[E^r_{2r}(r_0,\Theta,\Phi_0)]_h &= \frac{\omega\mu_0}{2\pi k_1}\cos\Phi_0 \\
&\quad\times\left[\frac{ik_2}{r_0}\left(\cos\Theta\sin\Theta - \frac{\cos\Theta\sin\Theta}{1+(k_1/k_2)\cos\Theta}\right.\right.\\
&\qquad\left. - \sin\Theta\cos\Theta + \frac{\sin\Theta\cos\Theta}{1+(k_1/k_2)\cos\Theta}\right)\\
&\qquad + \frac{k_1\sin^3\Theta}{k_2 r_0^2[1+(k_1/k_2)\cos\Theta]^3}\\
&\qquad\left. + \frac{k_1^2\sin^3\Theta\cos\Theta}{k_2^2 r_0^2[1+(k_1/k_2)\cos\Theta]^3}\right]e^{ik_2r_0}\\
&= \frac{\omega\mu_0}{2\pi k_2}\cos\Phi_0\,\frac{\sin^3\Theta}{r_0^2[1+(k_1/k_2)\cos\Theta]^2}\,e^{ik_2r_0},
\end{aligned}
\tag{18.11.12}
$$

$$
\begin{aligned}
[E^r_{2\Theta}(r_0,\Theta,\Phi_0)]_h &= \frac{\omega\mu_0}{2\pi k_1}\cos\Phi_0 \\
&\quad\times\left[\frac{ik_2}{r_0}\left(\cos^2\Theta - \frac{\cos^2\Theta}{1+(k_1/k_2)\cos\Theta}\right.\right.\\
&\qquad\left. + \sin^2\Theta - \frac{\sin^2\Theta}{1+(k_1/k_2)\cos\Theta}\right)\\
&\qquad\left. + \frac{k_1\sin^2\Theta[\cos\Theta-(k_1/k_2)\sin^2\Theta]}{k_2 r_0^2[1+(k_1/k_2)\cos\Theta]^3}\right]e^{ik_2r_0}\\
&= \frac{\omega\mu_0}{2\pi}\cos\Phi_0\\
&\quad\times\left[\frac{ik_2}{r_0}\left(\frac{\cos\Theta}{k_2+k_1\cos\Theta}\right)\right.\\
&\qquad\left. - \frac{k_1\sin^2\Theta\,[\sin^2\Theta-(k_2/k_1)\cos\Theta]}{k_2^2 r_0^2[1+(k_1/k_2)\cos\Theta]^3}\right]e^{ik_2r_0},
\end{aligned}
\tag{18.11.13}
$$

$$
\begin{aligned}
[B^r_{2\Phi}(r_0,\Theta,\Phi_0)]_h &= \frac{\mu_0}{2\pi}\cos\Phi_0\left[\frac{ik_2}{r_0}\left(\frac{k_2\cos\Theta}{k_2+k_1\cos\Theta}\right)\right.\\
&\qquad\left. - \frac{k_1\sin^2\Theta}{k_2 r_0^2[1+(k_1/k_2)\cos\Theta]^3}\right]e^{ik_2r_0}.
\end{aligned}
\tag{18.11.14}
$$

Note that $[E^r_{2r}(r_0,\Theta,\Phi_0)]_h$ has no $1/r_0$ terms, as required by the condition $\nabla\cdot\mathbf{E}=0$ in spherical coordinates.

The complete far field of the horizontal-wave or Beverage antenna as obtained from (18.8.1a–c) is

$$\begin{aligned} E^r_{2\Theta}(r_0, \Theta, \Phi_0) &= \frac{iI_x(0)(1 - e^{i(k_L - k_2 \sin\Theta \cos\Phi_0)h})}{k_L - k_2 \sin\Theta \cos\Phi_0} \cdot \frac{\omega\mu_0}{2\pi} \cos\Phi_0 \\ &\times \left[\frac{ik_2}{r_0} \left(\frac{\cos\Theta}{k_2 + k_1 \cos\Theta} \right) \right. \\ &\left. - \frac{k_1 \sin^2\Theta \, [\sin^2\Theta - (k_2/k_1)\cos\Theta]}{k_2^2 r_0^2 [1 + (k_1/k_2)\cos\Theta]^3} \right] e^{ik_2 r_0}, \end{aligned} \tag{18.11.15}$$

$$\begin{aligned} B^r_{2\Phi}(r_0, \Theta, \Phi_0) &= \frac{iI_x(0)(1 - e^{i(k_L - k_2 \sin\Theta \cos\Phi_0)h})}{k_L - k_2 \sin\Theta \cos\Phi_0} \cdot \frac{\mu_0}{2\pi} \cos\Phi_0 \\ &\times \left[\frac{ik_2}{r_0} \left(\frac{k_2 \cos\Theta}{k_2 + k_1 \cos\Theta} \right) \right. \\ &\left. - \frac{k_1 \sin^2\Theta}{k_2 r_0^2 [1 + (k_1/k_2)\cos\Theta]^3} \right] e^{ik_2 r_0}. \end{aligned} \tag{18.11.16}$$

The condition for a maximum effective length is with h given by h_m in (18.6.6).

It is of interest to compare the far-field formulas for the horizontal electric dipole close to the surface of the earth with the corresponding formulas from Section 4.2 for the vertical electric dipole on the surface of the earth. It is seen that the vertical-dipole formulas for $E^r_{2\Theta}(r_0, \Theta)$ and $B^r_{2\Phi}(r_0, \Theta)$ in (4.2.63) and (4.2.61) are like those for the horizontal dipole in (18.11.13) and (18.11.14) multiplied by the factor $(k_1/k_2)\sin\Theta$. Evidently, the upward radiation for the vertical dipole is much smaller because of the factor $\sin\Theta$, but the overall field with the same electric moment is greater by the factor k_1/k_2. Clearly, the vertical dipole is superior to the horizontal one for exciting a surface wave. However, the large directional gain that is achieved with the long traveling-wave arrangement of horizontal dipoles in the wave antenna cannot be obtained with vertical dipoles. This can easily be made greater than k_1/k_2.

Appendix A
Tabulation of Integrals

The general formulas for the components of the electromagnetic fields of vertical and horizontal electric dipoles near a boundary include a number of integrals that can be reduced to fairly simple forms. These are listed in a systematic manner together with their integrated values. The actual evaluations are carried out in Appendices B and C. Some of the integrals are required at nonzero values of z, others only at $z = 0$. These are listed in the form $I(\rho, z)$ or $I(\rho, 0)$ with $r = (\rho^2 + z^2)^{1/2}$ and, with $z^2 \ll \rho^2$, $r \sim \rho$ in amplitudes. Terms of higher order than $1/\rho^3$ are retained here, but are omitted in formulas throughout the book. In the integrals that are evaluated exactly, the approximation $r \sim \rho$ in amplitudes is not made.

$$I_1(\rho, z) = \int_0^\infty (k^2 - \lambda^2)^{-1/2}[J_0(\lambda\rho) + J_2(\lambda\rho)]e^{iz\sqrt{k^2-\lambda^2}}\, d\lambda$$

$$= -i\int_0^\infty (\lambda^2 - k^2)^{-1/2}[J_0(\lambda\rho) + J_2(\lambda\rho)]e^{-z\sqrt{\lambda^2-k^2}}\, d\lambda$$

$$= e^{ikz}\left(\frac{2}{k\rho} - \frac{iz}{k^2\rho^3} + \frac{1}{k^3\rho^3}\right) + ie^{ikr}\left(\frac{2i}{k^2\rho^2} + \frac{1}{k^3\rho^3}\right), \tag{A1a}$$

$$I_1(\rho, 0) = \frac{2}{k\rho} + \frac{1}{k^3\rho^3} + ie^{ik\rho}\left(\frac{2i}{k^2\rho^2} + \frac{1}{k^3\rho^3}\right), \tag{A1b}$$

$$I_2(\rho, z) = \int_0^\infty (k^2 - \lambda^2)^{-1/2}[J_0(\lambda\rho) - J_2(\lambda\rho)]e^{iz\sqrt{k^2-\lambda^2}}\, d\lambda$$

$$= -i\int_0^\infty (\lambda^2 - k^2)^{-1/2}[J_0(\lambda\rho) - J_2(\lambda\rho)]e^{-z\sqrt{\lambda^2-k^2}}\, d\lambda$$

$$= e^{ikz}\left(\frac{2iz}{k^2\rho^3} - \frac{2}{k^3\rho^3}\right) - ie^{ikr}\left(\frac{2}{k\rho} + \frac{i}{k^2\rho^2} - \frac{1}{4k^3\rho^3}\right), \tag{A2a}$$

$$I_2(\rho, 0) = -\frac{2}{k^3\rho^3} - ie^{ik\rho}\left(\frac{2}{k\rho} + \frac{i}{k^2\rho^2} - \frac{1}{4k^3\rho^3}\right), \tag{A2b}$$

$$I_3(\rho, z) = \int_0^\infty (k^2 - \lambda^2)^{-1/2}[J_0(\lambda\rho) + J_2(\lambda\rho)]e^{iz\sqrt{k^2-\lambda^2}}\lambda\, d\lambda$$

$$= -i\int_0^\infty (\lambda^2 - k^2)^{-1/2}[J_0(\lambda\rho) + J_2(\lambda\rho)]e^{-z\sqrt{\lambda^2-k^2}}\lambda\, d\lambda$$

$$= \frac{2}{\rho}\int_0^\infty (k^2 - \lambda^2)^{-1/2}J_1(\lambda\rho)e^{iz\sqrt{k^2-\lambda^2}}\, d\lambda$$

$$= \frac{2}{k\rho^2}\,(e^{ikz} - e^{ikr}), \tag{A3a}$$

$$I_3(\rho, 0) = \frac{2}{k\rho^2}\,(1 - e^{ik\rho}), \tag{A3b}$$

$$I_4(\rho, z) = \int_0^\infty (k^2 - \lambda^2)^{-1/2}[J_0(\lambda\rho) - J_2(\lambda\rho)]e^{iz\sqrt{k^2-\lambda^2}}\lambda\, d\lambda$$

$$= -i\int_0^\infty (\lambda^2 - k^2)^{-1/2}[J_0(\lambda\rho) - J_2(\lambda\rho)]e^{-z\sqrt{\lambda^2-k^2}}\lambda\, d\lambda$$

$$= -\frac{2}{k}\left[\frac{e^{ikz}}{\rho^2} + \left(\frac{ik}{r} - \frac{1}{\rho^2}\right)e^{ikr}\right], \tag{A4a}$$

$$I_4(\rho, 0) = -\frac{2}{k}\left[\frac{1}{\rho^2} + \left(\frac{ik}{\rho} - \frac{1}{\rho^2}\right)e^{ik\rho}\right], \tag{A4b}$$

$$I_5(\rho, z) = \int_0^\infty (k^2 - \lambda^2)^{-1/2}[J_0(\lambda\rho) + J_2(\lambda\rho)]e^{iz\sqrt{k^2-\lambda^2}}\lambda^2\, d\lambda$$

$$= -i\int_0^\infty (\lambda^2 - k^2)^{-1/2}[J_0(\lambda\rho) + J_2(\lambda\rho)]e^{-z\sqrt{\lambda^2-k^2}}\lambda^2\, d\lambda$$

$$= \frac{2}{\rho}\int_0^\infty (k^2 - \lambda^2)^{-1/2}J_1(\lambda\rho)e^{iz\sqrt{k^2-\lambda^2}}\lambda\, d\lambda$$

$$= 2\left[\frac{e^{ikz}}{k\rho^3} - \left(\frac{1}{\rho^2} + \frac{i}{2k\rho^3} + \frac{3}{8k^2\rho^4}\right)e^{ikr}\right], \tag{A5a}$$

$$I_5(\rho, 0) = 2\left[\frac{1}{k\rho^3} - \left(\frac{1}{\rho^2} + \frac{i}{2k\rho^3} + \frac{3}{8k^2\rho^4}\right)e^{ik\rho}\right], \tag{A5b}$$

$$I_6(\rho, z) = \int_0^\infty (k^2 - \lambda^2)^{-1/2}[J_0(\lambda\rho) - J_2(\lambda\rho)]e^{iz\sqrt{k^2-\lambda^2}}\lambda^2\, d\lambda$$

$$= -i\int_0^\infty (\lambda^2 - k^2)^{-1/2}[J_0(\lambda\rho) - J_2(\lambda\rho)]e^{-z\sqrt{\lambda^2-k^2}}\lambda^2\, d\lambda$$

$$= -\frac{4}{k\rho^3}\,e^{ikz} - 2\left(\frac{ik}{\rho} - \frac{3}{2\rho^2} - \frac{5i}{8k\rho^3} + \frac{1}{4k^2\rho^4}\right)e^{ikr}, \tag{A6a}$$

$$I_6(\rho,0) = -\frac{4}{k\rho^3} - 2\left(\frac{ik}{\rho} - \frac{3}{2\rho^2} - \frac{5i}{8k\rho^3} + \frac{1}{4k^2\rho^4}\right)e^{ik\rho}, \tag{A6b}$$

$$\begin{aligned}
I_7(\rho,z) &= \int_0^\infty (k^2-\lambda^2)^{-1/2}[J_0(\lambda\rho)+J_2(\lambda\rho)]e^{iz\sqrt{k^2-\lambda^2}}\lambda^3\,d\lambda \\
&= -i\int_0^\infty (\lambda^2-k^2)^{-1/2}[J_0(\lambda\rho)+J_2(\lambda\rho)]e^{-z\sqrt{\lambda^2-k^2}}\lambda^3\,d\lambda \\
&= \frac{2}{\rho}\int_0^\infty (k^2-\lambda^2)^{-1/2}J_1(\lambda\rho)e^{iz\sqrt{k^2-\lambda^2}}\lambda^2\,d\lambda \\
&= -2k\left(\frac{1}{r^2} + \frac{i}{kr^3}\right)e^{ikr},
\end{aligned} \tag{A7a}$$

$$I_7(\rho,0) = -2k\left(\frac{1}{\rho^2} + \frac{i}{k\rho^3}\right)e^{ik\rho}, \tag{A7b}$$

$$\begin{aligned}
I_8(\rho,z) &= \int_0^\infty (k^2-\lambda^2)^{-1/2}[J_0(\lambda\rho)-J_2(\lambda\rho)]e^{iz\sqrt{k^2-\lambda^2}}\lambda^3\,d\lambda \\
&= -i\int_0^\infty (\lambda^2-k^2)^{-1/2}[J_0(\lambda\rho)-J_2(\lambda\rho)]e^{-z\sqrt{\lambda^2-k^2}}\lambda^3\,d\lambda \\
&= -2k\left[\frac{ik}{r} - \frac{2}{r^2} - \frac{2i}{kr^3} - \frac{z^2}{r^2}\left(\frac{ik}{r} - \frac{3}{r^2} - \frac{3i}{kr^3}\right)\right]e^{ikr},
\end{aligned} \tag{A8a}$$

$$I_8(\rho,0) = -2k\left(\frac{ik}{\rho} - \frac{2}{\rho^2} - \frac{2i}{k\rho^3}\right)e^{ik\rho}, \tag{A8b}$$

$$\begin{aligned}
I_9(\rho,z) &= \int_0^\infty (k^2-\lambda^2)^{1/2}[J_0(\lambda\rho)+J_2(\lambda\rho)]e^{iz\sqrt{k^2-\lambda^2}}\,d\lambda \\
&= i\int_0^\infty (\lambda^2-k^2)^{1/2}[J_0(\lambda\rho)+J_2(\lambda\rho)]e^{-z\sqrt{\lambda^2-k^2}}\,d\lambda \\
&= 2\left[\left(\frac{k}{\rho} - \frac{1}{2k\rho^3} - \frac{iz}{2\rho^3}\right)e^{ikz} + \frac{ie^{ikr}}{k\rho^3}\right],
\end{aligned} \tag{A9a}$$

$$I_9(\rho,0) = 2\left(\frac{k}{\rho} - \frac{1}{2k\rho^3} + \frac{ie^{ik\rho}}{k\rho^3}\right), \tag{A9b}$$

$$\begin{aligned}
I_{10}(\rho,z) &= \int_0^\infty (k^2-\lambda^2)^{1/2}[J_0(\lambda\rho)-J_2(\lambda\rho)]e^{iz\sqrt{k^2-\lambda^2}}\,d\lambda \\
&= i\int_0^\infty (\lambda^2-k^2)^{1/2}[J_0(\lambda\rho)-J_2(\lambda\rho)]e^{-z\sqrt{\lambda^2-k^2}}\,d\lambda \\
&= 2\left[\left(\frac{iz}{\rho^3} + \frac{1}{k\rho^3}\right)e^{ikz} - \left(\frac{1}{\rho^2} + \frac{i}{2k\rho^3} + \frac{3}{8k^2\rho^4}\right)e^{ikr}\right],
\end{aligned} \tag{A10a}$$

$$I_{10}(\rho,0) = 2\left[\frac{1}{k\rho^3} - \left(\frac{1}{\rho^2} + \frac{i}{2k\rho^3} + \frac{3}{8k^2\rho^4}\right)e^{ik\rho}\right], \tag{A10b}$$

$$I_{11}(\rho,z) = \int_0^\infty (k^2-\lambda^2)^{1/2}[J_0(\lambda\rho)+J_2(\lambda\rho)]e^{iz\sqrt{k^2-\lambda^2}}\lambda\,d\lambda$$

$$= i\int_0^\infty (\lambda^2 - k^2)^{1/2}[J_0(\lambda\rho) + J_2(\lambda\rho)]e^{-z\sqrt{\lambda^2-k^2}}\lambda\, d\lambda$$

$$= \frac{2}{\rho}\int_0^\infty (k^2 - \lambda^2)^{1/2}J_1(\lambda\rho)e^{iz\sqrt{k^2-\lambda^2}}\, d\lambda$$

$$= 2k\left[\frac{e^{ikz}}{\rho^2} + \frac{ie^{ikr}}{kr^3}\left(1 + ikr\,\frac{z^2}{\rho^2}\right)\right], \tag{A11a}$$

$$I_{11}(\rho,0) = 2k\left(\frac{1}{\rho^2} + \frac{ie^{ik\rho}}{k\rho^3}\right), \tag{A11b}$$

$$I_{12}(\rho,z) = \int_0^\infty (k^2 - \lambda^2)^{1/2}[J_0(\lambda\rho) - J_2(\lambda\rho)]e^{iz\sqrt{k^2-\lambda^2}}\lambda\, d\lambda$$

$$= i\int_0^\infty (\lambda^2 - k^2)^{1/2}[J_0(\lambda\rho) - J_2(\lambda\rho)]e^{-z\sqrt{\lambda^2-k^2}}\lambda\, d\lambda$$

$$= -2k\left\{\frac{e^{ikz}}{\rho^2} + \left[\frac{1}{r^2} + \frac{2i}{kr^3} + \frac{z^2}{r^2}\left(\frac{ik}{r} - \frac{3}{r^2} - \frac{3i}{kr^3} - \frac{1}{\rho^2}\right)\right]e^{ikr}\right\}, \tag{A12a}$$

$$I_{12}(\rho,0) = -2k\left[\frac{1}{\rho^2} + \left(\frac{1}{\rho^2} + \frac{2i}{k\rho^3}\right)e^{ik\rho}\right], \tag{A12b}$$

$$I_{13}(\rho,z) = \int_0^\infty (k^2 - \lambda^2)^{1/2}[J_0(\lambda\rho) + J_2(\lambda\rho)]e^{iz\sqrt{k^2-\lambda^2}}\lambda^2\, d\lambda$$

$$= i\int_0^\infty (\lambda^2 - k^2)^{1/2}[J_0(\lambda\rho) + J_2(\lambda\rho)]e^{-z\sqrt{\lambda^2-k^2}}\lambda^2\, d\lambda$$

$$= \frac{2}{\rho}\int_0^\infty (k^2 - \lambda^2)^{1/2}J_1(\lambda\rho)e^{iz\sqrt{k^2-\lambda^2}}\lambda\, d\lambda$$

$$= 2k\left[\frac{e^{ikz}}{\rho^3} + ie^{ikr}\left(\frac{1}{\rho^3} + \frac{3i}{2k\rho^4}\right)\right], \tag{A13a}$$

$$I_{13}(\rho,0) = 2k\left[\frac{1}{\rho^3} + ie^{ik\rho}\left(\frac{1}{\rho^3} + \frac{3i}{2k\rho^4}\right)\right], \tag{A13b}$$

$$I_{14}(\rho,z) = \int_0^\infty (k^2 - \lambda^2)^{1/2}[J_0(\lambda\rho) - J_2(\lambda\rho)]e^{iz\sqrt{k^2-\lambda^2}}\lambda^2\, d\lambda$$

$$= i\int_0^\infty (\lambda^2 - k^2)^{1/2}[J_0(\lambda\rho) - J_2(\lambda\rho)]e^{-z\sqrt{\lambda^2-k^2}}\lambda^2\, d\lambda$$

$$= -\frac{4k}{\rho^3}\,e^{ikz} - 2k^2\left(\frac{1}{\rho^2} + \frac{7i}{2k\rho^3}\right)e^{ikr}, \tag{A14a}$$

$$I_{14}(\rho,0) = -\frac{4k}{\rho^3} - 2k^2\left(\frac{1}{\rho^2} + \frac{7i}{2k\rho^3}\right)e^{ik\rho}, \tag{A14b}$$

$$I_{15}(\rho,z) = \int_0^\infty (k^2 - \lambda^2)^{1/2}[J_0(\lambda\rho) + J_2(\lambda\rho)]e^{iz\sqrt{k^2-\lambda^2}}\lambda^3\, d\lambda$$

$$= i\int_0^\infty (\lambda^2-k^2)^{1/2}[J_0(\lambda\rho)+J_2(\lambda\rho)]e^{-z\sqrt{\lambda^2-k^2}}\lambda^3\,d\lambda$$

$$= \frac{2}{\rho}\int_0^\infty (k^2-\lambda^2)^{1/2}J_1(\lambda\rho)e^{iz\sqrt{k^2-\lambda^2}}\lambda^2\,d\lambda$$

$$= 2\left[\frac{ik^2}{r^3}-\frac{3k}{r^4}-\frac{3i}{r^5}-\frac{z^2}{r^2}\left(\frac{k^3}{r^2}+\frac{6ik^2}{r^3}-\frac{15k}{r^4}-\frac{15i}{r^5}\right)\right]e^{ikr}, \tag{A15a}$$

$$I_{15}(\rho,0) = 2\left(\frac{ik^2}{\rho^3}-\frac{3k}{\rho^4}-\frac{3i}{\rho^5}\right)e^{ik\rho}, \tag{A15b}$$

$$I_{16}(\rho,z) = \int_0^\infty [J_0(\lambda\rho)+J_2(\lambda\rho)]e^{iz\sqrt{k^2-\lambda^2}}\,d\lambda$$

$$= 2\left[e^{ikz}\left(\frac{1}{\rho}-\frac{iz}{2k\rho^3}\right)-\frac{z}{\rho}\frac{e^{ikr}}{k\rho^2}\right], \tag{A16a}$$

$$I_{16}(\rho,0) = \frac{2}{\rho}, \tag{A16b}$$

$$I_{17}(\rho,z) = \int_0^\infty [J_0(\lambda\rho)-J_2(\lambda\rho)]e^{iz\sqrt{k^2-\lambda^2}}\,d\lambda$$

$$= 2i\left(\frac{z}{\rho}\right)\left[\frac{e^{ikz}}{k\rho^2}-\left(\frac{1}{\rho}+\frac{i}{2k\rho^2}+\frac{3}{8k^2\rho^3}\right)e^{ikr}\right], \tag{A17a}$$

$$I_{17}(\rho,0) = 0, \tag{A17b}$$

$$I_{18}(\rho,z) = \int_0^\infty [J_0(\lambda\rho)+J_2(\lambda\rho)]e^{iz\sqrt{k^2-\lambda^2}}\lambda\,d\lambda$$

$$= \frac{2}{\rho}\int_0^\infty J_1(\lambda\rho)e^{iz\sqrt{k^2-\lambda^2}}\,d\lambda$$

$$= 2\left(\frac{e^{ikz}}{\rho^2}-\frac{z}{r}\frac{e^{ikr}}{\rho^2}\right), \tag{A18a}$$

$$I_{18}(\rho,0) = \frac{2}{\rho^2}, \tag{A18b}$$

$$I_{19}(\rho,z) = \int_0^\infty [J_0(\lambda\rho)-J_2(\lambda\rho)]e^{iz\sqrt{k^2-\lambda^2}}\lambda\,d\lambda$$

$$= -2\left[\frac{e^{ikz}}{\rho^2}+\left(\frac{z}{r}\right)\left(\frac{ik}{r}-\frac{1}{r^2}-\frac{1}{\rho^2}\right)e^{ikr}\right], \tag{A19a}$$

$$I_{19}(\rho,0) = -\frac{2}{\rho^2}, \tag{A19b}$$

$$I_{20}(\rho,z) = \int_0^\infty [J_0(\lambda\rho)+J_2(\lambda\rho)]e^{iz\sqrt{k^2-\lambda^2}}\lambda^2\,d\lambda$$

$$= \frac{2}{\rho}\int_0^\infty J_1(\lambda\rho)e^{iz\sqrt{k^2-\lambda^2}}\lambda\,d\lambda$$

$$= \frac{2}{\rho^3}\, e^{ikz} - 2\left(\frac{z}{\rho}\right)\left(\frac{k}{\rho^2} + \frac{3i}{2\rho^3}\right)e^{ikr}, \tag{A20a}$$

$$I_{20}(\rho, 0) = \frac{2}{\rho^3}, \tag{A20b}$$

$$I_{21}(\rho, z) = \int_0^\infty [J_0(\lambda\rho) - J_2(\lambda\rho)]e^{iz\sqrt{k^2-\lambda^2}}\lambda^2\, d\lambda$$

$$= -\frac{4}{\rho^3}\, e^{ikz} - 2k\left(\frac{z}{r}\right)\left(\frac{ik}{\rho} - \frac{3}{2\rho^2} - \frac{5i}{8k\rho^3} + \frac{1}{4k^2\rho^4}\right)e^{ikr}, \tag{A21a}$$

$$I_{21}(\rho, 0) = -\frac{4}{\rho^3}, \tag{A21b}$$

$$I_{22}(\rho, z) = \int_0^\infty [J_0(\lambda\rho) + J_2(\lambda\rho)]e^{iz\sqrt{k^2-\lambda^2}}\lambda^3\, d\lambda$$

$$= \frac{2}{\rho}\int_0^\infty J_1(\lambda\rho)e^{iz\sqrt{k^2-\lambda^2}}\lambda^2\, d\lambda$$

$$= -2k\left(\frac{z}{r}\right)\left(\frac{k}{r^2} + \frac{3i}{r^3} - \frac{3}{kr^4}\right)e^{ikr}, \tag{A22a}$$

$$I_{22}(\rho, 0) = 0, \tag{A22b}$$

$$I_{23}(\rho, z) = \int_0^\infty [J_0(\lambda\rho) - J_2(\lambda\rho)]e^{iz\sqrt{k^2-\lambda^2}}\lambda^3\, d\lambda$$

$$= 2k\left(\frac{z}{r}\right)\left(-\frac{ik^2}{r} + \frac{5k}{r^2} + \frac{12i}{r^3} - \frac{12}{kr^4}\right)e^{ikr}, \tag{A23a}$$

$$I_{23}(\rho, 0) = 0, \tag{A23b}$$

$$I_{24}(\rho, z) = \int_0^\infty [J_0(\lambda\rho) + J_2(\lambda\rho)]e^{iz\sqrt{k^2-\lambda^2}}\lambda^4\, d\lambda$$

$$= \frac{2}{\rho}\int_0^\infty J_1(\lambda\rho)e^{iz\sqrt{k^2-\lambda^2}}\lambda^3\, d\lambda$$

$$= -2k^2\left(\frac{z}{r}\right)\left(\frac{k}{\rho^2} + \frac{3i}{2\rho^3} - \frac{9}{8k\rho^4}\right)e^{ikr}, \tag{A24a}$$

$$I_{24}(\rho, 0) = 0, \tag{A24b}$$

$$I_{25}(\rho, 0) = \int_0^\infty J_0(\lambda\rho)\, d\lambda = \int_0^\infty J_1(\lambda\rho)\, d\lambda = \int_0^\infty J_2(\lambda\rho)\, d\lambda = \frac{1}{\rho}, \tag{A25}$$

$$I_{26}(\rho, z) = \int_0^\infty J_0(\lambda\rho)e^{iz\sqrt{k^2-\lambda^2}}\lambda\, d\lambda$$

$$= -\left(\frac{z}{r}\right)\left(\frac{ik}{r} - \frac{1}{r^2}\right)e^{ikr}, \tag{A26a}$$

$$I_{26}(\rho, 0) = 0, \tag{A26b}$$

$$I_{27}(\rho, z) = \int_0^\infty J_2(\lambda\rho)e^{iz\sqrt{k^2-\lambda^2}}\lambda\, d\lambda$$

$$= \frac{2}{\rho^2} e^{ikz} + \left(\frac{z}{r}\right)\left(\frac{ik}{r} - \frac{1}{r^2} - \frac{2}{\rho^2}\right) e^{ikr}, \tag{A27a}$$

$$I_{27}(\rho, 0) = \frac{2}{\rho^2}, \tag{A27b}$$

$$I_{28}(\rho, 0) = \int_0^\infty J_0(\lambda\rho)\lambda^2 \, d\lambda = -\frac{1}{\rho^3}, \tag{A28}$$

$$I_{29}(\rho, 0) = \int_0^\infty J_2(\lambda\rho)\lambda^2 \, d\lambda = \frac{3}{\rho^3}. \tag{A29}$$

Appendix B
Evaluation of Integrals with Radicals in the Integrand

The evaluation of the several integrals is often accomplished with one of the following first steps:

$$J_0(\lambda\rho) + J_2(\lambda\rho) = \frac{2}{\lambda\rho} J_1(\lambda\rho), \tag{B1}$$

$$J_0(\lambda\rho) - J_2(\lambda\rho) = \frac{2}{\rho} \frac{\partial}{\partial\lambda} J_1(\lambda\rho). \tag{B2}$$

The following integrals evaluated or listed in standard references are useful:

• Gradshteyn and Ryzhik [1980, p. 690, formula (6.571-1)],

$$\int_0^\infty [(\lambda^2 + \beta^2)^{1/2} \pm \lambda]^\mu J_\nu(\lambda\rho)(\lambda^2 + \beta^2)^{-1/2}\, d\lambda$$
$$= \beta^\mu I_m(\beta\rho/2) K_n(\beta\rho/2), \tag{B3}$$

where $m = (\nu \mp \mu)/2$, $n = (\nu \pm \mu)/2$, and $\operatorname{Re}\beta > 0$, $\rho > 0$, $\operatorname{Re}\nu > -1$, $\operatorname{Re}\mu < \frac{3}{2}$. [Note that formula (B3) gives the correct results when $\nu = 0$, $\mu = 2$, and $\nu = 1$, $\mu = 3$.]

• Watson [1958, p. 435, formulas (3) and (4)],

$$\int_0^\infty \frac{J_\nu(\lambda\rho)\, d\lambda}{(\lambda^2 + \beta^2)^{1/2}} = I_{\nu/2}(\beta\rho/2) K_{\nu/2}(\beta\rho/2); \qquad \operatorname{Re}\nu > -1, \tag{B4}$$

$$\int_0^\infty \frac{J_\nu(\lambda\rho)\lambda^{-\nu}\, d\lambda}{(\lambda^2 + \beta^2)^{\nu+1/2}}$$

$$= \frac{(2\rho/\beta^2)^\nu \Gamma(\nu+1)}{\Gamma(2\nu+1)} I_\nu(\beta\rho/2) K_\nu(\beta\rho/2); \qquad \operatorname{Re}\nu > -\tfrac{1}{2}, \tag{B5}$$

and p. 434, formula (2),

$$\int_0^\infty \frac{\lambda^{\nu+1} J_\nu(\lambda\rho)\, d\lambda}{(\lambda^2+\beta^2)^{\mu+1}} = \frac{\rho^\mu \beta^{\nu-\mu}}{2^\mu \Gamma(\mu+1)} K_{\nu-\mu}(\beta\rho);$$
$$-1 < \operatorname{Re}\nu < 2\operatorname{Re}\mu + \tfrac{3}{2}. \tag{B6}$$

• Gradshteyn and Ryzhik [1980, p. 719, formula (6.637.1)],

$$\int_0^\infty (\lambda^2+\beta^2)^{-1/2} e^{-z\sqrt{\lambda^2+\beta^2}} J_\nu(\lambda\rho)\, d\lambda$$
$$= I_{\nu/2}[\tfrac{1}{2}\beta(r-z)] K_{\nu/2}[\tfrac{1}{2}\beta(r+z)], \tag{B7}$$

where $r = (\rho^2+z^2)^{1/2}$, and $\operatorname{Re} z > 0$, $\operatorname{Re}\beta > 0$, $\rho > 0$, $\operatorname{Re}\nu > -1$.

• Watson [1958, p. 416, formula (4)] or Bateman [1953, Vol. 2, p. 95, formula (5)] with $b = \rho$, $t = \lambda$, $y = k$, $a = z$, ρ and z positive, and $k > 0$,

$$\int_0^\infty J_0(\lambda\rho) e^{-z\sqrt{\lambda^2-k^2}} (\lambda^2-k^2)^{-1/2} \lambda\, d\lambda$$
$$= e^{\mp ik\sqrt{\rho^2+z^2}} (\rho^2+z^2)^{-1/2} = \frac{e^{\mp ikr}}{r}; \qquad r = (\rho^2+z^2)^{1/2},$$

where $\arg(\lambda^2-k^2)^{1/2} = \pm\pi/2$ when $k > \lambda$ so that $(\lambda^2-k^2)^{1/2} = \pm i(k^2-\lambda^2)^{1/2}$. The singularity $\lambda = k$ is avoided by an indentation upward for the upper signs and downward for the lower signs. It follows that

$$\int_0^\infty J_0(\lambda\rho) e^{\mp iz\sqrt{k^2-\lambda^2}} \frac{\lambda\, d\lambda}{\pm i(k^2-\lambda^2)^{1/2}} = \frac{e^{\mp ikr}}{r}.$$

With $k = \beta + i\alpha$, the form required in all integrals in this book is with $\exp(i\gamma z)$, $\gamma = (k^2-\lambda^2)^{1/2}$, and $\exp(ikr)$. That is,

$$\int_0^\infty J_0(\lambda\rho) e^{iz\sqrt{k^2-\lambda^2}} \frac{\lambda\, d\lambda}{(k^2-\lambda^2)^{1/2}} = -\frac{ie^{ikr}}{r}. \tag{B8}$$

The integral $I_1(\rho, z)$ defined in (A1a) is

$$I_1(\rho,z) = \int_0^\infty (k^2-\lambda^2)^{-1/2} [J_0(\lambda\rho) + J_2(\lambda\rho)] e^{iz\sqrt{k^2-\lambda^2}}\, d\lambda. \tag{B9}$$

It can be evaluated with (B7) with $\nu = 0$ and $\nu = 2$. Thus, with $k = i\beta$,

$$I_1(\rho,z) = -i\int_0^\infty (\lambda^2+\beta^2)^{-1/2} [J_0(\lambda\rho) + J_2(\lambda\rho)] e^{-z\sqrt{\lambda^2+\beta^2}}\, d\lambda$$
$$= -i\{I_0[\tfrac{1}{2}\beta(r-z)] K_0[\tfrac{1}{2}\beta(r+z)] + I_1[\tfrac{1}{2}\beta(r-z)] K_1[\tfrac{1}{2}\beta(r+z)]\}, \tag{B10}$$

where $r = (\rho^2 + z^2)^{1/2}$. With $y_1 = \frac{1}{2}\beta(r - z)$ and $y_2 = \frac{1}{2}\beta(r + z)$, this becomes

$$I_1(\rho, z) = -i[I_0(y_1)K_0(y_2) + I_1(y_1)K_1(y_2)]. \tag{B11}$$

With the large-argument approximations of the Bessel functions, viz.,

$$I_0(z) \sim \frac{e^z}{\sqrt{2\pi z}}\left(1 + \frac{1}{8z} + \frac{9}{128z^2}\right) - \frac{ie^{-z}}{\sqrt{2\pi z}}\left(1 - \frac{1}{8z} + \frac{9}{128z^2}\right), \tag{B12}$$

$$K_0(z) \sim \sqrt{\frac{\pi}{2z}}\, e^{-z}\left(1 - \frac{1}{8z} + \frac{9}{128z^2}\right), \tag{B13}$$

$$I_1(z) \sim \frac{e^z}{\sqrt{2\pi z}}\left(1 - \frac{3}{8z} - \frac{15}{128z^2}\right) + \frac{ie^{-z}}{\sqrt{2\pi z}}\left(1 + \frac{3}{8z} - \frac{15}{128z^2}\right), \tag{B14}$$

$$K_1(z) \sim \sqrt{\frac{\pi}{2z}}\, e^{-z}\left(1 + \frac{3}{8z} - \frac{15}{128z^2}\right), \tag{B15}$$

it follows that

$$\begin{aligned}
&I_0(y_1)K_0(y_2)\\
&\quad = \left[\frac{e^{y_1}}{\sqrt{2\pi y_1}}\left(1 + \frac{1}{8y_1} + \frac{9}{128y_1^2}\right) - \frac{ie^{-y_1}}{\sqrt{2\pi y_1}}\left(1 - \frac{1}{8y_1} + \frac{9}{128y_1^2}\right)\right]\\
&\qquad \times \sqrt{\frac{\pi}{2y_2}}\, e^{-y_2}\left(1 - \frac{1}{8y_2} + \frac{9}{128y_2^2}\right)\\
&\quad = \frac{e^{y_1 - y_2}}{2\sqrt{y_1 y_2}}\left[1 + \frac{1}{8}\left(\frac{1}{y_1} - \frac{1}{y_2}\right) - \frac{1}{64y_1y_2} + \frac{9}{128}\left(\frac{1}{y_1^2} + \frac{1}{y_2^2}\right)\right]\\
&\qquad - \frac{ie^{-(y_1+y_2)}}{2\sqrt{y_1y_2}}\left[1 - \frac{1}{8}\left(\frac{1}{y_1} + \frac{1}{y_2}\right) + \frac{1}{64y_1y_2} + \frac{9}{128}\left(\frac{1}{y_1^2} + \frac{1}{y_2^2}\right)\right], \qquad \text{(B16)}\\
&I_1(y_1)K_1(y_2)\\
&\quad = \left[\frac{e^{y_1}}{\sqrt{2\pi y_1}}\left(1 - \frac{3}{8y_1} - \frac{15}{128y_1^2}\right) + \frac{ie^{-y_1}}{\sqrt{2\pi y_1}}\left(1 + \frac{3}{8y_1} - \frac{15}{128y_1^2}\right)\right]\\
&\qquad \times \sqrt{\frac{\pi}{2y_2}}\, e^{-y_2}\left(1 + \frac{3}{8y_2} - \frac{15}{128y_2^2}\right)\\
&\quad = \frac{e^{y_1 - y_2}}{2\sqrt{y_1 y_2}}\left[1 + \frac{3}{8}\left(\frac{1}{y_2} - \frac{1}{y_1}\right) - \frac{9}{64y_1y_2} - \frac{15}{128}\left(\frac{1}{y_1^2} + \frac{1}{y_2^2}\right)\right]\\
&\qquad + \frac{ie^{-(y_1+y_2)}}{2\sqrt{y_1y_2}}\left[1 + \frac{3}{8}\left(\frac{1}{y_1} + \frac{1}{y_2}\right) + \frac{9}{64y_1y_2} - \frac{15}{128}\left(\frac{1}{y_1^2} + \frac{1}{y_2^2}\right)\right]. \qquad \text{(B17)}
\end{aligned}$$

With $y_1y_2 = \frac{1}{4}\beta^2\rho^2$, $y_1 - y_2 = -\beta z$, and $y_1 + y_2 = \beta r$, (B16) and (B17) become

$$I_0(y_1)K_0(y_2) = e^{-\beta z}\left(\frac{1}{\beta\rho} + \frac{z}{2\beta^2\rho^3} + \frac{1}{2\beta^3\rho^3}\right)$$

$$- ie^{-\beta r}\left(\frac{1}{\beta\rho} - \frac{r}{2\beta^2\rho^3} + \frac{5}{8\beta^3\rho^3}\right), \tag{B18}$$

$$I_1(y_1)K_1(y_2) = e^{-\beta z}\left(\frac{1}{\beta\rho} - \frac{3z}{2\beta^2\rho^3} - \frac{3}{2\beta^3\rho^3}\right)$$
$$+ ie^{-\beta r}\left(\frac{1}{\beta\rho} + \frac{3r}{2\beta^2\rho^3} - \frac{3}{8\beta^3\rho^3}\right). \tag{B19}$$

Finally, the sum of (B18) and (B19) is

$$I_0(y_1)K_0(y_2) + I_1(y_1)K_1(y_2) = e^{-\beta z}\left(\frac{2}{\beta\rho} - \frac{z}{\beta^2\rho^3} - \frac{1}{\beta^3\rho^3}\right)$$
$$+ ie^{-\beta r}\left(\frac{2r}{\beta^2\rho^3} - \frac{1}{\beta^3\rho^3}\right). \tag{B20}$$

With $\beta = -ik$ and $z^2 \ll \rho^2$ so that $r \sim \rho$ in amplitudes, the final formula for $I_1(\rho, z)$ is

$$I_1(\rho, z) = e^{ikz}\left(\frac{2}{k\rho} - \frac{iz}{k^2\rho^3} + \frac{1}{k^3\rho^3}\right) + ie^{ikr}\left(\frac{2i}{k^2\rho^2} + \frac{1}{k^3\rho^3}\right). \tag{B21}$$

The integral $I_2(\rho, z)$ defined in (A2a) as

$$I_2(\rho, z) = \int_0^\infty (k^2 - \lambda^2)^{-1/2}\,[J_0(\lambda\rho) - J_2(\lambda\rho)]e^{iz\sqrt{k^2-\lambda^2}}\,d\lambda \tag{B22}$$

is identical to $I_1(\rho, z)$ except for the sign of $J_2(\lambda\rho)$. It follows that, with (B7) and $\nu = 0$ and $\nu = 2$,

$$I_2(\rho, z) = -i\int_0^\infty (\lambda^2 + \beta^2)^{-1/2}\,[J_0(\lambda\rho) - J_2(\lambda\rho)]e^{-z\sqrt{\lambda^2+\beta^2}}\,d\lambda$$
$$= -i\{I_0[\tfrac{1}{2}\beta(r - z)]K_0[\tfrac{1}{2}\beta(r + z)] - I_1[\tfrac{1}{2}\beta(r - z)]K_1[\tfrac{1}{2}\beta(r + z)]\}. \tag{B23}$$

With $y_1 = \frac{1}{2}\beta(r - z)$ and $y_2 = \frac{1}{2}\beta(r + z)$, this becomes

$$I_2(\rho, z) = -i[I_0(y_1)K_0(y_2) - I_1(y_1)K_1(y_2)]. \tag{B24}$$

With the large-argument approximations of the Bessel functions in (B12)–(B15) and (B18) and (B19), it follows that

$$I_0(y_1)K_0(y_2) - I_1(y_1)K_1(y_2)$$
$$= e^{-\beta z}\left(\frac{2z}{\beta^2\rho^3} + \frac{2}{\beta^3\rho^3}\right) - ie^{-\beta r}\left(\frac{2}{\beta\rho} + \frac{r}{\beta^2\rho^3} + \frac{1}{4\beta^3\rho^3}\right). \tag{B25}$$

With $\beta = -ik$, the final formula for $I_2(\rho, z)$ is

$$I_2(\rho, z) = e^{ikz}\left(\frac{2iz}{k^2\rho^3} - \frac{2}{k^3\rho^3}\right) - ie^{ikr}\left(\frac{2}{k\rho} + \frac{i}{k^2\rho^2} - \frac{1}{4k^3\rho^3}\right), \tag{B26}$$

where $r \sim \rho$ in amplitudes since $z^2 \ll \rho^2$.

With (B1), the integral $I_3(\rho, z)$ defined in (A3a) is

$$I_3(\rho, z) = \int_0^\infty (k^2 - \lambda^2)^{-1/2} [J_0(\lambda\rho) + J_2(\lambda\rho)] e^{iz\sqrt{k^2-\lambda^2}} \lambda \, d\lambda$$
$$= \frac{2}{\rho} \int_0^\infty e^{iz\sqrt{k^2-\lambda^2}} \frac{J_1(\lambda\rho)}{(k^2-\lambda^2)^{1/2}} \, d\lambda. \tag{B27}$$

With $k = i\beta$, $\beta = -ik$, and (B7) with $\nu = 1$, this becomes

$$I_3(\rho, z) = -\frac{2i}{\rho} \int_0^\infty e^{-z\sqrt{\lambda^2+\beta^2}} \frac{J_1(\lambda\rho)}{(\lambda^2+\beta^2)^{1/2}} \, d\lambda$$
$$= -\frac{2i}{\rho} \{ I_{1/2}[\tfrac{1}{2}\beta(r - z)] K_{1/2}[\tfrac{1}{2}\beta(r + z)] \}, \tag{B28}$$

where $r = (\rho^2 + z^2)^{1/2}$. With

$$I_{1/2}(y_1) = \frac{1}{\sqrt{2\pi y_1}} (e^{y_1} - e^{-y_1}); \qquad K_{1/2}(y_2) = \sqrt{\frac{\pi}{2y_2}} \, e^{-y_2}, \tag{B29}$$

it follows that

$$I_{1/2}(y_1) K_{1/2}(y_2) = \frac{1}{2\sqrt{y_1 y_2}} (e^{y_1 - y_2} - e^{-(y_1+y_2)}). \tag{B30}$$

With $y_1 = \frac{1}{2}\beta(r - z)$, $y_2 = \frac{1}{2}\beta(r + z)$, and $\sqrt{y_1 y_2} = \frac{1}{2}\beta\rho$,

$$I_{1/2}[\tfrac{1}{2}\beta(r - z)] K_{1/2}[\tfrac{1}{2}\beta(r + z)] = \frac{1}{\beta\rho} (e^{-\beta z} - e^{-\beta r}). \tag{B31}$$

It follows that, with $\beta = -ik$, (B28) becomes

$$I_3(\rho, z) = \frac{2}{k\rho^2} (e^{ikz} - e^{ikr}). \tag{B32}$$

The integral $I_4(\rho, z)$ as defined in (A4a) can be written as the sum of two integrals, viz.,

$$I_4(\rho, z) = \int_0^\infty (k^2 - \lambda^2)^{-1/2} [J_0(\lambda\rho) - J_2(\lambda\rho)] e^{iz\sqrt{k^2-\lambda^2}} \lambda \, d\lambda$$
$$= 2 \int_0^\infty e^{iz\sqrt{k^2-\lambda^2}} \frac{J_0(\lambda\rho)}{(k^2-\lambda^2)^{1/2}} \lambda \, d\lambda - I_3(\rho, z). \tag{B33}$$

With (B8) and (B32),

$$I_4(\rho, z) = -\frac{2ie^{ikr}}{r} - \frac{2}{k\rho^2} (e^{ikz} - e^{ikr})$$
$$= -\frac{2}{k} \left[\frac{e^{ikz}}{\rho^2} + e^{ikr} \left(\frac{ik}{r} - \frac{1}{\rho^2} \right) \right]. \tag{B34}$$

With (B1), the integral $I_5(\rho, z)$ as defined in (A5a) becomes

$$\begin{aligned} I_5(\rho, z) &= \int_0^\infty (k^2-\lambda^2)^{-1/2}[J_0(\lambda\rho)+J_2(\lambda\rho)]e^{iz\sqrt{k^2-\lambda^2}}\lambda^2\,d\lambda \\ &= \frac{2}{\rho}\int_0^\infty (k^2-\lambda^2)^{-1/2}e^{iz\sqrt{k^2-\lambda^2}}J_1(\lambda\rho)\lambda\,d\lambda \\ &= -\frac{2i}{\rho}\int_0^\infty (\lambda^2+\beta^2)^{-1/2}e^{-z\sqrt{\lambda^2+\beta^2}}J_1(\lambda\rho)\lambda\,d\lambda, \end{aligned} \tag{B35}$$

with $k = i\beta$. This integral can be evaluated with (B7) with $\nu = 0$. Thus, with $y_1 = \frac{1}{2}\beta(r-z)$, $y_2 = \frac{1}{2}\beta(r+z)$, and $r = (\rho^2+z^2)^{1/2}$,

$$\begin{aligned} &\frac{\partial}{\partial\rho}\int_0^\infty (\lambda^2+\beta^2)^{-1/2}e^{-z\sqrt{\lambda^2+\beta^2}}J_0(\lambda\rho)\,d\lambda \\ &\quad= \frac{\partial}{\partial\rho}I_0(y_1)K_0(y_2) \\ &\quad= -\int_0^\infty (\lambda^2+\beta^2)^{-1/2}e^{-z\sqrt{\lambda^2+\beta^2}}J_1(\lambda\rho)\lambda\,d\lambda \\ &\quad= -\frac{i\rho}{2}I_5(\rho, z) \end{aligned} \tag{B36}$$

since $(\partial/\partial\rho)J_0(\lambda\rho) = -J_1(\lambda\rho)\cdot\lambda$. Hence,

$$\begin{aligned} I_5(\rho, z) &= \frac{2i}{\rho}\frac{\partial}{\partial\rho}\int_0^\infty (\lambda^2+\beta^2)^{-1/2}e^{-z\sqrt{\lambda^2+\beta^2}}J_0(\lambda\rho)\,d\lambda \\ &= \frac{2i}{\rho}\mathcal{I}_1 \end{aligned} \tag{B37}$$

with

$$\mathcal{I}_1 = \frac{\partial}{\partial\rho}I_0(y_1)K_0(y_2). \tag{B38}$$

With the large-argument approximations of the Bessel functions in (B12) and (B13) and $I_0(y_1)K_0(y_2)$ in (B18), it then follows that

$$\begin{aligned} \mathcal{I}_1 &= \frac{\partial}{\partial\rho}I_0(y_1)K_0(y_2) \\ &= e^{-\beta z}\left(-\frac{1}{\beta\rho^2}-\frac{3z}{2\beta^2\rho^4}-\frac{3}{2\beta^3\rho^4}\right) \\ &\quad - ie^{-\beta r}\left(-\frac{1}{\beta\rho^2}+\frac{1}{\beta^2\rho^3}-\frac{15}{8\beta^3\rho^4}-\frac{1}{\rho}+\frac{1}{2\beta\rho^2}-\frac{5}{8\beta^2\rho^3}\right) \\ &= -e^{-\beta z}\left(\frac{1}{\beta\rho^2}+\frac{3z}{2\beta^2\rho^4}+\frac{3}{2\beta^3\rho^4}\right) \\ &\quad + ie^{-\beta r}\left(\frac{1}{\rho}+\frac{1}{2\beta\rho^2}-\frac{3}{8\beta^2\rho^3}+\frac{15}{8\beta^3\rho^4}\right), \end{aligned} \tag{B39}$$

where $r \sim \rho$ in amplitudes since $z^2 \ll \rho^2$. With $\beta = -ik$,

$$\mathcal{I}_1 = -e^{ikz}\left(\frac{i}{k\rho^2} - \frac{3z}{2k^2\rho^4} - \frac{3i}{2k^3\rho^4}\right) + ie^{ikr}\left(\frac{1}{\rho} + \frac{i}{2k\rho^2} + \frac{3}{8k^2\rho^3} - \frac{15i}{8k^3\rho^4}\right). \tag{B40}$$

It follows that the leading terms give

$$I_5(\rho, z) = \frac{2i}{\rho}\mathcal{I}_1 = 2\left[\frac{e^{ikz}}{k\rho^3} - e^{ikr}\left(\frac{1}{\rho^2} + \frac{i}{2k\rho^3} + \frac{3}{8k^2\rho^4}\right)\right]. \tag{B41}$$

The integral $I_6(\rho, z)$ defined in (A6a) can be expanded as follows with $k = i\beta$:

$$\begin{aligned} I_6(\rho, z) &= \int_0^\infty (k^2 - \lambda^2)^{-1/2}[J_0(\lambda\rho) - J_2(\lambda\rho)]e^{iz\sqrt{k^2-\lambda^2}}\lambda^2\, d\lambda \\ &= -2i\int_0^\infty (\lambda^2 + \beta^2)^{-1/2}J_0(\lambda\rho)e^{-z\sqrt{\lambda^2+\beta^2}}\lambda^2\, d\lambda \\ &\quad + i\int_0^\infty (\lambda^2 + \beta^2)^{-1/2}[J_0(\lambda\rho) + J_2(\lambda\rho)]e^{-z\sqrt{\lambda^2+\beta^2}}\lambda^2\, d\lambda \\ &= -2i\int_0^\infty (\lambda^2 + \beta^2)^{-1/2}J_0(\lambda\rho)e^{-z\sqrt{\lambda^2+\beta^2}}\lambda^2\, d\lambda - I_5(\rho, z). \end{aligned} \tag{B42}$$

The remaining integral can be expanded as follows:

$$\begin{aligned} \mathcal{I} &= \int_0^\infty (\lambda^2 + \beta^2)^{-1/2}J_0(\lambda\rho)e^{-z\sqrt{\lambda^2+\beta^2}}\lambda^2\, d\lambda \\ &= -\beta^2\int_0^\infty (\lambda^2 + \beta^2)^{-1/2}J_0(\lambda\rho)e^{-z\sqrt{\lambda^2+\beta^2}}\, d\lambda \\ &\quad + \int_0^\infty (\lambda^2 + \beta^2)^{1/2}J_0(\lambda\rho)e^{-z\sqrt{\lambda^2+\beta^2}}\, d\lambda. \end{aligned} \tag{B43}$$

The integrals on the right can be evaluated with (B7). Thus, with $\nu = 0$,

$$\begin{aligned} \mathcal{I}_1(\rho, z) &\equiv \int_0^\infty (\lambda^2 + \beta^2)^{-1/2}e^{-z\sqrt{\lambda^2+\beta^2}}J_0(\lambda\rho)\, d\lambda \\ &= I_0(y_1)K_0(y_2), \end{aligned} \tag{B44}$$

where $y_1 = \frac{1}{2}\beta(r - z)$, $y_2 = \frac{1}{2}\beta(r + z)$, and $r = (\rho^2 + z^2)^{1/2}$. Also,

$$\begin{aligned} \mathcal{I}_2(\rho, z) = \frac{\partial^2}{\partial z^2}\mathcal{I}_1(\rho, z) &= \int_0^\infty (\lambda^2 + \beta^2)^{1/2}e^{-z\sqrt{\lambda^2+\beta^2}}J_0(\lambda\rho)\, d\lambda \\ &= \frac{\partial^2}{\partial z^2}I_0(y_1)K_0(y_2). \end{aligned} \tag{B45}$$

From (B18),

$$\mathcal{I}_1(\rho, z) = e^{-\beta z}\left(\frac{1}{\beta\rho} + \frac{z}{2\beta^2\rho^3} + \frac{1}{2\beta^3\rho^3}\right)$$

$$- ie^{-\beta r}\left(\frac{1}{\beta\rho} - \frac{r}{2\beta^2\rho^3} + \frac{5}{8\beta^3\rho^3}\right). \tag{B46}$$

It follows that with $r = (\rho^2 + z^2)^{1/2} \sim \rho$ in amplitudes,

$$\begin{aligned}\frac{\partial}{\partial z}\mathcal{I}_1(\rho, z) = e^{-\beta z}&\left(-\frac{1}{\rho} - \frac{z}{2\beta\rho^3} - \frac{1}{2\beta^2\rho^3} + \frac{1}{2\beta^2\rho^3}\right)\\ &+ ie^{-\beta r}\left(\frac{1}{\rho} - \frac{1}{2\beta\rho^2} + \frac{5}{8\beta^2\rho^3}\right)\left(\frac{z}{\rho}\right).\end{aligned} \tag{B47}$$

Also,

$$\begin{aligned}\mathcal{I}_2(\rho, z) = e^{-\beta z}&\left(\frac{\beta}{\rho} + \frac{z}{2\rho^3} - \frac{1}{2\beta\rho^3}\right)\\ &+ ie^{-\beta r}\left[\left(-\frac{\beta}{\rho} - \frac{1}{2\rho^2} - \frac{5}{8\beta\rho^3}\right)\left(\frac{z^2}{\rho^2}\right)\right.\\ &\qquad\left.+ \frac{1}{\rho^2} - \frac{1}{2\beta\rho^3} + \frac{5}{8\beta^2\rho^4}\right].\end{aligned} \tag{B48}$$

When (B46) and (B48) are combined to give (B43) and the z^2/ρ^2 terms are neglected, the result is

$$\begin{aligned}\mathcal{I} &= -\beta^2\mathcal{I}_1(\rho, z) + \mathcal{I}_2(\rho, z)\\ &= e^{-\beta z}\left(-\frac{\beta}{\rho} - \frac{1}{2\beta\rho^3} + \frac{\beta}{\rho} - \frac{1}{2\beta\rho^3}\right)\\ &\quad + ie^{-\beta r}\left(\frac{\beta}{\rho} - \frac{1}{2\rho^2} + \frac{5}{8\beta\rho^3} + \frac{1}{\rho^2} - \frac{1}{2\beta\rho^3} + \frac{5}{8\beta^2\rho^4}\right)\\ &= -e^{-\beta z}\,\frac{1}{\beta\rho^3} + ie^{-\beta r}\left(\frac{\beta}{\rho} + \frac{1}{2\rho^2} + \frac{1}{8\beta\rho^3} + \frac{5}{8\beta^2\rho^4}\right).\end{aligned} \tag{B49}$$

This can be substituted in (B42) to give

$$I_6(\rho, z) = \frac{2i}{\beta\rho^3}\,e^{-\beta z} + 2e^{-\beta r}\left(\frac{\beta}{\rho} + \frac{1}{2\rho^2} + \frac{1}{8\beta\rho^3} + \frac{5}{8\beta^2\rho^4}\right) - I_5(\rho, z). \tag{B50}$$

With $\beta = -ik$ and $I_5(\rho, z)$ given by (B41),

$$\begin{aligned}I_6(\rho, z) &= -\frac{2}{k\rho^3}\,e^{ikz} + 2e^{ikr}\left(-\frac{ik}{\rho} + \frac{1}{2\rho^2} + \frac{i}{8k\rho^3} - \frac{5}{8k^2\rho^4}\right)\\ &\quad - \frac{2}{k\rho^3}\,e^{ikz} + 2e^{ikr}\left(\frac{1}{\rho^2} + \frac{i}{2k\rho^3} + \frac{3}{8k^2\rho^4}\right)\\ &\sim -\frac{4}{k\rho^3}\,e^{ikz} - 2e^{ikr}\left(\frac{ik}{\rho} - \frac{3}{2\rho^2} - \frac{5i}{8k\rho^3} + \frac{1}{4k^2\rho^4}\right).\end{aligned} \tag{B51}$$

The integral $I_7(\rho, z)$ defined in (A7a) can be expressed as a combination of the integrals $I_3(\rho, z)$ and $I_{11}(\rho, z)$ since

$$\lambda^2(k^2 - \lambda^2)^{-1/2} = k^2(k^2 - \lambda^2)^{-1/2} - (k^2 - \lambda^2)^{1/2}. \tag{B52}$$

Thus,

$$\begin{aligned} I_7(\rho,z) &= \int_0^\infty (k^2-\lambda^2)^{-1/2}[J_0(\lambda\rho)+J_2(\lambda\rho)]e^{iz\sqrt{k^2-\lambda^2}}\lambda^3\,d\lambda \\ &= k^2\int_0^\infty (k^2-\lambda^2)^{-1/2}[J_0(\lambda\rho)+J_2(\lambda\rho)]e^{iz\sqrt{k^2-\lambda^2}}\lambda\,d\lambda \\ &\quad -\int_0^\infty (k^2-\lambda^2)^{1/2}[J_0(\lambda\rho)+J_2(\lambda\rho)]e^{iz\sqrt{k^2-\lambda^2}}\lambda\,d\lambda \\ &= k^2 I_3(\rho,z) - I_{11}(\rho,z). \end{aligned} \tag{B53}$$

Formulas for $I_3(\rho, z)$ and $I_{11}(\rho, z)$ are, respectively, in (A3a) and (A11a) of Appendix A. The result is

$$\begin{aligned} I_7(\rho,z) &= \frac{2k}{\rho^2}\,(e^{ikz}-e^{ikr}) - 2k\left(\frac{e^{ikz}}{\rho^2}+\frac{ie^{ikr}}{kr^3}-\frac{z^2}{r^2}\frac{e^{ikr}}{\rho^2}\right) \\ &= -2k\left[\frac{1}{\rho^2}\left(1-\frac{z^2}{r^2}\right)+\frac{i}{kr^3}\right]e^{ikr} \\ &= -2k\left(\frac{1}{r^2}+\frac{i}{kr^3}\right)e^{ikr}. \end{aligned} \tag{B54}$$

The integral $I_8(\rho, z)$ defined in (A8a) as

$$I_8(\rho,z) = \int_0^\infty (k^2-\lambda^2)^{-1/2}[J_0(\lambda\rho)-J_2(\lambda\rho)]e^{iz\sqrt{k^2-\lambda^2}}\lambda^3\,d\lambda \tag{B55}$$

can be expanded with the identity in (B52) to give

$$\begin{aligned} I_8(\rho,z) = &-\int_0^\infty (k^2-\lambda^2)^{1/2}[J_0(\lambda\rho)-J_2(\lambda\rho)]e^{iz\sqrt{k^2-\lambda^2}}\lambda\,d\lambda \\ &+ k^2\int_0^\infty (k^2-\lambda^2)^{-1/2}[J_0(\lambda\rho)-J_2(\lambda\rho)]e^{iz\sqrt{k^2-\lambda^2}}\lambda\,d\lambda. \end{aligned} \tag{B56}$$

With Appendix A, this becomes

$$I_8(\rho,z) = -I_{12}(\rho,z) + k^2 I_4(\rho,z). \tag{B57}$$

The substitution of the explicit formulas for $I_{12}(\rho, z)$ in (A12a) and $I_4(\rho, z)$ in (A4a) gives

$$\begin{aligned} I_8(\rho,z) &= 2k\left\{\frac{e^{ikz}}{\rho^2}+\left[\frac{1}{r^2}+\frac{2i}{kr^3}+\frac{z^2}{r^2}\left(\frac{ik}{r}-\frac{3}{r^2}-\frac{3i}{kr^3}-\frac{1}{\rho^3}\right)\right]e^{ikr}\right\} \\ &\quad -2k\left[\frac{e^{ikz}}{\rho^2}+\left(\frac{ik}{r}-\frac{1}{\rho^2}\right)e^{ikr}\right] \\ &= 2k\left[-\frac{ik}{r}+\frac{1}{r^2}+\frac{2i}{kr^3}+\frac{z^2}{r^2}\left(\frac{ik}{r}-\frac{3}{r^2}-\frac{3i}{kr^3}\right)\right. \\ &\quad \left.+\frac{1}{\rho^2}\left(1-\frac{z^2}{r^2}\right)\right]e^{ikr}. \end{aligned} \tag{B58}$$

Hence,

$$I_8(\rho,z) = -2k\left[\frac{ik}{r} - \frac{2}{r^2} - \frac{2i}{kr^3} - \frac{z^2}{r^2}\left(\frac{ik}{r} - \frac{3}{r^2} - \frac{3i}{kr^3}\right)\right]e^{ikr}. \tag{B59}$$

The integral $I_9(\rho,z)$ defined in (A9a) has the form

$$I_9(\rho,z) = \int_0^\infty (k^2-\lambda^2)^{1/2}[J_0(\lambda\rho) + J_2(\lambda\rho)]e^{iz\sqrt{k^2-\lambda^2}}\,d\lambda. \tag{B60}$$

This can be expanded with $k = i\beta$ as follows:

$$\begin{aligned} I_9(\rho,z) &= 2\int_0^\infty (k^2-\lambda^2)^{1/2}J_0(\lambda\rho)e^{iz\sqrt{k^2-\lambda^2}}\,d\lambda - I_{10}(\rho,z)\\ &= 2i\int_0^\infty (\lambda^2-k^2)^{1/2}J_0(\lambda\rho)e^{-z\sqrt{\lambda^2-k^2}}\,d\lambda - I_{10}(\rho,z)\\ &= 2i\int_0^\infty (\lambda^2+\beta^2)^{1/2}J_0(\lambda\rho)e^{-z\sqrt{\lambda^2+\beta^2}}\,d\lambda - I_{10}(\rho,z). \end{aligned} \tag{B61}$$

The integral can be evaluated with (B7) differentiated twice with respect to z. Thus, from (B7) with $\nu = 0$,

$$\begin{aligned} &\frac{\partial^2}{\partial z^2}\int_0^\infty e^{-z\sqrt{\lambda^2+\beta^2}}\frac{J_0(\lambda\rho)}{(\lambda^2+\beta^2)^{1/2}}\,d\lambda\\ &\quad= \int_0^\infty (\lambda^2+\beta^2)^{1/2}J_0(\lambda\rho)e^{-z\sqrt{\lambda^2+\beta^2}}\,d\lambda\\ &\quad= \frac{\partial^2}{\partial z^2}I_0(y_1)K_0(y_2), \end{aligned} \tag{B62}$$

where

$$y_1 = \tfrac{1}{2}\beta(r-z); \qquad y_2 = \tfrac{1}{2}\beta(r+z); \qquad r = (\rho^2+z^2)^{1/2}. \tag{B63}$$

With the large-argument expansions of the Bessel functions in (B12) and (B13) and $r \sim \rho$ in amplitudes, it follows that

$$\begin{aligned} I_0(y_1)K_0(y_2) = e^{-\beta z}&\left(\frac{1}{\beta\rho} + \frac{z}{2\beta^2\rho^3} + \frac{1}{2\beta^3\rho^3}\right)\\ &- ie^{-\beta r}\left(\frac{1}{\beta\rho} - \frac{1}{2\beta^2\rho^2} + \frac{5}{8\beta^3\rho^3}\right), \end{aligned} \tag{B64}$$

$$\begin{aligned} \frac{\partial}{\partial z}I_0(y_1)K_0(y_2) &= e^{-\beta z}\left(-\frac{1}{\rho} - \frac{z}{2\beta\rho^3} - \frac{1}{2\beta^2\rho^3} + \frac{1}{2\beta^2\rho^3}\right)\\ &\qquad + ize^{-\beta r}\left(\frac{1}{\rho^2} - \frac{1}{2\beta\rho^3} + \frac{5}{8\beta^2\rho^4}\right)\\ &= -e^{-\beta z}\left(\frac{1}{\rho} + \frac{z}{2\beta\rho^3}\right)\\ &\qquad + ize^{-\beta r}\left(\frac{1}{\rho^2} - \frac{1}{2\beta\rho^3} + \frac{5}{8\beta^2\rho^4}\right), \end{aligned} \tag{B65}$$

$$\frac{\partial^2}{\partial z^2} I_0(y_1)K_0(y_2) = e^{-\beta z}\left(\frac{\beta}{\rho} + \frac{z}{2\rho^3} - \frac{1}{2\beta\rho^3}\right)$$
$$+ ie^{-\beta r}\left(\frac{1}{\rho^2} - \frac{1}{2\beta\rho^3} + \frac{5}{8\beta^2\rho^4}\right). \tag{B66}$$

Here, terms in z^2/ρ^2 are neglected. Thus, with $\beta = -ik$,

$$\int_0^\infty (\lambda^2 - k^2)^{1/2} J_0(\lambda\rho) e^{-z\sqrt{\lambda^2-k^2}}\, d\lambda$$
$$= e^{ikz}\left(-\frac{ik}{\rho} + \frac{z}{2\rho^3} - \frac{i}{2k\rho^3}\right) + ie^{ikr}\left(\frac{1}{\rho^2} - \frac{i}{2k\rho^3} - \frac{5}{8k^2\rho^4}\right). \tag{B67}$$

With $I_{10}(\rho, z)$ in (A10a), (B61) becomes

$$I_9(\rho, z) = 2i\left[e^{ikz}\left(-\frac{ik}{\rho} + \frac{z}{2\rho^3} - \frac{i}{2k\rho^3}\right) + ie^{ikr}\left(\frac{1}{\rho^2} - \frac{i}{2k\rho^3} - \frac{5}{8k^2\rho^4}\right)\right]$$
$$- 2\left[e^{ikz}\left(\frac{iz}{\rho^3} + \frac{1}{k\rho^3}\right) - e^{ikr}\left(\frac{1}{\rho^2} + \frac{i}{2k\rho^3} + \frac{3}{8k^2\rho^4}\right)\right]$$
$$= 2e^{ikz}\left(\frac{k}{\rho} + \frac{iz}{2\rho^3} + \frac{1}{2k\rho^3} - \frac{iz}{\rho^3} - \frac{1}{k\rho^3}\right) + 2e^{ikr}\left(\frac{i}{k\rho^3} + \frac{1}{k^2\rho^4}\right)$$
$$\sim 2\left[e^{ikz}\left(\frac{k}{\rho} - \frac{1}{2k\rho^3} - \frac{iz}{2\rho^3}\right) + \frac{ie^{ikr}}{k\rho^3}\right]. \tag{B68}$$

The integral $I_{10}(\rho, z)$ defined in (A10a) can be integrated as follows:

$$I_{10}(\rho, z) = \int_0^\infty (k^2 - \lambda^2)^{1/2}[J_0(\lambda\rho) - J_2(\lambda\rho)] e^{iz\sqrt{k^2-\lambda^2}}\, d\lambda$$
$$= i\int_0^\infty (\lambda^2 + \beta^2)^{1/2}[J_0(\lambda\rho) - J_2(\lambda\rho)] e^{-z\sqrt{\lambda^2+\beta^2}}\, d\lambda$$
$$= \frac{2i}{\rho}\int_0^\infty (\lambda^2 + \beta^2)^{1/2}\, \frac{\partial}{\partial\lambda} J_1(\lambda\rho) e^{-z\sqrt{\lambda^2+\beta^2}}\, d\lambda. \tag{B69}$$

Integration by parts, viz., $\int_0^\infty u\, dv = uv|_0^\infty - \int_0^\infty v\, du$, with

$$dv = \frac{\partial}{\partial\lambda}[J_1(\lambda\rho)]\, d\lambda; \quad v = J_1(\lambda\rho); \quad u = (\lambda^2 + \beta^2)^{1/2} e^{-z\sqrt{\lambda^2+\beta^2}},$$
$$du = \left(\frac{\lambda}{(\lambda^2+\beta^2)^{1/2}}\, e^{-z\sqrt{\lambda^2+\beta^2}} - z\lambda e^{-z\sqrt{\lambda^2+\beta^2}}\right) d\lambda, \tag{B70}$$

leads to

$$I_{10}(\rho, z) = -\frac{2i}{\rho}\left(\int_0^\infty (\lambda^2 + \beta^2)^{-1/2} J_1(\lambda\rho) e^{-z\sqrt{\lambda^2+\beta^2}} \lambda\, d\lambda\right.$$
$$\left. - z\int_0^\infty J_1(\lambda\rho) e^{-z\sqrt{\lambda^2+\beta^2}} \lambda\, d\lambda\right). \tag{B71}$$

The first integral in (B71) can be expressed in terms of $I_5(\rho, z)$. Thus,

$$\begin{aligned}\mathcal{I}_1 &= \int_0^\infty (\lambda^2+\beta^2)^{-1/2} J_1(\lambda\rho) e^{-z\sqrt{\lambda^2+\beta^2}} \lambda \, d\lambda \\ &= \frac{i\rho}{2} I_5(\rho, z) = \frac{i}{k\rho^2} e^{ikz} - \left(\frac{i}{\rho} - \frac{1}{2k\rho^2} + \frac{3i}{8k^2\rho^3}\right) e^{ikr}, \end{aligned} \tag{B72}$$

where $I_5(\rho, z)$ is evaluated in (B41). The second integral in (B71) can be evaluated with (B7) with $\nu = 0$. Thus,

$$\begin{aligned}&\frac{\partial}{\partial z}\int_0^\infty (\lambda^2+\beta^2)^{-1/2} e^{-z\sqrt{\lambda^2+\beta^2}} J_0(\lambda\rho)\, d\lambda \\ &\quad = -\int_0^\infty e^{-z\sqrt{\lambda^2+\beta^2}} J_0(\lambda\rho)\, d\lambda \\ &\quad = \frac{\partial}{\partial z}\{I_0[\tfrac{1}{2}\beta(r-z)] K_0[\tfrac{1}{2}\beta(r+z)]\} \end{aligned} \tag{B73}$$

and

$$\begin{aligned}\mathcal{I}_2 &= \int_0^\infty e^{-z\sqrt{\lambda^2+\beta^2}} J_1(\lambda\rho) \lambda \, d\lambda \\ &= -\frac{\partial}{\partial\rho}\int_0^\infty e^{-z\sqrt{\lambda^2+\beta^2}} J_0(\lambda\rho)\, d\lambda \\ &= \frac{\partial}{\partial\rho}\frac{\partial}{\partial z}\{I_0[\tfrac{1}{2}\beta(r-z)] K_0[\tfrac{1}{2}\beta(r+z)]\}. \end{aligned} \tag{B74}$$

With the large-argument approximations of the Bessel functions in (B12) and (B13) and $r \sim \rho$ in amplitudes,

$$\begin{aligned}\mathcal{I}_2 &= \frac{\partial}{\partial\rho}\frac{\partial}{\partial z}\left[e^{-\beta z}\left(\frac{1}{\beta\rho} + \frac{z}{2\beta^2\rho^3} + \frac{1}{2\beta^3\rho^3}\right)\right. \\ &\qquad\qquad \left. - ie^{-\beta r}\left(\frac{1}{\beta\rho} - \frac{1}{2\beta^2\rho^2} + \frac{5}{8\beta^3\rho^3}\right)\right] \\ &= \frac{\partial}{\partial\rho}\left[e^{-\beta z}\left(-\frac{1}{\rho} - \frac{z}{2\beta\rho^3} - \frac{1}{2\beta^2\rho^3} + \frac{1}{2\beta^2\rho^3}\right)\right. \\ &\qquad\qquad \left. + ize^{-\beta r}\left(\frac{1}{\rho^2} - \frac{1}{2\beta\rho^3} + \frac{5}{8\beta^2\rho^4}\right)\right] \\ &= e^{-\beta z}\left(\frac{1}{\rho^2} + \frac{3z}{2\beta\rho^4}\right) - ize^{-\beta r}\left(\frac{\beta}{\rho^2} + \frac{3}{2\rho^3} - \frac{7}{8\beta\rho^4} + \frac{5}{2\beta^2\rho^5}\right). \end{aligned} \tag{B75}$$

It follows that

$$\frac{2iz}{\rho}\mathcal{I}_2 \sim \frac{2iz}{\rho^3} e^{-\beta z} = \frac{2iz}{\rho^3} e^{ikz}. \tag{B76}$$

When (B76) and (B72) are substituted in (B71),

$$I_{10}(\rho, z) = 2\left[e^{ikz}\left(\frac{iz}{\rho^3} + \frac{1}{k\rho^3}\right) - e^{ikr}\left(\frac{1}{\rho^2} + \frac{i}{2k\rho^3} + \frac{3}{8k^2\rho^4}\right)\right]. \tag{B77}$$

The integral $I_{11}(\rho, z)$ in (A11a) can be evaluated as follows. With $k = i\beta$,

$$
\begin{aligned}
I_{11}(\rho, z) &= \int_0^\infty (k^2 - \lambda^2)^{1/2}[J_0(\lambda\rho) + J_2(\lambda\rho)]e^{iz\sqrt{k^2-\lambda^2}}\lambda\, d\lambda \\
&= \frac{2}{\rho}\int_0^\infty (k^2 - \lambda^2)^{1/2} J_1(\lambda\rho)e^{iz\sqrt{k^2-\lambda^2}}\, d\lambda \\
&= \frac{2i}{\rho}\int_0^\infty (\lambda^2 + \beta^2)^{1/2} J_1(\lambda\rho)e^{-z\sqrt{\lambda^2+\beta^2}}\, d\lambda. \qquad \text{(B78)}
\end{aligned}
$$

When (B7) with $\nu = 1$ is differentiated twice with respect to z, the result with (B31) and $\beta = -ik$ is

$$
\begin{aligned}
&\frac{\partial^2}{\partial z^2}\int_0^\infty (\lambda^2 + \beta^2)^{-1/2} J_1(\lambda\rho)e^{-z\sqrt{\lambda^2+\beta^2}}\, d\lambda \\
&= \int_0^\infty (\lambda^2 + \beta^2)^{1/2} J_1(\lambda\rho)e^{-z\sqrt{\lambda^2+\beta^2}}\, d\lambda \\
&= \frac{\partial^2}{\partial z^2}\{I_{1/2}[\tfrac{1}{2}\beta(r-z)]K_{1/2}[\tfrac{1}{2}\beta(r+z)]\}; \quad r = (\rho^2 + z^2)^{1/2} \\
&= \frac{\partial^2}{\partial z^2}\left(\frac{i}{k\rho}\,(e^{ikz} - e^{ikr})\right) \\
&= -\frac{1}{\rho}\frac{\partial}{\partial z}\left(e^{ikz} - \frac{z}{r}\, e^{ikr}\right) \\
&= -\frac{ik}{\rho}\, e^{ikz} + \frac{1}{\rho}\, e^{ikr}\left(\frac{\rho^2}{r^3} + ik\,\frac{z^2}{r^2}\right). \qquad \text{(B79)}
\end{aligned}
$$

Multiplication by $2i/\rho$ to give (B78) yields

$$
I_{11}(\rho, z) = 2\left[\frac{k}{\rho^2}\, e^{ikz} + \frac{i}{r^3}\, e^{ikr}\left(1 + ikr\,\frac{z^2}{\rho^2}\right)\right]. \qquad \text{(B80)}
$$

The integral $I_{12}(\rho, z)$ as defined in (A12a) can be expanded as follows:

$$
\begin{aligned}
I_{12}(\rho, z) &= \int_0^\infty (k^2 - \lambda^2)^{1/2}[J_0(\lambda\rho) - J_2(\lambda\rho)]e^{iz\sqrt{k^2-\lambda^2}}\lambda\, d\lambda \\
&= 2\int_0^\infty (k^2 - \lambda^2)^{1/2} J_0(\lambda\rho)e^{iz\sqrt{k^2-\lambda^2}}\lambda\, d\lambda \\
&\quad - \int_0^\infty (k^2 - \lambda^2)^{1/2}[J_0(\lambda\rho) + J_2(\lambda\rho)]e^{iz\sqrt{k^2-\lambda^2}}\lambda\, d\lambda \\
&= 2\int_0^\infty (k^2 - \lambda^2)^{1/2} J_0(\lambda\rho)e^{iz\sqrt{k^2-\lambda^2}}\lambda\, d\lambda - I_{11}(\rho, z). \qquad \text{(B81)}
\end{aligned}
$$

Here, the remaining integral can be evaluated with (B8) if this is differentiated twice with respect to z. Thus,

$$
\frac{\partial^2}{\partial z^2}\int_0^\infty (k^2 - \lambda^2)^{-1/2} J_0(\lambda\rho)e^{iz\sqrt{k^2-\lambda^2}}\lambda\, d\lambda
$$

$$= -\int_0^\infty (k^2-\lambda^2)^{1/2} J_0(\lambda\rho) e^{iz\sqrt{k^2-\lambda^2}} \lambda\, d\lambda$$

$$= \frac{\partial^2}{\partial z^2}\left(-\frac{ie^{ikr}}{r}\right)$$

$$= \left[\frac{k}{r^2} + \frac{i}{r^3} + \frac{z^2}{r^2}\left(\frac{ik^2}{r} - \frac{3k}{r^2} - \frac{3i}{r^3}\right)\right] e^{ikr} \tag{B82}$$

so that

$$2\int_0^\infty (k^2-\lambda^2)^{1/2} J_0(\lambda\rho) e^{iz\sqrt{k^2-\lambda^2}} \lambda\, d\lambda$$

$$= -2\left[\frac{k}{r^2} + \frac{i}{r^3} + \frac{z^2}{r^2}\left(\frac{ik^2}{r} - \frac{3k}{r^2} - \frac{3i}{r^3}\right)\right] e^{ikr}. \tag{B83}$$

With (B83) and (A11a), (B81) becomes

$$I_{12}(\rho, z)$$

$$= -2k\left\{\frac{e^{ikz}}{\rho^2} + \left[\frac{1}{r^2} + \frac{2i}{kr^3} + \frac{z^2}{r^2}\left(\frac{ik}{r} - \frac{3}{r^2} - \frac{3i}{kr^3} - \frac{1}{\rho^2}\right)\right] e^{ikr}\right\}. \tag{B84}$$

The integral $I_{13}(\rho, z)$ as defined in (A13a) is equivalent to

$$I_{13}(\rho, z) = \int_0^\infty (k^2-\lambda^2)^{1/2} [J_0(\lambda\rho) + J_2(\lambda\rho)] e^{iz\sqrt{k^2-\lambda^2}} \lambda^2\, d\lambda$$

$$= \frac{2}{\rho}\int_0^\infty (k^2-\lambda^2)^{1/2} J_1(\lambda\rho) e^{iz\sqrt{k^2-\lambda^2}} \lambda\, d\lambda. \tag{B85}$$

With $k = i\beta$, this integral becomes

$$I_{13}(\rho, z) = \frac{2i}{\rho}\int_0^\infty (\lambda^2+\beta^2)^{1/2} J_1(\lambda\rho) e^{-z\sqrt{\lambda^2+\beta^2}} \lambda\, d\lambda. \tag{B86}$$

This can be evaluated with (B7) with $\nu = 0$. The first step is to differentiate (B7) twice with respect to z to obtain

$$\frac{\partial^2}{\partial z^2}\int_0^\infty (\lambda^2+\beta^2)^{-1/2} J_0(\lambda\rho) e^{-z\sqrt{\lambda^2+\beta^2}}\, d\lambda$$

$$= \frac{\partial^2}{\partial z^2}\{I_0[\tfrac{1}{2}\beta(r-z)] K_0[\tfrac{1}{2}\beta(r+z)]\}, \tag{B87}$$

where $r = (\rho^2+z^2)^{1/2}$. The differentiation of the left side and the use of the large-argument forms of the Bessel functions in (B64)–(B66) give

$$\int_0^\infty (\lambda^2+\beta^2)^{1/2} J_0(\lambda\rho) e^{-z\sqrt{\lambda^2+\beta^2}}\, d\lambda$$

$$= \frac{\beta}{\rho}\left(1 + \frac{z}{2\beta\rho^2}\right) e^{-\beta z} + \frac{i}{\rho^2}\left(1 - \frac{1}{2\beta\rho}\right) e^{-\beta r}. \tag{B88}$$

The next step is the differentiation of both sides with respect to ρ to obtain the desired integral. Thus, with $\partial J_0(\lambda\rho)/\partial\rho = -J_1(\lambda\rho)\lambda$, the leading terms are

$$-\int_0^\infty (\lambda^2+\beta^2)^{1/2} J_1(\lambda\rho) e^{-z\sqrt{\lambda^2+\beta^2}}\lambda\, d\lambda$$
$$= -\frac{\beta}{\rho^2}\, e^{-\beta z} + ie^{-\beta r}\left(-\frac{\beta}{\rho^2}+\frac{1}{2\rho^3}-\frac{2}{\rho^3}\right), \tag{B89}$$

where $r \sim \rho$ in amplitudes. The substitution of (B89) into (B86) yields

$$I_{13}(\rho,z) = 2i\left[\frac{\beta}{\rho^3}\, e^{-\beta z} + ie^{-\beta r}\left(\frac{\beta}{\rho^3}+\frac{3}{2\rho^4}\right)\right]. \tag{B90}$$

With $\beta = -ik$,

$$I_{13}(\rho,z) = 2k\left[\frac{e^{ikz}}{\rho^3} + ie^{ikr}\left(\frac{1}{\rho^3}+\frac{3i}{2k\rho^4}\right)\right]. \tag{B91}$$

The integral $I_{14}(\rho,z)$ defined in (A14a) can be expanded into

$$\begin{aligned} I_{14}(\rho,z) &= \int_0^\infty (k^2-\lambda^2)^{1/2}[J_0(\lambda\rho)-J_2(\lambda\rho)]e^{iz\sqrt{k^2-\lambda^2}}\lambda^2\, d\lambda \\ &= 2\int_0^\infty (k^2-\lambda^2)^{1/2} J_0(\lambda\rho) e^{iz\sqrt{k^2-\lambda^2}}\lambda^2\, d\lambda \\ &\quad - \int_0^\infty (k^2-\lambda^2)^{1/2}[J_0(\lambda\rho)+J_2(\lambda\rho)]e^{iz\sqrt{k^2-\lambda^2}}\lambda^2\, d\lambda \\ &= 2\int_0^\infty (k^2-\lambda^2)^{1/2} J_0(\lambda\rho) e^{iz\sqrt{k^2-\lambda^2}}\lambda^2\, d\lambda - I_{13}(\rho,z). \end{aligned} \tag{B92}$$

Here, the first integral on the right can be evaluated with (B7) with $\nu = 0$, that is,

$$\mathcal{I} \equiv \int_0^\infty (\lambda^2+\beta^2)^{-1/2} e^{-z\sqrt{\lambda^2+\beta^2}} J_0(\lambda\rho)\, d\lambda$$
$$= I_0[\tfrac{1}{2}\beta(r-z)]K_0[\tfrac{1}{2}\beta(r+z)]. \tag{B93}$$

When both sides are differentiated twice with respect to ρ,

$$\frac{\partial^2\mathcal{I}}{\partial\rho^2} = \int_0^\infty (\lambda^2+\beta^2)^{-1/2} e^{-z\sqrt{\lambda^2+\beta^2}}\, \frac{\partial^2 J_0(\lambda\rho)}{\partial\rho^2}\, d\lambda$$
$$= \frac{\partial^2}{\partial\rho^2}\,\{I_0[\tfrac{1}{2}\beta(r-z)]K_0[\tfrac{1}{2}\beta(r+z)]\}. \tag{B94}$$

However,

$$\frac{\partial J_0(\lambda\rho)}{\partial\rho} = -\lambda J_1(\lambda\rho). \tag{B95}$$

A standard formula for Bessel functions is

$$J_1'(x) = J_0(x) - \frac{1}{x}\, J_1(x). \tag{B96}$$

Hence,

$$\frac{\partial^2 J_0(\lambda\rho)}{\partial\rho^2} = -\lambda\,\frac{\partial J_1(\lambda\rho)}{\partial\rho} = -\lambda^2 J_0(\lambda\rho) + \frac{\lambda}{\rho}\,J_1(\lambda\rho) \tag{B97}$$

so that

$$\begin{aligned}\frac{\partial^2\mathcal{I}}{\partial\rho^2} &= -\int_0^\infty (\lambda^2+\beta^2)^{-1/2}e^{-z\sqrt{\lambda^2+\beta^2}}J_0(\lambda\rho)\lambda^2\,d\lambda\\ &\quad+\frac{1}{\rho}\int_0^\infty (\lambda^2+\beta^2)^{-1/2}e^{-z\sqrt{\lambda^2+\beta^2}}J_1(\lambda\rho)\lambda\,d\lambda.\end{aligned} \tag{B98}$$

If both sides are differentiated twice with respect to z, the result is

$$\begin{aligned}\frac{\partial^2}{\partial\rho^2}\,\frac{\partial^2}{\partial z^2}\,\mathcal{I} &= -\int_0^\infty (\lambda^2+\beta^2)^{1/2}e^{-z\sqrt{\lambda^2+\beta^2}}J_0(\lambda\rho)\lambda^2\,d\lambda\\ &\quad+\frac{1}{\rho}\int_0^\infty (\lambda^2+\beta^2)^{1/2}e^{-z\sqrt{\lambda^2+\beta^2}}J_1(\lambda\rho)\lambda\,d\lambda\\ &= -\int_0^\infty (\lambda^2+\beta^2)^{1/2}e^{-z\sqrt{\lambda^2+\beta^2}}J_0(\lambda\rho)\lambda^2\,d\lambda - \frac{i}{2}\,I_{13}(\rho,z).\end{aligned} \tag{B99}$$

Thus,

$$\begin{aligned}I_{14}(\rho,z) &= 2i\int_0^\infty (\lambda^2+\beta^2)^{1/2}e^{-z\sqrt{\lambda^2+\beta^2}}J_0(\lambda\rho)\lambda^2\,d\lambda - I_{13}(\rho,z)\\ &= -2i\,\frac{\partial^2}{\partial\rho^2}\,\frac{\partial^2}{\partial z^2}\,\mathcal{I}\\ &= -2i\,\frac{\partial^2}{\partial\rho^2}\,\frac{\partial^2}{\partial z^2}\,\{I_0[\tfrac{1}{2}\beta(r-z)]K_0[\tfrac{1}{2}\beta(r+z)]\}.\end{aligned} \tag{B100}$$

From (B66),

$$\begin{aligned}F &\equiv \frac{\partial^2}{\partial z^2}\,\{I_0[\tfrac{1}{2}\beta(r-z)]K_0[\tfrac{1}{2}\beta(r+z)]\}\\ &= e^{-\beta z}\left(\frac{\beta}{\rho}+\frac{z}{2\rho^3}-\frac{1}{2\beta\rho^3}\right)+ie^{-\beta r}\left(\frac{1}{\rho^2}-\frac{1}{2\beta\rho^3}+\frac{5}{8\beta^2\rho^4}\right),\end{aligned} \tag{B101}$$

$$\frac{\partial F}{\partial\rho} = e^{-\beta z}\left(-\frac{\beta}{\rho^2}-\frac{3z}{2\rho^4}+\frac{3}{2\beta\rho^4}\right)-ie^{-\beta r}\left(\frac{\beta}{\rho^2}+\frac{3}{2\rho^3}-\frac{7}{8\beta\rho^4}\right), \tag{B102}$$

$$\frac{\partial^2 F}{\partial\rho^2} = e^{-\beta z}\left(\frac{2\beta}{\rho^3}\right)+ie^{-\beta r}\left(\frac{\beta^2}{\rho^2}+\frac{7\beta}{2\rho^3}\right), \tag{B103}$$

where terms smaller than $1/\rho^3$ have been neglected. It follows that, with $\beta = -ik$,

$$I_{14}(\rho,z) = -\frac{4k}{\rho^3}\,e^{ikz} - 2k^2\left(\frac{1}{\rho^2}+\frac{7i}{2k\rho^3}\right)e^{ikr}. \tag{B104}$$

The integral $I_{15}(\rho, z)$ as defined in (A15a) is

$$\begin{aligned} I_{15}(\rho, z) &= \int_0^\infty (k^2 - \lambda^2)^{1/2}[J_0(\lambda\rho) + J_2(\lambda\rho)]e^{iz\sqrt{k^2-\lambda^2}}\lambda^3\, d\lambda \\ &= \frac{2}{\rho}\int_0^\infty (k^2 - \lambda^2)^{1/2} J_1(\lambda\rho)e^{iz\sqrt{k^2-\lambda^2}}\lambda^2\, d\lambda. \end{aligned} \tag{B105}$$

This can be obtained from

$$\begin{aligned} I_{15}(\rho, z) &= -\frac{\partial^2}{\partial z^2} I_7(\rho, z) \\ &= -\frac{\partial^2}{\partial z^2}\frac{2}{\rho}\int_0^\infty (k^2 - \lambda^2)^{-1/2} J_1(\lambda\rho)e^{iz\sqrt{k^2-\lambda^2}}\lambda^2\, d\lambda \\ &= 2\frac{\partial^2}{\partial z^2}\left(\frac{k}{r^2} + \frac{i}{r^3}\right)e^{ikr}. \end{aligned} \tag{B106}$$

Performing the differentiations gives

$$\begin{aligned} \frac{\partial}{\partial z}\left(\frac{k}{r^2} + \frac{i}{r^3}\right)e^{ikr} &= \left(\frac{ik^2}{r^2} - \frac{k}{r^3} - \frac{2k}{r^3} - \frac{3i}{r^4}\right)\left(\frac{z}{r}\right)e^{ikr} \\ &= \left(\frac{ik^2}{r^3} - \frac{3k}{r^4} - \frac{3i}{r^5}\right)ze^{ikr}, \end{aligned} \tag{B107}$$

$$\begin{aligned} &\frac{\partial^2}{\partial z^2}\left(\frac{k}{r^2} + \frac{i}{r^3}\right)e^{ikr} \\ &= \left[\left(-\frac{k^3}{r^3} - \frac{3ik^2}{r^4} + \frac{3k}{r^5} - \frac{3ik^2}{r^4} + \frac{12k}{r^5} + \frac{15i}{r^6}\right)\left(\frac{z^2}{r}\right)\right. \\ &\qquad \left. + \frac{ik^2}{r^3} - \frac{3k}{r^4} - \frac{3i}{r^5}\right]e^{ikr} \\ &= \left[\frac{ik^2}{r^3} - \frac{3k}{r^4} - \frac{3i}{r^5} - \frac{z^2}{r^2}\left(\frac{k^3}{r^2} + \frac{6ik^2}{r^3} - \frac{15k}{r^4} - \frac{15i}{r^5}\right)\right]e^{ikr}. \end{aligned} \tag{B108}$$

Hence,

$$I_{15}(\rho, z) = 2\left[\frac{ik^2}{r^3} - \frac{3k}{r^4} - \frac{3i}{r^5} - \frac{z^2}{r^2}\left(\frac{k^3}{r^2} + \frac{6ik^2}{r^3} - \frac{15k}{r^4} - \frac{15i}{r^5}\right)\right]e^{ikr}. \tag{B109}$$

Appendix C
Evaluation of Integrals of Bessel Functions

Most of the integrals $I_{16}(\rho, z)$ to $I_{29}(\rho, 0)$ as listed in (A16a) to (A29) are straightforward integrals of Bessel functions that involve no radicals. Their evaluation is included here for convenience. Useful formulas are

- Watson [1958, p. 132, formula (1)],

$$\int z^{\nu+1} J_\nu(z)\, dz = z^{\nu+1} J_{\nu+1}(z); \tag{C1}$$

- Gradshteyn and Ryzhik [1980, p. 665, formula (6.511-1)],

$$\int_0^\infty J_\nu(\lambda\rho)\, d\lambda = \frac{1}{\rho}. \tag{C2}$$

The integrals $I_{16}(\rho, z)$ and $I_{17}(\rho, z)$ can be evaluated together since they differ by a sign. They are

$$\left.\begin{matrix} I_{16}(\rho, z) \\ I_{17}(\rho, z) \end{matrix}\right\} = \int_0^\infty [J_0(\lambda\rho) \pm J_2(\lambda\rho)] e^{iz\sqrt{k^2-\lambda^2}}\, d\lambda$$
$$= -\frac{\partial}{\partial z} \int_0^\infty (\lambda^2 + \beta^2)^{-1/2} e^{-z\sqrt{\lambda^2+\beta^2}} [J_0(\lambda\rho) \pm J_2(\lambda\rho)]\, d\lambda. \tag{C3}$$

Here, the integral on the right is (B7) with $\nu = 0$ and $\nu = 2$. Specifically, with $y_1 = \frac{1}{2}\beta(r - z)$, $y_2 = \frac{1}{2}\beta(r + z)$, and $r = (\rho^2 + z^2)^{1/2}$, the integrated forms are

$$\left.\begin{matrix} I_{16}(\rho, z) \\ I_{17}(\rho, z) \end{matrix}\right\} = -\frac{\partial}{\partial z} [I_0(y_1)K_0(y_2) \pm I_1(y_1)K_1(y_2)]. \tag{C4}$$

In carrying out the differentiation and evaluation, the following are involved: $dy_1/dz = -y_1/r$, $dy_2/dz = y_2/r$, $y_1 - y_2 = -\beta z$, $y_1 + y_2 = \beta r$, and $y_1 y_2 = \frac{1}{4}\beta^2\rho^2$. Thus,

$$\left.\begin{matrix} I_{16}(\rho,z) \\ I_{17}(\rho,z) \end{matrix}\right\} = -\left[-\frac{y_1}{r} I_0'(y_1)K_0(y_2) + \frac{y_2}{r} I_0(y_1)K_0'(y_2) \right. \\ \left. \pm \left(-\frac{y_1}{r} I_1'(y_1)K_1(y_2) + \frac{y_2}{r} I_1(y_1)K_1'(y_2) \right) \right]. \tag{C5}$$

The following relations are now needed:

$$I_0'(y_1) = I_1(y_1); \qquad K_0'(y_2) = -K_1(y_2), \tag{C6a}$$

$$I_1'(y_1) = -y_1^{-1}I_1(y_1) + I_0(y_1); \quad K_1'(y_2) = -y_2^{-1}K_1(y_2) - K_0(y_2). \tag{C6b}$$

With these,

$$\begin{aligned} \left.\begin{matrix} I_{16}(\rho,z) \\ I_{17}(\rho,z) \end{matrix}\right\} &= \frac{1}{r} [y_1 I_1(y_1)K_0(y_2) + y_2 I_0(y_1)K_1(y_2) \mp I_1(y_1)K_1(y_2) \\ &\qquad \pm y_1 I_0(y_1)K_1(y_2) \pm I_1(y_1)K_1(y_2) \pm y_2 I_1(y_1)K_0(y_2)] \\ &= \frac{1}{r}(y_1 \pm y_2)[I_1(y_1)K_0(y_2) \pm I_0(y_1)K_1(y_2)] \\ &= \frac{1}{r} \left\{ \begin{matrix} \beta r \\ -\beta z \end{matrix} \right\} [K_0(y_2)I_1(y_1) \pm K_1(y_2)I_0(y_1)]. \end{aligned} \tag{C7}$$

With the large-argument approximations of the Bessel functions in (B12)–(B15), these become

$$\begin{aligned} &\left.\begin{matrix} I_{16}(\rho,z) \\ I_{17}(\rho,z) \end{matrix}\right\} \\ &= \frac{1}{r} \left\{ \begin{matrix} \beta r \\ -\beta z \end{matrix} \right\} \left\{ \sqrt{\frac{\pi}{2y_2}}\, e^{-y_2} \left(1 - \frac{1}{8y_2} + \frac{9}{128y_2^2} \right) \right. \\ &\qquad \times \left[\frac{e^{y_1}}{\sqrt{2\pi y_1}} \left(1 - \frac{3}{8y_1} - \frac{15}{128y_1^2} \right) \right. \\ &\qquad\quad \left. + \frac{ie^{-y_1}}{\sqrt{2\pi y_1}} \left(1 + \frac{3}{8y_1} - \frac{15}{128y_1^2} \right) \right] \\ &\qquad \pm \sqrt{\frac{\pi}{2y_2}}\, e^{-y_2} \left(1 + \frac{3}{8y_2} - \frac{15}{128y_2^2} \right) \\ &\qquad \times \left[\frac{e^{y_1}}{\sqrt{2\pi y_1}} \left(1 + \frac{1}{8y_1} + \frac{9}{128y_1^2} \right) \right. \\ &\qquad\quad \left. \left. - \frac{ie^{-y_1}}{\sqrt{2\pi y_1}} \left(1 - \frac{1}{8y_1} + \frac{9}{128y_1^2} \right) \right] \right\} \\ &= \frac{1}{\rho r} \left\{ \begin{matrix} r \\ -z \end{matrix} \right\} \left[e^{-\beta z} \left(1 - \frac{1}{8y_2} - \frac{3}{8y_1} + \frac{3}{64y_1 y_2} + \frac{9}{128y_2^2} - \frac{15}{128y_1^2} \right) \right. \end{aligned}$$

$$+\,ie^{-\beta r}\left(1-\frac{1}{8y_2}+\frac{3}{8y_1}-\frac{3}{64y_1y_2}+\frac{9}{128y_2^2}-\frac{15}{128y_1^2}\right)$$
$$\pm\,e^{-\beta z}\left(1+\frac{3}{8y_2}+\frac{1}{8y_1}+\frac{3}{64y_1y_2}-\frac{15}{128y_2^2}+\frac{9}{128y_1^2}\right)$$
$$\mp\,ie^{-\beta r}\left(1+\frac{3}{8y_2}-\frac{1}{8y_1}-\frac{3}{64y_1y_2}-\frac{15}{128y_2^2}+\frac{9}{128y_1^2}\right)\Bigg]. \tag{C8}$$

It is now convenient to separate the formulas for $I_{16}(\rho, z)$ and $I_{17}(\rho, z)$. They are

$$I_{16}(\rho,z)=\frac{1}{\rho}\left[e^{-\beta z}\left(2+\frac{1}{4y_2}-\frac{1}{4y_1}+\frac{3}{32y_1y_2}-\frac{3}{64y_2^2}-\frac{3}{64y_1^2}\right)\right.$$
$$\left.+\,ie^{-\beta r}\left(-\frac{1}{2y_2}+\frac{1}{2y_1}+\frac{3}{16y_2^2}-\frac{3}{16y_1^2}\right)\right]$$
$$=\frac{1}{\rho}\left[e^{-\beta z}\left(2-\frac{z}{\beta\rho^2}\right)+ie^{-\beta r}\left(\frac{2z}{\beta\rho^2}-\frac{3zr}{\beta^2\rho^4}\right)\right] \tag{C9}$$

and

$$I_{17}(\rho,z)=-\frac{z}{\rho r}\left[e^{-\beta z}\left(-\frac{1}{2y_2}-\frac{1}{2y_1}+\frac{3}{16y_2^2}-\frac{3}{16y_1^2}\right)\right.$$
$$\left.+\,ie^{-\beta r}\left(2+\frac{1}{4y_2}+\frac{1}{4y_1}-\frac{3}{32y_1y_2}-\frac{3}{64y_2^2}-\frac{3}{64y_1^2}\right)\right]$$
$$=-\frac{z}{\rho r}\left[e^{-\beta z}\left(-\frac{2r}{\beta\rho^2}-\frac{3zr}{\beta^2\rho^4}\right)\right.$$
$$\left.+\,ie^{-\beta r}\left(2+\frac{r}{\beta\rho^2}-\frac{3}{8\beta^2\rho^2}-\frac{3r^2}{8\beta^2\rho^4}-\frac{3z^2}{8\beta^2\rho^4}\right)\right]. \tag{C10}$$

With $\beta = -ik$ and $z^2 \ll \rho^2$ so that $r \sim \rho$ in amplitudes, the leading terms are

$$I_{16}(\rho,z)=2\left[e^{ikz}\left(\frac{1}{\rho}-\frac{iz}{2k\rho^3}\right)-\frac{z}{\rho}\frac{e^{ikr}}{k\rho^2}\right], \tag{C11a}$$
$$I_{17}(\rho,z)=2i\left(\frac{z}{\rho}\right)\left[\frac{e^{ikz}}{k\rho^2}-e^{ikr}\left(\frac{1}{\rho}+\frac{i}{2k\rho^2}+\frac{3}{8k^2\rho^3}\right)\right]. \tag{C11b}$$

A useful check on $I_{17}(\rho, z)$ is obtained quite simply with (B2). Thus,

$$I_{17}(\rho,z)=\frac{2}{\rho}\int_0^\infty \frac{\partial}{\partial\lambda}\left[J_1(\lambda\rho)\right]e^{iz\sqrt{k^2-\lambda^2}}\,d\lambda. \tag{C12}$$

Integration by parts gives

$$I_{17}(\rho,z)=\frac{2}{\rho}\left(J_1(\lambda\rho)e^{iz\sqrt{k^2-\lambda^2}}\Big|_0^\infty\right.$$

$$+ iz \int_0^\infty (k^2 - \lambda^2)^{-1/2} J_1(\lambda\rho) e^{iz\sqrt{k^2-\lambda^2}} \lambda\, d\lambda \bigg). \tag{C13}$$

The integrated term vanishes at both limits and

$$\begin{aligned} I_{17}(\rho, z) &= izI_5(\rho, z) \\ &= 2i\left(\frac{z}{\rho}\right)\left[\frac{e^{ikz}}{k\rho^2} - \left(\frac{1}{\rho} + \frac{i}{2k\rho^2} + \frac{3}{8k^2\rho^3}\right)e^{ikr}\right], \end{aligned} \tag{C14}$$

which is exactly (C11b).

The integral $I_{18}(\rho, z)$ is

$$\begin{aligned} I_{18}(\rho, z) &= \int_0^\infty [J_0(\lambda\rho) + J_2(\lambda\rho)] e^{iz\sqrt{k^2-\lambda^2}} \lambda\, d\lambda \\ &= \frac{2}{\rho}\int_0^\infty J_1(\lambda\rho) e^{iz\sqrt{k^2-\lambda^2}}\, d\lambda \\ &= \frac{2}{\rho}\int_0^\infty J_1(\lambda\rho) e^{-z\sqrt{\lambda^2+\beta^2}}\, d\lambda, \end{aligned} \tag{C15}$$

where $\beta = -ik$. This can be integrated with (B7) with $\nu = 1$ and differentiation with respect to z. Thus,

$$\begin{aligned} &\frac{\partial}{\partial z}\int_0^\infty (\lambda^2 + \beta^2)^{-1/2} e^{-z\sqrt{\lambda^2+\beta^2}} J_1(\lambda\rho)\, d\lambda \\ &\quad = \frac{\partial}{\partial z}\{I_{1/2}[\tfrac{1}{2}\beta(r - z)]K_{1/2}[\tfrac{1}{2}\beta(r + z)]\}, \end{aligned} \tag{C16}$$

where $r = (\rho^2 + z^2)^{1/2}$. With (B31),

$$\begin{aligned} \int_0^\infty e^{-z\sqrt{\lambda^2+\beta^2}} J_1(\lambda\rho)\, d\lambda &= -\frac{\partial}{\partial z}\left(\frac{1}{\beta\rho}(e^{-\beta z} - e^{-\beta r})\right) \\ &= \frac{e^{-\beta z}}{\rho} - \frac{z}{r}\frac{e^{-\beta r}}{\rho}. \end{aligned} \tag{C17}$$

With $\beta = -ik$, it follows that

$$I_{18}(\rho, z) = 2\left(\frac{e^{ikz}}{\rho^2} - \frac{z}{r}\frac{e^{ikr}}{\rho^2}\right). \tag{C18}$$

The integral $I_{19}(\rho, z)$ is

$$\begin{aligned} I_{19}(\rho, z) &= \int_0^\infty [J_0(\lambda\rho) - J_2(\lambda\rho)] e^{iz\sqrt{k^2-\lambda^2}} \lambda\, d\lambda \\ &= -i\frac{\partial}{\partial z} I_4(\rho, z) \\ &= -2\left[\frac{e^{ikz}}{\rho^2} + e^{ikr}\left(\frac{z}{r}\right)\left(\frac{ik}{r} - \frac{1}{r^2} - \frac{1}{\rho^2}\right)\right]. \end{aligned} \tag{C19}$$

The integral $I_{20}(\rho, z)$ is evaluated in (B74) and (B75). It is

$$\begin{aligned} I_{20}(\rho, z) &= \int_0^\infty [J_0(\lambda\rho) + J_2(\lambda\rho)] e^{iz\sqrt{k^2-\lambda^2}} \lambda^2 \, d\lambda \\ &= \frac{2}{\rho} \int_0^\infty J_1(\lambda\rho) e^{iz\sqrt{k^2-\lambda^2}} \lambda \, d\lambda \\ &= 2e^{ikz} \left(\frac{1}{\rho^3} + \frac{3iz}{2k\rho^5} \right) - 2ze^{ikr} \left(\frac{k}{\rho^3} + \frac{3i}{2\rho^4} + \frac{7}{8k\rho^5} \right). \end{aligned} \tag{C20}$$

With (A6a), the integral $I_{21}(\rho, z)$ is

$$\begin{aligned} I_{21}(\rho, z) &= \int_0^\infty [J_0(\lambda\rho) - J_2(\lambda\rho)] e^{iz\sqrt{k^2-\lambda^2}} \lambda^2 \, d\lambda \\ &= -i \frac{\partial}{\partial z} I_6(\rho, z). \end{aligned} \tag{C21}$$

The differentiation of $I_6(\rho, z)$ gives

$$I_{21}(\rho, z) = -\frac{4}{\rho^3} e^{ikz} - 2k\left(\frac{z}{r}\right)\left(\frac{ik}{\rho} - \frac{3}{2\rho^2} - \frac{5i}{8k\rho^3} + \frac{1}{4k^2\rho^4}\right)e^{ikr}. \tag{C22}$$

The integral $I_{22}(\rho, z)$ is

$$\begin{aligned} I_{22}(\rho, z) &= \int_0^\infty [J_0(\lambda\rho) + J_2(\lambda\rho)] e^{iz\sqrt{k^2-\lambda^2}} \lambda^3 \, d\lambda \\ &= \frac{2}{\rho} \int_0^\infty J_1(\lambda\rho) e^{-z\sqrt{\lambda^2+\beta^2}} \lambda^2 \, d\lambda. \end{aligned} \tag{C23}$$

This can be integrated with (B8). Thus,

$$\frac{\partial}{\partial z} \int_0^\infty (\lambda^2 - k^2)^{-1/2} e^{-z\sqrt{\lambda^2-k^2}} J_0(\lambda\rho) \lambda \, d\lambda = \frac{\partial}{\partial z} \frac{e^{ikr}}{r}$$

so that

$$-\int_0^\infty e^{-z\sqrt{\lambda^2-k^2}} J_0(\lambda\rho) \lambda \, d\lambda = \left(\frac{ik}{r} - \frac{1}{r^2}\right)\left(\frac{z}{r}\right) e^{ikr}.$$

When both sides are differentiated with respect to ρ, this becomes

$$\begin{aligned} &-\frac{\partial}{\partial \rho} \int_0^\infty e^{-z\sqrt{\lambda^2-k^2}} J_0(\lambda\rho) \lambda \, d\lambda \\ &\quad = ze^{ikr} \left[ik\left(\frac{ik}{r^2} - \frac{1}{r^3}\right) - \frac{2ik}{r^3} + \frac{3}{r^4} \right] \left(\frac{\rho}{r}\right). \end{aligned}$$

Then, since $(\partial/\partial\rho) J_0(\lambda\rho) = J_0'(\lambda\rho) \cdot \lambda = -\lambda J_1(\lambda\rho)$,

$$\begin{aligned} &\int_0^\infty e^{-z\sqrt{\lambda^2-k^2}} J_1(\lambda\rho) \lambda^2 \, d\lambda \\ &\quad = -ze^{ikr} \left(\frac{k^2}{r^2} + \frac{ik}{r^3} + \frac{2ik}{r^3} - \frac{3}{r^4} \right) \left(\frac{\rho}{r}\right) \end{aligned}$$

$$= -ke^{ikr}\left(\frac{k}{r} + \frac{3i}{r^2} - \frac{3}{kr^3}\right)\left(\frac{\rho z}{r^2}\right).$$

Multiplication by $2/\rho$ to give (C23) yields

$$I_{22}(\rho, z) = -2k\, e^{ikr}\left(\frac{k}{r^2} + \frac{3i}{r^3} - \frac{3}{r^4}\right)\left(\frac{z}{r}\right). \tag{C24}$$

The integral $I_{23}(\rho, z)$ can be rearranged with (B2) to give

$$\begin{aligned} I_{23}(\rho, z) &= \int_0^\infty [J_0(\lambda\rho) - J_2(\lambda\rho)] e^{iz\sqrt{k^2-\lambda^2}} \lambda^3 \, d\lambda \\ &= \frac{2}{\rho}\int_0^\infty \frac{\partial}{\partial\lambda}[J_1(\lambda\rho)] e^{iz\sqrt{k^2-\lambda^2}}\lambda^3 \, d\lambda. \end{aligned} \tag{C25}$$

Integration by parts gives

$$\begin{aligned} I_{23}(\rho, z) = \frac{2}{\rho}\Bigg[& J_1(\lambda\rho) e^{iz\sqrt{k^2-\lambda^2}}\lambda^3 \Bigg|_0^\infty \\ & - \int_0^\infty J_1(\lambda\rho)\left(3\lambda^2 - iz\,\frac{\lambda^4}{(k^2-\lambda^2)^{1/2}}\right) e^{iz\sqrt{k^2-\lambda^2}}\, d\lambda\Bigg]. \end{aligned} \tag{C26}$$

Here, the integrated term vanishes at both limits and the integral can be expanded with the relation

$$\frac{\lambda^2}{(k^2-\lambda^2)^{1/2}} = \frac{k^2}{(k^2-\lambda^2)^{1/2}} - (k^2-\lambda^2)^{1/2} \tag{C27}$$

so that

$$\begin{aligned} I_{23}(\rho, z) = -\frac{2}{\rho}\Bigg(& 3\int_0^\infty J_1(\lambda\rho) e^{iz\sqrt{k^2-\lambda^2}}\lambda^2 \, d\lambda \\ & - izk^2 \int_0^\infty \frac{J_1(\lambda\rho)}{(k^2-\lambda^2)^{1/2}}\, e^{iz\sqrt{k^2-\lambda^2}}\lambda^2 \, d\lambda \\ & + iz\int_0^\infty (k^2-\lambda^2)^{1/2} J_1(\lambda\rho) e^{iz\sqrt{k^2-\lambda^2}}\lambda^2\, d\lambda\Bigg). \end{aligned} \tag{C28}$$

With Appendix A, this is equivalent to

$$\begin{aligned} I_{23}(\rho, z) &= -[3I_{22}(\rho, z) - izk^2 I_7(\rho, z) + izI_{15}(\rho, z)] \\ &= 2\Bigg[3ke^{ikr}\left(\frac{z}{r}\right)\left(\frac{k}{r^2} + \frac{3i}{r^3} - \frac{3}{kr^4}\right) - izk^3\left(\frac{1}{r^2} + \frac{i}{kr^3}\right)e^{ikr} \\ &\qquad - iz\left(\frac{ik^2}{r^3} - \frac{3k}{r^4} - \frac{3i}{r^5}\right)e^{ikr}\Bigg] \\ &= 2ke^{ikr}\Bigg[\left(\frac{z}{r}\right)\left(\frac{3k}{r^2} + \frac{9i}{r^3} - \frac{9}{kr^4}\right) + \left(\frac{z}{r}\right)\left(-\frac{ik^2}{r} + \frac{k}{r^2}\right) \\ &\qquad + \left(\frac{z}{r}\right)\left(\frac{k}{r^2} + \frac{3i}{r^3} - \frac{3}{kr^4}\right)\Bigg] \end{aligned}$$

$$= 2ke^{ikr}\left(\frac{z}{r}\right)\left(-\frac{ik^2}{r} + \frac{5k}{r^2} + \frac{12i}{r^3} - \frac{12}{kr^4}\right). \tag{C29}$$

The integral $I_{24}(\rho, z)$ is

$$I_{24}(\rho, z) = \int_0^\infty [J_0(\lambda\rho) + J_2(\lambda\rho)]e^{iz\sqrt{k^2-\lambda^2}}\lambda^4\, d\lambda$$
$$= \frac{2}{\rho}\int_0^\infty J_1(\lambda\rho)e^{iz\sqrt{k^2-\lambda^2}}\lambda^3\, d\lambda. \tag{C30}$$

It can be evaluated with the help of the formula

$$I_{24}(\rho, z)$$
$$= -i\frac{\partial}{\partial z}\int_0^\infty (k^2-\lambda^2)^{-1/2}[J_0(\lambda\rho) + J_2(\lambda\rho)]e^{iz\sqrt{k^2-\lambda^2}}\lambda^4\, d\lambda. \tag{C31}$$

This integral can now be expanded with the identity in (C27) so that

$$I_{24}(\rho, z)$$
$$= -i\frac{\partial}{\partial z}\left(k^2\int_0^\infty (k^2-\lambda^2)^{-1/2}[J_0(\lambda\rho) + J_2(\lambda\rho)]e^{iz\sqrt{k^2-\lambda^2}}\lambda^2\, d\lambda\right.$$
$$\left. - \int_0^\infty (k^2-\lambda^2)^{1/2}[J_0(\lambda\rho) + J_2(\lambda\rho)]e^{iz\sqrt{k^2-\lambda^2}}\lambda^2\, d\lambda\right). \tag{C32}$$

With Appendix A, this gives

$$I_{24}(\rho, z) = -i\frac{\partial}{\partial z}[k^2 I_5(\rho, z) - I_{13}(\rho, z)]$$
$$= -2i\frac{\partial}{\partial z}\left\{k^2\left[\frac{e^{ikz}}{k\rho^3} - \left(\frac{1}{\rho^2} + \frac{i}{2k\rho^3} + \frac{3}{8k^2\rho^4}\right)e^{ikr}\right]\right.$$
$$\left. - k\left[\frac{e^{ikz}}{\rho^3} + ie^{ikr}\left(\frac{1}{\rho^3} + \frac{3i}{2k\rho^4}\right)\right]\right\}$$
$$= -2ik^2\left[ik\frac{e^{ikz}}{k\rho^3} - \left(\frac{1}{\rho^2} + \frac{i}{2k\rho^3} + \frac{3}{8k^2\rho^4}\right)\left(\frac{ikz}{r}\right)e^{ikr}\right]$$
$$+ 2ik\left[\frac{ik}{\rho^3}e^{ikz} + i\left(\frac{1}{\rho^3} + \frac{3i}{2k\rho^4}\right)\left(\frac{ikz}{r}\right)e^{ikr}\right]$$
$$= 2k^2\left[\frac{e^{ikz}}{\rho^3} - \left(\frac{z}{r}\right)\left(\frac{k}{\rho^2} + \frac{i}{2\rho^3} + \frac{3}{8k\rho^4}\right)e^{ikr}\right.$$
$$\left. - \frac{e^{ikz}}{\rho^3} - \left(\frac{z}{r}\right)\left(\frac{i}{\rho^3} - \frac{3}{2k\rho^4}\right)e^{ikr}\right]$$
$$= -2k^2\left(\frac{z}{r}\right)\left(\frac{k}{\rho^2} + \frac{3i}{2\rho^3} - \frac{9}{8k\rho^4}\right)e^{ikr}. \tag{C33}$$

The integrals $I_{25}(\rho, 0)$ in (A25) follow from (C2). The combination of (A18a,b) and (A19a,b) gives (A26a,b), while (A27a,b) is obtained when

(A19a,b) is subtracted from (A18a,b). Similarly, (A28) is obtained by adding (A20b) to (A21b), and (A29) follows when (A21b) is subtracted from (A20b).

Appendix D
Evaluation of Integrals with Products and Quotients of Radicals

Integrals that contain quotients and products of the radicals $\gamma_1 = (k_1^2 - \lambda^2)^{1/2}$ and $\gamma_2 = (k_2^2 - \lambda^2)^{1/2}$ or the factors $\gamma_2 \exp(i\gamma_1 z)$ and $\gamma_2^{-1} \exp(i\gamma_1 z)$ occur in the evaluation of the components of the electric and magnetic fields. These can be integrated subject to the condition $|k_1^2| \gg |k_2^2|$. The required integrals are

$$\begin{aligned} I_A(\rho, z) &= \int_0^\infty \frac{\gamma_1}{\gamma_2} [J_0(\lambda\rho) + J_2(\lambda\rho)] e^{i\gamma_1 z} \lambda^3 \, d\lambda \\ &= \frac{2}{\rho} \int_0^\infty \frac{\gamma_1}{\gamma_2} J_1(\lambda\rho) e^{i\gamma_1 z} \lambda^2 \, d\lambda, \end{aligned} \tag{D1}$$

$$I_B(\rho, z) = \int_0^\infty \frac{\gamma_1}{\gamma_2} [J_0(\lambda\rho) - J_2(\lambda\rho)] e^{i\gamma_1 z} \lambda \, d\lambda, \tag{D2}$$

$$\begin{aligned} I_C(\rho, z) &= \int_0^\infty \gamma_1\gamma_2 [J_0(\lambda\rho) + J_2(\lambda\rho)] e^{i\gamma_1 z} \lambda \, d\lambda \\ &= \frac{2}{\rho} \int_0^\infty \gamma_1\gamma_2 J_1(\lambda\rho) e^{i\gamma_1 z} \, d\lambda, \end{aligned} \tag{D3}$$

$$\begin{aligned} I_D(\rho, z) &= \int_0^\infty \frac{\gamma_1}{\gamma_2} [J_0(\lambda\rho) + J_2(\lambda\rho)] e^{i\gamma_1 z} \lambda \, d\lambda \\ &= \frac{2}{\rho} \int_0^\infty \frac{\gamma_1}{\gamma_2} J_1(\lambda\rho) e^{i\gamma_1 z} \, d\lambda, \end{aligned} \tag{D4}$$

$$I_E(\rho, z) = \int_0^\infty \frac{\gamma_1}{\gamma_2} [J_0(\lambda\rho) - J_2(\lambda\rho)] e^{i\gamma_1 z} \lambda^3 \, d\lambda, \tag{D5}$$

$$I_F(\rho, z) = \int_0^\infty \frac{\gamma_2}{\gamma_1} [J_0(\lambda\rho) + J_2(\lambda\rho)] e^{i\gamma_1 z} \lambda \, d\lambda$$
$$= \frac{2}{\rho} \int_0^\infty \frac{\gamma_2}{\gamma_1} J_1(\lambda\rho) e^{i\gamma_1 z} \, d\lambda, \tag{D6}$$

$$I_G(\rho, z) = \int_0^\infty \frac{\gamma_2}{\gamma_1} [J_0(\lambda\rho) - J_2(\lambda\rho)] e^{i\gamma_1 z} \lambda \, d\lambda, \tag{D7}$$

$$I_H(\rho, z) = \int_0^\infty \gamma_1 \gamma_2 [J_0(\lambda\rho) - J_2(\lambda\rho)] e^{i\gamma_1 z} \lambda \, d\lambda, \tag{D8}$$

$$I_K(\rho, z) = \int_0^\infty \gamma_2^{-1} [J_0(\lambda\rho) - J_2(\lambda\rho)] e^{i\gamma_1 z} \lambda \, d\lambda, \tag{D9}$$

$$I_L(\rho, z) = \int_0^\infty \gamma_2 [J_0(\lambda\rho) + J_2(\lambda\rho)] e^{i\gamma_1 z} \lambda \, d\lambda$$
$$= \frac{2}{\rho} \int_0^\infty \gamma_2 J_1(\lambda\rho) e^{i\gamma_1 z} \, d\lambda, \tag{D10}$$

$$I_M(\rho, z) = \int_0^\infty \gamma_2^{-1} [J_0(\lambda\rho) + J_2(\lambda\rho)] e^{i\gamma_1 z} \lambda \, d\lambda$$
$$= \frac{2}{\rho} \int_0^\infty \gamma_2^{-1} J_1(\lambda\rho) e^{i\gamma_1 z} \, d\lambda, \tag{D11}$$

$$I_N(\rho, z) = \int_0^\infty \gamma_2 [J_0(\lambda\rho) - J_2(\lambda\rho)] e^{i\gamma_1 z} \lambda \, d\lambda, \tag{D12}$$

$$I_Q(\rho, z) = \int_0^\infty \gamma_2^{-1} [J_0(\lambda\rho) - J_2(\lambda\rho)] e^{i\gamma_1 z} \lambda^3 \, d\lambda, \tag{D13}$$

$$I_R(\rho, z) = \int_0^\infty \gamma_2^{-1} [J_0(\lambda\rho) + J_2(\lambda\rho)] e^{i\gamma_1 z} \lambda^3 \, d\lambda$$
$$= \frac{2}{\rho} \int_0^\infty \gamma_2^{-1} J_1(\lambda\rho) e^{i\gamma_1 z} \lambda^2 \, d\lambda, \tag{D14}$$

$$I_S(\rho, z) = \int_0^\infty \gamma_2 [J_0(\lambda\rho) + J_2(\lambda\rho)] e^{i\gamma_1 z} \lambda^3 \, d\lambda$$
$$= \frac{2}{\rho} \int_0^\infty \gamma_2 J_1(\lambda\rho) e^{i\gamma_1 z} \lambda^2 \, d\lambda, \tag{D15}$$

$$I_T(\rho, z) = \int_0^\infty \frac{\gamma_2}{\gamma_1} [J_0(\lambda\rho) + J_2(\lambda\rho)] e^{i\gamma_1 z} \lambda^3 \, d\lambda$$
$$= \frac{2}{\rho} \int_0^\infty \frac{\gamma_2}{\gamma_1} J_1(\lambda\rho) e^{i\gamma_1 z} \lambda^2 \, d\lambda. \tag{D16}$$

Integrals of this type are evaluated by expanding the integrands into a sum of three terms according to the following approximate representation of a function of three parameters and variables, viz., $f(\lambda, \beta, k_2)$ subject to

$$|\beta^2| \gg |k_2^2|. \tag{D17}$$

Let

$$\begin{aligned}
f_1(\lambda,\beta,k_2) &= f(\lambda\sim\beta,\beta,k_2),\\
f_2(\lambda,\beta,k_2) &= f(\lambda\sim k_2,\beta,k_2),\\
f_3(\lambda,\beta,k_2) &= f_1(\lambda\sim k_2,\beta,k_2) = f_2(\lambda\sim\beta,\beta,k_2).
\end{aligned}\tag{D18}$$

Then,

$$f(\lambda,\beta,k_2) \sim f_1(\lambda,\beta,k_2) + f_2(\lambda,\beta,k_2) - f_3(\lambda,\beta,k_2).\tag{D19}$$

Specifically, with $k_1 = i\beta$, $\gamma_1 = (k_1^2-\lambda^2)^{1/2} = i(\lambda^2+\beta^2)^{1/2}$, and $\gamma_2 = (k_2^2-\lambda^2)^{1/2} = i(\lambda^2-k_2^2)^{1/2}$,

$$\begin{aligned}
f(\lambda,\beta,k_2) &= \frac{\gamma_1}{\gamma_2}\,e^{i\gamma_1 z} = \left(\frac{\lambda^2+\beta^2}{\lambda^2-k_2^2}\right)^{1/2} e^{-z\sqrt{\lambda^2+\beta^2}}\\
&\sim \lambda^{-1}(\lambda^2+\beta^2)^{1/2}e^{-z\sqrt{\lambda^2+\beta^2}}\\
&\quad + \beta(\lambda^2-k_2^2)^{-1/2}e^{-\beta z} - \beta\lambda^{-1}e^{-\beta z},
\end{aligned}\tag{D20}$$

$$\begin{aligned}
f(\lambda,\beta,k_2) &= \gamma_1\gamma_2 e^{i\gamma_1 z} = -[(\lambda^2+\beta^2)(\lambda^2-k_2^2)]^{1/2}e^{-z\sqrt{\lambda^2+\beta^2}}\\
&\sim -[\lambda(\lambda^2+\beta^2)^{1/2}e^{-z\sqrt{\lambda^2+\beta^2}}\\
&\quad + \beta(\lambda^2-k_2^2)^{1/2}e^{-\beta z} - \beta\lambda e^{-\beta z}].
\end{aligned}\tag{D21}$$

With these approximations, the integrals (D1) through (D16) can be expanded and the resulting integrals expressed in terms of the formulas in Appendix A. Thus,

$$\begin{aligned}
I_A(\rho,z) &\sim \frac{2}{\rho}\biggl(\int_0^\infty (\lambda^2+\beta^2)^{1/2}J_1(\lambda\rho)e^{-z\sqrt{\lambda^2+\beta^2}}\lambda\,d\lambda\\
&\qquad + \beta e^{-\beta z}\int_0^\infty (\lambda^2-k_2^2)^{-1/2}J_1(\lambda\rho)\lambda^2\,d\lambda\\
&\qquad - \beta e^{-\beta z}\int_0^\infty J_1(\lambda\rho)\lambda\,d\lambda\biggr)\\
&= -iI_{13}(k_1;\rho,z) - ik_1e^{ik_1z}[iI_7(k_2;\rho,0) - I_{20}(\rho,0)]\\
&= -2ik_1\left[\frac{e^{ik_1z}}{\rho^3} + ie^{ik_1r}\left(\frac{1}{\rho^3}+\frac{3i}{2k_1\rho^4}\right)\right]\\
&\qquad - 2k_1\left(\frac{k_2}{\rho^2}+\frac{i}{\rho^3}\right)e^{ik_2\rho}e^{ik_1z} + 2ik_1\,\frac{e^{ik_1z}}{\rho^3}\\
&= -2k_1\left[\left(\frac{k_2}{\rho^2}+\frac{i}{\rho^3}\right)e^{ik_2\rho}e^{ik_1z} - e^{ik_1r}\left(\frac{1}{\rho^3}+\frac{3i}{2k_1\rho^4}\right)\right],
\end{aligned}\tag{D22}$$

$$\begin{aligned}
I_B(\rho,z) &\sim \int_0^\infty (\lambda^2+\beta^2)^{1/2}[J_0(\lambda\rho)-J_2(\lambda\rho)]e^{-z\sqrt{\lambda^2+\beta^2}}\,d\lambda\\
&\qquad + \beta e^{-\beta z}\int_0^\infty (\lambda^2-k_2^2)^{-1/2}[J_0(\lambda\rho)-J_2(\lambda\rho)]\lambda\,d\lambda
\end{aligned}$$

$$
\begin{aligned}
&- \beta e^{-\beta z} \int_0^\infty [J_0(\lambda\rho) - J_2(\lambda\rho)]\, d\lambda \\
&= -iI_{10}(k_1;\rho,z) - ik_1 e^{ik_1 z}[iI_4(k_2;\rho,0) - I_{17}(\rho,0)] \\
&= -2\left\{\left(-\frac{z}{\rho^3} + \frac{i}{k_1\rho^3}\right)e^{ik_1 z} - i\left(\frac{1}{\rho^2} + \frac{i}{2k_1\rho^3} + \frac{3}{8k_1^2\rho^4}\right)e^{ik_1 r}\right. \\
&\qquad \left. + \frac{k_1}{k_2}\left[\frac{1}{\rho^2} + \left(\frac{ik_2}{\rho} - \frac{1}{\rho^2}\right)e^{ik_2\rho}\right]e^{ik_1 z}\right\} \\
&= -2\left[\left(\frac{k_1}{k_2\rho^2} - \frac{z}{\rho^3} + \frac{i}{k_1\rho^3}\right)e^{ik_1 z} - i\left(\frac{1}{\rho^2} + \frac{i}{2k_1\rho^3} + \frac{3}{8k_1^2\rho^4}\right)e^{ik_1 r}\right. \\
&\qquad \left. + \frac{k_1}{k_2}\left(\frac{ik_2}{\rho} - \frac{1}{\rho^2}\right)e^{ik_2\rho}e^{ik_1 z}\right],
\end{aligned}
\tag{D23}
$$

$$
\begin{aligned}
I_C(\rho,z) &\sim -\frac{2}{\rho}\left[\int_0^\infty (\lambda^2+\beta^2)^{1/2} J_1(\lambda\rho) e^{-z\sqrt{\lambda^2+\beta^2}}\lambda\, d\lambda\right. \\
&\qquad \left. + \beta e^{-\beta z}\left(\int_0^\infty (\lambda^2 - k_2^2)^{1/2} J_1(\lambda\rho)\, d\lambda - \int_0^\infty J_1(\lambda\rho)\lambda\, d\lambda\right)\right] \\
&= iI_{13}(k_1;\rho,z) + k_1 e^{ik_1 z}[I_{11}(k_2;\rho,0) - iI_{20}(\rho,0)] \\
&= 2ik_1\left[\frac{e^{ik_1 z}}{\rho^3} + i\left(\frac{1}{\rho^3} + \frac{3i}{2k_1\rho^4}\right)e^{ik_1 r}\right] \\
&\qquad + 2k_1k_2\left(\frac{1}{\rho^2} + \frac{i}{k_2\rho^3}\, e^{ik_2\rho}\right)e^{ik_1 z} - 2ik_1\,\frac{e^{ik_1 z}}{\rho^3} \\
&= -2k_1\left[\left(\frac{1}{\rho^3} + \frac{3i}{2k_1\rho^4}\right)e^{ik_1 r} - k_2\left(\frac{1}{\rho^2} + \frac{i}{k_2\rho^3}\, e^{ik_2\rho}\right)e^{ik_1 z}\right],
\end{aligned}
\tag{D24}
$$

$$
\begin{aligned}
I_D(\rho,z) &\sim \int_0^\infty (\lambda^2+\beta^2)^{1/2}[J_0(\lambda\rho) + J_2(\lambda\rho)]e^{-z\sqrt{\lambda^2+\beta^2}}\, d\lambda \\
&\qquad + \beta e^{-\beta z}\int_0^\infty (\lambda^2 - k_2^2)^{-1/2}[J_0(\lambda\rho) + J_2(\lambda\rho)]\lambda\, d\lambda \\
&\qquad - \beta e^{-\beta z}\int_0^\infty [J_0(\lambda\rho) + J_2(\lambda\rho)]\, d\lambda \\
&= -iI_9(k_1;\rho,z) + k_1 e^{ik_1 z}[I_3(k_2;\rho,0) + iI_{16}(\rho,0)] \\
&= -2i\left[\left(\frac{k_1}{\rho} - \frac{1}{2k_1\rho^3} - \frac{iz}{2\rho^3}\right)e^{ik_1 z} + \frac{ie^{ik_1 r}}{k_1\rho^3}\right] \\
&\qquad + \frac{2k_1}{k_2\rho^2}\,(1 - e^{ik_2\rho})e^{ik_1 z} + \frac{2ik_1}{\rho}\, e^{ik_1 z} \\
&= e^{ik_1 z}\left(\frac{i}{k_1\rho^3} - \frac{z}{\rho^3} + \frac{2k_1}{k_2\rho^2}\right) + \frac{2e^{ik_1 r}}{k_1\rho^3} - \frac{2k_1}{k_2\rho^2}\, e^{ik_1 z}e^{ik_2\rho},
\end{aligned}
\tag{D25}
$$

$$\begin{aligned}
I_E(\rho,z) &\sim \int_0^\infty (\lambda^2+\beta^2)^{1/2}[J_0(\lambda\rho)-J_2(\lambda\rho)]e^{-z\sqrt{\lambda^2+\beta^2}}\lambda^2\,d\lambda \\
&\quad + \beta e^{-\beta z}\int_0^\infty (\lambda^2-k_2^2)^{-1/2}[J_0(\lambda\rho)-J_2(\lambda\rho)]\lambda^3\,d\lambda \\
&\quad - \beta e^{-\beta z}\int_0^\infty [J_0(\lambda\rho)-J_2(\lambda\rho)]\lambda^2\,d\lambda \\
&= -iI_{14}(k_1;\rho,z) + k_1 e^{ik_1z}[I_8(k_2;\rho,0)+iI_{21}(\rho,0)] \\
&= \frac{4ik_1}{\rho^3}\,e^{ik_1z} + 2ik_1^2\left(\frac{1}{\rho^2}+\frac{7i}{2k_1\rho^3}\right)e^{ik_1r} \\
&\quad - 2k_1k_2\left(\frac{ik_2}{\rho}-\frac{2}{\rho^2}-\frac{2i}{k_2\rho^3}\right)e^{ik_2\rho}e^{ik_1z} - \frac{4ik_1}{\rho^3}\,e^{ik_1z} \\
&= 2k_1\left[e^{ik_1r}\left(\frac{ik_1}{\rho^2}-\frac{7}{2\rho^3}\right) - e^{ik_1z}e^{ik_2\rho}\left(\frac{ik_2^2}{\rho}-\frac{2k_2}{\rho^2}-\frac{2i}{\rho^3}\right)\right], \qquad \text{(D26)}
\end{aligned}$$

$$\begin{aligned}
I_F(\rho,z) &\sim \frac{2}{\rho}\Bigg(\int_0^\infty (\lambda^2+\beta^2)^{-1/2}J_1(\lambda\rho)e^{-z\sqrt{\lambda^2+\beta^2}}\lambda\,d\lambda \\
&\quad + \beta^{-1}e^{-\beta z}\int_0^\infty (\lambda^2-k_2^2)^{1/2}J_1(\lambda\rho)\,d\lambda \\
&\quad - \beta^{-1}e^{-\beta z}\int_0^\infty J_1(\lambda\rho)\lambda\,d\lambda\Bigg) \\
&= iI_5(k_1;\rho,z) + k_1^{-1}e^{ik_1z}[I_{11}(k_2;\rho,0) - iI_{20}(\rho,0)] \\
&= 2i\left[\frac{e^{ik_1z}}{k_1\rho^3} - \left(\frac{1}{\rho^2}+\frac{i}{2k_1\rho^3}\right)e^{ik_1r}\right] \\
&\quad + \frac{2k_2}{k_1}\left(\frac{1}{\rho^2}+\frac{ie^{ik_2\rho}}{k_2\rho^3}\right)e^{ik_1z} - \frac{2i}{k_1\rho^3}\,e^{ik_1z} \\
&= -2ie^{ik_1r}\left(\frac{1}{\rho^2}+\frac{i}{2k_1\rho^3}\right) + \frac{2k_2}{k_1}\,\frac{e^{ik_1z}}{\rho^2} + \frac{2i}{k_1\rho^3}\,e^{ik_1z}e^{ik_2\rho}, \qquad \text{(D27)}
\end{aligned}$$

$$\begin{aligned}
I_G(\rho,z) &\sim \int_0^\infty (\lambda^2+\beta^2)^{-1/2}[J_0(\lambda\rho)-J_2(\lambda\rho)]e^{-z\sqrt{\lambda^2+\beta^2}}\lambda^2\,d\lambda \\
&\quad + \beta^{-1}e^{-\beta z}\int_0^\infty (\lambda^2-k_2^2)^{1/2}[J_0(\lambda\rho)-J_2(\lambda\rho)]\lambda\,d\lambda \\
&\quad - \beta^{-1}e^{-\beta z}\int_0^\infty [J_0(\lambda\rho)-J_2(\lambda\rho)]\lambda^2\,d\lambda \\
&= iI_6(k_1;\rho,z) + k_1^{-1}e^{ik_1z}[I_{12}(k_2;\rho,0) - iI_{21}(\rho,0)] \\
&= -\frac{4i}{k_1\rho^3}\,e^{ik_1z} - 2i\left(\frac{ik_1}{\rho}-\frac{3}{2\rho^2}-\frac{5i}{8k_1\rho^3}\right)e^{ik_1r} \\
&\quad - \frac{2k_2}{k_1}\left[\frac{1}{\rho^2}+\left(\frac{1}{\rho^2}+\frac{2i}{k_2\rho^3}\right)e^{ik_2\rho}\right]e^{ik_1z} + \frac{4i}{k_1\rho^3}\,e^{ik_1z}
\end{aligned}$$

$$= -\frac{2}{k_1}\left[\frac{k_2}{\rho^2}\,e^{ik_1z} + ie^{ik_1r}\left(\frac{ik_1^2}{\rho} - \frac{3k_1}{2\rho^2} - \frac{5i}{8\rho^3}\right) + e^{ik_1z}e^{ik_2\rho}\left(\frac{k_2}{\rho^2} + \frac{2i}{\rho^3}\right)\right], \tag{D28}$$

$$\begin{aligned}
I_H(\rho, z) \sim &- \int_0^\infty (\lambda^2 + \beta^2)^{1/2}[J_0(\lambda\rho) - J_2(\lambda\rho)]e^{-z\sqrt{\lambda^2+\beta^2}}\lambda^2\,d\lambda \\
&- \beta e^{-\beta z}\int_0^\infty (\lambda^2 - k_2^2)^{1/2}[J_0(\lambda\rho) - J_2(\lambda\rho)]\lambda\,d\lambda \\
&+ \beta e^{-\beta z}\int_0^\infty [J_0(\lambda\rho) - J_2(\lambda\rho)]\lambda^2\,d\lambda \\
&= iI_{14}(k_1;\rho, z) + k_1 e^{ik_1z}[I_{12}(k_2;\rho,0) - iI_{21}(\rho,0)] \\
&= -\frac{4ik_1}{\rho^3}\,e^{ik_1z} - 2ik_1^2\left(\frac{1}{\rho^2} + \frac{7i}{2k_1\rho^3}\right)e^{ik_1r} \\
&\quad - 2k_1k_2e^{ik_1z}\left[\frac{1}{\rho^2} + \left(\frac{1}{\rho^2} + \frac{2i}{k_2\rho^3}\right)e^{ik_2\rho}\right] + \frac{4ik_1}{\rho^3}\,e^{ik_1z} \\
&= -2k_1\left[ik_1\left(\frac{1}{\rho^2} + \frac{7i}{2k_1\rho^3}\right)e^{ik_1r} + \frac{k_2}{\rho^2}\,e^{ik_1z} + k_2\left(\frac{1}{\rho^2} + \frac{2i}{k_2\rho^3}\right)e^{ik_2\rho}e^{ik_1z}\right],
\end{aligned} \tag{D29}$$

$$\begin{aligned}
I_K(\rho, z) \sim &-i\int_0^\infty [J_0(\lambda\rho) - J_2(\lambda\rho)]e^{-z\sqrt{\lambda^2+\beta^2}}\,d\lambda \\
&- ie^{-\beta z}\int_0^\infty (\lambda^2 - k_2^2)^{-1/2}[J_0(\lambda\rho) - J_2(\lambda\rho)]\lambda\,d\lambda \\
&+ ie^{-\beta z}\int_0^\infty [J_0(\lambda\rho) - J_2(\lambda\rho)]\,d\lambda \\
&= -iI_{17}(k_1;\rho,z) + e^{ik_1z}I_4(k_2;\rho,0) + ie^{ik_1z}I_{17}(\rho,0) \\
&= 2\left(\frac{z}{\rho}\right)\left[\frac{e^{ik_1z}}{k_1\rho^2} - \left(\frac{1}{\rho} + \frac{i}{2k_1\rho^2} + \frac{3}{8k_1^2\rho^3}\right)e^{ik_1r}\right] \\
&\quad - \frac{2}{k_2}\,e^{ik_1z}\left[\frac{1}{\rho^2} + \left(\frac{ik_2}{\rho} - \frac{1}{\rho^2}\right)e^{ik_2\rho}\right],
\end{aligned} \tag{D30}$$

$$\begin{aligned}
I_L(\rho, z) \sim &\; i\int_0^\infty [J_0(\lambda\rho) + J_2(\lambda\rho)]e^{-z\sqrt{\lambda^2+\beta^2}}\lambda^2\,d\lambda \\
&+ ie^{-\beta z}\int_0^\infty (\lambda^2 - k_2^2)^{1/2}[J_0(\lambda\rho) + J_2(\lambda\rho)]\lambda\,d\lambda \\
&- ie^{-\beta z}\int_0^\infty [J_0(\lambda\rho) + J_2(\lambda\rho)]\lambda^2\,d\lambda \\
&= iI_{20}(k_1;\rho,z) + e^{ik_1z}I_{11}(k_2;\rho,0) - ie^{ik_1z}I_{20}(\rho,0)
\end{aligned}$$

$$= 2i\left[\frac{e^{ik_1z}}{\rho^3} - \left(\frac{z}{\rho}\right)\left(\frac{k_1}{\rho^2} + \frac{3i}{2\rho^3}\right)e^{ik_1r}\right]$$

$$+ 2e^{ik_1z}\left(\frac{k_2}{\rho^2} + \frac{ie^{ik_2\rho}}{\rho^3}\right) - \frac{2i}{\rho^3}\, e^{ik_1z}$$

$$= 2\left[\left(\frac{k_2}{\rho^2} + \frac{ie^{ik_2\rho}}{\rho^3}\right)e^{ik_1z} - \left(\frac{z}{\rho}\right)\left(\frac{ik_1}{\rho^2} - \frac{3}{2\rho^3}\right)e^{ik_1r}\right], \tag{D31}$$

$$I_M(\rho, z) \sim -i\int_0^\infty [J_0(\lambda\rho) + J_2(\lambda\rho)]e^{-z\sqrt{\lambda^2+\beta^2}}\, d\lambda$$

$$- ie^{-\beta z}\int_0^\infty (\lambda^2 - k_2^2)^{-1/2}[J_0(\lambda\rho) + J_2(\lambda\rho)]\lambda\, d\lambda$$

$$+ ie^{-\beta z}\int_0^\infty [J_0(\lambda\rho) + J_2(\lambda\rho)]\, d\lambda$$

$$= -iI_{16}(k_1; \rho, z) + e^{ik_1z}I_3(k_2; \rho, 0) + ie^{ik_1z}I_{16}(\rho, 0)$$

$$= -2\left[\frac{ie^{ik_1z}}{\rho} - \left(\frac{z}{\rho}\right)\frac{ie^{ik_1r}}{k_1\rho^2}\right] + \frac{2}{k_2\rho^2}\, e^{ik_1z}(1 - e^{ik_2\rho}) + \frac{2i}{\rho}\, e^{ik_1z}$$

$$= 2\left[\frac{e^{ik_1z}}{k_2\rho^2}(1 - e^{ik_2\rho}) + \left(\frac{z}{\rho}\right)\frac{ie^{ik_1r}}{k_1\rho^2}\right], \tag{D32}$$

$$I_N(\rho, z) \sim i\int_0^\infty [J_0(\lambda\rho) - J_2(\lambda\rho)]e^{-z\sqrt{\lambda^2+\beta^2}}\lambda^2\, d\lambda$$

$$+ ie^{-\beta z}\int_0^\infty (\lambda^2 - k_2^2)^{1/2}[J_0(\lambda\rho) - J_2(\lambda\rho)]\lambda\, d\lambda$$

$$- ie^{-\beta z}\int_0^\infty [J_0(\lambda\rho) - J_2(\lambda\rho)]\lambda^2\, d\lambda$$

$$= iI_{21}(k_1; \rho, z) + e^{ik_1z}I_{12}(k_2; \rho, 0) - ie^{ik_1z}I_{21}(\rho, 0)$$

$$= -\frac{4i}{\rho^3}\, e^{ik_1z} - 2ik_1\left(\frac{z}{r}\right)\left(\frac{ik_1}{\rho} - \frac{3}{2\rho^2} - \frac{5i}{8k_1\rho^3}\right)e^{ik_1r}$$

$$- 2k_2e^{ik_1z}\left[\frac{1}{\rho^2} + \left(\frac{1}{\rho^2} + \frac{2i}{k_2\rho^3}\right)e^{ik_2\rho}\right] + \frac{4i}{\rho^3}\, e^{ik_1z}$$

$$= -2\left\{ik_1\left(\frac{z}{r}\right)\left(\frac{ik_1}{\rho} - \frac{3}{2\rho^2} - \frac{5i}{8k_1\rho^3}\right)e^{ik_1r}\right.$$

$$\left. + k_2\left[\frac{e^{ik_1z}}{\rho^2} + \left(\frac{1}{\rho^2} + \frac{2i}{k_2\rho^3}\right)e^{ik_2\rho}e^{ik_1z}\right]\right\}, \tag{D33}$$

$$I_Q(\rho, z) \sim -i\int_0^\infty [J_0(\lambda\rho) - J_2(\lambda\rho)]e^{-z\sqrt{\lambda^2+\beta^2}}\lambda^2\, d\lambda$$

$$- ie^{-\beta z}\int_0^\infty (\lambda^2 - k_2^2)^{-1/2}[J_0(\lambda\rho) - J_2(\lambda\rho)]\lambda^3\, d\lambda$$

$$
\begin{aligned}
&+ ie^{-\beta z}\int_0^\infty [J_0(\lambda\rho) - J_2(\lambda\rho)]\lambda^2\,d\lambda \\
&= -iI_{21}(k_1;\rho,z) + e^{ik_1 z}I_8(k_2;\rho,0) + ie^{ik_1 z}I_{21}(\rho,0) \\
&= i\left[\frac{4}{\rho^3}\,e^{ik_1 z} + 2k_1\left(\frac{z}{r}\right)\left(\frac{ik_1}{\rho} - \frac{3}{2\rho^2} - \frac{5i}{8k_1\rho^3}\right)e^{ik_1 r}\right] \\
&\quad - 2k_2 e^{ik_1 z}\left(\frac{ik_2}{\rho} - \frac{2}{\rho^2} - \frac{2i}{k_2\rho^3}\right)e^{ik_2\rho} - \frac{4i}{\rho^3}\,e^{ik_1 z} \\
&= 2\left[ik_1\left(\frac{z}{r}\right)\left(\frac{ik_1}{\rho} - \frac{3}{2\rho^2} - \frac{5i}{8k_1\rho^3}\right)e^{ik_1 r}\right. \\
&\quad \left. - k_2\left(\frac{ik_2}{\rho} - \frac{2}{\rho^2} - \frac{2i}{k_2\rho^3}\right)e^{ik_2\rho}e^{ik_1 z}\right],
\end{aligned} \tag{D34}
$$

$$
\begin{aligned}
I_R(\rho,z) &\sim -i\int_0^\infty [J_0(\lambda\rho) + J_2(\lambda\rho)]e^{-z\sqrt{\lambda^2+\beta^2}}\lambda^2\,d\lambda \\
&\quad - ie^{-\beta z}\int_0^\infty (\lambda^2 - k_2^2)^{-1/2}[J_0(\lambda\rho) + J_2(\lambda\rho)]\lambda^3\,d\lambda \\
&\quad + ie^{-\beta z}\int_0^\infty [J_0(\lambda\rho) + J_2(\lambda\rho)]\lambda^2\,d\lambda \\
&= -iI_{20}(k_1;\rho,z) + e^{ik_1 z}I_7(k_2;\rho,0) + ie^{ik_1 z}I_{20}(\rho,0) \\
&= -\frac{2i}{\rho^3}\,e^{ik_1 z} + 2i\left(\frac{z}{\rho}\right)\left(\frac{k_1}{\rho^2} + \frac{3i}{2\rho^3}\right)e^{ik_1 r} \\
&\quad - 2e^{ik_1 z}\left(\frac{k_2}{\rho^2} + \frac{i}{\rho^3}\right)e^{ik_2\rho} + \frac{2i}{\rho^3}\,e^{ik_1 z} \\
&= -2\left(\frac{k_2}{\rho^2} + \frac{i}{\rho^3}\right)e^{ik_2\rho}e^{ik_1 z} + 2i\left(\frac{z}{\rho}\right)\left(\frac{k_1}{\rho^2} + \frac{3i}{2\rho^3}\right)e^{ik_1 r},
\end{aligned} \tag{D35}
$$

$$
\begin{aligned}
I_S(\rho,z) &\sim i\int_0^\infty [J_0(\lambda\rho) + J_2(\lambda\rho)]e^{-z\sqrt{\lambda^2+\beta^2}}\lambda^4\,d\lambda \\
&\quad + ie^{-\beta z}\int_0^\infty (\lambda^2 - k_2^2)^{1/2}[J_0(\lambda\rho) + J_2(\lambda\rho)]\lambda^3\,d\lambda \\
&\quad - ie^{-\beta z}\int_0^\infty [J_0(\lambda\rho) + J_2(\lambda\rho)]\lambda^4\,d\lambda \\
&= iI_{24}(k_1;\rho,z) + e^{ik_1 z}I_{15}(k_2;\rho,0) - ie^{ik_1 z}I_{24}(\rho,0) \\
&= -2ik_1^2\left(\frac{z}{r}\right)\left(\frac{k_1}{\rho^2} + \frac{3i}{2\rho^3} - \frac{9}{8k_1\rho^4}\right)e^{ik_1 r} \\
&\quad + 2\left(\frac{ik_2^2}{\rho^3} - \frac{3k_2}{\rho^4} - \frac{3i}{\rho^5}\right)e^{ik_2\rho}e^{ik_1 z},
\end{aligned} \tag{D36}
$$

$$
I_T(\rho,z) \sim \int_0^\infty (\lambda^2 + \beta^2)^{-1/2}[J_0(\lambda\rho) + J_2(\lambda\rho)]e^{-z\sqrt{\lambda^2+\beta^2}}\lambda^4\,d\lambda
$$

$$
\begin{aligned}
&+ \beta^{-1} e^{-\beta z} \int_0^\infty (\lambda^2 - k_2^2)^{1/2} [J_0(\lambda\rho) + J_2(\lambda\rho)] \lambda^3 \, d\lambda \\
&- \beta^{-1} e^{-\beta z} \int_0^\infty [J_0(\lambda\rho) + J_2(\lambda\rho)] \lambda^4 \, d\lambda \\
\sim\ & k_1^2 \int_0^\infty (\lambda^2 + \beta^2)^{-1/2} [J_0(\lambda\rho) + J_2(\lambda\rho)] e^{-z\sqrt{\lambda^2+\beta^2}} \lambda^2 \, d\lambda \\
&+ \int_0^\infty (\lambda^2 + \beta^2)^{1/2} [J_0(\lambda\rho) + J_2(\lambda\rho)] e^{-z\sqrt{\lambda^2+\beta^2}} \lambda^2 \, d\lambda \\
&+ \beta^{-1} e^{-\beta z} \int_0^\infty (\lambda^2 - k_2^2)^{1/2} [J_0(\lambda\rho) + J_2(\lambda\rho)] \lambda^3 \, d\lambda \\
&- \beta^{-1} e^{-\beta z} \int_0^\infty [J_0(\lambda\rho) + J_2(\lambda\rho)] \lambda^4 \, d\lambda \\
=\ & i k_1^2 I_5(k_1; \rho, z) - i I_{13}(k_1; \rho, z) + k_1^{-1} e^{ik_1 z} I_{15}(k_2; \rho, 0) \\
&- i k_1^{-1} e^{ik_1 z} I_{24}(\rho, 0) \\
=\ & 2ik_1^2 \left[\frac{e^{ik_1 z}}{k_1 \rho^3} - \left(\frac{1}{\rho^2} + \frac{i}{2k_1\rho^3} \right) e^{ik_1 r} \right] \\
&- \frac{2ik_1}{\rho^3} \left(e^{ik_1 z} + i e^{ik_1 r} \right) + \frac{2ik_2^2}{k_1 \rho^3} e^{ik_2\rho} e^{ik_1 z} \\
=\ & -2i \left[e^{ik_1 r} \left(\frac{k_1^2}{\rho^2} + \frac{3ik_1}{2\rho^3} \right) - e^{ik_2\rho} e^{ik_1 z} \frac{k_2^2}{k_1 \rho^3} \right].
\end{aligned}
\tag{D37}
$$

The first integral in $I_T(\rho, z)$ was expanded using the identity in (C27).

Appendix E
Evaluation of Integrals That Lead to Fresnel Integrals; $z^2 \ll \rho^2$

The following integrals occur in the formulas for the several components of the electric and magnetic fields and they all lead to the Fresnel integral.

Vertical Dipole in Region 1, Field in Region 1 (Chapter 3):

$$G_{1\rho}(\rho) = k_2^2 \int_0^\infty \gamma_1 \left(\frac{1}{N} - \frac{1}{N_0}\right) J_1(\lambda\rho)\lambda^2 \, d\lambda, \tag{E1}$$

$$G_{1\phi}(\rho) = k_2^2 \int_0^\infty \left(\frac{1}{N} - \frac{1}{N_0}\right) J_1(\lambda\rho)\lambda^2 \, d\lambda, \tag{E2}$$

$$G_{1z}(\rho) = k_2^2 \int_0^\infty \left(\frac{1}{N} - \frac{1}{N_0}\right) J_0(\lambda\rho)\lambda^3 \, d\lambda. \tag{E3}$$

Vertical Dipole in Region 1, Field in Region 2 (Chapter 3):

$$G_{2\rho}(\rho, z) = \int_0^\infty \gamma_1 \left(\frac{1}{N} - \frac{1}{N_0}\right) e^{-i\gamma_2 z} J_1(\lambda\rho)\lambda^2 \, d\lambda; \qquad z \le 0, \tag{E4}$$

$$G_{2\phi}(\rho, z) = \int_0^\infty \left(\frac{1}{N} - \frac{1}{N_0}\right) e^{-i\gamma_2 z} J_1(\lambda\rho)\lambda^2 \, d\lambda; \qquad z \le 0, \tag{E5}$$

$$G_{2z}(\rho, z) = \int_0^\infty \left(\frac{1}{N} - \frac{1}{N_0}\right) e^{-i\gamma_2 z} J_0(\lambda\rho)\lambda^3 \, d\lambda; \qquad z \le 0. \tag{E6}$$

Horizontal Dipole in Region 1, Field in Region 1 (Chapter 5):

$$H_{\rho 1}(\rho, 0) = -k_2^2 \int_0^\infty \gamma_1^2 \left(\frac{1}{N} - \frac{1}{N_0}\right) [J_0(\lambda\rho) - J_2(\lambda\rho)] \lambda \, d\lambda, \tag{E7}$$

$$\begin{aligned} H_{\phi 1}(\rho, 0) &= -k_2^2 \int_0^\infty \gamma_1^2 \left(\frac{1}{N} - \frac{1}{N_0}\right) [J_0(\lambda\rho) + J_2(\lambda\rho)] \lambda \, d\lambda \\ &= -\frac{2k_2^2}{\rho} \int_0^\infty \gamma_1^2 \left(\frac{1}{N} - \frac{1}{N_0}\right) J_1(\lambda\rho) \, d\lambda, \end{aligned} \tag{E8}$$

$$H_{z1}(\rho, 0) = -2k_2^2 \int_0^\infty \gamma_1 \left(\frac{1}{N} - \frac{1}{N_0}\right) J_1(\lambda\rho) \lambda^2 \, d\lambda, \tag{E9}$$

$$\begin{aligned} H_{B\rho 1}(\rho, 0) &= -k_2^2 \int_0^\infty \gamma_1 \left(\frac{1}{N} - \frac{1}{N_0}\right) [J_0(\lambda\rho) + J_2(\lambda\rho)] \lambda \, d\lambda \\ &= -\frac{2k_2^2}{\rho} \int_0^\infty \gamma_1 \left(\frac{1}{N} - \frac{1}{N_0}\right) J_1(\lambda\rho) \, d\lambda, \end{aligned} \tag{E10}$$

$$H_{B\phi 1}(\rho, 0) = -k_2^2 \int_0^\infty \gamma_1 \left(\frac{1}{N} - \frac{1}{N_0}\right) [J_0(\lambda\rho) - J_2(\lambda\rho)] \lambda \, d\lambda. \tag{E11}$$

In all of these integrals,

$$N = k_1^2 \gamma_2 + k_2^2 \gamma_1 = k_1^2 (k_2^2 - \lambda^2)^{1/2} + k_2^2 (k_1^2 - \lambda^2)^{1/2}, \tag{E12a}$$

$$N_0 = k_1^2 \gamma_2 = k_1^2 (k_2^2 - \lambda^2)^{1/2}, \tag{E12b}$$

and it is assumed that $|k_1^2| \gg |k_2^2|$. Since the significant contribution to all of these integrals occurs when λ is at or near k_2, it follows that in this range $k_1^2 \gg \lambda^2$ so that $\gamma_1 \sim k_1$.

For all but small arguments, the Bessel functions are well approximated by

$$J_0(\lambda\rho) \sim \frac{1}{(2\pi\lambda\rho)^{1/2}} \left(e^{i(\lambda\rho - \pi/4)} + e^{-i(\lambda\rho - \pi/4)}\right), \tag{E13}$$

$$J_1(\lambda\rho) \sim \frac{1}{(2\pi\lambda\rho)^{1/2}} \left(e^{i(\lambda\rho - 3\pi/4)} + e^{-i(\lambda\rho - 3\pi/4)}\right), \tag{E14}$$

$$\begin{aligned} J_2(\lambda\rho) &\sim \frac{1}{(2\pi\lambda\rho)^{1/2}} \left(e^{i(\lambda\rho - 5\pi/4)} + e^{-i(\lambda\rho - 5\pi/4)}\right) \\ &= -\frac{1}{(2\pi\lambda\rho)^{1/2}} \left(e^{i(\lambda\rho - \pi/4)} + e^{-i(\lambda\rho - \pi/4)}\right). \end{aligned} \tag{E15}$$

It follows that

$$J_0(\lambda\rho) - J_2(\lambda\rho) = 2J_0(\lambda\rho). \tag{E16}$$

With (E13)–(E16), it follows that each integral in (E1)–(E11) becomes the sum of two integrals. With

$$F_1(\lambda^2) \equiv \left(\frac{1}{N} - \frac{1}{N_0}\right); \quad F_2(\lambda^2) \equiv \left(\frac{1}{N} - \frac{1}{N_0}\right) e^{-i\gamma_2 z} \lambda^2, \quad z \le 0, \tag{E17}$$

and with $\gamma_1 \sim k_1$, (E1)–(E11) become

$$G_{1\rho}(\rho) = k_1 G_{1\phi}(\rho), \tag{E18}$$

$$G_{1\phi}(\rho) = k_2^2 \int_0^\infty \frac{F_1(\lambda^2)}{(2\pi\lambda\rho)^{1/2}} \, (e^{i(\lambda\rho - 3\pi/4)} + e^{-i(\lambda\rho - 3\pi/4)})\lambda^2 \, d\lambda, \tag{E19}$$

$$G_{1z}(\rho) = k_2^2 \int_0^\infty \frac{F_1(\lambda^2)}{(2\pi\lambda\rho)^{1/2}} \, (e^{i(\lambda\rho - \pi/4)} + e^{-i(\lambda\rho - \pi/4)})\lambda^3 \, d\lambda, \tag{E20}$$

$$G_{2\rho}(\rho, z) = k_1 G_{2\phi}(\rho, z), \tag{E21}$$

$$G_{2\phi}(\rho, z) = \int_0^\infty \frac{F_2(\lambda^2)}{(2\pi\lambda\rho)^{1/2}} \, (e^{i(\lambda\rho - 3\pi/4)} + e^{-i(\lambda\rho - 3\pi/4)}) \, d\lambda, \tag{E22}$$

$$G_{2z}(\rho, z) = \int_0^\infty \frac{F_2(\lambda^2)}{(2\pi\lambda\rho)^{1/2}} \, (e^{i(\lambda\rho - \pi/4)} + e^{-i(\lambda\rho - \pi/4)})\lambda \, d\lambda, \tag{E23}$$

$$H_{\rho 1}(\rho, 0) = -2k_1^2 k_2^2 \int_0^\infty \frac{F_1(\lambda^2)}{(2\pi\lambda\rho)^{1/2}} \, (e^{i(\lambda\rho - \pi/4)} + e^{-i(\lambda\rho - \pi/4)})\lambda \, d\lambda, \tag{E24}$$

$$H_{\phi 1}(\rho, 0) = -\frac{2k_1^2 k_2^2}{\rho} \int_0^\infty \frac{F_1(\lambda^2)}{(2\pi\lambda\rho)^{1/2}} \, (e^{i(\lambda\rho - 3\pi/4)} + e^{-i(\lambda\rho - 3\pi/4)}) \, d\lambda, \tag{E25}$$

$$H_{z1}(\rho, 0) = -2k_1 G_{1\phi}(\rho), \tag{E26}$$

$$H_{B\rho 1}(\rho, 0) = k_1^{-1} H_{\phi 1}(\rho, 0), \tag{E27}$$

$$H_{B\phi 1}(\rho, 0) = k_1^{-1} H_{\rho 1}(\rho, 0). \tag{E28}$$

Let the variable in the second integral be changed to $\lambda' = -\lambda$, $d\lambda' = -d\lambda$, $\lambda = \infty \to \lambda' = -\infty$. Thus, the second integral in (E19) becomes

$$-\int_0^{-\infty} \frac{F_1(\lambda'^2)}{i(2\pi\lambda'\rho)^{1/2}} \, e^{i(\lambda'\rho + 3\pi/4)} \lambda'^2 \, d\lambda'$$
$$= \int_0^{-\infty} \frac{F_1(\lambda'^2)}{(2\pi\lambda'\rho)^{1/2}} \, e^{i(\lambda'\rho - 3\pi/4)} \lambda'^2 \, d\lambda'. \tag{E29}$$

The second integral in (E20) is

$$\int_0^{-\infty} \frac{F_1(\lambda'^2)}{i(2\pi\lambda'\rho)^{1/2}} \, e^{i(\lambda'\rho + \pi/4)} \lambda'^3 \, d\lambda'$$
$$= \int_0^{-\infty} \frac{F_1(\lambda'^2)}{(2\pi\lambda'\rho)^{1/2}} \, e^{i(\lambda'\rho - \pi/4)} \lambda'^3 \, d\lambda'. \tag{E30}$$

The second integral in (E22) is

$$-\int_0^{-\infty} \frac{F_2(\lambda'^2)}{i(2\pi\lambda'\rho)^{1/2}} \, e^{i(\lambda'\rho + 3\pi/4)} \, d\lambda'$$

$$= \int_0^{-\infty} \frac{F_2(\lambda'^2)}{(2\pi\lambda'\rho)^{1/2}}\, e^{i(\lambda'\rho - 3\pi/4)}\, d\lambda'. \tag{E31}$$

The second integral in (E23) is

$$\int_0^{-\infty} \frac{F_2(\lambda'^2)}{i(2\pi\lambda'\rho)^{1/2}}\, e^{i(\lambda'\rho + \pi/4)}\lambda'\, d\lambda'$$
$$= \int_0^{-\infty} \frac{F_2(\lambda'^2)}{(2\pi\lambda'\rho)^{1/2}}\, e^{i(\lambda'\rho - \pi/4)}\lambda'\, d\lambda'. \tag{E32}$$

The second integral in (E24) is

$$\int_0^{-\infty} \frac{F_1(\lambda'^2)}{i(2\pi\lambda'\rho)^{1/2}}\, e^{i(\lambda'\rho + \pi/4)}\lambda'\, d\lambda'$$
$$= \int_0^{-\infty} \frac{F_1(\lambda'^2)}{(2\pi\lambda'\rho)^{1/2}}\, e^{i(\lambda'\rho - \pi/4)}\lambda'\, d\lambda'. \tag{E33}$$

The second integral in (E25) is

$$-\int_0^{-\infty} \frac{F_1(\lambda'^2)}{i(2\pi\lambda'\rho)^{1/2}}\, e^{i(\lambda'\rho + 3\pi/4)}\, d\lambda'$$
$$= \int_0^{-\infty} \frac{F_1(\lambda'^2)}{(2\pi\lambda'\rho)^{1/2}}\, e^{i(\lambda'\rho - 3\pi/4)}\, d\lambda'. \tag{E34}$$

When these are substituted in (E19), (E20), and (E22)–(E25), the resulting integrals are

$$G_{1\phi}(\rho) = k_2^2\left(\int_0^{\infty} + \int_0^{-\infty}\right)\left(\frac{1}{N} - \frac{1}{N_0}\right)\frac{e^{i(\lambda\rho - 3\pi/4)}}{(2\pi\lambda\rho)^{1/2}}\,\lambda^2\, d\lambda, \tag{E35}$$

$$G_{1z}(\rho) = k_2^2\left(\int_0^{\infty} + \int_0^{-\infty}\right)\left(\frac{1}{N} - \frac{1}{N_0}\right)\frac{e^{i(\lambda\rho - \pi/4)}}{(2\pi\lambda\rho)^{1/2}}\,\lambda^3\, d\lambda, \tag{E36}$$

$$G_{2\phi}(\rho, z) = \left(\int_0^{\infty} + \int_0^{-\infty}\right)\left(\frac{1}{N} - \frac{1}{N_0}\right)e^{-i\gamma_2 z}\,\frac{e^{i(\lambda\rho - 3\pi/4)}}{(2\pi\lambda\rho)^{1/2}}\,\lambda^2\, d\lambda; \quad z \le 0, \tag{E37}$$

$$G_{2z}(\rho, z) = \left(\int_0^{\infty} + \int_0^{-\infty}\right)\left(\frac{1}{N} - \frac{1}{N_0}\right)e^{-i\gamma_2 z}\,\frac{e^{i(\lambda\rho - \pi/4)}}{(2\pi\lambda\rho)^{1/2}}\,\lambda^3\, d\lambda; \quad z \le 0, \tag{E38}$$

$$H_{\rho 1}(\rho, 0) = -2k_1^2k_2^2\left(\int_0^{\infty} + \int_0^{-\infty}\right)\left(\frac{1}{N} - \frac{1}{N_0}\right)\frac{e^{i(\lambda\rho - \pi/4)}}{(2\pi\lambda\rho)^{1/2}}\,\lambda\, d\lambda, \tag{E39}$$

$$H_{\phi 1}(\rho, 0) = -\frac{2k_1^2k_2^2}{\rho}\left(\int_0^{\infty} + \int_0^{-\infty}\right)\left(\frac{1}{N} - \frac{1}{N_0}\right)\frac{e^{i(\lambda\rho - 3\pi/4)}}{(2\pi\lambda\rho)^{1/2}}\, d\lambda. \tag{E40}$$

Since the significant contributions to the integrals occur when $\lambda \sim k_2$ and since the zero of $N = k_1^2(k_2^2-\lambda^2)^{1/2}+k_2^2(k_1^2-\lambda^2)^{1/2} \sim k_1^2(k_2^2-\lambda^2)^{1/2}+k_2^2k_1$ occurs at $\lambda = k_1k_2/(k_1^2+k_2^2)^{1/2} \sim k_2 - k_2^3/2k_1^2$, the following change of variable is suggested:

$$\lambda = k_2(1+m\tau); \qquad d\lambda = k_2 m\, d\tau, \tag{E41}$$

where

$$m = \frac{k_2^2}{2k_1^2}; \qquad |m| \ll 1, \tag{E42}$$

and τ is a dimensionless variable of integration. It follows that, with $\sqrt{-1} = -i$,

$$\lambda^2 \sim k_2^2(1+2m\tau), \tag{E43a}$$

$$\gamma_2 = (k_2^2-\lambda^2)^{1/2} \sim k_2(-2m\tau)^{1/2} = \frac{k_2^2(-\tau)^{1/2}}{k_1} = -\frac{ik_2^2\tau^{1/2}}{k_1}. \tag{E43b}$$

In phases, (E41) is used; in amplitudes,

$$\lambda \sim k_2; \qquad \lambda^2 \sim k_2^2; \qquad \lambda^3 \sim k_2^3. \tag{E44}$$

With these values,

$$\begin{aligned} \frac{1}{N} - \frac{1}{N_0} &\sim \frac{1}{k_2^2k_1}\left(\frac{1}{\sqrt{-\tau}+1} - \frac{1}{\sqrt{-\tau}}\right) \\ &= \frac{1}{k_2^2k_1}\left(\frac{-1}{\sqrt{-\tau}\,(\sqrt{-\tau}+1)}\right) = \frac{1}{k_2^2k_1}\left(\frac{1}{\sqrt{\tau}\,(\sqrt{\tau}+i)}\right). \end{aligned} \tag{E45}$$

The lower limit, $\lambda = 0$, in the integrals (E35)–(E40) becomes $\tau = -1/m = -2k_1^2/k_2^2$, where $|2k_1^2/k_2^2| \gg 1$. It follows that the very large lower limit can be approximated by $-\infty$. This means that the second integrals in (E35)–(E40) become vanishingly small.

When (E41)–(E45) are used in the first integrals of (E35), (E36), (E39), and (E40) for the field components in Region 1, these are all found to contain the integral

$$I_{\phi 1}(R) = \int_{-\infty}^{\infty} \frac{e^{iR\tau}}{\sqrt{\tau}\,(\sqrt{\tau}+i)}\, d\tau, \tag{E46}$$

where

$$R = k_2\rho m = \frac{k_2^3\rho}{2k_1^2}. \tag{E47}$$

Alternatively,

$$I_{\phi 1}(R) = \int_0^{\infty} \frac{e^{iR\tau}}{\sqrt{\tau}\,(\sqrt{\tau}+i)}\, d\tau - \int_0^{\infty} \frac{e^{-iR\tau}}{\sqrt{\tau}\,(\sqrt{\tau}+1)}\, d\tau. \tag{E48}$$

These integrals can be evaluated. As a first step, let $\tau = \xi^2$, $d\tau = 2\xi\, d\xi$. Then, with $\sqrt{-1} = -i$,

$$I_{\phi 1} = 2\int_0^\infty \frac{e^{iR\xi^2}}{\xi + i}\, d\xi - 2\int_0^\infty \frac{e^{-iR\xi^2}}{\xi + 1}\, d\xi, \tag{E49}$$

$$\frac{dI_{\phi 1}}{dR} = 2i\int_0^\infty \frac{e^{iR\xi^2}}{\xi + i}\,\xi^2\, d\xi + 2i\int_0^\infty \frac{e^{-iR\xi^2}}{\xi + 1}\,\xi^2\, d\xi, \tag{E50}$$

and

$$\frac{dI_{\phi 1}}{dR} + iI_{\phi 1} = 2i\int_0^\infty [e^{iR\xi^2}(\xi - i) + e^{-iR\xi^2}(\xi - 1)]\, d\xi. \tag{E51}$$

Since

$$i\int_0^\infty e^{iR\xi^2} 2\xi\, d\xi = \frac{1}{R}\, e^u \Big|_0^\infty$$

and

$$i\int_0^\infty e^{-iR\xi^2} 2\xi\, d\xi = -\frac{1}{R}\, e^u \Big|_0^\infty,$$

it follows that

$$\begin{aligned} \frac{dI_{\phi 1}}{dR} + iI_{\phi 1} &= 2\int_0^\infty e^{iR\xi^2}\, d\xi - 2i\int_0^\infty e^{-iR\xi^2}\, d\xi \\ &= 2\,\frac{\sqrt{\pi}}{2\sqrt{-iR}} - 2i\,\frac{\sqrt{\pi}}{2\sqrt{iR}} \\ &= -\frac{\sqrt{\pi}}{\sqrt{R}}\,(e^{i\pi/4} + e^{i\pi/4}) = -\frac{2\sqrt{\pi}}{\sqrt{R}}\, e^{i\pi/4}. \end{aligned} \tag{E52}$$

Also,

$$\frac{d}{dR}\,(e^{iR} I_{\phi 1}) = e^{iR}\left(\frac{dI_{\phi 1}}{dR} + iI_{\phi 1}\right) = -2\sqrt{\pi}\, e^{i\pi/4}\,\frac{e^{iR}}{\sqrt{R}}. \tag{E53}$$

When this is integrated, the result is

$$\begin{aligned} \int_R^\infty \frac{d}{dR}\,[e^{iR} I_{\phi 1}(R)]\, dR = e^{iR} I_{\phi 1}(R)\Big|_0^\infty &= -e^{iR} I_{\phi 1}(R) \\ &= -2\sqrt{\pi}\, e^{i\pi/4}\int_R^\infty \frac{e^{iR}}{\sqrt{R}}\, dR. \end{aligned} \tag{E54}$$

Since $I_{\phi 1}(\infty) = 0$, it follows that

$$I_{\phi 1}(R) = 2\sqrt{\pi}\, e^{i\pi/4} e^{-iR}\int_R^\infty \frac{e^{it}}{\sqrt{t}}\, dt. \tag{E55}$$

The integral in (E55) is closely related to the Fresnel integral defined by

$$C_2(R) + iS_2(R) = \int_0^R \frac{e^{it}}{(2\pi t)^{1/2}}\, dt. \tag{E56}$$

Since

$$\int_0^\infty \frac{e^{it}}{(2\pi t)^{1/2}}\,dt = \tfrac{1}{2}(1+i), \tag{E57}$$

it follows that

$$I_{\phi 1}(R) = 2\sqrt{2}\,\pi e^{i\pi/4} e^{-iR}\mathcal{F}(R), \tag{E58}$$

where

$$\mathcal{F}(R) = \tfrac{1}{2}(1+i) - C_2(R) - iS_2(R). \tag{E59}$$

With (E45) and (E46), the G_1 integrals (H integrals) for the field in Region 1 due to vertical (horizontal) dipoles in Region 1 become

$$\begin{aligned} G_{1\phi}(\rho) &= \frac{k_2^3 m e^{-i3\pi/4}}{k_1(2\pi k_2\rho)^{1/2}}\, e^{ik_2\rho} I_{\phi 1}(R) \\ &= -\frac{ik_2^5}{k_1^3}\left(\frac{\pi}{k_2\rho}\right)^{1/2} e^{ik_2\rho}e^{-iR}\mathcal{F}(R), \end{aligned} \tag{E60}$$

$$\begin{aligned} G_{1\rho}(\rho) &= k_1 G_{1\phi}(\rho) \\ &= -\frac{ik_2^5}{k_1^2}\left(\frac{\pi}{k_2\rho}\right)^{1/2} e^{ik_2\rho}e^{-iR}\mathcal{F}(R), \end{aligned} \tag{E61}$$

$$\begin{aligned} G_{1z}(\rho) &= \frac{k_2^4 m e^{-i\pi/4}}{k_1(2\pi k_2\rho)^{1/2}}\, e^{ik_2\rho} I_{\phi 1}(R) \\ &= \frac{k_2^6}{k_1^3}\left(\frac{\pi}{k_2\rho}\right)^{1/2} e^{ik_2\rho}e^{-iR}\mathcal{F}(R), \end{aligned} \tag{E62}$$

$$\begin{aligned} H_{\rho 1}(\rho,0) &= -\frac{2k_1k_2^2 m e^{-i\pi/4}}{(2\pi k_2\rho)^{1/2}}\, e^{ik_2\rho} I_{\phi 1}(R) \\ &= -\frac{2k_2^4}{k_1}\left(\frac{\pi}{k_2\rho}\right)^{1/2} e^{ik_2\rho}e^{-iR}\mathcal{F}(R), \end{aligned} \tag{E63}$$

$$\begin{aligned} H_{\phi 1}(\rho,0) &= -\frac{2k_1k_2 m e^{-i3\pi/4}}{\rho(2\pi k_2\rho)^{1/2}}\, e^{ik_2\rho} I_{\phi 1}(R) \\ &= \frac{2ik_2^3}{k_1\rho}\left(\frac{\pi}{k_2\rho}\right)^{1/2} e^{ik_2\rho}e^{-iR}\mathcal{F}(R), \end{aligned} \tag{E64}$$

$$\begin{aligned} H_{z1}(\rho,0) &= -2k_1 G_{1\phi}(\rho) \\ &= \frac{2ik_2^5}{k_1^2}\left(\frac{\pi}{k_2\rho}\right)^{1/2} e^{ik_2\rho}e^{-iR}\mathcal{F}(R), \end{aligned} \tag{E65}$$

$$\begin{aligned} H_{B\rho 1}(\rho,0) &= k_1^{-1} H_{\phi 1}(\rho,0) \\ &= \frac{2ik_2^3}{k_1^2\rho}\left(\frac{\pi}{k_2\rho}\right)^{1/2} e^{ik_2\rho}e^{-iR}\mathcal{F}(R), \end{aligned} \tag{E66}$$

$$H_{B\phi 1}(\rho,0) = k_1^{-1} H_{\rho 1}(\rho,0)$$
$$= -\frac{2k_2^4}{k_1^2}\left(\frac{\pi}{k_2\rho}\right)^{1/2} e^{ik_2\rho} e^{-iR}\mathcal{F}(R). \tag{E67}$$

The G_2 integrals for the field in Region 2 of a vertical dipole in Region 1 are evaluated similarly. When (E41)–(E45) are used in the first integral of (E37) for $G_{2\phi}(\rho,z)$, this becomes

$$G_{2\phi}(\rho,z) = \frac{k_2 m e^{-i3\pi/4}}{k_1(2\pi k_2\rho)^{1/2}} e^{ik_2\rho} I_{\phi 2}(R,Z), \tag{E68}$$

where

$$I_{\phi 2}(R,Z) = \int_{-\infty}^{\infty} \frac{e^{iR\tau}}{\sqrt{\tau}\,(\sqrt{\tau}+i)} e^{-2Z\sqrt{\tau}}\, d\tau \tag{E69}$$

and

$$R = k_2\rho m = \frac{k_2^3\rho}{2k_1^2}; \qquad Z = \frac{k_2^2 z}{2k_1}, \quad z \le 0. \tag{E70}$$

Like $I_{\phi 1}(R)$, the integral $I_{\phi 2}(R,Z)$ can be expressed in terms of the Fresnel integral. As a first step, note that

$$\frac{1}{2}\frac{\partial}{\partial Z}(e^{-2iZ} I_{\phi 2}) = \left(\frac{1}{2}\frac{\partial I_{\phi 2}}{\partial Z} - iI_{\phi 2}\right) e^{-2iZ}$$
$$= e^{-2iZ}\int_{-\infty}^{\infty}\left(\frac{-1}{\sqrt{\tau}+i} - \frac{i}{\sqrt{\tau}\,(\sqrt{\tau}+i)}\right) e^{-2Z\sqrt{\tau}} e^{iR\tau}\, d\tau$$
$$= -e^{-2iZ}\int_{-\infty}^{\infty}\frac{1}{\sqrt{\tau}} e^{-2Z\sqrt{\tau}} e^{iR\tau}\, d\tau. \tag{E71}$$

This integral can be evaluated with the substitution $x^2 = \tau$, $2x\,dx = d\tau$. Also, $iR\tau - 2Z\sqrt{\tau} = iRx^2 - 2Zx = (\sqrt{iR}\,x - Z/\sqrt{iR})^2 - Z^2/iR$ so that

$$e^{-2iZ}\int_{-\infty}^{\infty}\frac{1}{\sqrt{\tau}} e^{-2Z\sqrt{\tau}} e^{iR\tau}\, d\tau$$
$$= 2e^{-2iZ} e^{iZ^2/R}\int_{-\infty}^{\infty} e^{(\sqrt{iR}\,x - Z/\sqrt{iR})^2}\, dx. \tag{E72}$$

With the substitutions

$$iy = \sqrt{iR}\,x - \frac{Z}{\sqrt{iR}}; \quad -y^2 = \left(\sqrt{iR}\,x - \frac{Z}{\sqrt{iR}}\right)^2; \quad i\,dy = \sqrt{iR}\,dx, \tag{E73}$$

the integral becomes

$$\frac{2i}{\sqrt{iR}} e^{-2iZ} e^{iZ^2/R}\int_{-\infty}^{\infty} e^{-y^2}\, dy = 2\left(\frac{\pi i}{R}\right)^{1/2} e^{-2iZ} e^{iZ^2/R}. \tag{E74}$$

It follows that

$$\frac{1}{2}\frac{\partial}{\partial Z}(e^{-2iZ}I_{\phi2}) = -2\left(\frac{\pi i}{R}\right)^{1/2} e^{-2iZ}e^{iZ^2/R}. \tag{E75}$$

This gives for $Z \leq 0$,

$$I_{\phi2}(R,Z) = -4\left(\frac{\pi i}{R}\right)^{1/2} e^{2iZ}\int_{\infty}^{-Z} e^{-2iZ'}e^{iZ'^2/R}\,dZ'. \tag{E76}$$

The evaluation of (E76) is carried out with the change of variable $Z' = Z'' - R$, $dZ' = dZ''$. Thus, with $I_{\phi2}(\infty, Z) = 0$,

$$I_{\phi2}(R,Z) = 4\left(\frac{\pi i}{R}\right)^{1/2} e^{2iZ}e^{-iR}\int_{R-Z}^{\infty} e^{iZ''^2/R}\,dZ''. \tag{E77}$$

Finally, let $t = Z''^2/R$, $dZ'' = \frac{1}{2}\sqrt{R/t}\,dt$ so that

$$I_{\phi2}(R,Z) = 4\left(\frac{\pi i}{R}\right)^{1/2} e^{2iZ}e^{-iR}\left(\frac{\pi R}{2}\right)^{1/2}\int_{P}^{\infty}\frac{e^{it}}{(2\pi t)^{1/2}}\,dt, \tag{E78}$$

where $P \equiv (R-Z)^2/R$ with $Z \leq 0$. The integral can be expanded as follows:

$$\mathcal{F}(P) \equiv \int_{P}^{\infty}\frac{e^{it}}{(2\pi t)^{1/2}}\,dt = \tfrac{1}{2}(1+i) - C_2(P) - iS_2(P), \tag{E79}$$

where

$$C_2(P) + iS_2(P) = \int_0^P \frac{e^{it}}{(2\pi t)^{1/2}}\,dt \tag{E80}$$

is the well-known Fresnel integral. In this notation,

$$I_{\phi2}(R,Z) = 2\sqrt{2}\,\pi e^{i\pi/4}e^{2iZ}e^{-iR}\mathcal{F}(P); \qquad Z \leq 0. \tag{E81}$$

With (E68) and (E81),

$$G_{2\phi}(\rho,z) = -\frac{ik_2^3}{k_1^3}\left(\frac{\pi}{k_2\rho}\right)^{1/2} e^{ik_2\rho}e^{iZ^2/R}e^{-iP}\mathcal{F}(P); \qquad Z \leq 0, \tag{E82}$$

where $P = (R-Z)^2/R$.

The integral for $G_{2z}(\rho,z)$ in (E38) differs from the integral for $G_{2\phi}(\rho,z)$ in the factor $\lambda e^{i\pi/2} \sim ik_2$. Thus, with (E82),

$$G_{2z}(\rho,z) = \frac{k_2^4}{k_1^3}\left(\frac{\pi}{k_2\rho}\right)^{1/2} e^{ik_2\rho}e^{iZ^2/R}e^{-iP}\mathcal{F}(P); \qquad Z \leq 0. \tag{E83}$$

Appendix F
Evaluation of Integrals That Lead to Fresnel Integrals; ρ and z Unrestricted

The evaluation of the several Fresnel-integral terms in Appendix E depends on the approximation (E41) that contributions to the integrals are limited to a narrow range defined by $\lambda = k_2(1+m\tau)$, where m is the small quantity $m = k_2^2/2k_1^2$. This is entirely adequate when the conditions $\rho^2 \gg |z|^2$ and $\rho^2 \gg d^2$ are valid. In order to determine the electromagnetic field for all values of ρ and z, a different approach is required. This is carried out in a generalized form for the integrals in Chapter 4. The correlation with the specific integrals in Chapter 4 and application to similar integrals in Chapter 3 are summarized at the end.

F.1 The Component $E_{2\rho}(\rho, z')$

Consider the integral for the component $E_{2\rho}(\rho, z')$ when the vertical dipole is at a height d in the air above the surface, as given by (4.2.2). With $z' = -z$, this is

$$E_{2\rho}(\rho, z') = \frac{i\omega\mu_0}{2\pi k_2^2} \int_0^\infty \left(\pm \frac{e^{i\gamma_2|z'-d|}}{2} - \frac{e^{i\gamma_2(z'+d)}}{2} \right.$$

$$+ \frac{k_1^2 \gamma_2 e^{i\gamma_2(z'+d)}}{N} \Bigg) J_1(\lambda\rho)\lambda^2\, d\lambda;$$

$$\begin{cases} z' > d \\ 0 \le z' \le d \end{cases}, \tag{F1}$$

where $N = k_1^2\gamma_2 + k_2^2\gamma_1$ and $\gamma_j = (k_j^2 - \lambda^2)^{1/2}$, $j = 1, 2$. The first two integrals are given exactly in (4.2.16) and (4.2.19). They are

$$\begin{aligned} E_{2\rho}^d(\rho, z') &= \frac{i\omega\mu_0}{4\pi k_2^2}\int_0^\infty \pm e^{i\gamma_2|z'-d|} J_1(\lambda\rho)\lambda^2\, d\lambda; \qquad \begin{cases} z' > d \\ 0 \le z' \le d \end{cases} \\ &= -\frac{\omega\mu_0}{4\pi k_2^2}\, e^{ik_2 r_1}\left(\frac{ik_2^2}{r_1} - \frac{3k_2}{r_1^2} - \frac{3i}{r_1^3}\right)\left(\frac{\rho}{r_1}\right)\left(\frac{z'-d}{r_1}\right), \end{aligned} \tag{F2}$$

$$\begin{aligned} E_{2\rho}^i(\rho, z') &= \frac{i\omega\mu_0}{4\pi k_2^2}\int_0^\infty e^{i\gamma_2(z'+d)} J_1(\lambda\rho)\lambda^2\, d\lambda \\ &= -\frac{\omega\mu_0}{4\pi k_2^2}\, e^{ik_2 r_2}\left(\frac{ik_2^2}{r_2} - \frac{3k_2}{r_2^2} - \frac{3i}{r_2^3}\right)\left(\frac{\rho}{r_2}\right)\left(\frac{z'+d}{r_2}\right), \end{aligned} \tag{F3}$$

where $r_1 = [\rho^2 + (z'-d)^2]^{1/2}$ and $r_2 = [\rho^2 + (z'+d)^2]^{1/2}$. Equation (F3) can be rearranged as follows:

$$\begin{aligned} E_{2\rho}^i(\rho, z') &= -\frac{i\omega\mu_0 k_2}{4\pi}\, e^{ik_2 r_2}\left(\frac{1}{k_2 r_2} + \frac{3i}{k_2^2 r_2^2} - \frac{3}{k_2^3 r_2^3}\right)\left(\frac{k_2\rho}{k_2 r_2}\right)\left(\frac{k_2(z'+d)}{k_2 r_2}\right) \\ &= \frac{i\omega\mu_0 k_2}{4\pi}\, I_0[k_2\rho, k_2(z'+d)]. \end{aligned} \tag{F4}$$

The remaining integral to be evaluated is

$$\begin{aligned} &2E_{2\rho}^i(\rho, z') + E_{2\rho}^s(\rho, z') \\ &\quad\equiv \frac{i\omega\mu_0 k_1^2}{2\pi k_2^2}\int_0^\infty \frac{\gamma_2 e^{i\gamma_2(z'+d)}}{N} J_1(\lambda\rho)\lambda^2\, d\lambda \\ &\quad= \frac{i\omega\mu_0 k_1^2}{2\pi k_2^2}\int_0^\infty \frac{(k_2^2-\lambda^2)^{1/2} e^{i(z'+d)(k_2^2-\lambda^2)^{1/2}}}{k_1^2(k_2^2-\lambda^2)^{1/2} + k_2^2(k_1^2-\lambda^2)^{1/2}} J_1(\lambda\rho)\lambda^2\, d\lambda. \end{aligned} \tag{F5}$$

Since the inequality $|k_1^2| \gg |k_2^2|$ is assumed, let the small quantity ϵ be defined by

$$\epsilon = \frac{k_2}{k_1}. \tag{F6}$$

Also let

$$\bar{z}' = k_2(z'+d); \qquad \bar{\rho} = k_2\rho; \qquad \bar{r}_2 = k_2 r_2 = (\bar{\rho}^2 + \bar{z}'^2)^{1/2}. \tag{F7}$$

The new variable of integration is $\zeta = \lambda/k_2$. With it,

$$E_{2\rho}(\bar{\rho}, \bar{z}') = E_{2\rho}^d(\bar{\rho}, \bar{z}') - E_{2\rho}^i(\bar{\rho}, \bar{z}') + \frac{i\omega\mu_0 k_2}{2\pi}\, I_\rho(\bar{\rho}, \bar{z}'), \tag{F8}$$

where

$$I_\rho(\bar{\rho},\bar{z}') = \int_0^\infty \frac{(1-\zeta^2)^{1/2} e^{i\bar{z}'(1-\zeta^2)^{1/2}}}{(1-\zeta^2)^{1/2} + \epsilon(1-\epsilon^2\zeta^2)^{1/2}} J_1(\zeta\rho)\zeta^2\, d\zeta. \tag{F9}$$

Since ϵ^2 is very small, let $(1-\epsilon^2\zeta^2)^{1/2} \sim 1$ [which is equivalent to approximating $\gamma_1 = (k_1^2 - \lambda^2)^{1/2}$ by k_1]. With this approximation,

$$I_\rho(\bar{\rho},\bar{z}') \sim \int_0^\infty \frac{(1-\zeta^2)^{1/2}}{(1-\zeta^2)^{1/2} + \epsilon} e^{i\bar{z}'(1-\zeta^2)^{1/2}} J_1(\zeta\bar{\rho})\zeta^2\, d\zeta. \tag{F10}$$

This integral is to be evaluated asymptotically for small ϵ. Complications arise from the fact that $\bar{z}'$ or $\bar{\rho}$ or both $\bar{z}'$ and $\bar{\rho}$ may be large. Consider the following:

(a) Neither $\bar{\rho}$ nor $\bar{z}'$ is large.

In this case, the integrand can be expanded in a power series in ϵ in the form

$$I_\rho(\bar{\rho},\bar{z}') = \sum_{n=0}^{\infty} (-\epsilon)^n I_n(\bar{\rho},\bar{z}'), \tag{F11}$$

where

$$I_n(\bar{\rho},\bar{z}') = \int_0^\infty (1-\zeta^2)^{-n/2} e^{i\bar{z}'(1-\zeta^2)^{1/2}} J_1(\zeta\bar{\rho})\zeta^2\, d\zeta. \tag{F12}$$

The contour of integration along the real axis is indented downward at $\zeta = 1$ so that

$$I_\rho(\bar{\rho},\bar{z}') = I_0(\bar{\rho},\bar{z}') - \epsilon I_1(\bar{\rho},\bar{z}') + \epsilon^2 I_2(\bar{\rho},\bar{z}') - \cdots, \tag{F13}$$

where

$$\begin{aligned} I_0(\bar{\rho},\bar{z}') &= \int_0^\infty e^{i\bar{z}'(1-\zeta^2)^{1/2}} J_1(\zeta\bar{\rho})\zeta^2\, d\zeta \\ &= -\frac{\bar{\rho}\bar{z}'}{\bar{r}_2^2}\left(\frac{1}{\bar{r}_2} + \frac{3i}{\bar{r}_2^2} - \frac{3}{\bar{r}_2^3}\right) e^{i\bar{r}_2}, \end{aligned} \tag{F14}$$

$$\begin{aligned} I_1(\bar{\rho},\bar{z}') &= \int_0^\infty (1-\zeta^2)^{-1/2} e^{i\bar{z}'(1-\zeta^2)^{1/2}} J_1(\zeta\bar{\rho})\zeta^2\, d\zeta \\ &= -\frac{\bar{\rho}}{\bar{r}_2}\left(\frac{1}{\bar{r}_2} + \frac{i}{\bar{r}_2^2}\right) e^{i\bar{r}_2}. \end{aligned} \tag{F15}$$

Here,

$$\frac{i\omega\mu_0 k_2}{2\pi} I_0(\bar{\rho},\bar{z}') = 2E_{2\rho}^i(\bar{\rho},\bar{z}'), \tag{F16}$$

where $E_{2\rho}^i(\rho,z')$ is defined in (F3) and (F4). When combined with $-E_{2\rho}^i(\bar{\rho}, \bar{z}')$, (F16) gives the positive image. It follows that (F8) is equivalent to

$$E_{2\rho}(\bar{\rho},\bar{z}') = E_{2\rho}^d(\bar{\rho},\bar{z}') + E_{2\rho}^i(\bar{\rho},\bar{z}')$$

$$+\frac{i\omega\mu_0 k_2}{2\pi}\,[-\epsilon I_1(\bar{\rho},\bar{z}')+\epsilon^2 I_2(\bar{\rho},\bar{z}')-\cdots]. \tag{F17}$$

This is valid when $\bar{\rho}$ and $\bar{z}'$ are of order 1. In general, $I_2(\bar{\rho},\bar{z}')$ is not expressible in terms of elementary functions.

The situation is especially simple when $\bar{z}'=0$. In this case, the integral $I_n(\bar{\rho},\bar{z}')$ in (F12) can be evaluated exactly with formula (51) of Bateman (1953, Vol. 2, p. 95). The result is

$$I_n(\bar{\rho},0)=\frac{i\pi}{2}\,\frac{1}{\Gamma(\frac{1}{2}n)}\,(\tfrac{1}{2}\bar{\rho})^{n/2-1}H^{(1)}_{n/2-2}(\bar{\rho});\quad n=2,\,3,\,\ldots,\,\infty. \tag{F18}$$

The first few terms in (F11) with $\bar{z}'=0$ are

$$I_\rho(\bar{\rho},0)=\frac{i\pi}{2}\left[-\epsilon\,\frac{2}{\pi}\,e^{i\bar{\rho}}\left(\frac{i}{\bar{\rho}}-\frac{1}{\bar{\rho}^2}\right)-\epsilon^2 H^{(1)}_1(\bar{\rho})-\epsilon^3\,\frac{2}{\pi}\,e^{i\bar{\rho}}\right.$$
$$\left.+\,\epsilon^4\,\frac{\bar{\rho}}{2}\,H^{(1)}_0(\bar{\rho})+O(\epsilon^5)\right]. \tag{F19}$$

This formula is instructive. With large values of $\bar{\rho}$, the first term on the right is of order $\epsilon\bar{\rho}^{-1}$, the second term is of order $\epsilon^2\bar{\rho}^{-1/2}$, the third term is of order ϵ^3, and the fourth term of order $\epsilon^4\bar{\rho}^{1/2}$. These terms clearly decrease rapidly when $\bar{\rho}$ is of the order 1 owing to the small parameter ϵ. On the other hand, when $\bar{\rho}$ is large of the order ϵ^{-2}, the first four terms are all of comparable magnitude. It is in the range $\bar{\rho}\sim\epsilon^{-2}$ that the argument $P=k_2^3\rho/2k_1^2$ of the Fresnel integral, when evaluated as in Appendix E, is of the order of 1.

(b) Either or both $\bar{\rho}$ and $\bar{z}'$ are large.

It is more difficult to determine the behavior of the integral $I_\rho(\bar{\rho},\bar{z}')$ when $\bar{\rho}$ or $\bar{z}'$ or both $\bar{\rho}$ and $\bar{z}'$ are large. For this purpose, the right side of (F10) can be rearranged to obtain an ordinary differential equation for $I_\rho(\bar{\rho},\bar{z}')$. With the identity

$$\frac{(1-\zeta^2)^{1/2}}{(1-\zeta^2)^{1/2}+\epsilon}=1-\frac{\epsilon}{(1-\zeta^2)^{1/2}+\epsilon}, \tag{F20}$$

it follows that

$$I_\rho(\bar{\rho},\bar{z}')=I_0(\bar{\rho},\bar{z}')-\epsilon\int_0^\infty\frac{e^{i\bar{z}'(1-\zeta^2)^{1/2}}}{(1-\zeta^2)^{1/2}+\epsilon}\,J_1(\zeta\bar{\rho})\zeta^2\,d\zeta$$
$$=I_0(\bar{\rho},\bar{z}')-\epsilon J_\rho(\bar{\rho},\bar{z}'). \tag{F21}$$

It is readily verified that the quantity

$$J_\rho(\bar{\rho},\bar{z}')\equiv-\frac{1}{\epsilon}\,[I_\rho(\bar{\rho},\bar{z}')-I_0(\bar{\rho},\bar{z}')] \tag{F22}$$

satisfies the differential equation

$$\frac{\partial J_\rho(\bar{\rho},\bar{z}')}{\partial\bar{z}'}=iI_\rho(\bar{\rho},\bar{z}'). \tag{F23}$$

Thus,

$$\begin{aligned}\frac{\partial J_\rho(\bar{\rho},\bar{z}')}{\partial \bar{z}'}\\ &= -\frac{1}{\epsilon}\left(\frac{\partial I_\rho(\bar{\rho},\bar{z}')}{\partial \bar{z}'} - \frac{\partial I_0(\bar{\rho},\bar{z}')}{\partial \bar{z}'}\right)\\ &= -\frac{1}{\epsilon}\frac{\partial I_0(\bar{\rho},\bar{z}')}{\partial \bar{z}'} + i\int_0^\infty \frac{(1-\zeta^2)^{1/2}}{(1-\zeta^2)^{1/2}+\epsilon}\, e^{i\bar{z}'(1-\zeta^2)^{1/2}} J_1(\zeta\bar{\rho})\zeta^2\, d\zeta\\ &\quad + \frac{1}{\epsilon}\frac{\partial I_0(\bar{\rho},\bar{z}')}{\partial \bar{z}'}\\ &= iI_\rho(\bar{\rho},\bar{z}').\end{aligned}$$

Equation (F23) with (F21) gives

$$\frac{\partial J_\rho(\bar{\rho},\bar{z}')}{\partial \bar{z}'} + i\epsilon J_\rho(\bar{\rho},\bar{z}') = iI_0(\bar{\rho},\bar{z}'). \tag{F24}$$

The solution of this equation is

$$J_\rho(\bar{\rho},\bar{z}') = -i\int_{\bar{z}'}^\infty e^{i\epsilon(\bar{z}-\bar{z}')} I_0(\bar{\rho},\bar{z})\, d\bar{z}. \tag{F25}$$

With the relation

$$\frac{\partial I_1(\bar{\rho},\bar{z}')}{\partial \bar{z}'} = iI_0(\bar{\rho},\bar{z}') \tag{F26}$$

[which follows directly from (F14) and (F15)], (F25) becomes

$$J_\rho(\bar{\rho},\bar{z}') = -\int_{\bar{z}'}^\infty \frac{\partial I_1(\bar{\rho},\bar{z})}{\partial \bar{z}}\, e^{i\epsilon(\bar{z}-\bar{z}')}\, d\bar{z}. \tag{F27}$$

Here, the integral is readily integrated by parts to give

$$J_\rho(\bar{\rho},\bar{z}') = I_1(\bar{\rho},\bar{z}') + i\epsilon\int_{\bar{z}'}^\infty I_1(\bar{\rho},\bar{z}) e^{i\epsilon(\bar{z}-\bar{z}')}\, d\bar{z}. \tag{F28}$$

Since $I_1(\bar{\rho},\bar{z}')$ is given in (F15), the integral becomes

$$J_\rho(\bar{\rho},\bar{z}') = I_1(\bar{\rho},\bar{z}') - \epsilon I_2(\bar{\rho},\bar{z}'), \tag{F29}$$

where

$$I_2(\bar{\rho},\bar{z}') = i\int_{\bar{z}'}^\infty \left(\frac{\bar{\rho}}{\bar{r}}\right)\left(\frac{1}{\bar{r}} + \frac{i}{\bar{r}^2}\right) e^{i\bar{r}} e^{i\epsilon(\bar{z}-\bar{z}')}\, d\bar{z} \tag{F30}$$

and

$$\bar{r} = (\bar{\rho}^2 + \bar{z}^2)^{1/2}. \tag{F31}$$

It follows from (F5) and (F8) that

$$2E_{2\rho}^i(\bar{\rho},\bar{z}') + E_{2\rho}^s(\bar{\rho},\bar{z}') = \frac{i\omega\mu_0 k_2}{2\pi}\, I_\rho(\bar{\rho},\bar{z}')$$

$$= \frac{i\omega\mu_0 k_2}{2\pi}\,[I_0(\bar{\rho},\bar{z}') - \epsilon I_1(\bar{\rho},\bar{z}') + \epsilon^2 I_2(\bar{\rho},\bar{z}')]. \tag{F32a}$$

With (F16), and in agreement with (F17), the surface-wave part is

$$E^s_{2\rho}(\bar{\rho},\bar{z}') = \frac{i\omega\mu_0 k_2}{2\pi}\,[-\epsilon I_1(\bar{\rho},\bar{z}') + \epsilon^2 I_2(\bar{\rho},\bar{z}')], \tag{F32b}$$

where $I_\rho(\bar{\rho},\bar{z}')$ as given by (F10) has been replaced by the alternative form (F21) with (F29); $I_0(\bar{\rho},\bar{z}')$ and $I_1(\bar{\rho},\bar{z}')$ are given in (F14) and (F15). It remains to evaluate $I_2(\bar{\rho},\bar{z}')$. This is given by (F30), rearranged here as

$$I_2(\bar{\rho},\bar{z}') = i\int_{\bar{z}'}^{\infty} \frac{\bar{\rho}}{\bar{r}^2}\left(1+\frac{i}{\bar{r}}\right)e^{i\bar{r}}e^{i\epsilon(\bar{z}-\bar{z}')}\,d\bar{z}. \tag{F33}$$

As has already been shown, the presence of the factor ϵ^2 multiplying $I_2(\bar{\rho},\bar{z}')$ makes its contribution to $I_\rho(\bar{\rho},\bar{z}')$ negligible when $\bar{r} = (\bar{\rho}^2+\bar{z}^2)^{1/2}$ is of the order of magnitude 1. It is, therefore, sufficient to evaluate $I_2(\bar{\rho},\bar{z}')$ approximately for $\bar{r}$ large. For this purpose, the method of stationary phase is appropriate. Equation (F33) can be rewritten in the form

$$I_2(\bar{\rho},\bar{z}') = ie^{i\bar{r}_2}\int_{\bar{z}'}^{\infty} e^{i\phi(\bar{z})}\,\frac{\bar{\rho}}{\bar{r}^2}\left(1+\frac{i}{\bar{r}}\right)d\bar{z}, \tag{F34}$$

where

$$\begin{aligned}\phi(\bar{z}) &= \epsilon(\bar{z}-\bar{z}') + \bar{r} - \bar{r}_2\\ &= \epsilon(\bar{z}-\bar{z}') + (\bar{\rho}^2+\bar{z}^2)^{1/2} - (\bar{\rho}^2+\bar{z}'^2)^{1/2}.\end{aligned} \tag{F35}$$

The phase has the properties that

$$\phi(\bar{z}') = 0 \tag{F36}$$

and that $\phi(\bar{z})$ is an increasing function of $\bar{z}$ when $\bar{z} > \bar{z}'$. Thus, the largest contribution to the integral $I_2(\bar{\rho},\bar{z}')$ comes from the region where $\bar{z}$ is close to $\bar{z}'$. The first two terms in the Taylor series expansion of $\phi(\bar{z})$ at $\bar{z}=\bar{z}'$ are

$$\phi(\bar{z}') = \left(\epsilon + \frac{\bar{z}'}{\bar{r}_2}\right)(\bar{z}-\bar{z}') + \frac{1}{2}\frac{\bar{\rho}^2}{\bar{r}_2^3}(\bar{z}-\bar{z}')^2 + O[(\bar{z}-\bar{z}')^3]. \tag{F37}$$

With $\bar{r}$ large, the quantity $(1+i\bar{r}^{-1})$ in (F34) can be approximated by 1 and $\bar{r}$ by $\bar{r}_2$ so that

$$I_2(\bar{\rho},\bar{z}') \sim \frac{i\bar{\rho}}{\bar{r}_2^2}\,e^{i\bar{r}_2}\int_{\bar{z}'}^{\infty}\exp\left\{i\left[\left(\epsilon+\frac{\bar{z}'}{\bar{r}_2}\right)(\bar{z}-\bar{z}') + \frac{1}{2}\frac{\bar{\rho}^2}{\bar{r}_2^3}(\bar{z}-\bar{z}')^2\right]\right\}d\bar{z}. \tag{F38}$$

Let

$$s = \frac{1}{\bar{r}_2^{1/2}}\left(\frac{\bar{\rho}}{\bar{r}_2}(\bar{z}-\bar{z}') + \frac{\bar{r}_2(\epsilon\bar{r}_2+\bar{z}')}{\bar{\rho}}\right); \qquad ds = \frac{\bar{\rho}}{\bar{r}_2^{3/2}}\,d\bar{z}.$$

Then,

$$I_2(\bar{\rho}, \bar{z}') = \frac{i\bar{\rho}}{\bar{r}_2^2}\, e^{i\bar{r}_2} e^{-iP_2} \int_{\bar{r}_2^{1/2}(\epsilon\bar{r}_2+\bar{z}')/\bar{\rho}}^{\infty} e^{is^2/2}\left(\frac{\bar{r}_2^{3/2}}{\bar{\rho}}\right) ds, \tag{F39}$$

where

$$P_2 = \frac{\bar{r}_2}{2}\left(\frac{\epsilon\bar{r}_2 + \bar{z}'}{\bar{\rho}}\right)^2. \tag{F40}$$

Now let

$$t = \tfrac{1}{2}s^2; \qquad dt = s\, ds. \tag{F41}$$

Then the integral becomes

$$I_2(\bar{\rho}, \bar{z}') = i\left(\frac{\pi}{\bar{r}_2}\right)^{1/2} e^{i\bar{r}_2} e^{-iP_2} \mathcal{F}(P_2), \tag{F42}$$

where

$$\mathcal{F}(P_2) = \int_{P_2}^{\infty} \frac{e^{it}}{(2\pi t)^{1/2}}\, dt = \tfrac{1}{2}(1+i) - \int_0^{P_2} \frac{e^{it}}{(2\pi t)^{1/2}}\, dt \tag{F43}$$

is the Fresnel integral. This is the desired form of the Fresnel-integral term. It is valid everywhere for $\bar{z}' \geq 0$ when $\epsilon = k_2/k_1$ is small. With $\bar{r}_2 = k_2 r_2$, $\bar{z}' = k_2(z'+d)$, $\bar{\rho} = k_2\rho$, and $\epsilon = k_2/k_1$, the final form for the complete surface-wave term (F32b) is

$$E_{2\rho}^s(\rho, z') = \frac{\omega\mu_0}{2\pi k_1}\, e^{ik_2 r_2}\left[\left(\frac{\rho}{r_2}\right)\left(\frac{ik_2}{r_2} - \frac{1}{r_2^2}\right) - T(\rho, z'+d)\right], \tag{F44}$$

where

$$T(\rho, z'+d) = \frac{k_2^3}{k_1}\left(\frac{\pi}{k_2 r_2}\right)^{1/2} e^{-iP_2}\mathcal{F}(P_2), \tag{F45}$$

$$r_2 = [\rho^2 + (z'+d)^2]^{1/2}; \qquad P_2 = S_2\left(\frac{S_2 + Z' + D}{R}\right)^2, \tag{F46}$$

$$S_2 = \frac{k_2^3 r_2}{2k_1^2}; \quad R = \frac{k_2^3\rho}{2k_1^2}; \quad Z' = \frac{k_2^2 z'}{2k_1}; \quad D = \frac{k_2^2 d}{2k_1}, \tag{F47}$$

and $\mathcal{F}(P_2)$ is defined in (F43).

F.2 The Component $B_{2\phi'}(\rho, z')$

The surface-wave integral for $B_{2\phi'}(\rho, z')$ is, from (4.2.21a,b),

$$B_{2\phi'}^s(\rho, z') = \frac{i\mu_0 k_1^2}{2\pi} \int_0^\infty \left(\frac{1}{N} - \frac{1}{N_0}\right) e^{i\gamma_2(z'+d)} J_1(\lambda\rho)\lambda^2\, d\lambda \tag{F48a}$$

and

$$B^s_{2\phi'}(\bar\rho,\bar z') = -\frac{i\mu_0 k_2^2}{2\pi}\left(\int_0^\infty \frac{e^{i\bar z'(1-\zeta^2)^{1/2}}}{(1-\zeta^2)^{1/2}} J_1(\zeta\bar\rho)\zeta^2\,d\zeta - \int_0^\infty \frac{e^{i\bar z'(1-\zeta^2)^{1/2}}}{(1-\zeta^2)^{1/2}+\epsilon} J_1(\zeta\bar\rho)\zeta^2\,d\zeta\right), \tag{F48b}$$

where $\bar z' = k_2(z'+d)$, $\bar\rho = k_2\rho$, and $\zeta = \lambda/k_2$. With (F15) and (F21), this is

$$B^s_{2\phi'}(\bar\rho,\bar z') = -\frac{i\mu_0 k_2^2}{2\pi}\,[I_1(\bar\rho,\bar z') - J_\rho(\bar\rho,\bar z')]. \tag{F49}$$

With (F29) and (F42),

$$\begin{aligned} B^s_{2\phi'}(\bar\rho,\bar z') &= -\frac{i\mu_0 k_2^2}{2\pi}\,\epsilon I_2(\bar\rho,\bar z') \\ &= \frac{\mu_0 k_2^3}{2\pi k_1}\left(\frac{\pi}{\bar r_2}\right)^{1/2} e^{i\bar r_2} e^{-iP_2}\mathcal{F}(P_2) \end{aligned} \tag{F50a}$$

and

$$\begin{aligned} B^s_{2\phi'}(\rho,z') &= \frac{\mu_0 k_2^3}{2\pi k_1}\left(\frac{\pi}{k_2 r_2}\right)^{1/2} e^{ik_2 r_2} e^{-iP_2}\mathcal{F}(P_2) \\ &= \frac{\mu_0}{2\pi}\, e^{ik_2 r_2} T(\rho, z'+d), \end{aligned} \tag{F50b}$$

where $\mathcal{F}(P_2)$ and P_2 are defined in (F43) and (F46). When this is combined with the direct and image terms,

$$\begin{aligned} B_{2\phi'}(\rho,z') = -\frac{\mu_0}{2\pi}\Bigg[&\frac{e^{ik_2 r_1}}{2}\left(\frac{\rho}{r_1}\right)\left(\frac{ik_2}{r_1}-\frac{1}{r_1^2}\right) + \frac{e^{ik_2 r_2}}{2}\left(\frac{\rho}{r_2}\right)\left(\frac{ik_2}{r_2}-\frac{1}{r_2^2}\right) \\ &- e^{ik_2 r_2}\,\frac{k_2^3}{k_1}\left(\frac{\pi}{k_2 r_2}\right)^{1/2} e^{-iP_2}\mathcal{F}(P_2)\Bigg]. \end{aligned} \tag{F51}$$

F.3 The Component $E_{2z'}(\rho, z')$

The surface-wave integral for $E_{2z'}(\rho, z')$ is given in (4.2.23a,b). It is

$$E^s_{2z'}(\rho,z') = \frac{\omega\mu_0 k_1^2}{2\pi k_2^2}\int_0^\infty e^{i\gamma_2(z'+d)}\left(\frac{1}{N_0}-\frac{1}{N}\right)J_0(\lambda\rho)\lambda^3\,d\lambda. \tag{F52}$$

This can be expanded in two ways. The first includes the two integrals in (F52). Thus,

$$E^s_{2z'}(\rho,z') = \frac{\omega\mu_0}{2\pi k_2^2}\left(\int_0^\infty \frac{e^{i\gamma_2(z'+d)}}{\gamma_2} J_0(\lambda\rho)\lambda^3\,d\lambda\right.$$

$$- k_1^2 \int_0^\infty \frac{e^{i\gamma_2(z'+d)}}{N} J_0(\lambda\rho)\lambda^3\,d\lambda\bigg). \tag{F53}$$

The second form makes use of the identity

$$\left(\frac{1}{N_0} - \frac{1}{N}\right) = \frac{k_2^2\gamma_1}{k_1^2\gamma_2 N} \sim \frac{k_2^2}{k_1\gamma_2 N}. \tag{F54}$$

The approximation $\gamma_1 = (k_1^2 - \lambda^2)^{1/2} \sim k_1$ is made as previously in this appendix because $|k_1^2| \gg |k_2^2|$ and because principal contributions to the integrals occur when λ is of the order of k_2. The second form is thus

$$E_{2z'}^s(\rho, z') = \frac{\omega\mu_0 k_1}{2\pi} \int_0^\infty \frac{e^{i\gamma_2(z'+d)}}{\gamma_2 N} J_0(\lambda\rho)\lambda^3\,d\lambda. \tag{F55}$$

With $\bar{z}' = k_2(z'+d)$, $\bar{\rho} = k_2\rho$, $\zeta = \lambda/k_2$, and $\epsilon = k_2/k_1$, the two forms become

$$E_{2z'}^s(\bar{\rho}, \bar{z}') = \frac{\omega\mu_0 k_2}{2\pi}\,[I_0(\bar{\rho}, \bar{z}') - I_z(\bar{\rho}, \bar{z}')], \tag{F56a}$$

where

$$I_0(\bar{\rho}, \bar{z}') = \int_0^\infty \frac{e^{i\bar{z}'(1-\zeta^2)^{1/2}}}{(1-\zeta^2)^{1/2}} J_0(\zeta\bar{\rho})\zeta^3\,d\zeta \tag{F56b}$$

and

$$I_z(\bar{\rho}, \bar{z}') = \int_0^\infty \frac{e^{i\bar{z}'(1-\zeta^2)^{1/2}}}{(1-\zeta^2)^{1/2} + \epsilon} J_0(\zeta\bar{\rho})\zeta^3\,d\zeta, \tag{F56c}$$

and

$$E_{2z'}^s(\bar{\rho}, \bar{z}') = \frac{\omega\mu_0 k_2}{2\pi}\,\epsilon J_z(\bar{\rho}, \bar{z}'), \tag{F57a}$$

where

$$J_z(\bar{\rho}, \bar{z}') = \int_0^\infty \frac{e^{i\bar{z}'(1-\zeta^2)^{1/2}}}{(1-\zeta^2)^{1/2}[(1-\zeta^2)^{1/2} + \epsilon]} J_0(\zeta\bar{\rho})\zeta^3\,d\zeta. \tag{F57b}$$

When the two forms are equated, it follows that

$$I_z(\bar{\rho}, \bar{z}') = I_0(\bar{\rho}, \bar{z}') - \epsilon J_z(\bar{\rho}, \bar{z}')$$

or

$$J_z(\bar{\rho}, \bar{z}') = -\frac{1}{\epsilon}\,[I_z(\bar{\rho}, \bar{z}') - I_0(\bar{\rho}, \bar{z}')]. \tag{F58}$$

It is readily shown that

$$\frac{\partial J_z(\bar{\rho}, \bar{z}')}{\partial \bar{z}'} = iI_z(\bar{\rho}, \bar{z}') = i[I_0(\bar{\rho}, \bar{z}') - \epsilon J_z(\bar{\rho}, \bar{z}')]$$

so that

$$\frac{\partial J_z(\bar{\rho}, \bar{z}')}{\partial \bar{z}'} + i\epsilon J_z(\bar{\rho}, \bar{z}') = iI_0(\bar{\rho}, \bar{z}'). \tag{F59}$$

This is the same differential equation obtained in (F24). It has the solution (F25), viz.,

$$J_z(\bar{\rho}, \bar{z}') = -i \int_{\bar{z}'}^{\infty} e^{i\epsilon(\bar{z}-\bar{z}')} I_0(\bar{\rho}, \bar{z})\, d\bar{z}. \tag{F60}$$

The integral for $I_0(\bar{\rho}, \bar{z})$ is given in (F56b). With this,

$$\begin{aligned} \frac{\omega\mu_0 k_2}{2\pi} I_0(\bar{\rho}, \bar{z}') &= \frac{\omega\mu_0}{2\pi k_2^2} \int_0^{\infty} \frac{e^{i\gamma_2(z'+d)}}{\gamma_2} J_0(\lambda\rho)\lambda^3\, d\lambda \\ &= -2E^i_{2z'}(\rho, z'). \end{aligned} \tag{F61}$$

As indicated, this is twice the negative image field, which is well known to be

$$-2E^i_{2z'}(\bar{\rho}, \bar{z}') = \frac{\omega\mu_0 k_2}{2\pi} I_0(\bar{\rho}, \bar{z}') \tag{F62a}$$

with

$$I_0(\bar{\rho}, \bar{z}') = -e^{i\bar{r}_2}\left[\frac{i}{\bar{r}_2} - \frac{1}{\bar{r}_2^2} - \frac{i}{\bar{r}_2^3} - \left(\frac{\bar{z}'}{\bar{r}_2}\right)^2 \left(\frac{i}{\bar{r}_2} - \frac{3}{\bar{r}_2^2} - \frac{3i}{\bar{r}_2^3}\right)\right]. \tag{F62b}$$

In the evaluation of the integral in (F60) for $E_{2\rho}(\rho, z')$ in (F25), only the leading $1/r_2$ term was retained since the entire term contributes negligibly to the field except when r_2 is large. Accordingly, the integral (F60) will be evaluated with

$$I_0(\bar{\rho}, \bar{z}) = -\frac{ie^{i\bar{r}}}{\bar{r}}\left(1 - \frac{\bar{z}^2}{\bar{r}^2}\right) = -\frac{i\bar{\rho}^2}{\bar{r}^3} e^{i\bar{r}}. \tag{F63}$$

That is,

$$J_z(\bar{\rho}, \bar{z}') = -\int_{\bar{z}'}^{\infty} \frac{\bar{\rho}^2}{\bar{r}^3} e^{i\bar{r}} e^{i\epsilon(\bar{z}-\bar{z}')}\, d\bar{z}. \tag{F64}$$

This is the same as $I_2(\bar{\rho}, \bar{z}')$ in (F33) when this is multiplied by the factor $i\bar{\rho}/\bar{r}$. Accordingly, the integrated form of (F64) is (F42) multiplied by this factor. Specifically,

$$J_z(\bar{\rho}, \bar{z}') = -\left(\frac{\pi}{\bar{r}_2}\right)^{1/2} \left(\frac{\bar{\rho}}{\bar{r}_2}\right) e^{i\bar{r}_2} e^{-iP_2} \mathcal{F}(P_2). \tag{F65}$$

With (F57),

$$E^s_{2z'}(\bar{\rho}, \bar{z}') = -\frac{\omega\mu_0 k_2^2}{2\pi k_1}\left(\frac{\pi}{\bar{r}_2}\right)^{1/2} \left(\frac{\bar{\rho}}{\bar{r}_2}\right) e^{i\bar{r}_2} e^{-iP_2} \mathcal{F}(P_2) \tag{F66a}$$

and

$$E^s_{2z'}(\rho, z') = -\frac{\omega\mu_0}{2\pi k_2} \frac{\rho}{r_2} e^{ik_2 r_2} T(\rho, z' + d), \tag{F66b}$$

with

$$T(\rho, z'+d) = \frac{k_2^3}{k_1}\left(\frac{\pi}{k_2 r_2}\right)^{1/2} e^{-iP_2}\mathcal{F}(P_2). \tag{F66c}$$

When this is combined with the direct and image fields, the complete formula is

$$\begin{aligned} E^s_{2z'}(\rho, z') = \frac{\omega\mu_0}{2\pi k_2}\Bigg\{ & \frac{e^{ik_2 r_1}}{2}\Bigg[\frac{ik_2}{r_1} - \frac{1}{r_1^2} - \frac{i}{k_2 r_1^3} \\ & - \left(\frac{z'-d}{r_1}\right)^2\left(\frac{ik_2}{r_1} - \frac{3}{r_1^2} - \frac{3i}{k_2 r_1^3}\right)\Bigg] \\ & + \frac{e^{ik_2 r_2}}{2}\Bigg[\frac{ik_2}{r_2} - \frac{1}{r_2^2} - \frac{i}{k_2 r_2^3} \\ & - \left(\frac{z'+d}{r_2}\right)^2\left(\frac{ik_2}{r_2} - \frac{3}{r_2^2} - \frac{3i}{k_2 r_2^3}\right)\Bigg] \\ & - e^{ik_2 r_2}\frac{k_2^3}{k_1}\left(\frac{\pi}{k_2 r_2}\right)^{1/2}\left(\frac{\rho}{r_2}\right) e^{-iP_2}\mathcal{F}(P_2)\Bigg\}, \end{aligned} \tag{F67}$$

where P_2 is defined in (F46) with (F47) and $\mathcal{F}(P_2)$ is given in (F43).

F.4 Application to Similar Integrals in Chapters 3 and 4

The integrals evaluated in this appendix can now be summarized for application to somewhat less general ones in Chapter 3. The two integrals involved are (F48a) and (F52), viz.,

$$e^{ik_2 r_0} T(\rho, z) = ik_1^2 \int_0^\infty \left(\frac{1}{N} - \frac{1}{N_0}\right) e^{i\gamma_2 z} J_1(\lambda\rho)\lambda^2\, d\lambda \tag{F68}$$

and

$$e^{ik_2 r_0}\frac{\rho}{r_0} T(\rho, z) = \frac{k_1^2}{k_2}\int_0^\infty \left(\frac{1}{N} - \frac{1}{N_0}\right) e^{i\gamma_2 z} J_0(\lambda\rho)\lambda^3\, d\lambda, \tag{F69}$$

where $r_0 = (\rho^2 + z^2)^{1/2}$ and

$$T(\rho, z) = \frac{k_2^3}{k_1}\left(\frac{\pi}{k_2 r_0}\right)^{1/2} e^{-iP_0}\mathcal{F}(P_0), \tag{F70}$$

$$P_0 = S_0\left(\frac{S_0 + Z}{R}\right)^2; \qquad S_0 = \frac{k_2^3 r_0}{2k_1^2}, \quad R = \frac{k_2^3 \rho}{2k_1^2}, \quad Z = \frac{k_2^2 z}{2k_1}. \tag{F71}$$

In (3.7.23) and (3.7.25), the following functions are defined:

$$G_{2\phi}(\rho, z) = \int_0^\infty \left(\frac{1}{N} - \frac{1}{N_0}\right) e^{-i\gamma_2 z} J_1(\lambda\rho)\lambda^2\, d\lambda; \qquad z \le 0, \tag{F72}$$

$$G_{2z}(\rho, z) = \int_0^\infty \left(\frac{1}{N} - \frac{1}{N_0}\right) e^{-i\gamma_2 z} J_0(\lambda\rho)\lambda^3\, d\lambda; \qquad z \le 0. \tag{F73}$$

A comparison of (F72) with (F68) and (F73) with (F69) shows that in the generalized form

$$\begin{aligned} G_{2\phi}(\rho, z) &= -\frac{i}{k_1^2}\, e^{ik_2 r_0} T(\rho, z) \\ &= -\frac{ik_2^3}{k_1^3}\left(\frac{\pi}{k_2 r_0}\right)^{1/2} e^{ik_2 r_0} e^{-iP_0} \mathcal{F}(P_0), \end{aligned} \tag{F74}$$

$$\begin{aligned} G_{2z}(\rho, z) &= \frac{k_2}{k_1^2}\frac{\rho}{r_0}\, e^{ik_2 r_0} T(\rho, z) \\ &= \frac{k_2^4}{k_1^3}\left(\frac{\pi}{k_2 r_0}\right)^{1/2}\left(\frac{\rho}{r_0}\right) e^{ik_2 r_0} e^{-iP_0} \mathcal{F}(P_0). \end{aligned} \tag{F75}$$

The corresponding integrals (4.2.21b) and (4.2.23b) in Chapter 4 are like (F74) and (F75) with z replaced by $z' + d$, r_0 by r_2, and ϕ by ϕ'. That is,

$$G_{2\phi'}(\rho, z' + d) = -\frac{ik_2^3}{k_1^3}\left(\frac{\pi}{k_2 r_2}\right)^{1/2} e^{ik_2 r_2} e^{-iP_2} \mathcal{F}(P_2), \tag{F76}$$

$$G_{2z'}(\rho, z' + d) = \frac{k_2^4}{k_1^3}\left(\frac{\pi}{k_2 r_2}\right)^{1/2}\left(\frac{\rho}{r_2}\right) e^{ik_2 r_2} e^{-iP_2} \mathcal{F}(P_2), \tag{F77}$$

where

$$P_2 = S_2\left(\frac{S_2 + Z' + D}{R}\right)^2 \tag{F78}$$

and

$$S_2 = \frac{k_2^3 r_2}{2k_1^2}; \quad R = \frac{k_2^3 \rho}{2k_1^2}; \quad Z' = \frac{k_2^2 z'}{2k_1}; \quad D = \frac{k_2^2 d}{2k_1}. \tag{F79}$$

Appendix G
Evaluation of Integrals for the Power in the Air and in the Earth Radiated by a Vertical Dipole in the Air above a Dielectric Earth

The two integrals J_{1a} and J_{2a} in (4.3.43) and (4.3.44) can be evaluated together. They are

$$\begin{aligned} J_{1a} + J_{2a} &= \operatorname{Re} \frac{3}{4k_2^3} \int_0^{k_2} \gamma_2^{-1}[|G_{21}|^2 - 2G_{21}]\lambda^3\, d\lambda \\ &= \operatorname{Re} \frac{3}{k_2^3} \int_0^{k_2} \gamma_2^{-1}\left(\frac{k_2^4\gamma_1^2}{N^2} - \frac{k_2^2\gamma_1}{N}\right)\lambda^3\, d\lambda \\ &= \operatorname{Re} \frac{3}{k_2} \int_0^{k_2} \gamma_2^{-1}\left(\frac{k_2^2\gamma_1^2 - \gamma_1(k_2^2\gamma_1 + k_1^2\gamma_2)}{N^2}\right)\lambda^3\, d\lambda \\ &= -\operatorname{Re} \frac{3k_1^2}{k_2} \int_0^{k_2} \frac{\gamma_1}{(k_1^2\gamma_2 + k_2^2\gamma_1)^2}\,\lambda^3\, d\lambda, \end{aligned} \tag{G1}$$

where $G_{21} = 2k_2^2\gamma_1 N^{-1}$ and $N = k_1^2\gamma_2 + k_2^2\gamma_1$. With the condition $k_2^2 \ll k_1^2$, it follows that in the range $0 \le \lambda \le k_2$, $\lambda^2 \ll k_1^2$. Hence, $\gamma_1 = (k_1^2 - \lambda^2)^{1/2} \sim k_1$. The integral, therefore, becomes

$$\begin{aligned} J_{1a} + J_{2a} &\sim -\operatorname{Re} \frac{3k_1^3}{k_2} \int_0^{k_2} \frac{\lambda^3\, d\lambda}{(k_1^2\gamma_2 + k_2^2 k_1)^2} \\ &= -\operatorname{Re} \frac{3k_1}{k_2} \int_0^{k_2} \frac{\lambda^3\, d\lambda}{[k_1(k_2^2 - \lambda^2)^{1/2} + k_2^2]^2}. \end{aligned} \tag{G2}$$

Let the variable be changed to

$$\zeta = \frac{(k_2^2 - \lambda^2)^{1/2}}{k_2}, \quad \lambda^2 = k_2^2(1-\zeta^2), \quad \lambda\, d\lambda = -k_2^2 \zeta\, d\zeta$$

so that

$$\begin{aligned} J_{1a} + J_{2a} &= -\operatorname{Re} \frac{3k_2}{k_1} \int_0^1 \frac{(1-\zeta^2)\zeta\, d\zeta}{(\zeta + k_2 k_1^{-1})^2} \\ &= -\operatorname{Re} \frac{3k_2}{k_1} \left(\int_0^1 \frac{\zeta\, d\zeta}{(\zeta + k_2 k_1^{-1})^2} - \int_0^1 \frac{\zeta^3\, d\zeta}{(\zeta + k_2 k_1^{-1})^2} \right). \end{aligned} \tag{G3}$$

These are elementary integrals with the integrated values

$$\begin{aligned} J_{1a} + J_{2a} &= -\frac{3k_2}{k_1} \left\{ \ln\left(\zeta + \frac{k_2}{k_1}\right) + \frac{k_2}{k_1(\zeta + k_2 k_1^{-1})} \right. \\ &\qquad - \left[\frac{1}{2}\left(\zeta + \frac{k_2}{k_1}\right)^2 - \frac{3k_2}{k_1}\left(\zeta + \frac{k_2}{k_1}\right) \right. \\ &\qquad\qquad \left.\left. + \frac{3k_2^2}{k_1^2} \ln\left(\zeta + \frac{k_2}{k_1}\right) + \frac{k_2^3}{k_1^3(\zeta + k_2 k_1^{-1})} \right] \right\}_0^1 \\ &= -\frac{3k_2}{k_1} \left[\left(1 - \frac{3k_2^2}{k_1^2}\right) \ln\left(1 + \frac{k_1}{k_2}\right) - \frac{3}{2} + \frac{3k_2}{k_1} \right]. \end{aligned} \tag{G4}$$

The third integral in (4.3.45) is

$$\begin{aligned} J_{3a} &= -\operatorname{Re} \frac{3}{k_2} \int_0^{k_2} \frac{\gamma_1}{\gamma_2 N} e^{2i\gamma_2 d} \lambda^3\, d\lambda \\ &\sim -\operatorname{Re} \frac{3}{k_2} \int_0^{k_2} \frac{1}{(k_2^2 - \lambda^2)^{1/2}} \frac{e^{2id(k_2^2 - \lambda^2)^{1/2}}}{k_1 (k_2^2 - \lambda^2)^{1/2} + k_2^2} \lambda^3\, d\lambda. \end{aligned} \tag{G5}$$

The approximation $\gamma_1 \sim k_1$ has been made because $\lambda^2 \ll k_1^2$ in the range $0 \le \lambda \le k_2$. With the same change of variable made in (G1), i.e., $\zeta = (k_2^2 - \lambda^2)^{1/2}/k_2$, the integral becomes

$$J_{3a} = -\operatorname{Re} \frac{3k_2}{k_1} \int_0^1 \frac{1-\zeta^2}{\zeta + k_2 k_1^{-1}} e^{2ik_2 d\zeta}\, d\zeta. \tag{G6}$$

Now let $m = k_2/k_1$ and $u = \zeta + m$ so that $\zeta = u - m$ and $1 - \zeta^2 = 1 - u^2 + 2um - m^2$. Then,

$$\begin{aligned} J_{3a} = -\operatorname{Re} 3m e^{-2ik_2 dm} &\left(\int_m^{m+1} (1 - m^2) \frac{e^{2ik_2 du}}{u}\, du \right. \\ &\left. - \int_m^{m+1} e^{2ik_2 du} u\, du + 2m \int_m^{m+1} e^{2ik_2 du}\, du \right). \end{aligned} \tag{G7}$$

Next, let $v = 2k_2du$ so that

$$J_{3a} = -\operatorname{Re} 3me^{-2ik_2dm}\left(\int_{2k_2dm}^{2k_2d(m+1)} (1-m^2)\,\frac{e^{iv}}{v}\,dv - \int_{2k_2dm}^{2k_2d(m+1)} \frac{e^{iv} v\,dv}{4k_2^2d^2} + \frac{m}{k_2d}\int_{2k_2dm}^{2k_2d(m+1)} e^{iv}\,dv\right). \tag{G8}$$

These integrals can be expressed in terms of the sine and cosine integrals, viz.,

$$\operatorname{Si} x = \int_0^x \frac{\sin u}{u}\,du; \qquad \operatorname{Cin} x = \int_0^x \frac{1-\cos u}{u}\,du. \tag{G9}$$

Thus,

$$\begin{aligned} J_{3a} &= -\operatorname{Re} 3me^{-2ik_2dm}\bigg\{(1-m^2)(\ln v - \operatorname{Cin} v + i\operatorname{Si} v) \\ &\qquad - \frac{1}{4k_2^2d^2}\,e^{iv}(-iv+1) - \frac{mi}{k_2d}\,e^{iv}\bigg\}_{2k_2dm}^{2k_2d(1+m)} \\ &= -\operatorname{Re} 3m\bigg\{e^{-2ik_2dm}(1-m^2)\bigg[\ln\left(\frac{1+m}{m}\right) \\ &\qquad - \operatorname{Cin} 2k_2d(1+m) + \operatorname{Cin} 2k_2dm \\ &\qquad + i\operatorname{Si} 2k_2d(1+m) - i\operatorname{Si} 2k_2dm\bigg] \\ &\qquad - \frac{1}{4k_2^2d^2}\,\{e^{2ik_2d}[1-2ik_2d(m+1)] - 1 + 2ik_2dm\} \\ &\qquad - i\,\frac{m}{k_2d}\,(e^{2ik_2d}-1)\bigg\}. \end{aligned} \tag{G10}$$

Rearrangement and substitution for $m = k_2/k_1$ give the final formula. It is

$$\begin{aligned} J_{3a} = -\frac{3k_2}{k_1}\Bigg(&\left(1-\frac{k_2^2}{k_1^2}\right)\Bigg\{\cos\frac{2k_2^2d}{k_1}\left[\ln\left(1+\frac{k_1}{k_2}\right) - \operatorname{Cin} 2k_2d\left(1+\frac{k_2}{k_1}\right) + \operatorname{Cin}\frac{2k_2^2d}{k_1}\right] \\ &+ \sin\frac{2k_2^2d}{k_1}\left[\operatorname{Si} 2k_2d\left(1+\frac{k_2}{k_1}\right) - \operatorname{Si}\frac{2k_2^2d}{k_1}\right]\Bigg\} \\ &+ \frac{1-\cos 2k_2d}{4k_2^2d^2} - \left(1-\frac{k_2}{k_1}\right)\frac{\sin 2k_2d}{2k_2d}\Bigg). \end{aligned} \tag{G11}$$

Since $J_{1g} + J_{2g} = -(J_{1a} + J_{2a})$, the only remaining integral is

$$J_{3g} = -\frac{3}{k_2} \operatorname{Im} \int_{k_2}^{\infty} \frac{e^{-2\nu_2 d}}{\nu_2} \frac{\gamma_1}{N} \lambda^3 \, d\lambda$$
$$= -\frac{3}{k_2} \operatorname{Im} \int_{k_2}^{\infty} \frac{(k_1^2 - \lambda^2)^{1/2} \, e^{-2d(\lambda^2 - k_2^2)^{1/2}} \lambda^3 \, d\lambda}{(\lambda^2 - k_2^2)^{1/2} [ik_1^2(\lambda^2 - k_2^2)^{1/2} + k_2^2(k_1^2 - \lambda^2)^{1/2}]}. \quad \text{(G12)}$$

In the range $\lambda > k_1$, $(k_1^2 - \lambda^2)^{1/2} = i(\lambda^2 - k_1^2)^{1/2} = i\nu_1$. Hence,

$$J_{3g} = -\frac{3}{k_2} \operatorname{Im}\left(\int_{k_2}^{k_1} \frac{\gamma_1 e^{-2d\nu_2} \lambda^3 \, d\lambda}{\nu_2(ik_1^2 \nu_2 + k_2^2 \gamma_1)} + \int_{k_1}^{\infty} \frac{i\nu_1 e^{-2d\nu_2} \lambda^3 \, d\lambda}{\nu_2(ik_1^2 \nu_2 + ik_2^2 \nu_1)} \right). \quad \text{(G13)}$$

Here, the second integral is real and, therefore, has no imaginary part. It follows that

$$J_{3g} = -\frac{3}{k_2} \operatorname{Im} \int_{k_2}^{k_1} \frac{\gamma_1(ik_1^2 \nu_2 - k_2^2 \gamma_1)}{\nu_2(-k_1^4 \nu_2^2 - k_2^4 \gamma_1^2)} \, e^{-2d\nu_2} \lambda^3 \, d\lambda. \quad \text{(G14)}$$

The integral that includes the second term in the numerator is real and has no imaginary part. Thus,

$$J_{3g} = \frac{3k_1^2}{k_2} \operatorname{Re} \int_{k_2}^{k_1} \frac{(k_1^2 - \lambda^2)^{1/2} e^{-2d(\lambda^2 - k_2^2)^{1/2}} \lambda^3 \, d\lambda}{k_1^4(\lambda^2 - k_2^2) + k_2^4(k_1^2 - \lambda^2)}. \quad \text{(G15)}$$

In this integral, let $\zeta = \lambda^2/k_2^2$ or $\lambda^2 = \zeta k_2^2$, $\lambda \, d\lambda = \frac{1}{2} k_2^2 \, d\zeta$. Then,

$$J_{3g} = \frac{3k_1^2}{2k_2} \operatorname{Re} \int_1^{k_1^2/k_2^2} \frac{(k_1^2 - k_2^2 \zeta)^{1/2} e^{-2k_2 d(\zeta - 1)^{1/2}} k_2^4 \zeta \, d\zeta}{k_1^4 k_2^2(\zeta - 1) + k_2^4(k_1^2 - k_2^2 \zeta)}. \quad \text{(G16)}$$

Here, the denominator is

$$k_2^2(k_1^4 - k_2^4)\zeta - k_1^2 k_2^2(k_1^2 - k_2^2) = k_2^2(k_1^4 - k_2^4)\left(\zeta - \frac{k_1^2}{k_1^2 + k_2^2}\right)$$

so that

$$J_{3g} = \operatorname{Re} \frac{3k_1^2 k_2}{2(k_1^4 - k_2^4)} \int_1^{k_1^2/k_2^2} \frac{(k_1^2 - k_2^2 \zeta)^{1/2} e^{-2k_2 d(\zeta - 1)^{1/2}} \zeta \, d\zeta}{\zeta - k_1^2(k_1^2 + k_2^2)^{-1}}. \quad \text{(G17)}$$

Now let $\tau^2 = \zeta - 1$ or $\zeta = \tau^2 + 1$, $d\zeta = 2\tau \, d\tau$. Also let $n^2 = (k_1^2/k_2^2) - 1$ and $m^2 = k_2^2/(k_1^2 + k_2^2)$. Then,

$$J_{3g} = \operatorname{Re} \frac{3k_1^2 k_2^2}{k_1^4 - k_2^4} \int_0^n \frac{(n^2 - \tau^2)^{1/2} e^{-2k_2 d\tau}(\tau^2 + 1)\tau \, d\tau}{\tau^2 + m^2}. \quad \text{(G18)}$$

Since $k_1^4 \gg k_2^4$, this reduces to the following two integrals:

$$J_{3g} = \operatorname{Re} \frac{3k_2^2}{k_1^2} (I_1 + I_2), \quad \text{(G19)}$$

where

$$I_1 = \mathrm{Re}\int_0^n \frac{(n^2-\tau^2)^{1/2}e^{-2k_2d\tau}\tau\,d\tau}{\tau^2+m^2}, \tag{G20a}$$

$$I_2 = \mathrm{Re}\int_0^n \frac{(n^2-\tau^2)^{1/2}e^{-2k_2d\tau}\tau^3\,d\tau}{\tau^2+m^2}. \tag{G20b}$$

Since $m^2 = k_2^2/(k_1^2+k_2^2) < k_2^2/k_1^2 \ll 1$, it can be neglected in I_2 but must be retained in I_1 because the integral becomes infinite when $m=0$.

The evaluation of I_1 is carried out in two parts obtained with $\tau = nt$. Thus, with $z = 2k_2dn$ and $a = m/n \sim k_2^2/k_1^2$,

$$I_1 = n\int_0^1 \frac{e^{-zt}(1-t^2)^{1/2}}{t^2+a^2}\,t\,dt = I_{11}+I_{12}, \tag{G21}$$

where

$$I_{11} = n\int_0^1 e^{-zt}\frac{t\,dt}{t^2+a^2}; \qquad I_{12} = -n\int_0^1 e^{-zt}\frac{1-(1-t^2)^{1/2}}{t^2+a^2}\,t\,dt. \tag{G22}$$

The integral I_{11} can be evaluated with formula (5.1.44) of Abramowitz and Stegun (1972, p. 230). With $x=-zt$ or $t=-x/z$,

$$\begin{aligned} I_{11} &= n\int_0^{-z} e^x\frac{x\,dx}{x^2+a^2z^2} \\ &= -\frac{n}{2}\left[e^{iaz}E_1(-x+iaz)+e^{-iaz}E_1(-x-iaz)\right]\Big|_0^{-z} \\ &= \frac{n}{2}\left[e^{iaz}E_1(iaz)+e^{-iaz}E_1(-iaz)-e^{iaz}E_1(z+iaz)\right. \\ &\qquad \left.-\,e^{-iaz}E_1(z-iaz)\right], \end{aligned} \tag{G23}$$

where $E_1(x)$ is defined in (5.1.1) of Abramowitz and Stegun (1972, p. 228). This can be expressed in terms of the more familiar exponential integral $E_i(-z) = -E_1(z)$ [see (5.1.2) on p. 228]. Also with $z(1\pm ia) = z[1\pm i(k_2^2/k_1^2)] \sim z$ (since $k_2^2 \ll k_1^2$), the desired formula is

$$I_{11} = -\frac{n}{2}\left[e^{iaz}E_i(-iaz)+e^{-iaz}E_i(iaz)+2E_1(z)\cos az\right]. \tag{G24}$$

Note that

$$\begin{aligned} E_i(\pm iaz) &= \mathrm{Ci}\,az \pm \left(\mathrm{Si}\,az - \frac{\pi}{2}\right) \\ &= \gamma + \ln az - \mathrm{Cin}\,az \pm i\left(\mathrm{Si}\,az-\frac{\pi}{2}\right), \end{aligned} \tag{G25}$$

where $\gamma = 0.5772$ and $\mathrm{Cin}\,x$ and $\mathrm{Si}\,x$ are defined in (G9). With this notation, the final formula for I_{11} is

$$I_{11} = -\frac{n}{2}\left\{e^{iaz}\left[\gamma+\ln az-\mathrm{Cin}\,az - i\left(\mathrm{Si}\,az-\frac{\pi}{2}\right)\right]\right.$$

$$+ e^{-iaz}\left[\gamma + \ln az - \operatorname{Cin} az + i\left(\operatorname{Si} az - \frac{\pi}{2}\right)\right]$$

$$\left. + 2E_1(z)\cos az\right\}$$

$$= -n\left[\cos az\,(\gamma + \ln az - \operatorname{Cin} az) + \sin az\left(\operatorname{Si} az - \frac{\pi}{2}\right)\right.$$

$$\left. + E_1(z)\cos az\right]. \tag{G26}$$

The integral I_{12} in (G22) can be evaluated as follows:

$$I_{12} = -n\lim_{\epsilon\to 0}\left(\int_0^1 e^{-zt}t^{-1+\epsilon}\,dt - \int_0^1 e^{-zt}t^{-1+\epsilon}(1-t^2)^{1/2}\,dt\right). \tag{G27}$$

Here, the first integral is given by formula (1) of Bateman (1953, Vol. 1, p. 255) with $c = 1+\epsilon$, $a = \epsilon$, and $x = -z$. The second integral is given by Gradshteyn and Ryzhik (1980, formula 1, p. 323) with $\mu = -z$, $\rho = \frac{3}{2}$, $\nu = \frac{1}{2}\epsilon$, and $u = 1$. The results are

$$I_{12} = -n\lim_{\epsilon\to 0}\left(\frac{\Gamma(\epsilon)}{\Gamma(\epsilon+1)}\,\Phi(\epsilon, 1+\epsilon; -z)\right.$$

$$- \frac{1}{2}\,B(\tfrac{1}{2}\epsilon, \tfrac{3}{2})\,{}_1F_2(\tfrac{1}{2}\epsilon;\,\tfrac{1}{2},\,\tfrac{1}{2}\epsilon+\tfrac{3}{2};\,\tfrac{1}{4}z^2)$$

$$\left. + \frac{z}{2}\,B(\tfrac{1}{2}\epsilon+\tfrac{1}{2},\,\tfrac{3}{2})\,{}_1F_2(\tfrac{1}{2}\epsilon+\tfrac{1}{2};\,\tfrac{3}{2},\,\tfrac{1}{2}\epsilon+2;\,\tfrac{1}{4}z^2)\right). \tag{G28}$$

The function Φ is defined in (1) of Bateman (1953, Vol. 1, p. 248). It is

$$\Phi(\epsilon,\,1+\epsilon;\,-z) = \sum_{k=0}^{\infty}\frac{(\epsilon)_k(-z)^k}{(1+\epsilon)_k k!} = 1 + \sum_{k=1}^{\infty}\frac{\epsilon}{k+\epsilon}\,\frac{(-z)^k}{k!}$$

$$\sim 1 + \epsilon\sum_{k=1}^{\infty}\frac{(-z)^k}{k(k!)} = 1 - \epsilon[\gamma + \ln z + E_1(z)]. \tag{G29}$$

The last step makes use of (5.1.11) of Abramowitz and Stegun (1972, p. 229). The Beta function is defined in (6.2.2) of Abramowitz and Stegun (1972, p. 258). It is

$$B(\tfrac{1}{2}\epsilon,\,\tfrac{3}{2}) = \frac{\Gamma(\frac{1}{2}\epsilon)\Gamma(\frac{3}{2})}{\Gamma(\frac{1}{2}\epsilon+\frac{3}{2})} \equiv \frac{2}{\epsilon}\,\frac{\Gamma(1+\frac{1}{2}\epsilon)\frac{1}{2}\sqrt{\pi}}{\Gamma(\frac{3}{2}+\frac{1}{2}\epsilon)}$$

$$= \frac{\sqrt{\pi}}{\epsilon}\,\frac{\Gamma(1)}{\Gamma(\frac{3}{2})}\,[1 + \tfrac{1}{2}\epsilon\,\psi(1) - \tfrac{1}{2}\epsilon\,\psi(\tfrac{3}{2})]$$

$$= \frac{2}{\epsilon}\,[1 - \tfrac{1}{2}\epsilon\,\gamma - \tfrac{1}{2}\epsilon(-\gamma + 2 - 2\ln 2)], \tag{G30}$$

where $\psi(z) = \Gamma'(z)/\Gamma(z)$ as defined in (6.3.1) of Abramowitz and Stegun (1972, p. 258), and with (6.3.2) and (6.3.4) on the same page. Thus,

$$B(\tfrac{1}{2}\epsilon,\ \tfrac{3}{2}) \sim \frac{2}{\epsilon} - 2 + 2\ln 2. \tag{G31}$$

Similarly,

$$B(\tfrac{1}{2}\epsilon + \tfrac{1}{2},\ \tfrac{3}{2}) = \frac{\Gamma(\frac{1}{2}\epsilon + \frac{1}{2})\Gamma(\frac{3}{2})}{\Gamma(\frac{1}{2}\epsilon + 2)}. \tag{G32}$$

With (6.1.18) and (6.1.15) of Abramowitz and Stegun (1972, p. 256),

$$\begin{aligned}\Gamma(\tfrac{1}{2}\epsilon + \tfrac{1}{2}) &= \frac{2^{1-\epsilon}\sqrt{\pi}\,\Gamma(\epsilon)}{\Gamma(\frac{1}{2}\epsilon)} \\ &= \frac{2^{1-\epsilon}\sqrt{\pi}\,\epsilon^{-1}\Gamma(\epsilon+1)}{2\epsilon^{-1}\Gamma(\frac{1}{2}\epsilon+1)} = \frac{\sqrt{\pi}\,\Gamma(\epsilon+1)}{2^{\epsilon}\Gamma(\frac{1}{2}\epsilon+1)}.\end{aligned} \tag{G33}$$

Also, $\Gamma(\frac{3}{2}) = \sqrt{\pi}/2$ and $\Gamma(\frac{1}{2}\epsilon + 2) = (\frac{1}{2}\epsilon + 1)\Gamma(\frac{1}{2}\epsilon + 1)$ so that

$$B(\tfrac{1}{2}\epsilon + \tfrac{1}{2},\ \tfrac{3}{2}) = \frac{\pi\Gamma(\epsilon+1)}{2^{1+\epsilon}(\frac{1}{2}\epsilon+1)\Gamma^2(\frac{1}{2}\epsilon+1)} \sim \frac{\pi\Gamma(1)}{2\Gamma^2(1)} \sim \frac{\pi}{2}. \tag{G34}$$

The hypergeometric series ${}_1F_2$ is defined as follows:

$$\begin{aligned}{}_1F_2(\tfrac{1}{2}\epsilon;\ \tfrac{1}{2},\ \tfrac{1}{2}\epsilon + \tfrac{3}{2};\ \tfrac{1}{4}z^2) &= 1 + \sum_{k=0}^{\infty} \frac{(\frac{1}{2}\epsilon)_{k+1}(\frac{1}{4}z^2)^{k+1}}{(\frac{1}{2})_{k+1}(\frac{1}{2}\epsilon + \frac{3}{2})_{k+1}(k+1)!} \\ &= 1 + \frac{\frac{1}{2}\epsilon}{\frac{1}{2}(\frac{1}{2}\epsilon + \frac{3}{2})} \sum_{k=0}^{\infty} \frac{(1 + \frac{1}{2}\epsilon)_k(\frac{1}{4}z^2)^{k+1}}{(\frac{3}{2})_k(\frac{1}{2}\epsilon + \frac{5}{2})_k(k+1)!}.\end{aligned} \tag{G35}$$

When higher-order terms in ϵ have been neglected, this reduces to

$${}_1F_2(\tfrac{1}{2}\epsilon;\ \tfrac{1}{2},\ \tfrac{1}{2}\epsilon + \tfrac{3}{2};\ \tfrac{1}{4}z^2) \sim 1 + \frac{2\epsilon}{3} \sum_{k=0}^{\infty} \frac{k!}{(\frac{3}{2})_k(\frac{5}{2})_k} \frac{(\frac{1}{4}z^2)^{k+1}}{(k+1)!}. \tag{G36}$$

The second series is

$$\begin{aligned}{}_1F_2(\tfrac{1}{2}\epsilon + \tfrac{1}{2};\ \tfrac{3}{2},\ \tfrac{1}{2}\epsilon + 2;\ \tfrac{1}{4}z^2) &= \sum_{k=0}^{\infty} \frac{(\frac{1}{2}\epsilon + \frac{1}{2})_k(\frac{1}{4}z^2)^k}{(\frac{3}{2})_k(\frac{1}{2}\epsilon + 2)_k k!} \\ &\sim \sum_{k=0}^{\infty} \frac{\frac{1}{2}(\frac{1}{4}z^2)^k}{(2)_k(\frac{1}{2} + k)k!}.\end{aligned} \tag{G37}$$

With these values, I_{12} becomes

$$I_{12} = -n \lim_{\epsilon\to 0}\left[\frac{1}{\epsilon}\{1 - \epsilon[E_1(z) + \gamma + \ln z]\}\right.$$

$$
\begin{aligned}
&\quad - \frac{1}{2}\left(\frac{2}{\epsilon} - 2 + 2\ln 2\right)\left(1 + \frac{2\epsilon}{3}\sum_{k=0}^{\infty} \frac{k!(\frac{1}{4}z^2)^{k+1}}{(\frac{3}{2})_k(\frac{5}{2})_k(k+1)!}\right) \\
&\quad + \frac{\pi z}{4}\sum_{k=0}^{\infty} \frac{(\frac{1}{4}z^2)^k}{(2)_k(1+2k)k!}\Bigg] \\
&= -n[-E_1(z) - \gamma - \ln z + 1 - \ln 2 - P(z)], \qquad \text{(G38)}
\end{aligned}
$$

where

$$
P(z) = \frac{2}{3}\sum_{k=0}^{\infty} \frac{k!(\frac{1}{4}z^2)^{k+1}}{(\frac{3}{2})_k(\frac{5}{2})_k(k+1)!} - \frac{\pi z}{4}\sum_{k=0}^{\infty} \frac{(\frac{1}{4}z^2)^k}{(2)_k(1+2k)k!}; \quad P(0) = 0, \qquad \text{(G39)}
$$

$$
\begin{aligned}
P'(z) &= \frac{2}{3}\,\frac{z}{2}\sum_{k=0}^{\infty} \frac{(\frac{1}{4}z^2)^k}{(\frac{3}{2})_k(\frac{5}{2})_k} - \frac{\pi}{4}\sum_{k=0}^{\infty} \frac{(\frac{1}{4}z^2)^k}{(2)_k k!} \\
&= \frac{z}{2}\,{}_1F_2(1;\, \tfrac{3}{2},\, \tfrac{5}{2};\, \tfrac{1}{4}z^2) - \frac{\pi}{4}\sum_{k=0}^{\infty} \frac{(\frac{1}{4}z^2)^k}{(k+1)!k!}. \qquad \text{(G40)}
\end{aligned}
$$

With formula (57) of Bateman (1953, Vol. 2, pp. 38 and 39) and formula (12) on p. 5, this becomes

$$
P'(z) = \frac{z}{3}\left(\frac{\Gamma(\frac{5}{2})L_1(z)}{2\pi^{-1/2}(\frac{1}{4}z^2)}\right) - \frac{\pi}{4}\cdot\frac{2}{z}\,I_1(z). \qquad \text{(G41)}
$$

Since $\Gamma(\frac{5}{2}) = \frac{3}{4}\Gamma(\frac{1}{2}) = \frac{3}{4}\sqrt{\pi}$,

$$
P'(z) = \frac{\pi}{2z}\,[L_1(z) - I_1(z)] \qquad \text{(G42)}
$$

and

$$
P(z) = \frac{\pi}{2}\int_0^z [L_1(z) - I_1(z)]\,\frac{dz}{z}. \qquad \text{(G43)}
$$

In these formulas, $L_1(z)$ is the modified Struve function and $I_1(z)$ is the modified Bessel function of order 1.

With $z = 2k_2nd \sim 2k_1d$,

$$
\begin{aligned}
I_{12} = \frac{k_1}{k_2}\Bigg(&E_1(2k_1d) + \gamma + \ln 2k_1d - 1 + \ln 2 \\
&+ \frac{\pi}{2}\int_0^{2k_1d} [L_1(z) - I_1(z)]\,\frac{dz}{z}\Bigg). \qquad \text{(G44)}
\end{aligned}
$$

With (G44) and (G26), (G21) becomes

$$
\begin{aligned}
I_1 &= I_{11} + I_{12} \\
&= -\frac{k_1}{k_2}\left[\cos\frac{2k_2^2d}{k_1}\left(\gamma + \ln\frac{2k_2^2d}{k_1} - \operatorname{Cin}\frac{2k_2^2d}{k_1}\right)\right.
\end{aligned}
$$

$$+\sin\frac{2k_2^2d}{k_1}\left(\mathrm{Si}\,\frac{2k_2^2d}{k_1}-\frac{\pi}{2}\right)-\left(1-\cos\frac{2k_2^2d}{k_1}\right)E_1(2k_1d)$$
$$-\gamma-\ln 2k_1d-\ln 2+1+\frac{\pi}{2}\int_0^{2k_1d}[I_1(z)-L_1(z)]\,\frac{dz}{z}\bigg]. \tag{G45}$$

The integral in (G45) can be rearranged with the recurrence relation

$$z^{-1}[I_1(z)-L_1(z)]=I_0(z)-L_0(z)-[I_1'(z)-L_1'(z)]. \tag{G46}$$

It follows that

$$\frac{\pi}{2}\int_0^{2k_1d}[I_1(z)-L_1(z)]\,\frac{dz}{z}$$
$$=\frac{\pi}{2}\left(\int_0^{2k_1d}[I_0(z)-L_0(z)]\,dz-[I_1(2k_1d)-L_1(2k_1d)]\right)$$
$$=\frac{\pi}{2}\,\{f_0(2k_1d)-[I_1(2k_1d)-L_1(2k_1d)]\}. \tag{G47}$$

The functions $f_0(2k_1d)$ and $I_1(2k_1d)-L_1(2k_1d)$ are tabulated in the range $0\le 2k_1d\le 5$ on p. 501 of Abramowitz and Stegun (1972). For large arguments,

$$\frac{\pi}{2}\int_0^{2k_1d}[I_1(z)-L_1(z)]\,\frac{dz}{z}$$
$$=\ln 2k_1d+\frac{\pi}{2}\,\{f_2(2k_1d)-[I_1(2k_1d)-L_1(2k_1d)]\}. \tag{G48}$$

The functions $f_2(2k_1d)$ and $I_1(2k_1d)-L_1(2k_1d)$ are tabulated on p. 502 of Abramowitz and Stegun in the range $0\le(2k_1d)^{-1}\le 0.2$ or $5\le 2k_1d\le\infty$.

Alternatively, an asymptotic formula can be derived for the integral in (G47) when k_1d is large. This is given in (4.3.53c). With it, (G45) becomes

$$I_1\sim-\frac{k_1}{k_2}\bigg[\cos\frac{2k_2^2d}{k_1}\left(\gamma+\ln\frac{2k_2^2d}{k_1}-\mathrm{Cin}\,\frac{2k_2^2d}{k_1}\right)$$
$$+\sin\frac{2k_2^2d}{k_1}\left(\mathrm{Si}\,\frac{2k_2^2d}{k_1}-\frac{\pi}{2}\right)$$
$$-\sum_{m=1}^{\infty}\frac{(2m)!(2m-1)!}{(m!)^2(4k_1d)^{2m}}+\frac{1}{2}\sum_{m=0}^{\infty}\frac{(-1)^m\Gamma(m+\frac{3}{2})}{\Gamma(\frac{1}{2}-m)(k_1d)^{2(m+1)}}\bigg]. \tag{G49}$$

The integral I_2 in (G20b) is evaluated with the help of (1) of Gradshteyn and Ryzhik (1980, p. 323) with $\mu=-2k_2d$, $\nu=1$, $\rho=\frac{3}{2}$, and $u=n$. The result, with $k_2nd\sim k_1d$, is

$$I_2=\tfrac{1}{2}B(1,\,\tfrac{3}{2})n^3\,{}_1F_2(1;\,\tfrac{1}{2},\,\tfrac{5}{2};\,k_1^2d^2)$$
$$-k_1dB(\tfrac{3}{2},\,\tfrac{3}{2})n^3\,{}_1F_2(\tfrac{3}{2};\,\tfrac{3}{2},\,3;\,k_1^2d^2). \tag{G50}$$

Here,

$$B(1,\,\tfrac{3}{2})=\frac{\Gamma(1)\Gamma(\frac{3}{2})}{\Gamma(\frac{5}{2})}=\frac{2}{3};\qquad B(\tfrac{3}{2},\,\tfrac{3}{2})=\frac{\Gamma(\frac{3}{2})\Gamma(\frac{3}{2})}{\Gamma(3)}=\frac{\pi}{8}.\tag{G51}$$

Hence,

$$I_2=n^3\left(\frac{1}{3}\,{}_1F_2(1;\,\tfrac{1}{2},\,\tfrac{5}{2};\,k_1^2d^2)-\frac{\pi k_1 d}{8}\,{}_1F_2(\tfrac{3}{2};\,\tfrac{3}{2},\,3;\,k_1^2d^2)\right).\tag{G52}$$

The hypergeometric series can be expanded in terms of the modified Struve function L and Bessel function I. Thus, with $(1)_k=k!$,

$$\begin{aligned}{}_1F_2(1;\,\tfrac{1}{2},\,\tfrac{5}{2};\,k_1^2d^2)&=1+\sum_{k=1}^{\infty}\frac{(k_1^2d^2)^k}{(\frac{1}{2})_k(\frac{5}{2})_k}=1+\sum_{k=0}^{\infty}\frac{(k_1^2d^2)^{k+1}}{\frac{1}{2}(\frac{3}{2})_k\times\frac{5}{2}(\frac{7}{2})_k}\\&=1+\tfrac{4}{5}k_1^2d^2\sum_{k=0}^{\infty}\frac{(k_1^2d^2)^k}{(\frac{3}{2})_k(\frac{7}{2})_k}\\&=1+\tfrac{4}{5}k_1^2d^2\,{}_1F_2(1;\,\tfrac{7}{2},\,\tfrac{3}{2};\,k_1^2d^2).\end{aligned}\tag{G53}$$

It follows with the first line on p. 39 of Bateman (1953, Vol. 2) that

$$\begin{aligned}{}_1F_2(1;\,\tfrac{1}{2},\,\tfrac{5}{2};\,k_1^2d^2)&=1+\frac{4}{5}\,k_1^2d^2\,\frac{L_2(2k_1d)\Gamma(\frac{7}{2})}{2\pi^{-1/2}k_1^3d^3}\\&=1+\frac{3\pi}{4k_1d}\,L_2(2k_1d).\end{aligned}\tag{G54}$$

Also, with formula (12) of Bateman (1953, Vol. 2, p. 5),

$$\begin{aligned}{}_1F_2(\tfrac{3}{2};\,\tfrac{3}{2},\,3;\,k_1^2d^2)&=\sum_{k=0}^{\infty}\frac{(k_1^2d^2)^k}{(3)_k k!}\\&=2\sum_{k=0}^{\infty}\frac{(k_1^2d^2)^k}{k!(k+2)!}=\frac{2}{k_1^2d^2}\,I_2(2k_1d).\end{aligned}\tag{G55}$$

It follows that

$$I_2=\frac{k_1^3}{k_2^3}\left(\frac{1}{3}-\frac{\pi}{4k_1d}\left[I_2(2k_1d)-L_2(2k_1d)\right]\right).\tag{G56}$$

This can be evaluated with the recurrence relations (12.2.4) and (9.6.26) of Abramowitz and Stegun (1972, pp. 498 and 376):

$$L_2(z)=L_0(z)-\frac{2}{z}\,L_1(z)-\frac{\frac{1}{2}z}{\sqrt{\pi}\,\Gamma(\frac{5}{2})}\tag{G57}$$

and

$$I_2(z)=I_0(z)-\frac{2}{z}\,I_1(z).\tag{G58}$$

With these formulas,

$$I_2(2k_1d) - L_2(2k_1d) = I_0(2k_1d) - L_0(2k_1d) - \frac{1}{k_1d}\,[I_1(2k_1d) - L_1(2k_1d)] + \frac{k_1d}{\sqrt{\pi}\,\Gamma(\frac{5}{2})}. \tag{G59}$$

The functions $I_0(2k_1d) - L_0(2k_1d)$ and $I_1(2k_1d) - L_1(2k_1d)$ are tabulated on p. 501 of Abramowitz and Stegun in the range $0 \leq 2k_1d \leq 5$ and on p. 502 in the range $5 \leq 2k_1d < \infty$.

Alternatively, use can be made of the power-series expansions of $L_2(2k_1d)$ and $I_2(2k_1d)$. These are convergent for all values of the argument. With $z = 2k_1d$, they are

$$L_2(z) = \left(\frac{z}{2}\right)^3 \sum_{k=0}^{\infty} \frac{(\frac{1}{2}z)^{2k}}{\Gamma(k+\frac{3}{2})\Gamma(k+\frac{7}{2})}, \tag{G60a}$$

$$I_2(z) = \left(\frac{z}{2}\right)^2 \sum_{k=0}^{\infty} \frac{(\frac{1}{2}z)^{2k}}{k!\,\Gamma(k+3)}. \tag{G60b}$$

When (G45) and (G56) are combined in (G19), the result is

$$\begin{aligned} J_{3g} &= \frac{3k_2^2}{k_1^2}\,(I_1 + I_2) \\ &= -\frac{3k_2}{k_1}\Bigg[\cos\frac{2k_2^2d}{k_1}\left(\gamma + \ln\frac{2k_2^2d}{k_1} - \operatorname{Cin}\frac{2k_2^2d}{k_1}\right) \\ &\qquad + \sin\frac{2k_2^2d}{k_1}\left(\operatorname{Si}\frac{2k_2^2d}{k_1} - \frac{\pi}{2}\right) - \left(1 - \cos\frac{2k_2^2d}{k_1}\right)E_1(2k_1d) \\ &\qquad - \gamma - \ln 2k_1d + 1 - \ln 2 + \frac{\pi}{2}\int_0^{2k_1d}[I_1(z) - L_1(z)]\,\frac{dz}{z}\Bigg] \\ &\quad + \frac{k_1}{k_2}\left(1 + \frac{3\pi}{4k_1d}\,[L_2(2k_1d) - I_2(2k_1d)]\right). \end{aligned} \tag{G61}$$

When $d = 0$,

$$J_{3g} = -\frac{3k_2}{k_1}\left(1 - \ln 2 - 2\ln\frac{k_1}{k_2}\right) + \frac{k_1}{k_2}. \tag{G62}$$

For large values of k_1d, asymptotic formulas for the integral in (G61) and for the term $L_2(2k_1d) - I_2(2k_1d)$ are, respectively, in (4.3.53c) and (4.3.53a,b). With these,

$$\begin{aligned} J_{3g} = -\frac{3k_2}{k_1}\Bigg[&\cos\frac{2k_2^2d}{k_1}\left(\gamma + \ln\frac{2k_2^2d}{k_1} - \operatorname{Cin}\frac{2k_2^2d}{k_1}\right) \\ &+ \sin\frac{2k_2^2d}{k_1}\left(\operatorname{Si}\frac{2k_2^2d}{k_1} - \frac{\pi}{2}\right) \end{aligned}$$

$$
- \sum_{m=1}^{\infty} \frac{(2m)!(2m-1)!}{(m!)^2(4k_1 d)^{2m}} + \frac{1}{2} \sum_{m=0}^{\infty} \frac{(-1)^m \Gamma(m+\frac{3}{2})}{\Gamma(\frac{1}{2}-m)(k_1 d)^{2(m+1)}}
$$

$$
- \frac{1}{4k_2^2 d^2} \sum_{m=0}^{\infty} \frac{(-1)^m \Gamma(\frac{3}{2}+m)}{\Gamma(\frac{3}{2}-m)(k_1^2 d^2)^m} \Bigg]. \tag{G63}
$$

Note that when $k_1 \to \infty$, $J_{3g} \to 0$.

Bibliography

Abramowitz, M., Stegun, I. A. (eds.): *Handbook of Mathematical Functions.* New York: Dover Publications, 1972.

Aharoni, J.: *Antennae.* Oxford: The Clarendon Press, 1946, pp. 223–229.

Attwood, S. S.: Surface-wave propagation over a coated plane conductor. *J. Appl. Phys.* **22**, 504–509 (1951).

Bannister, P. R., Dube, R. L.: Simple expressions for horizontal electric dipole quasi-static range subsurface-to-subsurface and subsurface-to-air propagation. *Radio Sci.* **13**, 501–507 (1978).

Baños, A., Jr.: *Dipole Radiation in the Presence of a Conducting Half-Space.* Oxford: Pergamon Press, 1966, pp. 151–158.

Baños, A., Jr., Wesley, J. P.: The horizontal electric dipole in a conducting half-space, Part I. Univ. of Calif. Marine Physical Lab., SIO Reference 53-33, 1953.

Baños, A., Jr., Wesley, J. P.: The horizontal electric dipole in a conducting half-space, Part II. Univ. of Calif. Marine Physical Lab., SIO Reference 54-31, 1954.

Barlow, H. M., Cullen, A. L.: Surface waves. *Proc. IEE (London)* **100**, Part III, 329–341 (1953).

Bateman Manuscript Project: *Higher Transcendental Functions, Vols. 1 and 2.* Erdélyi, A. (ed.). New York: McGraw-Hill Book Co., 1953.

Belrose, J. S., Litva, J., Moss, G. E., Stevens, E. E.: Beverage antennas for amateur communications. *QST* **67**, 22-27 (1983).

Beverage, H. H., Rice, C. W., Kellog, E. W.: The wave antenna. *Trans. AIEE* **42**, 215 (1923).

Boardman, A. D. (ed.): *Electromagnetic Surface Modes.* New York: John Wiley & Sons, Inc., 1982.

Brekhovskikh, L. M.: *Waves in Layered Media*, 2nd ed. New York: Academic Press, 1980.

Bremmer, H.: *Terrestrial Radio Waves.* New York: Elsevier Publishing Co., 1949.

Brown, M. F., King, R. W. P.: Shallow sounding of crustal regions using electromagnetic surface waves. *Radio Sci.* **21**, 831–844 (1986).

Brown, M. F., King, R. W. P., Wu, T. T.: Lateral-wave studies in a model lithosphere. *J. Appl. Phys.* **53**, 3387–3396 (1982).

Brown, M. F., King, R. W. P., Wu, T. T.: Experiments on the reflection of lateral electromagnetic waves. *J. Appl. Phys.* **55**, 3927–3933 (1984).

Bubenik, D. M.: A practical method for the numerical evaluation of Sommerfeld integrals. *IEEE Trans. Antennas Propag.* **AP-25**, 904–906 (1977).

Busch, H.: Theories der Beverage antenne. *Jahrb. Telegr. Telef.* **21**, 290–312 (1923).

Carver, K. R., Mink, J. W.: Microstrip antenna technology. *IEEE Trans. Antennas Propag.* **AP-29**, 2–24 (1981).

CCIR (International Radio Consultative Committee): *Propagation in Nonionized Media*, Vol. 5, Recommendations and Reports of the CCIR. Geneva, Switzerland: International Telecommunication Union, 1978, pp. 36 and 37.

Chave, A. D., Cox, C. S.: Controlled electromagnetic sources for measuring electrical conductivity beneath the oceans. I. Forward problem and model study. *J. Geophys. Res.* **87**, 5327–5338 (1982).

Collin, R. E.: *Field Theory of Guided Waves.* New York: McGraw-Hill Book Co., 1960, pp. 461–465 and 470–477.

Deibele, S., Beyer, J. B.: Measurements of microstrip effective relative permittivities. *IEEE Trans. Microwave Theory Tech.* **35**, 535–538 (1987).

Denlinger, E. J.: A frequency-dependent solution for microstrip transmission lines. *IEEE Trans. Microwave Theory Tech.* **19**, 30–39 (1971).

Dunn, J. M.: Electromagnetic Lateral Waves in Layered Media. Ph.D. Thesis, Harvard University, Cambridge, MA, 1984.

Dunn, J. M.: Lateral wave propagation in a three-layered medium. *Radio Sci.* **21**, 787–796 (1986).

Edwards, R. N., Nobes, D. C., DeLaurier, J. M.: Offshore electrical exploration of sedimentary basins: The effects of anisotropy in horizontally isotropic, layered media. *Geophysics* **49**, 566–576 (1984).

Edwards, R. N., Law, L. K., Wolfgram, P. A., Nobes, D. C., Bone, M. N., Triggs, D. F., DeLaurier, J. M.: First results of Moses experiment: Sea-sediment conductivity and thickness determination, Bute Inlet, British Columbia, by magnetometric offshore electrical sounding. *Geophysics* **50**, 153–160 (1985).

Fujiki, Y., Hayashi, Y., Suzuki, M.: Analysis of strip transmission lines by iteration method. *Electron. Commun. Japan* **55**, 74–80 (1972).

Getsinger, W. J.: Microstrip dispersion model. *IEEE Trans. Microwave Theory Tech.* **21**, 34–39 (1973).

Gradshteyn, I. S., Ryzhik, I. M.: *Table of Integrals, Series, and Products.* New York: Academic Press, 1980.

Gulla, D.: Surface Electric Field and Transient Signals Along IC Interconnects. Masters Thesis, Tufts University, Medford, MA, 1988.

Hansen, P. M.: The radiation efficiency of a dipole antenna located above an imperfectly conducting ground. *IEEE Trans. Antennas Propag.* **AP-20**, 766–770 (1972).

Hashimoto, M.: A rigorous solution for dispersive microstrip. *IEEE Trans. Microwave Theory Tech.* **33**, 1131–1137 (1985).

Hill, D. A., Wait, J. R.: Excitation of the Zenneck surface wave by a vertical aperture. *Radio Sci.* **13**, 969–977 (1978).

Hill, D. A., Wait, J. R.: Ground wave attenuation function for a spherical earth with arbitrary surface impedance. *Radio Sci.* **15**, 637–643 (1980).

Hoffmann, R. K.: *Handbook of Microwave Integrated Circuits.* Norwood, MA: Artech House, Inc., 1987, pp. 135–138.

von Hörschelmann, H.: Ueber die Wirkungsweise des geknickten Marconischen Senders in der drahtlosen Telegraphie. *Jahrb. drahtl. Telegra. Teleph.* **5**, 14–34 and 188–211 (1911).

Itoh, T., Mittra, R.: Spectral-domain approach for calculating the dispersion characteristics of microstrip lines. *IEEE Trans. Microwave Theory Tech.* **21**, 496–499 (1973).

Jahnke, E., Emde, F.: *Tables of Functions.* New York: Dover Publications, 1945.

Jansen, R. H.: The spectral-domain approach for microwave integrated circuits. *IEEE Trans. Microwave Theory Tech.* **33**, 1043–1056 (1985).

Kaufman, A. A., Keller, G. V.: *Frequency and Transient Soundings*, Vol. 16, Methods in Geochemistry and Geophysics Series. New York: Elsevier Publishing Co., 1983.

King, R. J.: Electromagnetic wave propagation over a constant impedance plane. *Radio Sci.* **4**, 255–268 (1969).

King, R. J., Wait, J. R.: Electromagnetic groundwave propagation—Theory and experiment. *Symposia Mathematica* **18**, 107–208 (1976); see Fig. 4.4.

King, R. J., Cho, S. H., Jaggard, D. L.: Height-gain experimental data for groundwave propagation, 2, Heterogeneous paths. *Radio Sci.* **8**, 17–22 (1973); see Figs. 3 and 4.

King, R. W. P.: *Theory of Linear Antennas.* Cambridge, MA: Harvard University Press, 1956.

King, R. W. P.: *Transmission-Line Theory*, 2nd ed. New York: Dover Publications, 1965, Chap. 4.

King, R. W. P.: Cylindrical antennas and arrays, Chap. 9 in *Antenna Theory, Part I.* Collin, R. E., Zucker, F. J. (eds.). New York: McGraw-Hill Book Co., 1969.

King, R. W. P.: New formulas for the electromagnetic field of a vertical electric dipole in a dielectric or conducting half-space near its horizontal interface. *J. Appl. Phys.* **53**, 8476–8482 (1982); erratum, **56**, 3366 (1984).

King, R. W. P.: The wave antenna for transmission and reception. *IEEE Trans. Antennas Propag.* **AP-31**, 956–965 (1983).

King, R. W. P.: On the reflection of lateral electromagnetic waves from perpendicular boundaries. *J. Appl. Phys.* **55**, 3916–3926 (1984).

King, R. W. P.: Scattering of lateral waves by buried or submerged objects. I. The incident lateral-wave field. *J. Appl. Phys.* **57**, 1453–1459 (1985).

King, R. W. P.: Scattering of lateral waves by buried or submerged objects. II. The electric field on the surface above a buried insulated wire. *J. Appl. Phys.* **57**, 1460–1472 (1985).

King, R. W. P.: Electromagnetic surface waves: New formulas and applications. *IEEE Trans. Antennas Propag.* **AP-33**, 1204–1212 (1985).

King, R. W. P.: Electromagnetic surface waves: New formulas and their application to determine the electrical properties of the sea bottom. *J. Appl. Phys.* **58**, 3612–3624 (1985).

King, R. W. P.: Properties of the lateral electromagnetic field of a vertical dipole and their application. *IEEE Trans. Geosci. Remote Sensing* **GE-24**, 813–825 (1986).

King, R. W. P.: Antennas in material media near boundaries with application to communication and geophysical exploration, Part I: The bare metal dipole. *IEEE Trans. Antennas Propag.* **AP-34**, 483–489 (1986).

King, R. W. P.: Antennas in material media near boundaries with application to communication and geophysical exploration, Part II: The terminated insulated antenna. *IEEE Trans. Antennas Propag.* **AP-34**, 490–496 (1986).

King, R. W. P.: Lateral electromagnetic pulses generated by a vertical dipole on a plane boundary between dielectrics. *J. Electrom. Waves Appl.* **2**, 225–243 (1988).

King, R. W. P.: The propagation of signals along a three-layered region: Microstrip. *IEEE Trans. Microwave Theory Tech.* **36**, 1080–1086 (1988).

King, R. W. P.: Wire and strip conductors over a dielectric-coated conducting or dielectric half-space. *IEEE Trans. Microwave Theory Tech.* **37**, 754–760 (1989).

King, R. W. P.: Lateral electromagnetic pulses generated on a plane boundary between dielectrics by vertical and horizontal dipole sources with Gaussian pulse excitation. *J. Electrom. Waves Appl.* **3**, 589–597 (1989).

King, R. W. P.: Lateral electromagnetic waves from a horizontal antenna for remote sensing in the ocean. *IEEE Trans. Antennas Propag.* **AP-37**, 1250–1255 (1989).

King, R. W. P.: Lateral electromagnetic waves and pulses on open microstrip. *IEEE Trans. Microwave Theory Tech.* **38**, 38–47 (1990).

King, R. W. P.: Electromagnetic field of a vertical dipole over an imperfectly conducting half-space. *Radio Sci.* **25**, 149–160 (1990).

King, R. W. P.: The electromagnetic field of a horizontal electric dipole in the presence of a three-layered region. *J. Appl. Phys.* **69**, 7987–7995 (1991).

King, R. W. P.: Electromagnetic field of dipoles and patch antennas on microstrip. *Radio Sci.* **27**, 71–78 (1992).

King, R. W. P., Brown, M. F.: Lateral electromagnetic waves along plane boundaries: A summarizing approach. *Proc. IEEE* **72**, 595–611 (1984).

King, R. W. P., deBettencourt, J. T., Sandler, B. H.: Lateral-wave propagation of electromagnetic waves in the lithosphere. *IEEE Trans. Geosci. Electron.* **GE-17**, 86–92 (1979).

King, R. W. P., Owens, M., Wu, T. T.: Properties of lateral electromagnetic fields and their application. *Radio Sci.* **21**, 13–23 (1986).

King, R. W. P., Prasad, S.: *Fundamental Electromagnetic Theory and Applications.* Englewood Cliffs, N.J.: Prentice-Hall, Inc., 1986.

King, R. W. P., Sandler, B. H.: Subsurface communication between dipoles in general media. *IEEE Trans. Antennas Propag.* **AP-25**, 770–775 (1977).

King, R. W. P., Sandler, B. H., Shen, L. C.: A comprehensive study of subsurface propagation from horizontal electric dipoles. *IEEE Trans. Geosci. Remote Sensing* **GE-18**, 225–233 (1980).

King, R. W. P., Sandler, B. H., Wu, T. T.: Cylindrical antennas immersed in arbitrary homogeneous isotropic media. *J. Appl. Phys.* **40**, 5049–5065 (1969); correction, **41**, 3559 (1970).

King, R. W. P., Smith, G. S.: *Antennas in Matter: Fundamentals, Theory, and Applications.* Cambridge, MA: The M.I.T. Press, 1981.

King, R. W. P., Wu, T. T.: Lateral waves: Formulas for the magnetic field. *J. Appl. Phys.* **54**, 507–514 (1983); erratum, **56**, 3365 (1984).

King, R. W. P., Wu, T. T., Shen, L. C.: The horizontal-wire antenna over a conducting or dielectric half-space: Current and admittance. *Radio Sci.* **9**, 701–709 (1974).

Kowalski, G., Pregla, R.: Dispersion characteristics of single and coupled microstrips. *Arch. Elek. Übertragung.* **26**, 276–280 (1972).

Kraichman, M. B.: *Handbook of Electromagnetic Propagation in Conducting Media.* Washington, D.C.: Superintendent of Documents, U.S. Government Printing Office, 1970.

Kraus, J. D.: *Antennas.* New York: McGraw-Hill Book Co., 1950, pp. 412 and 413.

Kuester, E. F., Chang, D. C.: An appraisal of methods for computation of the dispersion characteristics of open microstrip. *IEEE Trans. Microwave Theory Tech.* **27**, 691–694 (1979).

Leung, T., Balanis, C. A.: Pulse dispersion distortion in open and shielded microstrips using the spectral-domain method. *IEEE Trans. Microwave Theory Tech.* **36**, 1223–1226 (1988).

Lytle, R. J., Miller, E. K., Lager, D. L.: A physical explanation of electromagnetic surface wave formulas. *Radio Sci.* **11**, 235–243 (1976).

Mailloux, R. J., McIlvenna, J. F., Kernweis, N. P.: Microstrip array technology. *IEEE Trans. Antennas Propag.* **AP-29**, 25–37 (1981).

Myers, J. M.: Symmetry in Scattering by a Strip. Ph.D. Thesis, Harvard University, Cambridge, MA, 1962.

Norton, K. A.: The propagation of radio waves over the surface of the earth and in the upper atmosphere. *Proc. IRE* **24**, 1367–1387 (1936).

Norton, K. A.: The calculations of ground-wave field intensity over a finitely conducting spherical earth. *Proc. IRE* **29**, 623–639 (1941).

Ollendorf, F.: *Die Grundlagen der Hochfrequenztechnik.* Berlin: Springer-Verlag, 1925, pp. 576–583.

Pan, W.-Y.: Surface-wave propagation along the boundary between sea water and one-dimensionally anisotropic rock. *J. Appl. Phys.* **58**, 3963–3974 (1985).

Parkhomenko, E. I.: *Electrical Properties of Rocks.* New York: Plenum Press, 1967, pp. 87 and 134–136.

Perlmutter, P., Shtrikman, S., Treves, D.: Electric surface current model for the analysis of microstrip antennas with application to rectangular elements. *IEEE Trans. Antennas Propag.* **AP-33**, 301–311 (1985).

Rahmat-Samii, Y., Mittra, R., Parhami, P.: Evaluation of Sommerfeld integrals for lossy half-space problems. *Electromagnetics* **1**, 1–28 (1981).

Rokityansky, I. I.: *Geoelectromagnetic Investigation of the Earth's Crust and Mantle.* Berlin: Springer-Verlag, 1982.

Shafer, C. G.: Dominant mode of the microstrip transmission line. Cruft Laboratory Tech. Report No. 257, Harvard University, Cambridge, MA, 1957.

Shen, L. C., King, R. W. P., Sorbello, R. M.: Measured field of a directional antenna submerged in a lake. *IEEE Trans. Antennas Propag.* **AP-24**, 891–894 (1976); correction, **AP-26**, 872–873 (1978).

Siegel, M., King, R. W. P.: Electromagnetic fields in a dissipative half-space: A numerical approach. *J. Appl. Phys.* **41**, 2415–2423 (1970).

Siegel, M., King, R. W. P.: Radiation from linear antennas in a dissipative half-space. *IEEE Trans. Antennas Propag.* **AP-19**, 477–485 (1971).

Siegel, M., King, R. W. P.: Electromagnetic propagation between antennas submerged in the ocean. *IEEE Trans. Antennas Propag.* **AP-21**, 507–513 (1973).

Sommerfeld, A.: Electromagnetic waves near wires. *Wied. Annalen* **67**, 233–290 (1899).

Sommerfeld, A.: Propagation of waves in wireless telegraphy. *Ann. Phys. (Leipzig)* **28**, 665–737 (1909).

Sommerfeld, A.: Propagation of waves in wireless telegraphy. *Ann. Phys. (Leipzig)* **81**, 1135–1153 (1926).

Sommerfeld, A.: in *Die Differential und Integralgleichungen der Mechanik und Physik, Vol. II.* Frank, P., v. Mises, R. (eds.). Braunschweig, Germany: F. Vieweg and Son, 1935, pp. 932 and 933.

Sorbello, R. M., King, R. W. P., Lee, K.-M., Shen, L. C., Wu, T. T.: The horizontal-wire antenna over a dissipative half-space: Generalized formula and measurements. *IEEE Trans. Antennas Propag.* **AP-25**, 850–854 (1977).

Van der Pol, B.: On discontinuous electromagnetic waves and the occurrence of a surface wave. *IRE Trans. Antennas Propag.* **AP-4**, 288–293 (1956).

Van der Pol, B., Niessen, K. F.: Propagation of electromagnetic waves over a plane earth. *Ann. Phys. (Leipzig)* **6**, 273–295 (1930).

Vanyan, L. L.: *Electromagnetic Depth Soundings.* New York: Consultants Bureau, Plenum Publishing Corp., 1967.

Veghte, R. L., Balanis, C. A.: Dispersion of transient signals in microstrip transmission lines. *IEEE Trans. Microwave Theory Tech.* **34**, 1427–1436 (1986).

Wait, J. R.: The electromagnetic fields of a horizontal dipole in the presence of a conducting half-space. *Can. J. Phys.* **39**, 1017–1027 (1961).

Wait, J. R.: Theory of ground wave propagation, Chap. 5 in *Electromagnetic Probing in Geophysics.* Wait, J. R. (ed.). Boulder, CO: Golem Press, 1971, pp. 163–207; see also Baños (1966, pp. 240–241) for a list of relevant papers by J. R. Wait.

Wait, J. R., Hill, D. A.: Excitation of the HF surface wave by vertical and horizontal antennas. *Radio Sci.* **14**, 767–780 (1979).

Walter, C. H.: *Traveling-Wave Antennas.* New York: Dover Publications, 1965, pp. 315–318.

Watson, G. N.: *A Treatise on the Theory of Bessel Functions,* 2nd ed. London: Cambridge University Press, 1958.

Watt, A. D.: *VLF Radio Engineering.* Oxford: Pergamon Press, 1967.

Weiss, M. A.: Microstrip antennas for millimeter waves. *IEEE Trans. Antennas Propag.* **AP-29**, 171–174 (1981).

Wise, W. H.: Earthed condenser aerial radiation formulae. *Proc. IRE* **19**, 1684–1689 (1931).

Wu, T. T.: Theory of the microstrip. *J. Appl. Phys.* **28**, 299–301 (1957).

Wu, T. T., King, R. W. P.: Lateral waves: A new formula and interference patterns. *Radio Sci.* **17**, 521–531 (1982).

Wu, T. T., King, R. W. P.: Lateral waves: New formulas for $E_{1\phi}$ and E_{1z}. *Radio Sci.* **17**, 532–538 (1982); correction, **19**, 1422 (1984).

Wu, T. T., King, R. W. P.: Lateral electromagnetic pulses generated by a vertical dipole on the boundary between two dielectrics. *J. Appl. Phys.* **62**, 4345–4355 (1987).

Young, P. D., Cox, C. S.: Electromagnetic active source sounding near the East Pacific Rise. *Geophys. Res. Letters* **8**, 1043–1046 (1981).

Zenneck, J.: Propagation of plane electromagnetic waves along a plane conducting surface and its bearing on the theory of transmission in wireless telegraphy. *Ann. Phys. (Leipzig)* **23**, 846–866 (1907).

Index